Encyclopedia of Microbiology

Second Edition

Volume 3 L–P

Editorial Board

Editor-in-Chief

Joshua Lederberg

The Rockefeller University
New York, NY

Associate Editors

Martin Alexander
Cornell University
Ithaca, NY

Barry R. Bloom
Harvard School of Public Health
Boston, MA

David Hopwood
John Innes Centre
Norwich, UK

Roger Hull
John Innes Centre
Norwich, UK

Barbara H. Iglewiski
University of Rochester
   Medical Center
Rochester, NY

Allen I. Laskin
Laskin/Lawrence Associates
Somerset, NJ

Stephen G. Oliver
University of Manchester
Manchester, UK

Moselio Schaechter
San Diego State University
San Diego, CA

William C. Summers
Yale University School of Medicine
New Haven, CT

Encyclopedia
of
MICROBIOLOGY

Second Edition

Volume 3 L–P

Editor-in-Chief
Joshua Lederberg
The Rockefeller University
New York, NY

ACADEMIC PRESS
A Harcourt Science and Technology Company

San Diego   San Francisco   New York   Boston   London   Sydney   Tokyo

This book is printed on acid-free paper.

Copyright © 2000, 1992 by ACADEMIC PRESS

All Rights Reserved.
No part of this publication may be reproduced or transmitted in any form or by any
means, electronic or mechanical, including photocopy, recording, or any information
storage and retrieval system, without permission in writing from the publisher.

Requests for permission to make copies of any part of the work should be mailed to:
Permissions Department, Harcourt Inc., 6277 Sea Harbor Drive,
Orlando, Florida 32887-677

Academic Press
A Harcourt Science and Technology Company
525 B Street, Suite 1900, San Diego, California 92101-4495, USA
http://www.apnet.com

Academic Press
24-28 Oval Road, London NW1 7DX, UK
http://www.hbuk.co.uk/ap/

Library of Congress Catalog Card Number: 99-65283

International Standard Book Number: 0-12-226800-8 (set)
International Standard Book Number: 0-12-226801-6 Volume 1
International Standard Book Number: 0-12-226802-4 Volume 2
International Standard Book Number: 0-12-226803-2 Volume 3
International Standard Book Number: 0-12-226804-0 Volume 4

PRINTED IN THE UNITED STATES OF AMERICA
00 01 02 03 04 05 MM 9 8 7 6 5 4 3 2 1

Contents

L

M

N

O

P

Contents of Other Volumes

$$B$$

VOLUME 2

D

=== E ===

=== F ===

G

H

VOLUME 4

W

V

X

Y

Z

Contents by Subject Area

HISTORICAL

INFECTIOUS AND NONINFECTIOUS DISEASE AND PATHOGENESIS: HUMAN PATHOGENS

INFECTIOUS AND NONINFECTIOUS DISEASE AND PATHOGENESIS: IMMUNOLOGY

INFECTIOUS AND NONINFECTIOUS DISEASE AND PATHOGENESIS: PLANT PATHOGENS, BACTERIA

INFECTIOUS AND NONINFECTIOUS DISEASE AND PATHOGENESIS: PLANT PATHOGENS, FUNGI

STRUCTURE AND MORPHOGENESIS

SYSTEMATICS AND PHYLOGENY

TECHNIQUES

Preface

The scientific literature at large is believed to double about every 12 years. Though less than a decade has elapsed since the initiation of the first edition of this encyclopedia, it is a fair bet that the microbiology literature has more than doubled in the interval, though one might also say it has fissioned in the interval, with parasitology, virology, infectious disease, and immunology assuming more and more independent stature as disciplines.

According to the *Encyclopaedia Britannica,* the encyclopedias of classic and medieval times could be expected to contain "a compendium of all available knowledge." There is still an expectation of the "essence of all that is known." With the exponential growth and accumulation of scientific knowledge, this has become an elusive goal, hardly one that could be embraced in a mere two or three thousand pages of text. The encyclopedia's function has moved to becoming the first word, the initial introduction to knowledge of a comprehensive range of subjects, with pointers on where to find more as may be needed. One can hardly think of the last word, as this is an ever-moving target at the cutting edge of novel discovery, changing literally day by day.

For the renovation of an encyclopedia, these issues have then entailed a number of pragmatic compromises, designed to maximize its utility to an audience of initial look-uppers over a range of coherently linked interests. The core remains the biology of that group of organisms we think of as microbes. Though this constitutes a rather disparate set, crossing several taxonomic kingdoms, the more important principle is the unifying role of DNA and the genetic code and the shared ensemble of primary pathways of gene expression. Also shared is access to a "world wide web" of genetic information through the traffic of plasmids and other genetic elements right across the taxa. It is pathognomonic that the American Society for Microbiology has altered the name of *Microbiological Reviews* to *Microbiology and Molecular Biology Reviews.* At academic institutions, microbiology will be practiced in any or all of a dozen different departments, and these may be located at schools of arts and sciences, medicine, agriculture, engineering, marine sciences, and others.

Much of human physiology, pathology, or genetics is now practiced with cell culture, which involves a methodology indistinguishable from microbiology: it is hard to define a boundary that would demarcate microbiology from cell biology. Nor do we spend much energy on these niceties except when we have the burden of deciding the scope of an enterprise such as this one.

Probably more important has been the explosion of the Internet and the online availability of many sources of information. Whereas we spoke last decade of CDs, now the focus is the Web, and the anticipation is that we are not many years from the general availability of the entire scientific literature via this medium. The utility of the encyclopedia is no longer so much "how do I begin to get information on Topic X" as how to filter a surfeit of claimed information with some degree of dependability. The intervention of editors and of a peer-review process (in selection of authors even more important than in overseeing their papers) is the only foreseeable solution. We have then sought in each article to provide a digest of information with perspective and provided by responsible authors who can be proud of, and will then strive to maintain, reputations for knowledge and fairmindedness.

The further reach of more detailed information is endless. When available, many specific topics are elaborated in greater depth in the ASM (American Society of Microbiology) reviews and in *Annual Review of Microbiology*. These are indexed online. Medline, Biosis, and the Science Citation Index are further online bibliographic resources, which can be focused for the recovery of review articles.

The reputation of the authors and of the particular journals can further aid readers' assessments. Citation searches can be of further assistance in locating critical discussions, the dialectic which is far more important than "authority" in establishing authenticity in science.

Then there are the open-ended resources of the Web itself. It is not a fair test for recovery on a specialized topic, but my favorite browser, google.com, returned 15,000 hits for "microbiology"; netscape.com gave 46,000; excite.com a few score structured headings. These might be most useful in identifying other Web sites with specialized resources. Google's 641 hits for "luminescent bacteria" offer a more proximate indicator of the difficulty of coping with the massive returns of unfiltered ver-

biage that this wonderful new medium affords: how to extract the nuggets from the slag.

A great many academic libraries and departments of microbiology have posted extensive considered listings of secondary sources. One of my favorites is maintained at San Diego State University:

> http://libweb.sdsu.edu/scidiv/
> microbiologyblr.html

I am sure I have not begun to tap all that would be available.

The best strategy is a parallel attack: to use the encyclopedia and the major review journals as a secure starting point and then to try to filter Web-worked material for the most up-to-date or disparate detail. In many cases, direct enquiry to the experts, until they saturate, may be the best (or last) recourse. E-mail is best, and society or academic institutional directories can be found online. Some listservers will entertain questions from outsiders, if the questions are particularly difficult or challenging.

All publishers, Academic Press included, are updating their policies and practices by the week as to how they will integrate their traditional book offerings with new media. Updated information on electronic editions of this and cognate encyclopedias can be found by consulting www.academicpress.com/.

Joshua Lederberg

From the Preface to the First Edition

(Excerpted from the 1992 Edition)

For the purposes of this encyclopedia, microbiology has been understood to embrace the study of "microorganisms," including the basic science and the roles of these organisms in practical arts (agriculture and technology) and in disease (public health and medicine). Microorganisms do not constitute a well-defined taxonomic group; they include the two kingdoms of Archaebacteria and Eubacteria, as well as protozoa and those fungi and algae that are predominantly unicellular in their habit. Viruses are also an important constituent, albeit they are not quite "organisms." Whether to include the mitochondria and chloroplasts of higher eukaryotes is a matter of choice, since these organelles are believed to be descended from free-living bacteria. Cell biology is practiced extensively with tissue cells in culture, where the cells are manipulated very much as though they were autonomous microbes; however, we shall exclude this branch of research. Microbiology also is enmeshed thoroughly with biotechnology, biochemistry, and genetics, since microbes are the canonical substrates for many investigations of genes, enzymes, and metabolic pathways, as well as the technical vehicles for discovery and manufacture of new biological products, for example, recombinant human insulin. . . .

The *Encyclopedia of Microbiology* is intended to survey the entire field coherently, complementing material that would be included in an advanced undergraduate and graduate major course of university study. Particular topics should be accessible to talented high school and college students, as well as to graduates involved in teaching, research, and technical practice of microbiology.

Even these hefty volumes cannot embrace all current knowledge in the field. Each article does provide key references to the literature available at the time of writing. Acquisition of more detailed and up-to-date knowledge depends on (1) exploiting the review and monographic literature and (2) bibliographic retrieval of the preceding and current research literature. . . .

To access bibliographic materials in microbiology, the main retrieval resources are MEDLINE, sponsored by the U.S. National Library of Medicine, and the Science Citation Index of the ISI. With governmental subsidy, MEDLINE is widely available at modest cost: terminals are available at every medical school and at many other academic centers. MEDLINE provides searches of the recent literature by author, title, and key word and offers online displays of the relevant bibliographies and abstracts. Medical aspects of microbiology are covered exhaustively; general microbiology is covered in reasonable depth. The Science Citation Index must recover its costs from user fees, but is widely available at major research centers. It offers additional search capabilities, especially by citation linkage. Therefore, starting with the bibliography of a given encyclopedia article, one can quickly find (1) all articles more recently published that have cited those bibliographic reference starting points and (2) all other recent articles that share bibliographic information with the others. With luck, one of these articles may be identified as another comprehensive review that has digested more recent or broader primary material.

On a weekly basis, services such as Current Contents on Diskette (ISI) and Reference Update offer still more timely access to current literature as well as to abstracts with a variety of useful features. Under the impetus of intense competition, these services are evolving rapidly, to the great benefit of a user community desperate for electronic assistance in coping with the rapidly growing and intertwined networks of discovery. The bibliographic services of Chemical Abstracts and Biological Abstracts would also be potentially invaluable; however, their coverage of microbiology is rather limited.

In addition, major monographs have appeared from time to time—*The Bacteria, The Prokaryotes,* and many others. Your local reference library should be consulted for these volumes.

Valuable collections of reviews also include *Critical Reviews for Microbiology, Symposia of the Society for General Microbiology, Monographs of the American Society for Microbiology,* and *Proceedings of the International Congresses of Microbiology.*

The articles in this encyclopedia are intended to be accessible to a broader audience, not to take the place of review articles with comprehensive bibliographies. Citations should be sufficient to give the reader access to the latter, as may be required. We do apologize to many individuals whose contributions to the growth of microbiology could not be adequately embraced by the secondary bibliographies included here.

The organization of encyclopedic knowledge is a daunting task in any discipline; it is all the more complex in such a diversified and rapidly moving domain as microbiology. The best way to anticipate the rapid further growth that we can expect in the near future is unclear. Perhaps more specialized series in subfields of microbiology would be more appropriate. The publishers and editors would welcome readers' comments on these points, as well as on any deficiencies that may be perceived in the current effort.

My personal thanks are extended to my coeditors, Martin Alexander, David Hopwood, Barbara Iglewski, and Allen Laskin; and above all, to the many very busy scientists who took time to draft and review each of these articles.

Joshua Lederberg

Guide to the Encyclopedia

The *Encyclopedia of Microbiology, Second Edition* is a scholarly source of information on microorganisms, those life forms that are observable with a microscope rather than by the naked eye. The work consists of four volumes and includes 298 separate articles. Of these 298 articles, 171 are completely new topics commissioned for this edition, and 63 others are newly written articles on topics appearing in the first edition. In other words, approximately 80% of the content of the encyclopedia is entirely new to this edition. (The remaining 20% of the content has been carefully reviewed and revised to ensure currency.)

Each article in the encyclopedia provides a comprehensive overview of the selected topic to inform a broad spectrum of readers, from research professionals to students to the interested general public. In order that you, the reader, will derive the greatest possible benefit from your use of the *Encyclopedia of Microbiology*, we have provided this Guide. It explains how the encyclopedia is organized and how the information within it can be located.

ORGANIZATION

The *Encyclopedia of Microbiology* is organized to provide maximum ease of use. All of the articles are arranged in a single alphabetical sequence by title. Articles whose titles begin with the letters A to C are in Volume 1, articles with titles from D through K are in Volume 2, then L through P in Volume 3, and finally Q to Z in Volume 4. This last volume also includes a complete subject index for the entire

work, an alphabetical list of the contributors to the encyclopedia, and a glossary of key terms used in the articles.

Article titles generally begin with the key noun or noun phrase indicating the topic, with any descriptive terms following. For example, the article title is "Bioluminescence, Microbial" rather than "Microbial Bioluminescence," and "Foods, Quality Control" is the title rather than "Quality Control of Foods."

TABLE OF CONTENTS

A complete table of contents for the *Encyclopedia of Microbiology* appears at the front of each volume. This list of article titles represents topics that have been carefully selected by the Editor-in-Chief, Dr. Joshua Lederberg, and the nine Associate Editors. The Encyclopedia provides coverage of 20 different subject areas within the overall field of microbiology. Please see p. v for the alphabetical table of contents, and p. xix for a list of topics arranged by subject area.

INDEX

The Subject Index in Volume 4 indicates the volume and page number where information on a given topic can be found. In addition, the Table of Contents by Subject Area also functions as an index, since it lists all the topics within a given area; e.g., the encyclopedia includes eight different articles dealing with historic aspects of microbiology and nine dealing with techniques of microbiology.

ARTICLE FORMAT

In order to make information easy to locate, all of the articles in the *Encyclopedia of Microbiology* are arranged in a standard format, as follows:

- Title of Article
- Author's Name and Affiliation
- Outline
- Glossary
- Defining Statement
- Body of the Article
- Cross-References
- Bibliography

OUTLINE

Each entry in the Encyclopedia begins with a topical outline that indicates the general content of the article. This outline serves two functions. First, it provides a brief preview of the article, so that the reader can get a sense of what is contained there without having to leaf through the pages. Second, it serves to highlight important subtopics that will be discussed within the article. For example, the article "Biopesticides" includes subtopics such as "Selection of Biopesticides," "Production of Biopesticides," "Biopesticide Stabilization," and "Commercialization of Biopesticides."

The outline is intended as an overview and thus it lists only the major headings of the article. In addition, extensive second-level and third-level headings will be found within the article.

GLOSSARY

The Glossary contains terms that are important to an understanding of the article and that may be unfamiliar to the reader. Each term is defined in the context of the article in which it is used. Thus the same term may appear as a glossary entry in two or more articles, with the details of the definition varying slightly from one article to another. The encyclopedia has approximately 2500 glossary entries.

In addition, Volume 4 provides a comprehensive glossary that collects all the core vocabulary of microbiology in one A–Z list. This section can be consulted for definitions of terms not found in the individual glossary for a given article.

DEFINING STATEMENT

The text of each article in the encyclopedia begins with a single introductory paragraph that defines the topic under discussion and summarizes the content of the article. For example, the article "Eyespot" begins with the following statement:

> EYESPOT is a damaging stem base disease of cereal crops and other grasses caused by fungi of the genus *Tapsia*. It occurs in temperate regions world-wide including Europe, the USSR, Japan, South Africa, North America, and Australasia. In many of these countries eyespot can be found on the majority of autumn-sown barley and wheat crops and may cause an average of 5–10% loss in yield, although low rates of infection do not generally have a significant effect. . . .

CROSS-REFERENCES

Almost all of the articles in the Encyclopedia have cross-references to other articles. These cross-references appear at the conclusion of the article text. They indicate articles that can be consulted for further information on the same topic or for information on a related topic. For example, the article "Smallpox" has references to "Biological Warfare," "Polio," "Surveillance of Infectious Diseases," and "Vaccines, Viral."

BIBLIOGRAPHY

The Bibliography is the last element in an article. The reference sources listed there are the author's recommendations of the most appropriate materials for further research on the given topic. The bibliography entries are for the benefit of the reader and do not represent a complete listing of all materials consulted by the author in preparing the article.

COMPANION WORKS

The *Encyclopedia of Microbiology* is one of a series of multivolume reference works in the life sciences published by Academic Press. Other such titles include the *Encyclopedia of Human Biology, Encyclopedia of Reproduction, Encyclopedia of Toxicology, Encyclopedia of Immunology, Encyclopedia of Virology, Encyclopedia of Cancer,* and *Encyclopedia of Stress.*

Acknowledgments

The Editors and the Publisher wish to thank the following people who have generously provided their time, often at short notice, to review various articles in the *Encyclopedia of Microbiology* and in other ways to assist the Editors in their efforts to make this work as scientifically accurate and complete as possible. We gratefully acknowledge their assistance:

George A. M. Cross
Laboratory of Molecular Parasitology
The Rockefeller University
New York, NY, USA

Miklós Müller
Laboratory of Biochemical Parasitology
The Rockefeller University
New York, NY, USA

A. I. Scott
Department of Chemistry
Texas A&M University
College Station, Texas, USA

Robert W. Simons
Department of Microbiology and
    Molecular Genetics
University of California, Los Angeles
Los Angeles, California, USA

Peter H. A. Sneath
Department of Microbiology and Immunology
University of Leicester
Leicester, England, UK

John L. Spudich
Department of Microbiology and
    Molecular Genetics
University of Texas Medical School
Houston, Texas, USA

Pravod K. Srivastava
Center for Immunotherapy
University of Connecticut
Farmington, Connecticut, USA

Peter Staeheli
Department of Virology
University of Freiburg
Freiburg, Germany

Ralph M. Steinman
Laboratory of Cellular Physiology
    and Immunology
The Rockefeller University
New York, NY, USA

Sherri O. Stuver
Department of Epidemiology
Harvard School of Public Health
Boston, Massachusetts, USA

Alice Telesnitsky
Department of Microbiology and Immunology
University of Michigan Medical School
Ann Arbor, Michigan, USA

Robert G. Webster
Chairman and Professor
Rose Marie Thomas Chair
St. Jude Children's Research Hospital
Memphis, Tennessee, USA

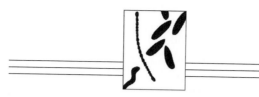

Lactic Acid Bacteria

George A. Somkuti

Eastern Regional Research Center and U.S. Department of Agriculture

I. Applications in Dairy Food Production
II. Applications in Milled Grain-Based Food Production
III. Metabolism of Lactic Acid Bacteria
IV. Production of Antimicrobial Substances
V. Plasmid Biology of LAB
VI. Genetic Development of LAB
VII. LAB in Human and Animal Health
VIII. Future Uses of LAB

promoter Nucleotide sequence involved in binding of RNA polymerase to initiate transcription.

secondary metabolite Microbial metabolic product not associated with energy generation and growth.

secretion element Nucleotide ("signal") sequence involved in secretion of a gene product.

starter culture Mixture of strains within a species, species of the same genus, or species of different genera, used to carry out dairy food fermentations on an industrial scale.

GLOSSARY

antigen Agent that induces synthesis of an antibody (immunoglobulin).

bacteriocin Ribosomally synthesized peptide, usually possessing a narrow range of antimicrobial activity.

bacteriophage Bacterial virus.

cloning vector DNA molecule for transporting new genetic material into host bacteria.

fusogen Chemical that induces the fusion of protoplasts resulting in genetic recombination.

heterologous gene Gene that is not native to the host bacterium.

insertion element Special nucleotide sequence that facilitates the chromosomal integration of foreign genes.

phenotype Characteristics of a microbe resulting from interaction of its genetic constitution with the environment.

plasmid Circular DNA molecule replicating independently from chromosomal DNA.

polymerase chain reaction Technique for producing additional copies of a nucleotide sequence.

primary metabolite Microbial metabolic product associated with energy generation and growth.

probe Labeled nucleic acid fragment useful in the detection of homologous nucleotide sequences.

probiotic Blend of live microbes exerting a beneficial effect on the host by improving properties of the native gastrointestinal microflora.

THE LACTIC ACID BACTERIA (LAB) are a group of nonspore-forming, gram-positive cocci and rods that are characterized by the production of lactic acid as a major metabolic end product of carbohydrate fermentation. They are also catalase negative, microaerophilic, acid tolerant, and lack cytochromes.

Since all LAB are fastidious in nutritional requirements, their most common ecological niches include nutrient-rich habitats, such as milk, meat, vegetables, and fruits, with plentiful simple sugars available for energy-generating metabolic pathways. The strictly fermentative LAB use either of two main pathways for carbohydrate dissimilation. If sugar metabolism proceeds via glycolysis and lactic acid is the primary end-product, the culture is said to be homolactic. On the other hand, if sugar metabolism proceeds via the 6-phosphogluconate/phosphoketolase pathway and end-products include CO_2, acetate, and alcohol, in addition to lactic acid, the culture is referred to as heterolactic. Traditionally, LAB included the genera *Lactococcus*, *Streptococcus*, *Lactobacillus*, and *Leuconostoc*. Expansion of earlier classification schemes

1

based on carbohydrate fermentation patterns, extreme temperature limits of growth, configuration of lactic acid produced, and the range of salt, acid, and alkaline tolerance, together with newer techniques, such as analysis of cell wall and fatty acid compositions, DNA–DNA homology, and ribosomal RNA sequencing data, has permitted the expansion of this core LAB group to include some 12 additional genera (Table I). The classification of LAB at both genus level and species level remains an evolving process and continues to be updated. The supplementation of historically applied criteria, including phenotypical and biochemical properties with DNA–DNA homology data, had an important role in helping to resolve classification problems. More recently, the introduction of new techniques that rely on rRNA probes (ribotyping) offers a significant refinement in the classification of LAB. The application of the PCR (polymerase chain reaction) technique has also become increasingly useful in LAB classification, as it requires a minimum amount of cells for extracting DNA or rRNA for amplification and sequencing. Although several features characteristic of all LAB cultures are discussed, this brief overview focuses primarily on the best studied group of LAB cultures which, through their metabolic activities, function as essential biocatalysts in the production of milk-based foods.

TABLE I
Recognized Genera of Lactic Acid Bacteria

Aerococcus
Alloiococcus
Carnobacterium
Dolosigranulum
Enterococcus
Globicatella
Lactobacillus
Lactococcus
Lactosphaera
Leuconostoc
Oenococcus
Pediococcus
Streptococcus
Tetragenococcus
Vagococcus
Weissella

TABLE II
Lactic Acid Bacteria Used as Starter Cultures in the Production of Fermented Dairy Foods

Homofermentative cocci:
Lactococcus lactis subsp. *lactis*
L. lactis subsp. *lactis* biovar *diacetylactis*
L. lactis subsp. *cremoris*
Streptococcus thermophilus

Heterofermentative cocci:
Leuconostoc lactis
Ln. mesenteroides subsp. *cremoris*

Homofermentative rods:
Lactobacillus delbrueckii subsp. *bulgaricus*
Lb. delbrueckii subsp. *lactis*
Lb. helveticus

Heterofermentative (facultative) rods:
Lactobacillus casei subsp. *casei*

Heterofermentative (obligate) rods:
Lactobacillus kefir

I. APPLICATIONS IN DAIRY FOOD PRODUCTION

In modern food production, the industrial-scale production of fermented dairy foods represents the most refined and best-controlled uses of lactic acid bacteria. LAB species used in the production of fermented dairy foods, such as cheeses and yogurt, are commonly referred to as starter cultures. Depending on the desired flavor and physical characteristics of the product manufactured, a starter culture may include selected strains of a single species or strains of different species from the same genus or different genera in specific combinations. Starter cultures may be supplemented by strains of other, non-LAB bacterial species such as *Propionibacterium or Brevibacterium*, in addition to fungi or yeast. The seemingly endless array of culture combinations, together with varying manufacturing conditions, allows the production of hundreds of varieties of fermented dairy foods. Species of LAB used primarily as components of dairy starter cultures are listed in Table II.

A. Lactococci

The homofermentative genus *Lactococcus* (formerly classified as Lancefield group N *Streptococcus*) includes five species, of which only one, *Lactococcus lactis*, is actually used in dairy fermentations. This mesophilic species is subdivided into two subspecies, designated as *L. lactis* subsp. *lactis* and *L. lactis* subsp. *cremoris*. Although both subspecies produce lactic acid from lactose and galactose, the capacity for maltose and ribose fermentation, growth in the presence of 4% NaCl and arginine hydrolysis characterize only *L. lactis* subsp. *lactis*. A variant of *L. lactis* subsp. *lactis* (biovar *diacetylactis*) can convert citrate to the aroma compound diacetyl, which imparts a desirable property to butter, buttermilk, and cottage cheese. In addition, carbon dioxide and traces of other compounds may be produced. Some strains of lactococci produce extracellular polysaccharides which are used in the commercial production of certain Scandinavian cultured dairy foods with a ropy texture.

B. Leuconostocs

The genus *Leuconostoc* includes two species, the mesophilic and heterofermentative *Leuc. mesenteroides* subsp. *cremoris* and *Leuc. lactis,* which are important in fermented dairy foods. Since they lack an adequate protease system, leuconostocs grow poorly in a milk environment. Nevertheless, they are frequently included in dairy starter cultures, for the purpose of transforming significant amounts of citrate to diacetyl, and to generate CO_2.

C. Streptococci

The reclassification groupings of LAB have left only one species of the genus *Streptococcus* with beneficial activities in association with dairy products, *Streptococcus thermophilus,* which is used as an essential component in starter culture mixes in the production of yogurt as well as Swiss- and Italian-style cheeses. Unlike mesophilic LAB, *S. thermophilus* tolerates temperatures up to 52°C but the scope of carbohydrate fermentation in most strains is limited to lactose, glucose, sucrose, and fructose. In dairy fermentations, as a rule, *S. thermophilus* is used in combination with other species of the genus *Lactobacillus*. On account of closely related DNA–DNA homology, the designation of *S. thermophilus* as a subspecies of *S. salivarius* was proposed. However, results of further DNA–DNA homology studies, in addition to the existing phenotypic differences, have led to confirmation of *S. thermophilus* as a separate species.

D. Lactobacilli

The genus *Lactobacillus* is a large, heterogeneous collection of genetically and physiologically diverse and highly acid-tolerant LAB species. *Lactobacillus* species with importance in dairy foods may be arranged in three main groups based on the presence or absence of fructose-1,6-diphosphate aldolase and phosphoketolase, enzymes with pivotal role in homo- and heterofermentative carbohydrate metabolism. Of the strictly homofermentative lactobacilli, *Lb. delbrueckii* subsp. *bulgaricus* and *Lb. helveticus* are used as components of starter cultures in the production of yogurt and Swiss- and Italian-style cheeses, while *Lb. delbrueckii* subsp. *lactis* is frequently a component of starter cultures used in hard cheese production. Another obligately homofermentative species is *Lb. acidophilus*, which is not used as a component of starter cultures but is added as a supplement to specialty dairy foods for its nutritional or "probiotic" benefits (e.g., "acidophilus milk"). A member of the facultatively heterofermentative group, *Lactobacillus casei* subsp. *casei*, is usually not a component of starter cultures but as a natural contaminant of cheese, it may play an important function in secondary fermentation during the product-ripening phase. Finally, *Lb. kefir*, an obligately heterofermentative LAB species characterized by the fermentation of carbohydrates to lactic acid, CO_2, and ethanol, is present only as a companion culture to yeast in kefir cultures, which are used for the cometabolism of lactose in the production of a fermented milk beverage. More recently, another obligately fermentative lactobacillus *Lb. reuteri*, which is a normal inhabitant of the intestinal tract of humans, pigs, and other animals, has been introduced, not as a starter, but as an adjunct culture in certain fermented dairy foods (yogurt) to impart a healthful ("probiotic") property to the product.

II. APPLICATIONS IN MILLED GRAIN-BASED FOOD PRODUCTION

The best characterized nondairy use of LAB is the sourdough fermentation process in breadmaking. The LAB participating in this process involves obligately homofermentative (*Lb. acidophilus, Lb. delbrueckii* subsp. *delbrueckii*), facultatively heterofermentative (*Lb. casei, Lb. plantarum, Lb. rhamnosus*), as well as obligately heterofermentative (*Lb. brevis, Lb. sanfrancisco, Lb. fermentum*) species, in combination with the yeast *Candida milleri*. The metabolic functions of LAB in the sourdough procedure contribute to the improved baking properties of rye bread, through partial hydrolysis of polysaccharides and proteins, and inhibit spoilage microbes by creating a low pH environment through the production of lactic and acetic acids, which also serve as flavoring agents in the finished product.

In addition to sourdough breads, LAB also play an important role in the fermentation of a large variety of staple food items, particularly in African countries. LAB identified in such fermented foods include lactobacilli, pediococci, leuconostocs, and lactococci.

III. METABOLISM OF LACTIC ACID BACTERIA

As mentioned earlier, LAB prefer an environment rich in simple carbohydrates, although a few species are known to degrade polysaccharides. Carbohydrate transport across the cytoplasmic membrane is driven by an ATP-dependent permease system in many lactobacilli, leuconostocs, and *S. thermophilus*. In all mesophilic lactococci and *Lb. casei*, transmembrane transport of sugars requires phosphorylation and the process is dependent on the phosphoenol pyruvate–phosphotransferase (PEP/PTS) system. For lack of an adequate level of galactokinase activity, certain thermotolerant lactobacilli (*Lb. delbrueckii* subsp. *bulgaricus, Lb. delbrueckii* subsp. *lactis, Lb. acido-*

philus) and *S. thermophilus* metabolize only the glucose moiety of lactose, while in lactococci, the galactose moiety of lactose is metabolized by the tagatose-6-phosphate pathway. Some LAB can also transport galactose by a permease system and process this hexose through the Leloir pathway.

The nutritionally fastidious LAB satisfy their requirements for peptides and amino acids as nitrogen sources through the activities of protease and peptidase enzymes which are accompanied by di- and tripeptide and amino-acid transport systems. In the best-studied lactococcal model, peptides generated by the membrane-bound proteinase system are translocated into the cell interior by peptide transport systems, where they are further hydrolyzed to amino acids by substrate-specific peptidases.

IV. PRODUCTION OF ANTIMICROBIAL SUBSTANCES

A. Primary Metabolites

The capacity of fermented dairy and other foods to resist invasion by spoilage bacteria has been known for centuries, and the antagonistic activity of lactic acid, the primary end-product of homofermentative carbohydrate metabolism in LAB, against a variety of food-borne pathogens has been studied in detail for decades. Similar preservative and antagonistic effects result from the catabolism of hexoses by heterofermentative LAB, during which equimolar amounts of lactic acid, acetic acid (or ethanol), and CO_2 are produced. This beneficial food preservative effect is caused by the low pH environment created during the course of carbohydrate fermentation. Certain LAB also produce, although in much smaller quantities, other antimicrobial substances that include hydrogen peroxide and diacetyl (with dual function as an aroma compound).

B. Bacteriocins

Bacteriocins are polypeptides produced by bacteria and have a narrow-range antimicrobial activity against closely related microbes. The best-characterized bacteriocin from LAB is the lactococcal nisin, which has found commercial application as a bioprotective agent, primarily against invasion of fermented dairy foods by deleterious *Clostridia,* and also as a preservative in canned foods. Bacteriocins have been the subject of intense studies in the past few years and new compounds have been discovered in practically all major groups of LAB, including lactobacilli, lactococci, leuconostocs, and *S. thermophilus.* The producing culture is usually endowed with resistance genes against the action of the bacteriocin and, at least in some cases, both production and immunity genes have been found on plasmids. The increase in research activity has been driven by the potential for industrial applications of bacteriocins, particularly those with broad-spectrum antimicrobial activity (e.g., pediocins of *Pediococcus acidilactici*), as bioprotective agents in various manufactured food products for controlling specific pathogens, such as *Listeria.*

V. PLASMID BIOLOGY OF LAB

The existence of extrachromosomal elements (plasmids) in LAB was first observed in the 1970s. As a rule, lactococci, lactobacilli, and leuconostocs harbor several plasmids, whereas most *S. thermophilus* strains are either plasmid-free or may harbor a single plasmid. The functionality encoded on many plasmids still remains unknown ("cryptic"). However, a copious number of metabolic functions in LAB have been linked to plasmids. These include key enzymes in primary metabolic activities, such as carbohydrate fermentation (lactose, sucrose, galactose, mannose, and xylose) and proteinase activity. The plasmids encoding these functions are usually large and vary in size between 17 kbp and ca. 50 kbp. On the other hand, the citrate permease gene, which is essential in citrate utilization in lactococci, is linked to a smaller, ca. 9 kbp plasmid. Several known DNA restriction and modification systems that protect LAB against bacteriophage attack are also encoded on plasmids. In addition, equally important phenotypic traits resulting from what may be viewed as secondary metabolism in LAB and which contribute to optimum culture performance, such as extracellular polysaccharide production (for improved

product consistency), and bacteriocin production (for controlling competing LAB strains or deleterious microbes), are frequently encoded on plasmids. Although continued interest in research on plasmid-encoded metabolic activities is justified, it must be kept in mind that with the exception of the above-mentioned enzyme activities, the bulk of essential carbon and nitrogen catabolic and biosynthetic pathways are chromosomally encoded in LAB.

VI. GENETIC DEVELOPMENT OF LAB

In the past, genetic engineering research on LAB centered around two main areas, one involving primary metabolic functions of LAB in fermented foods, such as hydrolytic enzymes, transmembrane transport systems, enzymes involved in flavor production, antimicrobials (bacteriocins) and biopolymers, while the second focused on the mechanism of DNA restriction/modification systems that defend LAB against bacteriophage attack. In more recent years, research on LAB has also expanded in novel areas concerned with health-related or probiotic activities, such as antitumor, antihypertensive, and immunostimulant effects, many of which still await confirmation by clinical data. There is also a growing interest in the development of LAB for nonfood applications, such as production systems for industrial and research enzymes (e.g., β-galactosidase, restriction endonucleases), bioactive peptides, and as carriers of antigens suitable for vaccination via mucosal immunization. Undoubtedly, the increased interest is attributable to the innocuous or "food-grade" status of LAB which, as edible components of cured cheeses and fermented dairy foods, may be consumed without any toxic effects in humans.

Genetic engineering of LAB for the purpose of developing strains with improved or novel metabolic capabilities requires efficient gene-transfer mechanisms, stable cloning vectors for carrying genes of interest, efficient promoter elements to facilitate gene expression in the new host, chromosomal integration factors to stabilize genes; secretion elements to promote the transport of gene products into the environment are desired.

A. Gene Transfer Mechanisms in LAB

The capacity for transformation via the uptake of DNA ("competence") is lacking in LAB. However, other gene-transfer systems, including conjugation, which requires cell-to-cell contact between donor and recipient cells, and phage-mediated transduction, have been described in various strains of LAB.

1. Conjugation

Conjugational transfer of DNA requires cell-to-cell contact, facilitated by the co-deposition of donor and recipient cells on a suitable membrane support. After several hours of contact, recipient cells with the new phenotype may be scored in a selective medium. The transfer of the lactose fermentation plasmid was achieved between strains of the same species (*Lactococcus*). Conjugation may also involve species of the same genus or different genera. Intrageneric and intergeneric conjugative transfer of the broad-host-range erythromycin resistance gene pAMβ1 (26.5 kbp) were achieved with lactococci, lactobacilli, and pediococci playing the donor/recipient role. Intergeneric *Lactococcus* to *Leuconostoc* transfer of the lactose fermentation plasmid has also been achieved.

2. Transduction

In transduction, temperate host-specific lysogenic bacteriophages mediate the transfer of genetic material. Using temperate phages, both chromosomal and plasmid-encoded genes may be transferred and, in lactococci, the transfer of proteinase and lactose metabolism genes was achieved. Transductional transfer of plasmids was also achieved in lactobacilli and *S. thermophilus*.

3. Protoplast Fusion and Transformation

The cell walls of many microbes may be removed by digestive enzymes, leading to the formation of protoplasts. In an osmotically protective environment (0.4 M sucrose, raffinose), protoplasts can fuse in the presence of a fusogen (polyethylene glycol), resulting in the recombination of both plasmid- and chromosome-encoded genes. After fusion, protoplasts may revert to normal cellular morphology in an osmoprotective medium. Protoplast fusion and transformation were achieved with both lactococci and lactobacilli. The conditions of protoplast fusion (ionic strength of protective buffers, polyethylene glycol molecular weight and concentration, length of treatment, stage of bacterial growth) have critical importance and must be worked out in each case.

4. Electrotransformation

Electrotransformation (electroporation) involves the momentary exposure of cells to a high electric field. The electric pulse (voltage setting varies with the intended host) induces polarization of cell membranes, resulting in a temporary breakdown of cell membrane integrity and increased cell permeability. During this transient and reversible phase, the temporary efflux, as well as influx, of macromolecules may occur, including DNA. In cells surviving the electric pulsing, normal cytoplasmic membrane architecture is reestablished and genetic transformants may be recovered in selective media. Electrotransformation as a rule requires larger amounts (microgram range) of DNA than conjugation or transduction, which may be accomplished with nanogram quantities of DNA. Although the electrotransformation technique is not universally applicable to all LAB strains, it has contributed significantly to progress in the genetic engineering of selected cultures. The electrotransformation-induced uptake of cloning vectors carrying chromosomal and plasmid-encoded genes has been achieved in all major groups of LAB, including lactococci, leuconostocs, pediococci, lactobacilli, and *S. thermophilus*.

B. Genetic Cloning Vectors

A large variety of cloning vectors have been developed for use in the genetic engineering of LAB. While some vectors have a narrow host range, several have been used successfully to deliver heterologous genes into lactococci, lactobacilli as well as *S. thermophilus*. In many cases, cloning vectors are constructed from one of the several structurally well-characterized cryptic plasmids of LAB.

Cloning vectors are frequently constructed to be bifunctional, i.e., to have the capacity to replicate in LAB and *Escherichia coli* (shuttle vectors). This is done for convenience, since the latter is readily transformable with nanogram amounts of DNA (e.g., ligation mixtures), and, frequently, serves as an intermediate developmental host for recombinant DNA structures before introduction into LAB. The stabilization and maintenance of cloning vectors in high copy numbers in new host systems frequently presents a challenge as deletion events and/or rearrangements resulting in the loss of vector functionality may occur.

Tracking genetically transformed LAB is usually based on antibiotic resistance genes built into cloning vectors. Since food application of such marker genes is undesirable, research has been intense on developing cloning vectors using specific metabolic ("food-grade") genes as traceable markers.

C. Promoters for Gene Expression in LAB

The expression of heterologous genes in LAB requires efficient promoter sequences. Since LAB strains frequently cannot recognize promoters of genes from foreign sources, replacement with a promoter sequence native to the host culture is often necessary. Promoter elements from various LAB have been isolated and used for driving the expression of a variety of heterologous genes in lactococcal and other LAB hosts.

D. Integration and Secretion Elements

Since several important enzyme systems (protease, β-galactosidase) in many LAB are encoded on unstable plasmids, integration of these, as well as heterologous genes, into the chromosome would be desirable. A variety of insertion elements and integration vectors have been used with success to integrate genes into the chromosome of lactococci and other LAB hosts.

There is ongoing research on the characterization and uses of secretion signal sequences. These DNA elements would be particularly useful in facilitating the amount of exported heterologous gene products by LAB strains, which would be essential to fulfill the promise of LAB as industrial production cultures in nonfood applications.

VII. LAB IN HUMAN AND ANIMAL HEALTH

Selected LAB strains and dairy foods fermented by or supplemented with them have, for a long time, enjoyed the reputation of having a positive effect on human and animal health. Although such "health-promoting" or "probiotic" effects were based on anecdotal observations, recent studies have substantiated some health-related claims.

Specific strains of LAB cultures (*Lb. acidophilus, Lb. casei, Lb. rhamnosus*) have been used in the alleviation of intestinal disorders, such as lactose maldigestion, enteric infections, infantile diarrhea, diarrhea induced by antibiotic therapy, and gastritis in humans. In addition, specific strains of *Lb. acidophilus* were shown to have a hypocholesterolemic effect. To be useful in these applications, the specific LAB strain must be highly resistant to acid and bile, and able to colonize, at least temporarily, the intestinal tract. Also, selected LAB strains have been reported to display antimutagenic, antitumor, and immune response-stimulating activity. Similar effects were also demonstrated in farm animals, for example, the incorporation of live cultures of the probiotic *Lb. reuteri* in the diet of chickens improved the control of salmonellosis and resulted in increased body weight.

VIII. FUTURE USES OF LAB

As biocatalysts, LAB will certainly continue to play an essential role in dairy food production. However, with the advancement, regulatory approval, and public acceptance of genetic technologies, the industrial application of new types of LAB with novel metabolic traits will undoubtedly yield milk-based foods with improved nutritional, health-promoting, and storage qualities. In addition to nisin, newer bacteriocins are anticipated to reach industrial production to be used as bioprotective agents against harmful bacteria in dairy and other types of food systems. Research on the potential of LAB as vaccine delivery systems in mucosal immunization is expected to result in products suitable for use in human and animal health care. In other nonfood industrial applications, genetic engineering research on LAB is expected to yield strains designed as efficient production systems for a variety of gene products.

See Also the Following Articles

BACTERIOCINS • DAIRY PRODUCTS • PLASMIDS

Bibliography

Dunny, G. M., Cleary, P. P., and McKay, L. L. (eds.). (1991). "Genetics and Molecular Biology of Streptococci, Lactococci, and Enterococci." American Society for Microbiology, Washington.

Gilliland, S. E. (ed.). (1985). "Bacterial Starter Cultures for Foods." CRC Press, Boca Raton, FL.

Grunberg-Manago, M., Buckingham, R. H., Dolnchin, A., Hershey, J., Söll, D., and Kaziro, Y. (eds.). (1988). "Lactic Acid Bacteria. I. Biochemistry and Physiology/Phages and Phage Resistance." *Biochemie* **70**, 303–460. Elsevier Science Publishers, Amsterdam, Netherlands.

Grunberg-Manago, M., Buckingham, R. H., Dolnchin, A., Hershey, J., Söll, D., and Kaziro, Y. (eds.). (1988). "Lactic Acid Bacteria. II. Genetics and Genetic Exchange Systems." *Biochemie* **70**, 461–590. Elsevier Science Publishers, Amsterdam, Netherlands.

Venema, G., Poolman, B., and Wouters, J. T. M. (eds.). (1990). "Lactic Acid Bacteria: Genetics, Metabolism and Applications." *FEMS Microbiol. Rev.* **87**, 1–188. Elsevier Science Publishers, Netherlands."

de Vos, W. M., Huls int't Veld, J. H. J., and Poolman, B. (eds.). (1993). "Lactic Acid Bacteria: Genetics, Metabolism and Applications." *FEMS Microbiol. Rev.* **12**, 1–272. Elsevier Science Publishers, Amsterdam, Netherlands.

Marth, E. H., and Steele, J. L. (eds.). (1998). "Applied Dairy Microbiology." Marcel Dekker, Inc., New York.

Salminen, S., and von Wright, A. (eds.). (1998). "Lactic acid bacteria: microbiology and functional aspects." Marcel Dekker, Inc., New York.

Tannock, G. W. (ed.). (1999). "Probiotics—A Critical Review." Horizon Press, Inc., Norfolk, England.

Venema, G., Veld, J. H. J. Huis int't, and Hugenholtz, J. (eds.). (1996). "Lactic Acid Bacteria: Genetics, Metabolism and Applications." Kluwer Academic Publishers, Dordrecht, Netherlands.

Lactic Acid, Microbially Produced

John H. Litchfield

Battelle Memorial Institute, Columbus, OH

GLOSSARY

batch process A technique in which all nutrients for the microorganism are added to the bioreactor only once and the products are separated in their entirety when the fermentation process is completed.

bioreactor or fermenter The culture vessel or system used for growing microorganisms and producing microbial metabolic products, such as lactic acid.

continuous process A technique in which nutrients are fed to the bioreactor and products are removed continuously during the process.

fed-batch or semicontinuous process A technique in which nutrients are fed to the bioreactor continuously and the products are removed intermittently, to maintain a fixed volume.

heterofermentative Referring to microorganisms that produce lactic acid, together with other organic acids, alcohols, aldehydes, ketones, and carbon dioxide as products of carbohydrate metabolism.

homofermentatative Referring to microorganisms that produce lactic acid as the sole product of carbohydrate metabolism.

immobilized cell bioreactor A technique in which microbial cells are entrapped in polymers or covalently bonded to or adsorbed onto inert materials in the bioreactor, which is operated in a continuous flow mode.

lactic acid bacteria A diverse group of bacteria that produces lactic acid as the primary metabolic product from the fermentation of carbohydrates.

LACTIC ACID, 2-hydroxypropanoic acid (molecular weight 90.08) was discovered in 1780 as "acid of milk" by Carl Wilhelm Scheele. Subsequently, in the nineteenth century, research by Louis Pasteur, Joseph Lister, and Max Delbruck established the lactic acid bacteria as the microorganisms involved in lactic acid production by fermentation.

Commercial manufacture of lactic acid by lactic acid bacterial fermentation of sugars for industrial uses first took place at the Avery Lactate Company in Littleton, Massachusetts, in 1883. Since that time, the lactic acid bacteria have been employed in manufacturing lactic acid from carbohydrate raw materials and as bacterial starter cultures in manufacturing cultured dairy products, such as cheeses, yogurt, and fermented milk products. Also, certain molds produce lactic acid from carbohydrate raw materials, such as sugars and starches. This article covers the microorganisms (bacteria and molds), processes, and product recovery methods that have been investigated for lactic acid production, and includes a discussion of selected commercial applications of this product. (See Benninga (1990) and Litchfield (1996) for more more detailed information.)

I. MICROORGANISMS FOR PRODUCTION

A. Chemical Forms of Lactic Acid

Lactic acid bacteria produce either of two optically active isomers (enantiomers) of lactic acid, L(+) or s(+) dextrorotary (Chemical Abstracts Service Registry No. (CAS) 79-35-4) or D(−) or R(−) levorotary

(CAS Registry No. 10326-41-7). As will be discussed subsequently, molds produce only the L(+) isomer. In some lactic acid bacteria, D- and L-lactic dehydrogenase enzyme activity forms racemic DL lactic acid (CAS Registry No. 598-82-3). The racemic form may also be produced by chemical synthesis from lactonitrile.

B. Bacteria

According to "Bergey's Manual of Determinative Bacteriology" (9th edition) (Holt *et al.,* 1994) the lactic acid bacteria are classified as follows:

1. gram-positive cocci: *Enterococcus* spp., *Lactococcus* spp., *Pediococcus* spp., *Saccharococcus* sp., *Streptococcus* spp.
2. endospore-forming gram-positive rods and cocci: *Bacillus* spp., *Sporolactobacillus* sp.
3. regular, nonsporing gram-positive rods: *Lactobacillus* spp.

From the standpoint of oxygen requirements, lactic acid bacteria are facultative anaerobes or microaerophilic and grow at low oxygen concentrations. Temperature and pH tolerance vary widely, 10–45°C and pH 3.5–9.6, respectively, depending upon the individual organism. They require complex nitrogen sources of amino acids and vitamins. From a metabolic standpoint, they are either homofermentative (produce lactic acid only) or heterofermentative (produce lactic acid and other metabolic products).

The homofermentative bacteria are of greatest interest for commercial lactic acid production, for obvious economic reasons. Sugars are metabolized by these bacteria to lactic acid through the Emden–Meyerhof pathway, with a theoretical conversion of 1 mol of glucose to 2 mols of lactic acid. Homofementative strains generally give >90% of the theoretical yield of lactic acid from glucose in both laboratory and commercial-scale fermentations. Some of the *Lactobacillus* spp. that are of interest for commercial-scale lactic acid production are facultatively heterofermentative, producing only lactic acid from glucose under anaerobic or microaerophilic conditions, but lactic and acetic acids, acetoin, and hydrogen peroxide under aerobic conditions.

Table I summarizes some of the key characteristics of bacteria that have been investigated for lactic acid production, including lactic acid isomer (enantiomer) produced, fermentation pattern, and typical raw materials (substrates). There has been recent interest in producing specific lactic acid isomers (L(+), D(−), DL) for synthesis of specialty polymers of lactic acid having desirable physical, chemical, and biodegradability characteristics. Consequently, the desired isomer is a criterion for selecting a suitable bacterial strain. Temperature, pH and lactic acid tolerances, substrate specificity, yields and lactic acid productivities (grams/liter/hour), and bacteriophage resistance are also important criteria for selecting suitable strains for commercial production.

Lactobacillus casei subsp. *rhamnosus* (formerly,

TABLE I

Characteristics of Selected Bacteria and Molds of Interest in Lactic Acid Production

Microorganism	Lactic acid isomer	Fermentation pattern	Raw material
Bacteria			
Lactobacillus amylophilus	L (−)	Homofermentative	starch
L. amylovorus	DL	Homofermentative	starch
L. casei subsp. *rhamnosus* (*L. delbrueckii,* NRRL B-445)	L (+)	Facultative hetero-fermentative	glucose, sucrose (molasses)
L. delbrueckii subsp. *bulgaricus*	D (−)	Homofermentative	cheese whey & permeate (lactose)
L. helveticus	DL	Homofermentative	cheese whey and permeate (lactose)
Molds			
Rhizopus arrhizus	L (+)	Heterofermentative	glucose, starch
R. oryzae	L (+)	Heterofermentative	glucose starch

Lactobacillus delbrueckii) has been widely used owing to its high yields of L(+)-lactic acid from glucose at 45°C. However, it cannot be used with cheese whey as a raw material because it is unable to utilize lactose, the sugar component of the whey. On the other hand, lactose is a suitable substrate for *L. delbrueckii* subsp. *bulgaricus* and *L. helveticus* that produce the D(−) and DL isomers, respectively, at 45°C. These strains can be used to produce the respective isomers from cheese whey. *L. helveticus* also tolerates pH values below 5.0.

Lactobacillus strains may also be susceptible to bacteriophage attack, which can result in cell lysis, disruption of the fermentation, and economic losses. Phages are specific for certain strains. For example, different phages are active against *L. delbrueckii* subsp. *bulgaricus* from those attacking *L. helveticus*. Consequently, phage resistance is an important criterion in selecting a production strain. In recent years, there has been considerable research on developing phage-resistant strains of lactic acid bacteria (Klaenhammer and Fitzgerald, 1994).

Recent advances in research on the molecular genetics of the lactic acid bacteria offer opportunities for future strain improvement. An increasing number of *Lactobacillus* genes have been cloned and sequenced. Where plasmids are present in organisms such as *L. casei* and *Lactococcus* spp., there is a potential for strain modification by genetic engineering/recombinant DNA techniques (Mercenier *et al.*, 1994).

C. Molds

Molds in the genus *Rhizopus* produce L(+)-lactic acid aerobically from glucose or sucrose. Also, *Rhizopus arrhizus* and *R. oryzae* convert starch or starch-based raw materials to L(+)-lactic acid since these species have amylolytic enzyme activity. Table I presents some of the important characteristics of *Rhizopus* spp. for lactic acid production. An advantage of *Rhizopus* spp. is that they are able to grow and produce lactic acid in synthetic media containing inorganic nitrogen sources, such as ammonium salts or nitrates and mineral salts, that simplify product recovery and purification with reduced production costs as compared with the complex nitrogen source production media required by *Lactobacillus* spp.

II. MICROBIOLOGICAL PROCESSES

A. Raw Materials and Process Systems

1. Raw Materials

Table I shows some of the typical carbohydrate raw materials that can be used to produce lactic acid by lactic acid bacteria and molds. These raw materials range from sugars, such as glucose, sucrose, or lactose, and waste materials containing them, to those containing complex carbohydrates such as starch or cellulose. Some *Lactobacillus* spp. (*L. amylophilus* and *L. amylovorus*) and *Rhizopus* spp. (*R. arrhizus* and *R. oryzae*) have amylolytic enzyme activity and are able to convert starch to glucose. Other organisms will require pretreatment of the starch with amylases to provide glucose for the lactic acid fermentation. Raw materials containing cellulose must be pretreated by chemical, physical, and/or enzyme methods to yield fermentable sugars. Acid hydrolysis of lignocellulosic materials may produce furfural from pentoses and hydroxymethylfurfural from hexoses that are inhibitory to many microorganisms.

In the United States, glucose produced by acid and enzyme conversion of corn starch by the corn wet milling industry has been the preferred raw material for commercial lactic acid production by microbial fermentation on the bases of cost, availability, and ease of product recovery and purification. In other regions, such as Europe and Brazil, sucrose from sugar beets or sugar cane has been used on cost and availability bases. Owing to the nutritional requirements of the lactic acid bacteria, glucose or sucrose must be supplemented with sources of amino acids and vitamins, such as corn steep liquor from the corn wet milling process, protein hydrolysates, and/or yeast extracts.

Since the 1930s, numerous processes have been investigated for microbial conversion of lactose in cheese whey to lactic acid. Annual production of cheese whey in the U.S. in the mid-1990s from dairy manufacturing operations is estimated as approximately 57×10^9 lb (25.9×10^9 kg) of liquid whey containing 4–4.5% lactose. Only approximately half of this whey is converted to 1.88×10^9 lb (8.5×10^8 kg) (dry basis) of whey products (dry and concentrated whey, whey protein concentrate, and lactose), with the remainder as waste. Also, cheese whey permeate from ultrafiltration processes that contains lactose is a potential raw material for lactic acid production.

Also, cheese whey and cheese whey permeate must be supplemented with sources of amino acids and vitamins such as corn steep liquor, protein hydrolysates, and/or yeast extracts, depending upon the requirements of the bacterial strain in concentrations ranging from 0.5–3%. A major problem in utilizing cheese whey and permeate for lactic acid production is the lack of microbiological stability of these raw materials. The costs associated with refrigerated transport and storage make it uneconomical to collect these dilute raw materials (7% solids) from widely dispersed cheese manufacturing plants. Also, owing its low solubility in water, lactose may crystallize out of whey under refrigeration. For commercial-scale production of lactic acid from cheese whey in the United States, the lactic acid fermentation plant should be located adjacent to a large cheese manufacturing plant that would produce an economically viable supply of cheese whey for production throughout the year, without incurring the costs and problems associated with collection, refrigerated storage and transportation of this raw material.

As mentioned previously, amylolytic *Lactobacillus* spp. and *Rhizopus* spp. can utilize starch as a raw material for lactic acid production. In the United States, the corn wet milling industry produces large quantities of starch which could serve as a raw material for lactic acid production using these microorganisms. Although microbial conversion of starch to lactic acid has been investigated on a laboratory scale, there is no commercial-scale process based on this raw material at the present time.

2. *Process Systems*

Fermenters for lactic acid production are preferably constructed of 316 low-carbon stainless steel, owing to the corrosive action of this acid. Also, stainless steel vessels are easier to clean and sanitize than are the wooden vessels used in the nineteenth and early twentieth centuries. Typically, fermenters used in processes based on lactic acid bacteria, that grow above 45°C, such as *L. casei* subsp. *rhamnosus,* are cleaned and steamed or treated with boiling water

and/or disinfectants to provide clean, aseptic, but nonsterile conditions and minimize contamination problems. However, the anaerobic butyric acid bacterium, *Clostridium butyricum,* has been the cause of contamination in nonsterile *Lactobacillus* fermentations operated under anaerobic rather than microaerophilic (low oxygen concentration) conditions. Processes based on *Rhizopus* spp., where the operating temperature and pH are 35°C and 5–6, respectively, require sterilization of both the production medium and fermenter to prevent contamination by bacteria, yeasts, and molds. In commercial-scale lactic acid bacterial processes, the cells are kept in suspension in the production medium by mixing with mechanical agitators or circulating the medium from the bottom to the top of the fermenter by pumping. It was pointed out previously that *Lactobacillus* spp. require low oxygen concentrations for growth and lactic acid production. Generally, agitation during the fermentation provides sufficient oxygen. However, *Rhizopus*-based processes are aerobic and oxygen must be provided by combining aeration with agitation in the fermenter.

B. Inoculum Development

Inoculum development is an important step in lactic acid production. Both the nutrient content of the inoculum medium and the size of the inoculum (number of colony-forming units (cfu) per ml or spores per ml) are important determinants of lactic acid concentrations and yields obtained. Published data based on laboratory-scale studies with *Lactobacillus* spp. indicate that supplementing the inoculum medium with yeast extract as a source of vitamins gives improved rates of lactic acid production. Also, the results of laboratory and small pilot plant studies indicate that initial cell counts of *Lactobacillus* spp. are usually in the range of 10^5–10^7 cfu/ml. In the case of *Rhizopus* processes, published values of inoculum spore counts are in the order of 10^8 per liter (L). There is a lack of published information on the inoculum conditions used in present commercial-scale lactic acid production processes. Several stages of inoculum growth can be employed starting with a small fermenter (1000–1500 L of medium), which is transferred after growth to a second stage (10,000–

20,000 L) and then used to inoculate 100,000 L of medium or greater volume in a production-scale fermenter.

C. Batch Processes

Generally, commercial-scale lactic acid fermentation processes have been operated in a batch free-cell mode, in which cells of lactic acid bacteria are maintained in suspension by agitation or mycelia of *Rhizopus* spp. in suspension by combined aeration and agitation. Table II summarizes some typical batch fermentations. With the widely used strain *of L. casei* subsp. *rhamnosus* (formerly, *L. delbrueckii* NRRL B445), typical glucose concentrations range from 15–20% (weight/volume basis), with final concentrations of lactic acid of 12–14%. Yields of lactic acid range from 0.70 to >0.90 g of lactic acid/g of glucose utilized. Factors limiting the concentration of lactic acid achievable in lactic acid bacteria-based processes are product inhibition, with increasing concentrations of undissociated lactic acid, and substrate inhibition of both growth and lactic acid formation, with increasing glucose concentrations. Typical lactic acid concentrations and yields with the mold *R. oryzae* from a 13% glucose concentration in the production medium are 9.3% and 0.74 g lactic acid/g glucose utilized, respectively. It is important to note that *Rhizopus* spp. are heterofermentative and conditions must be controlled to prevent excessive production of other organic acids, such as fumaric acid. Interest in cheese whey and starch as potential raw materials for fermentation production of lactic acid was mentioned previously.

Typical lactic acid yields from lactose in cheese whey in batch fermentations with *L. delbrueckii* subsp. *bulgaricus* are similar to those obtained in glucose-based fermentations (>0.90 g lactic acid/g lactose utilized). With starch, amylolytic *Lactobacillus* spp. and *Rhizopus oryzae* will also give high yields of lactic acid, >0.90 g/g and >0.72 g/g glucose equivalent utilized, respectively. Typical productivities of batch lactic acid fermentations are in the range of 3–5 g/L/hr.

D. Continuous Processes

Semicontinuous (fed batch) and continuous fermentation processes for lactic acid production have potential economic advantages over batch processes in terms of higher productivities (g lactic acid/L/hr). Free-cell continuous stirred-tank bioreactors (CSTR)

TABLE II
Lactic Acid Production by Batch Fermentation

Microorganism	Raw material, concentration (g/L)	Lactic acid productivity (P) (g/L/hr), concentration (C) (g/L)	Yield g/g substrate utilized
Bacteria			
L. amylophilus	corn, 45	P: 0.55, C: 31.2	0.68
	glucose, 20	P: 1.56, C: 21	0.93
L. amylovorus	starch (enzyme thinned), 120	P: 7.36 C: 96	0.94
L. casei subsp. *rhamnosus*	glucose	C: 120–135	0.80–0.90
L. delbrueckii subsp. *bulgaricus*	cheese whey & permeate, 50–150	P: 4.4–5 C: 60–115	0.91–0.99
L. helveticus	cheese whey permeate, 39.2–54	P: 3.7–3.8 C: 35–40	0.93
Molds			
R. arrhizus	glucose, 100	C: 79	>0.50
R. oryzae	glucose, 130	C: 93	0.72
	corn, 150	C: 53.2	0.44

can be used for both semicontinuous and continuous lactic acid processes. This type of bioreactor can also be combined with a flat, hollow fiber or crossflow membrane unit in a semiclosed loop for separating the product and cells and recyling the cells back to the bioreactor. Other continuous systems include immobilized cell bioreactors, in which the cells are trapped in or attached to an immobilization medium in a packed bed and dialysis membrane system, using a membrane permeable to lactic acid but impermeable to the cells, particularly electrodialysis units using an anion exchange membrane with an applied DC electric current.

Table III summarizes typical continuous systems for fermentation production of lactic acid. With free-cell CSTR systems, lactic acid productivities of 8.9–13 g/L/hr, as compared with batch values of 3–5 g/L/hr, have been obtained from glucose with lactic acid bacteria. Yields of lactic acid range from 0.70 to >0.90 g/g glucose utilized, similar to those obtained in batch fermentations. Lactic acid produc-

tivity is limited by lactic acid accumulation, allowing a maximum cell concentration of *Lactobacilli* to about 10^{10} cfu/ml, even when pH is controlled by continuous neutralization. With cheese whey and permeate, lactic acid productivities of 3–5 g/L/hr in CSTR systems with *L. delbrueckii* subsp. *bulgaricus* and *L. helveticus* are lower than those obtained from glucose with *L. casei* subsp. *rhamnosus* (*L. delbrueckii* NRRL B-445).

Coupling CSTR systems with membrane system to give a continuous membrane cell recycle bioreactor leads to substantial increases in lactic acid productivities. For example, a productivity of 76 g/L/hr from glucose by *L. casei* subsp. *rhamnosus* (*L. delbrueckii* NRRL 445) has been reported using a 100,000 molecular weight cutoff (MWCO) ultrafiltration membrane system at a cell concentration of 54 g (dry weight)/L. With cheese whey permeate, productivities as high as 84 g/L/hr have been reported with *L. delbrueckii* subsp. *bulgaricus* (cell concentration, 63 g/L) using a hollow fiber membrane system. A

TABLE III
Lactic Acid Production by Continuous Fermentations

Microorganism	Type of bioreactor, raw material	Lactic acid productivity (P) (g/L/hr), concentration (C), (g/L)	Yield g/g substrate utilized
L. amylovorus	Membrane cell recycle, starch	P: 8.4 C: 42	0.88–0.92
L. casei subsp. *rhamnosus*	CSTR, glucose	P: 8.93 C: 22–25	0.74
	Membrane cell recycle, glucose	P: 10–160 C: 34–59	0.95–0.99
	Immobilized cell, glucose	P: 3–20 C: 5–51	0.76–0.97
L. delbrueckii subsp. *bulgaricus*	CSTR, cheese whey	P: 1.77 C: 55	0.98
	Membrane cell recycle, cheese whey permeate	P: 84–85 C: 43–117	0.99
	Dialysis, cheese whey	P: 11.7 C: 80	0.97
L. helveticus	CSTR, cheese whey permeate	P: 3.7–5 C: 35–84.6	0.83–0.93
	Membrane cell recycle, cheese whey permeate	P: 15.8 C: 15	0.70
	Immobilized cell, cheese whey permeate	P: 2.6–13.5	0.50–0.95

problem with polymer membranes is that they cannot withstand the temperatures required for sterilization by heating. Also, membrane fouling requires cleaning procedures that ultimately weaken the membrane and leads to increased replacement costs. Although productivities are attractive, membrane cell recycle bioreactors have not been used for commercial-scale production, owing to the problems and costs associated with polymeric membranes. Ceramic membranes that can be heat-sterilized have been investigated on a laboratory scale in Japan and may offer an opportunity for future improvements to membrane cell recycle systems for lactic acid production.

Immobilized cell bioreactors using either lactic acid bacteria or molds do not require separating the cells from the fermentation medium. In studies with laboratory- and small pilot-scale units, cells of lactic acid bacteria have been entrapped in beads of natural polymers, such as agar, sodium or calcium alginates, or carrageenan. These polymeric beads are used in a packed column reactor configuration operated in a continuous flow mode. Natural polymer immobilization media are difficult to sterilize and keep free from contamination during operation. Also, these polymers may soften and leak cells from exposure to lactic acid. To avoid these problems, cells can be adsorbed as biofilms onto inert materials, such as sintered glass beads or Raschig rings, ceramic materials, and polypropylene, for examples. These materials are low cost, can be sterilized by heat, and develop biofilms readily on them without requiring a large cell population for initial immobilization.

As shown in Table III, reported productivities of immobilized cell systems are in the range of 5–20 g/l/hr and are lower than those reported for membrane cell recycle bioreactors. However, immobilized cell systems have been operated on a laboratory- or small-pilot plant scale for far longer periods than membrane recycle bioreactors, up to over 150 days with *Lactobacillus* spp., without contamination problems and with acceptable lactic acid yields in the range of 0.80–>0.90 g/glucose or lactose utilized. Again, there are no published reports of scale-up of immobilized cell systems for lactic acid production to a large pilot plant or commercial plant operation.

Dialysis fermentation systems for lactic acid pro-duction maintain low concentrations of lactic acid in the medium and, thereby, overcome its inhibitory effects on lactic acid bacteria. Also, higher purity lactic acid can be obtained with a dialysis membrane unit than in conventional batch fermentations requiring extensive purification steps. Electrodialysis has been investigated on a laboratory scale as a means for overcoming limitations in membrane diffusion rates in conventional dialysis systems and the large size of dialysis units that would be required on a commercial scale. In this case, lactate ion passes through the membrane and is removed in the anode compartment of the electrodialysis cell.

Electrodialysis units have been combined with CSTR-type fermenters and with immobilized cell bioreactors. With glucose as the substrate and *L. casei* subsp. *rhamnosus,* a typical reported lactic acid productivity value obtained in a combined electrodialysis immobilized cell bioreactor is 5 g/L/hr. An electrodialysis unit coupled to a CSTR-type fermenter has been reported to give a lactic acid productivity of 22 g/L/hr, a lactic acid concentration of 85 g/L, and a yield of 0.81 from cheese whey permeate using *L. helveticus*.

III. PROCESS CONTROL

Process controls in lactic acid production by fermentation include temperature, pH, substrate and lactic acid concentrations, and cell concentration. Temperatures during the fermentation can be controlled by cooling water coils immersed in fermenter vessels or by external cooling loops for continuous flow immobilized cell bioreactors. pH is a critical control factor in maximizing lactic acid concentrations in all types of bioreactors mentioned previously. For many years, pH during lactic acid production was monitored by measurements of periodic samples drawn manually and controlled by adding calcium carbonate to the production medium. Today, pH can be controlled continuously by computer-controlled systems, in which the lactic acid produced is neutralized at a designated value as it is formed during the fermentation.

Automated sampling and high performance liquid chromatography (HPLC) analytical systems have been developed for continuously measuring substrate and lactic acid concentrations. Also, on-line analytical systems such as flow injection analysis, near infrared (NIR) and Fourier transform infrared spectroscopy (FTIR), and fluorescence spectroscopy are available for determining glucose, lactic acid, and cell biomass concentrations. Furthermore, adaptive on-line optimizing control systems and fuzzy–expert knowledge-based systems have been employed to control lactic acid fermentations. There is a lack of published information on the applications of these methods to commercial-scale lactic acid fermentations.

IV. PRODUCT RECOVERY AND PURIFICATION

A major consideration in the commercial production of lactic acid is purifying the product to meet quality requirements for food, pharmaceutical, and medical materials applications. For these uses, lactic acid must be free of residual sugars, other organic acids, amino acids, complex organic nitrogen compounds, heavy metals, and Maillard reaction products between carbohydrates and aamino acids. It has been estimated that lactic acid recovery and purification costs may account for 50% of the final cost of manufacturing this product. Early commercial-scale lactic acid bacterial fermentation process employed neutralizing with calcium carbonate, heating to 82.2°C (180°F) to kill the lactic acid bacteria, filtering to separate calcium lactate, acidifying with sulfuric acid, filtering to remove calcium sulfate, concentrating the lactic acid solution by multiple vacuum evaporation, decolorizing with activated carbon, and precipitating heavy metals with sodium sulfide.

More recent methods for lactic acid recovery include centrifugation or membrane filtration, using either microfiltration (0.2 micrometer pore size) or ultrafiltration to separate the cells from the production medium. Heavy metals are removed by cation exchange resins. Lactic acid can be extracted from the fermentation medium using a strong anion exchanger, such as Amberlite IRA-420 in the carbonate form, and then converted to ammonium lactate by passing ammonium carbonate through the resin. The ammonium lactate is then converted to lactic acid by passing it through a strong cation exchange resin, such as Amberlite IR-120 in the hydrogen form. Also, reverse osmosis membrane systems or electrodialysis can be used to concentrate lactic acid as an alternative to multiple effect evaporation. Lactic acid can be purified by distillation of methyl, isobutyl, or 2-methyl-2-isobutyl lactate esters, followed by hydrolysis of the esters. Continuous countercurrent solvent extraction of lactic acid with isopropyl ether has been practiced on a commercial scale. Other extraction procedures that have been investigated include using tertiary amines, such as Adogen 364 in 60–70% isobutylheptylketone, quaternary ammonium salts such as Aliquat 336 at pH 5–6, or trioctyl phosphine oxide (TOPO) in kerosene. Extractive fermentation systems include a water immiscible phase in the fermenter to continuously remove lactic acid and reduce inhibition of the fermentation. Tertiary amines, such as Alamine 336, and solid phase polymers having teriary amine groups have been evaluated for extractive fermentation recovery of lactic acid. As far as it can be determined, none of these proceses has been practiced on a commercial scale.

V. COMMERCIAL APPLICATIONS

In the United States, the major use of lactic acid is in foods as an acidulant and preservative. The Food Chemicals Codex, 4th ed. (1996), sets the specifications for food-grade lactic acid. The 88% purity U.S.P. grade is used in food, pharmaceutical, and some industrial applications. Lactic acid and its salts are Generally Recognized As Safe (GRAS) food additives by the Food and Drug Administration (FDA) in the United States Code of Code of Federal Regulations (1998). Also, the FDA has cleared as food additives products of reactions of lactic acid with fatty acids, including lactylated fatty acid esters, calcium stearoyl-2-lactylate, and sodium stearoyl lactylate as dough improvers, emulsifiers, and plasticizers in foods.

Sodium lactate is used in parenteral and kidney dialysis solutions, while calcium and magnesium lactates can be used in treating mineral deficiencies. In addition, optically pure methyl, ethyl, and isopropyl lactate esters are used for synthesis of chiral molecules as pharmaceutical intermediates. In skin-care products, lactic acid is incorporated in formulations as an "alpha-hydroxy acid" for improving skin texture and appearance. Also, lactic acid and its sodium and calcium salts are used as humectants and lactic acid esters with fatty acids as emulsifiers, builders, and stabilizers in toiletries and personal care products. Owing to their lower toxicities, ethyl and butyl lactates have industrial applications as alternatives to glycol ethers. In particular, ethyl lactate can be used as a replacement for chlorinated hydrocarbon solvents for precision metal cleaning in the aerospace, electronics, and semiconductor industries. Other industrial applications for lactic acid and its compounds include tanning, textile finishing, and as an alkyd resin modifier and terminating agent for phenol–formaldehyde resins.

Polylactides, prepared from lactides, the dimers of lactic acid, and lactide–glycolide (glycolic acid dimer) copolymers are now in use as biodegradable and biocompatible polymers for surgical sutures and other medical materials. Also, biodegradable lactic acid polymers have been developed for applications such as disposable packaging and food service ware, diapers, personal hygiene products, and yard waste bags.

The global market for lactic acid in the late 1990s is estimated at 110 million pounds (49.9 million kilograms) per year. Estimates of the growth rate of lactic acid use are in the range of 3 to 5% per year.

See Also the Following Articles

Bioreactors • Dairy Products • Food Spoilage and Preservation • Industrial Fermentation Processes

Bibliography

Benninga, H. (1990). "A History of Lactic Acid Making." Kluwer Academic, Dordrecht, The Netherlands.

Code of Federal Regulations. (1998). Title 21, 172.844, 172.846, 172.848, 172.852, 184.1061, 184.1207, 184.1311, 184.1639, 184.1768. U.S. Government Printing Office, Washington, DC.

"Food Chemicals Codex" (4th ed.) (1996). National Academy Press, Washington, DC.

Holt, J. G., Krieg, N. R., Sneath, P. H. A., Staley, J. T., and Williams, S. T. (1994). "Bergey's Manual of Determinative Bacteriology" (9th ed.), pp. 528–528, 540, 566, 568. Williams & Wilkins, Baltimore, MD.

Klaenhammer, T. R., and Fitzgerald, G. F. (1994). Bacteriophage and bacteriophage resistance. *In* "Genetics and Biotechnology of Lactic Acid Bacteria" (M. J. Gasson and W. M. de Vos, eds.), pp. 106–168. Blackie Academic & Professional/Chapman & Hall, London, UK.

Litchfield, J. H. (1996). Microbiological production of lactic acid. *In* "Advances in Applied Microbiology," Vol. 42 (S. L. Neidleman and A. I. Laskin, eds.), pp. 45–95. Academic Press, San Diego, CA.

Mercenier, A., Pouwels, P. H., and Chassy, B. M. (1994). Genetic engineering of lactobacilli, leuconostocs and *Streptococcus thermophilus. In* "Genetics and Biotechnology of Lactic Acid Bacteria" (M. J. Gasson and W. M. de Vos, eds.), pp. 252–293. Blackie Academic & Professional/Chapman & Hall, London, UK.

Legionella

N. Cary Engleberg

University of Michigan Medical School, Ann Arbor

GLOSSARY

aerosol A colloidal suspension of microscopic particles in air.

alveolar macrophages Mononuclear phagocytes that are residents of the pulmonary airspaces.

autophagy A process of self-digestion by cells that are starved of amino acids, in which portions of the cytoplasm are circumscribed by endoplasmic reticulum and then catabolized as an internal supply of nutrients.

buffered charcoal yeast extract agar An enriched medium containing supplemental L-cysteine, iron pyrophosphate, and charcoal as an absorbant, which is used for the primary isolation and propagation of *Legionella* sp.

coiling phagocytosis An asymmetrical mode of particle uptake by phagocytic cells, in which pseudopods appear to form coils around the object being ingested.

dot "Defect in organelle trafficking"; *L. pneumophila* genes that are essential for evasion of phagosome–lysosome fusion and establishment of a replicative endosome after ingestion of bacteria by phagocytes. Several are also designated as "*icm*" genes.

icm "Intracellular multiplication"; a series of *L. pneumophila* genes required for survival and productive infection of mononuclear phagocytes. Several are also designated as "*dot*" genes.

γ-interferon A cytokine produced by T lymphocytes that activates macrophages, as well as stimulating a variety of other immune functions.

legionaminic acid The common designation for 5-acet-amidino-7-acetamido-8-O-acetyl-3,5,7,9-tetradeoxy-D-glycero-L-galacto-nonulosonic acid; a homopolymer of this sugar constitutes the serogroup-specific O-antigen of *L. pneumophila* serogroup 1 lipopolysaccharide.

lipopolysaccharide A major component of the gram-negative bacterial outer membrane which is composed of a lipid A moiety, a core polysaccharide, and repeats of short antigenic sequence of sugars.

Mip "Macrophage infectivity potentiator"; a legionella protein with activity as a *cis-trans* peptidyl prolyl isomerase that is needed for full expression of bacterial virulence.

monocytes Mononuclear phagocytes that circulate in the bloodstream but may differentiate into macrophages.

protozoa Free-living, unicellular, eukaryotic organisms which possess motility.

tumor necrosis factor-α A cytokine produced by T lymphocytes and macrophages that activates phagocytic cells and induces fever.

zinc metalloprotease A proteolytic enzyme that requires a molecule of zinc at the active site.

THE LEGIONELLACEAE are fastidious gram-negative bacteria that are common inhabitants of aquatic environments worldwide. In the natural environment, the Legionellaceae are intracellular parasites of free-living protozoa. These organisms may also inhabit manmade water distribution systems. A severe form of bacterial pneumonia, called Legionnaires' disease, occurs in susceptible humans who inhale infected aerosols from these contaminated water sources. In the human lung, *Legionella* are also intracellular pathogens, and infection of pulmonary macrophages is an essential step in pathogenesis. *Legionella pneumophila* is the most studied species within the family because of its potential to cause this pulmonary disease.

I. HISTORY

The Legionellaceae were unrecognized until 1976, when a large outbreak of severe pneumonia occurred at an American Legion Convention in Philadelphia (Fraser *et al.*, 1977; Tsai *et al.*, 1979). A total of 221 individuals who were exposed at the Bellevue–Stratford Hotel became ill, and 34 died within the first few weeks after the convention. The etiologic agent, *L. pneumophila*, was isolated first by inoculation of postmortem lung tissue into guinea pigs and then by subculture into a rich artificial medium (McDade *et al.*, 1977). Using an indirect immunofluorescent antibody assay, the investigators showed that >90% of the ill legionnaires had a fourfold or greater rise in titer against this organism. Using the same assay, it was also possible to screen sera that had been saved from previous epidemics of unexplained respiratory disease. Several previous outbreaks were found to have been associated with seroconversion to this organism, including a small outbreak of pneumonia at the same Philadelphia hotel during a convention 1974 (Table I). In addition, a retrospective examination of a "rickettsia-like" organism, isolated by guinea pig inoculation from the blood of a febrile patient in 1947, revealed that the organism was actually *L. pneumophila*. This organism is the earliest know isolate of *L. pneumophila* on record.

Seroconversion to *L. pneumophila* was also found to be associated with a large, unexplained outbreak of a nonfatal, flulike illness that occurred at the Department of Health in Pontiac, Michigan, in 1968.

At the time of the outbreak, the investigation included the exposure of sentinel guinea pigs to air in the implicated building. Reexamination of frozen lung samples from these guinea pigs 10 years after the outbreak confirmed the presence of *L. pneumophila*. By the early 1980s, two epidemic forms of disease associated with *L. pneumophila* were recognized and named: Legionnaires' disease, the potentially fatal pneumonitis that caused the 1976 American Legion outbreak and others, and "Pontiac Fever," the self-limited, nonpneumonic illness that caused the health department outbreak. Subsequent outbreaks have conformed to the epidemiologic behavior observed in the earliest investigations (Table I).

II. STRUCTURE AND METABOLISM

L. pneumophila is a gram-negative, nonencapsulated, aerobic bacillus. The organism usually measures $\geq 2\mu m$ in length and $0.3-0.9\mu m$ in width but may form long, filamentous forms in nutrient-deficient media. There is a typical gram-negative cell wall and a single, polar flagellum. Pili have been occasionally identified. The cell envelope contains a variety of branched-chain fatty acids and novel ubiquinones, which appear to be distinct in different *Legionella* species. Structural differences in these classes of compounds have been used to distinguish *Legionella* ssp. by gas–liquid chromatography.

The outer membrane includes a lipopolysaccharide that has been fully sequenced and found to have

TABLE I

Features of *Legionella pneumophila* Outbreaks of Prior to and Including the 1976 American Legion Convention[a]

Location	Year	No. ill	Est. attack rate	Case–fatality rate
Legionnaires' disease				
St. Elizabeth's Hospital, D.C.	1965	81	1.4%	17%
Benidorm, Spain	1973	89	–	3.4%
Odd Fellow's Convention, Philadelphia	1974	11	2.9%	10%
American Legion Convention, Philadelphia	1976	221	4.0%	17%
Pontiac fever				
Health Department, Pontiac, MI	1968	144	95%	0
James River, VA	1973	10	100%	0

[a] Adapted from Eickhoff, 1979.

several novel features which have pathophysiologic consequence. The LPS is markedly less endotoxic than enterobacterial LPS because it interacts poorly with the CD14 receptor on monocytes. This interaction may be inhibited by the long-chain fatty acids of *L. pneumophila* lipid A, some of which are twice the length of enterobacterial LPS. The repeating O antigen of *L. pneumophila* serogroup 1 LPS is a homopolymer of an unusual sugar, designated legionaminic acid. LPS is the immunodominant antigen of the Legionellaceae, and the O antigen is the determinant of serogroup specificity within the genus. The outer membrane also includes a single, predominant outer membrane protein which functions as a porin and as a target for human complement fixation.

The major secreted protein of *L. pneumophila* is a 38 kDa zinc metalloprotease. In purified form, this molecule possesses cytotoxic activity and is destructive to lung tissue *in vivo*. However, a site-directed mutation in this gene that abrogates proteolytic activity does not affect the virulence of *L. pneumophila* in the guinea pig model of infection.

L. pneumophila is a nonfermenter that utilizes amino acids and keto acids as primary carbon sources. *Legionella* ssp. do not grow on standard laboratory media but are usually cultivated on buffered charcoal yeast extract agar. This media is buffered to pH 6.85 to 6.95 and requires supplemental L-cysteine and ferric iron. Charcoal is added to the medium to prevent the generation of toxic impurities from agar during autoclaving. Typical iridescent colonies appear on this media about three days after streaking for isolation. A few biochemical tests can be used in the clinical microbiology laboratory to identify the Legionellaceae. For example, *L. pneumophila* hydrolyzes hippurate and is weakly catalase-positive but is negative for oxidase and urease. (For reviews on the clinical microbiology of the Legionellaceae, see Edelstein (1993) and Benson and Fields (1998)).

III. THE LEGIONELLACEAE AND THEIR NATURAL HABITAT

The family Legionellaceae consists of a single genus, *Legionella*. In addition to *L. pneumophila*, there

are 41 other species in the genus. Together, these species include 64 distinctive serogroups. The species and serogroups can be identified with a variety of phenotypic and genotypic methods. Most human legionellosis is associated with *L. pneumophila*, but *Legionella micdadei, L. bozemanii, L. dumoffii, L. longbeachae*, and others have also been associated with human disease, usually in immunocompromised individuals. Within the species *L. pneumophila*, human infection is caused primarily (but not exclusively) by a limited number of serogroups, e.g., serogroups 1, 4, and 6. Analysis of a large collection of *L. pneumophila* isolates using multilocus enzyme electrophoresis demonstrates that the species has a clonal structure. Thus, several clones of *L. pneumophila* have been isolated worldwide. In addition, clinical and environmental isolates were found in each of the clonal groupings, suggesting that all genetic lineages have the potential to be virulent for humans.

Legionella were first identified as parasites of amoebae in 1980 by Rowbotham (1980). Since then, *Legionella* have been observed to enter and to multiply within various species free-living protozoa, both in the laboratory and in the natural aquatic environment (for reviews, see Abu Kwaik *et al.* (1998) and Barabee *et al.* (1993)). Amoebae have also been isolated from potable water systems implicated as sources of legionellosis and shown to support the intracellular growth *in vitro* of *L. pneumophila* isolated from the same source. Moreover, *Legionella* ssp. can multiply in tap water when amoebae are also present. The isolation of *Legionella* ssp. from environmental sources can be enhanced by preincubation of water samples at 37°C when amoebae are also present in the sample. This maneuver dramatically enhances the recovery of *Legionella* ssp. compared with standard culture techniques. Like other aquatic bacteria, *Legionella* ssp. can exist in a viable, but noncultivable, state when incubated in pond water or within amoebae. Recently, several noncultivable intracellular bacterial pathogens of amoebae have been identified but not classified. Thus far, 16S RNA sequencing of some of these organisms suggests that they are previously unrecognized *Legionella* ssp. The role of these noncultivable "Legionella-like" bacteria in human pneumonia is still uncertain.

Since amoebae are common inhabitants of potable

water system and are nearly always present in systems contaminated with *Legionella* ssp, it has been suggested that they may serve as a "Trojan horse" to allow the bacteria to survive water treatment and as a nidus for growth of the bacteria once they colonize a water system. Intra-amoebic growth within a water system may also both amplify the numbers of bacteria and activate their virulence. Passage of *L. pneumophila* in co-culture with *Acanthamoeba* ssp. or *Hartmannella verimiformis* results in enhanced invasiveness for mammalian cells. It is also possible that the amoebae may play a more direct role as well. *Acanthamoeba* ssp. have been observed to release membrane-bound vesicles containing *L. pneumophila* that are small enough to be aerosolized and inhaled. Moreover, the addition of the common water-borne amoeba, *Hartmannella vermiformis,* to an intratracheal inoculum of *L. pneumophila* dramatically increases the intrapulmonary bacteria growth and virulence in a susceptible animal model. The introduction of amoebae that are preinfected with *L. pneumophila* into animal lungs has the same effect. Whether amoebae or amoebic products constitute part of the infectious inoculum in human Legionnaires' disease is still an open question.

IV. PATHOGENESIS

After the inhalation of contaminated water aerosols or the microaspiration of potable water, *L. pneumophila* may enter the small airways and pulmonary alveoli. Unlike other pathogens that cause bacterial pneumonia, *L. pneumophila* do not colonize the upper airway. Therefore, they must be inhaled as aerosol particles small enough to bypass the defenses of the upper respiratory tract, or they may be aspirated during drinking by individuals with defective swallowing mechanisms. In either case, the organism is acquired exclusively from an environmental source; transmission of the infection does not occur between humans or between animals and humans.

Having deposited in the lung, the bacteria are taken up by pulmonary alveolar macrophages, although they also have the capacity to infect type II alveolar epithelial cells. Nearly all of the growth of the bacteria in the lung occurs intracellularly. Indeed, intracel-

lular infection appears to be an absolute requirement for the development of productive infection and disease, and mutants of *L. pneumophila* that are impaired in intracellular survival or growth in macrophages cannot cause disease. It is, therefore, not surprising that the antibiotic treatment of infection is successful only when agents that accumulate within host cells (e.g., macrolides, quinolones, tetracyclines) are employed. Antimicrobials that are excluded by the plasma membrane (e.g., penicillins, cephalosporins, aminoglycosides) are not effective in Legionnaires' disease, even though they may be lethal to *L. pneumophila in vitro.*

A. The Intracellular Life Cycle of *L. pneumophila*

The sequential events that characterize the life cycle of *L. pneumophila* in phagocytic cells are summarized in Fig. 1. The uptake of the organism may be mediated by the CR1 and CR3 receptors after fixation of serum complement component C3 but may also occur in the absence of serum factors (Fig. 1A). "Coiling phagocytosis" is a prominent feature associated with the uptake of most strains; however, not all virulent strains produce this novel effect and other microorganisms as diverse as *Leishmania* ssp. and *Borrelia burgdorferi* have been observed to enter cells in this asymmetrical manner. During the process of ingestion, the plasma membrane surrounding the bacterium undergoes remodeling; certain membrane

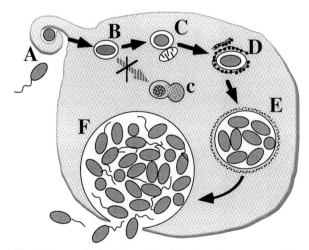

Fig. 1. Intracellular life cycle of *Legionella pneumophila.*

proteins, such as MHCI and II, are selectively excluded while others, such as CR3 and 5′-nucleotidase are retained (Fig. 1B). Two hours after uptake, the phagosome may be seen associating with mitochondria, smooth vesicles, or the nuclear membrane (Fig. 1C). Acidification of the phagosome does not occur, and the phagosomal membrane fails to acquire late endosomal markers, such as rab7 and LAMP-1. The consequence is a failure of phagolysosomal fusion that prevents exposure of the bacterium to intracellular killing. Mutants of *L. pneumophila* that cannot evade fusion are delivered to the lysosome within 30 minutes of uptake (Fig. 1C). In contrast, fully virulent organisms occupy an endosome that appears to be studded by ribosomes four hours after uptake (Fig. 1D). In fact, the endosome has become enveloped by rough endoplasmic reticulum, such as is seen in response to cellular amino acid starvation (i.e., autophagy). It is within this permissive endosome that bacterial multiplication proceeds (Fig. 1E) (Horwitz, 1992).

The intracellular bacteria undergo a major shift in protein expression, part of which is associated with a global stress response. At least one induced gene appears to be neither stress-related nor growth-phase related but plays a significant role in intracellular infection.

Later, as bacteria multiply to maximal numbers inside host cells, another phenotypic change occurs. The bacteria begin to express flagella, motility, cellular cytotoxicity, and sensitivity to physiologic concentrations of sodium chloride (Fig. 1F). Byrne and Swanson demonstrated that bacteria isolated from macrophages at this late point in the intracellular growth cycle are more likely to reinfect macrophages productively and to evade phagolysosomal fusion than bacteria isolated at earlier intracellular stages (Byrne and Swanson, 1998). In contrast, bacteria that are growing exponentially are aflagellate, nonmotile, noncytotoxic, resistant to sodium, and less capable of evading phagolysosomal fusion after uptake by macrophages. *In vitro* experiments show that this phenotypic switch can be simulated *in vitro* by growing bacteria to early stationary phase in broth. Moreover, it appears that the change occurs in response to amino acid depletion and is mediated by a rise in bacterial ppGpp concentrations, in a manner compa-

rable to the stringent response observed in *E. coli*. These findings are consistent with the observations of several investigators that spontaneous avirulent mutants of *L. pneumophila* are often persistently noncytotoxic or sodium-resistant, suggesting a link between these phenotypes and the capacity to infect macrophages successfully.

The cytotoxic potential of virulent *L. pneumophila* to kill cells *in vitro* at high multiplicity of infection is a contact-dependent phenomenon. This phenotype appears to depend on many of the same genes that mediate proper intracellular trafficking (see following) and causes necrotic death by osmotic lysis after insertion of 3 nm pores into the host cell membranes. The induction of this cytotoxicity at a point in intracellular growth when the hosts pool of amino acids have been spent may be a key mechanism to permit the bacteria to escape into the extracellular environment and to infect new host cells (Fig. 1F). Although cells may eventually undergo necrosis and release bacteria, recent evidence suggests that cells infected by smaller numbers of bacteria may undergo programmed cell death rather than necrosis. How the induction of apoptosis by the *L. pneumophila* might favor the parasitic cycle is not yet fully understood.

The life cycle of *L. pneumophila* within amoebae closely parallels the events described during macrophage infection (for a review, see Abu Kwaik *et al.*, 1998). Like macrophages, amoebae also ingest *L. pneumophila* by an asymmetric form of phagocytosis, although the uptake may be mediated by amoebic-specific receptors. Once ingested, the bacteria evade lysosomal fusion and localize to a membrane-vesicle surrounded by endoplasmic reticulum where bacterial growth occurs. Prior to release from the cell, *L. pneumophila* acquire motility as they do in human cells. These events are mediated by similar bacterial functions, since mutants that fail to grow in macrophages are inevitably defective in amoebic growth. The converse is not true, however, suggesting that additional bacterial functions may be required to infected amoebae successfully. These mechanistic similarities make a strong case that the factors that permit *L. pneumophila* to infect mammalian phagocytes and cause human disease evolved in protozoa. Since *L. pneumophila* are not transmitted between mammalian hosts or return to the natural environment by infected individuals, this hypothesis explains why this pathogen is adapted to intracellular life in the human lung without any apparent selection for pathogenic traits in this microenvironment.

B. Virulence Factors of *L. pneumophila*

The first virulence determinant of *L. pneumophila* was identified by making a site-directed mutation in the gene encoding Mip, a 24 kDa bacterial envelope protein. Mutation in this gene results in a 1.5 to 3 log reduction in infectivity for explanted alveolar macrophages, alveolar epithelial cells, and amoebae; it also results in a similar reduction in LD_{50} for guinea pigs. The sequence of the Mip gene revealed homology of the C-terminal half of the molecule with eukaryotic immunophilins of the peptidyl–prolyl isomerase (PPIase) type. Further study revealed that the Mip protein possesses PPIase activity; however, the role of this enzyme activity in virulence is uncertain, since single amino acids substitutions that abrogate PPIase still confer virulence. Subsequently, *mip* homologs in *Legionella micdadei* and *L. longbeachae* have been identified and specifically mutated; similar loss of infectivity for macrophages and amoebae have been demonstrated in these species as well.

The acquisition of iron is critical for all pathogenic bacteria. Byrd demonstrated that gamma-interferon restricts the growth of intracellular *L. pneumophila* exclusively by limiting the availability of intracellular iron. Gamma-interferon downregulates the expression transferrin receptors on macrophages and the cellular concentration of ferritin. It is known that bacterial growth is inhibited by these changes, since the inhibitory effect of gamma-interferon can be reversed by adding excess iron-loaded transferrin. Not surprisingly, *L. pneumophila* taken from iron-deficient cultures grown in a chemostat are defective in cellular infection. However, *L. pneumophila* has a *fur* homolog, which regulates the expression of an aerobactin synthetase homolog. Although *L. pneumophila* has not been shown to produce siderophores, mutation of this iron acquisition gene reduces intracellular infection by 80-fold. Several other genes associated with iron acquisition have been identified. Some of these genes are also required for infection of macrophages and amoebae.

By far, the most important advance in understanding the intracellular survival and trafficking of *L. pneumophila* has been gained from studying spontaneous mutants that fail to evade phagosome–lysosome fusion in macrophages. In a series of studies performed at Columbia and Tufts University, two chromosomal regions, encoding 18 and 7 genes each, have been found to encode functions that are essential for establishing intracellular infection (representative articles include Segal *et al.*, 1998; Segal and Shuman, 1998; Vogel *et al.*, 1998). The genes have been designated *icm* by one group (for intracellular multiplication) and *dot* (for defect in organelle trafficking) by the other. Figure 2 shows a map of the two regions with both nomenclatures. Mutation in almost any of these genes results in either a complete loss of cellular infectivity or attenuation. In all cases, the loss of infectivity is associated with a failure to evade phagolysosomal fusion, as well as a loss of the immediate contact cytotoxicity of *L. pneumophila* at high multiplicity of infections. Two lines of evidence strongly suggest that these genes encode a complex secretion system in the bacterial envelope. First, several of the genes in these two regions have homology to plasmid conjugation systems. *IcmP* and *icmO* (corresponding to *dotM* and *dotL*, respectively) have homology to genes on the R64 conjugative plasmid of *Salmonella* ssp. Similarly, *dotB* and *icmE* (*dotG*) have homology with two *vir* genes of the *Agrobacterium tumefasciens* T-DNA transfer system. *L. pneumophila* has also been shown to be capable of conjugally transferring plasmid RSF1010. However, mutations in several of the *dot–icm* genes—not just those with sequence homology to known conjugation genes—have been shown to abrogate plasmid transfer. The second line of evidence relates to the contact cytotoxicity phenotype. Mutations of icm/dot genes result in loss of this toxicity, which is known to be associated with the insertion of a pore into host cell membranes. It is likely that the multiplicity of proteins encoded by these loci either combine to form the pore or participate in its transfer or both.

Current theories about the *dot–icm* loci suggest that they encode a secretion complex that transfers as yet unknown factors to the host cell. Most likely, these factors mediate a change in the phagosomal membrane that excludes it from the normal endo-

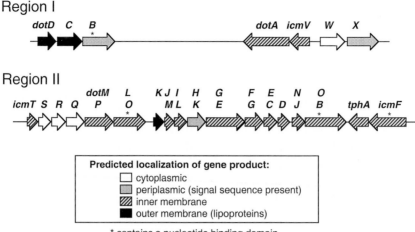

Fig. 2. Physical map of two regions on the chromosome of *Legionella pneumophila* that are required for bacterial survival and growth within macrophages. Both of the nomenclatures in current use (*dot* and *icm*) are given, and the putative localization of each gene product in the bacterial cell is indicated by the legend at the bottom of the figure.

cytic pathway, allowing the phagosome to adapt to a state permissive for bacterial growth. The pore that creates osmotic lysis at high multiplicity of infection may be a channel for transfer of virulence factors to the cell at low m.o.i. or a means to lyse the cell when the bacteria have spent the host cells' nutrient and reached stationary phase. At present, it remains to be determined which of the *dot–icm* gene products are the effectors and which comprise the secretion apparatus in this complex system.

V. IMMUNITY TO *L. PNEUMOPHILA*

In large outbreaks of Legionnaires' disease, the attack rate (i.e., the proportion of exposed individuals who become ill) is generally less than 5%. Those who become ill often have predisposing conditions, such as advanced age, immunosuppression, lung disease, or cigarette smoking (Hoge and Brieman, 1991). Young, healthy individuals are rarely affected, even with comparable exposures to an infectious source. These findings suggest that there may be active, innate resistance to legionellosis in healthy persons. The intracellular localization of the infection and the association with immunosuppression argue that this

innate immunity is primarily cellular. Thus, even relatively resistant animal species, such as rats and mice, may experience lethal infection after pretreatment with immunosuppressive drugs, such as corticosteroids or cyclophosphamide. In contrast, humoral immunity plays a minor role. *L. pneumophila* are resistant to serum complement, and bacteria opsonized with antibody and complement are resistant to killing by neutrophils and can replicate in macrophages. It appears that both innate and acquired immunity against *Legionella* infections are predominantly cell-mediated (Engleberg and Brieland, 1998).

In humans, legionellosis is associated with a preponderance of Th1-type cytokines (IFNγ and IL-12) in the circulation; whereas Th2-type cytokines (IL-4, IL-10) are not detected. *In vitro,* binding of *L. pneumophila* to mononuclear phagocytes results in the prompt induction of several monocytic cytokines, including tumor necrosis factor-α (TNFα), interleukin 1β, IL-6, GM-CSF, and macrophage inflammatory protein 2 (MIP-2). *In vivo,* an increase in pulmonary TNFα levels reaches a peak 24 hours after infection. This cytokine may be critical in the innate defense against infection because it has an autocrine effect on infected macrophages which limits intracellular infection. A study in rodents suggested that this effect of TNFα is partly mediated by nitric oxide (NO). However, a later study suggested that the elaboration of NO is an important stimulus for the expression of TNFα. Apart from its effect on macrophages. TNFα also induces neutrophils to become more bactericidal when *L. pneumophila* are ingested. This effect may help to limit the spread of infection once an acute inflammatory response has been triggered in the infected lung.

Gamma-interferon (IFNγ) is also induced during the course of primary infection and appears to augment host defense and to mediate bacterial clearance, although the specific cellular source of this cytokine is not known. IFNγ is bacteriostatic, since it acts by limiting essential iron. However, IFNγ also stimulates natural killer (NK) cells and endows them with an enhanced capacity to lyse *L. pneumophila*-infected cells. Moreover, IFNγ participates in the induction of nitric oxide in mononuclear phagocytes.

Treatment of animals with either anti-IFNγ or anti-TNFα antibodies enhances *L. pneumophila* infection, attesting to the importance of these cytokines in resistance to primary infection. In contrast, treatment with a Th2-type cytokine, IL-10, reverses the effects of IFNγ at the cellular level and impairs bacterial clearance.

The specific cellular components of the innate immune response are not precisely defined. Depletion of both CD4+ and CD8+ T cells impairs but does not entirely eliminate recovery in susceptible mice, suggesting that other inflammatory cells (perhaps NK cells) also have a critical role in the clearance of infection. The susceptibility of immunocompromised hosts to legionellosis may relate to defects in these cells or in their capacity to produce critically needed cytokines.

An acquired form of cellular immunity against legionellosis was first demonstrated by Horwitz in 1983. He showed that the supernate of mononuclear cells from patients convalescing from Legionnaires' disease incubated of *L. pneumophila* antigen contains factors that induce resistance to *L. pneumophila* in monocytes from a previously uninfected individual. Following this report, Horwitz and his colleagues used a guinea pig model to demonstrate that protective immunity can be induced by inoculation of sublethal doses of virulent *L. pneumophila*, inoculation

of avirulent *L. pneumophila*, *L. pneumophila* cell membrane fractions, or the purified metalloprotease. Although vaccination of humans against Legionnaires' disease is a theoretical possibility, it may not be practical, since the very population most in need of immunization (immunocompromised patients) are those who are least likely to respond to the vaccine or to possess intact immune effector mechanisms.

VI. SUMMARY

The Legionellaceae are parasites of free-living protozoa in the natural environment. To avoid phagolysosomal fusion and killing by host organisms, at least one species of *Legionella* (*L. pneumophila*) has acquired a complex secretion system that mediates the escape from the normal endocytic pathway and the trafficking of the organism to an endosome that is permissive for intracellular multiplication. The very same system mediates the survival and growth of these bacteria in macrophages, an essential requirement for the production of human disease. Other *Legionella* ssp. are also capable of infecting both protozoa and mammalian macrophages, but they are much less common causes of disease, and the mechanisms they use for intracellular survival have not been elucidated.

Humans acquire *Legionella* infection by inhalation of aerosols from contaminated water sources. Most immunocompetent individuals are probably capable of controlling the growth of the organism by innate immune mechanisms involving the secretion of IFN-γ and TNFα. Acquired immunity is cell-mediated via a CD4-dependent, Th1-predominant response. Human disease can be treated with antimicrobials that are active against *Legionella* ssp. and are taken up by host cells. The disease can be prevented by eliminating *Legionella* (and, perhaps, protozoa as well) from contaminated water systems.

See Also the Following Articles

CELLULAR IMMUNITY • LIPOPOLYSACCHARIDES • OUTER MEMBRANE, GRAM-NEGATIVE BACTERIA

Bibliography

Abu Kwaik, Y., Gao, L. Y., Stone, B. J., Venkataraman, C., and Harb, O. S. (1998). Invasion of protozoa by *Legionella pneumophila* and its role in bacterial ecology and pathogenesis. *Appl. Environ. Microbiol.* **64**, 3127–3133.

Barabee, J. M., Brieman, R. F., and DuFour, A. P. (1993). *"Legionella:* Current Status and Emerging Perspectives." American Society for Microbiology, Washington, DC.

Benson, R. F., and Fields, B. S. (1998). Classification of the genus *Legionella. Semin. Respir. Infect.* **13**(2), 90–99.

Byrne, B., and Swanson, M. S. (1998). Expression of *Legionella pneumophila* virulence traits in response to growth conditions. *Infection Immunity* **66**(7), 3029–3034.

Edelstein, P. H. (1987). The laboratory diagnosis of Legionnaires' disease. *Semin. Respir. Infect.* **2**, 235–241.

Edelstein, P. H. (1993) Legionnaires' disease. *Clin. Infect. Dis.* **16**, 741–749.

Eickhoff, T. C. (1979). Epidemiology of Legionnaires' disease. *Ann. Int. Med.* **90**, 499–502.

Engleberg, N. C., and Brieland, J. K. (1998). *Legionella,* infection and immunity. *In* "Encyclopedia of Immunology," 2nd ed., (P. J. Delves and I. Roitt, eds.), pp. 1542–1546. Academic Press, Ltd., London, UK.

Fraser, D. W., Tsai, T. R., Orenstein, W., Parkin, W. E., Beecham, H. J., Sharrar, R. G., Harris, J., Mallison, G. F., Martin, S. M., McDade, J. E., Shepard, C. C., and Brachman, P. S. (1977). Legionnaires' disease: Description of an epidemic of pneumonia. *N. Engl. J. Med.* **297**, 1189–1197.

Hoge, C. W., and Brieman, R. F. (1991). Advances in the epidemiology and control of *Legionella* infections. *Epidemiol. Rev.* **13**, 329–340.

Horwitz, M. A. (1983). Cell-mediated immunity in Legionnaires' disease. *J. Clin. Invest.* **71**, 1686–1697.

Horwitz, M. A. (1992). Interactions between macrophages and *Legionella pneumophila. Curr. Top. Microbiol. Immunol.* **181**, 265–282.

McDade, J. E., Shepard, C. C., Fraser, D. W., Tsai, T. R., Redus, M. A., and Dowdle, W. R. (1977). Legionnaires' disease: Isolation of a bacterium and demonstration of its role in other respiratory disease. *N. Engl. J. Med.* **297**, 1197–1203.

Rowbotham, T. J. (1980). Preliminary report on the pathogenicity of *Legionella pneumophila* for freshwater and soil amoebae. *J. Clin. Pathol.* **33**, 1179–1183.

Segal, G., Purcell, M., and Shuman, H. A. (1998). Host cell killing and bacterial conjugation require overlapping sets of genes within a 22-kb region of the *Legionella pneumophila* genome. *Proc. Natl. Acad. Sci. USA* **95**(4), 1669–1674.

Segal, G., and Shuman, H. A. (1998). How is the intracellular fate of *the Legionella pneumophila* phagosome determined? *Trends Microbiol.* **6**(7), 253–255.

Stout, J. E., and Yu, V. L. (1997). Legionellosis. *New Engl. J. Med.* **337**, 682–687.

Tsai, T. F., Finn, D. R., Plikaytis, B. D., McCauley, W., Martin, S. M., and Fraser, D. W. (1979). Legionnaires' disease: Clinical features of the epidemic in Philadelphia. *Ann. Intern. Med.* **90**, 509–517.

Vogel, J. P., Andrews, H. L., Wong, S. K., and Isberg, R. R. (1998). Conjugative transfer by the virulence system of *Legionella pneumophila. Science* **279**(5352), 873–876.

Leishmania

Gary B. Ogden

St. Mary's University and The University of Texas Health Science Center at San Antonio

Peter C. Melby

South Texas Veterans Health Care System and The University of Texas Health Science Center at San Antonio

GLOSSARY

amastigote The aflagelleted stage of *Leishmania*. Amastigotes are obligate intracellular parasites of mononuclear phagocytes of the vertebrate host.

delayed-type hypersensitivity reaction A cell-mediated immune reaction in the skin in response to foreign antigens, which is indicative of prior sensitization to those antigens.

metacyclogenesis Developmental and biochemical transformation, occurring within the sand fly midgut, of the noninfectious procyclic promastigote into the highly infectious metacyclic promastigote.

promastigote Flagellated insect stage of *Leishmania*. Promastigotes exist in the sand fly vector as either the replicative procyclic form (noninfectious) or the highly infectious (and nonreplicative) metacyclic form.

subclinical Referring to infection without any overt clinical signs of disease.

zoonosis An infectious disease in which the transmission cycle in nature is primarily within an animal population and humans are only incidentally infected.

LEISHMANIA are dimorphic intracellular protozoan parasites of vertebrates, which are transmitted by the bite of the phlebotomine sand fly vector. *Leishmania* belong to the kingdom Protista, are members of the order Kinetoplastida, and are one of nine genera in the family Trypanomastidae. Their complex life cycle dictates an interesting cell biology, as well documented cellular and biochemical changes must occur within these parasites as they prepare for the transition from life within the fly to survival in the vertebrate host.

As is the case for all Trypanomastidae, the molecular biology of *Leishmania* is fascinating and has helped to reveal novel mechanisms of RNA processing and uncommon mechanisms of gene regulation. *Leishmania* has also become an important model for studying the pathogenesis of intracellular parasitism and the subversion of the host's immune system for the benefit of the parasite. Multiple species of *Leishmania* are known to cause human diseases of the skin, mucosal surfaces, and organs of the reticuloendothelial system. The leishmaniases are a diverse set of zoonotic diseases in which humans are incidentally infected. Cutaneous leishmaniasis is usually mild, but may be cosmetically disfiguring. Visceral and mucosal diseases have a significant morbidity and mortality. The drugs used in treating these diseases are only partially effective, which underlines the importance of preventing exposure to sand flies.

I. CLASSIFICATION AND MORPHOLOGY

The classification of the genus *Leishmania* is shown in Table I. Characteristic of the order Kinetoplastida is the presence of a conspicuous Feulgen stain-positive (i.e., DNA-containing) kinetoplast (see following). All members of the family Trypanomastidae are parasitic for vertebrates or invertebrates and un-

TABLE I
Classification of the Genus *Leishmania*

Kingdom	Protista
Subkingdom	Protozoa
Phylum	Sarcomastigophora
Subphylum	Mastigophora
Class	Zoomastigophora
Order	Kinetoplastida
Suborder	Trypanosomatina
Family	Trypanomastidae
Genus	*Leishmania*
Subgenus	*Leishmania*
Subgenus	*Viannia*

dergo morphological changes during transition between stages of their life cycles. Of the nine genera in this family, only species of *Leishmania* and *Trypanosoma* are human pathogens. Two subgenera of *Leishmania*, *L. (Leishmania)* and *L. (Viannia)* are recognized. Conservatively, at least 14 *Leishmania* spp. are pathogenic for mammals, of which nine are recognized parasites of humans.

In mammals, amastigotes are an obligate intracellular parasite of mononuclear phagocytes. The promastigote form is found in female sand flies (genus *Phlebotomus* in the Old World and *Lutzomyia* and *Psychooptgus* in the New World), which are the only known vector for *Leishmania*. The promastigotes possess a single nucleus and are variable in length (15–25 μm) and shape (ellipsoid to slender). The most prominent features of stained promastigotes are the nucleus, the kinetoplast, and the flagellum (Fig. 1); the kinetoplast and the origin of the flagellum are located in the anterior region of the parasite. Amastigotes are round to oval in shape with a 2–10 μm diameter. This stage is aflagellar (i.e., the flagellum does not extend past the cell boundary), but the kinetoplast and nucleus remain visible in stained amastigotes.

II. LIFE CYCLE AND ECOLOGY

Leishmania spp. have a complex lifecycle (Fig. 2) with two morphologically distinct forms, the amasti-

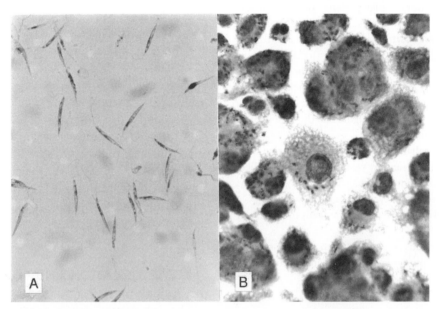

Fig. 1. Stages of the *Leishmania* parasite. In panel A, the elongated forms of the promastigotes, which parasitize the sand fly vector, are shown (stained with giemsa, 1000× magnification). The terminal flagella are not visible. In panel B, a macrophage is shown, which is infected with multiple intracellular amastigotes. Each of the darkly stained bodies within the cytoplasm of the macrophage is an amastigote (stained with giemsa, 600× magnification).

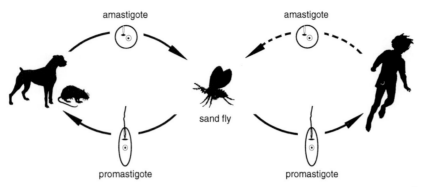

amastigote   amastigote

sand fly

promastigote   promastigote

Fig. 2. Life cycle of *Leishmania* spp. In this schematic diagram, transmission of the infectious promastigotes is shown to occur from the female phlebotomine sand fly vector to the two main reservoir animals (i.e., dogs and rodents). The promastigotes then infect mononuclear phagocytes and transform into amastigotes. The cycle is completed when the sand fly ingests amastigotes from the reservoir during a bloodmeal. As is typical in zoonotic transmission, humans are incidentally infected; however, as indicated by the broken arrow, humans rarely serve as reservoirs for the parasite. (We gratefully acknowledge the contribution of Hector Flores in constructing the diagram.)

gote and the promastigote. The female phlebotomine sand fly vector becomes infected during a bloodmeal, when the fly repeatedly bites an infected vertebrate host, creating a pocket of pooled blood. Upon drinking the bloodmeal, the fly ingests amastigotes, and within the bloodmeal itself, the amastigotes transform into the promastigote form. These early stage promastigotes, termed procyclics, actively replicate within a peritrophic membrane secreted by fly midgut epithelial cells (note that members of the *Viannia* subgenus replicate in the hindgut). About three days after being ingested by the fly, procyclic forms escape the peritrophic membrane and travel to the anterior portion of the midgut, where they attach to epithelial cells. This marks the early stage of metacyclogenesis, where procyclics transform into the nonreplicating, infectious, metacyclic promastigotes. Metacyclics are thinner and faster-moving than procyclics, but more importantly, about five days after the bloodmeal, biochemical changes occur in the surface coat of metacyclics (see LPG following), which cause their release from midgut epithelial cells. The newly released, and now highly infectious, metacyclic promastigotes are free to migrate anteriorly through the foregut, eventually reaching the fly buccal cavity and mouth parts. From the mouth parts, infectious promastigotes are dislodged into the wound created during the next

bloodmeal. Upon contaminating the wound of the vertebrate host, promastigotes are phagocytosed by tissue phagocytes (neutrophils and macrophages) responding to the damaged tissue. Within the macrophage phagolysosome, the promastigotes rapidly transform into nonmotile amastigotes. The amastigotes replicate within the acidic and hostile environment of the phagolysosome, eventually lysing the macrophage and releasing amastigotes to infect other macrophages. The cycle continues when a sand fly ingests free or amastigote-infected macrophages during a bloodmeal.

The transmission of *Leishmania* in most endemic areas is through a zoonotic cycle, where man is only incidentally infected. In general, the strains that cause cutaneous disease are maintained in rodent or other small mammal reservoirs, and the domestic dog is the usual reservoir for the strains that cause visceral disease. In most instances, more than one mammalian species has been found to be infected with a given parasite, but usually there is a principal reservoir and the other animals are incidental hosts that do not play a major role in the transmission cycle. The transmission between reservoir and sand fly is highly adapted to the specific ecology of the endemic region. For instance, the reservoirs for *L. major* in the Middle East are desert rodents and the sand fly vectors cohabit the rodent burrows, where they are protected from the extreme heat and low humidity of the desert environment. In the New World, the zoonotic cycle of parasites of the Viannia subgenus typically involves small mammals and sand flies that inhabit dense tropical and subtropical forests. Human infections occur when their activities bring them in contact with the zoonotic cycle. Anthroponotic transmission, where humans are the presumed reservoir, occurs with *L. tropica* in some urban areas of the Middle East and with *L. donovani* in India.

III. CELLULAR BIOLOGY

A. Plasma Membrane and Surface Molecules

1. Plasma Membrane

Leishmania possess a bi-leaflet plasma membrane similar to that found in metazoan cells. A membrane indistinguishable in appearance from the plasma membrane itself also surrounds the flagellum and, during development of promastigotes, these membranes undergo biochemical changes which modulate parasite attachment to host tissues (see discussion of LPG). The plasma membrane that lines the flagellar pocket actively acquires nutrients by pinocytosis, and pinocytotic vesicles have been observed to travel to the cytoplasm, where they fuse with lysosomes. The flagellar pocket is also active in secretion by exocytosis.

2. LPG

The entire surface of amastigotes and promastigotes, including the flagellum, is covered by a cell coat composed predominantly of lipophosphoglycan (LPG). LPG is anchored to the parasite via a phosphotidylinositol linkage, as are other surface molecules. This type of lipid anchor is unusual in the animal kingdom. In addition to the lipid anchor, LPG has three other domains: a glycan core, repeats of a saccharide–phosphate region, and an oligosaccharide cap. The glycan core and lipid anchor are conserved among all species of *Leishmania*, while the carbohydrates in the repeat region and the cap are highly variable in composition. LPG contributes to the parasite's virulence in several ways, protecting the parasite from oxidative damage and digestion in the phagolysosome, hindering macrophage chemotaxis, and playing a primary role in parasite attachment to macrophages. LPG enhances the infectivity of *Leishmania* by subverting the effects of serum complement. Infective *L. donovani* or *L. major* metacyclic promastigotes possess a thickened cell coat, due to an increase in the number of phosphorylated saccharides in the LPG. The LPG molecules in metacyclic promastigotes of *L. major* are long enough (almost double the length found in procyclics) to prevent the membrane attack complex (MAC) of the complement system, after forming on the parasite surface, from reaching the plasma membrane and lysing the parasite. Noninfective procyclic promastigotes have a thinner coat and are readily lysed by host complement.

Early in the development of *L. major* promastigotes in the sand fly, the terminal sugars of the LPG cap are predominantly galactose residues, but during metacyclogenesis, these residues are replaced by arabinopyranose. Because of these changes, procyclic promastigotes can attach to receptors on the epithelial layer of the sand fly midgut, allowing developing promastigotes to remain in the gut during digestion and excretion of the bloodmeal. Once development to the infectious metacyclic stage has taken place, alterations in the LPG allow the parasites to be released for passage to the mouth parts in order to infect the next host. Similar events occur in other species of *Leishmania*, although the specific changes in LPG structure may differ. Certain species of sand fly are competent hosts for particular species of *Leishmania*. Variations in LPG structure, which accompany varying ability to attach to sand fly epithelial cells, seem to control, at least in part, in which species of sand fly a particular parasite can successfully develop.

LPG plays a primary role in the attachment of complement-opsonized parasites to macrophages since it is the site of C3 deposition on the parasite surface. The macrophage CR1 and Mac-1 (CR3) complement receptors, which bind C3b and iC3b, respectively, mediate adhesion of the promastigotes to macrophages, but CR1 is primarily responsible for the phagocytosis of complement-opsonized metacyclic promastigotes. Consistent with these observations, *in vitro* experiments with *L. major* have demonstrated the increased intracellular survival of complement-fixed metacyclic promastigotes in macrophages.

3. GP63

The major protein found on the surface of all pathogenic species of *Leishmania* is a phosphotidyl-inositol-linked glycoprotein (gp), gp63. Biochemical and molecular analysis has identified gp63 as a type of zinc-endopeptidase. Like LPG, gp63 undergoes changes during the development of promastigotes, and an increase in the surface expression of gp63 during metacyclogenesis has been documented. Surface exposure of GP63, however, is largely blocked by the elongated metacyclic LPG. Gp63 may play a secondary role in the binding of *Leishmania* to the macrophage surface receptors, which subsequently mediate parasite entry by receptor-mediated endocytosis.

B. Flagellum

In addition to serving as a means of locomotion, electron microscopy reveals that the *Leishmania* flagellum sometimes functions as a tether, attaching the developing promastigotes to the tissues lining the midgut of the sand fly vector (see LPG, preceding). In the mammalian host, the functional role of the flagellum is curtailed, as the flagellum in amastigotes does not extend beyond the flagellar pocket. The flagellum has the typical 9 + 2 structure of metazoan cells, possessing nine microtubule doublets surrounding two unattached inner microtubules. The basal bodies at the base of the flagellum are structurally similar to centrioles found in animal cells.

C. Nucleus

The *Leishmania* nucleus is spherical or slightly ovoid in shape, with a diameter of approximately 1.5–2.5 μm. It is unusual in that the nuclear membrane remains intact during nuclear division. During division, the nucleus elongates and then constricts, splitting the nuclear membrane in two. Spindle microtubules are believed to elongate and push the dividing nucleus apart, but how the genetic material is partitioned is unknown. The nuclear membrane itself is typical in appearance, possessing a double unit membrane studded with 65–100 nm diameter nuclear pores. The nucleus contains chromatin, but the chromosomes do not condense or become visible during nuclear division.

D. Other Membrane-Bound Organelles

1. Mitochondrion and Kinetoplast

A single mitochondrion is present and functional in both the insect and mammalian forms of *Leishmania* (i.e., cyanide-sensitive respiration is found in both forms of the parasite). Within the mitochondrion is the kinetoplast, one of the defining morphological features of members of the order Kinetoplastida. This structure lies at the base of the flagellum, close to the basal bodies from which the flagellum arises. The kinetoplast is a large rod-shaped assemblage of kinetoplast DNA (kDNA), located within the mitochondrion. The kDNA itself is unusual, as spreads of purified kDNA visualized under the electron microscope appear predominantly as thousands of "minicircles" and a few "maxicircles" catenated (i.e., interlocked) into a large DNA network. This contrasts sharply with the mitochondrial DNA found in metazoan cells, which typically exists as a single circle of DNA.

2. Endoplasmic Reticulum and Golgi Apparatus

The endoplasmic reticulum (ER) is similar in appearance to that found in metazoan cells. It is contiguous with the nuclear membrane and may be rough (containing attached ribosomes) or smooth (lacking attached ribosomes) in appearance. Closely associated with the ER is the Golgi apparatus, which appears as flattened membrane structures, frequently stacked near the flagellar pocket (in addition to being a site of endocytosis, the flagellar pocket is active in exocytosis).

IV. MOLECULAR BIOLOGY AND CONTROL OF GENE EXPRESSION

A. Genomic Organization

Only recently has the development of pulsed-field gradient gel electrophoresis (PFGE) techniques allowed a determination of the number, size, and conformation of *Leishmania* chromosomes. The genome size has been estimated to be approximately $2.5–5.0 \times 10^7$ base pairs (bp), which is approximately 10 times the size of the genome of the bacterium *Escherichia coli*. PFGE karyotypes of *Leishmania* spp. have revealed a high degree of variability (some chromosomes are constant in size, while the length of others is not). The genomic material typically is divided among 25–30 linear chromosomes (ranging in size from 1.5×10^5 to 3.0×10^6 bp). *Leishmania* spp. appear to be asexual; sexual crosses between these organisms have not been successfully performed under laboratory conditions. Interestingly, however, most *Leishmania* chromosomes seem to be diploid, although this has been difficult to determine conclusively.

B. mRNA Processing

1. RNA Editing

Two of the most uncommon types of mRNA processing (i.e., RNA editing and trans-splicing) were first discovered in studies on trypanomastid protozoa. One of these, RNA editing, has changed our understanding of how genes are expressed in eukaryotes. The kDNA maxicircles, already mentioned, encode genes for ribosomal RNA and certain proteins necessary for oxidative phosphorylation and electron transport in the mitochondrion. The study of many of these genes revealed that the mature messenger RNA (mRNA) sequence was different from the genomic DNA sequence from which they were transcribed. Moreover, many of the genomic sequences did not code for a functional protein. These startling and controversial observations were reconciled when it was discovered that a posttranscriptional process, now termed RNA editing, resulted in the site-specific deletion or addition of uridine (U) residues in the mRNA, which created functional initiation sites for protein translation and/or created an open reading frame for protein translation. RNA editing, in effect, repairs "defects" in the DNA genome at the RNA level, allowing the expression of a functional protein from the mRNA. Although significant progress has been made, an agreement on the molecular events surrounding RNA editing has yet to be reached. However, guide RNAs (gRNAs) are know to be of critical importance to this process. The gRNAs carry the template for proper mRNA editing and may also contribute U residues to the editing process. In *L. tarentolae*, a lizard parasite, about 20 different classes of gRNAs are encoded in both mini- and maxicircles. What is most lacking in the study of RNA editing, however, is an explanation for the reliance of the trypanomastids on this unusual form of gene regulation.

2. Trans-Splicing

It is believed that, within any species of trypanomastid protozoa, all mRNAs possess an identical 5′ exon sequence. This 5′ sequence, called the "spliced leader" or "mini-exon," is not encoded by the gene corresponding to the mRNA transcript. The mechanism for linking the mini-exon to the mRNA, termed trans-splicing, was first discovered in *Trypanosoma brucei* and is now known to occur in nemotodes (in only 10–15% of their mRNAs) in addition to trypanomastids. The biochemistry of trans-splicing is similar to conventional cis-splicing, where one exon is fused to a second downstream exon by two sequential transesterification reactions, resulting in the joining of the two exons and the release of the intervening or "intron" sequence (in the form of a partially circular molecule, termed a "lariat structure") from the precursor mRNA. While both mechanisms use small nuclear RNAs in catalyzing this reaction, the crucial difference between cis- and trans-splicing is that the two exons in trans-splicing are not covalently linked. As a consequence of this, the trans-splicing reaction releases a branched or "Y-shaped" intron instead of a lariat. In all *Leishmania* spp. studied, the spliced leader RNA is transcribed from clusters of tandemly repeated genes (ranging from about 90 to 200 copies) on the genome. In *L. enrettii*, an 85-nucleotide (nt) spliced leader RNA is transcribed from a 440bp long gene. At the 5′ end of this transcript is the 35nt mini-exon sequence found at the 5′ end of all mature *L. enrettii* mRNAs.

C. Control of Gene Expression

For most eukaryotes and bacteria, gene expression is controlled primarily by regulating the initiation of transcription. In this paradigm, transcriptional promoters on the genome largely dictate where, and how often, transcription takes place. Promoter sequences have been identified in trypanosomes, but no promoter sequence has been clearly identified in any *Leishmania* spp. Transcription in *Leishmania* appears to be nonstringent, as promastigotes transfected with plasmids containing the gene for neomycin resistance (but containing no *Leishmania* sequences) were shown to actively transcribe the neomycin gene. All that was required was the incorporation of a signal for trans-splicing in the neomycin gene sequence. So far, other evidence for a true *Leishmania* promoter has been merely suggestive or inconclusive. Thus, gene regulation in *Leishmania* and other kinetoplastids seems to depend largely upon posttranscriptional mechanisms. This may be a consequence, again, of a peculiarity of trypanomastid molecular biology. Unlike most eukaryotes studied, trypanomastids produce polycistronic mRNAs. Translation of these transcripts, however, requires their conversion to monocistrons by trans-splicing at the 5′ end and polyadenylation at the 3′ end. The polyadenylation of an upstream gene is functionally coupled to trans-splicing of a downstream gene. Control of mature mRNA may then be controlled by regulating splicing and/or polyadenylation. The stability of the mRNA is also of importance, as the rate of mRNA degradation of a heat-shock protein, hsp83, is dependent upon 5′ and 3′ untranslated regions. In addition, gp63 transcripts have been shown to have a much shorter half-life in logarithmic phase cells, as compared to stationary phase cells.

D. DNA Transfection and Gene Targeting

A major contribution to the molecular analysis of *Leishmania* was the development of functional plasmid expression vectors in 1990. Independently, two groups used an *E. coli* plasmid in combination with the intergenic region of the α-tubulin gene or the flanking region of the *dhfr-ts* gene to construct a *Leishmania* plasmid. More recently, other plasmids have been made; several selectable markers are available for stable transfection (e.g., neomycin, hygromycin, bleomycin, and puromycin) and a number of phenotypic markers are available for

transient assays (e.g., β-galactosidase, luciferase, and β-glucuronidase). Negative selection is also possible using the thymidine kinase gene. Important questions about the pathogenesis of *Leishmania* have been answered using these plasmids. As mentioned previously, transfection of gp63-variant strains with gp63-expressing plasmids has helped to elucidate the role gp63 plays in infection. Similarly, functional complementation assays have been successfully employed in the study of LPG synthesis. Moreover, homologous recombination between a plasmid and the *Leishmania* genome may prove to be a powerful tool in studying the function of particular genes in these parasites. Researchers may now specifically target a gene for replacement, or eliminate the activity of a gene (i.e., knockout) by inserting a foreign gene in it. Although such gene knockout experiments are potentially powerful in studying *Leishmania*, it is important to remember that difficulties may arise, as *Leishmania* genes are usually diploid and may be present in tandem repeats on the chromosome.

V. PATHOGENESIS AND HOST RESPONSE

The *Leishmania* parasite has multiple ways in which it adapts to, and survives within, both the invertebrate and vertebrate hosts. Within the sand fly, promastigotes undergo metacyclogenesis to the infective stage (see preceding). When the sand fly takes a bloodmeal, the infective promastigotes are deposited in the host skin, which initiates the infection. As noted above, the surface LPG plays an important role in the entry and survival of *Leishmania* in the mammalian host by conferring complement resistance and by facilitating entry into the macrophage without triggering a respiratory burst.

The nature of the host immune response plays a critical role in the evolution and outcome of infection. There is extensive evidence from experimental animal models that cellular immune mechanisms mediate resistance to *Leishmania* infection. Human studies have confirmed, in a general sense, the findings of the experimental animal studies. The immunological mechanisms related to resistance and susceptibility are summarized in Fig. 3. Resistance to

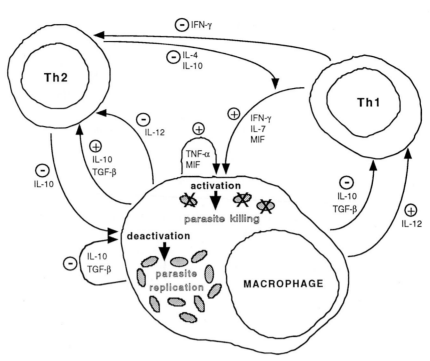

Fig. 3. Immunopathogenesis of *Leishmania* infection. The balance between the Th1 and Th2 cell response and the effect on macrophage activation and parasite killing is shown schematically. Inhibitory effects are noted by a (−) and stimulatory effects are noted by a (+). When Th1 cells are activated by *Leishmania* antigens, IFN-γ is produced, which downregulates the Th2 response and promotes macrophage activation. IL-7 and MIF may play a less prominent role in macrophage activation and parasite killing. TNF-α, produced by the macrophage, synergizes with IFN-γ to induce macrophage activation and parasite killing. If the *Leishmania* antigens induce a Th2 response with IL-4 and/or IL-10 production, the Th1 response is inhibited and the macrophage is deactivated, making it permissive for parasite replication. The infected macrophage itself may also produce IL-10 and TGF-β, which promote parasite survival by deactivating the macrophage.

leishmanial infection is associated with the capacity of CD4+ T cells (Th1 subset) to generate interferon-gamma (IFN-γ) in response to the parasite. Tumor necrosis factor-alpha (TNF-α) may contribute to protective responses by augmenting IFN-γ-mediated macrophage activation. Interleukin (IL)-12, which stimulates T cells and NK cells to produce IFN-γ, promotes Th1 subset expansion, and suppresses Th2 cells (IL-4), plays a critical role in the generation of a protective immune response. Other cytokines, such as interleukin-7 and macrophage migration inhibitory factor (MIF), have been shown to activate macrophages to eliminate intracellular *Leishmania,* but their role is less significant than that of IFN-γ. The

killing of intracellular *Leishmania* is mediated by the generation of reactive oxygen and nitrogen intermediates. The latter mechanism is prominent in rodent models of leishmaniasis, but it has not been shown to play a role in killing of *Leishmania* by human macrophages.

Susceptibility to leishmanial infection and progression of disease is associated with the production of IL-4 and IL-10 by Th2 cells and IL-10 and TGF-β by infected macrophages (see Fig. 3). IL-4 is a cross-regulatory cytokine that inhibits the expansion of the Th1 subset of T cells. IL-10 inhibits the production of cytokines by Th1 cells by suppression of accessory cell function and IL-12 production.

Infected macrophages have a diminished capacity to initiate and respond to an inflammatory response, thus providing a safe haven for the intracellular parasite. IL-10 and TGF-β, which are potent inhibitors of the antimicrobial activity of macrophages, are produced by T cells or infected macrophages at the site of infection. In addition to the production of these deactivating cytokines, infection of the macrophage can suppress its function. Leishmanial infection of macrophages has been reported to increase synthesis of the immunosuppressive molecules prostaglandin E$_2$ and TGF-β, blunt the IFN-γ-mediated upregulation of MHC Class II genes, decrease the capacity for IL-1 production, inhibit antigen presentation to T cells, and impair intracellular signaling in response to IFN-γ. Priming of monocytes with IFN-γ before infection augments IL-1 and TNF-α synthesis and increases leishmanicidal capacity.

Following inoculation of the *Leishmania* spp. by the sand fly vector, large numbers of T lymphocytes, macrophages, and polymorphonuclear leukocytes migrate into the site of infection. Host factors (genetic background, concomitant disease, nutritional status), parasite factors (virulence, size of the inoculum), and, possibly, vector-specific factors (sand fly genotype, immunomodulatory salivary constituents) influence the subsequent development and evolution of disease. Patients who develop localized cutaneous leishmaniasis (LCL, see following) generally exhibit strong protective anti-leishmanial T cell (Th1) responses and spontaneously heal within 3–6 months. Patients with mucosal leishmaniasis (ML; see following) exhibit a hyper-responsive cellular immune reaction which may contribute to the prominent tissue destruction seen in this form of the disease (see following). Patients with active disseminated disease (diffuse cutaneous leishmaniasis [DCL] or visceral leishmaniasis [VL]) demonstrate minimal or absent *Leishmania*-specific cellular immune responses.

VI. INFECTION AND DISEASE

In most instances, when *Leishmania* are transmitted to humans, the infection is controlled by the host response and overt disease does not develop. Field studies in endemic areas reveal people who have no history of disease, but evidence of infection by virtue of a positive Leishmanin (delayed-type hypersensitivity, DTH) skin test. In most endemic areas, subclinical infection (no overt disease) occurs more frequently than active disease. Epidemiological studies indicate that individuals with previous active cutaneous leishmaniasis or subclinical infection are usually immune to a subsequent clinical infection. Within endemic areas the prevalence of DTH skin test positivity (related to previous subclinical or clinical infection) increases with age and the incidence of clinical disease decreases with age, indicating that immunity is acquired in the population over time.

The leishmaniases (clinical disease) are estimated to affect 10–50 million people in endemic tropical and subtropical regions on all continents except Australia and Antarctica. Over the last decade, an increased number of cases have been reported throughout the world, either as new sporadic cases in endemic regions or new epidemic foci. Severe epidemics with more than 100,000 deaths from VL have occurred in India and Sudan. The disease has emerged in new foci because of the following factors: (i) the movement of susceptible people into existing endemic areas, usually because of agroindustrial development, (ii) an increase in the vector and/or reservoir populations due to agriculture development projects, (iii) heightened anthroponotic transmission due to rapid urbanization in some foci, and (iv) increased sand fly density because of scaled-back malaria vector control programs.

The different leishmaniases are distinct in their etiology, epidemiology, transmission, and geographical distribution. These features are summarized in Table II. With only rare exceptions, the *Leishmania* spp. that primarily cause cutaneous disease do not visceralize. (Cutaneous leishmaniasis may be localized to one or a few skin ulcers, which occur at the site of the sand fly bite [LCL] or, in rare instances, the parasite may disseminate to involve large areas of skin [DCL].) ML, an uncommon complication of LCL, is a destructive, disfiguring infection of the nasal and oropharyngeal mucosa. VL is the most serious form because there is progressive parasitism of the visceral reticuloendothelial organs (liver, spleen, and bone marrow).

LCL (also called oriental sore) can affect individu-

TABLE II
Major Species of *Leishmania* and Their Geographic Distribution

Parasite	Disease form	Reservoirs	Vectors	Distribution
Old World Leishmaniasis				
L. (L.) major	LCL	Desert rodents (Psammomys, Meriones, Gerbillus)	*Phlebotomus papatasi*	North Africa, the Middle East, central Asia, and the Indian subcontinent
L. (L.) tropica	LCL	Humans Rock hyraxes Unknown animals	*P. sergenti*	North Africa, the Middle East, central Asia, and the Indian subcontinent
L. (L.) ethiopica	LCL, DCL	Hyraxes	*P. pedifer* *P. longipes*	Ethiopian highlands, Kenya
L. (L.) infantum	VL, LCL	Domestic dog, wild canines	*P. perniciosus* *P. ariasi* *P. tobbi* *P. langeroni*	Mediterranean basin, Middle East, and central Asia
L. (L.) donovani	VL	Humans	*P. argentipes* *P. orientalis* *P. martini*	Kenya, Sudan, India, Pakistan, and China
New World Leishmaniasis				
L. (L.) mexicana	LCL, DCL	Forest rodents	*Lutzomyia olmeca olmeca* *Lu. Cruciata*	Southern Texas through Mexico and northern Central America
L. (L.) amazonensis	LCL, DCL	Forest spiny rats	*Lu. faviscutellata*	South America in the Amazon basin and northward
L. (L.) pifanoi	LCL, DCL	Probably rodents	Unknown	Venezuela
L. (L.) garnhami	LCL	Unknown	*Lu. youngi*	Venezuela
L. (L.) venezuelensis	LCL	Unknown	*Lu. olmeca bicolor*	Venezuela
L. (V.) braziliensis	LCL, ML	Forest rodents Opossums Sloths Domestic dogs Donkeys	*Ps. wellcomei* and others	South America from the northern highlands of Argentina and northward to Central America
L. (V.) panamensis	LCL, ML	Sloths	*Lu. trapidoi* *Lu. ylephiletor*	Panama, Costa Rica, Colombia
L. (V.) guyanensis	LCL	Sloths Lesser anteater	*Lu. umbratilis*	Guyana, Surinam, northern Amazon basin
L. (V.) peruviana	LCL	Unknown	*Lu. peruensis*	Peru, Argentinian highlands
L. (L.) chagasi	VL, LCL	Domestic dogs Foxes	*Lu. longipalpis*	Mexico (rare) through Central and South America

als of any age but children are the primary victims in many endemic regions. It typically presents as one or a few raised or ulcerated lesions located on skin that is exposed to the sand fly bite, e.g., the face and extremities. Cases of more than a hundred lesions have rarely occurred. The lesions develop at the site of the sand fly bite and increase in size and often ulcerate over the course of several weeks to months. The lesions remain localized to the skin and the patient generally has no systemic symptoms. Lesions caused by *L. major* and *L. mexicana* usually heal spontaneously after 3–6 months, leaving a depressed scar. Lesions on the ear pinna caused by *L. mexicana*, called chiclero's ulcer because they were common in chicle harvesters in Mexico and Central America, often follow a chronic, destructive course. In general, lesions caused by *L.* (*Viannia*) spp. tend to be larger and more chronic.

DCL is a rare form of leishmaniasis caused by organisms of the *L. mexicana* complex in the New World and *L. ethiopica* in the Old World. DCL manifests as large nonulcerating skin lesions, which often involve large areas of skin and may resemble the lesions of leprosy. The face and extremities are most commonly involved. Dissemination of the parasite from the initial lesion usually takes place over several years. It is thought that an immunological defect underlies this severe form of cutaneous leishmaniasis.

ML (also called espundia) is an uncommon but serious manifestation of leishmanial infection, resulting from spread of the parasite from the skin to the nasal or oropharyngeal mucosa from a cutaneous infection. ML is usually caused by parasites in the *L.* (*Viannia*) complex. ML occurs in less than 5% of individuals who have, or have had, skin lesions caused by *L.*(*V.*) *braziliensis*. Patients with ML most commonly have nasal mucosal involvement, which results in symptoms of nasal congestion, discharge, and recurrent nosebleeds. In severe cases, over the course of several years, there is marked soft tissue, cartilage, and even bone destruction, which lead to visible deformity of the nose or mouth, perforation of the nasal septum, and even obstruction of the airways.

VL (also called kala azar) typically affects children less than 5 years of age in the New World (*L. chagasi*) and Mediterranean region (*L. infantum*) and older children and young adults in Africa and Asia (*L. donovani*). Following inoculation of the organism into the skin by the sand fly, the child may have a completely asymptomatic infection (no clinical signs of disease) or 2–9 months later show signs of active VL. The classic clinical features of VL include high fever, enlargement of the liver and spleen, and severe cachexia (weight loss and muscle wasting). There is loss of bone marrow function, so the leukocyte and erythrocyte counts fall. Malnutrition is a risk factor for the development and more rapid evolution of active VL. Without the institution of drug therapy, death occurs in over 90% of patients.

VL has been increasingly recognized as an opportunistic infection associated with HIV infection. Most cases have occurred in southern Europe and Brazil, but there is potential for many more cases as the endemic regions for HIV and VL converge. Disease may result from reactivation of a longstanding subclinical infection.

VII. DIAGNOSIS, TREATMENT, AND CONTROL

A definitive diagnosis of leishmaniasis rests on the microscopic detection of amastigotes in tissue specimens or isolation of the organism by culture (performed only in specialized laboratories). A positive culture enables speciation of the parasite (usually by isoenzyme analysis by a reference laboratory). Serological tests are useful for the diagnosis of VL because of the very high level of anti-leishmanial antibodies, but not for ML or LCL, where the antibody response is minimal.

Specific anti-leishmanial therapy is not routinely needed for uncomplicated LCL caused by strains which have a high rate of self-healing (*L. major*, *L. mexicana*). Lesions that are extensive or located where a scar would result in disability or disfigurement should be treated. Cutaneous lesions suspected or known to be caused by members of the *Viannia* subgenus (New World) should be treated because of the low rate of spontaneous healing and the potential risk of developing mucosal disease. Likewise, patients with lesions caused by *L. tropica* (Old World), which are typically chronic and nonhealing, should be treated. All patients with VL or ML should receive therapy.

The pentavalent antimony compounds (sodium stibogluconate and meglumine antimoniate) have been the mainstay of anti-leishmanial chemotherapy for more than 40 years. These drugs are difficult to administer and have moderate toxicity. Using the recommended regimen, cure rates of 90–100% for

LCL, 50–70% for ML, and 80–100% for VL can be expected. Failure of antimony therapy is more common in some parts of the world. Relapses are common in patients who do not have an effective anti-leishmanial cellular immune response, such as patients who have DCL or are coinfected with HIV. These patients often require multiple courses of therapy.

Several alternative therapies have been used in the treatment of the leishmaniases, including the antifungal agent Amphotericin B and its related lipid-associated compounds, parenteral paromomysin (aminosidine) and pentamidine. Recombinant human interferon-gamma has been successfully used as an adjunct to antimony therapy in treatment of refractory cases of ML and VL. The oral antifungal drug ketoconazole has been effective in treating some patients with LCL.

Leishmaniasis is best prevented by avoidance of exposure to sand flies and use of insect repellents. Where transmission is occurring near homes, community-based insecticide spraying has had some success in reducing human infections. Control or elimination of infected reservoir hosts (e.g., seropositive domestic dogs) has had limited success. Where anthroponotic transmission is thought to occur, the early recognition and treatment of cases is essential. Because humans acquire longstanding immunity following infection with *Leishmania,* prevention of the disease through vaccination is a possible approach. Recent immunization studies involving experimental animal models are promising. Vaccination of humans, or domestic dogs to prevent transmission of VL, may have a role in the prevention of the leishmaniases in the future.

See Also the Following Articles

INSECTICIDES, MICROBIAL • SYMBIOTIC MICROORGANISMS IN INSECTS • ZOONOSES

Bibliography

Alvar, J., Canavate, C., Gutierrez-Solar, B., *et al.* (1997). *Leishmania* and human immunodeficiency virus coinfection: the first 10 years. *Clin Microbiol. Rev.* **10**, 198.

Berman, J. D. (1997). Human leishmaniasis: clinical, diagnostic, and chemotherapeutic developments in the last 10 years. *Clin. Inf. Dis.* **24**, 684.

Desjeux, P. (1996). Leishmaniasis. Public health aspects and control. *Clinics Derm.* **14**, 417–423.

Dye, C., Williams, B. G. (1993). Malnutrition, age, and the risk of parasitic disease: visceral leishmaniasis revisited. *Proc. R. Soc. Lond.* **254**, 33–39.

Cruz, A. K., and Tosi, L. R. O. (1996). Molecular biology. *Clin. Derm.* **14**, 533–540.

Grevelink, S. A., Lerner, E. A. (1996). Leishmaniasis. *J. Am. Acad. Dermatol.* **34**, 257–272.

Kreier, J. P., and Baker, J. R. (1987). "Parasitic Protozoa." Allen & Unwin, Boston.

McHugh, C. P., Melby, P. C., LaFon, S. G. (1996). Leishmaniasis in Texas: epidemiological and clinical aspects of human cases. *Am. J. Trop. Med. Hyg.* **55**, 547–555.

Molyneux, D. H., and Ashford, R. W. (1983). "The Biology of Trypanosoma and Leishmania, Parasites of Man and Domestic Animals." Taylor & Francis, London, UK.

Mosser, D. M., and Brittingham, A. (1997). *Parasitology* **115**, S9–S23.

Peters, W., and Killick-Kendrick, R. (1987). "The Leishmaniases in Biology and Medicine," Vols. 1 and 2. Academic Press, London, UK.

Sacks, D. L. (1992). The structure and function of the surface lipophosphoglycan on different developmental stages of Leishmania promastigotes. *Infect. Agents and Dis.* **1**, 200–206.

Wong, A. K. (1995). Molecular genetics of the parasitic protozoan Leishmania. *Biochem. Cell Biol.* **73**, 235–240.

Lignocellulose, Lignin, Ligninases

Karl-Erik L. Eriksson

University of Georgia, Athens

I. Introduction
II. Lignocellulose
III. Microorganisms Involved in Degradation of Lignocellulose
IV. Expression of Ligninolytic Enzymes—Physiological Demands
V. Ligninases

GLOSSARY

lignin(s) Highly stable polymers of mostly methoxylated phenyl–propanoic residues, synthesized as part of the cell wall of vascular plants, constituting the second most abundant organic polymer on earth after cellulose. The monomeric components in lignin(s) are p-coumaryl-alcohol (p-hydroxyphenyl), coniferyl-alcohol (guaiacyl), and sinapyl-alcohol (syringyl). Ligninases have here been used as a common name for the three extracellular phenol-oxidases (peroxidases) produced by white-rot fungi. These enzymes are lignin peroxidase (LiP), manganese peroxidase (MnP), and laccase. Each of these three enzymes participates in various ways in the degradation of lignin.

lignocellulose Plants (trees) in which the main components are cellulose, lignin, and hemicelluloses. The proportions among these three main components vary considerably in different plants (trees).

MICROBIAL DEGRADATION OF LIGNOCELLU-LOSIC MATERIALS is one of nature's most important biological reactions. Wood and other lignocellulosic materials are, in this way, converted to carbon dioxide, water, and humic substances.

I. INTRODUCTION

The oil crisis during the 1970s turned interest toward the utilization or renewable resources, and toward lignocellulosic materials in particular, to be used, instead of fossil fuels, for energy purposes. To release the solar energy stored in various lignocellulosic materials has been a prime target for research in laboratories around the world. Successful efforts have been made to utilize microorganisms, particularly wood-degrading fungi, and their enzymes in delignification of wood chips to allow for production of mechanical pulp with less energy expenditure. The cellulose- and hemicellulose-degrading enzymes have been used for saccharification of lignocellulosic materials to sugars for further conversion to ethanol. These enzymes are now also used for deinking of recycled papers. Protein production, based on microbial conversion of enzyme-derived sugars to protein, to feed the world's growing population is another important issue.

The focus of this article is on the white-rot fungi, the only microorganisms which, to any extent, can degrade all the lignocellulosic components, even the lignins. The production and characteristics of the three main extracellular enzymes employed for this purpose are described here in some detail.

II. LIGNOCELLULOSE

Lignocellulose is made up largely of cellulose, lignin, and hemicelluloses in various proportions. The most important lignocellulosic materials are wood and agricultural wastes, such as straw of various kinds and sugar cane bagasse. The microbiology of wood degradation has a long history of study and is now known in great detail. Presentation of this knowledge will, therefore, be the focus of this article.

Both the structure and chemical composition of wood greatly influence its degradation by microorganisms. The lignin content of angiosperms (hardwoods) varies between 20 and 25%, while this content in gymnosperms (softwoods) varies between 28 and 32%. There are also major differences in the structure of the hemicelluloses present in these two types of wood. The content of glucuronoxylan is high in hardwoods, while the dominating hemicellulose in softwood is glactoglucomannan. However, there is a great deal of variation among various woods, in their chemical composition and also in the composition of different types of cells in a tree. The types of lignins found in hardwoods and softwoods differs. Softwood lignin is almost entirely of guaiacyl-type structure, while hardwood lignins are of about 50/50 guaiacyl and syringyl types. The greatest concentration of lignin is in the middle lamella but it is also distributed throughout the secondary wall. Since the secondary wall is the largest portion of the total cell wall area, most of the lignin, 60–80%, is located here. Figure 1 shows a detailed composition of a softwood tracheid. Lignin is the last component to be synthesized in the fiber cell wall. It initially appears at the cell corners and in the middle lamella. When the deposition of cellulose microfibrils in the S-3 layer of the secondary wall has been completed, lignification proceeds extensively. Lignin is, thus, polymerized and deposited within a preformed cell-wall matrix. Accordingly, the space available for lignin deposition is defined from the outset. The lignin polymers are formed by oxidation of the lignin monomers, by the enzymes peroxidases and laccases, to their corresponding phenoxyradicals. These radicals polymerize spontaneously and, as far as is known, without the aid of any additional enzyme. The three-dimensional structure of lignin is unknown, although it is second only to cellulose in natural abundance and, thus, occupies an important position in the global carbon cycle. In the hemicellulosic polysaccharides,

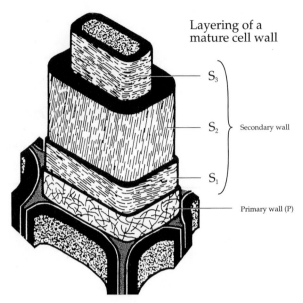

Fig. 1. Diagrammatic representation of a softwood tracheid (adapted from Côté, 1967).

side chains are present at more or less regular intervals, whose conformational preference determines how the lignin and the hemicellulose polymers interact.

Lignin serves several important functions in the cell-wall matrix of plants. It strengthens the cell-wall structure, thus lending mechanical support to the plant; it serves as a barrier to protect the plant against microbial attack; and it acts as a water-permeant seal for the xylem vessels, allowing for water transport from the roots to the top of plants and very high trees. Unfortunately, this same strength, stability, and hydrophobic contributions to plants and trees also impact upon mankind and the environment in many adverse ways: (1) in bleaching of wood pulp, residual lignin has been, and still is, removed from the polysaccharide components of wood using chlorine and clorine derivatives. This results in generation of toxic compounds constituting environmental hazards; (2) lignin is also an obstacle for efficient bioconversion of cell-wall polysaccharides in waste biomass to useful sugars and alcohols and limits its digestibility by cattle and other ruminants. However, degraded lignocellulosic materials, lignins, in particular, form the bulk of soil humus, without which life on earth could not be sustained.

III. MICROORGANISMS INVOLVED IN DEGRADATION OF LIGNOCELLULOSE

Lignocellulosic materials are decomposed in nature by microorganisms. However, the conditions must be conducive to microbial activity or the degradation process will not start or will be interrupted. Fungi, which by their hypha effectively penetrate wood, are the major decomposers of wood. Of the 1600–1700 species of wood-rotting fungi that have been identified in North America, 95% are white-rotters. Most of the white-rotters colonize hardwood trees with lower lignin content and higher hemicellulose content, but many also degrade softwoods. Since white-rot fungi are so dominating in wood degradation and also are the only ones that to any extent can degrade lignin, the focus here will be on this particular type of wood-rotting fungi.

The phenomena of wood decay and wood decomposition have been studied since the mid-nineteenth century. The three main types of wood rotting fungi are classified as white-rotters, brown-rotters, and soft-rotters. The latter two mainly degrade wood polysaccharides (Eriksson *et al.*, 1990). It is easy to distinguish between white- and brown-rotters by the color of the rotted wood. The ability of white-rotters to degrade lignin and the difference in color of advanced decay suggest that different enzymes are employed by these two types of wood-rotting fungi. Color formation around fungal mycelia, when phenols and tannins are added to a growth medium, is one way to distinguish between white-rot and brown-rot fungi. Only the white-rotters excrete phenoloxidases and, therefore, convert the added phenols to the more strongly colored quinones.

White-rot fungi commonly decay wood by attacking all the cell-wall components simultaneously (Fig. 2). However, there are also others who preferentially degrade the lignin component (Fig. 3). Originally, the term white-rot was used mostly for fungi which preferentially attacked lignin. Later, the white-rotters have been characterized as white-pocket, white-mottled, white-stringy rotters, etc., dependent upon the microscopic characteristics of the attack.

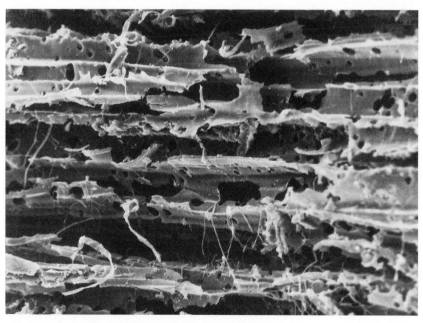

Fig. 2. Simultaneous attack on all the wood components by the white-rot fungus *Phanerochaete chrysosporium* (courtesy of T. Nilsson).

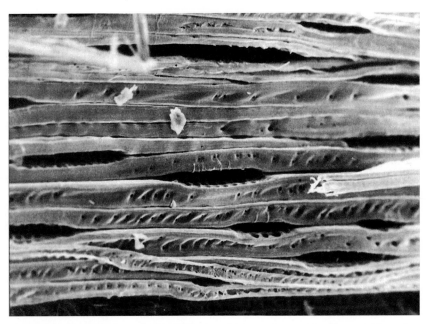

Fig. 3. Selective attack on the lignin by the white-rot fungus *Phellinus pini* (courtesy of R. A. Blanchette).

In the most interesting type of white-rot, the lignin is preferentially degraded within all cell-wall layers. Some species, such as *Phellinus pini, Ganoderma tsugae,* and *Ceriporiopsis subvermispora,* seem to always cause selective delignification (Fig. 3). When the middle lamellae are extensively attacked, it causes a separation of the cells. In a specific attack, lignin is then also degraded in the secondary wall but there are no visible lysis zones, erosion troughs, or thinned areas. In a selective attack of the lignin, the crystalline nature of cellulose is not destroyed. It is well known that degradation of crystalline cellulase takes place essentially only when there is a concerted action of both endo- and exoglucanases. In-depth studies of the plant cell-wall degrading enzymes produced by *C. subvermispora* demonstrated that this fungus did not produce an exoglucanase (cellobiohydrolase).

IV. EXPRESSION OF LIGNINOLYTIC ENZYMES— PHYSIOLOGICAL DEMANDS

White-rot basidiomycetes degrade lignin more rapidly and extensively than other groups of microorganisms. However, for lignin degradation to take place, white-rot fungi require an additional, more easily metabolizable carbon source. It has not been possible to demonstrate that lignin can serve as the sole carbon and energy source for any known microorganism (Eriksson *et al.,* 1990).

Phanerochaete chrysosporium has been the model organism for studies of lignin degradation by white-rot fungi (Kirk and Farrell, 1987). The ligninolytic system of *P. chrysosporium* is triggered mainly by nitrogen starvation, but it can also be triggered by carbon or sulfur starvation. It operates only under secondary metabolism (Buswell and Odier, 1987). These phenomena for triggering secondary metabolism are probably true for most white-rot fungi, although there are also examples of fungi which are not so strongly regulated by nitrogen starvation. Such fungi may be found in nitrogen-rich environment, such as in cattle dung piles, whereas in fungi growing in wood, where they encounter low nitrogen concentrations, lignin degradation would be repressed by a high nitrogen concentration.

Veratryl alcohol is a secondary metabolite in some white-rot fungi associated with their ligninolytic system, particularly in those producing lignin peroxidase (LiP). *P. chrysosporium* has been demonstrated to produce veratryl alcohol *de novo.* Veratryl alcohol is synthesized using the phenyl–alanine pathway. Addition of veratryl alcohol to cultures of *P. chrysosporium* has been demonstrated to induce LiP, and so does addition of lignin. Surprisingly, the increased LiP activity does not seem to give rise to a significant increase in the conversion of ^{14}C-labeled synthetic lignin to $^{14}CO_2$ in this fungus, which indicates that LiP is not the sole rate-limiting component in lignin metabolism.

Lignin degradation is an almost entirely oxidative process, which is why increased oxygen levels enhance lignin degradation considerably in various white-rot fungi (Kirk and Farrell, 1987). Cultures of *P. chrysosporium*, kept at an atmosphere of 5% O_2, released only 1% of totally available ^{14}C-ring-labeled carbon from synthetic lignin as $^{14}CO_2$ after 35 days of incubation. However, cultures maintained at 21 and 100% oxygen, respectively, generated approximately 47 and 57% of total ^{14}C-label as $^{14}CO_2$. The maximum rate of $^{14}CO_2$ evolution is approximately 3 times higher at 100% O_2-atmosphere compared to in air. This beneficial effect of O_2 on lignin biodegradation is probably applicable to white-rot fungi, in general.

It was originally reported that agitation of *P. chrysosporium* cultures completely repressed LiP production and lignin metabolism (^{14}C-lignin → $^{14}CO_2$). However, later conflicting results concerning agitation and lignin degradation appeared in the literature and it was reported that agitated cultures of *P. chrysosporium*, in which the mycelium had formed a single large pellet, readily produced $^{14}CO_2$ from ^{14}C-ring-labeled synthetic lignin. Production of LiP and also a complete oxidation of labeled lignin to $^{14}CO_2$ has later been demonstrated in agitated cultures of both wild-type and mutant strains of *P. chrysosporium*.

Lignin peroxidase (LiP) and manganese peroxidase (MnP) and the H_2O_2-generating systems seem to be the major components of the extracellular lignin degradation system in *P. chrysosporium*. Both LiP and MnP are regulated at the gene transcription level, i.e., by depletion of nutrient nitrogen. MnP activity is also dependent upon the presence of Mn(II) in the culture medium, and so is the production since the *mnp* gene transcription is also regulated by Mn(II).

V. LIGNINASES

A. Lignin Peroxidase

Lignin peroxidase (LiP) (ligninase, EC 1.11.14) was first discovered in 1983 in ligninolytic cultures of *Phanerochaete chrysosporium*, in which fungus it seems to constitute one of the major components of the ligninolytic system (Eriksson *et al.*, 1990). It was, for a long time, believed that the production in *P. chrysosporium* of both LiP and manganese peroxidase (MnP) was clear evidence that these two enzymes were necessary for lignin degradation. However, it has later been shown (Table I) that only about 40% of all studied white-rot fungi produce LiP. LiP catalyzes a large variety of reactions, such as cleavage of β-O-4 ether bonds and of C_α–C_β bonds in lignin. Cleavage of these bonds is basic for the depolymerization of lignin. The enzyme also catalyzes oxidation of aromatic C_α-alcohols to C_α-oxo compounds, hydroxylation, quinone formation, and aromatic ring cleavage (Eriksson *et al.*, 1990). LiP oxidizes its substrates by two conseutive 1-electron oxidation steps. Cation radicals are intermediates in these reactions. LiP has, compared to other phenol oxidases and peroxidases, an unusually high redox potential and can oxidize not only phenolic but also nonphenolic, methoxy-substituted lignin subunits. The importance of LiP in the degradation of lignin has been demonstrated in several studies. The enzyme can depolymerize dilute solutions of lignin. However, the net depolymerization is not that great, since phenoxyradicals are generated in the oxidation of phenolic substrates, as well as in demethoxylation and in ether cleavage reactions. These radicals readily repolymerize. LiP also oxidizes and degrades a variety of dimers and oligomers structurally related to lignin *in vitro* and catalyzes the production of activated oxygen species.

LiP has a catalytic cycle similar to that of horseradish peroxidase (Fig. 4). The native FeIII+ enzyme is first oxidized by H_2O_2 to compound I. One electron reduction of compound I with veratryl alcohol then takes place, or H_2O_2 oxidation results in compound II. Electron reduction of compound II by veratryl alcohol returns the enzyme to its native form, thus maintaining the catalytic cycle. However, in competi-

TABLE I
**Distribution of Ligninolytic Phenoloxidases
in White-Rot Fungi**

Organisms	LiP	MnP	Lac
Coriolopsis occidentalis	+	+	?[a]
Phlebia Brevispora	+	+	+
Phlebia radiata	+	+	+
Pleurotus ostreatus	+	+	+
Pleurotus sapidus	+	+	+
Trametes gibbosa	+	+	+
Trametes hirsutua	+	+	+
Trametes versicolor	+	+	+
Phanerochaete chrysosporium	+	+	+
Perenniporia medulla-panis	−	+	−
Trametes cingulata	−	+	−
Phanerochaete sordida	−	+	−
Bjerkandera sp.	−	+	+
Ceriporiopsis subvermispora	−	+	+
Cyathus stercoreus	−	+	+
Daedaleopsis confragosa	−	+	+
(Coriolus pruinosum)	+	+	?
Dichomitus squalens	−	+	+
Ganoderma volesiocum	−	+	+
Ganoderma colossum	−	+	+
Ganoderma lucidum	−	+	+
Grifola frondosa	−	+	+
Lentinus (Lentinula) edodes	−	+	+
Panus tigrinus	−	+	+
Pleurotus eryngii	−	?	+
Pleurotus pulmonarius	−	?	+
Rigidoporus lignosus	−	+	+
Stereum hirsutum	−	+	+
Stereum spp.	−	+	+
Trametes villosa	−	+	+
Pycnoporus cinnabarinus	−	−	+
Junghuhnia separabilima	+	−	+
Phlebia tremellosa (Merulius tremellosus)	+	?	+
Bjerkandera adusta	+	−	−
(Polyporus adustus)	−	+	+
	+	+	?
Coriolus consors	+	?	?

[a] ?, Information not given.

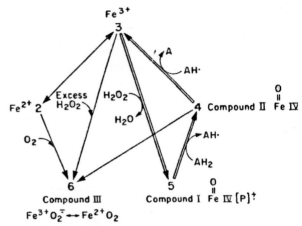

Fig. 4. The five redox states of lignin peroxidase (Renganathan and Gold, 1986).

tion with a reducing substrate, compound II can react with H_2O_2 and result in the formation of the catalytically less active compound III, which is stable but inactivated in the presence of H_2O_2. Compound III can transform back to the native enzyme and the cycle can get restarted.

It seems likely that the cation radicals of veratryl alcohol, the products of LiP catalysis, may mediate in the oxidation of lignin. These radicals may also assist in the reaction of LiP, compound II, with the reductant and, thereby, maintain the active peroxidase cycle.

Since their discovery, lignin peroxidases from various white-rot-fungi have been thoroughly studied. The LiP family contains multiple iso-enzymes with a molecular weight range of 38,000 to 43,000, with iso-electric points from 3.3 to 4.7. LiPs are glycoproteins of the oligomannose type with a number of possible O-glycosylation sites and with one or more N-glycosylation sites. It is not well understood why *P. chrysosporium*, and also other white-rot fungi, produce so many LiP iso-enzymes. One question has, therefore, been, are there specific roles, if any, for the individual iso-enzymes in lignin degradation? Also, do the different enzymes represent different posttranslational modifications of the product of a single gene, or are the iso-enzymes encoded by different genes? While there is no answer to the first question, all evidence indicates that each enzyme is encoded by a different gene. In addition to the molecular genetic studies of these lignin peroxidases, the x-ray crystal structure of LiP has recently been elucidated. These studies have, no doubt, led to a better understanding of the regulation and structure of the lignin-degrading enzyme system produced by *P.chrysosporium* and other white-rot fungi. Yet, with all of these advances, it has proven surprisingly difficult to demonstrate extensive ligninolytic activity using either isolated LiP or MnP. In fact, several investigators have reported polymerization, rather than depolymerization, of lignin preparation by LiP *in vitro*. So far, there has not been any application of this particular enzyme, which originally was thought to be an important breakthrough in the understanding of lignin degradation. To make matters even more confusing, an increasing number of studies have indicated that the value of *P. chrysosporium* as a model organism for lignin degradation might be limited since the majority of species within the group of white-rot fungi do not produce LiP.

B. Manganese Peroxidase

Manganese peroxidase (MnP) is another heme-containing extracellular fungal peroxidase. It was first identified in ligninolytic cultures of *Phanerochaete chrysosporium* as a Mn(II) and H_2O_2-dependent oxidase. It has now been purified and characterized from many other white-rot fungi. The mechanisms of MnP catalysis has been studied in detail. In a mixture with Mn^{2+} and H_2O_2 MnP oxidizes Mn^{2+} to Mn^{3+}, which oxidizes lignin, phenols, phenolic lignin model compounds, and high molecular weight chlorolignins. For Mn^{3+} to diffuse away from the enzyme and oxidize MnP substrates, it must be sufficiently stable and able to disassociate from the active site of the enzyme. Organic acids, metabolic products of white-rot fungi, form complexes with Mn^{3+}. These complexes are stable entities and allow for disassociation from the active site of the enzyme. Most of these chelators are carboxycilic acids such as mallonate, oxalate, and lactate. H_2O_2 also has the function to induce MnP gene transcription. In *P. chrysosporium*, MnP is induced 1.6-fold upon addition of Mn^{2+} and H_2O_2 compared to in their absence. However, induction of MnP by Mn^{2+} does not seem to be a general

trait in white-rot fungi. In *Phlebia radiata*, another white-rot fungus, high concentrations of Mn^{2+} had no influence on the induced levels of MnP, LiP, or laccase. It was also demonstrated that high Mn-containing cultures exhibited less efficient mineralization of synthetic lignin.

The crystal structure of MnP has been elucidated. The active site was found to consist of an His-ligand hydrogen bonded to an Asn residue and a distal side peroxide binding pocket formed by a catalytic His and Arg. The Mn^{2+} binding site was shown to be at the propionate end of the heme, Mn^{2+} being hexacoordinated by an Asp, two Glu-residues, a heme-propionate, and two water molecules.

Each MnP molecule was also found to contain 2 Ca ions, which are believed to have a structural role, and it has been shown that the thermal stability of the enzyme depends on the presence of these Ca^{2+} ions. Thermal inactivation appears to be a two-step process, and loss of Ca^{2+} decreases the enzyme stability; however, if excess Ca^{2+} is added to the medium, the enzyme can be reactivated. However, further inactivation, caused by the loss of the heme component, cannot be reversed.

Several other extracellular fungal enzymes are produced simultaneously with MnP and appear to work in accord with this enzyme. Laccase, which coexists in cultures of various fungi, was shown to work in concert with MnP in the degradation of lignosulfonates and solubilization of lignins. The highest degradation rates were obtained when the enzymes were working together. Similarly, an interaction between MnP and cellobiose dehydrogenase (CDH), both enzymes produced by *Trametes versicolor*, was also proposed. It is obvious that CDH can support MnP in different ways: (1) CDH oxidizes cellobiose to cellobionic acid, an efficient Mn^{3+} chelator; (2) CDH returns insoluble MnO_2 to the soluble Mn pool in the form of Mn^{2+} and Mn^{3+}. This reaction not only facilitates MnP production but also provides extra Mn^{2+} for MnP catalysis; (3) Quinones are reduced to their corresponding phenols by CDH. These phenols are substrates for MnP.

Both MnP and laccase can oxidize phenolic lignin substructures in various phenols but not nonphenolic lignin structures. However, both enzymes can attack nonphenolic lignin substructures in the presence of certain low molecular weight organic compounds, which act as mediators. It has, thus, been found that MnP, in the presence of glutathione (GSH), could efficiently oxidize veratryl alcohol, anisyl alcohol, and benzyl alcohol. The mechanism for this oxidation is that the formed Mn^{3+} oxidizes thiol to a thiyl radical, which abstracts a hydrogen from the substrate to form the corresponding aldehyde. This substrate oxidation was at least two fold higher under anaerobic conditions compared to under aerobic conditions.

It has also been demonstrated that, in the presence of Tween 80 or other lipids, $MnP + Mn^{2+}$ could oxidize a β-O-4 lignin model compound without the need for H_2O_2. This mechanism, by which lipid peroxy-radicals are generated, is called "MnP mediated lipid peroxidation." The peroxy-radicals easily abstract hydrogens from MnP substrates.

Several white-rot fungi, including *P. chrysosporium* and *T. versicolor,* have been shown to be able to degrade lignin and to bleach kraft pulp. There is a strong correlation between these abilities and the MnP activity in the culture solutions. The ability of MnP to increase brightness and decrease pulp kappa numbers has been well established. MnP purified from cultures of *T. versicolor,* was found to cause most of the demethylation and delignification of kraft pulp when compared to the effect of the complete, cell-free culture solution. It was also demonstrated that MnP bleached kraft pulp brown-stock, thereby releasing methanol. The bleaching effect was at its peak in cultures of *T. versicolor* when MnP production and activity were at their peak values.

C. Laccase

Laccase was first identified in the 1880s as a proteinaceous substance that catalyzed the lacquer curing process. With one of the defining reactions catalyzed by the enzyme, i.e., the ability to oxidize hydroquinone, the name "laccase" was implemented in the 1890s.

Laccases are so commonly found in the fungi that it is very likely that this enzyme may be ubiquitous in these organisms (Eriksson *et al.,* 1990). Laccases have also been found in a large variety of plant species, in insects, and, recently, also in a bacterium,

Azospirillum lipoferum. Laccases can function in different ways, such as participation in lignin biosynthesis (Dean and Eriksson, 1994), degradation of plant cell walls, plant pathogenicity, and insect sclerotization.

Laccases (p-diphenol: O_2 oxidoreductase; EC 1.10.3.2.) catalyze the oxidation of a variety of phenols, simultaneously reducing dioxygen to water. Substrate specificity of laccases are often quite broad and also vary with the source of the enzyme. Laccases are members of the blue copper oxidase enzyme family. Members of this family are characterized by having 4 cupric (Cu^{2+}) ions, where each of the known magnetic species (type 1, type 2, and type 3) is associated with a single polypeptide chain. The Cu^{2+} domain is highly conserved in the blue oxidases. While the crystallographic structure of a laccase has yet to be published, the crystallographic structure of ascorbate oxidase, another member of the blue copper oxidases, has been a good model for the structure of the laccase active site. Recently, the crystal structure of the Type-2 Cu depleted laccase from *Comprinus cinereus* at 2.2 Å resolution has been elucidated.

Not only p-diphenols, but also o-diphenols, aryldiamines, aminophenols, and hydroxyindols may be oxidized by laccases. To differentiate between laccase and other phenol oxidases is not a trivial matter, due to the relative nonspecificity of laccases in terms of their substrates. The combination, in white-hot fungi, of laccase with either lignin peroxidase (LiP) and/or manganese peroxidase (MnP) seems to be a more common combination of phenoloxidases than the LiP/MnP pattern found in *P. chrysosporium* (Table I).

Laccases produced by white-rot fungi are believed to participate mainly in the degradation of lignin, while laccases from other fungal species can serve different purposes. The fungal laccases can have very different characteristics, such as carbohydrate content, redox potential, substrate specificity, thermal stability, etc., depending upon the fungal species. The problem in assigning a role for laccase to substitute for the roles of either LiP or MnP in lignin degradation has been its low redox potential. The redox potentials of laccases studied so far have not been high enough to remove electrons from nonphenolic aromatic substrates. Laccases alone cannot oxidize these predominantly nonphenolic structures of lignin, which make up for 90% of lignin structures. However, it was demonstrated in the 1980s that laccase could oxidize a nonphenolic aromatic compound, rotenone, in the presence of chlorpromazine. It was further demonstrated that laccase, in the presence of syringaldehyde, could oxidize metoxylated benzyl alcohols. However, it was only when researchers at the Canadian Pulp and Paper Research Institute showed that two artificial laccase substrates, ABTS(2,2′-azinobis-[3-ethylbenzoline-6-sulfonate] and remazol blue could act as redox mediators, which enable laccase to oxidize also nonphenolic lignin model compounds, that the importance of laccase in lignin degradation was realized. When the same laboratory later demonstrated that kraft pulp could be partially delignified and demethylated by laccase from *T. versicolor* in the presence of ABTS, the importance of these findings was considerably extended. German researchers were then, in the mid-1990s, the first to apply the laccase mediator concept to pulp bleaching in pilot plant scale. For a redox mediator, they used 1-hydroxy benzoltriazol (1-HBT).

With the rapidly developing interest for laccase-based bleaching, investigations regarding the role played by laccases in lignin degradation by white-rot fungi were started at the University of Georgia. To identify white-rot fungi that produce large amounts of laccase, an extensive screening was undertaken. *Pycnoporus cinnabarinus*, found in the screening, was a white-rot fungus isolated from decaying pine wood in Queensland, Australia. This fungus turned out to be an ideal candidate for in-depth studies to investigate the importance of laccase in lignin degradation, since it produces only one isoelectric form of laccase, small amounts of an as yet unidentified peroxidase, and neither LiP nor MnP. The rate of lignin degradation by *P. cinnabarinus* is comparable to that of *P. chrysosporium*, despite the lack of both LiP and MnP. Contrary to what was first thought, *P. cinnabarinus* laccase had the same traits as practically all other laccases from white-rot fungi. The redox potential was not higher, the molecular mass, 76,500 Da, is comparable to other fungal laccases, spectroscopic characterization showed a typical laccase spectrum both in the UV/visible region and with EPR technique. These results confirm the presence of 4 Cu ions typical for an intact laccase active center. Glycosylation of the *P. cinnabarinus* was about 9%, which is just about average for fungal laccases. Comparison of the N-terminal protein sequence of this laccase with those of other fungal laccases showed the closest similarity to a laccase from *T. versicolor* (86%). High similarity was also found with laccases from other white-rot fungi, while, in contrast, the N-terminal sequences of laccases isolated from non-wood-rotting fungi were significantly different. These results seem to demonstrate that the lack of LiP and MnP does not exclude lignin degradation, even if the fungus produces only a laccase. These results were also taken as a support for the possible production by the fungus of its own laccase redox mediator system, allowing for the oxidation of nonphenolic lignin structures. Such a redox mediator system was also found and the redox mediator turned out to be 3-hydroxy antranillic acid (3-HAA). It could be demonstrated that *P. cinnabarinus* laccase, in the presence of 3-HAA, could oxidize also a nonphenolic lignin model dimer. This laccase redox mediator system, was also able to depolymerize synthetic lignin into low molecular weight oligomers.

The importance of laccase for lignin degradation by the white-rot fungus *P. cinnabarinus* was further demonstrated by production of laccaseless mutants of the fungus. It was shown that these laccaseless mutants were greatly reduced in their ability to metabolize ^{14}C-ring-labeled synthetic lignin. However, ^{14}CO$_2$, evolution could be restored in cultures of these mutants, to levels comparable to those of the wild-type cultures, by addition of purified *P. cinnabarinus* laccase. This is a clear indication that laccase is absolutely essential for lignin degradation by this white-rot fungus.

Although a lacccase mediator system could be both an interesting and a promising method for environmentally benign pulp bleaching, there are hurdles for such a system to be applied in pulp mills. The laccase mediators found so far are still too expensive, the effect of laccase mediator systems in pulp bleaching are still not satisfactory, and the mechanism for lignin degradation by the laccase/mediator system is yet unclear. To screen for more efficient laccase mediators, researchers at the University of Georgia developed a fast screening system. Monitoring the

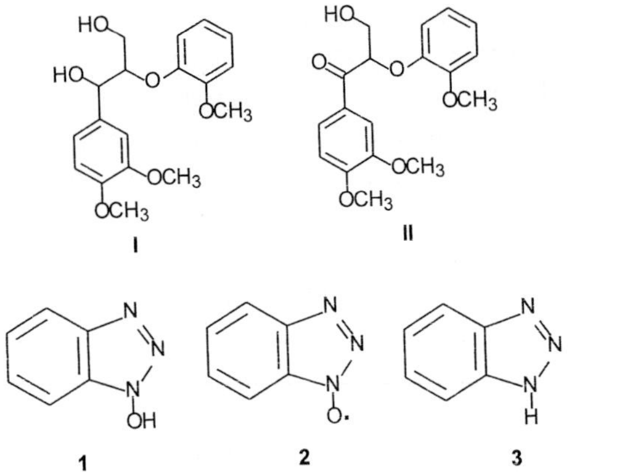

Fig. 5. Structures of mediator and lignin model compounds (reproduced with permission from K. Li *et al.*, 1998, *Biotechnology and Applied Boiochemistry* **27**, 239–243. © Portland Press Limited).

oxidation of compound I to compound II, Fig. 5, by HPLC was found to be useful for a fast screening of potentially effective laccase mediators, Scheme I. A lignin structure, such as the ketone II, is easily degraded by hydrogen peroxide under alkaline conditions. This would cause depolymerization of lignin macromolecules in pulp treated with an efficient laccase/mediator system, followed by treatment with an alkaline solution of hydrogen peroxide. This was also demonstrated to be true and substantial efforts to find effective laccase mediators are now made in many laboratories.

To investigate the importance, not only of the laccase mediators, but also of the laccases per se, several laccases were studied for the redox mediated oxidation of the nonphenolic lignin dimer I in Scheme I. In the presence of the redox mediators 1-HBT or

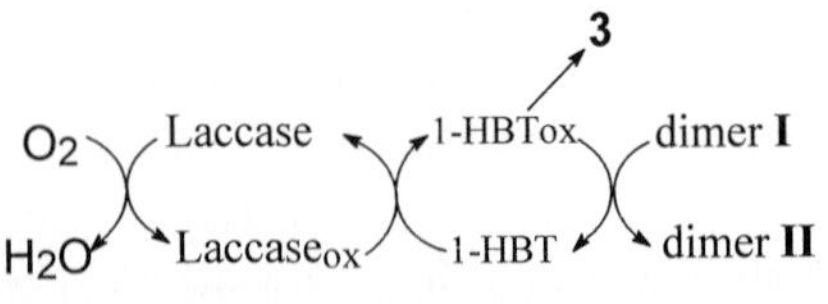

Scheme I. Proposed mechanism for the laccase-mediator oxidation of nonphenolic lignins. The number 3 in the scheme refers to compound 3 in Fig. 5 (reproduced with permission from K. Li *et al.*, 1998, *Biotechnology and Applied Biochemistry* **27**, 239–248,© Portland Press Limited).

violuric acid, the oxidation rates of dimer I by different laccases were found to vary considerably. In the oxidation of dimer I, both 1-HBT and violuric acid were consumed, to some extent. The redox mediators were simply converted to inactive components, such as benzoltriazol in the case of 1-HBT. Both 1-HBT and violuric acid inactivate the laccases. However, the presence of dimer I or any other lignin model compound in the reaction mixture, slows down this inactivation. The inactivation seems to be due mainly to the reaction of the redox mediator free-radicals, created by the laccases, with certain amino acids in the laccase molecule. With the present state of the art, it seems very possible that laccase plus an efficient redox mediator could be the environmentally benign pulp bleaching stage the pulp and paper industry is looking for.

See Also the Following Articles

ENZYMES, EXTRACELLULAR • PULP AND PAPER • TIMBER AND FOREST PRODUCTS

Bibliography

Buswell, J. A., and Odier, E. (1987). Lignin biodegradation. *CRC Crit. Rev. Biotechnol.* **6**, 1–60.

Côté, W. A., Jr. (1967). "Wood Ultrastructure—An Atlas of Electron Micrographs." Univ. of Washington Press, Seattle.

Dean, J. F. D., and Eriksson, K.-E. L. (1994). Laccase and the deposition of lignin in vascular plants. *Holzforschung* **48**, 21–24.

Eriksson, K.-E. L., Blanchette, R. A., and Ander, P. (1990). "Microbial and Enzymatic Degradation of Wood and Wood Components." Springer Verlag, Berlin.

Harvey, P. J., Gilardi, G.-F., Goble, M. L., and Palmer, J. M. (1993). Charge transfer reactions and feedback control of lignin peroxidase by phenolic compounds: Significance in lignin degradation. *J. Biotechnol.* **30**, 57–69.

Kirk, T. M., and Farrell, R. (1987). Enzymatic "combustion": The microbial degradation of lignin. *Ann. Rev. Microbiol.* **41**, 465–505.

Li, K., Helm, R. F., Eriksson, K.-E. L. (1998). Mechanistic studies of the oxidation of a non-phenolic lignin model compound by the laccase/I-HBT redox system. *Biotechnol. Appl. Biochem.* **27**, 239–243.

Renganathan, V., Gold, M, H. (1986). Spectral characterization of the oxidized states of lignin peroxidase, an extracellular heme enzyme from the white rot basidiomycete *Phanerochaete chrysosporium*. *Biochemistry* **25**, 1626–1631.

Lipases, Industrial Uses

Ching T. Hou

U.S. Department of Agriculture

GLOSSARY

enantioresolution   Separate molecules that are nonsuperimposable mirror image isomers.

interesterification   Interchange of esters between two triglycerides.

regioselectivity   The property of being able to select a specific position at triglyceride backbone.

stereospecificity   The property of being able to distinguish between the sn-1 and sn-3 position of triglyceride.

structured lipids   Triglycerides with tailor-made fatty acid composition.

transesterification   Interchange of esters.

triglyceride   Three fatty acids attached to a glycerol molecule.

LIPASES (TRIACYLGLYCEROL HYDROLASES E.C. 3.1.1.3)   are enzymes that catalyze the hydrolysis of triglycerides to glycerol and fatty acids. Microbial lipases are relatively stable and are capable of catalyzing a variety of reactions; they are potentially of importance for diverse industrial applications.

I. INTRODUCTION TO LIPASES

Lipase was first identified in pancreas by J. Eberle in 1834 and C. I. Bernard in 1856. Together with amylase and protease, it became one of the three major known digestive enzymes. However, because

of its difficulty in handling issues such as its water-insoluble substrate and a heterogeneous reaction system, lipase was rarely in the main stream of research. Plants, animals, and microorganisms all produce lipases. Animal lipases are found in several different organs, which contain, for example, pancreatic, gastric, and pregastric lipases. Earlier investigations were concerned mainly with enzymes which participated in lipid metabolism in animals. The most thoroughly studied has been the lipase from pancreas. Recently, increasingly more attention is being paid to lipases produced from bacteria and fungi.

In recent years, information on the mechanistic properties of lipases has become available. The notion that lipases have a catalytic triad, consisting of Ser-Asp-His, was confirmed by the structures of the human pancreatic lipase (HPL) and *Rhizomucor miehei* lipase (RML). For the *Geotrichum candidum* lipase (GCL), however, the catalytic triad was found to be Ser-Glu-His. In all three cases, the side chains of the active site amino acids form a configuration which is stereochemically very similar to that of serine proteases. In contrast to the proteases, the lipases share the common feature that the active site is buried in the protein. In the case of the HPL and RML, the active site is covered by a short amphipathic helix or "lid," whereas the active site of GCL seems to be covered by two nearly parallel amphipathic helices. The lid moves away upon interaction with the substrate. It has been proposed that this conformational change results in activation of these enzymes at an oil–water interface. However, this interfacial activation phenomenon does not appear in all lipases. For examples, lipases from *Pseudomonas aeruginosa* and *Candida antarctica B* do not have a "lid," and therefore lack interfacial activation. These enzymes constitute a bridge between lipases and esterases. Accordingly, lipases can be defined as esterases that are able to catalyze the hydrolysis of long-chain triacylglycerides.

II. SPECIFICITY

Lipases can be divided generally into the following four groups, according to their specificity in hydrolysis reactions.

1. Substrate Specific Lipases

Lipases are defined according to their ability to hydrolyze preferentially particular glycerol esters such as tri-, di-, monoglycerols, and phospholipids. For example, during digestion, the hydrolysis of triacylglycerol (TAG) is incomplete. The resulting diacylglycerol (DAG) is transformed into monoacylglycerol (MAG) but the hydrolysis of the latter is very slow. Accordingly, TAG is the favored substrate for most animal, plant, and microbial lipases. However, a few lipases, such as that from *Penicillium camembertii*, are reported to hydrolyze partial glycerides faster than TAG.

2. Regioselective Lipases

Regioselectivity is defined as the ability of lipases to distinguish between the two external positions (primary ester bonds) and the internal position (secondary ester bond) of the TAG backbone. During lipolysis of TAG substrates, 1,3-regioselective lipases preferentially hydrolyze the *sn*-1 and *sn*-3 positions over the *sn*-2 position. Examples of this type are lipases from pig pancreas, *Aspergillus niger*, *Rhizopus arrhizus* and *Mucor miehei*. True *sn*-2 regioselective lipase is very rare. The only lipase reported in this category is from *Candida antarctica*.

A number of lipases display little, if any, regioselectivity and hydrolyze all ester bonds in TAG. Examples of these random specificity lipases are numerous and include the lipases from *Penicillium expansum*, *Aspergillus sp.*, and *Pseudomonas cepacia*.

3. Fatty Acid Specific Lipases

Lipases can be specific for particular fatty acids or, more generally, for a class of fatty acids. Such lipases will hydrolyze glyceride esters of these acids, regardless of their position on the glycerol backbone. Examples of this type are lipases from *Penicillium roqueforti*, from premature infant gastric tissue (for short-chain fatty acids), and from *Geotrichum candidum* (for *cis*-9-unsaturated fatty acids).

4. Stereospecific Lipases

This type of specificity is defined as the ability of lipases to distinguish between the *sn*-1 and *sn*-3 positions of TAG. Reports of this type of specificity are relatively recent. Examples of this type are human lingual, *Candida antarctica B* and dog gastric lipases for *sn*-3; and *Humicola lanuginosa* and *Pseudomonas fluorescens* lipases for *sn*-1. Lipases are also able to differentiate between enantioisomers of chiral molecules. This ability has recently become very important in producing pure chiral isomers as intermediates for drug synthesis.

III. INDUSTRIAL USES OF LIPASES

The use of lipases in the bioconversion of oils and fats has many advantages over classical chemical catalysts. Lipases operate under milder reaction conditions over a range of temperatures and pressures that minimize the formation of side products. The usefulness of microbial lipases in commerce and research stems from their physiological and physical properties. In particular, (1) a large amount of purified lipase is usually available, (2) microbial lipases are generally more stable than animal or plant lipases, and (3) microbial lipases have unique characteristics compared with plant and animal lipases. Lipases catalyze, in either aqueous or nonaqueous systems, many reactions, such as hydrolysis, ester synthesis, transesterification, and enantioresolution of esters. Recently, a lipase-catalyzed reaction was demonstrated in supercritical fluid.

1. Hydrolysis of Esters

The Colgate–Emery Process operated at high temperature (250°C) and high pressure (50 atm) is used in the industrial production of fatty acids from tallow. There have been many attempts to replace this chemical process with a cleaner and milder bioprocess using lipases. However, as long as the cost of lipases is not reduced dramatically, there is no sign of using lipases in large-scale hydrolysis of tallow. Small quantities of high-quality fatty acids are currently produced by lipases in batch-wise processes. The biggest market for lipases for their hydrolysis function is in laundry detergent. Lipases used for this purpose include those from *Candida* and *Chromobacterium viscosum*. Since the early 1990s, much laundry powder detergent sold in the United States, Japan, and Europe contains microbial lipases. Dairy product flavors are produced by hydrolysis of milk fat with calf pregastric esterases. Microbial lipases have also been used to obtain specific flavor components by release of short- or medium-chain fatty acids from milk fat.

In leather manufacture, lipase is used in the processing of hides and skins to remove residual fats. It is now a common practice to utilize a mixture of lipases and proteases for this purpose.

In waste treatment, lipases are utilized in activated sludge and other aerobic waste processes, where thin layers of fats must be continuously removed from the surface of aerated tanks of permit oxygen transport.

Lipases are also used in the concentration of γ-linolenic acid from borage oil. Furthermore, a two-step enzymatic procedure for the isolation of erucic acid from rapeseed oil, based on chain-length discrimination by this lipase, has been developed.

2. Synthesis of Esters

For lipase-catalyzed synthesis and interesterification reactions, a specific water activity is needed to obtain good lipase activity and to minimize competitive hydrolysis reactions.

Monoacylglycerol (MAG) is one of the most used among a few FDA-approved emulsifiers. MAG is used in the food, cosmetics, and drug industries. Total world consumption is estimated at 60,000 tons annually. Currently, MAG is produced from tallow and glycerol in a continuous process using alkali catalyst at high temperature (220°C). The yield is 40–50%. Because of high temperature, the product MAG tends to have coloration and a burnt odor. There have been many reports on the production of MAG using lipases. Among them are: (a) by synthesizing MAG from fatty acid and glycerol using a high specific-

ity lipase from *Penicillium cummbertii*; (b) by first blocking two of the three hydroxy groups of glycerol with acetol and then reacting with fatty acid in the presence of lipase to produce MAG; and (c) by mixing tallow, glycerol and *Pseudomonas fluorecens* lipase at low temperature. The process (c) with 70–90% yields is the best one for industrial production of MAG so far. The ester synthesis activity of the commercially available lipases can be summarized, in general, as follows: bacterial lipases, such as those from *Pseudomonas sp.*, show higher ester synthesis activity, fungal lipases such as from *Mucor sp.* show lower, and yeast lipases such as *Candida* lipases show negligible ester synthesis activity. The reason for this is still unknown.

Lipases have been shown to have catalytic function in esterifying various alcohols with fatty acids. Lipase-catalyzed synthesis of sugar esters is also known.

3. Transesterification

1. Geraniol ester is one of the most important natural fragrances. Traditional methods, such as extraction from plant materials and direct biosynthesis by fermentation, are used for flavor and fragrance production. However, these methods exhibit a high cost of processing and a low yield of desired flavor component. Synthesis of geranyl acetate by lipase-catalyzed transesterfication (*Mucor miehei* lipase) in hexane was able to reach an 85% yield after 3 days of reaction.

2. Marcrocyclic lactones C_{14} to C_{16} are high-grade and expensive aromatic substances of decidedly musky fragrance. Chemical methods of synthesis used so far have been based on polycondensation of ω-polyhydroxycarboxylic acids to the respective polyesters and then on catalytic–thermal depolymerization and cyclization at high temperature and under vacuum. These methods are technically arduous because of the necessity of using high temperatures and high vacuum. Lipases have been used to produce macrocyclic lactones. For example: (1) lipase from *Pseudomonas sp.* and porcine pancreas catalyzed the lactonization of methylesters of ω-hydroxy acids with C_{12}–C_{16}, producing, with high yield, mono- and dilactones; (2) porcine pancreatic lipase lactonized various γ-hydroxy acid esters with

high yield and high enantioselectivity; (3) lipases from *Candida cylindracea, Pseudomonas sp.,* and porcine pancreas esterified dicarboxylic acids and diols with various carbon chain lengths to macrocyclic lactones; and (4) lipase from *Mucor javanicus* catalyzed the lactonization of 15-hydroxypentadecanoic and 16-hydroxyhexadecanoic acids to macrocyclic lactones.

3. Cocoa butter is the most expensive triglyceride. It consists of a saturated fatty acid, such as stearic, at its 1,3-positions and an unsaturated fatty acid at its 2-position (stearic–unsaturated fatty acid–stearic, or SUS). Normally, the starting materials are palm oil mid fraction (palmitic–oleic–palmitic, or POP) or olive oil (oleic–oleic–oleic, or OOO). The purpose of the lipase-catalyzed reaction is to introduce more stearate into the triglycerides by interesterification producing POS (palmitic–oleic–stearic) or SOS (stearic–oleic–stearic). The cocoa butter equivalent can be used in the chocolate and confectionery industry.

Following is a typical process: *Rhizopus chinensis* was grown on porous polyurethane particles. The mycelium-covered particles were collected, washed with acetone, and dried to form immobilized cells (lipase). These immobilized cells were used to produce (SOS) at 40% yield from olive oil and methyl stearate by interesterification. When this process was operated at less than 100 ppm water content, the half-life of the biocatalyst (lipase) was 1.7 months.

4. Synthesis of Structured Lipids

1,3-Regioselective lipases have been widely used on an industrial scale to obtain new fats with nutritionally improved properties. Human milk fat equivalent was synthesized by interesterifying tripalmitin with polyunsaturated fatty acid (PUFA). The result was a TAG rich in palmitic acid in position *sn*-2 and with PUFA on the 1,3-positions. Human milk fat substitute was also produced solely from vegetable sources for infant formula. The product with a saturated fatty acid (palmitic) at the *sn*-2 position and oleic acid at both 1 and 3 positions were produced from palm oil and oleic acid through interesterification using lipases.

A product with a saturated fatty acid in the position *sn*-2 was shown to improve digestibility and to enhance the absorption of other nutrients. Similarly, TAG with PUFA at the *sn*-2 position and medium-chain fatty acids at the 1,3-positions, which can be produced by interesterification using lipases, was shown to be more rapidly absorbed than the usual TAG. Short-chain preference lipases may be used in the production of low-calorie structured TAG. The short-chain specificity of *Candida antarctica* may prove useful in interesterification reactions to increase the ratio of medium-chain TAG in different oils. *Rhizomucor miehei* lipase was used in the synthesis of position-specific low-calorie structured lipids by interesterification of tristearin and tricaprin. The resulting structured lipids have their specific fatty acids at the *sn*-1,3 position.

Lipases were also applied to the interesterification of palm oil and soy oil in an effort to reduce the solid content of palm oil.

5. Improve PUFA Content in Fish Oil

The worldwide production of fish oil is around 1.6 million tons annually. Most of the fish oil is hydrogenated for use in margarine and shortening. However, fish oil has the highest ω-3 polyunsaturated fatty acids (ω-3 PUFA) content (20–30%) among all kinds of oils. Recently, ω-3 PUFA such as eicosapentaenoic (EPA) and docosahexaenoic (DHA) acids were reported to have many physiological activities (anti-blood-coagulation, enhancement of memory, etc.). There are physical methods known, such as solvent fractionation and winterization, to increase ω-3 PUFA content up to 40% in fish oil. With this limitation, many companies developed bioprocesses to concentrate ω-3 PUFA in fish oil. Examples are (1) A two-step process to concentrate ω-3 PUFA. Through selective hydrolysis by lipase, fish oil was converted to partial glycerides and free fatty acids. Partial glycerides were esterified with free PUFA by lipase to produce PUFA-enriched fish oil. (2) PUFA-enriched fish oil was produced by urea adduct and acidolysis or interesterification using lipases. The yield reached 50–60%. PUFA was also concentrated from fish oil by transesterification of fish oil with ethanol using *Pseudomonas lipase*.

The ability of *Mucor miehei* and *Candida cylindracea* lipases to discriminate against the ω-3 family of PUFA has been used for the selective harvesting of PUFA from fish oil.

6. Enantioresolution of Esters

One important specificity of lipases is their ability to differentiate between enantiomers of chiral molecules. Microbial lipases are increasingly used in kinetic resolution of chiral compounds that serve as synthons in the synthesis of chiral pharmaceuticals and agrochemicals. Hydroxy acids and their derivatives are major target molecules, because they constitute the framework of many chiral natural products and biologically active agents. Lipases have also been widely used for the resolution of racemic alcohols and carboxylic acids through asymmetric hydrolysis of the corresponding esters. The following are examples of industrial and potential industrial uses of lipases in the production of enantioselective isomers.

Lipase from *Candida cylindracea* catalyzes the acidolysis between racemic 2-methylalkanoates and fatty acids in heptane, with a preference for the (S)-configured esters. The enzymatic enantioselective resolution of 2-substituted propionic acids has been the subject of intense investigation. Much of this effort has centered on the production of (R)-2-chloro-propionic acid due to its high value as an intermediate in the synthesis of a number of commercially important herbicides. A substantial body of literature also exists on the production of (S)-2-arylpropionic acids, which are valuable as anti-inflammatory agents.

Lipases are used in the synthesis of chiral synthons as intermediates for the synthesis of paclitaxel (Taxol). Taxol, a complex polycyclic diterpene, exhibits a unique mode of action on microtubule proteins that are responsible for the formation of the spindle during cell division. Taxol has been used to treat various cancers, especially ovarian cancer. Currently, Taxol is produced from extracts of the bark of the Pacific yew tree by a cumbersome purification process. An alternative method to produce the chiral intermediate, 3R-cis-acetyloxy-4-pheny1-2-azetidinone, was developed using lipases from *Pseudomonas cepacia* and *Pseudomonas sp.* Both lipases achieve over 95% yield and 99.4% optical purity.

Lipases are used in the synthesis of a lactol, [3aS-(3aa,4a,7aa)]-hexahydro-4,7-epoxy-isobenzo-furan-1-(3H)-one, which is a key chiral intermediate for the total synthesis of a new cardiovascular agent, useful in the treatment of thrombolic disease. A *Pseudomonas fluorescens* and *Pseudomonas cepacia* lipase-catalyzed reaction achieved a greater than 85% yield and 97% optical purity of the chiral intermediate for the synthesis of a thromboxane A_2 antagonist.

Lipases are also used in the production of an intermediate for the synthesis of an antihypertensive drug. Captopril is designated chemically as 1-[(2S)-3-mercapto-2-methylpropionyl]-L-proline. Its S-configuration compound is 100 times more active than its corresponding R-enantiomer. Captopril prevents the conversion of angiotensin I to angiotensin II by inhibition of angiotensin-converting enzyme. Lipases from both *Pseudomonas cepacea* and a *Pseudomonas sp.* catalyzed the production of the S-isomer at greater than 32% yield and 96% optical purity. This reaction was conducted in methanol and with immobilized lipase.

Immobilized *Pseudomonas cepacia* lipase was used in organic solvent for the selective acylation of a key alcohol intermediate, which is used for the synthesis of Camptosa, a drug used in the treatment of ovarian cancer. A yield of greater than 46% conversion and 0.79 enantiomeric excess was achieved.

Lipases are also used in the production of a stereospecific isomer for the production of a β-blocker. β-blockers are a group of antihypertensive and cardiovascular drugs, which contain aryloxypropanolamine structure with an asymmetric carbon. There are over 24 drugs containing this type of moiety, with sales of over 3 billion dollars annually. Traditionally, racemic arylpropanolamine was used in the synthesis. Recently, due to social and economic demand, it is required to use only the physiologically active pure (S)-enantiomer for the synthesis of β-blockers. Both lipases and esterases are used in this respect.

Almost every pharmaceutical company has its own lipase process in producing synthons for the synthesis of their patented drugs. With the rapid progress in molecular biology and molecular modeling of protein three-dimensional structure, within a few years "tailor-made" lipases with improved properties in activity, stability, and designed specificity can be expected.

See Also the Following Articles

BIOCATALYSIS FOR SYNTHESIS OF CHIRAL PHARMACEUTICAL INTERMEDIATES • ENZYMES IN BIOTECHNOLOGY

Bibliography

Alberghina, L., Schmid, R. D., and Verger, R. (eds.) (1991). "Lipases: Structure, Mechanism and Genetic Engineering." VCH Publishers, Inc., New York.

Dordick, J. S., Patil, D. R., Parida, S., Ryu, K., and Rethswisch, D. G. (1992). Enzymatic catalysis in organic media. Prospects for the chemical industry. *In* "Catalysis of Organic Reactions" (W. E. Pascoe, ed.), pp. 267–292. Marcel Dekker, New York.

Foglia, T. A., and Villeneuve, P. (1997). Lipase specificities: Potential application in lipid bioconversion. *INFORM* 8, 640–650.

Gandhi, N. N. (1997). Application of lipases, a review. *Am. Oil Chem. Soc.* 74, 621–634.

Marcrae, A. R., and Hammond, R. C. (1985). Present and future applications of lipases. *Biotechnol. Genetic Eng. Rev.* 3, 193–217.

Yamane, T. (1987). Enzyme technology of lipids industry: An engineering overview. *J. Am. Oil Chem. Soc.* 64, 1657–1662.

Yamane, T. (1991). Reactions catalyzed by lipid-related enzymes. *Yukagaku* 40, 965–973.

Lipid Biosynthesis

Charles O. Rock

St. Jude Children's Research Hospital and the University of Tennessee

GLOSSARY

acyl carrier protein (ACP) A low molecular weight protein containing an active site sulfhydryl group attached to the protein via a 4′-phosphopantetheine linkage. ACP functions as the acyl group carrier for all of the intermediates in fatty acid biosynthesis.

Escherichia coli The organism that has been used for the vast majority of work on bacterial lipid metabolism.

fatty acids Long (12–20) saturated or unsaturated carbon chains that terminate in a carboxylic acid. Fatty acid composition defines the physical properties of the membrane bilayer.

genetic nomenclature The designation of genes using a four-letter code that is italicized, such as *fabH* and *plsB*. The corresponding protein products of the genes are abbreviated as FabH and PlsB.

phospholipids The major components of biological membranes that contain two fatty acids, esterified to the first two positions on the glycerol backbone, and a polar headgroup, attached to the third position via a phosphodiester bond.

type II fatty acid synthase The class of fatty acid synthases common to bacteria that consist of discrete, separable enzymes encoded by unique genes. Fatty acids are built by rounds of elongation that 2-carbon units to the growing acyl chain.

LIPIDS are crucial components of all bacteria. The fatty acids and phospholipid components are structural molecules that are responsible for the barrier properties of cell membrane. The major steps in the biosynthesis of fatty acids and phospholipids are known and the corresponding genes have been identified.

The bacterium most thoroughly understood at the molecular level is the gram-negative organism *Escherichia coli*, and lipid metabolism has been most extensively studied in this organism. In *E. coli*, the phospholipids are produced exclusively for use in membrane biogenesis and comprise 10% of the dry weight of the cell. Bacteria expend a great deal of energy to produce the fatty acid components of membrane lipids and their biosynthesis is tightly regulated. However, there are many other critical cellular components that are produced from intermediates and products of the fatty acid and phospholipid biosynthetic pathways. These components include lipoproteins, biotin, lipoic acid, and quorum-sensing molecules. This short review will draw from the wealth of information on *E. coli* lipid metabolism, but it is important to understand that the diversity of bacteria is enormous and that *E. coli* physiology cannot be considered typical.

I. FATTY ACID BIOSYNTHESIS

Most bacteria contain a type II, dissociable fatty acid synthase system. Each of the enzymes in the pathway are purified as discrete entities and are encoded by unique genes. There are many examples of isozymes that carry out the same basic chemical reaction, but due to their substrate specificity, fulfill different roles in producing the spectrum of fatty acid structures. A key feature of bacterial fatty acid synthesis is that the intermediates in the pathway are covalently attached to acyl carrier protein (ACP), a small (8.86 kDa) and highly soluble acidic protein. ACP is a necessary reactant in all steps of fatty acid biosynthesis including initiation, elongation and acyl transfer to the membrane bilayer. ACP is one of the most abundant proteins in cells constituting 0.25% of the total soluble protein in *E. coli*. The acyl groups are attached to the terminal sulfhydryl of the 4'-phosphopantetheine prosthetic group, which, in turn, is attached to a serine residue via a phosphodiester linkage. This prosthetic group undergoes metabolic turnover, and the apo-protein is inactive in fatty acid synthesis. ACP also has other func-

tions in bacteria. ACP is required for the synthesis of membrane-derived oligosaccharides and tightly associates with MukB, a protein involved in chromosome organization.

Acetyl-CoA carboxylase catalyzes the first step toward fatty acid synthesis (Fig. 1). The overall reaction can be divided into two half reactions: (a) the ATP-dependent carboxylation of biotin and (b) the transfer of the carboxyl group to acetyl-CoA. The biotin is covalently attached to a low molecular weight protein called biotin carboxy carrier protein. The biotin must be coupled to this protein to be active in the reaction. The overall reaction is catalyzed by a complex composed of four distinct gene products (*accA*, *accB*, *accC*, and *accD*), which dissociate into two dissimilar subunits that catalyze the half

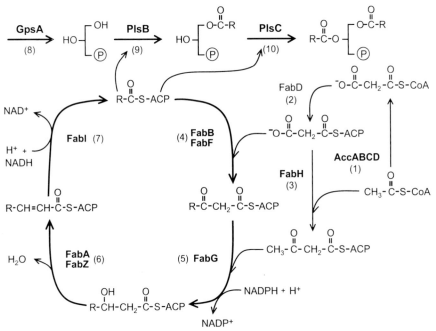

Fig. 1. Fatty and phosphatidic acid synthesis in *E. coli*. Malonyl-CoA is formed by the carboxylation of acetyl-CoA by the acetyl-CoA carboxylase complex (AccABCD) (1). The malonyl group is transferred to ACP by malonyl-CoA:ACP transacylase (FabD) (2) which provides the successive two carbon units to the growing acyl chain. Fatty acid synthesis is initiated by the condensation acetyl-CoA with malonyl-ACP by β-ketoacyl-ACP synthase III (FabH) (3). Subsequent rounds of elongation (steps 4–7) begin by the condensation of malonyl-ACP with acyl-ACP catalyzed by either β-ketoacyl-ACP synthase I (FabB) or II (FabF) (4). The resulting β-ketoacyl-ACP is reduced by the NADPH-dependent reductase, FabG (5). The β-hydroxyacyl-ACP is then dehydrated to enoyl-ACP by either of two dehydratases (FabA or FabZ) (6). The final step in each cycle is catalyzed by the NADH-dependent enoyl-ACP reductase (FabI) (7).

reactions when cell extracts are prepared. The acetyl-CoA carboxylase reaction supplies all of the malonyl groups that are needed for fatty acid elongation and is potentially a key regulatory step in the pathway. The expression of the acetyl-CoA carboxylase genes is growth-rate regulated, but the mechanisms that regulate the reaction at a biochemical level are less clear.

Acetyl-CoA functions as the primer for fatty acid synthesis (Fig. 1). Acetyl-CoA is directly incorporated into the first four carbon intermediate by β-ketoacyl-ACP synthase III (*fabH*), a condensing enzyme that catalyzes the condensation of acetyl-CoA with malonyl-ACP to form β-ketobutyryl-ACP. The malonyl groups used in fatty acid synthesis must first be transferred from CoA to ACP (Fig. 1). This is accomplished by the malonyl-CoA:ACP transacylase (*fabD*). This enzyme is essential, but highly active, and rapidly catalyzes the transacylation reaction.

The elongation cycle of fatty acid synthesis consists of a set of four reactions (Fig. 1). The first step is the elongation of the growing acyl chain by either of two condensing enzymes, β-ketoacyl-ACP synthases I (*fabB*) or II (*fabF*). This is an essentially irreversible step in the cycle. Not surprisingly, these condensing enzymes play an important role in regulating the product distribution in the pathway. The two condensing enzymes have overlapping substrate specificities, but both have unique, nonredundant functions. The *fabB* gene product is an essential component in unsaturated fatty acid synthesis, and the *fabF* gene product is responsible for the thermal regulation of membrane fatty acid composition.

Next, the β-ketoacyl-ACP is reduced by the NADPH-dependent β-ketoacyl-ACP reductase (*fabG*). This reaction is catalyzed by a single enzyme that is active on all chain lengths. The resulting β-hydroxyacyl-ACP is then dehydrated by one of two dehydratases (*fabA* or *fabZ*). Like the condensing enzymes, these two dehydratases have overlapping substrate specificities but play different roles in unsaturated fatty acid synthesis. *FabA* catalyzes the essential isomerization step in unsaturated fatty acid synthesis and *fabZ* functions to elongate the resulting unsaturated fatty acids. Both isozymes have overlapping specificity toward short-chain and saturated acyl-ACP intermediates. The final step in each cycle is catalyzed by the NADH-dependent enoyl-ACP reductase (*fabI*). There is only a single enoyl-ACP reductase that functions with all substrates. Rounds of elongation continue until the acyl-ACP chain length reaches 16–18 carbons. These chains are used as substrates by the glycerolphosphate acyltransferase in the first step in phospholipid formation (Fig. 1).

Unsaturated fatty acids are critical to maintaining the proper barrier properties of membranes. If the unsaturated fatty acid content falls below 15% of the total membrane fatty acids, *E. coli* stops growing and eventually lyses. Two enzymes are specifically required for unsaturated fatty acid biosynthesis. The FabA β-hydroxyacyl-ACP dehydratase carries out the key reaction at the branch point between saturated and unsaturated fatty acid synthesis. Not only does the FabA protein catalyze the dehydration reaction shown in Fig. 1, it also specifically isomerizes *trans*-2-decenoyl-ACP to *cis*-3-decenoyl-ACP. The FabB condensing enzyme is also essential for unsaturated fatty acid synthesis, and is probably specifically responsible for the elongation of the *cis*-3-decenoyl-ACP intermediate produced by FabA.

II. PHOSPHOLIPID BIOSYNTHESIS

Whereas the enzyme systems of fatty acid synthesis are soluble, the enzymes of phospholipid synthesis are membrane-bound. The glycerolphosphate acyltransferase (the *plsB* gene product) (Fig. 1) catalyzes the first step in phospholipid synthesis. This enzyme esterifies the acyl-ACP end-products of fatty acid synthesis to glycerolphosphate to form the first membrane associated lipid in the pathway, lysophosphatidic acid. A second acyltransferase forms the key intermediate phosphatidic acid (the *plsC* gene product), which is the precursor to all membrane phospholipids (Fig. 2). Both of these enzymes are tightly bound to the inner membrane and they can use either acyl-ACP or acyl-CoA as the acyl donors. These acyltransferases are responsible for the positional asymmetry in the fatty acid composition of phospholipids. The glycerolphosphate acyltransferase prefers saturated fatty acids as substrates and the lysophosphatidic acid acyltransferase prefers unsaturated fatty acid thioester substrates. This substrate specificity

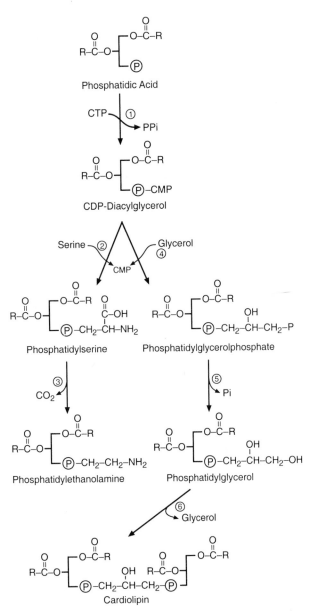

Fig. 2. Phospholipid biosynthesis in *E. coli*. The three major phospholipids of *E. coli* are synthesized from phosphatidic acid in a series of reactions catalyzed by six enzymes. (1) Phosphatidate cytidylyltransferase (*cds*); (2) phosphatidylserine synthase (*pass*); (3) phosphatidylserine decarboxylase (*psd*); (4) phosphatidylglycerolphosphate synthase (*pgsA*); (5) phosphatdiylglycerolphosphate phosphatase (*pgpA* or *pgpB*); and (6) cardiolipin synthase (*cls*).

accounts for the observed preference for saturated fatty acids at the 1-position and unsaturated fatty acids at the 2-position.

E. coli has a simple membrane phospholipid composition, consisting of phosphatidylethanolamine (75%), phosphatidylglycerol (15–20%), and cardiolipin (5–10%). The polar headgroups are attached to phosphatidic acid using six enzymes (Fig. 2). The first intermediate is CDP-diacylglycerol. The CDP-diacylglycerol reacts with either serine or glycerolphosphate to form either of two ephemeral intermediates, phosphatidylserine or phosphatidylglycerolphosphate. Phosphatidylserine is rapidly decarboxylated to the major membrane phospholipid, phosphatidylethanolamine. The second most abundant membrane component, phosphatidylglycerol, is produced by the dephosphorylation of phosphatidylglycerolphosphate. Cardiolipin is synthesized from the condensation of two phosphatidylglycerol molecules with the concomitant release of glycerol. All these enzymes are associated with the cytoplasmic membrane. The polar headgroup composition of the membrane is critical for proper membrane function and is thought to be regulated through feedback inhibition of the biosynthetic enzymes by the end-products of the pathway.

III. PHOSPHOLIPID TURNOVER

There are two predominant cycles of phospholipid turnover in *E. coli* that require ancillary enzymes for phospholipid synthesis. The polar headgroup of phosphatidylglycerol is metabolically active. The glycerolphosphate is transferred to a family of molecules called membrane-derived oligosaccharides. The resulting diacylglycerol is phosphorylated by diacylglycerol kinase (*dgk*) and the resulting phosphatidic acid, which reenters the phospholipid biosynthetic pathway (Fig. 2). In the overall cycle, only the glycerolphosphate portion of phosphatidylglycerol is consumed, while the lipid portion is recycled. The second cycle consists of the turnover of fatty acids at the 1-position of phosphatidylethanolamine. These fatty acids are removed from phosphatidylethanolamine by either transfer to lipoproteins (see following) or hydrolysis by phospholipase A$_1$. The accumulation of the lyso-phospholipid resulting from this metabolic turnover is deleterious to cells and is rapidly reacylated by an acyltransferase that utilizes only acyl-ACP as the acyl donor (*aas*). This acyltransferase is unique in that it activates a free fatty acid in the presence of ATP to an enzyme-bound ACP subunit. The Dgk and Aas ancillary enzymes preserve membrane integrity and conserve fatty acids that are energetically expensive to produce.

IV. ALTERNATE DESTINATIONS FOR FATTY ACIDS

Lipopolysaccharides are major components of the outer membrane of gram-negative bacteria and the destination of about 25% of the fatty acids. These lipids are composed of three domains. The region exposed to the cell exterior consists of a specific polysaccharide chain (O-antigen), which is the basis for the serological differences between strains. The O-antigen is attached to a core carbohydrate structure that is common to some groups of bacteria. This entire structure is linked to 2-keto-3-deoxyoctonate which, in turn, is attached to lipid A. Lipid A is a disaccharide that contains the only major fatty acid that is not a component of the phospholipid, 14-carbon fatty acid β-hydroxymyristate, attached by both ester and amide linkages. The β-hydroxymyristate is siphoned directly off the fatty acid biosynthetic pathway. Lipid A anchors the lipopolysaccharide molecules to the outer membrane and also functions as a host cell mitogen and endotoxin during bacterial infections.

A second destination for fatty acids is the acylation of membrane proteins. One of the most abundant proteins in *E. coli* is the outer membrane lipoprotein that contains an amino terminal cysteine residue with a diacylglycerol moiety attached to the sulfhydryl and an amide-linked fatty acid on the amino terminus. These fatty acids are not directly derived from the fatty acid biosynthetic pool, but rather are trans-acylated to the lipoprotein from the phospholipid pool. Fatty acylation is also necessary for the activa-tion of the *E. coli* virulence factor, prohemolysin, to the mature toxin. In both cases, the fatty acid modifications are required for the proper functioning of the proteins.

There are several essential small molecules that are derived, in part, from intermediates in fatty acid biosynthesis. Lipoic acid and biotin are two vitamins that have fatty acid moieties that are siphoned from the intermediates in fatty acid biosynthesis. Also, the synthesis of the quorum-sensing N-acyl-homoserine lactones requires intermediates from the fatty acid biosynthetic pathway.

V. REGULATION OF LIPID BIOSYNTHESIS

The control of the amount and composition of the phospholipids is critical to bacterial physiology and the pathways are tightly regulated at several points. Most of the enzymes of fatty acid and phospholipid synthesis are constitutively expressed and regulation of pathway activity is an intrinsic property of the proteins themselves. A clear example of this is temperature modulation of membrane phospholipid fatty acid composition, which is controlled by the activity of β-ketoacyl-ACP synthase II (the product of the *fabF* gene). This regulatory enzyme is critically important to maintaining the proper membrane function by ensuring that the majority of the phospholipids are in a fluid state at a given growth temperature. A second example is the interaction between the condensing enzymes and the glycerolphosphate acyltransferase, which governs the chain-length of fatty acids. However, there is also a transcriptional component to the regulation of fatty acid synthesis. The most interesting example of this is the reduced expression of an enzyme in fatty acid synthesis (*fabA*) coordinated with the increased expression of the enzymes of β-oxidation mediated by a single transcription factor (FadR) that functions either as a repressor or an activator, depending on the location of its DNA binding site in the promoter of the gene. Finally, fatty acid and fatty acid synthesis are tightly coupled and coordinately regulated with macromolecular synthesis. Thus, starvation for amino acids triggers the stringent response which leads to an abrupt cessation of both fatty acid and macromolecular synthesis. This coordination is critical to maintaining the proper ratio of proteins and lipids in the cellular membranes. Glycerolphosphate acyltransferase is a key regulator of phospholipid and fatty acid biosynthesis and responds to intracellular signals to coordinately modulate the rate of phospholipid synthesis. Fatty acid synthesis is regulated as a consequence of the accumulation of the acyl-ACP substrate for the glycerolphosphate acyltransferase, which act as feedback inhibitors of fatty acid initiate and elongation. The targets for acyl-ACP regulation of fatty acid synthesis are FabH, which governs the initiation of fatty acid synthesis, and FabI, which controls the rate of fatty acid elongation.

VI. FATTY ACID β-OXIDATION

Many bacteria are capable of degrading exogenous fatty acids as a carbon source for growth. A common transcriptional repressor coordinately regulates the genes required for the uptake and oxidation of fatty acids and the binding of this repressor to target DNA is mediated by the concentration of long-chain acyl-CoA. The biochemistry of the inducible enzymes of β-oxidation has been extensively studied and the pathway for degradation is defined. It is important to understand that acyl-CoAs serve as the substrates for the degradative β-oxidation enzymes, whereas the fatty acid biosynthetic enzymes use acyl-ACPs.

VII. INHIBITORS OF LIPID SYNTHESIS

The importance of fatty acid synthesis in bacterial physiology is underscored by the evolution of antibiotics that are produced by fungi to inhibit the growth of bacteria. Cerulenin is an irreversible inhibitor of the condensing enzymes and blocks the proliferation of a broad spectrum of bacteria. Thiolactomycin is a second antibiotic that inhibits the condensation reaction. Thiolactomycin does not inhibit the activity of type I fatty acid synthases found in mammals, whereas cerulenin does. The enoyl-ACP reductase component of type II fatty acid synthase is the target for the antituberculosis drug, isoniazid. The enzymes of fatty acid biosynthesis are emerging as important targets for the development of antibacterial drugs and it is anticipated that additional drugs will be developed in the future to exploit the dependence of bacteria on this essential pathway.

VIII. OTHER BACTERIAL LIPIDS

It needs to be emphasized that lipid metabolism in *E. coli* differs significantly from many other common bacteria. An extraordinary feature of bacterial lipids is their incredible diversity. Many bacteria do not make unsaturated fatty acids. Instead, they produce branched-chain fatty acids, using isobutyryl-CoA or 2-methylbutyryl-CoA as the primer. Some bacteria have a type I, multifunctional fatty acid synthase similar to the synthases found in eukaryotes. Obligate aerobes (i.e., *Bacilli*) synthesize unsaturated fatty acids by an oxygen-dependent reaction. Finally, a group of organisms once considered to be bacteria, but now recognized as a separate entity (the *Archaea*), have very unusual lipids build on mevalonic acid with complex lipids containing ether linkages.

See Also the Following Articles

Lipopolysaccharides • Outer Membrane, Gram-Negative Bacteria • Quorum Sensing in Gram-Negative Bacteria

Bibliography

Cronan, J. E., Jr., and Rock, C. O. (1996). Biosynthesis of membrane lipids. *In* "*Escherichia coli* and *Salmonella typhimurium*: Cellular and molecular biology" (F. C. Neidhardt, R. Curtis, C. A. Gross, J. L. Ingraham, E. C. C. Lin, K. B. Low, B. Magasanik, W. Reznikoff, M. Riley, M. Schaechter, and H. E. Umbarger, eds.), pp. 612–636. American Society for Microbiology, Washington, DC.

Raetz, C. R. H. (1996). Bacterial lipopolysaccharides: A remarkable family of bioactive macroamphiphiles. *In* "*Escherichia coli* and *Salmonella typhimurium*: Cellular and molecular biology" (F. C. Neidhardt, R. Curtiss, C. A. Gross,

J. L. Ingraham, E. C. C. Lin, K. B. Low, B. Magasanik, W. Reznikoff, M. Riley, M. Schaechter, and H. E. Umbarger, eds.), American Society for Microbiology, Washington, DC.

Raetz, C. R. H., and Dowhan, W. (1990). Biosynthesis and function of phospholipids in *Escherichia coli*. *J. Biol. Chem.* **265**, 1235–1238.

Rock, C. O., and Cronan, J. E., Jr. (1996). *Escherichia coli* as a model for the regulation of dissociable (type II) fatty acid biosynthesis. *Biochim. Biophys. Acta* **1302**, 1–16.

Rock, C. O., Jackowski, S., and Cronan, J. E., Jr. (1996). Lipid metabolism in procaryotes. *In* "Biochemistry of Lipids, Lipoproteins and Membranes" (D. E. Vance and J. E. Vance, eds.), pp. 35–74. Elsevier Publishing Co., Amsterdam, The Netherlands.

Lipids, Microbially Produced

Jacek Leman

Olsztyn University of Agriculture and Technology, Poland

GLOSSARY

biosurfactant lipid A surface-active molecule with a hydrophobic moiety that is a fatty acid and a hydrophilic moiety that is a carbohydrate, amino acid, peptide, or phosphate.

fatty acids Long-chain aliphatic carboxylic acids without (saturated) or with one (unsaturated) or more (polyunsaturated) double (–CH=CH–) bonds. Normally found esterified to glycerol (acylglycerols), but can be esterified to hydroxy compounds (waxes) or sterols.

lipid A molecule containing aliphatic or aromatic hydrocarbon. Can be categorized as polar lipids (with charged group, e.g., phospholipids, glycolipids, sulfolipids) and neutral lipids (with no charged group, e.g., acylglycerols, hydrocarbons, carotenoids, sterols) or as storage lipids (acylglycerols, wax esters), that are laid down as reserve of depot fat and source of energy when required, and structural lipids (phospholipids, lipoglycans), that structure the membranes of cells.

phospholipid Any lipid containing a phospho (H_2PO_3) group, but usually refers to lipids based on 1,2-diacylglycero-3-phosphate (phosphatidic acid).

single-cell oil A microbial lipid having the same triacylglycerol structure as plant oils and a similar distribution of fatty acids at the three positions of glycerol: saturated fatty acids are generally excluded from the central (*sn*-2) carbon atom; fatty acids are the same as either in plant oils (16:0, 18:1, 18:2) or fish and animal oils and fats (16:1, 20:4, 20:5, 22:6).

terpenoid lipids Lipids that are based on isopentenyl group ($CH_2=C(CH_3)CH=CH_2$) giving rise to sterols, carotenoids, polyprenols and the side chains of chlorophylls and quinons.

triacylglycerols A class of lipids in which the three alcohol groups of glycerol ($CH_2OH.CHOH.CH_2OH$) are esterified with a fatty acid. These, together with phospholipids, are the predominant lipid types in most eukaryotic cells.

MICROORGANISMS produce a wide array of lipids, sometimes quite complex, and can pragmatically be divided into oleaginous and nonoleaginous species, with respect to the ability for lipid accumulation, the former including the species accumulating lipid at more than 25% of cell dry weight. Microbial lipids are intracellular products, except glycolipids which, as biosurfactants, are extracellular. The lipids of microorganisms fulfill structural and functional roles in the cell and are of health importance for humans. The genetics and molecular biology of microbial lipids is not yet fully recognized, with the basis being laid by several major research groups worldwide. The entire subject of microbial lipids has been covered in a two-volume publication edited by Ratledge and Wilkinson (1988, 1989).

I. INTRODUCTION

A. Lipid Nomenclature

Lipids are molecules of molecular weight 100 to 500 that contain a substantial portion of aliphatic or aromatic hydrocarbon. (For the following, see Appendix.) Included are glycerolipids (I), sphingolipids (II), wax esters (III), terpenoids: sterols (IV) and carotenoids (V), ether lipids (VI), and more complex molecules, such as lipopolysaccharides and lipoproteins. Glycerolipids are the most common ones, including mono-, di-, and triacylglycerols, phospholipids, sulfolipids, and glycolipids. Hydrocarbon chains, containing a variable number of methylene ($-CH_2$) groups, that have a methyl ($-CH_3$) group at one end (the n or ω terminus) and a carboxylic ($-COOH$) group at the other end (the Δ terminus) are saturated fatty acids (e.g., $CH_3(CH_2)_nCOOH$). By the removal of pairs of hydrogen atoms from adjacent methylene groups, unsaturated fatty acids are formed, containing either one or more (polyunsaturated acids) double ($-CH=CH-$) bonds, mostly methylene interrupted ($CH_3(CH_2)_x(-CH_2-CH=CH-)_y(CH_2)_zCOOH$), but methylene noninterrupted (conjugated) ($CH_3(CH_2)_x$ $CH=CH-CH=CH-(CH_2)_yCOOH$) bonds can occur. The structure of fatty acids is represented using the following numerical system: $m:n$ $\Delta aZ(E)$, $bZ(E), \ldots cZ(E)$, describing an acid containing m carbon atoms with n double bonds being positioned at carbon atom a, b, $\ldots$ c from the Δ terminus and having *cis* (Z) or *trans* (E) geometrical configuration. Families of polyunsaturated fatty acids, in which the last double bond is respectively 3, 6, and 9 carbon atoms from the methyl end of the chain, are denoted as n-3, n-6, and n-9 fatty acids. Typical saturated fatty acids synthesized by microorganisms include lauric (12:0), myristic (14:0), palmitic (16:0), and stearic (18:0) acids. Representative examples of major polyunsaturated fatty acids are shown in Fig. 1.

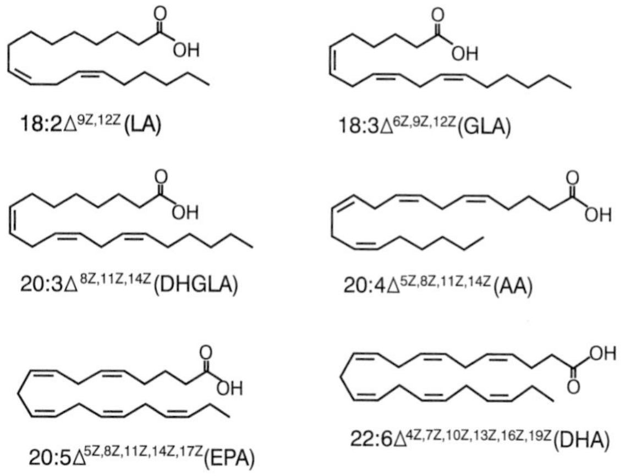

B. Lipid Biosynthesis

Saturated fatty acids are synthesized from acetyl-CoA produced during catabolism of carbohydrates. Two complex enzyme systems, i.e., acetyl-CoA carboxylase and fatty acid synthetase catalyze the reactions. Malonyl-CoA, derived from the acetyl-CoA carboxylase reaction is used to synthesize palmitic acid in a series of reactions, requiring NADPH and acetyl-CoA, catalyzed by the fatty acid synthetase complex. Further chain elongation with 2 carbon units occurs via Co-A derivatives, involving either acetate or malonate extender. Unsaturated fatty acids are most commonly synthesized by an aerobic path-

Fig. 1. Representative examples of polyunsaturated fatty acids synthesized by microorganisms.

way, in which the double bond of *cis* (Z) configuration is invariably introduced into the 9 position of the saturated fatty acid by specific monooxygenases in which NADPH is the coreductant. The desaturation produces n-9 acids, which may be converted to polyunsaturated fatty acids via pathways involving chain elongation and desaturation. Thus derived, two fatty acids, linoleic (18:2, n-6) and α-linolenic (18:3, n-3) are converted to two distinct families of polyunsaturated fatty acids through the n-6 and n-3 routes, respectively (Fig. 2), which can yield hydroxy-polyunsaturated fatty acids, such as prostaglandins, leukotrienes, and thromboxanes, collectively known as eicosanoids or lipid hormones. Triacylglycerols are synthesized from fatty acids and glycerol, which is a product of carbohydrate metabolism. Phosphoacylglycerols result from interaction of 1,2-diacylglycerols with phosphatidic acid or its esters with hydroxy compound. The biosynthesis pathways of other lipids (glycolipids, sulfolipids, waxes) have not been well established.

C. Lipid Importance

Lipids play both structural and functional roles in the microbial cell. The former assumes that lipids are the principal component of all membranes within the cell and the bacterial envelope, where the lipid is usually conjugated to a polysaccharide, e.g., lipid A in the outer membrane of some enterobacteria. The membrane lipids vary in nature and

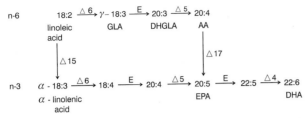

Fig. 2. Pathways for the biosynthesis of polyunsaturated fatty acids of the n-6 (linoleic) and n-3 (α-linolenic) series. Δ4, Δ5, Δ6, Δ15, and Δ17 are fatty-acid desaturases, and E is elongase. Desaturation occurs toward the carboxyl end of the molecule and chain elongation takes place at the carboxyl end, leaving the methyl end unaltered. GLA is γ-linolenic acid (6, 9, 12-*cis*-octadecatrienoic acid, 18:3 n-6); DHGLA is dihomo-γ-linolenic acid (8, 11, 14-*cis*-eicosatrienoic acid, 20:3 n-6); AA is arachidonic acid (5, 8, 11, 14-*cis*-eicosatetraenoic acid, 20:4 n-6); EPA is eicosapentaenoic acid (5, 8, 11, 14, 17-*cis*-eicosapentaenoic acid, 20:5 n-3); and DHA is docosahexaenoic acid (4, 7, 10, 13, 16, 19-*cis*-docosahexaenoic acid, 22:6 n-3). (Adapted from *Adv. Appl. Microbiol.* **43**, 1997, p. 210.)

fatty acid composition, mainly including phospholipids, in the bilayers of which other lipids, such as sterols or glycolipids, can be present. The fatty acid composition of membrane lipids is important since *cis*-unsaturation and the presence of branched-, cyclopropane-, or hydroxy acids ensure the required fluidity, permeability, and stability of the membrane, suitable for life processes. Functional roles of lipids include their being used as a storage material, and as essential elements of photosynthesis and respiratory chains. Lipids may also be produced by microorganisms to solubilize water-insoluble substrates to support the growth or affect adhesion of cells to hydrophobic phases. A more important function of lipids, particularly polyunsaturated fatty acids and phospholipids, is that they mediate cellular processes and ecological responses, including metamorphosis, reproduction, chemotaxis, water and solute transport, and immune function. Microbial lipids provide a unique class of polyunsaturated fatty acids for conversion into eicosanoids that regulate critical biological functions in humans, therefore attracting commercial interest in exploitation of microorganisms.

II. CHARACTERISTICS OF MICROBIAL LIPIDS

A. Archaeal Lipids

The *Archaea* lipids are unique and distinctly different from lipids found in all other living organisms. They are made up of repeating isopentanyl (VII) units that form the isopropanoid chains (also called phytanyl chains, VI) that are linked to glycerol via an ether bond ($-CH_2-O-CH_2-$) instead of an ester bond ($-CH_2-O-CO-$), and in the *sn* 2- and 3- positions (I) instead of the 1- and 2- positions, as found in other lipids. Typical lipids of *Archaea* that have been described include diether (VI) and tetraether lipids, phosphoglycolipids and sulfolipids, and cell wall-associated lipoglycans. The character of phytanyl lipids is determined by the presence of: (i) ether bonds that are much less susceptible to hydrolysis than are ester bonds, (ii) completely saturated alkyl chains with short side chains, which, while excluding oxidation, allows for close molecular packing and liquid crystallinity at ambient temperatures, (iii) pentacyclic rings in the biphanyl chains, ensuring fairly constant fluidity of the membranes in the extreme thermophiles (*Thermus, Caldariella, Sulfobous*). These characteristic features confer considerable stability to the membranes and cell wall of *Archaea*, explaining the resistance of many of these organisms to extremes of the environment, such as temperature, pH, and salinity.

B. Bacterial Lipids

The complex array of bacterial lipids varies from fatty acids through acylglycerols, fatty alcohols, glyco-, sulfo-, phospho-, and peptidolipids, to wax esters and hydrocarbons. Fatty acids synthesized by bacteria include straight-chain saturated fatty acids, ubiquitously found in nature, and unsaturated fatty acids of the cis-vaccenic acid (18:1 n-11) type or the less common oleic acid (18:1 n-9) type. Polyunsaturated fatty acids are not synthesized by the majority of bacteria, except in mycobacteria (e.g., phaleic acid, 36:5) with the double bonds not being methylene-interrupted (found in *Mycobacterium phlei* and *M. smegmatis*) and bacteria recovered from marine and freshwater sources (*Altermonas, Shewanella, Flexibacter, Vibrio*) that synthesize biologically important eicosapentaenoic acid (20 : 5), with the double bonds being methylene-interrupted. Branched-chain fatty acids with a methyl group positioned at different carbon atoms (VIII) are synthesized by *Bacillus, Clostridium, Listeria, Micrococcus, Sarcina, Staphylococcus, Flavobacterium, Bacteroides, Legionella, Vibrio, Desulphovibrio, Mycobacterium,* some *Pseudomonas,* and *Xanthomonas.* Cyclopropane fatty acids (19 : 0 and 17 : 0) are fairly common in bacterial lipids, with lactobacillic acid (IX) being most widely distributed in many organisms, including *Escherichia coli, Pseudomonas* spp., some *Rhizobium* and *Clostridium* spp., and, obviously, *Lactobacillus* spp. Hydroxy fatty acids include more common 3-hydroxy acids, synthesized by *E. coli,* and species of *Salmonella, Klebsiella, Yersinia, Vibrio, Pseudomonas, Xanthomomas,* and *Rhizobium,* and 2-hydroxy acids.

Hydroxy acids are constituents of the cell envelope of gram-negative bacteria (e.g., 3-hydroxy tetradecanoate (X), the most abundant type, present in the lipid A structure of some enterobacteria). Mycolic acids are very long chain, hydroxy, branched-chain fatty acids, occurring in *Mycobacterium, Nocardia,* and *Corynebacterium.* Some of them are the largest unsubstituted lipids (up to 180 carbon atoms long) that occur in any organism. The basic structure is that of a 2-alkyl, 3-hydroxy fatty acid (XI). Among typical lipids of bacteria, particularly when grown on hydrocarbons, there are biosurfactants that include glycolipids, e.g., rhamnose lipids from *Pseudomonas aeruginosa* and *P. fluorescens* (XII), lipoproteins, e.g., surfactin from *Bacillus subtilis,* and a similar biosurfactant produced by *B. licheniformis* (XIII), the polysaccharide-lipid complexes, e.g., emulsan from *Acinetobacter calcoaceticus.* Biosurfactants enable bacteria to grow on water-insoluble substrates (e.g., alkanes) and have antibiotic effect on various microbes. Common phospholipids include phosphatidylethanolamine, the methylated derivatives of which (XIV) together with phosphatidylserine are confined to the gram-negative bacteria. Waxes are synthesized by *Acinetobacter, Micrococcus,* and *Clostridium.* The polyester of β-hydroxybutyrate (PHB) (XV) is a storage lipid in bacteria, in contrast to triacylglycerols, the usual storage lipid of eukaryotic microorganisms. PHB is widely distributed, occurring in some 23 genera including *Alcaligenes, Azotobacter, Bacillus, Nocardia, Pseudomonas, Rhizobium,* and *Rhodococcus.* In some bacteria, PHB is replaced by poly-β-hydroxyalkanoate (PHA) (XV). PHB or PHA copolymers are biodegradable plastics being produced commercially using *Alcaligenes eutrophus.*

C. Eukaryote Lipids

The complex array of lipids occurring in bacteria is not seen in the yeasts. The range of fatty acids is limited to saturated and unsaturated fatty acids, usually of 16 and 18 carbon atoms, that account for more than 90% of the total fatty acids. Oleic acid (18 : 1) usually is the most abundant fatty acid. Yeasts produce polyunsaturated fatty acids, though *S. cerevisiae* is unable to produce either linoleic acid (18 : 2) or α-linolenic acid (18 : 3), which occur in most other yeasts. Hydroxy-polyunsaturated fatty acids (XVI) appear to be present in yeasts. All yeasts synthesize triacylglycerols, the composition of which is similar to plant oils. In oleaginous yeasts (see Section III), the amount of triacylglycerols that is accumulated can be up to 80% of the cell volume. The major phospholipids include phosphatidylcholine, -ethanolamine, -inositol, and -serine. Diphosphatidylglycerol (cardiolipin) and lysophosphatidylcholine are synthesized by *S. cerevisiae.* Sterols, most commonly, ergosterol and zymosterol, are present in yeasts, either free or esterified, in varying amounts, usually at no more than 1% of the cell dry weight, except in some species of *S. cerevisiae* which are able to accumulate as much ergosterol as up to 20% of the cell dry weight. A variety of carotenoids, either yellow (β-carotene) or red (astaxantin), are produced by species of *Rhodotorula, Phaffia,* and *Sporobolomyces.* Astaxantin is produced commercially using *Phaffia rhodozyma.* Sphingolipids and glycolipids synthesized by yeasts attract commercial interest; the former for some potential to the cosmetics and health care industries (e.g., N-stearolyphytosphingosine (XVII) produced commerically, using *Pichia ciferri*) and the latter as biosurfactants (e.g., sophorolipids (XVIII), produced by *Candida bombicola*) or perfumery intermediates (e.g., sophorolipid-derived macrocyclic lactone (XIX).

Mold lipids have a greater diversity than yeasts. Molds synthesize fatty acids from 12 to 24 carbon atoms long, including polyunsaturated and hydroxy fatty acids, the latter being found in association with fungal differentiation or particular stages of development, e.g., ricinoleic acid, 15HO-18:1, only present in sclerotical forms of *Claviceps* spp. Polyunsaturated fatty acids include biologically important n-3 and n-6 fatty acids and their hydroxy-derivatives (XVI). Triacylglycerols are accumulated extensively in some species, generally having the composition similar to those in yeasts or being rich in certain polyunsaturated fatty acids (see Section III.B). Phospholipids are of the same types as in yeasts. A wide range of sterols occurs in molds, including ergosterol, cholesterol, stigmasterol, and derivatives of these lipids. Carotenoids and gibberellins and diterpenes (plant growth hormones) are synthesized by most molds. Both groups of compounds have commercial value, though that of the former is limited because of cheaper chemical processes. Gibberellic acid (XX) is produced commercially using *Gibberella fujikuroi* (the telomorphic state of *Fusarium moniliforme*). Biosurfactant glycolipids, similar to those found in yeast but based on cellobiose or erythritol rather than sophorose, are synthesized by *Ustilago* spp.

D. Algal Lipids

Algae are photosynthetic organisms that are also able to ingest food from their surroundings in the absence of light and, thus, can be grown heterotrophically. In nature, they form the chief aquatic plant life, both in the sea and in fresh water. Blue-green algae (e.g., Anabena, Nostoc, Spirulina) contain chlorophyll characteristic of plants, in contrast to green and purple algae (*Rhodospirillaceae, Chromatiaceae, Chlorobiaceae*), the chlorophyll pigments of which differ in type and amount from the former. A wide variety of lipids produced by algae include neutral (storage) lipids, such as carotenoids (V), hydrocarbons, wax esters (III), and quinones (XXI), and polar (structural) lipids, such as galactolipids, lipopolysaccharide, sulfo-, and phospholipids. The main fatty acids in storage lipids are saturated fatty acids (e.g., palmitic acid 16:0) and unsaturated fatty acids (e.g., palmitoleic acid 16:1, oleic acid 18:1), whereas those in structural lipids are polyunsaturated fatty acids, i.e., "eukaryotic-like" substituents of acylglycerols. Among these, many have biologically important acids with 20 and 22 carbon atoms, e.g., arachidonic acid, eicosapentaenoic acid, and docosahexaenoic acid. The most studied algae in this respect include the species of *Euglena gracilis, Isochrysis galbana, Porphyridium cruentum, Chlorella minutissima, Spirulina platensis,* the latter containing γ-linolenic acid (18:3). The content of polyunsaturated fatty acids can be up to 30 to 60% of the total fatty acids, however, at the lipid content of cell not exceeding 15%. Some commercial interest has also been focused on algal carotenoids, particularly γ-carotene produced by *Dunaliella salina* which accumulates up to 10% of the biomass.

III. SINGLE-CELL OILS

Single-cell oil (SCO) is a triacylglycerol type of oil from oleaginous microorganisms, analogous to plant and animal edible oils and fats. Because of high costs of biotechnological production of SCO, commercial interests have centered on the highest value oils. The first major compilation of microbial oil production has been covered in the book edited by Kyle and Ratledge.

A. Biochemistry and Physiology of Oleaginicity

Oleaginous microorganisms are those accumulating triacylglycerols at more than 25% of their biomass. Oleaginicity is not a common feature among microorganisms, since only about 25 yeasts and molds (Table I) have been identified as oleaginous. Accumulation of lipid in the cell is typically a biphasic process, requiring an excess of carbon over other nutrients, particularly nitrogen. A C:N ratio of 40:1 is usually required. The first phase is characterized by rapid cell growth until the nitrogen is consumed, followed by the second phase, wherein the excess of carbon in the medium is converted to lipid. The rates of lipid synthesis in both phases is similar. The biochemical criterion of oleaginicity is the presence of cytosolic ATP–citrate lyase (ACL) that cleaves

TABLE I
Examples of Oleaginous Microorganisms with Oil Content above 50% of the Cell Dry Weight

Bacteria	Algae
Arthrobacter sp.	*Chlorella pyrenoidosa*
	Dunaliela salina
Yeasts	*Monalanthus salina*
Cryptococcus curvatus	
C. albidus	**Molds**
Endomyces vernalis	*Aspergillus fischeri*
Lipomyces lipofer	*Cunninghamella elegans*
L. starkeyi	*Entomophthora conica*
Rhodosporidium toruloides	*Fusarium bulbigenum*
Rhodotorula graminis	*Mortierella isabellina*
Trichosporon cutaneum	*M. vinacea*
T. pullulans	*Mucor circinelloides*
Yarrowia lipolytica	*Penicillium spinulosum*
	Pythium ultimum
	Ustilago zeae

citrate to yield acetyl–CoA units (Fig. 3). When cells have exhausted their supply of nitrogen in the medium, the intracellular concentration of AMP falls, preventing mitochondrial metabolism of citrate to α-ketoglutarate through the tricarboxylic acid cycle. Provided that carbon is still available in the medium, citrate begins to accumulate in the mitochondrion, from which it is transported out to the cytoplasm where is cleaved by ACL to acetyl–CoA units and oxalacetate. The latter, after being recycled to pyruvate via malic acid by malic enzyme, produces NADPH, needed for fatty acid biosynthesis. Since not all oleaginous yeasts and molds possess malic enzyme, an alternative means of generating NADPH and recycling the oxalacetate must exist in some species. A number of carbon sources, including hydrocarbons and most commonly available fermentation substrates, except for methanol and cellulose, can be utillized by oleaginous microorganisms. The yields of lipid from carbohydrate are generally from 15 to 20%, with the theoretical value being about 33%. Triacylglycerols account for up to 95% of the total lipids. The amount of lipid and its composition are affected by growth medium composition, pH, temperature, oxygen level and growth rate. Only temperature and nitrogen limitation appear to have

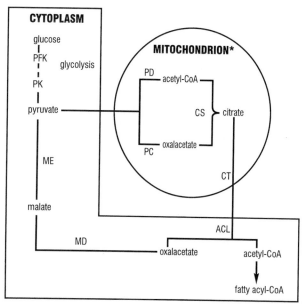

Fig. 3. Mechanism of fatty acid biosynthesis in oleaginous yeasts and molds. PFK, phosphofructokinase; PK, pyruvate kinase; PD, pyruvate dehydrogenase; PC, pyruvate carboxylase; CS, citrate synthetase; CT, citrate translocase; ACL, ATP citrate lyase; MD, malate dehydrogenase; ME, malic acid. *Mitochondrial citric acid cannot be metabolized through the reaction of the tricarboxylic acid cycle because, under N-limiting conditions, the conversion of isocitrate to α-ketoglutarate is prevented with a low intracellular concentration of AMP under these conditions, specifically required for isocitrate dehydrogenase activity. Thus, provided a supply of carbon is still available, the cells begin to accumulate citrate, when they have exhausted their supply of nitrogen, that is cleaved to yield acetyl-CoA units, the basic building block for fatty acid biosynthesis. (Adapted from *Adv. Appl. Microbiol.* **43**, 1997, p. 200.)

a definite effect on the lipid content, whereas the carbon source and temperature have a definite effect on the lipid content, whereas the carbon source and temperature have a definite effect on the lipid and fatty acid composition. The relationships between the availability of various nutrients and cell fatty acid content, cell oil and oil fatty acid contents, and cell concentration and lipid content are major economic issues.

B. Cocoa Butter Equivalents (CBE)

Cocoa butter equivalent should consist of two-thirds saturated fatty acids (i.e., 16 : 0 and 18 : 0) and one-third unsaturated fatty acid (18 : 1), to meet the ratio typical of cocoa butter. Oleaginous microorganisms accumulate lipids that generally have too high a content of oleic acid (40–50%), too low a content of palmitic acid (20–30%), and about four times lower content of stearic acid (5–10%), which, in cocoa butter, gives the product its characteristic hardness at ambient temperatures, with a sharp transition to the molten state at 30–32°C. Economic production of CBE requires that the microorganism accumulate more stearic acid at the expense of oleic acid and utilizes an inexpensive carbon source in the growth medium. Among the microorganisms employed for CBE imitation, yeast species predominate and *Cryptococcus curvatus* (formerly *Candida curvata*) has the greatest potential. Certain strains of *Mucor* also are capable of accumulating a high percentage (up to 27%) of stearic acid when grown on acetic acid as a sole carbon source. The substrates used for the yeast culturing are low-cost waste materials, such as sweet whey, prickly-pear juice, or monocarboxylic acids contained in wastes from the petrochemical process. Among the strategies for increasing the stearic acid content in yeast oil, those that rely on the inhibition or deactivation of Δ9-desaturase, responsible for the conversion of stearic acid to oleic acid, have appeared so far most efficient. To delete or partially inhibit Δ9-desaturase, mutation and spheroplast fusion methods are used, though an increase of stearic acid in wild *C. curvatus* cells can be achieved by limiting the oxygen uptake rate of the culture during the lipid-accumulating phase of cultivation.

C. Oils Rich in Polyunsaturated Fatty Acids

Potential microbial sources of oils rich in polyunsaturated fatty acids are phycomycete fungi, mainly *Mucorales*, and algae from the genera *Chlorophyceae, Cryptophyceae, Dimophyceae, Phaeophyceae,* and *Rhodophyceae*. The acids are typically present in triacylglycerols and phospholipids, which is important for their potential applications, such as fortified foods, nutritional supplements, health foods, and pharmaceuticals. A number of commercial processes are available worldwide, which use overproducing strains, selective inhibition of enzymes, and optimized cultivation conditions. Oils containing γ-linolenic acid, arachidonic acid, and docosahexaenoic acid are produced using *Mucor* spp., *Mortierella* spp., and *Pythium* spp. Phototrophic and heterotrophic algal systems employing *Chlorella* spp., *Spirulina* spp. or *Porphyridium* spp. operate for the production of arachidonic acid, eicosapentaenoic acid, and docosahexaenoic acid. Bacteria, such as *Rhodopseudomonas* spp. and *Shewanella* spp., and the mosses *Eurhynchium* spp. and *Rhytididiadelphus* spp. also are considered for the production of these acids. Although oils rich in polyunsaturated fatty acids often contain considerably higher levels of the desired acids than those from plant and animal sources (typically 9–80% w/w rather than 7–33% w/w), the total lipid contents of biomasses are usually much lower (2–12% fresh weight rather than 1.6–21%) and, therefore, biotechnological processes cannot yet compete with agricultural products. Improvements in the yields of lipid and acids are achievable by (i) genetic manipulation aiming at deleting one of the several desaturase enzymes, thereby causing the accumulation of the substrate fatty acids of desaturase, (ii) optimization of the cultivation conditions including those that affect enzymes necessary for NADPH supply, and (iii) technology development.

Appendix

$$^1CH_2OCO.R'$$
$$|$$
$$R''CO.O^2CH \qquad (I)$$
$$|$$
$$^3CH_2OCO.R'''$$

where R′ and R″ are fatty acyl groups, R‴ is fatty acyl group in triacylglycerols, carbohydrate in glycodiacylglycerols, phosphate ($-H_2PO_3$) or its hydroxy compound (choline, serine, ethanolamine) ester in phosphoacylglycerols, sulfate ($-SO_3H$) in sulfodiacylglycerols; 1, 2, and 3 are the stereospecific numbering (*sn*) for designating the position of fatty acyl groups.

$$CH_2O–X$$
$$|$$
$$CHNHCOR' \qquad (II)$$
$$|$$
$$R''CH=CH–CHOH$$

where R′ is fatty acyl group; R″ is 13 carbon atom acyl chain; X is H in ceramides, galactose, or glucose in cerebrosides.

$$\underset{\text{R}}{\text{HOCHCH}_2\text{CO.O}}(\underset{\text{R}}{\text{CHCH}_2\text{CO.O}})_n-\underset{\text{R}}{\text{CHCH}_2\text{CO.OH}} \quad \text{(III)}$$

where R is up to 12 carbon atom chain and n is up to 30,000.

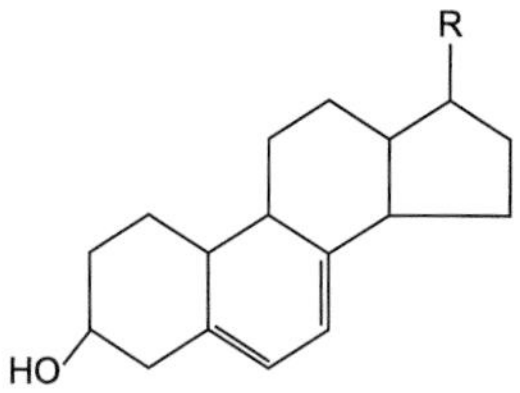

(IV)

where R is fatty acyl chain.

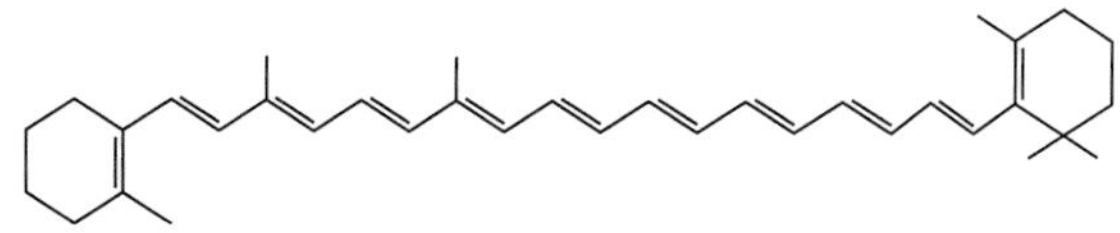

(V)

$$\begin{array}{c} \text{CH}_2\text{OSO}_3\text{H}^- \\ | \\ \text{O} \quad \text{CHOH} \\ \| \quad | \\ \text{CH}_2\text{OP}-\text{O}-\text{CH}_2 \\ | \\ \text{CHO}-\text{R}' \\ | \\ \text{CH}_2\text{O}-\text{R}'' \end{array} \quad \text{(VI)}$$

where R′ and R″ are 20 carbon atom phytanyl chains (⋏⌣).

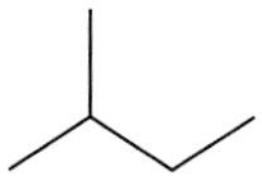

(VII)

$$\underset{\text{CH}_3}{\text{CH}_3\text{CH}(\text{CH}_2)_n\text{COOH}} \qquad \underset{\text{CH}_3}{\text{CH}_3\text{CH}_2\text{CH}(\text{CH}_2)_n\text{COOH}}$$

$$\underset{\text{CH}_3}{\text{CH}_3(\text{CH}_2)_2-\text{CH}(\text{CH}_2)_8\text{COOH}} \quad \text{(VIII)}$$

$$\underset{\text{CH}_2}{\text{CH}_3(\text{CH}_2)_5\text{CH}-\text{CH}(\text{CH}_2)_9\text{COOH}} \quad \text{(IX)}$$

$$\underset{\text{OH}}{\text{CH}_3(\text{CH}_2)_{10}\text{CH}-\text{CH}_2\text{COOH}} \quad \text{(X)}$$

$$\begin{array}{c} \text{OH} \\ | \\ \text{R}-\text{CH}-\text{CH}-\text{COOH} \\ | \\ (\text{CH}_2)_n\text{CH}_3 \end{array} \quad \text{(XI)}$$

n = 5–23, odd numbers only.

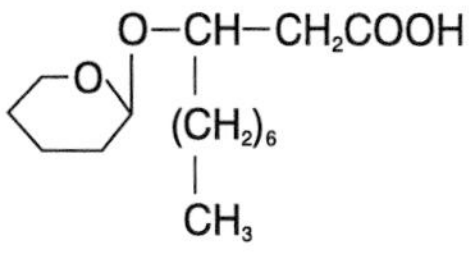

$$\begin{array}{c} \text{O}-\text{CH}-\text{CH}_2\text{COOH} \\ | \\ (\text{CH}_2)_6 \\ | \\ \text{CH}_3 \end{array} \quad \text{(XII)}$$

$$\begin{array}{c} \text{CH}_3)_2\text{CH}(\text{CH}_2)_8-\text{CHCH}_2\text{CO}-\text{Glu}-(\text{Leu})_2-\text{Val} \\ | \qquad\qquad\qquad | \\ \text{O} - \text{Ile} - \text{Leu} - \text{Asp} \end{array} \quad \text{(XIII)}$$

$$\begin{array}{c} \text{CH}_2\text{OCOR}' \\ | \\ \text{R}''\text{COO}-\text{CH} \quad \text{O} \\ | \qquad \| \\ \text{CH}_2\text{O}-\text{P}-\text{OCH}_2\text{CH}_2\text{N}(\text{CH}_3)_3 \\ | \\ \text{OH} \end{array} \quad \text{(XIV)}$$

$$\underset{\text{R}}{\text{HOCHCH}_2\text{CO.O}}(\underset{\text{R}}{\text{CHCH}_2\text{CO.O}})_n-\underset{\text{R}}{\text{CHCH}_2\text{COOH}} \quad \text{(XV)}$$

R = CH$_3$ in PHB
R = up to 12 carbon atom chain in PHA.

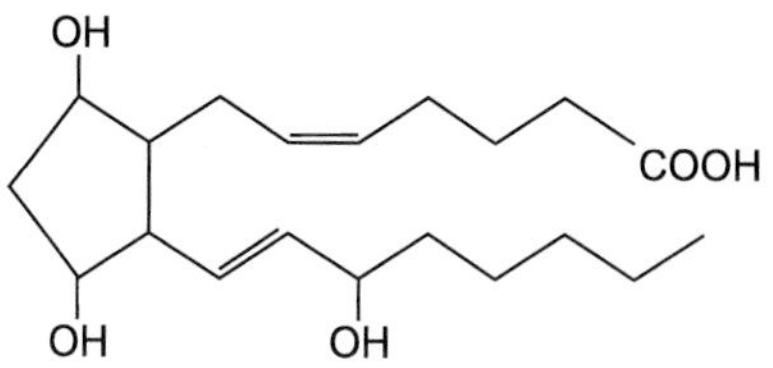

(XVI)

$$\begin{array}{c} \text{CH}_3(\text{CH}_2)_{13}\text{CH}-\text{CHCH}_2\text{OH} \\ | \quad | \\ \text{OH} \quad \text{NH} \\ | \\ (\text{CH}_2)_{17} \\ | \\ \text{CH}_3 \end{array} \quad \text{(XVII)}$$

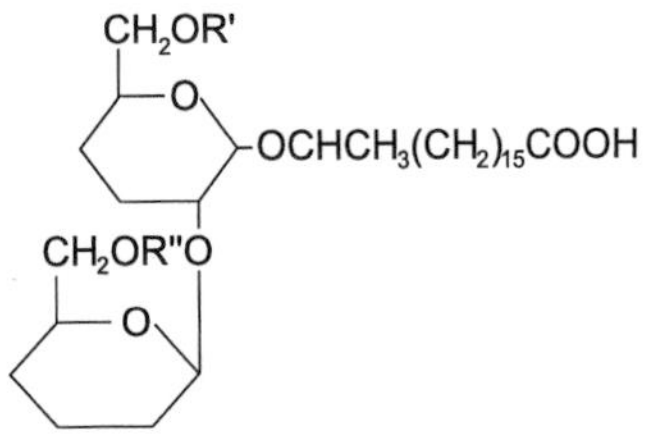

CH₂OR'

(XVIII)

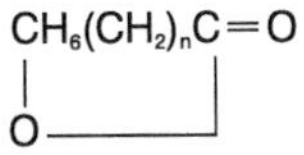

CH₆(CH₂)ₙC=O

(XIX)

COOH

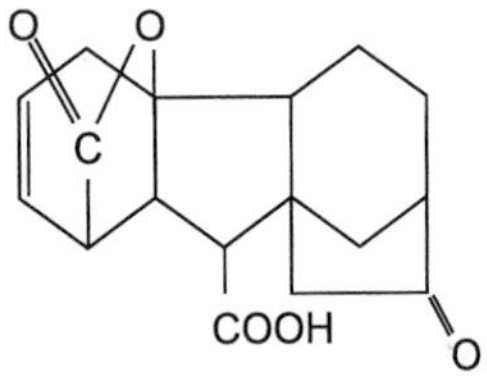

(XX)

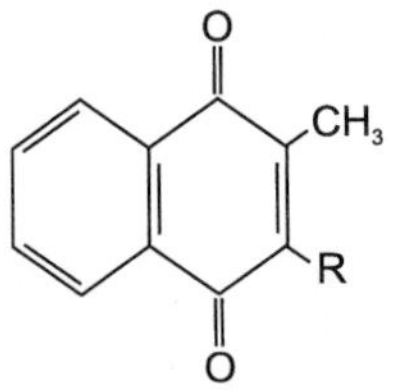

(XXI)

See Also the Following Articles

ARCHAEA • BIOSURFACTANTS • FRESHWATER MICROBIOLOGY • YEASTS

Bibliography

Horrobin, H., (ed.). (1998). "Polyunsaturated Fatty Acids" (Chemistry and Pharmacology of Natural Products Series). Cambridge Univ. Press, Cambridge, UK.

Kosaric, N., Cairns, L. W., and Gray, N. C. C. (eds.). (1993). "Biosurfactant and Biotechnology" (2nd ed.) Marcel Dekker Inc., New York.

Kyle, D. J., and Ratledge, C. (eds.). (1992). "Single Cell Oil." American Oil Chemists' Society, Champaign, IL.

Ratledge, C., and Wilkinson, S. G. (eds.). (1988, 1989). "Microbial Lipids," Vols. 1 and 2. Academic Press, London and New York.

Lipopolysaccharides

Chris Whitfield

University of Guelph, Ontario

ceae, Pseudomonadaceae, and *Vibrionaceae,* among others. S-LPS has a tripartite structure comprising lipid A, core oligosaccharide, and O-PS.

GLOSSARY

core oligosaccharide (core OS)   A branched and often phosphorylated oligosaccharide with varying glycan composition that is linked to lipid A. The inner core OS is more conserved and contains 3-deoxy-D-manno-otulosonic acid (Kdo) and L-glycero-D-manno-heptose (Hep) residues. The outer core OS is more variable in structure.

endotoxin   In gram-negative sepsis, the lipid A component of LPS may stimulate macrophages and endothelial cells to overproduce cytokines and proinflammatory mediators. This can lead to septic shock, a syndrome involving hypotension, coagulopathy, and organ failure.

lipid A   An acylated and phosphorylated di- or monosaccharide that forms the hydrophobic part of LPS.

lipooligosaccharide (LOS)   A form of LPS produced by many mucosal pathogens, including members of the genera *Neisseria, Haemophilus, Bordetella,* and others. LOS lacks O-PS but has oligosaccharide chains extending from the inner core OS. These chains are frequently antigenically phase-variable.

lipopolysaccharide (LPS)   An amphiphilic glycolipid found exclusively in gram-negative bacteria. LPS forms the outer leaflet of the outer membrane in the majority of gram-negative bacteria.

O polysaccharide (O-PS)   A glycan chain attached to the core OS. Structures of O-PSs vary considerably and give rise to O antigens that define O-serospecificity in serological typing.

R-LPS   A truncated LPS form that lacks O-PS and, in some cases, parts of the core OS.

S-LPS   A form of LPS common in the families *Enterobacteria-*

THE CELL ENVELOPE   of gram-negative bacteria is characterized by its outer membrane. The outer membrane is an asymmetric lipid bilayer, in which the inner leaflet contains phospholipids and the outer leaflet contains the unique amphiphilic glycolipid known as lipopolysaccharide (LPS).

There are estimated to be 2×10^6 LPS molecules per *Escherichia coli* cell. The distinctive structural features of LPS are crucial for the protective barrier properties of the outer membrane. In gram-negative sepsis, LPS molecules released from the bacterial surface stimulate macrophages and endothelial cells to overproduce cytokines and proinflammatory mediators, leading to the often fatal syndrome of septic shock. The involvement of LPS in this process is the reason that it is often referred to as endotoxin, and these biological effects have inspired a substantial part of LPS research. The complex structures of LPS molecules also provide fascinating research topics in the areas of synthesis and export of macromolecules, as well as in membrane biogenesis. The broad spectrum of LPS research is reflected in the activities of the International Endotoxin Society (http://www.kumc.edu/IES).

I. LIPOPOLYSACCHARIDE STRUCTURE

Early structural analyses of LPSs were driven, in part, by the need to resolve the identity of the molecule responsible for the endotoxic effect. One of the key breakthroughs in early LPS research came from the establishment of techniques for the extraction and isolation of LPS by O. Westphal and O. Lüderitz in the 1940s. The hot aqueous phenol method of Westphal and Lüderitz remains one of the most common and valuable extraction procedures in current use. This early work led to the understanding that the endotoxic phenomenon is attributable to LPS, and equally importantly, the finding that LPS molecules with similar compositions are found in different gram-negative bacteria. More recent application of nuclear magnetic resonance spectroscopy and mass spectroscopy analytical techniques has led to highly detailed and refined structures for LPS molecules from diverse bacteria. It is now clear that there are general structural features or themes that are highly conserved in LPSs from different sources, but that there is significant variation in the structural fine details.

Extensive research has been performed on the LPS molecules of *Salmonella enterica* serovar Typhimurium (hereafter, referred to as *Salmonella*) and *E. coli*, and these LPSs form a basis for comparative analysis of other LPSs. For the purpose of discussion, the LPSs of *E. coli* and *Salmonella* can be conveniently subdivided into three structural domains (Fig. 1). Lipid A is the hydrophobic part of the LPS molecule and forms the outer leaflet of the outer membrane. Extending outward from lipid A is the branched and often phosphorylated oligosaccharide known as the core oligosaccharide (core OS). The O antigen side-chain polysaccharide (O antigen; O-PS) is a polymer of defined repeat units, attached to the core OS. The O-PS extends from the surface to form a protective layer. This complete tripartite LPS structure is known as "smooth LPS" (S-LPS), taking its name after the "smooth" colony morphology displayed by enteric bacteria that have the complete molecule on their cell surfaces. Mutants with defects in O-PS or core OS assembly produce truncated LPS molecules. For example, the widely used *E. coli* K-12 strains carry a defect in O-PS biosynthesis. The resulting colonies lack the smooth character, and the truncated LPS is, therefore, widely known as "rough LPS" (R-LPS) (Fig. 1). Preparations of LPS from bacteria that produce S-LPS always contain a variable amount of trun-

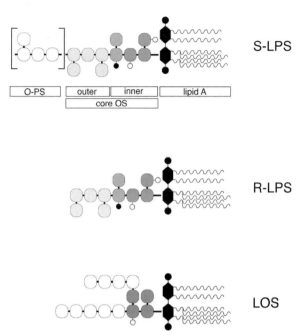

Fig. 1. Schematic diagram showing three different forms of LPS molecules. The tripartite S-LPS structure is typical of LPSs produced by members of the *Enterobacteriaceae*, *Pseudomonadaceae*, and *Vibrionaceae*. These bacteria also produce a variable amount of R-LPS that lacks O-PS and, in some cases, part of the core OS. Mucosal pathogens such as *Neisseria* and *Haemophilus* spp. lack O-PS but instead may have phase-variable oligosaccharide extensions attached to the core OS, to form LOS.

cated R-LPS. Some bacteria, particularly mucosal pathogens, naturally lack O-PS chains in their LPS. Their LPS contains oligosaccharide extensions attached to various points of a typical inner core OS, in a form of LPS known as lipooligosaccharide (LOS) (Fig. 1).

The LPS molecules from different bacteria typically show closer structural relationships in the cell-proximal lipid A and inner core OS regions and increasing diversity in the distal outer core OS and O-PS domains. The inner portions of the LPS molecule play important roles in establishing the essential barrier function of the outer membrane, and this likely places constraints on the extent of structural variation. The outer parts of the LPS molecule interact with environmental factors, such the host immune response. These selective pressures may have played a significant role in the diversification of outer LPS structures.

A. Lipid A

Free lipid A is not found on the bacterial cell surface but mild-acid treatment of most isolated LPSs releases lipid A by hydrolysis of the labile ketosidic linkage between the core OS and lipid A. Structural analysis of lipid A is hampered by its microheterogeneity, as well as its amphipathic properties. However, the lipid A components of LPSs from a variety of bacteria have now been resolved, revealing a family of structurally related glycolipids based on common architectural features. In enteric bacteria, the backbone of lipid A is formed by a disaccharide, comprising two glucosamine (GlcpN) residues joined by a β1→6-linkage. The disaccharide backbone is phosphorylated at the 1 (reducing) and 4′ (nonreducing) positions and is acylated with ester and amide-linked 3-hydoxyl saturated fatty acids. In *E. coli* and *Salmonella,* the fatty acyl chains on the nonreducing GlcpN residue are substituted by nonhydroxylated fatty acids, creating an asymmetric arrangement (Fig. 2).

Most lipid A variations in different species involve alterations in acylation or phosphorylation. For example, the acylation pattern is symmetrical in *Neisseria meningitidis,* whereas *Rhodobacter sphaeroides* lipid A (Fig. 1) is distinguished by amide-linked 3-oxotetradecanoic acid and the presence of unsaturated fatty acids. The structure of the *R. sphaeroides* lipid A is of particular importance since it results in an LPS lacking the normal biological activities

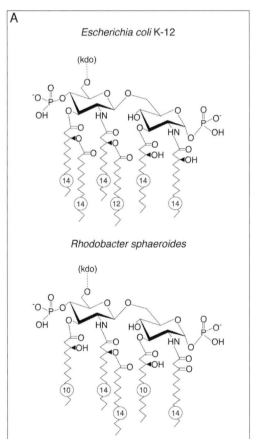
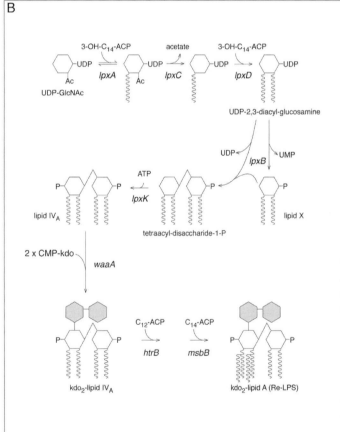

Fig. 2. Structure and biosynthesis of lipid A. Panel A shows the structure of the biologically active lipid A from *E. coli* and the naturally occurring nontoxic lipid A from *Rhodobacter sphaeroides.* Panel B shows the pathway for biosynthesis of lipid A from *E. coli* K-12. The genes encoding each enzyme in the pathway are indicated.

attributed to endotoxins. Some lipid A molecules (e.g., in the *Rhizobiaceae*) lack the 4′ phosphate. However, structural variations can be as extreme as the 2,3-diamino-2,3-dideoxy-D-glucose-containing monosaccharide backbone structure of lipid A molecules from *Pseudomonas diminuta* and *Rhodopseudomonas viridis*.

Microheterogeneity is evident in lipid A preparations isolated from a given bacterium. This can arise from nonstoichiometric modification of the phosphate residues (especially at the 4′-position) with substituents including 4-aminoarabinose (L-Arap4N), 2-aminoethyl phosphate (P-EtN) and, less frequently, GlcpN and arabinofuranose (D-Araf). In addition, there may be limited variation in the pattern of acylation. These structural variations, can be determined by the precise growth conditions used for the bacteria and can have a considerable impact on the function of LPS molecules (see following).

B. Core Oligosaccharides

For the purpose of discussion of structure–function relationships, the core OS is often divided into inner and outer core regions (Fig. 1). The inner core comprises characteristic residues of 3-deoxy-D-manno-octulosonic acid (Kdo) and L-glycero-D-manno-heptose (Hep). This region is highly conserved in enteric bacteria (Fig. 3A). The Kdo residues can be nonstoichiometrically modified by other sugars, or by 2-aminoethyl phosphate (P-EtN). The main chain Hep residues are also nonstoichiometrically decorated with phosphate, pyrophosphorylethanolamine (PP-EtN), or a side-branch Hep. In some nonenteric bacteria, the Kdo residue proximal to lipid A is phosphorylated. Others contain a derivative of Kdo, which may influence the lability of the linkage that is usually cleaved by mild-acid hydrolysis to release lipid A from the intact LPS molecule.

The outer core OS structure is quite variable. Within a species, these variations are quite limited and some may have a single outer core OS type. For many years, this was thought to be the case for *Salmonella* but a second core OS structure was reported recently. *E. coli* shows 5 distinct outer core OS types; all contain 5 glycose residues but they

differ in glycose content and organization (Fig. 3B). In addition to altering the antigenic epitopes and diagnostic bacteriophage receptors, variations in outer core OS structure give rise to altered sites for the attachment of O-PS.

C. O Polysaccharides

The O-PS is the most variable portion of the LPS molecule. A remarkable array of novel structures arises from alterations in constituent sugars, linkages, and both complete and partial substitutions with nonsugar residues. At their simplest, O-PSs are homopolysaccharides with a disaccharide repeat unit, consisting of a single monosaccharide component. Complex homopolysaccharides can result from larger repeat units defined by a specific sequence of glycosidic linkages. At their most complex, O-PSs can be heteropolysaccharides in which the repeat units contain several component sugars, together with nonsugar substituents, such as O-acetyl groups and amino acids. The structural diversity in O-PSs has been exploited in serological classification of isolates from a given bacterial species. In *E. coli*, there are approximately 170 distinct O serogroups. While the LPSs from most bacterial strains tend to have a single O-PS type, there is a growing number of bacteria where the lipid A-core serves as an acceptor for two or more different polymeric structures. A catalog of O-PS structures can be found online in the Complex Carbohydrate Structure Database (*www.ccrc. uga.edu/*).

The length of the O-PS attached to lipid A-core is heterogeneous but the distribution of chain lengths is both strain- and growth-condition dependent. This is best reflected in the patterns of LPS molecules revealed by SDS-PAGE analysis (Fig. 4).

D. Lipooligosaccharides

The LPS of mucosal pathogens such as *Neisseria* and *Haemophilus* spp. is smaller than S-LPS and takes a form that is known as lipooligosaccharide (LOS). These bacteria generally do not produce S-LPS. As an example, the LOS structure from *N. gonorrhoeae* is given in Fig. 3C. The oligosaccharide chains linked to the inner core OS provide distinct

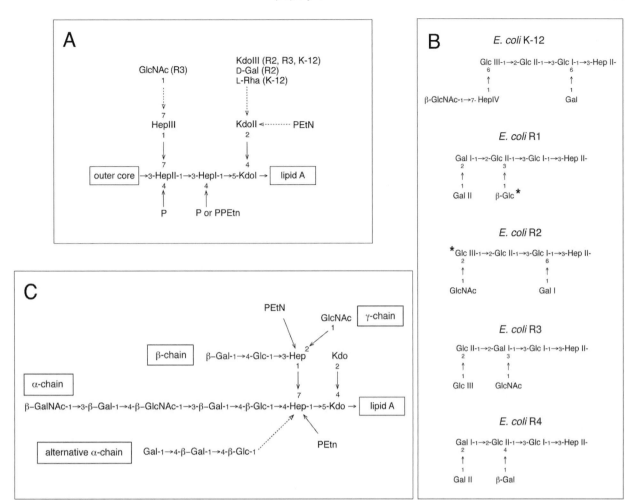

Fig. 3. Structure of the core OS region of LPS and LOS. Panel A shows the inner core region of the core OSs from *E. coli* K-12. The backbone structure is conserved but nonstoichiometric glycose modifications (dotted arrows) distinguish different core types (indicated in parentheses). Panel B shows the known outer core structures from the 5 core OSs of *E. coli*. They share a common overall structure, consisting of 5 glycoses linked to Hep II, but they differ in glycose content, glycose sequence, and side branch position. The site of O-PS ligation is not known for all of the core OS types but, where it has been determined, the attachment site is indicated by the asterisk. Panel C shows a representative LOS structure from *N. gonorrhoeae*. The LOS core is conserved among different strains but there are strain-specific differences in the attached oligosaccharide chains. The α- and β-chains are subject to antigenic variation. Unless otherwise indicated, all residues in the various structures are linked in the α-configuration.

antigenic epitopes, and a common feature in these bacteria is the phenomenon of phase variation, where the LOS epitopes are differentially expressed. This gives rise to difficulty in arriving at precise LOS structures, unless phase-locked variants are available. Some *Bordetella* spp. can form both S-LPS and LOS.

II. BIOSYNTHESIS AND ASSEMBLY OF LIPOPOLYSACCHARIDES

In *E. coli,* the minimal LPS molecule required for viablity consists of Kdo_2-lipid A (sometimes known as Re LPS; Fig. 2). Since *E. coli* can efficiently insert Re-LPS and larger LPS molecules on their cell sur-

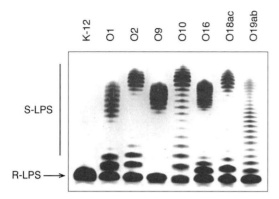

Fig. 4. Visualization of LPS molecules in silver-stained sodium dodecyl sulfate polyacrylamide gel electrophoresis (SDS-PAGE) gels. Shown are LPS samples from *E. coli* K-12 and from a variety of different *E. coli* O serogroups. *E. coli* K-12 derivatives harbor mutations that eliminate O-PS synthesis, leading to a fast migrating R-LPS band. Strains with S-LPS all produce a variable amount of R-LPS. The "ladders" of increasing molecular weight reflect LPS molecules in which the core OS is substituted by variable length O-PS. Each "rung" in the typical ladder contains lipid A-core OS substituted by an O-PS of defined chain length. Successive bands differ by an additional O-PS repeat unit in the chain. Note that the distribution of O-PS chain length is not uniform. Each strain has a "preference" for O-PSs of particular chain length. This gives rise to a "modal" distribution of O-PS lengths in a cluster of S-LPS bands.

faces, the LPS assembly pathways have generally been amenable to dissection by genetic approaches (mutant analysis). Assembly steps later than Re-LPS formation can be determined by a combination of structural analyses of mutant LPS structures and biochemical assays to establish enzymatic function. In contrast, the dependence on Re-LPS for viability has complicated studies on the biosynthesis steps leading to Re-LPS but, nevertheless, these have been systematically resolved by biochemical analyses. The assembly of LPS molecules is complicated by their tripartite structure. The lipid A-core portion is assembled at the cytoplasmic face of the plasma membrane and, once complete, it is transferred across the membrane to the periplasm. The long hydrophilic O-PS is assembled independently and exported to the periplasmic face of the plasma membrane prior to its ligation to preformed lipid A-core. Subsequent and, as yet, undefined steps translocate LPS to the cell surface

and assemble the nascent molecules into the outer membrane.

A. Synthesis of Lipid A

The pathway for biosynthesis of lipid A in *E. coli* (Fig. 2B) has been resolved primarily through the efforts of C. R. H. Raetz and co-workers and a detailed and historical perspective of this work is provided elsewhere. The acylated glucosamine derivatives that provide the halves of the lipid A backbone are formed from UDP-N-acetylglucosamine (UDP-GlcNAc) by sequential cytoplasmic reactions catalyzed by an *O*-acyltransferase (LpxA), a deacetylase (LpxC) and an *N*-acyltransferase (LpxD). The product of these reactions, UDP-2,3-diacylGlc*p*N, provides a direct precursor for the nonreducing diacyl-Glc*p*N derivative of the lipid A backbone. The reducing diacyl-Glc*p*N is generated by cleavage of a second UDP-2,3-diacylGlc*p*N molecule at the pyrophosphate bond, to form the monophosphoryl derivative known as lipid X. Lipid X was found to fortuitously accumulate in mutants with certain defects in phospholipid metabolism. Its isolation and structural elucidation provided an essential clue for unraveling the lipid A assembly pathway. The two halves of the lipid A backbone are then joined by the disaccharide synthase (LpxB). The 4′-kinase (LpxK) completes formation of the tetracyl-derivative known as lipid IV_A. This product has been useful for investigations of the biological activities of LPS. Full acylation of *E. coli* lipid A requires the prior addition of two inner core Kdo residues. The Kdo transferase (WaaA; formerly, KdtA) from *E. coli* is bifunctional and simultaneously adds two Kdo residues, providing substrates for the lauryl (HtrB) and myristoyl (MsbB) acyltransferases.

The essential features of the lipid A assembly pathway and its key enzymes appear to be conserved in different bacteria, although there are some subtle species-specific variations. There are clearly differences in the specifity of the acyltransferases. In the case of *P. aeruginosa*, late acylation enzymes have altered acceptor specificity and the reactions are not dependent on prior addition of Kdo to lipid IV_A. Some bacteria, for example, *H. influenzae*, have a monofunctional Kdo transferase, and the *Chlamydia* WaaA-homolog is trifunctional.

Due to the requirement for at least a minimal LPS in most bacteria, only conditional mutants of lipid A or Kdo synthesis have been isolated. This feature has been exploited in attempts to generate therapeutic antibacterial compounds, targeted against lipid A and CMP-Kdo synthesis. Structural analog of Kdo proved to be very effective inhibitors of CMP-Kdo synthetase *in vitro,* but their ultimate failure was due to the inability of most bacteria to take up the inhibitor. In the laboratory, this could be circumvented by synthesis of a dipeptide-linked prodrug which exploited the oligopeptide permease transporter for uptake. Unfortunately, this provides mutation of the transporter as a very simple route to drug resistance. Another attractive candidate for inhibitor development is the deacetylase reaction catalyzed by LpxC. Synthetic LpxC inhibitors have proven effective in animal challenge models for *E. coli* infections but not against *P. aeruginosa* and, again, this is likely a reflection of uptake problems. Nevertheless, such approaches show promise.

B. Synthesis of Core Oligosaccharide

The completed Kdo_2-lipid A provides an acceptor for glycosyltransferases that assemble the core OS (Fig. 5), as well as those that add the oligosaccharide chains of LOSs. These enzymes are peripheral membrane proteins that act at the cytoplasmic face of the plasma membrane. In *E. coli* and *Salmonella,* the structural genes for the core OS glycosyltransferases map together with the Kdo transferase gene (*waaA*), and genes required for modification of the Hep-region of the inner core OS (*waaPYQ*). These genes form the three separate operons in the chromosomal *waa*-region. Direct data for biochemical activities of individual core OS glycosyltransferase enzymes is unavailable in many cases. For example, while there is evidence for the assignment of the Hep I transferase (WaaC), studies of heptosyltransferases have generally been limited by the unavailability of the activated precursor, ADP-Hep. Interestingly, ADP-mannose will substitute for ADP-Hep as a substrate for the WaaC transferase but the product cannot be elongated further. Assignments of other glycosyltransfer-

ases have primarily been made by approaches where specific genes are individually mutated and the resulting LPS structure is resolved by chemical analysis. In this way, the sequence of assembly of the *E. coli* R1 core OS was established. Synthesis of the core OS backbone can be carried to completion in mutants lacking the modifications that decorate the Hep-region. As a result, the timing of these modification in the overall synthesis pathway is unknown. Modification of the Hep-region could follow immediately after the HepII residue is added (as shown in Fig. 5) or could occur once core OS elongation is completed.

C. Synthesis of O-Polysaccharides

Despite the diversity in O-PS structures, only three mechanisms are known for the formation of O-PS (Fig. 6). O-antigen synthesis begins at the cytoplasmic face of the plasma membrane with activated precursors (sugar nucleotides; NDP-sugars) and the process terminates with a nascent O-PS at the periplasmic face. The different pathways for assembly of O antigens vary primarily in the components required for polymerization, in the cellular location of the polymerization reaction, and in the manner in which material is exported across the plasma membrane. A carrier lipid, undecaprenyl phosphate (und-P), is involved in all three O-PS assembly pathways. The involvement of a carrier lipid scaffold may ensure fidelity in O-repeat unit structure.

The most prevalent pathways for O-PS synthesis are distinguished by the involvement (or not) of an "O-PS polymerase" enzyme, Wzy. The "Wzy-dependent" system (Fig. 6A) is the classical pathway first described in *S. enterica* serogroups A, B, D, and E. In the working model for this pathway, und-PP-linked O-repeat units are assembled by glycosyltransferase enzymes at the cytoplasmic face of the plasma membrane. The initial transferase is an integral membrane protein that transfers sugar-1-phosphate to the und-P acceptor. This is followed by sequential sugar transfers by additional peripheral glycosyltransferases to form an und-PP-linked repeat unit. The polymerization reaction occurs at the periplasmic face of the membrane and utilizes und-PP-linked O-PS-

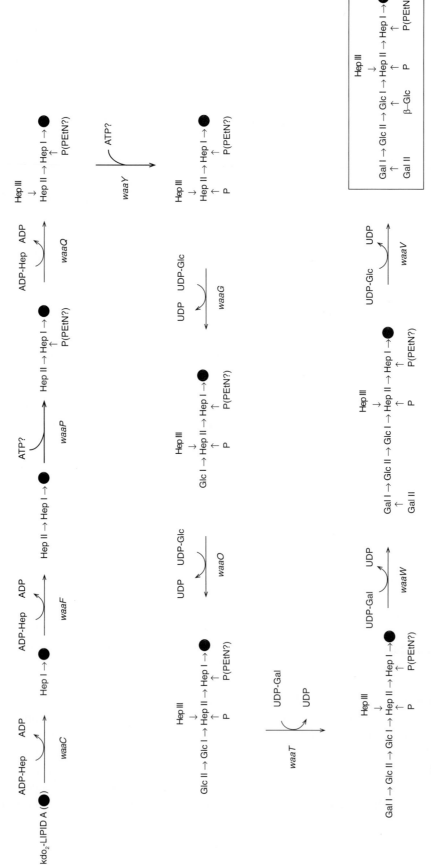

Fig. 5. Proposed pathway for the biosynthesis of the R1 type core OS from *E. coli*. The *waa* (core OS biosynthesis) gene encoding each enzyme in the pathway is shown. As drawn, the WaaPQY-mediated modification of the Hep region is shown occurring while extension of the core OS is still in progress. However, there is uncertainty surrounding the precise timing of the Hep-decoration. Extension of the core OS backbone is not dependent on these modifications, and they may equally well occur as a late step once the core OS backbone is complete.

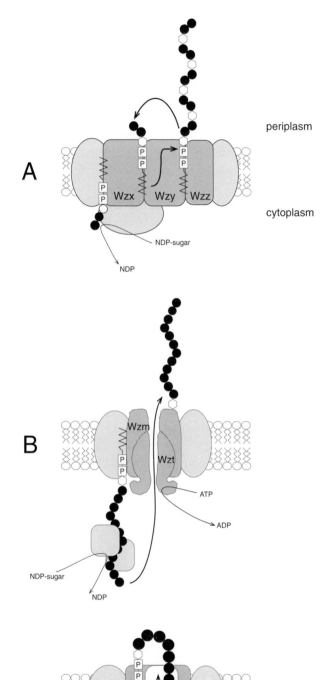

repeat units as the substrate. The individual und-PP-linked O-PS-repeat units must, therefore, be exported across the plasma membrane prior to polymerization, and preliminary biochemical analyses suggest that the likely candidate for this process is a multiple membrane-spanning protein, Wzx (formerly, RfbX). Polymerization of the O-PS repeat units minimally involves the putative polymerase (Wzy; formerly, Rfc) and the O-PS chain length regulator (Wzz, formerly Rol or Cld). A *wzy* mutant is unable to polymerize O-PS and its LPS comprises a single O-repeat unit attached to lipid A-core. In contrast, a *wzz* mutant makes S-LPS but loses the characteristic modal distribution of O-PS-chain lengths evident in SDS-PAGE analysis (Fig. 4). The O-PSs synthesized by this pathway are all heteropolymers and often have branched repeating unit structures.

The "ABC-2-transporter-dependent" pathway (Fig. 6B) is, so far, confined to O-PSs with relatively simple structures; these are generally linear homopolymers. As with the Wzy-dependent pathway, synthesis is initiated at the cytoplasmic face of the plasma membrane by an integral membrane glycosyltransferase enzyme, to form an und-PP-sugar. However, in this case, the initiating transferase acts once per O-PS chain. Additional peripheral glycosyltransferases then act sequentially and processively to elongate the und-PP-linked intermediate at the nonreducing terminus to form a fully polymerized und-PP-O-PS within the cytoplasm. The specificities of glycosyl transferases dictate the repeat-unit structure of the

Fig. 6. Models for the assembly of O-PS by the pathways termed Wzy-dependent (A), ABC-2-dependent (B), and synthase-dependent (panel C). All of the assembly systems begin at the cytoplasmic face of the plasma membrane and build the nascent O-PS in an undecaprenol pyrophosphoryl (und-PP)-linked form. Nucleotide diphospho (NDP)-sugars provide the activated precursors. The enzyme complexes include integral and peripheral proteins, and key components are indicated by name. The pathways differ in the mechanism and location of O-PS polymerization and in the manner by which O-PS (or O-repeat units) are transferred across the plasma membrane. Termination of O-PS synthesis occurs with ligation of the nascent O-PS to preformed lipid A-core OS at the periplasmic face of the membrane (not shown).

product. Only one und-PP acceptor is used per polymer and there is no equivalent of the polymerase (Wzy) or chain-length regulator (Wzz) enzymes. After polymerization, the nascent O antigen is transferred to the periplasmic face of the plasma membrane by an ATP-binding-cassette (ABC)-transporter, belonging to the ABC-2 family. The involvement of an ABC-transporter precludes the involvement of Wzx, since there is no requirement for export of individual und-PP-O-units. It remains unknown whether the O-PS retains its attachment to und-PP during export, whether it is removed from the lipid carrier for export, or, alternatively, if an alternative carrier molecule exists.

Recently, our laboratory described the "synthase-dependent" pathway for O-PS biosynthesis. This is currently confined to the homopolysaccharide O antigen (factor 54) of *S. enterica* serovar Borreze. The model for this pathway (Fig. 6C) proposes that the initiating glycosyltransferase forms an und-PP-sugar acceptor that is elongated by a single multifunctional synthase enzyme, in a manner analogous to eukaryotic chitin and cellulose synthases. There is no dedicated ABC-2 transporter or Wzx homologue in the O:54 system and all of the experimental evidence points to the synthase having dual transferase-export functions. It is not known whether this system requires the nascent polymer be linked to und-PP throughout the polymerization and export processes.

In most bacteria, the genes dedicated to O-PS biosynthesis are clustered on the chromosome. The O-PS gene clusters contain a predictable spectrum of genes encoding novel sugar nucleotide synthetases, glycosyltransferases, and the characteristic enzymes such as Wzx-Wzy-Wzz or an ABC-2 transporter. As a result, sequence data can give an accurate first evaluation of the O-PS biosynthesis pathway involved. As might be expected from the range of O-PS structures, the O-PS biosynthesis genetic loci are highly polymorphic. Genetic recombination within and between bacterial species has played a significant role in the diversification of O-PS. In some bacteria (e.g., *Shigella* spp.), phage-encoded genes provide additional determinants of O serotype specificity.

D. Terminal Reactions in LPS Assembly

Once complete, the nascent O-PS chain must be linked (ligated) to lipid A-core. All available evidence points to this reaction's occurring in the periplasm. Since lipid A-core is formed in the cytoplasm, it too must be exported to the ligation site at the periplasmic face of the plasma membrane. An ABC-transporter, known as MsbA, appears to be responsible for this step. The mechanism underlying the ligase reaction itself remains unknown. The *waaL* gene product, often referred to as "ligase," is currently the only known protein that is essential for ligation. Its assignment as the ligase is based only on mutant phenotypes and there is no supporting mechanistic information. Interestingly, the ligase from *E. coli* K-12 will ligate structurally distinct O-PSs formed by any of the three assembly pathways, indicating that the form in which nascent O-PS is presented for ligation is conserved.

Perhaps the most interesting open questions in LPS assembly surround the process by which the completed LPS is translocated through the periplasm and then inserted into the outer membrane. In most bacteria, the translocation pathway must have relaxed specificity, since it efficiently transfers a range of LPS molecules varying from S-LPS to Re-LPS to the cell surface.

III. FUNCTIONS AND BIOLOGICAL ACTIVITIES OF LIPOPOLYSACCHARIDES

A Lipopolysaccharides and Outer Membrane Stability

From the construction of precise mutants with LPS defects it is well established that the minimal LPS molecule required for survival of *E. coli* consists of Kdo_2-lipid A (Re-LPS). Although *E. coli* can assemble an outer membrane from Re-LPS, the barrier function of the resulting outer membrane is compromised. In *E. coli* and *Salmonella,* the phosphorylated Hep-region of the core OS is crucial for outer membrane stability by facilitating the cross-linking of adjacent LPS molecules by divalent cations or polyamines and by its interaction with positively charged groups on proteins. The inability to synthesize or incorporate Hep, or the loss of phosphoryl derivatives alone (i.e., a *waaP* mutant), gives rise to significant compositional and structural changes in the outer membrane. In *E. coli*, these mutants are known as "deep-rough" and their perturbed outer membrane structure leads to pleiotropic phenotypes. These include: (i) an increase in phospholipid : protein ratios and a depletion of in outer membrane porins, (ii) hypersensitivity to hydrophobic compounds, such as detergents, dyes, and some antibiotics; (iii) a "leaky" outer membrane which releases considerable amounts of periplasmic enzymes into the medium; (iv) induction of colanic acid exopolysaccharide synthesis; (v) reduced flagella expression; (vi) secretion of an inactive form of secreted hemolysin; and (vii) increased susceptibility to attack by lysosomal fractions of polymorphonuclear leukocytes and to phagocytosis by macrophages.

In most free-living bacteria, the important element in terms of outer membrane stability appears to be negative charges. However, phosphorylation is clearly not the only way to achieve a robust outer membrane, as some bacteria lack phosphorylation of the heptose region. For example, in the case of *Klebsiella pneumoniae*, glucuronic acid and Kdo residues in the core OS provide the only source of negative charges.

While Re-LPS is sufficient to allow growth of *E. coli* strains in the laboratory, it would not support growth in their natural environments. For such bacteria, assembly and phosphorylation of the core OS heptose region may, therefore, provide further avenues for therapeutic intervention. Perhaps the smallest wild-type LPS structure consists of only lipid A and a Kdo-trisaccharide, and is produced by members of the genus *Chlamydia*. Presumably, the intracellular growth environment for this organism, together with other features of the cell envelope, facilitate survival in the absence of a more complex LPS structure.

While it is generally considered that most gram-negative bacteria require at least some LPS in their outer membrane, there are exceptions. In one of the more intriguing recent observations, mutants of *N. meningitidis* with specific defects in lipid A assembly were found to be viable (Steeghs *et al.*, 1998). This organism normally produces LOS-form LPS but, after using a range of techniques, no LPS derivatives could be detected in the mutants. It is currently unclear whether this phenomenon extends beyond *N. meningitidis*, and its impact on the biology of this organism and on the ultrastructure of the resulting outer membrane remains to be established. A limited number of wild-type gram-negative bacteria are viable without LPS. In the case of *Sphingomonas paucimobilis*, no "typical" LPS is present in the outer membrane but, instead, the bacterium produces a glycosphingolipid—a modified ceramide-derivative containing glucuronic acid and an attached trisaccharide. This lipid functionally replaces LPS in the formation of a stable outer membrane.

B. O-Polysaccharides as a Protective Barrier

Molecular modeling of LPS structure and its organization in the outer membrane predict that the O-PS forms a significant layer on the cell surface. The O-PS partially lies flat on the cell surface, where the crossover of multiple chains forms a "felt like" network. Since the O-PS is flexible, it can extend a significant distance from the surface of the outer membrane. It is, then, not surprising that many properties attributed to the O-PS are protective. In particular, long-chain O-PS is often essential for resistance to complement-mediated serum killing and, therefore, represents a major virulence factor in many gram-negative bacteria.

The serum proteins in the complement pathway interact to form a membrane attack complex (MAC) that can integrate into lipid bilayers to produce pores, leading to cell death. The MAC can be formed via a "classical" pathway, where surface antigen–antibody complexes initiate MAC formation, or through the "alternative" pathway, where complement component C3b interacts directly with the cell surface in the absence of antibody to facilitate MAC formation. In gram-negative bacteria with S-LPS, resistance to the alternative pathway does not result from defects in C3b deposition. Instead, C3b is preferentially deposited on the longest O-PS chains, and the resulting MAC is unable to insert into the outer membrane. In addition to O-PS chain length, complement-resistance can also be influenced by the extent of coverage of the available lipid A core with O-PS. As is often the case, there are exceptions to such generalizations. For example, there are some *E. coli* strains with S-LPS that are serum-sensitive unless an additional capsular polysaccharide layer is present. Although R-LPS variants of *E. coli* and *Salmonella* are almost invariably serum-sensitive, other bacteria (including many with LOS) use alternate strategies to achieve resistance.

The bactericidal/permeability inducing protein (BPI) is an antibacterial product found in polymorphonuclear leukocyte-rich inflammatory exudates (Elsbach and Weiss, 1998). BPI binds LPS and may play a role in the clearance of circulating LPS but it also exhibits antimicrobial activity in the presence of serum. Resistance to BPI-mediated killing is also dependent on long chain O-PS.

C. Lipopolysaccharide and Gram-Negative Sepsis

One potential outcome of gram-negative infections is septic shock, a syndrome manifest by hypotension, coagulopathy, and organ failure. In the United States, gram-negative sepsis accounts for 50,000–100,000 deaths each year. Septic shock results from the liberation of LPS from the bacterial cell surface, a phenomenon that naturally ensues from the growth and proliferation of bacteria. Tissue damage is not a result of direct interaction between the host tissues and an LPS "toxin," but instead results from unregulated host production of cytokines and inflammatory mediators (including tumor necrosis factor (TNF-α) and a variety of interleukins) by stimulated macrophages and endothelial cells. Under normal circumstances, these components have beneficial effects and lead to moderate fever, general stimulation of the immune system, and microbial killing. In sepsis, however, their overproduction leads to tissue and vascular damage and the symptoms of sepsis. Since free LPS is required to initiate the process, treatment with antibiotics and the ensuing bacterial lysis may actually exacerbate the problem.

The last decade has seen significant advances in our understanding of the manner in which LPS interacts with animal cells and stimulates their production of mediator molecules. It was suspected for some time that lipid A was the component responsible for those biological activities that LPS exhibited in sepsis. Definitive proof came from the observations that some partial LPS structures and chemically synthesized lipid A derivatives display the same biological activities as the complete molecule. Importantly, other partial structures are not only biologically inactive but can act as antagonists of LPS molecules that are active. Well-studied antagonist LPS molecules include the precursor lipid IV_A and the naturally occurring *R. sphaeroides* lipid A molecule. Their structures share a common feature in that both are underacylated relative to the toxic *E. coli* hexaacyl lipid A (Fig. 2). These structures have directed the synthesis of potent synthetic LPS antagonists that are able to negate the effects of challenge with biologically active LPS.

Circulating LPS molecules naturally form micellar aggregates and a variety of host LPS binding proteins are important in mobilizing LPS monomers from such complexes. These include BPI (see previous discussion) and LPS-binding protein (LBP), a 60 kDa acute-phase protein produced by hepatocytes. One role of these proteins is to clear and detoxify LPS. For example, LBP is known to transfer LPS to high-density lipoprotein fractions. However, LBP is also a crucial component of the signaling pathway through which animal cells are stimulated to produce cytokines and inflammatory mediators.

The central pathway by which cells recognize low concentrations of LPS requires the participation of a receptor protein CD14. CD14-deficient cell lines, such as 70Z/3 pre-B lymphocytes, are less sensitive to LPS (i.e., responsive to nanomolar rather than picomolar levels), unless transfected with CD14. Consistent with these results, CD14-knockout mice have been shown to be 10,000-fold less sensitive to LPS *in vivo*. In myeloid cell lines, CD14 occurs as a 55 kDa glycosylphosphatidylinositol (GPI) anchored glycoprotein, attached to the membrane (mCD14 in Fig. 7). However, a variety of nonmyeloid (endothelial and epithelial) cells are also responsive to LPS, despite the fact that they lack mCD14. These require

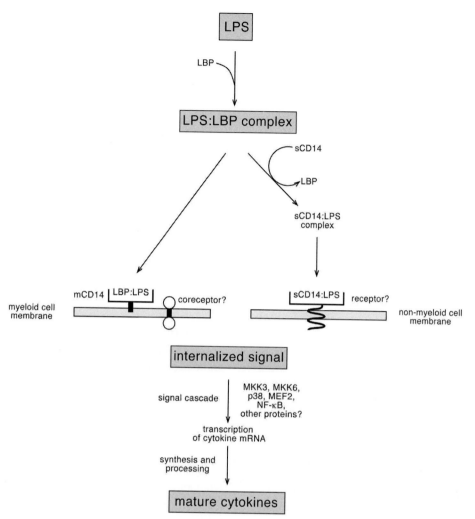

Fig. 7. Pathways involved in the stimulation of animal cells by LPS. The CD14-mediated central pathways are shown. In macrophages and other myeloid cell lines, the LPS binding protein (LBP) delivers LPS to the GPI anchored form of CD14 (mCD14). An unidentified coreceptor may participate in intracellular signal generation. In endothlial and other nonmyeloid cells, the soluble sCD14 activates the host cell through a route that may involve an additional unknown receptor. Other signaling pathways (not shown), such as that involving the β-integrin CD11/18, may play a role under specific circumstances. Once stimulated, a phosphorylation signal transduction cascade leads to transcription of genes encoding cytokines and inflammatory mediators. Elements of the pathways are also important for LPS clearing.

the participation of a serum protein, subsequently identified as a soluble form of CD14 (sCD14). Both sCD14 and mCD14 can bind LPS to form a complex, but the kinetics of binding are slow. LBP serves to overcome this rate-limiting step by delivering LPS to mCD14 or sCD14 (Fig. 7). At high LPS concentrations, CD14-independent stimulation is evident in CD14-deficient mice. This may reflect the use of non-physiological amounts of LPS, or may point to alternative pathways used under specific circumstances. For example, one possible alternate pathway involves the CD11/CD18 β-integrins and LBP. In comparison to CD14, the CD11/18 system is less responsive to LPS and the activation of a cellular response is slower. As predicted by the pathway shown in Fig. 7, blood taken from an LBP-knockout mouse shows a 1000-fold reduction in the ability to respond to LPS. However, the complexity of the system is reflected in the unexpected observation that these mice show no reduction in response to LPS *in vivo* (Fenton and Golenbock, 1998). One interpretation of the mouse phenotype is that one or more LBP-independent stimulatory pathways exist. These results exemplify the occasional difficulty in extrapolating data collected *in vitro,* often with purified components, to the situation in the animal.

While the LBP-CD14-mediated recognition of LPS has been studied in detail, the downstream steps in signal transduction remain obscure. It is generally accepted that an accessory coreceptor protein exists in a complex with CD14, to facilitate internalization of the signal generated by the interaction of LPS with CD14. The involvement of a putative transmembrane coreceptor helps explain several experimental observations. First, CD14 itself lacks a transmembrane domain to facilitate intracellular signaling. Second, CD14 alone is not able to discriminate between agonist and antagonist lipid A molecules (such as the *R. sphaeroides* lipid A), and the antagonist molecules do not appear to operate by blocking the ability of agonist LPS to bind to CD14. The antagonist and coreceptor might interact directly. In a related issue, the CD14-independent signal arising from administration of high concentrations of LPS could also reflect direct interaction of LPS with the putative coreceptor. The putative accessory receptor partner may be able to interact with other signaling pathways, for example, the one involving the CD11/CD18 β-integrins.

Once the relevant signal is transmitted across the animal cell membrane, a cascade of events leads to the release and overproduction of cytokines. The components of the latter stages of the response pathway are now beginning to be identified. The cascade involves rapid protein phosphorylation events, and isoforms of the p38 mitogen-activated protein (MAP) kinase family play a central role. LPS antagonists block phosphorylation of p38. pP38 is itself activated by the MAP kinase kinases, MKK3 and MKK6. Downstream, p38 has a number of substrates including the myocyte enhancer factor (MEF2) family of

transcription activators. The transcription factor NF-κB also plays an important regulatory role in the proinflammatory response, as well as in the development of LPS tolerance.

A variety of therapeutic approaches have been designed to interfere with specific steps in the process leading to septic shock. The numbers of mediators involved complicates strategies based on blocking cytokines themselves. Neutralizing an individual mediator would not be expected to be an effective therapy, as appears to be the case for antibody-neutralized TNF-α. Significant and well-publicized efforts have been directed to neutralizing the LPS signaling molecule by administering antibodies but, to date, attempts to develop therapies based on anti-endotoxin monoclonal antibodies have been disappointing. Two monoclonal antibodies have been the subject of clinical trials. Monoclonal antibodies E5 and HA1A were both raised by immunization with *E. coli* J5 and recognize the lipid A of this and other gram-negative bacteria. Unfortunately, the clinical trials provided no compelling evidence for the protective capacity of these antibodies. However, there are monoclonal antibodies that recognize the conserved inner core OS of *E. coli* and *Salmonella* and that show promise both *in vitro* and in animal models, suggesting alternate immunotherapeutic strategies (di Padova *et al.*, 1993). LPS neutralization could also be achieved by using proteins that bind LPS and both LBP and BPI are being pursued in this respect; BPI-based therapy is in clinical trials. Approaches that attempt to block LPS receptor pathways with synthetic LPS antagonists are proving effective in animal models (Wyckoff *et al.*, 1998). Equally promising is the application of anti-CD14 monoclonal antibodies that block the formation of CD14:LPS complexes and protect against LPS exposure in a rabbit model (Schimke *et al.*, 1998).

D. Adaptive Responses and Lipid A Modifications

It has been recognized for some time that the precise structure of lipid A determines resistance of the bacterium to polycationic polypeptides, proteins, and polycationic antibiotics, such as polymyxin. Modification of lipid A by addition of Ara*p*4N to the 4'-phosphate provides one important structural change that is correlated with resistance. The effect of this modification may be a dampening of the negative charges in lipid A, inhibiting the initial binding of polycations and their eventual perturbation of the outer membrane.

In *E. coli* and *Salmonella,* such modifications are nonstoichiometric, and the modifying enzymes (as well as polymyxin-resistance) are environmentally modulated (Guo *et al.,* 1997). In contrast, *Proteus mirabilis* and *Burkholderia cepacia* are constitutively resistant to polymyxin and their lipid A is extensively modified by Ara*p*4N. Work with *Salmonella* has established that the two-component environmental sensor comprising PhoP/PhoQ is required for transcription of genes essential for virulence in mice. The sensor appears to respond to microenvironments reflective of growth in mammalian cells, with the adaptive response optimizing the bacterial physiology for intracellular growth. Among the genes whose transcription is controlled by PhoP/PhoQ are *pmrA/B*, encoding an additional two-component sensory system. Together, these sensory systems modulate expression of enzymes that are thought to be involved in Ara*p*4N synthesis and its addition to lipid A. Such changes may provide an adaptive mechanism to counter host polycation defense molecules. However, an additional PhoP/PhoQ-controlled lipid A modification is the formation of a heptaacylated derivative. The resulting LPS has reduced biological activities, including diminished stimulation of TNF-α production in cell cultures. In this situation, LPS can be considered to be a dynamic, rather than static, molecule, whose structure and function can be modulated in response to cues from the host.

See Also the Following Articles

ABC TRANSPORT • LIPID BIOSYNTHESIS • OUTER MEMBRANE, GRAM-NEGATIVE BACTERIA

Bibliography

Di Padova, F. E., Brade, H., Barclay, G. R., Poxton, I. R., Liehl, E., Schuetze, E., Kochner, H. P., Ramsay, G., Schreier, M. H., McClelland, D. B. L., and Rietschel, E. (1993), A broadly cross-protective monoclonal antibody binding to *Esche-*

richia coli and *Salmonella* lipopolysaccharides. *Infect. Immun.* **61**, 3863–3872.

Elsbach, P., and Weiss, J. (11998). Role of the bactericidal/permeability-increasing protein in host defence. *Curr. Opin. Microbiol.* **10**, 45–49.

Fenton, M. J., and Golenbock, D. T. (1998). LPS-binding proteins and receptors. *J. Leukoc. Biol.* **64**, 25–32

Guo, L., Lim, K. B., Gunn, J. S., Bainbridge, B., Darveau, R. P., Hackett, M., and Miller, S. I. (1997). Regulation of lipid A modifications by *Salmonella typhimurium* virulence *genes phoP-phoQ. Science* **276**, 250–253.

Han, J., Bohuslav, J., Jiang, Y., Kravchenko, V. V., Lee, J. D., Li, Z. J., Mathison, J., Richter, B., Tobias, P., and Ulevitch, R. J. (1998). CD14 dependent mechanisms of cell activation. *Prog. Clin. Biol. Res.* **397**, 157–168.

Heinrichs, D. E., Yethon, J. A., and Whitfield, C. (1998). Molecular basis for structural diversity in the core regions of the lipopolysaccharides of *Escherichia coli* and *Salmonella enterica. Mol. Microbiol.* **30**, 221–232.

Joiner, K. A. (1988). Complement evasion by bacteria and parasites. *Ann. Rev. Microbiol.* **42**, 201–230.

Raetz, C. R. (1998). Enzymes of lipid A biosynthesis: Targets for the design of new antibiotics. *Prog. Clin. Biol. Res.* **397**, 1–14.

Raetz, C.R.H. (1996). Bacterial lipopolysaccharides: A remarkable family of bioactive macroamphiphiles. In "*Escherichia coli and Salmonella.* Cellular and Molecular Biology" (F. C. Niedhardt, ed.), pp. 1035–1063. American Society for Microbiology, Washington, DC.

Schimke, J., Mathison, J., Morgiewicz, J., and Ulevitch, R. J. (1998) Anti-CD14 mAb treatment provides therapeutic benefit after *in vivo* exposure to endotoxin. *Proc. Natl. Acad. Sci. USA* **95**, 13875–13880.

Steeghs, L., den Hartog, R., den Boer, A., Zomer, B., Roholl, P., and van der Ley, P. (1998). Meningitis bacterium is viable without endotoxin. *Nature* (Lond) **392**, 449–450.

Whitfield, C. (1995). Biosynthesis of lipopolysaccharide O-antigens. *Trends Microbiol.* **3**, 178–185.

Whitfield, C., Amor, P. A., and Köplin, R. (1997), Modulation of surface architecture of Gram-negative bacteria by the action of surface polymer:lipid A-core ligase and by determinants of polymer chain length. *Mol. Microbiol.* **23**, 629–638.

Whitfield, C., and Perry, M. B. (1998). Isolation and purification of cell surface polysaccharides from gram-negative bacteria. *Meth. Microbiol.* **27**, 249–258.

Wyckoff, T.J.O., Raetz, C.R.H., and Jackman, J. E. (1998), Antibacterial and anti-inflammatory agents that target endotoxin. *Trends Microbiol.* **6**, 154–159.

Low-Nutrient Environments

Richard Y. Morita

Oregon State University

GLOSSARY

anabiosis Return to life; sometimes referred to as cryptobiosis (latent life). The process may be brought about by cooling (cryobiosis), drying (anahydrobiosis), due to the lack of oxygen (anoxybiosis), and due to high salt concentration (osmobiosis).

copiotroph Organism that has the ability to grow at $100\times$ the concentration of organic matter that is found in the oligotrophic environment.

syntrophy Ecological relationship in which organisms provide nourishment for each other.

LOW-NUTRIENT ENVIRONMENTS, termed oligotrophic environments, mainly lack organic matter for the growth of heterotrophic bacteria. Other nutrients, such as phosphate and nitrogen (ammonium, nitrate, or nitrite), can also limit the growth of microbes in any ecosystem. This article concentrates on the lack of organic matter as a source of energy for the heterotrophic bacteria in any ecosystem.

I. OLIGOTROPHIC ENVIRONMENTS

Oligotrophic environments are created mainly by the microbes themselves. Because of the large numbers and various physiological types, the organic matter in any ecosystem becomes depleted over time, especially in places where the organic matter is not replenished mainly by photosynthetic organisms. The more labile organic matter is utilized rapidly, often as fast as it is supplied, eventually leaving the recalcitrant organic matter. Even the labile compounds may not be "free" since they may be complexed mainly to humus and clay, making them inaccessible to the microbes. Humus can also be complexed to clay-forming a clay–humate combination. Furthermore, the amount of "free" organic matter reported in the literature may be a result of the method of analysis. If all factors are favorable, the rate of decomposition of the organic matter is rapid. Generally, in the various ecosystems, all heterotrophic organisms are competing for the limited amount of organic matter present.

A. Levels of Organic Matter in Ecosystems

Each ecosystem has its own level of organic matter but it certainly is far less than that found in nutrient broth (3400 mg/l of dissolved carbon). The average values for soil are 25 mg organic C per gram for a fertile loam and 10 mg organic carbon per liter for aquatic systems, much of the carbon not being bioavailable to various microorganisms. Thus, the organic matter (energy) is severely limited for the indigenous microbial population. All estimates are that the amount of organic matter in ecosystems cannot support the active metabolism of the entire indigenous population; hence, the ecosystems are oligotrophic. It is rare when an ecosystem is eutrophic. As a result, most of the microbial population is inactive. Much of the organic carbon that enters the various ecosystems is cellulose, but not all microbes produce cellulase. As a result, syntrophy plays an important role in the microbiology of ecosystems.

B. Measurement of Activity

It is difficult to measure the microbial activity in any ecosystem sample. Although many different methods have been employed, none is accurate and dependable, mainly because perturbation takes place when the sample is taken. In addition, the surface (bottle or confinement) effect also hinders an accurate measurement. In many studies, the addition of small amounts of radioactive substrates to the ecological sample (called heterotrophic potential in aquatic samples and substrate-induced respiration) is employed, but this small amount of substrate may act as a primer for metabolic activity. For aquatic measurements, the samples are usually shaken, which compounds the error in the measurements. Furthermore, when very low concentrations (e.g., 10^{-12} M) of uniformly C-labeled glutamate is added to 72-day starved cells, none of the radioactivity is incorporated into cellular material. Chemostat studies show that when marine bacteria are placed in a chemostat employing natural-seawater concentrations of organic matter, the cells are washed out. In nature, many of the microorganisms may be in the dormant state, but during perturbation, addition of small amounts of radioactive substrate, or plating procedure, the dormant cells become active. For instance, the approximation on the turnover times of microbial biomass-C in soil ranges between 5.5 times per year (generation time of 66.36 days) and 0.4 times per year (generation time of 912.5 days), depending on the soil in question. The data, when extrapolated, indicate that maintenance energy demand surpasses the values known for the average C input into the soil. This indicates that the most microbes in soil are in a dormant (starvation-survival) state and the concept of "energy of maintenance" is invalid in the various ecosystems. The concept of energy of maintenance is not valid in the microbiology of ecosystems but applies only to a clone of bacteria in the laboratory. In the deep sea below the thermocline, the oxygen consumption has been estimated to be 0.002–0.004 ml/liter/yr for all organisms present.

The low concentration of organic matter in the deep sea is a "blessing in disguise" since a high concentration of organic matter would render the deep sea anoxic.

All data on the amount of organic matter (energy) in ecosystems definitely illustrate that oligotrophic environments are the norm, whereas eutrophic environments are rare. Most of the bacterial flora of any ecosystem are inactive, due to the lack of energy. The amount of organic matter in soil will not take care of the energy needs of the fungi present either. Thus, the indigenous microflora is in a starvation mode. Microbiologists recognize that spores are formed due to the lack of energy in the spore-forming bacteria. Recognizing the long period of time that microbes have inhabited the earth, the nonspore-forming bacteria (which predate the spore formers) must have also developed physiological mechanisms to survive periods of starvation, since green plant photosynthesis evolved much later than the bacteria. Unfortunately, microbiologists hardly ever deal with "survival of the species" or "survival of the fittest."

II. OLIGOTROPHIC, LOW-NUTRIENT, AND COPIOTROPHIC BACTERIA

The term "oligotrophic" was coined to describe the soil nutrient conditions in a German peat bog and later in limnology, where it was mostly employed. When applied to limnological conditions, the trophic state was determined by the phosphorus content, chlorophyll concentration, Secchi Disk depth (in meters), and primary production. No mention is made of the organic content because the energy in this situation was sunlight. The term was then adapted into microbiology without too much thought. The term oligotrophic bacteria still remains vague since it has been defined in the literature as (1) heterotrophic bacteria that have the capability to grow at minimal organic substrate concentrations (1–15 mg C/l), even though they can grow on richer media and (2) heterotrophs that have the ability to grow only at low-nutrient concentrations. The latter was sometimes termed "obligate" oligotrophic. Unfortunately, the obligate oligotrophs, when eventually tested under various conditions (various types of media, pH, Eh, temperature, etc.), were found to be able to adapt to higher concentrations of organic matter. However, if any are found in the future, they should be termed "oligocarbophilic," since priority is the rule in science. The term oligotrophic should be reserved to describe the environment, not the microorganisms.

Copiotrophic bacteria are defined as microorganisms that have the ability to grow in media containing $100\times$ the concentration of organic matter found in the oligotrophic environment from which they were isolated. Unfortunately, the term copiotrophic bacteria is a misnomer since *Escherichia coli* could be considered a copiotroph, because in the intestinal tract it reproduces only twice a day.

Originally, "low-nutrient bacteria" was used to describe organisms that were isolated on agar plates in which no nutrients were added to the agar. However, agar itself contains dissolved organic matter.

All the terms employed to describe heterotrophic bacteria that grow on low organic concentrations should be discarded until the organisms have undergone rigorous testing—a Herculean task.

III. STARVATION-SURVIVAL IN OLIGOTROPHIC ENVIRONMENTS

Due to lack of energy, microorganisms enter into the starvation-survival mode, and starvation-survival, a form of anabiosis, is defined as a physiological state resulting from an insufficient amount of energy for growth (increase in cellular biomass) and reproduction. This also occurs in chemolithotrophs. Bacteria do survive in ancient material and this has been attributed to anabiosis. The recent literature, mainly from the Russian microbiologists, on the occurrence of bacteria in permafrost attributes its presence to anabiosis (cryobiosis). The latest reports on the occurrence of bacteria in amber have reawakened microbiologists to the concept of viable cells in ancient material. However, the earliest report on the occurrence of bacteria in amber was in the early 1920s. In laboratory studies, *Pseudomonas syringae* subsp. *syringae* (initial concentration varied from 10^7 to 10^9 cells/ml, depending on the strain employed) was starved in distilled water for 24 years and viability in most strains was still retained (10^5

to 10^6 cells/ml). *E. coli, Klebsiella,* and *Pseudomonas cepacia,* when starved in sterile well water at room temperature and at a concentration of approximately 10^7 cells/ml, resulted in viable counts of 10^4–10^5 cells/ml after 624 days.

A. Starvation Process

Unfortunately, no data exist on the *in situ* starvation process in the various ecosystems except that there are many reports of bacteria in ancient material where the organic matter concentration is extremely low. All the laboratory studies dealing with starvation-survival start out with cells grown in energy sources at concentrations higher than normally found in any ecosystem. The cells are then harvested by centrifugation, washed (generally twice), and then placed in a starvation menstruum. The starvation-survival patterns determined by plate counts obtained by starving bacteria with time show that some species will (1) decline in numbers immediately, (2) produce no change in numbers, (3) increase in numbers followed by a decline and (4) increase in numbers. Unfortunately, most starvation studies deal with the third pattern. *Nitrosomonas cryotolerans* displays the second pattern, but insufficient studies on the various chemolithotrophs preclude making the statement that they all follow pattern 2. Very little is known about patterns 2 and 4.

When *Moritella* Ant-300, starting with an initial viable count of 3×10^7 cells/ml, was starved 98 days, a final count was obtained that was 0.3% of the initial count, but the total count remained level (Fig. 1) for cells grown at different dilution rates). (Cells grown at the lowest dilution rate resemble cells in nature much more than cells grown in batch culture or at higher dilution rates.) The doubling times for the dilution rate D = 0.0.5, D = 0.057, and D = 0.170 are 46.2 hr, 12.2 hr, and 4.1 hr, respectively. No cryptic growth could be demonstrated in this organism. Even after a long period of starvation, a few cells still remain viable so that the survival of the species is maintained. This concept is important in nature.

There appear to be three stages in the starvation-survival process, as depicted in Figs. 1 and 2 for *Moritella* Ant-300 and the time ranges were arbi-

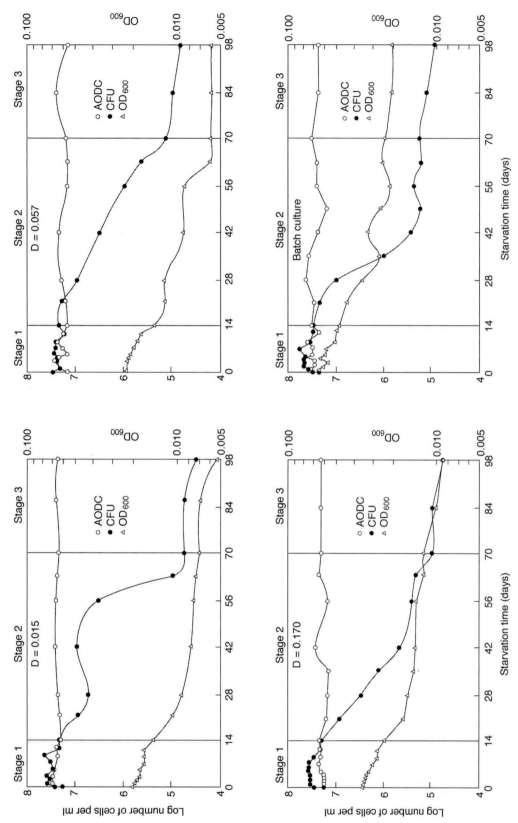

Fig. 1. Total cells (AODC, acridine orange direct count), viable cells (CFU, colony-forming units), and optical density (OD$_{600}$) of ANT-300 cells, with starvation time for cells from different dilution rates and batch culture. [From Moyer, C. L., and Morita, R. Y. (1989). *Appl. Environ. Microbiol.* **55**, 1122–1127.]

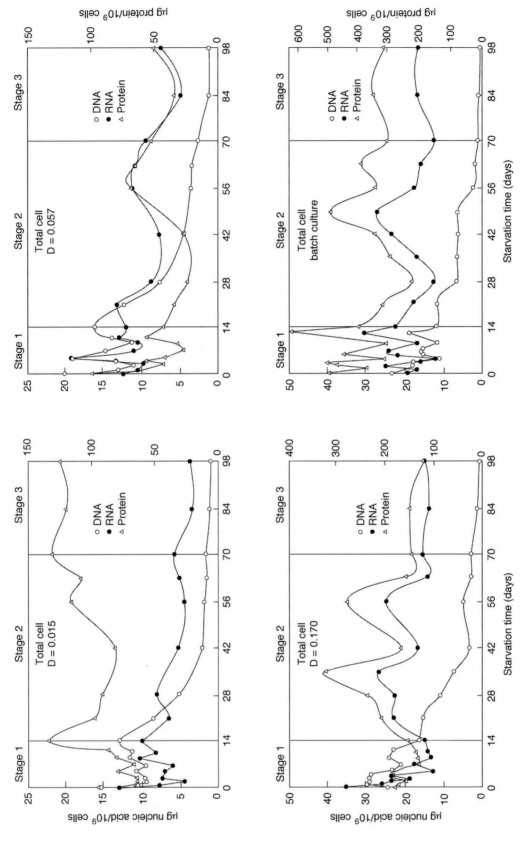

Fig. 2. DNA, RNA, and protein per total cell, with starvation times and stages for ANT-300 cells grown at different growth rates. [From Moyer, C. L., and Morita, R. Y. (1989). *Appl. Environ. Microbiol.* **55,** 2710–2716.]

trarily set from 0 to 14 days for the first stage, 14 to 70 days for the second stage, and greater than 70 days for the third stage.

Stage 1 represents a period where the total acridine orange direct counts (AODC) remain nearly level, but the OD_{600} drastically declines, indicating that the size of the cells must be small (see following). The viable count also increases during this period and the amount of proteins, DNA, and RNA greatly fluctuate. This is accompanied by rapid oxygen uptake, which takes place utilizing the reserve endogenous energy, and a decrease in phospholipid content. Chemotaxis, a mechanism for the cell to obtain energy, appears after 48 hr. The formation of fibrillar structures also occurs, needed for increased surface adhesion in some organisms. Adhesion aids the microbe in its initial stages of starvation because surfaces are known to attract organic matter (bottle, surface, or confinement effect).

In stage 2, the number of viable cells slowly decreases and certain proteins continue to be degraded, while others are not. During this period, some starvation stress proteins are synthesized (see following). In addition, the cells are "fine-tuning" themselves in preparation for stage 3.

During stage 3, the cells are in a state of metabolic arrest, analogous to the bacterial spore. Because no substrates are present, no metabolism takes place. How long cells can remain in this state remains a mystery, but for some organisms, this survival period can be a very long time, so that the cells can have some "biological" control over time.

Most of the bacteria found in the various ecosystems (soil, freshwater, and marine ecosystems) are ultramicrocells (dwarf cells, minicells, picoplankton, and nanoplankton).

Some of the ultramicrobacteria have the ability to pass through a 0.2-μm membrane and, subsequently, grow on media to a much larger size. The smaller the organism, the greater the surface/volume ratio, which permits the organism to better cope with obtaining nutrients from the environment. Thus, being the right size in the environments may be critical to its survival. When *Moritella* Ant-300 is grown in a chemostat, the size of the cells will depend upon dilution rate (Table I), where the slower the dilution rate, the smaller the cells. Accompanying the smaller cell size is the smaller amount of DNA/cell, hence, the smaller-nucleoid volume (Table I). The presence of ultramicrobacteria formed due to starvation is another factor suggesting that most ecosystems are oligotrophic.

B. Viable but Nonculturable

Starvation can induce the formation of viable but nonculturable cells. The first microorganism in which this property was demonstrated was *Escherichia coli*. In laboratory studies, when *Pseudomonus* sp. and *Moritella* Ant-300 were starved, the direct

---

TABLE I
Effect of Growth Rate on DNA Content in Relation to Average Cell Volume for Unstarved and Starved Ant-300 Cells

Culture	DNA/cell (fg ± SEM)	Cell volume (μm^3)	Estimated nucleoid volume (μm^3)	Nucleoid volume/cell volume
Unstarved batch	23.66 ± 0.01	5.94	0.33	0.06
$D = 0.170$ hr^{-1}	24.69 ± 0.01	1.16	0.35	0.30
$D = 0.057$ hr^{-1}	20.12 ± 0.06	0.59	0.28	0.48
$D = 0.015$ hr^{-1}	15.33 ± 0.27	0.48	0.21	0.45
Starved batch[a]	1.23 ± 0.17	0.28	0.017	0.06
$D = 0.170$ hr^{-1}	1.03 ± 0.35	0.19	0.014	0.08
$D = 0.057$ hr^{-1}	1.45 ± 0.02	0.18	0.020	0.11
$D = 0.015$ hr^{-1}	1.28 ± 0.02	0.05	0.018	0.40

[a] All cell samples were taken from stage 3 of starvation-survival.
[From Moyer, C. L., and Morita, R. Y. (1989). Effect of growth rate and starvation-survival on cellular DNA, RNA, and protein of a psychrophilic marine bacterium. *Appl. Environ. Microbiol.* **55**, 2710–2716.]

counts remained constant, the viable counts (CFU) readily decreased, but INT (reduction of tetrazolium salt) cell counts were 10X greater than the CFUs but not as high as the direct count. The reduction of INT indicates that cellular metabolism takes place, yet no CFU are formed and these would be termed viable but nonculturable. Viable but nonculturable does not mean that the cells will never be cultured. The question is whether or not these viable but noriculturable cells can be made to multiply and grow to form a colony on an agar plate. Research indicates that some of the bacteria can be resuscitated, but much further research is needed to determine the resuscitation methods for various organisms when they are rendered viable but nonculturable.

When determining the number of bacteria in any environmental sample, microbiologists realize that only a very small fraction of the total bacteria can be cultured. Furthermore, by molecular techniques, it is recognized that there are a great many other species of bacteria present which can not be cultured by the methods employed.

C. Stress Proteins

Stress proteins are formed when organisms are exposed to various stresses (hydrogen peroxide, heat, hydrostatic pressure, etc.). Starvation also induces their formation. Cross-protection has been noted among stress proteins. In various ecological situations, cells are generally starved before they are exposed to other stresses. Hence, the starved cells can handle the stress more successfully than can cells that are not starved.

IV. CONCLUSION

There can be no doubt that most ecosystems on Earth are oligotrophic. As a result, the microbes in nature are in the starvation-survival lifestyle most of the time, waiting for the right conditions (energy, pH, Eh, water content, etc.) to present themselves so that they can take their place in the cycling of matter on Earth.

See Also the Following Articles

Natural Selection, Bacterial • Starvation, Bacterial

Bibliography

Amy, P. S., and Haldeman, D. L. (1997). "The Microbiology of the Terrestrial Deep Subsurface," Lewis Publishing, Boca Raton, FL.
Iscobellis, N. W., and De Vay, J. E. (1986). Long-term storage of plant pathogenic bacteria in sterile distilled water. *Appl. Environ. Microbiol.* **52**, 388–389.
Keilin, D. (1959). The problem of anabiosis or latent life: history and current concepts. *Proc. Roy. Soc. of London, Ser. B* **150**, 149–191.
Kjelleberg, S. (1993). "Starvation in Bacteria." Plenum Press, New York.
Morita, R. Y. (1997). "Bacteria in the Oligotrophic Environment: Starvation–Survival Lifestyle," Chapman and Hall, New York.
Moyer, C. L., and Morita, R. Y. (1989). Effect of growth rate and starvation-survival on the viability and stability of a psychrophilic marine bacterium. *Appl. Environ. Microbiol.* **55**, 1122–1127.
Moyer, C. L., and Morita, R. Y. (1989). Effect of growth rate and starvation-survival on DNA, RNA and protein of a psychrophilic marine bacterium. *Appl. Environ. Microbiol.* **55**, 2710–2716.
Xu, H. S., Roberts, N., Singleton, F. L., Atwell, R. W., Grimes, D. J., and R. R. Colwell. (1982). Survival and viability of nonculturable *Escherichia coli* and *Vibrio cholerae* in estuarine and marine environment. *Microb. Ecol.* **8**, 313–323.

Low-Temperature Environments

Richard Y. Morita

Oregon State University

GLOSSARY

cryobiosis Anabiosis (latent life) due to freezing.

endolithotrophic Living inside rocks, usually sandstone.

homeophasic adaptation Adaptation of the cell membrane to maintain the lipids in a bilayer phase.

melt water Water melted by radiant energy in the polar regions.

permafrost Ground (ice, bedrock, soil) that remains frozen below 0°C for more than two years.

thermocline In the stratification of warm surface water over cold, deeper water, the transition zone of rapid temperature decline between the two layers.

upwelling Transport of water from the deep ocean to the surface, replacing the surface water that has moved offshore.

LOW-TEMPERATURE ENVIRONMENTS (sometimes referred to as the psychrosphere) dominate the Earth's biosphere. Although there are many types of low-temperature ecosystems on Earth, each has its distinct microbial community.

Because Mars and Europa's surface environments are cold, low-temperature environments, the cold-loving microbes are currently receiving much attention. The average temperature of the Earth is 15°C. It appears that life processes take place at temperatures above −10°C. As will be seen, the cold environments can be divided into two categories: psychrophilic (permanently cold) and psychrotrophic (seasonally cold or temperature fluxes into the mesophilic range) environments.

I. LOW-TEMPERATURE ENVIRONMENTS

A. Arctic and Antarctic Environments

The polar regions make up 14% of the Earth's surface. Although both polar regions are cold and lack sunlight during their respective winters, their differences lie in their land masses, land topographical features, large-scale water transport patterns, and the magnitude of nutrient supply. The latter is due to the major river inputs in the Arctic, which are lacking in the Antarctic. Both regions have different types of ecosystems, such as tundra, deserts, sea-ice fronts, snow, glaciers, continental ice sheets, mountains, lakes (many frozen the entire year), rivers, ice-bubble habitats, meltpools, and permafrost. In some ice-covered lakes, water may be present due to the geothermal heat. More than 29% of Earth's landmass is permafrost. Approximately 80% of Alaska, 50% of Russia (former USSR), 50% of Canada, and 20% of China is permafrost, and 80% of the Earth's biosphere is permanently cold. The lowest temperature recorded is −147°C at Lake Vostok, Antarctica. Microbes in these ecosystems must also contend with different salinities (caused by different ionic species), dehydration created by water freezing, different light intensities including seasonal darkness, and the lack of organic carbon such as found in most soils.

Great fluctuations in temperature may occur in

these regions due to solar radiation. Solar radiation hitting snow and ice, especially when dust particles are imbedded in the ice or snow, causes the formation of melt water, where microbial life increases. This melt water is the source of liquid water for the endolithotrophic bacteria as well as the water in certain lakes. It should also be realized that when melt water is formed, different salts (depending on location) dissolve into the water; hence, the growth of cold-loving bacteria that are also alkaliphilic may be enhanced.

B. Oceans

Approximately 71% of the Earth's surface is represented by the oceans and 90% by volume of it is colder than 5°C. The warmer surface water is separated from the cold water by the thermocline (discontinuity layer) and this thermocline becomes more shallow at the higher latitudes, until the thermocline is at the surface. This thermocline is an imperfect barrier that permits only a small amount of organic matter, produced by photosynthetic organisms in the photic zone, from sinking into the cold, deep sea. Unlike the land masses, this environment is rather constant in that the temperature and salinity do not vary greatly. In the deep sea, the main variable is the hydrostatic pressure, which increases approximately 1 atm for every 10 m depth, and there is no light. However, the light intensity at the surface of the cold ocean varies, depending on the season and the climatic conditions. Thus, we realize that many cold-living bacteria are also barophilic (pressure-loving).

In nearshore environments at lower latitudes, where temperatures are warmer, the water temperature may be suppressed due to upwelling of cold oceanic water.

C. Upper Atmosphere

At altitudes >10,000 m, the temperature is <10°C, while at altitudes >3000 m, the temperature is consistently <5°C. With increasingly higher altitudes, the temperature progressively decreases, and a temperature below −40°C has been recorded.

D. High Mountains

Where snow or ice remains year-round, low temperatures are always present, except on surfaces that receive solar radiation. At portions of the mountains where snow and ice melt, the temperature is cold part of the year. The same holds true for the mountain lakes, and if a thermocline exists in these freshwater bodies, then a permanent cold temperature exists.

II. LOW-TEMPERATURE ORGANISMS

A. Psychrophiles and Psychrotrophs

Microorganisms living in cold-temperature environments are known as psychrophiles, which have a maximum growth temperature at 20°C or lower, an optimum temperature at 15°C or lower, and a minimum growth temperature at 0°C or lower. The maximum growth temperature was set at 20°C simply because, in the U.S. laboratory, temperature is around 21°C to 22°C, which is definitely not cold. The cold-loving higher organisms are known as cryophiles but, because of common usage, the term psychrophiles has been retained for bacteria. For many years, psychrophiles were thought not to exist, mainly because investigators were working with psychrotrophs, which have their optimum and/or maximum growth temperature in the mesophilic range. As a result of this situation, many different terms were employed to describe the low-temperature organisms. Some of these terms were cyrophile, rhigophile, psychrorobe, thermophobic bacteria, Glaciale Bakterien, facultative psychrophile, psychrocartericus, psychrotrophic, and psychrotolerant. It was not until 1964 that the first true psychrophiles (*Vibrio* [*Moritella* gen. nov.] *marinus* MP-1 and *Vibrio* [*Colwellia* gen. nov.] *psychroerythrus*) were taxonomically described in the literature. Psychrotrophs (also known currently as psychrotolerant) are involved in the microbiology of foods and dairy products and are usually involved in spoilage processes. The amount of research done on psychrophiles is very small compared to the other thermal groups of bacteria. With renewed interest in life in outer space, more and more attention is being focused on these organisms.

All thermal groups of microorganisms can be found in low-temperature environments, even in the dry valleys of Antarctica. It is recognized that there is a continuum of cardinal temperatures among thermal groups, but the definition of psychrophiles is useful because of its ecological significance. That is, psychrophiles cannot be isolated from cold environments where temperature fluctuate into the mesophilic range, but psychrotrophs can be found in both environments and even thermophiles can be found in permafrost. Since cold temperature (cryobiosis) is a means by which organisms are preserved, it is only logical that all the various thermophiles and mesophiles can be found in low-temperature environments.

B. Biodiversity

From the old literature, various species of *Pseudomonas, Flavobacterium, Alcaligenes, Vibrio, Achromobacterium, Micrococcus, Microbacterium, Brevibacterium, Bacillus,* and *Clostridium* were considered to be psychrophiles, but many need to have their cardinal temperatures determined. *Polaromonas, Aquaspirillum, Arthrobacter, Bacteroides, Cytophaga, Marinobacter, Phonnidium, Planococcus, Psychroserpens, Gelidibacter,* and *Moritella* have been added. Genera that are both psychrophilic and barophilic are *Altermonas, Photobacterium, Colwellia,* and *Shewanella.* Thus, within the Eubacteria, psychrophiles appear to be widespread, being autotrophs or heterotrophs, aerobes or anaerobes, spore formers and non-spore formers, phototrophs and nonphototrophs. Isolation of a methanotroph has also been reported. Newly added to the list is an Archaea (*Methanogenium frigidum*) isolated from Ace Lake, Antarctica. It is a hydrogen-utilizing methanogen (max. temp. = 18°C; opt. temp. = 15°C; min. temp. = −2°C). Even a methyltroph has been described. Up to 34% of the procaryotic biomass in a coastal Antarctic surface waters (−1.5°C) is reported to be Archaea, identified by molecular phylogenic techniques. They remain to be isolated. How widespread the psychrophiles are within the Archaea remains to been seen.

Before the isolation of true psychrophiles, temperature precautions were not employed—mainly because investigators did not realize the abnormal thermosensitivity of psychrophiles. If the proper precautions are taken (taking the sample from a permanently cold environment; keeping the sample, media, pipettes, etc. cold), many other species of psychrophiles probably will be isolated. These precautions were not taken by early investigators trying to isolate psychrophiles. In addition, organisms that are psychrophilic as well as alkalophilic will probably be isolated from melt waters in the Dry Valley of Antarctica. As with many extremophiles, many require more than one extreme environmental condition.

III. PHYSIOLOGICAL AND BIOCHEMICAL CHARACTERISTICS

A. Growth

Microbes are found in permafrost where temperatures may be extremely low and microbes may have survived some millions of years. Yet, growth does require liquid water. At what temperature does the water in the cell and its immediate surroundings become frozen? This will depend on the composition of the cell and its immediate environment. The lowest temperature recorded for growth of a microbes is −11°C but the temperature of most of the permafrost, as well as other polar environments, is much less than −11°C. Because of these low temperature environments, no growth occurs until the temperature meets the minimum temperature of growth of the species in question. This means that it may be seasonal and growth thus occurs mainly in melt water. A good example of this seasonal growth is the case of the growth of the endolithotrophic bacteria in the Antarctic. It may be a "blessing in disguise" that growth in very cold ecosystems is for only short periods of time when melt water results from solar radiation because the energy source would become depleted. Recognizing that the environmental temperature is generally 5 to 20°C below the optimum growth temperature, growth is seldom at the maximal rate the organism is capable of in the environment, yet good growth can be obtained. In many of the psychrophiles, where the minimum temperature for growth has been determined, the minimum temperature for growth is less than 0°C (Fig. 1). When temperatures are below the minimum for growth, the organisms are in a state of cryobiosis. Thus, the time the microbes experience at temperatures above approximately −10°C is conducive to metabolism and growth.

In some of the psychrophiles, the maximum growth temperature is around 10°C (Fig. 1). Psychrophiles held a few degrees above the maximum for growth will expire and this is the reason why only psychrophiles are found in permanently cold environments and not in environments where temperature fluxes venture into the mesophilic temperature range.

B. Membrane

When psychophiles are exposed to temperatures above their maximum growth temperature, leakage of certain proteins (including enzymes) and amino acids, RNA, and DNA occurs. For example, *Moritella marinus* will leak cellular protein, malic dehydrogenase, glucose-6-phosphate, RNA, and DNA when exposed to a temperature a few degrees above its maximal growth temperature. Leakage of smaller amounts of the foregoing will also occur when cells are exposed to a temperature near the maximum.

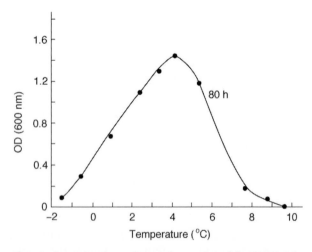

Fig. 1. Growth of an Antarctic psychrophile (AP-2-24), designated tentatively as a *Vibrio* sp. Incubation period as 80 hr in Lib-X medium employing a temperature gradient incubator. [From Morita, R. Y. (1975). *Bacteriol. Rev.* **39**, 144–157.]

Because of this situation, the membrane plays an important role in the viability of the psychrophile. However, it is not known how much leakage of intracellular material must take place before death results.

At the lower end of the growth temperature range, the cell must retain fluidity so that membrane-transport mechanisms can operate. In order to accomplish this, the psychrophile's membrane differs from the other thermal classes of microorganisms, mainly because it is composed of more polyunsaturated, short-chain, branched and/or cyclic fatty acids than the other thermogroups. Generally, when the temperature is decreased, the psychrophile responds with an increase in levels of (poly)unsaturated phospholipid and neutral lipids, which permit it to maintain membrane fluidity at low temperature. It has been reported that the acyl chain length of the phospholipids shortens when the temperature is lowered, but it takes several cell generations to occur. Psychrophiles have a higher portion of unsaturated fatty acids (especially hexadecenoic and octadecenoic acids than mesophiles. Psychrophiles grown at higher temperatures have more myristic acid and palmitic acid but, when grown at 0°C, there is an increase in docosahexaenoic acid content. Another paper reports the presence of eicosapentaenoic acid. *Cis/trans* isomerization of the double bonds of fatty acids is also found to be partly responsible for maintaining membrane fluidity. When *Moritella* ANT-300 was starved, there were induced qualitative and quantitative changes in fatty acids, with the major fatty acid (hexadecenoic) increased from 42 to 62.5%. This situation may be important, since psychrophiles do undergo cryptobiosis in their natural environment, which is generally proceeded by the process of starvation-survival. In another report, it was noted that psychrotrophs are able to alter their fatty-acid composition, but the psychrophiles are not, which may be important since psychrotrophs are found in environments where temperature fluxes go above 20°C.

Changes in the membrane-lipid desaturation are generally observed when the temperature decrease is brought about by the enzyme, desaturase, acting on the acyl chains of the membrane. As a result, there is an increase in unsaturated lipid, which helps maintain the membrane fluidity at the lower temperatures. This is followed by temperature-dependent changes in fatty-acid chain length and branching mediated by additional synthesis.

Thus, it appears that homeophasic adaptation does take place so that the membrane can carry out its important functions of uptake of nutrients and regulation of intracellular ionic composition, which are performed by carrier systems and ion pumps.

C. Enzymes

The most thermal-labile enzymes from psychrophiles that have been studied appear to be inactivated several degrees above the maximum growth temperature of the organisms. There are reports that this thermal lability is the cause of thermal death in psychrophiles, but this does not appear to be valid since RNA, DNA, and protein synthesis will occur 1° and 2°C above the maximum growth temperature. Nevertheless, these "psychrophilic" enzymes have the ability to function at psychrophilic temperature much better than do their "mesophilic" counterparts. The lowest temperature recorded for enzyme activity (a lipase) is −25°C.

IV. MICROBIAL ACTIVITY IN LOW-TEMPERATURE ENVIRONMENTS

Just how cold must the temperature become before the cycling of nutrients is halted? Definitely, there is much activity where ice (sea ice) is forming or melting. When ice is formed from water (mainly seawater) in the environment, it concentrates the various ions and organic molecules in the liquid phase so that the psychrophiles will have the organic matter necessary for growth. When the ice is formed, there are also small droplets of water within the ice that have not only higher concentrations of ions and organic matter, but also microbes. Freezing (like dehydration) does concentrate salts, which results in saline lakes in Antarctica, and *Halobacterium* sp. and *Halomonas* sp. have been isolated from these saline lakes but were found to be psychrotrophic. Where ice is melting, it releases these ions and organic matter so that they are readily available to the indigenous

microorganisms. Considerable microbial activity has been recorded in these environment. On the other hand, viable bacteria have been found in permafrost cores (ex. Halocene strata). However, at still lower temperatures, liquid water will cease to exist and the metabolic processes of microorganisms will also cease. The limit of life at the lower temperature is suggested to be −10°C, while the upper end, thus far, appears to be about 120°C for some of the hyperthermophiles. Both these limits are subject to change by future investigators.

In addition to the lack of liquid water, energy (especially organic matter for heterotrophs) is also lacking in many of the low-temperature environments. As a result, starvation-survival also takes place. As long as liquid water and nutrients (including energy) are present, metabolic processes will take place. If not, then some form of anabiosis will take place.

V. CONCLUSION

It is still not known how diverse the psychrophiles are, especially in the Archaea. The physiologies, biochemistries, and ecologies of this thermal group begs for more research, especially when we address psychrophiles that are also barophiles, with the probability that there may be psychrophiles that are also alkalophilic and/or halophilic.

Membranes of psychrophiles appear to be the main factor in psychrophily. Yet, much more research needs to be done along the lines of the physiologies, biochemistries, and ecologies of psychrophiles.

Many unanswered questions remain concerning the effect of low temperatures of −10°C on the viability and/or cryobiotic state of the microorganisms.

See Also the Following Articles

Alkaline Environments • Extremophiles • Low-Nutrient Environments

Bibliography

Baross, J. A., and Morita, R. Y. (1978). Life at low temperatures: Ecological aspects. *In* "Microbial Life in Extreme

Environments" (D. J. Kushner, ed.), pp. 9–71. Academic Press, London, UK.

Bowman, J. P., McCammon, S. A., Brown, M. V., Nichols, D. S., and McMeekin, T. A. (1997). *Shewanella gelidimarina* sp. nov. and *Shewanella frigidimarina* sp. nov., novel Antarctic species with the ability to produce eicosapentaenoic acid (20:5ω3) and grow anaerobically by dissimilatory Re(III) reduction. *Appl. Environ. Microbiol.* **63**, 3068–3078.

DeLong, E. F., Wu, K. Y., Prezelin, B. B., and Jovine, R. V. M. (1994). High abundance of Archaea in Antarctic marine picoplankton. *Nature* **371**, 695–697.

Gilichinsky, D. (1994). "Viable Microorganisms in Permafrost." Russian Academy of Sciences, Pushchino Scientific Center, Pushchino, Russia.

Franzmann, P. D., Liu, Y., Balkwill, D. L., Aldrich, H. C., De Macario, E. C., and Boone, D. R. (1997). *Methanogenium frigidum* sp. nov., a psychrophilic, H_2-using methanogen from Ace Lake, Antarctica. *Intern. J. System. Bacteriol.* **47**, 1068–1072.

Friedmann, E. I. (1993). "Antarctic Microbiology." Wiley-Liss, New York.

McMeekin, T. A., and Franzmann, P. D. (1988). Effect of temperature on the growth rates of halotolerant and halophilic bacteria isolated from Antarctic Lakes. *Polar Biol.* **8**, 281–285.

Morita, R. Y. (1975). Psychrophilic bacteria. *Bacteriol. Rev.* **30**, 144–167.

Russell, N. J., and Hamamoto, T. (1997). Psychrophiles. *In* "Extremophiles" (K. Horikoshi and W. D. Grant, eds.), pp. 25–45. Wiley-Liss, New York.

Vorobyova, E., Soina, V., Gorlenko, M., Minkovskaya, N., Zalinova, N., Mamukelashvili, A., Gilichinsky, D., Ravkina, E., and Vishnivetskaya, T. (1997). The deep cold biosphere: facts and hypothesis. *FEMS Microbiol. Rev.* **20**, 277–290.

Luteoviridae

Cleora J. D'Arcy and Leslie L. Domier

University of Illinois at Urbana–Champaign

GLOSSARY

accessory salivary gland A three- or four-celled organ that excretes a watery suspension into the salivary duct.

alate A winged individual of insect species that have both winged and wingless forms.

apterous Lacking wings.

cross protection The phenomenon in which plant tissues infected with one strain of a virus are protected from infection by other strains of the same virus.

ORF (open reading frame) A length of nucleotide sequence that can potentially be translated into a protein. An ORF usually begins with an AUG-initiation codon and ends with one of the three termination codons.

quasi-equivalence The state of a virus capsid in which all bonds between protein subunits are not strictly equivalent.

LUTEOVIRIDAE is a family of plant viruses that first were grouped because of their common biological properties. These properties include persistent transmission by aphid vectors and the induction of yellowing symptoms in many infected host plants. "Luteo" comes from the Latin *luteus*, which translates as "yellowish." Members of the *Luteoviridae* also cause other symptoms in infected hosts, including orange or red discolorations and leaf deformations, such as marginal rolling or enations, and stunting.

Members of the *Luteoviridae* cause economically important diseases in many food crops, including grains such as wheat and barley, vegetables such as potatoes and lettuce, and other crops, such as legumes and sugarbeets. These diseases were recorded decades and even centuries before the causal viruses were purified. For example, rolling of potato leaves was known since the 1700s, but was not recognized as a specific disease of potato until 1905 and as an insect-transmitted virus until a decade later. Epidemics of barley yellow dwarf in the United States were recorded for over 50 years before a virus was proposed as the cause in 1951. Stunted, deformed, and discolored plants in fields often were ascribed to abiotic factors, such as mineral excesses or deficiencies or stressful environmental conditions, or to other biotic agents, such as phytoplasmas.

All members of the *Luteoviridae* have small (ca. 25–28 nm diameter) icosahedral particles, composed of one major and one minor protein component and a single molecule of messenger-sense single-stranded RNA. The family is divided into three genera—Luteovirus, Polerovirus, and Enamovirus—based on the arrangements and sizes of the ORFs.

I. VIRION STRUCTURE AND COMPOSITION

Luteoviridae virus particles are approximately 25–28 nm in diameter, hexagonal in outline, and have no envelope (Fig. 1). They are composed of two proteins and a core of genomic ssRNA and contain no

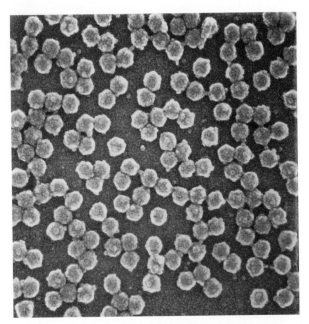

Fig. 1. Scanning electron micrograph of virus particles of barley yellow dwarf virus-PAV, magnified 200,000×. Virions are ca. 25 nm in diameter, hexagonal in appearance, and have no envelope.

lipids or carbohydrates. The structural proteins in the virions are a coat protein, encoded by ORF3, and a "readthrough" protein, which is a fusion of the products of the coat protein gene and the contiguous ORF5. Coat proteins range from 21 to 23 kDa, while the readthrough proteins are typically around 70 kDa. It is believed that most *Luteoviridae* virus particles consist of 180 protein subunits, arranged in a T = 3 icosahedron (see Section V.C.) *Luteoviridae* virions are moderately stable and are insensitive to treatment with chloroform or nonionic detergents, but are disrupted by prolonged treatment with high concentrations of salts. Virions of two genera, luteoviruses and poleroviruses, are insensitive to freezing.

II. GENOMIC ORGANIZATION

A. Genome Structure

The *Luteoviridae* have single-stranded messenger-sense RNA genomes, ranging in size from 5600 to about 5900 nucleotides (nts). Genome-linked pro-

teins (VPgs) are covalently attached to the 5′-termini of enamovirus and polerovirus genomic RNAs. However, the 5′-termini of the luteovirus genomic RNAs do not contain either a VPg or 5′-m⁷GTP cap. Unlike most messenger RNAs and the genomes of several other viruses, the genomic RNAs of all members of the *Luteoviridae* lack poly (A) tails.

B. Open Reading Frames

The genomic RNAs of the *Luteoviridae* contained five to seven open reading frames (ORFs; Fig. 2). ORFs 1, 2, 3, and 5 are shared among all members of the *Luteoviridae*. The luteoviruses lack ORF0. The enamoviruses lack ORF4. The luteovirus and polerovirus genomes contain two small ORFs downstream of ORF5, ORFs 6 and 7. In the enamoviruses and poleroviruses, ORF0 overlaps ORF1 by more than 600 nts, which, in turn, overlaps ORF2 by more than 600 nts. In the luteoviruses, the lengths of the overlaps of ORFs 1 and 2 are much shorter, e.g., ORF1 of barley yellow dwarf virus strain PAV (BYDV-PAV) overlaps ORF2 by just 15 nts. In all of the luteovirus and polerovirus genome sequences, ORF4 is contained within ORF3, which encodes the coat protein (CP). A single, in-frame amber (UAG) termination codon separates ORF5 from ORF3. The luteovirus genomes contain two small ORFs, 6 and 7, downstream of ORF5. The genomes of the poleroviruses potato leafroll virus (PLRV) and cucurbit aphid-borne yellows virus (CABYV) have an ORF6 contained within ORF5 and an ORF7 that represents the 3′-terminal region of ORF5.

C. Noncoding Sequences

The *Luteoviridae* have relatively short 5′ and intergenic noncoding sequences. The first ORF is preceded by just 21 nts in CABYV RNA and as much as 142 nts in soybean dwarf virus (SbDV) RNA. ORFs 2 and 3 are separated by 112 to 200 nts of noncoding RNA. There is considerable variation in the length of sequence downstream of ORF5, which ranges from 125 nts for cereal yellow dwarf virus-RPV (CYDV-RPV) to 650 nts for SbDV. The longer 3′ sequences contain ORF6 and, when present, ORF7. A short sequence located in the noncoding region just down-

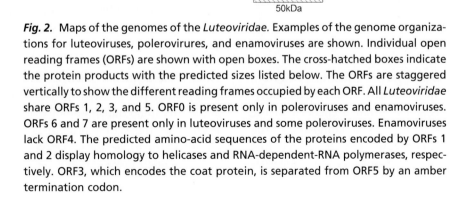

Fig. 2. Maps of the genomes of the *Luteoviridae*. Examples of the genome organizations for luteoviruses, poleroviruses, and enamoviruses are shown. Individual open reading frames (ORFs) are shown with open boxes. The cross-hatched boxes indicate the protein products with the predicted sizes listed below. The ORFs are staggered vertically to show the different reading frames occupied by each ORF. All *Luteoviridae* share ORFs 1, 2, 3, and 5. ORF0 is present only in poleroviruses and enamoviruses. ORFs 6 and 7 are present only in luteoviruses and some poleroviruses. Enamoviruses lack ORF4. The predicted amino-acid sequences of the proteins encoded by ORFs 1 and 2 display homology to helicases and RNA-dependent-RNA polymerases, respectively. ORF3, which encodes the coat protein, is separated from ORF5 by an amber termination codon.

stream of ORF5 in the BYDV-PAV genome has been shown to be a potent enhancer of cap-independent translation initiation.

D. Expression

ORFs 0, 1, and 2 are expressed directly from the genomic RNA. To express the downstream ORFs, members of the *Luteoviridae* transcribe shorter RNAs, called subgenomic RNAs (sgRNAs). The largest sgRNAs are from 2600 nts to 2900 nts in length (Fig. 2). Their sequences begin about 200 nts upstream of ORF3 in the end of ORF2 and extend to the 3' terminus of the genome. ORFs 3, 4, and 5 are expressed from sgRNA1. Luteoviruses, and at least some poleroviruses, produce a second sgRNA that expresses ORFs 6 and 7. Luteoviruses produce a third sgRNA, but it does not appear to encode a protein.

E. Functions of Encoded Proteins

1. ORF0

ORF0s are predicted to encode proteins of about 29 kDa. The functions of these proteins have not been determined unequivocally. The products of ORF0 have been predicted to be membrane-associated proteins that may play a role in host recognition and/or symptom development. Mutagenesis experiments with beet western yellows virus (BWYV) and PLRV have shown that the protein product of ORF0 is not required for replication, systemic movement, or aphid transmission of the viruses. However, viruses that did not express ORF0 usually produced less severe symptoms in inoculated plants than did the wild-type viruses. Therefore, the proteins are thought to be involved in the specificity of virus–host interactions.

2. ORFs 1 and 2

The predicted sizes of the proteins expressed from ORF1 distinguish the luteoviruses from the enamovirus and poleroviruses. The product encoded by the luteovirus ORF1 is about 40 kDa, compared to the 67 to 84 kDa proteins produced by the enamovirus and poleroviruses. For all members of the *Luteoviri-*

dae, nucleotide sequence analysis predicts that the ORF1 proteins include helicase functions required for separating RNA strands during replication. In addition, the ORF1-encoded proteins of enamovirus and poleroviruses contain the VPgs and a chymotrypsin-like serine protease responsible for the proteolytic processing of the ORF1 proteins. The protein products of ORF2s are expressed as a translational fusion with the product of ORF1. At a low, but significant, frequency during the expression of ORF1, translation continues into ORF2 through a -1 frameshift that produces a large protein containing sequences encoded by both ORFs 1 and 2 in a single polypeptide. All ORF2s encode proteins of 59 to 67 kDa that are very similar to known RNA-dependent RNA polymerases and, hence, likely represent the viral replicase.

3. ORFs 3 and 5

ORF3 of the *Luteoviridae* encodes the coat protein, which ranges in size from 21 to 23 kDa. The coat protein is required to form virus particles and for aphid transmission. ORF5s encode proteins of 29 to 56 kDa. However, ORF5s are expressed only as translational fusions with the products of ORF3s. About 10% of the time during the translation of ORF3, translation does not stop at the end of ORF3 and continues through to the end of ORF5. This results in a protein with the product of ORF3 at its amino terminus and the product of ORF5 at its carboxyl terminus. The ORF5 portion of this readthrough protein has been implicated in aphid transmission and virus stability. Experiments with PLRV and BYDV-PAV have shown that the amino terminal region of the ORF5 readthrough protein determines the ability of virus particles to bind to proteins produced by endosymbiotic bacteria of their aphid vectors. The interactions of virus particles with these proteins seem to be essential for the persistence of the luteoviruses and poleroviruses within aphids. Nucleotide sequence changes within ORF5 of the enamovirus pea enation mosaic virus-1 (PEMV-1) abolish its aphid transmissibility. The amino terminal portion of the ORF5 protein is highly conserved among *Luteoviridae;* the carboxyl termini of this protein are much more variable.

4. ORF4

Luteovirus and polerovirus genomes possess an ORF4 that is contained within ORF3 and encodes proteins of 17 to 21 kDa. Viruses that contain mutations in ORF4 are able to replicate in isolated plant protoplasts, but are deficient or delayed in systemic movement in whole plants. Hence, the product of ORF4 seems to be required for movement of the virus within infected plants. The 17 kDa protein encoded by ORF4 of PLRV has been shown to bind single-stranded nucleic acids *in vitro*, which is consistent with its proposed role in virus movement. This hypothesis also is supported by the observation that enamoviruses lack ORF4. While luteoviruses and poleroviruses are limited to phloem tissues, enamoviruses are able to move systemically through plants. The ability of enamovirus RNAs to move systemically in plants is imparted by pea enation mosaic virus-2 (PEMV-2), an umbravirus that copurifies with PEMV-1.

5. ORFs 6 and 7

Some luteovirus and polerovirus genomes contain small ORFs downstream of ORF5. No protein products have been detected from these ORFs in infected cells. As for ORF0, mutant BYDV-PAV genomes that do not express ORF6 are still able to replicate in protoplasts. The predicted sizes of the proteins expressed by ORFs 6 and 7 range from 4 kDa to 25 kDa. Based on mutational studies, it has been proposed that these genome regions may have a role in regulating transcription, possibly late in infection.

III. VIRUS–HOST INTERACTIONS

A. Host Range

Several members of the *Luteoviridae* have natural host ranges largely restricted to one plant family. For example, BYDV and CYDV infect many grasses, bean leafroll virus (BLRV) infects mainly legumes, and carrot red leaf virus (CtRLV) infects mainly plants in the *Umbelliferae*. Other members of the *Luteoviridae* infect plants in several or many different plant families. For example, BWYV infects more than 150 species of plants in over 20 families.

B. Effects on Hosts

Most members of the *Luteoviridae* cause yellow, orange, or red leaf discoloration, particularly on older leaves of infected plants. Leaves may become thickened, curled, or brittle, and plants are often stunted. The symptoms may persist, may vary seasonally, or may disappear soon after infection. Temperature and light intensity often affect symptom development, and symptoms can vary greatly with different virus isolates or strains and with different host cultivars.

BYDV and CYDV cause stunting when young plants are infected. Tillering is often reduced, but may increase in some barley cultivars. Loss of green color, beginning at the leaf tips, is usually the most obvious symptom. Various shades of yellow, orange, red, purple, or tan coloration occur, depending on the host species and cultivar.

BWYV induces interveinal yellowing, along with leaf thickening and brittleness, in most host plants. Symptoms develop acropetally, on oldest leaves first. Young leaves may never become affected.

PLRV causes persistent loss of chlorophyll in leaf tips and edges, which roll upward. Plants grown from infected tubers are stunted, with leaf rolling first appearing on the older leaves.

PEMV-1, together with the umbravirus, PEMV-2, causes enations, mosaic, and puckering of leaves, and stunting of infected plants. The symptoms are systemic and typically persist.

The external symptoms caused by luteoviruses and poleroviruses are the result of their infection of, and the subsequent collapse of, the plant's phloem tissue. Luteoviruses and poleroviruses are typically restricted to the phloem, where they are seen in sieve elements, companion and parenchyma cells. Virus particles are found in the cytoplasm, nuclei, and vacuoles of infected cells. Vesicles containing filaments and inclusions containing virus particles are common cytopathological effects of virus infection. Cell death results in phloem necrosis that spreads from inoculated sieve elements and causes symptoms by inhibiting translocation, slowing plant growth, and inducing the loss of chlorophyll. Thus, tissue specificity explains the symptoms caused by luteoviruses and poleroviruses. It also explains the relatively low virus titers obtained in purifications from infected plants, the lack of mechanical (sap) transmission, and the relatively long inoculation and transmission thresholds required for aphid transmission of these viruses.

PEMV-1 is able to replicate in plant protoplasts, but is unable to move systemically without the presence of functional PEMV-2 (an umbravirus). The complex of PEMV-1 and PEMV-2 is found in phloem and mesophyll tissues. Virions are seen in nuclei, cytoplasm, and cell vacuoles. The presence of the complex outside of the vascular tissue explains the mosaic symptoms and mechanical transmissibility of this virus.

It is difficult to get accurate estimates of yield losses due to members of the *Luteoviridae* because the symptoms are often overlooked or attributed to other agents. However, during the period 1951 to 1960, U.S. Department of Agriculture specialists estimated yield losses due to barley yellow dwarf, beet western yellows, and potato leafroll diseases were over $65 million. Yield losses attributable to infection by members of the *Luteoviridae* are most severe when young plants are infected.

IV. VIRUS–VECTOR INTERACTIONS

With the exception of the enamovirus PEMV-1, members of the *Luteoviridae* are transmitted from infected plants to healthy plants only by the feeding activities of specific species of aphids. There is no evidence for replication of the viruses within their aphid vectors. The most common aphid vector of the *Luteoviridae* that infect dicots is *Myzus persicae*. The members of the *Luteoviridae* that infect monocotyledenous plants (BYDV and CYDV) are transmitted by several different species of aphids. Particular virus strains are transmitted most efficiently by particular aphid species.

Members of the *Luteoviridae* are transmitted by aphids in a circulative manner. Aphids acquire the viruses from the vascular tissues of infected plants during feeding. The virus travels up the stylet through the food canal and into the gut. Virus particles are transported across the cells of this portion of the alimentary tract into the hemocoel (Fig. 3). While the BYDV, CYDV, and SbDV are transported across the hindgut, BWYV and PLRV are transported across the midgut. This transport process apparently involves receptor-mediated endocytosis of the viruses and the formation of tubular vesicles that transport the viruses through the epithelial cells and into the hemocoel. The viruses then travel through the hemocoel to the accessory salivary gland, where they again must pass through cell membranes by a receptor-mediated transport process. Once the virus reaches the salivary gland lumen, it is expelled with the saliva into the vascular tissue of host plants. The two membrane barriers of the gut and accessory salivary gland have different specificities/selectivities for the viruses that they will transport. The gut membranes are much less selective than are those of the accessory salivary glands. Viruses that are not transmitted by a particular species of aphid often will accumulate in the hemocoel, but not traverse the membranes of the accessory salivary gland. As has been mentioned, the interaction of the product of ORF5 and proteins expressed by aphid-borne symbiotic bacteria stabilizes virus particles so that they can persist in the hemocoel. Since large amounts of virus can accumulate in the hemocoels of aphids, they retain the ability to transmit virus for a long time, often for life.

Like the other members of the *Luteoviridae*, PEMV-1 is transmitted by aphids, although its interactions with aphid vectors are less well understood. In addition, it can be transmitted by rubbing sap from an infected plant onto a healthy plant. As has been mentioned, this difference in transmissibility is imparted by a second RNA, PEMV-2.

V. TAXONOMY

The current members of the *Luteoviridae* are listed in Table I. Criteria used to demarcate species of the family *Luteoviridae* include (1) differences in breadth and specificity of host range; (2) failure of cross protection in either one-way or two-way relationships; (3) differences in serological specificity with discriminatory polyclonal or monoclonal antibodies; (4) differences in amino-acid sequences of any gene product of greater than 10%.

Phylogenetic analysis of the predicted amino acid sequences of the polymerase (ORF2), coat protein

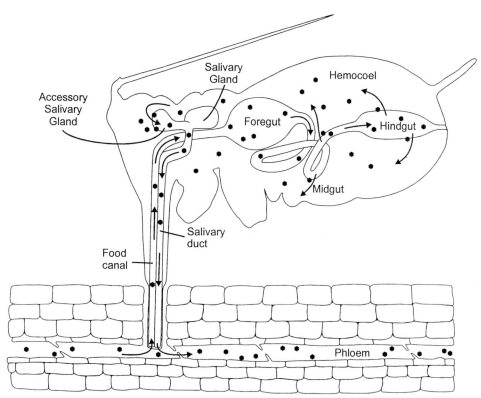

Fig. 3. Circulative transmission of luteoviruses and poleroviruses through their aphid vectors. Aphids (shown in longitudinal section) acquire luteoviruses and poleroviruses (filled hexagons) from phloem cells of infected plants during feeding. The viruses travel up the stylet, through the food canal, and into the gut, where they are transported via receptor-mediated endocytosis across the epithelial cells and into the hemocoel. BWYV and PLRV are transported across the midgut. BYDV, CYDV, and SbDV are transported across the hindgut. The viruses then travel through the hemolymph to the accessory salivary gland where, after passing through a second selective membrane barrier, they are expelled with the saliva into the vascular tissue of host plants.

(ORF3), and readthrough (ORF5) proteins separate the members of the *Luteoviridae* into three genera. Two barley yellow dwarf viruses (BYDV-PAV and BYDV-MAV) are in the genus luteovirus. Five species (potato leafroll virus, beet mild yellowing virus, beet western yellows virus, cereal yellow dwarf virus-RPV, and cucurbit aphid-borne yellowing virus) are in the genus polerovirus (a sigla from *potato leafroll*). A single virus, pea enation mosaic virus-1, is in the genus enamovirus (a sigla from pea *enation mosaic*).

A. Genus Luteovirus

The distinguishing features of this genus include the lack of ORF0 and the fact that the frameshift from ORF1 into ORF2 occurs at the termination codon of ORF1. The translation products of ORF1 and ORF2 form replication-related proteins, which are "carmoviruslike." The length of the noncoding sequence between ORF2 and ORF3 is about 100 nts. Luteovirus genomes contain an ORF6 downstream of ORF5. There is no evidence for the presence of a genome-linked protein and translation is by a cap-independent mechanism.

B. Genus Polerovirus

Polerovirus genome RNAs are linked to a protein (VPg), contain an ORF0, and have a noncoding region between ORF2 and ORF3 of about 200 nts.

TABLE I
Current Members of the *Luteoviridae*

Genus *Luteovirus*	
barley yellow dwarf virus-MAV	BYDV-MAV
barley yellow dwarf virus-PAV	BYDV-PAV
Genus *Polerovirus*	
beet mild yellowing virus	BMYV
beet western yellows virus	BWYV
cereal yellow dwarf virus-RPV	CYDV-RPV
cucurbit aphid-borne yellows virus	CABYV
potato leafroll virus	PLRV
Genus *Enamovirus*	
pea enation mosaic virus-1	PEMV-1
Unassigned species	
barley yellow drawf virus-GPV	BYDV-GPV
barley yellow drawrf virus-RMV	BYDV-RMV
barley yellow dwarf virus-SGV	BYDV-SGV
bean leafroll virus	BLRV
carrot red leaf virus	CtRLV
chickpea stunt disease associated virus	CpSDaV
groundnut rosette assistor virus	GRAV
Indonesian soybean dwarf virus	ISDV
soybean dwarf virus	SbDV
sweet potato leaf speckling virus	SPLSV
tobacco necrotic dwarf virus	TNDV

Frameshift from ORF1 into ORF2 occurs upstream of the termination of ORF1. Polerovirus genomes differ from those of enamoviruses, in that ORF4 is present within ORF3 and ORF5 is about 1400 nts long. Some polerovirus genomes contain an ORF6 within ORF5. Poleroviruses may invade nonphloem tissue to a limited extent in plants coinfected with other viruses, such as potyviruses.

C. Genus Enamovirus

The enamovirus genome lacks ORF4 (present in the luteoviruses and poleroviruses) and has an ORF5 of about 730 nts. Particles of the enamovirus PEMV-1 are approximately 28 nm in diameter and are proposed to be similar in structure to luteoviruses and poleroviruses (T = 3, 180 subunits). In nature, PEMV-1 occurs with PEMV-2, an umbravirus, which is proposed to have a 150 subunit arrangement lacking quasi-equivalence. This complex is found in mesophyll tissue and is mechanically transmissible.

PEMV-1 will multiply when inoculated to isolated leaf protoplasts, but there is no evidence that it can spread in plants.

D. Antigenic Properties

Luteoviruses and poleroviruses are strongly immunogenic. Species within a genus are more closely related serologically than are species in different genera. Serological relationships may be detected when comparing disrupted virus particles that are not detectable when intact virions are tested.

Virions produced in plants infected with PEMV-1 together with the umbravirus, PEMV-2, are moderately antigenic. No serological relationships have been reported between enamoviruses and either luteoviruses or poleroviruses.

E. Relationships with Other Viruses

Viruses in the family *Luteoviridae* have replication-related proteins which are sufficiently similar to those in other genera to suggest evolutionary relationships. The putative luteovirus polymerases resemble those of members of the family *Tombusviridae*. In contrast, polymerases of poleroviruses and enamoviruses resemble those of viruses in the genus *Sobemovirus*. These polymerase types are thought to be very distant in evolutionary terms. It has been suggested that these genomes originated through recombination between ancestral genomes containing the coat protein genes characteristic of the family *Luteoviridae* and genomes containing either of the two polymerase types.

VI. EPIDEMIOLOGY

Members of the family *Luteoviridae* have been reported from all parts of the world including temperate, subtropical, and tropical regions. Some of the viruses are found worldwide, such as BYDV, BWYV, CYDV, and PLRV. Others have more restricted distribution, such as tobacco necrotic dwarf virus (TNDV), which has been reported only from Japan, and groundnut rosette assistor virus (GRAV), which has been reported from four south Saharan countries in Africa. Some of the viruses are disseminated in infected planting material, such as PLRV in potato tubers. Extensive programs to produce clean stock are operated around the world to control these viruses.

Many other members of the *Luteoviridae* infect annual crops and are introduced each year by their aphid vectors. Since the viruses can be retained for days to weeks by their aphid vectors, these viruses can be spread more widely than those viruses that are transmitted nonpersistently. Cultivated crops, as well as weeds, play roles in virus epidemiology, both as hosts of the viruses and as hosts of the aphid vectors. The viruses must survive either in infected host plants or in aphid vectors. Alate aphid vectors may reproduce locally on infected cultivated, volunteer, or weed hosts and introduce virus into newly emerging crops. Mild weather conditions can favor aphid survival and increase the likelihood of disease epidemics.

Alternatively, alate aphids may be transported into cropping areas from distant locations on wind currents. These vectors may bring the virus with them or may first have to acquire virus from a locally infected host. The impact of migrating vectors depends on both meteorological events favoring long-distance transport and the susceptibility of the crop at the time of aphid arrival. Only aphid species that feed on a particular crop can transmit virus; short probes to determine plant suitability do not result in virus transmission. If a particular aphid species or morph does not feed on a crop, no primary infection will occur.

Secondary spread of the viruses is often primarily by apterous aphids in the crop, and is affected by the number of primary infection foci, the plant density, the occurrence of natural enemies and pathogens of the aphids, and the weather. Conditions that favor virus replication in host plants and aphid multiplication on these plants will favor rapid disease spread. Typically, in cooler and wetter weather, aphid multiplication is suppressed and slow spread of disease from foci of primary infection occurs. In warmer weather, aphids are more likely to multiply and to move rapidly from plant to plant, and secondary disease spread can be very rapid. The relative importance of primary introduction of virus by alate aphids and of secondary spread of virus by apterous aphids in disease epidemiology varies with the virus, the vector species, the crop species, and the environmental conditions.

A number of members of the *Luteoviridae* occur in complexes with other members of the family or with other plant viruses. For example, BYDV and CYDV are often found co-infecting cereals, and BWYV and SbDV are often found together in legumes. Some other plant viruses depend on members of the *Luteoviridae* for their aphid transmission, such as the groundnut rosette and bean yellow vein banding umbraviruses, which depend on groundnut rosette assistor (GRAV) and PEMV, respectively.

VII. DISEASE MANAGEMENT

A. Diagnosis

Infections by members of the *Luteoviridae* can result in significant losses in the quantity and quality in a wide range of crop products. An integral part of controlling infections by members of the *Luteoviridae* is the accurate diagnosis of infection. Infectivity assays were the first tests used to diagnose infections. They also have been used to identify species of vector aphids and vector preferences. For these assays, aphids are allowed to feed on infected plants and then transferred to sensitive indicator plants. These techniques are very sensitive, but can require 3 to 4 weeks to complete while researchers wait for symptoms to develop on indicator plants.

In spite of the limitations of very low virus titers in infected plants, serological tests have been developed for many members of the *Luteoviridae*. The viruses are purified from infected plants and then used as immunogens, typically in rabbits and mice. The antibodies produced by the animals in response to immunization with virus particles are used in serological tests, most commonly enzyme-linked immunosorbent assays, for the detection of infections. Serological assays are very sensitive and permit discrimination between different members of the *Luteoviridae* and sometimes between strains of a single virus.

More recently, techniques have been developed to detect the viral RNA from infected plant tissues. These techniques involve either nucleic-acid hybridization utilizing DNA probes synthesized from a cloned DNA copy of a segment of the viral genomes or polymerase chain reaction, which uses short oligonucleotide primers to specifically amplify short segments of the viral genomes for detection. Nucleic acid detection techniques have the highest sensitivity of the diagnostic techniques and can readily discriminate viruses and strains of a single virus species. However, they are the most expensive and labor-intensive of the diagnostic techniques.

B. Control

Once an infection is diagnosed, there are no methods currently available to directly treat plants infected by members of the *Luteoviridae*. Emphasis is placed on reducing losses through the use of tolerant or resistant plant varieties and, where feasible, reducing the spread of the aphids that transmit the viruses through the use of insecticides. Because many of the viruses are transmitted by migrating aphids and because these aphid movements often occur at similar times each year, it is sometimes possible to plant crops so that the young, highly sensitive plants are not in the field when the seasonal aphid migrations occur. Insecticides have been used in a prophylactic manner to prevent aphid populations from expanding in crops. Insecticide treatments may not prevent initial infections, but may greatly limit the secondary spread of aphids and viruses. In some instances biological control agents, such as predatory insects and parasites, have been effective in controlling aphid populations. Genes for resistance or tolerance to infection by members of the *Luteoviridae* have been identified in varieties of many crop plants infected by the viruses. For BYDV-PAV and PLRV, transgenic plants have been produced through DNA-mediated transformations that express portions of the virus genomes. In some cases, the expression of these virus genes in transgenic plants has conferred levels of virus resistance that are much higher than those afforded by normal plant resistance genes.

See Also the Following Articles

Plant Disease Resistance • Plant Virology, Overview • Potyviruses

Bibliography

D'Arcy, C. J., and Burnett, P. A. (eds.). (1995). "Barley Yellow Dwarf: 40 Years of Progress." APS Press. St. Paul, MN.

D'Arcy, C. J., Domier, L. L., and Mayo, M. A. (1999). *Luteoviridae.* VIIth Report of the International Committee on Taxonomy of Viruses. Academic Press, Inc., San Diego, CA.

Figueira, A. R., Domier, L. L., and D'Arcy, C. J. (1997). Comparison of techniques for detection of barley yellow dwarf virus PAV-IL. *Plant Dis.* 81, 1236–1240.

Garret, A., Kerlan, C., and Thomas, D. (1996). Ultrastructural study of acquisition and retention of potato leafroll luteovirus in the alimentary canal of its aphid vector, *Myzus persicae* Sulz. *Arch. Virol.* 141, 1279–1292.

Mayo, M. A., and Ziegler-Graff, V. (1996). Molecular biology of luteoviruses. *Adv. Virus Res.* 46, 413–460.

Miller, W. A., and Rasochova, L. (1997). Barley yellow dwarf viruses. *Annu. Rev. Phytopath.* 35, 167–190.

Peiffer, M. L., Gildow, F. E., and Gray, S. M. (1997). Two distinct mechanisms regulate luteovirus transmission efficiency and specificity at the aphid salivary gland. *J. Gen. Virol.* 78, 495–503.

Plant Viruses Online. Descriptions and lists from the VIDE database. *http://biology.anu.edu.au/Groups/MES/vide.*

Waterhouse, P. M., Gildow, F. E., and Johnstone, G. R. (1988). The luteovirus group. CMI/AAB Descriptions of Plant Viruses, No. 339. Association of Applied Biologists, Wellesbourne, Warwick, U.K.

Lyme Disease

Jenifer Coburn and Robert A. Kalish

Tufts-New England Medical Center, Boston

GLOSSARY

antigen A molecule that provokes an immune response.

arthritis Inflammation of joints.

host An organism in which a pathogen (or nonpathogenic species) can replicate and, potentially, cause disease.

pathogen An organism (parasite, fungus, bacterium) or virus that causes disease.

pathogenesis The study of pathogens and the processes by which they cause disease.

PCR The polymerase chain reaction, in which minute amounts of specific DNA sequences can be amplified by repeated cycles of DNA polymerization; can be used as an exquisitely sensitive method to diagnose infection.

reservoir A host species or inanimate source that supports the growth of a pathogen in nature, but is not necessarily susceptible to disease caused by the pathogen.

vector An organism that transmits a pathogen from one host to another.

virulence factor A protein or other molecule, or an inherent property, that contributes to the ability of a pathogen to cause disease.

LYME DISEASE is the result of infection by certain members of the *Borrelia* genus of spirochetes. The infection is transmitted by the bite of an infected tick and can affect multiple organ systems. Lyme disease is now the most common arthropod-borne infection in the United States and is widespread across Eurasia, as well. In the absence of antibiotic therapy, the infection can persist for years, despite the immune response mounted by the mammalian host. It has been proposed that, in some human patients with Lyme arthritis, the infection may trigger an autoimmune response that results in the persistence of symptoms long after apparent eradication of the spirochete by antibiotic therapy.

I. HISTORY OF LYME DISEASE

During the mid-1970s, a geographic cluster of what appeared to be juvenile rheumatoid arthritis that centered around Lyme, Connecticut, was brought to the attention of state health officials and rheumatologists at Yale Medical School. The arthritis typically involved only one or a few large joints, and many of the patients suffered recurrent bouts of painful, swollen joints interspersed with periods of remission. Some patients (or their parents) remembered an expanding skin rash that preceded the arthritis and other symptoms by a few weeks. Epidemiologic studies suggested that this epidemic of arthritis might be a vector-borne illness, and one patient remembered having had a tick bite. As the number of cases increased, a variety of other manifestations of this syndrome became apparent, and geographic associations of human disease and of a particular species of *Ixodes* tick were identified. Similarities to previously known syndromes in both Europe and

North America were recognized. Some of these syndromes were known to follow tick bites, and related *Ixodes* ticks were known to inhabit the relevant areas of Europe. It became clear that a tick-transmitted infectious agent was responsible for the collection of clinical manifestations that was termed Lyme disease.

Identification and subsequent isolation of spirochetes from patient samples in both North America and Europe, and the association of tick bites with certain skin lesions that had initially been reported in Europe in the late nineteenth or early twentieth century, confirmed the link between *B. burgdorferi* infection and both skin and joint manifestations and suggested that the infection occurred in human populations at least 100 years ago. Although Lyme disease has only been recognized in the United States for the last two decades, tick specimens in museum collections document the presence of *B. burgdorferi* on Long Island, New York, as early as the 1940s. The rising incidence is due not only to the identification and improved diagnosis of the infection, but also to an increase in the frequency of disease. In turn, this increased frequency is, in part, the result of population changes among the deer that are an essential factor in the life cycle of the tick vector in the northeastern United States and, in part, to lifestyle changes in the human population.

II. THE BIOLOGY OF *BORRELIA BURGDORFERI*

A. Agents of Lyme Disease

The three major *Borrelia* species that cause Lyme disease are *B. burgdorferi* (*sensu stricto*), *B. garinii*, and *B. afzelii*. Together, these three species are termed *B. burgdorferi* (*sensu lato*). These designations are the result of genetic typing of Lyme disease spirochetes that were isolated in diverse locations. What was originally thought to be a single species was found to comprise a group of closely related species, and additional subgroups of *B. burgdorferi* (*sensu lato*) have been identified. In addition, there are reports of a Lyme disease-like illness associated with an as yet unculturable spirochete in the southern United States, and *Borrelia* have been identified in

ticks and animals from geographic regions not yet recognized as endemic foci of Lyme disease. It is, therefore, possible that as the bacteria, their ecology, and the infection they cause are better understood, additional *Borrelia* species will formally be recognized as agents of Lyme disease. To date, however, *B. burgdorferi* (*sensu stricto*) is the only widely recognized agent of Lyme disease in North America. In Europe, *B. garinii* and *B. afzelii* predominate, but *B. burgdorferi* (*sensu stricto*) is also found. The three principal species of Lyme disease spirochetes are associated with similar, but not identical, disease syndromes in humans. It is likely that they share a number of virulence determinants, but that each species also has unique pathogenic properties. The distributions of the different *Borrelia* species that cause Lyme disease are likely to account for the observed geographic differences in disease manifestations. Many aspects of the biology of the genus *Borrelia,* and of the diseases caused by these spirochetes, have been reviewed previously (Barbour and Hayes, 1986).

B. Structure of *Borrelia* Species

The *Borreliae* are a group of vigorously motile, zigzag- to corkscrew-shaped bacteria, hence, the term "spirochetes." They are very long and thin, approximately 0.2–0.3 microns in diameter and 20–30 microns in length. Other pathogenic spirochetes include a number of *Borrelia* species that cause vector-borne relapsing fever, *Leptospiras* that can cause hepatitis and hemorrhagic fever in humans, and the *Treponemes,* which include the agent of syphilis. The spirochetes share a cellular structure that includes a double membrane (cytoplasmic and outer membranes) that, at a superficial level, has similarity to the membrane structure of gram-negative bacteria, although the constituents of the outer membranes of the two groups are very different. *Borrelia* species, for example, do not synthesize the classical LPS (lipopolysaccharide, or endotoxin), characteristic of gram-negative bacteria. The flagella, which are attached at both ends of the *Borrelia* cell, rotate much like those of other bacteria. In *Borrelia,* however, the flagella are located between the cytoplasmic and outer membranes. When confined to the periplasm, flagellar rotation appears to drive the rest of the cell around the long axis, generating the characteristic drill-like motility. The flagella may also be important in determining cell shape, as a spontaneous mutant strain of *B. burgdorferi* that does not make flagella is virtually straight.

Several experimental approaches suggest that the outer membranes of *T. pallidum* (the agent of syphilis) and *B. burgdorferi* contain few integral membrane proteins (Radolf *et al.,* 1995). This is an important observation, because both of these bacterial species cause chronic infections in the absence of antibiotic therapy, despite the immune response mounted by the infected host (e.g., human). It is, therefore, possible that, due to the scarcity of protein antigens, the outer surfaces of intact *B. burgdorferi* are not efficiently recognized by the immune system.

C. Cultivation of *B. burgdorferi*

In vitro cultivation of *B. burgdorferi* (*sensu lato*) requires very complex media. The base medium, which is normally used for culture of mammalian cells, must be supplemented with gelatin, bovine serum albumin, serum, and assorted other components to achieve growth of *Borrelia*. *B. burgdorferi* loses the ability to cause infection in animal models with continued laboratory culture (increased passage). In addition, protein expression and the genetic content of the bacteria change with prolonged culture. It should be noted that the bacterial strains used in some of these experiments were not clonal, so the *in vitro* genetic changes may, in part, reflect selection of variants that are better able to grow in the laboratory.

The growth rate of *B. burgdorferi* in laboratory media is very slow for a bacterium, with a doubling time of 12 to 24 hours, depending on the strain, the temperature, and the medium. The favored growth conditions are in liquid medium, in a sealed tube filled to approximately $\frac{3}{4}$ capacity, held stationary at between 32–34°C. These conditions reflect the microaerophilicity of the organism, i.e., the requirement for subatmospheric concentrations of oxygen. Colonies of *Borrelia* can be grown in medium supplemented with agarose at concentrations that are just sufficient to permit gelation. The plating efficiency,

however, can vary between strains and is generally lower for low-passage, infectious cultures.

D. Genetic Structure of *Borrelia burgdorferi*

The *Borrelia burgdorferi* genome is very unusual, in that the chromosome is linear, with covalently closed telomeres (ends). In addition to the chromosome, the Lyme disease spirochetes also carry a number of extrachromosomal genetic elements, or plasmids, some of which are also linear, some circular. It appears that some of the plasmids are indispensable and could therefore be thought of as mini-chromosomes. The chromosome and plasmid repertoire of a representative *B. burgdorferi* strain has been completely sequenced (Fraser *et al.,* 1997). The linear chromosome is just under one megabase in size, among the smallest bacterial chromosomes known. The plasmid content appears to differ somewhat between strains and may change with prolonged *in vitro* cultivation.

Several interesting clues to the biology of *B. burgdorferi* were revealed by the genome sequence. First, *B. burgdorferi* encodes very few recognizable biosynthetic enzymes, consistent with its requirement for a rich, complex medium for *in vitro* growth. Second, *B. burgdorferi* encodes on the order of ten times more lipidated secreted proteins than any other bacterium whose genome has been sequenced to date. This class of proteins is secreted across the cytoplasmic membrane and modified by the addition of a lipid moiety at the amino terminus. These "lipoproteins" are peripherally associated with the outer leaflet of either the cytoplasmic or outer membranes, or with the inner leaflet of the outer membrane. Despite their abundance, however, the biological significance of these proteins is largely unknown. Third, several groups of multiple genes encoding highly homologous proteins were identified. This potential for antigenic diversity might contribute to immune evasion during infection. Finally, as is the case for other bacterial genomes that have been sequenced in entirety, a large percentage of the predicted open reading frames encode proteins of unknown function. Further study of those that are shared with *T. pallidum* may shed light on the biology of these two pathogens, both of which must live within the mammalian environment while evading destruction by host defense mechanisms.

III. *BORRELIA BURGDORFERI* IN THE ENVIRONMENT

A. The Animal Reservoir of *B. burgdorferi*

Borrelia burgdorferi (*sensu lato*) is an obligate parasite of mammalian hosts; no free-living form has been identified. Because *B. burgdorferi* is transmitted to humans from an animal reservoir, it is often referred to as a "zoonotic" infectious agent. The normal host for the spirochete is generally a small mammal, depending on geographic location. In the northeastern United States, the host is the white-footed mouse, *Peromyscus leucopus*. On the U.S. west coast, the host is the dusky-footed woodrat (*Neotoma fuscipes*). These animals remain infected for life, but are apparently free of disease and do not show evidence of a significant inflammatory response to the spirochete. The level of infection in a population of the host mammal is maintained by transmission of the spirochetes to new generations by infected ticks, as transplacental transmission does not occur. For *Borrelia burgdorferi* to maintain itself, it must be carried by a sufficiently large percentage of the reservoir population that there is a good chance that feeding ticks will be infected. In addition, the density of the host population must be sufficiently high that infected ticks have a good chance of transmitting the bacteria to additional members of the reservoir species.

B. The Vectors of *B. burgdorferi* Infection

The tick vectors of *B. burgdorferi* (*sensu lato*) infections are members of the *Ixodes ricinus* complex, including *I. ricinus*, *I. dammini* (also known as *I. scapularis*), *I. pacificus*, and *I. persulcatus*. The particular species of tick that transmits *B. burgdorferi* varies with geographic location. It appears that only *Ixodes*

ricinus ticks will harbor the Lyme disease spirochetes and allow transmission to new hosts; other ticks feeding on the same infected animal either do not support bacterial survival or cannot transmit the spirochete to new hosts. *Ixodes* ticks are hard-shelled and feed slowly, remaining attached to the mammalian host for several days. This is an important factor in the transmission of *B. burgdorferi* to new mammalian hosts, as it appears that the spirochetes are introduced into the new host only after 24–48 hours of tick feeding.

The life cycle of the arthropod vector is also an important consideration in the biology of *Borrelia* species. The *Ixodes* ticks that carry *B. burgdorferi* mature and reproduce over a 2-year period, with only a single meal between each of the three stages of development: larva, nymph, and adult. The larvae and nymphs are very small; for example, the nymphal stage of *Ixodes dammini* is approximately 1 mm. The preferred host for both larval and nymphal stages of the tick *Ixodes dammini* in the northeastern United States is the white-footed mouse, the reservoir for *B. burgdorferi*. Newly hatched larvae become infected immediately if their first meal is taken from an infected mouse. After molting from the larval to the nymphal stage, the infected tick feeds again, thereby transmitting the spirochete to another member of the reservoir population. The nymphs, however, are not that choosy and occasionally feed on other mammalian hosts, e.g., humans and dogs, transmitting the bacteria to a host species that is essentially irrelevant to the biology of *B. burgdorferi*. On this basis, *B. burgdorferi* can be considered an accidental pathogen.

Following the nymphal meal, the tick molts again, giving rise to the adult stage of the tick. It is at this stage that deer become important in the ecology of *Borrelia* in the northeastern United States. The favored host for the adult tick is the white-tailed deer, upon which the ticks feed and reproduce. This accounts for the common name "deer tick" and for the rising incidence of Lyme disease in the northeastern United States over the last several decades. Due to conservation efforts and reforestation of former farmland, the deer population and range have increased. At the same time, the increased human population has fueled the development of suburban and rural areas. The high incidence of Lyme disease in the northeastern United States is, therefore, the result of the increasing of overlap in the "ranges" of humans and deer, as well as the complex interactions between the mammalian reservoir, the tick vector, and the spirochetal agent.

Analogous but geographically unique vector-host relationships occur in all regions in which Lyme disease is endemic. For example, on the west coast of the United States, the tick (*I. neotomae*) that is responsible for maintaining infection levels in the woodrat population does not feed on humans. It is only when a different tick, *I. pacificus,* occasionally feeds on an infected woodrat, then on a human, that the infection is transmitted to humans. Nymphal *I. pacificus,* however, prefers lizards, which do not support *B. burgdorferi* growth. The difference in tick behavior accounts for the difference in Lyme disease prevalence between the east and west coasts of the United States. *B. burgdorferi* (*sensu lato*) may exist in wildlife populations in many areas, but unless the tick vectors also feed on humans and the bacteria cause human disease, their existence may not garner much attention.

It is important to note here that, in certain geographic regions, other infectious agents can be transmitted to humans by the ticks that carry *B. burgdorferi.* These include a parasite, another bacterial pathogen, and viruses. *Babesia microti* is a parasite that replicates in red blood cells, resulting in lysis of the cells and hemolytic anemia. In the northeastern United States, the white-footed mouse is also the reservoir for *B. microti,* and field-collected *Ixodes* ticks may be infected with both *B. burgdorferi* and *B. microti.* The *Ehrlichia* species that can also be found in these ticks are obligate intracellular bacteria that infect granulocytes, which are essential to host defenses against infection. The tick-borne encephalitis agent and related viruses have also been identified in the *Ixodes* ticks that carry Lyme disease. It is, therefore, possible that a patient may be infected with more than one pathogen after a single encounter with an *Ixodes* tick and that the clinical picture of Lyme disease may be complicated by concomitant infections.

C. *B. burgdorferi* in the Tick

Studies on the biology of *B. burgdorferi* in the tick vector are challenging but the results so far have been informative in understanding certain aspects of the human infection. It is interesting that this bacterium, which is so fastidious in the laboratory, can apparently survive for long periods of time in a dormant state in the tick midgut. It has been shown that expression of at least two spirochetal proteins is altered while the tick is feeding (Schwan *et al.,* 1995). In the unfed (flat) tick, *B. burgdorferi* is found only in the midgut and expresses the well-characterized protein OspA. Osp (outer surface protein) A is a protein that is recognized by human patient sera (or infected mouse sera), but primarily at later stages of infection near the beginning of prolonged episodes of arthritis. Although its name suggests a surface localization, only a small fraction is actually on the outer surface of the intact spirochete. During the slow process of tick feeding, the spirochetes initially multiply in the midgut, then disseminate throughout the tick, presumably via the hemolymph (the tick version of blood). During this dissemination, OspA is no longer detectable on the great majority of spirochetes, but a different protein, OspC, is present. OspC, however, is not found on the spirochetes in unfed ticks. The functional significance of this result is unknown, but it does explain why sera from patients early after infection frequently recognize OspC but not OspA.

IV. FUNDAMENTAL ASPECTS OF THE IMMUNE RESPONSE

A. Innate and Acquired Immunity

Lyme disease is the result of infection by *B. burgdorferi* and the host's response to that infection. A number of textbooks provide much more detailed information on the incredible complexity of the immune system, but a very brief overview of those aspects that are relevant to Lyme disease will be given here. The mammalian (e.g., human) immune response can be divided into two fundamental components: innate and acquired. The innate response is not specific to any particular foreign entity and is, therefore, the first line of defense, able to respond to invasion by multiple pathogens. Components of innate immunity include complement, a series of blood-borne proteins that can form a complex on the surface of bacterial cells, resulting in disruption of the bacterial membranes and death of the invading bacteria. In addition, phagocytic cells, which ingest and kill most invading pathogens, normally patrol the tissues and are attracted in greater numbers to sites of infection. Two cell types, macrophages and neutrophils (also known as polymorphonuclear leukocytes or PMNs), are primarily responsible for these activities.

In contrast to the innate response, acquired immunity is developed in response to specific particular pathogens or antigens. Acquired immunity is mediated by B and T lymphocytes. Under normal conditions, B cells and T cells respond in a very specific manner only to particular antigens. B cells recognize and respond to antigens directly, but T cells require the formation of a complex between the antigen and a host-cell molecule termed MHC (major histocompatibility complex) by antigen-presenting cells (APCs). Only a small subset of either B or T lymphocytes will become active in fighting infection under normal circumstances. Both cell types can migrate to sites of infection in the tissues, promoting maximal stimulation by these antigens.

B cells multiply and secrete antibodies in response to specific antigens. Antibodies are proteins that can bind to the pathogen directly, increasing the efficiency of complement-mediated killing, or to products of the pathogen that facilitate its survival in the host environment. Two types of antibodies are relevant to the response against *B. burgdorferi*. Early in infection, antibodies of the IgM class are produced. As the infection proceeds, a switch to production of the IgG class occurs; this response is maintained for the duration of the infection. Antibodies are generally very specific, although if multiple pathogens share similar antigens, some cross-reactivity may occur.

The role of T cells in acquired immunity is more complex. Cytotoxic T lymphocytes (CTLs) kill other host cells that have been invaded by pathogens. T helper (Th) cells stimulate the activities of other immune system cells. Th1 cells activate macrophages and promote the inflammatory response, in which large numbers of immune system cells invade tissue at the site of infection. Th2 cells stimulate B cells to produce antibody and suppress the Th1 response. The type of T cell response mounted against an infection can, therefore, have a significant impact on the host as well as on the pathogen, since a predominantly inflammatory response may have greater functional consequences to the host than would a humoral (antibody) response. T cells, B cells, and phagocytic cells work together to limit or eradicate invading pathogens, and each cell type plays an important role in fighting infections.

Both innate and acquired immunity contribute to inflammation, which is a response to tissue injury, e.g., by bacterial infection. Initially, the inflammatory process may include the activation of chemical signals that directly affect the bacteria (e.g., complement) or that activate and attract cells of the immune system to sites of infection. Mixtures of phagocytes and lymphocytes then invade the tissue in response to the infection. These invading cells are referred to as an inflammatory infiltrate. If the immune response fails to eradicate the infecting bacteria, continued inflammation may result. From the point of view of the patient, inflammation results in redness, swelling, and pain in the affected area.

B. Measurement of the Immune Response

The laboratory-based methods of measuring the B and T cell responses of human patients and mice (the most commonly employed model of infection) to *B. burgdorferi* infection have been invaluable to the diagnosis and understanding of the development of Lyme disease. Two methods of measuring the antibody, or humoral response, are commonly used in the diagnosis of Lyme disease: ELISA and immunoblotting.

The ELISA (enzyme-linked immunosorbent assay) is useful in determining the level of antibody to a pathogen in patient sera, but it can be somewhat nonspecific. To increase the specificity in testing for infection by a particular pathogen, immunoblotting can also be performed. In this procedure (also known as "Western" blotting), particular proteins expressed by the pathogen are tested for reactivity with patient sera. Proteins to which a patient's serum reacts will appear as "positive bands" in the immunoblot. When used together, ELISA and immunoblot assays allow the measurement of both the level and the specificity of a patient's antibody response.

T cell responses can also be measured in the laboratory setting. In a "proliferation assay," T cells are cultured under conditions in which they can multiply only in response to the presence of antigen. In order to do so, they take up a radioactive precursor of DNA supplied in the medium. The cell-associated radioactivity, therefore, gives a measurement of the T cell response to the antigen. A second way of measuring T cell responses allows discrimination of the cytotoxic, Th1, and Th2 responses. In this assay, the type of cell that responds to an antigen and the characteristics of that response are measured. This system makes use of fluorescent antibodies directed against proteins expressed only by certain subsets of T cells or against products of T cells, e.g., cytokines.

The cytokine response is of particular interest in understanding diseases caused by infectious agents. Cytokines are small proteins that stimulate specific responses by target cells, e.g., cells of the immune system and the endothelial cells that line blood vessels. The Th1 cytokines, which include interferon-gamma (IFN-γ), generally stimulate the inflammatory response. The Th2 cytokines, such as interleukin-four (IL-4), promote the humoral response. The early host response, as measured by production of IFN-γ and IL-4, has been shown to be important in the development of Lyme disease in the mouse model (see following).

C. The Response to *B. burgdorferi*

In human Lyme disease patients and mice, the immune system appears to be important not only in controlling infection but also in the manifestations of infection that can be seen at the clinical level. As will be described in more detail, an inflammatory response is developed in response to the presence of *B. burgdorferi* in the joints and other tissues and is

responsible for the arthritis developed in both human patients and in the mouse model of infection. The humoral response appears to be important in decreasing the number of bacteria and, consequently, the inflammatory response in the joints, eventually reducing the severity of arthritis. Nevertheless, in many host species, e.g., humans and mice, even the multiple branches of the immune system working together are unable to clear the host of *B. burgdorferi* infection. *B. burgdorferi* must, therefore, be able to either avoid efficient recognition by the immune system or be able to evade clearance by the immune response mounted by the infected host. Long-term bacterial infection in a host that is able to mount an immune response (an immunocompetent host) is poorly understood.

In certain Lyme disease patients, it appears that the immune response can go awry, leading to recognition of the host's own proteins after infection by *B. burgdorferi*. This phenomenon, termed autoimmunity, is thought to be the cause of a number of chronic human diseases and may be responsible for the continued arthritis seen in some patients even after apparent eradication of the spirochete through antibiotic therapy.

V. CLINICAL MANIFESTATIONS OF LYME DISEASE

Lyme disease is frequently described as having three distinct clinical stages: early localized infection (stage 1), occurring days to weeks after the tick bite; early disseminated infection (stage 2), occurring during the first weeks to months of infection; and late or persistent infection (stage 3), which occurs months to years after disease onset (Table I) (Steere, 1989). The clinical stages of infection reflect the migration and dissemination of *B. burgdorferi*. It is important to note that, in reality, there may be overlap between the different stages of disease. Furthermore, any of the three stages of disease may be asymptomatic in a given individual. Indeed, it is estimated that approximately 5–10% of individuals who are infected with *B. burgdorferi* may have asymptomatic infection.

TABLE I
Summary of the Clinical Stages of Lyme Disease

Stage of disease	Time after tick bite	Disease manifestations
Early localized	Days to weeks	Erythema Migrans –Primary lesion
Early disseminated	Week(s) to months	Erythema Migrans –Secondary lesion(s) Flulike symptoms Joint and muscle pain –Migratory, transient Neurologic –Meningitis –Facial palsy –Peripheral neuropathy Cardiac –Heart block
Late	Month(s) to years	Arthritis Neurologic –Encephalopathy –Peripheral neuropathy Acrodermatitis

A. The Course of Human Lyme Disease

The first stage of disease, i.e., early localized infection, is manifest as an expanding skin rash, termed erythema migrans, surrounding the site of the tick bite (Fig. 1). This rash appears anywhere from 3 to 32 days (mean = 7) after the bite. Due to the small size of the tick, however, the bite goes unnoticed by many patients. The rash initially appears at the site of the tick bite as a small reddish area that may be flat or slightly raised, then gradually expands over a period of days to weeks to a final diameter of 3 to 68 cm (mean = 15). Some degree of central clearing generally occurs as the rash expands, yielding the characteristic ringlike or bullseye appearance of the mature erythema migrans skin lesion. The red outer edge follows the migration of the bacteria through the skin and is the site from which the spirochetes are most frequently cultivated. The rash is usually painless, but may also feature slight itching or burning sensations. Rarely, there is a more intense blistering of the rash. The erythema migrans lesion resolves fairly rapidly with antibiotic therapy or within a few weeks in untreated patients. It should

be emphasized that some patients do not develop erythema migrans. Others develop the rash in an area that is not easily visible, e.g., on the back side of the body or in areas covered by hair, and may therefore escape detection. During the early localized stage of the infection, patients generally feel well, although there may be some swelling of lymph nodes in the area of the rash and some patients may experience mild malaise or flulike symptoms.

The second stage of Lyme disease, early disseminated infection, occurs during the first weeks to months after the tick bite and is characterized by diverse symptoms involving multiple organ systems. This stage reflects the dissemination of the bacteria through the tissues and the bloodstream to multiple sites in the body. Many patients experience a flulike illness that may be severe, with high fevers, headache, and overwhelming fatigue. In contrast to many viral infections, upper respiratory symptoms, such as nasal congestion and cough, are uncommon in Lyme disease and are mild if they do occur. Migratory arthralgias (pain in multiple joints, but affecting any one for only a short time) and myalgias (muscle pain) are common and sometimes intense, but swelling of the joints at this stage of the disease is uncommon. Involvement of the nervous system at this stage is seen in approximately 20% of patients and may con-

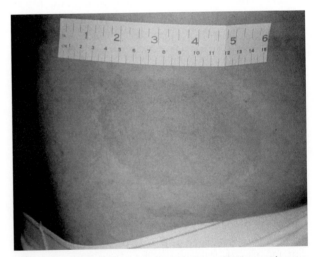

Fig. 1. An erythema migrans lesion. Photograph was kindly provided by Dr. Allen C. Steere; reprinted with permission from Steere *et al.* (1983) *Ann. Internal Med.* **99**, 76–82.

sist of meningitis, facial muscle paralysis (Bell's palsy), and inflammation of peripheral or spinal nerves. This inflammation causes numbness, feelings of pins and needles, or a sharp, shooting-type pain in the affected areas. A minority of patients (5%) develop an overt, transient carditis (inflammation of the heart muscle), which presents most commonly as heart block (interference of the endogenous pacemaker function, resulting in an unusually slow heart rate). This occasionally requires treatment with a temporary artificial pacemaker. Finally, secondary erythema migrans lesions at sites other than where the tick attached can also occur. Usually, the secondary lesions are smaller and fainter in color than the primary erythema migrans lesion.

Signs and symptoms of early disseminated disease gradually diminish over a period of weeks to months, even in patients not treated with antibiotic therapy. In some, there may be no further disease activity. Others, particularly if untreated, may develop manifestations of chronic infection several weeks to as long as several years later.

Late Lyme disease, the third stage of the illness, is characterized by arthritis, a variety of neurologic manifestations and, in Europe, a distinct type of skin lesion, acrodermatitis chronica atrophicans. The late manifestations of Lyme disease vary somewhat with the species of *Borrelia,* with *B. burgdorferi* (*sensu stricto*) being associated with both arthritis and neurologic disorders, *B. garinii* primarily with neurologic problems, and *B. afzelii* with acrodermatitis. At this stage, the bacteria can no longer be found in the circulation, and, in fact, are not abundant even in the tissues affected by the disease. Some patients remain asymptomatic for long periods of time before developing symptoms of late disease. Once they occur, the late manifestations of Lyme disease may persist or recur for years.

Lyme arthritis is rarely seen in patients treated adequately at the time of early infection, but occurs in approximately half of individuals who were asymptomatic at that time or those who were not treated at the time of early disease. The arthritis starts a mean of 6 months after initiation of infection and is characterized by marked swelling of one or a few large joints, with the knee eventually involved in nearly all cases. The attacks of arthritis initially tend to be brief, lasting days to a few weeks. With time,

however, episodes of arthritis may persist for months, or in a small percentage of cases, years. Antibiotic therapy is effective in eliminating arthritis in the majority of patients. Furthermore, data from patients who had arthritis prior to the discovery of a bacterial cause of Lyme disease and who, therefore, were not routinely treated with antibiotics, indicate that with each passing year, 10–20% of patients have spontaneous resolution of arthritis.

Approximately 10% of patients with Lyme arthritis do not respond to appropriate antibiotic therapy and continue to have active arthritis. Patients with what has been termed treatment-resistant chronic Lyme arthritis are more likely to have certain HLA-DRB1* alleles (which encode specific types of MHC molecules) and to have humoral and T cell responses to the OspA protein of *B. burgdorferi* (Kalish *et al.,* 1993). By PCR analysis, no *B. burgdorferi* DNA was detectable in the joint fluids of these patients following antibiotic therapy. These findings suggest a possible autoimmune response in the pathogenesis of this chronic, treatment-resistant arthritis. A candidate autoantigen, or human protein targeted by the immune response, has recently been proposed (Gross *et al.,* 1998). This protein is a subunit of a receptor expressed on lymphocytes that mediates adhesion of the cells to the antigen-presenting cells and to endothelial cells that line vessel walls. A small segment of the protein may be recognized by T cells that recognize a portion of OspA, accounting for the association of OspA reactivity with chronic Lyme arthritis.

Several distinct entities have been described for late neurologic involvement in Lyme disease. The peripheral neuropathy is characterized by predominantly sensory abnormalities, in which patients experience a waxing and waning pattern of numbness and tingling of their arms or legs or, less often, radiating from the spine. Muscle weakness as a result of the neuropathy is less common. Late neuropathy tends to be more subtle and have less of a pain component than the peripheral neuropathy seen during the early disseminated stage of disease. Electrophysiologic studies typically show abnormalities of the central axon of multiple nerves (axonal sensorimotor polyneuropathy).

Late Lyme encephalopathy is generally mild and not progressive, although it can potentially be quite disturbing. Symptoms of encephalopathy tend to wax and wane in severity and may include impairment of cognitive functions, such as short-term memory and language expression, and may be accompanied by fatigue, irritability, and sleepiness. Individuals may have difficulty performing mental tasks they previously accomplished with greater ease. Formal neuropsychological testing most often detects short-term memory impairment. Laboratory testing may reveal the presence of IgG antibodies to spirochetal proteins in the cerebrospinal fluid. In Europe, progressive *Borrelia* encephalomyelitis, a more disabling syndrome that can resemble multiple sclerosis, may occur after infection by *B. garinii*.

A considerable amount of confusion and controversy has surrounded the issue of the proper diagnosis of Lyme disease. Much of the problem stems from the overlap of the symptoms seen in Lyme encephalopathy with those seen in other disease states and in individuals experiencing stress, various degrees of depression or anxiety, or other emotional or physical difficulties. Several studies have documented the overdiagnosis of Lyme disease. In these studies, a significant number of patients instead had chronic fatigue syndrome or fibromyalgia, a chronic syndrome in which individuals suffer from fatigue, diffuse pain, and disrupted sleep patterns. A diagnosis of Lyme disease in the absence of *Borrelia burgdorferi* (*sensu lato*) infection may lead to a delay in the proper diagnosis and may subject patients to unnecessary treatments that have potential toxicity. Therefore, objective criteria should be used to make the diagnosis of Lyme disease. It is important to carefully consider the tick-exposure history, to look for signs and symptoms that have some degree of specificity for Lyme disease, and to be knowledgeable about the uses and pitfalls of laboratory tests for Lyme disease.

Acrodermatitis chronica atrophicans is a late skin lesion that occurs in about 10% of European patients. It occurs several months to years after the initial tick bite as an expanding, somewhat swollen, purplish area of skin. It is most often located on the limbs, usually on or near the hands or feet. Over a period of months to years, the area of involved skin gradually becomes atrophic (thinned) and may take on a characteristic "cigarette paper" appearance. *B. burgdorferi*

(*sensu lato*) has been cultured from the acrodermatitis rash, in some cases, years after the initial appearance of the lesion.

B. Laboratory Diagnosis of Lyme Disease

Proper use of the serologic tests for Lyme disease is essential. The guidelines established by the U.S. Centers for Disease Control should be followed at all times. The ELISA gives the best combination of sensitivity and specificity as a screening test and should be done first. IgM antibodies to *B. burgdorferi* are first detected 2 to 4 weeks after disease onset and generally (but not always) fade after several months. IgG antibodies appear within 4 to 8 weeks and may persist for many years in cases not treated at the time of early infection. As the infection continues, the spectrum of *B. burgdorferi* proteins recognized by IgG antibodies increases, so that many proteins are recognized. For individuals in whom symptoms have been present for less than one month, assays for both IgM and IgG antibodies to *B. burgdorferi* should be performed. If symptoms have been present for more than one month, only IgG antibodies need to be assessed.

A borderline or positive ELISA must be confirmed with a western blot assay, as there is a substantial risk of false positive results due to cross-reactive antibodies that are actually directed against non-*Borrelia* proteins. The U.S. Centers for Disease Control have recommended the following criteria for a western blot to be called positive. The IgM response must recognize at least two of three proteins termed p23, p39, and p41 (a flagellar protein of approximately 41 kilodaltons). IgG antibodies should recognize a minimum of five protein bands from among the following: p18, p23, p28, p30, p39, p41, p45, p58, p66, and p93.

It is important to be aware of the limitations of serologic testing for Lyme disease. Patients with early disease, particularly when localized, are often seronegative at presentation. In contrast, the sensitivity and specificity of testing for patients with late disease exceeds 90%. In addition, early antibiotic treatment may abrogate full development of the humoral immune response. Although early treatment is successful in the great majority of cases, a very small percentage of patients may not have a complete response to therapy and may later develop joint pain or chronic neurologic manifestations of the disease in the absence of a positive antibody response. Lyme arthritis is characterized by a highly expanded humoral immune response, with high titers of *B. burgdorferi*-specific antibodies and multiple bands on western blots, and should not be diagnosed in the absence of this response. Finally, a false positive test (more often ELISA than western blot) may be seen in the setting of other infectious or autoimmune diseases.

The polymerase chain reaction has been found to be quite sensitive in detecting *B. burgdorferi* DNA in the synovial fluid of untreated patients with arthritis and in biopsies of skin lesions. This technique can also be useful in cases of persistent arthritis following antibiotic therapy, in order to distinguish ongoing active infection in the joint (PCR positive) from an ongoing immune response in the absence of bacteria (PCR negative). In the latter case, the joint swelling and inflammation that continue after treatment suggest the possibility of an autoimmune response. The limitations of PCR include the potential for false positives if great care is not taken to minimize the possibility of laboratory DNA contamination, and the limited number of centers performing this assay.

Other tests have only limited use in the diagnosis of Lyme disease. A method to identify *B. burgdorferi* antigens in urine is available, but this test has not been adequately validated to the point that it can be recommended for diagnostic use. Direct cultivation is slow and labor-intensive and has good sensitivity only from the erythema migrans skin lesion, which can usually be recognized without the necessity of obtaining a skin sample for culture. The T cell proliferative assay remains primarily a research tool, as it is also time- and labor-intensive and is performed by only a few laboratories; its place in diagnosis remains unclear, given a wide range of differing experiences among different investigators with regard to its sensitivity, specificity, and utility.

C. Treatment of Lyme Disease

Antibiotic therapy is successful in the treatment of the great majority of Lyme disease cases. Multiple treatment studies have indicated a >90% success rate (in some studies, approaching 100%) when Lyme disease is treated in its early stages. Early Lyme disease is treated with a 2- to 4-week course of oral antibiotics; amoxicillin or doxycycline have been used most frequently and are probably equally effective. Late stage arthritis can also be treated with oral antibiotics; a 30-day course is recommended. Intravenous (IV) antibiotic therapy is indicated when neurologic disease is present (meningitis, peripheral nerve involvement, late encephalopathy) or when there are advanced degrees of heart block. Ceftriaxone is the drug of choice for IV therapy and requires administration only once daily. Penicillin is another option, but requires multiple doses per day. The duration of IV treatment ranges from 2 weeks to 1 month.

More prolonged courses or repeated courses of antibiotic therapy have not been shown to impart any additional benefit in the treatment outcomes of Lyme disease and increase the chance of an adverse drug effect. Trials investigating longer courses of antibiotics for patients who have ongoing symptoms (often consisting of fatigue, muscle and joint pain, and subjective cognitive difficulty) following Lyme disease are currently in progress. However, it is not clear that these symptoms are due to ongoing infection; rather, they may persist as part of post-Lyme syndrome, that resembles what is seen in patients with chronic fatigue syndrome or fibromyalgia, or following other infectious diseases.

VI. ANIMAL MODELS OF LYME DISEASE

Several animal models have been developed to study the course of *B. burgdorferi* infection, with hopes that this will shed light on the factors that are important in determining the course of human disease. The animals utilized include mice, rats, gerbils, hamsters, guinea pigs, dogs, and rhesus monkeys. Each model has its advantages and limitations, but the most widely used animal models are small rodents. Mice, in particular, allow analysis of the role of a well-characterized host genetic background in studies of infectious disease. The rhesus monkey, however, offers the opportunity to study aspects of *B. burgdorferi* infection that are seen in humans but not in mice, particularly erythema migrans and neuroborreliosis.

A. The Mouse Model

The most thoroughly characterized animal model is the inbred laboratory mouse (Barthold *et al.,* 1993, 1997). As is the case in human patients, *B. burgdorferi* infection results in a characteristic disease progression that affects multiple tissues. In addition, mice develop an early IgM response, then switch to a predominantly IgG response that expands as the infection continues. Inflammatory infiltrates and cytokine responses also appear to be similar in mice and humans. The major difference between mice and humans is that mice develop arthritis earlier in the course of infection and do not develop the chronic arthritis seen in some human patients. In addition, erythema migrans and neurologic manifestations of *B. burgdorferi* infection are not seen in mice. Finally, carditis is probably more frequent in mice than in human patients.

When laboratory mice are infected with *B. burgdorferi,* it appears that the spirochetes remain in the skin immediately surrounding the site of inoculation for a few days. The bacteria begin to disseminate widely within 4–7 days. At this stage of infection, the bacteria can be cultured reliably from a number of different tissues, including the blood, spleen, skin, and bladder, and can be detected by PCR in multiple sites.

Mice develop inflammatory arthritis and carditis within 2–3 weeks of infection. The arthritis affects the tibiotarsal joint (a major weight-bearing joint that is functionally analogous to the human knee, although, in mice, it is also called the ankle) most severely, but multiple joints are affected. Other tissues that comprise the musculoskeletal system, including tendons and ligaments, are also inflamed. Carditis is more difficult to assess without sacrificing the animal, but transient changes in heart rate can be observed. These disease manifestations reflect the fact that the spirochetes are relatively abundant in diverse tissues.

As the infection progresses from the disseminated phase into the chronic stage (starting 3–4 weeks after infection), the apparent number and tissue distribution of spirochetes detectable by either culture or PCR decreases. For example, *B. burgdorferi* is only rarely detected in the circulation at later stages of infection. In contrast, the spirochete is easily detectable in the skin, using either culture or PCR, as long as the animals are infected. In mice that are not treated with antibiotic therapy, *B. burgdorferi* persists for the life of the animals, although overt signs of infection disappear with time, even in mice that, at one time, had obvious arthritis. Some individual mice may suffer from transient, recurrent episodes of spirochetemia (spirochetes in the blood), arthritis, and carditis, although the inflammation seen in the affected tissues is much less severe than that seen at earlier stages of disease.

Several factors influence the development and progression of disease in mice. These factors include the infection route, the number of bacteria introduced into the animal, and the strain (genetic background) of the mice. In addition, younger animals mount a more vigorous inflammatory response and develop more significant arthritis than do older animals.

The infection route may be either by injection of spirochetes or by allowing infected ticks to feed on the animals. Tick-transmitted infection, of course, most closely mimics the conditions in which humans are infected, but is a more difficult system in which to work. Infection by injection of laboratory-cultivated *Borrelia* allows the comparison of the animals' responses to different strains or numbers of bacteria and facilitates studies in which large numbers of animals are required. Intradermal inoculation (i.e., into the skin) was found to be the most efficient route, requiring the lowest number of bacteria to establish infection. The intradermal route is that which most closely resembles tick-mediated transmission and, thereby, represents the most natural model of infection without the involvement of ticks. It should be noted that direct animal to animal transmission without surgical transplantation of infected tissue has never been observed, and that animal waste materials have never been found to be infectious.

When a mouse is infected by a low number of *B. burgdorferi* by intradermal inoculation, the animal's response resembles that seen when the infection occurs through a tick bite. At high doses of laboratory-cultivated organisms, however, the immune response mounted differs from that seen in the natural infection of either humans or mice. This response is largely directed against proteins, e.g., OspA and OspB, that are expressed by the spirochetes in culture, but not by bacteria in feeding ticks. *B. burgdorferi* can also be efficiently transmitted from one animal to another by transplantation of infected tissue, demonstrating that passage through a tick (or *in vitro* culture) is not required to establish an infectious state.

The severity of disease developed in response to *B. burgdorferi* infection of mice depends on the genetic background of the animals. The inbred mouse strains commonly used for studies of *B. burgdorferi* infection include BALB/c, C57BL/6, and C3H/He. All three mouse strains mount an antibody response that is reminiscent of that seen in humans. In general, C3H mice develop a somewhat higher level (higher titer) antibody response than do either C57BL/6 or BALB/c mice. This increased antibody titer might reflect the fact that more spirochetes are found in the tissues of C3H mice than of BALB/c or C57BL/6 mice.

The C3H mouse displays the greatest degree of inflammatory response, including arthritis and carditis, among the three aforementioned mouse strains. Arthritis in the C3H mouse is easily observed without sacrifice and dissection of the animal, and persists for longer periods of time. C57BL/6 mice are relatively resistant to the development of arthritis, regardless of the infectious dose, while in BALB/c mice, the severity of arthritis increases with the number of spirochetes. The C3H mouse model, therefore, provides the opportunity to study the inflammatory aspects of disease seen in humans, with the ability to manipulate the conditions of infection available only in animal models.

1. Histopathology of B. burgdorferi Infection in Mice

When sections of tissues from infected mice are examined under the microscope, it becomes apparent that *B. burgdorferi* can cause significant inflammation even if only a relatively small number of bacteria are present. In both mice and humans, *B. burgdorferi* is most abundant in the connective tissue matrix of all tissues or organs examined and is often associated with collagen bundles. There has been no suggestion in any of these studies of the presence of intact spirochetes within host cells. In sections of musculoskeletal structures from mice in the disseminated stage of infection, spirochetes are found in the synovial tissue that lines the joint and in the sites of attachment of muscles, tendons, and ligaments to bone. Spirochetes are also visible in cardiac tissues, particularly in the connective tissues associated with major blood vessels. The distribution of these bacteria within a single organ is not uniform; this is particularly true of the heart. As the infection progresses to the chronic phase in mice, spirochetes are found in connective tissues from a variety of sites but are rather scarce. In all stages of infection, *B. burgdorferi* can be seen in association with arterial walls from diverse sites, particularly in regions of inflammation of the vessel wall. In these arterial lesions, the spirochetes may be located in the connective tissue surrounding the vessel, within the vessel wall, or just underneath the endothelial cells that line blood vessels.

2. The Murine Immune Response to B. burgdorferi Infection

The ability to study and manipulate cytokine levels in the laboratory mouse has been important in understanding the progression of Lyme disease in this model. The levels of two cytokines, IFN-γ and IL-4, are thought to be important in determining the course of disease. IFN-γ is a "pro-inflammatory" cytokine and is an important activator of macrophages, resulting in increased clearance of pathogens by phagocytic killing. In contrast, IL-4 stimulates B cells, thereby activating the humoral or antibody-mediated response. IFN-γ is produced early in infection, although the levels differ between mouse strains. In C3H mice, the IFN-γ levels are initially lower than those observed in BALB/c mice, but climb steadily throughout the first 2 weeks of infection. In contrast, IFN-γ levels in BALB/c are constant throughout this time interval. BALB/c mice start producing IL-4 within 2 weeks after infection, while C3H mice produce no detectable IL-4. The severity of arthritis is similar in the two mouse strains in the first 2 weeks.

By 1 month postinfection, however, arthritis in BALB/c mice is largely resolved and IFN-γ levels have dropped significantly, while IL-4 levels remain high. In C3H mice, on the other hand, arthritis is severe, IFN-γ is still present, although at somewhat reduced levels, and IL-4 is undetectable. The differences between the two mouse strains can be attributed to the different cytokine responses to *B. burgdorferi* infection, because administration of antibodies that block the actions of IL-4 or IFN-γ alter the development of inflammatory disease.

The degree of the host response to the presence of *B. burgdorferi* in tissues can also be assessed microscopically. For example, in the joints, the synovial tissue, which is a thin lining of cells and connective tissue in the normal joint, displays overgrowth of cells and deposition of fibrous material. In addition, infiltration of large numbers of inflammatory cells (macrophages, neutrophils, and lymphocytes) is observed, with neutrophils predominating over macrophages. Large accumulations of fluid (edema) can be seen around the joints in some animals, but is not necessarily correlated with the degree of cellular infiltration (true arthritis). In the hearts of infected mice, infiltration of inflammatory cells is also seen, but in this site, macrophages are more numerous than neutrophils. The degree of inflammation in different portions of the heart and in the joints reflects the numbers of spirochetes quantitated either microscopically or by PCR. In the later stages of infection, arthritis and carditis in mice are largely resolved, although either may transiently reappear. During these recurrences, the numbers of spirochetes and the host inflammatory cells both increase in the affected tissues.

Mice carrying the *scid* mutation (severe combined immune deficiency) have been invaluable in deciphering the roles of the host immune system and the intrinsic properties of *B. burgdorferi* itself in the development and resolution of Lyme disease in the mouse model. *Scid* mice are unable to mount an acquired immune response, but are still able to clear the spirochetes from the circulation over a time course similar to that seen with wild-type mice. This result suggests that the acquired host response is not required for the decrease in spirochetemia as the infection progresses and that this is either a property

intrinsic to the bacteria in the mammalian environment or is the result of the actions of innate immunity. The development of arthritis in *scid* mice follows a time course similar to that seen in wild-type counterparts. In contrast to the situation in wild-type mice, however, arthritis in *scid* mice does not resolve at later stages of infection, and, in fact, grows progressively worse. These results demonstrate the importance of a specific immune response in the resolution of arthritis, which is consistent with the cytokine profiles observed in C3H and BALB/c mice.

Sera from *B. burgdorferi*-infected immunocompetent mice which have already resolved their arthritis can resolve the arthritis in infected *scid* mice, suggesting that arthritis resolution requires the production of specific antibodies. This result is unique to arthritis, however, as the *scid* mice remain infected. Carditis is not affected by transfer of immune serum. An interesting observation in these studies is that the arthritis-resolving antisera recognize *B. burgdorferi* in the joints, but not other tissues in infected *scid* mice. *B. burgdorferi* may, therefore, express one or more antigens specifically in the environment of the joint. It will be of great interest and significance to the understanding of how *B. burgdorferi* causes arthritis when the joint-specific antigen(s) is (are) identified.

Scid mice have also made possible experiments in which immune cells (B and/or T lymphocytes) or sera from *B. burgdorferi*-infected animals were tested to determine the roles of immunity mediated by these cell types in the prevention of infection. The key results of these experiments are that sera taken from mice infected with *B. burgdorferi* can be used to protect naive immunocompetent or *scid* mice from infection. This "passive immunization" is affected by the timing of serum harvest from the donor infected mice and of the transfer of sera relative to duration of infection in recipient mice.

Transfer of B and T cells from infected mice into naive *scid* mice have also been important in understanding the role of the host immune system in *B. burgdorferi* infection. B cells or mixtures of B and T cells from infected mice also protect naive *scid* mice and promote resolution of arthritis in infected mice. In contrast, T cells did not protect *scid* mice from infection and had no effect on arthritis. The results of these "adoptive transfer" experiments support the importance of the antibody response in protecting animals from infection and the resulting arthritis.

B. The Rhesus Monkey Model

When rhesus monkeys are infected by *B. burgdorferi*, either by the bite of an infected tick or by needle inoculation, the animals show manifestations of disease that closely parallel what is seen in humans (Roberts *et al.*, 1995). The humoral response, as assessed in immunoblots, is virtually identical to the human response. In humans, mice, and monkeys, infiltration of immune system cells into the walls of blood vessels and surrounding connective tissues is seen, and the invading cell types are similar in the different tissue sites examined.

The rhesus monkey frequently develops erythema migrans, which is not observed in mice. At both the gross and histologic levels, these skin lesions resemble those seen in human patients. Furthermore, secondary erythema migrans is seen in rhesus monkeys. Neurologic involvement is seen at the microscopic level, but functional deficits were not apparent. The rhesus monkey, however, provides a model in which to study invasion of nervous tissue by the spirochete and by the host's immune system cells, as these phenomena are not observed in mice. In the later (chronic) stages of infection, microscopic evaluation reveals arthritis in multiple joints in all animals, although externally visible joint swelling and limited range of motion of the affected joints were observed infrequently. The authors note that, in rhesus monkeys, significant changes at the microscopic level must occur before the animals show any signs of functional impairment. In contrast to humans, however, arthritis was not observed during the disseminated stage of disease. Carditis was also found in the rhesus model, although to a lesser extent than what is observed in the C3H mouse model.

The major disadvantage of the rhesus monkey is the expense of obtaining and housing the animals, but the development of erythema migrans and neurologic changes during *B. burgdorferi* infection in these animals provides a valuable model in which to study these aspects of Lyme disease. Furthermore, the opportunity to study the course of disease over a number of years in a primate might be important to understanding the chronic human disease.

VII. VIRULENCE MECHANISMS AND PATHOGENESIS OF *BORRELIA BURGDORFERI* INFECTION

In order for a bacterial species to be a pathogen, it must encounter a host species in which it can either do direct damage or induce a response that is detrimental to the host. *Borrelia burgdorferi* is considered a human pathogen because of the damage seen in human patients, resulting in the collection of clinical manifestations that we know as Lyme disease. In contrast, *B. burgdorferi* causes infection, but not disease, in its wild animal reservoirs, e.g., the white-footed mouse. As work in the inbred mouse model has shown, both bacterial and host factors appear to be important in the development of infection and disease. To date, however, the virulence mechanisms of *B. burgdorferi* are among the least understood in the field of pathogenic microbiology. Nevertheless, by analogy to other bacterial pathogens, potential virulence mechanisms might include adhesion to host cells and tissues, invasion of host tissues, or intoxication of host cells (Table II). Persistence within the immunocompetent host is essential to the survival of *Borrelia* and is likely to be an important factor in the development of disease.

A. Adherence to Host Cells

In order to gain entry into the mammalian host, *B. burgdorferi* employs the tick vector to do the physical work, namely, breaching the skin that normally serves as one of the innate barriers to infection. Upon introduction into the skin at the site of the tick bite, however, *B. burgdorferi* must continue the infection process on its own. Most bacterial pathogens employ at least one mechanism by which they can attach to the host and, in many cases, multiple bacterial molecules termed "adhesins" contribute to this process. Each adhesin recognizes a specific structure, or receptor, on the host cell. The complexity of adhesion mechanisms ensures that specific target tissues are colonized by the microorganism and that appro-

TABLE II
Virulence Mechanisms of *Borrelia burgdorferi*

Virulence mechanism	Host targets	Bacterial factors
Adherence	Heparin/Heparan sulfate proteoglycans	Unknown
	Chondroitin sulfate proteoglycans	Unknown
	Decorin (a subset of CS proteoglycans, preceeding)	Decorin-binding proteins
	Integrins	Unknown
	Fibronectin	Several candidate proteins
	Glycosphingolipids	Unknown
Intoxication of cells	Cell membranes	Hemolysin(s)
	Signaling mechanisms	Lipidated proteins
Invasion of host tissues	Extracellular matrix	Plasminogen-binding protein(s)
Antigenic variation	Evasion of immune clearance	*vlsE* Gene products others?

priate signals between the host and the pathogen are sent and received.

B. burgdorferi appears to employ multiple adhesion mechanisms, which are likely to enable the spirochete to establish the initial foothold at the site of the tick bite. In addition, the spirochete must interact with endothelial cells and the subendothelial matrix, both to gain entry into the circulation and to leave it to infect other tissues. Furthermore, persistent infection of certain tissues, including the joints, skin, nervous system, and heart, is likely to be, at least in part, the result of "tissue tropism" mediated by adhesion mechanisms. An additional aspect for this vector-borne pathogen is that adhesion to tick midgut cells and the down-regulation of this activity at the appropriate time are likely to be important in the transmission of the bacteria to new hosts. The *in vivo* significance of *B. burgdorferi* attachment to host remains to be formally tested, but, by analogy to other bacterial pathogens, this property is likely to be important to the ability of the spirochete to cause infection and disease.

Borrelia burgdorferi adheres to tick cells and to a variety of mammalian cell types in laboratory culture. In surveys of cell lines derived from several tick species, a spectrum of bacterial binding efficiency was found (Kurtti *et al.*, 1993). Most of the bacteria were bound to the tick cell surface by one tip. Tick cell membrane invaginations and coated pits were often associated with bound spirochetes. After extended coincubation, a few bacteria appeared to be

located in membrane-bound vacuoles in the cytoplasm, although it is possible that, due to the nature of the thin sections required for electron microscopy, this result reflects the plane of sectioning rather than true cellular invasion. An interesting observation made in these studies was that the tick cells showed morphological changes, primarily loss of extensions termed villi, in response to *B. burgdorferi* attachment. This system is a difficult one in which to work, however, because less is known of the receptors expressed by tick cells than of those expressed by human cells and of the biology of the tick cell responses to the engagement of these receptors.

Borrelia burgdorferi is able to bind to virtually every mammalian cell type tested, which probably reflects the ability of this bacterium to infect multiple tissues. As is the case with tick cells, *Borrelia* adhesion to mammalian cells is frequently tip-associated. Neutrophils and monocytes are able to bind, engulf, and kill the spirochete *in vitro,* supporting their potential roles in killing *B. burgdorferi* during infection. Glial cells, fibroblasts, endothelial cells, and epithelial cells may also be encountered by the spirochete in the course of infection, and all show some capacity to bind *Borrelia*. Many of these cell types have been employed to identify the bacterial and mammalian molecules that mediate these interactions.

At least two distinct classes of mammalian molecules, integrins and proteoglycans, have been characterized as receptors for *B. burgdorferi*. The bacterial molecules that mediate one of these binding pathways have been identified. Work in three laboratories independently identified proteoglycans as receptors for *B. burgdorferi*. Proteoglycans consist of glycosaminoglycan (GAG) chains linked to protein cores. The glycosaminoglycan chains are essentially disaccharide repeats of variable composition, length, and modification, generating a very heterogeneous mixture of molecules. Based on structures of the disaccharide repeats, these molecules are grouped into two broad families: the heparin/heparan sulfate and chondroitin sulfate glycosaminoglycans. Despite this heterogeneity, there is some specificity to GAG recognition by both microorganisms and by the host molecules with which they normally interact. Since glycosaminoglycans may be either associated with the cell or secreted by the cell into the extracellular milieu, this binding pathway can mediate attachment to extracellular matrix in tissues as well as to cell surfaces.

Several *B. burgdorferi* strains have been shown to bind heparin/heparan sulfate and chondroitin sulfate proteoglycans expressed by a variety of mammalian cell types in culture. Although the bacterial molecules that mediate attachment to all of these receptors have not yet been completely characterized, the proteoglycan repertoire of the mammalian cell and the spectrum of proteoglycans recognized by the bacterial strain both influence the efficiency of *Borrelia*-host cell interactions *in vitro* (Leong *et al.*, 1998).

One specific proteoglycan, decorin, has been more intensively studied as a receptor for *B. burgdorferi* attachment (Guo *et al.*, 1995). Decorin is a member of the chondroitin sulfate proteoglycan family and derives its name from the fact that it "decorates" collagen fibrils. Collagen is a major component of connective tissue and of the extracellular matrix. Given that *B. burgdorferi* has been seen associated with collagen in the tissues of infected humans and animals, this decorin-binding activity is very likely to play a role in the virulence and lifestyle of *B. burgdorferi* in the mammalian host.

Two decorin-binding proteins were identified in lysates of *B. burgdorferi* and subsequently cloned. Two other groups independently cloned the *dbp* (decorin binding protein) genes, using sera from infected mice to probe libraries of *B. burgdorferi* DNA. The two proteins are closely related and are encoded by a two-gene operon located on a linear plasmid.

Related molecules are also encoded by *B. afzelii* and *B. garinii*. Sera directed against DBP were protective against infection in mice. That the DBPs are expressed during *B. burgdorferi* infection increases the likelihood that decorin-binding might play a role in pathogenesis. It does not appear that the bacterial molecules that recognize decorin also recognize heparin, because heparin does not compete for *B. burgdorferi* binding to decorin. At least some *Borrelia* strains may, therefore, encode distinct mechanisms that mediate attachment to different classes of proteoglycans.

All three of the major species of Lyme disease spirochetes also bind to mammalian cell surface receptors termed integrins. Integrins are $\alpha + \beta$ heterodimeric protein receptors that are normally involved in a variety of cell–cell and cell–extracellular matrix interactions. The integrin family is considered to be vital to the signaling between the internal and external environments of mammalian cells. These receptors are functionally and structurally diverse, as there are multiple α and β subunits. Although integrins can be broadly grouped into subfamilies based on the β subunit, each particular heterodimer displays unique binding specificities. In some cases, the mammalian molecules that bind integrins (ligands) can be recognized by more than one receptor.

Borrelia burgdorferi (*sensu stricto*), *B. afzelii*, and *B. garinii* have all been shown to bind to integrins $\alpha_{IIb}\beta_3$, $\alpha_v\beta_3$, and $\alpha_5\beta_1$ (Coburn *et al.*, 1998). $\alpha_{IIb}\beta_3$, also known as the fibrinogen receptor, is a platelet-specific integrin. Integrins $\alpha_v\beta_3$ (the vitronectin receptor) and $\alpha_5\beta_1$ (the fibronectin receptor) are expressed by multiple cell types. Binding of *B. burgdorferi* to these receptors can be inhibited by reagents that inhibit binding of the normal mammalian integrin ligands, suggesting that the binding sites, at least partially, overlap. While binding to platelets appears to be mediated primarily by $\alpha_{IIb}\beta_3$ (Fig. 2), the proteoglycan and the integrin pathways both contribute to binding to epithelial and endothelial cells.

It should be noted here that several laboratories have shown that fibronectin, one of the mammalian integrin ligands, binds to *B. burgdorferi*. It is, therefore, possible that fibronectin serves as a "bridge" between the bacterium and certain integrins. It has

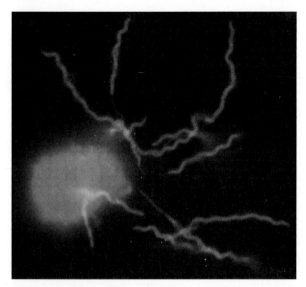

Fig. 2. *Borrelia burgdorferi* attached to human platelets. The spirochetes were illuminated by a green-fluorescent anti-Lyme spirochete antibody, the platelets with a red-fluorescent anti-integrin antibody. See color insert.

also been shown, however, that different *Borrelia* strains have distinct integrin preferences among the three receptors tested. This result is more consistent with variations in bacterial integrin ligands than with the scenario in which a mammalian molecule serves as a bridge to all integrins. The ability to bind fibronectin may, nevertheless, promote spirochetal attachment to mammalian cells and to extracellular matrix.

B. Invasion of Host Tissues

Work in three independent laboratories showed that *B. burgdorferi* binds plasmin(ogen). Plasmin is a mammalian protease responsible for degrading the fibrin component of blood clots; plasminogen is its inactive precursor (zymogen). The specificity of plasmin is, however, rather broad, and the enzyme can degrade many substrates, including extracellular matrix proteins. Under normal circumstances, the activity of plasmin is kept under tight regulation, but a number of bacterial pathogens can sequester plasmin(ogen) on their surfaces. If the zymogen is accessible to either host or bacterial plasminogen activators, active protease associated with the bacterial surface will be generated. This allows a bacterium with no proteolytic activity of its own to degrade

matrix proteins, thereby increasing the efficiency of tissue invasion. In the case of *B. burgdorferi,* binding of plasminogen, with subsequent activation by a plasminogen activator, generates active protease that resulted in increased ability of the spirochete to cross endothelial cell layers in culture.

Several *B. burgdorferi* proteins have been proposed to serve as receptors for plasmin(ogen). Plasmin(ogen) binds to many proteins that contain free amino groups, particularly the epsilon (ε) amino groups of lysine side chains. *B. burgdorferi* proteins tend to have a high lysine content when compared to other bacteria on a genome-wide basis. No consensus has yet been reached regarding the identity of the receptor(s) on intact bacteria, but a number of proteins in solubilized or detergent-extracted *B. burgdorferi* have been identified as candidates. It is possible that multiple proteins on the surface of *Borrelia* can bind plasmin(ogen), thereby ensuring that tissue invasion would not easily be lost to the organism.

Plasminogen-deficient mice (*plg*−) were employed to determine the role of plasmin(ogen) binding and activity in the pathogenesis of *B. burgdorferi* infection (Coleman *et al.,* 1997). This approach nicely circumvented two problems: the fact that no single spirochetal protein has been identified as a receptor for plasminogen and the lack of genetic tools with which to manipulate the genome of *Borrelia* species. After inoculation of *B. burgdorferi,* the *plg*− mice developed infection in all tissue sites tested, as did the wild-type controls. The number of spirochetes in the blood, however, was significantly less in *plg*− mice than in control mice. This result could reflect a decreased efficiency of spirochetal invasion of the circulatory system or an increased adherence of the bacteria to the vessel walls, perhaps in association with adherent platelets.

The most interesting results of this study were obtained when ticks were allowed to feed on *plg*− vs wild-type mice. Plasminogen and host plasminogen activators are both acquired during the tick feeding process. When infected ticks were fed on wild-type mice, plasmin(ogen) became associated with *B. burgdorferi* and remained on the spirochete during dissemination within the tick to the hemolymph and salivary glands. In contrast, when ticks were fed on *plg*− mice, spirochete dissemination to the hemolymph and salivary glands was dramatically reduced. Ticks fed on infected *plg−* mice acquired significantly fewer *B. burgdorferi* than did ticks fed on infected control mice. These results indicate that plasmin(ogen) acquisition by *B. burgdorferi* is important for the dissemination of the spirochetes, both in the mammalian circulatory system and in the tick, and is essential for efficient transmission of the spirochete in the reservoir/vector cycle.

C. Intoxication of Host Cells

Many bacterial pathogens synthesize proteins termed "toxins" that damage or kill host cells. Toxins can be broadly categorized into two groups: those that are active after delivery into the host cell cytosol and those that are active at the surface of the host cell. *Borrelia burgdorferi* has not been reported to produce any activities that resemble those of bacterial toxins that act inside the host cell, and no obvious homologies to toxins of this class were revealed by the genome sequence. *Borrelia burgdorferi* does, however, secrete a hemolytic activity. This activity, which is measured by lysis of red blood cells, is a property of certain surface-active toxins and defines the subset thereof termed "hemolysins." Despite the name, hemolysins may act on a variety of cell types to disrupt membrane integrity. This sort of intoxication may be responsible for the membrane perturbation and cell death observed in lymphocytes incubated with *B. burgdorferi in vitro.* A two-gene operon contained in one of the *B. burgdorferi* plasmids was cloned on the basis of conferring hemolytic activity to *E. coli,* but biochemical analyses suggested that the properties of the cloned hemolysin are somewhat different from those described earlier. The genome sequence revealed additional genes with homologies to other bacterial hemolysins, so it is possible that *B. burgdorferi* is armed with a diverse repertoire of hemolysins. The potential roles of the *B. burgdorferi* hemolytic activities have not yet been characterized *in vivo.*

Mammalian cells do respond to the presence of *B. burgdorferi,* but these effects do not appear to be the result of the action of a toxin in the classical sense. The morphological changes in tick cells after incubation with *B. burgdorferi* were mentioned earlier. It has been shown that endothelial cells in culture react to the presence of the spirochete, or to lipidated OspA and OspB, by increasing the expression of adhesion molecules and cytokines (Sellati *et al.,* 1996). In addition, outer surface proteins of *B. burgdorferi* can nonspecifically stimulate both B and T cells (Schoenfeld *et al.,* 1992; Simon *et al.,* 1995), which may exacerbate the host response to infection. The *in vivo* significance of these results is unclear, as OspA and OspB do not appear to be expressed until late in infection. If similar effects do occur *in vivo,* the presence of spirochetal outer surface lipoproteins would contribute to the inflammation seen in vessel walls and in neighboring tissues.

Lipidation of outer surface proteins, which occurs as the proteins are secreted by *Borrelia,* is required for the effects on B cells, T cells, and endothelial cells. It is, therefore, possible that other lipidated proteins (and there are many) may have similar effects on cells *in vivo.* These observations suggest the possibility that a vicious cycle may be initiated by *B. burgdorferi* during infection: if mammalian adhesion molecules are increased by the presence of the spirochete, the end result may be increased bacterial adherence to the cells. This hypothesis remains to be tested, as it has not been shown whether adhesion to cells by intact spirochetes is important in any aspect of the cellular response to *B. burgdorferi.*

D. Persistence of *Borrelia burgdorferi* in the Immunocompetent Host

Several mechanisms that might allow *B. burgdorferi* to survive in the mammalian host in the face of the immune response have been proposed. For example, because *B. burgdorferi* can cause chronic infection in the absence of appropriate antibiotic therapy, there has been a great deal of speculation regarding the possibility that the spirochete survives inside host cells, protected from the host immune response. Other bacterial pathogens adopt this sort of intracellular niche, but whether *B. burgdorferi* does so is an issue that awaits resolution. Experiments employing a variety of cultured cell types have suggested that *B. burgdorferi* might be taken up by the cells at low frequency. One caveat to this approach is that it is unclear how well the behavior of cells in culture reflects the biology of related cells in the host. Furthermore, the significance of the results suggesting a possible intracellular location in cultured cells is unclear, given that there is no evidence for an intracellular location in infected mice or humans. While the possibility of an intracellular location for the bacteria during chronic infection remains an important consideration, further work is needed to definitively answer this question.

One of the many alternative explanations for the persistence of *B. burgdorferi* in the mammalian host is that the spirochete is able to alter its outer surface, thereby evading clearance by the acquired immune response. This would be analogous to the situation with *Borrelia* species that cause relapsing fever, which frequently alter the sequence and expression of a family of major surface proteins. Recent work has identified a novel set of genes, *vls*, that generates antigenic variation in *B. burgdorferi* (Zhang *et al.*, 1997). Among this set of genes, only one, *vlsE*, is expressed at any given time from the single expression locus in the genome. The remaining related genes are not transcribed, but can be expressed after recombination into the expression site. The complex mechanism of recombination would, in time, allow almost infinite variation of the expressed sequence. Over the course of infection in the mouse model, considerable variation is observed, while spirochetes grown for a similar number of generations in laboratory culture media display no changes. This form of antigenic variation is not driven by the acquired immune response, however, as a similar degree of change occurs in *scid* mice. In addition, expression of VlsE is not required for *B. burgdorferi* to cause infection, as shown by the isolation of variants disrupted by premature stop codons and demonstration that these clones are still infectious. Further investigation of the phenomenon of antigenic variation may identify additional changes in the outer surface of *B. burgdorferi* during infection.

Identification of proteins and their corresponding genes that are expressed specifically during infection has become a common approach to understanding microbial pathogenesis in recent years. In many other bacterial species, this is done at the genetic level in the pathogen itself, but in *Borrelia* species, this strategy is not yet feasible. Instead, investigators have probed libraries that express *B. burgdorferi* genes in

E. coli with sera from human Lyme disease patients or from infected animals (Suk *et al.*, 1995). In addition to the utility of this method in cloning genes encoding antigens that are expressed *in vitro* and recognized by Lyme disease patient sera, this approach has allowed the identification of proteins that are not found, or at least are not abundant, in laboratory-cultivated spirochetes. Some examples of the genes that have been cloned and characterized using this approach are those encoding p93, p66, DBP, and various members of the Bmp, BapA/EppA, OspE, and OspF gene families. To date, however, the *dbp* gene products are the only proteins to which any functional significance has been ascribed.

E. Development of a Vaccine against *B. burgdorferi* Infection

Due to the potential for disabling manifestations of *B. burgdorferi* infection, and to the public fear of Lyme disease in endemic regions, considerable effort has been focused on the development of a vaccine. A number of *B. burgdorferi* proteins have been proposed as candidates for vaccine development, but most of the work has focused on OspA. Large-scale trials have been performed in humans, and OspA vaccines have recently been approved for human use.

Work in animal models has elucidated an interesting and novel mechanism by which the OspA vaccine works. When infected ticks are allowed to feed on animals that have been immunized with OspA, the *B. burgdorferi* in the midgut are killed early in the feeding process, before dissemination through the tick and transmission to the host can occur. It is likely that antibody-dependent complement killing is the primary mechanism in this process, but it has been known for some time that some anti-OspA antibodies can kill *B. burgdorferi* even in the absence of complement. This phenomenon would increase the efficiency of killing of the spirochete.

There are some potential drawbacks to the currently available OspA vaccine, however. First, although only slight heterogeneity exists in the *ospA* genes within *B. burgdorferi*, the *B. afzelii* sequences are somewhat different, and *B. garinii* shows considerable heterogeneity within that species. Second, reactivity to OspA has been associated with treatment resistant chronic Lyme arthritis, and there is some evidence that, in certain patients, OspA may trigger an autoimmune response to a human protein. The optimal vaccine candidate may not yet have been identified, but would be a highly conserved protein that does not have any association with potential autoimmune disease.

VIII. CONCLUSIONS AND FUTURE DIRECTIONS

Some of the most important directions for future work in the Lyme disease field will include the development of improved tools for the early diagnosis of infection. In addition, further standardization of diagnostic methods may eventually allow multiple laboratories, especially those not affiliated with large research centers, to perform the tests necessary for definitive diagnosis of *B. burgdorferi* infection.

It is also clear that the understanding of how *B. burgdorferi* causes infection and disease is still rudimentary when compared to many other bacterial pathogens. This is largely attributable to two factors: the complex nutritional requirements and slow growth rate of *B. burgdorferi* in the laboratory, and the lack of a well-developed genetic system. To date, no single *B. burgdorferi* molecule has been definitively shown to be important for pathogenesis, primarily because it is difficult to assess the importance of a bacterial product in the pathogenic process if mutants specifically deficient in that product cannot easily be generated. Nevertheless, progress is being made to alleviate some of these difficulties. First, the genome sequence might shed light on specific nutritional requirements, potentially allowing formulation of a simpler medium that allows faster growth. Second, the first steps toward manipulation of the genome in the laboratory have been taken (Tilly *et al.*, 1997). Mutations in the *gyrB* gene that confer resistance to the antibiotic coumermycin were selected by growing a wild-type, laboratory-adapted strain of *B. burgdorferi* in the presence of low concentrations of the antibiotic. This system has been exploited as a genetic tool, so that the mutant *gyrB* can be inserted into a copy of a different gene that is targeted for disruption in *B. burgdorferi*. This approach has been successfully used to generate mutations in *ospC* and in a noncoding portion of the plasmid that carries the *ospC* gene. As other antibiotic resistance markers are developed for use in *Borrelia* species, directed mutations will be generated more efficiently. Adaptation of these techniques to infectious *B. burgdorferi* will allow testing of the mutants for the ability to cause infection and disease in animal models. Finally, another approach to genetic manipulation of *B. burgdorferi* might be through the use of bacteriophages, viruses that infect bacteria. Although phagelike particles have been observed in *B. burgdorferi* cultures, phage as a genetic system are not yet available for *Borrelia*.

In order for *B. burgdorferi* to survive as a species, it must maintain infection in a mammalian host and be successfully transmitted to other members of the host species. At least some of the mechanisms that are employed by the spirochete to maintain itself in the wild are likely to be virulence factors in the context of the human host. Although considerable progress has been made over the last 20 years toward understanding the biology and pathogenic mechanisms of *Borrelia burgdorferi*, many of the most interesting questions remain unanswered and will be the focus of considerable effort in the future.

See Also the Following Articles

Adhesion, Bacterial • Flagella • Spirochetes • Vaccines, Bacterial

Bibliography

Barbour, A. G., and Hayes, S. F. (1986). Biology of *Borrelia* species. *Microbiol. Rev.* **50**, 381–400.

Barthold, S. W., de Souza, M. S., Janotka, J. L., Smith, A. L., and Persing, D. H. (1993). Chronic Lyme borreliosis in the laboratory mouse. *Am. J. Pathol.* **143**, 959–971.

Barthold, S. W., Feng, S., Bockenstedt, L. K., Fikrig, E., and Feen, K. (1997). Protective and arthritis-resolving activity in sera of mice infected with *Borrelia burgdorferi*. *Clin. Inf. Dis.* **25**(suppl. 1), S9–19.

Coburn, J., Magoun, L., Bodary, S. C., and Leong, J. M. (1998). Integrins $\alpha_v\beta_3$ and $\alpha_5\beta_1$ mediate attachment of Lyme disease spirochetes to human cells. *Infect. Immun.* **66**, 1946–1952.

Coleman, J. L., Gebbia, J. A., Piesman, J., Degen, J. L., Bugge, T. H., and Benach, J. L. (1997). Plasminogen is required for efficient dissemination of *B. burgdorferi* in ticks and for enhancement of spirochetemia in mice. *Cell* **89,** 1111–1119.

Fraser, C. M. *et al.* (1997). Genomic sequence of a Lyme disease spirochete, *Borrelia burgdorferi. Nature* **390,** 580–586.

Gross, D. M., Forsthuber, T., Tary-Lehmann, M., Etling, C., Ito, K., Nagy, Z. A., Field, J. A., Steere, A. C., and Huber, B. T. (1998). Identification of LFA-1 as a candidate autoantigen in treatment-resistant Lyme arthritis. *Science* **281,** 703–706.

Guo, B. P., Norris, S. J., Rosenberg, L. C., and Hook, M. (1995). Adherence of *Borrelia burgdorferi* to the proteoglycan decorin. *Infect. Immun.* **63,** 3467–3472.

Kalish, R. A., Leong, J. M., and Steere, A. C. (1993). Association of treatment-resistant chronic Lyme arthritis with HLA-DR4 and antibody reactivity to OspA and OspB of *Borrelia burgdorferi. Infect. Immun.* **61,** 2774–2779.

Kurtti, T. J., Munderloh, U. G., Krueger, D. E., Johnson, R. C., and Schwan, T. G. (1993). Adherence to and invasion of cultured tick (Acarina: Ixodidae) cells by *Borrelia burgdorferi* (Spirochaetales: Spirochaetaceae) and maintenance of infectivity. *J. Med. Entomol.* **30,** 586–596.

Leong, J. M., Wang, H., Magoun, L., Field, J., Morrissey, P. E., Robbins, D., Tatro, J. B., Coburn, J., and Parveen, N. (1998). Different classes of proteoglycans contribute to the attachment of *Borrelia burgdorferi* to cultured endothelial and brain cells. *Infect. Immun.* **66,** 994–999.

Radolf, J. D., Goldberg, M. S., Bourell, K. W., Baker, S. I., Jones, J. D., and Norgard, M. V. (1995). Characterization of outer membranes isolated from *Borrelia burgdorferi,* the Lyme disease spirochete. *Infect. Immun.* **63,** 2154–2163.

Roberts, E. D., Bohm Jr., R. P., Cogswell, F. B., Lanners, H. N., Lowrie, R. C. Jr., Povinelli, L., Piesman, J., and Phillip, M. T. (1995). Chronic Lyme disease in the rhesus monkey. *Lab. Investig.* **72,** 146–160.

Schwan, T. G., Piesman, J., Golde, W. T., Dolan, M. C., and Rosa, P. (1995). Induction of an outer surface protein on *Borrelia burgdorferi* during tick feeding. *Proc. Natl. Acad. Sci.* **92,** 2909–2913.

Schoenfeld, R., Araneo, B., Ma, Y., Yang, L., and Weis, J. J. (1992). Demonstration of a B-lymphocyte mitogen produced by the Lyme disease pathogen, *Borrelia burgdorferi. Infect. Immun.* **60,** 455–464.

Sellati, T. J., Abrescia, L. D., Radolf, J. D., and Furie, M. B. (1996). Outer surface lipoproteins of *Borrelia burgdorferi* activate vascular endothelium *in vitro. Infect. Immun.* **64,** 3180–3187.

Simon, M. M., Nerz, G., Kramer, M. D., Hurtenbach, U., Schaible, U. E., and Wallich, R. (1995). The outer surface lipoprotein A of *Borrelia burgdorferi* provides direct and indirect augmenting/costimulatory signals for the activation of CD4+ and CD8+ T cells. *Immunol. Lett.* **45,** 137–142.

Steere, A. C. (1989). Lyme disease. *New Engl. J. Med.* **321,** 586–596.

Suk, K., Das, S., Sun, W., Jwang, B., Barthold, S. W., Flavell, R. A., and Fikrig, E. (1995). *Borrelia burgdorferi* genes selectively expressed in the infected host. *Proc. Natl. Acad. Sci.* **92,** 4269–4273.

Tilly, K., Casjens, S., Stevenson, B., Bono, J. L., Samuels, D. S., Hogen, D., and Rosa, P. (1997). The *Borrelia burgdorferi* circular plasmid cp26: Conservation of plasmid structure and targeted inactivatio of the *ospC* gene. *Mol. Microbiol.* **25,** 361–373.

Zhang, J.-R., Hardham, J. M., Barbour, A. G., and Norris, S. J. (1997). Antigenic variation in Lyme disease spirochetes by promiscuous recombination of VMP-like sequence cassettes. *Cell* **89,** 275–285.

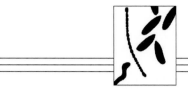

Malaria

Kostas D. Mathiopoulos

University of Patras, Greece

GLOSSARY

basic reproduction rate   The estimated number of secondary malaria infections potentially transmitted within a susceptible population from a single nonimmune individual.

continuous fever   Elevated fever with a fluctuation of not more than one degree Celsius. Compare to "remittent" and "intermittent" fevers.

endemic malaria   A constant incidence of cases over a period of many successive years.

epidemic malaria   A periodic or occasional sharp increase of the malaria incidence in a given indigenous population.

hypnozoite   The resting stage of the parasite in the hepatocytes. This situation occurs only with *P. vivax* or *P. ovale*. Reactivation of these parasites causes the relapse of the disease.

incidence   A measure of the number of *new* cases of disease that develop in a population of individuals at risk during a given time interval. Compare to "prevalence."

intermittent fever   Elevated fever with temperature returning to normal one or more times during a 24-hour period. Compare to "continuous" and "remittent" fevers.

morbidity/mortality rate   Morbidity rate is the incidence rate of nonfatal cases in the total population at risk during a specified period of time. Mortality rate expresses the incidence of death in a particular population during a period of time.

prevalence   A measure of the proportion of individuals in a population who have the disease at a specific instant and an estimate of the probability (risk) that an individual

will be ill at a particular point in time. Compare to "incidence."

recrudescence   A recurrence of symptoms in a patient whose bloodstream infection has previously been at such a low level as not to cause symptoms. Compare to "relapse."

relapse   A recurrence that takes place after complete initial clearing of the erythrocytic infection; implies reinvasion of the bloodstream by parasites from the exoerythrocytic stages. Compare to "recrudescence."

remittent fever   Elevated fever with a fluctuation of more than one degree Celsius. Compare to "continuous" and "intermittent" fevers.

splenomegaly   Enlargement of the spleen.

vectorial capacity   A mathematical formula that expresses the malaria transmission risk, the "receptivity" to malaria of a defined area.

MALARIA   is a serious, acute, and chronic relapsing infection in humans that is characterized by periodic attacks of chills and fever, anemia, splenomegaly, and often fatal complications. The disease is also observed in apes, rats, birds, and reptiles. It is caused by various species of protozoa (one-celled organisms) called sporozoans that belong to the genus Plasmodium. The parasites are transmitted by the bite of various species of mosquitoes. Human malaria is transmitted by mosquitoes of the genus Anopheles.

Malaria occurs throughout the tropical and subtropical regions of the world and is one of the leading causes of morbidity and mortality (Fig. 1). About 300–500 million cases of malaria worldwide result in 1,000,000 deaths per year, mostly among children in Africa under 5 years of age. Malaria has been estimated to represent 2.3% of the overall global

Encyclopedia of Microbiology, Volume 3
SECOND EDITION

131

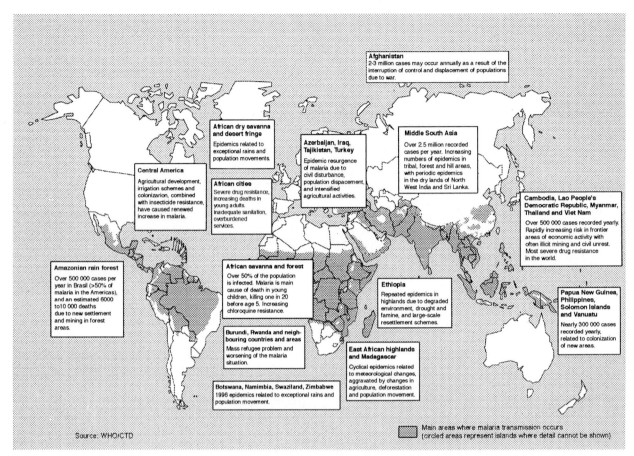

Fig. 1. Malaria distribution and problem areas. Despite intensive control efforts, malaria remains one of the leading causes of morbidity and mortality, afflicting more than 500 million individuals, mainly in the tropical zone. (Reproduced from WHO/CTD TDR, Progress Report 1995–1996.)

disease burden and 9% of that in Africa, ranking third among major infectious disease threats, after pneumococcal acute respiratory infections (3.5%) and tuberculosis (2.8%). The disease is often linked to the movement of refugees and populations seeking work and to environmental change, including agriculture, forestry, mining, and water development projects. In Africa alone, the estimated annual direct and indirect costs of a malaria were $800 million in 1987 and over $1800 million in 1995. Malaria is also common in Central and the northern half of South America and in South and Southeast Asia. The disease also occurs in countries bordering on the Mediterranean, in the Middle East, and in East Asia.

I. HISTORICAL OUTLINE

Malaria is one of the most ancient infections known (Bruce-Chwatt, 1988). Descriptions of typical malaria fevers are found in ancient Sumerian and Egyptian writings dating before 2000 B.C. and in the Chinese medical classic of 2700 B.C., the Nei Ching. The Greek "Father of Medicine" Hippocrates, in the fifth century B.C., tried to give a scientific explanation to the disease, differentiating malarial fevers into three types according to their time cycles and associating them with swampy areas. The term "malaria" (which, in Italian, means "bad" or "evil air," presumably deriving from the association of the disease with the dank effluvium rising from the marshes) was not used until much later. The first reference was probably made by the English writer Horace Walpole, who wrote from Rome in 1740 of "a horrid thing called mal'aria that comes to Rome every summer and kills one." It is not known when malaria first made its appearance in the Americas, but it is highly probable that it was a post-Columbian importation. Some rather severe epidemics were first noted in 1493.

The association between swampy or marshy areas and the disease has long been recognized, but the roles of the mosquito and of the malarial parasite were not known until the very end of the nineteenth century. In 1880, the French army surgeon Alphonse Laveran described the malarial parasite in the blood of a patient in Algeria and recognized it as the cause of malaria. In 1897–1898, the British physician Sir Ronald Ross proved that bird malaria is transmitted by Culex mosquitoes, and he described the entire life cycle of that parasite in the mosquito. In 1898, the Italian investigators Amico Bignami, Giovanni Battista Grassi, and Giuseppe Bastianelli first infected humans with malaria by mosquitoes, described the full development of the parasite in humans, and noted that human malaria is transmitted only by anopheline mosquitoes.

During the twentieth century, much research was devoted to malaria control. Larvicides in the form of oil or Paris green were introduced for preventing the breeding of mosquitoes in various types of waters. Such methods had outstanding success in controlling malaria and yellow fever in Cuba and the Panama Canal Zone. The ravages of malaria experienced during World War I and the difficulties of securing cheap supplies of quinine (the only antimalarial known until then) stimulated a line of research in Germany and then in France, England, and the United States, that resulted in a series of synthetic antimalarials. In the meantime, the use of pyrethrum sprays against adult mosquitoes, which managed to reduce the amount of malaria in rural areas of South Africa, revolutionized malaria control approaches.

Successful programs resulting in the eradication of *Anopheles gambiae* from Brazil in 1934–1949 and from Egypt in 1948, and in the interruption of ma-

laria transmission in Sardinia, Cyprus, Greece, Guyana, and Venezuela, made worldwide eradication of malaria through vector control seem a feasible goal. Despite the opposition of several malariologists, and with the insecticide DDT as the first-line weapon, such a "malaria eradication" program was justifiable for two main reasons. From an economic point of view, it seemed that the cost to eradicate malaria was much lower than the damage the disease did to the world economy. From a practical point of view, the development of DDT resistance in malaria vectors in several countries in the early 1950s prompted the effort to eliminate the disease before resistance made control impossible. In 1955, the eighth World Health Assembly in Mexico City launched the malaria eradition initiative; application of DDT within households was the primary stratagem. The goal was to sufficiently reduce infected vector populations feeding on humans to interrupt transmission of parasites rather than to eradicate all vectors. Africa south of the Sahara was essentially excluded from the project because of the magnitude of malaria transmission and the lack infrastructure in those countries. In the 1960s, however, it was progressively made clear that this objective was unattainable and, gradually, the various financing agencies began reducing their contributions. The eradication program was formally abandoned by WHO at the World Health Assembly of 1969. The following year the Assembly recommended "malaria control" schemes within the general health systems of countries still afflicted by malaria and supported the intensification of both basic and applied research, which had been nearly halted during the eradication campaign years.

At the time of the interruption of the malaria eradication program, about 80% of the world population had been involved in the campaign, eradicating malaria in 36 nations. The world population that was not under malaria risk was estimated in 1968 as 37.6%. Now, malaria is still endemic in 91 countries and presents a risk for over 2 billion people, i.e., about 40% of the world population.

Eradication was most successful in the temperate zones, where transmission was epidemic and unstable (e.g., Tunisia, Morocco) and on islands. Indoor spraying of DDT that had a residual action was the last successful tool in a long series of antimalarial interventions that contributed to the destabilization of malaria transmission, reducing the vector density and forcing the mosquitoes to move outdoors. In temperate zones that are characterized by a rigid winter climate, the outdoor temperature was below the necessary threshold required by the parasite to complete its life cycle in the mosquito, thus interrupting transmission. Residual insecticide did not have the same success in tropical countries. The outdoor temperatures are, in general, favorable to vector survival all year long, as well as to the development of the plasmodium inside the mosquito. The transmission level of the disease that, in several zones in sub-Saharan Africa, reaches 100 infective bites per person per year, is such that even a 90% reduction of transmission would not have any significant effect in malaria diffusion. Ironically, in such zones the impact of economical progress associated with agricultural development, irrigation of arid zones, deforestation and desalinization of coastal zones, interventions that have a great impact on the ecosystem, combined with socioeconomic instability and dubious health service systems, result in increase of malaria.

Given this situation, the only possibilities for intervention should address the reduction of mortality through the reinforcement of the basic sanitary services, improvement of the diagnosis techniques and a better accessibility to antimalarial drugs. In fact, the present Global Strategy for Malaria Control of WHO, since the beginning of 1990s, is based on the precocious diagnosis and immediate therapy, on the planning and application of prevention methods that are adequate for the local conditions, and on the constant surveillance and reinforcement of local capacities of the research and administration of the problem. In the 51st World Health Assembly in May of 1998, the Director General of WHO presented the "Roll Back Malaria" project that proposes the better utilization of the existent control systems in the afflicted countries, through reinforcement of the sanitary infrastructure and socioeconomic development. The project aims to gain the financial support of the World Bank and the industrial world in order to decrease the malaria mortality to 50% by the year 2010.

II. THE DISEASE

A. Disease Manifestation

Human malaria is caused by four species of Plasmodium: *Plasmodium falciparum, P. vivax, P. mariae,* and *P. ovale.* The primary attack usually begins with several days of malaise, anorexia, headache, myalgia and low-grade fever, followed by the classic malarial paroxysms of chills, fever, and sweating. After several such paroxysms, symptoms tend to assume a periodicity characteristic of each species of Plasmodium. The bouts may be accompanied by nausea, vomiting, diarrhea, and the development of anemia and splenomegaly. Between paroxysms, the patient may feel well except for fatigue.

Falciparum malaria (malignant tertian malaria) is characterized by febrile paroxysms every 36–48 hours. Falciparum infections are more severe and may cause the death of up to 25% of adults not previously exposed to the organism. Complications seen in these severe infections include renal insufficiency and failure, pulmonary edema, neurologic symptoms, and severe hemolytic anemia. Pregnant women and their fetuses are particularly vulnerable to the effects of *P. falciparum* and infections during pregnancy may result in abortion, stillbirth, or lower-than-normal birth weight. In children, repeated malarial infections may cause chronic anemia and splenomegaly. In children and nonimmune adults, infection may provoke the accumulation of parasitized red blood cells in the capillaries of internal organs, thus block their function. When this clogging occurs in the brain capillaries, it causes "cerebral malaria" the principal cause of death from malaria. *P. falciparum* is found predominantly in tropical areas and accounts for 50% of all malaria cases and as many as 95% of all malaria deaths in the world. Vivax malaria (benign tertian malaria) is also characterized by febrile attacks about every 48 hours. A particularity of vivax malaria (and also of ovale) is its ability to relapse even after several months, an issue that will be discussed. *P. vivax* is found predominantly in Asia and accounts for at least 43% of malaria cases in the world. Quartan malaria is caused by *P. mariae,* characterized by febrile paroxysms every fourth day (72 hours) and by a very long duration, at times, up to 40 years. Malariae malaria and ovale malaria combined contribute fewer than 7% of malaria cases worldwide.

1. The Double Life of Plasmodium

The malarial parasite has a complicated double life cycle, with a sexual reproductive cycle while it lives in the mosquito and an asexual reproductive cycle while in the human host. The life cycle of Plasmodium, in both human and mosquito, is shown in Fig. 2. When the female mosquito bites an infected person, she draws into her stomach blood that may contain male and female gametocytes. In the mosquito, the male or microgametocyte undergoes a process of maturation, resulting in the production of a number of microgametes. The extrusion of these delicate spindle-shaped gametes has been termed "exflagellation." At the same time, the female or macrogametocyte matures to become a macrogamete, after which it may be fertilized by the microgamete, forming a zygote. The zygote becomes elongated and active and is called an ookinete. The ookinete penetrates through the stomach wall of the mosquito and rounds up just beneath the outer covering of that organ to become an oocyst. Growth and development of the oocyst result in the production of a large number of slender, threadlike sporozoites, which break out and wander throughout the body of the mosquito. Length of the developmental cycle in the mosquito depends not only on the species of Plasmodium, but also upon the mosquito species and the ambient temperature. It may range from as short a time as 8 days in *P. vivax* to as long as 35 days in *P. malariae*. The final destination of the sporozoites is the salivary glands of the female mosquito. With each bite, the female mosquito secretes saliva at the site, which contains anticoagulant factors whose role is to prevent blood clotting and facilitate the blood meal. But, along with the saliva, she also inoculates sporozoites into the skin vessels of the human host.

Sporozoites injected into the bloodstream leave the blood vascular system within a period of 40 minutes and invade the parenchymal cells of the liver. Subsequent development of *P. falciparum* and *P. malariae* differs from that seen in *P. vivax* and *P. ovale*. In all four species, an asexual multiplication takes place within the liver cells, but, in *P. vivax* and

P. ovale, a proportion of the infecting sporozoites are believed to enter a resting stage before undergoing asexual multiplication, while others undergo this multiplication without delay. The resting stage of the parasite is known as hypnozoite. After a period of weeks or months, reactivation of the hypnozoite initiates asexual division. Hypnozoite reactivation is thought to bring about the relapse characteristic of *P. vivax* and *P. ovale*. Hypnozoites are not found in the development of *P. falciparum* and *P. malariae*, and relapses do not occur in disease caused by these species. (Note: "Relapse" and "recrudescence" have special meanings when applied to malaria; see Glossary.)

Asexual multiplication in the liver results in the production of thousands of tiny merozoites in each infected cell. Rupture of these infected cells releases merozoites into the circulation. In the bloodstream, an asexual cycle takes place within the red blood cells. This process, known as schizogony, results in the formation, within a period of 48 to 72 hours, of from 4 to 36 new parasites in each infected cell. At the end of the schizogonic cycle, the infected blood cells rupture, liberating merozoites, which, in turn, infect new red blood cells. Ruptured red blood cells liberate products of metabolism of the parasites and of the red blood cells, and it is thought that if large numbers of infected cells rupture simultaneously, the volume of toxic materials thrown into the bloodstream may be sufficient to bring about a malarial paroxysm. Generally, in the initial stages of infection, rupture of the infected cells is not synchronous, so that fever may be continuous or remittent, rather than intermittent. It is theorized that the fever peaks may have a regulatory effect on the development cycle, speeding up those that are out of phase, so that after a number of days, the febrile cycle develops a 48- or 72-hour periodicity, depending upon the species of the parasite. In mixed infections with two or more species, or in the early stages of infection with one species, there may be daily (quotidian) paroxysms or even double paroxysms in one day.

Some time after the asexual parasites first appear in the bloodstream, usually not until after the patient has become clinically ill, gametocytes appear in the red blood cells. These forms, derived from merozoites similar in appearance to those that continue the

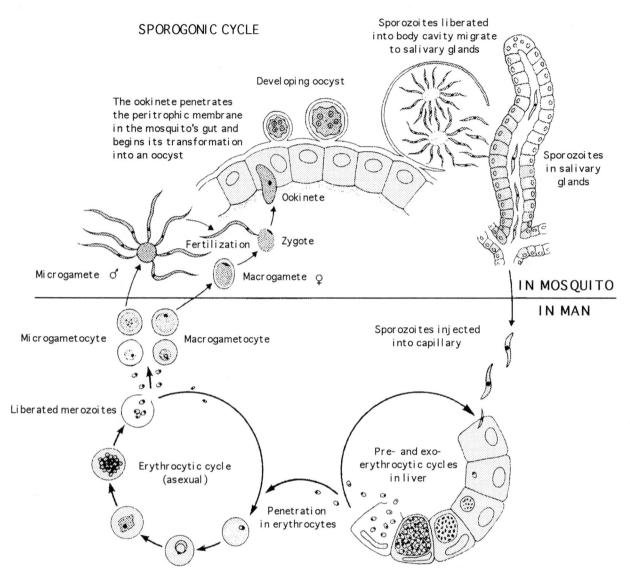

SPOROGONIC CYCLE

Sporozoites liberated into body cavity migrate to salivary glands

Developing oocyst

The ookinete penetrates the peritrophic membrane in the mosquito's gut and begins its transformation into an oocyst

Sporozoites in salivary glands

Ookinete

Fertilization   Zygote

Microgamete ♂

Macrogamete ♀

IN MOSQUITO

IN MAN

Microgametocyte   Macrogametocyte

Sporozoites injected into capillary

Liberated merozoites

Erythrocytic cycle (asexual)

Pre- and exo-erythrocytic cycles in liver

Penetration in erythrocytes

Fig. 2. The life cycle of malaria parasites in the mosquito and in the human host. Female anopheline mosquitoes, while they ingest a blood meal, inoculate sporozoites into the bloodstream that rapidly invade hepatic cells. There, they develop as intracellular hepatic schizonts. After about six days, the schizont ruptures, releasing about 20,000 merozoites within a hepatic cell. Merozoites rupture from hepatic cells and pour into the bloodstream to invade erythrocytes. There, they undergo the asexual erythrocytic cycle, where they mature from rings to schizonts in 48 to 72 hours, the time varying with the malaria species. Merozoites are released to invade other erythrocytes. Some red blood cell merozoites differentiate to sexual gametocytes. These male and female gametocytes (micro- and macrogametocytes, respectively) are the only forms able to survive in the mosquito midgut following a blood meal. Within the mosquito midgut, gametocytes transform into gametes and fertilize; the newly formed zygote transforms into an ookinete, a motile stage that crosses the midgut wall and develops into an oocyst on the outer side of the gut. The oocyst grows over 10 to 14 days, producing sporozoites that migrate to the salivary glands, ready to enter the vetebrate bloodsteam during a subsequent blood meal.

asexual cycle, grow but do not divide, and, finally, form the male and female gametocytes. Gametocytes continue to circulate in the bloodstream for some time and, if ingested by a mosquito of the genus Anopheles, undergo the a sexual cycle (gametogony) and subsequent development (sporogony) in the mosquito.

2. Diagnosis

Thick and thin blood films are traditionally used for the diagnosis of malaria. Proper stains and staining techniques are important to successful identification of the parasites. Of similar importance is the timing of blood examination. Depending upon the stage of the developmental cycle of the parasite at which the blood sample is taken, various stages of the parasite will appear in the blood. The thin blood film is always fixed with methyl alcohol, either before it is stained or as a part of the staining process. This enables the parasite to be observed with minimum distortion and within its host erythrocyte. On the other hand, the thick smear is not fixed, as the erythrocytes need to be lysed during staining to allow light to penetrate through the smear during microscopical examination. Consequently, the erythrocytes cannot be detected and the parasites frequently undergo some distortion due to the nonisotonicity of the staining solution. Thin blood films, therefore, reveal the parasite in greater morphological detail but the thick smear has the advantage that parasites are detected with greater frequency. Also, various serologic tests are available (indirect hemagglutination test, ELISA, and indirect fluorescent antibody tests), mostly based on antibodies to the asexual blood stages. More recently, PCR-based assays have been developed that are capable of identifying the different Plasmodium species from tiny amounts of blood.

3. Pathophysiology

The primary pathologic effects of a malarial infection are the result of hemolysis of infected and noninfected red blood cells, the liberation of metabolites of the parasite followed by the immunologic response of the host to this antigenic material, and the formation of malarial pigment.

The various species of malarial parasites differ in their ability to infect red blood cells. Merozoites of *P. vivax* and *P. ovale* are able to invade only reticulocytes, whereas those of *P. malariae* are limited to the senescent cells nearing the end of their lifespans. Thus, there is a natural limit placed on these infections: reticulocytes compose less than 2% of the total number of red blood cells and those suitable for infection by *P. malariae* even less *P. falciparum* is able to invade all ages of red blood cells indifferently and, consequently, infections by this parasite result early in considerable degrees of anemia.

Mechanisms of red blood cell invasion have been studied by means of electron microscopy (Hadley *et al.*, 1986). Briefly, the apical end of the parasite touches the surface of the erythrocyte, which invaginates slightly. A junction forms between merozoite and erythrocyte. The erythrocyte becomes ringlike and expands as the parasite invades the red blood cells and then, in turn, contracts to seal off the parasite within its vacuole. Initial recognition and attachment may require specific determinants on the surface of the red blood cell. For example, in the case of *P. vivax*, the determinant seems related to the Duffy blood group antigen. Glycophorin A is apparently involved in the attachment process of *P. falciparum*.

Rupture of the infected red blood cells brings on the malarial paroxysm. Lysis of numerous uninfected cells during the paroxysm, plus enhanced phagocytosis of normal cells, in addition to the cell remnants and other debris produced by schizogony, leads to both anemia and enlargement of the spleen and liver. The enlarged spleen becomes palpable during the first few weeks of infection, during which time it is soft and subject to rupture. If the infection is treated, the spleen returns to normal size, but, in chronic infections, it continues to enlarge and becomes hard. Malarial pigment (hemozoin) collects to give the organ a grayish to dark brown or black color. The liver becomes congested, and the Küpffer cells are packed with hemozoin, which, indeed, is seen throughout the viscera. Hemozoin is derived from the hemoglobin of the infected red blood cell and, as it is insoluble in plasma, its formation depletes the iron stores of the body, thus adding to the anemia.

P. falciparum is the most virulent of the four malaria parasites that infect humans and this is attributed to its ability to adhere to endothelial cells lining small blood vessels in internal organs and, in some cases, to noninfected erythrocytes, thus forming rosettes. When large numbers of parasites accumulate in the brain, small vessels may become plugged by masses of parasitized red blood cells, thus contributing to life-threatening cerebral malaria. A number of endothelial cell receptors (ICAM-1, CD-36, V-CAM, E-selectin, chondroitin-4-sulphate, and thrombospondin) are now known to have the potential to bind infected erythrocytes.

4. Immunology

Both acquired and natural immunity play a role in resistance to malarial infections. Acquired immunity is particularly important in areas with a high level of malaria transmission. Infants born to native parents in the regions of African in which malaria is hyperendemic (and presumably other areas of high endemicity) are almost completely free of malaria during the first few months of life. This is presumably due to the passive transfer of maternal immunity. Antibodies to blood stages of the parasites have been detected in such infants, as well as antisporozoite antibodies to *P. falciparum* in similar titers in mothers and babies, but the babies lose their titers by the age of 6 months. Antisporozoite antibodies (stage- and species-specific for both *P. falciparum* and *P. vivax*) in adults are seen in hyperendemic areas. Children exposed to repeated infections in hyperendemic areas develop a high degree of immunity, though this immunity does not imply an eradication of the infection but, rather, a balance between parasite and host.

Natural immunity to malaria infections is also very important. West African blacks and their descendants in the Americas and elsewhere possess a relative immunity to *P. vivax* and also to experimental infection with *P. cynomogli* (a monkey malaria agent of the Oriental Region, able to infect humans in the laboratory and a perfect model for human vivax malaria), since their red blood cells lack the appropriate parasite receptor. This explanation of this remarkable resistance was postulated when it was noted that Duffy-blood-group-negative human erythrocytes (Fy/Fy) are resistant to invasion *in vitro* by *P. knowlesi*, a species of monkey malaria (Miller *et al.,* 1976). Duffy-positive erythrocytes (Fy^a or Fy^b) were readily infected. It was subsequently observed that, among black and white volunteers who had been exposed to the bites of *P. vivax*-infected mosqui-

toes, only the Duffy-negative blacks were immune to infection. This suggested that the high prevalence of Duffy-negativity in West Africans and of approximately 70% of American blacks accounts for the resistance of these individuals to *P. vivax.* Instead of *P. vivax, P. ovale* is seen commonly in West Africa. In East Africa, where a higher percentage of persons are Duffy-positive, *P. vivax* is seen.

Other genetic deficiencies have been shown to confer resistance to malaria. For example, it was noted that glucose-6-phosphate dehydrogenase (G6PD)-deficient individuals are several times more resistant to *P. falciparum* infection (Luzzato and Bienzle, 1979). G6PD-deficiency involves three alleles, Gd^A, Gd^B, and Gd^{A-}. Clinically, one sees no increased resistance in hemizygous (Gd^{A-}) males (the gene involved is on the X chromosome) or homozygous (Gd^{A-}/Gd^{A-}) females, but it is seen in heterozygous (Gd^{A-}/Gd^B) females. The precise mechanism whereby this mosaicism offers protection is uncertain but it certainly involves the erythrocyte invasion process, since G6PD-deficient cells are from 2 to 80 times more resistant to invasion by *P. falciparum* than are normal erythrocytes. Various hemoglobinopathies, and especially the presence of hemoglobin S, have also been found to be related to increased resistance to *P. falciparum* infection, apparently because the parasite cannot develop normally in red blood cells that contain hemoglobin S (Pasvol *et al.,* 1978). Malaria mortality in patients with the sickle-cell trait has been found to be less than one-twentieth of that expected. These hemoglobinopathies (as well as thalassemia and hemoglobin C or E, also considered by some to be associated with increased resistance to *P. falciparum* infection) have a geographic distribution such that their presence could be readily explained if they confer a selective advantage to their possessors in a hyperendemic malaria zone.

B. Disease Control

1. Chemotherapy: From the Bark of the Cinchona Tree to Synthetic Antimalarials to the Bark of Qinghaosu Tree

Prior to World War I, quinine and its related alkaloids were the only specific antimalarials. The expeditions to the New World and the subsequent transatlantic exchanges brought to Europe the bark of a Peruvian tree known to cure people from intermittent fevers. The tree was given the generic name cinchona by Linnaeus in 1735 from the name of the countess that, according to the legend, was the first European to be cured from an intermittent fever by drinking infusions of its bark. Since World War I, a large number of synthetic drugs were produced, some of which have become mainstays of malaria therapy. Antimalarial drugs may be used for the treatment of clinical cases, for prophylaxis, and for the prevention of disease transmission. The principal families of antimalarial drugs are quinine, 4-aminoquinolines (e.g., chloroquine and amodiaquine), 8-aminoquinolines (primaquine), folic acid antagonists (e.g., pyrimethamine and chlorguanide), sulfones and sulfonamides, antibiotics (e.g., tetracycline), and quinoline methanols (e.g., mefloquine). Resistance was noted early to certain antimalarials (chlorguanide, pyrimethamine), but this was of little clinical significance until chloroquine was found ineffective in the treatment of certain strains of *P. falciparum* in South America and Southeast Asia. Frequently, resistance to chloroquine means resistance to other synthetic antimalarials as well, and successful treatment may require the use of quinine. Unfortunately, some strains have developed a partial resistance to quinine also and require the use of combination therapy. Rational use of the antimalarial drugs is based on a knowledge of their effects upon the various stages of the life cycle.

Quinine, chloroquine and its derivatives, and the acridine dye quinacrine are generally effective against the asexual erythrocytic stages of the human malarias but are ineffective against their exoerythrocytic stages or the gametocytes of *P. falciparum*. The mechanism of action of this group of drugs is thought to involve inhibition of the enzymatic synthesis of DNA and RNA. Mefloquine hydrochloride has been found effective for treatment of both chloroquine-sensitive and multiply drug-resistant strains of *P. falciparum,* when administered in a single oral dose. The mechanism of its action is not known. It is blood schizonticide and does not affect the exoerythrocytic stages. Resistance to mefloquine has become widespread in recent years.

The 8-aminoquinoline drugs, of which primaquine is the least toxic, are most effective against the exo-

erythrocytic stage and gametocytes of the various species of human malaria. They are relatively ineffective against the asexual erythrocytic parasites. Their mechanism of action is unknown but it is thought that the site of action involves the plasmodial mitochondria.

Chlorguanide and the structurally related diaminopyrimidine, pyrimethamine, are relatively slow-acting drugs of limited value in the treatment of the sexual erythrocytic stages of malaria. Both of these drugs inhibit metabolism of folic acid and affect nuclear division of the parasites. *P. falciparum* strains resistant to both these drugs are now widespread in Africa.

The sulfonamides and sulfones have been know for many years to have antimalarial properties, but they were not extensively tested until the emergence of resistant strains of *P. falciparum* created an acute need for new and effective antimalarials. Both the longer-acting sulfonamides and the sulfone dapsone have been clinically tested and found to have some schizonticidal activity, especially against *P. falciparum*. The sulfonamides and sulfones are PABA antagonists; their effect on the malarial parasite apparently derives from their ability to block the synthesis of folic acid from para-aminobenzoic acid. Dapsone and other sulfones inhibit intraerythrocytic–parasite glycolysis and the utilization of adenosine, apparently by blockage at the red blood cell membrane.

A number of antibiotics, including the tetracyclines, rifampin, clindamycin, erythromycin, and chloramphenicol, have some effect on the erythrocytic cycle. When used with quinine, tetracycline and clindamycin increase the cure rate in chloroquine-resistant *P. falciparum* infections.

Many new drugs are being tested for antimalarial activity. Some of those showing promise include a phenanthrenemethanol, halofantrine, which is effective against multiply drug-resistant *P. falciparum* and requires a short treatment course, and two types of herbal extracts. Quassinoids, derived from plants of the family Simoroubaceae, include glaucarubin, which has anti-amebic activity. Qinghaosu, or artemisinin, is an extract of *Artemesia annua,* used for centuries in China for treatment of malaria. Recently, it has been purified and its chemical structure determined. A number of derivatives have proved particularly effective in the treatment of severe and multidrug-resistant falciparum malaria. Although most information on qinghaosu compounds concerns the treatment of adults, there is sufficient data to conclude that children tolerate the drugs very well and that the therapeutic response is similar to that observed in adults. Its efficacy, the lack of apparent toxicity, its rapid action, and its low cost make artemisinin and its derivatives particularly promising. Although there are still no drug resistant cases reported, resistance to artemisinin has been demonstrated *in vitro*.

2. *Malaria Vaccine Development*

The development of an effective malaria vaccine represents one of the most important approaches to providing a cost-effective intervention for addition to currently available malaria control strategies. At the same time, it also presents a major scientific challenge for several reasons. Host immunity clearly develops after malarial infections, as demonstrated by the lower parasite count in the blood and the milder clinical symptoms of the individuals in endemic zones. This immunity, however, is lost when the individual moves to a nonendemic area, since, upon return, he or she could suffer from severe disease. The reason of this loss is due to either the poor immunogenicity of the parasite (as far as protective immunity is concerned) or to the fact that immunity may be strain-specific, or both. Over the last decade, there has been considerable progress in the understanding of immune mechanisms involved in conferring protection to malaria and the identification of vaccine candidates and their genes. However, due to the complexity and cost of malaria vaccine development, and the relative lack of commercial interest, few vaccine candidates have, so far, progressed to clinical trials. Most of these candidates represent unmodified parasite antigens in traditional adjuvants, such as alum. Moreover, work over the last 15 years, which has focused on the identification of vaccine candidate antigens, is now being augmented by new adjuvants soon to be available for human use, and new approaches such as DNA vaccines and structural modification of antigens, to circumvent some of the strategies the parasite uses to avoid the host's immune response.

There are three main types of vaccine: asexual blood-stage vaccine; transmission-blocking vaccines; and pre-erythrocytic vaccines (Fig. 3). Asexual blood-stage antigens may constitute "antidisease" vaccines in that such vaccines would reduce severe morbidity and mortality from malaria, but would not necessarily provide sterile immunity in the vaccinated individual. Indeed, it is extraordinarily rare for sterile immunity ever to be developed in nature. A transmission-blocking vaccine would function by blocking fertilization of the female gamete or by inhibiting the development of the parasite in the mosquito. Because this vaccine would block transmission of the disease but would not provide protection against parasitemia or clinical illness, it probably would always be used in combination with one of the other two types of vaccines. An effective pre-erythrocytic vaccine would prevent infection of the hepatocytes following a mosquito bite, so that the immunized patient would not develop either parasitemia or symptoms. A patient receiving this vaccine would also be incapable of infecting additional mosquitoes, and transmission of the disease would, therefore, be interrupted. If the vaccine were not completely effective, however, and a few organisms escaped to invade liver cells, parasitemia and clinical illness would result, as in a naive individual.

Asexual Blood-Stage Vaccines

In 1993, a WHO-sponsored task force evaluated some 20 asexual blood-stage candidate *P. falciparum* antigens and prepared a strategy for their development, leading to clinical testing and field trials. Merozoite Surface Protein-1 (MSP-1) was determined to be a leading candidate antigen, and several recent clinical trials in humans have involved vaccines with portions of this antigen, including the multicomponent asexual blood-stage vaccine candidate SPf66 developed in Colombia (Patarroyo *et al.*, 1988). This is a synthetic cocktail peptide vaccine against *P. falciparum* malaria, which was demonstrated to give partial protection in Aotus monkeys and, subsequently, in humans. This is the first multicomponent blood-stage vaccine to undergo extensive field trials throughout the world. Despite mixed results, the many field trials conducted to date have had a major impact on the thinking and design of

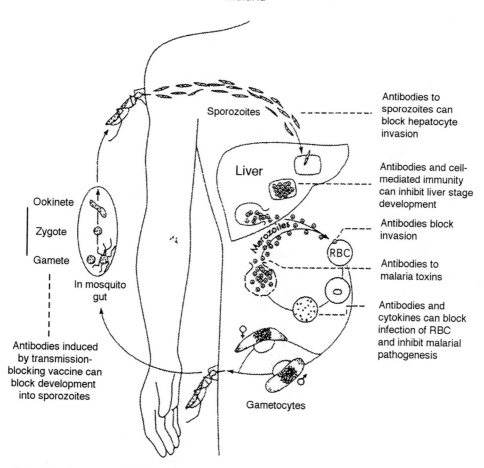

Sporozoites

Liver

Ookinete

Zygote

Gamete

In mosquito gut

RBC

Merozoites

Gametocytes

Antibodies to sporozoites can block hepatocyte invasion

Antibodies and cell-mediated immunity can inhibit liver stage development

Antibodies block invasion

Antibodies to malaria toxins

Antibodies and cytokines can block infection of RBC and inhibit malarial pathogenesis

Antibodies induced by transmission-blocking vaccine can block development into sporozoites

Fig. 3. Vaccine targets in the malaria parasite life cycle. RBC, red blood cells. [Modified and redrawn from Hadley, T. J., Klotz, F. W., and Miller, L. H. (1986). *Annu. Rev. Microbiol.* **40**, 451–477.]

vaccine trials in naturally exposed populations. Another candidate is an engineered, attenuated vaccinia virus, multistage, multicomponent *P. falciparum* vaccine, which includes a transmission-blocking vaccine candidate Pfs-25 (see following), together with six additional leading candidate antigens (three pre-erythrocytic proteins and three asexual blood stage antigens). The engineered, attenuated virus vaccine approach, if successful, would provide a particularly cost-effective means of delivering multiple antigens in one vaccine formulation designed to elicit both humoral and cellular immune responses. Indeed, the concept of priming with an attenuated viral vaccine, followed by a booster injection with a recombinant antigen—the "prime boost" principle—has considerable support in the HIV-vaccine development field.

Natural boosting may be very relevant in the case of malaria.

Transmission Blocking Vaccines

An effective transmission-blocking vaccine would help deal with the emergence of drug-resistant parasites or potential escape variants selected by partially effective pre-erythrocytic stage of asexual blood-stage vaccines. It may also be effective at eliminating transmission in areas of low endemicity. However, it is considered unlikely that a transmission-blocking vaccine would be used on its own in areas of high endemicity, but, rather, would be used in combination with effective pre-erythrocyte and asexual blood-stage antigens. Pfs-25, the leading *P. falciparum* transmission-blocking vaccine candidate under development, has recently entered clinical trials. Preliminary results suggest that the vaccine is safe and induces antibodies, but that the antibodies lack functional transmission-blocking activity in the *in vitro* membrane feeding assay. Additional preclinical studies are testing new, more potent adjuvants and antigen combinations. In addition, Pfs-25 is included in studies designed to test the prime boost concept for generating functional transmission-blocking immunity that has been mentioned.

Pre-erythrocytic Vaccines

Pre-erythrocytic malaria vaccine development has been driven by the observation that solid immunity in humans can be achieved following immunization via mosquito bites with large numbers of radiation-attenuated *P. falciparum* sporozoites. Until recently, all attempts at reproducing this generation of immunity using immunization of volunteers with purified proteins or peptides had been relatively unsuccessful. There is now a promising vaccine candidate, RTS,S, that consists of sporozoite coat proteins expressed in the hepatitis B surface coat and formulated in a novel adjuvant, that has demonstrated protection in 6 of 7 individuals against challenge with five *P. falciparum*-infected mosquitoes. Additional pre-erythrocytic candidate antigens can also be combined with the basic vaccine to form a multicomponent vaccine, with the idea of overcoming a potential immune-selection of antigenic variants of the malaria parasite.

Second Generation and Naked DNA Vaccines

Most of the above-mentioned candidate malaria vaccines share the common feature of being directed toward a limited set of antigens and, in some instance, toward certain epitopes. However, many of these epitopes show considerable antigenic diversity between different isolates, suggesting the need for strategies aimed at circumventing this strain variation. One approach is the development of multistage–multicomponent vaccines, aimed at covering all the possible variants. This may be possible for certain antigens which appear to exhibit limited variation, but probably will not be feasible for antigens showing multiple variation and many gene copies, unless it can be shown that pathogenic strains constitute only a small proportion of all possible variants. Another approach is the elicitation of immune responses to nonvariant (sometimes nonimmunogenic) portions of the candidate antigens. This can be achieved through creation of hapten–carrier conjugates carrying the specific epitopes in question or through amino acid substitution of such epitopes. Such "second generation" vaccines hold considerable promise for inducing effective immunity against the malaria parasite.

As more and more genes coding for malaria antigens and proteins (including the above-mentioned variant forms) are identified and elucidated, the acquisition rate of malaria sequences into the database has followed an exponential curve. Based on genetic engineering technology, encouraging preliminary results have been obtained using a promising new method for vaccination against malaria, which involves the use of a DNA plasmid encoding various malaria genes, the so-called "naked DNA" vaccine strategy. This multicomponent approach would appear to have several advantages, including stimulating T cell immunity and providing a cost-effective means of presenting several malaria antigens from different stages of the life cycle (and, possibly, cytokines/adjuvants) by a single injection. In addition, the method is ideally suited for use in screening for promising candidate antigens, as studied in various animal models.

III. THE MOSQUITO VECTOR

A. Natural History

1. Life Cycle

Malaria transmission is connected to the bite of mosquitoes of the genus *Anopheles* and, for this reason, the malaria environment has historically been associated with marshes or stagnant waters. Mosquito eggs are deposited in water and, after two to three days, they hatch, liberating an aquatic larva (Fig. 4). The oviposition site is characteristic for the different mosquito species: from animal hoofprints filled with rainwater, small pools, and manmade water collections, to ponds, lakes, streams, and swamps. Mosquito species also vary in their preference for

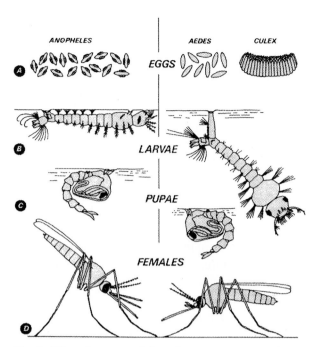

Fig. 4. Life cycle stages of three important genera of mosquito [according to Marshall, J. F. (1938). Brit. Mus. Nat. His., London, UK, republished in 1966 by Johnson Reprint Co., New York]. (A) Shape of eggs, which are laid on the water's edge (Aedes, Culex) or on the water itself (Anopheles); (B) respiring larvae at the water surface; (C) respiring pupae; (D) sitting females: note the characteristic 45° angle of the resting Anopheline female.

shaded or unshaded, fresh or brackish, and still or slow-moving waters. During the 7 to 10 days of their aquatic life, mosquitoes go through four larval developmental stages and a pupal stage. Absence of an air tube permits the larva to lie against the surface, often in the line of intersection formed by air, water, and emergent vegetation, and this sequestration appears to protect against predators. The larva feeds mainly upon material contained in the surface film. Pollen may be an important nutrient. The pupal stage lasts about one to two days, during which the mosquito does not feed. Males emerge first and they start hovering over the breeding site in search of a newly emerging female to copulate with. Mating takes place in flight and only once in the life of the female mosquito, after which the male sperm is placed in the female's spermatothece and is used to fertilize the eggs that the female later deposits. Males and females recognize each other through a series of signals, among which is the frequency of the wing beat. Both males and females require frequent intake of nectar, honeydew, or similar carbohydrate nutrients. However, female mosquitoes also need a blood meal that provides them with the necessary protein contribution for egg maturation and, perhaps, hibernation. Though most mosquito species are zoophilic, feeding preferentially on animals, some species are anthropophilic, having the blood meal from humans. After the meal, the mosquito enters a phase of relative inactivity that permits the digestion of the blood meal and the maturation of the eggs. Within two to three days, the female mosquito is ready to deposit 100 to 300 eggs and to start looking for a new blood meal source. The cycle from the mosquito bite, the maturation of the eggs and the oviposition takes from 2 to 5 days, depending on the mosquito species and the environmental conditions, such as temperature and humidity.

2. Ecology

Although Anopheles mosquitoes are most frequent in tropical or subtropical regions, they are found in temperate climates and even in the Arctic during the summer. As a rule, Anopheles are not found at altitudes above 2000–2500 m. Not all anopheline mosquitoes are malaria vectors. From over 450 species known in the world, only 50 are able to transmit malaria parasites, and of those, less than 30 are considered efficient malaria vectors. Among the main malaria vectors in different geographical areas, one can mention *A. freeborni* and *quadrimaculatus* in North America, *A. albimanus* in Central and South America, *A. atroparvus* in the North Eurasian zone, *A. labranchiae* and *sacharovi* in the Mediterranean basin, *A. gambiae* and *funestus* in the African continent, and *A. culicifacies, stephensi,* and *dirus* in the East Asian zones. Worth noting also is a large area of the Pacific Ocean, roughly bounded by New Zealand, the Galapagos islands, Hawaii, Fiji and New Caledonia, that is free from Anopheles mosquitoes and where there is no indigenous malaria.

Anopheline mosquitoes, in general, are adapted to relatively more permanent aquatic habitat than, for example, Aedes mosquitoes. The water must remain for several weeks or more, the surface must be free of mold, and particular vegetation may be required.

In Africa, for example, *A. gambiae* exploits open sunlit pools, such as the brick pits used for making mud-bricks for home construction. Because mud-brick buildings require constant repair, the brick pits are located close to home, and villages tend to develop where the water table is high enough to flood the pits. Maximum breeding densities are reached soon after the onset of seasonal rains. A companion African vector, *A. funestus,* becomes most abundant with onset of the dry season. Sunlit pools containing emergent marginal vegetation provide the main breeding sites. Thus, clearing of forests near developing villages renders human populations vulnerable to risk of perennial malaria, even where rainfall is seasonal. Combined transmission by *A. gambiae* and *A. funestus* results in seasonally constant transmission patterns.

Similar constant transmission patterns due to a combination of vectors characterize the Indian subcontinent. In that region, *A. culicifacies* exploits irrigated fields and the ditches that drain them. The rainy season is the period of maximum productivity of that vector, while drying conditions promote maximum populations of *A. stephensi,* a mosquito that breeds in water puddled in the beds of streams. Similar perennial transmission can occur where multiple vectors are present.

Such ecological generalities, however, cannot automatically be applied across broad regions. For example, Sri Lankan *A. culicifacies* breeds in water-filled hollows left in the sandy beds of receding rivers. Thus, in contrast to the situation in nearby India, *A. culicifacies*-transmitted malaria in Sri Lanka follows drought. This vector increases as a kind of mirror image of irrigation resulting from increased pudding when river water is diverted to irrigated fields. Neither the fields themselves nor the ditches that drain them support *A. culicifacies* in Sri Lanka.

Clearly, understanding these complexities of vector ecology is very important for malaria epidemiology and control. One can easily imagine that considering issues like endophily vs exophily or anthropophily vs zoophily could have an extremely significant impact on the choice of a control program. Yet another discovery that was the result of a detailed vector analysis proved to be of paramount importance in malaria control. That was the discovery of morphologically identical mosquito species complexes. A characteristic example is that of the *A. gambiae* complex of the African malaria vectors (Coluzzi, 1992). The *A. gambiae* complex contains six members, two brackish-water species present on opposite coasts (*A. melas* and *A. merus,* on the west and the east African coasts, respectively), three freshwater species (*A. gambiae* sensu strictu, *A. arabiensis,* and *A. quadriannulatus*), and a species that breeds in mineral-rich water in a restricted location in Uganda (*A. bwambae*). Two of the freshwater forms, *A. gambiae* and *A. arabiensis,* are strongly anthropophilic and are the most effective vectors of malaria. *A. quandriannulatus,* on the other hand, is mainly zoophilic and has insignificant impact on human malaria transmission. Therefore, in various areas of southeastern Africa where *arabiensis* and *quadriannulatus* share the same habitat (zones of sympatry), it would be greatly misleading for a malariologist if he or she were not able to distinguish between these two species, a vector one and a nonvector one. It would be less misleading, in fact, pooling *A. arabiensis* and *A. funestus,* the other major African malaria vector, two species that both show a high vectorial capacity resulting from convergence in biting and resting behavior. The *A. gambiae* sibling species cannot be distinguished by conventional morphological means. Hybrids are nonfertile, a fact that provided the original clue pointing towards a species complex. The component populations are distinguished by banding characteristics of their polytene chromosomes. More recently, PCR-based molecular diagnosis has proved useful in distinguishing members of the complex, even in field conditions. Such cryptic species, which represent genetically discrete interbreeding units, seem to be the rule among mosquito populations. Examples multiply with increasing knowledge of vector biology.

3. Vectorial Capacity

What makes a mosquito an efficient vector? The first indispensable condition that would constitute an efficient vector is its susceptibility to infection. Some anopheline species do not transmit malaria simply because they are refractory to Plasmodium. In these cases, the malaria parasites, although ingested by the mosquito with the blood meal, either die or are incapable of completing their sporogonic cycle and reaching the salivary glands. The phenomenon is not clearly understood but it is certainly related to biochemical processes in the body of the mosquito (see following). Other obvious factors that determine whether a particular species of Anopheles is an important vector are the frequency of its feeding on humans (anthropophily), the mean longevity, and its density in relation to humans. Vectorial capacity synthesizes those variables that affect the ability of a vector to transmit disease. Patterns of transmission of vector-borne disease depend on intensity of vector–host contact. Intuitive and numerical assessment of this contact requires an understanding of vector–parasite–host ecology and methods for estimating relevant parameters. To transmit a parasite, a vector population must be competent to deliver an infectious dose as well as be in an ecological situation favoring host contact.

In the 1950s, Macdonald developed a predictive model for malaria incidence. This work stimulated field studies, which eventually proved its usefulness and practicality. Macdonald's model, as it is often called, represents the basis of epidemiological entomology (Macdonald, 1957). Based on a few parameters, the model calculates the number of secondary cases a certain mosquito population will generate from a single case introduced into a region. This convention represents the basic reproduction rate. If the equation is simplified to consider the number of potentially infective bites, only entomological parameters need be considered, thus greatly simplifying field work. This simplified equation was labeled vectorial capacity:

$$VC = \frac{ma^2 p^n}{-\ln p}$$

In epidemiological terms, this equation may be translated as follows: when a person is bitten ma times per day (ma is the man-biting rate, where m represents the relative density of female mosquitoes and a is the man-biting habit) and a proportion p^n of the vector population survives the incubation period of the malaria parasite in the mosquito, and if this proportion is expected to live for another $(1/-\ln p)$ days, during which they bite another person a times a day, then we may estimate the rate of potentially infective bites. This is the measure of transmission which determines the endemic level

of malaria in given conditions. Values of vectorial capacity of over 2.5 indicate stability, values below 0.5 indicate unstable conditions of transmission. Probably the most important contribution of Macdonald's equation is that it permits us to rank the various components of vectorial capacity in a hierarchy important to epidemiological studies and control operations. Most powerful is the vector's longevity, which contributes exponentially (p is the probability of survival through one day and p^n the probability of vector's survival through sporogonic period of parasite). The duration of the sporogonic cycle is about 9 days for *P. vivax* and 12 days for *P. falciparum* at 26°C. Obviously, any mosquito that lives less than this period will not be able to transmit the infection. The next most important factor is the vector's narrowness of host range (a) which contributes as the square; and the vector abundance (m), which provides a linear contribution. It becomes clear that vector larvicides or other measures will reduce m, screening of houses will reduce a, while residual spraying will also greatly reduce the factor p.

A demonstration of the usefulness of the above analysis comes from the interruption of malaria transmission from European countries. Malaria eradication from the southern European countries, achieved in the early 1950s, and its maintenance ever since is generally attributed to effective vector control by indoor insecticide spraying. However, the success of the eradication campaign that was achieved in Europe at the time was a combination of an ecoepidemiological situation that was marginal for malaria (particularly for *P. falciparum* transmission), an infrastructure that was well in place and functioning for years, and the use of DDT that was able to lower vectorial capacity below its critical value. First, malaria transmission in southern Europe was characterized by marked seasonal transmission and by vectorial capacity that was not much greater than the critical value, mainly due to the moderate temperatures of the Mediterranean. Second, malaria control activities were in place for some time that had already resulted in reduction of transmission by the use of larviciding, elimination of vector-breeding sites, improved housing and health care, more general use of antimalarial drugs, house mosquito proofing and personal protection, zooprophylaxis, and the introduction of larvivorous fish. Finally, it was the arrival of DDT that gave the last essential reduction to the already diminished vectorial capacity. DDT has a strong irritant effect on the mosquito vector, mosquitoes would either avoid treated surfaces or would die after coming into contact with them. Thus, indoor spraying would decrease vector–human contact, decrease vector longevity, and increase length or interrupt the sporogonic cycle. However, all this was possible because malaria transmission in these countries was largely house-dependent. The same efficacy would not be expected in tropical climates, where epidemiologically stable malaria can be transmitted outdoors and where higher temperatures maintain a higher mosquito longevity and a faster sporogonic cycle.

B. Vector Control

1. Insecticides

The greatest part of vector control efforts has, until now, been focused on the use of insecticides. The first insecticides used in the 1930s were based on the extracts of pyrethrum flowers. Residual spraying with pyrethrum would reduce malaria transmission, even in certain tropical setups. Pyrethrum efficiency was, however, limited since large quantities were required in order to obtain killing of the mosquito vectors, while its effect would be very short-lasting. With the start of World War II, the Americans placed a great emphasis on synthetic insecticides that could be used in military activities, a decision that was guided by the fact that pyrethrum production was mainly in the hands of the Japanese. None imagined that the result of these researches would yield a product that already existed in the market as a Suiss insecticide against plant parasites and lice. The product that is famous by now by the name of DDT (dichlorodiphenyltrichloroethane) was chemically synthesized first in 1874 and its insecticide activity against both the adult and the larval forms of several insects was well documented since 1939. However, it was not until 1944 that became the main component of the malaria eradication campaign launched in Europe by WHO. Indoor DDT spraying in the 1950s and 1960s greatly contributed to eradicating malaria from Europe and in greatly reducing malaria

prevalence in almost all parts of Asia. DDT was insect-specific, had a lasting residual action, and, very importantly, had a low cost, making it affordable for the less-developed countries. However, its extensive and often exaggerated use led to the development of insecticide resistance all over the world only a few years later. But it was not until the publication of Silent Spring (Carson, 1962) that its toxic effect on domestic animals and wildlife, such as raptorial birds, made the authorities reconsider the benefits from its use and, finally, discontinue its application in most countries.

Generally, four principal groups of pesticides have been used for malaria control: pyrethrum and the synthetic pyrethroids, chlorinated hydrocarbons (e.g., DDT and dieldrin), organophosphates (e.g., malathion and fenitrothion), and carbamates (e.g., propoxur). These compounds may be used as aerosols (space spraying) and/or may be applied to walls and other surfaces where mosquitoes rest (residual spraying). Space spraying is most appropriate in densely populated areas and is usually reserved for situations in which a rapid reduction of the mosquito population is desired in a limited geographic area. In large malaria control programs, residual spraying is a more effective and economical approach.

Pyrethrum is used primarily as a space spray because of its rapid knockdown effect and low toxicity to humans. It has no residual action. In contrast, the chlorinated hydrocarbons, organophosphates, and carbamates, as well as several of the newer synthetic pyrethroids, retain their toxicity for mosquitoes for up to 6 months on sprayed surfaces. When a mosquito rests on a surface sprayed with one of these compounds, it picks up minute amounts of the insecticide on its body. As a result of this exposure, the mosquito's life span is shortened and both the chance of disease transmission and the mosquito's reproductive capacity are reduced. Some pesticides, such as the carbamates, also have a fumigant action that kills mosquitoes even without direct contact.

Vector resistance to insecticide is a major problem in malaria control programs. An estimated $5 billion are spent each year on insecticides, 90% of which are used in agriculture and 10% in public health. Resistance to insecticides emerges because of selection pressure by an insecticide on a population of mosquitoes. The result of this pressure is that mosquitoes that are genetically more resistant survive a normally lethal dose of the insecticide. If exposure to the pesticide continues, the proportion of resistant mosquitoes will continue to increase. Some mosquitoes may also avoid contact with sprayed surfaces (behavioral resistance). In such cases, malaria transmission may continue, even though sensitivity testing shows that the mosquito is susceptible to the insecticide. By 1985, at least 117 mosquito species had been reported as resistant to one or more insecticides, with 67 of these in the genus Anopheles. Not only are the most widely used insecticides (DDT, dieldrin) now exhibiting reduced effectiveness, but resistance to replacement insecticides (organophosphates and carbamates) is also developing in some areas. Currently, the areas with the most serious insecticide resistance problems are parts of India, where *A. stephensi* and *A. culicifacies* are resistant to the chlorinated hydrocarbons and organophosphate compounds, and Central America and Turkey, where *A. albimanus* and *A. sacharovi*, respectively, have also developed resistance to the carbamates. Relief from this situation is not forthcoming since manufacturers are generally reluctant to invest money in the development of new insecticides for public health and because the cost of the second-line insecticides is prohibitive for many developing countries.

2. Insecticide-Treated Bednets and Curtains

Bed nets are widely used all over the world against nuisance mosquitoes. Since most Anopheles species bite at night, it has long been assumed that nets must be useful against malaria. The first mention of bed nets is probably by Herodotus in the fifth century B.C., who described its use by the Egyptian fishermen (who were using them to catch fish during the day and protect them from "gnat" bites at night). However, the impregnation of the nets in pesticide is a more recent practice that not only protects the individual from the mosquito bites but also has a very important mass effect. By placing the pesticide on a surface which mosquitoes are bound to encounter in their attempts to bite sleeping people, one targets more precisely the malaria vector population and increases the exposure of the mosquito, thus making its killing more efficient. Even mosquitoes

that have managed to enter the net and have bitten a sleeping individual are likely to rest on the net at some point and, thus, take a lethal dose of the insecticide. Results from the use of insecticide-treated bed nets are very encouraging. In a pilot study in the Gambia in 1990–1991, there was documented a 63% reduction in mortality in children under 5 sleeping under insecticide-treated bed nets, compared with a control group. This stimulated the Gambian government to initiate a countrywide program of net-impregnation in villages with a primary health care clinic. Since then, larger-scale studies and interventions carried out in several African countries have confirmed the important impact on all-cause child mortality in differing epidemiological situations, with observed reductions of between 16 and 33% in children under 5 years of age. These results suggest that approximately 500,000 African children might be saved each year from malaria-related mortality if the nets, treated with a biodegradable pyrethroid insecticide, were widely and properly used.

3. Larval and Biological Control

Malaria control measures directed against the aquatic stages of anopheline mosquitoes have been in use since the beginning of the century. Antilarval measures can be either chemical or biological. Chemical larvicides range from petroleum oils and Paris green (copper acetoarsenite) to pesticides, such as fenthion and temephos, which are also active against the adult mosquito. Biologic methods of larval control include the use of larvivorous fish or insect pathogens. The North American fish *Gambusia affinis* was successfully used to reduce malaria incidence in Italy and Greece, where malaria transmission was unstable. Introduction of *G. affinis* to 3800 wells in India reduced the number of wells containing anopheline larvae by about 75%. Environmental problems resulting from competition of *G. affinis* with indigenous fish have led to investigations of many other fish species as potential control agents. The bacterial endospore toxins produced by various strains of *Bacillus sphaericus* and *Bacillus thuringiensis israelensis* have been used as larvicidal agents in several situations. Finally, pathogens and parasites, including viruses, fungi, nematodes, and protozoa, have occasionally been used for malaria control.

4. Environmental Management

Since the flight range of most anopheline mosquitoes is thought to be 1–3 km, source reduction, i.e., the elimination of mosquito breeding sites close to areas of human habitation, may significantly reduce mosquito density and the risk of infection. Environmental management includes filling or draining of collection of water, strengthening and clearing of streams and rivers to eliminate vegetation and pockets of standing water, and intermittent drying of irrigated fields. Breeding sites may also be rendered unsuitable to anopheline mosquitoes by changes in the amount of shade, the degree of water salinity or the water level in impounded waters. This kind of source reduction has been used successfully in Malaysia, Europe, the U.S.S.R., Cuba, and the southern United States. Although the initial cost of these projects may be high, in the long run, they are among the most cost-effective forms of malaria control since they may result in permanent elimination of mosquito breeding sites. Source reduction is particularly cost-effective in densely populated areas, such as the outskirts of large cities, where a limited number of moderate to large breeding places can be identified. This technique is not feasible with species of Anopheles that breed in small or occasional collections of water that appear and disappear throughout the rainy season, as is in many parts of sub-Saharan Africa.

Agricultural policies or other economic activities that change land use, such as the creation of dams, irrigation schemes, and commercial tree cropping and deforestation, can have a tremendous impact on the transmission of vector-borne diseases. Land-use change is invariably financed following appraisal of the economic benefit and cost of the development activity. Health outcomes are often not taken into account in this appraisal, due to the lack of convincing quantitative information on how much the risk of disease is uniquely attributable to the land-use change and the difficulties of valuing health benefits and costs in financial terms. In Southeast Asia, for example, the deforestation of the natural habitat of the malaria vector *Anopheles dirus* has recently forced the mosquito to adapt to new commercial tree plantations (orchards, rubber, or teak plantations). Several practices have increased the incidence of malaria on those plantations, as, for example, the discarding of coconut shells, used for the collection of latex on rubber plantations, which has provided good breeding sites for *A. dirus,* or the harvesting in orchards at a time that coincides with the peak transmission season for malaria.

5. Genetic Manipulation of Vector Populations

Several recent developments in the genetic and molecular analysis of the malaria vector have made scientists consider that their appropriate application could be used as a control strategy for malaria or other vector-borne diseases. Up until now, genetic control programs for insects have been based on the reduction or elimination of the population by the introduction of sterility factors, usually by release of sterilized males. Genetic control programs for insects have been quite successful in at least two cases, that of the screw worm and the Mediterranean fruit fly. In both cases, the target insect had a relatively low intrinsic rate of increase and the high program cost was nonetheless lower than the tangible economic benefit. Initial experiments with the mass release of *Aedes aegypti* males were unsuccessful, probably because of the use of crude and debilitating chemo- and radio-sterilants that impaired reproductive competitiveness in the released males. Small genetic control projects targeting Anopheles mosquitoes did not meet any particular success, mainly due to problems associated with the very high intrinsic rate of increase of the mosquitoes and the high cost of the intervention.

Current genetic control strategies are based not on elimination or reduction of the vector population, but on modification of the capacity of the natural vector to support parasite development. The idea is to introduce a genetic construct into the existing vector population that carries one or more parasite-inhibiting genes, without affecting the vector population's fitness. Such a control strategy would target not the mosquito itself, but, rather, the parasite within the mosquito.

This approach faces at least four problems. One is the mechanism that could stably and functionally introduce the genes of interest into the mosquito vector. The second is the availability of appropriate genes and the understanding of their mode of action. The third regards issues of successful release of these transgenic insects in nature, their survival, maintenance, and eventual equilibrium with the natural population. And the fourth regards ethical issues associated with these releases.

First, techniques for introducing DNA into other diptera (e.g., drosophila or medfly) are already well established and are mainly based on the use of transposable elements (*P* or *Minos*). Development of such tools for mosquitoes is under way, with a number of candidate systems under investigation but none, as yet, established. Second, the search for appropriate antiparasite genes is taking several different routes. Mosquito antiparasite genes could, for example, be involved in the recognition of receptors in the salivary glands by the sporozoites. In this case, blocking of these surface receptors in the mosquito would render the plasmodium incapable of completing its sporogonic cycle. Other antiparasite genes could be those involved in the mosquito's own defense system. For example, insects typically use to encapsulate and melanize invading parasites and parasitoids by a phenoloxidase-mediated pathway. In fact, a laboratory strain of *A. gambiae* has been selected that can encapsulate ookinetes when they reach the basement membrane (Collins *et al.*, 1986). These mosquitoes lay down a melaninlike deposit around the parasite, parasite development is aborted, and the encapsulated oocysts do not give rise to sporozoites. These selected mosquitoes show different degrees of refractoriness to several different strains of Plasmodium parasites. Third, successful maintenance of any gene or construct after the release of a transgenic mosquito in nature needs a drive mechanism. Two types of drive mechanisms have been suggested. One is meiotic drive, where a given chromosome is transmitted to more than the expected 50% of offspring. Any desirable genes linked to the driven chromosome would eventually approach fixation, even with the release of relatively few individuals. Meiotic drive occurs during hybrid dysgenesis in Drosophila. The second type of drive mechanism is the exploitation of genetic traits that reduce heterozygote fitness. For example, the gene to be driven could be introduced into a translocation chromosome, such that homozygotes formed are viable and fertile, whereas heterozygotes display reduced fertility of viability. Efficiency of the drive mechanism could be improved by providing the released individuals with some form of temporary advantage. For example, insecticide resistance

could be incorporated into the genome and then insecticide applied. Ideally, the insecticide resistance gene would be fused to the desirable gene and introduced as a unit to prevent disruption of useful combinations by meiotic recombination.

Less emphasis is often given to ethical issues that would accompany any effort to establish transgenic malaria-incompetent mosquitoes in a populated site. Such mosquitoes would be peridomestic and anthropophilic. They would, therefore, be vectors of other anthroponoses, such as lymphatic filariasis and certain arboviral infections. These mosquitoes would certainly be pests in numbers that, at least initially, would be much higher than those of the local mosquito populations. Experiences such as the violent reaction of the local Indian residents to a mass release of laboratory-reared *Culex pipiens* should not be underestimated.

IV. THE EVOLUTION OF THE MALARIA–MOSQUITO– HUMAN RELATIONSHIP

The disease and its distribution has always been influenced by the specific relationships among the major actors, humans, mosquitoes, and plasmodia. Due to the aquatic dependence of the mosquito, the disease has followed the evolution of the human race from its origins to recent times. The evolutionary history of the Plasmodium probably precedes that of humans by several millions of years. Plasmodia probably evolved from intestinal parasites of vertebrate animals that were propagated through cyst production, excreted by the feces, and capable of resisting the external environment. These were then adapted to live in internal organs (e.g., liver) and then the blood, an adaptation that required a blood-feeding vector that would transmit and propagate the disease. About 200 species of Plasmodia are known that infect humans and other primates, rodents, birds, and reptiles. The four human Plasmodia are probably the result of distinct evolutionary routes, as is confirmed by recent molecular studies. *P. vivax*, *P. malariae* and *P. ovale*, responsible for the benign forms of malaria, have accompanied our species throughout human evolution, although there is also the hypothesis that these species were differentiated in Asia about a million years ago with the appearance of *Homo erectus*. *P. falciparum*, on the other hand, the most malignant of all human plasmodia, has been considered the least adapted to humans, probably acquired by a recent host switch, perhaps from (domestic) birds as recently as the onset of agriculture some 5000 to 10,000 years ago (Waters *et al.,* 1991). Other studies, however, place *P. falciparum* phylogenetically closest to the chimp plasmodium *P. reichenowi* and indicate that it may be the oldest of the human plasmodia (Ayala *et al.,* 1998). In that case, a plausible hypothesis is that a relatively benign form of *P. falciparum* was left for millions of years in some restricted geographic area of the African continent and, during more recent environmental transformations, a more malignant form was selected and diffused. These environmental changes were probably connected to the development of agriculture about 10,000 years ago, during which human populations increased due to increased food resources and agricultural activities changed the environment, favoring the diffusion of mosquitoes. In fact, the evolution of certain characters of the *A. gambiae* and *A. funestus* African vector complexes is probably associated with agricultural activities of about that time. Particularly, the anthropophily, the preference for indoor, climatically protected resting sites, and the choice of manmade larval breeding sites in irrigated zones were all results of a close association with humans from the larval to the adult stage. Given this human–mosquito relationship and combined with the higher density of mosquitoes and humans, *P. falciparum* transmission was made possible, despite its high virulence for humans.

See Also the Following Articles

GLOBAL BURDEN OF INFECTIOUS DISEASES • INSECTICIDES, MICROBIAL • PESTICIDE BIODEGRADATION • PLASMODIUM

Bibliography

Ayala, F. J., Escalante, A. A., Lal, A. A., and Rich, S. M. (1998). Evolutionary relationships of human malaria parasites. *In* "Malaria: Parasite Biology, Pathogenesis and Protection" (I. W. Sherman, ed.), pp. 285–300. ASM Press, Washington, DC.

Bruce-Chwatt, L. J. (1988). History of malaria from prehistory to eradication. *In* "Malaria: Principles and Practice of Malariology (W. H. Wernsdorfer and I. McGregor, eds.), pp. 1–59. Churchill Livingstone.

Carson, R. (1962). "Silent Spring." Houghton Miffin Company, Boston, MA.

Collins, F. H., Sakai, R. K., Vernick, K. D., Paskowtiz, S., Seeley, D. C., Miller, L. H., Collins, W. E., Campbell, C., and Gwadz, R. W. (1986). Genetic selection of a *Plasmodium*-refractory strain of the malaria vector *Anopheles gambiae*. *Science* **234**, 607–610.

Coluzzi, M. (1992). Malaria vector analysis and control. *Parasitology Today* **8**, 113–118.

Desowitz, R. S. (1991). "The Malaria Capers." Norton, New York.

Hadley, T. J., Klotz, F. W., and Miller, L. H. (1986). Invasion of erythrocytes by malaria parasites: A cellular and molecular overview. *Annu. Rev. Microbiol.* **40**, 451–477.

Luzatto, L., and Bienzle, U. (1979). The malaria/G.-6-P.D. hypothesis. *Lancet* **2**, 1183–1184.

Macdonald, G. (1957). "The Epidemiology and Control of Malaria." Oxford Univ. Press.

Miller, L. H., Mason, S. J., Clyde, D. F., and McGinniss, M. H. (1976). The resistance factor to *Plasmodium vivax* in blacks: The Duffy-blood group genotype, FyFy. *N. Engl. J. Med.* **295**, 302–304.

Pasvol, G., Weatherall, D. J., Wilson R. J. M. (1978). Cellular mechanisms for the protective effect of haemoglobin S against *P. falciparum* malaria. *Nature* **274**, 701–703.

Patarroyo, M. E., Amador, R., Clavijo, P., Moreno, A., Guzman, F., Romero, P., Tascon, R., Franco, A., Murillo, L. A., Ponton, G. *et al.* (1988). A synthetic vaccine protects humans against challenge with asexual blood stages of *Plasmodium falciparum* malaria. *Nature* **332**, 158–161.

Waters, A. P., Higgins, D. G., and McCutchan, T. F. (1991). *Plasmodium falciparum* appears to have arisen as a result of lateral transfer between avian and human hosts. *Proc. Natl. Acad. Sci. USA* **88**, 3140–3144.

Mapping Bacterial Genomes

J. Guespin-Michel
Université de Rouen, France

F. Joset
LCB, CNRS, France

GLOSSARY

bacterial artificial chromosomes (BACs) Vectors based on plasmid F (from *E. coli*) origin of replication. Such vectors, which can carry 24 to 100 kbp, are stable in *E. coli*.

contigs Adjacent or partially overlapping clones.

fine mapping Ordering intragenic mutations or very tightly linked genes.

fingerprinting assembly Ordering a restriction map via comparison of the restriction patterns of overlapping clones.

gene encyclopedia (or ordered gene library) Library of cloned DNA fragments, ordered in a sequence reconstructing the gene order on the corresponding chromosome.

genetic map Linkage (or chromosomal) map showing ordered genes, independently of the method used for ordering.

linking clones Clones which, when used as probes, hybridize with two adjacent macrorestriction fragments or two clones from a library.

polarized chromosomal transfer Sequential transfer of (part of) a bacterial chromosome from a donor to a recipient cell, mediated by a conjugative plasmid.

pulsed field gel electrophoresis (PFGE) Electrophoretic device allowing separation of very long (up to Mb) DNA molecules.

physical (macrorestriction) mapping Positioning landmarks, such as restriction sites, along a bacterial chromosome (or any DNA molecule).

rare-cutter endonucleases (or rare-cutters) Endonucleases (mostly Class II restriction enzymes) that have only a limited number of recognition/cutting sites on a whole chromosome.

yeast artificial chromosomes (YACs) Shuttle yeast–*E. coli* vectors, possessing all requirements (telomeric and centromere regions), allowing their reproduction and segregation in *S. cerevisiae*. Such vectors can carry 75–2000 kbp.

POSSESSING GENOMIC (MOSTLY CHROMO-SOMAL) GENETIC MAPS, i.e., ordered genes on the different elements of a genome, is an important tool for geneticists and, thus, an early aim when studying a new species. The level of achievement of this work has been closely related to the available techniques.

Until the mid-1980s, genome mapping relied on the classical concept of recombination linkage and, thus, could be achieved only in strains for which suitable natural *in vivo* DNA transfer processes were available. Possible drawbacks due to heterogeneity of recombination frequencies could not be avoided. More or less extensive chromosomal maps had thus been constructed for about 15 species. A major breakthrough has been the possibility to construct so-called physical maps, i.e., to position landmarks such as restriction sites (restriction, or physical, maps) along the DNA molecules. The first, still incomplete, physical map was published in 1987. Localization of genes along this physical map, i.e., its transformation into a genetic map, can be achieved, more or less

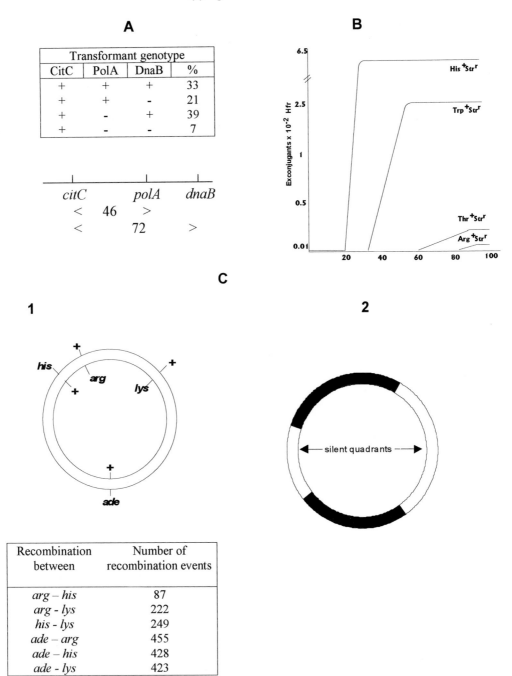

Fig. 1. Survey of the various approaches for *in vivo* chromosome mapping in bacteria. (A) Fine mapping in *Bacillus subtilis* (the so-called 3-point test). Three loci were differently marked in a donor (*citC⁺*, *dnaB⁺*, and *polA⁻*) and a recipient (*citC⁻*, *dnaB⁻*, and *polA⁺*) strain; single *citC⁺* transformants were selected and their genotype determined for the other two characters (panel I); the least frequent class (7%) must correspond to a requirement for 4 crossovers, yielding the map shown, with relative distances calculated as frequencies of recombination (panel 2). (B) Chromosomal mapping by conjugation in *E. coli*. A fully prototrophic, streptomycin resistant (Str^r) Hfr strain was mated with a polyauxotrophic, Str^s F⁻ one; mating was interrupted as a function of time, and the proportion of each possible type of exconjugants determined; extrapolation towards the abscissa

precisely, by several methods. Extant chromosomal maps obtained by *in vivo* approaches (the case of *E. coli* is particularly enlightening) or molecular techniques (gene identification through partial sequencing or via hybridization with, hopefully, conserved genes from gene banks) greatly facilitate the work. Thus, each new map construction is further facilitated by comparison with all maps or genetic information available. More than 100 such maps, which widely differ in the number of genes positioned and the precision of their localization, are presently available, and, in principle, there is no cultivable strain which cannot thus be mapped. Therefore, even though the traditional genetic methods are still fully valuable for strain constructions or gene function analysis, they should no longer be useful for mapping purposes. Whole genome sequencing has been the next advance, with the first one published in 1995. Sixteen fully sequenced genomes from Bacteria and Archaea are now published, and nearly 50 more are in progress. This, however, does not render obsolete physical and genetic maps, since genome sequencing will not, for some time yet, be performed on as many species and strains as the former, which also quite often constitute useful requisites for sequencing.

I. *IN VIVO* GENETIC MAPPING OF BACTERIAL CHROMOSOMES

As soon as genetic methods became available for bacteria, they have been readily used for gene mapping, mostly applied to chromosomes, as opposed to plasmids. Various portions of chromosomes have thus been defined, depending on the techniques available. Fine mapping, used for ordering intragenic mutations or very close loci, was opposed to broad mapping, allowing the treatment of (almost) whole chromosomes. The latter is sometimes taken as the only case of genetic chromosomal mapping.

Fine mapping is based on the analysis of recombination data, assuming a direct relationship between the distance separating two markers and the recombination frequency between them (Fig. 1A). Although it allows a reliable and precise estimation of the order of the markers, this approach is hampered due to heterogeneity of the recombination frequencies along the DNA molecule. Recombination events are initiated at particular sites (sequences) along DNA molecules, known as chi (χ) sites, which are usually not distributed regularly; thus, construction of a genetic map on recombination frequencies may be biased. Fine mapping is essentially performed via transformation (natural or artificial) or transduction. This explains why it applies only to small portions (usually a few percent) of chromosomes. It is, however, available for most species, provided sufficient effort is devoted to finding the relevant DNA transfer tools.

Chromosome mapping in gram-negative bacteria deduces gene order from the relative time required for their transfer during a sequential (polarized or oriented) transfer of the (whole) chromosome, after so-called interrupted mating experiments (Fig. 1B). Transfer occurs between two (donor and receptor)

yields the minimal time required for transfer of each character and, thus, their relative distances (in minutes) assuming the rate of transfer is constant along the whole chromosome. (C) Mapping in *Streptomyces*. (1) Four-marker mapping in *Streptomyces coelicolor*. Doubly marked strains (parent 1: *his ade;* parent 2: *arg lys*) were crossed, recombinants selected and analyzed for their complete genotype; the recombination frequencies between pairs of markers allowed the gene arrangement to be deduced; (2) The 1992 version of the map, obtained by similar methods, carried about 130 loci (short lines outside the circle), 5 accessory elements (insertion sequences, phage insertion sites), and the two almost completely "silent" quadrants. The actual linear structure of the chromosome was later established. A circular linkage map had been arrived at because, in merodiploïds as used for genetic mapping, double crossovers are required between linear chromosomes to generate recombinants harboring complete chromosomes. From this situation ensues the presence of both ends of the chromosome from the same parent in the recombinant molecule, which leads to apparent linkage between these ends and, thus, to the erronous interpretation of a circular structure. [A and B: Adapted from Joset, F., and Guespin-Michel, J. (1993). *In* "Prokaryotic Genetics: Genome Organization, Transfer and Plasticity." Blackwell Science, Oxford. C: From Smokvina *et al.* (1988). *J. Gen. Microbio.* **134**, 395–402. D: Adapted from Kieser, H. M., Kieser, T., and Hopwood, D. A. (1992). *J. Bacteriol.* **174**, 5496–5507.]

suitably marked bacteria, the process being controlled by conjugative plasmids integrated in the chromosome and defining the polarized transfer direction. Recombination of the transferred material (i.e., marker exchange) is necessary for its stabilization, a prerequisite for its detection. However, the recombination frequencies are high enough to insure a high rate of integration, and, thus, this requirement does not bias the overall outcome. Conjugation was first discovered in *E. coli,* in which plasmid F performs the transfer (see Section VI). Plasmid F can function in only a few related gram-negative bacteria. So chromosomal mapping could be performed in other species, either via endogenous conjugative plasmids, when existing (e.g., in *Pseudomonas aeruginosa*) or via broad host-range conjugative plasmids (often of the *incP* group), engineered so as to be able to integrate more or less randomly into the desired host chromosome.

In gram-positive bacteria, the situation depends on the species. No system for polarized chromosomal transfer is available in species with low GC content, such as *Bacillus* or *Streptococcus*. Partial *in vivo* chromosomal mapping has been performed in *B. subtilis,* thanks to a very large transducing phage (genome size ≈ 2% of the host chromosome). In *Streptomyces,* conjugative plasmids are abundant, but there is no evidence for progressive chromosome transfer, and no kinetic mapping is possible. Therefore, final recombination frequencies are used for *in vivo* mapping. Analyses of the progeny of various four-factor crosses between doubly auxotrophic parents were first performed in *S. coelicolor* strain A3(2) (Fig. 1C). A compiled analysis of the resultant partial maps was used to deduce the order of the markers. The resulting circular linkage map consisted of two well-marked regions, separated by two very long "silent" quadrants. Analysis of recombinants issuing from matings between appropriately marked parents showed very limited linkage between the two marked regions, thus defining the unusual length, in terms of crossover units, of these silent quadrants. Their relative lengths were later better estimated by statistical analysis of the heterozygous regions of merodiploids in a population of heteroclones (colonies arising from partially diploid cells). Linkage maps were similarly established for other *Streptomyces* species.

II. PHYSICAL (MACRORESTRICTION) MAPS

Physical mapping can readily be performed for small plasmids by comparing the sizes of fragments obtained after treatment with sets of restriction enzymes, since only a limited number of such fragments are formed. To obtain the same kind of results on a larger scale, e.g., for a chromosome, one must be able to cut the DNA into a reasonable number of pieces (about 20 is optimal). This implies the formation of large fragments (most of them larger than 50 kbp), for which resolution of their size is not possible via classical agarose gel electrophoresis. Two conditions, i.e., the availability of enzymes with very few cutting sites on whole chromosomes and tools allowing size resolution for large fragments, opened the era of chromosomal physical, or macrorestriction, mapping.

A. Rarely Cutting Site-Specific Endonucleases

Physical mapping requires the formation of reproducible fragments and, thus, precludes random breaks due to shearing during extraction procedures, particularly frequent for large DNA molecules, such as bacterial chromosomes (or megaplasmids). Prevention of random breakage is achieved by trapping the cells, before any other treatment, into small plugs of agarose. Cell lysis is performed inside the agarose. Small molecules diffuse out while the DNA remains entrapped. The endonucleases can penetrate into the agarose network and the cleaved DNA fragments are electroeluted from the plug.

Several types of enzymes have only a limited number of recognition/cutting sites on whole chromosomes. Ten or so restriction enzymes (all belonging to Class II) are "rare-cutters" for many genomes, due to their 8-bp recognition sequence (Table I). *Not*I is the most extensively used. But extreme GC contents or reduced representations of particular sets of nucleotides of certain genomes allow adding a few restriction enzymes with six-base recognition sequences to the list of possible rare-cutters (Table I). Works are in progress to modify the recognition specificity of some frequently cutting restriction endonucleases so

TABLE I
Examples of Endonucleases with Rare-cutting Frequency in Bacterial Genomes

Recognition particularity	Enzyme	Recognition sequence
8-nucleotide recognition sequence	NotI	GC/GGCCGC
	Sse83871	CCTGCA/GG
	SwaI	ATTTAAAT
	PacI	TTAATTAA
	PmeI	GTTTAAAC
	SgrAI	CACCGGCG
	SrfI	GCCCGGGC
	SgfI	GCGATCGC
	FseI	GGCCGGCC
	AscI	GGCGCGCC
Hyphenated 8-nucleotide recognition sequence	SfiI	GGCCNNNN/NGGCC
Rare-cutters in G + C rich genomes	AseI	ATT/AAT
	DraI	TTT/AAA
	SspI	AAT/ATT
Rare-cutters in T + A rich genomes	SacII	CCG/CGG
	SmaI	CCC/GGG
	RsrII	CG/GNCCG
	NaeI	GCC/GGC
Overlap on TAG, a rare stop codon in prokaryotes	SpeI	A/CTAGT
	XbaI	T/CTAGA
Overlap on GC/CG or CC/GG sequences, relatively rare in prokaryotes	AvrII	C/CTAGC
	NheI	G/CTAGC

as to increase the size of their recognition site (this has been recently achieved for *Eco*RV). Proteins which participate in intron processing have a recognition sequence of 18 to 26 bp and have often proved valuable as rare-cutters. One of these, largely used for bacterial chromosome mapping, is protein I-*Ceu*I, produced by a chloroplast intron of *Chlamydomonas eugametos*. It recognizes a 26 bp sequence, usually present only in bacterial-type 23s rRNA genes. Several devices have been used to protect part of the cutting sites of frequently cutting enzymes. "Peptide nucleic acid clamps" (bis-PNAs), that bind strongly and sequence-specifically to short homopyrimidine stretches, shield overlapping methylation/restriction sites and, thus, reduce the number of accessible sites for the corresponding enzyme. A strategy called "Achilles' heel cleavage" (AC) consists in introducing into a genome a unique site (such as the phage lambda *cos* site) that can be cleaved only by a specialized enzyme.

B. Pulsed Field Gel Electrophoresis (PFGE)

In 1982–1984, Schwartz and Cantor devised a method allowing the separation of DNA fragments ranging from 35 to 2000 kbp. The molecules were electrophoresed in an agarose gel subjected to electric fields alternately orientated at roughly right angles, hence, the name of the method, "pulsed-field gel electrophoresis," or PFGE. The rationale of the method is that the longer a DNA molecule, the slower it reorientates at each change of direction of the electric field. This, accordingly, slows down its overall migration speed along the average direction of migration. Various types of PFGE may be chosen, depending on the sizes of the fragments to separate. This also allows distinguishing circular plasmids, which migrate in a very special way in PFGE.

The reliability of this method depends on the identification of the complete set of fragments generated by the restriction treatments. Its main drawbacks are that: (i) nonambiguous resolution of the digestion fragments may be hindered if the fragments are too numerous or if two (or more) have very close lengths; (ii) fragments too small to be detected or so small that they would have eluted from the gel before it was examined, may be generated.

C. Construction of Macrorestriction Maps

Several strategies (a few examples will be described) can be used to reconstruct the alignment of the fragments along the chromosome. Only rarely will one strategy be sufficient.

1. The fragments obtained after digestion with one endonuclease are individually digested by a second enzyme, and reciprocally. Comparisons of the sizes of the resulting doubly digested fragments allow locating the different cleavage sites, as for plasmid mapping (Fig. 2A). The method is optimized by 2D-electrophoresis. The main drawbacks, again, are the existence of several fragments with the same length and possible elution of small fragments from the plug during the preparation of the gel. The interest of a physical map with large intervals is limited, so further cutting with other enzymes can be pursued. The method is generally accurate if the total number of secondary fragments does not exceed 20–25, thus hampering simultaneous treatment with two enzymes.

2. Other approaches allow obtaining more precise genomic location of large numbers of fragments, i.e., to link adjacent fragments. A linking clone is a clone which contains a site for a rare-cutting enzyme and overlaps two adjacent macrorestriction fragments. Adjacent segments are called contigs, referring to their contiguous positions on the chromosome. Hybridization of a labeled restriction fragment with a whole DNA library allows the detection of the two contigs flanking the corresponding restriction site. Linking clones can be obtained from a DNA library by selecting clones which contain rare-cutter sites. The latter can be tagged for instance by insertion of a marker. Thus, the *Not*I sites of *Listeria monocytogenes* chromosome were individually labeled with a Km^r cassette (a DNA sequence originating from a

transposon, containing a gene for resistance to an antibiotic; in this case, kanamycin), the Km^r clones from the corresponding *Eco*RI libraries, selected in *E. coli,* represented the linking clones for the *Not*I restriction fragments (Fig. 2B). Known open reading frames (ORFs) can be similarly used as markers (Fig. 2C). Willems *et al.* (1998) have sequenced the 58 *Not*I/*Sau*3A fragments of the *Coxiella burnetii* genome. Checking in databases whether chance partial reading frames present at the *Not*I side of some fragments could correspond to registered ORFs shared by two fragments allowed linking 10 out of the 29 *Not*I fragments. Amplification by polymerase chain reaction (PCR) was then performed on the whole chromosome, using random pairs of primers directed towards the *Not*I sites of the remaining *Not*I/*Sau*3A fragments. Amplification meant that the two corresponding fragments were adjacent, i.e., were contigs.

3. A method derived from that devised by Smith and Birnstiel (1976) has been applied to *Pseudomonas aeruginosa* mapping (Heger *et al.*, 1998). Fragments formed after partial digestion by a frequent-cutter are separated by PFGE and hybridized with end probes from each of the rare-cutter fragments of the same genome (Fig. 2D). Hybridization of a single frequent-cutter fragment with two probes identifies the rare-cutter contigs. In addition, comparison of the sizes of the partial digests can be used to establish a restriction map.

The presence of repetitive sequences (multifamily genes) or copies of mobile elements may lead to false alignments biased by erroneous apparent identity of the corresponding regions. Other methods, such as fingerprinting assembly (comparison of the restriction patterns of supposedly overlapping clones), have been devised to check previous results or to construct contig charts. The main limit to a list of available approaches is the imagination of the workers, which should be applied to devise any technique or combination of techniques that allow overcoming a suspected problem.

The first physical chromosomal map, that of *E. coli* (strain K12), was published in 1987 (see Section VI). Since then, an exponentially growing number of macrorestriction maps have been issued. These maps, however, are not very precise, and most do

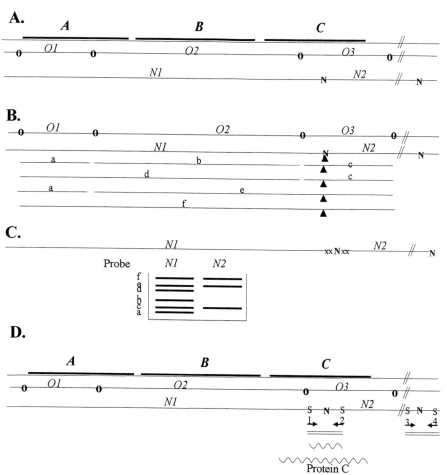

Fig. 2. Construction of linking clones. Application to a portion of chromosome with genes A, B, and C. (A) Comparison of the sizes of fragments obtained after separate or double digestions by pairs of rare-cutter endonucleases (O and N) can allow positioning of the cleavage sites. *O1, O2, O3, N1,* and *N2* represent the fragments obtained after single digestion by enzymes O or N, respectively. (B) Insertion of a cassette (▲) in an N site and identification of the fragments obtained after partial digestion with another enzyme (O) carrying the cassette indicates that fragment *O3* links fragments *N1* and *N2* (i.e., fragments *N1* and *N2* are contigs). (C) Partial digests of the whole chromosome with enzyme O are electrophoresed and hybridized, with each of the N fragments, labeled at their extremities (xx). Only the fragments encompassing region *O3* give a signal (heavy line) with *N2,* while all are recognized by *N1;* thus *O3* overlaps (links) *N1* and *N2.* (D) If end-sequencing of N fragments indicates the presence of reading frames, primers (→) directed towards the N sites are designed. If the sequence of the fragment amplified on whole DNA from a pair of primers (here, S1 and S2) reconstitutes a continuous reading frame showing homology with a known protein, the N1 and N2 fragments are probably contigs.

not carry any genetic information. A further step was to provide actual ordered cloned libraries (so-called encyclopedias), which are useful tools for subsequent genetic mapping.

III. ORDERED CLONED DNA LIBRARIES, OR ENCYCLOPEDIAS

An encyclopedia consists of a library of DNA fragments, cloned into a vector, ordered so as to reconstruct with sufficient overlap the order on the chromosome. The first one produced, again, was for *E. coli*, in 1987 (see Section VI).

A. Choosing the Vector for an Ordered Cloned Library

The available vectors cover a large range of possible sizes of the inserts they carry: λ-based vectors (10–25 kbp), cosmids (5–50 kbp), P1-based vectors (90 kbp), yeast artificial chromosomes (YACs) (75–2000 kbp), and bacterial artificial chromosomes (BACs) (20–100 kbp). Due to their lack of stability in *E. coli*, YACs have been used for bacterial genome mapping only for *B. subtilis*, *Myxococcus xanthus*, and *Pseudomonas aeruginosa*. The BACs, based on the

E. coli F plasmid origin of replication, are much more stable, and have been extensively used for eukaryotic gene libraries. Their first use in bacteria was for the construction of a *Mycobacterium tuberculosis* encyclopedia in 1998.

B. Assembling the Encyclopedia

Constructing an encyclopedia implies finding overlapping clones, in order to define contigs. One must start with a number of clones 10- to 20-fold in excess over the number corresponding to the length of the chromosome. This ratio depends on the portion of the insert required for detection of linkage (the minimal detectable overlap, MDO). The methods used to order the clones are similar to those described for the construction of macrorestriction maps. The work can be strongly facilitated by available knowledge, such as macrorestriction or genetic maps, or sequenced regions. Then, a minimal overlapping map may be proposed. Thus 420 BAC clones (20–40 Mbp) have allowed the construction of the 4,4 Mb map of the circular chromosome of *M. tuberculosis*, but the minimal overlapping set (or miniset) requires only 68 unique BAC clones (Fig. 3) (Brosch *et al.*, 1998).

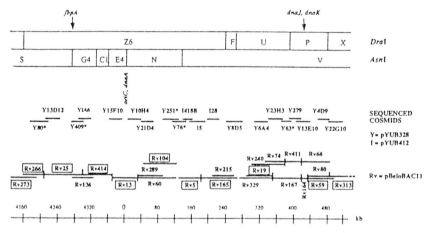

Fig. 3. The BAC ordered library for *M. tuberculosis,* superimposed on the physical and genetic maps. About 1/4 of the whole chromosome is shown, starting at the arbitrary 0 (*oriC*) site. Two top lines: restriction maps for *Dra*I and *Asn*I, showing a few gene locations. Cosmids, cosmids sequenced during the sequencing project; BACs, a representative set of BAC clones positioned relatively to the cosmids by end-sequencing and restriction mapping; kb, length of the chromosome. [Adapted from Brosch, R. *et al.* (1998). *Infection and Immun.* **66**, 2221–2229.]

IV. CONVERSION OF PHYSICAL MAPS INTO GENETIC MAPS: POSITIONING GENES ALONG THE PHYSICAL MAP

Restriction maps are converted into genetic ones by locating genes with reference to the restriction sites. Localization can be achieved by DNA hybridization or by sequence comparison when available. The limiting factor of this conversion lies in the number of known genes available (i.e., cloned or sequenced) and in the precision of the physical map. In favorable cases (for instance, *E. coli*), the restriction pattern of a cloned fragment encompassing a given gene is sufficient to localize the gene on the fragment, and thus, on the chromosome. Difficulties may arise if one or several cutting sites are protected as part of an overlapping site for a different modification system in the original host, but no longer so when subcloned and amplified in the cloning host (usually *E. coli*).

The genes used as probes can originate from the host itself or from a heterologous host. In the latter case, the strains should be sufficiently related and possess the same function, so that DNA sequence conservation can be expected to allow efficient annealing. PCR probes using primers covering conserved protein or DNA regions have also been thoroughly used for widely distributed genes. For instance, the cleavage site for I-*Ceu*I, which is specific of 23S rRNA genes, allows localizing these rRNA loci. A very efficient method using transposons with rare-cutting sites was also developed. When a mutation in a known gene has been obtained with such a transposon, the corresponding macrorestriction pattern displays the replacement of one fragment by two fragments, and subsequently allows a precise location of the transposon, hence, of the gene. The transposon Tn5 naturally harbors a single *Not*I site, but other transposons have been engineered so as to harbor similar single rare-cutter sites. In addition, transposon insertions can be intraspecifically transferred by conventional genetic methods, thus allowing easy comparison of chromosomal organization of related strains.

Comparisons of genetic with physical maps have shown a general good agreement with regard to gene order, but less so to distances between genes. This reflects biases introduced by nonregular distribution of recombination hotspots along the chromosome

in the course of genetic mapping. Thus, restriction mapping of the chromosome of *Streptomyces* showed the estimates obtained by *in vivo* mapping to have been rather accurate. Surprisingly, however, the *Streptomyces* chromosome has turned out to be linear instead of circular.

V. GENOME SEQUENCES

Whole genome sequences are presently, and will more readily in the future be, available. Will this render the approaches via physical mapping obsolete, as the genetic methods have mostly become? This does not seem likely for several reasons. The most obvious is that, even though genome sequencing becomes cheaper, it is still time consuming. Thus, it does not seem likely that the 4000 or so presently cultivable bacterial species will have their genome sequenced, whereas a physical map is more easily feasible. It should also be recalled that a physical map, even more as an ordered library, is often a prerequisite for a genome-sequencing project, mainly for the larger bacterial genomes.

Furthermore, sequencing a genome means sequencing the chromosome of a representative strain (and possibly isolate) of a species. There is growing evidence of a large plasticity of the genomes. Thus, physical maps and/or encyclopedias will remain the easiest way to approach this problem. For instance, *Spe*I-restricted fragments of 97 strains of *Pseudomonas aeruginosa* isolated from clinical and aquatic environments have been hybridized with YACs carrying 100 kb inserts (about 3% of the chromosome) from three chromosomal regions of the well-known strain PAO. At this scale, little genomic diversity was detected in these representatives of the species. In contrast, a study of 21 strains of the same species isolated from cystic fibrosis patients, and analyzed by several rare-cutting endonucleases, revealed that blocks of up to 10% of the genome could be acquired or lost by different strains.

The problem in translating a genomic sequence into a genetic map is the identification of the genes, i.e., of the encoded functions. Sequence homology with a known element (at the DNA, RNA, or protein level), be it of prokaryotic or eukaryotic origin, allows postulating a function with some confidence.

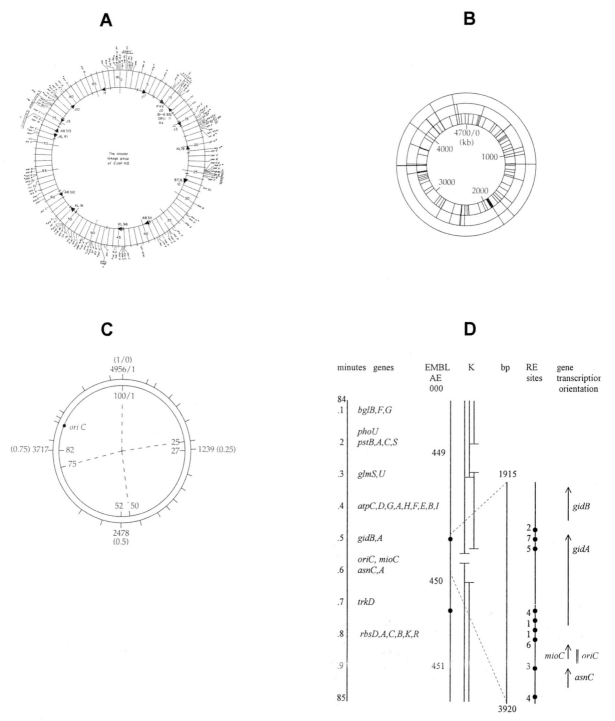

Fig. 4. Successive chromosomal maps of *E. coli* K12.(A) The 1967 issue of the circular linkage map, showing 166 loci (outer circle) and Hfr origins and orientations of transfer (arrows, inner circle) used for mapping. The total length (90 minutes) had not yet been normalized to 100 units. (B) The Kohara's physical map, showing the restriction sites for only 3 endonucleases (each circle); the outer one is for the rare-cutter *Not*I. (C) Comparison of the physical (*Not*I, outer circle) and normalized genetic lengths (in minutes), with the *thr* gene arbitrarily assigned at position 100/1: several genetic distances have been distorted to match the physical map. (D) The present complete genetic map indicating

However, about 30% of the potential ORFs of all sequenced genomes do not show homology to any known element. Their identification will constitute the main challenge of what is now referred to as the "after-sequencing," or post-genomic, molecular biology.

VI. THE *E. COLI* K12 CHROMOSOMAL MAP

The building of the chromosomal map of an *E. coli* K12 strain represents, historically, a typical textbook example, since all available techniques have successively been applied until the complete molecular information was reached with the sequencing. It, thus, allows summarizing, for just one strain, all the steps in genome mapping. A genetic linkage map showing 99 genes (first edition in 1964), then 166 (1967 edition), was based on conventional mapping procedures, using recombination after conjugation or transduction by phage P1 (Fig. 4A). This approach also provided the first proof of the circularity of a bacterial chromosome. Further accumulation of information via the same approach during the next 10–15 years led to a more complete map positioning about 1500 genes, representing 20–25% of the whole chromosome.

During the early 1980s, the introduction of molecular techniques allowed the cloning and sequencing of one-third of these genes and, thus, also provided their complete restriction maps. In 1987, Kohara and coworkers prepared a genomic library, now called an encyclopedia, using as cloning vector a modified λ phage. The whole library was contained in 1056 clones, each carrying 15–20 kbp-long inserts. From this library, an almost complete physical map showing restriction sites for eight endonucleases was obtained, using PFGE procedures and adapted computer programs (Fig. 4B). It took another 2304 clones and the use of newly published information on restriction or sequencing of local regions to deal with most of the remaining ambiguities or gaps, yielding a 4700 kbp-long molecule. Simultaneously, Cantor's group, using similar approaches, proposed a 4600 kbp-long chromosome, of which all ambiguities but one were solved. This whole map was covered by 22 NotI fragments. Correlation between these restriction maps and the known linkage map was excellent in terms of gene order. Some distortions of the genetically estimated distances, however, were necessary for an optimal alignment with the restriction profile (Fig. 4C). One cause for these discrepancies probably lies in unequal crossing-over frequencies along the chromosome.

The present Kohara's phage λ ordered library covering the whole genome is commercially available as a set of 476 phages, immobilized on a nylon hybridization membrane. A computer program was developed by Danchin's group to ease the experimental work of mapping a new character, by first working out its most probable location(s) through comparisons of restriction profiles. Another program (1995) allowed localizing a cloned fragment by simply determining the sizes of the hybridizing fragments obtained by the 8 restriction endonucleases used by Kohara. This allowed localizing a locus within 7 kb.

The complete sequence, published in 1997, describes a 4,639,221-bp-long chromosome, a figure very close to those reached by the previous physical data. Compilation of all data derived from genetic and molecular approaches, or predicted from sequence analyses or comparisons, has yielded what could be the complete set of information on this chromosome (see Fig. 4D for an example of this map). It is interesting to note that, even though location and precise length of a gene are now defined on a base level, the standard coordinate scale using

---

the minute coordinates (only the portion around *oriC* is shown) aligned along the corresponding fully sequenced map; the latter shows, successively, the sequence entries in the GenBank/EMBL/DDBJ data libraries (EMBL), the Kohara miniset clones (K), the base position (bp), restriction segments for eight enzymes (RE), and the identified genes (genes) with their orientation. [A: Adapted from Hayes, W. (1968). "The Genetics of Bacteria and Bacteriophages." Blackwell Science, Oxford. B and C: Adapted from Joset, F., and Guespin-Michel, J. (1993). Op. cit. D: Adapted from Berlyn, M. K. B. (1998). *Microbiol. Mol. Biol. Rev.* **62,** 814–984, and Rudd, K. E. (1998). *Microbiol. Mol. Biol. Rev.* **62,** 985–1019.]

a 0–100 arbitrary units (so-called minutes), derived from the times of transfer via conjugation, has been maintained. The available sequenced genome has now provided a new, easier means to localize any cloned fragment, by sequencing a small part of this fragment and aligning it by computer methods on the whole map.

The knowledge available, mostly as restriction data, from various *E. coli* strains provides growing evidence for a very low level of polymorphism in this species. Most of it may be due to movement of mobile elements (ISs, transposons, prophages, pathogenicity islands). As a consequence, the K12 map gains a wider validity than its use for the specific specimen strain from which it was constructed. This, however, is known not to be a universal situation (e.g, Streptomycetes, *Nesseria,* and to a lesser extent, *Pseudomonas*).

VII. CONCLUSION: THE INTEREST OF MAPPING BACTERIAL GENOMES

The publication of the first physical maps of bacterial genomes has started a new field in molecular biology called genomics, i.e., the study of integral genome structures. Although genomics has reached its full significance with the study of genome sequences, physical maps have provided, besides the initial tools for genome sequencing, numerous important results, such as original genomic structures (linear or multiple chromosomes, very large plasmids, and linear plasmids).

Comparison of related strains or species has been and will be of paramount importance to study the plasticity of genomes, i.e., their capacities of variations in chromosome organization, gene sequence, gene content per species, etc. Examples of this are the existence of large deletions in *Streptomyces* spp. genomes, the presence of inversions, deletions, or additions of genes or regions (e.g., in *Neisseria, Pseudomonas,* and *Bacillus*). To this purpose, PFGE, gene encyclopedia, or genome sequences of one or a few well-known strains serve to compare other strains of the same or related species.

See Also the Following Articles

DNA RESTRICTION AND MODIFICATION • DNA SEQUENCING AND GENOMICS • TRANSDUCTION: HOST DNA TRANSFER BY BACTERIOPHAGES • TRANSFORMATION, GENETIC

Bibliography

Berlyn, M. K. B. (1998). Linkage map of *Escherichia coli* K-12, edition 10: The traditional map *Microbiol. Mol. Biol. Rev.* **62**, 814–984.

Cole S. T., and Saint, Giron, I. (1994). Bacterial genomics. *FEMS Reviews* **14**, 139–160.

Fonstein, M., and Haselkorn, R. (1995). Physical mapping of bacterial genomes. *J. Bacteriol* **177**, 3361–3369.

Joset, F., and Guespin-Michel, J. (1993). Construction of genomic maps. *In* "Prokaryotic Genetics: Genome Organization, Transfer and Plasticity," pp. 369–387. Blackwell Science, Oxford.

Leblond, P., and Decaris, B. (1998). Chromosome geometry and intraspecific genetic polymorphism in Gram positive bacteria revealed by pulse field gel electrophoresis. *Electrophoresis* **19**, 582–588 (and, in general, this whole special issue on genome mapping).

Rudd, K. E. (1998). Linkage map of *Escherichia coli* K-12, edition 10: The physical map. *Microbiol. Mol. Biol. Rev.* **62**, 985–1019.

Meat and Meat Products

Jerry Nielsen

Oklahoma State University–Oklahoma City

GLOSSARY

bacteriocin A protein produced by a bacterium that is inhibitory to other similar species of bacteria.

lactic acid bacteria A group of bacteria belonging to a diversity of genera used to effect fermentation in meats to produce lactic acid as the primary end product of carbohydrate metabolism.

psychrotroph A bacterium that can grow at 7°C or less, irrespective of its optimum growth temperature.

serovar Strains of different antigenic forms of a pathogen.

starter culture A microbial strain or mixture of strains, species, or genera used to effect fermentation in meat and bring about desirable changes in the finished product.

MEAT has always been a staple in the diet of most people around the world. Fossil records indicate that early man was a meat eater. Fresh meat spoils easily so early meat eaters had to consume it fresh until methods were discovered to preserve it.

Early methods of meat preservation included drying, salting, or smoking, alone or in combination. At some point in history, lactic acid bacteria grew in or on the meat, aiding in the preservation process. When refrigeration was developed, fresh meat could be stored longer. However, many spoilage microorganisms and pathogens are psychrotrophs, so refrigeration did not always ensure quality and safety.

Spoilage microorganisms and most pathogens are common in nature so there is always a potential for contamination. As the amount of processing increases, so does the potential for contamination. Desirable microorganisms, such as lactic acid bacteria, impart a tangy character as well as subtle flavors to processed meats, thus increasing the value of poor cuts. Spoilage microorganisms make meat inedible and pathogens can cause illness or death. Many techniques are presently used to ensure that meat is wholesome and safe and additional methods are being developed.

I. COMPOSITION AND MICROBIAL CONTENT OF MEAT

Meat is both a versatile and nutritious part of a balanced diet. Its protein content varies from 15 to 22% of the wet weight. Lipid content may vary from a low of 0.5 up to 37%, depending on the cut of meat and how it is prepared. Easily absorbed low-molecular weight compounds, such as amino acids and simple sugars, make up from 1 to 3.5% of the wet weight.

The meat from a healthy animal contains very few, if any, bacteria at the time of slaughter. Lymph nodes may contain bacteria that have been filtered from the tissue. Some common genera of bacteria isolated include *Staphylococcus sp.*, *Streptococcus sp.*, *Clostridium sp.*, and *Salmonella sp.* Typically, lymph nodes are removed at the time of slaughter or early in processing. Bacteria are present in large numbers on hides and hooves of animals such as cattle, sheep, and swine. Feathers and skin of poultry also contain large numbers of bacteria. The intestinal tracts of all those animals contain large numbers of potentially harmful bacteria. Initial contamination of meat can occur during sticking and bleeding, skinning, and evisceration of the animal. Additional sources of microorganisms include knives, air, workers, carts, containers, and tables. Processed meats may be contaminated by workers, grinders, stuffers, and even by spices. One contaminated carcass may contaminate other carcasses if proper sanitation procedures are not employed as workers go from carcass to carcass. Additionally, rinse water may be a source of contamination for poultry carcasses.

II. DESIRABLE MICROORGANISMS

Microorganisms that cause a predictable and desirable change in meat are used for preservation. Most of these microorganisms produce lactic acid, which lowers the pH, preventing the growth of undesirable microorganisms. Other, subtle flavors are also produced by bacteria which, in turn, gives the fermented meat a richer flavor that cannot be duplicated by direct acidification. The cheaper, less desirable cuts of meat are used for fermented products, thus increasing the marketability of the whole carcass. Common curing bacteria include pediococci, lactobacilli, micrococci, and streptococci.

A. Sources

Curing microorganisms are part of the normal flora associated with food products. Originally, the normal flora were allowed to grow in the meat product. Typically, lactobacilli were present. If sufficient numbers of lactic acid bacteria were present, fermentation occurred in timely manner and the meat did not spoil. Product flavor and quality was inconsistent from batch to batch. When a good batch of sausage was produced, subsequent batches were "back inoculated" with the good sausage. Occasionally, toxin-producing bacteria, such as *Staphylococcus aureus,* were present and able to grow long enough to produce toxins before growth was inhibited by lactic acid. Most commercial meat processors now use concentrated starter cultures for their fermented products. The number of microorganisms is consistent

so there is uniformity in quality from batch to batch. Also, the fermentation process occurs rapidly, thus preventing the growth of undesirable microorganisms. Most commercial starter cultures for meat products contain pediococci.

B. Products

Most fermented sausages contain beef, pork, or a combination of the two meats. Occasionally, poultry meat is used. Starter cultures produce lactic acid, which lowers the pH so undesirable microorganisms cannot grow. Lactic acid also is responsible for the tangy flavor. Some common sausages are summer sausage, thuringer, cervelat, Lebanon bologna, pepperoni, and salami. These products are different because of the combination of spices, fat content, meat combinations, texture, and cooking methods. Some salamis are further cured by inoculating the surface with other microorganisms that add subtle but specific flavors. Rather than use starter cultures, some processors add lactic or acetic acid. The proper pH can be attained but flavor components produced by bacteria are absent.

III. UNDESIRABLE MICROORGANISMS

Any microorganism that causes a change in meat that would adversely alter the appearance, texture, flavor, odor, or safety of the product is considered undesirable.

A. Spoilage Microorganisms

The first sign of spoilage is normally an off-odor, which is slightly sour or putrid. End products, usually from the breakdown of glucose, cause this odor. As the undesirable microorganisms continue to grow, color may change to gray or green, due to the breakdown of hemoglobin and ribose. The slick film is produced when microorganisms polymerize glucose to dextran. Proteolysis is normally confined to the surface so destruction is more extensive in ground meat. Breakdown of proteins and lipids results in very offensive odors. Since a large percentage of common spoilage microorganisms are psychrotrophs, spoilage can occur slowly in the refrigerator but much more rapidly as the holding temperature increases.

1. Sources

Pseudomonads are the most common spoilage bacteria. Species include *Pseudomonas aeruginosa*, *P. fragi*, and *P. lundensis,* which is fluorescent. Pseudomonads are common in soil and water and are found on surfaces of most plants. They can break down such a large variety of substrates, they not only survive, but grow just about anywhere. Additionally, pseudomonads are psychrotrophs, so they are commonly found in foods that spoiled in the refrigerator.

Other bacteria associated with the spoilage of fresh refrigerated meat are *Acinetobacter sp., Moraxella sp., Aeromonas sp., Alcaligenes sp.,* and *Micrococcus sp.* Yeasts and molds can also spoil fresh meat. Common members of this group include *Cladosporum sp., Geotrichum sp., Mucor sp., Thamnidium sp., Candida sp.,* and *Rhodotorula sp.* Spoilage bacteria found in processed meats include *Lactobacillus sp., Acinetobacter sp., Bacillus sp., Micrococcus sp., Serratia sp.,* and *Staphylococcus sp.* Yeasts and molds include *Aspergillus sp., Penicillium sp., Rhizopus sp., Thamnidium sp.,* and *Candida sp.* The bacteria are part of the normal surface flora of plants and animals and may also be found in soil, water, and air. Yeasts and molds may also be found in all these places. Since fungal spores are generally more resistant to environmental extremes than bacteria, they are more likely to be transported by air-circulating equipment.

2. Effect of Microorganisms on Meat

Spoilage microorganisms break down glucose first. Growth continues on low-molecular weight compounds, such as amino acids and other simple sugars. Many microorganisms also have proteases and lipases, which break down proteins and lipids once the smaller molecules are depleted. The degradation of proteins causes softness of tissue, while degradation of lipids results in off-odors, stickiness, and rancid flavors in the cooked meat. Aerobic spoilage by bacteria results in surface slime on meat, caused by polymerization of glucose to dextran. Ground meat may be slick or slimy throughout. From this type of spoilage, organisms such as *Pseudomonas sp.,*

Acinetobacter sp., Alcaligenes sp., and *Bacillus sp.* may be routinely isolated.

Aerobic yeasts and molds cause sticky surfaces on meat as well as fuzzy spots. Black spots are usually caused by *Cladosporum sp.,* white spots by *Sporotrichum sp.,* and green spots by *Penicillium sp.* Molds may also cause off-flavors that are usually described as "moldy." In most cases, surface contamination can be trimmed off solid cuts of meat with little or no undesirable effect.

Anaerobic spoilage can also occur in vacuum-packaged fresh or processed meat. Souring, which results in a soured odor, may be caused by a number of lactic acid bacteria or some species of *Clostridium.* Lactic acid bacteria, especially lactobacilli, often grow on vacuum-packaged processed meats without causing detectable changes in flavor or aroma. Growth is noticeable because the small amount of fluid in the package changes from clear to cloudy and white. Lactobacilli are also responsible for the greening on the surface of cured meats, such as bacon and ham. Again, there are usually no detectable changes in flavor or aroma. Putrefaction results from the decomposition of protein and can be caused by *Clostridium sp., Alcaligenes sp.,* or *Proteus sp.*

B. Pathogens

Meat provides an excellent environment for the growth of pathogens. This is confirmed by the number of food-borne illnesses each year attributed to meat. Many pathogens are psychrotrophs and multiply during refrigerated storage. Others may flourish when meat is held at an elevated temperature or receives an insufficient heat treatment during preparation. Also, the possibility of cross-contamination exists because most food handlers in restaurants and in the home have too little knowledge of food safety.

1. Sources

Pathogens are common and are routinely isolated from the environment. *Aeromonas sp.* is common in water, while *Listeria sp., Clostridium sp., E. coli, Streptococcus sp.,* and *Salmonella sp.,* are common in soil. Infected animals may transmit *Listeria sp., Salmonella sp.,* and *Campylobacter sp.* Healthy animals, as well as humans, carry *E. coli* as part of the normal intestinal flora, while *Staphylococcus sp.* and *Streptococcus sp.* are part of the normal flora of the upper respiratory tract.

2. Pathogenesis

a. Food-Borne Intoxication

Food-borne intoxication results when a person consumes food in which bacteria have grown and released toxic by-products. The bacteria do not survive to reach the intestinal tract. In most cases, the food has been improperly processed, stored, or prepared for consumption.

i. Gastroenteritis The most common form of food-borne intoxication is caused by *Staphylococcus aureus,* which produces an enterotoxin. The microorganism may be salt- and nitrate-tolerant and be able to grow on cured meat. Usually, no changes occur in the food, so growth is not detected. The most common source of contamination is the food handler, because *S. aureus* is part of the normal flora of the skin and upper respiratory tract. As it grows, the microorganism produces a toxin that is heat-stable and is not destroyed by normal cooking. Many meat products are involved in outbreaks of staph food poisoning.

After consumption of contaminated food, symptoms occur in 2–6 hours. These include nausea, vomiting, abdominal cramps, and diarrhea. Also, headache, muscle cramps, sweating, chills, prostration, weak pulse, shock, and shallow respiration may occur in severe cases. Duration of the illness is brief, and mortality is less than 1%. Usually, no treatment is given with the exception of replacing fluids and electrolytes. Most cases are probably not reported to medical authorities.

Clostridium perfringens may grow in the intestinal tract but the toxin it produces is the cause of gastroenteritis. The microorganism is traced to meat products more often than to other foods. Foods that have been cooked and allowed to cool slowly are most likely to support growth of this microorganism. The sublethal heat treatment stimulates germination of endospores, and the microorganism multiplies as the food cools. The vegetative cells grow in the food or gut for a short time and, during sporulation, release a toxin that results in excessive fluid release into the intestine. Symptoms occur in 8–24 hours, depending on the original number of cells. The most common symptoms are abdominal pain, diarrhea, and gas. Fever, nausea, and vomiting are uncommon. Recovery is usually uneventful.

Another, less common form of food-borne intoxication is caused by *Bacillus cereus.* This microorganism can grow in the gut as well but its toxin is what causes discomfort. This microorganism has been implicated in outbreaks of gastroenteritis traced to meat loafs. Symptoms of this illness occur within 8–16 hours after ingestion. They include nausea, abdominal cramps, watery diarrhea, and some vomiting. Recovery is uneventful, and most cases are not reported.

ii. Systemic Intoxication The most serious food-borne intoxication is botulism. *Clostridium botulinum* grows anaerobically in low-acid foods that have been improperly processed. It may occur in canned foods that have had insufficient heat treatment to destroy the endospores. Most outbreaks of botulism have been traced to home-canned foods. Some vacuum-packaged foods, such as cold-cuts, ham, and bacon, can support the growth of *C. botulinum* if no inhibiting agent has been used. The microorganism is proteolytic and produces a toxin during growth. A lethal amount of toxin can be produced before growth of the bacteria is noticeable. The toxin is heat-sensitive and is easily destroyed by the cooking process.

After consumption, the toxin is absorbed through the small intestine. Symptoms, which occur in 12–36 hours, include nausea, vomiting, diarrhea, fatigue, dizziness, double vision, and headache. Constipation occurs later. Involuntary muscles become paralyzed and death is usually due to respiratory failure. The mortality rate for botulism is almost 20%. An antitoxin is available, and the earlier it is administered, the more effective it is.

b. Food-Borne Infection

A food-borne infection occurs when a person consumes food containing some pathogen. The food serves as the vehicle of entry into the body and the pathogen then grows inside the host, causing illness.

i. Gastroenteritis Salmonellosis is a leading cause of food-borne infection. It accounts for more than one-third of the known causes of food-borne illness. There are many serovars of *Salmonella,* and it can be isolated from a variety of sources.

Normally, salmonellae are eliminated by proper processing of meat. Sources of post-processing contamination may be humans, animals, or equipment. Salmonellae may grow in food without producing detectable changes. Most outbreaks occur because of mishandling of food or cross-contamination. Because of this, most outbreaks are traced to restaurants or cafeterias, and, in many cases, more than one food is found to be contaminated. Poultry is the most common source of salmonellae.

After ingestion of contaminated food, symptoms occur in 12–36 hours and include nausea, vomiting, abdominal pain, and diarrhea. These may be preceded by headache, chills, fever, weakness, prostration, and drowsiness. Mortality is less than 1%, however, because of the large number of cases of salmonellosis, about 1/4 of all deaths from food-borne illnesses are caused by salmonellae. Duration is normally 2–3 days, with an uneventful recovery. Some people may become asymptomatic carriers of salmonellae.

In recent years, *Campylobacter jejuni* has been recognized as a major cause of food-borne infection. The Food and Drug Administration recently reported that *C. jejuni* is the leading cause of diarrhea in the United States. The microorganism is associated with warm-blooded animals. It is microaerophilic and requires low levels of oxygen and elevated levels of carbon dioxide for optimum growth, so it is very difficult to grow routinely in the lab. As culturing methods have improved, campylobacters have been isolated from patients with gastroenteritis more often than salmonellae. Symptoms of the illness occur in 2–3 days and include abdominal pain, cramps, diarrhea, headache, fever, and sometimes bloody stools. Cross-contamination to other products is possible; however, *C. jejuni* is sensitive to heat and is easily destroyed by cooking.

Several strains of *E. coli* are recognized as pathogens and are referred to collectively as enterovirulent *E. coli.* One strain in particular, *E. coli* 0157:H7 (enterohemorrhagic *E. coli*) has been implicated in outbreaks of hemorrhagic colitis, which were traced to undercooked ground beef. The microorganism is commonly found in the gut of cattle and probably gets on the carcass during slaughtering. It is a psychrotroph so it is able to grow while the carcass is chilling. When contaminated beef is ground, bacteria from the carcass get on the grinder and contaminate the rest of the beef going through. The bacteria survive during frozen storage. Only a small number need to survive the cooking process in order to cause disease. When ground beef is cooked thoroughly, the bacteria are destroyed. In addition to gastroenteritis, this bacterium can cause systemic illnesses, such as acute kidney disorders. The fatality rate can be high, depending on the age and physical condition of the victim and how soon treatment is administered.

A relatively uncommon, but potentially problematic, type of gastroenteritis is caused by *Yersinia enterocolitica.* It is associated with warm-blooded animals and is routinely isolated from a small number of samples of beef and pork in retail outlets. Symptoms of the illness occur within 24–36 hours and include acute abdominal pain resembling appendicitis, fever, headache, diarrhea, vomiting, nausea, and chills. *Yersinia enterocolitica* is a psychrotroph, so its presence on meat could pose a problem if the meat is not properly cooked.

ii. Systemic Infection *Lysteria monocytogenes* is a common pathogen that can use food as a vector into humans. Rather than gastroenteritis, the bacterium can cause abortion, meningitis, encephalitis. Those most susceptible to this bacterium are pregnant women, fetuses, and those with compromised immune systems. Although few people are infected, the mortality rate is about 30%. The microorganism is a psychrotroph and has been shown to survive in other extremes, such as high-temperature processing and low pH.

Although *Trichinella spiralis* is a nematode, it can be transmitted to humans by pork if it is not cooked to an internal temperature of 175°F. It is not as common now since most swine are raised in confinement and do not have much opportunity to become infected with the organism. Once inside the host, the adult nematodes release larvae into the digestive tract, where they invade the mucus mem-

TABLE I
Comparison of the Types of Food-Borne Illnesses

	Food-borne intoxication	*Food-borne infection*
Cause	Ingestion of preformed toxin which irritates intestinal tract.	Ingestion of pathogen which colonizes intestinal tract causing irritation.
Onset time after ingestion	2–12 hours	12–36 hours
Common signs and symptoms	Nausea, vomiting, abdominal cramps, diarrhea, and headache	Headache, fever, nausea, vomiting, abdominal pain, and diarrhea
Treatment	Replace fluids and electrolytes	Replace fluids and electrolytes
Duration of illness	12–24 hours	1–3 days
Mortality rate	<1%	<1%
Common causative organisms	*Staphylococcus aureus, Bacillus cereus, Clostridium perfringens*	*Salmonella enteritidis, Campylobacter jejuni, E. coli*

brane and mature. Onset of symptoms varies from a few days to several weeks. Symptoms include nausea, vomiting, diarrhea, profuse sweating, colic, and loss of appetite, which may last for several days. The larvae may travel through blood vessels and lymphatics to muscle tissue, where they encyst. Later symptoms are related to muscle soreness and swelling. Prevention is the best control. Trichinae can be destroyed by freezing fresh pork at −15°C for 20 days or longer, cooking to an internal temperature of 175°F, or using low doses of ionizing radiation. A comparison of food-borne intoxication and food-borne infection is found in Table I.

IV. CONTROL OF UNDESIRABLE MICROORGANISMS

Many of the procedures used in producing meat and meat products for the consumer are concerned with controlling microorganisms. Much of the expense of production is dedicated to ensuring that a safe, wholesome product is available to consumers.

A concept called Hazard Analysis and Critical Control Points (HACCP) has been incorporated into meat processing as well as other industries. The system identifies places in the process where contamination or other problems might occur. These are called critical control points. When the hazards are identified, the critical control points are established, and samples are taken before the product goes beyond that point. More samples are taken but less expense is involved, because the problem is solved in an unfinished product rather than a finished product, thus making rework or disposal an easier task.

A. Physical

The best way to assure the safety of meat is to prevent contamination. Many procedures are used to limit the number of microorganisms in a meat-processing operation. Some processors spray animals with water before slaughter to remove manure and soil from the hair. All equipment in the slaughter area is thoroughly cleaned and sanitized before and after use. Knives and saws are frequently scalded to prevent cross-contamination, and boots and aprons are rinsed often to remove residue. Once microorganisms attach to meat, they are very difficult to remove.

A high-pressure water spray is used to rinse carcasses before they are chilled. This removes some microorganisms but may add others such as *Pseudomonas sp.* if they are present in the water. The low number of microorganisms, plus the cold storage temperature, keeps most microorganisms in check.

Once the meat is cut or processed, other methods are used to inhibit microorganisms. Refrigeration extends the shelf life by slowing the growth of many microorganisms. However, *Aeromonas sp., B. cereus, E. coli, Listeria sp.,* and *Yersinia sp.* are all psychrotrophs. Freezing stops the growth of microorganisms and severely limits the action of enzymes. The number of viable microorganisms decreases with time in the frozen state but some microorganisms survive.

High temperatures denature proteins and inactivate enzymes, thus destroying most vegetative cells. Canning can result in complete sterilization and can be used to prevent spoilage. Microwave energy is sometimes used to destroy microorganisms. The killing effect is due to the heat generated and not to microwaves.

Some meat products are dried as a means of preservation. Drying inhibits growth of microorganisms by limiting or removing the water necessary for growth. With products such as jerky, dry sausage, and salamis, drying is used in combination with salt, spices, and smoking for preservation. Other products are freeze-dried and sealed in packages, which extends the shelf life indefinitely.

Radiation is another method of inactivating microorganisms. Ultraviolet radiation is absorbed by pyrimidines, resulting in lethal mutations. Its effectiveness is determined by time and intensity of light but it does not penetrate, so only the surfaces of objects are sanitized. Ultraviolet light also irritates the eyes very quickly and causes reddening of skin, so its use is limited to times when no one is present. It is sometimes used to limit the growth of microorganisms on meat that is aged at higher-than-normal refrigeration temperature.

Ionizing radiation causes major changes in the DNA of bacterial cells, which results in death of the organisms. Gamma rays are produced from cobalt-60 or cesium-137. They have the ability to penetrate up to 20 cm into foods. Cathode rays emit beta particles. The penetrating ability is poor but the rays are directional, so they can be used more efficiently than can gamma rays.

X-rays produced by high-energy electrons are more efficient than gamma rays from cobalt-60. They penetrate better and the dose distribution is more uniform. Monitoring the dose of radiation is easier, so a more standard application is possible.

Large doses of ionizing radiation can sterilize meat. Low doses can extend shelf life. Very low doses can inactivate trichinae. Bacterial enzymes such as lipases and proteinases, which are already present in meat, are not inactivated, so some undesirable changes can still occur.

Ionizing radiation may cause secondary changes in meat products, such as an increase in pH, increase in H_2S, and methyl mercaptan. Natural antioxidants may be destroyed in fats, resulting in oxidation and reduction in the levels of some vitamins.

B. Chemical

Additives used in meat-processing vary from the very simple to very complex. High concentrations of sugar and salt cause plasmolysis of bacterial cells. Sodium and potassium nitrates and nitrites were used originally because they were converted to nitrosomyoglobin, which preserved the red color of meats. They were later found to inhibit the growth of *C. botulinum*. More recently, these compounds have been shown to be carcinogenic in laboratory tests at high concentration, but, because of the antibacterial activity, the benefit outweighs the danger so their use is still permitted.

Other preservatives include methyl and propyl parabens, which inhibit bacterial growth. Sorbates inhibit yeasts and molds and are used in many products in addition to meats. Meat has been smoked over wood fires for centuries. It adds flavor and also improves the color and finish of meat. Smoke, or the heat from generating it, has a drying effect and also adds preservatives, such as formaldehyde, phenols, and cresols. Other compounds include low-molecular weight organic acids, ketones, and aldehydes. These compounds are inhibitory to vegetative cells.

Modified atmosphere packaging is used to extend the shelf life of some meats. At CO_2 concentrations of 20% or more, *Pseudomonas sp.*, *Moraxella sp.*, and *Acinetobacter sp.*, as well as other spoilage microorganisms, are inhibited. Lactobacilli may continue to grow, but their growth is slow and the by-products of metabolism are not as offensive. One drawback of modified atmosphere packaging is that pathogens such as *Clostridium sp.* and *Campylobacter sp.* may be stimulated to grow.

C. Biological

Lactic acid bacteria preserve foods by using available carbohydrates to produce lactic acid, which is inhibitory to many other bacteria. More recently, other compounds, produced in small quantities, have been found to have an antagonistic effect on other microorganisms. Lactobacilli produce hydrogen peroxide but cannot break it down, so it accumulates in quantities sufficient to inhibit *S. aureus* and *Pseudomonas sp*. Lactobacilli have been inoculated onto the surfaces of cured hams and other large cuts of meat to inhibit other microorganisms during refrigerated storage. Growth of lactobacilli in the refrigerator is slow if it occurs at all, so no detectable changes occur.

Many lactic acid bacteria produce protein-containing, bactericidal molecules called bacteriocins. These compounds are produced by a very diverse group of bacteria. Typically, bacteriocins can control *Listeria monocytogenes, C. botulinum,* and sometimes *S. aureus.*

Nisin, produced by *Lactococcus lactis,* was first recognized in 1928 as an inhibitory substance. Since then, it has been extensively investigated and characterized. It is approved by the Food and Drug Administration as a food additive to prevent growth of many gram-positive bacteria, including *C. botulinum* and *L. monocytogenes.*

Many of the pediococci, which are used in sausage fermentations, also produce bacteriocins. Production of bacteriocins is plasmid-controlled, and the antimicrobial spectrum is narrower than that of nisin. Of importance is the ability of these bacteriocins to inhibit *L. monocytogenes.* These bacteriocins are not approved for use in a pure form as food additives so the bacteriocin-producing starter culture has to be used. Thus, its use is limited to cultured products.

See Also the Following Articles

Bibliography

Cross, H. R., and Overby, A. J. (eds.). (1988). "Meat Science, Milk Science and Technology." Elsevier Science Publishers B.V., New York.

Egan, A. F., and Roberts, T. A. (1987). Microbiology of meat and meat products. *In* "Essays in Agricultural and Food Microbiology" (J. R. Norris and G. L. Pettiphen, eds.), pp. 167–198, John Wiley & Sons, New York.

Frazier, W. C., and Westhoff, D. C. (1988). "Food Microbiology" (4th ed.). McGraw-Hill, New York.

Gilliland, S. E. (ed.) (1985). "Bacterial Starter Cultures of Foods." CRC Press, Boca Raton, FL.

Mercury Cycle

Tamar Barkay

Cook College, Rutgers University

GLOSSARY

biomagnification Process by which an increase in the concentration of mercury in biological tissues/matter occurs when organisms are consumed by others at higher trophic levels.

broad-spectrum mercury resistance Bacterial resistance to both Hg(II) salts and organomercury compounds.

carbanion Negatively charged carbon atom, as in a methyl group, that is involved in methylation of positively charged Hg(II).

corrinoid coenzyme A coenzyme group that participates in methyl-transfer reactions of negatively charged methyl groups, as in the methylation of Hg(II).

degassing Natural process by which Hg^0 is transported from soils and waters to the atmosphere due to its volatility.

detoxification Microbial transformations of mercury that result in formation of volatile mercury species that are removed from the immediate environment of the cell.

disproportionation A mechanism whereby two molecules of an intermediate redox status interact, resulting in the production of a more reduced and a more oxidized molecule, as in the case of two molecules of Hg(I) producing Hg^0 and Hg(II).

hydroperoxidases Enzymes whose major function is the decomposition of peroxides and that are involved in the oxidation of Hg^0 to Hg(II) by some bacteria.

mer operon Structural organization of genes that encode the various functions needed for the reduction of Hg(II) and dissimilation of organomercury compounds. Often carried on plasmids and transposons.

narrow-spectrum mercury resistance Bacterial resistance that is limited to Hg(II) salts; resistance to organomercurial compounds is not evident.

oxidative demethylation Degradation of $CH_3Hg(I)$ by anaerobic bacteria that results in the production of CO_2 rather than CH_4.

photodegradation Light-induced breakdown of $CH_3Hg(I)$.

protonolysis Mechanism by which the C-Hg bond is cleaved by the enzyme organomercurial lyase to produce Hg(II) and a reduced organic moiety.

sulfate-reducing bacteria Bacteria that use SO_4^{2-} as a terminal electron acceptor in anaerobic respiration; play an important role in the geochemical cycling of mercury in both formation and degradation of $CH_3Hg(I)$.

THE MERCURY CYCLE describes fluxes of mercury between different compartments in the biosphere. Microbial (and physical-chemical) transformations play a key role in this cycle by altering the chemical form of mercury compounds. Mercury as a toxicant is of most concern in its methylated form in aquatic environments, where it is available for bioaccumulation by aquatic organisms. Thus, transformations of mercury that favor methylmercury formation and its accumulation have far-reaching effects on human and animal health.

I. INTRODUCTION

A. Origin, Properties, and Uses of Mercury

Deposits of mercury (throughout this article, the precise chemical nomenclature for mercury species is followed; the term "mercury" is used to describe the element or when the chemical form is unknown) in the Earth's crust are predominantly in the form

TABLE I
Physicochemical Properties of Mercury

Atomic number	80
Atomic weight	200.59
Density	13.5 g ml^{-1} at 20°C
Boiling point	356.9°C
Condensation temperature	−130 to −140°C
Diffusion of vapor	0.1124 cm^2 s^{-1}

of cinnabar (HgS) and are of primary magmatic origin. The content of mercury in the Earth's crust is estimated to be $1 - 30 \times 10^{-6}$ weight percentage. Mercury is a heavy metal (density of 13.5 g ml^{-1} at 20°C) and is the only metal to exist as liquid at room temperature. Additional physical properties of mercury are listed in Table I. Mercury is produced by roasting cinnabar-enriched ore at 600–700°C in the presence of air and condensing gaseous mercury that is stripped from the reaction. Industrial uses of mercury include production of paints, batteries, instruments, and the simultaneous generation of chlorine and caustic soda by electrolysis, using a flowing cathode of Hg^0. The latter process is becoming obsolete as it is being replaced by new technologies. Anthropogenic sources of mercury are divided into five categories: burning of fossil fuels, mining and smelting activities, metal processing, production of goods, and consumption of goods (including the incineration of waste). Relative to natural sources, anthropogenic sources account for 25–50% of the mercury input to the environment (estimates vary greatly).

B. Mercury Toxicity

Four unique properties make mercury a potent toxicant: (1) a strong affinity of Hg(II) and organomercurial compounds for thiol groups, (2) a tendency to form covalent bonds with organic molecules, (3) the high stability of the Hg–C bond that results from a low polarity and a low affinity for oxygen, and (4) a strong tendency to maximize bonding to two ligands in a linear stereochemistry. Mercury in nature exists in three major forms: as elemental (Hg^0), mercuric ion (Hg[II]), and methylmercury (CH$_3$Hg[I]). The most abundant form of Hg, HgS, is a ligand of Hg(II) with sulfide. Whereas both Hg(II) and CH$_3$Hg(I) are known toxicants, the later is of most concern to public health. Methylmercury readily crosses the walls of the gastrointestinal tract due to its fast transport through biological membranes, accumulating in the envelopes of nerve cells to cause neurological damage. Methylmercury in association with tissue is stable and is neither degraded nor excreted from the body at significant rates. As a result, CH$_3$Hg(I) is accumulated by organisms through their lifetime, and the concentration of CH$_3$Hg(I) is further magnified through the trophic interactions of the food web. Predatory fish that occupy top compartments in the food chain may contain mercury levels (essentially all tissue-associated mercury is in the methylated form) that are magnified by a factor of 10^7 relative to mercury concentrations in the water. Consumption of fish thus tainted with CH$_3$Hg(I) may cause poisoning. Public health regulations in the United States prohibit consumption of fish with tissue mercury concentration of >1 ppm (Canada and some states regulate at the 0.5 ppm level). Risk is especially severe for young children and fetuses whose nerve system is in a highly active developmental stage. Mercury poisoning is manifested in various neuromuscular malfunctions, ranging from numbing of the extremities to the loss of eyesight, paralysis, and even death.

Other chemical forms of mercury are of less concern. Mercuric mercury is excreted from the body within days of exposure, but chronic exposure may cause renal damage. The inert nature of HgS, a consequence of its insolubility, assures low toxicity. Exposure to Hg^0 is of concern in industries such as mining and smelting, where volatile Hg^0 is emitted in closed spaces (concentrations of >0.05 mg m^{-3} are considered hazardous). Elemental mercury may cross the gastrointestinal and pulmonary walls, and, once in the bloodstream, it is oxidized to Hg(II). The ingestion of Hg^0 from amalgams in dental fillings has been shown to cause renal damage in laboratory animals.

II. THE MERCURY CYCLE IN THE BIOSPHERE

Cycling of mercury in the biosphere (Fig. 1) is governed by the physicochemical properties of the

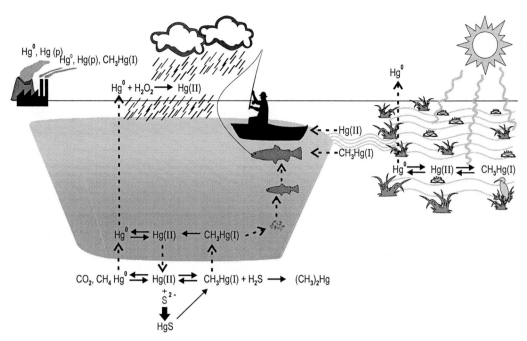

Fig. 1. The mercury cycle in the biosphere. Only major processes and chemical species are depicted. Dashed arrows indicate fluxes of mercury; full solid arrows indicate chemical and microbial transformations. The sizes of the solid arrow heads depict the relative magnitude of the indicated transformations.

participating mercury species (Table II). The relatively inert mercury in HgS is solubilized in alkaline and acidic solutions. The photosolubilization of HgS to Hg(II) occurs at wavelengths of <650 nm and is enhanced by oxygen and electrolytes. Microbial methylation of Hg(II) from HgS has been observed, albeit at extremely slow rates.

Mercuric mercury is a highly active species and is subject to physical interactions and chemical and microbiological transformations. It forms complexes with organic and inorganic ligands and readily adsorbs to surfaces of particulates. The latter plays a significant role in aquatic environments, where settling particulates carry Hg(II) to the sediment. In

TABLE II
Physicochemical Properties of Mercurial Compounds That Govern Mercury Cycling in the Biosphere

Hg^0	• High vapor pressure	Henry's constant, H = 0.3 dimensionless
	• Low aqueous solubility	6×10^{-6} g 100^{-1} ml water at 25°C
Hg(II)	• Covalent bonding	
	• Affinity to thiol groups	
	• Ligand formation	
	• Low vapor pressure	$H < 10^{-4}$ dimensionless
	• High aqueous solubility for most salts	$7.4 \times$ g 100^{-1} ml cold water ($HgCl_2$)
Alkylmercury	• Stability of C–Hg bond	
	• High solubility in water and hydrocarbons	
	• Rapid transport through biological membranes	
	• Low vapor pressure	$H < 10^{-4}$ dimensionless for $CH_3Hg(I)$
HgS	• Very insoluble	Log K_s = −0.51 mol L^{-1} at 25°C.

sediments and other anaerobic environments that contain S= (a product of microbial sulfate reduction), the formation of insoluble HgS is favored (log K_s = −0.51 mol L^{-1} at 25°C), dramatically decreasing the availability of mercury for further geochemical cycling.

Chemical and microbial reactions for Hg(II) methylation have been described (see following). The major product of these reactions is $CH_3Hg(I)$, a stable species with high solubility in water and hydrocarbons and a low vapor pressure (Henry's constant, H <10^{-4} dimensionless). The persistence of $CH_3Hg(I)$ is governed by the high stability of the C−Hg bond. This stability is due to a low polarity that, together with a low affinity of mercury to oxygen, decreases susceptibility of the bond to hydrolytic cleavage. One mechanism for the breakdown of $CH_3Hg(I)$ is a protonolytic attack on the C−Hg bond. This reaction is an extremely slow chemical process, but it is accelerated 10^8-fold when mediated by an enzyme, organomercurial lyase (see following). Bacteria in sediments, waters, and soils produce this enzyme, and they degrade $CH_3Hg(I)$ and other organomercury compounds *in situ*. Photodegradation is a newly discovered mechanism for the degradation of $CH_3Hg(I)$.

The importance of dimethylmercury ($[CH_3]_2Hg$) in the biosphere is not clear. Early work indicated that, in anoxic sediments under alkaline conditions, its formation was favored, whereas $CH_3Hg(I)$ was the prevalent product of the methylation reaction at low pH. Trace amounts of $(CH_3)_2Hg$ have been detected in air samples. However, highly sensitive methods for speciation and analysis of mercury compounds in environmental samples failed to detect this species in most natural waters, fish tissue, or rainwater.

Reduction of Hg(II) to Hg^0 is mediated by both microbial and chemical mechanisms (see following) and results in transport of Hg^0 from soils and waters to the atmosphere, due to its high vapor pressure (air in equilibrium with Hg^0 contains 13.2 ng cm^{-3} at 20°C, H = 0.3 dimensionless) and its low water solubility (6×10^{-6} g 100 ml^{-1} at 25°C). Volatilization (degassing) of Hg^0 from soils and waters is a natural process that is accelerated by volcanic and geothermal activities. The total input of mercury to the atmosphere has been estimated at 10^{10} g yr^{-1}, with anthropogenic sources accounting for 10–50% of this flux.

Bacteria in pure culture have been shown to oxidize Hg^0 to Hg(II) (as will be discussed). This reaction might play a role in returning Hg^0 to the pool of Hg(II) that is available for methylation in lakes, where the production of Hg^0 exceeds its transport rate to the atmosphere, resulting in Hg^0 supersaturation in the water. The oxidation of Hg^0 in natural waters has not been investigated to date.

Atmospheric mercury exists as gas or is adsorbed to particulates (Hg$_{[p]}$). The gaseous fraction is comprised of Hg^0, $CH_3Hg(I)$, and trace concentrations of Hg(II), $(CH_3)_2Hg$, and chlorinated forms. The residence time of mercury in the atmosphere depends on its physical and chemical properties. Thus, Hg$_{(p)}$ is subjected to dry deposition at short distances from its source, and Hg(II) and $CH_3Hg(I)$ are returned to the surface locally with rain (due to their high water solubility). Elemental mercury, on the other hand, is globally distributed and has a long atmospheric residence time (of about 1 yr). Oxidation of Hg^0 in the atmosphere by H_2O_2 (formed by the reaction of H_2O with O_3) to Hg(II) may proceed under acidic conditions and results in the subsequent wet deposition of mercury as Hg(II).

III. MICROBIAL TRANSFORMATIONS OF MERCURY

Microorganisms mediate four transformations of mercury; they reduce Hg(II) to Hg^0, they break down $CH_3Hg(I)$ (and other organomercury compounds), they methylate Hg(II), and oxidize Hg^0 to Hg(II) (Fig. 1). All four transformations are also mediated by chemical reactions in the environment. A summary of known mechanisms of mercury transformations is presented in Table III. The following sections are a summary of the current state of knowledge regarding the molecular, biochemical, and genetic mechanisms of microbial mercury transformations. For comparison, chemical transformations are briefly described.

A. Mercury Resistance

Because other articles in this book cover microbial resistance to mercury, this section briefly summa-

TABLE III
Mechanisms of Mercury Transformations

	Hg(II) reduction	*CH₃Hg(I) degradation*	*Methylation*	*Hg(O) oxidation*
Biological	• Enzymatic–mercuric reductase	• Enzymatic–organomercurial lyase	• Methyl-group transfer by corrinoid coenzyme (bacteria)	• Hydroperoxidases (e.g., catalase)
	• Indirect–reduced metabolites	• Oxidative demethylation–anaerobic bacteria	• "Incorrect" methionine synthesis (fungi)	
Abiological	• Free radicals associated with humic substances	• Photodegradation	• Photochemical	• H_2O_2-mediated under acidic conditions in the atmosphere
	• Disproportionation of Hg(I)		• Humic and fulvic acid induced	
			• Transformation of $CH_3Hg(I)$ to $(CH_3)_2Hg$ in presence of H_2S	

rizes this knowledge and places it in the context of microbial tolerance to mercury among natural microbial communities.

Mercury-resistant bacteria are widely distributed and quite ubiquitous in nature. Resistant organisms account for as many as 1–10% of aerobic heterotrophic bacteria, and they can be isolated without prior enrichment. However, resistant strains are more abundant in mercury-polluted environments, where 50% may grow on fully supplemented media in the presence of as high as 50 μM Hg(II) (sensitive strains can tolerate concentrations at the 1 μM range, on these media). The mechanism of resistance has been investigated in representatives of numerous aerobic bacterial species (Table IV) and was found to be due to the reduction of Hg(II) and its subsequent spontaneous volatilization as Hg^0. Among strains resistant to Hg(II), approximately 10–30% also tolerate organomercury compounds. Genetic determinants for mercury reduction are divided into two classes: narrow spectrum, those encoding tolerance and reduction of Hg(II), and broad spectrum, those encoding tolerance and reduction of organomercury compounds, as well as of Hg(II) (Fig. 2). Broad-spectrum mercury resistance is conferred by degradation of organomercury to Hg(II) and a reduced organic moiety. Thus formed, Hg(II) is subsequently reduced to Hg^0. Because Hg^0 is volatile, the reduction of Hg(II) and the break down of organomercury lead to the removal of mercury from the immediate environment of the organism. Thus, these are detoxification mechanisms. In this regard, mercury resistance is unique among mechanisms of metal tolerance. Other metals may be chemically transformed by microbial activities, yet tolerance is commonly mediated by mechanisms that restrict accessibility of metals to sensitive

TABLE IV
Eubacteria Shown to Tolerate Hg(II) by Its Reduction to Hg^0

Gram-negative
Enteric bacteria

Serratia marcescens	*Rhodococcus*
Klebsiella spp.	*Yersinia* spp.
Proteus spp.	*Paracoccus*
Providencia spp.	*Aeromonas* spp.
Shigella spp.	*Alcaligenes* spp.
Escherichia coli	*Azotobacter*
Salmonella spp.	*Citrobacter*
Enterobacter spp.	*Erwinia*
Bacteroids fragilis	*Flavobacterium*
	Morganella
	Xanthomonas
	Pseudomonas spp.

Gram-positive
Staphylococcus aureus
Group B *streptococcus*
Streptomyces spp.
Bacillus spp.
Mycobacterium spp.
Clostridium perfringens
Planococcus

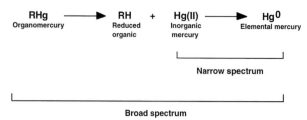

Fig. 2. Classification of mercury resistance determinants. Narrow spectrum encodes tolerance to Hg(II) salts; broad spectrum encodes tolerance to a variety of organomercury compounds in addition to resistance to Hg(II).

targets in the interior of the cell. An exception to the universality of Hg(II) reduction as a mechanism of mercury resistance among aerobes was described for *Enterobacter aerogenes,* where a loss of a plasmid-encoded membrane protein resulted in a decreased tolerance and increased uptake of Hg(II). This mechanism has not been demonstrated in other bacteria, nor has its distribution among environmental isolates been investigated. A protective role of intra- and extracellular reduced glutathione against mercurials and other metals has also been proposed.

Little information is available regarding resistance to Hg(II) and organomercury among anaerobes. It is believed that metal toxicity is alleviated in anaerobic environments by interactions with reduced sulfides. Yet, the susceptibility of anaerobic processes, such as methanogenesis and sulfate reduction, to the presence of metals indicates that metal tolerance among anaerobic microbes is a topic that needs to be addressed. Clinical and sewage isolates of *Bacteroides ruminicola* and *Clostridium perfringens* reduced Hg(II) to Hg^0, yet this activity was not inducible and was not plasmid-mediated, as is commonly found among aerobic bacteria. In *Clostridium cochlearium,* resistance to $CH_3Hg(I)$ was due to formation of Hg(II) and its subsequent sequestering by $S^=$, and in the case of a highly $CH_3Hg(I)$-tolerant *Desulfovibrio desulfuricans,* the formation of $(CH_3Hg)_2S$ and the subsequent slow disproportionation to the volatile $(CH_3)_2Hg$ and HgS was proposed as a resistance mechanism. This mechanism would suggest that methylation, if it results in the formation of the volatile $(CH_3)_2Hg,$ may be considered a detoxification mechanism. Indeed, a vitamin B_{12}-requiring auxotroph of an anaerobic bacterium was more sensitive

to Hg(II) than its wild-type strain. Because vitamin B_{12} it thought to be the methyl-donating agent in methylation (see following), this observation supports the role of Hg(II) methylation in detoxification. However, $CH_3Hg(I)$, rather than $(CH_3)_2Hg$, is commonly identified as the product of microbial methylation, and $CH_3Hg(I)$ is at least as toxic as Hg(II) to microorganisms.

B. Reduction of Hg(II)

1. *Mechanisms of Hg(II) Reduction*

Microorganisms reduce Hg(II) directly by metabolic functions that evolved specifically to cope with mercury toxicity or indirectly by means of oxidizable microbial metabolites.

a. Direct Reduction by *mer*-Specified Reactions

The reduction of Hg(II) and organomercury compounds to Hg^0 is specified by the mercury (*mer*)-resistance operon. The following is a brief description of this operon, its mode of regulation, and the activities of *mer*-encoded enzymes.

The *mer* operon is composed of genes that encode the transport of mercurials into the cell, genes encoding transformations of mercurials, and genes regulating the operon's expression (Fig. 3). Two *mer*-specified functions are involved in the active transport of Hg(II) into the cell. The scavenging of Hg(II) in the periplasmic space is carried out by MerP (the gene product of *merP*) and the transfer of Hg(II) across the cytoplasmic membrane, which is mediated by MerT. Transmembrane transfer of Hg(II) is commonly encoded by *merT* but, in some organisms, other genes, *merC* or *orfF*, encode for proteins with a similar function. Once in the cytoplasm, Hg(II) is reduced by the mercuric reductase (the *merA* gene product), a pyridine nucleotide–disulfide oxidoreductase. Organomercurial lyase (the gene product of *merB*), the enzyme that splits the Hg–C bond in a broad range of organomercurial compounds, is located in the cytoplasm of bacteria with a broad spectrum resistance to mercury. The action of the *mer*-specified functions is dependent on the sequential interaction of Hg(II) with cysteine residues of the respective Mer polypeptides in a process that has been likened to "a bucket brigade" operation. Thus,

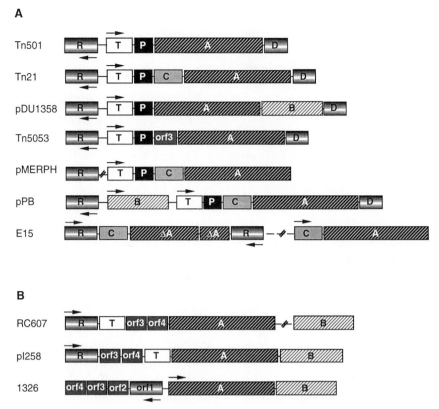

Fig. 3. The relatedness among several *mer* operons originating in gram-negative (A) and gram-positive (B) bacteria. Genes are depicted as boxes with solid filling for genes involved in transport functions, stippled filling for genes specifying mercury transformations, and gradient fillings for regulatory genes. Arrows show direction of transcription and horizontal lines show spaces between genes. Large spaces between genes indicate promoter regions. Genes separated by broken horizontal lines with vertical slashes indicate that there is no physical continuity between those genes and the rest of the operon. The depicted systems were originally isolated from: Tn501–*Pseudomonas aeruginosa*; Tn21–*Shiegella flexneri*; pDU1358–*Serratia marcescens*; Tn5053–*Pseudomonas fluorescens*; pMERPH–*Pseudomonas putrifaciens*; pPB–*Pseudomonas stutzeri*; E15–*Thiobacillus ferrooxidans*; RC607–*Bacillus* sp.; pI258–*Staphylococcus aureus*; 1326–*Streptomyces lividans*. Prepared after Osborn *et al.*, 1997.

Hg(II) is transferred from MerP through MerT to the reductase in the cytoplasm. The product of the reduction, Hg^0, diffuses out of the cell due to its low aqueous solubility and high vapor pressure (see preceding). The mode by which organomercury is transported into the cell is, at present, unknown. At least in one case, that of the soil bacterium, *Pseudomonas* sp.K62, MerP and MerT were involved in transport of phenylmercuric acetate but not of $CH_3Hg(I)$.

The gene products of two genes, *merR* and *merD*, act as the regulators of *mer* operons. MerR binds to the operator/promoter region of the operon, together with RNA polymerase, to form a complex that, in the absence of Hg(II), bends the DNA in an angle that does not allow transcription of the operon's structural genes. When present in the cell cytoplasm, Hg(II) induces the formation of a tricoordinate metal-bridged MerR by interacting with cysteine residues and this binding causes a conformational change leading to a realignment of the -10 with the -35 regions of the DNA and the commencement of transcription. Thus, *mer* is positively and negatively regulated by MerR. MerR also represses its own transcription. The product of the *merD* gene acts as a negative regulator of the *mer* operon.

The structures of several *mer* operons are now known and their comparative analysis shed light on the diversity and evolution of mercury resistance. The most important conclusions from these analyses stress that *mer* evolution is governed by recombinations of genes that are transferred horizontally among bacteria, a proposition that is supported by the frequent association of *mer* with conjugal plasmids and transposable elements, and that this trait emerged early during microbial evolution. Generally, *mer* operons contain an organizational variation of the genes and regulatory elements that have been described (Fig. 3). The presence or absence of *merC* is the basis for the classification of *mer* among gram-negative operons. The amino-acid sequences of regions critical for the activity of Mer functions (e.g., cysteine pairs involved in Hg[II] binding, the active site of the mercuric reductase) show a strong homology among all operons. Findings on the homology and diversity among *mer* operons have led to our present view of the operon as having a mosaic structure where (at least among gram-negative bacteria) a core structure consisting of *merR*, *merT*, *merP*, and *merA*, where homology is well preserved, is accompanied by auxiliary genes, *merC*, *merC*-like, and *merB*, whose amino acids and DNA sequences are more diverse. These auxiliary genes are inserted at either 3′ or 5′ to *merA*.

b. Indirect Microbial Reduction of Hg(II)

Some algae reduce Hg(II) to Hg⁰. Volatilization by *Chlorella pyrenoidosa* was dependent on light and was inhibited by 3-(3,4-dichlorophenyl)-1,1-dimethyl urea, a specific inhibitor of photosynthesis, a finding that could be explained if organic acids that are produced during photosynthesis were the reducing agents (the extracellular nature of this process was not investigated). However, loss of Hg(II) as Hg⁰ from growing cultures of *Chlamydomonas reinhardi* was the same when incubations were carried out under aerobic/light or anaerobic/dark conditions and the activity was destroyed by treatments with trichloroacetic acid. Thus, various microbial metabolites indirectly reduce Hg(II).

2. Significance of Microbial Hg(II) Reduction in the Geochemical Cycling of Mercury

Bacteria that reduce Hg(II) by the *mer*-mediated mechanism are ubiquitous in nature, yet their role may be limited to locations highly contaminated by anthropogenic inputs of mercury. Concentrations of Hg(II) in most natural environments are too low to effectively induce the *mer* operon (e.g., natural waters contain mercury concentrations at the 10^{-12}–10^{-15} M range). However, strong evidence indicates that biological activities contribute significantly to the flux of Hg⁰ to the atmosphere. The nutrient-enriched waters near the equatorial upwelling in the Pacific Ocean are supersaturated with Hg⁰, and latitudinal transects showed Hg⁰ concentrations to be positively correlated with biological productivity. Biologically formed Hg⁰ in natural water with low Hg(II) concentrations may be microbially mediated in an indirect process driven by reduced microbial metabolites (see preceding).

Physicochemical means of Hg(II) reduction include free radicals associated with humic substances and the disproportionation of 2Hg(I). Organic acids, even weak ones such as those amply present in natural waters and soils, can reduce Hg(II) to 2Hg(I), which then disproportionates spontaneously to Hg⁰ and Hg(II):

$$2\text{Hg(I)} \rightleftarrows \text{Hg}^0 + \text{Hg(II)}$$

Disproportionation may be an important step in the mechanism by which reduced microbial metabolites affect conversion of Hg(II) to Hg⁰.

C. Degradation of Organomercury Compounds

1. Microbial Degradation of CH₃Hg(I)

Bacteria degrade organomercury compounds (among them CH₃Hg[I]) by the activity of the organomercurial lyase, which is specified by the *merB* gene. The enzyme cleaves the Hg–C bond in a broad range of organomercury compounds by a protonolytic attack that results in the production of a thiolated Hg(II) and a reduced organic moiety (CH_4 in the case of $CH_3Hg[I]$). The thiolated Hg(II) serves as a substrate for the mercuric reductase, resulting in evolution of Hg^0. Phenylmercuric acetate acts as an inducer of broad spectrum *mer* operons (i.e., containing *merB*), and the carboxy terminal of MerR, from such operons, is thought to facilitate interaction with organomercury.

Sediment samples that are incubated with $^{14}CH_3Hg(I)$ emit a mixture of $^{14}CO_2$ and $^{14}CH_4$, indicating the existence of an oxidative demethylation mechanism. Furthermore, the proportion of CO_2 is increased when the sediments are incubated anaerobically. This observation suggests that oxidative demethylation occurs under anaerobic conditions, while, under aerobic conditions, the *mer*-specified reaction is predominant. Experiments with specific inhibitors and amendments with electron acceptors indicate that both methanogens and sulfate-reducing bacteria contribute to oxidative demethylation. It has been proposed that pathways analogous to monomethylamine degradation and acetate oxidation by methanogens and sulfate-reducing bacteria, respectively, are responsible for oxidative demethylation. The role in nature of bacteria with broad-spectrum *mer* operons in the breakdown of organomercury is probably limited to mercury-contaminated environments where concentrations are high enough to efficiently induce the operon.

2. *Abiological Mechanisms of CH₃Hg(I) Degradation*

The photodegradation of CH3Hg(I) was recently described. Illumination of lake waters containing ng L^{-1} concentrations of $CH_3Hg(I)$ resulted in degradation rates that were directly related to $CH_3Hg(I)$ concentrations and to the intensity of photosynthetically active radiation. The nature of the degradation products and the specific wavelength that is responsible for this activity are, at present, unknown so that a mechanism for photodegradation cannot be proposed at present. Photodegradation is most likely the major mechanism of $CH_3Hg(I)$ degradation in clear freshwater lakes because incubation of water samples in the dark results in a 350-fold decrease in the

turnover of $CH_3Hg(I)$ relative to light incubations. In darker waters and sediments where light intensity is abolished, microbial activities are likely the dominant mechanism of $CH_3Hg(I)$ degradation.

D. Methylation of Mercury

1. *Microbial Mechanisms of Methylation*
a. *Corrinoid-Mediated Methylation*

Many species of bacteria and fungi can methylate Hg(II). This biotransformation requires transfer of a methyl group to Hg(II). Three coenzyme groups are known to participate in methyl transfer reactions: (1) N^5-methyltetrahydrofolate, (2) *S*-adenosylmethionine, and (3) methylcorrinoids, such as methylcobalamine (CH_3–B_{12}). Of the three, CH_3–B_{12}, coenzymes are the only coenzymes that transfer a carbanion (in this case, a negatively charged CH_3) and are, therefore, the only coenzymes that could be directly involved in biological methylation of positively charged Hg(II). This rationale guided early efforts aimed at the isolation of methylating factors from bacteria that produce CH_3–B_{12} under anaerobic conditions. Thus, a Hg(II)-methylating factor in cell-free extracts of a methanogen that had been isolated from sewage sludge was described. Methanogens generate methane and reduced B_{12} (B_{12-r}) by reacting CH_3–B_{12} with hydrogen. It was suggested that $CH_3Hg(I)$ was formed by the spontaneous transfer of CH_3 from the CH_3–B_{12} and the regeneration of CH_3–B_{12} by a methyl-transfer reaction, using N^5-tetrahydrofolate as a donor. Based on these observations, methanogenic bacteria were implicated in Hg(II) methylation in anaerobic sediments. However, in the mid-1980s, experiments with anaerobic sediments and specific inhibitors of methanogens (2-bromoethane sulfonic acid) and of sulfate-reducing bacteria (SRB) (MoO_4^{2-}) suggested that, *in situ,* the later rather than the former methylate Hg(II). Several pure cultures of Hg(II)-methylating SRB were subsequently isolated from saline and freshwater sediments. Additional experiments suggested that methylation by *Desulfovibrio desulfuricans* involved the biosynthetic pathway of acetyl coenzyme A (acetyl-CoA). The presence of Hg(II) inhibited acetyl-CoA synthesis by cellfree extracts of the active bacterium and propyl iodide, an inhibitor of transmethylation

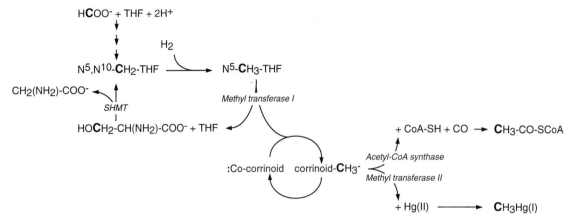

Fig. 4. A proposed mechanism for the methylation of Hg(II) *by Desulfovibrio desulfuricans* LS. The carbon that is incorporated into CH$_3$Hg(I) is indicated by a bold letter. Enzyme names are italicized; SHMT, serine hydroxymethyl transferase; CoA, Coenzyme A; THF, tetrahydrofolate. Prepared after Choi *et al.*, 1994.

by CH$_3$–B$_{12}$, inhibited both acetyl-CoA and CH$_3$Hg(I) formation. Based on these observations, it was proposed that the biosynthesis of CH$_3$Hg(I) proceeds by a transfer of CH$_3^-$, that originates in either foramate or serine, from N^5-CH$_3$-tetrahydrofolate to a corrinoid coenzyme and a subsequent enzymatically mediated methyl transfer to Hg(II) (Fig. 4).

Mercury methylation by SRB presents a paradox. Sulfate is required as an electron acceptor in the metabolism of SRB, yet the product of SO$_4^{2-}$ reduction, S$^=$, reacts with Hg(II) to form HgS, limiting availability of the substrate for the methylation reaction. It is conceivable that *in situ* Hg(II) methylation occurs within a narrow range of SO$_4^{2-}$ concentrations, allowing SRB metabolism but not inhibiting methylation.

b. Other Mechanisms of Methylation

Microorganisms not known to form CH$_3$–B$_{12}$ can methylate mercury. Thus, in *Neurospora crassa*, methylation serves to detoxify mercury and it is carried out by an "incorrect" synthesis of methionine (B$_{12}$ is not produced by *N. crassa*). Resistant mutants accumulated CH$_3$Hg(I), and this accumulation was stimulated by homocysteine or L-cysteine. According to the proposed mechanism, the transmethylase involved in the conversion of homocysteine to methionine uses Hg-homocysteine as a substrate.

2. Abiological Mechanisms of Methylation

Methylmercury may be formed abiologically in soils, sediments, and natural waters. Two mechanisms for methylation of Hg(II) and a mechanism for the formation of (CH$_3$)$_2$Hg from CH$_3$Hg(I) have been described (Table III). Humic and fulvic compounds, particularly those with low molecular weight, methylate Hg(II). Methylation rate increases with temperature and Hg(II) concentration and is optimal at pH <5. The reaction is stimulated by the presence of other metals (Fe, Cu, Mn, and Al) that act as catalysts.

Researchers have described a photochemical methylation of Hg(II) that occurs in the presence of methyl group donors, such as methanol or ethanol, but most effectively with acetic acid. Methylation is optimal at light of wavelengths <420 nm and pH <5. It is stimulated by absence of O$_2$, but methylation does occur under aerobic conditions and in the presence of colored compounds that act as photosensitizers. This may be one mechanism that is responsible for the production of CH$_3$Hg(I) in sunlight-irradiated sediments.

Formation of the volatile (CH$_3$)$_2$Hg by a reaction between CH$_3$Hg(I) and H$_2$S was observed in anoxic sediments. A possible mechanism to explain this observation proposed the formation of volatile ([CH$_3$Hg]$_2$S). This compound is unstable and disproportionates to $(CH_3)_2Hg$ and HgS. The significance of this reaction in nature is unclear because of the rarity of $(CH_3)_2Hg$ in natural waters and sediments (see preceding).

3. Significance of Microbial Methylation in the Mercury Cycle

Although several abiological mechanisms for Hg(II) methylation have been described and conditions for these reactions commonly exist, quantitative analysis indicated that, at least in anoxic sediments, biotransformations account for 90% of the methylmercury formed. Methylation is believed to be a slow process (no reliable methods for rate determinations are currently available) and concentrations of $CH_3Hg(I)$ in natural waters are extremely low (in the ng L^{-1} range). Even in mercury-contaminated environments having high concentrations of total mercury, $CH_3Hg(I)$ levels are low; however, these low concentrations are sufficient to cause accumulation in aquatic biota.

Methylation is widely accepted as an anaerobic process that occurs in anoxic environments; however, methylation also proceeds in aerated sediments and waters, although with lower yields. Methylation under aerobic conditions may be due to the existence of anaerobic niches in aerated environments or to activities of aerobic microorganisms that methylate Hg(II).

D. Elemental Mercury Oxidation

The ability of bacteria to oxidize Hg^0 to Hg(II) was recently described in *Escherichia coli* and two soil bacteria, *Bacillus* and *Streptomyces*. This transformation was attributed to hydroperoxidases (e.g., catalase) and other, as yet unidentified, enzymes in *E. coli*. A role for this transformation in the geochemical cycling of mercury remains to be established.

See Also the Following Articles

HEAVY METAL POLLUTANTS: ENVIRONMENTAL AND BIOTECHNO-LOGICAL ASPECTS • HEAVY METALS, BACTERIAL RESISTANCES

Bibliography

Choi, S.-C, Chase, Jr., T., and Bartha, R. (1994). Metabolic pathways leading to mercury methylation in *Desulfovibrio desulfuricans* LS. *Appl. Environ. Microbiol.* **60**, 4072–4077.

Marvin-Dipasquale, M. C., and Oremland, R. S. (1998). Bacterial methylmercury degradation in Florida Everglades peat sediments. *Environ. Sci. Technol.* **32**, 2556–2563.

Osborn, A. M., Bruce, K. D., Strike, P., and Ritchie, D. A. (1997). Distribution, diversity and evolution of the bacterial mercury resistance (*mer*) operon. *FEMS Microbiol. Rev.* **19**, 239–262.

Sellers, P., Kelly, C. A., Rudd, J. W. M., and MacHutchon, A. R. (1996). Photodegradation of methylmercury in lakes. *Nature* **380**, 694–697.

Silver, S., and Phung, L. T. (1996). Bacterial heavy metal resistance: New surprises. *Annu. Rev. Microbiol.* **50**, 753–789.

Smith, T., Pitts, K., McGarvey, J. A., and Summers, A. O. (1998). Bacterial oxidation of mercury metal vapor, Hg(0). *Appl. Environ. Microbiol.* **64**, 1328–1332.

Summers, A. O. (1992). Untwist and shout: A heavy metal-responsive transcriptional regulator. *J. Bacteriol.* **174**, 3097–3101.

Weber, J. H. (1993). Review of possible paths for abiotic methylation of mercury (II) in the aquatic environment. *Chemosphere* **26**, 2063–2077.

Metal Extraction and Ore Discovery

James A. Brierley
Newmont Metallurgical Services, Englewood, CO

GLOSSARY

autotroph Microorganism capable of using CO_2 as a sole source of carbon for biosynthesis and obtaining its energy from either sunlight (photolithotroph) or oxidation of inorganic elements and compounds (chemolithotroph).

biogenic Produced or formed by organisms, such as the formation of some metal sulfide deposits.

bioleaching Solubilization of metals from minerals by microbially mediated oxidation or reduction reactions.

biosorption Binding of anionic or cationic metals and metalloids to living or dead biomass.

peptidoglycan Bacterial cell-wall saclike polymer composed of glycan chains cross-linked by short peptide units; also called mucopeptide or murein.

DIVERSE MICROORGANISMS interact with minerals and metals to either solubilize metal species or concentrate metals from solution. The microbial activities that lead to either metals' mobilization or sequestration include enzymatic-catalyzed processes, organic compound complexation, precipitation, and physico-chemical adsorption on cell surfaces.

Metals' solubilization occurs via aerobic respiration processes, in which minerals are oxidized for an energy source. Anaerobic respiration processes also cause metals' solubilization. The mineral serves as an electron acceptor and the metal is solubilized under reducing conditions. Metals are also mobilized with organic chelating compounds produced by microorganisms. This activity is associated with providing to bacteria metal nutrients, such as iron, which are not otherwise available unless chelated.

Metals' concentration is a result of transport of the metal to the interior of the cell for enzymatic reduction, a process sometime associated with detoxification of metals. The cell-wall structure of many microorganisms has functional groups that adsorb and concentrate metal species. Bacteria also precipitate metals, either by direct enzymatic reduction or by generation of sulfide via anaerobic reduction of sulfate. Sulfide is a powerful precipitant of heavy metals.

I. BASIC REACTIONS OF MICROBIAL SOLUBILIZATION OF METALS

Metals' solubilization from mineral substrates occurs as a consequence of oxidation (respiration) of the mineral, to couple the available energy with the cells' anabolic metabolism. One group of microorganisms, including bacteria and Archaea, is able to oxidize sulfide minerals, such as pyrite (Eq. 1) and chalcocite (Eq. 2) for energy. The iron and copper of the respective

$$4FeS_2 + 15O_2 + 2H_2O \rightarrow 2Fe_2(SO_4)_3 + 2H_2SO_4 \quad (1)$$

$$2Cu_2S + 5O_2 + 2H_2SO_4 \rightarrow 4CuSO_4 + 2H_2O \quad (2)$$

minerals become solubilized, or leached, as metal sulfates. This aerobic respiration process is common to the chemolithotrophic autotrophic bacteria, that also utilize CO_2 as their carbon source, concurrent with oxidation of iron and sulfur in minerals. The bacteria commonly associated with mineral oxidation include the mesophilic *Thiobacillus ferrooxidans, Leptospirillum ferrooxidans,* moderately thermophilic *Sulfobacillus thermosulfidooxidans,* and thermophilic *Acidianus brierleyi, Metallospheara sedula,* and strains of the genus *Sulfolobus.* Biooxidation of sulfur (Eq. 3) is also an aerobic respiration process associated

$$S^0 + 1\tfrac{1}{2}O_2 + H_2O \rightarrow H_2SO_4 \qquad (3)$$

with *Thiobacillus thiooxidans* and other bacteria. This metabolic process leads to acidification of an environment or an industrial process in which these types of bacteria function. Acidic conditions, which can be pH 2.0 or less, dissolve numerous minerals species, resulting in leaching of metals, metalloids, and alkaline earth elements.

Conversely, bacteria capable of anaerobic respiration can also solubilize metals. In this case, the metal acts as the electron acceptor for the bacterial oxidation of either organic or inorganic substrates. Several types of bacteria grow using iron or manganese oxide minerals as electron sinks for oxidation of organic substrates. Organisms of the genera *Shewanella* and *Geobacter* reduce and solubilize the iron and manganese from their mineral forms (Eqs. 4 and 5). These genera

$$\tfrac{1}{2}MnO_2 + 2H^+ + e^- \rightarrow \tfrac{1}{2}Mn^{2+} + H_2O \qquad (4)$$
$$FeOOH + 3H^+ + e^- \rightarrow Fe^{2+} + 2H_2O \qquad (5)$$

oxidize organic substrates during the course of reduction of iron or manganese. Enzymes catalyze these reactions. Dissimilatory ferric-iron reductases, which reduce Fe^{3+} to Fe^{2+} (Eq. 5), are membrane-bound systems induced by anoxic conditions. Dissimilatory manganese reductase activity, solubilizing manganese as reduced Mn^{2+} (Eq. 4), is also membrane-bound, but little is known about this system.

II. MICROBIAL EXTRACTION OF METALS FROM MINERALS

Microbial extraction of metals from minerals, a process termed bioleaching, has been applied on an industrial scale for recovery of copper, uranium, and gold. Heavy metals (cadmium, copper, iron, molybdenum, nickel, and zinc) are solubilized from their sulfide–mineral matrices as a consequence of mineral biooxidation. The bioleach process is also a consequence of acid produced from oxidation of sulfur and mineral sulfide. The sulfuric acid produced solubilizes a wide range of metals and metalloids. These are aerobic metal-mobilization processes, a function of microbial respiration activity.

Microbial processes also play a role in metals' solubilization in anaerobic conditions, using anaerobic respiration as has been discussed. Acidophilic bioleaching bacteria, *T. ferrooxidans* and *S. thermosulfidooxidans,* promote metal solubilization in oxygen-depleted environments. Ferric iron serves as the electron acceptor for oxidation of sulfide minerals by *T. ferrooxidans. Sulfobacillus* uses either ferric iron or iron oxide minerals (e.g., ferric hydroxide, jarosite, and goethite) as electron acceptors for mineral oxidation. Likely, anaerobic bioleaching is of less importance for commercial mineral biooxidation processes, compared with aerobic bioleaching, as oxygen limitation in these processes reduces the activity of bioleaching of mineral sulfides.

III. BIORECOVERY OF METALS FROM SOLUTION

Biorecovery of metals from solution, a process referred to as "biosorption," occurs by either active or passive mechanisms. Active metal transformation processes require viable microbes, enzymatically catalyzing the alteration of the metal, leading to sequestration or concentration. Three mechanisms are involved in active metal transformations: (1) precipitation, (2) intracellular accumulation and complexation, and (3) oxidation/reduction. Passive transformations are physical–chemical processes that do not require active viable microorganisms, but do involve products of the microorganism (Fig. 1). Mechanisms of passive biosorption include (1) metal chelation by extracellular microbial products and (2) physical–chemical binding of metals to the surface structures of microorganisms.

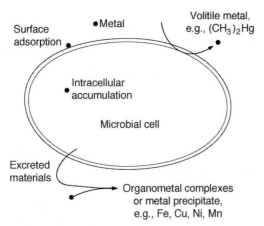

Fig. 1. Diagrammatic relationship among the processes involved in microbial transformation of metals.

A. Precipitation

Metals are precipitated as highly insoluble sulfides resulting from the metabolic activity of anaerobic sulfate-reducing bacteria. There are several genera of sulfate-reducing bacteria, including *Desulfovibrio* and *Desulfotomaculum*. Other genera also capable of reducing sulfate to sulfide are *Desulfococcus*, *Desulfobacter*, *Desulfobulbus*, *Desulfosarcina*, and *Desulfonema*. Bacterial sulfate-reduction, using an oxidizable organic substrate, by these bacteria is typified by Eq. (6). Sodium

$$2CH_3CHOHCO_2Na$$
$$+ MgSO_4 \rightarrow 2CH_3CO_2Na$$
$$+ CO_2 + MgCO_3 + H_2O \quad (6)$$

lactate is oxidized with the concurrent reduction of sulfate. Elemental sulfur also serves as an electron acceptor for some bacteria and forms sulfide. Some of the sulfate-reducing bacteria grow mixotrophically, oxidizing hydrogen (Eq. 7) and using organic compounds as a carbon source.

$$SO_4^{2-} + 4H_2 \rightarrow S^{2-} + 4H_2O \quad (7)$$

Precipitation of heavy metals by the activity of sulfate-reducing bacteria is used for treatment of water from mining and milling operations. Purposely built wetlands, in which the microbial sulfate-reduction activity is primarily responsible for metal precipitation, treat metalliferous mine drainage. The metal–sulfide precipitates accumulate in the anaerobic sediment layers of the wetland. Heavy metal-contaminated ground water is also treated using sulfate-reducing bacteria, growing on either methanol or hydrogen in large tank-reactors. The contaminated water is pumped to large anaerobic bioreactor tanks, where the bacteria generate sulfide for precipitation of the metals in the form of a recoverable sludge. Residual sulfide is biologically oxidized to sulfur, a salable end-product.

Sulfate-reducing bacteria probably participated in the formation of economic deposits of sulfur. The world's largest known sulfur deposits, in Texas and Louisiana, are believed to be the result of biogenic processes attributable to participation of the sulfate-reducing microorganisms. Some sulfide ore deposits of iron, copper, and lead are also the consequence of metal precipitation with biogenic sulfide.

Bacteria of the genus *Citrobacter* precipitate metals by the reaction of a phosphatase enzymes cleaving glycerol 2-phosphate, liberating HPO_4^{2-}. The phosphate precipitates metals, which accumulate on the bacterial cell surface. Cadmium, lead, and uranium precipitate by this process.

B. Binding of Metals to Microbial Surfaces

Biosorption of soluble metals to bacterial cell surfaces is considered a metal-extraction process by which metals are concentrated from solutions. Biosorption mechanisms include ion-exchange, precipitation, and complexation. Surface accumulation of metals may be a process by which microorganisms protect themselves from the toxic effects of metals. The peptidoglycan of the cell-wall layer of bacteria is an active site for the metal-accumulation reactions. This layer is most effective in accumulation of heavy metal cations (e.g., Cd^{2+}, Cu^{2+}, Pb^{2+}, Zn^{2+}). Oxyanions (AsO_4^{2-}, CrO_4^{2-}, MoO_4^{2-}) are poorly accumulated by the peptidoglycan layer.

An unusual property of some bacteria and bacterial endospores is an ability to precipitate soluble gold. The initial precipitation reaction nucleates further deposition of gold. This mechanism has been hypothesized as the process that formed some of today's gold deposits (Fig. 2). Some filiform gold from placer

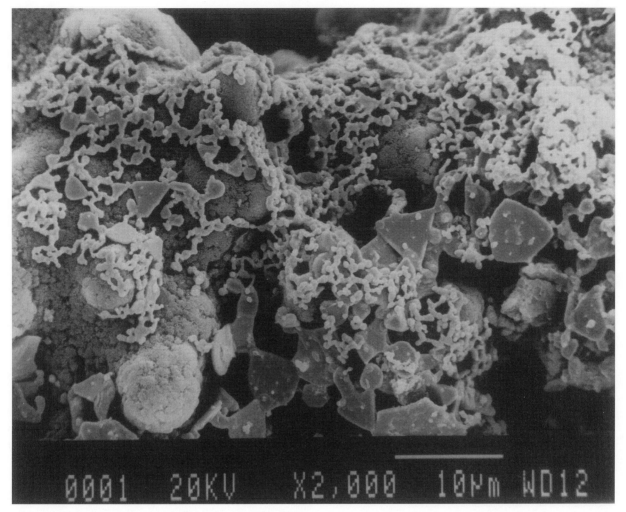

Fig. 2. Filiform or "mossy" gold deposit, magnified 2000×, similar to those deposits hypothesized to be of possible microbial origin (courtesy of Newmont Gold Company).

gold particles is interpreted as pseudomorphs of *Pedomicrobium*-like budding bacteria (Watterson, (1991)).

Nonliving bacteria, such as *Bacillus subtilis,* other bacterial species, fungi, and algae can be immobilized to form granular products that accumulate heavy-metal contaminants from process and wastewaters (Fig. 3). Immobilized nonliving biomass provides an alternative to conventional ion-exchange resins for recovery of metalloids and heavy and precious metals from dilute streams. Living biomass can also be immobilized in a nontoxic matrix for metal removal. A metal precipitation process requiring metabolically active microorganisms, as described above for *Citro-bacter,* would use immobilized living bacteria. How-ever, practical application of either nonliving or viable immobilized biomass for metals' removal has yet to be fulfilled on a large scale.

C. Intracellular Accumulation and Complexation

A wide array of organic ligands, produced by microorganisms, complex metal ions. Complexation leads to significant biochemical and geochemical transformations. Organic complexing agents are a means by which microorganisms obtain inorganic micronutrients, under conditions in which the inorganic elements are otherwise insoluble. For example, siderophores, such as ferrioxamine B, rhodotorulic

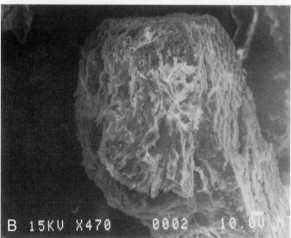

Fig. 3. *Bacillus subtilis* cells immobilized into granule form for use in removal and recovery of heavy metals from dilute wastewater solutions. (A) Granules used in column reactor systems. (B) SEM of single-granule product of *B. subtilis* cells magnified 470×; individual cells no longer discernible as a result of the granulization process.

acid, or enterobactin, complex iron for transport into the cell to fulfill biochemical needs. Siderophores are produced by a number of bacteria, including species of *Pseudomonas, Actinomyces, Azotobacter,* and *Arthrobacter.*

Metals accumulate within the cells of microorganisms. Diagnostic procedures for identification of bacteria include a microbe's ability for intracellular metals' accumulation. Oxyanions of the metalloids selenium and tellurium are internally reduced and precipitated by some bacteria. To what extent this impacts metal-transformation processes in nature is unknown, but the potential exists for biomass-related concentration of metals.

IV. ORE DISCOVERY: MICROBIAL PROSPECTING

Some microbiological techniques offer the potential for discovering ore bodies, but none is accepted or practiced on a commercial scale. Conventional geological prospecting techniques still remain the methods of choice by commercial mining firms for discovery of new ore bodies. However, these microbiological techniques offer some insight into the nature of mineral and microbe associations.

The bacteria associated with biooxidation of minerals, *T. ferrooxidans, L. ferrooxidans,* or *S. thermosulfidooxidans,* are indicators of the presence of sulfide mineralization, perhaps of sufficient size to classified as an ore body. The exclusive requirement of these bacteria for oxidizable mineral substrates restricts their occurrence to those conditions with metal sulfides. However, their occurrence does not necessarily indicate that an economic "ore body" is present.

A specific bacterium, *Bacillus cereus,* is also proposed as an indicator organism for occurrence of a metalliferous ore body. This microbe, or its spore form, is present in greater than background concentrations in soils associated with metal ore deposits.

Modern molecular techniques for measuring microbial density, diversity, and phylogeny are not fully developed as tools for bioexploration. However, further developments may result in their use as specific biosensors and microprobes in bioprospecting.

V. CONCLUSIONS

As a group, many microorganisms remarkably affect metals, leading to their extraction, either from insoluble minerals or from solutions. The latter process leads to concentration of metals. Metal solubilization is a function of either aerobic or anaerobic respiratory metabolism. The consequences can be environmentally deleterious, with an increase in soluble heavy metals, or the same processes can be beneficially applied in a controlled manner for economical extraction and recovery of metal values.

"Extraction" of metals from solution by microbes is accomplished by active metabolism or by sequestration on a cell's surface. These microbial activities are ongoing in cycling of metals between soluble and insoluble states. These activities lead to mineral deposits or they are used for water treatment.

Use of microorganisms in mineral exploration is not yet practical. Association of select groups of bacteria with a given habitat is useful as an indicator of biogeochemical processes ongoing in a particular defined environment.

See Also the Following Articles

Autotrophic CO₂ Metabolism • Heavy Metal Pollutants • Ore Leaching by Microbes

Bibliography

Banfield, J. F., and Nealson, K. H. (eds.) (1997). "Geomicrobiology: Interactions between Microbes and Minerals." Mineralogical Society of America, Washington, DC.

Beveridge, T. J., and Doyle, R. J. (eds.) (1989). "Metal Ions and Bacteria." Wiley, New York.

Eccles, H., and Hunt, S. (eds.) (1986). "Immobilisation of Ions by Bio-Sorption." Ellis Horwood, Chichester, UK.

Ehrlich, H. L. (1996). "Geomicrobiology." (3rd ed., revised and expanded). Marcel Dekker, New York.

Ehrlich, H. L., and Brierley, C. L. (eds.) (1990). "Microbial Mineral Recovery." McGraw-Hill, New York.

Nealson, K. H., and Saffarini, D. (1994). Iron and manganese in anaerobic respiration: Environmental significance, physiology, and regulation. *Annu. Rev. Microbiol.* 48, 311–343.

Hughes, M. N., and Poole, R. K. (1989). "Metals and Microorganisms." Chapman and Hall, New York.

Smith, R. W., and Misra, M. (eds.) (1991). "Mineral Bioprocessing." The Minerals, Metals & Materials Society, Warrendale, PA.

Volesky, B. (ed.) (1990). "Biosorption of Heavy Metals." CRC Press, Boca Raton, FL.

Watterson, J. R. (1991). Preliminary evidence for the involvement of budding bacteria in the origin of Alaskan placer gold. *Geology* 20, 315–318.

Methane Biochemistry

David A. Grahame and Simonida Gencic

Uniformed Services University of the Health Sciences, Bethesda, MD

I. Carbon Dioxide Reduction to CH_4
II. Methylated Substrate Reduction to CH_4
III. Acetate Cleavage to Carbon Dioxide and CH_4

GLOSSARY

acetyl-CoA decarbonylase/synthase (ACDS) Multienzyme complex responsible for synthesis and cleavage of acetyl-CoA in methanogens.

anaerobes Microorganisms capable of growth in the complete absence of molecular oxygen. Obligate anaerobes are those able to grow only under conditions in which oxygen is completely excluded. They are usually killed by oxygen.

antiporter Membrane transporter protein catalyzing coupled, simultaneous counter-transport across the membrane of two solutes in opposite directions.

archaea A group of prokaryotes that constitute a phylogenetically separate domain of organisms distinct from Eubacteria and Eukaryotes. Archaea are often found (although not exclusively) living in extreme terrestrial or aquatic environments, such as hot springs, undersea thermal vents, or under conditions of high alkalinity and/or concentrated saline.

chemiosmosis Process by which energy stored in the form of a transmembrane electrochemical gradient (generated by enzymatic reactions or electron transport) is coupled to the formation of ATP.

CO dehydrogenase Enzyme catalyzing redox interconversion of carbon monoxide and carbon dioxide.

coenzyme M (CoM) Small, two-carbon substrate (HS–CH_2CH_2–SO_3H), which acts as a characteristic methyl group carrier in all methanogenic pathways.

corrin Substance representing the core ring structure in vitamin B_{12}, that serves to coordinate the central cobalt atom. It is a cyclic tetrapyrrole ring system, in which three of the pyrrole rings are present in the dihydro form and the fourth is reduced to the tetrahydro level. Two of the pyrroles are bonded together directly, whereas methine group bridges provide the other three links between adjacent pyrrole rings.

electrochemical gradient Transmembrane free energy potential made up of two components, the concentration gradient (osmotic component) and the voltage difference across the membrane (electrical component).

ferredoxin A low molecular weight protein, containing one or more iron–sulfur clusters, that functions to transfer electrons from one enzyme system to another.

hydrogenase Enzyme capable of redox activation of molecular hydrogen.

methanogens Microrganisms that produce methane as metabolic end-product. Methanogens constitute a large group within the Archaea. They are strict anaerobes that utilize and are limited to growth on a relatively narrow range of simple one-carbon compounds and acetate.

methanogenesis The process of methane formation by methanogens.

methyl-CoM reductase Primary enzyme responsible for methane formation in all methanogens.

methylotrophic methanogens Methanogens capable of utilizing a variety of simple methylated compounds, which, in most cases, are able to serve as sole sources of carbon and energy.

porphin Parent compound of the porphyrins, a cyclic tetrapyrrole structure in which the pyrrole rings are connected by one-carbon methine groups.

THE BIOCHEMISTRY OF METHANE formation by methanogenic Archaea is both fascinating and unusual. Discoveries of novel cofactors and unique enzymatic mechanisms have turned up repeatedly in studies that have now defined several different methane-producing metabolic pathways. All of these pathways join together in the final step of methane formation, carried out by the unique enzyme methyl-coenzyme M reductase. Nearly 25 years after the discovery of coenzyme M (CoM) as the substrate of this reductase, we are now able to comprehend the enzymatic mechanism of methane production, and to understand many of the other critical enzymatic steps in these pathways, involving the activation and dissimilation of a number of methanogenic substrates including methanol, methylamines, methylthiols, H_2+CO_2, formate, and acetate.

The methanogens are obligate anaerobes that play an important role in the terminal stages of microbial decomposition in a wide range of anaerobic habitats. Methanogens growing on a variety of simple carbon substrates contribute approximately one-billion tons of methane to the yearly production of this gas on Earth; and, in the metabolic process of methanogenesis, they obtain essentially all of their energy requirements. The ecological role of methanogens is to remove the metabolic byproducts and reducing equivalents produced by other anaerobic organisms during decomposition of complex organic substrates. This allows other members in the anaerobic microbial ecosystem to proceed with reactions (e.g., H_2 formation) that would be otherwise thermodynamically unfavorable. Methane does not undergo further anaerobic reaction; thus, it serves well as the ultimate repository for metabolically generated reducing power in such environments. Since methane is a gas, the reducing equivalents are readily dissipated by simple diffusion into the aerobic world, where about two-thirds of it is then oxidized by methanotrophic bacteria, and the rest is released to the atmosphere. Because of their important ecological contribution, it is no surprise that methanogens are found in numerous types of anaerobic environments, including the digestive tracts of animals and humans, marine sediments, and fresh water sediments of rivers, lakes, marshes and swamps, and in industrial settings used for biomass conversion and sewage treatment. Depending on the ecosystem, different substrates may be available to support the growth of methanogens. Carbon dioxide and molecular hydrogen are common, and nearly all methanogens are able to obtain energy using H_2 in the reduction of CO_2 to methane. Nonetheless, nearly two-thirds of the methane

formed in the living world is derived from the cleavage of acetic acid—a methanogenic pathway that becomes increasingly thermodynamically favorable as H_2 levels in the environment decrease. A limited number of species, the methylotrophic methanogens, are also able to disproportionate a variety of methylated substrates (e.g., methylamines, methanol, methylsulfides) to methane and carbon dioxide. The following sections describe details of these three different biochemical pathways, originating from $H_2 + CO_2$, methylated substrates, and acetate, and culminating in the formation of methane. These transformations can be considered to operate in separate stages; they are presented here in a modular format, with each block defining a principal enzyme system and/or characteristic class of coenzymes.

I. CARBON DIOXIDE REDUCTION TO CH₄

The discovery of coenzyme M in the early 1970s was the initiation point from which rapid progress developed in unraveling the biochemistry of methanogenesis. Coenzyme M (2-mercaptoethanesulfonic acid) is the smallest of a number of unique and/or unusual coenzymes that participate in methane formation. Coenzyme M acts as the common acceptor for methyl groups, forming methyl-CoM ($CH_3-S-CH_2CH_2SO_3^-$) in all methanogenic pathways. The various enzyme cofactors utilized in these pathways are categorized, generally, as either electron carriers or one-carbon group carriers, and are described in Table I, with structures shown in Scheme I.

Reduction of carbon dioxide during growth on $H_2 + CO_2$ proceeds according to the stoichiometric relationship shown in Eq. (1). The first step involves

$$CO_2 + 4H_2 \rightarrow 2H_2O + CH_4 \qquad (1)$$

methanofuran (MFR, Table I) which acts as carbon dioxide-reduction cofactor in fixation of CO_2 at the formyl level. Reductive formylation of MFR by the enzyme formylmethanofuran dehydrogenase generates N-formyl-MFR, which is the first stable one-carbon bound species in the CO_2 reduction pathway, as shown in Fig. 1. Formyl-MFR dehydrogenase from

TABLE I
Methanogen Cofactors Acting as Electron and/or One-Carbon Carriers

Name	Molecular characteristics	Carrier function
Methanofuran (MFR)	2,4-disubstituted furan ring system, bearing an aminomethyl group at position 2 and connected to a complex side chain containing as many as five free carboxylate groups.	C-1 carrier. N-formyl attachment occurs on the furan 2-aminomethyl group. Formyl-MFR is the first stable product of CO_2 reduction to methane.
Tetrahydromethanopterin (H_4MPt)	Analog to tetrahydrofolate. A modified tetrahydropteridine ring system is attached to "methaniline" as side chain. Methaniline differs in several respects from the side chain of tetrahydrofolate, including linkage through an electron-donating methylene moiety in place of the electron withdrawing amide formed with *p*-aminobenzoic acid.	C-1 carrier. Similar to H_4 folate with functional groups N-5 and N-10. Carries one-carbon species at formyl, methenyl, methylene, and methyl oxidation states. The closely related tetrahydrosarcinapterin (H_4SPt) carries out similar functions in species of *Methanosarcina* and others.
Coenzyme F_{420} (Factor 420, F_{420})	8-hydroxy-5-deazaflavin derivative with intense blue-green fluorescence in the oxidized state and characteristic absorbance at 420 nm.	Electron carrier. Accepts electrons from H_2 catalyzed by F_{420}-dependent hydrogenase. Reduced by F_{420}-dependent formate dehydrogenase in methanogens growing on formate. Acts as hydride donor in reduction of methenyl-H_4MPt and methylene-H_4MPt.
Methanophenazine	2-hydroxyphenazine derivative with a polyisoprenoid side chain.	Electron carrier. Membrane-bound mediator of electron transport.
Coenzyme M (CoM)	2-mercaptoethanesulfonic acid. Simple two-carbon compound with one sulfonate and one sulfhydryl group.	C-1 carrier. Methylation of the thiol group forms thioether CH_3-S-CoM, the substrate directly reduced to methane.
Component B (CoB, HS–HTP)	7-mercaptoheptanoylthreonine phosphate. Free SH group present at the end of a seven-carbon fatty acyl chain; amide linkage to threonine phosphate forms a hydrophilic head	Electron carrier. Electron donor for methyl-CoM reduction to methane, oxidized to form mixed disulfide with CoM (CoM-S-S-HTP).
Coenzyme F_{430} (F_{430})	Nickel-containing porphinoid, yellow and nonfluorescent, highly reduced tetrapyrrole ring system with absorption maximum at 430 nm.	Both electron and C-1 carrier. Methyl-CoM reductase cofactor which accepts electrons from HS-HTP and a CH_3 group donated by methyl-CoM to form a methyl-Ni(II)-F_{430} species, capable of releasing methane.
Factor III (B_{12}–HBI)	Corrinoid (B_{12}) cofactor containing 5-hydroxybenzimidazole as α axial ligand to cobalt.	C-1 carrier. Acts as methyl transfer cofactor in activation of all methylotrophic growth substrates, methylates H_4SPt during cleavage of acetate, and participates in formation of CH_3-S-CoM from CH_3-H_4MPt (or one of its analogs) in nearly all methanogens.

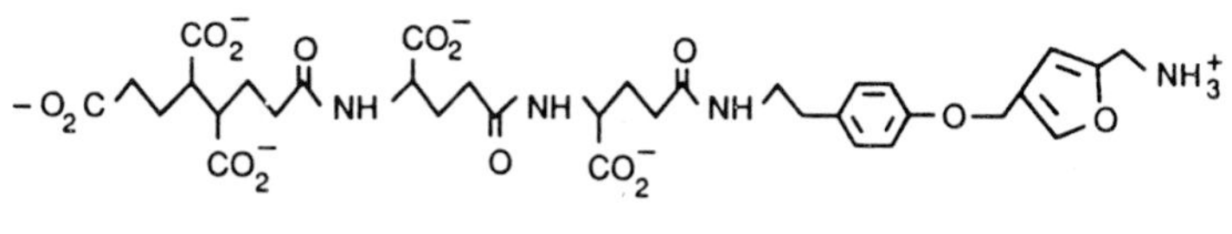

Methanofuran

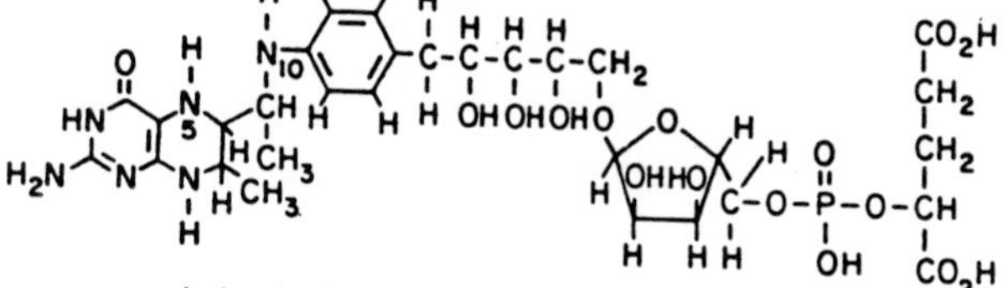

tetrahydromethanopterin (H_4MPT)

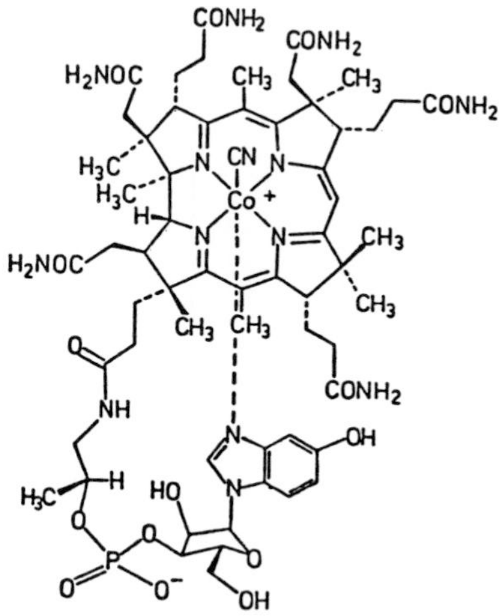

Co ß-cyano-5'hydroxybenzimidazolyl-cobamide
(factor III)

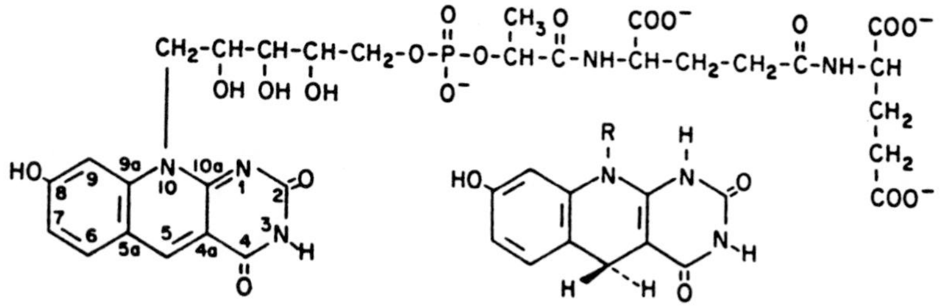

Oxidized

Reduced ($F_{420}H_2$)
coenzyme F_{420}

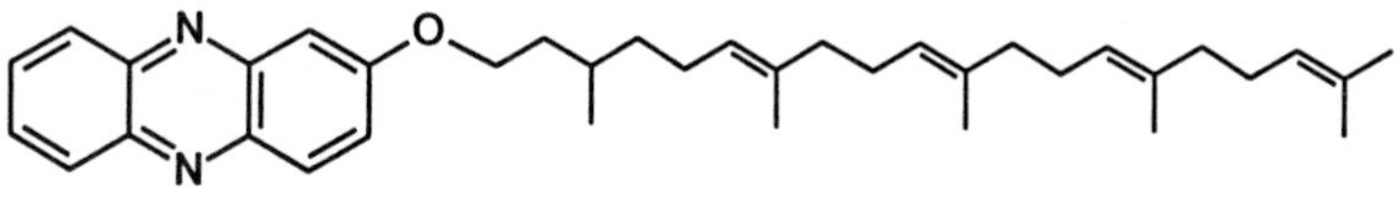

$HS-CH_2-CH_2-SO_3^-$

coenzyme M (HS-CoM)

$CH_3-S-CH_2-CH_2-SO_3^-$

CH_3-S-CoM

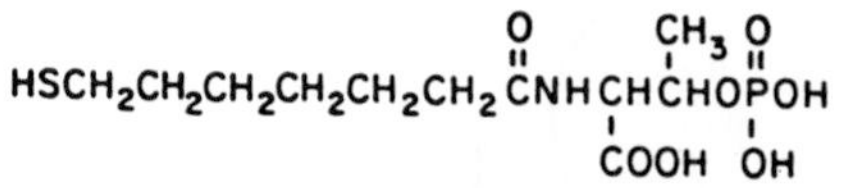

Methanophenazine

$HSCH_2CH_2CH_2CH_2CH_2CNHCHCHCHOPOH$
7-mercaptoheptanoylthreonine phosphate
(HS-HTP)

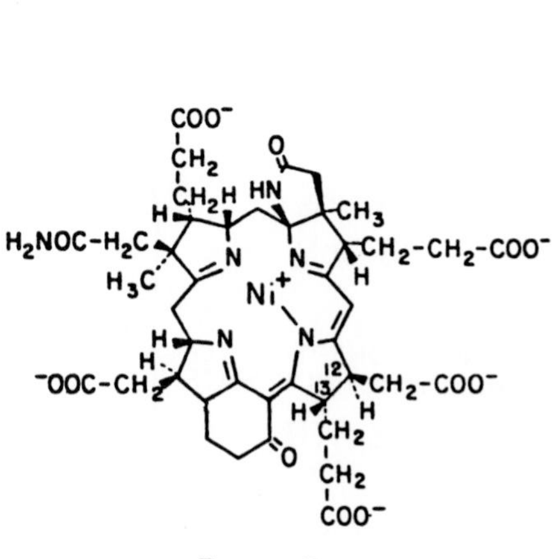

Factor F_{430}

Scheme I. Structures of methanogen cofactors.

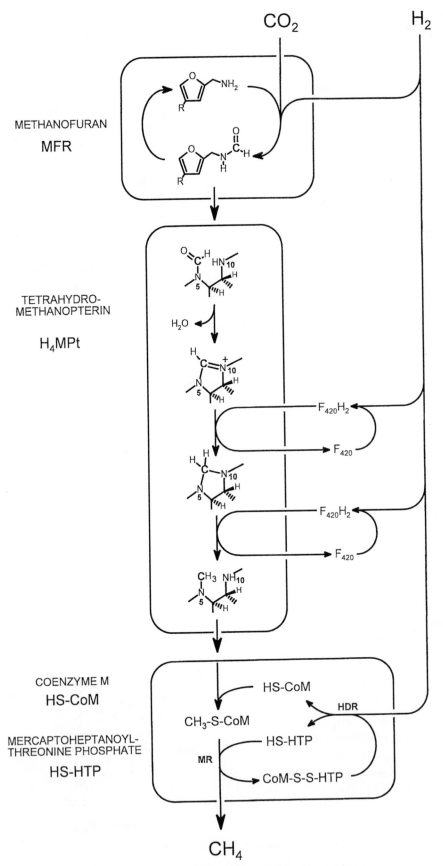

Fig. 1. Metabolic pathway of CO_2 reduction to methane.

Methanosarcina barkeri contains at least 5 different subunits, and harbors non-heme iron and acid-labile sulfur, in addition to one atom of molybdenum present in a molybdopterin–guanine dinucleotide cofactor. Catalysis presumably occurs via reduction of intermediate methanofuran carbamate, present in rapid equilibrium with methanofuran and CO_2. A tungsten-containing isozyme is also found, notably in certain hyperthermophilic methanogen species, such as *Methanococcus jannaschii*.

The next step in the pathway is the transfer of the formyl group from formyl-MFR to an analog of tetrahydrofolate known as tetrahydromethanopterin (H_4MPt). The N^5-formyl-H_4MPt so formed is then converted to N^5, N^{10}-methenyl-H_4MPt ($CH\equiv H_4MPt^+$) by a cyclohydrolase enzyme. Thereafter, reduction of $CH\equiv H_4MPt^+$ to the methylene level, and then to form methyl-H_4MPt, is carried out in two successive 2-electron steps. One of the enzymes involved, found widely distributed in hydrogenotrophic species, is a novel hydrogenase that brings about direct hydrogenation of $CH\equiv H_4MPt^+$ by H_2. The enzyme is unique among hydrogenases in that it contains no redox active metal cofactors. Its reaction mechanism most likely involves activation of $CH\equiv H_4MPt^+$, by out-of-plane distortion of the methinyl cation, to form a superelectrophilic species capable of hydride abstraction from H_2. All methanogens contain an enzyme also capable of reducing $CH\equiv H_4MPt^+$ to $CH_2=H_4MPt$ in a reaction dependent on the reduced deazaflavin coenzyme F_{420} ($F_{420}H_2$). Coenzyme $F_{420}H_2$ is used in all species for the next reaction, reduction of $CH_2=H_4MPt$ to form N^5-methyl-H_4MPt. The required $F_{420}H_2$ is provided by reduction of F_{420} with H_2 by the enzyme F_{420}-reducing hydrogenase. By comparison, $F_{420}H_2$ generated in the pathway of formate conversion to methane, is furnished by F_{420}-dependent formate dehydrogenase. Formate enters the pathway only after its oxidation to carbon dioxide; it then follows the same route as CO_2 reduction.

Energy in the form of ATP is generated by a chemiosmotic mechanism. Free energy is conserved in a transmembrane electrochemical gradient, generated, in part, at the next step in the pathway, an exergonic transfer of the methyl group from CH_3-H_4MPt to coenzyme M. Methyl group transfer is catalyzed in *Methanobacterium thermoautotrophicum* by a membrane-bound protein complex containing eight different subunits, ranging from 12–34 kDa, many of which are predicted to contain membrane-spanning domains. A similar enzyme is present in other species, but may differ in the number of subunits. The reaction occurs in two steps. The first involves transfer of the methyl group from CH_3-H_4MPt to a corrinoid cofactor bound specifically to the 23 kDa subunit. The reaction is then completed in the second step in formation of CH_3–S–CoM by methyl group transfer from the corrinoid cofactor to HS–CoM. The reaction is dependent on Na^+ ion, and, during transfer of the methyl group, the complex acts to pump sodium ions across the membrane. A separate Na^+/H^+ antiporter serves to convert the sodium ion potential to a transmembrane proton potential, thus driving the synthesis of ATP.

Methyl-coenzyme M reductase acts in the final reaction of methanogenesis to convert methyl-CoM to methane, using reducing equivalents supplied by its thiol compound substrate 7-mercaptoheptanoylthreonine phosphate (HS–HTP). Formation of the reduced product CH_4 is accompanied by production of an oxidized product, the mixed disulfide of CoM and HS–HTP, namely, CoM–S–S–HTP. The reduction of CoM–S–S–HTP is a second process coupled with electrochemical energy conservation, and is brought about by the enzyme heterodisulfide reductase. Electrons used in this reduction are likely donated directly by reduced methanophenazine, a 2-hydroxyphenazine derivative with a polyisoprenoid side chain associated in the membrane. The crystallographic structure of the 300 kDa $\alpha_2\beta_2\gamma_2$ methyl-CoM reductase reveals two structurally identical active sites. Each active site contains a molecule of the Ni-porphinoid coenzyme F_{430} bound at the bottom of a long channel that extends 30 Å out to the surface of the protein. The HS–HTP substrate fits into the channel, with its aliphatic, thiol-bearing heptanoyl side chain in extended conformation directed toward the active site. Because the bulky hydrophilic threonine phosphate head group can not enter the channel (which is only 6 Å wide at its narrowest part), steric limitations prevent the HS–HTP thiol from reaching all the way to the Ni atom in F_{430}. Nonetheless, proximity to coenzyme M allows for formation of the product CoM–S–S–HTP mixed disulfide. The unusual mechanism of this enzyme involves transfer of the methyl group from methyl-CoM to nickel of the F_{430} in the Ni(I) state. A Ni(II)-methyl intermediate, formed in the reaction, undergoes irreversible protonolysis to release methane. Regeneration of Ni(I) from Ni(II) involves radical species formed as a result of a process initiated either by the decomposition of a transiently formed methyl-Ni(III) adduct or by an initial homolytic cleavage of the CH_3–S bond of CH_3–S–CoM. This intriguing mechanism, found in no other organisms, has been adopted for methane formation by all species of methanogens.

II. METHYLATED SUBSTRATE REDUCTION TO CH$_4$

Methylated compounds, such as methanol and methylamines, are produced in decay of aquatic animals and plants and serve as the growth substrates for methylotrophic methanogen species, such as those found in the family *Methanosarcinaceae*. Methylotrophic growth generally involves the disproportionation of a methylated substrate such that, for every 3 methyl groups reduced to methane (each by two electrons), one methyl group undergoes 6-electron oxidation to CO_2. Although the oxidation pathways remain poorly described, a number of enzymes en route to CH_4 have been identified as participating in methyl group transfer and activation of methylated compounds, including mono-, di-, tri-, and tetramethylamine, dimethylsulfide, methylmercaptopropionate, and methanol. Fermentation of methanol is described according to Eq. (2). All methylotrophic

$$4CH_3OH \rightarrow 3CH_4 + CO_2 + 2H_2O \qquad (2)$$

pathways, including methanol conversion, follow variations of a two-step theme involving, (1) substrate activation with methyl transfer to a corrinoid (B_{12}) protein and (2) transfer of the methyl from the corrinoid protein to coenzyme M. Corrinoid proteins play prominent roles in methylotrophic pathways. Cells grown on methylated substrates are rich in corrinoid compounds, and, if accidentally exposed

to air, cells and cell extracts turn from dark brown to red, due to oxidation of corrinoid coenzymes to the Co^{3+} state. The pathway of methanol conversion is one of the more thoroughly studied, and is shown in Fig. 2. Initially, methanol is activated by an enzyme designated methanol:5-hydroxybenzimidazolyl methyl-transferase (MT1), which transfers the methyl group to a corrinoid protein specific for the methanol pathway. The reaction is remarkable since it results in the breakage of a stable carbon–oxygen bond, with formation of a much less stable carbon–cobalt bond. The methyl group is then transferred to coenzyme M by the enzyme methylcobamide:coenzyme M methyltransferase (MT2). In order for the initial methyl group activation to proceed, the B_{12} cofactor must be in the highly reduced Co^{1+} oxidation state, in which the central cobalt atom is capable of acting as a supernucleophile in attack on the methyl moiety. Oxidation to the Co^{2+} level results in loss of function and, to guard against this, methanogens contain a separate system for reductive reactivation that employs that ATP and an activating protein, designated methyltransferase activating protein.

A different set of MT1 and corrinoid proteins is employed in the metabolism of each different methylated growth substrate. Several MT2 proteins also exist, and exhibit metabolic specificity, in that different MT2 isozymes function exclusively in different pathways. One isozyme acts in methanol dissimilation, whereas a second form of MT2 functions in the conversion of various methylamines. A third MT2 protein takes part in metabolism of methylsulfides and is found tightly associated as a subunit of a 480 kDa corrinoid-containing protein complex, known as methylthiol:coenzyme M methyltransferase. The MT2 reaction is analogous to the second half reaction of methionine synthase, which acts to form methionine from homocysteine, inasmuch as both enzymes catalyze transfer of the methyl group from a methyl-B_{12} protein to a low molecular weight thiol acceptor. Methyl-CoM is the common product formed from the various methylotrophic substrates, as well as from all other methanogenic substrates. Reduction of methyl-CoM to methane proceeds as described under the pathway for CO_2 reduction, but

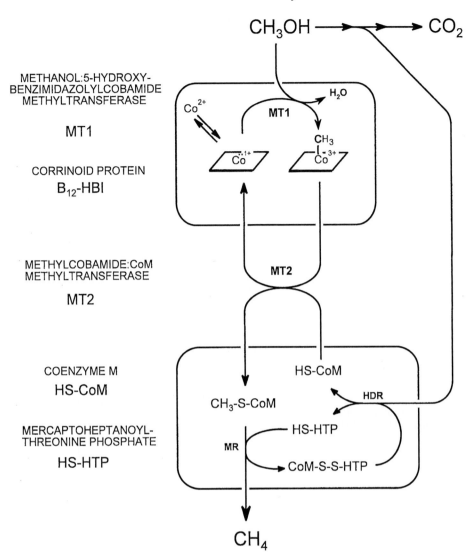

Fig. 2. The pathway of methanogenesis from methanol.

the major electron source most often originates from concurrent methyl group oxidation instead of from hydrogen. The electrons serve to reduce CoM–S–S–HTP, which acts, as it does in all methanogenic pathways, as the penultimate oxidant, continuously being regenerated by the action of methyl-CoM reductase. Evidence indicates that energy available from electron transport is conserved via a chemiosmotic mechanism; however, individual components that make up the electron transport chain in this pathway and others still remain to be identified and characterized.

III. ACETATE CLEAVAGE TO CARBON DIOXIDE AND CH₄

Acetic acid is a significant product of decomposition of complex organic material in anaerobic environments, and the further degradation of acetate accounts for the major portion of the total methane produced biologically. Acetic acid is also oxidized by other anaerobic microorganisms; however, these convert both carbon atoms of acetate to CO_2, which requires a separate oxidizing substrate (e.g., sulfate,

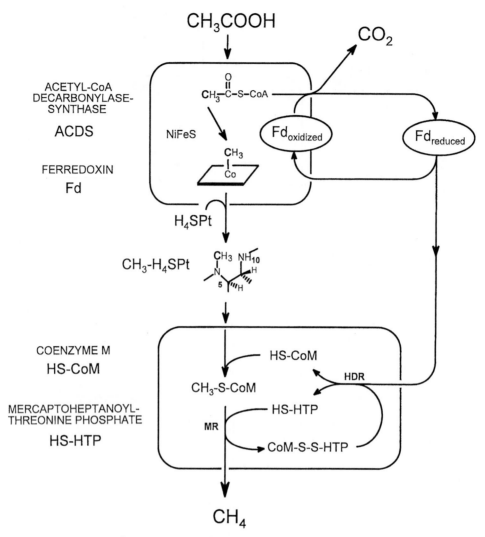

Fig. 3. Acetate pathway leading to CH_4 and CO_2.

nitrate). In contrast, methanogens are capable of disproportionating acetate by a process in which oxidation of the carboxyl group to CO_2 is coupled to reduction of the methyl group to methane, according to Eq. (3).

$$CH_3-COOH \rightarrow CO_2 + CH_4 \qquad (3)$$

The central event in the pathway of acetate conversion (Fig. 3) is the cleavage of the acetate carbon–carbon bond. This reaction occurs following the activation of acetate to form acetyl-CoA by the enzymes acetate kinase and phosphotransacetylase. There-

after, cleavage of acetyl-CoA takes place in a highly unusual reaction, in which two types of bonds are broken, the thioester C–S and the acetate C–C bond. The reaction is carried out by a multienzyme complex of a molecular mass around two million daltons, roughly the size of a ribosome. The complex is designated as acetyl-CoA decarbonylase/synthase (ACDS) and was formerly called the CO dehydrogenase–corrinoid enzyme complex. It is composed of five different subunits, ranging in molecular mass from 20 to 90 kDa; evidence suggests that it contains eight copies of each type. A variety of cofactors are present, including a corrinoid moiety, and non-heme iron and acid labile sulfide arranged in several different types of metal clusters, some of which are also associated with nickel. High rates of methanogenesis are needed to meet the energy needs of methanogens growing on various carbon substrates, and growth on acetate is no exception. When acetate is present as the sole source of carbon, high methane formation rates impose an equally high demand for acetate cleavage. Cells of *Methanosarcina thermophila* growing on acetate invest substantial resources to produce the quantities of ACDS needed—this accounts for as much as 20–22% of the buffer soluble protein in exponentially growing cells. The 240 Å diameter ACDS complex particles are readily visible in electron micrographs following hypotonic lysis directly on specimen grids of *M. thermophila* cells grown under conditions in which they lack outer cell-wall layers. Crude extracts analyzed by SDS PAGE display eight prominent protein bands, three of which correspond to the subunits of methyl-CoM reductase, with the other five belonging to the ACDS complex. In the action of the complex on its substrate acetyl-CoA, cleavage of the carbon–carbon bond is accompanied by immediate oxidation of the carbonyl group to CO_2, producing electrons that are donated directly to reduce ferredoxin. The corrinoid cofactor serves as an intermediate methyl group carrier, acting to remove the methyl group from the site of cleavage and to donate it finally to the substrate tetrahydrosarcinapterin (H_4SPt), forming N^5-methyl-tetrahydrosarcinapterin (CH_3–H_4SPt). The process of carbonyl group excision is not fully understood, but is believed to involve metal-based decarbonylation of an acetyl–enzyme species, formed as a result of interaction of acetyl-CoA at a reduced metal center in the enzyme active site. The overall reaction is reversible and, in environments containing sufficient reducing power, the ACDS complex is used in the reverse direction to synthesize acetyl groups needed for autotrophic growth of methanogens on single-carbon substrates. In this process, carbon fixed in the form of acetyl-CoA is then further reduced to generate pyruvate for entry into biosynthetic pathways forming other cellular components and macromolecules, i.e., proteins, polysaccharides, lipids, and nucleic acids.

Energy produced in acetate conversion to methane and carbon dioxide is conserved in the formation of a transmembrane electrochemical gradient coupled to ATP synthase, as in other methanogenic pathways. Exergonic transfer of the methyl group from CH_3–H_4SPt to coenzyme M, such as in the pathway of CO_2 reduction, results in the translocation of sodium ions. Sufficient energy to support a second site of ion translocation would be provided from electron transport from reduced ferredoxin via membrane-bound carriers to methanophenazine, and, ultimately, to reduce the CoM–S–S–HTP heterodisulfide. However, further research is needed to define the nature of the electron transport carriers and the identity of other ion translocating membrane proteins.

In summary, outlines of the biochemistry are now established for three types of methane-forming metabolic pathways, organized in a modular fashion, which utilize the following different classes of growth substrates: (1) H_2 + CO_2 and formate, (2) a variety of simple methylated compounds, and (3) acetate. A collection of unique coenzymes is involved that function as co-catalysts in a number of fascinating enzyme systems. Some of the enzymes employ catalytic mechanisms that are seldom, if ever, found in other types of organisms. The enzyme responsible for producing methane, methyl-CoM reductase, connects the divergent pathways at the level of methyl-CoM. Energy is conserved in methanogens by a chemiosmotic process, with an unusual mechanism of forming a transmembrane ion gradient coupled to an exergonic methyltransfer reaction. Key players involved in the activation and transfer or methyl groups in the different pathways are the methanopterin and methanofuran-related enzymes in CO_2 reduction, corrinoid proteins and methyltransferases in methylotrophic metabolism, and the acetyl-CoA decarbonylase/synthase complex carrying out oxidative cleavage of acetyl-CoA in the major methane-producing pathway of acetate conversion. Although studies on the methanogens have often been complicated by factors such as the oxygen sensitivity of the cells and their components, nevertheless, rewards along the trail to understand the novel biochemistry of methanogenesis have been and will continue to be great.

See Also the Following Articles

ECOLOGY, MICROBIAL • METHANOGENESIS

Bibliography

DiMarco, A. A., Bobik, T. A., and Wolfe, R. S. (1990). Unusual coenzymes of methanogens. *Annu. Rev. Biochem.* **59**, 355–394.

Ermler, U., Grabarse, W., Shima, S., Goubeaud, M., and Thauer, R. K. (1997). Crystal structure of methyl-coenzyme M reductase: The key enzyme of biological methane formation. *Science* **278**, 1457–1462.

Ferry, J. G. (1992). Biochemistry of methanogenesis. *Crit. Rev. Biochem. Mol. Biol.* **27**, 473–503.

Ferry, J. G. (ed.). "Methanogenesis: Ecology, Physiology, Biochemistry & Genetics. (1993). Chapman & Hall, New York.

Reeve, J. N., Nölling, J., Morgan, R. M., and Smith, D. R. (1997). Methanogenesis: Genes, genomes, and who's on first? *J. Bacteriol.* **179**, 5975–5986.

Thauer, R. K. (1998). Biochemistry of methanogenesis: A tribute to Marjory Stephenson. *Microbiology* **144**, 2377–2406.

Methane Production/Agricultural Waste Management

William J. Jewell

Cornell University

I. Methane Fermentation of Agricultural
   Wastes—Background
II. Methane Production System Designs and Limitations

GLOSSARY

agricultural wastes All photosynthetically fixed by-products of food production, including animal wastes, crop residues, food-processing wastes (both liquid and solid), and food wastes.

ammonia toxicity Dissolved ammonia can exist in two forms in chemical equilibrium with each other—ionized ammonium ion (NH4+) and unionized ammonia (NH$_3$). The majority of dissolved ammonia is in the ionized form in healthy digesters, but an exponential increase in the unionized form occurs as the pH increases above 7.0. It is the unionized ammonia form that inhibits or kills bacteria by passing freely through biological membranes. Inhibitory concentrations are around 50 mg/l of NH$_3$–N, but slow-rate digesters have been reported to acclimate to concentrations in excess of 10 times this value.

anaerobic digestion, anaerobic methane fermentation Microbial decay process that converts organic carbon to biogas in the absence of oxygen; most often observed at temperatures above 25°C, but commonly occurs in nature at temperatures down to zero and above 65°C.

biogas Naturally evolved gas above saturation values in an anaerobic digester mainly of methane (CH$_4$) and carbon dioxide (CO$_2$) but with traces of composed carbon monoxide, hydrogen, hydrogen sulfide, ammonia, and water vapor. Methane always comprises more than 50% of the biogas when the digester is operating in a stable condition.

renewable energy Carbon-based energy that is derived from recently fixed carbon dioxide from the atmosphere.

MORE THAN 30 MICROBIAL SPECIES are involved in an intimate microbial ecosystem that breaks down complex organic matter to methane, carbon dioxide, other trace elements, and water. Uncontrolled losses of methane from such conversions release $20 billion dollars per year of methane into the atmosphere. Processes have been developed that are capable of converting to energy very low concentrations of organic matter in water (0.1 g per liter) at temperatures approaching 0°C, as well as concentrated organic matter in animal wastes and landfills (up to 500 gram per liter), and at temperatures over 65°C.

Almost all applications of methane conversion are at 35°C, due to limited understanding of microbial interactions at higher and lower temperatures. No psychrophillic (<15°C temperature optimum) methanogen has been identified, even though much of the world's background methane production occurs at this temperature. One of nature's most abundant natural polymers, lignin, is completely nonbiodegradable under methanogenic conditions. An understanding of anaerobic microbial systems and the chemical composition of agricultural wastes results in reliable predictions of biogas generation potential and process limitations.

I. METHANE FERMENTATION OF AGRICULTURAL WASTES—BACKGROUND

There are more than 100 operating anaerobic digesters on U.S. farms and probably about that many applied to other agricultural wastes. Methane contribution to the atmosphere from all organic waste, just in the United States, equals 43 million metric tons and has an energy value of $8 billion per year, and, worldwide, total release of methane into the atmosphere is more than 10 times this amount. Most of this methane originates from uncontrolled microbial conversion of agricultural wastes, either on farms or in landfills.

A. The Potential

Anaerobic conversion of natural organic matter may be the most efficient energy conversion process on Earth (not including the relatively low solar energy capture efficiency of photosynthesis). Properly designed anaerobic digesters can harvest over 90% of the "available carbon" in complex organic matter in a few weeks' exposure time, and 90% of this energy can be harvested as substitute natural gas. For example, a field of corn may have a harvestable crop yield containing 420 kJ/ha-yr (160×10^6 Btu/ac-yr) of energy. Anaerobic digestion of this harvested corn biomass could generate 380 kJ/ha-yr (144×10^6 Btu/ac-yr) of substitute natural gas and 2.2 mt/ha-yr (1 ton/ac-yr) of humus, containing most nutrients consumed in biomass production.

B. Historical Overview

Many full-scale digesters have operated for decades in the United States. Over 10,000 large-scale anaerobic digesters are in operation at sewage treatment plants, and more then twenty million small-scale anaerobic reactors (known as "septic tanks") are in use. A number of concentrated industrial wastewaters, mainly breweries and meatpacking wastes, are routinely treated by anaerobic processes. Simple flexible liner-covered anaerobic lagoons are used as reactors, and biogas is recovered and utilized in some of these facilities. However, few biomass and agricultural wastewater digesters have been constructed.

II. METHANE PRODUCTION SYSTEM DESIGNS AND LIMITATIONS

Animal waste solids conversion efficiencies vary over a wide range—from a low of around 10% of total dry organic matter with dairy cow wastes to a high of over 75% with fresh poultry wastes. All animal wastes, food wastes, and fiber wastes can be successfully digested. Because of differing environmental requirements of the complex microbial system, and varying composition of waste organics, fermentation requirements vary widely.

A. Fate of Organic Matter in Anaerobic Digestion

Organic matter is characterized in terms of mass including water (wet mass) and mass excluding water (dry mass) (see Fig. 1). Dry matter consists of two components—ash and organic matter. Organic matter (volatile solids measured at 550°C) is further divided into two fractions—biodegradable, or "available" organics, and refractor/organics. Only the biodegradable organics are available to methanogenic microbial systems for conversion to methane.

Since lignin is nonbiodegradable under anaerobic conditions, refractory organics are enriched in lignin by a factor of 2 to 3 times as biodegradable carbon is removed. Nitrogen is initially divided among three components—dissolved ammonia, urea, and organic nitrogen. The accumulation of ammonia during digestion can result in microbial inhibition or toxicity.

B. Predicting Methane Production from Anaerobic Digestion of Organic Matter

Efficiencies of organic waste conversion by anaerobic methane fermentation can range over an eightfold amount, from 90% of the dry matter to a low of about 10% of the dry matter. When the composition of a substrate is known, it is possible to predict gas volume generation, gas composition, and microbial growth characteristics.

Organic matter lignin content can be used as an indicator of nonbiodegradability

$$BVS = -2.82\ L_s + 83,$$

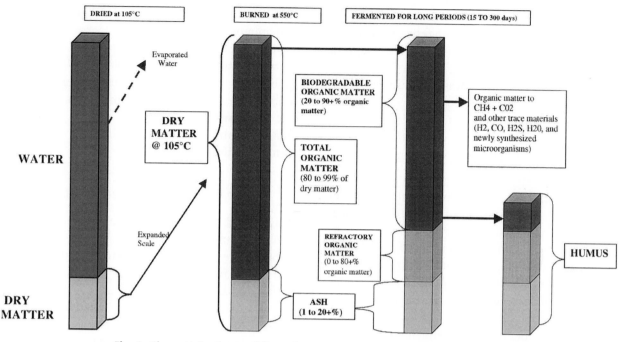

Fig. 1. Characterization and fate of organic matter in microbial conversions.

where BVS is volatile solids that can be converted to methane, expressed as percentage of total dry volatile solids, and L_s = lignin content of organic matter, expressed as percentage of dry volatile solids using the 72% sulfuric acid method.

Methane fermentation that results in destruction of 1 g chemical oxygen demand (COD) produces 0.35 liters of CH_4 at standard temperature and pressure (STP) (5.6ft^3 CH_4/pound COD destroyed). At 35°C, one g COD converted to CH_4 generates 0.395 L CH_4 (6.3 ft^3 CH_4/lb COD destroyed).

A simplistic way of estimating gas production relates methane energy content to energy content of organic matter and its biodegradability. Since most of the energy is converted to methane, this enables a rapid technique, useful in making gas estimates. Energy content of methane is 30,000 kJ/m^3 (1000 Btu/ft^3) and organic matter contains between 11,600 kJ/kg to 21,000 kJ/kg (5000 and 9000 Btu/lb. and averages about 8000 Btu/lb.). Thus, when a pound of organic matter is converted to biogas during fermentation, methane will be produced with an en-

ergy content of 8000 Btu, or 8 ft^3 CH_4 (@ STP) generated per pound organic matter converted.

C. Kinetics, Bioreactor Sizes, Effluent Biomass Composition, and Limitations

Maximum anaerobic digestion conversion efficiencies with most animal waste occur at organic loading rates of between 0.1 and 0.2 pounds of organic matter as volatile solids per cubic foot of reactor per day (lb. VS/ft^3 $_r$-d) (1.2 to 2.4 g VS/l-d). Since the equivalence of organic mass (VS) to methane is about 0.51 CH_4/gΔVS-d, maximum conversion rates can be described by

methane production rate (MPR) = [OLR][f$_B$][0.5]

where MPR is the methane production rate in liters of methane per liter of reactor per day, OLR is the organic loading rate (gTVS/l$_r$-d, 0.5 is a "typical" methane to mass conversion equivalence), and f$_B$ is the biodegradable fraction of organic matter (TVS).

Depending on diet, dairy cow wastes (mixture of urine, manure, waste feed, bedding) biodegradability varies from 25 to 50% VS, with a typical value of $f_B = 0.4$.

Thus, if we choose a dairy waste bioreactor loading rate of 1.8 g TVS/l_r-d (0.15 lb TVS/ft^3-d), the estimated methane production rate would be

$$MPR = 1.8\,(0.4)(0.5) = 0.36\ l\ CH_4/l_r\text{-d}.$$

Process "efficiency" of waste treatment and fate of nutrients, mainly dissolved nitrogen, limit applications of this technology. "Efficiency of waste treatment" can mean odor reduction, aesthetic change, biodegradable matter destruction, and mass reduction. Anaerobic digestion is applied to achieve odor reduction, and little concern is directed to other process efficiency definitions.

Of all the factors that affect fermentation of more concentrated organic wastes (i.e., animal waste, garbage, etc.), ammonia concentration is most important because of its toxicity to methanogenic bacteria. Of the two forms of ammonia (NH_3 unionized or "free" ammonia and NH_4^+), unionized ammonia is most toxic.

The amount of unionized ammonia is a function of total ammonia concentration, pH, and temperature of the reactor as described by the equation

$$NH_4^+ \leftrightarrow NH_3 + H^+,\ pK = 9.3\ @\ 35°C.$$

Upper limit on free ammonia concentration was thought to be around 50 mg/l for a healthy methanogenic fermentor, but slower rate reactors can adapt to concentrations three or more times this concentration.

Note that, in cow manure digestion, free ammonia at 35°C and a pH of 7.5 are below inhibitory limitations (around 90 mg/l NH_3). However, if the pH shifts to 8.0, the unionized ammonia concentration increases to around 300 mg/l, and the unionized ammonia is many times the inhibitory level. Thus, while mesophilic digestion of dairy cow manure may operate uninhibited, higher temperature applications may result in ammonia toxicity, due to higher pH

levels that cause higher concentrations of unionized ammonia.

High concentrations of nitrogen in swine and poultry wastes cause significant problems with anaerobic applications unless the ammonia concentration is controlled below toxic levels. Dilution water can enable these wastes to be treated, however, large volumes of dilution water create additional waste management limitations.

D. Wastewater Treatment

Significant advancements have been made in anaerobic treatment of organic wastewaters using microbial process variations that enable better control over the slow-growing anaerobic microbial mass. These processes use microbes in a form that are easily, rapidly, and efficiently separated from treated wastes. This is accomplished by increasing their settleability (e.g., with granular sludge in upflow microbial sludge blankets) or in immobilized biofilms (e.g., in moving particle reactors, such as expanded beds). Efficiency of biodegradable organic removals from wastewaters can exceed >85% at hydraulic retention times of only a few hours with new innovative anaerobic microbial reactors.

See Also the Following Articles

BIOREACTORS • LIGNOCELLULOSE, LIGNIN, LIGNINASES • WASTEWATER TREATMENT, INDUSTRIAL

Bibliography

Chynoweth, D. P., and Isaacson, R. (eds.). (1997) "Anaerobic Digestion of Biomass." Elsevier Science. London, UK.

Chynoweth, D. P., Wilke, A. C., and Owens, J. M. "Anaerobic Processing of Piggery Wastes: A Review." Paper presented at the 1998 ASAE Meeting, Orlando, FL.

Gibbs, M. J. (ed.). (1998). "Costs of Reducing Methane Emissions in the United States." U.S. Environmental Protection Agency Report. Methane and Utilities Branch, Atmospheric Pollution Prevention Division, Office of Air and Radiation. Washington, DC.

Hill, D. T., and Cobb, S. A. (1996). Simulation of process steady-state in livestock waste methanogenesis. *Trans. ASAE* 39(2), 565–573.

Jewell, W. J., Cummings, R. J., and Richards, B. K. (1993). Methane fermentation of energy crops: Maximum conversion kinetics and in situ biogas purification. *Biomass Bioenergy* **5**, 261–278.

Literature Review. (1996). "Anaerobic Processes" pp. 480–497. Water Environment Research, Water Environment Federation, Alexandria, VA.

Lusk, P. (1998). "Methane Recovery from Animal Manures: The Current Opportunities Case Book." U.S. Department of Energy Report: NREL/SR-580-25145.

Speece, R. E. (1996). "Anaerobic Biotechnology: For Industrial Wastewaters." Archae Press, Nashville, TN.

Stafford, D. A., Wheatley, B. I., and Hughes, D. E. (1979). "Anaerobic Digestion." Applied Science Publishers.

Zhang, R., and Felmann, D. J. (1997). "Animal Manure Management—Agriculture Scoping Study." The Epri Agricultural Technology Alliance—Electric Power Researach Institute, C109139.

Methanogenesis

Kevin R. Sowers

University of Maryland Biotechnology Institute

I. Historical Overview
II. Diversity and Phylogeny
III. Habitats
IV. Physiology and Biochemistry
V. Molecular Genetics
VI. Summary

GLOSSARY

acetotrophic methanogen   Methanogen that decarboxylates acetate to CH_4 and CO_2.

coenzyme M   Coenzyme that serves as the carbon carrier during the final reduction of bound CH_3 to CH_4.

DNA/RNA polymerases   Enzymes that catalyze the synthesis of DNA or RNA, respectively, from a DNA template.

H_2-consuming acetogen   Microorganism that synthesizes acetate from H_2 and CO_2

H_2-producing acetogen   Microorganism that generates H_2 by oxidizing short-chain fatty acids or alcohols

hydrogenotrophic methanogen   Methanogen that generates methane by the reduction of CO_2 with H_2.

interspecies H_2 transfer   Mutually beneficial transfer of H_2 from H_2-producing microorganisms to H_2-consuming microorganisms in an anaerobic ecosystem.

methanofuran   Coenzyme that serves as a carbon carrier during the initial reduction of CO_2 in the methanogenic pathway.

methylotrophic methanogen   Methanogen that generates methane by reduction of methyl groups.

obligate syntroph   Microorganism that must interact mutualistically with one or more other microorganisms to metabolize a specific substrate; refers specifically to interspecies H_2 exchange in this article.

restriction endonuclease   Enzyme that protects organisms by degrading ("restricting") foreign DNA that enters the cell.

tetrahydromethanopterin   Coenzyme that serves as the carbon carrier during sequential reduction of bound CO_2 to CH_3.

BIOLOGICAL METHANOGENESIS   has a significant role in the global carbon cycle. This process is one of several anaerobic degradative processes that complement aerobic degradation by utilizing alternative electron acceptors in habitats where O_2 is not available.

The hierarchy of electron acceptors is $O_2 > NO_3^- > Fe^{+3} > Mn^{+4} > SO_4^{-2} > CO_2$, where reduction of CO_2 to form CH_4 is the ultimate terminal process catalyzed by degradative microbial consortia in the absence of all other electron acceptors. Methanogenesis is catalyzed exclusively by prokaryotic single-cell microorganisms, classified as methanogenic *Archaea*. Although these microorganisms require highly reduced, anaerobic conditions for growth, methanogenesis is ubiquitous in the environment and occurs in many habitats, including sewage digestors, waste landfills, marine and freshwater sediments, algal mats, wet wood of living trees, mammalian gastrointestinal tracts, ruminants, termite guts, geothermal vents, and deep subsurface rock (Fig. 1). CH_4 generated by methanogens is ultimately converted to CO_2 by CH_4-oxidizing bacteria or released into the atmosphere where it accumulates as a greenhouse gas. The combined processes of aerobic and anaerobic degradation, which include methanogenesis, ensures that CO_2 "fixed" as cell carbon by photosynthetic organisms, and consumed by higher organisms in the food chain, such as herbivores and carnivores, is eventually restored to the atmosphere as CO_2, thus completing the global carbon cycle.

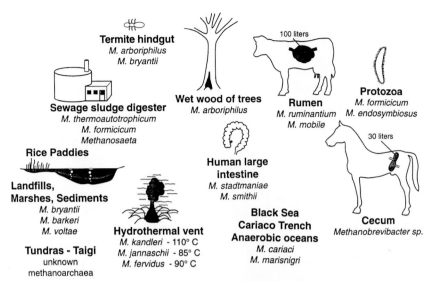

Fig. 1. Habitats of the methanogenic Archaea. [Reproduced, with permission, from Wolfe, R. S. (1996). *ASM News* **62,** 529–534. ASM Press.]

I. HISTORICAL OVERVIEW

The generation of combustible gas, presumably CH₄, has been reported by Pliny as early as during the Roman Empire. Legendary manifestations of methanogenesis include the will-ó-the-wisp, hypothesized to have resulted from the spontaneous combustion of marsh gas, and fire-breathing dragons, conjectured to have resulted from the accidental ignition of gas from CH₄-belching ruminants. The close association between decaying plant material and the generation of "combustible air" was first described by the Italian physicist Alessandro Volta in 1776, when he reported that gas released after disturbing marsh and lake sediments produced a blue flame when ignited by a candle. Bechamp, a student of Pasteur, was the first to establish that methanogenesis was a microbial process, which was corroborated by others throughout the remainder of the nineteenth century and early twentieth centuries. Because of the methanogens' requirement for strict anaerobic conditions, the first isolates were not reported until the 1940s. The approach used for isolation, the shake culture, involved adding microorganisms to molten agar growth medium containing a reductant, such as pyrogallol–carbonate, to prevent O_2 from diffusing into the agar. However, this approach was not suitable for isolating and maintaining methanogens in pure culture for long periods of time, as the medium was not sufficiently anaerobic. It was not until 1950 that a simple, effective technique was developed that provided the rigorous conditions required for routine isolation and culturing of methanogenic *Archaea*. The technique, referred to as the "Hungate Technique," employs gassing cannula, O_2-free gases, and cysteine–sulfide reducing buffers to prepare a highly reduced, O_2-free medium. Boiling initially deoxygenates medium and a cannula connected to an anaerobic gas line, such as N_2 or CO_2, is inserted into the vessel to displace air as the medium cools. The medium is dispensed into culture tubes or serum vials while purging with anaerobic gas and then the medium is sealed with a rubber stopper or septum. The vessel containing reduced medium and anaerobic gas effectively becomes an anaerobic chamber for culture growth. The development of the anaerobic glove box has further simplified culturing of methanogens by providing a means for colony isolation in petri plates containing anaerobic medium (Fig. 2). Inoculated plates are then transferred to an anaerobe jar that is purged with anaerobic gas and hydrogen sulfide to create conditions necessary for growth.

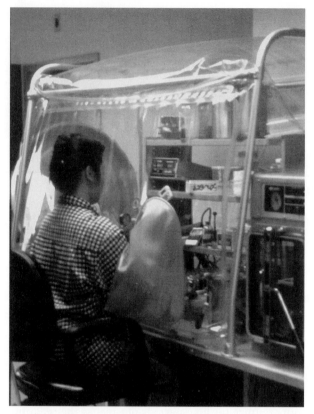

Fig. 2. Anaerobic glove box used for growing methanogens. Materials are introduced into the glove box through an airlock located on the side.

II. DIVERSITY AND PHYLOGENY

The methanogens are members of the *Archaea*, one of three domains of life proposed by C. Woese on the basis of 16S rRNA sequence, which also include the *Bacteria* and *Eukarya* (Fig. 3). *Archaea* have morphological features that resemble the *Bacteria*; they are unicellular microorganisms that lack a nuclear membrane and intracellular compartmentalization. In contrast, several molecular features of the *Archaea* have similarity to the *Eukarya*; these feature include histone-like DNA proteins, a large multicomponent RNA polymerase, and eukaryal-like transcription initiation. Despite the similarities to the other domains, *Archaea* also have unique characteristics that distinguish them from the *Bacteria* and *Eukarya*. These distinguishing features include membranes composed of isoprenoids ether-linked to glycerol or carbohydrates, cell walls that lack peptidoglycan, syn-

thesis of unique enzymes, and enzyme cofactor molecules. An additional unifying characteristic among the *Archaea* is their requirement for extreme growth conditions, such as high temperatures, extreme salinity, and, in the case of the methanogens, highly reduced, O_2-free anoxic environments.

Although the methanogens are a phylogenetically coherent group and have a limited substrate range, they are morphologically and physiologically diverse. They include psychrophilic species from Antarctica that grow at 1.7°C to extremely thermophilic species from deep submarine vents that grow at 113°C; acidophiles from marine vents that grow at pH 5.0 to alkaliphiles from alkaline lake sediments that grow at pH 10.3; species from freshwater lake sediments that grow at saline concentrations below 0.1 M to extreme halophiles from solar salterns that grow at nearly saturated NaCl concentrations; autotrophs that use only CO_2 for cell carbon and heterotrophs that utilize reduced carbon compounds. Despite the range and diversity of growth habitats where methanogens are found, methanogens have one common attribute: they all generate CH_4 during growth.

There are currently over 60 described species of methanogens in five orders within the archaeal kingdom *Euryarchaeota* (Fig. 3). Characteristics of methanogenic *Archaea* are described in Table I. The order *Methanococcales* includes marine autotrophs that grow exclusively by CO_2 reduction with H_2. Morphologically, these species form irregularly shaped cocci. Instead of a rigid cell wall, typical of most *Bacteria*, these species form an S-layer, composed of a paracrystalline array of glycoproteins, and are subject to osmotic lysis at NaCl concentrations below seawater. This order includes several mesophilic species, the moderate thermophile *Methanococcus thermolithotrophicus*, and the extreme thermophiles *Methanococcus igneus*, *Methanococcus jannaschii*, and *Methanococcus infernus*.

The order *Methanobacteriales* is composed predominantly of rod-shaped cells that grow by CO_2 reduction with H_2. The exception is the genus *Methanosphaera*, which grows as cocci and uses H_2 to reduce methanol instead of CO_2. Cells have a rigid cell wall approximately 15–20 nm thick and, when stained for thin-section electron microscopy, resemble the electron dense mono-layer cell wall of

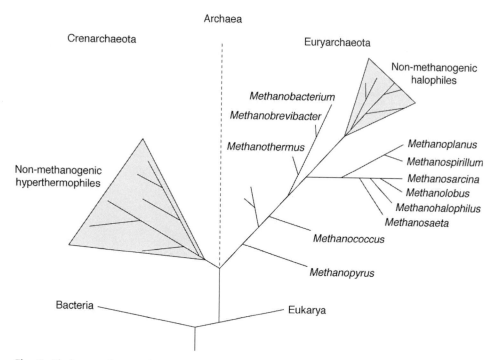

Fig. 3. Phylogenetic tree based on 16S rRNA sequence showing genera of methanogens within the archaeal kingdom Euryarchaeota.

gram + bacteria. These archaeal cells walls are composed of pseudomurein, which is chemically distinguishable from bacterial murein by the substitution of N-talosaminuronic acid for N-acetylmuramic acid and substitution of $\beta(1,3)$ for $\beta(1,4)$ linkage in the glycan strands, and substitution of D-amino acids for L-amino acids in the peptide cross-linkage. *Methanothermus* species also have an additional cell-wall layer, composed of glycoprotein S-layer that surrounds the pseudomurein. *Methanobrevibacter* species are all mesophilic, *Methanobacterium* includes mesophiles and moderate thermophiles with optimal growth temperatures are high as 75°C, *Methanothermobacter* species are exclusively moderate thermophiles, and *Methanothermus* species are extreme thermophiles, with maximum growth temperatures as high as 97°C.

The order *Methanomicrobiales* contains genera that are diverse in morphology and physiology. Most species grow as cocci and rods. In addition, *Methanoplanus* forms flat platelike cells with characteristically angular ends. Another species, *Methanospirillum hungatei*, forms a helical spiral. The cell walls in this order are composed of a protein S-layer and are sensi-

tive to osmotic shock or detergents. In addition to the S-layer, *M. hungatei* also has an external sheath that is composed of concentric rings stacked together. Species are generally slightly halophilic and include mesophiles, moderate, and extreme thermophiles. Most species grow by CO_2 reduction with H_2, but some species also use formate or secondary alcohols as electron donors for CO_2 reduction.

The order *Methanosarcinales* is the most catabolically diverse species of methanogens. In addition to growth and methanogenesis by CO_2 reduction with H_2, some species grow by the dismutation, or "splitting," of acetate and by methylotrophic catabolism of methanol, methylated amines, pyruvate, and dimethylsulfide. While some species of *Methanosarcina* can grow by all three catabolic pathways, *Methanosaeta* species are obligate acetotrophs and all other genera are obligate methylotrophs. All species have a protein S-layer cell wall and most species grow as irregularly shaped cocci. However, several species of *Methanosarcina* also synthesize a heteropolysaccharide matrix external to the S-layer. This external layer can be up to 200 nm thick and is composed primarily of a nonsulfonated polymer of N-acetylgalactosamine

TABLE I
Description of Methanogenic Archaea[a]

Taxonomic epithet	Morphology	Substrates[b]	Optimum growth conditions[c]		Isolation source
			pH	Temp (°C)	
Order *Methanobacteriales*					
Family *Methanobacteriaceae*					
Genus *Methanobacterium*					
alcaliphilum	rod	H	8.4	37	alkaline lake sediment
bryantii	rod	H,2P,2B	6.9–7.2	37–39	sewage digestor
defluvii	rod	H,F	6.5–7.0	60–65	methacrylic waste digestor
espanolae	rod	H	5.6–6.2	35	kraft mill sludge
formicicum	rod	H,F,2P,2B	6.6–7.8	37–45	sewage digestor
ivanovii	rod	H	7.0–7.4	45	sewage digestor
palustre	rod	H,F,2P,CP	7.0	37	peat bog
subterraneum	rod	H,F	7.8–8.8	20–40	deep granitic groundwater
thermoaggregans	rod	H	7.0–7.5	65	cattle pasture
thermoautotrophicum	rod	H	7.2–7.6	65–70	sewage digestor
thermoflexum	curved rod	H,F	7.9–8.2	55	methacrylic waste digestor
thermophilum	rod	H	8.0–8.2	62	digestor methane tank
uliginosum	rod	H	6.0–8.5[d]	40	marsh sediment
wolfei	rod	H	7.0–7.7	55–65	sewage/river sediment
Genus *Methanobrevibacter*					
arboriphilicus	coccobacillus	H,F	7.8–8.0	30–37	cotton wood tree
curvatus	curved rod	H	7.1–7.2	30	termite hindgut
cuticularis	rod	H,F	7.7	37	termite hindgut
filiformis	filamentus rod	H	7.0–7.2	30	termite hindgut
oralis	coccobacillus	H	6.9–7.4	36–38	human subgingival plaque
ruminantium	coccobacillus	H,F	6.3–6.8	37–39	bovine rumen
smithii	coccobacillus	H,F	6.9–7.4	37–39	sewage digestor
Genus *Methanosphaera*					
cuniculi	coccus	H/Me	6.8	35–40	rabbit rectum
stadtmanae	coccus	H/Me	6.5–6.9	36–40	human feces
Family *Methanothermaceae*					
Genus *Methanothermus*					
fervidus	rod	H	6.5	83	solfataric hot spring
sociabilis	rod	H	6.5	88	solfataric mud
Order *Methanococcaales*					
Family *Methanococcaceae*					
Genus *Methanococcus*					
"aeolicus"	irreg. coccus	H,F	nr	nr	nr
deltae	irreg. coccus	H,F	nr	37	river delta sediment
igneus	irreg. coccus	H	5.7	88	marine hydrothermal vent
infernus	irreg. coccus	H	6.5	85	marine hydrothermal vent
jannaschii	irreg. coccus	H	6.0	85	marine hydrothermal vent
maripaludis	irreg. coccus	H,F	6.8–7.2	35–39	marine marsh sediment
thermolithotrophicus	irreg. coccus	H,F	6.5–7.5	65	thermal coastal sediment
vannielii	irreg. coccus	H,F	7.0–9.0[d]	36–40	marine sediment
voltae	irreg. coccus	H,F	6.7–7.4	32–40	estuarine sediment
Order *Methanomicrobiales*					
Family *Methanomicrobiaceae*					
Genus *Methanoculleus*					
bourgensis	irreg. coccus	H,F	7.4	37	tannery waste digestor
marisnigri	irreg. coccus	H,F,2P,2B	6.2–6.6	20–25	marine sediment
oldenburgensis	irreg. coccus	H,F	7.5–8.0	45	river sediment
olentangyi	irreg. coccus	H	nr	37	river sediment
palmolei	irreg. coccus	H,F,2P,2B,CP	6.9–7.5	40	palm oil wastewater reactor
thermophilicus	irreg. coccus	H,F	7.0	55	thermal marine sediment
Genus *Methanomicrobium*					
mobile	curved rod	H,F	6.1–6.9	40	bovine rumen
Genus *Methanolacinia*					
paynteri	irreg. rod	H,F,2P,2B,CP	6.6–7.2	40	marine sediment
Genus *Methanospirillum*					
hungatei	sheathed spiral	H,F	6.6–7.4	30–37	sewage sludge

continues

Continued

Taxonomic epithet	Morphology	Substrates[b]	Optimum growth conditions[c]		Isolation source
			pH	Temp (°C)	
Genus *Methanogenium*					
cariaci	irreg. coccus	H,F	6.8–7.3	20–25	marine sediment
frigidum	irreg. coccus	H,F	6.5–7.9	15	antarctic lake
frittonii	irreg. coccus	H,F	7.0–7.5	57	lake sediment
organophilum	irreg. coccus	H,F,E,1P,2P,2B	6.4–7.3	30–35	marine sediment
Genus *Methanofollis*					
liminatans	irreg. coccus	H,F,2P,2B	7.0	40	industrial wastewater
tationis	irreg. coccus	H,F	7.0	37–40	solfataric hot pool
Genus *Methanocalculus*[e]					
halotolerans	irreg. coccus	H,F	7.6	38	oil field
Family *Methanocorpusculaceae*					
Genus *Methanocorpusculum*					
aggregans	irreg. coccus	H,F	6.4–7.2	35–37	sewage digestor
bavaricum	irreg. coccus	H,F,2P,2B,CP	7.0	37	sugar plant wastewater
labreanum	irreg. coccus	H,F	7.0	37	tar pit lake
parvum	irreg. coccus	H,F,2P,2B	6.8–7.5	37	whey digestor
sinense	irreg. coccus	H,F	7.0	30	distillery wastewater
Family *Methanoplanaceae*					
Genus *Methanoplanus*					
endosymbiosus	irreg. disk	H,F	6.6–7.1	32	marine ciliate
limicola	plate	H,F	7.0	40	drilling swamp
petrolearius	plate	H,F,1P	7.0	37	offshore oil field
Family *Methanosarcinaceae*					
Genus *Methanosarcina*					
acetivorans	irreg. coccus, pseudosarcina	AC,ME,MA,DMS,MMP	6.5–7.5	35–40	marine sediment
barkeri	irreg. coccus, pseudosarcina	H,AC,ME,MA	6.5–7.5	30–40	sewage digestor
mazei	irreg. coccus, pseudosarcina	AC,ME,MA	6.5–7.2	30–40	sewage digestor
siciliae	irreg. coccus	AC,ME,MA,DMS,MMP	6.5–6.8	40	marine sediment
thermophila	irreg. coccus pseudosarcina	AC,ME,MA	6.0	45–55	sewage digestor
vacuolata	irreg. coccus pseudosarcina	H,AC,ME,MA	7.5	40	methane tank sludge
Genus *Methanolobus*					
bombayensis	irreg. coccus	ME,MA,DMS	7.2	37	marine sediment
taylorii	irreg. coccus	ME,MA,DMS	8.0	37	estuarine sediment
tindarius	irreg. coccus	ME,MA	6.5	37	lake sediment
vulcani	irreg. coccus	ME,MA	7.2	37	submarine fumarole
Genus *Methanococcoides*					
"burtonii"	irreg. coccus	ME,MA	7.7	23.4	antarctic saline lake
methylutens	irreg. coccus	ME,MA	7.0	30–35	marine sediment
Genus *Methanohalophilus*					
halophilus	irreg. coccus	ME,MA	7.4	26–36	marine cyanobacterial mat
mahii	irreg. coccus	ME,MA	7.4	35–37	saline lake sediment
oregonense	irreg. coccus	ME,MA,DMS	8.6	35	saline alkaline aquifer
portucalensis	irreg. coccus	ME,MA	6.5–7.5	40	solar salt pond
zhilinae	irreg. coccus	ME,MA,DMS	9.2	45	alkaline lake sediment
Genus *Methanohalobium*					
evestigatum	irreg. coccus	ME,MA	7.4	50	salt lagoon sediment
Genus *Methanosaeta*					
concilii	sheathed rod	AC	7.1–7.5	35–40	pear waste digestor
thermophila	sheathed rod	AC	6.7	55–60	thermophilic sludge digestor
Order *Methanopyrales*					
Family *Methanopyraceae*					
Genus *Methanopyrus*					
kandleri	sheathed rod	H	6.5	98	geothermal marine sediment

[a] Type strain descriptions.

[b] H, hydrogen/carbon dioxide; F, formate; AC, acetate; ME, methanol; MA, methylamines; H/ME, methanol reduction with hydrogen; E, ethanol; 1P, 1-propanol; 2P, 2-propanol; 2B, 2-butanol; CP, cyclopentanol; DMS, dimethylsulfide; MMP, methylmercaptopropionate

[c] nr = not reported.

[d] Only a range reported.

[e] Family epithet currently uncertain.

and D-glucuronic or D-galacturonic acids. The matrix is called methanochondroitin because of its chemical similarity to a mammalian connective tissue component, known as chondroitin. At freshwater NaCl concentrations, *Methanosarcina* spp. that synthesize methanochondroitin grow in multicellular aggregates rather than as single cells, but when grown at marine salt concentrations or with high concentrations of divalent cations, such as Mg^{+2}, they no longer synthesize methanochondroitin and grow as single cells. *Methnaosaeta* species have an external sheath that appears similar in structure to that previously described for *M. hungatei*.

The order *Methanopyrales* is the most deeply branching methanogenic archaeon and presently includes only one species, *Methanopyrus kandleri*. This species is an obligate hydrogenotroph and grows as a rod with a pseudomurein cell wall surrounded by a protein S-layer, similar to that described for *Methanothermus*.

III. HABITATS

A. Interspecies H_2 Transfer

As has been described, methanogens utilize a limited number of simple substrates. In most habitats, they depend on other anaerobes to convert complex organic matter into substrates that they can catabolize. Therefore, unlike aerobic habitats, where a single microorganism can catalyze the mineralization of a polymer by oxidation to CO_2, degradation in anaerobic habitats requires consortia of interacting microorganisms to convert polymers to CH_4. These interactions are dynamic with the methanogens affecting the pathway of electron flow, and, consequently, carbon flow, by a process called interspecies H_2 transfer. In this association, the H_2-utilizing methanogens maintain a low H_2 partial pressure that allows certain reactions to be thermodynamically favorable. One physiological group of microorganisms affected by this process is the H_2-producing acetogens. The reactions carried out by these microorganisms for growth are not thermodynamically favorable (i.e., $+\Delta G^{0'}$ under physiological growth conditions), as the H_2 they generate accumulates and growth is

subsequently inhibited. However, in association with H_2-consuming microorganisms, such as the methanogens, the H_2 partial pressure is maintained at levels low enough to make the reaction thermodynamically favorable (i.e., $-\Delta G^{0'}$ under physiological growth conditions). Because of their dependence on the H_2-consuming microorganisms, the H_2-producing acetogens are often referred to as obligate syntrophs.

Another physiological group of microorganisms affected by interspecies H_2 transfer is the fermentative anaerobes that synthesize a hydrogenase. In many of these microorganisms, substrate oxidation is linked to the electron carrier nicotinamide adenine dinucleotide (NAD), which has higher redox potential (-320 mv) than H_2 (-414 mv) under standard conditions. However, if the H_2 partial pressure is maintained at a low level (<10 Pa) by H_2-utilizing microorganisms, then H_2 production from NAD becomes thermodynamically favorable. This enables the fermentative microorganisms to reoxidize NADH by reducing protons to form H_2 rather than reducing pyruvate to form dicarboxylic acids and alcohols. This synergistic process enables the fermentor to conserve ATP by synthesizing more acetate and less reduced products. In addition, the net products, acetate and H_2, serve as substrates for growth by the methanogen. The result is that carbon and electron flow is directed towards more efficient degradation by the consortium to the gases CH_4 and CO_2. CO_2 reenters the carbon cycle directly and CH_4 is either oxidized by methylotrophs to CO_2 as it diffuses into the aerobic zone or it enters the atmosphere.

Synergistic interspecies substrate transfer has also been reported to occur with formate and acetate. Removal of formate by formate-utilizing methanogens and the removal of acetate by acetotrophic methanogens have been shown to create a thermodynamic shift that favors butyrate and propionate degradation in syntrophic co-cultures of acetogenic *Bacteria* and methanogenic *Archaea*.

B. Freshwater Sediments

As organic matter accumulates on lake and river basins, the sediment becomes anaerobic, because oxygen is depleted by the activities of aerobic microorganisms. In eutrophic environments with high

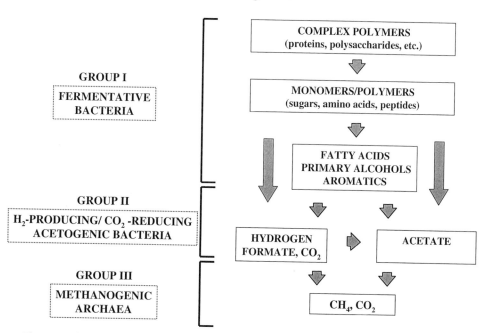

Fig. 4. Carbon flow in an anaerobic microbial consortium from freshwater sediments.

organic loading, the anaerobic region occurs immediately below the sediment surface and even into the water column if the activities of the aerobic microorganisms exceed the rate of oxygen diffusion. Once oxygen and alternative electron acceptors such as NO_3^-, Fe^{+3}, Mn^{+4}, and SO_4^{-2} are depleted, methanogenesis becomes the ultimate degradative process. The flow of carbon in anaerobic freshwater sediments is shown in Fig. 4. Fermentative bacteria that synthesize hydrolytic enzymes, such as cellulases, proteases, amylases, and lipases, catalyze degradation of complex polymers to soluble monomers. The fermentative bacteria then ferment the soluble products to H_2, CO_2, simple alcohols, and fatty acids, including significant generation of acetate, due to interspecies H_2 transfer. H_2-producing acetogenic bacteria then catalyze the oxidation of alcohols and fatty acids to H_2, CO_2, and acetate. The third consortium member, the methanogenic *Archaea*, utilizes the simple substrates H_2, CO_2, and acetate generated by the fermentative and acetogenic bacteria, to generate CH_4. The net result of this consortium is that carbon and electrons are directed toward the synthesis of CH_4 and CO_2, which then reenter the global carbon cycle. CH_4 is oxidized by aerobic methanotrophic bacteria as it diffuses through the aerobic regions of

the water column and reenters the atmosphere as CO_2. However, some CH_4 escapes from shallow water bodies, such as rice fields, and through the vascular systems of aquatic plants and enters the atmosphere.

C. Anaerobic Bioreactors

Methanogenic bioreactors are used for the conversion of organic wastes to CH_4 and CO_2. These anaerobic digestors are found in nearly all sewage treatment plants where wastes in the form of sewage sludge, polymeric particulate material generated by settling of raw sewage, are converted to CH_4 and CO_2 by a consortium of microorganisms in freshwater sediments similar to that previously described. In contrast to most sediment environments, microbial metabolism is higher in a bioreactor because of greater rates of organic loading and higher temperatures (35–40°C) generated by heating the reactor, often by combusting the CH_4 generated in the degradative process. As a result, the rate-limiting factor in a bioreactor is the slow growth rates of the acid-consuming H_2-producing acetogens and the acetate-utilizing methanogens, which require retention times of 14 days or greater, to compensate for their slow metabolic activities. Another critical factor is the requirement for H_2-utilizing methanogens to maintain H_2-partial pressures below 10 Pa. Perturbations in the process that result in inhibition and subsequent "washout" of either of these metabolic groups will result in a drop in pH, resulting in acid accumulation and subsequent inhibition of the entire reactor process. Perturbations can include sudden overloading with a readily fermented organic substrate that results in rapid accumulation of H_2 and fatty acids or introduction of a toxic compound that disrupts the microbial balance.

Anaerobic bioreactors have also been tested as a low-cost method for treatment of other types of particulate organic wastes, including animal manure and crop wastes. Nonparticulate industrial wastes, including many food processing by-products, and organic solvents, such as chlorinated aliphatic and aromatic compounds, are often too dilute to be economically treated by standard bioreactor configurations, which would require long retention times. In order to decrease the hydraulic retention time of the

waste without washing out the biomass, high-rate anaerobic bioreactor configurations have been developed. Examples of these reactors include fixed-film reactors, in which biofilms of microbial consortia are retained in the vessel on solid supports, such as plastic or ceramic matrixes, glass beads, or sand grains. Other designs exist, such as the anaerobic upflow sludge blanket process, in which biomass is immobilized by the aggregation of microbial consortia into distinct granules (Fig. 5). The granules usually consist of three discrete layers of microorganisms that include acetate-utilizing methanogens in the interior, H_2-producing acetogens and H_2-utilizing methanogens in the middle layer, and fermentative microorganisms in the outermost layer. Settler screens separate the granules from the treated water and gas is collected at the top of the reactor. In contrast to particulate waste reactors, in which microbial biomass is generated at the same rate of hydraulic washout, requiring retention times as long as several days, growth of biomass in these

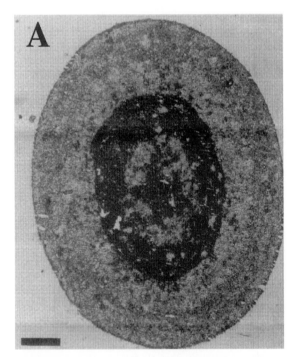
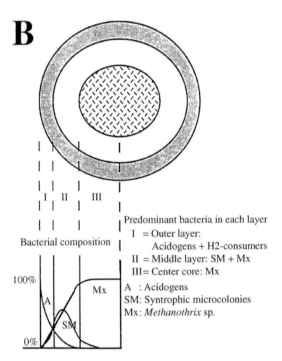

Fig. 5. An anaerobic granule from a brewery wastewater digestor. Panel A shows a thin-section photomicrograph of a granule under a bright-field microscope. Bar, 0.25 mm. Panel B illustrates the bacterial composition of a granule. Methanothrix former generic epithet for Methanosaeta. [Reproduced from Fang, H. H. P., Chui, H. K., and Li, Y. Y. (1994). *Water Sci. Tech.* **30**, 87–96, and Fang, H. H. P., Chui, H. K., and Li, Y. Y. (1995). *Water Sci. Tech.* **31**, 129–135. Copyright 1994 and 1995, respectively, with permission from Elsevier Science.]

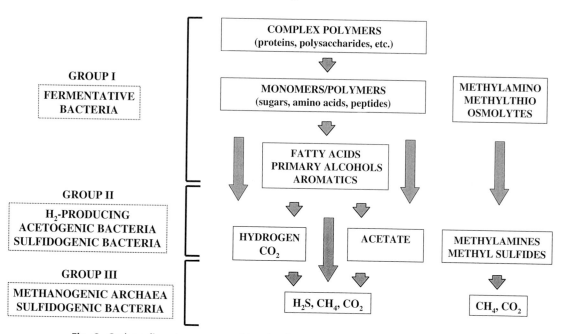

Fig. 6. Carbon flow in an anaerobic microbial consortium from marine sediments.

high-rate reactors is uncoupled from retention time of the waste and can have retention times as short as 1 hr. Hydraulic retention times have also been reduced in anaerobic fermentors that operate at temperatures of up to 60°C, since the metabolic rates of thermophilic microorganisms, including acetate-utilizing methanogens, are greater than those of mesophiles.

D. Marine Habitats

In marine habitats where substrates are limited, sulfate-reducing bacteria outcompete methanogens as the terminal members of the anaerobic consortium (Fig. 6). Their predominance is a result of the availability of the electron acceptor sulfate in seawater (ca. 30 mM). Their lower K_s for substrate utilization enables the sulfate-reducing bacteria to use low concentrations of H_2 and acetate at rates that are greater than those of methanogens. Methanogenesis is predominant in environments where SO_4^{-2} has been depleted. These environments include the lower depths of sediments and elevated coastal marshes, where the rate of SO_4^{-2} reduction is greater than the rate of diffusion from seawater, and also sediments that receive large amounts of organic matter, such as eutro-

phic coastal regions and submarine trenches. In these regions, the three-member methanogenic consortium is similar to that in freshwater environments, but it is composed of halophilic and halotolerant species. Although the acetotrophic methanogens *Methanosarcina* and *Methanosaeta* have been isolated from marine methanogenic enrichments, isotope studies performed in sediment suggest that most of the acetate is oxidized by a H_2-producing syntroph rather than by splitting to CH_4. Methanogens also generate CH_4 from methylated amines and thiols, which are readily available in the marine environment as metabolic osmolytes. Since methylated amines are not used by SO_4^{-2} reducing bacteria, this class of compound is "noncompetitive" and can be used by methanogens in habitats that contain high SO_4^{-2} concentrations.

CH_4 generated in sediments is often consumed as it diffuses through the SO_4^{-2}-reducing region of the sediments, before reaching the aerobic regions. However, the microbes that catalyze this anaerobic oxidation have not been described. Although much of the CH_4 generated in sediments is consumed in the SO_4^{-2}-reducing regions, the water column in the open ocean is supersaturated with CH_4, compared with the atmospheric concentration. This may result from a combination of unoxidized CH_4 that escapes from sediments, and methanogenic activity in the gastrointestinal tracts and fecal material of marine animals. Biologically generated CH_4 in some organic-rich buried sediments can accumulate as gas deposits. Since natural gas deposits generated abiotically are used as indicators of petroleum, these biologically generated CH_4 deposits can act as false indicators of petroleum deposits during oil exploration. Under high hydrostatic pressures generated in deep ocean sediments, biogenic CH_4 can also accumulate as solidified CH_4 hydrates.

E. Ruminant Animals

Ruminant animals include both domestic (cows, sheep, camels) and wild (deer, bison, giraffes) animals. These animals have a large chamber, called the rumen, before the stomach, in which polymers, such as cellulose, are fermented by bacteria to short-chain fatty acids, H_2, CO_2, and CH_4. The rumen is similar to an anaerobic bioreactor, except that the short retention time created by swallowing saliva is less than the generation time of H_2-producing fatty-acid oxidizers and acetate-utilizing methanogens (Fig. 7). As a result, acetate, propionate, and butyrate are not degraded by the consortium, but are absorbed into the bloodstream of the host animal as carbon and energy sources. The volatile gases (e.g., CH_4, CO_2) are removed from the animal by belching. Acidification of the system by the acids is prevented by bicarbonate in the saliva of the animal. Carbon diverted to CH_4 and belched into the atmosphere represents a loss of energy to the animal. Ruminant nutritionists have been attempting to increase feed efficiency by adding methanogenic inhibitors, such as monensin, to feed, thereby, diverting carbon flow to metabolites that can be utilized by the animal.

F. Xylophagous Termites

All known termites harbor a dense microbial community of anaerobic bacteria, and, in the case of lower termites, they also contain cellulolytic protozoa that catalyze the digestion of lignocellulose from wood. As in the rumen, these microorganisms have a synergistic relationship with the termites by converting polymers to short-chain fatty acids that are used as carbon and energy sources for their host. The carbon flow in the hindgut of soil-feeding and fungus-cultivating termites is similar to that in rumen, but in wood- and grass-eating termites, most H_2–CO_2 is

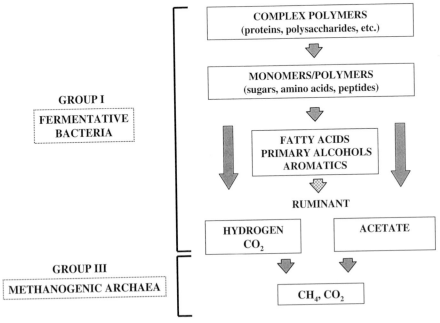

Fig. 7. Carbon flow in an anaerobic microbial consortium from the rumen.

converted to acetate instead of CH_4. Generally, methanogenesis outcompetes H_2–CO_2 acetogenesis and the factors that cause the predominance of acetogenesis in some termites in not known. Two species of H_2-utilizing *Methanobrevibacter* have been isolated from the hindgut of the subterranean termite that exhibit catalase activity and are particularly tolerant to oxygen exposure. Oxygen tolerance may be important for recolonization by bacterial consortia after expulsion of the hindgut contents during molting. Reinoculation is achieved by transfer of hindgut contents from other colony members, which are exposed to air during the process.

G. Human Gastrointestinal Tract

The human colon serves as a form of hindgut where undigested polymers and sloughed off intestinal epithelium and mucin are dewatered, fermented by bacterial consortia, treated with bile acids, and held until defecation. Fatty acids generated by fermentation are absorbed into the bloodstream and can provide approximately 10% of human nutritional needs. Methanogenesis occurs in 30–40% of the human population, with the remaining population producing H_2 and CO_2 instead. Acetogenesis and SO_4^{-2} reduction from H_2–CO_2 also occur in the human colon, but studies with colonic bacterial communities suggest that these activities are only prevalent in the absence of methanogenic activity. The level of CH_4 produced by an individual corresponds to the population levels of *Methanobrevibacter*, but factors controlling the occurrence of methanogens in the human population are not known. Diet does not appear to have a significant role in determining whether an individual harbors an active methanogenic population, but hereditary and individual physiological factors may have a role. For example, methanogens may be absent from individuals that excrete higher levels of bile acids, which are inhibitory to methanogens.

H. Protozoan Endosymbionts

Methanogens are present as endosymbionts in many free-living marine and freshwater anaerobic protozoa, where they are often closely associated with hydrogenosomes, organelles that produce H_2, CO_2, and acetate from the fermentation of polymeric substrates. The products of the hydrogenosomes are substrates for methanogenesis. It is conceivable that the methanogens have a synergistic role by lowering the H_2 partial pressure to create a favorable thermodynamic shift in the protozoan's fermentation reaction. Also, evidence suggests that excretion of undefined organic compounds by the methanogen provides an advantage to the protist host. Endosymbionts are also found in flagellates and ciliates that occur in the hindgut of insects, such as termites, cockroaches, and tropical millipeds. Although rumen ciliates do not harbor endosymbionic methanogens, many have ectosymbionic methanogens that may have an analogous function.

I. Other Habitats

Habitats that have a source of organic carbon and a high water content can become anaerobic as a result of respiratory depletion of oxygen, and, subsequently, support methanogenic communities. Examples include soils waterlogged by heavy rainfall, marshes, rice paddies, rotting heartwood of trees, and landfills. Most of the CH_4 generated in these habitats is released into the atmosphere. However, many landfills are now vented and collected to prevent a buildup of potentially combustible CH_4 underground or in nearby dwellings, and some communities harvest the vented biogas for heat and energy.

Methanogenic habitats are also found in geohydrothermal outsources, such as terrestrial hot springs and deep-sea hydrothermal vents. Methanogens from these environments use geothermally generated H_2–CO_2 for methanogenesis and most are hyperthermophilic, requiring growth temperatures as high as 110°C. In terrestrial sites, methanogens are usually associated with microbial mats composed of photosynthetic and heterotrophic consortia. However, in deep sea hydrothermal vents, where there is no light available for photosynthetic production of organic carbon, the methanogens and other autotrophic bacteria serve as primary producers of cell carbon for a complex community of heterotrophic microbes and animals that accumulates in the vicinity of a hydrothermal vent.

Methanogenesis also occurs in high saline environments, such as Great Salt Lake, Utah, Mono Lake, California, and solar salt ponds. Methanogens from these environments generate CH_4 from methylated amines and dimethylsulfide, which are synthesized by animals and plants as osmolytes. Methanogenesis has been detected in subsurface aquifers, where, it has been proposed, that H_2 generated by an abiotic reaction between iron-rich minerals in basalt and ground water is used as a substrate for methanogenesis. Methanogenesis has also been detected in deep subsurface sandstone, where it is hypothesized that methanogenic consortia use organic compounds that diffuse from adjacent organic-rich shale layers.

IV. PHYSIOLOGY AND BIOCHEMISTRY

A. Catabolic Pathways

All methanogens generate CH_4 during growth by three basic catabolic pathways: autotrophic CO_2 reduction with H_2, formate, or secondary alcohols; acetotrophic cleavage of acetate; or methylotrophic dismutation of methanol, methylated amines, or methylthiols. The common reactions for methanogenesis are shown in Table II. Six new coenzymes were discovered that serve as carbon carriers in the methanogenic pathway (Fig. 8). Methanofuran (MFR) is an analog of molybdopterins, which occur in enzymes that catalyze similar reactions in the *Bacteria* and *Eukarya*. Tetramethanopterin (H_4MPT) is an analog of tetrahydrofolate (H_4THF), which is also a one-carbon carrier in bacterial and eukaryal systems. Although H_4MPT was initially found to be unique to methanogens, it has since been detected in other *Archaea* and, more recently, enzymes catalyzing the methyl transfer for MFR and H_4MPT have been found to coexist with the H_4THF pathway in the CH_4-utilizing methanotrophs. Methyl coenzyme M (CoM–SH), 7-mercaptoheptanoylthreonine phosphate (HS–HTP) and cofactor F_{430} are currently unique to the methanogens.

The methanogenic sequence is initiated by a two-electron reduction of CO_2 and methanofuran by formyl-MFR dehydrogenase (a) to form formyl-MFR (Fig. 9). The formyl group is then transferred to H_4MPT by formyl-MFR:H_4MPT formyltransferase (b) yielding formyl-H_4MPT. A homolog of H_4MPT, tetrahydrosarcinapterin (H_4SPT), found in *Methanosarcina* spp., differs by an additional glutamyl moiety in the substituted R group. Formyl-H_4MPT cyclization to methenyl-H_4MPT is catalyzed by N5,N10-methenyl-H_4MPT cyclohydrolase (c). N5,N10-methylene-H_4MPT dehydrogenase (d) and N5,N10-methylene-H_4MPT reductase (e) catalyze

TABLE II
Reactions and Free Energy Yields from Methanogenic Substrates

Substrate	Reaction	$\Delta G^{0'}$ (kJ/mol CH_4)
Hydrogen/carbon dioxide	$4H_2 + HCO_3^- \rightarrow CH_4 + 3H_2O$	−135
Formate	$4HCOO^- + 4H^+ \rightarrow CH_4 + 3CO_2 + 2H_2O$	−145
Carbon monoxide	$4CO + 5H_2O \rightarrow CH_4 + 3HCO_3^- + 3H^+$	−196
Ethanol[a]	$2CH_3CH_2OH + HCO_3^- \rightarrow 2CH_3COO^- + H^+ + CH_4 + H_2O$	−116
Hydrogen/methanol	$CH_3OH + H_2 \rightarrow CH_4 + H_2O$	−113
Methanol	$4CH_3OH \rightarrow 3CH_4 + HCO_3^- + H_2O + H^+$	−105
Trimethylamine[b]	$4CH_3NH^+ + 9H_2O \rightarrow 9CH_3 + 3HCO_3^- + 4NH_4 + 3H^+$	−76
Dimethylsulfide[c]	$2(CH_3)_2S + 3H_2O \rightarrow 3CH_4 + HCO_3^- + 2H_2S + H^+$	−49
Acetate	$CH_3COO^- + H_2O \rightarrow CH_4 + HCO_3^-$	−31
Pyruvate	$4CH_3COCOOH + 2H_2O \rightarrow 5CH_4 + 7CO_2$	−31

[a] Other short chain alcohols are utilized.
[b] Other methylated amines are utilized.
[c] Other methylated sulfides are utilized.

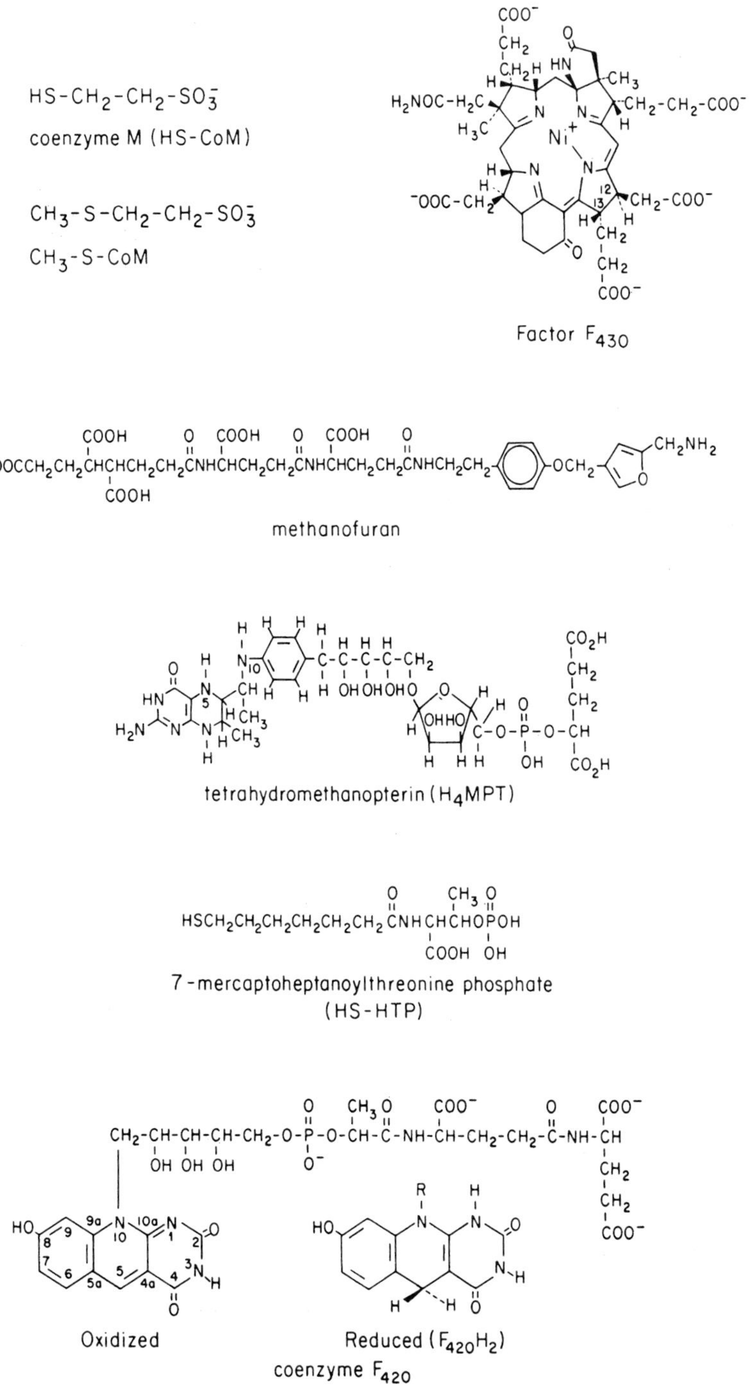
HS-CH2-CH2-SO3-
coenzyme M (HS-CoM)
CH3-S-CH2-CH2-SO3-
CH3-S-CoM
Factor F430
methanofuran
tetrahydromethanopterin (H4MPT)
7-mercaptoheptanoylthreonine phosphate
(HS-HTP)
Oxidized
Reduced (F420H2)
coenzyme F420

Fig. 8. Structures of coenzymes that participate in the methanogenic pathway. [Reproduced, with permission, from Rouviere, P. E., and Wolfe, R. S. (1988). *J. Biol. Chem.* **263,** 7913–7916. American Society for Biochemistry and Molecular Biology, Inc.]

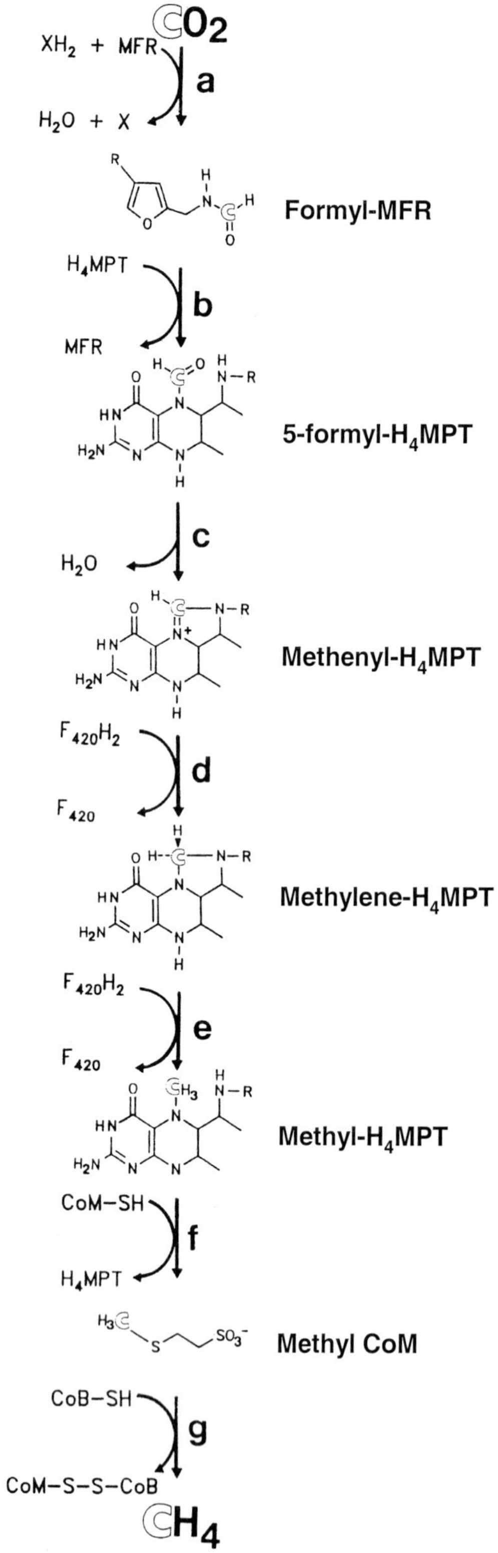

the sequential reduction of methenyl-H_4MPT by the electron carrier coenzyme F_{420} to methylene-H_4MPT and methyl-H_4MPT. The methyl group is then transferred to CoM-SH by N5-methyl-H_4MPT : CoM-SH methyltransferase (f) forming methyl-S-CoM. Methyl CoM reductase (g) catalyzes the terminal reduction of methyl-S-CoM by two elections from HS–HTP to CH_4. CoM–SS–HTP is the product of the terminal reaction, which is subsequently reduced by heterodisulfide reductase to regenerate the reduced forms CoM–SH and HTP–SH.

Methylotrophic catabolism of methanol and methylated amines requires three polypeptides. Methanol is catabolized by transfer of its methyl group to a corrinoid-binding protein, which is methylated by a substrate-specific methyltransferase, methanol : 5-hydroxybenzinidazolyl (MT1). The methyl group is then transferred from the corrinoid protein to coenzyme HS–CoM by methylcobamide : CoM methyltransferase (MT2). Trimethylamine, dimethylamine, and monomethylamine each require a distinct corrinoid-binding protein, which is methylated by a substrate-specific methyltransferase. The methyl group is then transferred from the corrinoid protein to coenzyme HS–CoM by a common MT2 homolog. In contrast, catabolism of the methylthiols dimethylsulfide and methylmercaptopropionate is catalyzed by only two polypeptides: a corrinoid-binding protein tightly bound to a methylcobamide : CoM methyltransferase homolog of MT2. Methyl-S-CoM generated from methanol, methylated amines, and methylthiols is reduced to CH_4 in the methanogenic pathway, as has been described. A portion of the methyl groups generated from methylotrophic catabolism is oxidized in reverse sequence in a pathway identical to the CO_2 reduction pathway, after what appears to be a direct transfer of the methyl groups to H_4MPT. However, the mechanism of this transfer is not yet known. This oxidative sequence generates

Fig. 9. Methanogenic pathway in which CO_2 is sequentially reduced as coenzyme-bound intermediates to form CH_4. Details are described in the text. [Reproduced, with permission, from Weiss, D. S., and Thauer, R. K. (1993). *Cell* **72,** 819–822. Cell Press.]

electrons for the reduction of CoM–S–S–HTP in the methyl-S-CoM reductase system.

The acetotrophic pathway for acetate catabolism proceeds by initial "activation" of acetate by formation of acetyl CoA. *Methanosarcina* spp. synthesize acetyl CoA by sequential activities of phosphotransacetylase and acetyl kinase. In contrast, activation of acetate to acetyl CoA by *Methanosaeta* spp. is catalyzed in a single step by acetyl-CoA synthase. In both genera, cleavage of the C–C and C–S bonds of acetyl-CoA is then catalyzed by the acetyl-CoA decarbonylase/synthase complex, yielding enzyme-bound methyl and carbonyl groups. The complex contains CO:acceptor oxidoreductase, Co-β-methyl-cobamide : tetrahydropterine methyltransferase, and acetyl-CoA synthase activities. The five subunit complex consists of a two polypeptide CO-oxidizing nickel/iron–sulfur component, a two-polypeptide corrinoid/iron–sulfur component, and a single polypeptide of unknown function. The nickel/iron–sulfur component catalyzes the cleavage of acetyl-CoA, the oxidation of the bound carbonyl to CO_2, and methyl transfer to the corrinoid/iron–sulfur component. The methyl group is sequentially transferred to H$_4$SPT by a currently unknown process and to HS–CoM by H$_4$MST : CoM-SH methyltransferase. Methyl-S-CoM is reductively demethylated to CH_4 by methylreducase as previously described. The enzyme-bound carbonyl group is oxidized to CO_2. *Methanosarcina* spp. and autotrophic methanogens growing on H$_2$/CO$_2$ synthesize acetate, presumably by reversing the direction of the acetyl-CoA decarbonylase/synthase complex in a reaction analogous to acetyl CoA synthase in acetate-utilizing *Clostridia*.

B. Bioenergetics

The methanogenic *Archaea* derive their metabolic energy from autotrophic CO_2 reduction with H$_2$, formate or secondary alcohols, cleavage of acetate, or methylotrophic dismutation of methanol, methylated amines, or methylthiols (Table II). Currently, there is little evidence to convincingly support substrate level phosphorylation. Evidence that redox reactions in the catabolic pathways are catalyzed, in part, by membrane-bound enzyme systems and are dependent upon electrochemical sodium ion or proton gradients indicates that electron transport phosphorylation is responsible for ATP synthesis. Both gradients generate ATP for metabolic energy via membrane ATP synthases. Several reactions in the methanogenic pathway are sufficiently exergonic to be coupled with energy conversion, including reduction of methyl CoM (-29 kJ/mol) and methyl transfer from H$_4$MPT/H$_4$SPT to HS-CoM (-85 kJ/mol). In addition, the oxidation of formyl-MF (-16 kJ/mol) during methyl oxidation in the methylotrophic pathway and CO oxidation (-20 kJ/mol) in the acetotrophic pathway are also exergonic.

During growth on H$_2$–CO$_2$ or H$_2$–methanol, the H$_2$:heterodisulfide oxidoreductase system catalyzes the H$_2$-dependent reduction of CoM–S–S–HTP to HS–CoM (Fig. 10). Electron translocation across the membrane generates a proton gradient for ATP synthesis via an A$_1$A$_0$ ATPase. During methylotrophic growth of methanol or methylamines, heterodisulfide dehydrogenase is linked to F$_{420}$ dehydrogenase instead of hydrogenase for CoM–S–S–HTP reduction and generation of a proton gradient. Reduced F$_{420}$ is generated by methylene-H$_4$MPT dehydrogenase and methylene-H$_4$MPT reductase. There is evidence for a third type of heterodisulfide oxidoreductase system in the acetotrophic pathway that is linked to acetyl CoA decarbonylase via ferridoxin.

In addition to proton gradients, methanogens also use sodium ion gradients to drive endogonic reactions and generate ATP. During growth on H$_2$–CO$_2$ or acetate, vectoral Na$^+$ translocation is coupled to methyl-H$_4$MPT:CoM-SH methyltransferase. The Na$^+$ gradient generates ATP via a F$_1$F$_0$ ATPase. During growth on methanol and methylamines, a sodium ion gradient is formed by a Na$^+$/H$^+$ antiporter and the methyl-H$_4$MPT:CoM-SH methyltransferase sodium pump is used in reverse to drive the endergonic methyl transfer from methyl CoM to methyl-H$_4$MPT, for subsequent oxidation of the methyl groups.

Formyl-MFR dehydrogenase is involved in the bioenergetics of autotrophic growth on H$_2$–CO$_2$ or methylotrophic growth on methanol. During hydrogenotrophic growth, the exergonic CO_2 reduction to formyl-MF is driven by a hydrogenase-generated H$^+$ or Na$^+$ gradient. The reaction is likely to be reversible

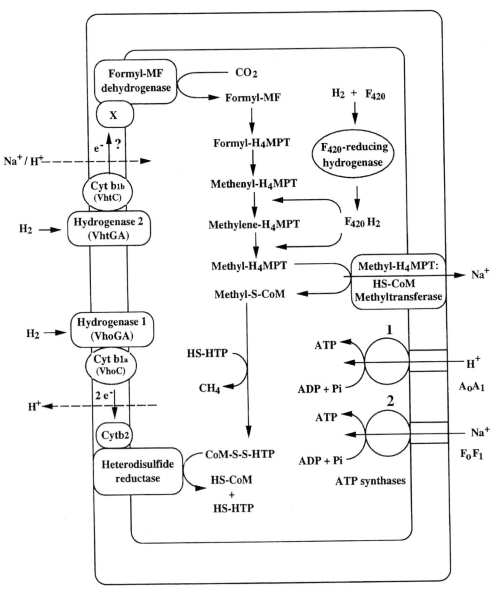

Fig. 10. Proposed mechanisms of energy conservation during methanogenesis from H_2–CO_2 by *Methanosarcina*. 1, A_0A_1 ATPase; 2, F_0F_1 ATPase; Cyt $b1_b$, cytochrome $b1_b$; Cyt $b1_a$, cytochrome $b1_a$; Cyt b2, cytochrome b2; Vho and Vht, membrane-bound hydrogenases; X and question mark denote unknown mechanism of electron transfer. [Reproduced, with permission, from Deppenmeier, U., Möller, and Gottschalk, G. (1996). *Arch. Microbiol.* **165**, 149–163. Springer-Verlag, Gmbh.]

during CO_2 generation by methylotrophs, resulting in a net H^+ or Na^+ gradient.

C. Biosynthetic Pathways

Most methanogens assimilate carbon by CO_2 fixation or acetate uptake and current evidence indicates that assimilation occurs via acetyl-CoA synthesis in a pathway analogous to the Ljungdahl–Wood pathway in acetotrophic clostridia. This conclusion is supported (1) by labeling studies; (2) by the presence of acetyl CoA synthase in autotrophic methanogens; and (3) by the absence or partial absence of enzymes required for other routes of CO_2 fixation, including the Calvin cycle, reverse tricarboxylic acid cycle, the serine pathway, and ribulose monophosphate cycle of the methylotrophs and the hydroxypropionate pathway of *Chloroflexus*. Acetotrophic methanogens likely utilize acetyl CoA directly synthesized by acetate kinase and phosphotransacetylase in the catabolic acetate pathway. In the autotrophic methanogens, current evidence suggests that methyl-H_4MPT, formed by CO_2 fixation in the H_4MPT reductive pathway, likely donates a methyl group to a Fe/S corrinoid protein where it is subsequently transferred to HS–CoA by acetyl CoA synthase. This pathway is a reversal of the acetotrophic pathway for acetate catabolism. During growth on methanol or methylated amines, methyl-H_4MPT, formed by direct methyl transfer to H_4MPT, is also the likely precursor for acetyl CoA synthesis in methylotrophs.

Acetyl CoA is reductively carboxylated by pyruvate oxidoreductase and enters the predominant biosynthetic pathways as pyruvate via an incomplete tricarboxylic acid cycle (TCA). Both reductive and oxidative partial TCA cycles are detected in methanogens and their distribution among species is based on the bioenergetics of α-ketoglutarate synthesis. The oxidative incomplete TCA pathway requires a second acetyl CoA and is restricted to acetotrophic methanogens, which readily synthesize acetyl CoA from acetate via acetate kinase and phosphotransacetylase. In contrast, the reductive incomplete TCA pathway requires two additional reductive reactions and a reductive carboxylation to synthesis of α-ketoglutarate. This pathway is commonly found in autotrophic methanogens, which have a steady reducing potential from H_2.

Studies based on labeling and identification of specific enzymatic activities in selected species, thus far, indicate that biosynthesis of the aromatic aspartate, glutamate, histidine, pyruvate, and serine families of amino acids occurs by pathways similar to those described in the *Bacteria* and *Eukarya*. Biosynthesis of hexoses, required as precursors for cells walls and reserve polysaccharides, such as glycogen and trehalose, proceeds via gluconeogenesis from phosphoenolpyruvate. Pentoses are formed from fructose 1,6-phosphate and glyceride 3-phosphate via transketolase and transaldolase in *Methanobacterium* and *Methanococcus*. In contrast, *Methanospirillum* forms pentoses by oxidation of glucose-6-phosphate. Labeling studies in *Methanospirillum* and *Methanobacterium* indicate that pyrimidine and purines are synthesized by the expected pathways and they also have a purine salvage pathway. The polar core lipids, which include diphytanyl glycerol ethers and diphytanyl diglycerol tetraethers, are synthesized via the mevalonate pathway for isoprenoids.

V. MOLECULAR GENETICS

A. Genome Structure

The methanogens a have single circular chromosome that ranges from 1.6×10^6 to 3.0×10^6 base pairs in size. The percentage guanosine + cytosine (%G + C) ranges from 23 to 62% and high %G + C content is not always associated with high growth temperatures. It is not known whether replication starts at a single origin, as in *Bacteria*, or at multiple origins, as in *Eukarya*. The relatively small genome sizes and short generation times suggest that the mechanism to control timing of DNA replication may be similar to that found in the *Bacteria*. Adenosine–thymidine intergenic spacer regions have been detected in species of *Methanococcus*, but the role of these regions is unknown. Insertion elements have been detected in chromosomal DNA from methanogens. *Methanobrevibacter smithii* harbors a 1381 base-pair element that is flanked by 29 base-pair terminal repeat sequences and a 1501 base-pair sequence flanked by 6 base-pair repeats has been detected in *Methanobacterium thermoautotrophicum* and *Methanobacterium thermoformicium*. Both elements have an open reading frame hypothesized to be a transposase. Extrachromosomal DNA has been detected in methanogens in the form of plasmids, viruses, and virus like particles. At this time, eleven plasmids, ranging in size from 4.4 to 20 kilobase pairs, have been detected in four genera of methanogens. Plasmids from *M. thermoautotrophicum* encode a type II DNA restriction–modification system, and open reading frames in a plasmid from *Methanosarcina acetivorans* show homology to site-specific recombinases and RepA-like replication initiation proteins. The functions of the remaining plasmids are cryptic

TABLE III
Molecular Features of the Three Phylogeneic Domains

Feature	Eukarya	Bacteria	Archaea
Genome	Multiple linear	Single circular[a]	Single circular[b]
Chromatin	Histone-mediated	No conserved mechanism	Histone-mediated
Extrachromosomal DNA	Plasmids, viruses, IS units	Plasmids, viruses, IS units	Plasmids, viruses, IS units
DNA polymerase	Families A[c], B and X	Families A, B and C	Families B and X
RNA polymerase	Three classes, complex	One class $\beta\beta'\alpha_2\sigma$	One class, complex[d]
Gene structure	Single gene	Multiple gene operon	Multiple gene operon
Transcription promoter	TATA box	$-35/-10$ sequence	TATA box
Transcription terminator	AAUAAA sequence	intrinsic, σ-dependent	intrinsic, oligo T
Ribosomal RNA	28S-5.8S/18S	23S-5S/16S	23S-5S/16S
Translation initiation	5′ cap	Ribosomal binding site	Ribosomal binding site
Initiator transfer RNA	Methionine	Formylmethionine	Methionine

[a] Other short chain alcohols are utilized.
[b] Some species have multiple copies of the same genome.
[c] Mitochondrial.
[d] AB′B″C based on homology to eukaryal RNA polymerase II.

at this time. The mechanism of plasmid replication in methanogens is not known, but RepA proteins are associated with plasmids and phage that replicate by a rolling circle mechanism. Lytic viruses have been detected in three strains of *Methanobacterium* and one strain of *Methanobrevibacter smithii*. These viruses show a varying range of host specificities. A temperate virus-like particle has been isolated that can integrate into the chromosome of *Methanococcus voltae*.

In contrast to the similarities of DNA structure in methanogenic *Archaea* and *Bacteria*, DNA-modifying proteins in the methanogens share features common to both the *Bacteria* and *Eukarya* (Table III). All cells must package their genomic DNA within the limited space of the cells. In contrast to the *Bacteria*, which do not appear to have a conserved mechanism for DNA packaging, *Archaea* and *Eukarya* chromosomal DNAs are compacted by histones into defined structures called nucleosomes, which are further assembled to form chromatin. The small, basic histonelike proteins from *Methanothermus fervidus* have homology to *eukaryal* histones and appear to have conserved the minimal structure for creating positive superturns, required to build a nucleosome. Archaeal histone-like proteins increase the melting temperature of linear DNA *in vitro* by as much as 25°C and likely protect DNA from heat denaturation *in vivo*. Reverse gyrase, also detected in *M. fervidus*, may

contribute to heat resistance of DNA by creating stable positive supercoils in inter-histone regions. Histone-like proteins isolated from mesophilic *Methanosarcina* cause concentration-dependent inhibition or stimulation of gene transcriptions *in vitro*, which suggests that they may also have a role in gene regulation.

B. DNA Replication, Repair, Modification, and Metabolism

Although our understanding of DNA replication and repair in the *Bacteria* and *Eukarya* is well advanced, comparatively little is known about these processes in the methanogenic *Archaea*. Studies on DNA replication show that aphidicolin, a specific inhibitor of eukaryal DNA replication, inhibits DNA polymerases from *Methanococcus vannielii* and *Methanococcus voltae*, but does not inhibit DNA polymerase from *Methanobacterium thermoautotrophicum*. Sequences of genes encoding both the aphidicolin-sensitive and -insensitive methanogen DNA polymerases reveal they are homologous to family B and X DNA polymerases from the *Bacteria* and *Eukarya*. *Methanococcus jannaschii* contains a single gene with two inteins that encode a B-type DNA polymerase. In contrast, *M. thermoautotrophicum* has two polymerases: a B-type DNA polymerase composed of two polypeptides and an X-type DNA polymerase.

DNA repair mechanisms have also been identified in methanogens. A photoreactivation system in *M. thermoautotrophicum* is mediated by class II photolyase with homology to metazoan photolyases. Genes encoding putative eukaryal DNA repair proteins RAD2 and RAD51, and bacterial DNA repair proteins uvrABC, have been identified in genomic sequences of *M. jannaschii* and *M. thermoautotrophicum*. Several type II restriction endonuclease–methyltransferase systems have been identified and four methanogen endonucleases are available commercially. In addition, putative type I restriction–modification enzymes have been identified by sequence annotation of the *M. thermoautotrophicum* genome. Despite their sensitivity to oxygen, a Fe-superoxide dismutase has been characterized from *M. thermoautotrophicum* and catalase activity has been detected in species of *Methanobrevibacter*. Putative genes encoding for related DNA replication and repair proteins, such as helicases, ligases, topoisomerases, endonucleases, recombinases, and replication factors, have been identified in genomic sequences of *M. jannaschii* and *M. thermoautotrophicum*, but their function has not yet been confirmed.

C. Gene Structure and Transcription

The organization of methanogenic genes in tightly linked clusters is similar to the operon configuration found in the *Bacteria*. As in the *Bacteria*, the archaeal operons are transcribed from an upstream promoter into polycistronic RNAs. However, archaeal genes are transcribed by a multicomponent RNA polymerase that is structurally homologous to eukaryal RNA polymerase and recognizes a TATA promoter with high sequence homology to the consensus motif for the *Eukarya*. Unlike the variable upstream distance of eukaryal promoters, the archaeal promoter element is located at a consistent distance, approximately 20–30 bp upstream from the transcription initiation site, a range similar to the conserved −35 bp region observed in the *Bacteria*. Site-directed deletion studies conducted with methanogenic *Archaea* by *in vitro* techniques indicate that efficient transcription and start-site selection is dependent upon a TATA promoter. This arrangement closely resembles the core structure of RNA polymerase II promoters

in the *Eukarya*. Purified archaeal RNAPs from *Methanobacterium thermoautotrophicum* and *Methanococcus voltae* fail to initiate site-specific transcription without the addition of TATA binding protein (TBP) and transcription factor TFIIB. Yeast and human TATA-binding proteins can substitute for TBP in a *Methanococcus*-derived archaeal cell-free transcription system, indicating that they are functionally homologous. Additional genes that encode the eukaryal-like transcription factors TFIIIC and TFIIS have been putatively identified in the genomic sequences of methanogenic *Archaea*. Although the spatial and temporal nature of these transcription mechanisms is not yet known, the results suggest that a DNA-protein recognition site, analogous to that in the *Eukarya*, is required for site-specific RNAP recognition in the *Archaea*. This mechanism would involve sequence recognition by TFIIB–TBP and recruitment of other transcription factors to form a recognition complex that would be recognized by polymerase and initiate transcription. The mechanisms of transcriptional regulation in highly regulated genes has not yet been determined, but *in vitro* studies on *Methanococcus maripaludis* reveals that point mutations in a palindrome located downstream of the *nif*H transcription start site results in derepression of expression. This suggests that a bacterial-type repressor may mediate gene expression in some highly regulated methanogen genes. Generally, transcription is terminated following an inverted repeat sequence located downstream of methanogen genes. Transcription termination sites are similar to the ρ-independent terminators in the *Bacteria*, and likely form a stem–loop secondary structure to mediate termination. A second type of transcription terminator, which consists of a single or several tandemly arranged oligo-T sequences, is found in hyperthermophilic methanogens. The occurrence of this terminator in hyperthermophiles suggests that the stem–loop structures characteristic of σ-independent promoters may be unstable at higher growth temperatures.

D. RNA Structure and Translation

The stable RNAs transcribed by methanogen genes have been investigated in some detail. Methanogen ribosomes resemble bacterial ribosomes. They are composed of two protein subunits of 30S and 50S and three rRNA components of 23S, 16S, and 5S, which, assembled, yield a ribosome of 70S. Archaeal and bacterial ribosomal proteins are functionally homologous and can be interchanged to create an active ribosome *in vitro*. Genes encoding rRNA are arranged in the order 16S–23S–5S and the number of operon copies varies in number from one to four. The organization and order of genes encoding methanogen ribosomal proteins also resembles that found in the *Bacteria*. Methanogen tRNAs contain the sequence 1-methylψψCG substituted for the sequence TψCG, typically found in the arm of bacterial and eukaryal tRNAs. Introns have been detected in genes encoding tRNA.

Some methanogen mRNAs have poly-A$^+$ tails, but, as in the *Bacteria,* they average only 12 bases in length. Protein-encoding genes employ the same genetic code as *Bacteria* and *Eukarya*, and codon preferences reflect the overall base composition of the genome and the level of gene expression. The codon ATG is used frequently as an initiation codon, as well as GTG and TGG. A ribosomal-binding site located upstream of structural genes, when transcribed, is complementary to the 3′ terminal sequence of methanogen 16S rRNA. There is little information on the mechanisms of translation based on biochemical experimentation. However, genome analysis reveals that methanogens possess protein initiation factors that share homology with both bacterial and eukaryal IF proteins. Inteins have been identified by genomic sequencing of *M. jannaschii,* which suggests that protein-splicing occurs in methanogens. In addition, evidence has been found for phosphorylation of proteins at a tyrosine residue, which is a mechanism of post-translational control in the *Bacteria* and *Eukarya.*

E. Genomics and Gene Function Analysis

The 1.66 megabase pair genome of *M. jannaschii* and 1.8 megabase pair genome of *M. thermoautotrophicum* have been completely sequenced and annotated. For the thermophile, *M. thermoautotrophicum,* 46% of the total open reading frames had significant

similarity to database sequences with assigned function. In contrast, only 38% of the total open reading frames in *M. jannaschii* could be assigned a function based on sequence similarity with all proteins, and only 19% had similarity with *M. thermoautotrophicum* genes, which suggests that the majority of proteins in this hyperthermophile are uniquely adapted for growth at high temperatures. Overall, the majority of archaeal open reading frames with similarity to bacterial sequences include genes for small molecule biosynthesis, intermediary metabolism, transport, nitrogen fixation, and regulatory functions. Archaeal open reading frames with similarity to eukaryal sequences include genes for DNA metabolism, transcription, and translation. The *M. thermoautotrophicum* has several genes not identified in *M. jannaschii*. These include a *grp*E–*dna*J–*dna*K heat shock operon, a proteasome–chaperonin, additional DNA repair proteins, DNA helicases, nitrogenase subunits, ribonucleotide reductase, and proteases. The presence of Cdc6 homologs and three histones suggests that DNA replication initiation and chromosome packaging is eukaryal, but detection of *fts*Z suggests that bacterial-type cell division occurs. Additional unique features include an archaeal B-type DNA polymerase with two subunits, putative RNAP A′ subunits that suggest that possibility of additional mechanisms for gene selection, and two introns in the same tRNA^{Pro} (CCC) gene, which establishes a new precedent.

The availability of genome sequences of methanogens and current activities to sequence others make the development of archaeal gene-transfer systems essential for confirmation of gene function. However, advances in the genetics of methanogens have been limited to studies *in vitro* or using heterologous systems, such as *E. coli,* because of the lack of tractable gene transfer systems for these microorganisms. Two gene-transfer systems have recently been developed for species of *Methanococcus* and *Methanosarcina.* Both systems utilize hybrid shuttle vectors derived from native archaeal plasmids. Since plasmids occur in low copy number in the methanogens, the DNA to be transferred is ligated into the vector, which is then amplified in *E. coli* with ampicillin as a selectable marker. The modified plasmid is then transferred into the methanogen by polyethylene glycol-mediated (*Methanococcus* sp.) or liposome-mediated (*Methanosarcina*) transformation. A second selectable marker in the plasmid, such as *pac* (puromycin acetyl transferase) controlled by a highly expressed methanogen promoter *mcr* (methyl CoM reductase), provides selection in the methanogen. Both autonomously replicating plasmids for introducing specific phenotypes and integration plasmids for disrupting genes by homologous recombination have been designed. These systems are highly efficient, yielding 10^7 to 10^9 transformants μg^{-1} DNA. Methods for transposon-mediated mutagenesis have also been developed. A transducing phage has been isolated from *M. thermoautotrophicum,* but it is not currently useful for gene transfer because of its limited burst size (~6/cell). Conjugation has not been observed in methanogens. Although gene-transfer systems are somewhat limited at this time, development of new and more sophisticated systems is ongoing. Impediments that need to be overcome include a lack of understanding of mechanisms for vector replication, retention and segregation, and the limited number of selectable and phenotypic markers that are currently available for the methanogens.

VI. SUMMARY

The methanogenic *Archaea* have a pivotal role in the global carbon cycle by complementing aerobic processes that ultimately lead to the oxidation of organic carbon to CO_2. However, a steady increase in the levels of atmospheric CH_4 that has coincided with the increase in the human population is a cause for concern, since CH_4 is a greenhouse gas. Methane's contribution to global warming results from its high infrared absorbance and its role in complex chemical reactions in the stratosphere that affect the levels of ozone. Increased waste disposal activities, such as landfills, are a significant source of atmospheric CH_4. Another significant source results from agricultural activities, such as the increased use of domesticated ruminants for production of meat and dairy products and the increased development of rice paddies. Understanding the properties of methanogens and the roles they have in the global carbon cycle will have important implications in addressing the issue of global warming as the human population increases.

The application of methanogens in biotechnology has been largely limited to waste management, which is often coupled to limited biogas production. Although the petroleum crisis of the 1970s led to an interest in methanogenic biogas production by fermentation of sources ranging from agricultural products to marine kelp, cost-effective technologies were never fully developed and interest has since waned with the drop in petroleum prices. Other potential applications for methanogens include the production of novel pharmaceuticals, corrinoids, and thermostable enzymes. To date, methanogens have yielded only a few restriction endonucleases as commercial products. However, significant advances in our understanding of the physiology and biochemistry of methanogens over the past two decades, combined with the recent developments in gene transfer systems and advances in genome sequencing, make the application of methanogens for biotechnology more plausible in the near future.

See Also the Following Articles

ARCHAEA • FRESHWATER MICROBIOLOGY • RUMEN FERMENTATION • TRANSCRIPTIONAL REGULATION IN PROKARYOTES

Bibliography

Bult, C. J., White, O., Olsen, G. J., Zhou, L. X., Fleischmann, R. D., Sutton, G. G., Blake, J. A., Fitzgerald, L. M., Clayton, R. A., Gocayne, J. D., *et al.* (1996). Complete genome sequence of the methanogenic archaeon, *Methanococcus jannaschii. Science* **273,** 1058–1073.

Daniels, L. (1992). Biotechnological potential of methanogens [Review]. *Biochem. Soc. Symp.* **58,** 181–193.

Deppenmeier, U., Muller, V., and Gottschalk, G. (1996). Pathways of energy conservation in methanogenic archaea. *Arch. Microbiol.* **165,** 149–163.

Ferry, J. G. (1993). Methanogenesis. *In* "Microbiology Series" (C. A. Reddy, A. M. Chakrabarty, A. L. Demain, and J. M. Tiedje, eds.), p. 536. Chapman and Hall, New York.

Forterre, P. (1997). Archaea: What can we learn from their sequences? *Curr. Opin. Genet. Develop.* **7,** 764–770.

Kates, M., Kushner, D. J., Matheson, A. T., and eds. (1993). "The Biochemistry of Archaea (archaebacteria)," Vol. 26 D. J. K. M. Kates, A. T. Matheson, eds. Elsevier, Amsterdam and New York.

Kotelnikova, S., and Pedersen, K. (1997). Evidence for methanogenic Archaea and homoacetogenic Bacteria in deep granitic rock aquifers. *FEMS Microbiol. Rev.* **20**, 339–349.

Smith, D. R., DoucetteStamm, L. A., Deloughery, C., Lee, H. M., Dubois, J., Aldredge, T., Bashirzadeh, R., Blakely, D., Cook, R., Gilbert, K., *et al.* (1997). Complete genome sequence of Methanobacterium thermoautotrophicum Delta H: Functional analysis and comparative genomics. *J. Bacteriol.* **179**, 7135–7155.

Sowers, K. R., and Schreier, H. J. (1999). Gene transfer systems for the archaea. *Trends Microbiol.* **7**, 212–219.

Sowers, K. R., and Schreier, H. J. (1995). Methanogens. *In* "Archaea: A Laboratory Manual" (F. T. Robb, K. R. Sowers, S. DasSharma, A. R. Place, H. J. Schreier, and E. M. Fleischmann, eds.), p. 540. Cold Spring Harbor Laboratory Press, Cold Spring Harbor.

Thauer, R. (1997). Biodiversity and unity in biochemistry. *Anton Leeuwenhoek Int. J. Gen. M.* **71**, 21–32.

Thomm, M. (1996). Archaeal transcription factors and their role in transcription initiation. *FEMS Microbiol. Rev.* **18**, 159–171.

Method, Philosophy of

Kenneth F. Schaffner

George Washington University

GLOSSARY

cause A condition that is necessary in the circumstances for bringing about an event.

model An idealized, usually causal, mechanism or collection of mechanisms that account for a biological outcome.

scientific method A collection of general principles of scientific inquiry, test, and evaluation.

strong inference A systematic attempt to devise mutually exhaustive alternative hypotheses and subject them to crucial experimental tests.

theory Generally, a collection of related principles that explain a domain.

PHILOSOPHY OF METHOD is a branch of the philosophy of science that investigates scientific methodology. Though "scientific method" is largely implicit in microbiology, microbiologists do implement a number of methods of scientific inquiry that have been analyzed by philosophers, such as the methods of difference and comparative experimentation. In addition, these methods are put into practice against a backdrop of more general philosophical assumptions concerning scientific progress, scientific realism, scientific inference, and the nature of empirical evidence. The philosophy of method is best described using particular microbiological examples illustrating the application of specific methods of experimental inquiry.

I. SCIENTIFIC METHOD IN GENERAL

The search for a method that would articulate and codify the principles of scientific inquiry has a long history. Philosophers, in particular, have proposed a number of different approaches to scientific method, beginning with Aristotle. In more modern times, Descartes and Francis Bacon wrote on the subject, but the study of scientific method attained its most systematic treatments in the work of the nineteenth-century philosophers and scientists William Whewell, Stanley Jevons, and John Stuart Mill. It was the last who forcefully re-presented the methods of agreement, difference, concomitant variation, and other methods which continue to have an influence among contemporary philosophers. Later in this article, some of the more current positions of such philosophers as Karl Popper and Thomas Kuhn will also be cited as contributing in important ways to this subject.

Though some philosophers studying scientific method, such as Francis Bacon, hoped to furnish a method of *discovery*, the majority of philosophical thinking in this area has tended to follow Popper's view that states

The initial stage, the act of conceiving or inventing a theory, seems to me neither to call for logical analysis or be susceptible of it. The question how it happens that a new idea occurs to a man—whether it is a musical theme, a dramatic conflict, or a scientific theory—may be of great interest to empirical psychology; but it is irrelevant to the logical analysis of scientific knowledge. The latter is concerned not with *questions of fact* but only with questions of *justification* or *validity*.

This view of Popper's has come under attack by a number of recent philosophers and artificial intelligence theorists. These proponents of a logic of scientific discovery have been able to develop discovery programs in a number of domains, including organic chemistry and mathematics, and active research in this area is continuing. Advances in this area are quite technical and typically domain-specific, and, thus, it will not be possible in the context of the present article to discuss in any detail these various attempts to develop a logic of discovery; this selection will primarily be concerned with the logic of testing and of proof.

It should be noted that "scientific method" has been largely *implicit* in the writings of microbiologists. There are some occasional reflections on methodological issues, for example, Koch's development of his "postulates" for assessing the causation of disease by a bacteriological agent and Duesberg's recent appeals to methodological issues in connection with the cause(s) of HIV disease. Another interesting methodological debate is the recent controversy about the causative agent—putatively a prion—of the spongiform encephalopathies, such as "mad cow disease." Though these explicit debates are the exception, microbiologists do put into practice various philosophies of method, and it is the function of the present article to make explicit what figures in a largely silent way in microbiology.

II. SCIENTIFIC METHOD IN BIOLOGY; OBJECTIVE TRUTH AND SCIENTIFIC PROGRESS

A. Theory and Experiment in Physics Compared and Contrasted with Biology

Biology in general, and microbiology in particular, share a number of common methodological assumptions with all the natural sciences, including the physical sciences. The life sciences, however, also possess some special methodological features which it will be useful to distinguish. Because each of the

natural sciences seeks reliable general knowledge, the methods mentioned briefly in Section I, such as the methods of agreement and difference, are widely employed in the life and nonlife sciences. These methods can be thought of as attempting to discern the causal structure of the world, and, in their application, scientists endeavor to identify possible confounding factors which can lead to spurious inferences about causes and effects. Thus, all natural scientists attempt to control for interfering and extraneous factors, frequently by setting up a control comparison or a control group. Such controls are a direct implementation of what Mill termed the method of difference and Claude Bernard the method of comparative experimentation, which will be reviewed in some detail in Section III.

In the biological sciences, including microbiology, some added complexity is frequently encountered due to biological diversity and the number of systems which strongly interact in living organisms. It is the backdrop of evolutionary theory that allows us to understand why there can be both extensive and subtle variation in organisms and mechanisms, as well as why there may be narrowly defined precise mechanisms that are (nearly) biologically universal, such as the genetic code. Variation due to meiosis, mutation, and genetic drift, for example, predict that extensive variation should occur most frequently in evolving populations where strong selection pressures toward precise and universal mechanisms are not present. Alternatively, where variations would almost certainly be lethal, there exist strong pressures toward the fixation of (nearly) universal mechanisms. Thus, evolutionary theory, at a very general level, explains some of the specific and general features of other theories and models in microbiology.

Because broad and subtle variations may be encountered in microbiology, special attention frequently needs to be given to ensuring the (near) identity of the organisms under investigation, except for those differences which are the focus of the scientist's inquiry. Accordingly, the development of special strains of organisms and the identification and classification of populations (and subpopulations) assume an urgency that frequently can be ignored in the physical sciences, where, for example, all electrons are identical. It is in connection with the satisfaction of these urgent needs that the development of the techniques of pure culture become so vitally important.

Organism and subsystem variation do not only influence the experimental arena, but also figure in what constitutes the biological analog of "theory" as well. Genetic and environmentally produced variation frequently results in the need of biologists (and microbiologists) to focus on "model" organisms and on prototypical systems, which are highlighted against a backdrop of similar but different organisms and mechanisms (see Schaffner, 1998). Biologists, thus, often find themselves practicing what a 1985 National Academy of Sciences report called "many–many modeling" in a complex "biomatrix." The report, which re-presented the results of a series of workshops directed by Harold Morowitz, introduced a notion of the "matrix" of biomedical knowledge: "The workshops demonstrated that the results of biomedical research can be viewed as contribution to a complex body, or matrix, of interrelated biological knowledge built from studies of many kinds of organisms, biological preparations, and biological processes at various levels." From within this multidimensional matrix, many–many modeling occurs, in which analogous features at various levels of aggregation are related to each other across various taxa. The report notes: "An investigator considers some problem of interest—a disease process, some normal physiological function, or any other aspect of biology or medicine. The problem is analyzed into its component parts, and for each part and at each level, the matrix of biological knowledge is searched for analogous phenomena . . . Although it is possible to view the processes involved in interpreting data in the language of one-to-one modeling, the investigator is actually modeling back and forth onto the matrix of biological knowledge."

B. Method and Scientific Progress

In his historical monograph on the development of microbiology, Collard distinguished four historical eras. The period of *speculation* comprised the epoch from about 5000 B.C. until the work of the microscopist Leeuwenhoek around 1675 ushered in the era of *observation*. Beginning in the mid-nineteenth cen-

tury, the era of *cultivation* began with Pasteur's studies of fermentation and Koch's development of pure culture methods employing solid media. Collard suggests that the modern physiological era, which is dominated by the elucidation of bacteriological and biochemical mechanisms, commenced about 1900. Implicit in Collard's chronology is the inference that the more recent methodologies are more scientifically sound and objective, in contrast to the earlier, more speculative inquiries in microbiology.

It may initially appear odd to scientifically trained microbiologists to even raise the question of whether there *is* any "objective truth" in science. Over the past 25 or so years, however, serious questions have been raised about the nature of scientific "truth" and "progress" by the influential work of Kuhn, some philosophers of science (e.g., Feyerabend), and several sociologists of science (e.g., Latour and Woolgar), and a brief discussion of these issues may be helpful.

C. Truth and Progress in Science

Throughout this century, philosophers of science have engaged in vigorous disputes about the nature of scientific truth. An examination of the history of science in general, and microbiology in particular, would lead one to the conclusion that there have existed many "good" scientific theories that have not survived to the present day. Kuhn's characterization of scientific revolutions provides a superb (if ultimately misleading) introduction to examples of these discarded theories. Such theories have gone through the stages of discovery, development, acceptance, rejection, and extinction. Further examination of extinct theories, however, would show that they possessed a number of beneficial consequences for science.

Incorrect and literally falsified theories have several explanatory functions, and have systematized data, stimulated further inquiry, and led to a number of important practical consequences. For example, the false Ptolemaic theory of astronomy was extraordinarily useful in predicting celestial phenomena and served as the basis for oceanic navigation for hundreds of years. Newtonian mechanics and gravitational theory, which is incorrect from an Einsteinian and quantum mechanical perspective, similarly served both to intelligibilize the world and to guide its industrialization. In the biological sciences, the false evolutionary theory of Lamarck systematized and explained significant amounts of species data, and in microbiology, Pasteur's false nutrient-depletion theory of the immune response, nonetheless, served as the background for the development of the anthrax vaccine. Such examples lead one toward what has been termed an *instrumentalistic* analysis of scientific theories (or hypotheses).

The basic idea behind such a position is to view theories and hypotheses as *tools* and not as purportedly true descriptions of the world. For a thoroughgoing instrumentalist, the primary function of scientific generalizations is to systematize known data, to predict new observational phenomena, and to stimulate further experimental inquiry. Such an approach bears strong analogies to the "constructivist" program of several sociologists of science, such as Latour and Woolgar, who conceive of many biomedical entities (e.g., neuroendocrine releasing factors) as being "constructed" rather than as "discovered." To constructivists, such putatively "real" microbiological substances are actually *constructed socially* (as conceptualized entities), as part of a complex give-and-take among laboratory machine readings and discourse among laboratory scientists.

Though such positions as Kuhnian relativism, instrumentalism, and social constructivsm are prima facie attractive, they are inconsistent with other features of scientific inquiry. The relation of scientific theory to "observations" in the laboratory is exceedingly complex. In spite of the weblike connections between theoretical postulation and laboratory experiments, however, multiple empirical constraints on theoretical speculations generate a stability and "stubbornness" of both data and theory, to use Galison's terms. For example, scientists view the distinction between what they term "direct" and "indirect" evidence as important. Even though, as we shall see below, the distinction is relative, it is, nonetheless, significant to note that scientists *behave* as if the distinction is important, and that "direct evidence" would seem to support a more *realistic* analysis of scientific theories (or hypotheses). A realistic type of alternative to the instrumentalist position would characterize scientific theories as *candidates* for *true* descriptions of the world. Though not denying the importance of theories' more instrumentalistic functions, such as prediction and fertility, the realist views these features as partial indications of a theory's *truth*. The history of recent philosophy of science and the more general discipline of "science studies" has seen an oscillation between these realist and instrumentalist positions, as well as the development of some interesting variants of these positions. The relativist reviews have been strongly criticized, especially by realism-defending scientists, the give-and-take of which has been described as the 1990s "science wars" [compare Collins and Pinch (1994) with Gross and Levitt (1994)]. This, however, is a subject which cannot be pursued within the limitations of this article.

Suffice it to say that, in spite of the variation in positions that scientists and philosophers of science have taken about the nature of ultimate scientific truth, these varying positions do *not* disagree about the need to report accurately and faithfully what the scientist has observed or reasoned to in his or her investigations.

III. SPECIFIC BUT PHILOSOPHICALLY GENERAL METHODS OF EXPERIMENTAL INQUIRY

This section provides a more systematic and in-depth review of the classical methods of experimental inquiry that were very briefly introduced in Section I. We begin from John Stuart Mill's characterization of these experimental methods, since Mill's analysis is *provisionally* adequate for our purposes and is also well known.

A. "Mill's" Methods

Mill's first experimental method is known as the "method of agreement." This was stated by Mill as follows:

If two or more instances of the phenomenon under investigation have only one circumstance in common, the circumstance in which alone all the instances agree, is the cause (or effect) of the given phenomenon.

It should immediately be pointed out that this method is most difficult to satisfy in microbiological inquiry because of the complexity of organisms. To vary every relevant character save one, is largely unrealizable, though it may occasionally be done in very narrowly circumscribed investigations in molecular biology, for example, in genetic codon analysis.

The applicability of the method of agreement is also suspect in complex biological organisms for two other general reasons, originally pointed out by Mill in his *System of Logic*. These general reasons were referred to by Mill as the *plurality of causes* and the *intermixture of effects*. In the former, Mill noted that in those complex cases where the same consequence could be the result of different jointly sufficient antecedents, the method of agreement yielded *uncertain* conclusions. In complex and adaptable biological organisms, such a situation is often likely. The problem of the "intermixture of effects" produced even greater difficulties for the method of agreement. This occurred in cases where effects were "intermixed," i.e., were either not separable into clearly defined components, or else the result was emergent and incalculable on the basis of the causal components. As examples of these two species of the intermixture of effects, Mill cited the vector addition of forces in mechanics and the production of water from oxygen and hydrogen. Mill also believed that biological properties were emergent with respect to the physicochemical, with attendant difficulties. Suffice it for now to note that the method of agreement possesses serious imperfections in its application to complex circumstances, as are found in biological organisms.

Mill's methods also include, beside the better known method of agreement and the method of difference, the method of residues and the method of concomitant variations. The latter, according to Mill, may be stated in canonical form as

Whatever phenomenon varies in any manner whenever another phenomenon varies in some particular manner, is either a cause or an effect of that phenomenon, or is connected with it through some fact of causation.

This method has important uses in biology and medicine, and an illustration of it will be considered in the following section on specific implementation of the methods. Again, as in the case of the method of agreement, the complexity of biological organisms makes a simple and straightforward application of the method of concomitant variations uncertain. When we are attempting to determine a major causal role which, say, an entity plays in a system, a simple recording of concomitant variations is almost always insufficient.

The method of residues is likewise suspect in its simple application to biological and microbiological causation. This method, in which one "*subduct[s] from any phenomenon such part as is known by previous inductions to be the effect of certain antecedents*," and considers "*the residue of the phenomenon . . . [to be] the effect of the remaining antecedents*," is similarly difficult to apply in complex systems in which the interfering residues are likely to be very large, both in number and in their interactions.

The method of difference supplemented in a number of cases with a statistical interpretation, which bears certain analogies to Mill's methods of concomitant variation, is often the most appropriate method of experimental inquiry to establish empirically and directly scientific claims. The method of difference was stated by Mill in the following manner:

If an instance in which the phenomenon under investigation occurs, and an instance in which it does not occur, have every circumstance in common save one, that one occurring only in the former, the circumstance in which alone the two instances differ is the effect or the cause, or an indispensable part of the cause of the phenomenon.

The application of the method of difference, like the other methods, is not automatic in any experimental situation and, as has been observed by a number of thinkers, *presumes* an analysis of the situation into all relevant factors which can be examined one at a time.

This rather demanding requirement can be ameliorated in a way that is reasonably faithful to Mill's approach by following some suggestions of the great nineteenth-century physiologist and philosopher of method, Claude Bernard. In his influential monograph *An Introduction to the Study of Experimental Medicine*, Bernard noted that these presumptions of an antecedent analysis of the experimental situation and the ability to examine factors one at a time were not necessarily valid in experimental medicine, and he urged that the method of difference be further distinguished into (i) the method of "counterproof" and (ii) the method of "comparative experimentation." According to Bernard, in the method of counterproof, one assumes that a complete analysis of an experimental situation has been made, i.e., that all complicating and interfering factors have been identified and controlled. Subsequently, one eliminates the suspected cause and determines if the effect in which one is interested persists. Like Karl Popper in this century, Bernard believed that experimental medical investigators avoided counterproof as a method, since they feared attempts to disprove their own favored hypotheses. Bernard defended a strong contrary position and maintained that counterproof was essential to avoid elevating coincidences into confirmed hypotheses. He argued, however, that those entities which fell into the province of biology and medicine were so complex that any attempt to specify *all* of the causal antecedents of an effect was completely unrealistic. As a remedy for this problem, he urged the consideration of *comparative experimentation:*

Physiological phenomena are so complex that we could never experiment at all rigorously on living animals if we necessarily had to define all the other changes we might cause in the organism on which we were operating. But fortunately it is enough for us completely to isolate the one phenomenon on which our studies are brought to bear, separating it by means of comparative experimentation from all surrounding complications. Comparative experimentation reaches this goal by adding to a similar organism, used for comparison, all our experimental changes save one, the very one which we intend to disengage.

Bernard referred to comparative experimentation as "the true foundation of experimental medicine."

B. Strong Inference

The use of counterproof, crucial experiments, and comparative experimentation finds unified application in what Platt termed "strong inference." Platt contended that rapid progress could be made in the biological sciences if well-formulated, alternative hypotheses were subjected to crucial experiments, designed to eliminate most of the alternatives. He characterized this approach as involving the following steps:

1. Devising alternative hypotheses;
2. Devising a crucial experiment (or several of them) with alternative possible outcomes, each of which will, as nearly as possible, exclude one or more of the hypotheses;
3. Carrying out the experiment so as to get a clean result;
1.′ Repeating the procedure, making subhypotheses or sequential hypotheses to refine the possibilities that remain; and so on.

In Section IV, we shall see some examples of how this "strong inference" approach is coupled with the methods of experimental inquiry in several recent examples in microbiology.

IV. MICROBIOLOGY: ITS SCOPE AND SUBJECT AREAS; REPRESENTATIVE ILLUSTRATIONS OF THE PHILOSOPHY OF METHOD

The scope of microbiology is astonishingly broad. The most recent edition of Zinsser's well-known textbook comments on this subject, stating that:

With each passing year the term *microbiology* becomes a less satisfactory umbrella for the many disciplines that it attempts to cover. Bacteriology, immunology, virology, mycology, and parasitology have each long since become separate and independent disciplines.

Zinsser provides a continuing rationale for the subject area, however, noting that these numerous specialties "are treated together . . . simply because they deal with the agents that cause infectious diseases and with the mechanisms by which hosts defend against them" (1992, ix). Be that as it may, any analysis of the philosophy of method of a domain that includes infection, bacteria, spontaneous generation and origin of life, the germ theory of disease, viruses, immunity, and chemotherapy must, of necessity, be selective. In the following paragraphs, several examples are developed that are representative of methodology in microbiology. These illustrations will be used to provide a specific and accessible approach to the somewhat abstract topic of the philosophy of method in microbiology.

A. The Example of Koch's Postulates and AIDS Virology

As noted in the discussion earlier, self-conscious, explicit methodology is rare in the writings of microbiologists. One prominent and influential exception to this general rule are the Koch's postulates—more accurately termed the Henle–Koch postulates—developed to assess bacteriological causation of disease. About 10 years ago, these postulates returned to prominence at the center of a potentially momentous controversy in AIDS virology. The postulates can also be used as a framework for investigating another controversial example of disease causation—prions—as well as to point toward the limits of "direct evidence" in microbiology. A brief account of the postulates' history and the philosophical presuppositions will facilitate an understanding of their role in the contemporary debate to be discussed later in this article.

The postulates have their proto-origin in Henle's work in 1840 and were then further developed by his pupil, the great microbiologist Robert Koch, in lectures in 1884 and 1890. There are several slightly different formulations of the postulates (see Evans, who follows Rivers as well as Zinsser), but the essence can be summarized as follows:

1. The infectious agent occurs in every case of the disease in question and under circumstances which can account for the pathological changes and the clinical course of the disease.
2. It must be possible to isolate the agent from all cases of the disease and to grow the agent in pure culture.
3. After being fully isolated from the diseased animals (or humans) and repeatedly grown in pure culture, it can induce the disease anew by inoculation into a suitable host.

In addition, some formulations add a fourth postulate:

4. The infectious agent occurs in no other disease as a fortuitous and nonpathogenic agent.

These are stringent conditions, and even at the time that Koch enunciated them, a number of etiological agents were known which did not fully meet the criteria, including bacteria isolated from typhoid fever and cholera. In the ensuing 100 years since the articulation of the postulates, several types of hitherto unrecognized forms of disease, including chronic diseases, cancer, and a number of viral diseases, have required the modification and extension of these postulates. In 1976, Evans attempted to bring together a number of these different themes into a "unified" epidemiological conception of disease causation.

It is, however, the more traditional and stringent form of the postulates which was cited over the course of the past decade in an attempt by molecular virologist Peter Duesberg to question the causal relationship between HIV and AIDS. The theses of Duesberg's work were that (1) HIV (the AIDS virus in either its HIV-1 or HIV-2 form) is neither necessary nor sufficient for AIDS, though HIV is a good "marker" for American AIDS, and (2) AIDS is actually an autoimmune disease caused by as yet unknown pathogens, probably acting in concert with a variety of environmental insults, including hard psychoactive as well as medical drugs (AZT) and blood transfusions (particularly in the case of hemophiliacs). In the eyes of most scientists, Duesberg's principal arguments have been effectively answered, though Duesberg has continued to criticize the standard view in several recent books (see Duesberg, 1996).

One of Duesberg's major arguments is that HIV fails to satisfy Koch's postulates. He wrote concerning the first of Koch's postulates that "there is no free virus in most—and very little in some—persons with AIDS, or in asymptomatic carriers." Though he admits that various viral elements (and, by definition of the disease, antibody to HIV) can be found in most or all of AIDS patients, he contended that HIV is not present in ways that can account for the loss of T cells or for the clinical course of the disease, which lags 8 or more years behind infection. Further, with respect to Koch's second postulate, Duesberg contends that there is an "often over 20% failure rate in isolation of HIV from AIDS patients." Finally, citing Koch's third postulate, Duesberg notes that HIV does not produce AIDS in any animal model (chimpanzees had been inoculated unsuccessfully), and the data from accidental inoculations of humans (through donor semen, laboratory accidents involving HIV researchers, and blood transfusions) is not consistent with a role of HIV as the cause of AIDS.

The response to Duesberg's arguments has been threefold. First, a number of commentators have pointed out that Koch's postulates in the form cited above fail to be satisfied in a number of *other* diseases, primarily of the viral and immunological type we encounter in AIDS, and that the postulates need to be taken as "guidelines" and supplemented by broader epidemiological evidence. Second, when a broader, more epidemiological conception of causation is implemented using epidemiological data, that the evidence for HIV as *the* cause of AIDS is overwhelming. Finally, molecular investigations of HIV pathophysiology have deepened in recent years. New coreceptors for the virus have been indentified that explain some resistance to HIV and may also help explain its oddly delayed pathogenesis. These receptors have also permitted better animal models of HIV/AIDS to be created. Duesberg has rejoined to this criticism, arguing that appeals to epidemiological considerations do not make the case that the proponents of HIV think they do and continues to aggressively defend his own theory (Duesberg, 1996).

It is not the purpose of this article to delve further into the details of the ongoing controversy regarding HIV and AIDS, but several points need to be highlighted that illustrate the role of the different philosophies of method involved. First, problems with the failure of HIV to satisfy Koch's first postulate have led to further speculations about the mechanism(s) of T cell depletion. These speculations include syncytia formation, "co-factor" pathogens, HIV as the trigger for an autoimmune disease, and, most recently, the proposal of a "super-antigen" theory. Investigators generally believe that the evidence for these hypotheses are "indirect," at best, and that more "definitive" or "direct proof" is needed. Thus, the Henle–Koch postulates provide a framework and the failure to satisfy them in any simple way should provide the motivation for additional investigation, which

may move the subject more in the direction of "direct" evidence for explanatory models of AIDS pathogenesis. In the next illustration of the philosophy of method, another example of how ambiguous results that do not fully meet the requirements of Koch's postulates is examined, which raises the question of whether fully conclusive evidence is ever forthcoming in microbiology. Our last example of just how "direct" evidence can be obtained will partially answer this question affirmatively, with the continuing caveat, however, that in science, even direct proof continues to remain somewhat conditional.

B. Prions and the Question of the Causative Agents of the TSEs

Modifications needed to extend Koch's postulates to take into account later microbiological developments are also a useful framework for a consideration of the cause of the transmissible spongiform encephalopathies (TSEs). These diseases include Creutzfelt–Jakob disease (CJD) in humans, scrapie in sheep, and bovine spongiform encephalopathy (BSE), or "mad cow" disease, in cattle. A number of investigators have attempted to determine the etiological agents and the details of the pathogenesis of the TSEs. At present, TSEs are generally thought to be caused by a novel infectious agent, a "prion." "Prion" is a term coined by Stanley Prusiner, to represent a "proteinaceous infectious particle." Prusiner received a Nobel Prize in 1997 for his accomplishments in this area and was cited for his "discovery of 'Prions—a new biological principle of infection'." That prions are the fundamental cause of TSEs is still somewhat controversial, and an alternative viral causation hypothesis continues to have its defenders.

If one followed Koch's postulates, as outlined previously, in the TSE area, one would attempt (1) to show the (various) infectious agent(s) occurs in every case of the diseases and accounts for the pathogenesis, (2) to isolate the agent and to purify it, and (3) to use the purified agent(s) to induce the disease(s) in new hosts. In addition, one might want (4) to show the agent does not occur in any other disease as a nonpathological agent. Though Prusiner does not explicitly use Koch's postulates in his various writings on prions and TSEs, the postulates can function as a framework to approach his various arguments. Prusiner's early work was directed at the sheep TSE, scrapie. Following work by Alper that suggested the infectious agent was a protein that did not include nucleic acid, Prusiner was able to show that scrapie prions contained a protein he called PrP, short for "prion protein." These proteins are also unusually resistant to degradation by protease enzymes. Working with Leroy Hood, the protein was then partially sequenced, and the sequence used to construct a probe that could identify the genetic source of the protein. Remarkably, it turned out that the PrP gene can be found in hamsters, mice, and humans, but, most of the time, cells that make PrP are not pathological. An unorthodox, unprecedented, but possible, explanation was that PrP could exist in two different tertiary forms with the same amino-acid sequence: one normal, called "cellular PrP" (PrP^c) and the other associated with scrapie, termed "scrapie PrP" (PrP^{Sc}). This heretical idea was subsequently confirmed and considerable work has been done that examines various conformational changes in PrP and their pathological effects. (For an in-depth discussion of Prusiner's struggle with the unorthodoxy of the prion concept and its relations to the "central dogma of protein synthesis," see Keyes, 1999.)

Prusiner and his group initially attempted (implicitly) to conform to Koch's second postulate by using the PrP gene to generate pure copies of PrP. They thought they could then implement Koch's third postulate and, as they wrote, "inject the protein molecules into animals, secure in the knowledge that no elusive virus was clinging to them. If the infections caused scrapie, we would have shown that protein molecules could, as we had proposed, transmit the disease." But as Prusiner adds, "by 1986, we knew the plan would not work" (Prusiner, 1995). It was difficult to induce the gene to make the high levels of PrP needed, and, in addition, the PrP that was made was the benign "cellular" type.

An alternative strategy was then devised, that looked to inherited, rather than transmissible, forms of the spongiform encephalopathies. Gestmann–Straussler–Scheineker disease is a very rare familial human SE that Prusiner and other investigators were able to link with a point mutation in the PrP gene.

This served as the backdrop for the creation of genetically altered mice carrying a mutated PrP gene. One could first examine whether the mutated gene in the transgenic mice produced scrapie and then whether brain tissue from these mice could cause scrapie in "healthy" mice, albeit only healthy transgenic mice expressing a low level of mutant PrP. These events do, in fact, occur and are claimed by Prusiner to be "solid evidence that the protein encoded by the mutant gene had been solely responsible for the transfer of the disease" (1995). Let us step back and examine this claim more carefully and also consider a critic's view of this "solid" or "persuasive" evidence.

Prima facie, the experiment seems to satisfy Koch's first postulate, since *all* of the transgenic mice making high levels of PrP did develop mouse scrapie. Transplanted infected brain tissue, however, caused only *many* (not all) of the recipients to develop the disease. A proponent of the competing viral hypothesis, Bruce Chesebro, finds the transgenic mouse experiment "inconclusive" (Chesebro, 1998). He notes that the transgenic mice differ from all known TSE models, because no abnormal protease-resistant PrP—a hallmark of scrapie PrP—is detectable in the diseased brain tissue. Also, disease occurs only when the mutant gene is overexpressed, not when it is found in its normal site as a single copy. Finally, transmission by infected brain tissue is only successful in transgenic mice—the low-level PrP producers—and not in nontransgenic animals. Chesebro suspects that "other molecules" from the diseased brains could be causative of the disease, and, further, doubts the utility of the transgenic model, because none of the transmitted materials has been demonstrated to have typical properties of TSE agents, such as resistance to inactivation by heat. Other investigators also believe that additional TSE experimental findings point toward something more than prions as the cause of these diseases, perhaps a virus.

The prion hypothesis and the competing viral hypothesis are, thus, not unambiguously and definitively tested by this transgenic model. But this is only one of a number of lines of evidence supporting the prion hypothesis, and this article cannot delve into the multiple arguments and counterarguments in support of and against the prion view, which continue to evolve (see, for example, the very recent report by Safar *et al.* in *Nature Medicine,* that provides data that different protein conformations are responsible for variations in prions' "strains"). This fragment of the story, however, does suggest that Koch's postulates can be a useful framework within which to analyze possible etiological agents of disease, albeit one that needs extensions. Such extensions include both new understandings of molecular biological processes never dreamed of by Koch, as well as powerful new experimental techniques, such as the use of transgenic and knockout animals. But the story also indicates how difficult it is to obtain conclusive or "direct" evidence of a causal hypothesis in microbiology, particularly in the still comparatively early stages of an investigation of a novel model of infection, even with this new powerful molecular knowledge and associated techniques.

C. The Example of the Isolation of the Repressor

Our third example illustrating the application of methods of experimental inquiry is drawn from molecular genetics and involves progress in clarifying the nature of the mechanism of genetic control in bacteria, more specifically, the development and testing of the operon model of genetic regulation. The model, which is principally due to the work of Jacob and Monod (1961) developed over a number of years. The initial form of the mechanism dates from 1960–1961 and has had a most important impact on both experiment and theory construction in molecular biology (and in embryology). It has been extraordinarily well corroborated by a variety of experiments. It was further developed in the 1960s, 1970s, and 1980s.

The model proposed the existence of a new class of genes, termed regulator genes, which were responsible for the synthesis of a substance, later determined to be a protein of about 160,000 molecular weight, which was termed a repressor. The repressor, in inducible systems, such as the *lac* region of *Escherichia coli,* binds specifically with a DNA region termed the operator locus. This operator has adjacent to it several structural genes, which are under the operator's control. When the repressor is bound to the operator, the associated structural genes are not

transcribed into messenger RNA, and, accordingly, no proteins or enzymes associated with those structural genes are synthesized. An inducer can specifically interact with the repressor, altering its three-dimensional structure, and thus render it incapable of binding to the operator. The enzyme RNA polymerase, which attaches to a DNA sequence called a promoter in the presence of a protein called CAP and which can transcribe the structural genes, is then no longer prevented from initiating transcription of the structural genes adjacent to the operator, and mRNA synthesis and then protein synthesis commence. Repressible systems employ a similar regulatory logic, only, in that case, the initially ineffective repressor is aided in its operator-binding capacity by an exogenously added corepressor. It has been often noted by biologists that such a regulatory system is most useful to the organism since it allows unnecessary genes to be turned off when the proteinaceous enzyme products of those genes are not required for a metabolic process. This results in energy saving and is conceived of as evolutionarily useful to the organism.

In spite of the breadth of evidence for the operon model when it was first proposed in 1961, it can be said that the evidence, at that time, was primarily of an *indirect* genetic type. Monod in his Nobel lecture suggested, for example, that in the early 1960s the repressor—the crucial regulatory entity—seemed "as inaccessible as the matter of the stars." The difficulty of being able to provide more direct biochemical evidence for the existence of the repressor and its postulated properties stimulated at least one alternative theory of genetic regulation, involving a mechanism different from that of the operon model.

In 1966, however, Walter Gilbert and Benno Müller-Hill reported that they had been able to isolate the elusive repressor, and, in 1967, the same authors were able to demonstrate that the repressor bound directly to the operator DNA. These experiments were most important for the field of regulatory genetics, and they were confirmed by Mark Ptashne's essentially simultaneous isolation of the repressor in the phage λ system and the discovery that it also bound to DNA. These experiments, and the analyses of the experimental results, illustrate both the application of Mill's methods and Platt's "strong inference."

Gilbert and Müller-Hill noted in their paper reporting their results that though they felt the Jacob and Monod model to be the simplest, other models would also "fit the data" available. "Repressors," they continued, "could have almost any target that would serve as a block to any of the initiation processes required to make a protein. A molecular understanding of the control process has waited on the isolation of one or more repressors." Isolating the repressor was most difficult because of the hypothesized very low concentration of repressor in the *E. coli* cells. Gilbert and Müller-Hill solved this problem by looking for and finding a mutant form of the bacteria (i^t or tight-binding), which produced a repressor with about a 10-fold increased affinity for the inducer. As an inducer, they used the chemical IPTG (isoprophylthiogalactoside), which induces the *lac* operon but which, in contrast to lactose, is not metabolized by the induced enzyme. The repressor was detected by placing an extract of the i^t mutant in a dialysis sac in a solution of radioactive IPTG (labeled to facilitate detection of very low amounts of it), and looking for an increased concentration of IPTG inside the sac after about thirty minutes (at 4°C).

Using the increased concentration as a sign of the presence of repressor, Gilbert and Müller-Hill purified the extract and analyzed the binding component. They discovered that the enzymes which would attack DNA and RNA would not destroy the inducer binding ability of the presumed repressor, but that an enzyme which attacked proteins would. Heating (above 50°C) also destroyed the ability to bind inducer. This was rather direct evidence (though still conditional on the hypotheses (1) that the enzymes possessed anti-DNA/anti-RNA capacity and (2) that heat denatures proteins) that the repressor *was* a protein. The repressor (labeled with the radioactive inducer) was centrifuged and sedimented on a glycerol gradient. The "profile" of the sedimented repressor indicated that it had a molecular weight of about 150,000–200,000.

In their paper, Gilbert and Müller-Hill cited what they termed "negative and positive controls" on their supposition that they had isolated the repressor product of the *i* gene. The most important "negative controls" involved examining the extracts from i^s (the super repressed) and i$^-$ (constitutive) mutants. The former type (i^s) presumably synthesizes a repressor which has lost the ability to bind inducer, and the latter (i$^-$) had been hypothesized to produce an (significantly) incomplete repressor. For both of these mutants, no binding of IPTG was observed. (Gilbert and Müller-Hill noted that "the various mutant forms of the *i* gene were put into identical genetic backgrounds, so that unknown variations from strain to strain could not confuse the issue.")

As "positive controls," Gilbert and Müller-Hill plotted graphs representing the binding constants of the wild type and i^t mutant to demonstrate the essential identity of the repressors. An examination of the binding ability of the repressor for various other substances, including galactose and glucose (and yielding a decreasing *sequence* of affinities), "gave further support for the [thesis that the isolated] material . . . [was] the repressor."

In the following year, Gilbert and Müller-Hill (1967) used this partially purified radioactive repressor to test for *in vitro* binding to the operator of *lac* DNA. In this experiment, Gilbert and Müller-Hill used strains of bacteriophage which carried different forms of the *lac* operon in their DNA.

By sedimenting mixtures of the various strains of the phage DNA with the radioactive repressor, Gilbert and Müller-Hill found that:

1. In the absence of inducer, the radioactive repressor binds to the DNA.

2. In the presence of IPTG, which releases the repressor from the DNA, no binding is observed.

3. The effect of the IPTG is specific, for substitution by a chemical ONPF (ortho-nitro-phenyl-fucoside; a substance which binds to the DNA but does not induce) has no effect.

4. Use of two mutant phage o^c strains, in which the affinity of the operator for the repressor is expected to be factors of 10 and 200 respectively weaker, yielded sedimentation profiles in rough quantitative conformity with these binding expectations.

These experiments, which appear to utilize the methods of comparative experimentation or difference (the "negative control") and concomitant variation (the "positive controls"), have been characterized as offering "direct" evidence and "proving" the hypotheses of the Jacob–Monod model. Clearly, they do so, however, with the aid of auxiliary hypotheses involving, among others (a) the effects of growth in radioactive media so as to specifically label the "observed" entities and (b) the analysis of centrifugation and sedimentation techniques in density gradients. The experiments require a chain of reasoning. There is, in addition, (c), very general negative auxiliary assumption that what is observed is not an artifact. It must be further emphasized that the direct proof of the Jacob–Monod operon model is contingent on the assumption that other, and perhaps more important and overriding factors, are not involved, i.e., some competing, but as yet unarticulated, hypothesis which could account for all these results. The proposal of a novel mechanism of control which would account for the genetic data as well as Gilbert and Müller-Hill's findings (perhaps by showing some of them to be artifacts), but which would disagree with basic tenets of the operon model cannot be *logically* excluded. The existence of Gilbert and Müller-Hill's experiments are thus a *conditional* direct proof, the conditions depending on (1) the truth of auxiliary hypotheses and (2) the nonexistence of an inconsistent competing theory or model which would account for the same data as the operon model.

Let us now generalize from this example. The "directness" and force of the evidence is obtained by choosing one or more *central* (roughly equivalent to "essential") property(ies) of the hypotheses in the theory or model to be tested. In the operon model's case, these properties were initially suggested in Monod's work, namely, the repressor's "recognition" of the inducer and the operator.

In addition, plausible competitor theories and models are scanned to determine if the test condition(s), if realized, would support the model under test while falsifying the competitors. Recall that *genetic* evidence was insufficient to do this, whereas the Gilbert–Müller-Hill experiments (as well as Ptashne's) were. This exhibits Platt's "strong inference" in specific ways.

The evidence was considerably strengthened by providing both positive and negative controls. These controls are essentially equivalent to the well-known Mill's methods of difference and concomitant variation, discussed previously.

These three particular examples, the use of the Henle–Koch postulates to assess the causal role of HIV and the TSEs, and the employment of a strategy of strong inference implementing methods of difference and concomitant variation to identify and characterize the repressor, illustrate the application of several methodological themes found in microbiology. Each of the examples indicates that the application of such methods requires creative insight coupled with detailed knowledge of the subject domain, as well as exemplary technique, in the case of laboratory investigations. In addition, the examples illustrate that the path from speculation and indirect evidence for scientific hypotheses to direct proof is difficult and frequently depends on background assumptions that themselves could, at some time, be called into question. Such a process of continued questioning of received assumptions, coupled with the proposal of new, precisely formulated hypotheses subject to experimental test and evaluation by rigorously applied scientific method, is the essence of excellent science, and though this process is typically implicit in microbiology, it is no less an important constituent of scientific progress in the subject.

See Also the Following Articles

AIDS, HISTORICAL • HISTORY OF MICROBIOLOGY • PRIONS

Bibliography

Chesebro, B. (1998). BSE and prions: Uncertainties about the agent. *Science* 279, 42–43.

Collard, P. (1976). "The Development of Microbiology." Cambridge Univ. Press, Cambridge.

Collins, H. M., and Pinch T. (1994). "The Golem: What Everyone Should Know About Science." Cambridge Univ. Press, New York.

Duesberg, P. H. (1991). AIDS epidemiology: Inconsistencies with Human Immunodeficiency Virus and with infectious disease. *Proc. Natl. Acad. Sci. USA* 88, 1575–1579.

Duesberg, P. H. (1996). "Inventing the AIDS Virus." Regnery Publishing Co., Washington, DC.

Evans, A. S. (1989). Does HIV cause AIDS? An historical perspective. *J. Acq. Immun. Def. Synd.* 2, 107–113.

Gilbert, W., and Müller-Hill, B. (1966, 1967). Isolation of the

Lac repressor: The *Lac* operator is DNA. *Proc. Natl. Acad. Sci. USA* **56**, 1891–1898 and **58**, 2415–2521.

Gross, P. R., and Levitt, N. (1994). "Higher Superstition: The Academic Left and Its Quarrels With Science." Johns Hopkins Univ. Press, Baltimore.

Holmes, F. L. (1974). "Claude Bernard and Animal Chemistry: The Emergence of a Scientist." Harvard Univ. Press, Cambridge, MA.

Keyes, M. (1999). The prion challenge to the 'Central Dogma' of molecular biology. 1965–1991. Part I: Prelude to prions; Part II: The problem with prions. *Stud. Hist. Phil. Biol. Biomed. Sci.* **30C/1**, 1–19; **30C/2**, 181–218.

Kuhn, T. S. (1970). "The Structure of Scientific Revolutions" (2nd ed.) Univ. of Chicago Press, Chicago.

Laudan, L. L. (1981). "Science and Hypothesis: Historical Essays on Scientific Methodology." Kluwer Inc., Boston.

Morowitz, H. (1985). "Models for Biomedical Research: A New Perspective." National Academy of Sciences Press, Washington, DC.

Platt, J. R. (1964). Strong inference. *Science* **146**, 347–353.

Popper, K. R. (1959). "The Logic of Scientific Discovery." Free Press, New York.

Prusiner, S. (1995). The prion diseases. *Sci. Amer.* January, 48–57.

Safar, J., Wille, H., Itri V., Groth, D., Serban, H., Torchia, M., Cohen, F. E., and Prusiner, S. B. (1998). Eight prion strains have PrP(Sc) molecules with different conformations. *Nat. Med.* October, 1157–1165.

Schaffner, K. (ed.) (1985). "Logic of Discovery and Diagnosis in Medicine." Univ. of California Press, Berkeley, CA.

Schaffner, K. (1986). Exemplar reasooning about biological models and diseases: A relation between the philosophy of medicine and philosophy of science. *J. Med. Philos.* **11**, 63–80.

Schaffner, K. (1998). Model organisms and behavioral genetics: A rejoinder. *Phil. Sci.* **65**, 276–288.

Wolfgang, K., Joklik, H. P., Willett, D., and Amos, B. (eds.) (1992). "Zinsser Microbiology," 20th ed. Appleton and Lange, Norwalk, CT.

Methylation of Nucleic Acids and Proteins

M. G. Marinus

University of Massachusetts Medical School

GLOSSARY

bacteriophage Virus specific for particular strain of bacteria.

epigenetic Producing a phenotypic, but not genotypic, change.

methyltransferase Enzyme methylating particular residues in a macromolecule.

modification Methylation of DNA bases.

promoter Sequence at which RNA polymerase initiates transcription.

S-adenosyl-l-methionine (SAM) Universal methyl donor used by methyltransferases.

restriction Inactivation of DNA by nucleolytic cleavage.

METHYLATION of polymers, such as DNA, RNA, and protein, occurs at the postsynthetic level. That is, methylation occurs after formation of the polymer rather than incorporation of methylated precursors during synthesis. Methylation results from the action of enzymes, methyltransferases, which are very specific not only for the particular polymer but also for specific residues in defined sequences in the polymer.

In some cases, the biological role for methylated residues of polymers has been elucidated. This article focuses on methylation of macromolecules in *Escherichia coli*, since it is in this organism that most information on the biochemistry and genetics of methylation is available.

I. DNA, RNA, AND PROTEIN METHYLTRANSFERASES

Methyltransferases catalyze the transfer of methyl groups from S-adenosyl-l-methionine (SAM) to specific residues in the polymer. For example, DNA adenine methyltransferase (DAM) from *E. coli* catalyzes the formation of N6-methyladenine at adenines in the sequence 5'-GATC-3'. Only the N6 position is modified and adenine residues in other sequences are not methylated nor are adenines in RNA. In double-stranded nucleic acids, methylation substrates are symmetric (5'-GATC-3'/3'-CTAG-5') and two independent methyltransferase binding events are required.

All methyltransferases acting on polymers use SAM as the methyl donor, which results in the formation of S-adenosylhomocysteine (SAH) after catalysis. SAH is a powerful inhibitor of methyltransferase action and has a physiological role in limiting the action of this class of enzymes. Amino acid sequence comparisons of the DNA methyltransferases suggests that they evolved from a common ancestral protein by gene duplication, followed by extensive divergence. DNA cytosine methyltransferases, however, retain conserved amino acid sequence motifs, suggesting a similar basic mechanism of catalysis. This mechanism involves the transient and covalent binding of the enzyme through nucleophillic attack of cysteine-177 to the number 4 carbon atom of the pyrimidine ring. This leads to the activation of the number 5 carbon to receive the methyl group from SAM. After completion of methyl group transfer, SAH is released and the enzyme dissociates from the substrate sequence. A similar mechanism of catalysis has been proposed for RNA methyltransferases.

The atomic structures of several DNA methyltransferases have been solved recently and, in some instances, the structure of the enzyme bound to DNA. These structures have shown that the enzyme rotates the target nucleotide about 180° into a concave catalytic pocket. This rotation is accomplished without any breakage of chemical bonds and has been termed "base flipping." Although the number of demonstrated examples of base flipping is currently limited, this mechanism is surprisingly elegant and simple. It is tempting to speculate that this may be the basic mechanism of action, not only for DNA methyltransferases in general, but for RNA methyltransferases as well. At present, there are no atomic structures for either RNA or protein methyltransferases.

II. DNA METHYLATION

Three methylated bases have been identified in the DNA of bacteria: N6-methyladenine, 5-methylcytosine, and N4-methylcytosine. At maximal, these modified bases are present at 1–2% the level of the normal bases. The distribution of these methylated bases is widespread in bacteria and also can occur in some Archaea. It is unclear from the literature if N4-methylcytosine has been misidentified as 5-methylcytosine in some cases. There are also reports of bacteria lacking methylated bases but it is possible that such results may reflect the insensitivity of the techniques used. In *E. coli* K-12, strains lacking all endogenous DNA methyltransferases have been constructed by genetic means and these have no detectable methylated bases in DNA. Such strains are viable as long as their ability to recombine DNA is not impaired. Since endogenous DNA methylation in *E. coli* blocks the action of some restriction endonucleases used for recombinant DNA technology, plasmid DNA is often prepared from these methylation-deficient strains to avoid this problem.

A. Restriction and Modification of DNA

DNA transfer of plasmid, bacterial virus or chromosomal DNA is thought to occur at a low level between bacterial strains and species. Most bacteria have surveillance systems that allow them to differentiate "foreign" DNA from their own DNA. The basis of one of these systems is the possession of an endonuclease (e.g., *Eco*RI) that recognizes a specific sequence in DNA (5′-GAATTC-3′) and cleaves it only when unmethylated. A corresponding DNA methyltransferase (M. *Eco*RI) recognizes the identical sequence and modifies it to 5′-GAmeATTC-3′, where the second A is methylated. Therefore, bacteria possessing the *Eco*RI methyltransferase will have all susceptible sequences methylated and, thus, be immune to *Eco*RI endonuclease action. Introduction into the bacterium of homologous or heterologous DNA lacking the specific *Eco*RI modification leads to recognition and cleavage of such DNA and subsequent degradation.

A very large number of restriction enzymes have been isolated from bacteria, each having a unique DNA recognition sequence. The availability of this battery of enzymes has been a major contributing factor to the success of recombinant DNA technology in allowing the mapping of genes and their isolation from chromosomes.

One other interesting aspect of restriction/modification is that it can be the basis of epigenetic behavior. For example, bacterial viruses containing DNA not modified at *Eco*RI sites will not kill bacteria, since their DNA is degraded upon entering the bacteria. However, at a low frequency, a few viruses survive and, when these are tested again, they now infect and kill the same bacteria at high efficiency. The virus has not mutated in any way; the only difference is that its DNA has become methylated and, thus, is now resistant to *Eco*RI endonuclease action. Since methylation is dependent on the host in which the virus was last propagated, it is not a heritable change but an epigenetic one.

A second system that bacteria possess to protect themselves from foreign DNA is the ability to detect "foreign" modifications on such DNA and degrade it, using specific endonucleases. For example, in the early days of recombinant DNA technology, it was difficult to isolate genes from animals and plants in the standard bacterial host, *Escherichia coli* K-12. We now know that this failure was due to the destruction of the DNA because it contained methylated cytosine and was recognized by an endonuclease called Mcr (for methylcytosine restriction). This enzyme recognizes and cleaves any methylcytosine-containing DNA that does not have the *E. coli* K-12 signature cytosine methylation. Currently, *E. coli* strains lacking all known restriction systems are used to successfully isolate human and plant genes.

B. Dam Methylation

In some bacteria (and all eukaryotes), most methylated bases are not part of restriction/modification systems. The Dam and Dcm methylation systems of *E. coli* K-12 are the best-characterized of these. The *dam* and *dcm* genes encode DNA adenine (Dam) and DNA cytosine methyltransferases (Dcm), respectively. Dam modifies the adenine residue in 5′-GATC-3′ sequences to N6-methyladenine and Dcm modifies the second cytosine in 5′-CCAGG-3′ or 5′-CCTGG-3′ sequences to 5-methylcytosine. The combined action of these two methyltransferases is responsible for about 99% of the total methylated bases in DNA. The methyltransferase (M. *EcoK*) that is part of the modification/restriction system contributes the remaining small percentage of the total, owing to the infrequent occurrence of the DNA sequence it recognizes.

1. Dam-Directed Mismatch Repair

The concentration of Dam in the cell is regulated to be less than that needed for rapid methylation of all available sites in the DNA. This results in under-methylation of the newly synthesized DNA chain relative to the parental strand, which is fully methylated. This difference in methylation state is exploited by a DNA repair system (Dam- or methyl-directed DNA mismatch repair) that removes errors generated by the replication machinery from the newly synthesized strand. A specific endonuclease, MutH, cleaves mismatched hemimethylated DNA only in the unmethylated strand, thereby targeting repair to that strand. Inactivation of this mismatch repair pathway increases the spontaneous mutation frequency almost a 1000-fold relative to the wild-type strain, indicating its importance in proofreading newly replicated DNA.

2. Initiation of Chromosome Replication

Chromosome replication in *E. coli* begins at a sequence called *oriC* (origin of replication), which has about ten-fold more GATCs than expected on a random basis. Once chromosome duplication begins, it is important to suppress further initiation until the cell requires it. It appears that *dam* methylation is involved in the suppression mechanism. Upon replication, the *oriC* region becomes hemimethylated and is thought to be bound by specific proteins and sequestered in the cell membrane. At some later stage in the cell cycle, the complex is released from the membrane and dissociates. The free *oriC* region is then methylated by Dam, thereby preparing it for the next initiation event. Experimental evidence supporting this model includes the inability of hemimethylated DNA molecules to replicate in *E. coli dam* mutants.

3. Regulation of Gene Expression

Dam-methylation can influence gene expression by two different mechanisms. First, there are several promoter sequences that have 5′-GATC-3′ in one of two critical regions: the "−10" and the "−35" hexamers. Methylation at these sites can increase, decrease, or have no effect on transcription initiation, depending on the specific promoter. The promoter regulating transposition of the tetracycline-resistance element, Tn*10,* is active only in a hemimethylated state, which occurs during the brief period when it is replicated. Transposition is, thereby, coordinated with the cell cycle.

The second mechanism by which Dam methylation can control gene expression relies on competition by Dam and other regulatory proteins for the overlapping sites on the chromosome. Methylation of the sequence then locks gene expression into a particular mode. For example, Cap (cyclic AMP (cAMP) binding protein), with cAMP, binds to specific sequences upstream of many promoters and acts to increase the basal level of transcription. The binding site for Cap–cAMP can overlap with the Dam recognition sequence. Methylation of the sequence can cause interference with Cap–cAMP binding, thus preventing high level expression of the gene. High-level gene expression is achieved only if Cap–cAMP is continuously bound to the site, preventing access to Dam and subsequent modification. In practice, such sites are identified in chromosomes by their lack of Dam methylation. Note that expression of such genes constitutes epigenetic behavior.

4. Control of Bacteriophage DNA
   Packaging into Virions

Bacteriophage P1 encodes its own *dam* gene and Dam methylation is essential for its survival. The critical step at which Dam methylation acts is the introduction of bacteriophage DNA into the head (capsid) of maturing virus particle. The DNA is synthesized in long pieces that need to be cut into smaller pieces, in order to fit properly into the head. This process requires that the DNA be fully methylated, since it is only in this state that the DNA can be cleaved by the appropriate enzyme. Obviously, lack of methylation prevents successful encapsidation of DNA and the production of complete virus particles.

C. Dcm Methylation

Unlike Dam methylation, there is no solid evidence for a physiological role of 5-methylcytosine. Given its widespread occurrence in bacteria, however, there must be some evolutionary advantage in retaining it. 5-methylcytosine in DNA is spontaneously converted to thymine at a low level. This conversion produces a mutation and *E. coli* and other bacteria have evolved a repair system that converts the thymine back to cytosine and then to 5-methylcytosine. This repair system will infrequently convert normal thymine into cytosine and is probably responsible for the overabundance of 5′-CCAGG-3′ pentanucleotide sequences in the *E. coli* genome relative to 5′-CTAGG-3′. If so, this low level of spontaneous mutagenesis may be advantageous to evolving bacteria.

III. RNA METHYLATION

In contrast to DNA, ribosomal and transfer RNA have a wide variety of modified bases, including at least 20 methylated ones. It has been estimated that there must be at least 75 RNA modification enzymes in *E. coli* and, of these, at least 20 are RNA methyltransferases. Only a few of these have been characterized to date. There are no reports of methylated bases in messenger RNA but this may be difficult to measure, given the inherent instability of this class of molecules.

Although mutant strains deficient in certain ribosomal RNA methyltransferases are viable, subtle effects on messenger RNA translation by ribosomes have been detected. For example, in one such mutant strain (*ksgA*), ribosomal subunits have decreased affinity for each other and an increased requirement for a translation initiation factor. It may be that loss of one ribosomal RNA modification can be compensated for by the translation machinery. This may not be the case in eukaryotes since disruption of the homologous *ksgA* gene in yeast is lethal. Mutant strains lacking ribosomal methylation are often resistant to antibiotics (kasugamycin, viomycin), again suggesting subtle alterations in the translation apparatus.

The TrmA transfer RNA methyltransferase, which methylates uracil at position 54 of transfer RNA, is essential for viability, since *E. coli* lacking it are not viable. However, since it appears from other experiments that such methylation is not essential for growth, it has been proposed that the TrmA enzyme may have another essential function. Other mutant strains affecting tRNA methylation are viable but biochemical experiments suggest that methylated bases have subtle effects on the coupling of transfer RNA to its cognate amino acid and interaction of transfer RNA with messenger RNA (codon–anticodon interaction) on the ribosome.

IV. PROTEIN METHYLATION

Methylation can occur at side-chain nitrogens in arginine, lysine, *iso*-aspartic acid, histidine, and glutamic acid residues in proteins. Only a few protein methyltransferases, however, have been purified and characterized.

Bacteria exhibit chemotactic behavior toward certain chemicals to which they are either attracted or repelled. Receptors in the cell membrane are used to monitor the concentration of such chemicals and to signal the flagella motor to turn on or off, which, in a cumulative manner over time, results in movement of the cell either toward or away from the chemical. Methylation/demethylation at four glutamic acid residues on the CheR membrane receptor protein is part of the signaling cascade. A methyltransferase and a methylesterase (which removes the methyl groups) have been implicated in this process. Inactivation of either one of these enzymes leads to loss of chemotactic behavior. Presumably, changes in the cell's environment alter the pattern of methylation/demethylation at the four glutamic acid residues and the subsequent response of the flagellar motor. CheR methylation/demethylation is the only known instance of a reversible methylation involved in regulating a protein's activity.

Flagella are responsible for the bacterium's movement and are composed of a structural protein called flagellin. The FliB protein methyltransferase produces N-methyllysine in flagellin. The physiological role for this methyltransferase, however, is unclear, since *E. coli* bacteria missing this enzyme are not impaired in flagellar locomotion. Similarly, mutant strains of *E. coli* lacking methylated lysines and glutamic acids in ribosomal proteins show no impairment of growth. Spontaneous chemical degradation of aspartic acid to *iso*-aspartic can occur and subsequent methylation of the *iso*-aspartic residue is part of a protein repair pathway. Mutations eliminating the methyltransferase, however, appear to have no effect on *E. coli* viability.

Pili (fimbriae) are rodlike appendages the bacterial cell uses to attach to various other cells. For example, pathogenic bacteria require pili to attach to host cells in order to infect them. Pili are composed of pillin protein, and type 4 pillin has been reported to contain N-methylphenylalanine at its amino-terminal end. However, there are no reports in the literature on how this methylated amino acid is formed, although the action of a specific methyltransferase would seem probable.

The translation elongation factor, EF-Tu, is monomethylated at lysine-56 in growing bacteria but is gradually converted to dimethyllysine as *E. coli* cells enter the stationary phase of growth. The physiological significance of this observation currently remains unknown.

In summary, postsynthetic methylation of macromolecules allows the cell to impose an additional regulatory step on various biological processes. The cell can easily switch between methylated and unmethylated states of a macromolecule by specific enzymatic methylation or inhibiting it. Recognition of either state by proteins or other molecules allows for specific signals to be generated. The epigenetic nature of the modification offers an alternative in switching between states that is not achievable by traditional genetic means.

See Also the Following Articles

BACTERIOPHAGES • CHROMOSOME REPLICATION AND SEGREGATION • DNA RESTRICTION AND MODIFICATION • FIMBRIAE, PILI • FLAGELLA

Bibliography

Björk, G. R. (1996). Stable RNA modification. *In* "*Escherichia coli* and *Salmonella*: Cellular and Molecular Biology" (2nd ed.) (F. C. Neidhardt *et al.* eds.), pp. 861–886. ASM Press Washington, DC.

Chiang, P. K., Gordon, R. K., Tal, J., Zeng, G. C., Doctor, B. P., Pardhasaradhi, K., and McCann, P. (1996). s-Adenosylmethionine and methylation. *FASEB J.* **10**, 471–480.

Clarke, S. (1985). Protein carboxyl methyltransferases: Two distinct classes of enzymes. *Annu. Rev. Biochem.* **54**, 479–506.

Marinus, M. G. (1996). Methylation of DNA. *In* "*Escherichia coli* and *Salmonella*: Cellular and Molecular Biology" (2nd ed.) (F. C. Neidhardt *et al.*, eds.), pp. 782–791. ASM Press, Washington, DC.

Redaschi, N., and Bickle, T. A. (1996). DNA restriction and modification systems. *In* "*Escherichia coli* and *Salmonella*: Cellular and Molecular Biology" (2nd ed.) (F. C. Neidhardt *et al.*, eds.), pp. 773–781. ASM Press, Washington, DC.

Roberts, R. J., and Cheng, X. D. (1998). Base flipping. *Annu. Rev. Biochem.* **67**, 181–198.

Methylotrophy

J. Colin Murrell and Ian R. McDonald

University of Warwick, Coventry, UK

GLOSSARY

c_1 compounds Compounds more reduced than carbon dioxide containing one or more carbon atoms, but no carbon–carbon bonds.

methane monooxygenase The key enzyme for the oxidation of methane to methanol in bacteria.

methanotroph A methylotrophic bacterium with the ability to grow on methane as its sole carbon and energy source (methane-oxidizing bacteria).

proteobacteria A kingdom of bacteria divided into five groups based on their 16S ribosomal RNA sequences; a group of bacteria diverse in their morphology, physiology and lifestyle.

METHYLOTROPHY refers to the ability of microorganisms to utilize one-carbon compounds more reduced than CO_2 as sole energy sources and to assimilate carbon into cell biomass at the oxidation level of formaldehyde. Methylotrophic organisms must synthesize all cellular constituents from methylotrophic compounds, such as methane, methanol, methylated amines, halogenated methanes, and methylated sulfur species. A diverse range of both aerobic and anaerobic prokaryotes and eukaryotes can utilize methanotrophic substrates for growth.

I. HISTORICAL PERSPECTIVE

Methylotrophs were first discovered in 1892 by Leow, who described a pink bacterium growing on methanol, methylamine, formaldehyde, and also on a variety of multi-carbon compounds. This organism was called *Bacillus methylicus* and was almost certainly what is now know as the pink-pigmented facultative methylotroph (PPFM) *Methylobacterium extorquens. Bacillus methanicus* was the first methane-oxidizing bacterium, reported by Söhngen in 1906. This methanotroph was isolated in pure culture from aquatic plants. This isolate was subsequently lost but was reisolated in 1956 by Dworkin and Foster and renamed *Pseudomonas methanica*. At around that time, the PPFM *Pseudomonas* AM1 was isolated on methanol by Quayle and colleagues. This was to become the "workhorse" organism for many of the biochemical and molecular biological studies on the metabolism of methanol and has now been renamed as *Methylobacterium extorquens* AM1. In 1970, Whittenbury and colleagues isolated over 100 new strains of methane-oxidizing bacteria. The characterization of these organisms was carried out and the scheme proposed still remains the basis for current classification schemes for methanotrophs.

II. SIGNIFICANCE

A. Global Carbon Cycle

Methane is the most abundant organic gas in the atmosphere. It is a very potent greenhouse gas and it absorbs infrared radiation considerably more efficiently than CO_2 and, therefore, makes a significant contribution to global warming. Current understanding of the global methane budget suggests that methane-oxidizing bacteria play an important role in oxidizing a large proportion of the methane produced by methanogenic bacteria in environments such as wetlands, ricefields, tundra, and the marine environment. Therefore, these bacteria are a significant sink for methane in the environment in modulating net emissions of methane and may provide an important negative feedback on future methane increases in wetland and soil environments. It is, therefore, important to learn more about the role of methanotrophs in the global carbon cycle.

B. Biotechnology

Both methane and methanol are relatively cheap feedstocks for fermentation processes and methylotrophs have received considerable attention for a number of biotechnological applications. Initial work on the production of single-cell protein (SCP) was carried out with methanotrophs. Growth yields of these organisms on methane were high but two drawbacks included the high oxygen demand in the fermentation process and the explosive nature of their substrates methane and oxygen. Methanol-utilizing bacteria, e.g., *Methylophilus methylotrophus,* have been successfully used for SCP production in very large-scale fermentation processes (ICI Pruteen Process). However, due to the fall in price of agricultural protein products, such as soya protein, over recent years, SCP from methanol has not been particularly competitive on a commercial scale.

Methanol-utilizing bacteria have also been exploited for the production of vitamins, polymers, and amino acids and these processes may be more economically viable. For example, auxotrophs of thermophilic gram-positive methanol utilizers can excrete relatively large amounts of lysine and other amino acids, which can then be used for animal feedstock supplements. The possibilities of genetically engineering methylotrophs for the overproduction of amino acids is also being explored.

Methanotrophs have also been investigated for the production of bulk chemicals, such as propylene oxide. Methane monooxygenase is able to insert oxygen into a number of aliphatic and aromatic compounds other than methane. These co-oxidation properties of methanotrophs, particularly those that contain soluble methane monooxygenase (sMMO), are unique and also unusual, since these organisms appear to derive no benefit from this process. sMMO is able to co-oxidize propylene to propylene oxide, a valuable compound in organic synthesis. However, due to the toxic nature of propylene oxide, alternatives to the use of whole cells, such as immobilized enzymes, together with the problem of regenerating reductant for sMMO may need to be considered.

C. Bioremediation

Methanotrophs have received considerable attention for their potential use in bioremediation processes. The enzyme soluble methane monooxygenase not only oxidizes methane but will co-oxidize a wide variety of aliphatic, substituted aliphatic, and aromatic compounds. sMMO is able to degrade several pollutants, including vinyl chloride, trichloroethylene, and other halogenated hydrocarbons that contaminate soil and groundwater. Challenges facing the use of methanotrophs *in situ* in bioremediation processes include ensuring supply of the substrates methane and oxygen and/or reductant and also overcoming problems associated with the negative regulation of sMMO by copper ions. Methylotrophs containing specific dehalogenases may also be useful in clean-up of industrial solvent-contaminated sites. For example, dichloromethane is metabolized by some *Methylobacterium* species using a dehalogenase.

More recently, methylotrophs with the ability to degrade methyl chloride and methyl bromide have been isolated. Degradation of methyl bromide, a potent ozone-depleting gas currently used as a pesticide in agriculture, is a particularly interesting trait for methylotrophs and may be useful in mitigating methyl bromide loss to the atmosphere during soil fumigation processes.

Methylotrophic bacteria can also grow on some methylated sulfur compounds found in toxic wastes, such as paper mill effluents. Others can degrade aliphatic sulfonates and, therefore, may be important in degradation of detergents and related compounds in the environment. Research is under way to investigate the metabolism of halogenated methanes and one-carbon compounds containing sulfur, in order to be able to explore the bioremediation potential and to exploit the properties of these novel methylotrophs.

D. Expression Systems

The methylotrophic yeast *Pichia pastoris* is now becoming one of the best hosts for the production of foreign proteins because of the presence of the strong methanol-inducible promoter AOX1. This allows high level expression of a large number of biotechnologically and pharmaceutically important proteins in a controlled fashion during growth of a yeast on a relatively cheap substrate.

III. HABITATS AND ECOLOGY

Methane is produced by methanogenic bacteria in a number of diverse environments in the biosphere. Methane-oxidizing bacteria, which require both methane and oxygen for growth, are generally found on the fringes of anaerobic environments and are probably responsible for oxidizing much of the methane derived from methanogens, before it escapes to the atmosphere. They appear to be ubiquitous in nature and have been isolated from many different environments, including freshwater and lake sediments, rivers, groundwater aquifers, seawater, marine sediments, rice paddies, sewage sludge, decaying plant material, acidic peat bogs, and alkaline lakes. Psychrophilic representatives may also be isolated from Arctic and Antarctic tundra and thermophilic methanotrophs growing at temperatures as high as 70°C have recently been obtained from hot springs. They can also be isolated from polluted environments. Methanotrophs also appear to exist as symbionts, for example, in the gill tissue of marine mussels or tube worms. The carbon assimilated by these methanotrophic endosymbionts probably supplies much of the organic carbon necessary for growth of these marine organisms. Methanotrophs from certain environments, for example, the putative symbionts already described, do not respond well to conventional enrichment and isolation techniques and molecular ecology experiments employing phylogenetic and functional gene probes suggest the presence of many new, and as yet uncultivated, methanotrophs in the environment.

Methanol is also a relatively abundant substrate for the growth of methylotrophs in the environment. Methanol is released during the decomposition of plant lignins and pectins and other compounds that contain methoxy groups. Bacteria that utilize methanol are frequently found in association with methane-oxidizing bacteria, presumably growing on the methanol excreted by methanotrophs. Methanol utilizers such as *Methylobacterium* are frequently found on the leaves of aquatic and terrestrial plants and on decaying plant material. Methanol-utilizers have also been isolated from the marine environment. Bacteria that utilize methyamine, many of which also grow on methanol, are also widespread in nature. Methylated amines are the products of degradation of some pesticides, lecithin and carnitine derivatives, and of trimethylamine oxide. Methylamine-utilizing bacteria are common in both terrestrial and marine environments. Methylated sulfur compounds, such as dimethylsulfoxide (DMS), dimethylsulfide, and dimethyldisulfide, are capable of supporting the growth of certain methylotrophs, such as *Hyphomicrobium* and *Thiobacillus*. DMS is the most abundant organic sulfur gas in the environment, produced in the marine environment from the cleavage of dimethylsulfoniopropionate, an algal osmoregulator. DMS is oxidized in the upper atmosphere to sulfur dioxide and the C_1 sulfur compound methanesulfonic acid (MSA). This MSA falls to earth by wet and dry deposition and, recently, it has been demonstrated that terrestrial and marine bacteria can grow methylotrophically on this C_1 substrate. Halogenated methanes are widely used as industrial solvents in the chemical industry and some, e.g., dichloromethane, have been shown to be methylotrophic substrates for some strains of *Methylobacterium* in polluted soils. Methyl bromide and methyl chloride are natural products released into the biosphere in large amounts from marine phytoplankton, algae, and wood-decaying fungi and these are also methylotrophic substrates for newly isolated bacteria.

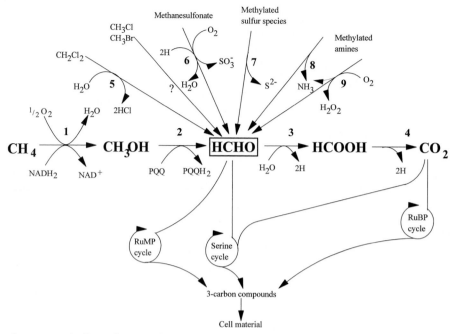

Fig. 1. Metabolism of one-carbon compounds by aerobic methylotrophic bacteria. Enzymes: 1, methane monooxygenase; 2, methanol dehydrogenase; 3, formaldehyde dehydrogenase; 4, formate dehydrogenase; 5, dichloromethane dehalogenase; 6, methanesulfonic acid monooxygenase; 7, methylated sulfur dehydrogenases or oxidases; 8, methylated amine dehydrogenases; 9, methylamine oxidase.

IV. METHANOTROPHS (METHANE UTILIZERS)

A. Physiology and Biochemistry

Methanotrophs grow by oxidizing methane to methanol using the pathway shown in Fig. 1. Methanol is further metabolized to formaldehyde by the pyrolloquinoline quinone- (PQQ-) linked enzyme methanol dehydrogenase (MDH), an enzyme found in all gram-negative methylotrophic bacteria. Approximately half of the formaldehyde produced is further oxidized to yield carbon dioxide, resulting in generation of reducing power for biosynthesis and the initial oxidation of methane. The carbon dioxide is not fixed into cell carbon in significant amounts but is lost to the atmosphere. The remainder of the formaldehyde is assimilated into cell carbon by one of two pathways. In Type I methanotrophs, formaldehyde is condensed with ribulose phosphate into hexulose-6-phosphate in the ribulose monophosphate pathway. Type II methanotrophs utilize the serine pathway for the incorporation of formaldehyde into the cell.

The initial reaction in the RuMP pathway involves the addition of 3 mol formaldehyde to 3 mol ribulose 5-phosphate to produce 3 mol hexulose 6-phosphate. Rearrangement reactions similar to those in the Calvin cycle for carbon dioxide fixation result in the production of glyceraldehyde 3-phosphate and the regeneration of 3 mol of ribulose 5-phosphate. Glyceraldehyde 3-phosphate is used for the synthesis of cell material. The serine pathway is very different from other formaldehyde assimilation pathways. There are no enzymatic reactions in common with the RuMP or methylotrophic yeast assimilatory pathways. In the serine pathway, 2 mol formaldehyde are condensed with 2 mol glycine to form 2 mol serine. The serine is converted to 2 mol of 2-phosphoglycerate. One phosphoglycerate is assimilated into cell material while the other is converted to phosphoenolpyruvate (PEP). The carboxylation of PEP yields oxaloacetate that is subsequently converted to 2 mol glyoxylate. Transamination of glyoxylate with serine as the amino donor regenerates the two glycine acceptor molecules. The phosphoglycerate, which is assimilated, undergoes transformations in central metabolic routes to provide carbon backbones for the synthesis of all cell materials.

Methane oxidation is carried out by the enzyme methane monooxygenase (MMO). A membrane-bound, particulate methane monooxygenase (pMMO) appears to be present in all methanotrophs grown in the presence of relatively high concentrations of copper ions. pMMO appears to consist of at least 3 polypeptides of approximately 46, 23, and 20 kDa, and contains a number of copper clusters. Further characterization of pMMO at the biochemical level is ongoing in a number of research groups.

In some methanotrophs, a second form of MMO, a cytoplasmic, soluble form (sMMO) is synthesized in growth conditions when the copper-to-biomass ratio is low. This sMMO is structurally and catalytically distinct from the pMMO and has a broad substrate specificity, oxidizing a wide range of aliphatic and aromatic compounds. The sMMO enzymes of *M. capsulatus* (Bath) and *M. trichosporium* OB3b both consist of three components—A, B, and C. Protein A is the hydroxylase component of the enzyme complex and contains a binuclear iron–oxo center, believed to be the reactive center for catalysis. Protein B, is a single polypeptide, contains no metal ions or cofactors, and functions as a regulatory, or coupling, protein. Protein C is a reductase containing 1 mol each of FAD and a 2Fe2S cluster which accepts electrons from NADH and transfers them to the diiron site of the hydroxylase component. The x ray crystal structure of the hydroxylase component of sMMO is now known, a fact which has further stimulated research on the mechanism of oxidation of methane by this unique enzyme.

B. Molecular Biology

The genes encoding sMMO complex have been cloned from several methanotrophs. The gene cluster contains genes encoding the α, β, and γ subunits of Protein A (*mmoX, Y,* and *Z*), Protein B (*mmoB*), and Protein C (*mmoC*). Derived polypeptide sequences of sMMO components from three methanotrophs showed a high degree of identity, highlighting the conserved nature of this enzyme complex. Amino acid sequences within the α subunit of Protein A align well with the four helix iron coordination bundle of the R2 protein of ribonucleotide reductase and is characteristic of a family of proteins that contains a catalytic carboxylate-bridged diiron center.

Differential expression of sMMO and pMMO is regulated by the amount of copper ions available to the cells; sMMO is expressed at low copper-biomass ratios, whereas pMMO is expressed at high copper-biomass ratios. The transcriptional regulation of the sMMO gene cluster appears to be under the control of a copper-regulated promoter. Transcription of the sMMO gene cluster is negatively regulated by copper ions. Activation of *pmo* transcription by copper ions is concomitant with repression of sMMO gene transcription in both methanotrophs, suggesting that a common regulatory pathway may be involved in the transcriptional regulation of sMMO and pMMO.

C. Molecular Ecology

Difficulties in using traditional culture-based techniques to study the ecology of methanotrophs (e.g., slow growth of methanotrophs and scavenging of nonmethanotrophs on agar plates) has hampered studies. Application of molecular biology techniques to methanotrophs has been aided considerably by the sequencing of a number of 16S rRNA genes of methanotrophs and methylotrophs and the cloning of several methanotroph-specific genes. Hanson and colleagues, (1991, 1996) have used 16S rRNA sequence data to examine phylogenetic relationships within genera of methanotrophs and methylotrophs.

16S rRNA data coupled with PCR technology have also been used successfully in analyzing methanotrophs in the marine environment. Seawater samples were enriched for methanotrophs by adding essential nutrients and methane. Changes in composition of the bacterial population was then monitored by analysis of 16S rDNA libraries. The dominant 16S rRNA sequence that was present in samples after enrichment on methane was found to show a close phylogenetic relationship to *Methylomonas*, indicating that novel methanotrophs related to extant *Methylomonas* spp. were present in enrichment cultures.

Genes unique to methanotrophs include those encoding sMMO and pMMO polypeptides. The high degree of identity between sMMO genes has enabled the design of PCR primers which specifically amplify sMMO genes directly from a variety of different freshwater, marine, soil, and peat samples. Results suggest that there is considerable diversity of methanotrophs in these environments.

Another functional gene probe for methanotrophs is one based on the pMMO, present in all extant methanotrophs. Sequence data on *pmo* and *amo* (ammonia monooxygenase—a related enzyme found in nitrifying bacteria) genes has allowed the design of

degenerate PCR primers which will specifically amplify DNA genes encoding *pmoA* or *amoA* from many different methanotrophs and nitrifiers. Analysis of the predicted amino acid sequences of these genes from representatives of each of the phylogenetic groups of methanotrophs (α and γ Proteobacteria) and ammonia oxidizing nitrifiers (β and γ Proteobacteria) suggests that the particulate methane monooxygenase and ammonia monooxygenase may be evolutionarily related enzymes.

Another potentially useful marker is *mxaF*, encoding the large subunit of methanol dehydrogenase, which is present in virtually all gram-negative meth-

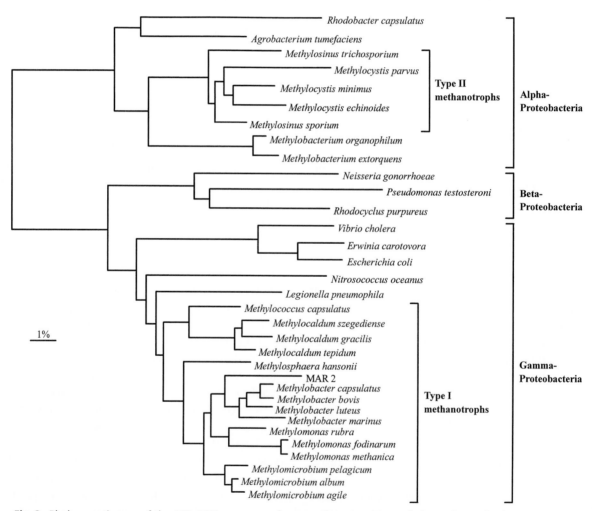

Fig. 2. Phylogenetic tree of the 16S rRNA sequences from methanotrophs, methylotrophs, and other representative bacteria belonging to the α, β, and γ subdivisions of the Proteobacteria. MAR 2 is a marine methanotroph isolate.

TABLE I
Characterization of Methanotrophs

Genus	Phylogeny	Membrane type	Major PLFA	Formaldehyde assimilation pathway	Enzyme type[a]	Mol % G + C content of DNA
Methylobacter	Gamma Proteobacteria	Type I	16:1	RuMP	pMMO	49–54
Methylocaldum	Gamma Proteobacteria	Type I	16:1	RuMP/Serine	pMMO	57–59
Methylococcus	Gamma Proteobacteria	Type I	16:1	RuMP//Serine	pMMO/sMMO	59–66
Methylomicrobium	Gamma Proteobacteria	Type I	16:1	RuMP	pMMO/sMMO	49–60
Methylomonas	Gamma Proteobacteria	Type I	16:1	RuMP	pMMO/sMMO	51–59
Methylosphaera	Gamma Proteobacteria	Type I	16:1	RuMP	pMMO	43–46
Methylocystis	Alpha Proteobacteria	Type II	18:1	Serine	pMMO/sMMO	62–67
Methylosinus	Alpha Proteobacteria	Type II	18:1	Serine	pMMO/sMMO	63–67

Note. RuMP, ribulose monophosphate pathway; pMMO, particulate methane monooxygenase; sMMO, soluble methane monooxygenase. All strains grow on CH_4; some will also grow on CH_3OH.

[a] All strains possess pMMO; however, some also possess the sMMO.

ylotrophs. *mxaF* is highly conserved and is, therefore, a good indicator of the presence of these organisms in the natural environment.

D. Phylogeny and Taxonomy

Methanotrophs are all gram-negative bacteria and can be classified into two groups. Type I methanotrophs of the genera *Methylomonas, Methylobacter, Methylomicrobium,* "*Methylothermus,*" *Methylococcus, Methylosphaera,* and "*Methylocaldum*" utilize the ribulose monophosphate (RuMP) pathway for the assimilation of formaldehyde into cell carbon, possess bundles of intracytoplasmic membranes, and are members of the γ-subdivision of the purple bacteria (class Proteobacteria). Type II methanotrophs, such as *Methylosinus* and *Methylocystis,* utilize the serine pathway for formaldehyde fixation, possess intracytoplasmic membranes arranged around the periphery of the cell, and fall within the γ-subdivision of the Proteobacteria. Classification schemes, based on

pheno- and chemo-taxonomic studies have been strengthened as a result of the nucleotide sequencing of both 5S and 16S ribosomal RNA (rRNA) from a large number of methanotrophs and methylotrophs (Fig. 2). The key features of representative genera of methanotrophs are summarized in Table I. Properties such as mol % G + C content of DNA, membrane fatty acid composition, nitrogen fixation, and some morphological features, can be used to discriminate between Type I and Type II methanotrophs.

V. AEROBIC METHYLOTROPHS

A. Methanol Utilizers

The bacteria capable of growth on methanol are more diverse than those capable of growing on methane. They include a variety of gram-negative and gram-positive strains and include both facultative and obligate methylotrophs (Table II). The metha-

TABLE II
Characteristics of Aerobic Methylotrophic Bacteria

Group	Growth substrates	Major assimilation pathway	GC content (mol%)
Obligate gram-negative methylotrophs			
Methylobacillus	CH_3OH, CH_3NH_2	RuMP	50–55
Methylophaga	CH_3OH, CH_3NH_2, fructose	RuMP	38–46
Methylophilus	CH_3OH, CH_3NH_2, fructose[a]	RuMP	50–55
Facultative gram-negative methylotrophs			
Acidomonas[b]	CH_3OH, multi-carbon compounds	RuMP	63–66
Hyphomicrobium	CH_3OH, CH_3NH_2[a], DMSO[a], DMS[a], CH_3Cl[a], some 2-C and 4-C compounds[a]; denitrification	Serine	60–66
Methylobacterium	CH_3OH, CH_3NH_2, CH_3Cl[a], multi-carbon compounds	Serine	65–68
Methylosulfonomonas	MSA, CH_3OH, CH_3NH_2, HCOOH, poor growth on multi-carbon compounds	Serine	60–62
Microcyclus	CH_3OH, multi-carbon compounds	RuBP	65–67
Paracoccus	CH_3OH, CH_3NH_2, multi-carbon compounds, denitrification	RuBP	66
Rhodopseudomonas	CH_3OH[a], multi-carbon compounds; photosynthesis	RuBP	62–72
Thiobacillus	CH_3OH[a], CH_3NH_2[a], H_2S, S_2O_3, multi-carbon compounds[a]	RuBP	52–69
Xanthobacter	CH_3OH, multi-carbon compounds	RuBP	67–69
Facultative gram-positive methylotrophs			
Amycolatopsis[c]	CH_3OH[a], multi-carbon compounds,	RuMP	ND
Arthrobacter	CH_3NH_2[a], multi-carbon compounds,	RuMP	ND
Bacillus	CH_3OH[a], CH_3NH_2[a], multi-carbon compounds	RuMP	60–70
Mycobacterium	CH_4[a], CH_3OH[a], multi-carbon compounds	RuMP	65–69

[a] Some strains.
[b] Formerly *Acetobacter methanolicus*.
[c] Formerly *Nocardia methanolica*.

nol-utilizing bacteria can be divided according to their carbon assimilation pathway. The methanol-utilizers that contain the serine cycle for formaldehyde assimilation include *Methylobacterium* and *Hyphomicrobium* strains. Most of the gram-negative methanol utilizers that contain the RuMP cycle are obligate methylotrophs, with the exception of the facultative *Acidomonas* strains. The gram-positive methanol utilizers, which all contain the RuMP cycle, are facultative methylotrophs. All of the known gram-negative methanol- and methane-utilizing bacteria contain an enzyme for oxidizing methanol called methanol dehydrogenase. This enzyme, which oxidizes primary alcohols, contains the cofactor pyrroloquinoline quinone (PQQ). Methanol dehydrogenase is one of a family of PQQ-linked enzymes known as quinoproteins. The electrons from the oxidation of methanol are transferred from the PQQ cofactor to a specific soluble cytochrome *c* and, from there, through to other carriers to the terminal oxidase. Methanol dehydrogenases are highly conserved throughout the gram-negative methylotrophic bacteria. Studies of the molecular genetics of this system have revealed that the synthesis of a fully active methanol oxidizing pathway requires a total of at least 32 genes, among which are the *pqq* genes involved in cofactor biosynthesis and a comprehensive set of *mxa mxb, mxc,* and *mxd* genes, some of which encode the structural enzymes and others which encode proteins involved in regulation of gene expression and protein activity.

Different types of methanol dehydrogenases occur in the gram-positive methylotrophs. A methylotrophic *Amycolatopsis* species contains an unusual quino alcohol dehydrogenase, and the methylotrophic *Bacillus* species contain a methanol dehydrogenase enzyme that is not PQQ linked, but instead is linked to NAD.

Recently, an aerobic methylotrophic bacterium, *Methylobacterium extorquens,* was found to contain a cluster of genes that are predicted to encode some of the enzymes from methanogenic and sulfate-reducing Archaea involved in C_1 transfer, thought to be unique to this group of strictly anaerobic microorganisms. Enzyme activities were also detected in *M. extorquens* and mutants defective in some of these genes were unable to grow on C_1 compounds, sug-gesting that the archaeal enzymes also function in aerobic C_1 metabolism. Thus, methylotrophy and methanogenesis involve common genes that cross the bacteria/archaeal boundaries.

Some bacteria, such as *Paracoccus, Thiosphaera,* and *Xanthobacter,* can grow on methanol by oxidizing this methylotrophic substrate to carbon dioxide but then fixing this carbon into cell biomass using the enzyme ribulose bisphosphate carboxylase/oxygenase (RuBISCO). These types of organisms are, therefore, considered as autotrophic methylotrophs.

B. Utilization of Methylated Amines

Many of the bacteria that grow on methanol are also capable of utilizing methylated amines. Some bacteria have been isolated that are capable of growth on methylated amines but do not grow on methanol or methane, such as *Pseudomonas aminovorans* and *Arthrobacter* P1. Several species of methylotrophic bacteria are able to utilize methylamine as a sole source of carbon and energy. Three different systems for the oxidation of primary amines are known. These are methylamine dehydrogenase, found in some gram-negative methylotrophs; amine oxidase, found in gram-positive methylotrophs; and indirect methylamine oxidation via N-methyl-glutamate dehydrogenase, found in the remaining gram-negative methylotrophs.

The methylamine dehydrogenases (MADH) are periplasmic proteins, consisting of two small and two large subunits. Each small subunit has a covalently bound prosthetic group called tryptophan tryptophylquinone (TQQ). MADHs can be divided into two groups, based on the electron acceptors that they use. The MADHs from restricted facultative methylotrophic bacteria belonging to the genus *Methylophilus,* use a *c*-type cytochrome as a electron acceptor, whereas all other MADHs use blue copper proteins called amicyanins. A group of genes called the *mau* genes are responsible for the synthesis of MADH. The *mau* gene cluster of *Paracoccus denitrificans* consists of 11 genes, 10 of which encode the structural proteins or proteins involved in cofactor biosynthesis and which are transcribed in one direction, whereas the 11th regulatory gene (*mauR*) is located upstream and is divergently transcribed.

C. Utilization of Halomethanes

Certain *Methylobacterium* and *Methylophilus* species can grow on dichloromethane as sole carbon and energy source (Fig. 2). The key enzyme in the aerobic degradation of CH_2Cl_2 is dichloromethane dehalogenase, which catalyzes the glutathione (GSH)-dependent dehalogenation of CH_2Cl_2 to formaldehyde and chloride ions. The formaldehyde is subsequently assimilated via the serine pathway. Some *Hyphomicrobium* and *Methylobacterium* species also grow on methyl chloride, although the exact mechanism for utilization of this C_1 compound is not yet know. Methyl bromide also appears to be a C_1 substrate for certain bacteria and is currently receiving considerable attention as a potential methylotrophic substrate. Again, the bacteria responsible have not been fully characterized but it is likely that these novel bacteria are widespread in the terrestrial and marine environment and play an important role in cycling of halogenated methanes.

D. Utilization of Methylated Sulfur Species

There are some methylotrophic bacteria which are capable of utilizing methylated sulfur compounds, such as dimethylsulfoxide (DMSO), dimethylsulfide (DMS), and dimethyldisulfide (DMDS). Most of these are of the genus *Hyphomicrobium* but certain *Thiobacillus* strains have been reported. DMSO and DMS metabolism has not been studied in great detail in methylotrophic bacteria but it is believed that in *Hyphomicrobium* DMSO is reduced to DMS, which is, in turn, converted to formaldehyde and methanethiol. The methanethiol may then be converted by an oxidase to formaldehyde and H_2S with the concomitant production of hydrogen peroxide. The formaldehyde produced would then be assimilated into cell carbon by the serine pathway in *Hyphomicrobium*. Methanesulfonic acid (MSA) is also a C_1 source for certain methylotrophs, which appear to be ubiquitous in the environment. The terrestrial strain *Methylosulfonomonas* and the marine strain *Marinosulfonomonas* oxidize MSA to formaldehyde and sulfite, using a methanesulfonic acid monooxgenase. The formaldehyde is subsequently assimilated into the cell via the serine pathway. Some strains of *Methylobacterium* and *Hyphomicrobium* can also utilize MSA. Most of the MSA-utilizers also grow well on other C_1 compounds such as methanol, methylamine, and formate, but not methane. Certain *Hyphomicrobium* species can grow on monomethylsulfate.

VI. ANAEROBIC METHYLOTROPHS

All extant methanotrophs are obligate aerobes. However, there is now good biogeochemical and biological evidence that methane oxidation occurs in marine environments, such as sulfate-rich sediments, alkaline soda lakes, and some freshwater lakes. However, to date, no anaerobic bacteria that will grow on or oxidize methane have been isolated from these environments and cultivated in the laboratory. It is not known if such bacteria are true methanotrophs or if a consortium of bacteria is involved in these processes. One hypothesis is that sulfate is the terminal electron acceptor for anaerobic methane oxidation but further experimental work is required here. Some studies indicate that, although methane is oxidized to carbon dioxide, the methane carbon is not assimilated into cell biomass.

Methanol can also be oxidized by facultatively anaerobic bacteria of the genus *Hyphomicrobium*, using nitrate as a terminal electron acceptor. Some acetogenic and methanogenic bacteria, which are strict anaerobes, are capable of growth on C_1 compounds, such as methanol and methylamines. During anaerobic growth, these bacteria convert such C_1 compounds to methane, acetate, or butyrate, rather than carbon dioxide. Carbon from the original C_1 substrate is not assimilated via formaldehyde but their methyl groups are incorporated into acetyl CoA, a precursor for cellular constituents. Therefore, following the original definition of methylotrophy, these obligate anaerobes that utilize C_1 compounds, such as methanol and methylamine, are not normally considered as methylotrophs.

VII. METHYLOTROPHIC YEASTS

The ability of some yeasts to grow on methanol as a source of carbon and energy has been discovered only relatively recently. They can be isolated from soil, rotting fruits, and vegetables, or plant material, again suggesting that methanol derived from methoxy groups in wood lignin or pectin is an important factor in the ecology of these yeasts. Methylotrophic yeasts belong to the fungi perfecti, form ascospores that are hat-shaped and homothallic. They are members of the genera *Hansenula, Pichia,* and *Candida* and they metabolize methanol via alcohol oxidases in peroxisomes. Assimilation of formaldehyde is accomplished by the xylulose monophosphate cycle. Yeast cultures that use methane as a sole carbon and energy source have also been described. These strains were slow growing and have received very little attention over the past 20 years.

See Also the Following Articles

BIOREMEDIATION • ECOLOGY, MICROBIAL • INDUSTRIAL FERMENTATION PROCESSES • METHANE BIOCHEMISTRY

Bibliography

Anthony, C. (1982). "The Biochemistry of Methylotrophs." Academic Press, London, UK.

Conrad, R. (1996). Soil microorganisms as controllers of atmospheric trace gases (H$_2$, CO, CH$_4$, OCS, N$_2$O, and NO). *Microbiol. Rev.* **60**, 609–640.

Hanson, R. S. (1991). The obligate methanotrophic bacteria *Methylococcus, Methylomonas, Methylosinus* and related bacteria. *In* "The Prokaryotes" (A. Balows, H. G. Truper, M. Dworkin, W. Harder, and K. H. Schleifer, eds.). Vol. 1, pp. 2350–2365. Springer-Verlag, New York.

Hanson, R. S., and Hanson, T. E. (1996). Methanotrophic bacteria. *Microbiol. Rev.* **60**, 439–471.

Large, P. J., and Bamforth, C. W. (1988). "Methylotrophy and Biotechnology." Longman Scientific and Technical, New York.

Leak, D. J. (1992). Biotechnological and applied aspects of methane and methanol utilizers. *In* "Methane and Methanol Utilizers" (J. C. Murrell and H. Dalton, eds.), pp. 245–279. Plenum Press, New York.

Lidstrom, M. E. (1991). Aerobic methylotrophic bacteria. *In* "The Prokaryotes" (A. Balows, H. G. Truper, M. Dworkin, W. Harder, and K. H. Schleifer, eds.). Vol. 1, pp. 431–445. Springer-Verlag, New York.

Lidstrom, M. E., and Tabita, F. R. (1996). "Microbial Growth on C$_1$ Compounds." Kluwer Academic, Dordrecht, The Netherlands.

Murrell, J. C., and Dalton, H. (1992). "Methane and Methanol Utilizers." Plenum Press, New York.

Murrell, J. C., and Kelly, D. P. (1993). "Microbial Growth on C$_1$ Compounds." Intercept Press, Andover, UK.

Murrell, J. C., and Kelly, D. P. (1996). "The Microbiology of Atmospheric Trace Gases: Sources, Sinks and Global Change Processes." NATO ASI Series, Springer Verlag.

Murrell, J. C., McDonald, I. R., and Bourne, D. G. (1998). Molecular methods for the study of methanotroph ecology. *FEMS Microbiol. Ecol.* **27**, 103–114.

Oremland, R. S. (1993). "Biogeochemistry of Global Change." Chapman and Hall, New York.

Microbes and the Atmosphere

Ralf Conrad

Max-Planck-Institut für Terrestrische Mikrobiologie, Germany

I. Composition of the Atmosphere
II. Cycling of Atmospheric Trace Gases

GLOSSARY

abiontic soil enzymes Free enzymes and those bound to dead or nonproliferating cells.

bunsen solubility coefficient (α_B) A dimensionless index of gas solubility, equivalent to Henry's law constant (K_H) [$\alpha_B = 22.414\ K_H$] at 1 atm (101.3 Kpa) pressure. α_B is the ratio between aqueous and gaseous concentrations at standard gas pressure.

chemolithotrophic microorganisms Microorganisms that oxidize reduced inorganic compounds (e.g., H_2, H_2S, NH_4^+, Fe^{2+}) to obtain energy for growth.

methanogenic archaea Obligately anaerobic microorganisms, belonging to the domain of *Archaea*, which form methane as their metabolic end-product.

methylotrophic bacteria Bacteria that grow aerobically with C_1 compounds as their energy and carbon source, e.g., methane, methanol, methylamine, dimethylsulfide, formate.

mixing ratios (of trace gases in the atmosphere) Commonly given in units of volume percent (%) or parts per million (ppmv) or parts per billion (ppbv) by volume. In contrast to concentrations (mass per volume), mixing ratios are independent of temperature and pressure.

trace gases Atmospheric constituents with mixing ratios in the ppmv range and less. The atmospheric mixing ratio is the result of simultaneous operation of processes which release (sources) or remove (sinks) the trace gas. The atmospheric budget is given by the sum of all sources (annual release rates), balanced by the sum of all sinks. Imbalance between total sources and sinks results in either increasing or decreasing atmospheric mixing ratios. The lifetime of a trace gas is given by its total mass present in the atmosphere divided by the budget.

upland soils Soils that, in contrast to wetland soils, are usually well aerated and generally oxic, but may contain anoxic microniches, especially when the water content is higher than field capacity, i.e., the water content can be retained in soil against gravity.

EARTH'S ATMOSPHERE is to a large part the result of microbial activity. This applies to the minor atmospheric constituents, the so-called trace gases, as well as to the abundance of O_2 in our modern atmosphere. Without microbial life, the atmosphere of our planet would definitely be different, with many consequences for habitability and life.

I. COMPOSITION OF THE ATMOSPHERE

A. Evolution of Oxygen

Probably the most drastic change in the habitability of Earth was the evolution of the oxic atmosphere. The early composition of the atmosphere after Earth had formed about 4.5 Gyr (4.5 billion years) ago was probably similar to that of our neighboring planets Mars and Venus, i.e., CO_2, H_2O, and N_2. The condensation of H_2O into the oceans allowed the dissolution of CO_2, so that the early atmosphere consisted mainly of N_2. CO_2 was also present but was in dissolution equilibrium with the oceans similar to today. Reduced gases (H_2, CH_4, H_2S, NH_3) were only present at very low mixing ratios (<1%), and O_2 was also only present at traces <1 ppmv. The generally anoxic conditions on Earth gradually changed with the evolution of oxygenic photosynthesis, i.e., mainly with the evolution of cyanobacteria. The formation of photosynthetic O_2 started at about 3.5–3.8 Gyr ago, but most of the O_2 formed was scavenged by ferrous iron and sulfide and resulted in the precipitation of sulfate and iron oxides. The latter can today be retrieved in the so-called banded iron rocks. Oxygen reached significant levels (about 0.2%) in the atmosphere probably about 2 Gyr ago, a time when ferrous iron started to get oxidized in continental rock formations. About 700 million years ago, the atmospheric O_2 mixing ratios reached about 2%. This mixing ratio resulted in the formation of stratospheric ozone, which shielded Earth's surface from detrimental UV radiation and, thus, allowed the beginning of evolution of life on the continents. The contemporary atmospheric O_2 mixing ratio of 21% was probably reached about 350 million years ago, when evolution of plants and animals boomed on the continents. Hence, most of the change of an anoxic to the contemporary oxic atmosphere was driven by photosynthesis in the oceans, which was dominated, at least at the beginning, by cyanobacteria. The enormous mass of O_2 that has been produced by aquatic microorganisms can be calculated from the equivalent amount of organic matter that has been buried in sediments. Sedimentary stone contains, on average, about 0.5% organic carbon, which is equivalent to a total of 3.2×10^{22} g O_2. Only about 2–5% of this amount is found in the contemporary atmosphere; the rest has been used to oxidize ferrous iron or sulfide in the ancient ocean. The major constituents of our contemporary atmosphere are N_2 (78.08%), O_2 (20.95%), and Ar (0.93%).

B. Atmospheric Trace Gases

The most abundant trace gas is CO_2 (360 ppmv), followed by Ne (18.2 ppmv), He (5.24 ppmv), and CH_4 (1.75 ppmv). We know little about the dynamics of trace gases in the ancient atmosphere during the early history of Earth, but are better informed about changes during the last 150,000 years, i.e., back to the last interglacial period. This information is obtained from analysis of gas bubbles in ice cores retrieved from Arctic and Antarctic glaciers. For example, the mixing ratio of CH_4 was only about 300 ppbv during the glacial period and 600 ppbv during the interglacial period, while it is presently about 1750 pbbv and still increasing at about 0.5% per year (recent observations indicate that the CH_4 increase may now be slowing down, and a new steady state may soon be reached). The increase is believed to be caused by human activities, such as agriculture, industry, and mining. Most of the other trace gases are increasing as well, and this observation concerns scientists and the public because of many possible adverse effects on the habitability of our planet. With the exception of H_2, which seems to have no important effects, all the other trace gases listed in Table I play important roles for the functioning of our environment. Many trace gases, such as CO, CH_4, and NO, are key compounds in the chemistry of the troposphere and regulate the oxidizing capacity of our atmosphere. Some, like CH_4 and N_2O, act as greenhouse gases next in importance to CO_2. Others affect the chemistry of the stratosphere either

TABLE I
Contribution of Microbes in the Soil and the Ocean to the Global Budgets of Atmospheric Trace Gases

Trace gas	Mixing ratio (ppbv)	Lifetime (days)	Total budget (Tg yr^{-1})	Contribution (%) to budget by Soil	Contribution (%) to budget by Ocean	Importance
H_2	550	1,000	90	95 (sink)	<4 (source)	Insignificant
CO	100	100	2,600	15 (sink)	<4 (source)	Atmospheric chemistry
CH_4	1,750	4,000	540	60 (source)	2 (source)	Greenhouse gas; atmospheric chemistry
OCS	0.5	1,500	2.3	25 (source)	30 (source)	Formation of stratospheric aerosol
DMS	<0.1	<0.9	38	<0.1 (source)	95 (source)	Cloud formation
N_2O	310	60,000	15	70 (source)	25 (source)	Greenhouse gas; atmospheric chemistry
NO	<0.1	1	60	20 (source)	<1 (source)	Atmospheric chemistry
NH_3	<1	5	66	4 (source)	15 (source)	Acidification of ecosystems

by forming radiation-reflecting aerosols (OCS) or by destroying the ozone shield (N_2O). Cloud formation seems to be greatly influenced by dimethyl sulfide, which gives rise to condensation nuclei by aerosol formation. Aerosols can also be formed from trace gases like NO or NH_3, which are finally deposited and may result in acidification of ecosystems. Any change in the dynamics of these trace gases on either the regional or the global scale may, therefore, have important effects for the environment and for human beings. Hence, it is important to know by which mechanisms the trace gas cycles are driven and controlled and to find out those which are possibly affected by human activity.

II. CYCLING OF ATMOSPHERIC TRACE GASES

A. Implications of the Oxic Atmosphere

The availability of O_2 in the atmosphere allowed the evolution of aerobic chemolithotrophic and methylotrophic bacteria, many of which nowadays play important roles in the cycling of reduced gases such as H_2, CO, CH_4, or NH_3 (see following). However, it should be emphasized that reduced gases are thermodynamically unstable in an oxic atmosphere and would be oxidized chemically (e.g., by ozone or OH in the atmosphere) over geological timescales. In thermodynamic equilibrium, the atmospheric mixing ratios of these gases would be on the order of 10^{-45}, i.e., many orders of magnitude lower than they are in our real world (Table I). Hence, there must be sources of these gases which eventually form the contemporary mixing ratios. Indeed, many of these sources are of biological origin and microorganisms in soils and oceans are responsible for many of these biological reactions (Table I). For the overall global cycles of most of the atmospheric trace gases, biogenic sources actually are more important than biogenic sinks. The contribution of soils and oceans to the global trace gas budgets is mostly on the side of sources (Table I). Only in the cycles of H_2 and CO do microbial sink reactions play a greater role than microbial source reactions. However, for these

gases, major sources are the chemical oxidation of CH_4 and other volatile hydrocarbons in the atmosphere, which, in turn, are predominantly produced by biological reactions. Hence, the conclusion that the *net contribution* of microorganisms to trace gas budgets is mainly on the side of trace gas production, rather than trace gas consumption, is also true for these gases. It is also true that the global cycles of most of the trace gases listed in Table I are substantially influenced by microbial metabolism in soils and oceans, and that the contributions by vegetation, animals, biomass burning, and oxidation of biological compounds further emphasizes the importance of biological processes. Thus, it is obvious that the biosphere plays the central role in the global trace gas cycles.

B. Microbial Processes Involved in Trace Gas Cycling

Before a trace gas can be released into the atmosphere, it must be produced in the microbial habitat, e.g., the soil or the aquatic environment. Trace gas production results in an increase in its local concentration so that a gradient is established between the production site and the atmosphere, thus resulting in a flux. In general, this flux is attenuated by simultaneous consumption reactions which take place in the same habitat. Hence, the net flux of a trace gas is the result of simultaneous production and consumption reactions. Thus, a more or less intensive cycling of the produced gases goes on within the soil and the water even before they become trace gases in the atmosphere. In other words, the emission rates of trace gases into the atmosphere would be much higher than they are, and, consequently, so would the steady state mixing ratios of the trace gases in the atmosphere, if microbial consumption reactions were not operating. Table II summarizes some typical ranges of gas concentrations in the aqueous phase of soil and in the ocean water body, in comparison to the concentrations which would be expected from equilibrium dissolution with respect to the mixing ratios in the atmosphere. The extremely large range of concentrations illustrates that soil and water can as well be undersaturated as oversaturated with re-

TABLE II
Typical Trace Gas Concentrations in the Aqueous Phase of Upland and Wetland Soils and in Ocean Water, in Comparison to the Concentrations Which Are in Equilibrium with the Atmospheric Partial Pressure at 25°C

Trace gas	Bunsen solubility coefficient (25°C)	Concentration in equilibrium with the atmosphere (nM)	Concentration in the aqueous phase of soil (nM)	Concentration in ocean water (nM)
H_2	0.017	0.42	<0.2–100	0.5–4
CO	0.021	0.09	<0.01–100	0.1–5
CH_4	0.032	2.43	<0.3–1,400,000	2–4
OCS	0.468	0.01	<0.01–0.1	<0.01–0.1
DMS	13.2	<0.06	?	2–5
N_2O	0.561	7.76	<25	7–8
NO	0.044	<0.002	<20	<0.002–0.04
NH_3	313	<14	$<10^6$	30–4600

spect to the atmosphere. However, oversaturation is the general rule, thus explaining why soils and oceans act predominantly as sources for trace gases (Table I). The production and consumption reactions of trace gases which are relevant in soils and oceans are summarized in Table III.

C. Production Reactions of Trace Gases

Hydrogen is produced by microorganisms through the catalysis by the enzyme hydrogenase. Hydrogen production is typical for fermenting bacteria which, thereby, regenerate redox carriers during the fer-

TABLE III
Major Trace Gas Producing and Consuming Reactions That Are Relevant in Soils and Oceans

Trace gas	Production in soil	Production in water	Consumption in soil	Consumption in water
H_2	N$_2$-fixing microbes; fermenting bacteria	N$_2$-fixing microbes; fermenting bacteria	abiontic hydrogenases; Knallgas bacteria; anaerobic microbes	abiontic hydrogenases (?); Knallgas bacteria
CO	chemical reactions	photochemical reactions	ammonium oxidizers; anaerobic microbes (?)	ammonium oxidizers
CH_4	methanogenic archaea	methanogenic archaea	methanotrophic bacteria	methanotrophic bacteria
OCS	microbes with thiocyanate hydrolase; unknown reactions	photochemical reactions	microbes with carbonic anhydrase; unknown reactions	microbes with carbonic anhydrase; unknown reactions
DMS	sulfide methylation (?)	DMSP degradation	thiobacilli; methylotrophic bacteria	thiobacilli; methylotrophic bacteria
N_2O	nitrifying bacteria; denitrifying bacteria	nitrifying bacteria; denitrifying bacteria	denitrifying bacteria	denitrifying bacteria
NO	nitrifying bacteria; denitrifying bacteria	nitrifying bacteria; photochemical reactions	heterotrophic bacteria; denitrifying bacteria	denitrifying bacteria (?); heterotrophic bacteria (?)
NH_3	urease; animal excretions	animal excretions	nitrifying bacteria; assimilation by microbes and plants	nitrifying bacteria; assimilation by microbes and algae

Note. A question mark (?) indicates that the role of this reaction is unclear.

mentation of organic substrates under anaerobic conditions. Fermentation is the most important H_2-production process in anoxic environments, such as flooded rice fields, wetlands, and aquatic sediments, but it is unclear to what extent fermentation can support the H_2 production in oxic environments, such as upland soils and ocean water. Hydrogen is also produced by the nitrogenase reaction. Thus, H_2 production is observed in legume fields or in aquatic environments inhabited by N_2-fixing cyanobacteria.

Carbon monoxide can be produced by microorganisms, e.g., during degradation of heme, flavonoids, and aromatic compounds, but microbial production is apparently not relevant in soils and oceans. Instead, CO is produced by chemical reactions from organic matter (including dead microbial biomass) in soil and by photochemical reactions from dissolved organic compounds in ocean water. Likewise, CO production in plants also seems to be dominated by photochemical rather than metabolic reactions. Thus, CO is mainly produced abiologically, though by reactions with biologically produced substrates.

Methane is produced by methanogenic archaea, which are obligately anaerobic microbes. Methanogenesis is of great importance in all anoxic environments as soon as other oxidants such as nitrate, Mn(IV), Fe(III), and sulfate have been reduced and the redox potential has decreased to values <50 mV. Therefore, wetlands, flooded rice field soils, and landfills are major sources in the global CH_4 cycle. Methanogenic archaea are also abundant in the intestinal tracts of many animals, so that ruminants and termites are also significant sources of atmospheric CH_4. Upland soils and oceans, on the other hand, are too oxic to be ideal habitats for methanogens. Therefore, these environments emit only little CH_4. Nevertheless, some CH_4 production does take place, most probably by methanogenic strains that are relatively O_2-tolerant.

Carbonyl sulfide seems to be produced mainly from substances of biological origin, such as cysteine, methionine, and thiocyanate. The production mechanisms are largely unknown. An OCS-producing thiocyanate hydrolase has been described in bacteria. However, OCS is also produced in sterilized soil. In ocean water, OCS is produced by both photochemical and nonphotochemical reactions. In principle, the production processes resemble those of CO.

Dimethylsulfide is produced by several metabolic reactions. In ocean water, the most important production process is the degradation of dimethylsulfonium propionate (DMSP = $(CH_3)_2S^+CH_2CH_2COOH$), which is biosynthesized as an osmolyte in algae (e.g., *Emiliana huxleyi, Phaeocystis* sp.). The DMSP is then degraded to DMS plus acrylic acid by the enzyme DMSP lyase, which is found in many aquatic bacteria and algae. DMS is also produced and emitted from *Sphagnum*-wetlands. There, DMS seems to be produced either from dimethylated precursors, such as S-methylmethionine or from methoxyaromatic compounds, e.g., syringic acid, which are found in peat. DMS production from S-methylmethionine is known to occur in plant metabolism. DMS production from syringic acid has been described for *Holophaga foetida,* a homoacetogenic bacterium belonging to the kingdom *Acidobacteria. H. foetida* methylates sulfide to DMS by using the methoxy groups of syringic acid.

Nitrous oxide is a metabolic product of nitrification and denitrification. Ammonia-oxidizing nitrifiers produce N_2O by reduction of their product nitrite when O_2 is limiting (nitrifier-denitrification). Since O_2 limitation is a widespread condition in soils and oceans, nitrification is one of the most important processes of N_2O formation. After organic nitrogen is mineralized to ammonium, aerobic nitrifying bacteria oxidize ammonium to nitrate. In many environments, the supply of nitrate and, thus, nitrate reduction is limited by nitrification. Next to nitrification, denitrification is also important for the production of N_2O, which is an obligatory intermediate in the sequential reduction of nitrate to N_2. Denitrification is suppressed at atmospheric O_2 partial pressures. Under completely anoxic conditions when nitrate concentrations are relatively low, nitrate is usually completely reduced to N_2, with little concomitant N_2O production. Production of N_2O by denitrification is more pronounced when nitrate concentrations are relatively high or when conditions are microaerobic with limited O_2 suppy, e.g., in moist soils above field capacity.

Nitric oxide is also produced by the same reactions and bacteria that produce N_2O. Only the quantities are different, since other enzymes are involved. Thus,

NO is produced by nitrite reductase and further reduced to N_2O by NO reductase. The N_2O is then reduced by N_2O reductase to N_2. The production of N_2O seems to be more O_2-sensitive than the production of NO. Therefore, NO production is more common in dry soils, whereas N_2O production is more common in wet soils. In most soils, NO production seems to be quantitatively more important than N_2O production. Little is known about NO production in oceans, but rates seem to be generally low. In water, NO seems to be chemically produced from photolysis of nitrite, which is produced by ammonium-oxidizing nitrifiers.

Ammonia is produced from the hydrolysis of urea and of other N-compounds excreted by animals. The hydrolysis is accomplished by the enzyme urease, which is produced by bacteria and is ubiquitous in soil and water. The production and subsequent volatilization of NH_3 is controlled by the pH, which affects the chemical equilibrium between the NH_4^+ ion and NH_3 ($pK_a = 9.3$ at 25°C).

D. Consumption Reactions of Trace Gases

Hydrogen is consumed by many different bacteria. In anoxic environments, such as sediments or rice fields, H_2 is mainly consumed by methanogenic archaea and by bacteria reducing sulfate or ferric iron. In oxic environments, such as soil and ocean water, the decomposition process is less clear, but abiontic hydrogenases, rather than metabolically active bacteria, seem to be involved. These abiontic hydrogenases are able to rapidly oxidize H_2 at atmospheric concentrations, whereas aerobic H_2-oxidizing bacteria (Knallgas bacteria) are only able to utilize H_2 at elevated concentrations, e.g., in the vicinity of H_2-producing organisms. It is assumed that soil organisms are the primary source for abiontic soil hydrogenases.

Carbon monoxide is also consumed by many different bacteria. Some bacteria (the so-called carboxydotrophs) are able to grow chemolithotrophically by oxidizing CO as their sole energy source, but only at elevated CO concentrations. At the low atmospheric concentrations, CO consumption in soil and ocean water is most probably accomplished by ammonium-oxidizing nitrifiers, which have a similar half-saturation constant for CO oxidation as that determined in soil and water samples directly. The oxidation of CO in nitrifiers is catalyzed by the ammonia mono-oxygenase reaction, which utilizes CO as a fortuitous substrate instead of NH_3.

Methane is consumed by methanotrophic bacteria, i.e., a group of methylotrophs which are able to aerobically grow with CH_4 as their sole energy and carbon source. There are two major groups of methanotrophs (type I and II), which are mainly distinguished by their phylogenetic affiliation (γ- and α-*Proteobacteria*) and their carbon assimilation pathways (ribulose monophosphate and serine pathway, respectively). The responsible enzyme is methane monooxygenase, which exists as a membrane-bound or a cytoplasmic enzyme, the latter only in type II species. The ecological niche of these methanotrophs seems to be in wetlands and aquatic habitats, where they attenuate the CH_4 flux from the anoxic methanogenic sites through the oxic surface layer into the atmosphere. The CH_4 concentration in these habitats is in the micromolar range and, thus, sufficiently high to be utilized by the methanotrophs. However, the low atmospheric concentrations of CH_4 are in the nanomolar range. Atmospheric CH_4 is consumed in upland soils by still unknown bacteria, which are characterized by a high affinity for CH_4 (nanomolar range half-saturation constants), in contrast to the methanotrophs described previously, which exhibit half-saturation constants in the micromolar range.

Carbonyl sulfide can be utilized by sulfur-oxidizing bacteria, such as thiobacilli. These bacteria hydrolyze OCS to CO_2 and H_2S and then grow by oxidizing H_2S as their energy source. The hydrolytic step seems to be catalyzed by the enzyme carbonic anhydrase, which reacts with OCS as well as with CO_2. Hence, it is likely that bacteria that contain carbonic anhydrase, e.g, to facilitate assimilation of CO_2, will also be able to hydrolyze OCS. However, the role of these bacteria for OCS turnover in the environment is not yet clear.

Dimethylsulfide is utilized by aerobic S-oxidizers (e.g., *Thiobacillus thioparus*) and aerobic methylotrophs (e.g., *Hyphomicrobium sp.*, *Methylophaga sulfidovorans*), which gain energy by oxidizing DMS via methylsulfide to H_2S plus CO_2. Methyl groups are sequentially eliminated as formaldehyde, in reactions catalyzed by DMS monooxygenase and methanethiol oxidase. In the thiobacilli, H_2S is oxidized to sulfate and allows the generation of additional energy. Microbial DMS utilization significantly attenuates the flux of DMS from the ocean into the atmosphere. Chemical photooxidation of DMS to dimethyl sulfoxide contributes to flux attenuation. Under anaerobic conditions, e.g., in sediments and microbial mats, DMS may be utilized by sulfate-reducing bacteria, phototrophic bacteria, and methanogenic archaea.

Nitrous oxide is consumed by denitrification through the enzyme N_2O reductase, which catalyzes the reduction of N_2O to N_2. Most denitrifiers are able to grow with N_2O instead of nitrate as an oxidant. Nitrous oxide reductase is an O_2-sensitive enzyme and is only induced in the absence of O_2, so consumption of N_2O is only possible in anoxic microniches in soil and water. The role of N_2O consumption for attenuation of N_2O emission from soil and water into the atmosphere is still unclear. However, there is growing evidence that soils may occasionally even act as a small net sink for atmospheric N_2O.

Nitric oxide is consumed by denitrification through the enzyme NO reductase, which reduces NO to N_2O. However, growth on NO has not been proven irrefutably. Similarly as for N_2O, reduction of NO requires anoxic conditions. However, NO can also be consumed by various heterotrophic bacteria under aerobic conditions through fortuitous oxidation to nitrate. Growth of the bacteria is not stimulated by the oxidation of NO. In soil, this reaction, nevertheless, seems to be of greater importance than the reduction by denitrifiers. Methanotrophic bacteria are also able to consume NO both in pure culture and in soil. Microbial NO consumption causes a rapid turnover of NO in soils. Depending on the relative rates of NO production and consumption in the soil, and on the local concentration of NO in the ambient atmosphere, soils may act as either a source or a sink for atmospheric NO. Except in polluted areas, however, soils are usually net sources.

Ammonia is the energy substrate of aerobic chemolithotrophic nitrifiers (*Nitrosomonas* sp., *Nitrosospira* sp., *Nitrosococcus* sp.), which oxidize it to nitrite with the enzymes ammonia monooxygenase and hydroxylamine oxidoreductase. Ammonia is also con-

sumed by bacteria, algae, and plants, which use it as a nitrogen source for biosynthesis of cell material. The role of microbial NH_3 consumption for attenuation of NH_3 emission is unclear.

See Also the Following Articles

ECOLOGY, MICROBIAL • EXOBIOLOGY • METHANOGENESIS • NITROGEN CYCLE

Bibliography

Andreae, M. O., and Schimel, D. S. (eds.). (1989). "Exchange of Trace Gases between Terrestrial Ecosystems and the Atmosphere. Dahlem Konferenzen." John Wiley, Chichester, UK.

Bouwman, A. F. (ed.). (1990). "Soils and the Greenhouse Effect." John Wiley, Chichester, UK.

Bouwman, A. F., Lee, D. S., Asman, W. A. H., Dentener, F. J., VanderHoek, K. W., and Olivier, J. G. J. (1997). *Global Biogeochem. Cycles* **11**, 561–587.

Buat-Ménard, P. (ed.). (1986). "The Role of Air–Sea Exchange in Geochemical Cycling." Reidel, Dordrecht, The Netherlands.

Conrad, R. (1988). *Adv. Microb. Ecol.* **10**, 231–283.

Conrad, R. (1996). *Microbiol. Rev.* **60**, 609–640.

Fenchel, T., King, G. M., and Blackburn, T. H. (1998). "Bacterial Biogeochemistry. The Ecophysiology of Mineral Cycling." Academic Press, San Diego.

Holland, H. D. (1984). "The Chemical Evolution of the Atmosphere and Oceans." Princeton Univ. Press, Princeton, NJ.

"Intergovernmental Panel on Climate Change (IPCC) Reports" 1990, 1992, 1994, 1995. Cambridge Univ. Press, Cambridge, UK.

Kelly, D. P., and Smith, N. A. (1990). *Adv. Microb. Ecol.* **11**, 345–385.

L'Air Liquide (1976). "Gas Encyclopaedia." Elsevier, Amsterdam, The Netherlands.

Lovelock, J. E. (1979). "Gaia: A New Look at Life on Earth." Oxford Univ. Press, Oxford, UK.

Murrell, J. C., and Kelly, D. P. (eds.). (1996). "Microbiology of Atmospheric Trace Gases: Sources, Sinks and Global Change Processes." Springer, Berlin.

Oremland, R. S. (ed.). (1993). "Biogeochemistry of Global Change. Radiatively Active Trace Gases." Chapman & Hall, New York.

Prinn, R. G. (ed.). (1994). "Global Atmospheric–Biospheric Chemistry." Plenum, New York.

Raynaud, D., Jouzel, J., Barnola, J. M., Chappellaz, J., Delmas, R. J., and Lorius, C. (1993). *Science* **259**, 926–934.

Rogers, J. E., and Whitman, W. B. (eds.). (1991). "Microbial

Production and Consumption of Greenhouse Gases: Methane, Nitrogen Oxides, and Halomethanes." American Society for Microbiology, Washington, DC.

Schidlowski, M. (1980). *In* "The Handbook of Environmental Chemistry," Vol. 1A, (O. Hutzinger, ed.) pp. 1–16. Springer, Berlin.

Schlesinger, W. H. (1997). "Biogeochemistry—An Analysis of Global Change" (2nd ed.). Academic Press, San Diego.

Slanina, S. (ed.). (1997). "Biosphere–Atmosphere Exchange of Pollutants and Trace Substances." Springer, Berlin.

Warneck, P. (1988). "Chemistry of the Natural Atmosphere." Academic Press, San Diego.

Microscopy, Confocal

Guy A. Perkins and Terrence G. Frey

San Diego State University

GLOSSARY

Airy disk The diffuse central spot of light with concentric rings of light of decreasing intensity, formed by the microscope imaging a point source of light.

charge-coupled device (CCD) An imaging device that converts light into electrical charges which vary in proportion to the amount of incident light and a two-dimensional array of CCDs can thus serve as an accurate electrical representation of the image.

chromatic aberration Aberration arising from the lens material that brings light of different wavelength to focus at different points along the optical axis.

deconvolution A mathematical technique that uses an empirical measure of the blurring of an image caused by the point spread function to deblur an image by quantitatively redistributing out-of-focus intensity.

depth of field The slice of the specimen that contributes to the in-focus image at a particular focus level.

epi-illumination Illumination of the specimen by light from above, usually by light emerging from the objective lens.

fixed specimen A mode of preservation intended to minimize swelling, shrinkage, evaporation, decay, and all forms of distortion of cellular components. Fixation is usually done by chemical action to stabilize or fix the cell contents so that they may be subjected to further treatment that would otherwise damage them.

fluorescence A molecule is fluorescent if it absorbs light and, after a delay of less than one-millionth of a second, re-emits light of a longer wavelength (lower energy).

fluorophore A molecule that will fluoresce when excited by light of a certain wavelength.

image contrast The ability to distinguish various components of the structure of the object by different intensity levels in the image.

numerical aperture (N.A.) A measure of the resolving power of the objective lens, equal to the product of the refractive index of the medium in front of the lens and the sine of the angle between the outermost ray entering the lens and the optical axis.

objective lens The lens system that is closest to the object that forms an image of the object.

ocular lens The eyepiece (lens) of an optical microscope.

optical sectioning Imaging sections of the specimen by stepping the focal plane through a specimen and recording 2-D images on each focal plane.

phase Light waves oscillate and, hence, are periodic in nature. A light wave has two components, amplitude and phase. The phase specifies the particular stage in the periodic cycle of the wave with respect to a specific origin.

photobleaching To extinguish the fluorescence by the action of light incident on the specimen, chemically altering the fluorophore.

photomultiplier tube (PMT) A very sensitive detector of light that amplifies the initial photocurrent by a factor of 10^8 or more.

point spread function (PSF) The relationship between a point object and the image of it after passing through the objective of the microscope simply put, the image-forming characteristics of an objective lens.

polarized light The direction of the electric field, which is in plane for plane-polarized light and rotates in a circle for circularly polarized light.

refractive index (index of refraction) The effect of a material object on the incident wave, defined as the ratio of the velocity of light in a vacuum to that in the object.

resolution   The ability to distinguish objects that are close together; defined as the minimum distance between points or parts of the object which appear as distinct points in the image.

spherical aberration   Aberration arising from rays which enter a lens at different distances from the optical axis and are focused at different distances along the axis when they are refracted by the lens.

THE FIELD OF OPTICAL MICROSCOPY has been revolutionized in the past 10 years by the widespread use of confocal microscopes. Confocal microscopy can complement conventional optical microscopy principally in (1) observing submicron objects in thick specimens, e.g., in slices throughout a cell, (2) detailing living specimens, (3) immunofluorescence imaging, and (4) 3-D optical imaging. Although the first confocal microscope was invented in the late 1950s, their use was limited to only a handful of labs until the introduction of commercial microscopes a decade ago. In essence, the confocal microscope offers the capability, not easily accomplished with conventional optical microscopes, of imaging thick microorganisms or biological tissue in three dimensions (3-D). Confocal microscopes are commonly used in two modes, fluorescence and epi-illumination bright field, which provides reflection contrast. Fundamental differences exist between conventional optical and confocal microscopes.

I. HOW DOES CONFOCAL MICROSCOPY DIFFER FROM CONVENTIONAL OPTICAL MICROSCOPY?

The conventional microscope is essentially a parallel-processing instrument that images the entire object field simultaneously. In a conventional optical microscope operated in the reflection contrast mode, the specimen is illuminated uniformly through the objective lens by an incoherent light source, typically, a filament lamp. The objective lens produces an image that can be viewed through the ocular lens. When the image is in focus, its features appear sharp, but when the image is defocused, the features become blurred without change in the total light intensity. Imaging the wide object field places a severe requirement on the performance of the optical components, in order to avoid introducing distortions that affect the image quality. In epifluorescence microscopy of microbes larger in size than bacteria and of thick sections of tissue (50–100 μm thick), there are typically many fluorescent labels distributed throughout the volume. The illumination of the specimen with light at the excitatory wavelength excites fluorescence emissions throughout the whole depth of the specimen and not just at the focal plane. There is usually an undesirably high influence of the background in the focal plane that is caused both by light coming from above and below the focal plane and also from stray light from the same focal plane. This high background, which can seriously degrade the contrast and sharpness of the image, is called the out-of-focus blur and is largely eliminated in confocal microscopy. While it is possible to remove most of this blur from conventional optical images by computational mechanisms, the confocal microscope presents an alternative method, in which almost all the light contributing to the out-of-focus blur is prevented by an optical mechanism from contributing to the image.

With confocal microscopes, all that is asked of the optics is that a good image of one point be provided. The price paid is that scanning must be performed to build up an image of the entire field. Depending on the application, this price is often worth paying because the principal advantage of point imaging is that defocused images disappear instead of becoming blurred, as in conventional optical microscopy. In contrast to conventional optical microscopy, confocal microscopes possess very narrow depths of field. This feature allows for optical sectioning, which permits the viewing of different layers in a specimen without interference from out-of-focus layers. Optical sections obtained from a confocal microscope build up a three-dimensional image of the object, in which each section has significantly improved contrast, compared to images from a conventional microscope. Furthermore, the use of pinholes in both the illumination and detection components of a confocal microscope produces a sharper point spread function, so that the lateral resolution can be as much as 1.4 times better than a conventional microscope with comparable optics. The most common mode of confocal microscopy in biological applications is fluorescence using fluorophores conjugated to antibodies, in order to map the positions of specific molecules, or using fluorophores which report the pH, membrane potential, ion concentrations, etc. of their environment. Lasers are commonly used sources of excitation light for fluorescence confocal microscopes, as they provide an intense beam of highly coherent light which is scanned across the specimen. Thus, the most common type of confocal microscopy used in biology is termed laser scanning confocal microscopy or LSCM.

II. OPTICAL SECTIONING

Optical sectioning occurs because confocal microscopes illuminate one small spot on the specimen at a time by focusing the image of a pinhole onto the specimen. Light reflected from the specimen is imaged by the objective lens and sent back to a matching pinhole in front of the detector, as shown schematically in Fig. 1. The out-of-focus information does not strike the detector pinhole and so does not pass through it. Scanning the illumination spot or specimen in a raster pattern forms an image of the specimen. Because the optical path from the illuminating pinhole to the specimen is the same as from the detector pinhole to the specimen, the microscope is said to be confocal. Even though the depth of field is very narrow, the confocal microscope can be operated as though it had an infinite depth of field, as a result of the confocal rejection of out-of-focus information. Hence, a complete in-focus stack of projection images down the z-axis may be recorded by progressively stepping the focal plane through the entire 3-D specimen (axial z optical sectioning) while performing successive x,y scans in the focal plane. Vertical sectioning may also be achieved by scanning in x and z at a particular y position. This z-scanning can be undertaken by a precision-motor drive on the microscope's focus knob or by coupling a piezoelectric z translator to the specimen stage. In this way, optical sections parallel to the optical axis can be generated, something impossible to obtain by con-

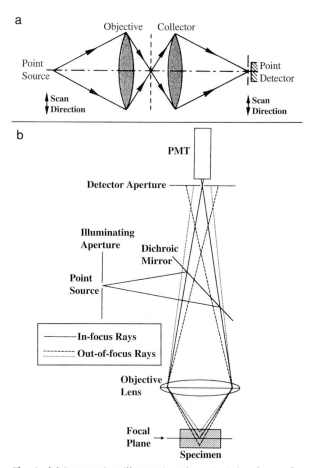

Fig. 1. (a) Ray tracing illustrating the principle of a confocal scanning microscope. Light from the point source probes a very small region of the object (vertical dashed line) and the point detector ensures that only light from that same small area is detected. The point source and point detector are moved synchronously as the object is scanned. (b) Schematic diagram showing how out-of-focus light is rejected and how depth discrimination, which allows for optical sectioning, is achieved. A laser light source provides excitation illumination that is reflected by a dichroic mirror through an objective lens and brought to a focus in the specimen. The focused light is scanned in a horizontal (*xy*) plane. The fluorescent light emitted by fluorophores in the specimen is collected by the same lens and then passed back through the dichroic mirror to impinge on the detector aperture. Only light from the plane of focus is focused through this aperture into the PMT detector.

ventional optical microscopy, except by resampling a complete deblurred 3-D image.

In general, it is preferred to use as small a pinhole as possible to maximize the lateral resolution and to sharpen the image plane being imaged. However, one may need to sacrifice this increase in resolution when there is motion in the specimen or with weakly fluorescing microorganisms or faint signals of antibodies conjugated to fluorophores. By opening the pinhole iris, but only to an extent not much greater than the Airy disk diameter of the objective lens PSF, the advantage of higher resolution is lost; but much of the capability for eliminating out-of-focus blur is retained. It is important to realize that with fluorescence microscopy, the spatial intensity distribution of the fluorophore is measured. This measurement is interpreted to reveal the spatial distribution of the target. However, not all targets, usually antigens, are equally accessible. Confocal microscopy can be used to determine the penetration of the fluorophore, as well as to determine differential localization of the target in 2-D planes throughout a 3-D object. Through the appropriate use of emission filters and dichroic beam splitters, multiwavelength LSCM can be achieved.

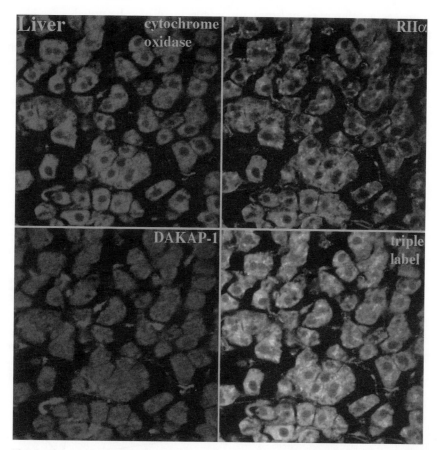

Fig. 2. Immunohistochemistry performed by triple-antibody labeling of a 70-μm slice of liver tissue. LSCM was used to record images of a 3-μm optical section near the surface of the slice. The cytochrome c antibodies specifically label mitochondria and, hence, are not in the nucleus (the circular regions inside each cell). The RIIα antibodies produce a typical Golgi labeling pattern inside the hepatocytes. DAKAP-1 is a unique anchoring protein for regulatory subunits of protein kinase A (PKA), which are involved in cAMP-mediated cell signaling pathways. It is unique in its anchoring of both RIα and RIIβ regulatory subunits of PKA. Triple labeling shows the superposition of the three antibody labeling patterns. See color insert.

Current LSCMs have the capability of simultaneously imaging three or more wavelengths by employing an argon/krypton mixed-gas laser producing spectral lines at 488 nm, 568 nm, and 647 nm to selectively excite three different fluorophores. The advantage of multiwavelength LSCM is the ability to compare the distributions of three different antibodies in colocalization studies (Fig. 2). One benefit of multiwavelength LSCM, compared to conventional fluorescence optical microscopy, is that the images in the different channels are in exact spatial register, since the corresponding pixels in each channel are measured simultaneously during the illumination of a single point in the specimen. Thus, the superposition of pseudocolor images, a different color for each channel, allows one to easily locate regions of colocalized fluorophores based on the resultant mix of colors. For example, if equal intensities of red, green, and blue occur in the various channels at an image point, then the resultant color will be white in the superimposed image (Fig. 2). Of course, the same advantage holds in serial optical sectioning, simplifying any subsequent image processing involving stacks of images. Sometimes, the LSCM is equipped with a nonconfocal detector to simultaneously record a conventional transmitted scanned image, e.g., using the Nomarski configuration. One channel would then have the Nomarski image and the other channel would have confocal epifluorescence images of the same field of view. In this manner, the distribution of the fluorophores may be displayed in suitable pseudocolors superimposed on the monochromatic transmission image, and the location of the fluorophores in relation to the structural features in the specimen may be noted.

Confocal microscopes may vary in their illumination sources between white light and laser light and in their mode of scanning (Fig. 3). Direct observation is possible with designs such as the tandem scanning microscope, which will be described, but with laser

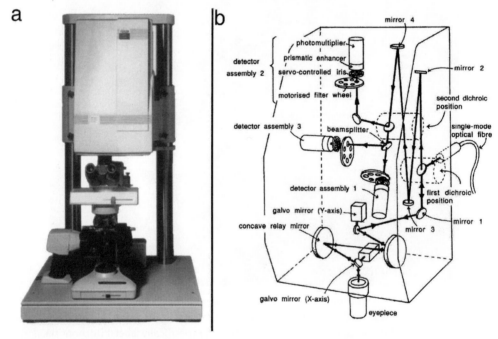

Fig. 3. (a) Image of a popular laser-scanning confocal microscope. The scan head can be attached to almost any make of upright or inverted microscope (courtesy of Bio-Rad Laboratories, Hercules, CA). (b) Schematic diagram of the laser scanning system found in the box attached to the top of the microscope shown in (a) (from *Handbook of Biological Confocal Microscopy,* Plenum Press, New York, NY).

scanning, optical detectors, typically photomultiplier tubes, must be combined with image processing to create an image. An inevitable consequence of this approach is that the image is acquired in an electrical format. This facilitates collection of digital images and certain forms of image enhancement, the simplest being contrast enhancement. In laser-scanning confocal microscopes (LSCMs), either the beam or the specimen must be scanned to obtain a complete image. An advantage of these microscopes is that they are typically built around conventional optical microscopes. Specimen scanning is the least complicated optical arrangement and eliminates off-axis aberrations in the objective lens. This approach provides space-invariant imaging, ensuring that the contrast and resolution are identical across the entire field of view and are decoupled from the magnification. Thus, it is possible to continuously vary the magnification attainable with a single objective lens and zoom in and out of a field of view. In turn, this means that low-magnification images can be taken with high numerical aperture objectives. This situation is a particular advantage in fluorescence imaging, where a higher numerical aperture achieves a greater level of signal collection, important for reducing photobleaching and for detecting weak fluorescent signals. But mechanical rastering of the specimen suffers from the drawback of slow scanning speed, on the order of 10 s per scan, which is too slow for many fluorescence applications common in microbiology.

In contrast, most beam-scanning microscopes operate at 1–2 scans per second and use either acousto-optical modulators or galvanometer-driven mirrors to deflect the laser beam (Fig. 3). The image is then displayed on a computer monitor. For scanning at TV rates (30 scans per second), rotating mirror wheels or galvanometer-driven mirrors offer the speed required for this high scan rate. The optical microscope is coupled to a laser light source, a high-resolution CCD camera, and a precision-drive motor mounted on the z-axis of the microscope. Laser light reflected from the sample travels back along the same optical path as the incident laser beam and is recorded by a high-resolution CCD camera. The laser-generated image is then displayed on a standard TV monitor. The drawbacks to beam-scanning microscopes are

the increased complexity of the optical system and non-space-invariant imaging, which may be important at the edge of the field and when image enhancement is performed. To limit edge distortions, some LSCMs utilize only one-third of the field width. With LSCMs, the magnification is coupled to the resolution and so it is necessary to change the objective lens in order to cover the entire magnification range. Therefore, care must be exercised in comparing images recorded with different numerical aperture and magnification objective lenses. Since significant rejection of out-of-focus blur can be obtained with detector apertures larger than required for true confocal imaging, it is advantageous to equip the micro-

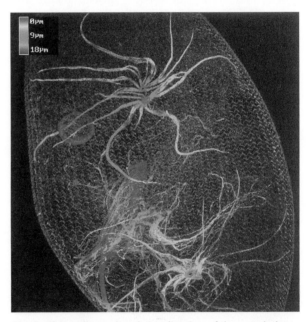

Fig. 4. Three-dimensional visualization of the cytoskeleton of *Paramecium caudatum*. *P. caudatum* were labeled immunocytochemically with anti-alpha-tubulin antibodies and detected by cy3-antibody conjugates. By use of the confocal principle, 1-μm thick sections (perpendicular to the z-axis) were recorded to reconstruct a three-dimensional image by serial optical sectioning. This *P. caudatum* was captured at the beginning of the division process. Labeled microtubules make visible the contractile vacuole complex, each with 2 pores due to doubling, collecting canals, cortex, oral apparatus, and micronucleus. (Courtesy of Theo Brigge, from Naturwissenschaften, Vol. 85, Aug. 1998, Springer-Verlag, Heidelberg.)

scope with a detector aperture with a variable size opening. Because confocal image quality depends on a number of empirical variables, such as the numerical aperture, the size and shape of the detector aperture, the fluorophores used, the wavelength of the exciting laser light, and the nature of the specimen itself, it is often required to optimize the opening of the detector aperture for each particular application.

Higher resolution along the z-axis can be achieved by combining confocal microscopy with deconvolution techniques. One first records 30-nm optical sections with a precision-drive motor and computer software reconstructs the 3-D image, followed by deconvolution to correct for the PSF. Deconvolved 3-D confocal images can attain 100-nm resolution, which is comparable to the best lateral resolving power, and allows for vertical measurements and noncontact surface profiles. This depth discrimination has applications in microbiology in the study of microbe surfaces in defined solutions, for example, and may be displayed either as a HEIGHT image, in which lighter areas correspond to points closer to the observer, or as a 3-D height profile plot (Fig. 4).

III. TANDEM SCANNING MICROSCOPE

Depending on the application, the fact that a computer or television screen is required in order to see an image with an LSCM may be a disadvantage. With dynamic processes and fast-moving microbes, it is convenient to be able to view a confocal image directly by eye in real time. The tandem scanning microscope (TSM) is a confocal microscope that was designed for real-time viewing, if we accept the definition of real time to mean much faster than video or frame repetition rates. Tandem scanning means scanning the illumination and detection simultaneously. The basic idea is to use a large array of pinholes on a rotating disk instead of a single pinhole so that many points on the specimen are simultaneously scanned by many focused light beams. The pinholes are separated from each other by a distance large enough so that there is little interaction between the images formed by the individual pinholes. The complete image is formed by rapidly rotating the array of pinholes so as to fill in the image space

between them. TSMs have the advantage of not needing a laser source. Incoherent white light or mercury arc lamps can be used, generating an apparently continuous optical image that may be viewed conventionally through the ocular lens or captured by an electronic imaging camera. They can operate at rates up to 700 frames/s, usually reported for a class of TSMs called RSOMs (real-time scanning optical microscopes).

TSMs work well with reflection contrast (epi-illumination). The beauty of such a microscope is that it permits the examination of unlabeled, unstained living microbes. The one drawback, compared to the LSCM, is the low transmitted light level, because the rotating disk of pinholes, called the Nipkow disk, dramatically reduces the amount of light available for imaging. However, with a sensitive recording modality, such as an image-intensifying TV camera, high-speed frame store, cooled CCD detector, or videodisk, the quality of the images can be better than that of a LSCM. Coupled with video microscopy, TSMs can be configured for interference microscopy, such as phase, polarization or differential interference contrast (Nomarski), that is ideal for the dynamic, high-resolution observations of living specimens and for detailing the behavior of macromolecular assemblies without the need to label them fluorescently. In contrast, LSCMs are difficult to configure for interference microscopy. TSMs can image multiple-wavelength fluorescence. With an appropriately corrected objective lens that has no significant longitudinal chromatic aberration, fluorescence excitation will give rise to a multicolored image in which all wavelengths are imaged simultaneously, from the near-infrared through to the ultraviolet. This means that imaging in real color, as well as real time, can be accomplished; however, recording in the infrared or ultraviolet ranges requires a suitable TV camera and filters. With fluorescent signals, image intensifiers are generally required because of the low level of light transmitted through the Nipkow disk.

Applications of confocal microscopy to microbiological imaging are boundless and so only a few representative examples will be presented here. Operated in bright-field mode, confocal microscopy offered substantial improvement in signal-to-noise to enable a study of chromosomal material during division in *E. coli.* A number of microorganisms are well suited for confocal reflection microscopy; among them are scaly flagellates and other microbes with naturally reflective coats, fungal colonies, and spores. Other microbes, such as *Euglena,* are ideal for Nomarski confocal microscopy because they have a large phase transition just beneath their surface. However, when multiple layers of phase transition exist close to each other inside a specimen, interference fringes can result that degrade the image quality. Interference patterns can be seen when imaging the staminal hairs of *Tradescantia* and the water-imbibed spores of the moss *Dawsonia superba.* When these spores are observed with a conventional optical microscope, their cells walls are almost invisible, but in confocal reflection microscopy, the fine wall ornamentation is strongly highlighted. Naturally fluorescent microbes can be studied in 3-D using confocal fluorescence microscopy; examples include the algae *Cyanoderma* and *Trichophilus,* found in the hair of sloths; algae in mussels, such as *Conchocelis* and *Ostreobium*; the alga *Stomatochroon,* found in the substomatal space in leaves; and *Sacculina,* a parasitic crustacean in crabs.

IV. ABERRATIONS TO WATCH FOR IN MICROBIOLOGICAL IMAGING

Various types of aberrations can exist in confocal microscopes that may degrade the imaging process. One aberration can surface when optical sectioning is performed deep within aqueous specimens because high numerical aperture, oil-immersion objective lenses are corrected for observation of specimens close beneath the coverslip. Thus, the image quality may be compromised by the effects of spherical aberration far from the surface of the lens. If serial optical sectioning of aqueous specimens is performed, it is best to use a high numerical aperture, water-immersion objective lens and to position it close to the specimen. This arrangement has the additional advantage that the physical change in focal position is equivalent to the depth moved through the specimen. Otherwise, the focal position would require correction by the ratio of the refractive index of the immersion oil relative to that of the aqueous medium and specimen.

Another aberration that is manifest in multiwavelength fluorescence imaging occurs in the presence of significant chromatic aberration. For example, in off-axis imaging, used in scanned-beam microscopes, chromatic aberration will cause an axial shift in the optimal confocal position of the detector aperture in relation to the illuminating aperture. This axial shift has two components, one that is wavelength dependent and another that is dependent on the scan position. When undertaking triple-label epifluorescence imaging, for example, the exciting wave and the three emission waves need to be brought to the same focal point by the objective lens. The presence of significant chromatic aberration in this situation will result in one or more of the fluorescence emissions being collected from focal planes different from that being illuminated. As a result, the fluorescence intensity is diminished and the resolution in optical sectioning is reduced. Moreover, the emission images are no longer in register along the z-axis, having been collected from different z-planes. The effects of chromatic aberration in the objective lens may be eliminated to a great extent by reducing the effective numerical aperture of this lens. In practice, the diaphragm at the back focal plane of the objective lens, or in a conjugate image plane, is closed down to allow less light through. The obvious drawback is the concomitant reduction in image brightness.

V. SPECIMEN PREPARATION FOR CONFOCAL MICROSCOPY

A few aids to specimen preparation will be presented for the three most commonly used confocal preparative techniques in microbiology: fluorescence, immunogold, and surface replicas. The preparation of specimens for epifluorescence confocal imaging will be treated first because of its prominence in microbiology. When first used, fluorescence confocal imaging appears to produce images that are less bright than those produced by conventional fluorescence imaging, although with an enhanced signal-to-noise ratio. Thus, it is worthwhile to label specimens more intensely than would be suitable for conventional fluorescence imaging. Higher concentrations of reagents are generally used to increase the signal. However, one must take care to optimize the appropriate blocking procedure and to perform negative controls. To avoid spurious cross-reactivity, it is recommended that the antisera used be affinity-purified and that the number of secondary antibody–fluorophore conjugates per IgG molecule be optimized. Introduction of antifade media, such as DABCO or gelvatol, is always recommended with fixed specimens to reduce the attenuation of the fluorescence signal upon illumination, especially when the specimen is to be scanned many times to collect a stack of optical sections. However, with live microorganisms or cultured cells, these antifade media cannot be used. Instead, the excitation intensity is reduced by lowering the laser power or by inserting neutral density filters into the beam path. It is recommended that, if serial optical sections are to be collected, the sections be collected front to back so that the inevitable progressive fading, in this case, causing the back of the object to appear fainter than the front, can be used as an appropriate depth marker. In this way, corrections for both the photobleaching and the signal attenuation, as a function of specimen depth, can be accomplished by a computational procedure.

Colloidal gold labels can be visualized in confocal microscopes configured in the reflection contrast mode. These labels are commonly conjugated to antibodies and used in immunoelectron microscopy because gold is electron-dense and so is contrasted sharply in electron micrographs. Since the reflection signal from colloidal gold is strongly detected in the confocal set-up and is resistant to photobleaching, colloidal gold labeling may be preferable to immunofluorescence labeling where multiple scans are needed, as in stacks of optical sections. Colloidal gold particles as small as 5-nm diameter can be detected with signal averaging under optimal conditions. While immunogold confocal microscopy works admirably for fixed specimens mounted in a high refractive index medium, such as glycerol, it does not lend itself with the same facility when living microbes and cells are used. Part of the problem is the low refractive index of commonly used culture media, which contributes to the reflec-

tion signal. The other part of the problem is the reflections from the cells, particularly from the plasma membrane. With this inherent difficulty, only gold particles greater than 40-nm diameter can be visualized.

Another application of confocal reflection imaging is with surface replicas. Microbes or cells are preserved by rapid freezing, followed by either freeze-fracture, deep-etch, or spray-dry techniques. A replica is then made by coating with either platinum or platinum–carbon. These methods have been used for many years with the transmission electron microscope. Surface replica imaging has also proved to be feasible with the confocal microscope configured in the reflection mode, enabling the 3-D topographical detail of the specimen surface to be examined. One advantage of this technique is the possibility of correlative light and electron microscopy. For example, living bacteria may first be imaged by immunofluorescence confocal microscopy after labeling specific surface antigens. The bacteria may then be rapidly frozen, freeze-fractured, and surface replicated. The ultrastructural detail can be studied by electron microscopy, including the positions of surface immunogold particles. Afterwards, the 3-D topography of the replica may be determined by immersing it in oil and imaging with the confocal microscope. The information obtained by confocal microscopy complements that from electron microscopy because, while electron microscopy provides high-resolution detail, global 3-D topography is not easily garnered. With suitable image processing, information from the different types of images may be combined.

VI. RECENT TECHNICAL DEVELOPMENTS

Recent technical developments will be briefly catalogued here, especially those that impact microbiological imaging. By modifying the Nipkow disk used in real-time TSM, Nomarski or Differential interference contrast imaging, which is a mainstay in microbiological imaging, can be achieved in the confocal mode. Thus, the advantages of optical sectioning afforded by confocal microscopes can be used to study the boundary features, such as membrane surfaces, that are well highlighted in the Nomarski microscope. Another technical advance has been made through the coupling of a CCD array camera with a rapid scan rate confocal microscope. Very high optical scan rates can be achieved because the CCD is an integrating device, which means that CCD readout rate can be independent of the specimen scan rate. The CCD has additional advantages over PMTs as manifested by a higher quantum efficiency and better red sensitivity for multiwavelength fluorescence imaging. A further development is the marriage of confocal microscopy with spectroscopy, in which a polychromator grating is placed in the beam path after the light has passed through a slit detector (instead of a pinhole detector). The spectral dispersion is then recorded using a 2-D CCD array producing an (x,λ) image (λ = wavelength), in which the second dimension contains the spectra of light emitted at each point, x. In practice, the microscope collects a single x,y image at a fixed wavelength. By repeating this measurement at different wavelengths, a series of wavelength-specific images that comprise an (x,y,λ) 3-D data set for each focal z-plane can be generated. Applications of confocal spectroscopy include studies of certain halobacteria, which contain integral membrane proteins that have incorporated retinal prosthetic groups, capable of photocycling between states characterized in the visible spectrum.

Recent advances in confocal super-resolution fall into two categories. The first involves a new confocal configuration that permits the independent collection of amplitude and phase information from the light passing through the pinhole detector. Subsequent digital filtering of the signal permits a doubling of the spatial resolution. The second involves the collection of off-axis light that is normally excluded by the confocal detector aperture. An array of detectors is used in place of the single detector. Computer algorithms, similar to those employed in deconvolution microscopy, process the additional information, leading to a doubling of the lateral resolution. These two super-resolution confocal microscopes are important in fluorescence imaging as they are used to minimize photobleaching by reducing the excitation illumination or to reduce the level of labeling with cytotoxic fluorophores. By using these microscopes at "half-performance," which would still produce the resolution limit characteristic of conventional confocal microscopes, one can still collect sufficient signal to form a satisfactory image.

Other recent technical developments, such as two-photon excitation, 4-Pi confocal microscopy, and Theta confocal microscopy, merit their own section and will be described next.

VII. TWO-PHOTON EXCITATION, 4-PI CONFOCAL MICROSCOPY, AND THETA CONFOCAL MICROSCOPY

Two-photon excitation refers to the simultaneous absorption of two photons with wavelengths λ_1 and λ_2, where λ_1 may equal λ_2, that produces a molecular excitation. The two photons combine their energies to effect the same excitation that would otherwise require a single photon of higher energy, described by the shorter wavelength, $\lambda_1\lambda_2/(\lambda_1+\lambda_2)$. Two-photon laser-scanning microscopy (TPLSM) does not require confocal (pinhole) detection, even though it is commonly used. But since one of its advantages is optical sectioning, TPLSM is often described in relation to confocal microscopy (Fig. 5). TPLSM provides all of the advantages of a conventional (one-photon) confocal microscope in fluorescence microscopy plus two features important to microbiological imaging. One feature is the absence of out-of-focus photobleaching. One-photon confocal microscopy excites the fluorophores throughout the depth of the specimen and then collects only the signal derived from the in-focus section. Thus, photobleaching occurs in all out-of-focus planes, and not just in the in-focus plane, and may be severe during the time it takes to collect a stack of optical sections. TPLSM, on the other hand, almost completely limits the excitation to the immediate axial region of the focal plane. This narrow excitation depth is brought about because the two-photon excitation rate is proportional to the square of the incident intensity (just as the rate of a chemical reaction, $2A + B \rightarrow C$, depends on the square of the concentration of A), which is high enough to produce fluorescence emission only where the waist of the excitation beam is smallest,

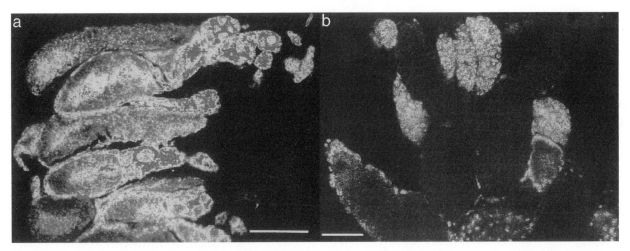

Fig. 5. Two-photon confocal microscopy. (a) DAPI-labeled ovary of uniparental *Aphytis yanonensis*. Microorganisms are evident in the young oocytes and possibly in the nurse cells. (b) DAPI-labeled ovary of biparental *Aphytis melinus*. Host nuclei are brightly stained, but no microorganisms were indicated (courtesy of Einat Zchori-Fein, from *Current Microbiology* **36**, 1–8).

i.e., at the plane of focus. Since the fluorescence excitation of all out-of-focus planes is avoided, no photobleaching occurs in these planes and, perhaps equally important, no pinhole detector is needed. A by-product of the narrow excitation depth is the high axial (z) resolution attained in TPLSM, equivalent to that of the best confocal microscope. Because photobleaching can be greatly reduced, TPLSM works well for dynamic imaging of fluorescence distributions in living cells.

The other feature of TPLSM of interest to microbiologists is the imaging of ion or other biochemical indicators that require ultraviolet excitation. For example, two photons of red light from the laser can excite an ultraviolet-absorbing fluorophore, e.g., Indo-1 or Fura-2, which are sensitive to the concentration of free Ca^{++}, which then emits blue fluorescence. A significant advantage is the avoidance of the need for UV optics or the poor chromatic corrections for most microscope lenses in the ultraviolet range. In the majority of cases, two-photon excitation occurs whenever there is one-photon absorption at a wavelength corresponding to half the wavelength of the excitation photons. This holds not only for the ultraviolet-excited fluorophores, such as Fura-2 or Indo-1, but also for the visible-excited fluorophores, such as fluorescein.

The instrumentation in TPLSM is different in two

aspects from LSCM—the laser and the detection system—and will be briefly described. Because of the large excitation intensities required for TPLSM, mode-locked lasers capable of generating intense, ultrashort pulses of red or infrared light less than 100 femtoseconds duration at high repetition rates (~100 MHz) are used. The use of ultrashort pulses and high repetition rates is essential to acquire scanned images in a reasonable amount of time, while not heating the specimen to "biologically intolerable" levels. Because of excellent optical sectioning by excitation alone, there is more flexibility in the design of TPLSM detectors, which, in turn, allows greater fluorescence collection efficiency compared to the LSCM. Since the resolution and optical sectioning are defined by the excitation alone, the emitted light does not have to be focused. The principal distinction in detectors is whether the emitted light passes back through the scanning mirrors, called descanned detection, or whether the detector can collect the emitted light from the entire image area at all times, called whole-area detection. Descanned detection relies on a pinhole that is several times larger than the optimal diameter used in LSCM and allows the use of detectors with small entrance apertures, such as avalanche photodiodes. This set-up is analogous to confocal detection and is sometimes termed "confocal TPLSM." In general, the pinhole cannot be made small enough to produce a substantial increase in lateral resolution, as is possible in a LSCM, because of a large drop in detection efficiency, due to the smaller diffraction-limited volume produced by the nonlinear optics employed in TPLSM. Whole-area detection uses a dichroic mirror between the scanner and the microscope to pass the light through a transfer lens that is positioned so the back aperture of the objective lens is conjugate to a PMT. One drawback of whole-area detection is contamination from ambient room light, which is eliminated in descanned detection by the pinhole. Instead of detecting with a PMT, it is possible to focus the light onto a CCD camera. This substitution is equivalent to wide-field (nonconfocal) fluorescence microscopy, but with only a thin slice rather than the whole thickness of the sample being excited. A variation of whole-area detection is called external detection because the emitted light bypasses the objective lens altogether. External detection is particularly useful in the ultraviolet range where many objective lenses become opaque.

Two-photon excitation can also be used with either 4Pi or Theta confocal fluorescence microscopies. In 4Pi-confocal microscopy, a second objective lens is placed opposite the first. This arrangement increases the effective aperture angle and, thus, provides enhanced resolution along the optical (z) axis. Two-photon excitation further enhances the resolution along this axis. In Theta confocal microscopy, the specimen is viewed along two different axes and the resolution in the axial direction is forced to be the same as the lateral resolution. When used in the two-photon fluorescence mode, Theta confocal microscopy can improve the axial resolution by approximately four times.

The application of TPLSM to microbiology has been demonstrated in several instances. In parasitic *Hymenoptera*, the vertical transmission of symbiotic bacteria, *Wolbachia*, from nurse cells to developing oocytes was documented using TPLSM. The usefulness of TPLSM in following the symbionts from the posterior pole to regions surrounding nuclei throughout the embryo in later stages of embryogenesis is an example of the potential of this technique for dynamic imaging. The ability to image the embryo through several cycles of division, because of TPLSM's reduced photodamaging mechanism, allowed results to be acquired from a single specimen that would otherwise require painstaking effort using many different specimens. Another application is the measurement of free calcium concentrations as a function of both space and time using the indicator Indo-1 in motile cells. Great interest is currently building for using two-photon excitation for photorelease ("uncaging") applications in microbiology. For example, "caged" fluorophores attached to drugs being developed against pathogenic bacteria can be activated (uncaged) by two-photon excitation and then followed to study 3-D diffusional dynamics.

See Also the Following Articles

Diagnostic Microbiology • Microscopy, Optical

Bibliography

Pawley, J. B. (1995). "Handbook of Biological Confocal Microscopy." Plenum Press, New York.

Shotton, D. (1993). "Electronic Light Microscopy." Wiley-Liss, New York.

Wilson, T. (1990). "Confocal Microscopy." Academic Press, San Diego, CA.

Microscopy, Electron

Susan F. Koval

University of Western Ontario, London, Ontario

Terrance J. Beveridge

University of Guelph, Guelph, Ontario

GLOSSARY

conventional thin sectioning A technique in which cells are chemically fixed, heavy metal-stained, dehydrated, embedded in plastic, and sectioned on an ultramicrotome for viewing by TEM.

electron energy loss spectroscopy (EELS) A technique in which the change in energy and wavelength of a transmitted electron can be detected and related to an element for compositional analysis.

electron spectroscopic imaging (ESI) A technique similar to EELS, but in which the electron energy lines are separated and used for imaging to produce a TEM image of the distribution of a desired element.

energy dispersive x-ray spectroscopy (EDS) A technique in which the x-rays which are emitted from the surface of a specimen are detected and separated into their unique energy lines, so that compositional analysis can be performed.

environmental scanning electron microscope (ESEM) A new type of SEM which allows the specimen to be viewed at partial pressure and humidity, so that its natural topography can be seen.

freeze-etching and fracturing A technique in which cells are rapidly frozen, fractured while frozen, etched under vacuum, and a platinum–carbon replica made. Especially good for viewing the insides of membranes.

freeze-substitution A technique similar to conventional thin sectioning except the cells are physically frozen to preserve structure and then chemically fixed at a very low temperature.

negative staining A technique in which a particulate sample is surrounded by a solution of a heavy metal salt and allowed to dry. Especially useful for small particles, such as viruses.

rotary-shadowing A technique in which a particulate sample is rotated during exposure to a directed metal vapor so that the metal builds up around the particles. Especially useful for viewing DNA.

scanning electron microscope (SEM) An electron microscope that is used to discern the topography of a sample through the emission of secondary electrons, once the specimen has been energized by an electron beam.

scanning transmission electron microscope (STEM) An electron microscope which combines many of the operational principles of both a SEM and a TEM. A transmitted image is formed.

selected area electron diffraction (SAED) A technique in which a TEM is converted into an electron diffractometer, so that the lattice spacings of small crystalline objects can be seen.

shadow-casting The same as rotary shadowing without specimen rotation. In this way, a metal vapor is deposited at an angle on a particulate specimen so that it can be seen by TEM.

transmission electron microscope (TEM) An electron microscope that produces an image from the electrons which have passed through a (thin) specimen. The image depends on the electrons which are scattered by the specimen.

LIGHT MICROSCOPY was essential to the discovery of microorganisms and was used during the first half of the twentieth century to study the shape and general makeup of these cells. With the introduction of the electron microscope in the 1940s, scientists were provided with an optical instrument which extended the range of observable structure far beyond the limits of those imposed by light optics. The visualization of the structure of extremely small particles was now possible (such as individual viruses and, even, individual cellular components from bacteria, fungi, and protozoa). Electron microscopy could also be used to further distinguish between the two basic types of cells: prokaryotes and eukaryotes. Of course, this distinction is a fundamental aspect of microbiology because it involves the study of both prokaryotic (Archaea and Bacteria) and eukaryotic (algae, protozoa, and fungi) unicellular microorganisms.

Currently, there is widespread interest in the isolation of microorganisms with new and unusual physiological capabilities, especially those microbes from so-called extreme environments. Size and form are not always useful characteristics in determining the taxonomic affiliation of a microorganism and electron microscopy must be done to determine cellular organization of a new isolate. Electron microscopy is also a tool for biochemists and molecular biologists who wish to determine the localization of expressed foreign gene products or to examine the ultrastructure of mutants. The association of microorganisms with various kinds of surfaces (both biological and inert) can also be studied by electron microscopy. Indeed, electron microscopy is one of the few techniques which can be used to visualize the macromolecular components of cells, their interdigitation into physical structures within cells, and their surface associations with either animate or inanimate interfaces.

I. TRANSMISSION ELECTRON MICROSCOPY (TEM)

A. Basic Principles

The essential design of an electron microscope requires a source of accelerated electrons, a high vacuum system, and a series of electromagnetic lenses. Physicists in the 1920s discovered that accelerated electrons in a vacuum travel in straight lines and have a wavelength thousands of times shorter than that of visible light. Because resolving power is attributable to wavelength, point-to-point resolutions of ~0.1 nm are possible with ideal specimens. Thus, the resolving power of a typical TEM is on the order of 1000 times greater than that for a light microscope. Because biological matter is often thick and friable (unstable) once in a TEM, resolutions of only ~1.0 nm are frequent with microbiological samples. Although lanthanum hexaboride (LaB_6) and field-emission sources are available, the source of electrons at the top of the instrument is, typically, a pointed tungsten filament, which is encased in a Wehnelt cylinder (commonly, the entire unit is called the "electron gun"). The electrons are accelerated by the anode (which, being at "ground," is much more positive than the filament) through a small hole in the center of the Wehnelt cylinder and down the microscope column.

Physicists also discovered in the 1920s that magnetic fields have an analogous effect on electrons to that of glass lenses on visible light. Thus, electron microscopes use electromagnetic lenses in the condenser and imaging systems. The condenser lens focuses the electron beam onto the specimen and is also used to control the brightness of the illumination. A proportion of these electrons interact with atoms within the specimen as they pass through and are scattered, thereby eventually forming an image. The objective lens produces an image of the specimen, formed by electrons which pass *through* the specimen (hence, the term *transmission* electron microscopy). A series of other lenses—usually, diffraction, intermediate, and projector lenses—are then used to provide a greater magnification of the image on a fluorescent screen. Because the entire electron path, from filament to screen, has to be under vacuum ($\sim 1 \times 10^{-6}$ torr; otherwise, the electrons would collide with air molecules and be absorbed), the final image has to be viewed through a thick glass window in the projection chamber. Usually, the image is recorded on photographic film because electrons have the same influence on certain photographic material as light. Some modern TEMs contain slow-scan CCD (charge-coupled device) cameras, which allow the easy digitization of an image, so that image enhancement by computers is made easier. But since these CCD systems must have exquisite resolution, they are very expensive. In electron microscopes with such digital imaging systems (or, sometimes, even TV cameras), the image can be also viewed on an external monitor. The need for a high vacuum system, coupled with the series of electromagnetic lenses that are needed to focus this type of radiation, results in a much larger, more complex, expensive instrument than the light microscope. An electron microscope requires a dedicated room to house it and one or more rooms to process specimens before they are visualized.

B. Specimen Preparation

1. Thin Sectioning

This technique is commonly used by microbiologists to view the internal structure of cells at relatively high resolution. Even though high accelerating voltages (i.e., 60–120 kV) are used in TEMs, specimens have to be very thin to allow the electrons to penetrate them so that an image is formed. For conventional thin sectioning, microorganisms must be chemically fixed (to cross-link and stabilize structures), washed (to remove unbound toxic fixatives), and dehydrated (for embedment in a water-incompatible plastic resin), so that the specimen can be solidified for the cutting of very thin sections (~60 nm thick). The initial fixation step is one of the most crucial steps in this process. The purpose of fixation is to preserve cellular structure with minimal alteration during the subsequent steps. Penetration of fixative by diffusion can be a problem with relatively large blocks of tissue, but specimen size is usually not a problem with unicellular microorganisms. Various fixation protocols are available, but a common method that works well with bacteria is sequential fixation in glutaraldehyde (to cross-link protein) and osmium tetroxide (to cross-link lipids). Because osmium has a high atomic number, it easily scatters electrons and, accordingly, osmium tetroxide acts as both a fixative and an electron dense stain. Cells can also be stained with uranyl acetate (uranium also has a high atomic number) prior to dehydration. It is best to "enrobe" the cells in purified agar, usually prior to the osmium tetroxide and uranyl acetate treatments, to consolidate the small cells so that they mimic a 1 mm³ block of tissue that can be easily transferred from solution to solution. This makes dehydration and plastic infiltration prior to embedding easier. The agar blocks containing chemically fixed cells are usually dehydrated in an

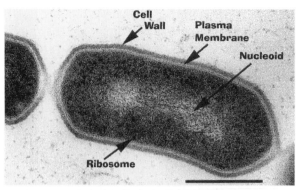

Fig. 1. Conventional thin section of the gram-positive bacterium, *Corynebacterium parvum*. Bar = 500 nm.

alcohol or acetone series and then embedded with a plastic resin in a small capsule. Epon 812 (an epoxy resin), LR White (a methacrylate), or Spurr's medium (a vinyl resin) are good embedding plastics for routine purposes. The cured or hardened blocks are trimmed and thin sections cut with a very sharp glass or diamond knife using an ultramicrotome. Sections are usually post-stained with uranyl acetate and lead citrate to enhance contrast before viewing. Figures 1 and 2 provide examples of conventional thin sections of gram-positive and gram-negative bacteria, respectively, where the general structural features of prokaryotic cells are clearly demonstrated, i.e., cell wall, plasma membrane, nucleoid, and ribosomes. Most bacteria with specialized physiology (e.g., phototrophs, nitrifiers, methylotrophs) usually have more sophisticated internal membrane systems (Fig. 3).

2. Negative Staining

This technique had its origins in the classical India ink stain used in light microscopy. The principle is to immerse the specimen in a thin aqueous film of a dilute heavy metal salt (the "negative stain") and then dry down the suspension on a suitably prepared copper TEM grid. Because the microorganisms are dispersed within the negative stain (which scatters electrons), they appear white against the dark background of the stain. Most material does not require chemical fixation prior to negative staining. Thus, this can be the quickest and simplest technique used in electron microscopy but neophytes can be assured that it takes experience and dedication. A variety of negative stains are available: ammonium molybdate,

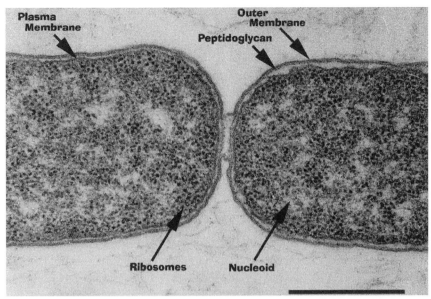

Fig. 2. Conventional thin section of the gram-negative bacterium, *Escherichia coli.* This shows a cell that is in the late stages of division. Bar = 500 nm.

uranyl acetate, and sodium or potassium phosphotungstate are commonly used.

Negative staining has been instrumental since 1959 in the elucidation of the structure of viruses, including bacteriophages (Fig. 4). Even though bacteria are the smallest of cells, they are too thick for a microscopist to gain much high-resolution information by negative staining. It is more useful to break up microorganisms and to visualize their subcellular particles, such as ribosomes, cell wall (Fig. 5) or membrane fragments, and isolated flagella (Fig. 6). Once these have been defined, it is sometimes possible to mentally piece the cell back together again. Even though bacteria are large by TEM standards, negative stains of whole cells do provide information on cell shape, placement of flagella, pili (fimbriae)

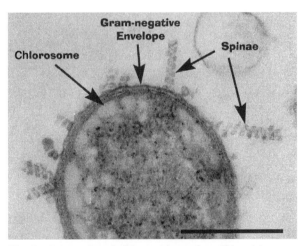

Fig. 3. Conventional thin section of a *Chlorobium* sp., showing the photosynthetic organelles (chlorosomes) and spinae emanating from the gram-negative cell wall. Bar = 500 nm.

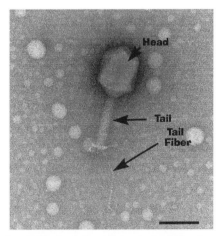

Fig. 4. A negatively stained T₂ bacteriophage from *E. coli,* which shows the icosahedral "head" attached to a "tail." Tail fibers are attached to the bottom of the tail. Bar = 50 nm.

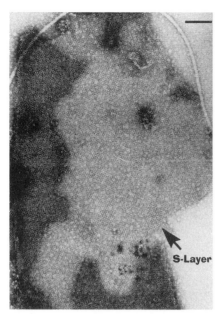

Fig. 5. Negatively stained image of an S-layer (a cell wall component) of *Desulfotomaculum luciae*, showing the tetragonal arrangement of protein subunits. Bar = 100 nm.

and spinae, presence of a capsule or S-layer (Fig. 5), and cell–cell associations

Although negative staining is one of the simplest and fastest electron microscopic techniques, patience is often required to obtain the best possible negatively stained image. The distribution of stain and particles on the support film may vary and several attempts may be required, using different combinations of staining and particle concentrations, before accurate TEM analyses can be made. In addition, care must

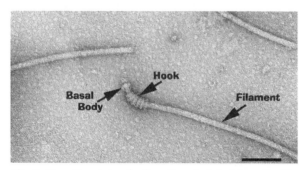

Fig. 6. Negative stain of a flagellum isolated from *Bacillus circulans*. Bar = 100 nm.

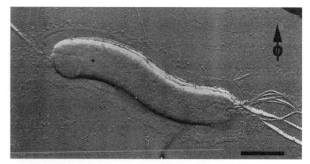

Fig. 7. Shadow-cast of an *Aquaspirillum metamorphum* cell, showing the tufts of flagella at each pole. Bar = 1 μm.

be taken in the interpretation of these images, as they are subject to artifacts produced by drying of the stain around the specimen, by the surface tension forces directed against cells during drying, by stain–specimen interactions, and by damage in the electron beam.

3. Shadow-Casting (Metal Shadowing)

This technique was one of the first used to reveal biological structure with the electron microscope. Shadow-casting involves the directed coating of a specimen at a specific angle with an electron-dense metal vapor (e.g., platinum or tungsten) while in a high vacuum. The metal builds up against the sides of the specimen (or contours on the specimen), whether they be cells or subcellular structures, providing enough electron density to form a faithful replica of the specimen's surface and, thus, is useful for determining topography (Fig. 7). Scanning electron microscopy (SEM) can also reveal the topography of surfaces, but usually at a lower resolution. When the metal is deposited at an angle (unidirectionally), the specimen height can be determined by measuring the shadow length and calculating the height, geometrically, from the known shadow angle. Specimens can also be metal shadowed on all sides by rotary-shadowing. This technique has had excellent success with thin fibers, such as nucleic acids (e.g. chromosomal and plasmid DNA and mRNA), since it artificially thickens them with the heavy metal deposits on their sides. Essential to these studies is the method of spreading the nucleic acids on the air:fluid interface (i.e., the Kleinschmidt technique) prior to transfer to a TEM grid and subsequent rotary-shadowing, usually with platinum or palladium.

4. Freeze-Etching

Freeze-etching provided an entirely different way of preserving the native structure of microorganisms. Instead of using the harsh toxic chemical fixatives that are used in conventional thin-sectioning, freeze-etching relies on the rapid freezing of cells to physically "fix" their structural attributes *in situ*. To do this, entirely new protocols and equipment had to be designed; the freezing had to be so rapid that crystalline ice was not formed. Typically, if normal refrigerator freezers are used, small particles within water nucleate the formation of ice crystals, which grow in size as freezing proceeds. Since microorganisms are small particles, they make excellent nucleating agents for ice crystal development; as these crystals grow, they shatter the cells and deform cellular structure. In the 1960s, rapid freezing for freeze-etching was performed by manually immersing small suspensions of microorganisms (attached to gold or copper planchets for rapid heat transfer) into Freon 22 held at liquid nitrogen temperature ($-196°C$). Freon has an excellent thermal transfer index and rapidly drains the microbial cells of their heat, freezing them so rapidly that an amorphous or vitreous ice is formed. This physical fixation is so good and the cells so naturally preserved that, when thawed, the microorganisms come back to life.

Rapid freezing alone is not enough manipulation to make the cells suitable for TEM observation. The cell pellet is much too thick and the cells are unstained. To overcome these problems, the frozen pellet is subjected to three separate manipulations within a freeze-etching device; freeze-fracture, freeze-etching, and replica formation. Here, the frozen pellet is kept at $-196°C$ while a high vacuum (1×10^{-6} torr) is generated around it. Next, the pellet is *freeze-fractured* and, since it is still frozen, the fracture plane follows the regions of least bond energy (i.e., the easiest route through the pellet); fractures occur over cell surfaces, through cell cytoplasms, and (most interestingly) through the hydrophobic domains of membrane bilayers. Since there are hundreds of microbial cells exposed during the fracturing of the pellet, the entire surface of the fracture exposes cells which have been cleaved through a full range of structural attributes. In this way, external and internal structures are exposed, but this fresh fracture surface is extremely flat and featureless. Now, the surface is *freeze-etched*. The pellet is heated to $-100°C$ and a "cold-finger" (cooled to $-196°C$) is held over the fracture surface for 30–120 s (actually, this cold-finger is the knife which was used to fracture the pellet). Because of the vacuum in the freeze-etching device and the temperature differential between the frozen pellet ($-100°C$) and the cold-finger ($-196°C$), the amorphous ice of the fracture surface sublimes onto the cold-finger. The structural attributes of the cells, because they are made of more solid materials, become exposed as the surface ice is etched away. Freeze-etching confers topography to the fractured surface. Structures made up of minute particles, such as bacterial S-layers (Fig. 8), require short etching times, whereas larger structures, such as extracellular polysaccharide matrices and whole cells, require longer periods (cf. Fig. 9).

Even though the sample has been fractured and etched, it is still too thick to be put into a TEM; a thin *replica* of the fractured, etched surface must be produced. For this, the surface is shadow-cast at a defined angle (usually with platinum vapor at an angle of 10–25°) and then entirely covered by a thin coating of carbon. (Platinum and carbon evaporators

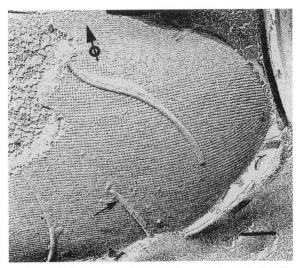

Fig. 8. Freeze-etching of the surface of *Aneurinibacillus thermoaerophilus*, showing its square-packed S-layer. Bar = 100 nm. (By permission of P. Messner, Zentrum für Ultrastrukturforschung, Universität für Bodenkultur, Vienna.)

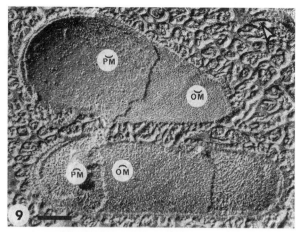

Fig. 9. Freeze-etching of two cells of *E. coli,* showing the concave and convex fractures that are possible through the outer (OM) and plasma membranes (PM). Bar = 100 nm.

are built into the freeze-etching device). This produces a thin, platinum-shadowed carbon replica (or mask) of the surface. Next, the vacuum is broken, the cell pellet melts and the replica floats onto the surface of distilled water in a small petri dish. Because cellular debris clings to the internal face of the replica, it is washed with concentrated sulfuric acid, bleach, and water before TEM observation. Since one of the preferential fracture planes occurs through lipid bilayers, freeze-etching is an excellent technique to study proteinaceous aggregates within cell membranes (Fig. 9).

5. Freeze-Substitution

It is well recognized that the use of chemical fixatives during the conventional thin-sectioning technique does not accurately preserve delicate structures. Freeze-etching with the use of rapid freezing provides a more accurate representation of microbial cells, but this technique produces relatively low-resolution replicas and is subject to the whim of where the fracture plane occurs. Several laboratories have developed a new technique over the last decade that combines the best attributes of both thin-sectioning and freeze-etching; it is called *freeze-substitution.*

Here, thin layers of microorganisms are rapidly frozen to obtain exquisite physical fixation. Since we can no longer use Freons, propane held at −196°C

is often the cryogen of choice. To ensure even more rapid freezing times, the microorganisms are "plunged" by specialized equipment into the cryogen so that vitreous ice is assured. Even more sophisticated equipment either "slams" the cells onto an ultracold, polished gold surface or "jet freezes" the sample between jets of super cooled gas. Once frozen, the specimen is put into a freeze-substitution mixture (held at −100°C) containing chemical fixatives (usually osmium tetroxide) and heavy metal stains (usually uranyl acetate) dissolved in an organic solvent (usually acetone) for 48 hr. Since the mixture is held at such a low temperature, the specimen remains well-frozen and preserved. Gradually, the cells become chemically fixed and heavy-metal stained. At the same time, the ice is substituted by the organic solvent and the cells are, therefore, dehydrated. (Often, a molecular sieve is added to the freeze-substitution mixture to entrap the water removed from the specimen.) Once the process is complete, the specimen can be warmed to room temperature, removed, and embedded in a suitable plastic, so that thin-sectioning can be done with an ultramicrotome. Freeze-substituted bacterial cells are much better preserved than their conventionally fixed counterparts. Their nucleoids intermingle with most of the cytoplasm, their ribosomes are larger (but often more difficult to distinguish), and their surfaces show more infrastructure (Fig. 10).

Other cryotechniques (such as the use of thin frozen films, cryo-ultramicrotomy, and low temperature plastics for embedding) have also been used on microorganisms with various degrees of success, but these are highly specialized methods requiring dedicated equipment and advanced expertise. These techniques will not be discussed.

II. SCANNING ELECTRON MICROSCOPY

A. Basic Principles

Scanning electron microscopy (SEM) is entirely different from TEM, even though they both use a high-energy, columnated beam of electrons. The SEM does not produce a transmitted image and that which is produced provides only a surface view of

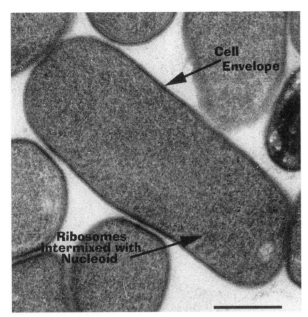

Fig. 10. Thin section of freeze-substituted *E. coli* cells, revealing that the nucleoid is woven throughout the cytoplasm and is intermixed with the ribosomes. The cell envelope consists of the outer and plasma membranes, and an electron-dense periplasm is sandwiched between them. Bar = 500 nm.

the specimen; this microscope is best at topographical imaging.

The SEM consists of an electron gun, 1–2 condenser lenses, and a specimen stage (or platform); all are held under high vacuum ($\sim 1 \times 10^{-6}$ torr). These components can be considered as similar to the upper one-half of a TEM; the electron gun emits electrons which are forced into a thin columnated beam by the condensers as the beam is directed towards the specimen. Accelerating voltages are typically at ~ 50 kV. In addition, the SEM column is equipped with a series of electron deflectors (called scanning coils), which can precisely control the location of the beam as it strikes the specimen. These are computer-controlled, so that the beam can be scanned back and forth so that a designated surface area of the specimen is illuminated in a raster. (This is similar to the manner in which the electron gun in a television forms an image on the TV screen.) As the electron beam strikes the specimen, it penetrates a few micrometers into the surface and (through a complex series of atomic interactions in-

volving orbital electrons) liberates low-energy secondary electrons. These are trapped by a secondary electron detector to produce a signal that is amplified and conveyed to a TV screen. The same computer which directs the scanning of the electron beam is also in charge of image production, so that the image is "painted" onto the screen simultaneously as the specimen is scanned. Since the image is ultimately derived from secondary electrons emitted from the specimen's surface, only topography can be discerned. The higher regions of the specimen flood the detector with many secondary electrons and are, therefore, "whiter" on the TV screen, whereas those electrons from the lower regions are not as easily detected and are darker. This produces a sense of 3-dimensionality for the SEM image (Fig. 11).

For a number of reasons, high resolution is difficult with an SEM. The diameter of the electron beam is important because its illumination point is the area from which the secondary electrons are derived. (Typically, the emission area for secondary electrons is $2\times$ greater than the electron beam illumination area. If the illumination point is 25 nm in diameter, the emission area will be ~ 50 nm, and this will be the nominal resolution of the SEM under these conditions.) When the electron beam strikes a sample, heat and electrical charge can build up, which will distort the image unless it is rapidly conducted away. Because biological matter is a notoriously poor conductor, distortion is a serious problem, especially when attempting to view small particles, such as

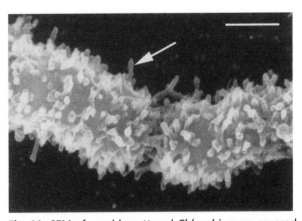

Fig. 11. SEM of a gold-sputtered *Chlorobium* sp. covered with special surface appendages called "spinae" (arrow). Bar = 500 nm.

microorganisms. A thin film of a conducting metal (usually gold) is often deposited (or "sputtered") onto a specimen's surface before viewing by SEM so that charge and heat is rapidly dissipated. This metal film is nanometers thick and also reduces resolution. In microbiology, SEMs are usually used to determine the shapes of microorganisms, their interaction with other cells or surfaces, and the arrangements of relatively large appendages such as flagella and spinae (Fig. 11).

Newer SEMs can be equipped with "field-emission" systems that are capable of directing extremely narrow (~0.5 nm), coherent beams of electrons onto specimens, so that higher resolutions are possible. There are also available low-voltage systems (~5–10 kV) that reduce the heating and charging problems of conventional SEMs, so that uncoated specimens can be more easily observed, thereby giving better resolution. Even cryospecimen stages can be obtained, which keep the specimen frozen during observation. Maybe the most exciting aspect of SEM is the recent production of environmental scanning electron microscopes (ESEMs) which, through a breakthrough in aperture and pumping designs, allows partial atmospheric pressures in the specimen chamber. For the first time, biological specimens can be viewed by ESEM under a humidity that ensures natural hydration. As one would expect, field-emission guns, low-voltage systems, and ESEMs are expensive and require expert operators; they are not frequently found in microbiology departments. But they are gradually being found in EM facilities that provide campuswide service or in regional centers; most of these are quite accessible to researchers with a worthy project.

B. Specimen Preparation

Unlike specimen preparation for TEM, that for SEM is relatively uncomplicated. As long as a researcher realizes that the microbial sample must be subjected to a high vacuum and the energy of an electron beam, suitable preparatory protocols are straightforward. A high vacuum dictates that the sample must be completely dry, and (as stated previously) that heating and charging by the electron beam are dissipated by the deposition of thin metal films on the specimen's surface.

1. Drying

Since microorganisms are comparatively small cells, the easiest way of drying is to leave thin films of cells open to the air. Unfortunately, this simple method subjects the cells to high surface tension as the surrounding fluid evaporates, which tends to flatten and distort them. Often, too, microorganisms are surrounded by capsules and extracellular polymers which, during drying, condense and aggregate on the cells, obscuring their surfaces. Appendages such as flagella and pili can also condense onto the surface and, in doing so, become distorted. Better methods of drying, which maintain the extension of appendages and external matrices are, therefore, required.

Freeze-drying is one of these. Here, cells are rapidly frozen (to reduce ice crystallization) and are put into a vacuum chamber, where the ice sublimes from them. Because water is not liquified during freeze-drying, surface tension does not collapse external structures. SEMs that are equipped with a cryo-stage can have the frozen (hydrated) material directly placed on the specimen stage and the vacuum pumps of the microscope can be used to freeze-dry the specimen.

Critical point-drying is another technique which avoids the problem of surface tension during the evaporation of a fluid. For this, the natural water in a specimen is gradually displaced with an organic solvent that is compatible with water (e.g., ethanol). Once the water has been removed by the solvent, the specimen is put into a critical point dryer that can withstand high pressure. The pressure is raised to a point that will convert gaseous CO_2 to liquid CO_2. Once the microorganisms are equilibrated to this pressure, the chamber is continually flushed with CO_2, until all ethanol is removed (ethanol and carbon dioxide dissolve in one another), and the cells are completely saturated in liquid CO_2. Now, as the pressure is gradually decreased, the liquid CO_2 (both inside and outside the cells) gradually enters its gaseous phase and bleeds off. Once all CO_2 has vaporized, the cells are completely dry.

2. Metal Film Deposition

Although thin metal films can be evaporated onto the surfaces of SEM specimens by the sophisticated evaporators used for shadow-casting or freeze-etching preparatory to TEM, a less expensive device is normally used. In this device, only a partial vacuum is drawn and a precious metal, such as gold, is heated to evaporation temperature, so that the vaporized metal completely coats the specimen. As the evaporation proceeds, there is a "spitting" sound and "sparking" is seen at the electrodes where the metal is vaporized; this technique is called "sputtering" (and the equipment, a sputtering device). Because gold is often the metal of choice, the SEM sample is now said to be gold sputtered. The thickness of the gold film can be followed by monitoring increased electrical conductance as the film is formed.

Sometimes, especially for cohesive films of microorganisms (e.g., biofilms), pretreatment of SEM specimens with osmium tetroxide can infiltrate them with enough metal for conductance so that gold sputtering is not necessary. When an ESEM is used, the samples remain hydrated and this water acts as a natural conducting agent; "charging" is not a problem. And, as previously mentioned for low-voltage SEM (LV-SEM), the same is true.

III. OTHER TYPES OF ELECTRON MICROSCOPY

This section is not meant to be all inclusive and the techniques mentioned are only described because they have been valuable at one time or another in the microscopy of microorganisms. These techniques are not easily available and they usually require dedicated equipment and expertise. Scanning probe microscopies (SPM), such as atomic force microscopy (AFM) and scanning tunneling microscopy (STM), have occasionally been used to decipher microbial structure (for more information, see Firtel and Beveridge, 1995); since these techniques are not considered to be bona fide electron microscopies, they will not be considered further.

1. Scanning Transmission Electron Microscopy (STEM)

STEM is a partnership between TEM and SEM. Scanning coils are inserted into a TEM column, so that the electron beam can be scanned over a thin specimen. The beam actually penetrates the specimen so that a weak transmitted image is produced, which must be amplified and displayed on a TV screen. As with SEM, the scanning of the electron beam must be synchronized with image production on the TV screen and resolution (even if it is a transmitted image) is still dependent on beam size. Dedicated STEMs with field emission systems can generate electron beams approaching 0.1 nm, but extremely thin specimens (less than 600 nm thick) are necessary to achieve resolutions below 1.0 nm. Accordingly, fragments of microorganisms are more useful for STEM than intact cells. Indeed, ribosomal subunits and enzyme complexes have been some of the particles best viewed. STEM's great advantage over TEM is that, as the specimen is scanned by the electron beam, only extremely small regions of the specimen are subjected to the illumination and only for small units of time; the energy load on the specimen is greatly reduced and the chances of viewing unaltered material substantially increased. But with price-tags of ~$1,000,000, dedicated STEMS are rarely encountered by microbiologists.

2. Electron Energy Loss Spectroscopy (EELS), Energy Dispersive X-ray Spectroscopy (EDS), and Electron Spectroscopic Imaging (ESI)

EELS and EDS are techniques which monitor certain signals emanating from the specimen as the electron beam passes through it in a TEM or STEM. For EELS, some of the electrons which pass through have lost energy, since they have strongly interacted with the constituent atoms of the specimen. The energy loss of each electron is directly equatable to the atomic makeup of the cells, so that elemental analysis can be done. EELS is especially useful for detecting relatively low atomic number elements such as nitrogen, oxygen, and carbon.

EDS monitors the spectrum of x-rays which are emitted from the specimen's surface as the electron

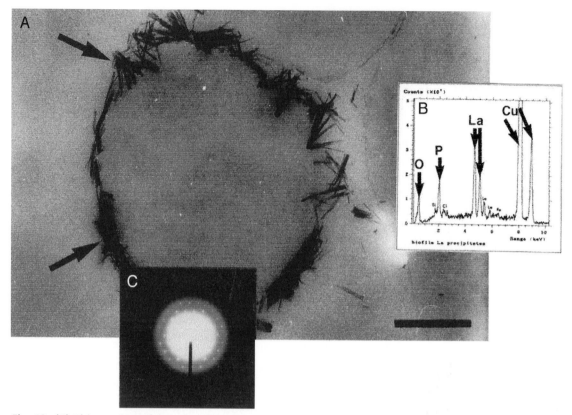

Fig. 12. (A) Thin section of *Pseudomonas aeruginosa* covered with a crystalline lanthanum mineral on its surface. Bar = 500 nm. (B) EDS spectrum of the minerals showing high lanthanum peaks (the copper peaks are due to the copper TEM grid). (C) SAED of a single lanthanum crystal, revealing diffraction spots equitable to the atomic lattice.

beam passes through the sample. Each x-ray is a quantum of energy that is given off whenever an electron passes so close to an atom that its atomic orbital electrons need to be reshuffled. Each x-ray energy line of an EDS spectrum is representative of an element in the specimen (Fig. 12A and B). EDS is particularly sensitive for detecting high atomic number elements and is useful for analyzing toxic heavy metals and minerals associated with microorganisms (Fig. 12A).

ESI, unlike EDS and EELS, is an actual imaging technique and it can exhibit the cell according to its elementary composition. An electron energy filter is built into (or under) the column of a TEM, so that the transmitted signal can be separated into distinct energy lines and distinct lines chosen for imaging. In this way, only those lines indicative of a certain element can be used to form the TEM image. For

example, if a magnetotactic bacterium is imaged and the energy lines for iron chosen, the magnetosomes can be clearly imaged since they contain magnetite (Fe_3O_4). For isolated ribosomal subunits, the rRNA can be seen by following the phosphorus energy lines, since nucleic acids are phosphate-rich.

3. Selected Area Electron Diffraction (SAED)

SAED is a technique which utilizes a TEM as a diffractometer. This is done by turning off all of the imaging lenses in the microscope except the diffraction lens (under normal operation, this is one of the magnifying lenses). If the specimen possesses regions of crystallinity, each will act as a diffraction grating and will diffract the electron beam. Exact spacings can be calculated because there is an inverse relationship, which follows Bragg's Law, between the diffraction spacing and the crystal lattice spacing (i.e., the larger the diffraction spacing, the smaller the crystal lattice). SAED can be used to determine subunit spacings in bacterial S-layers and the intramolecular folding of the S-protein monomers which make up each subunit (Stewart *et al.,* 1985). But, admittedly, SAED analysis of organic crystals is difficult because of their inherent friability during the long beam-exposure times required for diffraction; too much mass is lost from the crystals during diffraction to keep the crystal lattice intact. For this reason, SAED is a technique that is better used on small inorganic crystalline objects, such as fine-grained minerals which are associated with microorganisms (Fig. 12c). Certainly, other diffraction techniques are available to researchers, such as X-ray diffraction and neutron diffraction, but these require large macroscopic crystals or large volumes of suspended particles. SAED is virtually the only technique which can study small individual crystalline particles (~100–250 nm in diameter) and this makes it especially useful when studying microorganisms. Both EDS and SAED have been instrumental in elucidating how bacterial surfaces nucleate the development of fine-grained minerals and have, therefore, helped define the new scientific field of microbial biogeochemistry.

See Also the Following Articles

BACTERIOPHAGES • CRYSTALLINE BACTERIAL CELL SURFACE LAYERS • EXTREMOPHILES • FREEZE-DRYING OF MICROORGANISMS

Bibliography

Beveridge, T. J. (1998). The ultrastructure of gram-positive cell walls. *In* "Gram-Positive Pathogens" (V. Fischetti, R. Novick, J. Ferretti, D. Portnoy, and J. Rood, eds.), pp. 3–10. American Society for Microbiology, Washington, DC.

Beveridge, T. J. (1995). The periplasmic space and the concept of periplasm in gram-positive and gram-negative bacteria. *ASM News* **61**, 125–130.

Beveridge, T. J. (1989). The structure of bacteria. *In* "Bacteria in Nature: A Treatise on the Interaction of Bacteria and their Habitats," Vol. 3 (E.R. Leadbetter and J.S. Poindexter, eds.), pp. 1–65. Plenum Pub. Co., New York.

Beveridge, T. J., and Graham, L. L. (1991). Surface layers of bacteria. *Microbiol. Rev.* **55**, 684–705.

Beveridge, T. J., Popkin, T. J., and Cole, R. M. (1994). Electron Microscopy. *In* "Methods for General and Molecular Bacteriology" (P. Gerhardt, ed.) pp. 42–71. American Society for Microbiology, Washington, DC.

Beveridge, T. J., Hughes, M. N., Lee, H., Leung, K., Poole, R. K., Savvaidis, I., Silver, S., and Trevors, J. T. (1996). Metal-microbe interactions: Contemporary approaches. *Adv. Microbial Physiol.* **38**, 178–243.

Firtel, M. and Beveridge, T. J. (1995). Scanning probe microscopy in microbiology. *Micron* **26**, 347–362.

Firtel, M., Southam, G., and Beveridge, T. J. Electron microscopy. Protocols. *In* "Archaea—A Laboratory Manual" (K. R. Sowers, ed.), pp. 123–140. Cold Spring Harbor Laboratory Press, NY.

Hayat, M. A. (1986). "Basic Techniques for Transmission Electron Microscopy," pp. 1–411. Academic Press, New York.

Holt, S. C., and Beveridge, T. J. (1982). Electron microscopy: Its development and application to microbiology. *Can. J. Microbiol.* **28**, 1–53.

Paul, T. R., Graham, L. L., and Beveridge, T. J. (1993). Freeze-substitution of medically-important bacteria. *Rev. Med. Microbiol.* **4**, 65–72.

Robards, A. W., and Sleytr, U. B. (1985). "Low Temperature Methods in Biological Electron Microscopy." Elsevier, NY.

Sprott, G. D., and Beveridge, T. J. (1993). Microscopy. *In* "Methanogenesis: Ecology, Physiology, Biochemistry and Genetics" (J. G. Ferry, ed.), pp. 81–127. Chapman & Hall, New York.

Stewart, M., Beveridge, T. J., and Sprott, G. D. (1985). Crystalline order to high resolution in the sheath of *Methanospirillum hungatei*. *J. Mol. Biol.* **183**, 509–515.

Microscopy, Optical

Guy A. Perkins and Terrence G. Frey

San Diego State University

GLOSSARY

aperture iris A diaphragm that can be adjusted to vary the diameter of an aperture to regulate the amount of light admitted to a lens.

birefringent structures Two refractive indices at right angles to one another in a material.

condenser lens The lens in a lens system that gathers light from the source and directs it onto an object.

depth of field The slice of the specimen that contributes to the in-focus image at a particular focus level.

diffraction The spatial distribution of waves that are scattered from an object.

epi-illumination The illumination produced when the specimen is illuminated from above by light emerging from the objective lens.

fixed specimen A specimen preserved to minimize swelling, shrinkage, evaporation, decay and all forms of distortion of cellular components. Fixation is usually done by chemical action to stabilize or fix the cell contents so that they may be subjected to further treatment that would otherwise damage them.

fluorescence The property of absorbing light and after a delay of less than one-millionth of a second reemiting light of a longer wavelength (lower energy).

image contrast The property allowing one to distinguish various components of the structure of the object by different intensity levels in the image.

Kohler illumination The illumination used with light sources of irregular form or brightness, hence with most conventional optical microscopes. The image of the light filament is made large enough to fill the aperture iris and is focused on the condenser lens, which is focused so that the image of the aperture iris is in focus with the specimen.

long-pass filter An optical filter that blocks wavelengths shorter than a characteristic cut-off wavelength and transmits longer wavelengths.

objective lens The lens in a lens system that is closet to the object that forms an image of the object.

ocular lens The eyepiece lens of an optical microscope.

optical path The product of the distance traveled by a ray of light and the index of refraction of the various materials it passes through.

optical sectioning Imaging sections of the specimen by stepping the focal plane through a specimen and recording 2D images on each focal plane.

phase The particular stage in the periodic cycle of a light wave with respect to a specific origin. The other component of a light wave is amplitude.

photobleaching Extinguishing the fluorescence by the photochemical action of light incident on the specimen chemically, altering the fluorophore.

point spread function (PSF) The relationship between a point object and the image of it formed by the objective lens of the microscope; simply put, the image-forming characteristics of an objective lens.

polarized light Light affected by an electric field, in the direction of a plane for plane-polarized light and rotating in a circle for circularly polarized light.

refraction The change in the direction of propagation of a wave when it passes from one material to another in which the index of refraction is different.

refractive index (index of refraction) The effect of a

material object on the incident wave, defined as the ratio of the velocity of light in a vacuum to that in the object.

resolution   The ability to distinguish objects that are close together and defined as the minimum distance between points or parts of the object that appear as distinct points in the image.

short-pass filter   An optical filter that blocks wavelengths longer than a characteristic cut-off wavelength and transmits shorter wavelengths.

OPTICAL MICROSCOPES   all operate on the same basic principle: Light is projected toward an object, such as a microorganism, and is transmitted through the object with varying amounts of scattering and absorption or is reflected off the object. The scattered, absorbed, or reflected light can be collected by a lens or set of lenses to form an image that is detected with a sensing device such as the eye, photographic film, or a video system. The image produced by the interaction of this light with the object may reveal size, shape, form, and other structural characteristics such as compartmentalization, distribution, and colocalization of subcellular components. Because of the nature of microbiology, the optical microscope, also called the light microscope, has been of crucial importance to the reliable identification and understanding of microorganisms. Until the late seventeenth century when van Leeuwenhoek constructed the optical microscope, the presence of microorganisms smaller than the human eye could see was only a matter of speculation. Since then, much of what is currently known about these organisms has been discovered with various microscopes.

I. INTRODUCTION

In all optical microscopes, the formation of an image results from the interference of direct and diffracted light. Most microscopes exploit this interference to improve image contrast. When one is confronted with the need to use an optical microscope to identify or characterize a microorganism, the first question to ask is, "How does one choose the type of microscope that best allows for the visualization of what one wants to see?" Subsequent questions might be, "Can I view live specimens, or must they be fixed? Will I need to use a stain, label, or some other marker? Can I see different features by using complementary types of microscopes?" To answer these questions one must not only be aware of the different types of optical microscopes, but one must also understand the distinctive images produced by each, which may yield different aspects of microbial morphology. Hence, it is good practice to use at least two different types of optical microscopy, where possible, when examining any specimen. The advantages and limitations of the six most commonly used types of optical microscopes—bright-field, dark-field, phase-contrast, Nomarski, polarization and fluorescence—will be described along with the applications to microbiology that illustrate the power of each technique. In addition, advances in optical techniques, such as video microscopy, deconvolution microscopy and near-field scanning microscopy, will be briefly treated here.

II. BRIGHT-FIELD MICROSCOPY

The most common optical microscope is the bright-field microscope (Fig. 1), named because the image of an object is dark against a brighter background. The bright-field microscope is a compound microscope because it is a multilens system. Kohler illumination of the specimen is the basic mode of operation and is sufficient for many microbial specimens as long as they are stained or contain inherent color or contrast. This illumination is achieved with the condenser lenses and its purpose is, first, to achieve an evenly lit field of view against which the specimen can be recognized and, second, to obtain the maximum resolution of fine detail by having a wide cone of radiation. Kohler illumination is produced by first positioning a lens in front of the light source so that its image is not in the specimen plane. A second condenser lens places an image of the surface of the first lens on the specimen with as short a focal length as possible to obtain as wide a cone of light as possible to illuminate the specimen. An aperture iris is located between the light source and the condenser lenses to control the amount of light entering the condenser system. Improved contrast

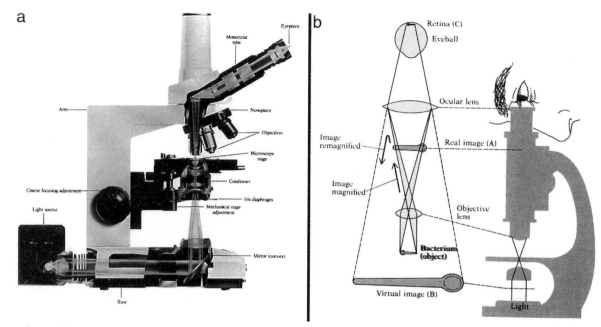

Fig. 1. (a) Compound bright-field microscope. The main parts are labeled and the cut-away shows the optical path (courtesy of Nikon, Inc., Rochester, NY). (b) Schematic of the image-formation properties of a bright-field microscope. Light passes through the objective lens forming an inverted real image (A). The ocular lens remagnifies the image and forms a virtual image (B). The lens system of the eye forms an image of this virtual image (C) (courtesy of Edward Alcamo, "Fundamentals of Microbiology," 5th ed. (1997). Benjamin/Cummings, Menlo Park, CA).

can be achieved by closing down the iris, and the temptation exists to do this with low-contrast specimens. However, a loss of fine-detail resolution occurs, and as the aperture becomes smaller spurious boundaries develop around fine features as a result of diffraction phenomena. After passing through the specimen, the light is collected by the objective lens to form a magnified image, which is further magnified by the ocular lens.

It is important to realize that bright-field microscopy, as do all the microscopies discussed here, allows for the observation of a magnified image of the specimen and not the specimen itself. The image formed by the objective lens is called the "real image" because it can be projected onto a screen. This image is not seen by the observer. The real image is the object for the ocular lens, which magnifies it to create a "virtual image" that can be seen only by the observer. The total magnification is the product of the objective-lens and ocular-lens magnifications. The

highest magnifying objective lens is usually 100× and the ocular lens typically has a magnification of 10×. Thus, the maximum magnification attained is usually 1000×, which is sufficient to discern the finest details that can be resolved in the light microscope.

When visualizing microbes, it is important to understand how the microscope resolution governs the detail of what can be seen and what the maximum achievable resolution is. The minimum distance (d) between two objects that can be resolved as distinct objects by the lens system is governed by this equation

$$d = 0.5\lambda/\text{N.A.},$$

where λ is the wavelength of light used for illumination and N.A. is the numerical aperture, which is characteristic of the objective lens and the medium between the objective lens and the specimen, usually air or immersion oil. N.A. $= n \sin(\phi)$, where

n is the index of refraction of this medium and ϕ is the half-angle of the cone of light. Higher resolution can be achieved with immersion oil because it has a higher refractive index (~ 1.5, which is the same for glass) than air, which means that many light rays that would not enter the objective lens in air due to reflection and refraction at the surfaces of the objective lens and slide will now enter. The resolution limit for normal operation of optical microscopes is defined for light with the shortest wavelength at the blue end of the visible spectrum, about 500 nm, and with an oil-immersion objective with a numerical aperture of about 1.25. With these values in the equation, $d = 200$ nm or 0.2 μm. For example, if two bacterial cells are separated by 0.2 μm, they can be distinguished; but if they are closer than 0.2 μm, they will appear as blurred single cell. A common misconception that has propagated through textbooks for a number of years is that particles smaller than the resolution limit cannot be visualized. Individual particles can indeed be visualized, often with the aid of a label, but their size will be exaggerated by the point-spread function (PSF) of the objective, and clumps of these particles may not be separately resolvable. For example, distributions of viruses, which have sizes well below the resolution limit of optical microscopes, can be easily observed when they are fluorescently labeled. Under bright-field microscopy, most microbes are invisible because they don't have high concentrations of molecules that absorb visible light. This is also true of living single cells or cell monolayers, which means that they cannot alter the incident light by absorption enough to produce much contrast. There are two ways to improve the low contrast. One way is to use simple or differential staining. For example, differential staining can make bacterial structures such as capsules, endospores, and flagella visible. The second way is to use different optical modalities that take advantage of differences in optical density or refractive index inside the specimen. Examples are the phase-contrast microscope and the Nomarski microscope (see Sections IV and V). For bright-field microscopy, slightly closing the condenser aperture may

aid in the detection of certain microbes such as fungi and protozoa because the light will be incident on the edges of the object at a sharper angle, which increases the contrast.

III. DARK-FIELD MICROSCOPY

In dark-field microscopy oblique illumination is used. With no specimen present, no light from the condenser reaches the objective-lens system, producing a dark image. However, if a specimen is placed in the path of this oblique illumination, its reflective structures will cause the incident light to be reflected at all angles. Some of this light will reach the objective and so will appear bright. In practice, the light below the condenser lens system is blocked in a central circle so that only light from the outer ring impinges on the object at a sharp angle. Thus, undeviated rays are not collected by the objective lens, and only light that has been reflected or refracted by the object is collected by the objective lens and used to form an image (see Figs. 2a and 2b). Hence, the background field surrounding the object appears black, whereas the object appears bright. Because the dark-field image is an image produced by deviated light, it is in effect a map of the discontinuities of refractive index and other light-diffracting features of the object. Considerable internal structure is often visible inside living unstained cells and microorganisms by dark-field microscopy. Moreover, dark-field microscopy is the preferred technique in the diagnosis caused by spiral bacteria, such as the distinctively shaped *Treponema pallidum*, which causes syphilis, because these organisms are small and thin (~ 0.15 μm in diameter) and do not stain well, and so cannot be visualized with bright-field microscopy. Furthermore, observation of their characteristic motility is important for presumptive identification. Even substructures much smaller than the resolving power of the microscope can be observed with dark-field microscopy. For example, bacterial flagella, which are only ~ 0.02 μm in diameter can be visualized because they diffract enough light for their outline to be observed; however, their size appears larger because of the effects of the PSF.

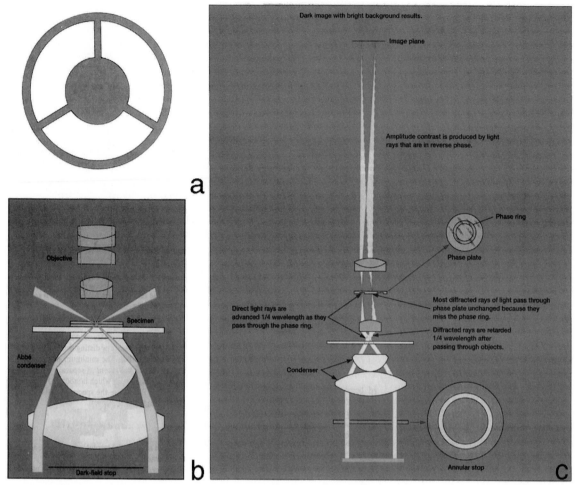

Fig. 2. Dark-field microscopy. A bright-field microscope can be converted into a dark-field microscope by placing a beam stop (a) underneath the condenser lens system (b). A hollow cone of light is produced by the condenser so that only light that has been reflected or refracted by the specimen enters the objective. (c) Phase-contrast microscopy (courtesy of Lansing Prescott, "Microbiology," 2nd ed., (1993). Wm. C. Brown, Dubuque, IA with permission of the McGraw-Hill Companies).

IV. PHASE-CONTRAST MICROSCOPY

If the specimen has insufficient contrast for bright-field microscopy, as is the case with unstained micro-organisms, then phase-contrast microscopy may be a good choice. These types of specimens can be described as phase objects because they do not significantly reduce the transmitted light intensity; that is, they do not absorb much light, yet they differ in refractive index from the background. Hence, the transmitted beam can be physically separated according to differences in the optical path traversed through the specimen as compared to the back-ground. Interference between the reunited components of the beam that passed through the specimen and through the background produces detectable differences in intensity in the image (see Fig. 2c). The image produced is effectively a map of the optical-path differences between the features in the specimen and between the specimen and the background.

The condenser of the phase-contrast microscope has an opaque beam stop in it, surrounded by a thin transparent ring. This ring lets pass a hollow cone of light that is refracted to different extents by variations in the refractive index the specimen. The phase retardation produced by biological material is ~1/4

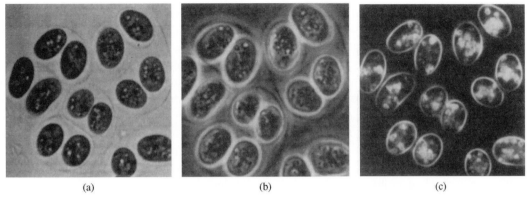

Fig. 3. Visualization of the cyanobacterium *Gleocapsa* using three types of optical microscopy: (a) bright-field microscopy, (b) phase-contrast microscopy, and (c) dark-field microscopy. Each type of microscopy affords a different view of the internal structure (courtesy of Edward Alcamo, "Fundamentals of Microbiology," 5th ed. (1997). Benjamin/Cummings, Menlo Park, CA).

wavelength. This diffracted wave is again retarded by ~1/4 wavelength when it strikes a phase plate in the objective-lens system. Thus, the wave diffracted by the specimen and the incident light that missed the specimen will be ~1/2 wavelength out of phase and so will interfere destructively. The combination of the two light waves produces an image that resembles one produced by absorption contrast (i.e., like bright-field microscopy). The background is bright and the phase object appears dark and well-contrasted (see Fig. 3). The contrast is further heightened through the use of color filters, green being the most common because the human eye is most

sensitive to this wavelength. Phase-contrast microscopy has remarkable sensitivity, capable of distinguishing differences of 7/360 of a wavelength. However, it has a disadvantage in that the image of each feature is surrounded by a halo of bright light, and this halo degrades the resolution of small objects. Phase-contrast can be useful for weakly stained specimens (see Fig. 4), but it is not suitable for heavily stained or thick specimens because the interference from multiple diffractions is uninterpretable.

The uses of phase-contrast microscopy in microbiology are many and varied. Individual bacteria can be easily observed and quantified, and their move-

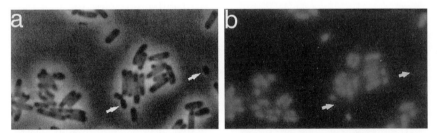

Fig. 4. Combination phase-contrast and fluorescence microscopy used to find anucleate cells. (a) *Escherichia coli* cells deficient in RecA/RecD proteins were stained with the fluorescent dye DAPI, which binds specifically to DNA. The nucleoids (chromosomal DNA) shine brightly with light-blue fluorescence when observed in a fluorescence microscope. In a system with dual phase-contrast and fluorescence imaging capabilities, the nucleoids are light blue and the other cytoplasmic portions are dark blue, which allows the nucleoid to be clearly visible in the cell shape. A small portion of these cells is anucleate (arrows), which is not visible in (b), which shows the same field in fluorescent mode only (courtesy of Judith Zyskind, (1992). *Microbiol. Molec.* **6**, 2525–2537).

ment can be studied. Nuclei and endospores can be measured for both *z*-dimension and refractive index, which provide insight into their density. The fine structures of yeast, protozoa, and molds can be studied. Because the refractive indices of protein solutions are related to their concentrations, the phase-contrast microscope can map the concentration or mass of proteins inside the cytoplasm or other subcellular compartments. Phase-contrast microscopes can also detect inclusion bodies containing sulfur, poly β-hydroxybutyrate, or polymetaphosphate because the refractive indices of these compounds are quite different from that of water.

V. NOMARSKI MICROSCOPY

Nomarski or differential interference contrast (DIC) microscopy is superb for looking at unstained living cells because of the high contrast produced along the edges of biological structures, such as the membranes and cell walls of microbes (see Fig. 5). In a Nomarski microscope, the specimen is illuminated with two beams of plane-polarized light, which differ by 90° in their polarization planes. One beam is called the measuring beam and the other is called

Fig. 5. Nomarski microscopy of oocysts of *Cryptosporidium parvum*. This microbe is an enteric apicomplexan capable of causing moderate to severe diarrhea in immunocompetent individuals and life-threatening disease in immunocompromised individuals. Each oocyst is ~5.2 ×4.6 μm in dimension (courtesy of S. J. Upton, Kansas State University).

the comparison beam. After passing through the objective, the measuring and comparison beams are displaced laterally relative to each other by a small amount. The distance between both beams is kept at about the resolution limit of the microscope, although it is sometimes set below the resolution limit because the interference pattern can be detected at higher resolution. The beam passing through a given point of the object interferes with a beam passing through in the immediate vicinity. As a result, local differences in optical thickness are visualized (hence the name differential interference contrast). The gradients of optical-path differences are made visible largely by the shear effect on one of the beams at the boundary of discrete biological structures, such as the cell membrane. Hence, as with phase-contrast microscopy, Nomarski microscopy detects abrupt changes of refractive index to provide increased contrast. The images produced with a Nomarski microscope appear three-dimensional because one side of a boundary appears lighter than the other side, as if light were falling on it and casting shadows. It is important to understand that the suggested heights and depths due to this three-dimensional effect are simply representations of the gradient of optical-path differences through the specimen, which may or may not correspond with real relief structure. With Nomarski microscopy, it is easy to be deceived into believing something is real that is not. In other words, it is not possible to say to what extent these optical-path differences are due to a longer path traversed by one beam or a higher refractive index, or a combination of the two. It is advisable to always focus up and down repeatedly to determine the relationship of one structure to another.

Because the comparison beam is not a constant reading of the background, but rather a heterogeneous bundle of light that has passed through the entire specimen, no measurement of optical-path differences is possible. However, there exists a different interference microscope, called the Jamin–Lebedeff interference microscope, which can quantify optical-path differences. In the Jamin–Lebedeff microscope, the comparison beam is made to pass around the specimen before interfering with the measuring beam. By measuring the optical-path difference, such parameters as film thickness or mass of a nucleus, even in a living cell, can be calculated. It is even possible to calculate concentrations and mass of small components, such as a nucleic acids inside nuclei.

Although the Nomarski microscope does not produce halos around structures, as seen with the phase-contrast microscope, it is inferior when looking at thin objects because certain fine details are not seen as sharply and distinctly, due to rather small optical-path differences produced in thin specimens. Hence, phase-contrast microscopy is preferred when observing wet mounts of most bacteria. On the other hand, Nomarski microscopy can be superior to phase-contrast microscopy with thick specimens because optical sectioning can be achieved, which is also one of the attractive features of confocal microscopy. For example, real-time optical sectioning with a Nomarski microscope can allow for the observation of spiral chloroplast movement in filamentous algae. Unlike with phase-contrast microscopy, weakly stained specimens can be beautifully contrasted with Nomarski microscopy, because a certain amount of absorption and interference effects can add to the interference pattern, which is immune from the light-scattering problems this sort of specimen causes with phase-contrast modalities. Sometimes, rotating one of the polarizing prisms can enforce only those contrasting elements that arise by local differences in absorption. This technique has been applied to weakly stained histochemical preparations and has resulted in a relatively high-resolution mapping.

VI. POLARIZATION MICROSCOPY

Palarized-light microscopy uses either plane- or circular-polarized light to create images that map changes in the polarization of light or changes in the amount of polarized light transmitted or scattered by a specimen. Its use in microbiology is mainly focused on birefringent structures, which indicate the presence of some form of molecular order in the specimen. Examples of birefringent structures are chloroplasts, mitotic spindles, collagen, elastin, myofibrils, myelin sheaths, starch, and cellulose.

In addition to the features found in a conventional bright-field microscope, a typical polarization microscope has two linear polarizers, one placed in the substage condensing system and another placed above the objective-lens system. Each linear polarizer transmits plane-polarized light, and for routine purposes they are adjusted so that the plane-polarized light passing through each is mutually perpendicular or crossed. By using crossed polarizers the background intensity is suppressed to virtual darkness; that is, in the absence of a birefringent object, no light is transmitted through the top polarizer. Images of birefringent objects are high-contrast, brilliant, and often in an array of colors that provides information on the relative orientation of the ordered objects. It is sometimes helpful to rotate the polarizers slightly from the crossed position to make the non-birefringent material visible by absorption. This is a simple mechanism for locating the birefringent structures in a stained specimen without totally extinguishing the illumination of the amorphous parts of the specimen, allowing for a better mapping of birefringence relative to the overall view of the specimen. In this way, large-scale polarizaion studies of the quantitative and qualitative changes in the mitotic spindle during mitotic division in different cell types were performed. Similarly, rearrangements of chromatin in cells undergoing differentiation were visualized. An exciting application is the combination of well-characterized dyes with differential polarization microscopy to examine the arrangement of proteins, lipids, polysaccharides, nucleic acids, and porphyrins in cells. An example is the use of specific stains to monitor changes in RNA structure that can be differentiated from those of DNA *in situ* in a differential polarization field of view.

VII. FLUORESCENCE MICROSCOPY

The optical microscopes considered thus far create an image from light that passes through a specimen. An object can also be seen because it emits light. Fluorescence microscopy takes advantage of the imaging possibilities afforded by the fluorescent light emitted by these objects and has emerged as the most used type of optical microscopy for identification of many microorganisms in both diagnostic and research laboratories. The importance of epifluorescence imaging in microbiology has vastly increased because of the provision of a multitude of highly specific fluorescent probes. Moreover, it is possible to obtain fluorescence images noninvasively at light levels that are not damaging to living cells. Fluorescent molecules are called fluorophores or fluors and emit light in a specific, narrow band of wavelength when excited by incident light of shorter wavelength (higher energy).

To achieve its full potential, the fluorescence microscope must fulfill two principal objectives. First, the fluorophore in the specimen must be excited effectively. This exciation is best accomplished with wavelengths close to the fluorophore's absorption peak. Second, the microscope's lens system must collect as much of the relatively weak fluorescent light as possible. Five main components are essential for the observation of fluorescence—the light source, excitation and emission filters, fluorophores, lenses, and a fluorescence-detecting unit. The fluorescence can be observed visually through the microscope's ocular and photographed or recorded with a photomultiplier or digital camera. Commonly, the light source is a mercury vapor arc lamp, which produces an intense beam in the ultraviolet and visible spectrum. The most common microscope arrangement is dark-field, which provides a black background against which the fluorescent objects are brightly lit, with the color dependent on the fluorophore used. The image quality depends largely on the image brightness and contrast, as well as on the microscope optics. The contrast is defined as the ratio of fluorophore emission to the background light. Optimal image contrast, however, may result in a dim image due, for example, to narrow-band excitation filters that have low-transmission signatures. Thus, in most applications, a compromise is made between the image contrast and the level of background fluorescence permitted in order to allow easy visualization and measurement, and for photography with reasonably short exposure times.

Fluorescence microscopy universally uses epi-illumination, which has the advantages that unabsorbed light is directed away from the observer or recording system, and that the alignment is simplified because the objective lens acts as its own condenser (i.e., the same lens system acts as objective and con-

denser). Consequently, focusing the lens system on the specimen results in the correct alignment of this lens. Because the illuminated field is the field of view, the observer can be aware of which areas of the specimen may be undergoing photobleaching. Moreover, epi-illumination permits a straightforward changeover between the fluorescence configuration and the transmitted-light configuration, which is useful when one wants to obtain an overview of the specimen with phase-contrast microscopy in combination with immunofluorescence microscopy.

To observe fluorescence, it is necessary to prevent the exciting light from mixing with the emitted fluorescent light that forms the image. By using filters, one can select an appropriate wavelength for the excitation of the fluorophore and a second filter can eliminate light of wavelengths other than that emitted by the fluorophore, which will always emit photons of longer wavelength (lower energy) than absorbed. The second filter is very important, as it removes the much more intense excitation light, so that the emitted signal is detected in the absence of the excitation light and other stray light. Fluorescence microscopy has extreme sensitivity to the rather low intensities of fluorescent light, specific detection of one among many fluorophores that may be present, and an excellent signal-to-noise ratio. Because a specimen can be stained with a finite quantity of fluorescent dye, which decays or photobleaches when illuminated, the microscope must be configured to efficiently excite and detect the limited number of emitted photons before complete bleaching has occurred. For this reason, the filters for excitation and emission should be carefully chosen for the application of interest. For example, the choice of filters depends on the quantum yield of the fluorophore (i.e., the ratio of emitted light to light absorbed) and on its spectral characteristics, such as its emission peak and the broadness of its emission band. Band-pass filters, which transmit one particular region (band) of the light spectrum with high transmission (~90%) are normally used to filter the excitation light, while a long-pass filter is commonly used as an emission filter. In order to extend the range of observation to two-color fluorescence, a short-pass filter is added for sequential observation of two fluorophores in the same specimen. The long-pass filter is chosen to pass the fluorescence emission from one fluorophore with a characteristic peak wavelength and block the shorter-wavelength emission from the second fluorophore. The short-pass filter is chosen to pass the emission from the second fluorophore with shorter characteristic-peak wavelength while blocking the emission from the first, if minimal overlap of the characteristic spectral bands for the two fluorophores exists.

Microorganisms can be studied by fluorescence microscopy in five ways.

1. Some naturally occurring substances such as albumen, cellulose, starch, and many fats fluoresce. Naturally occurring fluorescence is called primary fluorescence.

2. Microbes can be coated with fluorophores, one of the most common being fluorescein, which causes them to fluoresce. This is an example of secondary fluorescence.

3. Fluorophores can be inserted inside cells, where they bind to specific structures. A large number of fluorophores interact with structures containing nucleic acids (see Fig. 4), some of the most common of these being acridine orange, ethidium bromide, propidium iodide, and DAPI (diamidinophenylindole).

4. Fluorescent microscopy can also be used to monitor intracellular compartments through the use of fluorescent indicator dyes that report on their local environments. Some of the fluorescent dyes useful for microscopy are membrane-potential indicators such as JC-1; using this probe, it was shown that heterogeneity in mitochondrial membrane potential exists between different mitochondria in the same cell. Other fluorescent dyes are ion indicators that can report that pH or concentrations of calcium, chloride, magnesium, potassium, and sodium ions either qualitatively or quantitatively.

5. Fluorescence microscopy is used to detect fluorescently labeled antibodies bound to their antigens *in situ*.

In all cases, it is advisable to observe and store the specimen in a way that reduces fading and quenching. Quenching produces a decrease in fluorescence due to the presence of oxidizing agents, heavy metals, or other fluorophores. To minimize quenching and fading, it is best to store specimens in the dark at 4°C and to add an antifade component to the mounting medium, such as gelvatol, *N*-propylgallate or DABCO.

Immunofluorescence light microscopy is gaining prominence in fluorescence microscopy because the most specific detectors of biological molecules are antibodies that bind tightly to antigens against which they are directed. Visible flags for the presence of an antigen involves the use of an antibody labeled with a fluorophore. Antibodies do not necessarily need to be directed against specific proteins. They can also be produced against entire classes of pathologic agents that share common antigens, or against specific components, such as the carbohydrate found in the cell walls of specific bacteria. Immunofluorescence can be performed by either direct or indirect antibody labeling. Direct immunofluorescence requires the conjugation of a primary antibody with a fluorophore, commonly rhodamine and fluorescein. Indirect immunofluorescence, on the other hand, uses a fluorophore conjugated secondary antibody that was raised against the immunoglobulin type of the primary antibody. For example, if the primary antibody is raised in a rabbit, the secondary antibody might be a goat anti-rabbit antibody. Indirect immunofluorescence is the more common of the two techniques because making a fluorophore conjugate of each primary antibody is time-consuming and can decrease the affinity or specificity for the antigen. Furthermore, the indirect technique may be more sensitive because more than one molecule of the secondary antibody may bind to a molecule of the primary antibody, creating an amplification of the signal. Indirect immunofluorescence has motivated the development of a large number of commercially available secondary antibodies conjugated to various fluorophores, so in order to do immunofluorescence microscopy, one need only raise and characterize the primary antibody and then purchase the secondary antibody.

A key consideration in immunofluorescence microscopy is confirmation of the specificity of both the fluorophore and antibodies. In order for immunofluorescence results to be validly interpreted, one must demonstrate that the secondary antibody recognizes only the primary antibody and that the primary antibody recognizes only the antigen. Although antibodies can be very specific in their high-affinity binding, they can also exhibit lower affinity nonspecific binding. For this reason, it is usually necessary to treat the specimen with an excess of proteins, such as bovine serum albumin, fish gelatin, or normal goat (or donkey) serum, in order to coat surfaces with proteins that bind with low affinity that can be pushed aside by the high-affinity binding of antibodies. This blocks nonspecific binding sites, reducing background fluorescence. Secondary antibodies must not cross-react with intrinsic proteins and must not recognize each other. It is also necessary to establish whether autofluorescence from compounds intrinsic to the tissue can mimic the appearance of fluorophores. This test is easily done by examining an unstained specimen using the same excitation and emission filters intended for use with the fluorophore.

The applications of fluorescence microscopy to microbiology are many and varied. For example, *Mycobacterium tuberculosis,* the bacterium that causes tuberculosis, can be easily identified after staining with fluorophores or after labeling with fluorescent antibodies. With the appropriate choice of fluorescent labels and proper equipment, one can do multilabeling (sometimes called multicolor) to localize two or more molecular species using fluorophores with different excitation or emission wavelengths. An example is the use of DiOC6 to label mitochondria and ethidium bromide to label mitochondrial DNA in *Euglena gracilis* cells.

Immunomicroscopy can still be realized in the absence of a fluorescent microscope. An example of this is the immunoperoxidase stain method, in which horseradish peroxidase takes the place of the fluorophore conjugated to an antibody. Horseradish peroxidase produces an orange-brown precipitate that is easily visible in bright-field microscopy. The immunoperoxidase method has been used to detect virus antigens. Another example is the biotin-avidin-enzyme stain method. Antimicrobial antibodies can be bound to the small molecule biotin, which has a strong affinity for the protein avidin. Avidin has four binding sites and so can be easily conjugated to a color-producing enzyme and used to label antibodies bound to antigens that can be detected with bright-field microscopy by staining for enzyme activity. Commercial products are available that use the biotin-avidin-enzyme method to detect viruses and microorganisms such as Epstein-Barr virus, cytomegalovirus, hepatitis B Virus, herpes I and II viruses, adenovirus 2, and *Chlamydia.* Direct immunofluorescence methods have been developed to identify antigens found on group A streptococci, enteropathogenic *E. coli, Haemophilius influenza b, Neisseria meningitidis, Salmonella typhi, Listeria moncytogenes,* and *Shigella sonnei,* among other pathogens. Indirect immunofluorescence is used extensively in clinical settings to detect the presence of antibodies in serum after a person's exposure to microbes. For example, indirect immunofluorescence kits are available commercially to identify the presence of *Treponema pallidum* antibodies in the diagnosis of syphilis. As more monoclonal antibodies are developed, more conjugated labels will be produced for the rapid, direct detection of pathogens in clinical material by fluorescence microscopy. In the field of microbial ecology, microorganisms can be observed in a relatively undisturbed ecological niche by tagging them with the fluorophore acridine orange, which allows for their detection in a soup of other particles.

When one uses fluorescence microscopy on live cells, care must be taken to avoid disrupting the function of the antigen that affects normal physiology by antibody binding. Monovalent Fab fragments directly conjugated to the fluorophore may alleviate this problem. At times, however, the intentional disruption of antibody function is useful for studying the effects on cellular function. Surface antigens are straightforward to label. Internal antigens can be labeled with minimal cell trauma by microinjection, electroporation, fusion with liposomes, or by a biolistic gene gun. Specific sequences of nucleic acids can be localized by *in situ* hybridization using a particular oligonucleotide probe that is fluorescently labeled. By using vital dyes and fluorescence ratio imaging, it is possible to study dynamic processes inside cells, such as fluctuations in intracellular pH and free-calcium concentrations in distinct cellular compartments, such as mitochondria versus cytoplasm. Another powerful technique is fluorescence analog cytochemistry. First a normal protein or lipid is biochemically purified and conjugated to a fluorophore. Its native properties are verified and then it is reincorporated into living cells. Once the cells have been checked for viability, fluorescence microscopy is used to study its redistribution and function in response to pharmacological or physiological stimuli.

A further application to microbiology is time-resolved fluorescence microscopy (TRFM), fluorescence lifetime spectroscopy coupled to microscopic recordings. This technique provides information on the dynamics of fluorescent probes and how their biological environments influence them. The coupling of spatial and positional information with temporal information provides a new way to identify and study inhomogeneous milieu inside cells. For example, methanogenic bacteria are known to be autofluorescent (primary fluorescence) because a constituent enzyme, coenzyme F420, fluoresces. The overlap of its fluorescent signature with those from organic and inorganic compounds within samples containing this class of bacterium has prevented its detection using conventional fluorescence microscopy. However, TRFM has been successfully used to resolve the various decay times of the different fluorophores and thus to monitor these bacteria.

A. FRAP

Fluorescence recovery after photobleaching (FRAP) contributed to the establishment of the fluid-mosaic model of cellular membranes through its capability to measure protein and lipid mobility and continues to have an impact in topics of microbe membrane dynamics. The basic idea is that a living cell is incubated with a fluorescent ligand that binds selectively to an exposed epitope of a membrane protein or lipid. Intense light from a mercury lamp or from a laser will irreversibly photobleach the portion of the membrane that is being irradiated in a short amount of time. After photolysis of the fluorescent tag, the fluorescence is measured as a function of time on the bleached portion and on an unirradiated area, and compared. Fluorescence in the bleached area "recovers" as photolyzed molecules diffuse out of the bleached area and unphotolyzed molecules diffuse into it from neighboring areas. The fluorescence curves are fitted to theoretical curves,

and lateral diffusion coefficients and flow velocities are determined. If the fluorescence recovery is measured in a polarizing microscope, rotational mobility can be monitored. Some photobleaching techniques do not involve a fluorescence recovery, and are included under the umbrella term fluorescence microphotolysis (FM). Continuous photobleaching FM is used to couple the measurement of diffusion with reaction products. Photobleaching methods were used to measure transport across membranes, for instance of ions. These methods need not be limited to membranes. For example, photobleaching techniques have been used to study cytoskeleton dynamics, the mobility of macromolecules between the nucleus and the cytoplasm, and even the mobility between neighboring cells.

In the conventional photobleaching instrument, a continuous-wave laser is coupled to an epifluorescence microscope, a photon counter, and a computer. There is also a device that can rapidly switch the laser power between a low level, used in fluorescence measurement, and a high level, used for photolysis. In the 1990s, it has become more common to use a laser-scanning microscope (LSM). The LSM operates by focusing the laser beam down to the diffraction minimum and scanning the object point by point. The principal difference between the conventional instrument and the instrument employing the LSM is that the illuminated area in the former is typically 1–2 μm, but in LSM is about 0.2 μm. The consequence is that the LSM instrument achieves a significantly higher resolution in x, y, and z. By scanning in x- and y-planes and stacking x-and y-scans, the LSM instrument permits imaging in extended areas and volumes.

The photobleaching experiment proceeds as follows. The specimen is initially scanned at low laser power to select the area to be photolysed. The selected area is then photolysed. Repetitive scans at low laser power follow to monitor the time development of the fluorescence distribution. This method is called "pulsed photolysis" and is based on a number of assumptions or simplifications that may or may not hold for any given specimen. First, it is assumed that fluorescence can be measured without photolysis. Second, it is assumed that photolysis can be accomplished instantaneously. Quite often, these simplifying conditions can be well approximated by specimens containing large numbers of fluorophores that yield a strong signal. However, in many cases, only a small number of fluorophores is present and the signal is weak. For this situation, continuous FM was developed that allows for photolysis during fluorescence measurements. The principal difference in operating procedure is that the laser does not alternate between power levels, but instead is kept at a constant intermediate level. The consequence is that both photolysis and diffusion occur at the same time in the illuminated area. Inside the illuminated area with time, a ring-like appearance develops with a dark center and a bright rim because fresh fluorophores enter the illuminated area from adjacent areas as fluorophores are continuously photolysed. A biphasic decay accurately describes this system if the fluorescence in the illuminated area is integrated. The initial steep decay represents the fast photolytic conversion of those fluorophores that are present in the area at zero time. The shallow decay reflects the diffusion of fresh fluorophores from adjacent regions into the illuminated area and their photolysis.

The measurement of transport across membranes can also be performed by either pulsed photolysis or continuous photolysis. It requires compartments with a closed surface membrane, such as intact cells, organelles, or vesicles. The goal is to deplete the internal compartment(s) of fluorescence and then to directly monitor the influx of fluorescent-tagged molecules into the compartment by transport across the membrane. A confocal microscope is usually used to insure spatial selectivity by optical sectioning. The confocal arrangement bestows spatial selectivity with the emitted signal, but not with the excitation light because the regions along the laser-beam path above and below the compartment of interest are being excited. Even better would be two-photon microscopy, which bestows both excitation and emission spatial selectivity. It is important to realize that photolysis triggers diffusion external and internal to the compartment, as well as membrane transport. However, if the diffusion rate is many times higher than the rate of transport across the membrane, then both internal and external regions will be virtually in diffusional equilibrium and the observed fluorescence recovery is a measure of membrane transport.

The applications of photobleaching techniques to microbiology provide clues to basic aspects of membrane biology, for instance, the formation of microdomains in the plasma membrane. The influence of the membrane potential on lateral protein mobility can be measured. Changes in mobility during aging, viral transformations, or cell-fusion events can be monitored. An example of the importance of this technique comes from a study of the parasite *Hymenolopis diminuta* exposed to complement. Regional elimination of membrane fluidity was observed by exposure to normal rat serum and rat C reactive protein.

VIII. DECONVOLUTION MICROSCOPY

Three-dimensional imaging using optical microscopy revolves around two approaches, confocal microscopy and deconvolution microscopy. The major limitation of 3D imaging with conventional optical microscopes is out-of-focus blur. Confocal microscopy removes the out-of-focus blur by optical methods. Deconvolution microscopy removes the out-of-focus blur by computational means (see Fig. 6). For example, in the case of fluorescence microscopy, the illumination of the entire field of view with intense light excites fluorescence emissions throughout the

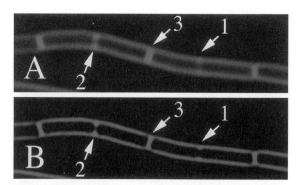

Fig. 6. Deconvolution microscopy distinguishes between partial and complete septa. Exponentially growing wild-type (PY79) bacteria were stained with FM 4-64. (A) The original medial section in color. (B) Following deconvolution, partial septa (arrows 1 and 2) are clearly distinguished from complete septa (arrow 3) (courtesy of Kit Pogliano, University of California, San Diego).

depth of the specimen. Significant portions of the emitted light coming from regions of the specimen above and below the focal plane pass through the objective lens and contribute to the image as out-of-focus blur, reducing significantly the sharpness and contrast of the image. When imaging in 3D, the depth of field and resolution in the third dimension (i.e., the direction of the optical axis) must be considered. For optical microscopes, the depth of field is large compared to the in-plane resolution limit, but small compared to the typical specimen thickness. Thus, true projection images cannot be obtained. This means that a multiple-tilt method cannot be used without additional information or corrections to construct a 3D image, such as used in electron microscope tomography or CAT scans. Furthermore, because of the relatively small depth of field, contamination of the image by out-of-focus contributions from adjacent levels of the specimen adds to the image at a given focal plane, increasing the background and reducing contrast. Thus, the image is a poor representation of a section through the specimen. This problem of degraded contrast is severe for epifluorescence imaging because it is a dark-field mode and, with normal levels of fluorescence labeling, the out-of-focus components build up to give a large flare. This means that the in-focus detail is seen as a relatively small modulation against the large out-of-focus flare.

The first step in deconvolution microscopy is to gain as full an understanding as possible of the imaging system. To achieve accurate computational image deblurring, the 3D PSF must be determined. One can think of the image obtained as the object convoluted with the PSF. Deblurring then involves deconvolving the PSF from the 3D image, producing a more accurate representation of the object. The PSF can be determined empirically by imaging sub-resolution point test objects (e.g., small fluorescent beads) using the same objective lens and imaging conditions that will be used to collect the specimen data. Alternatively, one may attempt to calculate the PSF from the imaging characteristics of the objective lens, but this approach is less satisfactory. However, accurate knowledge of the PSF is not enough to calculate an exact solution to deblur an image plane. This is because to determine the out-of-focus contri-

butions to an image plane, the in-focus images at other focal planes must be determined. But the in-focus images of the other planes in the 3D specimen required for this calculation are themselves unknown because these planes also can be imaged only in the presence of out-of-focus blur.

Three procedures have been developed to overcome this dilemma, which use the blurred images at the various focal planes as initial approximations to the corrected in focus images. A mathematical description of the deconvolution algorithms is beyond the scope of this article; the procedures will only be discussed in general terms. One method uses the complete stack of blurred images from which a projection of the 3D specimen is computed down the optical axis onto a plane. This projection image is deblurred by deconvolving the PSF. Other projections are computed down axes that are displaced a few degrees from the optical axis and likewise deblurred. In this way, stereoscopic pairs of projections can be created that provide some 3D information. However, stereoscopic pairs are not a complete deblurred 3D data set. A second method, which does create a complete deblurred 3D set of images, uses information from the sections immediately above and below the image to be deblurred. This "nearest-neighbor" deblurring is only approximate, but has the advantage of quick computation. However, with the increasingly fast processors in current-generation workstations, the time required when using the entire set of optical sections to deblur each image is no longer an impediment. This more computationally intensive procedure is the third method, which is now the preferred approach because of the improved image quality attainable over the other two approaches. There are two methods commonly used in this approach: (1) 3D inverse filtering, and (2) constrained iterative deconvolution. Three-dimensional inverse filtering is the more direct method because all that is required is an inverse filter operation, usually a Wiener filter. In practice, each z-axis line is first expanded by linearly interpolating between the first and last values to remove discontinuities in the transform, as the fast Fourier transform is cyclical. Next it is Fourier-transformed in z, filtered, and inverse transformed in z. Those sections that correspond to the original stacks are finally transformed in the *xy* plane. The Wiener filter is an example of a linear restoration method. Such linear methods have the drawback of negative ripples around strong feature that can obscure or distort adjacent weaker features. Nonlinear restoration methods that possess a posivity constraint can eliminate the problem with ripples and may provide a more realistic restoration by using positivity or other a priori information, such as spatial boundaries. These nonlinear methods are of necessity iterative, and so tests for convergence must be employed. The strategy is to find a constrained solution that when convolved with the known PSF will regenerate the observed data. This requires a first guess at the solution and, with each iteration, the pixel-by-pixel differences between the observed data and the convolved guess are used to update the next guess. Convergence is defined when the mean difference falls below a defined threshold and is usually reached in 5–10 iterations.

As with all 3D imaging schemes that rely on volume reconstruction by sections, the *z*-dimension resolution is not as high as the in-plane resolution. Improvement in *z*-resolution can be made by collecting two series of optical sections along orthogonal axes, that is, by rotating the specimen 90° after collecting the first series. Hence, iterative optimization must be applied using the section data to avoid introducing artifacts. Even though confocal and deconvolution microscopies are complementary techniques for deblurring images, the algorithms developed for deconvolution imaging can be applied to confocal images to remove residual distortions. This distortion is present because even for minimum pinhole (best case) there is still considerable anisotropy in the PSF resulting in lower resolution along the optical axis. Whereas confocal images possess an image quality similar to deconvolved images, deblurring confocal images yields a small but significant increase in image clarity.

Because both deconvolution and confocal microscopies are designed to remove out-of-focus blur, a brief comparison of the advantages of each will help microbiologists choose the appropriate instrument for the application at hand. There are three advantages of deconvolution microscopy: (1) the ability to use arc lamps to excite the full range of typical fluorophores, including those excited by ultraviolet light. (Laser-scanning confocal microscopy is limited to fluorophores excited by specific laser lines, although with the wide range of excitation wavelengths to choose from with the currently available fluorophores, the few laser lines available are adequate for typical applications), (2) simultaneous parallel data collection from all regions of the specimen, which results in rapid image acquisition (typical laser-scanning confocal microscopy uses serial data collection), and the (3) collection of all the light that enters the objective lens (confocal microscopy passes through light only from the focal plane). In essence, the deconvolution method "restores" the image by returning the out-of-focus light to its proper place, making more efficient use of light and minimizing exposure to light that can damage the specimen, such as photobleaching of fluorophores. There are four advantages of confocal microscopy: (1) improved focal sectioning because of a better optical transfer function, (2) simplicity of use (the deconvolution algorithms are not always straightforward to use), (3) the display of images within a second or two, without having to pass through a program, (4) better time resolution if one selects a small region or even a single line of the image in order to study intracellular changes of conditions (e.g., at Ca^{2+} concentration, pH, and membrane potential).

IX. OPTICAL VIDEO AND DIGITAL MICROSCOPY

Optical video microscopy involves the acquisition of images from optical microscopes with a video camera. It is important to realize that these video images are recorded with analog signals, which can be subsequently converted to digital signals. Optical digital microscopy acquires optical images with non-video-rate digital cameras that are of solid-state design (CCD cameras) and can record at high resolution with high sensitivity, absolute photometric accuracy, and with a large dynamic range. Because video microscopy increasingly uses digital techniques, the distinction between it and digital microscopy more purely defined is becoming blurred. The older nomenclature "video microscopy" is still used as an umbrella term for both video and digital microscopies when applied to the recording of dynamic processes. Perhaps a more comprehensive term is electronic light microscopy (ELM) because all forms of this type of microscopy involve the electronic encoding of an image produced by a light microscope. But ELM is far more than the electronic recording of conventional optical microscope images for convenience or economy, as an alternative to cinematography. These recording techniques coupled with image processing have become invaluable for studying the dynamics of living cells because of the demand for high resolution, fast acquisition, and nanometer precision. Combined with the fastest computer processors available, ELM can be performed in real time at video-rate processing and analysis speeds.

One powerful application of ELM is the study of dynamic cellular processes, including the movement of structures as much as an order of magnitude smaller than the diffraction-limited resolution of the optical microscope. These include individual microtubules, transport vesicles, and colloidal gold particles conjugated to antibody molecules and used to label specific cellular structures. In order to image intracellular structures, one normally employs one of the phase-contrast modes, usually Nomarski (DIC). In the resulting image, object structures are convoluted with the PSF of the objective lens, increasing the apparent sizes of very small structures. For example, the images of 25-nm diameter microtubules are approximately 200 nm, the diffraction-limited resolution of the best optical microscopes. The signals from these subresolution structures are very small and are superimposed on a bright background whose inhomogeneities are of the same magnitude as the signal; thus, they cannot be seen by eye. Recording the images wih a video camera with a good light sensitivity allows one to increase dramatically the contrast of the image so that the dynamics of small objects can be visualized at video rates (30 frames per second).

First the optics of the microscope are adjusted to maximize the phase contrast of the specimen. This generally results in an increased signal from the specimen superimposed on a background too bright to be studied by eye. The video camera, however, has a much larger dynamic range than the human eye and can measure small fluctuations on a high background. In order to record these small changes with the full dynamic range of the video tube, the dark level is adjusted to bring the background down and the gain is increased to maximize the signals from the specimen. These adjustments are analogous to adjusting the brightness and contrast, respectively, on a television set. The result of these operations is usually still contaminated by background fluctuations of the same magnitude as the signal. These are caused by nonuniform illumination of the specimen plus contributions from various elements of the optical system. This contamination can be removed by digitally subtracting a background image obtained without a specimen, yielding an image of the specimen with a greatly enhanced contrast. One limitation of video cameras is their relatively low resolution, typically 525 or 625 horizontal lines, depending upon the video standard employed. But by employing an additional stage of image magnification, the image can be projected onto the tube of the video camera at a resolution high enough for the optical resolution of the microscope to be adequately sampled in the video image. In case, the effect of the limited resolution of the video camera is a narrow field of view.

An example of the power of contrast-enhanced ELM is the study of microtubule dynamics and of organelle and vesicle transport along microtubules. Microtubules are approximately 25 nm in diameter, but can be readily observed by ELM following contrast enhancement. Observations of the simultaneous growth and shrinkage of different microtubules in the same field of view have helped confirm a model for the assembly and disassembly of these cytoskeletal elements. ELM has also demonstrated the rapid movement of vesicles and organelles pulled by motor molecules along microtubules. Another example is the ability to image collodial gold labels more than an order of magnitude smaller than the resolution limit of the optical microscope. The use of colloidal gold particles as specific labels for microscopic studies was previously the domain of immunoelectron microscopy, whereas their use in optical microscopy has been nanovid microscopy. Stationary gold particles as small as 5 nm have been visualized and gold particles as a small as 20 nm have been tracked inside living cells. Because the images of these particles have a diameter of about 200 nm, the microscope's resolution limit, if two such particles lie closer than 200 nm they cannot be separately resolved. However, by measuring the image intensity it may be possible to determine how many particles are in close proximity. The capacity to coat colloidal gold with any chosen antibody and then follow this probe in a living cell opens up avenues of correlative light and electron microscope studies. For example, direct correlations can be made between dynamic processes observable with nanovid microscopy in the living cell and specific macromolecular relationships revealed only by electron microscopy.

Video cameras are also widely used in fluorescence light microscopy to record weak fluorescent signals and to measure changes in fluorescence that monitor movement of particles or changes in ion concentrations. In this case, the characteristics of the image and the corresponding requirements of the video camera are different than those employed with phase-contrast microscopy. In fluorescence microscopy, one is measuring a weak signal against a dark background, and it is often advantageous to employ a very sensitive camera such as a silicon intensifier target (SIT) tube or intensified SIT (ISIT), which are two to three orders of magnitude more sensitive than a conventional video camera. By using a very sensitive camera, one can study faint images resulting from a low level of labeling used to minimize the disruption of cellular processes or a low level of excitation illumination used to minimize photobleaching. These very sensitive cameras suffer from relatively poor resolution, poor photometric accuracy, geometric distortion, and low signal-to-noise (S/N) ratio. S/N can be improved dramatically by digital signal averaging, that is, by acquiring and averaging multiple video frames with a computer and video digitizing hardware. However, by frame averaging reduces the rate at which images are acquired, so the measurement of rapid processes may not be possible. More recently, cameras employing 2D CCD arrays have become the instrument of choice for low-light-level imaging. When cooled to reduce the dark current to nearly zero, these CCD arrays offer high resolution (up to 3,000 × 3000 pixels per frame) with very high sensitivity, large dynamic range, high S/N, and photometric accuracy (a linear response to photons). They are also inherently integrating detectors, so sensitivity can be increased simply by allowing the image to accumulate for a longer period of time before reading it out (increasing the exposure). In most respects, cooled CCD array cameras are ideal image-recording devices. Their one significant weakness is that they are relatively slow; they are often referred to as slow-scan CCD cameras. This is due to the fact that reading out the data from a large CCD array takes some time even with fast computers, and achieving the highest sensitivity requires relatively long exposures. Recent improvements have increased the rate of image acquisition to well below a second, so cooled CCD cameras will continue to replace video cameras in many applications.

If one wants to measure movement within tens of nanometers of activities such as microorganisms swimming or diffusion of macromolecules on a cell surface, which have velocities in the micron per second range, then sample rates of at least the video rate of 30 frames per second are required. Some transient fluorescence activities may require sampling at rates of over 100 frames per second. These fast rates have started to yield temporal information in toxicology studies on pathogenic microorganisms. In the past, toxicologists have been able to determine at what concentration and time microorganisms die after the introduction of toxic chemicals. Now, questions can be answered detailing at what concentrations and times microbes begin to deviate from their normal behavior, and the morphology and response can be monitored at sublethal concentrations of toxic agents. Thus, insight into the mechanism or site of attack of a toxic agent's action can be obtained.

X. NEAR-FIELD SCANNING OPTICAL MICROSCOPY

Near-field scanning optical microscopy (NSOM) is a relatively new subfield of optical microscopy that allows visualization with subwavelength resolution, also called superresolution. NSOM is a combination of optical microscopy with scanning probe techniques, like those used in atomic force microscopy.

The principle is based on a probe consisting of an aperture smaller than the wavelength of light that is positioned in close proximity to the specimen. By laterally scanning the specimen with near-field optics, one can generate an image at a resolution dependent on only the probe size and the probe-to-specimen separation, each of which can be pushed into the 50- to 100-nm regime. Thus, NSOM extends light microscopy to a higher spatial resolution than is currently available with confocal microscopes and other sophisticated microscopes.

To understand the probe's operation, imagine a hole in an opaque material. When light impinges on this hole, it can pass through only if the hole's diameter is larger than half the light's wavelength. For example, if a hole is made with a diameter of 50 nm, visible and near-UV light cannot pass through it. However, the electrical-field associated with the light will protrude into the hole. The electrical-field intensity decreases exponentially as the light source is moved from the hole. Holes of this size can now be prepared on the sharp tip apex of an optical fiber by coating it with aluminum. When laser light is coupled to the tip and brought to within 10 nm of the specimen, its electrical field causes the specimen to emit light from a small region. Usually, a NSOM scan head is attached to a conventional upright fluorescence microscope that can be used in confocal mode. This setup enables the visualization of a large field of view by bright-field microscopy zoom-in by confocal microscopy, and final zoom-in and topographical recording by NSOM. By scanning the specimen in the same way that laser-scanning confocal microscopy (LSCM) images are scanned, an optical image is constructed. For an accurate optical image, the distance of the tip above the sample must be kept constant. This is accomplished by using a shear-force mechanism in which the tip oscillates horizontally above the specimen. The oscillation is damped by atomic shear forces when the oscillating tip approaches the specimen. The damping translates to a measure of the distance between tip and specimen. In this way,

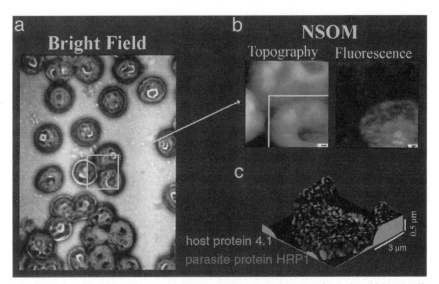

Fig. 7. Colocalization of host–parasite proteins in malaria-infected red blood cells by NSOM. (a) Bright-field image. The NSOM tip was positioned above the three cells in the framed area. (b) NSOM imaging in topography mode (left) and fluorescence mode (right). Although all three cells were imaged in topography mode, only the one in the lower right corner was infected and made visible in fluorescence mode. (c) 3D zoom around the infected cell. The fluorescence signal was superimposed (please see the reference for a color figure) to the topography to show the distribution of HRP1 in the erythrocyte membrane. Courtesy of Magowan, C. *PNAS* **94,** 520–525. Copyright 1997 National Academy of Sciences, U.S.A.

the probe provides both optical and topographical data through the same measurement.

However, because the achievable resolution degrades with increasing distance from the probe, it is possible to obtain superresolution information in three dimensions only within a few tens of nanometers from the surface. NSOM therefore falls into the class of surface-probe microscopies (e.g., atomic force microscopy). Superresolution surface features can be obtained from thick as well as thin specimens by filtering, although subsurface structures are usually better studied with complementary 3D microscopies, such as confocal or deconvolution microscopies. The surface-penetrating power is nonetheless sufficient to map cytoskeletal structures in addition to cell membranes and membrane proteins. Resolutions as fine as 30–50 nm have been obtained on biological specimens using NSOM. Another important aspect of NSOM is the possibility of sensitive mapping of dynamic events between two fluorophores, connected by a short DNA strand by using fluorescence energy transfer (FRET).

An advantage of NSOM over microscopies that offer higher resolution, such as scanning transmission electron microscopy, scanning tunneling microscopy, and atomic force microscopy, is the wide range of contrast mechanisms available. Contrast in NSOM can be accomplished by absorption, polariza-tion, refractive index, fluorescence, and reflectivity. From the perspective of microbiology, fluorescence may be the most powerful contrast mechanism because a wide variety of fluorescent probes have previously been developed for light microscopy. A powerful application of NSOM coupled with fluorescence was demonstrated through the mapping of parasite proteins inside human red blood cells (erythrocytes). For example, after erythrocytes are invaded by a human malaria parasite, NSOM coupled with immunofluorescence labeling shows colocalization of erythrocyte cytoskeletal proteins and parasite proteins expressed during intraerythrocytic growth and transported to the erythrocyte cell membrane (see Fig. 7). The high-resolution information provided by NSOM between proteins in biological membranes has wide-spread applicability in many areas of current microbiology.

See Also the Following Article

Microscopy, Confocal

Bibliography

Cherry, R. (1991). "New Techniques of Optical Microscopy and Microspectroscopy." CRC Press, Boca Raton, FL.

Lacey, A. J. (1989). "Light Microscopy in Biology: A Practical Approach." IRL Press, Oxford.

Shotton, D. (1993). "Electronic Light Microscopy". Wiley-Liss, New York.

Mutagenesis

Richard H. Baltz

CognoGen Biotechnology Consulting

GLOSSARY

amplification   The tandem reiteration of a segment of DNA.

deletion   The loss of a segment of DNA from the genome. Spontaneous deletions can range in size from several nucleotides to multiple contiguous genes. Deletions can also be generated by recombinant DNA methods.

duplication   The addition of a second copy of a segment of DNA in the genome. Spontaneous duplications usually encompass one to several genes, and generally occur in tandem. Targeted duplications of specific genes can be generated by recombinant DNA methods.

frameshift mutation   The deletion or addition of a small number of nucleotides (but not three or a multiple of three) in an open reading frame that shifts the translational register out of frame, causing the production of a protein containing an altered amino acid sequence downstream of the frameshift.

gene disruption   The physical disruption of the integrity of a gene by insertion mutagenesis or by deletion, usually of an internal segment of the gene. Gene disruptions are commonly constructed by recombinant DNA methods to study gene function.

insertion mutation   A mutation caused by the insertion of a segment of DNA into another segment of DNA. Spontaneous insertions are often caused by the transposition of IS elements or transposons. Insertions can be generated in the laboratory by transposons or by the use of recombinant DNA methods.

insertion sequence (IS) element   Elements that encode the ability to transpose from one site to another in the genome. IS elements differ from transposons in that they generally do not encode selectable markers such as antibiotic resistance genes. IS elements can be converted into transposons by inserting selectable markers into them naturally or by using recombinant DNA methods.

inversion   The physical breakage and rejoining of a segment of DNA that places the segment in opposite orientation relative to its normal orientation.

mutagen   A physical or chemical agent that induces mutations.

mutation   A heritable change in the nucleotide sequence of an organism. Mutations include base-pair substitutions, deletions, duplications, transpositions (translocations), and inversions.

transition   A type of base-pair substitution mutation that substitutes a purine for a purine or a pyrimidine for a pyrimidine. There are two types of transition mutations, G-C to A-T and A-T to G-C.

transposition (translocation)   The movement of a segment of DNA from one location in the genome to another. Transpositions are usually mediated by transposons or IS elements.

transposon   A DNA element that is capable of transposing from one DNA site to another. Transposons generally encode the ability to transpose (transposase and sometimes resolvase) and encode a selectable marker such as an antibiotic resistance gene.

transversion   A type of base-pair substitution mutation that substitutes a purine for a pyrimidine or a pyrimidine for a purine. There are four types of transversion mutations, G-C to C-G, G-C to T-A, A-T to T-A, and A-T to C-G.

MUTAGENESIS   is fundamental to the evolution and diversity of microbes. The ability to undergo mutagenesis is inherent in the chemical structure and properties of DNA, and in its mechanisms of replication and repair. Mutations can arise spontaneously and can also be induced by certain chemicals and by radiation.

The rates of spontaneous mutation in viruses and bacteria are controlled by DNA replication and repair enzymes. Induced mutation rates are also under genetic control and are often mediated by repair enzymes. Mutagenesis has played an important role defining gene function and will continue to play a prominent role as multiple bacterial genome sequences become available for functional analysis. Mutagenesis has also played a critically important role in industrial microbiology, particularly for the generation of highly productive strains of actinomycetes used to produce antibiotics and other secondary metabolites.

I. MUTATION AND EVOLUTION OF SPECIES

Mutation coupled with recombination has been the driving force in the evolution of microbial species. It has been argued that in the early days of life on Earth, the progenitors of the three kingdoms, the Bacteria, the Eukarya, and the Archaea, probably experienced very high rates of mutation and horizontal gene transfer. As time progressed, enzyme systems evolved to regulate mutation and recombination rates, species evolved to compete in specialized niches, and mutation and recombination rates generally declined. The species that we encounter today generally express highly regulated mutation rates that can change in response to chemical, physical, and nutritional stimuli.

II. SPONTANEOUS MUTATION

A. Spontaneous Mutation and Genome Size

Spontaneous mutation is a somewhat archaic term, predating Watson and Crick's discovery of the structure of DNA. Spontaneous implies that these mutations just happen naturally; they might be considered as "green" mutations in some political circles, and therefore probably good or at least acceptable. It is now known that spontaneous mutation rates are under genetic control and vary inversely with genome size in haploid microbes. Thus, microbes carefully titrate their spontaneous mutation rates to balance the selective advantages afforded by being able to mutate and adapt to different environments with the disadvantages of deleterious or lethal mutations. It makes no sense to accumulate both advantageous and deleterious mutations in the same cell if the deleterious mutations cannot be readily recombined out. It stands to reason that microbes with large genomes must have relatively lower mutation rates per base pair than their smaller cousins to avoid this dilemma. Microbes with very small genomes, including a number of infectious viruses, can afford to have very high mutation rates. This poses a constant challenge to the pharmaceutical industry because certain viruses can mutate readily to become resistant to antiviral agents. Spontaneous mutation to antibiotic resistance is less of a problem in bacterial infections. Most antibiotic-resistance determinants in pathogenic bacteria reside on self-transmisible plasmids or transposons. A notable exception is *Mycobacterium tuberculosis,* the causative agent for tuberculosis. In *M. tuberculosis,* resistance to different combinations of isoniazid, rifampin, pyrazinamide, ethambutol, and streptomycin arise by sequential spontaneous mutations. Effective treatments are obtained when three agents are used in combination over the appropriate period of time, thus eliminating the possibility of selecting rare mutants containing three spontaneous mutations.

B. Sources of Spontaneous Mutation

1. Deamination of Cytosine

Although as a first approximation, spontaneous mutations do "just happen naturally," there are chemical and enzymatic underpinnings that control the rates of mutation. Some mutations arise from the inherent chemical properties of the components of DNA. One component, cytosine (C), undergoes heat- and acid-induced deamination to yield uracil (U) at physiologically relevant rates. If the U is not repaired, it will code as thymine (T) during the next round of replication, resulting in a C to T transition mutation. Different microbes may modulate mutation rates by this pathway differently, depending on how efficiently they express uracil N-glycosylase, a DNA-

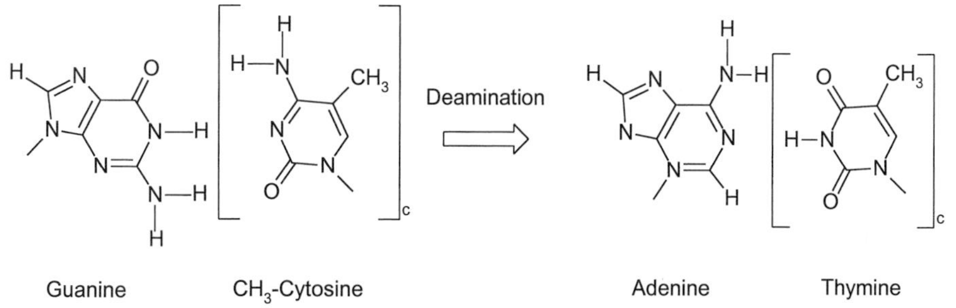

Fig. 1. Spontaneous deamination of 5-methylcytosine to thymine (in brackets). During DNA replication, adenine is incorporated across from thymine, resulting in a G-C to A-T transition mutation. Cytosine deamination is also induced by heat and acid.

repair enzyme that cleaves the U from DNA, leaving an apyrimidinic site that can be further processed by repair enzymes to reinsert C. If C happens to be methylated, then deamination yields T directly (see Fig. 1). In this case, the uracil N-glycosylase cannot repair the damage, so the mutation will be fixed if a mismatch repair system does not correct the damage. Most bacteria express restriction and modification systems. Often the modification is methylation of cytosine within a precise context of DNA, so the existence of methylcytosine in bacterial DNA is not uncommon.

2. Depurination

Depurination, the loss of guanine (G) or adenine (A) bases from DNA, is also induced by heat and acid at physiologically relevant rates. Depurination can be repaired by error-free or error-prone repair systems, so the relative levels of expression of these systems could be used to modulate the rates of mutation by this pathway.

3. Oxidative Damage

Oxidative damage to dGTP and to G in DNA can generate 8-hydroxyguanine (7,8-dihydro-8-oxoguanine) or GO. This is an inherent "spontaneous" chemical reaction that occurs in oxidizing environments. GO can be incorporated into DNA from 8-oxo-dGTP across from A or C. GO in DNA can be replicated, incorporating C or A on the opposite strand. GO can also be eliminated from 8-oxo-dGTP or from DNA by the enzymes encoded by *mutT* and *mutM* genes, respectively. Another enzyme, the product of the *mutY* gene, can remove a mispaired A from

the strand across from a GO residue. Mutations in any of these three genes in *Escherichia coli* cause enhanced spontaneous mutation rates. The levels of expression of these genes or genes with similar functions, and the modulation of the oxidizing environment by catalases, peroxidases, superoxide dismutases, and anaerobiosis, for instance, could lead to different rates of spontaneous mutation by this mechanism.

4. DNA Repliction Errors

Spontaneous mutation rates are also controlled at the level of DNA replication. Bacterial and some bacteriophage DNA polymerases contain both polymerase and exonuclease activities. The polymerase controls the choice of dNTP, but the accuracy is limited to an error frequency of ~1 in 10^4. The exonuclease activity acts as a proofreading mechanism by cleaving misincorporated bases. The combination of polymerase and exonuclese proofreading give an overall accuracy of spontaneous misincorporation of ~1 in 10^8. Subtle changes in the fidelity of the DNA polymerase or exonuclease functions can modulate spontaneous mutation rates.

Having incorporated some wrong bases during DNA replication, certain bacteria have yet another opportunity to correct the errors by mismatch repair. In *E. coli* and *Salmonella typhymurium*, mismatch repair employs three genes, *mutH*, *mutS*, and *mutL*. This DNA-repair system recognizes a variety of different kinds of mismatches, including small insertions or deletions of up to three nucleotides. It distinguishes the newly replicated DNA strand from the template strand by the state of methylation of C at

GATC sites. It initiates a single-strand break in the newly replicated DNA across from a methylated template site, and other exonucleases cleave a string of bases in the DNA (which can be extensive) to just beyond the mismatch. The single-strand region is then repaired by DNA polymerase and DNA ligase, restoring the wild-type sequence to the site of the original mismatch. Mutations in any of the mismatch-repair genes in *E. coli* lead to elevated levels of spontaneous mutation. Because *mutS* and *mutL* genes have been observed in a number of other bacteria, mismatch repair is likely to play a significant role in modulating spontaneous mutation rates.

5. Deletions and Amplifications

Unlike many single-cell prokaryotes that contain circular chromosomes, the industrially important filamentous *Streptomyces* species have linear chromosomes with teleomeres with the potential to form secondary structures. These linear chromosomes are very large (about 8 Mb, or nearly twice the size as the *E. coli* chromosome), and the DNA near the ends is prone to high-frequency spontaneous deletion and amplification. The deletions can be as large as 1 Mb, and the mutants remain prototrophic (ie., able to grow on minimal medium) in some cases. Segments of DNA of varying unit sizes can also amplify several hundredfold, adding well over 1 Mb to the mutant strains. This high spontaneous deletion and amplification rate is likely to also occur in many other actinomycetes, the most important group of industrial microorganisms. In some cases, deletions or amplifications occur across repeated sequences. Genetically unstable actinomycetes can be stabilized by mutation and selection for stable production of secondary metabolites.

6. Transposable Elements

Bacteria commonly harbor insertion sequence (IS) elements and transposons. Transposable elements have the ability to translocate, often site-specifically, from one site to another in the genome. Many transposable elements carry outward-reading promoter sequences. If a transposable element with promoter activity inserts upstream of a gene, it can activate transcription or change the level of transcription, thus causing a spontaneous mutation. If a transposable element inserts into an open reading frame, it can disrupt the function of the gene, causing a mutation. Some transposable elements leave small deletions when they exit a site. Thus the number, target specificity, and relative activity of transposable elements can influence the overall rate of spontaneous mutation in microbes.

III. INDUCED MUTATION

Mutations can be induced by radiation, by certain chemicals and by other forms of environmental stress. Ultraviolet radiation (UV) causes a variety of different kinds of base-pair substitutions, frameshift mutations, and deletions. Some common chemical mutagens are hydroxylamine (HA), methyl methanesulfonate (MMS), ethyl methanesulfonate (EMS), 4-nitroquinoline-1-oxide (NQO), and *N*-methyl-*N*^{1}-nitro-*N*-nitrosoguanidine (MNNG). Of these, MNNG is generally the most potent. These chemical mutagens induce G-C to A-T transition mutations predominantly. The mutagenicity of at least some of these agents depends on how the premutational lesion is dealt with by error-free and inducible error-prone (SOS) DNA-repair systems. Mutation rates can also be modulated by an inducible adaptive repair system that can remove methyl groups from *O*-6-methylguanine, a major mutagenic analog generated by alkylation. Mutation rates can be increased by 1000- to 10,000-fold by MNNG treatment of growing bacterial cells.

IV. APPLICATIONS OF MUTAGENESIS

A. Basic Research

Mutagenesis is an important tool for basic research in microbiology. Much of what is known about biosynthetic and catabolic pathways in bacteria has been supported by having mutants defective in steps in the pathways of interest. Likewise, mutations in other cellular functions give the corresponding gene a phenotype that can help define the function of the gene. Much has been deduced about gene a function using chemically induced or radiation-induced mutants of bacteria. A drawback of this approach is that some phenotypes are the result of more than one mutation. This is more likely when very potent mutagens are used. A more rigorous approach is to delete an internal segment of a gene in frame so that the phenotype of the resulting mutant can be assigned unambiguously to a single gene. A number of genetic engineering methods are now available to facilitate these strain constructions. As more bacterial genome sequences become available, the development of gene disruption and in frame-deletion methods for a variety of microbes will be critical for the analysis of gene function. This will be particularly important in the identification of virulence factors and novel targets for antibiotic development in bacterial pathogens.

B. Industrial Strain Improvement

Mutagenesis has been an indispensable tool for process development in the fermentation industry. Many antibiotics and other secondary metabolites, such as antitumor agents, antifungal agents, insecticides, and anthelminthic agents, are produced by the filamentous actinomycetes. Actinomycetes isolated from soil generally produce very low amounts of secondary metabolites in laboratory media. However, product yields can be increased dramatically by sequential mutagenesis and screening. In carrying out mutagenesis for strain improvement, it is critical to maximize the level of mutagenesis to give the highest frequency of strains containing only one mutation affecting the secondary metabolite yield. This is necessary because many mutations cause reduced yield and mutations that increase product yield are rare. Thus the coupling of two or more mutations affecting yield in the same strain will invariably lead to lower product yields. A model based on Poisson statistics has been developed that allows this process to be optimized. In practice, only the most potent mutagenic agents such as MNNG can induce levels of mutagenesis in actinomycetes that approach the optimum mutational multiplicity.

See Also the Following Articles

Diversity, Microbial • DNA Repair • Strain Improvement • Transposable Elements

Bibliography

Baltz, R. H. (1986). Mutagenesis in *Streptomyces* spp. *In* "Manual of Industrial Microbiology and Biotechnology" (A. L. Demain and N. A. Solomon, eds.), pp. 184–190, ASM Press, Washington, DC.

Baltz, R. H. (1998). New genetic methods to improve secondary metabolite production in *Streptomyces. J. Indust. Microbiol. Biotechnol.* **20**, 360–363.

Baltz, R. H. (1999). Mutagenesis. *In* "Encyclopedia of Bioprocess Technology: Fermentation, Biocatalysis and Bioseparation" (M. C. Flickenger and S. W. Drew, eds.), pp. 1819–1822. Wiley, New York.

Berg, D. E., and Howe, M. M. (1989). "Mobile DNA." ASM Press, Washington, DC.

Blanchard, J. S. (1996). Molecular mechanisms of drug resistance in *Mycobacterium tuberculosis. Annu. Rev. Biochem.* **65**, 215–239.

Drake, J. W. (1991). Spontaneous mutation. *Annu. Rev. Genet.* **25**, 125–146.

Drake, J. W. (1993). Rates of spontaneous mutation among RNA viruses. *Proc. Natl. Acad. Sci. U.S.A.* **90**, 4171–4175.

Friedberg, E. C., Walker, G. C., and Siede, W. (1995). "DNA Repair and Mutagenesis." ASM Press, Washington, DC.

Humayun, M. Z. (1998). SOS and mayday: Multiple inducible mutagenic pathways in *Escherichia coli. Mol. Microbiol.* **30**, 905–910.

King, G., and Murray, N. E. (1994) Restriction enzymes in cells, not eppendorfs. *Trends Microbiol.* **2**, 465–469.

Michaels, M. L., and Miller J. H. (1992). The GO system protects organisms from the mutagenic effect of the spontaneous lesion 8-hydroxyguanine (7,8-dihydro-8-oxoguanine). *J. Bacteriol.* **174**, 6321–6325.

Miller, J. H. (1983). Mutational specificity in bacteria. *Annu. Rev. Genet.* **17**, 215–238.

Modrich, P., and Lahue, R. (1996). Mismatch repair in replication fidelity, genetic recombination, and cancer biology. *Annu. Rev. Biochem.* **65**, 101–133.

Volff, J. N., and Altenbuchner, J. (1998). Genetic instability of the *Streptomyces* chromosome. *Mol. Microbiol.* **27**, 239–246.

Woese, C. (1998). The universal ancestor. *Proc. Natl. Acad. Sci. U.S.A.* **95**, 6854–6859.

Mycobacteria

John T. Belisle and Patrick J. Brennan

Colorado State University

GLOSSARY

cell wall   One of the outer (protective) layers of the bacterial cell; the most characteristic feature of *Mycobacterium* in its structure, biogenesis, and genetics.

genomes   The complete genetic material of an organism; the *M. tuberculosis* has been fully sequenced and those for *M. leprae* and *M. avium* are under investigation.

leprosy   A wasting disease in humans characterized by lesions in the extremities; fewer than one million sufferers worldwide.

metabolism   The chemical and physical processes occurring within a living organism that are involved in maintaining normal life status and function. The general metabolism of *Mycobacterium* is not unusual, other than in biogenesis of cell wall components.

Mycobacterium leprae   Causative agent of human leprosy.

Mycobacterium tuberculosis   Causative agent of human tuberculosis.

tuberculosis   A chronic infectious condition affecting the lungs and other organs; a disease of vast proportions.

MYCOBACTERIUM is a bacterial genus containing at least 30 different species, including *M. tuberculosis* and *M. leprae,* the causative agents of tuberculosis and leprosy. Members of this genus are also opportunistic pathogens and are common causes of infection in AIDS patients. Although this group of bacteria is positive by Gram stain, the bacteria are not true gram-positive organisms, but are classified as acid-fast bacilli, a staining characteristic that is attributed to the unique cell wall of mycobacteria. This article discusses the physiology and genetics of *Mycobacterium* spp., as well as the history and pathogenesis of common mycobacterial diseases.

I. GENERAL CHARACTERISTICS OF *MYCOBACTERIUM*

The genus *Mycobacterium* is composed of at least 30 different species and many subspecies and is commonly subdivided into two groups, slow-growing and rapid-growing species (Table I). All members of this genus are nonmotile, nonspore forming chemoheterotophs that possess genomes with a high (55 to 70%) G+C content. Most *Mycobacterium* spp. are aerobic; however, some species exhibit the ability to survive in both aerobic and microaerophilic environments. Mycobacterial cells are generally observed as straight or curved rods, but branched, coccobacillary, or filamentous morphologies are observed for some species. Phylogenetic classification of mycobacteria demonstrates a closer relationship to gram-positive than gram-negative bacteria, and, in fact, these bacilli are positive for the Gram stain. The cell wall of mycobacteria is its most distinguishing feature and is characterized by the covalent linkage of mycolyl–arabinogalactan to the peptidoglycan backbone. This lipophilic structure provides a protective barrier for the cells and is responsible for the acid-fast nature of the bacilli. The majority of *Mycobacterium* spp. are saprophytic organisms, commonly found in water and soil. However, this group of bacteria also encompasses several pathogens and opportunistic pathogens, the most noted of which are *M. tuberculosis* and *M. leprae,* the causative agents of tuberculosis and leprosy, respectively, and the *M. avium* complex,

TABLE I
"Approved" Mycobacterial Species

Slowly Growing

M. tuberculosis (N)	*M. bovis* (N)	*M. africanum* (N)
M. microti (N)	*M. kansasii* (P)	*M. marinum* (P)
M. simiae (P)	*M. asiaticum* (P)	*M. gordonae* (S)
M. scrofulaceum (S)	*M. szulgai* (S)	*M. avium* (N)
M. intracellulare (N)	*M. lepramurium* (N)	*M. paratuberculosis* (N)
M. malmoense (N)	*M. haemophilum* (N)	*M. farcinogenes* (N)
M. triviale (N)	*M. terrae* (N)	*M. nonchromogenicum* (N)
M. ulcerans (N)	*M. gastri* (N)	*M. xenopi* (N)
M. shimoidei (N)	*M. celatum* (N)	

Rapidly Growing

M. chelonae (N)	*M. fortuitum* (N)	*M. phlei* (S/P)
M. smegmatis (S/P)	*M. aurum* (S)	*M. gadium* (S)
M. neoaurum (S)	*M. vaccae* (S)	*M. chitae* (S)
M. duvalii (S)	*M. flavescens* (S)	*M. gilvum* (S)
M. komossense (S)	*M. senegalense* (N)	*M. parafortuitum* (S)
M. thermoresistibile (S)	*M. diernhoferi* (N)	*M. rhodesiae* (S)
M. aichiense (S)	*M. porcinum* (S)	*M. sphagni* (S)
M. austroafricanum (S)	*M. obuense* (S)	*M. shinshuense* (S)
M. pulveris (S)	*M. agri* (N)	*M. chubuense* (S)
M. tokaiense (S)	*M. fallax* (S)	

Noncultivable (or of very fastidious growth)

M. leprae	*M. genavense*	*M. interjectum*
M. confluentis	*M. intermedium*	

Note. Key to pigmentation: N, nonchromogen; P, photochromogen; S, scotochromogen.

important opportunistic infections in the advanced stages of AIDS. Other species of particular interest are *M. bovis*, responsible for bovine TB, and *M. africanum*, probably a subspecies of *M. bovis*; *M. paratuberculosis* of the extended *Mycobacterium avium* complex (MAC) family, responsible for Johne's disease and, perhaps, implicated in Crohn's disease; *M. marinum*, responsible for swimming pool granuloma; and *M. ulcerans*, responsible for Buruli ulcers and the only species with a clearly defined polyketide toxin.

II. HISTORY AND STATUS OF MAJOR MYCOBACTERIAL DISEASES

It can be said that the field of mycobacteriology was born in 1882, when Robert Koch reported the isolation of the causative agent of tuberculosis. However, Armauer Hansen had described *Mycobacterium leprae*, the first recognized bacterial human pathogen, in 1873. It was nearly one and a half decades later before the genus *Mycobacterium* was adopted to encompass the bacterial species responsible for tuberculosis and leprosy, as well as other acid-fast bacilli that were known as nonhuman pathogens.

A. Human Tuberculosis

Tuberculosis originated as an endemic disease in animals during the Paleolithic period, long before any known disease in man. In fact, modern genetic analyses of ancient human remains demonstrate that human tuberculosis was present only in agricultural communities, thus supporting the hypothesis that the domestication of animals permitted this disease to evolve into an infectious human pathogen. In recent history, the spread of tuberculosis reached epidemic proportions with the emergence of cities at the start of the Industrial Revolution. Poor socioeconomic conditions, overcrowding, and improper management of refuse have contributed to a tuberculosis pandemic that is prevalent to this day. It is estimated that, at its peak in Europe and the United States, tuberculosis accounted for one out of seven deaths and nearly 33% of all deaths for people in their most productive years. The implementation of patient management resulted in a decline of tuberculosis cases, and the introduction of antituberculosis agents, such as streptomycin, pyrazinamide, and isoniazid in the middle of this century, dramatically decreased the mortality rate of tuberculosis patients, but only in the so-called developed/industrialized countries. Over the last decade, a resurgence of tuberculosis was noted in the United States and some Western European countries. This increase was due to the reactivation of the disease in the elderly, the spread of AIDS, and the immigration of individuals from endemic areas of the world, as well as to a decline in socioeconomic conditions in metropolitan areas and the degradation of the infrastructure for the diagnosis and treatment of the disease. Associated with this increase in the incidence of tuberculosis was the emergence of multi-drug-resistant strains of *M. tuberculosis*. A rapid response to this resurgence in tuberculosis by state and federal agencies in the United States has resulted in a continual decline in the incidence of tuberculosis since 1993. However, the global outlook for tuberculosis is less promising. The most recent statistics by the World Health Organization indicate that one-third of the population of the world is infected with tuberculosis. Of these individuals, it is estimated that 8–10 million develop active disease and 3 million die from tuberculosis each year. It is estimated that tuberculosis is responsible for the deaths of more youths and adults than any other infectious disease, and tuberculosis is the single greatest cause of deaths among women of reproductive age; over 900 million women are currently infected with *M. tuberculosis*, one million will die, and 2.5 million will develop active disease.

B. Leprosy

The book of Leviticus has descriptions of leprosylike diseases. Early accounts of a leprosylike disease are found in the *Charaka Samhita,* an Indian text written between 600 and 400 B.C. It is believed that leprosy was brought from India to Greece in the fourth century B.C. by the army of Alexander the Great. In the New Testament, there is the well-known encounter between Jesus and the 10 lepers on the road to Galilee. In the first English translation of the *Vulgate* in 1384, Wycliffe translated *lepra* as *leprosy.* People with leprosy have been reviled, ostracized, and isolated since time immemorial. In 1948, the International Leprosy Association recommended the term *leprosy patient* instead of *leper,* and, indeed, in many of the South American countries, the term *Hansen's Disease* is preferred.

The WHO Expert Committee, in its fifth report, in 1977, estimated the prevalence of leprosy at 12 million cases, and the WHO Study Group on the Epidemiology of Leprosy in Relation to Control, in 1983, referred to an estimate of 11,525,000 cases. In 1991, the WHO Study Group on Chemotherapy for Leprosy Control recommended multiple drug therapy (MDT) for regimens for both multibacillary and paucibacillary leprosy, which were subsequently universally implemented and are very effective and well tolerated. Due to the substantial progress in leprosy control through MDT, the World Health Assembly in 1991 was prompted to call for the elimination of leprosy as a public health problem by the year 2000, defining elimination as attaining a level of prevalence below one case per 10,000.

At the beginning of 1999, out of 122 countries which were considered endemic in 1995, 94 had reached the elimination target, and the leprosy prevalence had been reduced 85% since 1985. Thus, the figure for global prevalence was 834,988 registered cases, with a 1.4 rate per 10,000—excellent but still well short of the WHO goal. Moreover, the numbers of new cases detected has continued to increase over the past decade; the number at the beginning of 1999 was 795,117 (13.3 per 100,000). The reason for this is not clear; WHO attributes it to the very active ongoing leprosy elimination campaigns. However, we know little about the transmission of leprosy or the incidence of leprosy or the possibility of an environmental source or reservoir. Clearly, leprosy will continue to be a problem for years to come.

C. Bovine Tuberculosis

The dimensions of global bovine tuberculosis are unknown but are considered to be vast, even in comparison with human tuberculosis. The extent of bovine tuberculosis in the water buffalo and lion herds of Kruger National park in South Africa is but one example. Other notable instances are found in Great Britain and New Zealand, where bovine tuberculosis is maintained as an endemic infection in wild animals. Over the past 10 years, Great Britain has experienced a rising incidence of TB in cattle herds, especially in the southwest of England and in South Wales, and the likeliest source of infection is the badger (*Meles meles*), whose natural habitat often lies near or within cattle pasture areas. Much the same problem occurs in Ireland. In those countries, the badger is a protected and much loved animal, leading to a very vigorous debate among conservationists, the farming industry. and public health workers. *M. bovis* is closely related to *M. tuberculosis.* In the United Kingdom in 1934, about 40% of cows in dairy herds were infected with bovine TB, and *M. bovis* was estimated to cause about 2000 human deaths annually. However, the introduction of milk pasteurization and regular tuberculin testing of cattle and compulsory slaughter of reactors have had a dramatic beneficial effect on human and cattle health. In 1995, less than 0.5% of cattle in Great Britain had TB and only 32 cases of human TB were attributed to *M. bovis.* The same measures, notably the draconian implementation of test and slaughter, in the United States has virtually wiped out bovine TB, except along the southern border. Bovine tuberculosis in developing countries is one of the great silent epidemics of which we know little.

Bovine paratuberculosis (Johne's disease) is also of vast proportions, particularly in the northern hemisphere, and its relationship to Crohn's disease is also a matter of vigorous debate but of little concrete research.

D. Atypical Mycobacterioses

Although once thought of as saprophytic organisms, many *Mycobacterium* spp. classified as atypical mycobacteria are now associated with a wide array of diseases that involve lymphatic, pulmonary, skeletal, cutaneous, and severe disseminated infections. Atypical mycobacteria were originally defined as a group of acid fast bacilli that were avirulent in guinea pigs and did not produce rough colony morphotypes.

These bacteria have also been referred to as Mycobacteria Other Than Tuberculosis (MOTT). In the first part of the twentieth century, many mycobacterial isolates from patients with pulmonary disease were misidentified as *M. tuberculosis,* and mycobacteria from other sources were disregarded or poorly described. However, during the 1940s and 1950s, it was recognized that some atypical mycobacteria were important human pathogens. The most noted of the atypical *Mycobacterium* species, *Mycobacterium avium,* was originally described as the causative agent of avian tuberculosis, and, in 1942, the recovery of "avium-like" bacilli from a man suffering silicotuberculosis was reported. Several years later, an intracellular acid-fast organism, named *Nocardia intracellularis,* was isolated from a disseminated infection. This organism and an acid-fast bacillus, termed the *Batty bacillus,* were finally recognized as one and designated *M. intracellulare.* It is now clear that *M. avium* and *M. intracellulare* are closely related, and these two species form what is now termed the *M. avium* complex (MAC). *M. avium, M. intracellulare, M. scrofulaceum,* and *M. lepraemurium* are often called the MAIS serocomplex, sharing the same type of glycopeptidolipid (GPL) antigens. The MAC are a source of major medical health problems, in that members of the complex are responsible for the most common opportunistic bacterial infections associated with AIDS patients. Other atypical mycobacteria that are known pathogens or opportunistic pathogens are *Mycobacterium kansasii, Mycobacterium scrofulaceum, M. marinum, Mycobacterium ulcerans, Mycobacterium fortuitum, Mycobacterium simiae, Mycobacterium szulgai,* and *Mycobacterium xenopi.*

III. PHYSIOLOGY OF *MYCOBACTERIUM*

A. General Metabolism

The two dominant physiological features of mycobacteria are: (a) slow growth (even the fastest-growing mycobacteria take 3–4 days to grow in the laboratory, compared to 16–18 hrs for *E. coli*) and (b) structure and associated metabolism of the cell envelope, unique among prokaryotes. There is still no satisfactory answer for the exceptionally slow growth rate of mycobacteria: whether it is the relative impermeability of the envelope or aberrations in nucleic acid synthesis or the fact that the very slow-growing mycobacteria have only one set of rRNA genes.

Mycobacteria are obligate aerobic, but, *in vivo,* they are oxygen-limited in their metabolism. Instracellular mycobacteria repress their respiratory metabolism in response to the low amount of O_2 available, and, in situations where tissue necrosis has occurred, *M. tuberculosis* may be completely deprived of O_2 but can survive by depressing all oxidative activities and entering a dormant state. In this situation, the anaerobic glyoxolate shunt may provide a basal metabolic rate. Sporulating type genes may be activated, and the cell wall may thicken. Chemotherapy of the dormant bacillus is an important research topic.

In broad terms, the metabolism of mycobacteria is not exceptional, i.e., carbohydrate metabolism, energy production, and the biosynthesis of low molecular weight metabolites and macromolecules follow the routes of other prokaryotes. Mycobacteria can utilize a wide range of carbon compounds for growth in the laboratory and, presumably, in host tissue. They are also able to assimilate a wide range of lipids and proteins through the products of extracellular lipases and proteases. *M. tuberculosis* relies principally on carbohydrates as a source of metabolic energy. The Embden–Myerhoff and the pentose phosphate pathways generate the usual range of intermediates for macromolecular biosynthesis and also lead to the formation of acetyl CoA and oxaloacetate leading to citric acid, the reactions of the tricarboxylic acid cycle, and the major source of $NADH/FADH_2$, from which oxidative phosphorylation will lead to ATP generation, and, through formation of α-oxoglutarate and NH_3 assimilation, will lead to amino acids and purine and pyrimidine bases. However, the *in vivo Mycobacterium* is surrounded by an abundance of mixed preformed nutrients, such as lipids, carbohydrates, and proteins. Since these can be readily and simultaneously assimilated, the result will be repression of the enzymes for *de novo* synthesis. However, since these *de novo* pathways will operate when the organisms are grown in the laboratory, we know much about them and little about the *in vivo*-grown *Mycobacterium*. Since *M. leprae* is derived from living tissues,

a comparison of the genome and proteome of *M. tuberculosis* and *M. leprae* will prove fascinating in demonstrating the essential requirements for *in vivo* growth.

Iron acquisition, metabolism, and storage in mycobacteria are one basic metabolic process with special unique features. Mycobacteria have evolved their own class of unique siderophores, the exochelins, the extracellular siderophores, and the mycobactins, the intracellular siderophores. Both the exochelins and the mycobactins of *M. tuberculosis* are of the one class of structure but differ in the nature of the alkyl side chains—the exochelins, the agents of iron acquisition, have short alkyl chains, whereas those of the mycobactins have substantially longer chains, ensuring their retention within the cell envelope and the cellular transmission of iron. There is only limited available information on subsequent events in the uptake of the ferri-siderophore complex. The physical receptors for the transport of the entire complex across the cell wall and membrane, the iron regulated envelope proteins (IREPs), have been recognized. Bacterioferritin, the iron storage protein, at least for *M. leprae*, has been fully studied. However, the level of knowledge is not yet to be compared to such as the Fur system of *E. coli*.

B. The Mycobacterial Cell Wall— Structure and Biosynthesis

1. Architecture of the Mycobacterial Cell Wall

The cell wall of mycobacteria is a cornucopia of novel structures and biosynthetic pathways. Present-day understanding of the architecture of the mycobacterial cell envelope arises from electron-microscopic studies, which demonstrate alternating zones of electron density and transparency reminiscent of gram-negative cells. This evidence is in accord with chemical models of a substantial lipid-permeability barrier, a staggered pseudo-outer membrane contributed in one-half by the ca. C_{56} meromycolate and ca. C_{24} α chains of mycolic acids and, in the other, by an assortment of free lipids with medium-chain (C_{30}, e.g., the mycocerosates) and short-chain fatty acids. This concept has received credibility from recent permeability, x-ray diffraction, calorimetry, and biochemical studies. Thus, the outer half of this lipid barrier and the variable glycocalyx zone beyond it contain the many extracellular polysaccharides and peculiar species-specific glycolipids of mycobacteria, such as the glycopeptidolipids (GPLs) in the case of the *M. avium* complex, the lipooligosaccharides (LOSs) in the case of *M. kansasii, M. xenopi, M. fortuitum,* and many other species, and the phthiocerol-containing phenolic glycolipids in the case of *M. leprae* and other mycobacteria. The non-specific lipids, such as the nonglycosylated phthiocerol-containing lipids (phthiocerol dimycocerosate; phenol–phthiocerol dimycocerosate), the trehalose mycolates (e.g., cord factor), and the sulfolipids are presumably also located in this region of the cell wall. The inner half of this lipid barrier is composed of the mycolic acids anchored to the arabinogalactan–peptidoglycan complex. Lipoarabinomannan (LAM), either mannose-capped (ManLAM) or uncapped (AraLAM), important immunomodulators, is probably primarily anchored in the cytoplasmic membrane and extends through the cell wall proper. Mycobacteria are also endowed with distinctive phospholipids, the phosphatidylinositol mannosides, some of which are precursors of LAM. These are also located primarily in the plasma membrane, although there is also evidence for their existence in the outer lipid environment.

2. Structure of the Cell Wall Core— The Mycolylarabinogalactan– Peptidoglycan Complex

A spate of intensive investigations from 1950–1975 allowed the definition of the insoluble cell wall matrix as a cross-linked peptidoglycan linked to arabinogalactan and esterified at the distal ends by the mycolic acids. Peptidoglycan consists of alternating units of N-acetylglucosamine (GlcNAc) and modified muramic acid (Mur) residues, to which are attached the peptide side chains. These tetrapeptide side chains consist of L-alanyl-D-isoglutaminyl-meso-diaminopimelyl-D-alanine (L-Ala-D-Glu-A$_2$pm-D-Ala) with the Glu being further amidated. This type of peptidoglycan is one of the most common found in bacteria. However, mycobacterial peptidoglycan differs in two ways: Some or all of the Mur residues

are N-acylated with glycolic acid (MurNGly), and the cross-links include a proportion between two A$_2$pm residues as well as between A$_2$pm and D-Ala.

The arabinogalactan is also unique, not only in its elemental sugars, but, unlike most bacterial polysaccharides, it lacks repeating units, composed instead of a few distinct structural motifs: (i) the arabinose (Ara) and galactose (Gal) residues are in the furanose (f) form; (ii) the nonreducing termini of arabinan consists of the structural motif [β-D-Ara$f$-(1→2)-α-D-Ara$f$]$_2$-3,5-α-D-Ara$f$-(1→5)-α-D-Ara$f$; (iii) the majority of the arabinan chains consist of 5-linked α-D-Ara$f$ residues with branching introduced by 3,5-α-D-Ara$f$; (iv) the arabinan chains are attached to C-5 of some of the 6-linked Gal$f$ residues, and approximately 2–3 arabinan chains are attached to the galactan core; (v) the galactan regions consist of linear alternating 5- and 6-linked β-D-Gal$f$ residues; (vi) the galactan region of arabinogalactan is linked to the C-6 of some of the MurNGly residues of peptidoglycan via a special diglycosyl-P bridge, α-L-Rha$p$-(1→3)-D-GlcNAc-(1→P); the mycolic acids are located in clusters of four on the terminal hexaarabinofuranoside, but only two-thirds of these are mycolated; (vii) the arabinan chains consist of a tricosarabinoside (23-mer), with three such chains attached to the galactan unit; (viii) the galactan also consists of 23 Gal residues of the repeating linear structure [β-D-Gal$f$-(1→5)-β-D-Gal$f$-(1→6)]$_n$, devoid of any branching; (ix) the points of attachment of the arabinan chains are close to the reducing end of galactan, itself linked to peptidoglycan *via* the linker disaccharide-P.

3. Biosynthesis of the Linkage Unit, Galactan and Arabinan of Arabinogalactan (Fig. 1)

The Gal$f$ residues of cell wall galactan originate in UDP-Gal$f$ by way of UDP-Gal$p$ and the respective mutase, and the Ara$f$ residues of arabinan originate in decaprenol-P-Ara$f$ (C$_{50}$-P-Ara$f$), probably formed *via* the non oxidative pentose shunt, probably as P–Ara–P–P *via* P–Ribose–P–P. The biosynthesis of the linkage unit and attached galactan has recently been established. Membranes from *M. smegmatis* and *M. tuberculosis* catalyzed the incorporation of radioactivity from UDP-[^{14}C]GlcNAc into two polyprenol-

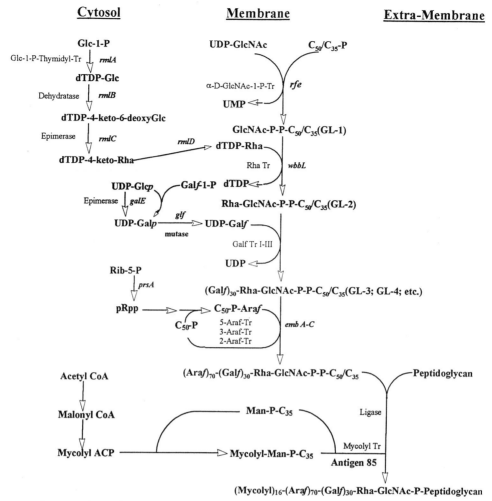

Fig. 1. Pathway for the synthesis of the mAGP complex of mycobacterial cell walls.

containing glycolipids (GL 1 and GL 2), polyprenol–P–P–GlcNAc (GL 1). Incorporation of [^{14}C]Rha from dTDP- [^{14}C]Rha took place only into GL 2, demonstrating that GL 2 is a polyprenol–P–P–GlcNAc–Rha. The inclusion of UDP-[^{14}C]Gal*p* resulted in exclusive labeling of GL 3 (polyprenyl–P–P–GlcNAc–Rha–Gal) and GL 4 [polyprenyl–P–P–GlcNAc–Rha–(Gal)$_2$], and higher versions (GL 5, 6, and so on), indicating growth of the galactan chain on the polyprenol–P–P–GlcNAc–Rha unit, eventually resulting in a polymer which possessed the characteristics of lipid-linked galactan with some attached Ara*f* residues (Fig 1). The final steps in the assembly of the cell wall core are not known, such as the transfer of the arabinan galactan-linker unit to peptidoglycan, further modification of the arabinan, and attachment of mycolic acids.

4. Structure and Biosynthesis of Mycolic Acids: Action of INH

The mycolic acids are β-hydroxy-α-alkyl long chain (ca. C-80) fatty acids that are the most distinguishing feature of members of the *Mycobacterium* genus. The parent long chains are called meromycolate chains, whereas the α-branched chain is simply called the α-chain. *M. tuberculosis* contains α mycolates, characterized by cyclopropane groups; the methoxyl mycolates contain a methoxy group in the meromycolate chain and the keto mycolates contain a keto group. The complete structures of these and other mycolates from other mycobacteria are well known and well described.

Much of the biosynthesis and underlying genetics of mycolate synthesis has been described only recently (Fig. 2). Synthesis occurs through repetitive cycles of condensation, keto reduction, dehydration, and enoyl reduction, catalyzed by four enzymatic activities: β-ketoacyl synthase (KAS), β-ketoacyl reductase (KR), β-hydroxyacyl dehydrase (DE), and enoyl reductase (ER). The saturated fatty acid produced during a round of such reactions is used as the substrate for another round of extensions, until a terminal extension value is reached. The genes and activities responsible for the insertion of the cyclopropane groups and other adducts have been identified. The mechanism of the final condensation of the α-chain and meromycolate is not fully understood.

The association between catalase/peroxidase, INH susceptibility/resistance, and mycolic acid biosynthesis had been known for years. It is now known that INH susceptibility and resistance are mediated by the *katG* gene, which encodes an 80 kDa protein containing heme and structural motifs characteristic of several bacterial catalase/peroxidases. The *M. tuberculosis katG* gene restored sensitivity to INH in a resistant mutant of *M. smegmatis*, and deletion of the gene resulted in INH resistance in *M. tuberculosis*. Clearly, INH is a pro-drug that requires the *katG* product for activation, and it has been proposed that AhpC, a homolog of the thioredoxin-dependent alkyl hydroperoxide reductase, can interact directly with activated INH, compensating for the loss of KatG peroxidase in INH resistance.

The primary effect of INH is on mycolic acid synthesis, and, recently, a novel gene, *inhA*, was identified, which, through point mutations within the 5′-regulatory region, conferred resistance to both INH and ETH. The InhA protein has now been shown to catalyze the NADH-specific reduction of long-chain (C_{12}–C_{24}) 2-trans-enoyl ACP intermediates involved in fatty-acid elongation, consistent with its involvement in the early stages of mycolic acid biosynthesis. This important work served to reopen the debate on the polymorphic nature of INH targets and mechanism of resistance and provided a clearer understanding of the biochemistry of the chain elongation

phases of mycolic acid synthesis. For instance, Barry and colleagues proposed that, in *M. tuberculosis*, INH specifically inhibits the insertion of a $\Delta5$-double bond into a C_{24} fatty acid, and, thus, that the target in the case of *M. tuberculosis* was a $\Delta5$-desaturase, rather than a β-keto enoyl reductase; the latter may be the primary target in *M. smegmatis*. Thus, apparently contrasting views and information on the action of INH, also prevalent in the 1960s and 1970s, can be reconciled by invoking either difference of action in different mycobacteria or pleotrophic effects. Nevertheless, the interest in INH has precipitated a deeper understanding of mycolic acid synthesis.

5. Biosynthesis of LAM

LAM biosynthesis is initiated by two α-mannosyl-transferases, which utilize the donor GDP-Man and the acceptor phosphatidylinositol (PI) to yield phosphatidylinositol dimannoside (PIM$_2$), which is then extended by the mannosyl donor C_{35}/C_{50}–P–Man to yield linear $\alpha(1\rightarrow6)$ lipomannan (LM), a step which is sensitive to amphomycin, a lipopeptide antibiotic that specifically inhibits polyprenol–P-requiring translocases. This linear LM then serves as a template for further additions of Man from either GDP–Man or C_{35}/C_{50}–P–Man to form mature LM, which is then the recipient of single Ara units or polymerized arabinosides, similar to the arabinosylation of the galactan of AG. The genes encoding the rich array of glycosyl transferases involved in LAM and AG synthesis are now being recognized, cloned, and function attributed to them.

IV. GENETICS OF MYCOBACTERIA

The genome size of most *Mycobacterium* spp. range between 3.1×10^9 to 4.5×10^9 daltons; however, the genomes of *M. tuberculosis*, *M. leprae*, and *M. bovis* are considerably smaller. The genomes of these three species are approximately 2.8×10^9, 2.5×10^9, and 2.2×10^9 daltons, respectively. The G+C content of mycobacteria is relatively high, with most species possessing genomes with 64 to 70% G+C. Studies utilizing DNA–DNA hybridization demonstrate that distinct clusters of closely related species exist, such as *M. bovis*, *M. africanum*, and *M. tubercu-*

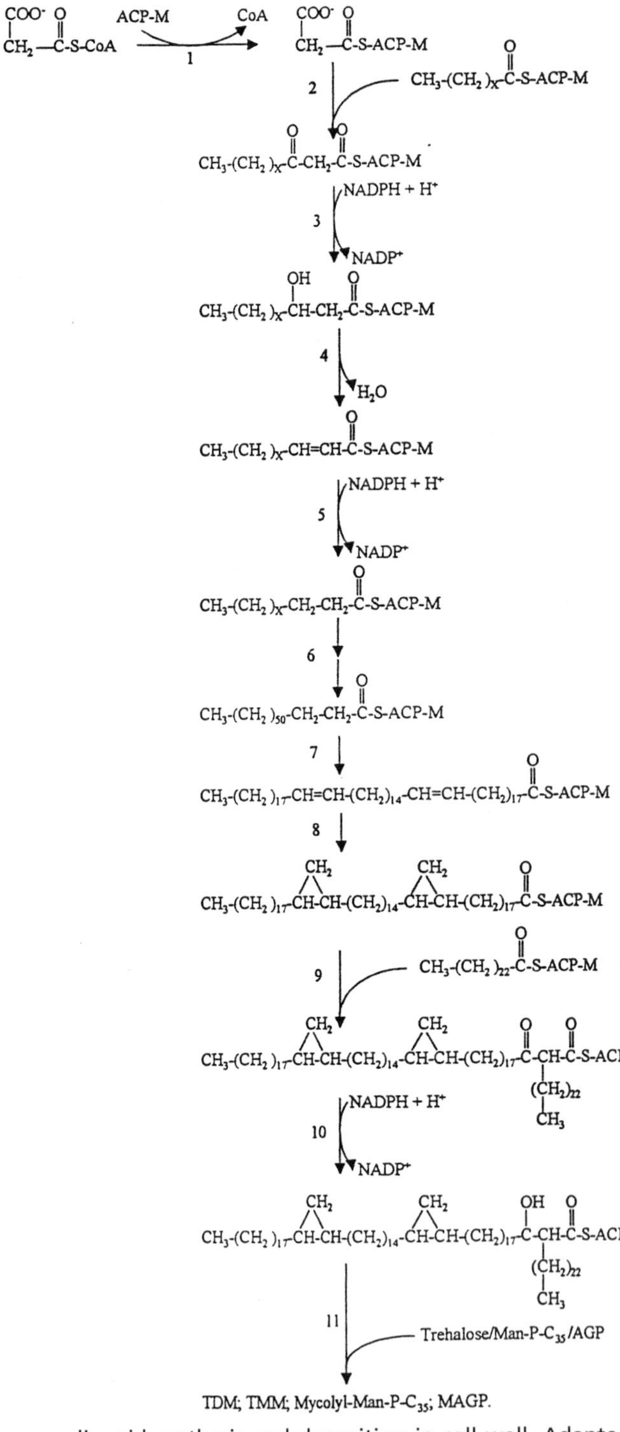

Mycobacteria

Fig. 2. Possible pathway for mycolic acid synthesis and deposition in cell wall. Adapted from Barry *et al.* (1998). *Prog. Lipid Res.* **37,** 143–179. (1) FabD, a malonyl–CoA–ACP transacylase; (2) KAS, β-ketoacyl ACP synthase; (3) KR, β-ketoacyl ACP reductase; (4) DE, β-hydroxyacyl ACP dehydratase; (5) EA, enoyl ACP reductase; (6) many repetitions of steps 2–5; (7) possible desaturation, as discussed by Barry *et al.* (1998); (8) introduction of cyclopropane groups, as discussed by Barry *et al.* (1998); (9) addition of the α-chain to form the 3-keto mycolate intermediate; (10) reduction of the 3-keto group; (11) transfer of mature mycolates from one carrier to either trehalose or a polyprenyl-P-Man or the final AGP complex.

losis, and *M. avium* and *M. intracellulare.* However, outside these clusters, a dramatic divergence exists between species of the *Mycobacterium* genus. The use of highly conserved genes also provides markers for defining the genetic relationship among *Mycobacterium* spp. Analysis of the nucleotide sequence of a number of 16s rRNA genes demonstrates a distinct separation of fast - and slow-growing mycobacteria. On average, the slow growers share 96% sequence homology, while the sequence homology of the slow growers to fast growers is significantly lower. The minor sequence variations that exist between the rRNA genes of *Mycobacterium* spp. have been successfully exploited for identification of *M. tuberculosis, M. avium,* and *M. intracellulare* in clinical specimens.

A. The *M. tuberculosis* Genome

The most significant advances in understanding the genetics of mycobacteria have come from sequencing the entire genome of *M. tuberculosis,* and undoubtedly, this knowledge will grow with the completion of the *M. leprae* and *M. avium* genomes. The genome of *M. tuberculosis* is 4,411,529 base pairs and encodes for 3924 putative open reading frames. Interestingly, a large number of these genes (approximately 250) encode enzymes involved in fatty-acid metabolism; an obvious reflection of the complexity of the cell wall and the myriad of fatty acids and other lipids produced by this organism. Additionally, 10% of the open reading frames are assigned to two large families of acidic, glycine-rich proteins, the PE and PPE families. The PE gene products possess a Pro–Glu motif near the N terminus and the PPE proteins possess a Pro–Pro–Glu motif. Many of the PE and PPE proteins also possess repetitive amino-acid sequences that are based on polymorphic G+C rich sequences (PGRS) and major polymorphic tandem repeats (MPTRs), respectively. The precise function of these proteins is unknown; however, the large number of gene products encompassed by these two families is indicative of biological importance. A third major group of sequences found in the genome of *M. tuberculosis* are insertion sequences. In *M. tuberculosis* strain H37Rv, 16 copies of IS6110 and 6 copies of IS1081 were detected. Additionally, the genome of this organism possesses 32 different IS elements, the majority of which were not described prior to the sequencing of the genome.

B. Insertion Sequences

Early analyses of the genomes of *Mycobacterium* spp. included a search for unique repetitive sequences that could be used for diagnostic and epidemiological purposes. This undertaking ultimately resulted in the discovery of mycobacterial insertion (IS) elements (Table II). Thus far, about 23 separate IS elements have been defined in *Mycobacterium* spp. Of these, Tn610 of *M. fortuitum* is the only naturally occurring transposon isolated from a *Mycobacterium* spp. This transposon possesses flanking IS6100 elements and harbors a gene encoding resistance to sulfonamide (*sul3*). Two atypical insertion sequences have been isolated: IS900 from *M. paratuberculosis* and IS901 from *M. avium.* These IS elements are atypical in that they lack inverted repeats and they do not duplicate the target site upon insertion. The IS901 along with IS6110 of *M. tuberculosis* have been widely used for epidemiological evaluation and molecular diagnosis of *M. avium* and *M. tuberculosis* infections. Two IS elements of *M. smegmatis* (IS6120 and IS1096) are known to actively participate in transposition. Both of these insertion sequences possess inverted repeats and generate 9 and 8 bp duplications of target DNA. More importantly, IS6120 and IS1096 have been developed as tools for transposon mutagenesis of mycobacterial genomes, an important application for the study of pathogenicity and gene function.

C. Extrachromosomal and Mycobacteriophage DNA

As with most prokaryotic organisms, *Mycobacterium* spp. possess extrachromosomal DNA in the form of plasmid and prophage. These two forms of DNA have received extensive attention, due to their value as epidemiological markers, their encoding of phenotypic traits, and, more recently, their utility as vectors for genetic manipulation of mycobacteria. The actual isolation of mycobacterial plasmids was first reported in 1979, with the analysis of several *M.*

TABLE II
Selected Insertion Sequences of *Mycobacterium* spp.

Insertion sequence	Family	Host	Size	IR[a]	DR[b]
IS986	IS3	M. tb. complex	1358	Yes	Yes
IS987	IS3	M. tb. complex	1355	Yes	Yes
IS900	IS116/900	M. paratuberculosis	1451	No	No
IS901	IS116/900	M. avium	1472	No	No
IS1081	IS256	M. tb. complex	1324	Yes	No
IS1096		M. smegmatis	2275	Yes	Yes
IS1110	IS116/900	M. avium	1457	No	No
IS1137	IS3	M. smegmatis	1361	Yes	Yes
IS1141	IS3	M. intracellulare	1596	Yes	Yes
IS1245	IS256	M. avium	1313	Yes	
IS1395	IS256	M. xenopi	1323	Yes	
IS1407	IS256	M. celatum	1325	Yes	
IS1408	IS256	M. branderi	1325		
IS1512	IS256	M. gordonae	1712		
IS1547	IS900	M. tb. complex	1351	Yes	No
IS1549	IS4	M. smegmatis	1634	Yes	Yes
IS1601	Composite	M. avium	1462	Yes	Yes
IS1652		M. kansasii	947	No	Yes
IS2404	ISRm3	M. ulcerans	1274	Yes	Yes
IS2606	ISRm3	M. ulcerans	1404	Yes	No
IS6100	IS6	M. fortuitum	880	Yes	No
IS6110	IS3	M. tb. complex	1361	Yes	Yes
IS6120	IS256	M. smegmatis	1486	Yes	Yes

[a] Inverted repeats.
[b] Direct repeats.

avium strains. Although plasmids have been isolated from several *Mycobacterium* spp., specific biological functions can be attributed to only a small number of these. In *M. scrofulaceum,* mercury reductase activity was associated with either a 177- or 240-kb plasmid, and the presence of the 177-kb plasmid also correlated with resistance to copper. Additionally, the ability of *M. avium* complex strains to respond to high levels of oxygenation was associated with the presence of a plasmid.

Several laboratories have attempted to correlate plasmids of *M. avium* with virulence; however, the results of such studies are conflicting. It should also be noted that no plasmids have been isolated from *Mycobacterium* spp. that are true pathogens. The most noted mycobacterial plasmid is pAL500, which was isolated from *M. fortuitum.* This plasmid has been used in the construction of numerous mycobacterial

vectors and its origin of replication has been sequenced. Only two of the five open reading frames of this plasmid (*repA* and *repB*) are necessary for replication. Additionally, a 435-bp fragment (*inc*) is required in cis orientation for replication and influences copy number. This region possesses several repeated motifs and confers incompatibility.

The earliest report of a mycobacteriophage infection was made in 1947. It is now evident that *Mycobacterium* spp. possess both lytic and lysogenic bacteriophage, and, like their hosts, these phages contain DNA of a relatively high G+C content (60 to 70%). Additionally, most of the mycobacteriophage are sensitive to organic solvents, suggesting that associated lipids influence the infectivity or integrity of the phage particles. As for mycobacterial cell surface receptors, little is known. However, the D4 phage, which is lytic for *M. smegmatis,* has been demonstrated to use the terminal methylated rhamnose of the nonspecific GPL as a receptor. The majority of mycobacteriophages are defined by their ability to produce lytic infections. This trait was exploited by epidemiologists for the typing of mycobacterial isolates but has largely been replaced by DNA probes. The most thoroughly studied of the mycobacteriophages are L5 and D29. Mycobacteriophage D29 produces lytic infections in both fast- and slow-growing *Mycobacterium* spp. In contrast, the L5 phage is a temperate phage that forms stable lysogens in *M. smegmatis*. It also infects slow-growing *Mycobacterium* spp.; however, this occurs only in the presence of high concentrations of Ca^{2+}. Sequencing of the genomes of both of these phages has been completed and reveals that they are closely related and possess similar gene organization. A major difference between the genomes is the presence of a 3.6 kb region in L5 phage and the absence of this region in the genome of the D29 phage. This 3.6-kb fragment possesses a repressor gene, and the lack of this gene in the D29 phage accounts for its inability to form lysogens.

D. The *M. leprae* Genome

In the 1980s, Dr. Stewart Cole constructed an ordered library of *M. leprae* cosmids, and, subsequently, a fingerprinting technique was used to order the cosmids, which allowed their assembly into blocks of contiguous DNA sequences. At present, the *M. leprae* chromosome can be represented as four contigs, and, by summing their sizes, one can estimate a size of about 2.8 megabases. A genome-sequencing project was initiated by Dr. Cole. The lessons from the sequence of the first cosmid to be sequenced (B1790) were surprising. Twelve candidate genes were found. However, the most surprising feature was a scarcity of genes with only ~50–60% of the potential coding (gene density) capacity being used, compared to about 85% in the case of *E. coli*. Later, the sequencing of another nine cosmids again demonstrated that genes were relatively scarce and that noncoding regions were extensive. The presence of large numbers of pseudogenes in *M. leprae* probably accounts for many of the phenotypic differences between the leprosy and tubercle bacilli, and this is

exemplified by their respective responses to isoniazid. In *M. tuberculosis*, *katG* encodes catalase–peroxidase, a heme-containing enzyme that mediates the toxic effect of isoniazid. Comparison of the corresponding regions of *M. leprae* and *M. tuberculosis* revealed the presence of numerous mutations in the *M. leprae* gene that abolished activity. This undoubtedly explains why the leprosy bacillus produces no catalase–peroxidase and displays high-level resistance to isoniazid. Until the sequence of *M. leprae* is completed, the precise number of genes will remain unknown. Preliminary estimates suggest that the proteome may contain as few as 1600 proteins. The genome of *M. leprae* is roughly 1.4 Mb smaller than that of *M. tuberculosis,* and its G+C content (57%) is significantly lower than those of all other mycobacterial genomes. Although deletion of coding sequences almost certainly led to some genome shrinkage, two other factors may also have contributed to the size difference. First, *M. leprae* contains very few members of the PE and PPE gene families which account for ~450 kb of the *M. tuberculosis* chromosome. Second, traces of far fewer insertion sequences (IS) and bacteriophages have been found in *M. leprae* than in *M. tuberculosis* H37Rv, where they contribute over 120 kb. Although the genome sequence is not yet complete, it is already clear that *M. leprae* contains at least 50 genes that do not have counterparts in *M. tuberculosis*. Their products could find application as immunodiagnostics.

V. PATHOGENESIS OF MYCOBACTERIAL DISEASES

A. Tuberculosis

Tuberculosis remains one of the most prevalent and deadly of all infectious diseases. An additional concern with the current tuberculosis problem is the recent emergence of multiple drug resistant (MDR) strains (strains of *M. tuberculosis* that are resistant to two or more of the first line antibiotics). On a global scale, only 20% of tuberculosis cases involve MDR strains. However, hot spots of tuberculosis associated with MDR strains occur. For instance, in 1998, about 50% of all cases in Latvia were due to

MDR strains of *M. tuberculosis*. An additional problem is the damage caused by strains of enhanced transmission and perhaps virulence, such as the multi-drug- resistant *M. tuberculosis* strain W family, responsible for much of the increased tuberculosis in New York City in the early 1990s, and the *M. tuberculosis* CDC 1551 strain, responsible for a large outbreak of tuberculosis in areas of Tennessee and Kentucky in which the index patient, the source patient, and one other patient were responsible for extensive transmission of the organism to over 70% of 429 contacts.

1. Pulmonary Tuberculosis

Infection by *M. tuberculosis* generally occurs via the inhalation of a small number of viable bacilli contained in small (1–10 μm) aerosolized droplets. Particles of this size are able to reach alveolar spaces and are not trapped by the ciliated epithelium of the bronchia. Once the bacilli reach the alveolus, they are ingested by alveolar macrophages and transported to regional lymph nodes. The bacilli will replicate or be destroyed, depending on the microbicidal activity of the macrophage and virulence of the bacilli. Infection of the macrophage induces the release of chemokines that attract additional immune effector cells. Additionally, some macrophages will be overwhelmed by the replicating bacilli and eventually die, releasing the bacteria and other chemo attractants. During the second and third weeks of infection, the number of bacilli are increasing logarithmically and continue to infect other macrophages in the primary foci of infection. By the fourth week of infection, the cell-mediated immune response has been primed and antigen-specific T cells activate killing of the bacilli by the macrophage. Additionally, the death of the macrophages increases, leading to the formation of a granuloma or tubercle. Activated macrophages continue to enter the granuloma to control bacterial growth. In most cases, the granuloma is calcified and the infection remains quiescent. If the initial infection is not controlled or if reactivation occurs after several years, liquification of the granuloma occurs. This provides the bacteria with an excellent environment for bacterial growth. The addition of a strong DTH response to the infection exacerbates the disease by facilitating cavity formation and erosion of the bronchial walls. At this point, the tuberculosis patient develops a productive cough and infectious materials can be transmitted through respiratory excrements.

Reactivation of disease or the development of post-primary tuberculosis can result several years after the initial infection. The cause of reactivation is not fully understood, but evidence indicates a strong correlation with the loss or weakening of the cell-mediated immune response and reactivation. Thus, factors that can contribute to reactivation include aging, alcoholism, drug abuse, illness, malnutrition, or immunodeficiency disorders. The association of tuberculosis with HIV infection is worth noting. HIV infects and damages those cells of the immune system that are responsible for an acquired immune response to the tubercle bacilli. Thus, individuals infected with HIV have a much greater chance of developing tuberculosis from a primary infection with *M. tuberculosis* as compared to the general population. Moreover, AIDS patients that were infected with *M. tuberculosis* prior to contracting AIDS are at a much higher risk of reactivation.

During the primary infection with *M. tuberculosis*, the bacteria can be disseminated throughout the body via the blood and lymphatic systems. Presumably, the bacilli are carried out of the site of the primary infection by infected macrophages. Additionally, the development of cavitary tuberculosis can also result in the spread of the bacilli and the formation of lesions in other tissues to include the nervous system, cervical lymph nodes, the kidneys, the bone marrow, and skin. Disseminated disease, termed miliary tuberculosis, is a severe form of extra pulmonary tuberculosis. This form of the disease is well noted in individuals that are immunocompromised.

2. Diagnosis of Tuberculosis

The diagnosis of an *M. tuberculosis* infection can be made using a number of techniques: evaluation of clinical symptoms, examination of sputum or other body fluids by light microscopy, chest x-rays to detect granulomas, culturing of sputum or other biological fluids, and immunological means. In many of the developing countries where tuberculosis is endemic, the primary means of diagnosis is the evaluation of sputum smears by light microscopy and, where available, radiological examination. However, microscopic examination of sputum samples identifies only about 50% of pulmonary tuberculosis cases. Sensitivity of detection is greatly increased when culturing of the sputum is performed. Several diagnostic kits for the serodiagnosis of tuberculosis are available; however, the sensitivity and specificity of these kits have been questioned. In the United States, one of the primary means of identifying tuberculosis patients is by skin-testing, or DTH reactivity to purified protein derivative (PPD). Skin-testing has been a mainstay of the tuberculosis control program in the United States for several decades and has been useful in controlling tuberculosis in the country. Drawbacks to skin-testing are its inability to distinguish between active and inactive disease, cross-reactivity with other mycobacterial infections, anergy in patients with well-developed cavitary tuberculosis, and negative reactivity in immunocompromised individuals.

Culturing remains the most reliable method for the diagnosis of tuberculosis and detection of the bacilli. The methodology for the culturing of *M. tuberculosis* from clinical specimens (primarily sputum) are well established and commonly use sodium hydroxide and N-acetyl-L-cysteine to decontaminate the sample. Specimens are then plated onto medium, such as Lowenstein–Jensen or Middlebrook 7H11, which provide good support for the growth of *M. tuberculosis*. Specific antibiotics which do not affect *M. tuberculosis* are commonly incorporated into the agar to prevent the growth of other bacteria. In support of culturing methods, a number of high-throughput techniques are available. One widely used method for the detection of mycobacteria is the BACTEC AFB System (Becton Dickinson Diagnostic Instruments, Sparks, Maryland) This methodology uses a medium containing a combination of antibiotics that inhibit growth of nonmycobacterial organisms and the specific growth of *Mycobacterium* spp. is detected by the β-oxidation of [^{14}C]palmitate and release of [^{14}C]carbon dioxide. The detection time for *M. tuberculosis* averages 9 to 14 days and may be less than 7 days for some fast-growing strains. A variation to this is the Mycobacteria Growth Indicator Tube (MGIT) System, which uses a fluorescent indicator dye that detects the consumption of oxygen. In addition to rapid culturing systems, a number

of analytical and molecular techniques have been developed for the rapid identification of *M. tuberculosis*. Among these is the Microbial Identification System (Microbial ID, Inc., Newark, DE) that is capable of identifying over 26 *Mycobacterium* species through HPLC analysis of fatty-acid profiles, especially mycolic acids. The first molecular-based system was the Gen-Probe Rapid Diagnostic System (Gen-Probe Corp., San Diego, CA) and was based on the sequence of the 16s rRNA genes and capable of distinguishing between *M. avium*, *M. intracellulare*, and *M. tuberculosis*. The principle of the test is that the rRNA released from lysed bacteria is hybridized with an [^{125}I]-labeled single-stranded DNA probe. This system has also been modified to use none isotopic acridinium ester labeled DNA probes (AccuProbes; Gen-Probe Corp., San Diego, CA).

The middle part of the twentieth century saw the use of the first chemotherapeutic agent (streptomycin) for the treatment of tuberculosis. Since that time, several other antibiotics have been introduced. Currently, the first-line drugs used for the treatment of tuberculosis are isoniazid, pyrazinamide, rifampin, and ethambutol or streptomycin. In general, treatment begins with a 2-month treatment with isoniazid, rifampin, and pyrazinamide, followed by treatment for 4 months with isoniazid and rifampin. It is also recommended that ethambutol or streptomycin be included in the initial regimen, until drug susceptibility testing of the isolate is completed. The emergence of MDR strains of *M. tuberculosis* is not due to a single mutation or the activity of a MDR efflux system, but is the result of multiple mutations in the genes encoding the targets of the various drugs. It is believed that such mutation arose from tuberculosis patients' failing to adhere to drug treatment programs and that mutations and selection for these mutations occurred over an extended period of time as new antibiotics were prescribed to control infections that were not fully resolved and that were now resistant to the drugs used in previous treatments.

B. Leprosy

We are not certain of the route of transmission of leprosy. The shedding of the large number of bacilli in the nasal discharge of lepromatous leprosy patients and deposition in the nasal mucosa of susceptible individuals is being accepted as the likely route. The organism is well adapted and completely dependent on the host, and, according to the genome sequence, has shed most virulence factors, and so the disease is slowly evolving and unique in its clinical and immunopathological spectrum. Leprosy patients can be classified within a six-stage clinical immunopathological spectrum, according to the Ridley–Jopling classification. Indeterminate leprosy often goes unrecognized and consists of a single or a few hypopigmented lesions, perhaps with some sensory loss and minimal histological changes. Spontaneous healing usually occurs, but some patients may progress to one of the established forms of leprosy. In tuberculoid leprosy (TT), individuals may still have single or few maculas or raised hypopigmented lesions, but often one of the nerves in the region is enlarged or thickened. Histopathologic examination reveals highly ordered granulomas with epithelioid cells, multinucleated Langhans giant cells, and a discrete organization of CD4$^+$ and CD8$^+$ cells, but few acid-fast bacilli. Individuals with TT manifest a strong delayed-type hypersensitivity (DTH) to antigens but produce relatively low levels of antibody. Nerve damage in TT may be due to cell-mediated immunity (CMI), to *M. leprae* antigens in sites adjacent to nerves. In borderline tuberculoid leprosy (BT), the skin lesions resemble those of TT patients but are more numerous. The number of *M. leprae* vary in lesions from undetectable to a few. In borderline leprosy (BB), there are numerous erythematous or hyperpigmented lesions of varying sizes, containing epithelioid cells together with macrophages and T cells but no multinucleated giant cells. Bacteria are readily detected. Borderline lepromatous (BL) leprosy resembles lepromatous disease except that at least some of the skin lesions are selectively anesthetic. Due to the host CMI, peripheral nerve trunk involvement is more widespread than in LL and proliferation of bacilli in the mucous membranes is less than in LL. Skin lesions in BL leprosy contain predominantly macrophages with far more lymphocytes than LL. Polar lepromatous leprosy in the multibacillary (MB) form is the widespread anergic form of the disease. Proliferation of *M. leprae* results in a variety of skin lesions, notably nodules or lepromas.

The skin lepromas in advanced LL may contain 10^{10} *M. leprae* per gram of tissue. Characteristically, the cooler parts of the body are involved: the dermis and peripheral nerves, anterior of the eye, and the nasal mucosa. LL lesions are characterized by a massive accumulation of macrophages filled with large numbers of acid-fast bacilli and often containing large amounts of lipid, which are believed to be mostly bacterial in origin, phenolic glycolipid (PGL-I) and phthiocerol dimycocerosate (PDM). There is a potent antibody response in LL to PGL-I, LAM, arabinogalactan, and the protein antigens, and this response has some applicability in diagnosis. There is also a clearly identified progressive anergy in T-cell response that is limited to specific antigens of the leprosy bacillus, a phenomenon that has never been adequately explained. LL lesions display a striking paucity of T cells with a preponderance of the CD8$^+$ suppressor/cytotoxic subset yielding a CD4$^+$/CD8$^+$ ratio of 0.6:1. In TT lesions, more T cells are evident and exhibit mostly CD4$^+$ helper/inducer cells with a CD4+/CD8$^+$ ratio of 1.9:1. The distribution of these cells in the microenvironment of the granuloma is also remarkable. In TT lesions, CD4$^+$ are distributed throughout the lesion in close proximity to the granuloma macrophages, whereas cytotoxic CD8$^+$ are located on the periphery. In contrast, in the LL lesion, CD8$^+$ cells are dispersed throughout, close to the *M. leprae*-laden macrophages. It is thought that the CD8$^+$ T-suppressor cells down-regulate macrophage activation and CMI. The first human demonstration of the Th1/Th2 paradigm has been reported in human leprosy where the TT lesion demonstrates a Th1-type response, producing principally IFN-γ, IL-2, and lymphotoxin, largely responsible for activation of macrophages. Conversely, the LL lesion demonstrates distinct Th2 response by producing IL-4, IL-5, and IL-10, which promote antibody production.

So-called reactions in leprosy are clinically apparent, immunologically mediated inflammatory conditions occurring during the course of the disease in about 50% of patients. These manifestations of leprosy are due to the immunologic response of the host to the bacilli. They are basically of two types. Type 1 or reversal reactions are generally agreed to be a result of DTH and affect patients with borderline to tuberculoid leprosy. They are characterized by edema and erythema of preexisting lesions and a tendency for the overall disease classification to upgrade. Type 2 or erythema nodosum leprosum (ENL) lesions have long been thought to be manifestations of an Arthus type of hypersensitivity reaction. They are seen in BL and LL patients and are characterized by the development of crops of tender, erythematous skin nodules and fever. Both types of reaction can involve any tissue containing antigens of the leprosy bacillus; there, lesions of ENL are not confined to the skin, but can involve the eye, joints, nasal mucosa, etc.

Type 1 reactions histologically consist of edema on a BT or TT background initially. If the outcome of the type 1 reaction is upgrading, there may be an early increase in the number of lymphocytes. In servere type 1 reactions, caseation necrosis may occur. Type 2 reactions are characterized by an influx of neutrophils on a BL or LL background. A vasculities, involving arterioles or venules, is demonstrable in about half of the cases in type 2 reactions.

The extraordinary tropism of *M. leprae* for Schwann cells has now been explained in molecular terms. Laminan-2 in the basal lamina of the Schwann cell–axon unit serves as an initial target for *M. leprae* interaction with peripheral nerves. Laminan 2 anchors to Schwann cells through laminin receptors, specifically α-dystroglycan, and thereby provides a conduit for *M. leprae* into the Schwann cell. The ligand on *M. leprae* responsible for this interaction is not known; PGL-I is a likely candidate on account of its specificity for *M. leprae*.

The diagnosis of leprosy has not changed in a century, still relying on visual examination and assessment of tactile response to pin-prick or heat. With increasing numbers of new patients being identified, many with single lesions and thus of unsure diagnosis, there is a desperate need from research of tests of greater discriminatory value.

See Also the Following Articles

AEROSOL INFECTIONS • CELL WALLS, BACTERIAL • SURVEILLANCE OF INFECTIOUS DISEASES

Bibliography

Adler, J. J., and Rose, D. N. (1996). Transmission and pathogenesis of tuberculosis. *In* "Tuberculosis" (W. N. Rom and S. Garay, eds.), pp. 129–140. Little, Brown, and Company, Boston.

Barry, C. E., III, Lee, R. E., Mdluli, K., Sampson, A. E., Schroeder, B. G., Slayden, R. A., and Yuan, Y. (1998). Mycolic acid: structure, biosynthesis and physiological functions. *Prog. Lipid Res.* 37, 143–179.

Baulard, A. R., Besra, G. S., and Brennan, P. J. (1999). The cell wall core of *Mycobacterium*: structure, biogenesis, and genetics. *In* "Mycobacteria: Molecular Biology and Virulence" (C. Ratledge and J. Dale, eds.), pp. 240–259. Blackwell Science, Oxford.

Brennan, P. J., and Nikaido, H. (1995). The envelope of mycobacteria. *Annu. Rev. Biochem.* 64, 29–63.

Cole, S. T. *et al.* (1998). Deciphering the biology of *Mycobacterium tuberculosis* from the complete genome sequence. *Science* 393, 537–544.

Dale, J. W. (1995). Mobile genetic elements in mycobacteria. *Eur. Respir. J.* 8 Suppl. 20, 633s–648s.

Grange, J. M. (1996). The biology of the genus *Mycobacterium*. *Soc. Appl. Bacteriol. Symp. Ser.* 25, 15–95.

Krebs, J. R., Anderson, R. M., Clutten-Brock, T., Donnelly, C. A., Frost, S., Morrison, W. I., Woodroffe, R., and Young, D. (1998). Badgers and bovine TB: Conflicts between conservation and health. *Science* 279, 817–818.

Krahenbuhl, J. L., and Adams, L. B. *Mycobacterium leprae* as an opportunistic pathogen. (1999). *In* "Opportunistic Intracellular Bacteria and Immunity" (L. J. Paradise, ed.), pp. 75–90. Plenum Press, New York.

Modlin, R. L. (1994). Th1–Th2 paradigm: insights from leprosy. *J. Invest. Dermatol.* 102, 828–832.

Parsons, L. M., Jankowski, C. S., and Derbyshire, K. M. (1998) Conjugal transfer of chromosomal DNA in *Mycobacterium smegmatis*. *Mol. Microbiol.* 28, 571–582.

Ramaswamy, S., and Musser, J. M. (1998). Molecular genetic basis of antimicrobial agent resistance in *Mycobacterium tuberculosis*: 1998 update. *Tubercle Lung Dis.* 79, 3–28.

Rambukkana, A., Yamada, H., Zanazzi, G., Mathus, T., Salzer, J. L., Yurchenco, P. D., Campbell, K. P., and Fischetti, V. A. (1998). Role of α-dystroglycan as a Schwann cell receptor for *Mycobacterium leprae*. *Science* 282, 2076–2079.

Shinnick, T. M., and Good, R. C. (1994). Mycobacterial taxonomy. *Eur. J. Clin. Microbiol. Infect. Dis.* 13, 884–901.

Valway, S. E., Sanchez, M. P., Shinnick, T. F., Orme, I., Agerton, T., Hoy, D., Jones, J. S., Westmoreland, H., and Onorato, I. M. (1998). An outbreak involving extensive transmission of a virulent strain of *Mycobacterium tuberculosis*. *New Engl. J. Med.* 338, 633–639.

Wheeler, P., and Ratledge, C. (1994). Metabolism of *Mycobacterium tuberculosis*. *In* "Tuberculosis" (B. R. Bloom, ed.) pp. 353–385. ASM Press, Washington, DC.

World Health Organization. "TB. WHO Report on the tuberculosis epidemic 1997. Global TB Programme." WHO, Geneva, Switzerland.

Mycorrhizae

Michael F. Allen

University of California, Riverside

GLOSSARY

arbuscule A structure produced by a fungus responsible for nutrient and C exchange between a plant and fungus.

hartig net Fungal hyphae that weave between plant-root cortical cells.

mutualism A mutually beneficial association of organisms.

mycorrhiza A mutualistic symbiosis that has coevolved between a host plant and a fungus, localized in the root or root-like structure of a plant.

parasitism An association of organisms in which one partner gains and the other loses.

symbiosis Two organisms living in initimate contact.

MYCORRHIZAE are mutualistic associations between plants and fungi that are the most widespread of terrestrial symbioses. These symbioses are receiving increasing attention from land managers for the production of agricultural and forest products, and for the preservation of natural areas. Because of the increasing recognition that careful management of all components must be an integral part of any sustainable ecosystem, whether designed for humans or nature, understanding of the key role of mycorrhizae becomes important for all microbiologists and plant scientists. Here an overview is provided of what mycorrhizae are and what general roles they play.

I. DEFINITION

Mycorrhizae were described in the 1880s (Frank, 1885). During the next 20 years, the types of mycorrhizae and their widespread distribution were surprisingly well documented. Further research into the basic types and functioning of mycorrhizae has led to our extensive understanding of the relationships between plants and fungi.

A mycorrhiza is a mutualistic symbiosis coevolved between a host plant and a fungus localized in the root or root-like structure of a plant. This definition uses the original concepts of de Bary, in which a symbiosis is defined as organisms intimately living together, and mutualism a mutually beneficial association. It further incorporates the idea that a mycorrhiza is not a transient interaction. This definition is consistent with the many studies demonstrating net gain in growth by a plant and the dependence of the mycorrhizal fungus on direct transport of C from the plant. It is also consistent with early research in which a mycorrhiza was determined by the absence of observable necrosis in the plant tissue when invaded by mycorrhizal fungi, as contrasted with parasites (Gallaud, 1905). Clearly there are difficulties with this definition (see Section III), but these really do not detract from this basis for understanding the structure and functioning of mycorrhizae.

The first issue is to recognize that a mycorrhiza is not an ephemeral relationship, but one that has coevolved such that both partners have benefited. Although coevolution is difficult to prove, we assess mycorrhizae based on an accumulation of data and historical evidence that the two organisms have developed structures that have been experimentally shown to provide benefits to both partners. It is important to distinguish mycorrhizae from rhizosphere inhabitants, which may mineralize nutrients or provide toxins against a third party and which provide a short-term gain but for which there are no measurable benefits in fitness over multiple generations.

II. STRUCTURE

Mycorrhizae are distinguished by morphological structures that have been related to mutualistic functioning by numerous studies. The basic premise is that nutrients flow from the substrate (generally soil) to the plant via the fungus, and C from the plant to the fungus. This reciprocal transport is mediated by particular structures designed to facilitate resource transfer. Plant roots are rather good at finding those regions of soil with water and, subsequently, efficient at transporting C below ground to find those regions. However, because of the large diameter of the roots, they are not well suited for intensively searching soil for nutrients bound to organic matter and clay, or for competing with the myriad of soil microorganisms. Fungal hyphae, on the other hand, are narrow (the tips of the absorbing hyphae can be as small as 2 μm in diameter) allowing them to compete with other microorganisms for those nutrients, but cannot fix C in the amounts necessary for growth and reproduction. Thus, a mycorrhiza must have a structure designed for nutrient exchange from fungus to plant and for C exchange from plant to fungus, and must have mechanisms designed to facilitate those transfers.

There are two major categories of mycorrhizae, endomycorrhizae and ectomycorrhizae. These revolve around the type of structure in the root that facilitates the transport of resources between symbionts (Fig. 1). Endomycorrhizae penetrate the cell wall of the plant (but not the membrane). Within the host cell, the hyphae form specialized structures that increase the surface area of contact and facilitate exchange. There are several types of these structures, including arbuscules and coils. Ectomycorrhizae remain outside the cell wall, but produce an extensive network of intercellular hyphae designed to maximize the surface area of contact, known as a hartig

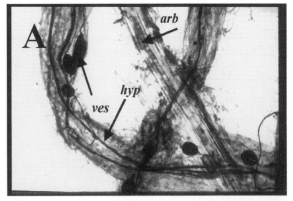

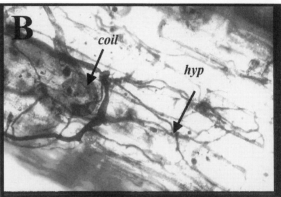

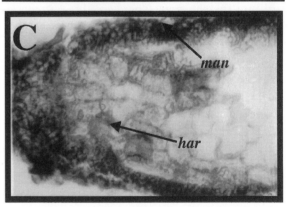

Fig. 1. Structures for the transport of resources between plan and fungal symbionts. These are differentiated between endomycorrhizae forming (A) arbuscules (*arb*), vesicles (*ves*), and hypae (*hyp*); (B) coils (*coil*) and internal hyphae (*hyp*); and (C) ectomycorrhizae that do not penetrate the cells, but that have extensive surface area for exchange between cells, with a mantle (*man*) and hartig net (*har*).

TABLE I
Differentiation of Mycorrhizal Types[a]

	Arbuscular (AM)	Ecto (EM)	Ectendo-	Arbutoid	Monotropoid	Ericoid	Orchid
Fungi							
septate	−	+	+	+	+	+	+
aseptate	+	−	−	−	−	−	−
Intracellular	+	−	+	+	+	+	+
Fungal sheath	−	+	+	+	+	−	−
Hartig net	−	+	+	+	+	−	−
Fungi	Zygomycetes	Basidiomycete/ Ascomycetes	Basidiomycete/ Ascomycetes	Basidiomycete	Basidiomycete	Ascomycete	Basidiomycete
Plants	Most oders, herbaceous and woody	Mostly woody plants	Mostly woody plants	Ericales	Monotropaceae	Ericales, Bryophytes	Orchidaceae

[a] From Smith and Read (1997).

net. All mycorrhizae form one or both of these structures and the particular combination plus the associated fungi determines the type of mycorrhiza (Table I).

The fungal mycelium extends from the root outward into the substrate. There are three categories of mycelia that form in a mycorrhiza. The least extensive is one in which the fungus forms only a small amount of hyphae extending into the surrounding substrate, for example, Ascomycetes that form ericoid mycorrhizae. In temperate to arctic habitats, these associations are found in conditions in which there is a ready diffusion of nutrients, such as bogs and in sandy and nutrient-deficient soils of Australia and South Africa. The fungi forming ericoid mycorrhizae are Ascomycetes that have been found to have the ability to break down complex organic compounds transporting limiting nutrients, especially N. The next group are the arbuscular mycorrhizae (AM) formed by the Glomales (Zygomycetes). These have runner hyphae that extend large distances outward, infecting multiple roots from many plants. In addition, these have a finely branched absorbing hyphal network that intensively searches the soil surrounding individual roots for nutrients. These fungi appear to have minimal enzymatic capacity but their intensive exploration increases the surface area for uptake of elements, especially those that are not moved by mass flow (e.g., HPO_4^-, and NH_4^+). Their

hyphae also directly attach to soil and excrete compounds that aggregate soil. Finally, ectomycorrhizae (EM) and related types of mycorrhizae (such as arbutoid and ectendo-) are structured to take up local or distant resources. Some of these can form rhizomorphs (basidiomycetous hyphae that wrap around each other, forming a cord) that allow the mycelium to extend long distances in search of resources and new plants, and highly branched hyphae (both Basidiomycetes and Ascomycetes) that can take up elements directly or break down organic compounds transporting organic resources, especially amino acids. They also can excrete organic acids that alter nutrient availability. These fungi are highly variable in their taxonomy and in their external structure. Some Zygomycetes (Endogonales) form EM, but little work on their external hyphal structure and functioning has been published. Many Ascomycetes and Basidiomycetes forming EM have a wide variation in their external architecture. Ectendo- and arbutoid mycorrhizae are generally considered forms of EM containing some endomycorrhizal structures, such as coils.

The other types that need further study are the monotropoid and orchid mycorrhizae. In the monotropoid mycorrhiza, the fungi appear largely to be the same fungi forming EM, but they apparently transport enough C to the achlorophyllous (nonchlorophyll-containing) plant to support its C needs. This

C is largely shunted from the surrounding forest trees. A similar mechanism probably also characterizes an achlorophyllous orchid, although no evidence has been presented. Importantly, most orchids are largely achlorophyllous at the protocorm stage, but become green and fix their own C after forming the first green leaf. After photosynthesis, the direction of C flow shifts from fungus-to-plant to plant-to-fungus.

III. AMBIGUITIES

As in most biological relationships, the mycorrhizal relationships between organisms at any one time represent a continuum from positive to negative. Thus, any two organisms can interact anywhere along that continuum, depending on the organisms themselves and the environmental conditions in which they find themselves at the moment. In particular, four alternative interactions involving potential mycorrhizal fungi and plants always need to be separated from a known mycorrhiza. These are apparent or temporary parasitism, phylogenic mycotrophy, multiple partnerships, and undescribed mycorrhizae.

A. Apparent or Temporary Parasitism

Of these, apparent parasitism is the most controversial. Johnson and Graham (1997) have discussed the notion of parasitism by mycorrhizal fungi. However, the data rarely show a continuous parasitism (throughout a life cycle) by a fungus against all hosts in that community. In a multispecies community, there are real preferences among species that supports the hypothesis that coevolution is occuring. Further, mycorrhizal fungi often change the allocation of C in a host. This can appear to be a growth reduction when only one parameter, such as shoot growth, is measured because shoot production may increased with one plant species, but root production may increase in the other. Finally, during different stages of the life cycle of both symbionts, there are times when little reciprocity is measured as increased mass of the host.

One exception may be the mycorrhizae formed by achlorophyllous plants. Although the plants obtain C and nutrients, and often various growth-promotion substances, it remains unclear what the fungus receives. The relationship tends to be rather specific and likely coevolved. However, it is still not known what the fungus receives for its investment.

B. Phylogenic Mycotrophy

Stahl (1900) grouped plants into three mycorrhizal categories, nonmycotrophic, facultatively mycotrophic, and obligately mycotrophic. The last two represent a gradient and are difficult to clearly distinguish. The first, nonmycotrophic, was based on family groupings. He noted that plants of several families such as Chenopodiaceae, Brassicaceae, Cyperaceae, and Amaranthaceae did not appear to form mycorrhizae. Most of these either predominated in disturbed soils where the mycorrhizal inoculum had been reduced, or in specialized habitats in which mycorrhizae might be expected to provide lesser benefits. Many subsequent researchers have cataloged mycorrhizal associations based on plant families. Work on some of these, such as annuals in the Chenopodiaceae, indicated that mycorrhizal fungi caused both a necrotic response similar to plant pathogens, and growth and fitness reductions in a field situation in which a mycorrhizal fungus was paired with a nonmycotrophic host. Importantly, although subsequent work has shown that plant family may not be an adequate level on which to characterize mycotrophy status, clearly there are plants that either do not form mycorrhizae, or that demonstrate fidelity to a particular group of fungi or type of mycorrhiza.

C. Multiple Partnerships

Communities comprise many species of fungi and plants interacting through changing intra-and interannual conditions. Mycorrhizal fungal clones extend across an area with many plant species, and each plant may associate simultaneously with many fungal species. Competitive outcomes between plants are influenced by the presence or absence of mycorrhizae and by the local fungal species composition. Because these associations have coevolved over long time periods, it should be expected that there are temporal reductions in the performance of some plants when they encounter different fungal species (and vice versa). Further, if the fungus is not providing a limiting resource to the host, the fungus will still attempt to obtain C to improve its own fitness. Thus, in complex communities, there will be winners and losers. This does not equal parasitism, in which a parasite has evolved to invade a host, leading to a consistent reduction in the growth and fitness of that host.

D. Undescribed Mycorrhizae

Although surveys have been undertaken for over a century, we still know little about the ecological relationships in many ecosystems. There are likely to be new types of mycorrhizae about which we know little. Examples include canopy epiphytes and wetlands (in which the mycorrhizae, except for ericoid associations, have been largely neglected). Further, plants may form types of mycorrhizae other than those commonly known and in unusual conditions. Examples include arbuscular mycorrhizae in Pinaceae, EM in the Ericaceae, and AM in Ericaceae. These often lead to a new understanding of plant–fungus relationships.

IV. EVOLUTION

Although the general functioning of mycorrhizae is relatively simple (carbon goes from plant to fungus; nutrients go from fungus to plant), there is a wide variety in the manner of the flow and the characteristics that regulate those flows. This is largely a result of the evolutionary history of each type of association. Mycorrhizae have been around for a very long time and the many independent phylogenic relationships among plant–fungus combinations show that this symbiosis has evolved many times. For this reason, we should expect and not be surprised at the diversity of mycorrhizal responses measured.

Fungi, like all heterotrophs, must find an existing source of organic C. There are two sources, detritus and living material. In soil, the predominant forms of C are roots (or rhizoids) and decaying plant material. Those fungi that use decaying material must be able to readily disperse to find simple C compounds, or expend energy-producing enzymes that break

down complex carbohydrates. Those fungi that invade roots to extract C have access to simple carbohydrates, but must be able to avoid plant resistance mechanisms that have evolved to keep out pathogens.

AM fungi are extremely ancient. The roots of early land plants (400 million years ago) had AM fungal structures that are unchanged morphologically from those found today. At that time, the Zygomycetes were the most advanced group of fungi. There were fungi that lived saprobically, and those that formed lichens and mycorrhizae. There is controversy as to whether the members of the group forming AM are mono- or polyphyletic, but they probably diverged from a common ancestor more than 350 million years ago. They may, in fact, be closely related to *Geosiphon*, a fungus forming an endophytic symbiosis with *Nostoc*, a cyanobacterium. This is clearly enough time to foster structural and functional differences. Every order of plants that has subsequently appeared have (or have had) some or all members forming AM associations, except for the orchids.

Fungi of one order of Zygomycetes (Endogonales), and many families of Basidiomycetes and Ascomycetes fungi form the remaining types of mycorrhizae (Table 1). These fungi have arisen independently from many different fungal groups. For example, *Hysterangium* and *Trappea,* genera closely related to the Phallales, a typically saprobic group, form EM. The real question here is, did the EM habit develop from saprobes, or saprobes from EM?

The orchid mycorrhizae represent a separate challenge. These are evolutionarily recent plants that have associated with a unique group of evolutionarily advanced fungi. These fungi are Basidiomycetes that do not form sexually reproductive structures or clamp connections. They are classified as *Rhizoctonia* because of their imperfect status. Systematic work is proceeding rapidly due to the introduction of molecular methods and I anticipate a much fuller treatment of the mycorrhizal associates in the near future.

V. FUNCTIONING

The basic role of mycorrhizae is to increase resource acquisition, primarily by exploring the soil volume for those resources. This has been known for many decades. However, the mechanisms whereby those resources are accessed are probably as varied as the habitats in which they exist and the genetic and developmental characteristics of the organisms involved. It is this variation that provides the basis for exciting research and opens a need for interactions between microbiologists, plant biologists, and land managers. There is a very large amount of literature on this topic. Here I will address four topics: environmental limits, the interface, the soil–fungus–root–leaf continuum, and community interactions–multiple associations.

A. Environmental Limits

Mycorrhizae affect plant production and the composition of the plant community. The association is dramatically affected by decisions of land managers in human-dominated ecosystems. Anything that moves or buries soil affects mycorrhizae. In nature, volcanoes, landslides, and glaciers remove or bury topsoil. On a smaller scale, animals such as gophers mix the soil, disrupt the hyphal networks, and sever the roots and hyphae. Even collembola and mites will feed on or clip hyphae emanating from roots, decreasing the number of mycorrhizal connections. Further, anything that affects photosynthesis alters mycorrhizae. The grazing on photosynthetic tissue by mammals or insects reduces the C allocation to the plant roots and, subsequently, to the mycorrhizal fungi. Similarly, human activities also alter mycorrhizal activity. Topsoil removal or redistribution, erosion, and vegetation removal by overgrazing, agriculture, forest-harvest practices, or land development can reduce mycorrhizae.

If these disturbances are small-scale relative to a larger undisturbed landscape, it is likely that the fungi can reinvade, either by hyphal expansion (up to a few dm/year) or dispersal by animals, wind, or water. If the surrounding matrix is also disturbed, there is a low chance for invasion either because the vectors have been eliminated or because there is no surrounding source of inoculum. In these cases, artificial inoculation techniques must be used. The point is that neither the plants or the fungi readily tolerate severe disturbances and, in many cases, once these are eliminated mycorrhizae do not necessarily return.

B. The Interface

An intriguing aspect of mycorrhizae is that one organism, a fungus, invades another, a plant, yet the necrotic reaction that occurs against known pathogens is not observed. Or is it? In fact, the interface between plant and fungus is very dynamic and important for characterizing the symbiosis.

The control appears to start before there is any direct connection in that fungal germination, growth rates, and direction are influenced by the presence of a plant root. Mycorrhizal fungi have hyphae designed, at least in part, to search long distances for host roots. In Basidiomycetes, these may be in the form of rhizomorphs, or cords, that extend across space. When they encounter a root, they unwind and form a net mycelium that can make multiple invasion points. In AM fungi, although no true rhizomorphs are formed, multiple hyphae also unwind when approaching a root, initiating multiple infection points (Fig. 2). Also, single hyphae can invade as a single infection point. Directional growth and extensive hyphal elongation from the germ tube appear to be largely controlled by CO_2 and by flavonoids.

After contact, the first direct connection between plant and fungus is established. A mycorrhizal fungus is likely to attempt to penetrate all roots with which it comes in contact. Because a mycorrhizal fungus is an obligate symbiont, it must gain C from a host. It

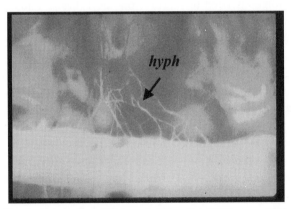

Fig. 2. Unwinding AM hypahe (*hyph*) initiating multiple infection points for mycorrhizae.

is unlikely that it will pass up a root. At that point, the recognition system probably determines if a mycorrhiza develops and if the relationship is mutualistic. If the plant is not genetically conditioned to be mycorrhizal, it reacts as it would to a plant pathogen. In some cases, such as with the nod($-$) and myc($-$) *Pisum* mutant, further hyphae expansion into the root is stopped. In the case of *Salsola kali,* a nonmycotrophic chenopod, a massive browning reaction results, with subsequent root death and sometimes plant death. Although this is a negative interaction, the fungus can gain enough C to reproduce. If a plant is genetically adapted to be mycorrhizal, the genes for pathogen protection remain disorganized. A host of genes for regulating symbiotic action then appear to take over. These include such activities as hormone balance, inhibition of root-hair elongation, phosphate-transport genes, and the H$^+$ATPase. These are only now being discovered and the array of symbiosis-regulating genetic interactions should be an important research topic for some time to come.

C. The Soil–Fungus–Root–Leaf Continuum

A unicellular alga is bathed in a nutrient solution from which it extracts nutrients necessary to fix C for growth and reproduction. A terrestrial plant must explore the soil for those same nutrients. About 450 million years ago, plant rhizoids were rather inefficient for obtaining those resources. More would be gained by associating with a fungus. Although the plant might lose some C, it would gain those resources needed to increase photosynthesis. Thus, a mycorrhiza developed and became the dominant lifestyle for most plants.

In most terrestrial environments, plants are rarely C limited. In areas with adequate precipitation, open stomates draw water through the soil–plant–atmosphere continuum, taking up water and nutrients such as nitrates or salts that move by mass flow. In these ecosystems, nutrients that are not transported by mass flow are especially limiting, such as HPO$_4^-$ and NH$_4^+$. Mycorrhizal hyphae are efficient at taking up immobile nutrients because they explore soil cracks and penetrate organic layers taking up these nutrients. Nutrients can be taken up for several centimeters to several meters, depending on the

fungus. In some cases, these hyphae also produce organic acids that modify nutrient availability or siderophores to make iron available for uptake and use. In other cases, the fungi may produce an array of enzymes that break down organic N.

In areas with low or highly seasonal precipitation, the fungus plays an even larger role. The mycelium continues to explore the soil volume for nutrients as long as it has a positive growth rate. In addition, the hypha provides both structure and glue, binding soil together and soil to roots. As soil–water potentials decline below -0.5 MPa, the largest resistance to water flow becomes the root surface as the root shrinks back from the soil. Mycorrhizal hyphae stretch across these boundaries and hold soil particles together. Thus, as water stress increases, the acquisition of soil resources dependent on mass flow continues. This is especially critical during the period between high soil moisture and extreme drought, the "ecological crunch" periods. Finally, two organisms, one inside the other, alter almost all aspects of the host physiology in very complex manners. These range from altered hormonal balance (either through direct production and transport by the fungus or by altered resource allocation) to changing stomatal behavior (e.g., rate of closure).

The acquisition of resources results in a plant C drain to build the fungal mycelium (some of which the fungus siphons off for reproduction). This generally represents a small C drain over the life of a plant with the result of an increase in C acquisition due to the enhancement of the other limiting resources. Nevertheless, there is a wide variation in the estimates of C allocation to mycorrhizae on an individual plant basis. This is probably due to both difficulties in measurement technologies and in the diversity in mycorrhizal symbioses. This can be virtually unnoticed ($<1\%$ of the daily C gain) to the point at which there can be a net C drain.

D. Community Interactions— Multiple Associations

The vast majority of mycorrhizal studies have concentrated on the interactions between a single plant and a single fungus. This, in part, has led to the view that these organisms should not be studied as independent organisms, but as a "dual organism."

However, in the field, each plant is associated with multiple fungi, and each fungus forms mycorrhizae with more than a single plant. Each plant, and indeed differing roots of the same plant, compete for soil resources, and thus for mycorrhizal partners.

What are the extent of these interactions? No one really knows. The extent of a root system is reasonably well described. In an annual cropping system, roots extend outward from the plant base a few centimeters to a few meters. At the other end, a clonal plant such as aspen will extend across many kilometers. A fungal clone probably ranges from a few centimeters (in an ericoid mycorrhiza) to many meters in EM and AM. It is also important to remember that a fungal clone will contain many different nuclei in the same connected mycelium.

The fungus may persist through time. Once a mycelial network is established, it remains for an unknown period. Theoretically, barring a major disturbance, a mycelial clone could last indefinitely. In addition, these clones probably interact with many plants. AM fungi are present in the roots of both a C3 and a C4 grass and developed different mycorrhizal structures at different seasons, depending on the host photosynthesis. As these plants were interspersed, it is highly likely that they were interconnected by the same network.

Understanding this clonal structure is especially important to the understanding of mycorrhizae. In an annual cropping system or in greenhouse pots (the most common study method), the mycelium is disrupted, resulting in the death of most of the mycelium. That mycelial network must be rebuilt, requiring a large amount of C each time. In a perennial system such as a forest or grassland, the mycelium can continue to expand and a plant can tap into an existing network. As the mycelial network is the organ that acquires nutrients, understanding its dynamics is key to describing mycorrhizal benefits.

VI. THE FUTURE—CHANGING GLOBAL CONDITIONS

Due to the pressures of the increasing human population and utilization of the environment, mycorrhizal symbioses will play a larger role in the sustainability of the environment. These pressures can take several forms. Land use, elevated CO_2, terrestrial eutrophication, and changing climates all affect mycorrhizae. Because mycorrhizae affect C fixation by plants, atmospheric C source and sink relationships are affected.

As discussed, virtually all land manipulations affect mycorrhizae, largely negatively, by disrupting the hyphal network or directly killing the fungus. Techniques for growing a few of the mycorrhizal fungi have been developed and used for specialized reintroductions. New techniques for reinoculation of a diverse suite of fungi for the different conditions should be a high priority for microbiological research. This is especially important in developing regions where fertilizers are too expensive to maintain high food and fiber production.

Elevated atmospheric CO_2 is a clear consequence of human activity. However, mycorrhizal associations largely evolved under conditions of higher atmospheric CO_2 than is present today (Allen 1996). If anything, mycorrhizae largely increase when nutrients are limited, including when CO_2 is even less limiting to plant growth. However, terrestrial eutrophication is coincidentally occuring with elevated CO_2. N deposition in response to urbanization (NO_x) and agriculture (NH_4^+) puts a resource into the soil for which plants are largely dependent on mycorrhizae. With increasing N and P in soils, a reduction in mycorrhizae is often noted. Interactions between elevated CO_2 and N are not well studied, but the few studies (for example, work by Klironomos and colleagues) do not show promising results for sustainability.

The dynamics of mycorrhizae under changing climatic regimes is not understood. In AM, it appears that plants move faster than fungi and probably adapt to the local fungus. Many EM associations are rather specific, and both plant and fungus must move together. More work on the ability of these organisms to codisperse is needed, along with information on the ability of both plants and fungi to adapt to changing conditions.

Mycorrhizae are widespread terrestrial symbioses. The vast majority of plants are highly dependent on fungi for their soil resources in the field. A large number of fungi are mycorrhizal (including most of the highly desirable edible mushrooms) and depend on specific hosts for their survival. These associations are threatened by many land-use practices. The mycorrhizal benefits already documented under highly varying conditions demonstrate a need to maintain mycorrhizae. Importantly, human-made and catastrophic natural disturbances often have the same basic effects on mycorrhizae, and many practices for restoring mycorrhizae can be based on successional processes that occur naturally. Developing new techniques for reintroducing mycorrhizae that mimic natural successional processes in lands that are already destroyed will be essential both for sustainable agriculture and forestry, and for restoring natural areas.

See Also the Following Articles

Ecology, Microbial • Heterotrophic Microorganisms • Microbes and the Atmosphere • Rhizoctonia • Soil Microbiology

Bibliography

Allen, M. F. (1988). Re-establishment of mycorrhizae following severe disturbance: Comparative patch dynamics of a shrub desert and a subalpine volcano. *Proc. Royal Soc. Edinburgh* **94 B**, 63–71.

Allen, M. F. (1991). "The Ecology of Mycorrhizae." Cambridge University Press, New York.

Allen, M. F. (ed.) (1992). "Mycorrhizal Functioning." Chapman Hall, New York.

Allen, M. F. (1996) The ecology of arbuscular mycorrhizae: A look back into the 20th century and a peek into the 21st. *Mycolog. Res.* **100**, 769–782.

Frank, A. B. (1885) Uber die auf Wurzelsymbiose beruhende Ernährung gewisser Baume durch unterirdische Pilze. *Berichte der Deutsche Botanische Gesellschaft* **3**, 128–145.

Gallaud, I. (1905). Etudes sur les mycorrhizes endotrophs. *Revue generale de botanique* **17**, 5–500.

Gianinazzi-Pearson, Dumas-Gaudot, V. E., Gollotte, A., Tahiri-Alaqui, A., and Gianinazzi, S. (1996). Cellular and molecular defense-related root responses to invasion by arbuscular mycorrhizal fungi. *New Phytologist* **133**, 45–57.

Harley, J. L. (1969). "The Biology of Mycorrhizae," 2nd ed. Leonard Hall, London.

Janos, D. P. (1993). Vesicular-arbuscular mycorrhizae of epiphytes. *Mycorrhiza* **4**, 1–4.

Johnson, N. C., Graham, J. H., and Smith, F. A. (1997). Functioning of mycorrhizal associations along the mutualism-parasitism continuum. *New Phytologist* **135**, 575–586.

Martin, F., Laurent, P., De Carbalho, D., Burggess, T., Murphy, P., Nehls, U., and Tagu, D. (1995). Fungal gene expression during ectomycorrhizal formation. *Canadian J. Botany* **73**, S541–S547.

Smith, S. E., and Read, D. J. (1997). "Mycorrhizal Symbiosis," 2nd ed. Academic Press, San Diego, CA.

Stahl, E. (1900). Der Sinn der Mycorrhizenbildung. *Jahrbuch fuer wissenschaftliche Botanik* **34**, 539–668.

Mycotoxicoses

Stan W. Casteel and George E. Rottinghaus

University of Missouri, Columbia

GLOSSARY

mycotoxicoses Diseases induced in animals and humans by toxic mold metabolites.

mycotoxins Secondary mold metabolites that impair human health and animal production through disease and reduced efficiency.

toxigenic molds Species of fungi capable of producing mycotoxins.

MYCOTOXINS are a group of structurally diverse, mold-elaborated compounds that induce diseases known as mycotoxicoses in humans and animals. An estimated 25% of the world's food crops are thought to be contaminated with mycotoxins. Ingestion of sufficient quantities of mycotoxin-contaminated material induces acute or, more commonly, chronic intoxication. Clinical signs and lesions of acute mycotoxicoses are generally referable to the most sensitive target organ. In contrast, most mycotoxin-associated health problems present as obscure chronic conditions related to depressed growth, emaciation, or increased susceptibility to infectious disease. Epidemiologic studies conducted in Africa and Asia indicate that a particular mycotoxin, aflatoxin, is a human liver carcinogen.

I. INTRODUCTION

Mycotoxicoses are distinct from the mycoses, which involve the invasion of living tissue by actively growing fungi. As secondary mold metabolites, mycotoxins contaminate animal feed and human food ingredients in the absence of intact fungal elements. Conversely, the presence of toxigenic fungi in food and feed commodities does not guarantee the presence of mycotoxins. In many instances, the processing of feedstuffs may mask the presence of mold growth without a concomitant destruction of mycotoxin. Mycotoxin contamination can only be established by appropriate chemical analysis. Visibly contaminated sources of animal feed and human food are consumed worldwide during times of famine and through ignorance of the adverse health effects. The epidemiologic investigations designed to determine the effects of such consumption are few. Data from numerous animal studies indicate that consumption of mycotoxin-contaminated food has a high potential for inducing a variety of human diseases.

II. FACTORS AFFECTING MYCOTOXIN PRODUCTION

Approximately 100 species of fungi produce mycotoxins, at least a dozen of which are associated routinely with naturally occurring mycotoxicoses in humans and animals. Fungi capable of producing mycotoxins (i.e., toxigenic fungi) frequently contaminate food consumed by humans and animals. Common toxin-producing genera include *Aspergillus*, *Claviceps*, *Fusarium*, and *Penicillium*. Genera of lesser importance include *Alternaria*, *Byssochlamys*,

Chaetomium, Cladosporium, Diplodia, Phoma, Phomopsis, Pithomyces, Rhizopus, Sclerotinia, Stachybotrys, and *Trichothecium.* Mycotoxin formation in appropriate substrates requires the presence of specific strains of these filamentous fungi or molds and is profoundly influenced by environmental factors such as aeration, humidity, temperature, drought stress, and insect damage. Because of this preference for substrates and the unique conditions for toxin formation, there often is considerable regional and seasonal variation associated with the elaboration of specific mycotoxins. Most feeds and foods are susceptible to invasion by toxigenic fungi during some phase of production, transport, or storage. Environmental stress and subsequent reduced vigor predispose crop plants to infestation, colonization, and contamination by toxigenic fungi. Fungal populations in mold-infested grains inadvertently incorporated into animal feeds may be destroyed when they are exposed to the unusual conditions of temperature and pressure during milling. The absence of recognizable mold infestation, however, does not preclude the presence of relatively heat-stable mycotoxins in processed foods and feeds. Mycotoxins have been isolated from diverse foods and commodities, including wheat, barley, oats, corn, rice, peanuts, cottonseed, milk, eggs, cheese, spices, dried beans, apples, apple juice, English walnuts, moldy hay, mixed feeds, pelleted livestock feed, and commercial dog food.

Fungal growth and mycotoxin production result from an interaction of the fungus, the substrate, and the environment. The correct combination of these factors determines the infestation and colonization of the substrate, and the type and amount of mycotoxin produced. Fungal spores are dispersed by wind currents, rain, and insects. Different strains of toxigenic molds vary in growth rate, virulence, competitiveness with other fungi, toxigenic potential, and the quantity and type of mycotoxin produced. In some cases, different fungi produce the same mycotoxin, whereas in other situations a single species may produce more than one mycotoxin.

Not only is mycotoxin formation a problem for humans and animals ingesting such material, but fungal proliferation in the host often results in systemic disease in the plant itself, as well as the destruction of feed and food commodities. An exception to fungal-induced plant pathology by a mycotoxigenic fungus involves the stress and disease resistance conferred on certain grasses infected with endophytic fungi.

The precise factors that initiate mycotoxin production are not well defined. Mycotoxins often contaminate crops in the field prior to harvest. Toxigenic species of the *Claviceps, Aspergillus,* and *Fusarium* genera are known for their ability to elaborate mycotoxins on crops standing in the field. Postharvest contamination occurs when crop drying is delayed, allowing the moisture content to exceed critical levels required for mold growth. The conditions of storage associated with mold growth and mycotoxin production include condensation within metal bins, high-moisture grain, leaky roofs, and prolonged storage time. *Aspergillus* species generally present a problem in improperly stored commodities such as corn, cottonseed, and peanuts. Mold infestation and mycotoxin production on crops is often restricted to certain geographical areas and is closely associated with environmental factors such as relative humidity, ambient temperature, drought stress, and insect damage. In addition, crop growth stage, mechanical-harvesting damage, and other fungal growth competition affect mold growth and toxin production. These conditions vary significantly among geographical locations. Because grain commodities are shipped to a variety of distant locations, mycotoxin-related health problems may lose their geographic identity.

III. IMPACT OF MYCOTOXINS ON HUMAN HEALTH

Human exposure to mycotoxins commonly occurs through the consumption of foods contaminated with metabolites of fungal growth. Such exposure is unavoidable because fungal growth in unprocessed foods is ubiquitous and difficult to prevent. Historically, the first documented case of human mycotoxicosis occurred in the Middle Ages, St. Anthony's Fire, as it was commonly called, obtained epidemic proportions in central Europe from the ninth to the eighteenth centuries. Recognized as ergotism by the

1850s, this malady was characterized by delirium, hallucinations, convulsions, and gangrene of the extremities. Studies led to the identification of *Claviceps purpurea* as the fungal agent invading rye, oats, and wheat. Mass outbreaks of apparent mycotoxicoses are now unusual in humans, but relatively common in animals. Developing countries are at increased risk because of food shortages, improper cultivation and storage methods, and the absence of mycotoxin-monitoring and -control systems. Such occurrences are unlikely to occur in the United States because of market controls on contaminated food. Concern continues about the possible adverse effects of long-term exposure to low levels of mycotoxins in the food supply.

Interest in human mycotoxicoses also stems from the isolation and identification of aflatoxin, elaborated by the mold *Aspergillus flavus,* in Ugandan peanuts. The consumption of these aflatoxin-contaminated peanuts by ducks and turkeys in England caused several hundred thousand animal deaths in the early 1960s. Further research in laboratory animals revealed aflatoxins as potential carcinogens in the human food chain in many parts of the world. Some human population studies in Africa and Asia have reported positive associations between exposure to aflatoxins and the incidence of liver cancer. Hepatitis B virus is an additional risk factor that may play a role in determining the relative susceptibility of humans to the carcinogenic effects of aflatoxin. Such correlations have not been demonstrated in the United States. Similar investigations in other countries have yielded mixed results.

Additional human health problems associated with mycotoxins include alimentary toxic aleukia caused by *Fusarium* toxins, and human nephropathy (kidney disease) induced by ochratoxin A. These problems, because they are geographically restricted to other areas of the world, are of little practical significance in the United States. The possibility cannot be ruled out that exposure to these toxins may contribute to the incidence of human cancer. Therefore, the U.S. regulatory system of mycotoxin monitoring and control must be maintained to assure that exposure to mycotoxins in the United States remains low. Regrettably, economic restrictions preclude this essential regulation in developing countries.

A. Ergot Alkaloids

Ergot alkaloids are the products of *Claviceps* spp. fungi that invade and replace the developing ovary of certain grasses and cereal grains. Rye is the most susceptible to this invasion, with spores carried by insects or the wind to the ovaries where they germinate into hyphal filaments. As the hyphae penetrate into the ovary of the grain, a dense mass of filaments forms, eventually consuming the entire substance of the grain and hardening into a purple curved body, called a sclerotium. This sclerotium, or ergot body as it is commonly called, is a rich source of ergot alkaloids. Ergot alkaloids consist of a nucleus of lysergic acid with the addition of side chains dividing the group into the peptide and amine alkaloids. Ergotamine is representative of the first group and ergonovine of the second group. Numerous other pharmacologically active alkaloids have been isolated, but the overall pharmacological effect of the group is smooth muscle contraction. Vasoconstrictive, oxytocic, and adrenergic-blocking activities are common to all of the ergot alkaloids. The therapeutic uses of the ergot alkaloids and their synthetic derivatives include vascular-headache (i.e., migraine) treatment, uterine atonia, hyperprolactinemic infertility, and adjunctive treatment of Parkinsonism. Of the toxigenic fungi, the *Claviceps* genus is unique because the infection of cereal grains occurs only during inflorescence and seed-maturation processes, with the associated mycotoxins confined to macroscopic sclerotia. Cereal grains destined for commercial sale are subject to government inspection and are rejected if they contain more than 0.05%, 0.1%, or 0.3% (by weight) ergot bodies in wheat, barley, or oats, respectively.

Human exposure results in two types of ergotism, convulsive and gangrenous. The necrotic gangrenous form of ergotism is due to the sustained peripheral constriction of blood vessels, which results in loss of sensation, swelling, and inflammation of limbs, with necrosis and the eventual loss of fingers, toes, and, in the most severe instances, feet or limbs. The convulsive form of ergotism is characterized by numbness and a tingling sensation in the hands and feet with severe muscle contractions, cramps, vomiting, and hallucinations.

Epidemics of ergotism in Europe during the Middle Ages (St. Anthony's Fire) were due to the consumption of ergotized rye bread. Although ergotism has been known to be caused by *Claviceps purpurea* since the 1850s and public health regulations have been established, a number of isolated incidences of ergotism have occurred in Russia, Ireland, France, Ethiopia, and India in the twentieth century. Agricultural practices and strict grain inspection procedures in the United States have eliminated the possibility of significant ergot contamination in our food supply, but surveys of Canadian grain foods have revealed that rye, wheat, and triticale products contain low concentrations of the ergopeptine alkaloids produced by *Claviceps purpurea*.

B. Aflatoxins

Aflatoxins are a group of potent liver toxins produced mainly by strains of *Aspergillus flavus* and *Aspergillus parasiticus,* which coexist with and grow on almost any crop or food. Aflatoxin is a problem because of its extensive preharvest contamination of corn, cottonseed, Brazil nuts, pistachios, and peanuts, and because residues from contaminated feed may appear in milk. Preharvest aflatoxin contamination is associated with high ambient temperatures ($>70°F$), insect damage, and severe and prolonged drought stress. In addition, areas of excessive moisture in storage facilities may result in foci of high aflatoxin production. Because of its carcinogenic potential and extreme liver toxicity, even low concentrations of aflatoxin are believed to pose a risk to human health. Consequently, its presence in major food commodities, as well as in milk in minute amounts (>0.5 ppb), is a perceived food-safety issue and has caused serious economic losses to producers, food handlers, and food processors.

Aflatoxin refers to a group of toxins designated aflatoxin B_1, B_2, G_1, and G_2. The most abundant and toxic member of this group under natural contamination conditions is aflatoxin B_1. Aflatoxin M_1 refers to the hydroxylated metabolite found in the milk of dairy cows exposed to aflatoxin B_1. The average excretion of aflatoxin M_1 in milk is approximately 0.9% of the total aflatoxin intake. Aflatoxin levels regulated by the U.S. Food and Drug Administration are dictated by the presumption that it cannot be completely eliminated from foodstuffs. Current U.S. guidelines limit the total aflatoxin concentration in feed or foodstuffs for human or dairy cattle use to 20 ppb (or the equivalent ng aflatoxin/g of feed or foodstuff). The one U.S. regulatory exception for food for human use applies to milk, which has an action level of 0.5 ppb for aflatoxin M_1. Clearly, children are the population segment of concern because they are more sensitive to adverse effects and ingest more milk on a body-weight basis than adults. A European commission has recently amended their regulation on the maximum levels of aflatoxins allowed in foodstuffs. Cereals and processed products for human consumption or as an ingredient in foodstuffs have a regulatory limit of 2 ppb for aflatoxin B_1 or 4 ppb for total aflatoxins. The maximum permitted level for aflatoxin M_1 in milk was set at 0.05 ppb.

Human health consequences of aflatoxin ingestion range from acute liver toxicity to chronic disease such as liver cancer. Although suspected aflatoxin-related effects have been reported from developing countries, similar problems have not been documented in humans in the United States, where the food supply is well regulated. Acute aflatoxicosis in humans from Taiwan and Uganda was characterized by vomiting, abdominal pain, pulmonary edema, and fatty infiltration and necrosis of the liver. In 1974, a documented outbreak of aflatoxicosis occurred in western India. Unseasonal rains and crop failures led to the consumption of heavily molded corn. Corn samples from the area contained from 6–16 ppm (or the equivalent mg aflatoxin/kg of food) aflatoxin. Ninety-seven fatalities resulted, mostly from gastrointestinal hemorrhage. Microscopic examination of liver sections revealed extensive damage compatible with acute aflatoxicosis.

The biological reactivity of aflatoxin B_1 is well known, although the precise mechanisms of acute toxicity and carcinogenicity continue to be elucidated. The hepatocarcinogenicity of aflatoxin B_1 is correlated with its biotransformation to a highly reactive, electrophilic epoxide, which reacts with DNA, RNA, and protein to form covalent adducts. The formation of DNA adducts is thought to be the initiating event in the pathogenesis of liver disease.

Because aflatoxin B_1 is a potent liver carcinogen in some experimental animals, it has been classed as a probable human carcinogen by the International Agency for Research on Cancer. Epidemiological studies show a positive correlation between aflatoxin ingestion and human liver cancer. Regions of the world with the highest incidence of liver cancer correspond to areas where climatic conditions favor aflatoxin contamination of food supplies. Confounding factors that limit the validity of these studies include hepatitis B viral infections, malnutrition, and parasitic infections. In addition, higher primates are relatively less sensitive to the chronic effects of aflatoxins. For example, monkeys are considerably more resistant than are rats to the carcinogenic effects of aflatoxins. Although human susceptibility is not precisely known, biochemical and metabolic similarities suggest that humans, like monkeys, are relatively less sensitive to the chronic effects of aflatoxins. Many questions concerning aflatoxin's carcinogenic potential in humans remain to be answered.

C. Trichothecenes

These mycotoxins have a cyclic sesquiterpene structure with an epoxy group at positions 12 and 13 and a double bond at 9 and 10. The epoxy group and double bond are important for biological activity. Trichothecenes are potent inhibitors of protein synthesis in eukaryotic cells and interfere with the initiation, elongation, and termination phases of this anabolic process. The trichothecenes are stable in storage and their thermal stability is remarkable, often withstanding the rigors of cooking. The *Fusarium* genus is the most important producer of trichothecenes because of its worldwide distribution and ability to produce significant quantities of these mycotoxins. Less important genera include *Cephalosporium*, *Myrothecium*, *Stachybotrys*, and *Trichoderma*.

The incidence of trichothecene-contaminated agricultural commodities is second only to aflatoxin. In Japan, high concentrations of dexynivalenol (DON) and nivalenol have been identified in corn, wheat, and barley. Deoxynivalenol is frequently found in U.S. and Canadian wheat at levels greater than 1 ppm, particularly when cool wet growing and harvesting conditions favor the growth of *F. graminearum*. Lower concentrations of DON have been detected in barley, rice, and rye. The U.S. Food and Drug Administration has issued advisory levels for DON of 1 ppm in finished wheat products for human consumption. Diacetoxyscirpenol (DAS) and T-2 toxin are of importance in other parts of the world. From a human health standpoint, T-2 appears to be the only toxin of concern from this group. Trichothecene-producing fungi grow at relatively low temperatures and toxins may be produced close to the freezing point. *Fusarium* species are generally more common than toxigenic *Aspergillus* in the middle and northern temperate regions of the world.

A presumed outbreak of T-2 intoxication occurred in the former Soviet Union during the winter of 1944. Precise identification of T-2 was not performed at this time; however, later experimentation with cats produced a similar syndrome. World War II disrupted agricultural production and resulted in millet, wheat, and barley being overwintered in the field. The consumption of these moldy grains induced a fatal syndrome resembling radiation poisoning in about 10% of the exposed population. This fatal T-2 induced disease in humans was labeled alimentary toxic aleukia (ATA). Clinical signs of ATA are divided into four progressive stages beginning 3–9 days following exposure, with skin inflammation, sore throat, abdominal pain, salivation, headache, weakness, and a rapid heart rate. Stage 2 is characterized by the suppression of blood-forming elements (bone marrow depression), inducing anemia and depressed platelet and white blood cell (aleukia) counts. Stage 3 is characterized by bleeding from the gums, nose, and gastrointestinal tract, as well as increased susceptibility to infectious agents. Death often occurs in this stage. The hallmark of stage 4 is an improvement of bone marrow function, denoted by rising white blood cell counts. Patients reaching this stage usually recover with no permanent damage.

Concern about the trichothecenes has focused on their alleged use in chemical warfare. "Yellow rain," as it is popularly labeled, has been used to describe the release of yellow particles from low-flying aircraft on rural civilians in Laos, Cambodia, and Afghanistan. Victims reported painful skin sores, dizziness, respiratory distress, a rapid onset of hemorrhaging, and death in some individuals. Survivors of the acute syndrome allegedly develop a radiomimetic illness consistent with T-2 poisoning. A cause–effect relationship was not established, despite the identification of T-2, HT-2, and DAS in the liver, kidney, and intestines of casualties.

D. Ochratoxin

Ochratoxins are a group of seven mycotoxins produced by genera of *Aspergillus* and *Penicillium*. Structurally, they are isocoumarin derivatives linked by an amide bond to phenylalanine. Of the group, ochratoxin A is of the most toxicological importance. Ochratoxin is most commonly detected in small grains of northern Europe and the Balkans, where it is thought to be the cause of an endemic nephropathy (kidney disease) in the human population. This malady is a chronic kidney disease of rural Bulgaria, Romania, and Yugoslavia. The pathological similarities between Balkan endemic nephropathy and ochratoxin-induced porcine nephropathy suggest a common origin. Circumstantial evidence supporting this link includes an increased incidence of deaths from nephropathy during periods that favor fungal growth, and the frequent isolation of *Penicillium* species from indigenous food samples.

Ochratoxin contamination of barley, corn, wheat, oats, rye, green coffee beans, and peanuts has been reported in the United States, Canada, Denmark, Poland, Sweden, Yugoslavia, and the United Kingdom. Milk residues of ochratoxin A have been detected in cows given high doses, but these milk residues are not considered a threat to human health. Ochratoxin A was detected in the kidney, liver, and muscle of chickens receiving ochratoxin-amended experimental diets, but the residues were cleared within 18 hours of removal of the contaminated diet. The residues found in the kidney, liver, fat, and skeletal muscle of hogs slaughtered in Denmark have resulted in the condemnation of carcasses. Surveys of foodstuffs in the United States indicate that human exposure to significant levels of ochratoxin in this country is unlikely. According to surveys of food and feedstuffs, concentrations of ochratoxin above 5 ppm are rare and the highest concentration ever identified was 27.5 ppm. The minimum temperature and water content required in corn for ochratoxin production by *P. viridicatum* are 4°C and 18.5%.

E. Fumonisin

The worldwide occurrence of the fungus *Fusarium moniliforme* on field and stored foodstuffs, including corn, small grains, wheat, rice, sorghum, and citrus fruits, is well documented. This ubiquitous fungus is involved in several human and animal diseases associated with the consumption of contaminated grain. Its incidence in grain has been associated with human esophageal cancer in Transkei, Africa, and China. Cultures of *F. moniliforme* on corn also have been shown to induce liver cancer in rats.

A family of structurally related mycotoxins are produced by toxic strains of *F. moniliforme* and have been found in measurable levels in corn throughout the world. The most important, fumonisin B_1, has cancer-promoting activity in rats and may prove to be a primary carcinogen in humans. Several toxic strains of *F. moniliforme* exhibit cancer-promoting activity in a short-term cancer initiation–promotion bioassay in rats. Because of fumonisin's possible cancer-causing properties, additional information is required before the scientific and economic recommendations for tolerance levels can be established for fumonisin in corn products destined for human consumption. Switzerland is the only country that has established a tolerance level for fumonisin of 1 ppm (FB_1 plus FB_2) in corn for human consumption. The results of long-term fumonisin-feeding studies currently being conducted by the National Center for Toxicological Research will provide data for quantitative risk assessment. It is felt that a level of 1 ppm fumonisin will be appropriate in the United States, but that lower tolerance levels may be required in countries where corn is a staple part of the diet.

IV. IMPACT OF MYCOTOXINS ON ANIMAL HEALTH

Mycotoxins induce a variety of adverse health effects in humans and animals. The economic consequences of reduced productivity, increased susceptibility to infectious agents due to immune suppression, insidious damage to vital organs and tissues, and decreased reproductive capacity are many times greater than that of overt clinical intoxication in livestock. Unquestionably, mycotoxicoses of animals occur, affecting their health and production efficiency. The subtle effects of mycotoxin ingestion on livestock performance are difficult but not impossible to assess in terms of economic losses. However, the subtle effects on the human population are unknown, with the best estimates based on extrapolations from animal mycotoxicoses or animal experimentation and data from epidemiological investigations. Because of the involvement of large volumes of stored feed concentrates, mycotoxicoses are particularly important in mass groups of animals such as feedlot cattle, swine, poultry, and dairy cows.

A. Ergot Alkaloids

The ingestion of ergot-contaminated grain and grass seed induces ergotism. This condition can take one of two forms in livestock. Gangrenous ergotism is induced by alkaloids produced by *C. purpurea*, whereas nervous ergotism is induced by *C. paspali* alkaloids. The gangrenous form is most commonly seen in cattle ingesting ergotized grain or seed-containing hay. The ingestion of rations containing from 0.3–1.0% by weight ergot bodies for several weeks has been associated with gangrenous ergotism. Hind-limb lameness, fever, and increased pulse and respiratory rates in affected cattle are frequent clinical findings. Close examination discloses swelling and tenderness of the fetlock joint and pastern. Within about a week of onset, the affected part loses sensation and an indented line appears at the junction between the normal and distal gangrenous tissue. Eventually, the affected feet, the tip of the tail, and ears become gangrenous and slough off. The toxic mechanism is related to the vasoconstrictive action on smooth muscle of peripheral arterioles. Sustained effects lead to injury of vascular endothelium. Initially, these actions reduce the peripheral blood flow, followed by stasis and terminal necrosis of the extremities due to thrombosis. A reduction in blood flow to the extremities is exacerbated by cold weather.

During warm weather, a summer hyperthermia syndrome is associated with cattle ingesting ergot-contaminated feedstuffs. The clinical signs include general poor-doing with increased respiratory rate, excessive salivation, increased water intake and urination, and extended periods spent standing in shaded areas or in water. The symptoms are similar to those that arise from livestock grazing tall fescue (*Festuca arundinacea*) infected with the endophyte fungus *Acremonium coenophialum* (reclassified as *Neotyphodium* genus). Tall fescue provides the primary groundcover on an estimated 25 million acres in the United States. Although the agronomic characteristics of a particular variety, Kentucky-31, are excellent, animal performance is inconsistent and clinical intoxication does occur. Four distinct syndromes are recognized in livestock, fescue foot, summer slump, lactation failure, and fat necrosis. These problems are attributed to the diazophenanthrenes (perloline and perlolidine), the pyrrolizidines (N-acetyl and N-formyl lolines), and the ergopeptine alkaloids found in *Acremonium*-infected fescue. Ergovaline accounts for 84–97% of the total ergopeptine alkaloid fraction and is most closely correlated with fescue toxicosis. Fescue foot is a dry gangrene of the extremities of cattle with the onset occurring in cold weather. The summer slump syndrome is most common in cattle, but also can be seen in horses and sheep. Signs include fever, increased respiratory rate, rough hair coat, excessive salivation, and heat intolerance. Lowered reproductive performance and lactation failure due to prolactin inhibition are seen in both horses and cattle. Fat necrosis occurs sporadically in cattle grazing tall fescue and often is detected during routine rectal examinations of normal animals. These hard masses of fat are usually located in the omentum of the abdominal cavity and are occasionally associated with intestinal strangulation when looped around a length of bowel.

Nervous ergotism is most frequently seen in cattle pastured on Dallis grass (*Paspalum dilatatum*) parasitized by *Claviceps paspali*. Occasionally, cases are seen in cattle grazing other forages, notably tobosa grass and rye grass. Ergot alkaloids are not completely responsible for this clinical syndrome. In addition to alkaloids, this fungus also produces several indole diterpene tremorgens referred to as paspali-nine and the paspalitrems. The clinical syndrome in cattle is characterized by hyperexcitability, belligerence, and tremors that intensify with excitement or forced movement. Undisturbed animals stand with their rear legs extended. When forced to move, cows exhibit exaggerated flexure of the forelegs and incoordination that leads to stumbling and collapse. Cows usually remain down until the tremors subside, at which time they may stand and resume grazing. More severely affected animals may show extensor rigidity and clonic convulsions when recumbent. The clinical signs may appear within 2–7 days of initial exposure, but usually subside a week after their removal from infected grasses.

The effects of ergot alkaloid ingestion in pregnant sows include the premature delivery of piglets and diminished or absent lactation (agalactia). Ergocryptine will curtail or prevent the prolactin surge necessary for the maintenance of lactation in cattle, sheep, goats, swine, and rats. Because of their high-grain diet, sows are particularly at risk for this effect. Ergot levels of 0.3% of the diet in late gestation have induced agalactia in sows.

B. Aflatoxins

Aflatoxins depress the carbohydrate metabolism, decrease protein synthesis, impair lipid transport and key enzyme systems, and reduce natural defense mechanisms. Generally, young animals are more susceptible to the toxic effects than are mature animals. Poultry are more sensitive to the adverse effects of aflatoxins than are mammals. Among domestic mammals, the approximate order of sensitivity from most to least is dog > young swine > pregnant sows > calves > fattening pigs > mature cattle > sheep. The relative resistance of mature ruminants is a result of rumen microbial-detoxification mechanisms.

Clinical aflatoxicosis is primarily a reflection of the adverse effects of aflatoxin on liver function. Subacute aflatoxicosis in swine is characterized by decreased feed-conversion efficiency, depressed growth, toxic hepatitis, icterus, toxic nephrosis, and hemorrhagic enteritis. Daily exposure to aflatoxin for more than 7–10 days results in liver lesions of fibrosis, edema of the gall bladder, centrilobular hemorrhage, fatty change, necrosis, and biliary hyperpla-

sia. In swine rations, dietary levels of 2–4 ppm are associated with acute fatal toxicosis, whereas rations containing 260 ppb for several weeks cause reduced growth rate. A protein-deficient diet enhances the toxicity of aflatoxin, whereas a high-protein diet is somewhat protective. There is extensive evidence that aflatoxin depresses cell-mediated immune function, thus lowering the resistance of several animal species to bacterial, fungal, and parasitic infections.

Cattle, sheep, and other ruminants appear less susceptible to aflatoxin than are monogastric mammals or poultry. Calves, which are functionally monogastrics, are more susceptible than are mature cattle. Generally, rations containing 1–2 ppm aflatoxin fed to mature cattle for a few weeks results in reduced weight gain and depressed milk production. Rations with as little as 1 ppm aflatoxin were lethal to steers within 60 days. Aflatoxicosis in cattle is characterized by depression, anorexia, reduced growth, decreased milk production, subnormal body temperature, and dry muzzle. As in several other species, lesions in the liver include fatty degeneration, vacuolated liver cells, liver necrosis, bile-duct proliferation, and diffuse fibrosis.

The U.S. Food and Drug Administration (FDA) has established recommended levels in animal feed ingredients that take into consideration animal safety and the subsequent residue levels in meat, eggs, and milk. The recommended aflatoxin levels apply to the total aflatoxins present in the feedstuff. Guidance aflatoxin levels are finishing beef cattle, <300 ppb; beef cattle, swine, and poultry, <300 ppb; finishing swine over 100 pounds, 200 ppb; breeding cattle, swine, and mature poultry, <100 ppb; and immature and dairy animals, <20 ppb.

C. Trichothecenes

The major concern in this group has centered around three toxins known as T-2 toxin, deoxynivalenol (synonymous with DON and vomitoxin), and diacetoxyscirpenol (DAS). Trichothecenes have been isolated from at least five genera of fungi, the most important from a veterinary standpoint being *Fusaria*. Although *Fusaria* are common contaminants in field and storage units, their mycotoxin products are produced predominantly during field infection rather than post-harvest elaboration. The conditions favoring trichothecene formation include cool wet weather, especially during harvest. T-2 toxin is effectively elaborated at 8–15°C, and production can occur at temperatures near freezing. Mycotoxin screening of several thousand grain samples submitted to the Veterinary Medical Diagnostic Laboratory at the University of Missouri-Columbia from 1981–1998 has not revealed detectable concentrations of T-2 and DAS. This contrasts with the frequent detection of significant quantities of DON in cereal grains and mixed feeds. Based on results from Missouri and from laboratories in surrounding states, clinical problems in livestock associated with T-2 and DAS have not been adequately documented in the United States.

In general, the problems associated with trichothecene exposure of livestock include poor feed conversion, reduced feed intake, feed refusal, diarrhea, dermatitis, hemorrhage, immunosuppression, coagulopathy, decreased egg and milk production, and decreased reproductive performance. Microscopic lesions include hemorrhage and necrosis of rapidly proliferating tissues of the intestinal mucosa, bone marrow, and lymphoid system, such as spleen and Peyer's patches. The induction of these lesions is consistent with the primary mechanism of action of T-2 toxin in particular, which is protein-synthesis inhibition. Field reports of trichothecene toxicosis have been based on the association of *Fusarium*-contaminated feedstuffs consumed by affected livestock, but purified toxins have been used in the laboratory to mimic field diseases. Many *Fusarium* species produce more than one toxin when colonizing cereal grains, thus complicating the reproduction of the disease under experimental circumstances.

The ingestion of T-2 or DAS-amended feed by swine caused the erosion of the oral mucosa and esophagus, along with necrosis on the snout due to the corrosive nature of these mycotoxins. Mild diarrhea occurred in calves given 0.08 mg T-2/kg of body weight for 9 days, whereas those receiving 0.64 mg/kg on a daily basis developed bloody diarrhea. Ulcerations were seen in the abomasum and rumen in the absence of depressed leukocyte count, hemorrhage, or bone marrow lesions. Broiler chicks fed rations containing 4.8–16 ppm T-2 toxin for

3 weeks developed circumscribed, proliferative, caseous-like plaques in and around the mouth. In addition, the growth rate was depressed, together with necrosis of lymphoid tissue. Lymphoid depletion and necrosis in the thymus, spleen, mesenteric lymph nodes, and Peyer's patches are consistent lesions induced by T-2 in all the animal species studied.

Deoxynivalenol is commonly associated with dose-dependent feed refusal in swine field cases. Swine are the species most sensitive to the effects of DON, with ruminants, horses, and poultry much less so. Swine rations contaminated with 5–10 ppm DON may cause a 50% reduction in feed intake or vomiting and feed refusal. No adverse effects were detected in cattle fed rations containing 10.5 ppm DON for 18 weeks. Concentrations of 1 ppm or less are not usually associated with adverse effects on swine performance. Other complaints connected with naturally occurring DON-contaminated swine rations include decreased reproductive performance, reduced weight gain, diarrhea, death, and vomiting. Deoxynivalenol levels exceeding 10 ppm in complete rations consistently induce vomiting in pigs. Concentrations of this magnitude are seldom seen under natural conditions. The FDA has suggested guidance levels for deoxynivalenol in animal feeds. Levels of DON are not to exceed 10 ppm (not to exceed 50% of the diet) in ruminating beef and feedlot cattle older than 4 months; 10 ppm (not to exceed 50% of diet) in poultry; 5 ppm (not to exceed 20% of diet) in swine; and 5 ppm (not to exceed 40% of diet) in all other animals.

D. Ochratoxin

Ochratoxin is nephrotoxic in all species tested, and there is some evidence of immunotoxicity as well. Ochratoxin is teratogenic in mice, rats, hamsters, and the developing embryos of chickens. Large doses induce fetal resorption. Repeated exposure to low doses of ochratoxin A is necessary to induce kidney toxicity in experimental animals. Renal excretion is the primary route of elimination of ochratoxin A. The mechanism of renal toxicity is not completely understood, but it may be related to the mistaken recognition of ochratoxin A by the kidney as an essential amino acid due to the presence of the phenylalanine moiety. There is some evidence that ochratoxin may be carcinogenic in mice.

Ochratoxicosis has been reported in poultry and swine in Sweden and Denmark and in turkeys and chickens in the United States. Ochratoxin A is recognized as the primary cause of porcine nephropathy in Sweden and Denmark. The incidence of this disease in Denmark is estimated to be 10–20 cases/10,000 pigs/year.

Swine fed diets containing 1 ppm ochratoxin A develop renal disease within a few days, characterized by increased frequency and volume of urination, compensatory increase in water consumption, stunted growth and reduced feed conversion. The clinical course may persist for several weeks in pigs exposed to low concentrations of ochratoxin A in feed with signs often reversible when contaminated feed is withdrawn. Pigs consuming diets containing 25–50 ppm ochratoxin A have a more rapid onset of depression, reduced feed intake, and loss of body weight followed by diarrhea, polyuria, polydipsia, dehydration, and death in less than a week. The no-effect concentration of ochratoxin A in swine rations is less than 0.2 ppm.

Prolonged exposure to ochratoxin-contaminated diets induces gross lesions of pale white to tan, fibrotic kidneys. Microscopic findings include diffuse interstitial fibrosis and decreased numbers of renal cortical tubules. Some tubules may be collapsed, whereas others are dilated and lined by squamous epithelium. Renal tubular epithelial cells are often degenerative to necrotic. Degenerative to necrotic changes occasionally may be seen in the liver.

E. Fumonisin

A number of mycotoxins have been isolated from *Fusarium moniliforme*. These include gibberellin and its analogs, plant-growth hormones; moniliformin; fusarin C, a mutagen; the fusariocins, known cytotoxins; and fusaric acid and its analogs. However, none has been shown to be responsible for the swine pulmonary edema syndrome associated with consumption of contaminated rations. A new class of mycotoxins, the fumonisins, have been isolated and characterized from cultures of *F. moniliforme* grown on corn. Four fumonisins were isolated and given the trivial names fumonisin A_1, A_2, B_1, and B_2. The fumonisins are structurally similar to sphinganine and their mechanism of action involves the disruption of sphingolipid biosynthesis. Fumonisin has been shown to inhibit ceramide synthase and, in all species tested, exposure to fumonisin resulted in increased sphinganine/sphingosine ratios in serum and tissue. This rapid increase in the sphinganine/sphinosine ratio can be used as a very sensitive indicator of fumonisin exposure. Fumonisin B_1 in the diet (0.1%) of rats has cancer-promoting activity in addition to the induction of an insidious and progressive toxic liver disease.

An acute fatal respiratory distress syndrome in swine was linked to the consumption of fumonisin-contaminated corn screenings from the 1989 corn harvest. The clinical history included the consumption of corn screenings for 3–10 days prior to the onset of acute respiratory difficulty and death. All ages of swine were affected with a mortality rate of 10–40%. Removal of the corn screenings from the diet had positive effects within 24 hours. Lesions included the massive accumulation of fluid in and around the lungs. Corn screenings from 12 farms with pigs similarly affected contained from 20–330 ppm fumonisin B_1. Six of eight pigs fed contaminated corn screenings developed respiratory difficulty and died within 5 days. The two survivors developed subacute liver disease. Necropsies revealed lesions of orange to yellow livers. Fumonisin is also the cause of equine leukoencephalomalacia (ELEM), a fatal brain disorder of horses and other equides. Field outbreaks of ELEM, synonymous with moldy corn poisoning or cornstalk disease, occur sporadically in many countries. The disease is most often characterized by an acute and severe onset of central nervous system derangement, but also can produce a hepatic syndrome manifested by icterus, generalized systemic hemorrhaging, and edema. The first signs are usually intermittent anorexia, depression, and cachexia with incoordination, aimless walking or walking in circles, blindness, and head pressing. Death usually ensues within 48 hours of onset.

Feeding studies have shown fumonisin has a low order of toxicity for poultry and ruminants. Purified fumonisin (80 ppm) did not affect broiler performance. Cattle fed 15, 31, and 148 ppm fumonisin showed only mild liver changes with increased liver enzymes in serum at the highest level.

Recommended levels for fumonisin in animal feeds are horses, <5ppm; swine, <10 ppm; and poultry and cattle, <50 ppm.

F. Zearalenone

Zearalenone is a resorcyclic lactone produced primarily by *Fusarium roseum*. This mycotoxin has been detected in corn, wheat, barley, oats, sorghum, rice, sesame, commercial rations, corn silage, corn meal, and corn flakes. Natural contamination with zearalenone is detected most commonly, and at the highest levels, in Midwest corn. Maximal fungal growth occurs when ambient temperature is about 25°C and grain-moisture content is greater than 23%. Much lower temperatures (10–15°C) are required to induce enzymes responsible for zearalenone production. Optimal conditions for zearalenone production occur with alternating periods of these temperature extremes. As a result of these requirements, zearalenone is found most often in high-moisture ear corn stored over the winter, especially in northern climates. Concentrations of several hundred parts per million have been detected in stored corn, whereas corn in the field rarely has more than 10 ppm. Zearalenone induces estrogenic effects (i.e., endocrine disruption) in many species. Organs normally receptive to estrogenic compounds include the tubular organs of the female reproductive tract, the ovaries, and the mammae. Prepubertal female swine are most sensitive to these effects, with as little as 1–2 ppm in the diet inducing vulvar enlargement and precocious mammary development within 7 days of continuous exposure. Unabated, this syndrome often progresses to a high incidence of rectal and vaginal prolapses. Other reported problems in female swine include anestrus, conception failure, pseudopregnancy, decreased pigs per litter, and abortion. Zearalenone induces feminization in immature male swine, characterized by testicular atrophy, and swelling of the prepuce and mammary glands. Mature male swine are highly resistant to the effects of zearalenone. Unusually high concentrations of zearalenone (>20

ppm) in cattle rations are associated with infertility and udder enlargement.

IV. DIAGNOSIS OF MYCOTOXICOSES

Diagnostic confirmation of a mycotoxicosis depends on a compilation of evidence. This means compatible clinical signs and lesions, together with chemical identification of the appropriate mycotoxin in the food or feed available at the time the problem occurred. The presence of toxigenic molds is not a sure indicator of mycotoxin contamination.

In acute mycotoxicoses, feed and food sources are often readily available for chemical confirmation of mycotoxin contamination. Problems associated with chronic low-level mycotoxin ingestion are ill-defined and often difficult to trace to a specific contaminated feed or food commodity. The insidious onset of a chronic mycotoxicosis often precludes procurement of representative food and feed samples necessary for chemical confirmation. Contaminated food sources are often exhausted by the time the problem is finally recognized. Securing a representative sample from a contaminated food source is a major problem because of the uneven distribution of mycotoxins in large grain stores. Contamination in a storage unit can vary widely or, in the case of aflatoxin, even among kernels on the same ear of corn.

V. CONTROL OF MYCOTOXINS

The continued surveillance of the feed and food supply for mycotoxin contamination using chemical analyses, surveys, and epidemiologic studies is necessary to minimize the risk to human and animal health. In an effort to minimize the mycotoxin contamination of the food supply, methods are being developed to control the formation of mycotoxins and to detoxify or decontaminate the tainted commodity. A variety of adsorbent materials have been shown to bind aflatoxin. One compound in particular, hydrated sodium calcium aluminosilicate, effectively binds aflatoxin in swine and poultry rations, thereby reducing its bioavailability. Ammoniation of feeds results in a significant degradation of aflatoxin in peanuts, cottonseed meals, and corn. A mycotoxin management system is important for the detection, prevention, and detoxification of these compounds, together with improved methods of food production, harvesting, storage, and processing, to promote the effective control of mycotoxin-contaminated commodities.

See Also the Following Articles

FOOD-BORNE ILLNESSES • SMUTS, BUNTS, AND ERGOT

Bibliography

Bacon, C. W., and Hill, N. S. (1997). "Neotyphodium/Grass Interactions." Plenum Press, New York.

Beasley, V. R. (1989). "Trichothecene Mycotoxicosis: Pathophysiologic Effects," Vol. I–II. CRC Press, Boca Raton, FL.

Cheeke, P. R. (1998). "Natural Toxicants in Feeds, Forages, and Poisonous Plants," 2nd ed. Interstate Publishers, Danville, IL.

Cole, R. J., and Cox, R. H. (1981). "Handbook of Toxic Fungal Metabolites." Academic Press, New York.

Council for Agricultural Science and Technology (1989). "Mycotoxins. Economic and Health Risks," Task Force Report No. 116. Council for Agricultural Science and Technology, Ames, IA.

Ellenhorn, M. J. (1997). Mycotoxins. *In* "Ellenhorn's Medical Toxicology: Diagnosis and Treatment of Human Poisoning" 2nd ed., pp. 1876–1879. Williams & Wilkins, Baltimore, MD.

Jackson, L. S., DeVries, J. W., and Bullerman, L. B. (1996). "Advances in Experimental Medicine and Biology, Vol. 392, Fumonisins in Foods." Plenum Press, New York.

Leeson, S., Diaz, G. J., and Summers, J. D. (1995). "Poultry Metabolic Disorders and Mycotoxins." University Books, Guelph, Canada.

Marasas, W. F. O., and Nelson, P. E. (1987). "Mycotoxicology: Introduction to the Mycology, Plant Pathology, Chemistry, Toxicology, and Pathology of Naturally Occurring Mycotoxicoses in Animals and Man." Pennsylvania State University Press, University Park, PA.

Miller, J. D., and Trenholm, H. L. (1994). "Mycotoxins in Grains: Compounds Other Than Aflatoxin." Eagan Press, St. Paul, MN.

Mycotoxins (1989). *Arch. Environ. Contamination Toxicol.* **18.**

Osweiler, G. D., Carson, T. L., Buck, W. B., and Van Gelder, G. A. (1985). Biotoxins. *In* "Clinical and Diagnostic Veterinary Toxicology," pp. 409–426. Kendall/Hunt, Dubuque, IA.

Quisenberry, S. S., and Joost, R. E. (eds.) (1990). "Proceedings of the International Symposium on *Acremonium*/Grass Interactions." Louisiana Agricultural Experiment Station, Baton Rouge, LA.

Residues of natural toxins in feedstuffs and food. (1993). *J. Animal Sci.* **71.**

Smith, J. E., and Henderson, R. S. (1991). "Mycotoxins and Animal Foods." CRC Press, Boca Raton, FL.

U.S. Department of Agriculture. (1990). "A Perspective on Aflatoxin in Field Crops and Animal Food Products in the United States, A symposium." Agricultural Research Service, ARS-83. USDA, Washington, DC.

Myxobacteria

David White

Indiana University

GLOSSARY

A motility genes   Genes required for single-cell motility.

A-signal   A mixture of amino acids produced by certain myxobacteria that is required for an early stage of fruiting body formation. It is a cell-density signal.

cooperative feeding   The phenomenon by which populations of myxobacteria secrete enzymes that degrade extracellular nutrient and share the products of digestion.

C-signal   An intercellular signal required by certain myxobacteria for fruiting-body formation.

fruiting body   A multicellular structure housing resting cells, called myxospores. The myxobacteria are the only bacteria that form fruiting bodies.

gliding motility   A creeping movement on a solid surface.

myxospore   A resting cell formed by myxobacteria. It is resistant to desiccation.

pheromone   A lipoidal signal required by certain myxobacteria for fruiting body formation.

rippling   Tiny concentric waves of gliding cells in a swarm.

slime   Secreted material consisting of polysaccharide and protein. It is found in the slime trails.

slime trails   Secreted polysaccharides and proteins left behind as a trail as myxobacteria glide on solid surfaces. Gliding is more efficient in slime trails.

S motility genes   Genes required for group motility.

sporangiole   The cyst-like structure on the tops of stalks of fruiting bodies. The sporangioles house the myxospores.

swarm   A colony of myxobacteria that expands because of gliding motility.

MYXOBACTERIA are aerobic free-living rod-shaped gliding gram-negative bacteria that grow primarily in soil, on tree bark, on herbivore animal dung, and on decaying vegetation. Their distinguishing characteristic is that they form multicellular structures called fruiting bodies. Myxobacteria are predators. They feed on other microorganisms, such as other bacteria and yeast, with secreted hydrolytic (digestive) enzymes and use the products of lytic digestion as a source of nutrients. Hence they are usually found in niches rich in microbial life. Myxobacteria may also use molecules released from decaying vegetation, such as proteins and peptides, as nutrients. In the laboratory, most genera can be grown on bacteria, yeast or yeast extract, proteins, peptides, or mixtures of amino acids. One genus (*Sorangium*) can be grown on glucose and can degrade cellulose. Myxobacteria use secreted hydrolytic enzymes to break down larger molecules, such as proteins, to smaller molecules on which they grow.

I. INTRODUCTION

A. Why Myxobacteria Are Studied

Myxobacteria signal each other during growth and development, and they are studied in part to learn more about such communication and how it affects gene activity, cell movements, and morphogenesis. This is of widespread interest among microbiologists

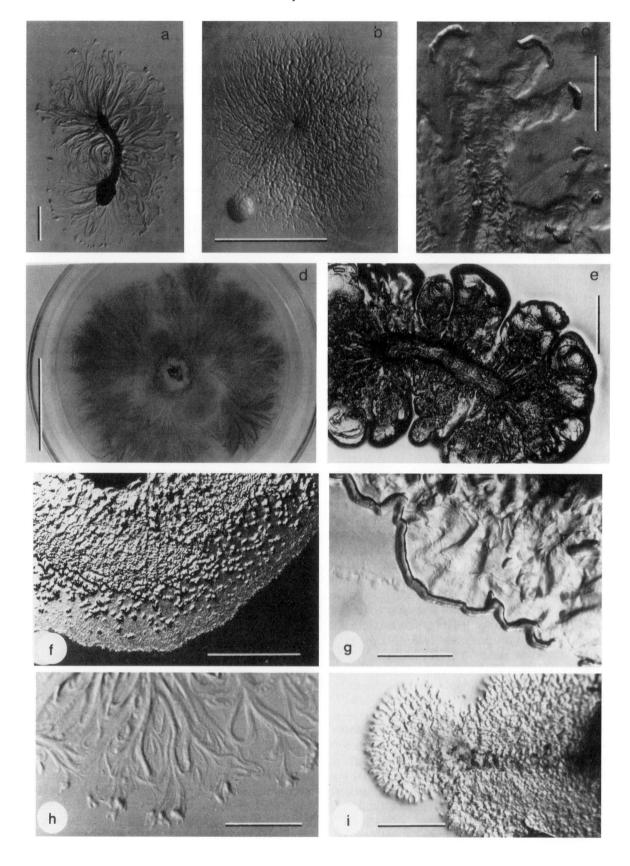

because it is now understood that intercellular signaling among bacteria is quite common. (See Shapiro and Dworkin, 1997.)

A second reason that myxobacteria are studied is that they produce compounds (e.g., antibiotics) that inhibit the growth of other organisms. Indeed, some strains are of commercial importance because of this. For example, myxobacteria are known to produce antibacterial, antifungal, antiviral, and, to a lesser extent, insecticidal compounds. This has been reviewed by Reichenbach and Höfle (1993). Many such compounds have been detected by screening methods, and their structures are being elucidated. It has been estimated that at least 50% of all myxobacterial strains tested produce substances with growth-inhibiting activity against bacteria, yeast, and filamentous fungi. Examples of two very important antibiotics are myxothiazol and stigmatellin, which are electron-transport inhibitors widely used in research.

Here an introduction to and broad overview of the myxobacteria is presented. Interested readers should consult reviews of myxobacteria (Reichenbach and Dworkin, 1992; Rosenberg, 1984; Dworkin and Kaiser 1993; Dworkin, 1996; Shimkets and Dworkin, 1997; Koch and White, 1998) and the remarkable films of myxobacteria movement and fruiting body formation made by Reichenbach, Huenert, and Kuczka (1965a, 1965b).

II. GLIDING MOTILITY AND SWARMING

In their natural habitat, myxobacteria live as dense populations on solid surfaces such as soil particles or decaying vegetation, rather than dispersed as single cells or a film of cells in water such as lakes or streams. A population of myxobacteria on a solid surface is called a swarm because the edge of the colony spreads, similar to a swarm of insects seeking new territory. Figure 1 shows typical swarms of different myxobacteria on agar surfaces. A few myxobacteria placed in the center of an agar surface will grow and move out to cover most of the agar in a petri dish within a few days (Fig. 1d). This differs from most bacteria, which generally remain as a small colony in the center of the plate. During swarming, the edge of the colony continually expands outward, but the visible swarm itself does not leave one area and go to another. However, microscopic observations reveal that cells can leave the main body of the swarm and move as a group away from the edge. They generally return to the main swarm or are joined by the leading edge of the swarm as it expands outward (Fig. 2).

The expansion of a colony of myxobacteria (swarming) is due to a particular type of cell motility called gliding motility. (Gliding motility occurs in certain bacteria other than myxobacteria.) This has been described as a creeping motion in either the forward or backward direction on a solid surface. There are no known locomoter organelles, such as flagella, and the mechanism of gliding motility is poorly understood. The cells glide in slime trails that they produce (Fig. 2b).

As discussed later, an important feature of myxobacteria movement is the coordinate behavior between cells, which is based on sophisticated intercellular signaling systems. The coordinated gliding movement allows the edges of the swarms to send out streams of cells called flares that shift directions, which allows the cells to find and feed on other microorganisms and sometimes separate transiently from the edge of the swarm. A model explaining how this might occur has been recently formulated by

---

Fig. 1. Myxobacteria swarm colonies. (a) *Stigmatella erecta.* Bar = 1 mm. (b) *Myxococcus xanthus.* Bar = 0.5 mm. (c) *Polyangium* sp. Bar = 1 mm. (d) *Cystobacter violaceus.* Cells were inoculated in the center and the swarm spread over most of the agar surface in the petri dish. Bar = 30 mm. (e) *Nannocystis exedens.* Bar = 5 mm. (f) *Corallococcus coralloides.* Young fruiting bodies are seen throughout the swarm. Bar = 5 mm. (g) *Polyangium* sp. Bar = 1 mm. (h) *Stigmatella erecta.* Bar = 1 mm. (i) *Nannocystis exedens.* Bar = 2 mm. (From Reichenbach, H., and Dworkin, M. (1992). The Myxobacteria, pp. 3416–3487. *In* "The Prokaryotes" Second edition. Vol. IV. (A. Balows *et al.*, eds.), 2nd ed., Vol. 4, pp. 3416–3487. Springer Verlag, Berlin.)

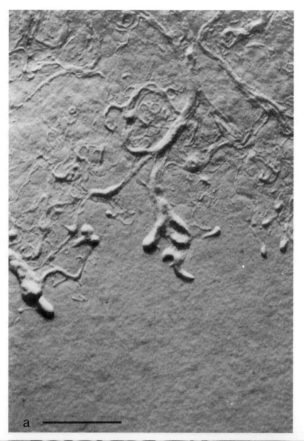

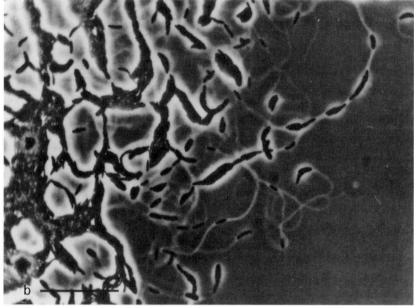

Fig. 2. The edge of a swarm. (a) *Stigmatella aurantiaca.* Bar = 290 μm. (b) *Myxococcus fulvus,* showing individual cells and slime trails. Bar = 275 μm. (From Reichenbach, H. (1984). Myxobacteria: A most peculiar group of social prokaryotes. *In* "Myxobacteria: Development and Cell Interactions" (E. Rosenberg, ed.), pp. 1–50. Springer-Verlag.)

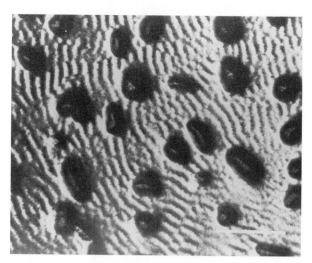

Fig. 3. Ripples formed by *Myxococcus xanthus* during fruiting body formation on agar. The dark masses are immature fruiting bodies. Bar, 400 microns. (From Shimkets, L. J., and Kaiser, D. (1982). Induction of coordinated movement of *Myxococcus xanthus* cells. *J. Bacteriol.* **152,** 451–461.)

Koch (1998). Another consequence of coordinate movements and intercellular signaling is rippling. During rippling, the swarms form tiny advancing waves that resemble the ripples in a body of water (Fig. 3). Cooperative movement also results in a unique form of multicellular development called fruiting body formation (see Section III), when nutrient supplies are depleted.

III. FRUITING BODY AND MYXOSPORE FORMATION

Myxobacteria have one outstanding characteristic that sets them apart from all other known bacteria. When swarms of myxobacteria are faced with nutrient depletion, the individual cells glide into aggregation centers to form hundreds of tiny (approximately 0.1–0.2 mm) multicellular structures called fruiting bodies. A scanning electron micrograph of fruiting bodies made by the myxobacterium *Stigmatella aurantiaca* is shown in Fig. 4, and the morphological stages of fruiting body formation for that organism are shown in Fig. 5.

Within the fruiting bodies, the cells convert to resting cells called myxospores, whose biological properties have been studied in *Myxococcus xanthus* and to some extent in *S. aurantiaca*. Such studies have shown that the myxospores are resistant to desiccation, higher temperatures, and ultraviolet radiation. In certain genera the process of myxospore formation entails cellular morphogenesis (i.e., a shortening and thickening of the cell) and can be quite dramatic; for example, the conversion of a rod-shaped cell to a refractile spherical cell occurs in the genus *Myxococcus*. In the myxobacterium *M. xanthus*, the formation of myxospores can be induced in cell suspension by adding 0.5 M glycerol to the medium. The morphogical stages of glycerol-induced myxospore formation are the same as myxospore formation in fruiting bodies, and these are shown in Fig. 6. The fruiting bodies consist of 10^4 to 10^6 myxospores, are generally brightly pigmented (e.g., shades of orange, yellow, and red), and, as mentioned, are quite small (less than 0.2 mm).

The shapes of the fruiting bodies vary with the type of myxobacterium (Figs. 7–8). Some are simply mounds of myxospores in loose or hardened slime, whereas others consist of stalks supporting walled-

Fig. 4. Fruiting bodies of *Stigmatella aurantiaca*. Scanning electron micrograph; both a side view and a top view are shown. The individual stalk is seen to support several sporangioles, which house the resting cells called myxospores. The fruiting bodies are approximately 0.1 mm tall.

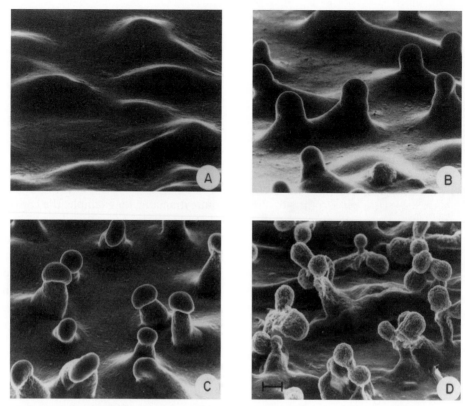

Fig. 5. Scanning electron micrographs of different stages of fruiting body formation of *Stigmatella aurantiaca*. (A) Early aggregates. (B) Early stalk formation. (C) Late stalk formation. (D) Mature fruiting bodies. Bar = 20 μm. (From Qualls, G. T., Stephens, K., and White, D. (1978). Morphogenetic movements and multicellular development in the fruiting myxobacterium, *Stigmatella aurantiaca. Dev. Biol.* **66,** 270–274.)

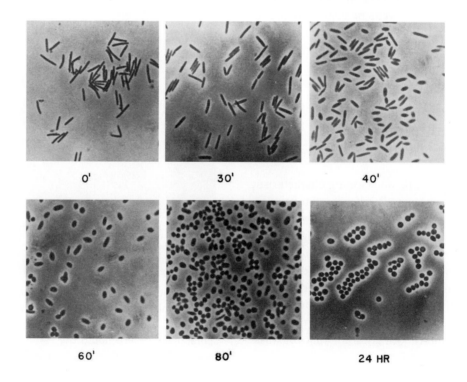

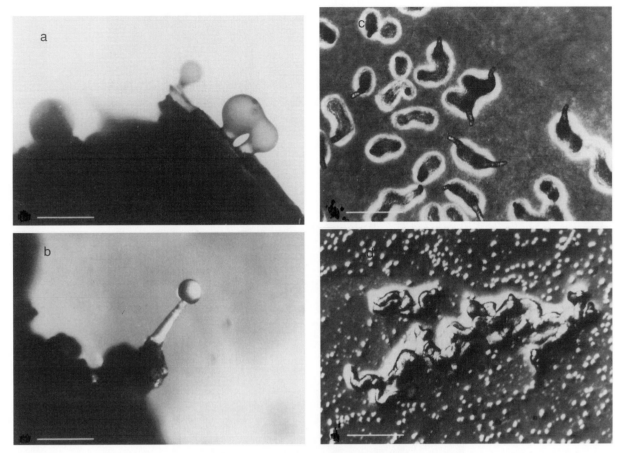

Fig. 7. Shapes of fruiting bodies of different genera of myxobacteria. (a) *Myxococcus fulvus* on rotting wood. Bar = 125 µm. (b) *Myxococcus stipitatus* on soil. Bar = 125 µm. (c) *Corallococcus coralloides* on agar. Bar = 215 µm. (d) *Archangium serpens* on agar. Bar = 215 µm. (From Reichenbach, H. (1984). Myxobacteria: A most peculiar group of social prokaryotes. *In* ''Myxobacteria: Development and Cell Interactions'' (E. Rosenberg, ed.), pp. 1–50. Springer-Verlag, Berlin.)

sporangioles within which the myxospores are housed. Myxobacteria fruiting bodies have been described since the early nineteenth century. However, the early reports did not recognize these organisms as bacteria, but classified them as fungi because of their stalks supporting what appeared to these early mycologists and botanists to be conidia. It remained for the American botanist Roland Thaxter, who in the late nineteenth and early twentieth centuries

studied the life cycle of these creatures to recognize that they were bacteria and not fungi.

IV. LIFE CYCLE

Myxobacteria can be grown either in cell suspension in liquid culture or on an agar surface. Growth on an agar surface is required in order for swarms to

Fig. 6. Morphological changes during glycerol-induced myxospore formation in *Myxococcus xanthus*. Glycerol at a final concentration of 0.5 M was added to a growing culture. At various times after the addition of the glycerol, the cells were photographed. The average length of the cells before the shape change (0') was 5 µm.

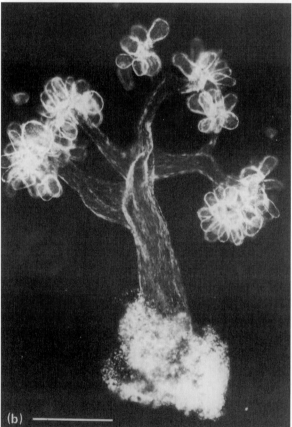

Fig. 8. (a) *Chondromyces apiculatus.* Bar = 105 μm. (b) *Chondromyces crocatus.* Slide mount, dark field. Bar = 130 μm. (From Reichenbach, H. (1984). Myxobacteria: A most peculiar group of social prokaryotes. *In* ''Myxobacteria: Development and Cell Interactions'' (E. Rosenberg, ed.), pp. 1–50. Springer-Verlag.)

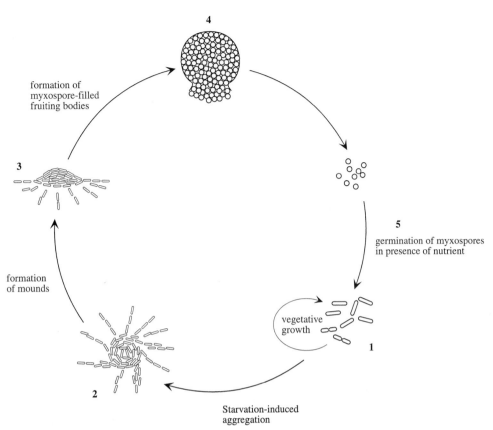

Fig. 9. Life cycle of *Myxococcus xanthus*. 1. Vegetative growth; cells grow in dense populations called swarms on solid surfaces. 2. Aggregation; when nutrients are depleted, cells glide into aggregation centers, each one consisting of many thousands of cells. 3. Mound formation; each aggregation center becomes a mound of cells as bacteria continue to accumulate. 4. The mound develops into a fruiting body when the cells differentiate into resting cells called myxosopores; each myxospore is surrounded by a coat (capsule). 5. When nutrients become available once more, the myxospores germinate into vegetative cells, returning the population to a new growth phase. Once the cells deplete the supply of nutrients, they can aggregate and form myxospore-filled fruiting bodies within 24 hr. (From White, D. (1999). "The Physiology and Biochemistry of Prokaryotes." Oxford University Press, Oxford.)

form and the population of myxobacteria to complete the life cycle. The life cycle of myxobacteria can best be described as consisting of two stages: (1) a vegetative (growth) stage and (2) a fruiting (developmental) stage (Fig. 9).

A. Vegetative Stage

During the vegetative stage, the myxobacteria consist of a feeding and growing swarm of cells (Fig. 9). In nature, myxobacteria use other microorganisms (e.g., other bacteria and yeast) as a source of nutrients and can be grown in the laboratory on these food sources. However, they can also be grown in the laboratory on proteins, peptides, mixtures of amino acids, and, in some cases, carbohydrates.

1. Cooperative Feeding

It has been concluded that myxobacteria in the swarm feed cooperatively when growing on other microorganisms in nature. This conclusion is based on the finding that a single myxobacterial cell has relatively poor reproductive ability when placed on an agar surface, separate from a swarm, that has as the sole source of nutrients an undigested protein such as casein. It has been assumed that such would be the case in nature if a single cell became separated from the swarm and was left to feed on its own. Because a swarm of myxobacteria grows and reproduces quite well on other microorganisms, it has been concluded that the hydrolytic enzymes secreted by the population of myxobacteria is sufficient to provide nutrients for the swarm. In other words, the myxobacteria within the swarm share the products of digestion. This has been called cooperative feeding, and indeed myxobacteria swarms have been likened to wolf packs devouring their prey.

B. Fruiting Stage

After the depletion of nutrients, the cells in the swarm move into hundreds of aggregation centers and construct a fruiting body in each aggregation center (Figs. 5, 9). Within the fruiting bodies, the cells convert to myxospores. The conversion of a vegetative cell to a myxospore often entails a change in the shape of the vegetative cell, and frequently the synthesis of a capsule surrounding the cell. Depending on the genus of myxobacterium, the shape change may involve a shortening and thickening of the cell, or the conversion of the rod-shaped vegetative cell into a spherical cell. Fig. 6 depicts the shape changes that occur when *Myxococcus xanthus* rods convert to spherical myxospores. Such conversions of *M. xanthus* occur either within fruiting bodies or when vegetative cells are incubated with 0.5 M glycerol in the laboratory. In addition, myxospores may differ physiologically from their progenitor vegetative cells. For example, myxospores are often dormant and resistent to desiccation. Indeed, myxospores within dried fruiting bodies may remain viable for several years.

1. Survival Advantage to Forming Fruiting Bodies

Fruiting bodies are physically dispersed, perhaps by sticking to the bristles of insects that roam through a "field" of fruiting bodies and then randomly falling off the insects in new location. The cooperative feeding behavior of the swarms offers a clue to the advantage of forming multicellular fruiting bodies. When the fruiting bodies are dispersed to new locations where nutrients are available, the myxospores germinate and each fruiting body produces one or more small swarms, ready to feed, grow, and expand in search of new prey.

V. CLASSIFICATION BASED ON DIFFERENT SHAPES OF FRUITING BODIES, VEGETATIVE CELLS, AND MYXOSPORES

There are 12 genera and approximately 40 species of myxobacteria. The genera of myxobacteria are classified according to the shapes of their fruiting bodies, as well as the shapes of the vegetative cells and myxospores. A detailed description of the taxonomy of the myxobacteria can be found in the articles by Reichenbach and Dworkin (1992) and Reichenbach (1993). Examples of myxobacteria fruiting bodies are shown in Figs. 7–8. Some myxobacteria vegetative cells are long thin rods with tapered ends, whereas others are shorter, thicker cylindrical rods with round ends. Myxospores also vary in their size and shape. Some are spherical and refractile, some are simply shortened thick rods, and others resemble their progenitor vegetative cells.

A. Fruiting Bodies That Are Myxospores Encased in Slime

Myxobacteria belonging to the genus *Myxococcus* form fruiting bodies that are simply mounds of myxospores encased in soft slime (Fig. 7a,b). There is usually a constriction at the base, and sometimes a small stalk. If the slime is hard, then then genus is usually *Coralococcus* or *Archangium* (Fig. 7c,d.) The shapes of the latter two are nonspherical.

B. Myxospores Encased in Walled Structures

Several genera of myxobacteria house their myxospores in walled structures called sporangioles, which can be sessile or borne on stalks, depending on the genus (Fig. 8). There is no soft slime within the sporangioles. *Nannocystis* and *Haploangium* make fruiting bodies consisting of single sporangioles that are sessile (on or in the substrate). *Angiococcus, Cystobacter, Polyangium, Sorangium,* and *Nannocystis* form groups of sessile sporangioles (on or in the substrate). *Melittangium* and *Stigmatella erecta* form single sporangioles supported by a stalk. *S. aurantiaca* and *Chondromyces* spp. form clusters of sporangioles supported by an unbranched stalk (Figs. 4, 8a). *C. crocatus* forms clusters of sporangioles on top of a branched stalk (Fig. 8b).

VI. ISOLATION AND CULTIVATION

A. Most Common Habitats

Myxobacteria are generally viewed as the predators of microorganisms and colonize habitats rich in microbial growth. They can be isolated primarily from neutral or slightly alkaline top soil that is rich in organic material, such as compost or garden soil, decaying plant material in the woods (e.g., rotting woods), bark from (both living and dead) trees, and the dung of herbivorous animals such as rabbit, deer, and sheep. A general description of their isolation and purification follow; consult Reichenbach and Dworkin (1992) for detailed instructions.

B. Isolation

1. Isolation from Dung, Bark, or Rotting Wood

Specimens thought to contain myxobacteria may be placed in petri dishes lined with two to three layers of moist filter paper. The paper should be kept moist because if the atmosphere is too dry, the material will be overgrown with mold. If it too wet, fruiting bodies will not develop. The plates are placed in an incubator at 30°C, usually kept moist by a container of water. They can also be incubated at room temperature; however, the lower temperature slows the development of fruiting bodies. After a few days to two weeks, fruiting bodies develop on the samples.

2. Isolation from Soil

A petri dish is filled with soil that may contain myxobacteria. The soil is moistened with water, and autoclaved dung pellets from wild rabbits are placed in the soil. The dung has bacteria that serve as "bait" for the myxobacteria. Over a period of days, myxobacteria migrate from the soil to the dung pellets, grow there, and form fruiting bodies.

3. Isolation from Swarms

A third way to isolate myxobacteria is to place the source (e.g., soil) on streaks of microorganisms. The microorganisms may be bacteria or yeast, which are streaked in large amounts on nonnutrient agar. The plates are incubated at 30°C and inspected every day. The myxobacteria in the soil will grow and lyse the food microorganisms. Careful observations with a dissecting microscope will reveal one or more thin swarms that move away from the lysed microorganisms. The edges of these swarms frequently have only myxobacteria. The myxobacteria can be isolated by transferring from the edge of the swarm to a fresh plate. Eventually, fruiting bodies may develop, and these also can be transferred.

C. Purification

It is sometimes possible to isolate a pure culture by carefully transferring a fruiting body to a fresh nutrient plate consisting of peptides and yeast extract, or autoclaved yeast. This is generally done using a dissecting microscope and a drawn-out pasteur pipette to move the fruiting body. Usually, however, it is necessary to purify the strain by an enrichment technique. To do this, fruiting bodies or cells from the edge of the swarm are transferred to streaks of a food microorganism on a nonnutrient plate. After swarms appear, new cultures are made from their edges.

VII. RESEARCH

Myxobacteria are an attractive group of organisms for research, in part because molecular genetic techniques such as transposon mutagenesis, gene cloning, restriction mapping, and methods to study the expression of specific genes during development are available, especially for *M. xanthus* and *S. aurantiaca*.

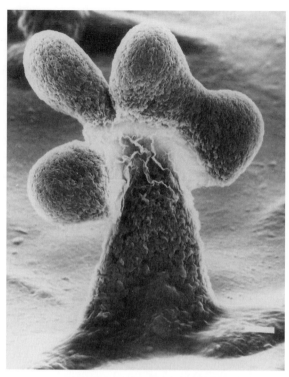

Fig. 10. Fruiting body of *Stigmatella aurantiaca*. Scanning electron micrograph. The fruiting body cover (coat) is missing, revealing the cells in the sporangioles and the stalk. Bar = 8 μm. (From Stephens, K., and White, D. (1980). Scanning electron micrographs of fruiting bodies of the myxobacterium *Stigmatella aurantiaca* lacking a coat and revealing a cellular stalk. *FEMS Microbiol. Lett.* **9,** 189–192.)

A. Morphogenetic Movements during Fruiting Body Formation

There have been careful microscopic studies of aggregation and fruiting body formation in several myxobacteria. As an example, fruiting body formation in *S. aurantiaca* will be described. This myxobacterium forms a fruiting body consisting of a stalk supporting several sporangioles (Fig. 10). The stalk is cellular, at least in the early stages. Stages in fruiting body formation by *S. aurantiaca* are shown in Fig. 5. Aggregates form (Fig. 5A), which grow taller as the stalk grows (Fig. 5B). In an intermediate stage, a globular population of cells forms on top of the stalk (Fig. 5C), and the structure resembles a mushroom. Later, the globular portion of cells on top of

the stalk develops into several sporangioles housing myxospores (Fig. 5D).

B. Intercellular Signaling

The coordinate behavior of myxobacteria during vegetative swarming and fruiting body formation is dependent on intercellular signaling. Intercellular signaling has been analyzed both genetically and biochemically in the myxobacterium *M. xanthus*. Some of the signaling influences motility, *per se,* and some is required for multicellular development and the formation of myxospores.

1. *Intercellular Signaling and Gliding Motility*

M. xanthus glides both as single cells and as groups of cells within the swarms, and these two patterns of motility are determined by separate genes. The two sets of genes are system A genes (for *adventurous*), which are required for single-cell motility, and system S genes (for *social*), which are required for group motility. Mutants in system A genes (A^-S^+) can glide only as groups. Mutants in system S genes (A^+S^-) glide only as single cells. Both single-cell and group motility are significantly stimulated when the cell density is increased, reflecting the fact that cell-to-cell communication stimulates gliding motility in both systems. The signaling systems are not understood, but appear to require cell–cell contact in both the A and S motility systems. For example, certain A^-S^- cells will be stimulated to move when mixed with wild-type (A^+S^+) cells but will not move if the wild-type and mutant cells are separated by a space on the agar. There are five multigene loci involved in A motility system signaling. If complementary pairs of mutants are mixed, they both move, indicating that each one of the pair provides a signal not produced by the other member of the pair. Because of this, it has been suggested that there are five signals for the A motility system. The signaling molecules for the A system have not been identified, but presumably they are bound to the cell surface or to fibrils that extend from the cell surface. A single mutation in the S motility system can be complemented by wild-type cells, reflecting an S signal lacking in the mutant cells.

2. Intercellular Signaling and Development

Early evidence for cell-to-cell signaling involved in development came from the isolation of *Myxococcus* mutants that were unable to form normal fruiting bodies and myxospores. These cells had mutations in five genes and were able to form normal fruiting bodies and myxospores when mixed with wild-type cells. They would also form fruiting bodies when mixed with each other in complementary pairs, indicating the existence of five different signals. Two of these signals have been isolated and identified. One, the A-signal, is a mixture of specific amino acids. Mutants unable to produce the A-signal fail to aggregate, suggesting that the A signal acts early in development. When the concentration of A-signal amino acids exceeds a certain threshold (10 μM), the signal informs the population of cells that their cell density has reached the requisite number required to build fruiting bodies. Thus, the A-signal is a quorum signal analogous to other quorum signals, including acylated homoserine lactones produced by a wide range of gram-negative bacteria. The C-signal has also been characterized. Mutants defective in making C-signal construct abnormal aggregates and only after a significant delay. It is believed that the C-signal acts at a developmental stage after the A-signal. The C-signal appears to be synthesized by a membrane-associated protein that has been localized to the surface of the cells.

The myxobacterium *S. aurantiaca* secretes a lipoidal pheromone. The pheromone induces cells to aggregate and form fruiting bodies but the mechanism by which this occurs is not known.

Thus, the myxobacteria communicate with a variety of extracellular signaling molecules of different chemical compositions. Some are freely diffusible in the medium and increase in concentration during development, at least one is lipoidal, and others are attached to the cell surface. An important area of research is to identify these molecules and to elucidate the signaling pathways that result in multicellular behavior and cellular differentiation.

VII. SUMMARY

Bacteria are generally responsive to external signals that may be secreted from other cells, including other bacteria, and they respond to environmental factors such as oxygen, temperature, and osmotic pressure. The signals usually alter gene expression. For example, lipids called acylated homoserine lactones are secreted by various gram-negative bacteria and serve as quorum sensors. This means that certain genes in these bacteria are expressed only when the population of cells reaches a threshold cell density, and the signal that the cell density has been reached is the accumulation in the extracellular medium of an acylated homoserine lactone secreted by the cells. Examples of genes whose expression is stimulated by acylated homoserine lactones include the luminescence genes in *Vibrio fischeri*, the genes required for conjugation in *Agrobacterium tumefaciens*, virulence genes in *Pseudomonas aeruginosa*, and virulence genes in the plant pathogen *Erwinia carotovora*. Peptides are often secreted by gram-positive bacteria for intercellular signaling. For example, *Bacillus subtilis* uses secreted peptides to activate genes in the sporulation and competence pathways. From that point of view, myxobacteria are not unique among bacteria simply because they signal each other to alter gene expression and behavior. What sets myxobacteria apart from other bacteria is that the signaling, which relies on both secreted molecules as well as cell–cell contact, results in complex patterns of movement in the swarms, culminating under starvation conditions in the formation of fruiting bodies and myxospores. This is what one might expect of eukaryotic organisms that have multicellular stages during their life cycle (e.g., the cellular slime molds). To find such signaling and consequent complex cell behavior in a bacterium is indeed remarkable. Small wonder that myxobacteria were originally mistaken for fungi! The identification of the chemical components in the myriad signaling pathways, learning how the signals are transduced into directed cell movements, and determining developmental gene expression are major goals in myxobacteria research.

See Also the Following Articles

DEVELOPMENTAL PROCESSES IN BACTERIA • QUORUM SENSING IN GRAM-NEGATIVE BACTERIA • SECONDARY METABOLITES

Bibliography

Dunny, G. M., and Leonard, B. A. B., (1997). Cell-cell communication in gram-positive bacteria. *Annu. Rev. Microbiol.* **51**, 527–564.

Dworkin, M. (1996). Recent advances in the social and developmental biology of the myxobacteria. *Microbiol. Rev.* **60**, 70–102.

Dworkin, M., and Kaiser, D. (eds.) (1993). "Myxobacteria II." American Society for Microbiology, Washington, DC.

Fuqua, C., and Greenberg, E. P. (1998). Self perception in bacteria: Quorum sensing with acylated homoserine lactones. *Curr. Opinion in Microbiol.* **1**, 183–189.

Koch, A. L. K. (1998). The strategy of *Myxococcus xanthus* for group behavior. *Antonie van Leevwenhoek* **73**, 299–313.

Koch, A. L., and White, D. (1998). The social lifestyle of Myxobacteria. *BioEssays* **20**, 1030–1038.

Reichenbach, H. (1993). Biology of the myxobacteria: Ecology and taxonomy. *In* "Myxobacteria II" (M. Dworkin and D. Kaiser, eds.), pp. 13–62. American Society for Microbiology, Washington, DC.

Reichenbach, H., and Dworkin, M. (1992). The myxobacteria. *In* "The Prokaryotes" (A. Balows *et al.*, eds.), 2nd ed., Vol. 4, pp. 3416–3487. Springer-Verlag, Berlin.

Reichenbach, H., Heunert, H. H., and Kuczka, H. (1965a). "Schwarmentwicklung und Morphogeneses bei Myxobakterien- *Archangium, Myxococcus, Chondrococcus, Chondromyces*." Encylcop. Cinematogr. Film C893. Institute of Wissen Film, Göttingen.

Reichenbach, H., Heunert, H. H., and Kuczka, H. (1965b). "*Myxococcus* spp. (Myxobacterales)—Schwarmentwicklung und Bildung von Protocysten." Encylcop. Cinematogr. Film E779. Institute of Wissen Film, Göttingen.

Reichenbach, H., and Höfle, G. (1993). Production of bioactive secondary metabolites. *In* "Myxobacteria II" (M. Dworkin and D. Kaiser, eds.), pp 347–397. American Society for Microbiology, Washington, DC.

Rosenberg, E. (ed.) (1984). "Myxobacteria: Development and Cell Interactions." Springer-Verlag, Berlin.

Shapiro, J. A., and Dworkin, M. (eds.) (1997). "Bacteria as Multicellular Organisms." Oxford University Press, New York.

Shimkets, L. J., and Dworkin, M. (1997). Myxobacterial multicellularity. *In* "Bacteria as Multicellular Organisms" (J. A. Shapiro and M. Dworkin, eds.), Oxford University Press, New York.

Myxococcus, Genetics

N. Jamie Ryding and Lawrence J. Shimkets

University of Georgia

GLOSSARY

bacteriophage A bacterial virus.

cosmid A plasmid-cloning vector containing bacteriophage lambda *cos* (cohesive end) sites incorporated into the plasmid DNA.

ectopic expression Artificial expression of a gene under conditions in which it would not normally be expressed.

electroporation The use of electrical shock to introduce small and temporary holes in the bacterial cell membrane for the purpose of introducing DNA.

fruiting body A macroscopic multicellular structure produced by myxobacteria in response to environmental stress.

gliding motility Locomotion on solid surfaces in the absence of flagella.

homologous recombination A process by which two nearly identical genetic elements are brought together for the purpose of combining them or separating them.

megabase One thousand kilobases of nucleic acid.

myxospore An asexual dormant cell produced in the fruiting body.

operon A cluster of genes whose expression is controlled by a single promoter.

reporter A gene product whose activity is readily recognized by selection or screening.

transduction The transfer of host genes from one cell to another via the agency of a bacteriophage.

transposon A genetic element that has the ability to move (transpose) from one site on a DNA molecule to another. In addition to genes involved in transposition, it carries other genes, conferring selectable phenotypes such as antibiotic resistance.

MYXOCOCCUS is a rod-shaped gram-negative soil bacterium that exhibits gliding motility. It secretes proteases and other hydrolytic enzymes to degrade other bacteria and so provide peptides and amino acids to be used as a nutrient source. It displays a level of cooperation during feeding, as individuals benefit from the activity of enzymes secreted by their neighbors. To preserve the integrity of a swarm of cells during the resting phase, many thousands of individuals aggregate to form a fruiting body, within which a portion of cells form spores. The purpose of this section is to describe the genetic tools and strategies that have been used to try to understand the biology of this organism.

I. IDENTIFICATION OF GENES OF INTEREST

Myxococcus xanthus has a life cycle that can be depicted in terms of two interlocking circles describing vegetative growth and fruiting body development. Cells reproduce by transverse binary fission and continue vegetative growth as long as nutrients are available (Fig. 1). Amino acids serve as the principal carbon and energy source. When cells encounter nutritional stress, tens of thousands aggregate together to form multicellular fruiting bodies containing dormant asexual myxospores. *M. xanthus* has a superb genetic system and can be manipulated genetically, as can any other member of the Proteobacteria such as *Escherichia coli.* Genetic analyses of *M. xanthus* have focused primarily on the developmental cycle. As development is an alternative to growth, it is possible to isolate and maintain develop-

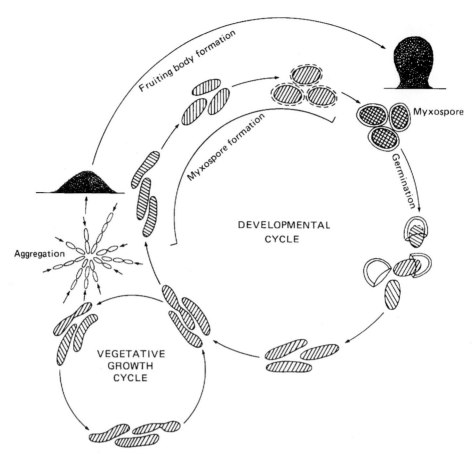

Fig. 1. Life cycle of *M. xanthus.* The cells replicate by transverse binary fission during the vegetative growth cycle. When cells experience amino acid limitation, they move to an aggregation center where they build a fruiting body. Inside the fruiting body, they form asexual spores. The fruiting body is not drawn to scale; it is a few hundredths of a millimeter in diameter, in contrast to vegetative cells, which are about 5–7 $\times$ 0.7 μm. (Adapted from Dworkin, M. (1985). "Developmental Biology of Bacteria." Benjamin/Cummings, Menlo Park, CA.)

mental mutants in a viable and healthy state. The purpose of this article is to describe genetic tools that are available for work with *M. xanthus. Myxobacteria II* (Dworkin and Kaiser, 1993) contains much information on the biology, genetics, and biochemistry of these organisms, including an extensive chapter on established genetic techniques.

A. Mutagenesis

The study of *M. xanthus* has relied on mutagenesis to reveal the genetic loci that are important for certain aspects of the organism's biology. In early studies,

mutagenesis was carried out by the treatment of whole cells with a chemical mutagen or UV light, but more recent mutant searches have involved transposon mutagenesis using Tn5 or a derivative. Although the kanamycin resistance gene encoded by Tn5 may be successfully used as a selectable marker in *M. xanthus*, variants have been engineered to encode resistance to tetracycline (Tn5-*132*) and trimethoprim (Tn5-*Tp*), to allow other genetic elements that require kanamycin selection to be introduced into a transposon-containing strain. Transposons are delivered into *M. xanthus* by means of a variant of an *E. coli* bacteriophage P1 in which the transposon has previously inserted. Because P1 cannot itself be stably maintained in *M. xanthus,* antibiotic-resistant strains should only result from transposition of the Tn5 onto the *M. xanthus* chromosome. Brian Julien has constructed mini-Tn5 derivatives, which employ a bacteriophage P4 delivery system. The mini-Tn5 derivatives transpose at a 50- to 100-fold higher frequency than wild-type Tn5.

The single-base changes that are created by chemical mutagenesis may result in phenotypes that would not be seen when using a transposon (which often causes a complete loss of function of the gene into which it has inserted). These may include mutations in essential genes that subtly alter the function of their product and temperature-sensitive mutations. However, the use of a transposon has an advantage over chemical mutagenesis, in that by its nature the mutation is directly linked to a selectable marker that may be used to transfer the mutation to other strains by means of generalized transduction using the *M. xanthus* bacteriophages Mx4 or Mx8, or to select for the transposon and its flanking chromosomal DNA when cloning into an *E. coli* vector. Approximately 10% of the transposon mutants examined contained two copies of the transposon, making it generally advisable to transfer a newly discovered insertion to a "clean" genetic background by transduction before proceeding with further analysis. Transposons appear to be stable during generalized transduction.

Work with *M. xanthus* genetics has focused on the isolation of mutants in fruiting body development and motility. Several developmental mutants have later been remutagenized to allow the selection of suppressor mutations that at least partially restore development by counteracting in some way the effect of the original mutation. Suppressor mutations are useful in that they may allow the identification of genes whose loss or modification would be undetectable in a wild-type strain, and genes that are able to suppress the original defect because their products interact in some way with that of the primary gene.

B. Reporter Gene Activity

One strategy that has been used to identify classes of genes that are transcribed under particular condi-

tions or whose products are localized to a subcellular compartment has been to use transposons that contain a reporter gene. The reporter genes that have been used in *M. xanthus* are *lacZ* (encoding β-galactosidase) and *phoA* (encoding alkaline phosphatase).

The transposon Tn5 *lac* contains a promoterless copy of *lacZ* that has a ribosome-binding site, allowing transcriptional fusions to be formed when the transposon inserts into the chromosome downstream of a promoter. This technique was used by Kroos, Kuspa, and Kaiser (1986) to identify genes that were uniquely activated by starvation, which they estimated to be approximately 8% of the genes on the *M. xanthus* chromosome or approximately 750 genes, assuming an average gene size of 1 kb. Only 0.3% of the insertions caused a noticeable defect in development, suggesting that fewer than 30 development-specific genes are essential for fruiting body morphogenesis. This particular approach would not have identified genes that were essential for both growth and development.

The *phoA* gene within Tn5 *phoA* lacks both a promoter and ribosome-binding site. The insertion of this transposon in frame with an existing gene results in the expression of a chimeric protein made up of part of the endogenous gene product and alkaline phosphatase. This transposon was used by Kalos and Zissler (1990) to identify gene products that were exported to the periplasm and beyond, as alkaline phosphatase is inactive within the cytoplasm. About 0.3% of the insertions produced alkaline phosphatase activity. Assuming that only one of six insertions is in the correct reading frame, one can calculate that *M. xanthus* exports about 200 proteins. Among the proteins that were identified were several Cgl proteins that play an essential role in adventurous gliding motility and that can be transferred between cells upon contact.

C. Sequence Homology

The nucleotide sequences of genes encoding proteins that have evolved from a common ancestor tend to show conservation in areas that are important to the protein's function. This conservation has been used to identify members of interesting classes of proteins in *Myxococcus,* which have previously been identified in other organisms. Homologs have been detected by using either the gene from another organism directly as a heterologous probe (as in the case of the sigma 54 homolog *rpoN*), or by designing opposing pairs of oligonucleotides that match regions that are conserved within the family of proteins and using these to amplify a portion of the gene from *Myxococcus* chromosomal DNA by polymerase chain reaction (PCR). PCR has been used to isolate genes from *M. xanthus* that encode eukaryotic-like protein serine–threonine kinases.

II. CLONING METHODS IN *M. XANTHUS*

A. Cloning Strategies

M. xanthus DNA is routinely cloned in *E. coli* vectors. The strategies that may be used to clone the wild-type copy of a gene from *M. xanthus* that has been identified by mutagenesis depend on the method of mutagenesis that has been employed. Genes that have been mutated by a chemical mutagen or by UV irradiation may be cloned by identifying a fragment of wild-type DNA from a genomic library that restores the mutant to a wild-type phenotype. Although this method of genetic complementation could also be used for transposon mutants, it is more common for chromosomal DNA flanking a transposon to be cloned by either transforming *E. coli* with a library of chromosomal DNA from the transposon-containing strain and selecting directly for the resistance gene, or by carrying out inverse PCR. In the case of inverse PCR, chromosomal DNA from the strain containing the transposon is digested with a restriction enzyme that cuts frequently in *Myxococcus* DNA, ligated under conditions that favor recircularization of the restriction fragments, and used as the template in a PCR reaction in which the primers are complementary to the ends of the transposon, reading outward. Either method results in the isolation of a fragment of DNA that contains chromosomal DNA flanking the point of insertion of the transposon, which may then be used to identify clones containing the wild-type copy from a library of *M. xanthus* DNA in either a lambda phage or cosmid library

by hybridization. The recovery of the transposon by selection for its marker has been taken one step further by the construction of a variant of Tn5, called TnV, in which an the origin of replication from *E. coli* plasmid pSC101 has been inserted into the transposon. In this case, chromosomal DNA from an *M. xanthus* strain containing TnV may be digested with a restriction enzyme, recircularized by ligation, and used to transform *E. coli* directly, selecting for the marker on the transposon and thus a self-replicating plasmid containing DNA that had been flanking the insertion point in *M. xanthus*.

B. Reintroduction of DNA into *M. xanthus*

Although genes may most easily be cloned and manipulated in *E. coli*, further study often requires DNA to be reintroduced into *Myxococcus*. Two means of delivering DNA into *Myxococcus* have been used, the electroporation of *M. xanthus* cells in the presence of DNA and the transduction of plasmids from *E. coli* by the bacteriophage P1. Plasmid DNA may be packaged as a random event during the lytic phase of P1 infection in *E. coli*, presumably as multimers, as at least 40 kb of linear DNA is required for packaging to take place. More efficient transfer can be achieved by using a plasmid vector that carries the incompatibility region (*inc*) of P1. The vector plasmid may then only be maintained in a strain lysogenic for P1 as part of a cointegrate. Packaging of the cointegrate during lytic phase yields a high proportion of phage that contains vector DNA flanked by P1 sequence, which resolves after infection of *M. xanthus*, allowing maintenance of only the vector portion.

No native plasmids have been found in *Myxococcus* species that could be modified to act as vectors, but an *E. coli* plasmid (pACYC184) has been mutagenized to allow it to replicate in *M. xanthus*. DNA may also be maintained in *M. xanthus* by integration into the chromosome, either by using a site-directed integration system from the bacteriophage Mx8, or by homologous recombination between the cloned DNA and the chromosomal copy. Integration of Mx8 into the chromosome of *M. xanthus* occurs by a site-specific recombination event between the *attP* site on the phage and the *attB* site on the chromosome.

High-frequency insertion of plasmids into the *M. xanthus* chromosome requires only a 2.9-kb fragment that contains *attP* within the C terminus of the coding region of a gene called *intP*, whose predicted product shows homology to site-specific recombinases of the integrase family.

III. GENETIC MAPPING

A. Short-Range Mapping and Strain Construction

Short-range mapping is accomplished by generalized transduction using the bacteriophages Mx4 or Mx8, which package 56- and 62-kb DNA, respectively, or about 0.6% of the genome. Individual genes are transduced at a frequency in the range of 10^{-5} to 10^{-7} per phage particle. If two genetic markers are close enough on the bacterial chromosome, they may be packaged together and coinherited in the recipient. The cotransduction frequency of two markers is inversely proportional to the distance between them, allowing one to derive the physical distance between two markers from the cotransduction frequency.

In this era of genome sequencing, the value of short-range genetic mapping has diminished greatly: It is far more useful to locate a mutation by DNA sequencing. But a powerful application for generalized transduction, namely the ability to construct otherwise isogenic strains, is an essential component of the *Myxococcus* genetics toolbox. If the sole genetic difference between two strains is a mutation in a single gene, then it follows that any biochemical or behavioral abnormalities observed *in vivo* are a consequence of that mutation. This approach has been instrumental in establishing the genetic networks involved in the social and developmental behaviors of *M. xanthus*.

B. Long-Range Mapping

Myxobacteria contain a single circular chromosome with sizes ranging from 9.2–10.0 mb, depending on the species. Long-range mapping in *M. xanthus* is performed by DNA–DNA hybridization

to a series of partially overlapping *M. xanthus* clones contained in yeast artificial chromosomes (YAC). The physical and genetic map is shown in Fig. 2 (see also Table I). We are exploring strategies to sequence the *M. xanthus* genome.

IV. GENE ANALYSIS

A. Gene Replacements

It is frequently desirable to perform gene replacements in order to determine whether a particular gene, or a base within that gene, is essential for a biological process. If the gene was identified by hybridization, rather than mutagenesis, gene replacement is necessary to assess the importance of the gene in the life cycle of the organism. For example, the *pkn-1* gene encoding a protein serine–threonine kinase was identified in a library of cloned genes on the basis of hybridization to a PCR product made by amplifying *M. xanthus* DNA with primers to highly conserved regions of this family of kinase genes. Replacement of the wild-type version of this gene with a disrupted version resulted in low spore yield, thereby demonstrating that *pkn-1* is indeed required for fruiting body development.

In the case of an operon, a transposon insertion in an upstream gene usually blocks the transcription of all downstream genes. This makes it difficult to determine whether the mutant phenotype is due to the mutation in the gene harboring the insertion or a polar effect on a downstream gene. In this case, it is advantageous to construct a mutation that is less polar on downstream genes. In-frame deletions remove a portion of the coding sequence while retaining the translational start and stop signals, and are usually less polar on downstream genes than transposon insertions because they allow the ribosome to remain engaged to the transcript as it shifts from one translational unit to the next. In-frame deletions in the pilin gene *pilA* and an upstream gene *pilS* were constructed by Wu and Kaiser (1996) using a plasmid integration–excision strategy facilitated by *sacB* from *B. subtilis*, which confers sucrose sensitivity on *M. xanthus*. Versions of *pilA* and *pilS* engineered to contain in-frame deletions were cloned in

100 kbp

a vector containing a kanamycin resistance gene and *sacB,* and then used to direct integration into the *M. xanthus* chromosome by homologous recombination. After approximately 15 generations under nonselective conditions, the strains were plated onto media containing sucrose, thus selecting for strains that had undergone a second recombination event that had resulted in the loss of the vector sequences. In cases in which the second recombination event had taken place on the opposite side of the deletion from the initial event, the wild-type gene was effectively replaced in the chromosome by the mutant version containing the deletion. It was found that the *pilA* mutant lacked pili and social motility, whereas the upstream *pilS* mutant was normal for both properties, thereby demonstrating the utility of this approach.

B. Reporter Genes

Transposons containing *lacZ* have been used to identify genes in the *M. xanthus* chromosome that are transcribed under developmental conditions. However, transcriptional fusions such as these have not only allowed the identification of new genes, but also facilitated the detailed study of the temporal and spatial patterns of expression of the transcriptional unit in which the transposon is inserted. For example, a transcriptional fusion of the *devTRS* operon to *lacZ* was useful in determining that the

promoter is highly expressed in only a portion of developing cells, providing one line of evidence that cells can have different developmental fates. Reporter–gene fusions may also be transferred to strains that contain other mutations to examine the transcriptional dependence of a promoter on those genes.

C. Ectopic Expression

The ability to manipulate the concentration of a gene product *in vivo* has a number of applications, including the visualization of gene products, analysis of regulatory pathways, and determination of the role of the protein in the life cycle of the organism. Four promoters have been used for this purpose. The *mglA* and *lonD* (*bsgA*) promoters are constitutive and are expressed at reasonably high concentrations during vegetative growth and development. The *tac* promoter is induced about eight-fold with isopropyl-β-D-thiogalactoside (IPTG) in *M. xanthus.* The light inducible *carQRS* promoter increases expression about 47-fold in the presence of blue light.

Crawford and Shimkets observed that the *socE* gene was likely to be essential for growth because they were unable to isolate a *socE* mutant by conventional gene replacement technology. To examine the molecular basis of the lethality they placed the *socE* gene under the control of the light-inducible *carQRS* promoter and discovered that growth and macromo-

Fig. 2. A linearized physical and genetic map of the circular *M. xanthus* DK1622 chromosome. The three sets of long boxes present a composite physical map consisting of the *Ase*I (upper) and *Spe*I (lower) restriction map in each set. The alphabetical name of each restriction fragment is shown in the appropriate box along with the numerical size in kilobases. The right end of the top segment is contiguous with the left end of the middle segment, and the right end of the middle segment is contiguous with the left end of the lower segment. The *Ase*I sites in the far upper left and the far lower right of the figure are the identical site, shown twice for clarity. The short vertical lines below the physical map represent the distance in megabases from the left end of the top segment. The short horizontal lines below the physical map represent the locations and sizes of specific (numbered) yeast artificial chromosome (YAC) clones. To make the map more concise, only the informative, nonredundant, YACs are shown. The names and locations of mapped genes and transposon insertions are shown above the long lines. The solid black boxes on the long lines represent the genetic loci named above them, with the width of the box reflecting the limits of the exact location, not the size of the locus represented. Thus, the rRNA operon *rrnA* (upper left hand corner) is located somewhere on YAC 1142. Omega (Ω) with the alphanumeric sequence that follows it designates a genomic site identified by a particular insertion of transposon Tn5. For tracking the genealogy of a particular transposon insertion, ΩDK4414 (lower right), for example, indicates the site of a transposon insertion originally isolated in strain DK4414. The transposon itself is not actually present in the DK1622 genome that this map represents. Mxalpha is an 80-kb prophage.

TABLE I
Genetic Markers of *Myxococcus xanthus* DK1622[a]

Gene symbol	Mnenomic	Homologous YAC clones[b]	Map position (kbp)[c]	Phenotypic trait[d]
agIB	Adventurous gliding	1280, 1702	8723	Adventurous motility
asgB	A signal	665, 799	3662	Transcriptional regulator
bsgA	B signal	933, 1465	5161	ATP-dependent protease (Ion)
carC	Carotenoid synthesis	336	1350	Unknown
carE	Carotenoid synthesis	336	1350	Unknown
carQ	Carotenoid synthesis	1305	5322	Activator of carQRS
carR	Carotenoid synthesis	1305	5322	Membrane bound inhibitor of carQRS
carS	Carotenoid synthesis	1305	5322	Required for light activation of carB
cgIB1	Contact gliding	557, 1120, 1183	3844	Contact-stimulated A motility
cgIF1; cgIF2	Contact gliding	452, 1520	3198	Contact-stimulated A motility
csa-1604	Cell surface antigen	912, 1360	4421	Cell surface antigen
csgA	C signal	1409	2212	Short-chain alcohol dehydrogenase
devR	Development	198, 1135, 1751	9171	Required for development
devS	Development	198, 1135, 1751	9171	Required for development
DK4401		935	2848	Dev. regulated promoter
DK4406		470, 912	4321	Dev. regulated promoter
DK4408		336, 458	1433	Dev. regulated promoter
DK4427		1400	6865	Dev. regulated promoter
DK4445		797, 1167	5447	Dev. regulated promoter
DK4455		336	1375	Dev. regulated promoter
DK4457			2407	Dev. regulated promoter
DK4469		1142	110 or 126	Dev. regulated promoter
DK4473		198, 1135, 1751	9171	Dev. regulated promoter
DK4474		263, 1479	6447	Dev. regulated promoter
DK4491		776, 1183	3914	Dev. regulated promoter
DK4494		423, 1690	2185	Dev. regulated promoter
DK4497		1181	4944	Dev. regulated promoter
DK4500		780	1982	Dev. regulated promoter
DK4511		470, 912	4321	Dev. regulated promoter
DK4521		812, 1038	5592	Dev. regulated promoter
DK4529		617, 933, 1465	5211	Dev. regulated promoter
DK4530		1683	1021	Dev. regulated promoter
DK4531		1412	8152	Dev. regulated promoter
dsgA	D signal	1080, 1683	895	Translation initiation factor IF-3
dsp	Dispersed	486, 1383	8538	Social motility
ER304		897, 1306, 1566, 1868	7642	Autocide-dependent development
ER419		263, 1479	6392	Antibiotic TA biosynthesis
ER1310			5079	Antibiotic TA biosynthesis
ER1912			5039	Antibiotic TA biosynthesis
ER6118			5085	Antibiotic TA biosynthesis
fprA	Flavoprotein	1409	2212	Pyridoxine 5′-phosphate oxidase
frzA	Frizzy	797, 1305	5372	CheW-like coupling protein
frzB	Frizzy	797, 1305	5372	Motility regulation
frzCD	Frizzy	797, 1305	5372	MCP
frzE	Frizzy	797, 1305	5372	Histidine kinase/response regulator
frzF	Frizzy	797, 1305	5372	MCP methyl transferase
frzG	Frizzy	797, 1305	5372	MCP methyl esterase
gyrB	Gyrase	931, 1730	588	DNA gyrase
hemG	Heme biosynthesis	1409	2212	Protoporphyrinogen oxidase
LS234		1181	4951	Dev. regulated promoter
LS237		263, 1479	6447	Dev. regulated promoter
LS255		486, 1383	8502	Dev. regulated promoter
LS257		912, 1360	4475	Dev. regulated promoter

continues

Continued

Gene symbol	Mnenomic	Homologous YAC clones[b]	Map position (kbp)[c]	Phenotypic trait[d]
LS259		406, 1343	2751	Dev. regulated promoter
LS263		461, **799**	3621	Dev. regulated promoter
LS267		486, **1383**	8532	Dev. regulated promoter
LS409		617, 1465	5041	Dev. regulated promoter
LS420		1409	2255	Dev. regulated promoter
LS421		933, **1465**	5118	Dev. regulated promoter
LS441		739, **1416**	8900	Dev. regulated promoter
LS442		819	4182	Dev. regulated promoter
LS444		1416	8873	Dev. regulated promoter
mbhA	Myxobacterial Hemagglutinin	209, 1213	9109	Lectin
mgIA	Mutual gliding	105, 984	2503	G protein
mgIB	Mutual gliding	105, 984	2503	Stimulates mgIA
Mx8 attB	Mx8 attachment	**780**	1973	Bacteriophage Mx8 integration site
Mx alpha	Mx alpha prophage	429	2442–2522	Tandem copies of Mx alpha
ops	Other protein S	504, 798	7075	Dev. regulated gene
oriC	Origin	1387, 1619	225	Chromosomal origin of replication
PH1215		**1120**, 557	3842	Aventurous motility
PH1222		486, **1383**	8512	Social motility
PH1255		486, **1383**	8542	Social motility
PH1258		**1619**	125–177	Social motility
PH1272		**1120**, 557	3802	Adventurous motility
PH1284		263, **1479**	6387	Adventurous motility
PH1302		131, 1049, 1134	6172	Adventurous motility
PH1329		**1566**	7476	Social motility
rif	Rifampicin resistance	557, 1120, 1183	3855	High-level rifampicin resistance
rrnA	rRNA	1142	75	rRNA operon
rrnB	rRNA	1569	1673	rRNA operon
rrnC	rRNA	461, 799, 1242	3542	rRNA operon
rrnD	rRNA	131, 1049	6168	rRNA operon
sgI-3119	Social gliding	921	8494	Social motility
socA	Suppressor of C	535, **1400**	6354	CsgA-like short chain alcohol dehydrogenase
socB	Suppressor of C	535, **1400**	6354	Membrane anchor protein
socC	Suppressor of C	535, **1400**	6354	Negative regulator of socABC
socD	Suppressor of C		8812	Histidine protein kinase
socE	Suppressor of C	1683	845	C signal suppressor
socF	Suppressor of C	907	1045	C signal suppressor
spo-406	Sporulation	776, **1183**	3911	Essential for sporulation
spo-417	Sporulation	617	5231	Essential for sporulation
spo-418	Sporulation	402, 454	6357	Essential for sporulation
spo-422	Sporulation	131, 1049, 1134	6188	Essential for sporulation
spo-423	Sporulation	430, 939	4701	Essential for sporulation
spo-510	sporulation	276	6654	Essential for sporulation
stk-1907	Sticky	**1036**	4634	Cell adhesion
tagA-H	Temp. sensitive Aggregation	1690, 1419	2107	34°C aggregation system
tgIA		557, 1120, 1183	3861	Contact-stimulated S motility
tps	True protein S	504, 798	7075	Spore-coat protein S
uraA	Uracil	897, 1306, **1566**	7537	Orotidine-5′-phosphate decarboxylase

[a] A transposon linked to the gene of interest was mapped to a particular *Ase*I restriction fragment by restriction enzyme analysis. Where hybridization data to the ordered YAC library is available, a more precise map location is given along with the names of the homologous YAC clones.

[b] Bold YAC numbers refer to linkages inferred from the map structure, but not proven by hybridization.

[c] Location is relative to the *Ase*I fragment J-G junction which is 1/9232 kbp; average range ±30 kbp.

[d] MCP, methyl-accepting chemotaxis protein.

lecular synthesis ceased when *socE* expression was disrupted.

D. Genetic Analysis of *Myxococcus* Behavior

Genetic approaches to the analysis of *M. xanthus* have focused primarily on the multicellular motility and development behaviors.

Genetic analysis of gliding motility has revealed that there are two large multigene systems controlling cell behavior (for review see Dworkin, 1996; Brun and Shimkets, 1999). The A (or adventurous) system controls the movement of individual cells, whereas the S (or social) system controls the movement of groups of cells. Many of the S-system genes have been cloned and sequenced, and used to determine the machinery of S motility. At least three cell-surface appendages are involved, including type IV pili, lippolysaccharide, and fibrils, which are long thin appendages of polysaccharide and protein that extend away from the cell surface to interact with other cells. It has been suggested that the retraction of the pili provides force for cell translocation across a solid surface.

The analysis of development has focussed on the genes involved in the transfer of signals between cells (for review see Shimkets, 1999). A set of conditional developmental mutants was isolated, which failed to develop alone but could develop when mixed with wild-type cells or cells in a different complementation group. Presumably a molecule passes from the wild-type to the mutant, which restores the capacity to develop. The change is only transient in that the germinated spores still retain the mutant genotype and phenotype. Five complementation groups, designated A–E, have been identified and all the genes represented by the mutants have now been cloned and sequenced. In several cases, the signaling system was revealed when the appropriate biochemistry was performed. In several other cases, the signaling system remains mysterious. For example, the sole gene in the B complementation group encodes an intracellular ATP-dependent protease, and the sole gene in the D complementation group encodes translation

initiation factor IF2. In neither case is it obvious how intercellular signaling occurs.

V. PROSPECTUS

A fine suite of genetic tools is available for the analysis of *Myxococcus*. The remarkable behaviors exhibited by growing and developing cells are proving amenable to genetic analysis. The *Myxococcus* genome sequence is expected to further simplify genetic approaches to the analysis of this fascinating creature.

Acknowledgments

This work was supported by National Science Foundation grant MCB9601077.

See Also the Following Articles

DNA SEQUENCING AND GENOMICS • MUTAGENESIS • POLYMERASE CHAIN REACTION • TRANSDUCTION: HOST DNA TRANSFER BY BACTERIOPHAGES

Bibliography

Brun, Y., and Shimkets, L. J. (ed.). (1999). "Prokaryotic Development." American Society for Microbiology Press, Washington, DC, in press.

Dworkin, M. (1996). Recent advances in the social and developmental biology of the myxobacteria. *Microbiol. Rev.* 60, 70–102.

Dworkin, M., and Kaiser, D. (eds.) (1993). "Myxobacteria II." American Society for Microbiology Press, Washington, DC.

He, Q., Chen, H.-W., Kuspa, A., Cheng, Y., Kaiser, D., and Shimkets, L. J. (1994). A physical map of the *Myxococcus xanthus* chromosome. *Proc. Natl. Acad. Sci. U.S.A.* 91, 9584–9587.

Kalos, M., and Zissler, J. (1990). Transposon tagging of genes for cell-cell interactions in *Myxococcus xanthus*. *Proc. Natl. Acad. Sci. U.S.A.* 87, 8316–8320.

Kroos, L., Kuspa, A., and Kaiser, D. (1986). A global analysis of developmentally regulated genes in *Myxococcus xanthus*. *Dev. Biol.* 117, 252–266.

Shimkets, L. J. (1999). Intercellular signaling during fruiting body development of *Myxococcus xanthus*. *Annu. Rev. Microbiol.* 53, 525–549.

Wu, S. S., and Kaiser, D. (1996). Markerless deletions of *pil* genes in *Myxococcus xanthus* generated by counterselection with the *Bacillus subtilis sacB* gene. *J. Bacteriol.* 178, 5817–5821.

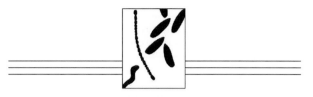

Natural Selection, Bacterial

Daniel E. Dykhuizen

State University of New York at Stony Brook

GLOSSARY

fitness The average contribution of one allele or strain to the next generation compared to other alleles or strains.

genetic drift Random changes in the frequencies of two or more alleles or genotypes due to sampling from one generation to the next.

mutation Spontaneous change in the sequence of bases in a DNA molecule.

NATURAL SELECTION is formally defined as the differential survival and reproduction of distinct phenotypes. The differences in phenotype are caused at least in part by differences in genotype. The selective environment acts on the phenotypic variation to produce the differential survival and reproduction. This differential is called natural selection.

sure of the portion of the phenotypic variation that caused by genetic variation, and it varies from zero (when all the phenotypic variation is caused by environmental variation) to one (when all the phenotypic variation is caused by genetic variation). Heritability is a relative term. It increases with an increase in the amount of genetic variation in a population and decreases with an increase in the amount of variation in the epigenetic environment. Differential survival and reproduction can change gene frequency. The change in gene frequency causes adaptation of the population to the local environment. Thus, natural selection increases adaptation to a particular environment at a particular time, but does not create global perfection.

When humans create the differential surival and reproduction by picking certain organisms to reproduce, as in animal and plant breeding, it is referred to as artifical selection. In microbiology, almost all the selection is mediated through the environment and, therefore, is natural selection. For example, we select antibiotic resistance by plating a bacterial culture on agar plates containing a particular antibiotic. We have also selected antibiotic resistance in nature through our extensive use of antibiotics in medicine and in agriculture, to the extent that for many pathogenic species we have little or no option in antibiotic treatment.

I. DEFINITION OF NATURAL SELECTION

Figure 1 gives an extended model of natural selection. The portion of the environment that interacts with the genotype to produce the phenotype is called the epigenetic environment. Heritability is the mea-

II. OTHER FORCES CHANGING GENE AND GENOTYPE FREQUENCY

There are three forces that change gene frequency in a population in addition to natural selection. These are mutation, migration, and genetic drift. Mutation

Copyright © 2000 by Academic Press.
All rights of reproduction in any form reserved.

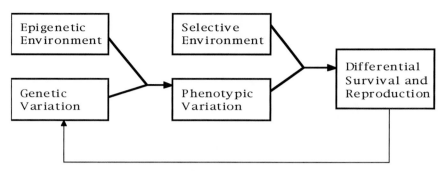

Fig. 1. Pictorial model of natural selection.

is a weak force, doing little to change gene frequency, but it is critically important in providing new variation. Migration and the consequent gene flow is required to prevent populations from diverging from each other. However, if there is enough migration between populations, on the order of a single migrant per generation, all the populations of a single species can be roughly considered a single population. If this is the case, migration does not have to be considered when partioning the causes that change gene frequency within a species over time. Genetic drift describes the effect of chance in changing gene frequency as genes pass from one generation to the next. If the population is small, drift is a strong force. As the population size increases to infinity, the effect of drift approaches zero. However, a new mutation, which by definition is present in only a single copy, is likely to be lost by chance even in a very large population. The chances a new mutation with a selective advantage of 2%, which is considered a very large advantage, will be fixed in the population is only about 2%.

Mixis, or lateral transfer of genes, changes genotype frequencies and can introduce new genes or new alleles from sources outside the species. In bacteria, all reproduction is clonal, and sex or genetic exchange involves the transfer of a piece of DNA between lineages. The size of this piece of DNA ranges from a small (200–600 bp) to a large part of the genome. The size of the transfers in nature is poorly known, but I would guess that most are in the range of 10–40 kb or 10–40 genes. In bacteria, in contrast to animals, the transfer of pieces of DNA from other species is expected. Thus, species definitions in bacteria cannot use genetic isolation. Rather,

some definition involving natural selection in particular environments will have to be developed.

III. SELECTIVE FATE OF A NEW MUTATION

A new mutation can be detrimental, advantageous, or selectively neutral. A segment of DNA transferred into the species from outside can be considered in the same terms as a new mutation. A detrimental mutation is eliminated both by selection and by genetic drift. Most selectively neutral mutations are also lost by drift. The probability that a selectively neutral mutation will be fixed is $1/N$, where N is the population size. An advantegous mutation will be fixed, with the probability and time to fixation dependent on the selective advantage.

A. The Neutral Theory of Molecular Evolution

This theory assumes that (almost) all new mutations are either detrimental or selectively neutral. This is a reasonable assumption if organisms are well adapted. Few mutations make on organism better, adopted but many of make it worse, and many more don't matter. A mutation needs to change the phenotype in a way that effects longevity and fecundity to be selected; otherwise it will be selectively neutral. A synonymous mutation, one that changes the nucleotide but not the amino acid sequence, is likely to be neutral. A mutation that changes the amino acid but does not change the function of the enzyme is likely to be neutral. A mutation that changes the function of an enzyme but does not change the flux through the pathway is likely to be neutral. A mutation that changes the flux through the pathway but does not change the physiology is likely to be neutral. All this is conditioned on the normal environment. A mutation that is neutral in one environment may not be neutral in a different environment. By neutral, we do not mean a mutation has absolutely no fitness effects, but rather one for which the fitness effects in the usual environments of the species are so small that genetic drift is a stronger force than selection. Thus, drift rather than selection dominates the fixation process. Because the effect of drift is a function of population size, mutations that are selectively neutral in a species with a small population size can be selected (or, usually, selected against) in species with a large population size.

This theory makes a number of predictions. Positions, which are the homologous nucleotide or amino acid sites across the sequences under study, will be selectively neutral if they vary within a species or change between species. Positions that are monomorphic within a species and unchanging across species will probably be those positions for which change is detrimental. Consequently, these unchanging positions are the functionally important sites.

The neutral theory can also be used to investigate selection. The neutral theory predicts that the nucleotide polymorphism within a species is μN, where μ is the neutral mutation rate and N the population size; and that the divergence between two related species is $2\mu T$ where T is the time of divergence and μ is again the neutral mutation rate. The ratio of polymorphism to divergence is $N/2T$, which should be the same for all genes in a particular species. The effects of selection can be determined by looking at this ratio. A ratio that is significantly smaller than expected for a particular gene indicates that this gene has been under directional selection, whereas a ratio that is significantly larger than expected indicates the presence of balancing selection.

B. Advantageous Mutations

Not all mutations will be detrimental or neutral. For adaptation to proced, at least a few mutations have to be adaptive. An adaptive mutation will hap-

pen in a particular bacterial cell. This cell will contain a particular variant for every polymorphic locus in the species. If the species is purely clonal, as the advantageous mutation sweeps through the species, all variation will be lost. The alleles in the cell with the advantageous mutation, even the detrimental ones, will be fixed. This phenomenon is called periodic selection. The more general term for the fixation of an advantageous mutation is selective sweep.

Consider a species of bacteria with lateral transfer of genetic material. The region containing the advantageous mutation will be transferred into other backgrounds and other regions will be transferred into the background of the clone containing the advantageous mutation. Thus, the genetic variation in the species will be eliminated only in a region surrounding the adaptive mutation. The size of this region will depend on the amount of lateral transfer, the size of the pieces transferred, and the selective advantage of the adaptive mutation. The size of this region will be smallest when the transfer is frequent, the piece size is small, and the selection is weak.

A region surrounding the *gapA* and *pabB* genes in *E. coli* shows evidence of a recent slective sweep. The polymorphism level is 5-fold lower than expected given the amount of polymorphism at this gene in *Salmonella* and the ratio of polymorphism at other genes for *E. coli* and *Salmonella*, suggesting that selection has reduced the diversity in this region of the *E. coli* chromosome. Also, the DNA sequences from

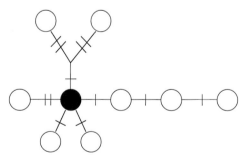

Fig. 2. A network showing the relationships between the alleles of the glyceraldehyde 3-phosphate dehydrogenase (*gapA*) gene of *Escherichia coli*. Each circle represents a different nucleotide sequence and the hatch marks on the lines between the circles represent the number of base changes. The solid circle is the presumed ancestor. This figure is redrawn from Guttman and Dykhuizen (1994).

multiple *gapA* alleles form a network with many short branches coming directly off the sequence that was swept to fixation. (Fig. 2). The ancestral sequence is still part of the population.

IV. TYPES OF SELECTION MAINTAINING POLYMORPHISM

A. Balancing Selection

Balancing selection creates a stable equilibrium such that two or more alleles are maintained in the population, much longer than if they were neutral, often for a very long time. This requires that the alleles not be easily interconvertible by mutation. The selective differences can be due to a single mutational change or to many. Because the alleles are maintained in the population for a long time, various neutral mutations accumulate in the linked region of DNA around the functional changes, creating a chromosomal region of high polymorphism. The size of this region will depend on the recombination rate; the higher the rate, the smaller the region.

B. Escape Selection

Escape selection is like balancing selection except that the mutations that convert one allele into another are common. Neutral variation does not build up around the selected sites. This selection gives more rapid amino acid changes compared to synonymous nucleotide change.

C. Frequency-Dependent Selection

There are two forms of balancing and escape selection—frequency-dependent selection and diversifying selection. In frequency-dependent selection, the rare alleles have a selective advantage. This type of selection is frequently found in proteins on the cell surface of pathogens.

An example of frequency-dependent balancing selection is the *ospC* gene of *Borrelia burgdorferi*, the spirochete that causes Lyme disease. The outer-surface protein C is the major protein on the surface of *B. burgdoferi* as the bacteria are transferred from their tick vector into the skin of a mammal. It is also a protective antigen. Immunization with the OspC protein provides resistance to infection with homologous strains, strains having the same *ospC* allele as the one used for the antigen, but not to heterologous ones, strains having very different *ospC* alleles. If the sequences of *ospC* alleles are compared, they are either very similar, with less than 2% sequence variation, or very different, with greater than 8% sequence variation. The median sequence variation of those that are different is about 15%. These alleles shared a common ancestor a very long time ago and have been maintained in the population by balancing selection. The proposed mechanism is that after an infection with one *ospC* allele, the host becomes resistant to other infections with similar *ospC* alleles, but remains sensitive to infection by strains carrying the divergent alleles. If an ospC allele is common in a population of *Borrelia*, there is a high chance that any mammal infected by a strain carrying this allele will already be resistant to it, causing the infection to fail. But if an allele is rare, then there is a much lower chance that mammal infected will be resistant and a much higher chance that the infection will be established. Thus, there is selection against common alleles and selection for rare alleles. Any time that an allele becomes rare by chance, selection will increase its frequency. The allele frequency distribution supports this mechanism. In a small population of 72 strains from ticks obtained from a single sampling site, we found 11 distinct alleles. The frequency distribution of these alleles is more even than expected by chance, showing the presence of balancing selection.

D. Diversifying Selection

In diversifying selection functionally different alleles are selected in different enviroments. An example of diversifying escape selection is the *fimH* gene of *Escherichia coli*. The FimH protein is a lectin-like protein found at the tip of type 1 or mannose-sensitive fimbriae. Within *E. coli*, this gene shows little diversity, unlike the *ospC* of *B. burgdorferi*, with most alleles showing around 99% similarity. However, there are functionally two classes. About 80% of the commensal fecal strains have a FimH, which binds only to tri-mannose structures, whereas 70% of the strains isolated from urinary infections will bind both mono- and tri-mannoses. The strains that bind mono- and tri-mannoses bind to uroepithelium significantly better and have a 15-fold colonization advantage in the bladder in mice when compared to the FimH that only binds to tri-mannoses. However, these strains that bind to both types of mannoses are inhibited by the mono-mannoses in the saliva and do not bind well to the mucosa of the oropharynx. Binding to the mucosa of the oropharynx has been shown to be important in the colonization of the intestinal tract, probably because each cell that binds to the mucosa can grow there awhile and therefore get more chances of getting through the acid bath of the stomach. Thus, the FimH that binds both mono- and tri-mannoses have an advantage in colonization and growth in the bladder, but a disadvantage in colonization of the intestinal tract.

When the *fimH* alleles are sequenced, most are different from one another by a single amino acid change. When the alleles are connected as part of a network, the *fimH* alleles that bind only tri-mannoses are on the interior nodes, whereas the alleles that bind both mono- and tri-mannoses are at the tips (Fig. 3). This pattern shows that the *fimH* alleles that bind only tri-mannoses are ancestral and the others are descendants, each containing mutations selected to survive in the urinary tract. Each mutation changes a different amino acid, showing that it is easy to convert from the ancestral type to the uropathogenic type. Thus, there is escape selection as *E. coli* adapts to the urinary tract.

V. ANTIBIOTIC RESISTANCE

Antibiotic resistance is so common in pathogenic bacteria that many people have concluded that the age of antibiotics is quickly drawing to a close. Much work has been done on the genetics and biochemistry of antibiotic resistance, but almost nothing has been done on the natural selection. We have assumed that the natural selection of resistance is simple: resistance is selected when antibiotics are present and sensitivity is selected when they are absent. As we have seen here, natural selection is not just a dualism of for and against, but it arises out of a complex interaction of environment and genetics. The direction and intensity of selection is very sensitive to the environment. Models are now being developed to understand how human behavior can promote or retard the selection for resistance while we use antibiotics as our major defense against bacterial diseases. For example, the initial models suggest that cycling, using one antibiotic for a while and then another, will not extend the effective life of these antibiotics, compared to using one until it becomes useless and then using another.

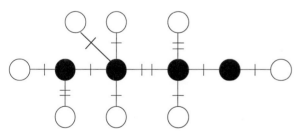

Fig. 3. A network showing the relationship between the alleles of the adhesin of type 1 fimbriae (FimH) from *Escherichia coli*. Each circle represents a different amino acid sequence and the hatch marks on the lines between the circles represent the number of amino acid changes. The solid circles are the alles that only bind tri-mannoses and are normally found in the intestinal tract. The open circles are the alleles that bind both mono- and tri-mannoses and are mostly found in the urinary tract. This figure is redrawn from Sokurenko *et al.* (1998).

Bibliography

Baumberg, S., Young, J. P. W., Wellington, E. M. H., and Saunders, J. R. (eds.) (1995). "Population Genetics of Bacteria," 52nd Symposium of the Society for General Microbiology. Cambridge University Press, Cambridge.

Dykhuizen, D. E., and Dean, A. M. (1990). Enzyme activity and fitness: Evolution in solution. *Trends Ecol. Evol.* **5**, 257–262.

Guttman, D. S., and Dykhuizen, D. E. (1994). Detecting selective sweeps in naturally occurring *Escherichia coli*. *Genetics* **138**, 993–1003.

Hartl, D. L. (1988). "A Primer in Population Genetics," 2nd ed. Sinauer Associates, Sunderland, MA.

Hudson, R. R. (1990). Gene genealogies and the coalescent process. *Oxford Surveys Evol. Biol.* **7**, 1–44.

Kimura, M. (1983). "The Neutral Theory of Molecular Evolution." Cambridge University Press, Cambridge.

Sokurenko, E. V., Chesnokova, V., Dykhuizen, D. E., Ofek, I., Wu, X. -R., Krogfelt, K. A., Struve, C., Schembri, M. A., and Hasty, D. L. (1998). Pathogenic adaptation of *Escherichia coli* by natural variation of the FimH adhesin. *Proc. Natl. Acad. Sci. U.S.A.* **95**, 8922–8926.

Nitrogen Cycle

Roger Knowles

McGill University

GLOSSARY

ATP Adenosine triphosphate, the energy-rich molecule synthesized in energy-yielding reactions and consumed in biosynthetic processes.

biofilm A community of microorganisms on a solid surface. They are usually embedded in polysaccharide slime produced by members of the community.

diazotroph A bacterium that can fix gaseous N_2.

heterotroph An organism deriving both its energy and its carbon from organic compounds.

lithotroph An organism deriving its energy from the oxidation of inorganic compounds.

megaplasmid A very large extrachromosomal circle of DNA present in some cells in addition to the chromosome.

periplasm The space between the inner and outer membranes of gram-negative bacteria. It may contain certain enzymes and cytochromes.

phylogeny A classification of microorganisms that represents their putative evolutionary relationships.

repression The inhibition of synthesis of an enzyme or other gene product, often brought about by an end product or other molecule that makes the enzyme reaction or the pathway in which it occurs unnecessary.

DESPITE THE VERY LARGE AMOUNT OF COMBINED NITROGEN COMPOUNDS present in the biosphere, lithosphere, and hydrosphere, over 99.9% of all global nitrogen is in the atmosphere in the form of dinitrogen. A small amount of this nitrogen gas is converted to oxides during electrical storms in the atmosphere and is washed out in rainfall. However, most transformations of nitrogen are catalyzed by microorganisms, which are thus of critical importance in the control of nitrogen availability for the growth of crop plants, in the treatment of waste, and in the conversion to forms that are leachable and result in contamination of groundwaters.

I. INTRODUCTION

Some aspects of the biochemistry and microbiology of each nitrogen cycle process are presented here, followed by ecological and environmental implications. The most reduced forms of nitrogen are the organic nitrogen compounds, such as amino acids, and the first product of their decomposition, ammonia (NH_4^+ or NH_3) (Fig. 1). The most oxidized form is nitrate (NO_3^-), and between ammonia and nitrate are compounds of different oxidation states: the gases dinitrogen (N_2), nitrous oxide (N_2O), nitrogen monoxide (NO or nitric oxide), and nitrogen dioxide (NO_2); and the nonvolatile hydroxylamine (NH_2OH) and nitrite (NO_2^-). Oxidative reactions (leading to the right in Fig. 1) tend to occur under aerobic (oxygenated) conditions, and reductive reactions leading to the left occur mostly under low-oxygen or anaerobic conditions. Environmental conditions dictate the type and abundance of microorganisms occurring in nature and they thus can critically control the kinds of nitrogen transformations that are likely to occur.

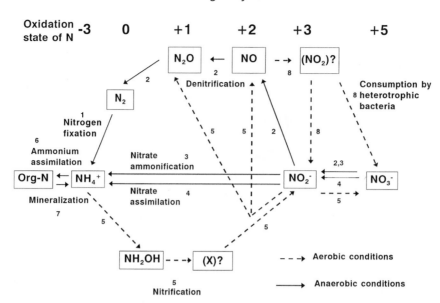

Fig. 1. Relationships among the microbiological processes in the N cycle. The oxidation state of N is indicated, and aerobic and anaerobic processes are shown as dashed and solid arrows, respectively. 1, Nitrogen fixation; 2, denitrification; 3, nitrate ammonification; 4, nitrate assimilation; 5, nitrification; 6, ammonium assimilation; 7, mineralization; and 8, consumption of NO by heterotrophic bacteria.

II. DINITROGEN FIXATION

A. Biochemistry and Microbiology

The ability to fix (assimilate) gaseous nitrogen is restricted to the archaea and the eubacteria (together, the prokaryotes) and is found in higher life forms (eukaryotes) only in association with N_2-fixing (diazotrophic) bacteria. The fixation of dinitrogen is catalyzed by an enzyme complex, nitrogenase, consisting of two subunits of dinitrogenase reductase (an iron protein), two pairs of subunits of dinitrogenase (an iron–molybdenum enzyme), and an iron–molydbdenum cofactor (FeMoCo). Dinitrogenase reductase uses ATP as the energy source to transfer energized electrons to the dinitrogenase enzyme. The very strongly triple-bonded molecule of dinitrogen is successively reduced on the surface of the FeMoCo cofactor, yielding two molecules of ammonia as the first detectable product. The ammonia is then used in the synthesis of amino acids and proteins (Section III). During dinitrogen reduction, some electrons and ATP are used to reduce protons to dihydrogen (H_2),

a wasteful process. The overall stoichiometry of these reactions is

$$N_2 + 8e^- + 8H^+ + 16ATP \rightarrow 2NH_3 + 16ADP + 16P_i + H_2$$

Some (mainly aerobic) bacteria have adapted by using a recycling (or uptake) hydrogenase to reoxidize the H_2, returning electrons to the electron-transport chain and regaining valuable ATP. Nevertheless, nitrogen fixation remains a very energy-expensive process and the availability of energy and reducing power is frequently a major limiting factor.

Another limiting factor is the presence of oxygen (O_2). Because the nitrogenase complex is extremely sensitive to O_2 inactivation, it is active only in cells or environments with very low O_2 concentrations. One species of *Azotobacter* possesses a third protein in the nitrogenase complex, the FeS II or Shethna protein, which affords some protection from O_2 inactivation.

An interesting property of the nitrogenase complex

is that in addition to being able to reduce dinitrogen to ammonia, it can also reduce a number of other small molecules that are somewhat similar in size and conformation to N_2. For example, it was discovered in the late 1960s that nitrogenase could reduce acetylene to ethylene ($H-C{\equiv}C-H$ to $H_2C{=}CH_2$) and this reaction provided a very simple, inexpensive, and very sensitive assay for nitrogenase activity that could be applied to systems varying in complexity from enzyme preparations to soil samples. The sample to be assayed is exposed to C_2H_2 and the product C_2H_4 is readily detected by gas chromatography. In addition, nitrogenase reduces hydrogen cyanide ($H-C{\equiv}N$), hydrogen azide ($H-N^{-}-N^{+}{\equiv}N$), nitrous oxide ($N \equiv N^{+}-O^{-}$), and other compounds. All of these reactions consume ATP, so the reduction of N_2O by nitrogenase is different from the reaction catalyzed by the N_2O reductase of denitrifying bacteria, which contributes to ATP synthesis (Section VA), and which, interestingly, is inhibited by (but does not reduce) C_2H_2.

About 20 catalytic proteins, cofactors, and other components are necessary for N_2 fixation. They are coded for by genes (*nif*) that may be tightly clustered, as in *Klebsiella pneumoniae* (Fig. 2), or scattered throughout the bacterial chromosome, or even present on very large megaplasmids (as in some of the rhizobia that infect the roots of legumes and cause the formation of N_2-fixing root nodules). The *nif* system is tightly regulated to conserve resources for the cell, and it is repressed by both O_2 and NH_4^+.

Microorganisms possessing *nif* genes and able to fix N_2 are taxonomically and phylogenetically very diverse, but because the nitrogenase structural genes are highly conserved, some believe that the process is very ancient in the evolutionary sense. Table I shows some selected N_2-fixing genera in major phylogenetic groups, indicating the very varied ecological types represented.

It should be noted that two other, rather rare, nitrogenases are known, one containing vanadium instead of molybdenum, and the other containing only iron. They are produced only in molybdenum-deficient environments and their ecological significance is not clear.

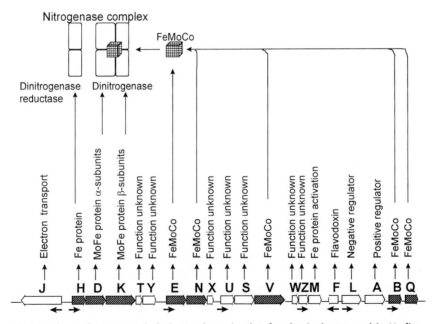

Fig. 2. The *nif* genes and their products in the facultatively anaerobic N_2 fixer *Klebsiella pneumoniae*. Structural genes are shaded, and products and functions shown. The direction of transcription is shown, as is the composition of the nitrogenase complex.

TABLE I
Phylogenetic Distribution of Selected Genera of N$_2$-Fixing Organisms[a]

Groups/genera	Comment	Groups/genera	Comment
Archaea (7)		Firmibacteria (gram positive) (3)	
Methanococcus (5)	Methanogen	*Bacillus* (3)	Fac. aerobes
Proteobacteria, alpha subdivision (24)		*Clostridium* (16)	Anaerobes
Azospirillum (4)	Plant-associative	*Desulfotomaculum* (3)	SO$_4^-$ reducer
Bradyrhizobium (1)	Soybean nodules	Green sulfur bacteria (4)	
Methylosinus (2)	CH$_4$ oxidizer	*Chlorobium* (3)	H$_2$S oxidizer
Rhizobium (5)	Legume nodules	Actinomycetes (4)	
Rhodobacter (5)	Photosynthetic	*Frankia*	Non-legume nodules
Proteobacteria, beta subdivision (8)		Heliobacteria (3)	
Alcaligenes (3)	Common het'troph	*Heliobacterium* (3)	
Thiobacillus (1)	H$_2$S oxidizer	Cyanobacteria (>27)	Blue-green bacteria
Proteobacteria, gamma subdivision (17)		*Anabaena*	
Azotobacter (6)	Aerobic het'troph	*Nostoc*	
Klebsiella (4)	Env. and path.	*Oscillatoria*	
Methylococcus (4)	CH$_4$ oxidizer	*Spirulina*	
Pseudomonas (1)	Common het'troph	*Trichodesmium*	Marine
Vibrio (4)	Aquatic het'troph	Campylobacter (1)	
Proteobacteria, delta subdivision (2)		*Campylobacter* (1)	Some pathogens
Desulfovibrio (7)	SO$_4^-$ reducer		

[a] Numbers in parentheses after group names indicate the number of genera in the group, and those after genus names indicate the number of species in the genus with N$_2$-fixing ability. Abbreviations: Env., environmental organism; Fac., facultative (grows aerobically or anaerobically); het'troph, heterotroph; Path., pathogenic strains exist.

B. Ecology of N$_2$ Fixation

The very high ATP requirement and the high O$_2$ sensitivity of the nitrogenase system means that N$_2$-fixing organisms are active only in certain environments.

1. Environments with high C and energy availability, such as readily decomposable high-C organic matter, high-C root exudates (e.g., N$_2$ fixers associated with sugar cane) or photosynthate (e.g., cyanobacteria, and root nodules formed on legumes or other plants).

2. Cells or environments with low O$_2$ concentrations. Anaerobes such as *Clostridium* spp. naturally must avoid O$_2$ to grow, whether the environment is nitrogen-deficient (i.e., they need to fix N$_2$) or not. Facultative anaerobes such as *Klebsiella* spp. fix N$_2$ only when O$_2$ is absent, but some obligate aerobes such as *Azotobacter* can fix N$_2$ under aerobic conditions by using a type of "respiratory protection" in which they develop a special branch of their electron-transport chain that permits them to consume O$_2$ (and therefore also organic carbon) at a high rate to maintain reducing conditions in the cytoplasm where the nitrogenase is located. Indeed, *Azotobacter* has the highest respiration rate per unit mass of any known organism.

Many free-living diazotrophs have no specific mechanisms of O$_2$ protection. When they must fix N$_2$ they become microaerophilic and grow only in environments having O$_2$ concentrations less than about one-twentieth of that in the atmosphere, that is, $\leq 25\ \mu$M dissolved O$_2$. Such bacteria are members of the genera *Azospirillum, Aquaspirillum, Magnetospirillum, Pseudomonas,* and the diazotrophic methanotrophs. Some of these organisms are found inhabiting the rhizosphere of many plants, including rice.

Cyanobacteria such as *Anabaena* produce specialized cells called heterocysts about every 10 or 12 cells in a chain of photosynthetic cells. Such heterocysts lack the O_2-producing photosystem II and support nitrogenase with carbohydrate translocated from the neighboring photosynthetic cells. Most *Anabaena* and related spp. are active in the free-living state in rice paddies and other aquatic systems. However, an *Anabaena* sp. lives symbiotically in specialized cavities in the fronds of the aquatic fern *Azolla*, which can contribute large amounts of fixed nitrogen in rice paddies.

Other cyanobacteria do not develop heterocysts. Species of *Trichodesmium* are filamentous and fix N_2 only in the center of colonial aggregates where O_2-producing photosynthesis does not occur. They are significant in, for example, the N-deficient Caribbean Sea, where, however, wave action can disrupt aggregates and allow O_2 to inhibit N_2 fixation. The single-celled *Gloeocapsa* has evolved another strategy in which it photosynthesizes during the day and fixes N_2 at night. Yet other cyanobacteria, with fungi, form symbioses called lichens that have a great capacity to colonize rocks and other inhospitable substrates. They are therefore important primary colonizers in nature, accumulating organic matter and allowing other forms of life (such as mosses, ferns, and higher plants) to become established.

Rhizobia of the genera *Rhizobium, Bradyrhizobium,* and *Sinorhizobium* are soil bacteria that can infect the roots of plants of the leguminosae (alfalfa, peas, beans, vetches, and soybean). Others can lead to the formation of stem nodules on some tropical plants such as *Sesbania*. To initiate infection, they attach to root hairs, exchange signal molecules with the plant, pass into the root-hair cell by reorientating its growth and causing an invagination of the cell wall, and promote the synthesis of an infection thread that infects cells in the root cortex. Bacteria are released from this thread and differentiate into specialized cells called bacteroids, in which the *nif* and associated *fix* genes that support nodule N_2 fixation are derepressed. The bacteroids use plant dicarboxylic acids to fuel the reduction of N_2 to ammonia, which is converted to amino acids or ureides, mainly by host enzymes, for transportation to the rest of the plant.

Oxygen is required by the bacteroids and is supplied at a very low free concentration but at a high flux by leghemoglobin, a carrier having a very high affinity for O_2.

Nodules formed on the roots of nonlegumes are called actinorhiza. The causal agents are species or strains of the genus *Frankia,* filamentous actinomycetes that can infect often commercially important trees and shrubs of genera such as the temperate *Alnus* (alder) and the semi-tropical *Casuarina* (horsetail pine). Such genera are important ecologically as colonizers of nutrient-poor soils and moraines, and are used for stabilization and windbreak purposes.

III. ASSIMILATION AND AMMONIFICATION

Most microbes in nature can not fix N_2 and must obtain their nitrogen supply in the form of ammonia (NH_4^+), nitrate (NO_3^-), or free amino acids. Ammonia is assimilated by the glutamate dehydrogenase or (in N-deficiency) the glutamine synthetase–glutamate synthase (GS-GOGAT) system. Nitrate is reduced by assimilatory nitrate and nitrite reductases to NH_4^+ (Fig. 1), a system that is repressed by NH_4^+, but is unaffected by O_2. Growth may be limited by C or N availability, and in agricultural soils supplied with residues having a high ratio of C to N, such as straw, the N-starved microorganisms may be serious competitors with the plant roots for available nitrogen. This suggests the advisability of reducing the C content by composting plant material before its application to soils.

Nitrogen tied up in microbial biomass components is released only on death or lysis of the cells. Other microbes then produce a variety of proteolytic enzymes and deaminases that degrade proteins, as well as other enzyme systems that attack nucleic acids and wall components. The major ultimate product of these reactions is NH_4^+ and the process is termed ammonification or mineralization (Fig. 1).

The ecological implication of the assimilation and ammonification reactions is that the microbial biomass in terrestrial and aquatic systems is in a state of turnover, with the two processes often being in a steady state such that there is no marked change in the concentration of NH_4^+. Perturbations of the ecosystem can upset the steady state, high C inputs resulting in net assimilation and high N inputs resulting in net ammonification. A great variety of microbes carry out these two processes and activity is likely in both aerobic and anaerobic environments.

IV. NITRIFICATION

The oxidation of NH_4^+ through NO_2^- to NO_3^- is carried out mainly by two highly specialized groups of lithotrophic bacteria that use the oxidation reactions as their sole source of energy and reducing power to fix CO_2 to the level of cellular organic C components. They can thus grow in completely inorganic environments, providing NH_4^+, CO_2, and O_2 are available. They are of great environmental importance because they convert the relatively immobile NH_4^+ to the anionic NO_3^- which is mobile; can be leached into lakes, rivers, or groundwaters; and is a major substrate for denitrification. The nitrifying bacteria are also of significance in aerobic secondary sewage treatment, in which NH_4^+ from ammonification is converted to NO_3^-. NO_3^- in drinking water obtained from contaminated lakes, rivers, or aquifers can be reduced to NO_2^- in the human gut and cause the conversion of hemoglobin to methemoglobin, with a great loss in O_2-carrying capacity. This condition is referred to as methemoglobinemia (in infants, called blue babies). The NO_2^- can also react with amines forming carcinogenic nitrosamines.

A. Biochemistry and Microbiology

1. NH_4^+ Oxidation

The oxidation of ammonia (NH_3 is the actual substrate) is catalyzed by an ammonia monooxygenase (AMO) in an inner membrane. O_2 is an obligate requirement because one of the atoms is incorporated into the substrate, the other is reduced to water, and hydroxylamine (NH_2OH) is produced. Reducing

power is supplied from the ubiquinone pool (Fig. 3). The NH_2OH is oxidized to NO_2^- by means of a hydroxylamine oxidoreductase that donates the electrons released to a cytochrome 554, and then through a cytochrome 552 to the terminal oxidase that reduces O_2 to water on the inner face of the inner membrane (Fig. 3). The overall reactions are

$$NH_3 + O_2 + 2H^+ + 2e^- \rightarrow NH_2OH + H_2O$$
$$NH_2OH + H_2O \rightarrow NO_2^- + 5\,H^+ + 4e^-$$

The redox reactions in the membrane are associated with the extrusion of protons into the periplasm, thus creating a proton-motive force and supporting ATP synthesis. ATP and reducing power are needed for growth and fixation of CO_2 by the Calvin cycle using ribulosebisphosphate carboxylase–oxygenase (RuBisCO). The ammonia monooxygenase genes (*amo*) show significant homology with those of methane monooxygenase (pMMO), providing some evidence that the genes may be evolutionarily related.

During the oxidation of NH_4^+, small amounts (about 0.3% of the N oxidized) of NO and N_2O are produced, the amounts increasing in O_2 deficiency to 10% or more of the N oxidized. It is known that the NO is produced from NO_2^- by a copper-containing nitrite reductase (CuNir, Fig. 3), but the mechanism by which N_2O and, in a few cases, small amounts of N_2 are produced is not known. An NO reductase has not been isolated or characterized from these organisms.

In *Nitrosomonas* sp., the genes for AMO, *amo*A and *amo*B, coding for the two subunits, are duplicated, there are three copies of *hao*, and some of the cytochrome genes are present in more than one copy. There is evidence that these genes are more scattered on the chromosome than are those of the N_2 fixation and denitrification systems.

Ammonia-oxidizing bacteria are limited to six genera that are members of the beta subdivision of the Proteobacteria, except for *Nitrosococcus oceanus*, which is in the gamma subdivision (Table II). Some are found commonly in soils; others in aquatic environments. Many have arrays of cytomembranes arising from infoldings of the cytoplasmic membrane and these are believed to contain important components of the NH_4^+-oxidizing system. However, much

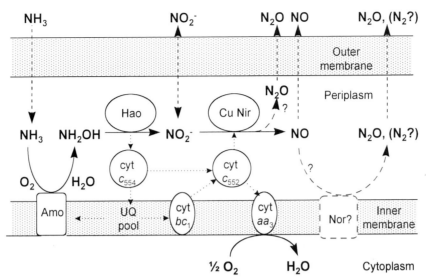

Fig. 3. The cellular topography of the NH_4^+-oxidizing system in *Nitrosomonas europaea*. The components shown are ammonia monooxygenase (AMO), NH_2OH oxidoreductase (Hao), copper NO_2^- reductase (CuNir), a putative NO reductase (Nor), and several cytochromes with location and direction of electron transport (dotted lines). Solid lines indicates reactions; dashed lines, diffusion of substrates and products. The mechanism of production of N_2O is not known, but it is speculated that it could be a minor product of the CuNir or a product of an undescribed Nor. The production of N_2 is reported for one strain but the mechanism is unknown.

remains to be discovered about the detailed biochemistry of this system.

2. NO_2^- Oxidation

The oxidation of NO_2^- appears to be a one-step process catalyzed by a nitrite oxidoreductase, an enzyme containing iron and molybdenum that may be located in the membrane. The reaction is

$$NO_2^- + H_2O \rightarrow NO_3^- + 2H^+ + 2e^-$$

It is not yet clear how a proton-motive force is generated for the support of ATP synthesis. The nitrite

TABLE II
Phylogenetic Distribution of Genera of Nitrifying Bacteria[a]

Ammonia oxidizers		Nitrite oxidizers	
Proteobacteria, beta subdivision		Proteobacteria, alpha subdivision	
Nitrosomonas (1)	Peripheral memb.	*Nitrobacter* (2)	Polar cap of memb.
Nitrosospira (5)	Acid soils	*Nitrococcus* (2)	Marine, tubular memb.
Nitrosovibrio (2)		*Nitrospira* (1)	Marine, halophilic
Nitrosolobus (2)		*Nitrospina* (1)	Marine, no memb.
Nitrosococcus (1)	Marine		
Proteobacteria, gamma subdivision			
Nitrosococcus oceanus	Marine, equatorial stack of cytomembranes		

[a] Numbers in parentheses indicate the number of species in the genus (note that the two spp. of *Nitrosococcus* are in different subdivisions of the Proteobacteria). Memb. indicates arrays of cytomembranes that are often characteristic of the genus.

oxidoreductase can operate in the reductive direction and some strains of NO_2^- oxidizers are reported to be able to grow anaerobically. However, both the NO_2^- oxidizers and the NH_4^+ oxidizers are essentially aerobic organisms, requiring O_2 for growth and activity. The fixation of CO_2 is by means of ribulose bisphosphate carboxylase–oxygenase.

Nitrite-oxidizing bacteria are confined to four genera (Table II) in the alpha subdivision of the Proteobacteria. All are lithotrophic but some can grow mixotrophically (that is, they can grow using a combination of lithotrophic and heterotrophic pathways) and one, *Nitrobacter hamburgensis*, can grow heterotrophically on organic carbon alone.

B. Ecology of Nitrification

The obligatory requirement of the nitrifiers for O_2 means that nitrification occurs only in aerobic environments. Ammonia must come from ammonification, either in the same location or by diffusion from neighboring anaerobic environments where excess NH_4^+ accumulates (Fig. 4). Nitrification provides NO_3^-, the most important nitrogen source for plant growth, and occurs readily in most agricultural and other soils, especially at near-neutral pH values, and the production by nitrifying bacteria of NO and smaller amounts of N_2O makes such environments important global sources of these trace gases. NO plays important roles in tropospheric chemistry and is a factor in acid precipitation and ozone turnover. It was recently recognized that NO from nitrification or other sources can be consumed under aerobic conditions by many heterotrophic bacteria that convert it to NO_3^- as the major product. This process can greatly modify the net flux of NO to the atmosphere. N_2O is radiatively active, absorbing infrared radiation from Earth and thus acting as a greenhouse gas. When N_2O diffuses to the stratosphere, it is converted photochemically to NO, which then catalyzes the conversion of ozone to O_2.

Based on the behavior of nitrifying bacteria in pure culture, it was believed that nitrification would not occur in environments with pH values ≤ 6.5. However, it is now known that at least some of the nitrification that can occur in acidic forest soils is brought about by NH_4^+ oxidizers such as *Nitrosospira* spp.

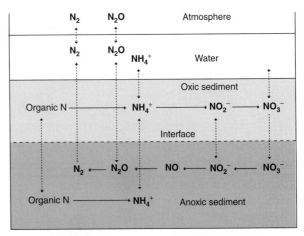

Fig. 4. The stratification of N-cycle processes that occurs in aquatic sediment. Solid lines indicate reactions; dotted lines, diffusion of substrates and products. Nitrification occurs in an upper aerobic (oxic) layer and denitrification occurs in a lower anaerobic (anoxic) layer to the extent that NO_2^- or NO_3^- penetrate this layer. Mineralization of organic N occurs throughout. Ionic N oxides diffuse into the water column, and N_2O and N_2 diffuse into the atmosphere. Denitrification may be supplied with NO_3^-, either from the oxic layer nitrification or from the water column. A similar system can be established in biofilms or in water-saturated soils or soil aggregates.

growing in colonial aggregates that are surrounded by less acid-sensitive NO_2^- oxidizers. Such aggregates are observed in aquaculture and other aquatic systems in which nitrification is an important process reducing the potential problem of ammonia toxicity and providing the denitrification substrate NO_3^-.

In agriculture, the potential losses of fertilizer nitrogen through leaching or denitrification of the nitrification product NO_3^- is considered to be a serious problem. A number of "nitrification inhibitors," such as nitrapyrin (sold as N-Serve by the Dow Chemical Company) are commercially available and usually specifically inhibit the ammonia monooxygenase enzyme, thus preventing the conversion of anhydrous ammonia (NH_3) or cationic NH_4^+ fertilizers to the leachable and denitrifyable anion NO_3^-.

Sewage treatment and certain other waste-treatment processes promote nitrification in a secondary activated sludge reactor in which biologically available organic carbon (often monitored as BOD, biochemical oxygen demand) is mineralized and converted to microbial biomass and CO_2. The NO_3^- that is produced, if fed into a river or lake as receiving water, can stimulate eutrophication and the growth of algae and macrophytes (aquatic higher plants), so it is preferably subject to tertiary or nitrogen-removal treatment in which denitrification is encouraged (Section V).

C. Heterotrophic Nitrification

Although it is clear that most of the global nitrification is catalyzed by the lithotrophic nitrifiers, there are some heterotrophs that can produce NO_2^- and NO_3^- from reduced nitrogen compounds or from ammonia by mechanisms that are not yet clear.

$$RNH_2 \rightarrow RNHOH \rightarrow RNO \rightarrow RNO_2 \rightarrow NO_3^-$$
$$NH_4^+ \rightarrow NH_2OH \rightarrow NOH \rightarrow NO_2^- \rightarrow NO_3^-$$

In the heterotrophic nitrifier–denitrifier *Thiosphaera pantotropha* (recently reclassified as a strain of *Paracoccus denitrificans*), there do not appear to be genes homologous to the *amo* or *hao* of lithotrophic nitrifiers, so the nitrification mechanism would seem to be different. Other heterotrophic nitrifiers include fungi such as *Aspergillus flavus*, relatively acid insensitive and perhaps responsible for much of the nitrification in acidic forest soils, or bacteria such as *Arthrobacter* sp. However, there is no evidence that any of these organisms can conserve energy from the oxidations that they catalyze, and so the evolutionary significance of this process is not known.

It is difficult to distinguish between lithotrophic and heterotrophic nitrification activity, but in laboratory experiments, use can be made of the fact that acetylene (in low ppmv concentrations) inhibits the Amo of the former but not the oxidation system of the latter, and chlorate inhibits lithotrophic NO_2^- oxidation but apparently not heterotrophic NO_3^- production.

V. DENITRIFICATION

When oxygen becomes limiting, some microorganisms, mainly aerobic bacteria, have the ability to switch to the use of the nitrogen oxides NO_3^-, NO_2^-, NO, and N_2O as terminal acceptors of electrons in their metabolism. This process is known as denitrification, and it permits organisms to continue what is essentially a form of aerobic respiration in which the end product is dinitrogen. However, intermediates sometimes accumulate. Denitrification is of major importance because it closes the global nitrogen cycle, maintaining a balance in atmospheric dinitrogen; it is responsible for significant losses of fertilizer nitrogen in agriculture; it is a critical process in the nitrogen-removal component of modern tertiary wastewater-treatment plants; leakage of the intermediate N_2O adds to the greenhouse gas load of the troposphere and also indirectly causes catalytic destruction of stratospheric ozone; and it is possible that the long-term accumulation of nitrous oxide in closed systems may cause human or animal health problems.

A. Biochemistry and Microbiology

Most of the denitrifying organisms of significance in nature are bacteria, and they have an aerobic type of electron-transport (cytochrome) chain leading to O_2 as the terminal electron acceptor. However, when O_2 is deficient or absent, a cellular regulator (Anr or Fnr) switches on the synthesis of a series of reductase enzymes that successively reduce NO_3^-, NO_2^-, NO, and N_2O to N_2. This type of respiration is less efficient than O_2 respiration, and so growth of the microorganisms is slower under anaerobic conditions. The NO_3^- reductase (Nar) is a membrane-bound molybdo-enzyme with the active site on the inner face of the membrane (Fig. 5). Also, a soluble NO_3^- reductase enzyme (Nap) was reported to be located in the periplasm (between the inner and outer membranes) of some gram-negative bacteria. NO_2^- reductases (Nir) are of two types, copper-containing and cytochrome cd_1-containing, and are periplasmic. NO reductase (Nor) is membrane-bound, and N_2O reductases (Nos) are also mostly periplasmic Cu enzymes, except in the gliding bacterium *Flexibacter canadensis*, in which it is in the membrane.

Nitrate is the major N oxide substrate for denitrification. It is taken up through the inner membrane by means of an antiport (at least in some bacteria) that is O_2 sensitive, and thus NO_3^- uptake is inhibited by O_2. The NO_2^- product from reduction of NO_3^-

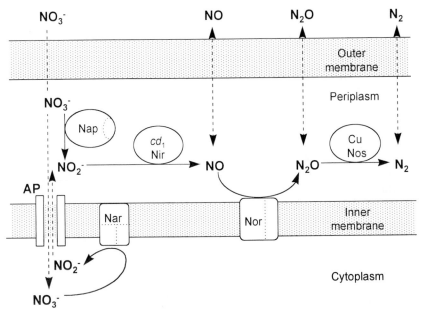

Fig. 5. The cellular topography of the denitrification system in *Paracoccus denitrificans*. The components shown are a $NO_3^- - NO_2^-$ antiport (AP); membrane-associated NO_3^- reductase (Nar) and NO reductase (Nor); and periplasmic NO_3^-, NO_2^-, and N_2O reductases (Nap, cd_1Nir, and CuNos, respectively). For simplicity, cytochromes and electron-transport pathways are not shown.

exits via the $NO_3^- - NO_2^-$ antiport and undergoes successive reduction by the periplasmic NO_2^- reductase (Nir), the membrane-bound NO reductase (Nor), and the periplasmic N_2O reductase (Nos) (Fig. 5). Electrons for these reductions are supplied via NADH dehydrogenase, the ubiquinone pool, a cytochrome bc_1 complex in the inner membrane, and other cytochromes in the periplasm (not shown in Fig. 5). These membrane redox reactions result in the extrusion of protons (H^+) into the periplasm and the generation of a proton motive force that allows the cell to make ATP.

The genes encoding the denitrification system are generally chromosomal and are tightly linked, as shown in Fig. 6 for *Pseudomonas stutzeri*. The NO_2^- and NO reductase genes, *nir* and *nor*, are coregulated, presumably to avoid the possible toxicity of NO, because NO could accumulate if Nor activity was deficient. Both low O_2 and the presence of a nitrogen oxide are necessary for the transcription of the denitrification genes. Organisms exist that possess only certain subsets of the four reductases; thus the nitro-

gen oxide substrates used and the final products can vary with the species (Fig. 7).

Phylogenetically, bacteria that are able to denitrify are found widely distributed in very diverse groups (Table III), and this has led to much speculation about the evolutionary significance of this distribution. A few fungi (for example, *Fusarium oxysporum* strains) were recently shown to denitrify, and they produce N oxide reductases with some similarities to those in bacteria. However, there is no evidence that such fungi are important denitrifiers in nature. Most denitrifiers are typical heterotrophs, such as *Pseudomonas*, *Alcaligenes*, and *Ralstonia* spp., using organic carbon as a source of reducing power.

B. Ecology of Denitrification

1. Primary Controlling Factors

Denitrifiers are ubiquitous, growing best aerobically, so some potential for denitrification exists in many habitats. The actual activity depends on three primary factors—limitation of O_2, availability of N

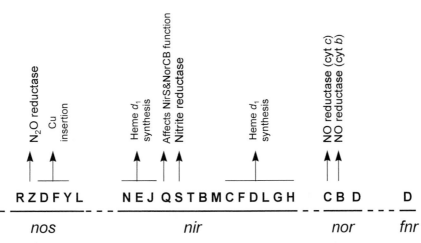

Fig. 6. The chromosomal arrangement of some of the denitrification genes in *Pseudomonas stutzeri*. The *nir, nor,* and *nos* genes are shown to be neighboring clusters in this bacterium. In others, they may be more scattered. Products and functions are shown.

oxides, and availability of reductant (mainly as organic carbon).

a. Oxygen

Depletion of O_2 occurs in environments in which its consumption by biological activity is greater than its rate of supply by mass transport or diffusion. Thus, aquatic sediments, waterlogged or irrigated soils, and the centers of large water-saturated soil aggregates are some of the habitats in which the genes for N oxide reductases of denitrifiers will be derepressed and the enzymes synthesized. If respiratory activity is high and O_2 is supplied by molecular diffusion (as in biofilms), the gradients of O_2 concentration can be very steep, conditions becoming nearly anaerobic within hundreds of micrometers of an air-saturated boundary layer. Although O_2 represses the synthesis of the N oxide reductases and inhibits the

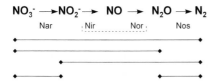

Fig. 7. Schematic diagram showing that the presence of different subsets of the denitrification-pathway enzymes in different organisms can affect the substrates used and the products released.

activity of the enzymes, some denitrifying systems are reported to be relatively insensitive to O_2. However, studies using microelectrodes have shown significant denitrification only in the anaerobic or nearly anaerobic zone. The N_2O reductase is often reported to be the most sensitive to O_2, so exposure of a denitrifying system to O_2 may result in the release of N_2O rather than N_2. Indeed, the Nos enzyme seems to be generally more sensitive than the other reductases to unfavorable conditions. Thus, low pH values and inhibitory components in the environment can cause the release of N_2O. The marked inhibition of N_2O reduction by acetylene, coupled with the measurement of N_2O accumulation by gas chromatography, is used as a sensitive assay for denitrification.

b. N Oxides

In environments in which O_2 depletion occurs, the denitrifiers become energy limited unless reducible N oxides are present. Nitrification is ultimately the major source of these oxides, but they also come from precipitation and fertilizers. As mentioned in Section IVB, nitrification, the oxidation of NH_4^+ to NO_2^- and NO_3^-, is an obligately aerobic process that often occurs close to denitrification zones, on opposite sides of aerobic–anaerobic interfaces, and so these processes are frequently tightly coupled (Fig. 4). Such interfaces can occur in soils, but they are

Groups/genera		Groups/genera	
Archaea (4)		Proteobacteria, delta subdivision	
Halobacterium	Halophilic	*Beggiatoa*	H_2S oxidizer
Proteobacteria, alpha subdivision		*Nitrosococcus*	NH_4^+ oxidizer
Agrobacterium	Soil heterotroph	*Pseudomonas*	Soil heterotroph
Aquaspirillum	Aquatic heterotroph	Proteobacteria, epsilon subdivision	
Azospirillum	N_2 fixer	*Campylobacter*	Some pathogens
Bradyrhizobium	Soybean nodules	*Wolinella*	Grows on fumarate + H_2
Hyphomicrobium	Methylotroph	Firmibacteria (gram-positive)	
Magnetospirillum	Magnetic bacterium	*Bacillus*	Some fac. anaerobes
Paracoccus		*Jonesia*	
Rhodobacter	Photosynthetic	Cytophagales, green S bacteria	
Rhodopseudomonas	Photosynthetic	*Cytophaga*	Gliding bacteria
Sinorhizobium	Legume nodules	*Flexibacter*	Gliding bacteria
Proteobacteria, beta subdivision			
Alcaligenes	Common heterotroph		
Neisseria	Pathogens		
Nitrosomonas	NH_4^+ oxidizer		
Ralstonia	(Alcaligenes)		
Thiobacillus	H_2S oxidizer		

very important in aquatic sediments, where both the sediment and the water column may be sources of NO_3^-. The relative importance of sediment nitrification and the water column in providing NO_3^- for denitrification may depend on the depth of O_2 penetration, and thus on the steepness of the NO_3^- diffusion gradient from the surface to the zone of denitrification. Sediment nitrification is likely to be the major source of NO_3^- for denitrification in cases in which O_2 penetrates deeply, the NO_3^- gradient from the surface to the interface is less steep, and the nitrification potential occurs close to the aerobic–anaerobic interface.

c. Reductants

In environments with low organic-matter content, the supply of electron donors (and thus energy) may be a critical factor. Denitrification capacity may be highly correlated with water-soluble soil organic carbon, and denitrifiers are reported that degrade various aromatic compounds anaerobically. Such bacteria are of interest in the bioremediation of contaminated environments. Other bacteria can use di-

hydrogen, elemental sulfur, or oxidizable sulfur compounds as electron donors, but it is not known whether this is of significance in nature.

2. Accumulation of Intermediates and Effects of Plants

Transient nitrite accumulation occurs, at least in the cultures of some denitrifiers, when NO_3^- concentrations above about 0.3 mM inhibit the reduction of NO_2^-. The NO_2^- is reduced once the NO_3^- concentration decreases sufficiently. As mentioned earlier, O_2 exposure and low pH values can cause the escape of N_2O from a denitrifying system, but the escape of NO appears to be minimal because of the very high apparent affinity of NO reductase for NO. Thus, denitrification is not as significant a contributor of NO to the atmosphere as nitrification.

Active plant roots are strong sinks for O_2 and NO_3^- and are sources of oxidizable organic carbon released in exudates. The roots of aquatic macrophytes act as sources of O_2. Therefore, in either case, they have the potential to both stimulate and inhibit denitrification in the rhizosphere. The actual outcome seems to depend on the availability of NO_3^- and carbon in the O_2-depleted part of the rhizosphere soil.

3. Dissimilatory Reduction of Nitrate to Ammonia (Nitrate Ammonification)

In highly reducing habitats, such as sludges and the rumen, where the ratio of available organic carbon to electron acceptor (NO_3^-) is high, the NO_3^- is reduced not by denitrification to gaseous products but via specific NO_3^- and NO_2^- reductases to NH_4^+ (Fig. 1). Both enzymes are soluble and cytoplasmic, and the process acts mainly as an electron sink. It is found in some fermentative organisms such as *E. coli* and some species of *Clostridium*.

VI. SUMMARY AND GLOBAL IMPLICATIONS

The microorganisms involved in the many N-cycle processes respond to changes in their environment and in the availability of their substrates. Human influences, especially since the 1930s, have greatly changed both environments and substrates. Increases in biological N_2 fixation in legume crops and in chemical production of inorganic fertilizers have roughly doubled the total N inputs to the terrestrial N cycle. The retention of N has not been sufficient to prevent losses of inorganic N and other nutrients to rivers, lakes, and oceans, and it is now known that ecosystems such as wetlands along shorelines are important in restricting (through assimilation and denitrification) such input of NO_3^- to rivers and lakes. The conversion of N forms has increased volatile losses of NH_3, NO, and N_2O to the atmosphere, and the long-range transport and deposition of these molecules or their oxidation products. This has resulted in the "N-saturation" of many previously N-limited natural terrestrial and aquatic ecosystems, with consequent changes in flora, decreased diversity, and increased acidity both from rainfall of nitric acid (from the oxidation of NO) and from the nitrification of NH_4^+. Other effects on the atmosphere itself include the contribution of NO to photochemical smog and of N_2O to both the greenhouse effect and the destruction of stratospheric ozone.

See Also the Following Articles

BIOFILMS AND BIOFOULING • CARBON AND NITROGEN ASSIMILATION, REGULATION OF • HETEROTROPHIC MICROORGANISMS • INDUSTRIAL EFFLUENTS: SOURCES, PROPERTIES, AND TREATMENTS

Bibliography

Cole, J. (1996). Nitrate reduction to ammonia by enteric bacteria: Redundancy, or a strategy for survival during oxygen starvation? *FEMS Microbiol. Lett.* **136**, 1–11.

Harper, L. A., Mosier, A. R., Duxbury, J. M., and Rolston, D. E. (eds.) (1993). "Agricultural Ecosystem Effects on Trace Gases and Global Climate Change." Special publication no. 55. American Society of Agronomy, Madison, WI.

Kuenen, J. G., and Robertson, L. A. (1988). Ecology of nitrification and denitrification. *In* "The Nitrogen and Sulphur Cycles" (J. A. Cole and S. J. Ferguson, eds.), pp. 161–218. Cambridge University Press, Cambridge.

Paul, E. A., and Clarke, F. E. (1989). "Soil Microbiology and Biochemistry." Academic Press, San Diego, CA.

Prosser, J. (ed.) (1986). "Nitrification." IRL Press, Oxford.

Revsbech, N. P., and Sørensen, J. (eds.) (1990). "Denitrification in Soil and Sediment." Plenum Press, New York.

Sprent, J. I. (1987). "The Ecology of the Nitrogen Cycle." Cambridge University Press, Cambridge.

Stacey, G., Burris, R. H., and Evans, H. J. (eds.) (1992). "Biological Nitrogen Fixation." Chapman & Hall, London.

Vitousek, P. M., Arbar, J. D., Howarth, R. W., Likens, G. E., Matson, P. A., Schindler, D. W., Schlesinger, W. H., and Tilman, D. G. (1997). Human alterations of the global nitrogen cycle: Sources and consequences. *Ecol Appl.* **7**, 737–750.

Zumft, W. G. (1997). Cell biology and molecular basis of denitrification. *Microbiol. Molec. Biol. Rev.* **61**, 533–616.

Nitrogen Fixation

L. David Kuykendall
USDA
Robert B. Dadson
University of Maryland, Eastern Shore

Fawzy M. Hashem
University of Maryland, Eastern Shore
Gerald H. Elkan
North Carolina State University

I. Significance of Biological Nitrogen Fixation
II. The Biological Nitrogen Fixation Process
III. Free-Living Nitrogen-Fixing Bacteria
IV. Associative Nitrogen-Fixing Bacteria
V. Symbiotic Nitrogen-Fixing Bacteria

GLOSSARY

cyanobacteria Photosynthetic prokaryotes of some species that fix nitrogen, either free living or symbiotically with a broad range of lower plant taxa.

diazotroph An organism capable of biological nitrogen fixation.

fixed nitrogen Chemically reduced or biologically fixed nitrogen that is in a chemical form that can be assimilated readily by most microbes or plants.

frankia Symbiotic nitrogen-fixing actinomycetes; a broad spectrum of shrubs and trees are hosts.

megaplasmid Extremely large ($\geq$500 kb) extra-chromosomal DNA molecules that encode the symbiotic nodulation and nitrogen-fixation abilities of certain legume-nodulating bacteria.

nif gene A bacterial gene coding for either nitrogenase enzyme or its subunits (i.e., *nifHDK*), or for proteins in the regulation of nitrogenase expression.

nitrogenase A highly conserved enzyme complex that catalyzes the conversion of atmospheric dinitrogen to ammonia; nitrogenase consists of two components, dinitrogenase reductase (an iron protein) and dinitrogenase (an iron molybdenum protein).

nitrogen fixation Reduction of inert atmospheric N_2 gas to a metabolizable active form by prokaryotic organisms (eubacteria and archaebacteria).

nod factors Lipochitooligosaccharides that are chemical signals produced and released by either *Rhizobium*, *Bradyrhizobium*, *Azorhizobium*, or *Mesorhizobium*. They are essential for the bacterial capacity to induce nodule organogenesis.

nod genes The genes in legume-nodulating bacteria that code for the development of nodules on the proper legume host. There are two types—a common *nod* region, which consists of a structurally and functionally conserved cluster of genes; and host-specific nodulation genes, which cannot complement nodulation defects in other species.

rhizobiophage Specific bacterial viruses (bacteriophages) virulent for *Rhizobium* and related genera.

Rhizobium, Bradyrhizobium, Azorhizobium, Sinorhizobium, Mesorhizobium, and Allorhizobium Six genera, representing four distinct families, of nitrogen-fixing bacteria that form nodules on the roots or stems of various legume hosts. The complex relationships they form with a limited number of host legumes vary both in host range and efficiency of symbiotic nitrogen fixation according to genetically programmed two-way communication between bacteria and host via chemical messengers.

symbiosis Mutually beneficial sharing between two distinct organisms, perhaps the best known examples of which are legumes and their bacterial microsymbionts.

NITROGEN FIXATION refers to that property of some taxa of eubacteria or archaebacteria to enzymatically reduce atmospheric N_2 to ammonia. The ammonia produced can then be incorporated by means of other enzymes into cellular protoplasm. Nitrogen fixation only occurs in prokaryotes and not in higher taxa.

I. SIGNIFICANCE OF BIOLOGICAL NITROGEN FIXATION

Although nitrogen is an essential nutrient for life, little available nitrogen is present in mineral form. Above every hectare of soil at sea level, there are 78 million kg of inert N_2 (dinitrogen) gas. Plants, as eukaryotic autotrophs, need either an oxidized or reduced form of nitrogen for anabolism. Only certain prokaryotic organisms "fix" N_2 at physiological temperatures. Humans can mimic biological nitrogen fixation using the Haber–Bosch process of chemically reducing nitrogen at high temperatures. However, the process consumes precious fossil fuels, contributes to global warming, and pollutes the environment. Worldwide industrial production of NH_3 is annually about 100 million metric tons, three-fourths of which is manufactured for fertilizer to grow crop plants.

Thus, humans produce about 75 million mt/year of nitrogenous fertilizer, whereas the annual contribution of reduced nitrogen from the biological process called nitrogen fixation is roughly estimated to be about two to three times that amount. As shown in Table I, nitrogen fixation occurs in every natural environment, including the sea. The energy cost for nitrogen fixation is significant. Perhaps 10% of the available fossil fuel energy is used for the production of fertilizer nitrogen, whereas Hardy (1980) estimated that between 1 and 2 billion tons of plant carbohydrates derived from photosynthesis fuel the biological process of nitrogen fixation.

TABLE I
Estimated Annual Amount of N_2 Fixed Biologically in Various Systems

System	N_2 fixed mt/year $\times 10^6$
Legumes	35
Nonlegumes	9
Permanent grassland	45
Forest and woodland	40
Unused land	10
Total land	139
Sea	36
Total	175

TABLE II
Families and Genera of Nitrogen-Fixing Eubacteria, Excluding Cyanobacteria

Family and genus	Family and genus
Acetobacteriaceae	Methanomonadaceae
Acetobacter	*Methylobacter*
Azotobacteraceae	*Methylococcus*
Azomonas	*Methylocystis*
Azotobacter	*Methylomonas*
Azotococcus	*Methylosinus*
Beijerinckia	Pseudomonodaceae
Derxia	*Pseudomonas*
Xanthobacter	Rhizobiaceae
Bacillaceae	*Allorhizobium*
Paenibacillus (Bacillus)	*Rhizobium*
Clostridium	*Sinorhizobium*
Desulfotomaculum	Hypomicrobiaceae
Baggiatoaceae	*Azorhizobium*
Baggiatoa	Phyllobacteriaceae
Thiothrix	*Mesorhizobium*
Vitreoscilla	Bradyrhizobiaceae
Chlorobiaceae	*Bradyrhizobium*
Chlorobium	Rhodospirallaceae
Pelodictyron	*Rhodomicrobium*
Chloroflexaceae	*Rhodopseudomonas*
Chloroflexus	*Rhodospirillum*
Chromatiaceae	Spirillaceae
Amoebobactor	*Aquaspirillum*
Chromatium	*Azospirillum*
Ectothiorhodospira	*Campylobacter*
Thiocapsa	*Herbaspirillum*
Thiocystis	Streptomycetaceae
Corynebacteriaceae	*Frankia*
Anthrobacter	Thiobacteriaceae
Enterobacteriaceae	*Thiobacillus*
Citrobacter	Vibrionaceae
Enterobacter	*Vibrio*
Erwinia	Uncertain family
Escherichia	*Alcaligenes*
Klebsiella	*Desulfovibrio*

There are many genetically diverse nitrogen-fixing eubacteria falling into 27 families and 80 genera, and there are at least three thermophilic nitrogen-fixing genera of Archaebacteria. These nitrogen-fixing families and genera are listed in Table II; the "blue-green algae" or cyanobacteria are listed in Table III. With the exception of the Azotobacteraceae, there is no

TABLE III
Families and Genera of Cyanobacteria with Nitrogen-Fixing Species

Family and genus	Family and genus
Chroococcaceae	Oscillatoriaceae
Chlorogloea	*Lyngbya*
Chroococcidiopis	*Microcoleus*
Gloeothece	Oscillatoria
Syenchococcus	*Phormidium*
Mastigocladaceae	*Plectonema*
Mastigocladus	Pleurocapsaceae
Michrochaetaceae	*Pleurocapsa*
Michrochaete	Rivulariaceae
Nistocaceae	*Calothrix*
Anabaena	*Dichothrix*
Anabaenopsis	*Gleotrichia*
Aphanizomenon	Scytonemataceae
Aulosira	*Scytonema*
Cylindrospermum	*Tolypothrix*
Nodularia	Stigonematacaeae
Nostoc	*Fischerella*
Pseudanabaena	*Hapalosiphon*
Raphidiopsis	*Stigonema*
Richelia	*Westelliopsis*

genus (or family) whose species are all nitrogen-fixing. The potential amount of nitrogen fixed by these bacteria depends largely on the ecosystem in which the organisms are active, as will be discussed in later sections. The amount of nitrogen fixed ranges from only trace amounts for some free-living soil bacteria to 584 kg/ha/year for the tropical tree legume *Leucaena*.

II. THE BIOLOGICAL NITROGEN FIXATION PROCESS

Dinitrogen gas is both chemically inert and very stable, requiring much energy to break the triple bond and reduce the $N \equiv N$ to ammonia in an endothermic reaction: $3H_2 + N_2 \rightarrow 2 NH_3$. This can be accomplished chemically by the Haber–Bosch process or biologically by prokaryotic organisms using adenosine triphosphate (ATP) energy to initiate the bond-breaking reaction. The prokaryotes able

to fix nitrogen are extremely heterogeneous, with representatives that are autotrophic, heterotrophic, aerobic, anaerobic, photosynthetic, single-celled, filamentous, free-living, and symbiotic. Phylogenetically, these organisms are extremely diverse and yet the nitrogen-fixing process and enzyme system are similar in all of these organisms and depend on a nitrogenase enzyme complex, a high-energy requirement and availability (ATP), anaerobic conditions for nitrogenase activity, and a strong reductant.

Nitrogenase has been purified from all known nitrogen-fixing eubacteria. It consists of two components, dinitrogenase reductase (an iron protein) and dinitrogenase (an iron–molybdenum protein). Both enzymes are needed for nitrogenase activity. Nitrogenase reduces N_2 and H^+ simultaneously, using 75% of the electron flow in the reduction of nitrogen and 25% in H^+ reduction. A key characteristic of this enzyme complex is that both components are quickly and irreversibly lost on interaction with free oxygen, regardless of the oxygen requirements of the microbe. The protein usually has a molecular mass of 57,000–72,000 Da and consists of two identical subunits coded for by the *nifH* gene. It has a highly conserved amino acid sequence. The iron–molybdenum protein has a molecular mass of about 220 kDa and has four subunits, which are pairs of two different types. The α-subunit is coded for by the *nifD* gene and is 50 kDa in size. The β-subunit is coded for by the *nifK* gene and has a mass of 60 kDa. These enzymes have been sequenced and demonstrate considerable similarity. The nitrogenase complex is large and may amount to 30% of total cell protein.

Theoretically, as much as 28 moles of ATP are consumed in the reduction of 1 mole of N_2. Depending on the method used and the nature of the organism in question, *in vitro* studies show that the energy requirements vary between a minimum of 12–15 moles and 29 moles of ATP/N_2. This energy is not only required for the reduction process, but also to maintain the anaerobic conditions needed for the reaction.

Various strategies are used to exclude oxygen from the reaction in nonanaerobic bacteria. Facultative organisms fix nitrogen only under anaerobic conditions. Aerobic organisms exhibit a wide range of methods for protecting the enzyme complex from oxygen. Many grow under microaerophilic conditions accomplished by scavenging free oxygen for metabolism or sharing the ecosystem with other organisms that consume the excess O_2. Some evidence indicates that some free-living aerobes, such as *Azotobacter,* can change the conformation of the nitrogenase protein to form a less oxygen-sensitive protein. Photosynthetic aerobes such as cyanobacteria can form special cells called heterocysts in which nitrogen fixation occurs and which lack the O_2-evolving mechanism that is part of photosynthesis. The best-known protective mechanisms are found in symbiotic nitrogen-fixing associations (i.e., *Rhizobium*–legume), in which the nitrogenase system in the endophyte is protected from excess oxygen by a nodule component, leghemogloblin. Nonleguminous nodules (with *Frankia* or cyanobacteria) are probably protected by other, as yet undescribed, oxygen-restrictive mechanisms. There are other protective mechanisms as well, but all of these require the diversion of energy from the nitrogen-fixation process itself to maintain a favorable environment for fixation.

In addition to the limitations to biological nitrogen fixation (BNF) caused by free oxygen, generally the presence in the environment of combined nitrogen, such as ammonia or nitrate, strongly inhibits both nodulation and N_2 fixation. Thus, it is unnecessary to use chemical fertilizer in large quantities for legume crop production.

III. FREE-LIVING NITROGEN-FIXING BACTERIA

There are many diverse nitrogen-fixing prokaryotes, representing 22 families and 52 genera of eubacteria (excluding the cyanobacteria), as well as three thermophilic genera of archaebacteria. Most of these genera fix nitrogen as free-living diazotrophs. These were first described by Winogradsky in 1893 (*Clostridium pasteurianum*), and Beijerinck in 1901 (*Azotobacter*). The discovery of other free-living nitrogen-fixing bacteria lagged until the availability of ^{15}N stable isotopic and acetylene-reduction techniques became common. Thus, most of the free-living nitrogen-fixing bacteria listed in *Bergey's Manual of Systematic Bacteriology* have been discribed using these techniques. It is generally accepted that diazotrophs obtain their carbon and energy supplies from root exudates and lysates, sloughed plant cell debris, and organic residues in soil and water. They are found "completely free-living" or in loose associations as a result of root or rhizosphere colonization (the associative bacteria are discussed separately). The quantity of nitrogen fixed is a matter of some controversy. Russian workers have estimated that *C. pasteurianum* or *Azotobacter* contribute perhaps 0.3 kg N_2/ha/year, compared with about 1000 times that amount provided by a good leguminous association. Associative organisms such as *Azospirillum* have been estimated to fix from trace amounts to 36 kg/ha. The limitations of the terrestrial BNF system is due not only to the difficulty in obtaining sufficient energy and reductant but also because of the need to divert substrate for respiratory protection of nitrogenase. The cyanobacteria can overcome the environmental constraints faced by most other nitrogen fixers. Being photosynthetically active, these prokaryotes use sunlight to fix CO_2 and are thus independent of external energy needs. The families and genera of nitrogen-fixing cyanobacteria are listed in Table III.

The free-living cyanobacteria are distributed widely in humid and arid tropical surface soils. Extensive studies, especially under rice paddy conditions, have been conducted in India, Japan, and the Phillippines. Of 308 isolates from Philippine paddy soils, most were identified as *Nostoc* or *Anabaena*. Reports of fixation rates ranged from 3.2 to 10.9 kg N/ha/year. Reports of 15–20 kg N/ha/year in rice in the Ivory Coast, 44 kg N/ha/year in Lake George in Uganda, and 80 kg N/ha/year in paddy fields in India have been published. Under temperate conditions, BNF by cyanobacteria has been a major problem in the eutrophication of lakes.

About 125 strains of free-living cyanobacteria representing 10 families and 31 genera in all taxonomic groups have been shown to fix nitrogen, but the extent to which fixation occurs and conditions needed for fixation vary greatly. There are four types of nitrogen-fixing cyanobacteria—heterocystous filamentous, nonheterocystous filamentous, unicellular reproducing by binary fission or budding, and unicellular reproducing by multiple fission. Heterocysts are thickened, specialized cells occurring at regular intervals in some filamentous cyanobacteria. These cells lack the oxygen-evolving component of the photosynthetic apparatus. Heterocysts appear to be the only cells capable of fixing nitrogen aerobically as well as anaerobically. The main function of the heterocysts seems to be to compartmentalize nitrogen fixation because there is little or no photosynthesis in the heterocysts, and all of the energy translocated is available for nitrogen fixation. The other groups of cyanobacteria need to be examined as well because they appear quite active as fixers under anaerobic conditions. Nitrogen fixation in paddy rice could thus be improved.

One should not overlook the potential for the improvement of nitrogen fixation in unique environments. There are many reports of intimate associations between nitrogen-fixing bacteria and animals (e.g., termites and ruminants). Because *nif* genes can be transferred and expressed between the enteric bacteria *Klebsiella* and *Escherichia coli*, it may be possible to increase nitrogen fixation by rumen bacteria. Perhaps ultimately in ruminants, it will be practical to substitute engineered diazotrophic enterics for the plant protein now required by these animals.

IV. ASSOCIATIVE NITROGEN-FIXING BACTERIA

Since the 1960s, another type of plant–bacterial interaction has been described as "associative." This interaction was shown to result from the adhesion of the bacteria to the root surfaces of wheat, corn, sorghum, and other grasses. The major associative nitrogen-fixing systems are summarized in Table IV. *In vitro* studies show that many of these bacteria can achieve high rates of N_2 fixation under optimum conditions. However, in the rhizosphere, the ability to survive, grow, and colonize plant roots is a precondition limiting the potential for nitrogen-fixation. The characteristics required for an organism to flourish in the rhizosphere are the ability to withstand the changing physical and chemical soil environments, grow well and obtain all the energy needed from carbon and mineral supplies in the root zone, and

TABLE IV
Major Associative Diazotrophs

Plant species	Principle microorganism
Rice (*Oryza sativa*)	*Achromobacter*, Enterobacteriaceae, *Azospirillum brasilense*
Sugarcane (*Saccharum* spp.)	*Azotobocter*, *Beijerinckia*, *Klebsiella*, *Derxia*, *Vibrio*, *Azospirillum*, Enterobacteriaceae, Paenibacillaceae
Pearl millet (*Penniseturn galucum*) and sorghum (*Sorghum bicolor*)	*Azospirillum*, *Paenibacillus polymyxa*, *P. azotofixans*, *P. macerans*, *Klebsiella*, *Azotobacter*, *Derxia*, *Enterobacter*
Maize (*Zea mays*)	*Azospirillum lipoferum*, *Azotobacter vinelandii*
Grasses	
Paspalum notatum var. *batatais*	*Azotobacter paspali*
Panicum maximum	*Azospirillurn lipoferum*
Cynodon dactylon	*Azospirillum lipoferum*
Digitaria decumbens	*Azospirillum lipoferum*
Pennisetum purpureum	*Azospirillum lipoferum*
Spartina alterniflora Loisel	*Campylobacter*
Wheat	
Triticum spp.	*Paenibacillus polymyxa*

compete successfully with other rhizosphere organisms for the limited energy and nutrients available. Estimates of *in vivo* fixation are extremely variable and give rise to a recurring question as to whether energy substrates in the rhizosphere are sufficient to support growth and nitrogen fixation by these associative bacteria at sufficient levels. The nitrogenase system is repressed by bound nitrogen. Therefore, in the presence of nitrogen fertilizer, BNF by free-living or associative bacteria is reduced. Thus, a mixed BNF-nitrogen fertilizer system would work better under field conditions if the microorganisms could be genetically engineered so that the expression of nitrogenase is derepressed when bound nitrogen is present.

V. SYMBIOTIC NITROGEN-FIXING BACTERIA

Only three groups of nitrogen-fixing bacteria have evolved mutually beneficial symbiotic associations with higher plants: (1) the filamentous bacterium *Frankia*, forming root nodules with a number of plants such as alder, *Purshia*, and Russian olive; (2) heterocystous cyanobacterium with a number of diverse hosts from *Cycads* to *Azolla*; and (3) *Rhizobium* and allied genera with legumes.

A. *Frankia*

The genera of dicotyledonous plants found to be nodulated by *Actinorhizae (Frankia)* are listed in Table V. These include more than 180 species in more than 20 genera, representing at least eight families and seven orders of plants. There is no obvious taxonomic pattern among these hosts. Most of the species

TABLE V
Genera of Nitrogen-Fixing Plant Genera with *Frankia* Symbiosis

Genus	Number of nodulated species
Alnus	34
Casuarina	25
Ceanothus	31
Cercocarpus	4
Chamaebatia	1
Colletia	3
Coriaria	14
Cowania	1
Discaria	6
Dryas	3
Elaeagnus	17
Hippophae	1
Kentrothamnus	1
Myrica	26
Purshia	2
Rubus	1
Satisca	2
Shepherdia	3
Talguena	1
Trevoa	2

described are shrubs or trees and are found in temperate climates, but they have a wide growth range and could be grown in the tropics. In fact, *Purshia tridento* already is an important rangeland forage crop in Africa, and other *Purshia* species are harvested for firewood. *Casuarina*, a vigorous nitrogen fixer, has been planted in Thailand where it can be harvested for construction lumber after 5 years of growth. Almost all the hosts are woody, ranging from small shrubs to medium-sized trees.

Frankia, the nitrogen-fixing endophyte found in nodules of these nonleguminous plants is a genus of prokaryotic bacteria closely related to the actinomycetes. In pure culture, these endophytes behave as microaerophilic, mesophilic, or heterotrophic organisms, usually with septate hyphae that develop sporangia. Isolates vary morphologically and nutritionally. Most strains can fix atmospheric nitrogen in pure culture. Nitrogenase genes are highly conserved, and *Frankia* nitrogenase enzymes closely resemble those of other nitrogen-fixing bacteria. *Frankia* was first reliably isolated in 1978. The organism grows slowly, requiring 4–8 weeks for visible colonies to be formed in culture. *Frankia* is similar to other aerobic actinomycetes, producing a separate filamentous mycelium that can differentiate into sporangia and vesicles. The cells are routinely gram-positive, but unlike other gram-positive bacteria, *Frankia* has a discontinuous membranous layer. Molecular methods for taxonomy such as DNA–DNA hybridization have demonstrated considerable genetic diversity between isolates within this family, but only a limited number of isolates have been analyzed.

Although the majority of actinorhizal microsymbionts apparently infect their plant hosts via a root hair-mediated mechanism, some of them can infect by the direct penetration of the root by means of the intercellular spaces of the epidermis and cortex. As a result of infection, a root meristem is induced as the *Frankia* hyphae penetrate the cells, but the hyphae remain enclosed by a host-produced polysaccharide layer. *Frankia* grows in the nodule occupying a major part of the host-produced nodular tissue. Root nodule tissue usually composes 1–5% of the total dry weight of the plant. *Frankia* can form vesicles whose walls evidently adapt to O_2 in such a way that they can fix nitrogen at atmospheric O_2 levels, but the precise mechanism is obscure.

It appears that the infective abilities of each *Frankia* strain are limited to one or a few plant genera. Currently, four host-specificity groups have been identified: (1) strains infective on *Alnus, Comptonia,* and *Myrica;* (2) strains infective on Casuarinaceae; (3) flexible strains infective on species of *Elaeagnaceae* and the promiscuous species of *Myrica* and *Gymnostoma;* and (4) strains infective only on species of the family Elaeagnaceae. Strains in specificity groups 1 and 2 infect the host plant via the root hair, whereas group 4 isolates infect by intercellular penetration. Some strains can use both modes of infection, and these are in the "flexible" specificity group 3.

Because these organisms were only recently isolated and cultured, only in the 1990s have taxonomic, physiological, and genetic studies been started, and not all isolates have been successfully cultured. In contrast to *Rhizobium,* it is possible to routinely obtain nitrogen fixation *ex planta* in pure culture. It has been possible to inoculate the host plant with pure cultures and obtain effective nodule formation.

B. Cyanobacteria

Cyanobacteria form symbioses with the most diverse group of hosts of any other nitrogen-fixing system known. Interestingly, whereas *Rhizobium* can form associations with highly evolved plants, cyanobacteria associate with more primitive plants, such as lichens, liverworts, a pteridophyte (*Azolla*), gymnosperms (e.g., *Cycads*), and angiosperms (i.e., *Gunnera*). The cyanobacterium *Nostoc* forms symbioses with all of the taxa other than the ferns, indicating a potential for genetic manipulation to increase host range. The habitat range is also wide, from tropics to arctic, and includes freshwater, soil, saltwater, and hot springs.

As previously discussed, cyanobacteria are an important source of biologically fixed nitrogen. Whereas the free-living cyanobacteria fix up to 80 kg N_2/ha, the *Azolla–Anabaena* symbiosis can produce three times that amount. This value is based on multiple cropping, but rates in rice paddies have been reliably reported from 1.4–10.5 kg N/ha/day as daily averages over the whole growing season.

Nitrogen fixation by cyanobacteria was first reported in 1889 shortly after the *Rhizobium*–legume symbiosis was described. Until the 1970s, research with this system was intermittent, but, partly due to the recently recognized importance of these eubacteria in agriculture and the environment, research has been steadily accelerating. Cyanobacteria often have been classified by botanists using mainly morphological and anatomical criteria rather than by bacteriologists, who use molecular methods. The recognized families and genera of cyanobacteria containing nitrogen-fixing species are summarized in Table III. Because morphological characteristics can vary greatly depending on growth conditions, only limited physiological, biochemical, and molecular genetic studies have been conducted. These indicate that there is considerably more diversity in this taxon than is accounted for by the traditional taxonomy.

Cyanobacteria possess the requirements for the higher plant type of photosynthesis—water is the ultimate source of reductant and oxygen is evolved with CO_2 fixation via the Calvin cycle. The photosynthetic pigments are located in the outer cell regions.

Among filamentous, heterocyst-forming cyanobacteria, the ability for BNF appears universal. In nonheterocystous filamentous forms and unicellular forms, the ability to fix nitrogen is much less common. The heterocysts, which are found spaced along the cell filaments, appear colorless. These cells can not fix CO_2 or evolve O_2, but can generate ATP by photophosphorylation and can fix nitrogen aerobically. This is apparently due to a modified thickened cell wall, which interferes with oxygen diffusion. In these organisms then, photosynthesis and nitrogen fixation occur in different cells, which protects nitrogenase from excess oxygen stress.

As stated previously, cyanobacteria associate with almost every group of the plant kingdom, forming symbiotic nitrogen-fixing associations. These associations, however, occur with the more primitive plants. Specificity is less tight than it is with either *Rhizobium* or *Frankia*. Generally, cyanobacteria can fix nitrogen and grow independently of their plant partner. However, when they live symbiotically, photosynthesis is often diminished in favor of increased nitrogen fixation, sufficient for both partners of the symbiosis.

The cyanobacteria usually invade existing normal morphological structures in the host plant, such as leaf cavities, rather than evoking specialized structures such as the nodules caused by *Rhizobium* or *Frankia*, although in some cases, as with the roots of cycads, infection is followed by morphological change.

One reason for the renewed interest in this group of bacteria is the *Azolla–Anabaena* symbiosis. *Azolla* is a free-floating fern commonly found in still waters in temperate and tropical regions and often found in rice paddies. There are seven species. *Azolla* is a remarkable plant and under suitable conditions can double in weight in about four days. The plant forms a symbiotic relationship with the cyanobacterium *Anabaena azollae*. *Azolla* provides nutrients and a protective leaf cavity for *Anabaena*, which in turn provides nitrogen for the fern. Under optimum conditions, this symbiosis results in as much or perhaps more nitrogen fixed than does the legume–*Rhizobium* symbiosis. If inoculated into a paddy and intercropped with rice, this symbiosis can satisfy the nitrogen requirement for the rice. In Southeast Asia, where there is a wet and dry season, *Azolla* is recommended for intercropping with paddy rice, but there are problems because *Azolla* is an extremely efficient scavenger of nutrients and will compete with the rice for phosphate, so careful management is required.

Anabaena azollae symbiotically fixes about triple the amount of nitrogen fixed by free-living *Anabaena*. The high photosynthetic rate of *Azolla* no doubt supplies energy in this symbiosis. The *Anabaena* forms more heterocysts when growing in association with *Azolla* than when growing alone. Similarly, when free-living heterocystic cyanobacteria are grown in a nitrogen-starved environment, they form extra heterocysts. The hypothesis, then, is that, when sufficient energy and nutrients are present for metabolism, *Anabaena* can increase its nitrogen-fixation capacity to maximize growth. It has also been reported by several researchers that nitrogen fixation in this symbiotic system is not repressed by the presence of bound nitrogen.

Although *Azolla* has been cultivated in China and Vietnam for centuries, its use represents a new technology for most areas. Even at a relatively low technological level, there is a great potential for the *Azolla–Anabeana* system as a green manure supplement or animal feed. Little is known about the plant–bacterial interactions, but the fact that *Anabaena* can be induced to increase nitrogen fixation gives promise that the system can be optimized. The lack of repression of nitrogenase by bound nitrogen of *Anabaena* nitrogenase works in favor of a crop rotation following legumes.

C. The *Rhizobium*–Legume Symbioses

With worldwide distribution, the Leguminosae is one of the largest plant families. It consists of about 750 genera and an estimated 18,000 species. Members of the Leguminosae have traditionally been placed into three distinct subfamilies based on floral differences—Mimosoideae, Caesalpinoideae, and Papilionoideae. Although only about 20% of the total species have been examined for nodulation, these species are representative of all three subfamilies of legumes. Virtually all species within the Mimosoideae and Papilionoideae are nodulated, but about 70% of the species in the subfamily Caesalpinoideae are not nodulated. It is important to note that *Bradyrhizobium* strains of the cowpea type have been shown to form effective symbioses with the nonlegume *Parasponia*, which is a member of the *Ulmaceae*. This is the only verified nitrogen-fixing association between a *Rhizobium* and a nonlegumous plant.

The soil-improving properties of legumes were recognized by ancient agriculturalists. For example, Theophrastus (370–285 BC) in his "Enquiry into Plants" wrote, "Of the other leguminous plants the bean best reinvigorates the ground," and in another section, "Beans are not a burdensome crop to the ground: they even seem to manure it." However, it was only in 1888 that Hellriegel and Wilfarth established positively that atmospheric nitrogen was assimilated by root nodules. This was quickly followed by the experiments of Beijerinck, who used pure-culture techniques to isolate the root-nodule bacteria and proved that they were the causative agents of dinitrogen assimilation. He initially called these organisms *Bacillus radicicola*, but they were subsequently named *Rhizobium leguminosarum* by Frank (1889).

Early researchers considered all "rhizobia" to be a single species capable of nodulating all legumes. Extensive cross-testing on various legume hosts led to a taxonomic characterization of these special bacteria based on bacteria–plant cross-inoculation groups, which were defined as "groups of plants within which the root-nodule organisms are mutually interchangeable." The concept of cross-inoculation groupings as taxonomic criteria held for a very long time and has only gradually fallen into disfavor, although much of this philosophy is retained in the taxonomic scheme. There is a wide range in the efficiency of the *Rhizobium*–legume symbiosis. Estimates for the amounts of nitrogen fixed are summarized in Tables VI and VII.

As cells without endospores, bacteria of the family *Rhizobiaceae* are normally rod-shaped, motile, with one polar or subpolar flagellum or two to six peritrichous flagella, aerobic, and gram-negative. Considerable extracellular slime is usually produced during growth on carbohydrate-containing media with many carbohydrates used. Some *Rhizobium* and *Agrobacterium* evidently overlap; their DNA homology is very high. 16S RNA sequence analysis indicates very similar molecular phylogeny. Also, there is an almost complete lack of distinguishing characteristics other

TABLE VII
Nitrogen Fixed by Tropical and Subtropical Forage and Browse Plants, Green Manure, and Shade Trees[a]

Plant	Average	Range
Centrosema pubescens	259	126–395
Desmodium intortum	897	—
Leucaena glauca	277	74–584
Lotonosis bainesii	62	—
Sesbania cannabina	542	—
Stylosanthes sp.	124	34–220
Phaseolus atropurpurea	291	—
Mikanea cordata	120	—
Pueraria phaseoloides	99	—
Enterolobium saman	150	—

[a] In kg N/ha/year.

than those that are carried on extrachromosomal elements or plasmids. Some bacterial taxonomists are proposing the amalgamation of *Agrobacterium*, *Allorhizobium*, *Rhizobium*, and *Sinorhizobium*.

Traditionally, legume-nodulating bacteria have been recognized as falling into two major phenotypic groups according to growth rate. The term "fast growers" commonly refers to strains associated with alfalfa, clover, bean, and pea because, in culture, these organisms grow much faster (less than one-half the doubling time of slow growers, or < 3 hr) than the "slow growers" exemplified by soybean and cowpea rhizobia (generation time > 6 hr). Although there is phenotypic and genotypic diversity within these major groupings, and some overlap, numerous studies demonstrated the validity of this approach. *Mesorhizobium* strains, however, have intermediate growth rates between 3 and 6 hr (See Table VIII.)

The relative fastidiousness of the slow growers has been substantiated by recent studies. Although the major biochemical pathways seem to be similar, evidence suggests that the preferred pathway may be different. 16S RNA analysis of the fast- and slow-growing symbionts have confirmed that these groupings indeed represent very distinct genetic phyla because the similarity coefficient (S_{AB}) of the RNA is only 0.53. Thus, with modern gene analysis, the fast- and slow-growing *Rhizobium* fall into widely separate groups. Jordan (1982) transferred the slow-growers

TABLE VI
Nitrogen Fixed by Pulses[a]

Plant	Average	Range
Vicia faba (faba beans)	210	45–552
Pisum sativum (peas)	65	52–77
Lupinus spp. (lupines)	176	145–208
Phaseolus aureus (green gram)	202	63–342
Phaseolus aureus (mung)	61	—
Cajanus cajan (pigeon pea)	224	168–280
Vigna sinensis (cowpea)	198	73–354
Canavalia ensiformis (jack bean)	49	—
Cicer arietinum (chickpea)	103	—
Lens culinaris (lentil)	101	88–114
Arachis hypogaea (peanut)	124	72–124
Cyamopsis tetragonolobus (guar)	130	41–220
Calopogonium mucunoides (calapo)	202	370–450

[a] In kg N/ha.

TABLE VIII
Differences between Fast- and Slow-Growing Rhizobia[a]

| Characteristic | Rhizobial type[b] | |
	Fast-growing	Slow-growing
Generation time	< 3 hr	> 6 hr
Carbohydrate substrate	Uses pentoses, hexoses, and mono-, di- and trisaccharides	Uses pentoses and hexoses solely
Metabolic pathways	EMP[2], low activity Strain-specific ED, main pathway TCA, fully active PP present	EMP, low activity ED, main pathway TCA, fully active Hexose cycle present
Flagellation type	Peritrichous	Subpolar
Symbiotic gene location	Plasmids	Chromosome only
Nitrogen-fixing gene location	*nif H, D,* and *K* on same operon	*nif D, K,* and *H* on separate operons
Intrinsic antibiotic resistance	Low	High

[a] *Mesorhizobium* are intermediate phenotypically and phyllogenetically.

[b] ED, Entner–Parners pathway; EMP, Embden–Meyerhoff; PP, pentose phosphate pathway; TCA, tricarboxylic acid cycle.

to the new genus *Bradyrhizobium*. Recent findings using numerical taxonomy, carbohydrate metabolism, antibiotic susceptibilities, serology, DNA hybridization, and RNA analysis all demonstrate the validity of the fast- and slow-growing groupings. *Bradyrhizobium* is transferred to the new family *Bradyrhizobiaceae* in the new edition of *Bergey's Manual of Systematic Bacteriology*. A summary of some of the differences is found in Table VIII. Thus, whereas in the first edition of *Bergey's Manual*, the slow-growing strains were placed in the new genus *Bradyrhizobium*, they now fall into their own family along with close relatives such as *Afipia* and *Nitrobacter*. The genus *Bradyrhizobium* now comprises three species, *B. japonicum, B. elkanii,* and *B. liaoningense*, all of which nodulate soybean (Table IX). Other bradyrhizobia are known to occur (e.g., the peanut bradyrhizobia) but these have not been classified. Researchers suggest that until further taxa within the genus are proposed, these should be described with

TABLE IX
Current Taxonomic Classification of the Rhizobia

Recognized genus	Recognized species
Allorhizobium	*Allorhizobium undicola*
Azorhizobium	*Azorhizobium caulinodans*
Bradyrhizobium	*B. japonicum, B. elkanii, B. liaoningenes*
Mesorhizobium	*M. amorphae, M. ciceri, M. huakuii, M. loti, M. mediterraneum, M. plurifarium, M. tianshanese*
Rhizobium	*R. etli, R. galegae, R. gallicum, R. giardinii, R. hainanense, R. huautlense, R. leguminosarum, R. mongolense, R. tropici*
Sinorhizobium	*S. fredii, S. medicae, S. meliloti, S. saheli, S. terangae, S. xinjiangansis*

the appropriate host plant given in parentheses [i.e, the peanut rhizobia–*Bradyrhizobium* sp. (*Arachis*)].

The fast-growing legume-nodulating bacteria (sometimes still called "rhizobia,") were all originally placed within the genus *Rhizobium*. There were only a few species, *R. leguminosarum*, *R. meliloti*, *R. loti*, *R. galegae*, and *R. fredii*. The first three species, *R. phaseoli*, *R. trifolii*, and *R. leguminosarum* were amalgamated into the single type species *R. leguminosarum* as biovars. The biovar *phaseoli* had tremendous genetic diversity and a number of new bean-nodulating species have been named, *R. tropici*, *R. etli*, *R. gallicum*, *R. giardinii*, or *R. mongolense*. *Rhizobium fredii* was the first of a series of species consisting of fast-growing rhizobacteria that effectively nodulate Chinese soybean cultivars, originally thought to be nodulated only by *B. japonicum*. *R. fredii* was reassigned to a new genus, *Sinorhizobium*, in 1988. This controversial new genus also contains *S. meliloti* and several close relatives, which all share a very close phylogenetic relationship with the type species, *S. fredii*. Alfalfa plants are nodulated by *S. fredii*, *S. meliloti*, and *S. medicae*. Soybean is nodulated by *B. japonicum*, *B. elkanii*, *B. liaoningense*, *S. fredii*, and *Mesorhizobium tianshanense*. The latter new genus and species are closely related to *Rhizobium loti*, which is now appropriately named *Mesorhizobium*, along with other newly described close relatives because they are intermediate, both in growth rate and molecular phylogeny, between *Bradyrhizobium* and *Rhizobium*. Thus, they now belong to the family Phyllobacteriaceae. *Allorhizobium* is a newly proposed genus for the microsymbiont of an aquatic legume, *Neptunia natans*. The taxonomic scheme is summarized here in a list of the recognized species of legume-nodulating rhizobacteria.

1. *Allorhizobium undicola* fixes nitrogen with *Acacia* spp., *Faidherbia* spp., and *Lotus arabicus*. Most *Allorhizobium undicola* strains are closely related to *Agrobacterium*.

2. *Azorhizobium caulinodans* forms stem nodules on *Sesbania rostrata* and readily fixes nitrogen *ex planta* when microaerobic conditions are provided. It belongs to the family Hypomicrobiaceae.

3. *Bradyrhizobium japonicum* forms root nodules on species of *Glycine* (soybean) and on *Macroptilium*

atropurpureum (siratro). Some strains of *B. japonicum* express hydrogenase activity with the soybean host and are hence more efficient in symbiotic nitrogen fixation.

4. *Bradyrhizobium elkanii* normally forms root nodules on species of *Glycine* (soybean), the "nonnodulating" *rj1rj1* mutant soybean that fails to nodulate with *B. japonicum*, black-eyed peas (*Vigna*), mung bean, and *Macroptilium atropurpureum* (siratro). Unlike *B. japonicum*, *B. elkanii* often produces rhizobitoxine-induced chlorosis on sensitive soybean cultivars. Strains of *B. elkanii* often are hydrogenase positive on *Vigna* but not on *Glycine*, suggesting that they possess more symbiotic affinity or compatibility with the former than the latter.

5. *Bradyrhizobium liaoningense* is an extra-slow-growing, soybean-nodulating *Bradyrhizobium* isolated from alkaline Chinese soils.

6. *Mesorhizbium amorphae* was isolated from nodules of *Amorpha fruticosa* growing in North China.

7. *Mesorhizobium ciceri* was isolated from chickpeas (*Cicer*) grown in uninoculated fields over a wide geographic range, including Spain, the United States, India, Russia, Turkey, Morocco, and Syria.

8. *Mesorhizobium huakuii* was isolated from *Astragalus sinicus*, a green manure crop grown in rice fields in southern parts of China, Japan, and Korea.

9. *Mesorhizobium loti* nodulates *Lotus corniculatus*, *Lotus tenuis*, *Lotus japonicum*, *Lotus krylovii*, *Lotus filicalius*, and *Lotus schoelleri*.

10. *Mesorhizobium mediterraneum* is exclusively a *Cicer*-nodulating bacterium.

11. *Mesorhizobium tianshanense* isolates were obtained from *Glycyrrhiza pallidiflora*, *G. uralensis*, *Glycine max*, *Sophora alopecuroides*, *Swainsonia salsula*, *Caragara polourensis*, and *Halimodendron holodendron* growing in Xinjiang Region of China. Most of the host plants are wild and indigenous to that region, except *Glycine max*, which is of course a cultivated crop that originated in northeastern Asia.

12. *Mesorhizobium plurifarium* nodulates *Acacia senegal*, *A. tortilis*, *A. nilotica*, *A. seyal*, *Leucaena leucocephala*, and *Neptunia oleracea*, but most strains do not nodulate *Sesbania rostrata*, *S. pubescens*, *S. grandiflora*, *Ononis repens*, or *Lotus corniculatus*.

13. *Rhizobium etli* nodulates and fixes nitrogen in

association with *P. vulgaris* exclusively; it includes nonsymbiotic strains.

14. *Rhizobium galegae* nodulates *Galega orientalis* and *Galega officinalis* and is specific to this plant genus.

15. *Rhizobium gallicum* nodulates and fixes nitrogen in association with *Phaseolus* spp., *Leucaena leucocephala, Macroptilium atropurpureum,* and *Onobrychis viciifolia.*

16. *Rhizobium giardinii* nodulates *Phaseolus* spp., *Leucaena leucocephala,* and *Macroptilium atropurpureum.*

17. *Rhizobium hainanense* is found in nodules of *Desmodium sinuatum* or *Stylosanthes guianensis, Centrosema pubescens, Desmodium triquetrum, D. gyroides, D. sinatum, D. heterophyllum, Tephrosia candida, Acacia sinicus, Arachis hypogaea, Zornia diphylla, Uraria crinita,* and *Macroptilium lathyroides.*

18. *Rhizobium huautlense* nodulates *Sesbania herbacea, S. rostrata,* and *Leucaena leucocephala.*

19. *Rhizobium leguminosarum* nodulates with some, but not necessarily all *Pisum* spp., *Lathyrus* spp., *Vicia* spp., *Lens* spp., temperate species of *Phaseolus* (*P. vulgaris, P. angustifolius,* and *P. multiflorus*), and *Trifolium* spp.

20. *Rhizobium mongolense* was recently isolated from *Medicago ruthenica,* but it also nodulates *Phaseolus vulgaris.* It is a very close relative of *Rhizobium gallicum.*

21. *Rhizobium tropici* forms nodules on *Phaseolus vulgaris* and Leucaena spp. The type strain, CFN299, nodulates *Amorpha fruticosa.*

22. *Sinorhizobium fredii* effectively nodulates *Glycine max cv.* "Peking," *Glycine soja, Vigna unguiculata,* and *Cajanus cajan.* Also nodulates alfalfa. In 1985, Dowdle and Bohlool reported new strains that are symbiotically competent with North American cultivars of soybean. Their molecular phylogeny, however, has not been determined yet.

23. *Sinorhizobium medicae* is alfalfa-nodulating, with a close phylogenetic relationship to *S. meliloti.*

24. *Sinorhizobium meliloti* forms nitrogen-fixing nodules on *Melilotus, Medicago,* and *Trigonella.*

25. *Sinorhizobium saheli* is found in nodules of *Sesbania* spp. growing in the Sahel and can nodulate *Acacia seyal, Leucaena leucocephala,* and *Neptunia oleracea.*

26. *Sinorhizobium terangae* is also found in nodules of *Acacia* spp., *Sesbania* spp., *Leucaena leucocephala,* and *Neptunia oleracea.*

27. *Sinorhizobium xinjiangense* nodulates soybean and is a close relative of *S. fredii.*

The taxonomy of the nitrogen-fixing bacteria is in a dynamic state of change. As molecular information accumulates, the cataloging of phenotypic data has not kept pace and further revision, and "reversions" will be necessary. New approaches to classification are needed because the scheme, unfortunately, does not function to allow the identification of isolates without DNA-sequence analysis.

1. Nitrogen Fixation

The nitrogenase complex is a highly conserved enzyme system and, as stated earlier, is basically common to all of the dinitrogen-fixing prokaryotes. The evidence is conclusive that there are differences in location of nitrogenase genes of *Rhizobium* and *Bradyrhizobium.* In all of the fast-growing *Rhizobium* species, which have a chromosome size of about 3500 kb, the structural *nifH, D,* and *K* genes are localized on extremely large plasmids or megaplasmids. In *Bradyrhizobium, nif* genes have been mapped on the 8700-kb chromosome. In *Mesorhizobium,* on the other hand, the *nif* genes are located on the chromosome in some species and on megaplasmids in other species.

Plasmids and megaplasmids, present in a wide variety of the nitrogen-fixing bacteria, control many phenotypic and genetic characteristics of the bacterial cells. Those in *Rhizobium, Sinorhizobium,* and *Azorhizobium* species carry genes controlling symbiotic functions, which are clustered on a single large plasmid termed symbiotic plasmid, "pSym." Two or more plasmids that carry genes controlling symbiotic functions, *nod, fix* and the nitrogenase structural (*nifHDK*) genes, have been found in certain strains of the nitrogen-fixing species. In addition to symbiotic plasmid(s), rhizobia strains may carry 1–10 plasmids that range in size from 30 to more than 1000 MDa. These plasmids are highly stable and have beneficial roles in the soil environment, and plasmid profile analysis is sometimes used to discriminate among *Rhizobium* strains.

The analysis and comparison of *nif* DNA in the fast- and slow-growing "rhizobia" have established the affinity coefficient (S_{AB}) for the nucleotide sequence (*nifH, nifD,* and *nifK*) from the two groups. The same analysis was done comparing amino acid sequences of the nitrogenase Fe and FeMo protein polypeptides. A considerable sequence conservation reflects the structural requirements of the nitrogenase proteins for catalytic functions. The S_{AB} *nif* values (based on *nifH* sequences) between fast- and slow-growing organisms indicated that these are almost as distant from each other as they are from other gram-negative organisms. The results suggest that *nif* genes evolved in a manner similar to the bacteria that carry them rather than by a more recent lateral distribution of *nif* genes among microorganisms. Again, the phylogenetic difference between fast and slow groups is apparent. Although this general system is common to all of nitrogen-fixing prokaryotes, several concomitant alternative systems have been described; but these evidently lack biological significance and they have not been shown to be present or active in the "rhizobia."

2. Nodule Formation

Nodule initiation and subsequent maturation is an interactive process involving the eukaryotic host legume and the prokaryotic *Rhizobium*. The process is complex, resulting in biochemical and morphological changes in both symbionts and leading to the capacity to reduce atmospheric nitrogen. Initially, the proper *Rhizobium* species proliferates in the root zone of a temperate leguminous plant and becomes attracted and attached to the root hair. A chemotactic response attracts the bacteria to the root surface. At the surface, the bacteria secrete Nod factors, which are certain chitolipooligosaccharides (CLOS) that alter the growth of epidermal root hairs so that they are deformed. CLOS molecules chemically induce nodule organogenesis in extremely low concentrations ($\sim 10^{-10}$ M). In some tropical legumes such as peanuts (*Arachis*), root hairs are not the primary invasion sites. However, the alternative invasion process, "crack entry," has been well-documented because infection occurs at the site of lateral-root emergence.

The root hair infection process consists of several events leading to nodule formation: (1) recognition by the "rhizobia" of the legume, (2) attachment to the root hair, (3) curling of the root hair, (4) root-hair infection by the bacteria, (5) formation of an infection thread, (6) nodule initiation, and (7) transformation of the vegetative cells in the nodules to enlarged pleomorphic forms, called bacteroids, which fix nitrogen.

Based on morphology, there are two kinds of nodules, determinative and indeterminative. In general, indeterminative nodules are formed by fast-growing "rhizobia" and are characterized by a defined meristem during nodule growth. Determinative nodules arise from cortex tissue. Legumes nodulated by *Bradyrhizobium* form determinative nodules close to the endodermis, which is near the xylem poles in the root.

The formation of nodules on legumes is the result of a coordinated development involving many plant and bacterial genes. Studies of the nodulation (*nod*) genes of rhizobia have depended on the development of molecular genetic tools. Many of the genes involved in the nodulation process have been located and identified.

Legume roots grown axenically do not appear as morphologically distinctive from other plant roots, so that the abilities of these plants to respond to microbial signals and then alter their metabolism to form nodules are not explained by morphology alone. It has been conclusively demonstrated that genetic information from both symbionts controls nodulation and the host range of nodulation by a *Rhizobium* species. Metabolically, there are three types of nodules (often termed effective, inefficient, and ineffective). Effective nodules contain a high density of bacteria actively fixing dinitrogen. Inefficient nodules may contain a similar density of the bacteria, but only a relatively low level of fixed dinitrogen results from the symbiosis. Ineffective symbiosis occurs with bacteria not able to nodulate or fix nitrogen normally. Because the regulatory roles of the plant and bacterial genes in nitrogen fixation have not been generally elucidated, the reasons for differential nitrogen-fixing ability of nodules continues to be unclear. There is, however, an already observed compatibility in legume host–*Rhizobium* interaction that can be technologically exploited to enhance dinitrogen fixation. Such an interaction makes it possible to optimize dinitrogen fixation under field conditions in a cultivar through inoculation with an effective *Rhizobium* strain.

Nodulation genes are defined by their effect on the bacteria's ability to generate the nodulation process on the proper legume host. Because most individual species of legume-nodulating rhizobacteria can each nodulate a limited number of host legumes and plant genes also limit the symbiosis, it follows that a recognition exists between bacteria and host. Thus, there are two types of *nod* genes, a common *nod* region, which consists of a structurally and functionally conserved cluster of genes; and host-specific nodulation genes, which cannot complement nodulation defects in other genera or species. *Nod* genes have been studied in varying degrees in different species (usually in the fast-growing rhizobia). In these organisms, four genes have been identified in two transcription units (*nodD* and *nodABC*). Two additional genes, apparently on the same transcriptional unit, *nodI* and *nodJ,* have been identified. Genetic maps of the common *nod* cluster, drawing together the information from many sources, have been published. The *nodABC* appears to be functionally interchangeable among all rhizobia, and mutations in these genes cause complete nodulation failure. These genes are involved in cell division and root-hair deformation. The *nodIJ* genes cause a delay in the appearance of nodules.

The second group of *nod* genes are termed "host-specific." These genes are not conserved because alleles from various rhizobia cannot substitute for each other on different hosts. Bacteria carrying mutations in these host-specific genes cause abnormal root-hair reactions. Many genetic *nod* loci have been identified in a variety of the symbiotic nitrogen-fixing species. The list includes at least 15 *nod* genes. In many cases, these have been cloned and sequenced and the gene product associated with a step in nodulation. Although the amino acid sequences of many of the nodulation-gene products have been described, the biochemical functions of these genes have not been fully determined. Possible exceptions are the *nodD* genes, which are positive gene regulators.

The centenary of this first demonstration of biological nitrogen fixation occurred during 1986. During that period, many papers were published expanding the knowledge base, both basic and applied. In 1991, the U.S. Department of Agriculture patented an improved soybean inoculant *Bradyrhizobium japonicum* strain that results in a significant increase in growth and soybean yield. Since then, this improved strain has been commercially produced. Currently, the subject of BNF is of great practical importance because of the use of fossil fuels in the manufacture of nitrogenous fertilizers. The increased scarcity and higher costs of fossil fuel have made it important to optimize biological nitrogen fixation as an alternative to chemical nitrogen. In addition, the increasing usage of nitrogen fertilizer has resulted in unacceptable levels of water pollution, which occurs only to much lesser extent when the biologically fixed forms of nitrogen are used. With the additional research capabilities resulting from the developing field of biotechnology, it is evident that interest in this field will continue and that we may reach a level of accumulated knowledge that will allow full use of BNF as an alternative to the Haber–Bosch chemical industrial process.

3. Rhizobiophages

Rhizobiophages occur commonly in the rhizosphere of legumes and are often associated with susceptible *Bradyrhizobium, Rhizobium,* or *Sinorhizobium* strains. They reduce rhizobial populations in soils and negatively affect the nitrogen-fixing abilities of these bacteria with the host legume plant. Rhizobiophages can be used to distinguish between rhizobial strains through "phage typing." Furthermore, rhizobiophages are potential biocontrol agents useful for reducing the number of susceptible rhizobial cells in the soils, thus decreasing nodule occupancy by the undesirable indigenous bacterial strain, and thereby increasing the nodule occupancy by superior strains used as inoculant. The use of specific bacterial viruses or bacteriophages as biocontrol agents requires the identification of symbiotically competent, rhizobiophage-resistant *Rhizobium* or *Bradyrhizobium* strains that have a demonstrated ability to promote the growth and yield of their specific legume hosts.

See Also the Following Articles

CYANOBACTERIA • NODULE FORMATION IN LEGUMES • RHIZOSPHERE

Bibliography

Boland, G. J., and Kuykendall, L. D. (eds.) (1998). "Plant-Microbe Interactions and Biological Control." Marcel-Dekker, New York.

Dilworth, M. J., and Glenn, A. R. (eds.) (1991). "Biology and Biochemistry of Nitrogen Fixation." Elsevier, Amsterdam.

Elkan, G. H. (ed.) (1987). "Symbiotic Nitrogen Fixation Technology." Marcel Dekker, New York.

Elkan, G. H., and Upchurch, R. G. (eds.) (1997). "Current Issues in Symbiotic Nitrogen Fixation." Kluwer, Dordrecht, The Netherlands.

Elmerich, C., Kondorsi, A, and Newton, W. E. (eds.) (1998). "Biological Nitrogenase Fixation for the 21st Century." Kluwer, Dordrecht, The Netherlands.

Gresshof, P. M. (ed.) (1990). "Molecular Biology of Symbiotic Nitrogen Fixation." CRC Press, Roca Raton, FL.

Hennecke, H., and Verma, D. P. S. (eds.) (1990). "Advances in Molecular Genelic of Plant-Microbe Interaction." Kluwer, Dordrecht, The Netherlands.

Somasegaran, P., and Hoben, H. J. (eds.) (1994). "Handbook for Rhizobia." Springer-Verlag, New York.

Spaink, H. P., Kondorosi, A., and Hooykaas, P. J. J. (eds.) (1998). "The *Rhizobiaceae*: Molecular Biology of Model Plant-Associated Bacteria." Kluwer, Dordrecht, The Netherlands.

Sprent, J. I., and Sprent, P. (eds.) (1990). "Nitrogen Fixing Organisms." Chapman and Hall, London.

Stacy, G., Burris, R., and Evans, H. J. (eds.) (1992). "Biological Nitrogen Fixation." Chapman and Hall, New York.

Nodule Formation in Legumes

Peter H. Graham
University of Minnesota

GLOSSARY

bacteroids Cells of rhizobia in the nodule that have undergone surface change, are often swollen and irregular in shape, and express nitrogenase.

cross-inoculation The ability of bacterial strains from two or more different legumes to each produce nodules on one another's host(s).

endocytosis A process releasing rhizobia from the infection thread, but surrounding them with plant-derived membrane material limiting the host defense responses.

ineffective Limited in the ability to fix nitrogen (N_2); refers to host–rhizobial combinations.

infection thread A plant-derived tube through which rhizobia move down the root hair or between cells in the root cortex.

inoculation The application of artificially cultured rhizobia to legume seed or soil in order to improve nodulation.

nodule A gall-like structure that develops on the root or stem of legumes following infection by compatible rhizobia.

nodulins Gene products expressed in host tissue during nodulation and N_2 fixation.

peribacteroid membrane A host-derived membrane that surrounds rhizobial cells following their release into the cells of the host.

promiscuity The ability of some rhizobia to nodulate with many legumes.

symbiosis An association between two organisms in which each derives benefit.

SYMBIOTIC N_2-FIXING BACTERIA, collectively known as rhizobia, infect legumes, forming stem or root nodules. The bacteria derive energy from the host for growth and N_2 fixation and are protected from external stresses; the host has access to a form of N it cannot otherwise use. Rates of N_2 fixation vary, but commonly range from 50–200 kg N_2 fixed ha^{-1} year^{-1}, reducing the plant's need for soil or fertilizer N. Collectively, nodulated crop plants fix 32–53 Tg N year^{-1}, with significant additional N_2 fixation by legumes in natural terrestrial ecosystems.

I. INTRODUCTION

Nitrogen needs in agriculture can only grow. If the emphasis is toward fertilizer N, increases to around 134 Tg N fertilizer produced year^{-1} can be anticipated by the year 2020. N deposition and environmental N pollution will be widespread, and fossil fuel will be consumed at an alarming rate. Further, developing countries will be at a disadvantage because many subsistence farmers will not be able to afford significant N use. A more balanced approach is to use fertilizer N mainly for high-value crops, and to place greater emphasis on N_2 fixation for other crop and pasture systems. In Brazil, the emphasis on symbiotic N_2 fixation has permitted a major increase in soybean production, while saving more than $1.8 billion in fertilizer N costs annually. Field application of some recent advances in inoculation, nodulation, and N_2 fixation could benefit many crops and ecosystems.

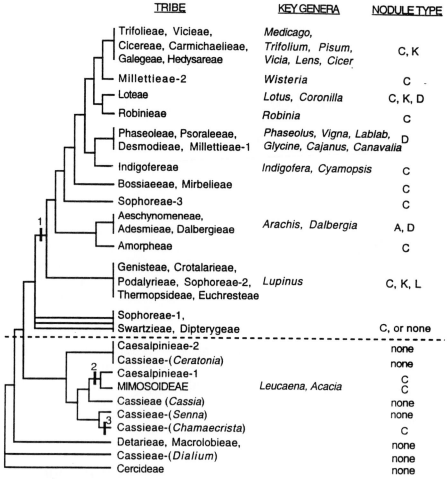

Fig. 1. Phylogeny of the Leguminosae based on *rbc* L DNA-sequence data. (Modified from Doyle and Doyle [1997] and printed with permission.) Tribes appearing more than once on the tree are polyphyletic. The numbered rectangles indicate stages in evolution beyond which most species are nodulated. C and K, A and D, and L refer to indeterminate, determinate, and lupinoid (collar) nodule shapes, respectively.

II. EVOLUTION IN THE NODULATION OF LEGUMES

Nodulation and symbiotic N_2 fixation are restricted to a clade of plants that includes both legume and actinorhizal species. Not all legumes bear nodules; only 23% of Caesalpinioideae are nodulated, though 90% of Mimosoideae and 97% of Papilionoideae nodulate. Nodulation in legumes could have arisen independently on several occasions in the evolution of legumes, including in the genus *Chamaecrista* (see Fig. 1). This genus has also attracted attention be-

cause some species retain the rhizobia within infection threads during symbiosis, whereas in others they are released into cells of the host. Because legumes such as bean nodulate with several distinct species of rhizobia, other studies have concluded that multiple evolution in symbiosis is unlikely. Instead, a continuum is proposed from nonnodulation, through rare or mixed nodulation capacity in each subfamily, to the abundant nodulation of most Papilionoideae and Mimosoideae. Bryan *et al.* (1996) suggest that even unnodulated legumes may fix N_2 and describe cells resembling bacteroids within the roots of *Adenan-*

thera, Ceratonia, Bauhinia, Gleditsia, and *Pelto-phorum.*

The ability of humans to manipulate both the host and the rhizobia has accelerated evolutionary change in some symbioses. Thus, trade in *Phaseolus* beans during the colonial period introduced this crop to the more acid soils of Brazil and eastern Africa, where the survival of the normal bean microsymbiont *R. etli* was affected. A result is that beans in such acid soils are now often associated with the more acid-tolerant *R. tropici.*

Microorganisms producing root and stem nodules on legumes are divided into five genera (*Rhizobium, Azorhizobium, Mesorhizobium, Sinorhizobium,* and *Bradyrhizobium*), with creation of a sixth genus (*Allo-rhizobium*) likely. Twenty-four species are recognized, but rhizobial taxonomy is far from stable, and species epithets are constantly under change. An updated list of species is maintained at *http://www.rhizob ium.umn.edu.*

III. MECHANISMS OF INFECTION

A. Types of Infection

Rhizobia can infect their hosts and induce root- or stem-nodule formation by several mechanisms, the most common of which are:

1. Infection through the root hairs with infection threads carrying the rhizobia into the root cortex (Hirsch, 1992), as seen in medics and clovers.

2. Entry via wounds or through lesions caused by emergence of secondary roots (Boogerd and van Rossum, 1997), as occurs in peanut and *Stylosanthes.* Rhizobia spread intercellularly or cause infected cells to collapse, colonizing the space.

3. Infection via cavities surrounding adventitious root primordia on the stems of *Sesbania, Aeschyno-mene, Neptunia,* or *Discolobium* (Boivin *et al.,* 1997).

Rhizobia can also nodulate the nonlegume *Paras-ponia,* a process involving the formation of callus-like prenodules, intercellular spread of rhizobia via infection threads, and merging of prenodule tissue with a modified lateral root.

The same rhizobial isolate may infect one host (e.g., *Macroptilium*) via root hair penetration, but another (e.g., *Arachis*) through crack entry and intercellular spread. Similarly, one organism may produce both stem and root nodules.

B. Visible Changes during Root-Hair Infection

The Fahraeus slide technique (Fahraeus, 1957) and the root-tip marking procedure (Bhuvaneswari *et al.,* 1981) are seminal to our knowledge of root-hair infection. One allowed the repeated observation of the infection process in small-seeded legumes such as clovers; the other showed differences in the susceptibility of immature and mature root hairs to infection, and allowed research to focus on those parts of the root where infection was most common.

Compatible rhizobia begin to attach to root hairs of their host within minutes of inoculation, and attachment increases over time. Attached cells cap the root-hair tip, and are often oriented end-on to their host. Adhesion is initially mediated by the calcium (Ca)-binding protein rhicadhesin, or by plant lectins, with subsequent bonding via production of cellulose fibrils.

Rhizobia cause localized hydrolysis of the root-hair cell wall, and promote invagination of the host plasma membrane, with additional plant-cell material deposited about them as they infect. The enzymes involved in hydrolysis are cell bound and difficult to quantify, and several differ from those normally associated with plant infection. Rhizobial penetration causes root-hair growth at the point of infection to cease, and leads to root-hair curling, first visible some 6–18 hr after inoculation. The proportion of root hairs infected is low, and the percentage of these giving rise to nodules, highly variable. An electron micrograph taken at this stage in infection shows the root hair curled into a shepherd's crook, penetration by several rhizobia, and the beginnings of an infection thread (see Fig. 2a). Rhizobia, still encased within a plant-derived infection thread, move down the root hair to the root cortex (Fig. 2b). Cell division in the root cortex precedes their arrival and gives rise to the nodule primordia, and in some legumes to an uninfected meristematic region. The spread

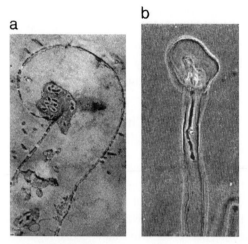

Fig. 2. (a) Root hair deformation, curling, and infection—an early stage in the nodulation of clover by *R. leguminosarum bv trifolii.* (From Sahlman and Fahraeus, 1963 with permission.) (b) Rhizobia contained in the infection thread move down the root hair in the direction of the root cortex. (From Fahraeus, 1957, with permission.)

of the infection thread among cells of the nodule primordium follows, with rhizobia released into their host by endocytosis. Rhizobia never gain free intracellular access. They are initially confined by the infection thread and later surrounded by the host-derived peribacteroid membrane.

Nodulation is usually evident 6–18 days after inoculation, but this varies with the strain and cultivar used, the inoculant density, and temperature. Initially, nodulation is heaviest in the crown of the root, with secondary flushes of nodules on lateral roots as the first-formed nodules senesce.

C. The Molecular Basis for Nodule Formation

Molecular studies of infection received impetus from the demonstration that most nodulation genes in *Rhizobium* were plasmid-borne. Mapping studies followed, with the sequencing of the entire symbiotic plasmid from the promiscuous strain NGR234 only completed in 1997.

Further studies showed that some nodulation genes occur in all rhizobia (with genes from one strain correcting mutations in another), whereas others only occurred in strains nodulating a particular

host(s). Further, only one gene, *nod*D, was always expressed, with the others only turned on in the presence of their host. A search for host factors triggering *nod*-gene expression followed, with flavonoids from the root shown to be responsible for triggering the *nod*-genes, legume species shown to vary in the mix of flavonoids they produced, and rhizobia shown to be able to respond differently to specific flavonoids. Thus naringenin and genistein stimulated *nod* gene expression in *B. japonicum*, whereas luteolin was required for *S. meliloti*. Because flavonoids may potentiate rhizobial infection, but be limiting in specific cultivars or at low temperature, some flavonoid preparations are now sold commercially.

More than 50 nodulation genes have been identified in rhizobia. Some are involved in nod-gene regulation, or protein secretion, but most are involved in the synthesis of lipochitooligosaccharide "nod factors" (Long, 1996). These molecules all have the same core structure (coded for by the common nodulation genes), but vary in the side chains each carries, affecting host range (see Fig. 3). Strains may produce several nod factors differing slightly in composition. They act as powerful morphogens that at 10^{-9}–10^{-11} M can deform root hairs and initiate the cortical cell division typical of nodule initiation. Table I lists some of the changes effected by these substances and the time frame in which they occur. High-affinity legume-receptor molecules for nod factors have been postulated but not identified; John *et al.* (1997) discuss the broader role of this class of compounds in plant growth regulation.

Roles for plant lectins, rhizobial extracellular polysaccharides (EPS), and lipopolysaccharides (LPS) in nodule formation continue to be explored. Lectins are nonenzymatic carbohydrate-binding proteins occurring in different forms throughout the plant. Cross-bridging between root-hair lectins and the rhizobial surface has been proposed for many years but is not fully accepted. More definitive evidence of lectin involvement in nodulation was the recent demonstration of host-range modification in white clover plants transformed with the gene for pea seed lectin. The possibility that lectins bind rhizobial nod factors has been discounted, and a more indirect effect of lectins on cell responsiveness to nod factors of all types has been proposed.

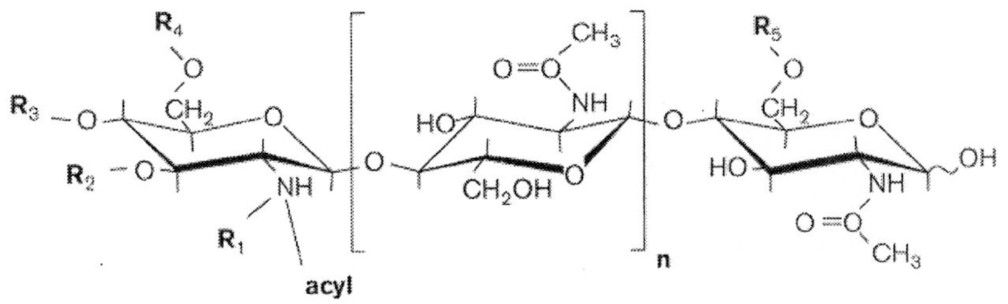

Fig. 3. The general structure of Nod factors elicited by rhizobia. In this structure *n* = 2 or 3, and the substitutions possible at each indicated position are R₁ for H or methyl; R₂ and R₃ for H, carbomoyl, or 4-O methyl carbomoyl; R₄ for H, acetyl, or carbomoyl; R₅ for H, sulfate, acetyl 2-O-methyl fucose, 4-O-sulfo-2-O-methylfucose, 3-O-acetyl-2-O-methyl-fucose, 4-O-acetyl-fucose, or D-arabinose; and Acyl for C16-C20. Modified from Schultze and Kondorosi (1996) and used with permission.

Rhizobial EPS mutants produce nodules on alfalfa devoid of rhizobia and ineffective. In contrast, hosts giving rise to determinate nodules (see Section IV.A) generally nodulate normally with such mutants. An exception is soybean, in which EPS mutant strains are delayed in nodule formation and induce a marked plant defense response. For pea and alfalfa rhizobia defective in EPS synthesis, μM quantities of wild-type EPS, or coinoculation with a *nod⁻ eps⁺* strain, can restore some degree of invasive ability. Because the quantity of EPS needed for nodulation is small, it has been suggested that it acts as a signal molecule for infection-thread growth beyond the epidermal cell layer.

Rhizobial LPS production is essential for normal nodule development, but reponses obtained with LPS mutants again vary with plant species. LPS mutants of *R. leguminosarum* failed to properly colonize the nodule tissue in peas, induced marked host defense responses, and were ineffective. Similar mutants of *B. japonicum* and *R. etli* were deficient in infection-thread formation, and formed nodules that were small, white, and devoid of rhizobia. In both cases, a role for rhizobial LPS in masking the elicitors of plant defense response has been proposed.

D. Nodule-Specific Gene Expression

The interaction of host and rhizobia is accompanied by the expression of nodule-specific proteins or nodulins, originally defined as specific to infected root hairs or nodule tissue, and needed for nodule formation and function. This definition has blurred over time. Most nodulins also occur in other tissues; a number have also been identified in actinorhizal or mycorrhizal symbioses. Pea mutants that neither nodulate or form mycorrhizal associations are a further indicator of overlap in symbiosis. Some nodulins may be expressed following inoculation with either Nod factor or rhizobia; others are only expressed in the presence of rhizobia.

Nodulin expression can vary temporally and spatially. Early nodulins are involved in infection or nodule development, and may be expressed within 6 hr of inoculation. Expression may be transient or,

TABLE I
Effects of Purified Nod Factor on Legumes

Plant response	Time after inoculation	Minimum concentration (M)
Depolarization of plasma membrane	15 s	10^{-11}
Increase in root hair pH	15 s	10^{-10}
Ca²⁺ spiking in root hairs	10 min	10^{-9}
Expression of early nodulin genes	6–24 hr	10^{-12}
Root hair deformation	~18 hr	10^{-12}
Cortical cell divisions	days	10^{-9}
Empty nodule formation	weeks	10^{-9}

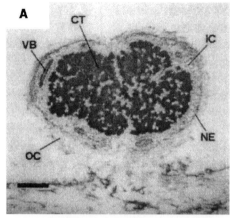

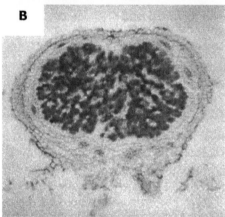

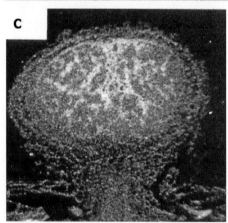

Fig. 4. Sections of mature bean nodules following staining with toluidine blue and *in situ* hybridization with leghemoglobin or uricase-II probes. (A) Bright-field micrograph stained with toluidine blue, showing the determinate nodule structure, central tissue (CT), vascular bundles (VB), nodule endodermis (NE), and inner (IC) and outer (OC) cortex. Bar equals 200μm. (B) Bright-field micrograph showing *in situ* localization of the leghemoglobin transcript. Hybridization signals are evident as dark dots in

as with nodulins associated with new cell infection, may extend over the life of the nodule. Well-studied early nodulins include Enod 40, first detected in the root pericycle opposite incipient nodule primordia and postulated to function in hormone perception; Enod 12, expressed in cells adjacent to growing infection threads; and Enod 2, located in the parenchyma and postulated to have a role in regulation of oxygen supply to bacteroids.

Later nodulins are expressed just before or with the onset of N_2 fixation and are related to nodule function, carbon and nitrogen metabolism, and O_2 transport. Examples include leghemoglobin, PEP carboxylase, and enzymes involved in allantoic acid synthesis and ammonia assimilation (see Fig. 4). More detailed information is provided by Pawlowski (1997).

E. Plant Defense Response and Symbiosis

Plants respond to invasion by rapid localized tissue necrosis, synthesis of phytoalexins or hydrolytic enzymes, and by lignification of plant tissue. Rhizobia usually avoid such responses, though enzymes involved in the flavonoid and isoflavonoid pathway leading to phytoalexin synthesis have been found in a number of legumes soon after inoculation, and glyceollin I levels in soybean show transient increases soon after infection. Chalcone synthase levels in root hairs rise soon after inoculation, but decline thereafter, whereas those in the root increase. This response appears to follow the rhizobia into the root cortex and to be a part of normal nodule development. Enhanced flavonoid levels may be needed for hormone production in the nodule, to induce cytoskeletal rearrangement, or to influence *nod* gene expression in the rhizosphere. In contrast, transient phytoalexin production during nodulation could stem from the brief exposure of the rhizobia during entry, with the host response subsequently muted as

invaded cells. (C) Dark-field micrographs of the *in situ* localization of uricase-II transcripts. Silver grains are visible as white dots in uninvaded nodule cells. From Tate *et al.* (1994), with permission.

the rhizobia become enclosed within the infection thread and peribacteroid membrane. A more typical host defense response with elevated phytoalexin levels has been noted in specific soybean–*Bradyrhizobium* combinations, and following the abortion of infection threads or inoculation with EPS-defective mutants. The peribacteroid membrane could serve to protect rhizobia from host recognition, especially in organisms that spread intercellularly. Deterioration of this membrane could be a factor contributing to declining N_2 fixation in senescent nodules. Salicylic acid accumulation is an additional defense response shown in alfalfa.

IV. CHARACTERISTICS OF THE LEGUME NODULE

A. Nodule Shape

Nodule shape is a characteristic of the host plant and is regulated by the pattern of cortical cell division following infection. Rounded, determinate nodules, such as occur on *Phaseolus* and *Glycine,* lack a persistent nodule meristem (see Fig. 4), whereas nodules on *Medicago* and *Trifolium,* in which the meristem is persistent, are elongate, sometimes lobed and indeterminate (see Fig. 5). Hirsch (1992) contrasts the

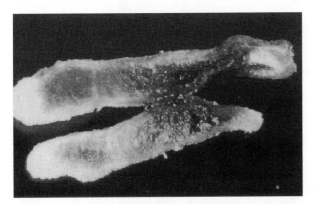

Fig. 5. The internal organization of an indeterminate nodule from *Medicago sativa,* showing a white uninfected meristematic area on the left, with progression through stages of infection by *S. meliloti,* active N_2 fixation, and leghemoglobin production, to the beginnings of nodule senescence on the right. Photo courtesy of C. P. Vance, with permission.

points of structure and function in each nodule type. More rarely, collar nodules, as found in lupin, can completely encircle the stem.

B. Nodule Number and Size

The number of nodules per plant is affected by the host species and rhizobial strain, by the number and efficiency in nodulation of the inoculant rhizobia, by the presence of existing nodules, and by nutritional and soil stresses (see Section VI).

In classic studies with subterranean clover, Nutman (1967) noted that variety and strain affected both the time to nodule formation and number of nodules per plant, with the host effects usually much greater than for the strain. In crosses between lines differing in number of nodules per plant, abundant nodulation was dominant over sparse nodulation, but abundantly nodulated lines bore a higher proportion of their nodules on lateral roots.

Subsequent studies have shown a definite time limit for the infection of immature, initially susceptible root hairs. For many species, inoculation delays can mean that such root hairs mature and are not infected, and that when nodules are produced, they form further down the taproot or on lateral roots.

Autoregulation is evident in the predominantly crown nodulation of well-inoculated plants and in the irregular distribution of nodules along the root. Even young nodules that have yet to fix N_2 can limit further nodulation in the adjacent root area. Thus, when the same root-tip marked plants were inoculated twice at intervals of 15 hr, each inoculation gave rise to similar numbers of new infections, but few of the infections initiated following the second inoculation developed into nodules. A very high rate of inoculation can also depress nodulation, though the basis for this response is not known. Supernodulating plants with single-gene mutations, such as have been described for soybean and pea, no longer regulate nodulation and give rise to many nodules, often at the expense of plant growth. Autoregulation depends in some degree on the N status of the plant. Thus, a plant with many small ineffective nodules may continue to nodulate, whereas one with fewer effective nodulates will not. For this reason, the number of nodules number per plant is only likely to have meaning under conditions that affect inoculant numbers or interfere with nodulation per se.

Mechanisms involved in the regulation of nodule formation have yet to be clearly defined, but involve both shoot and root factors. A role for ethylene is also possible. Legumes evolve ethylene for up to 6 days after infection, and ethylene can inhibit nodule formation. Conversely aminoethoxyvinyl glycine (AVG), an inhibitor of ethylene biosynthesis, stimulates nodulation in some plants. A hypernodulating pea mutant that is resistant to the effects of ethylene and soybean cultivars that nodulate normally following repeated treatment with ethylene have been described.

V. SPECIFICITY IN NODULATION

Given the complex signaling involved, specificity in nodulation is to be expected. Many caesalpinioid legumes never form nodules; individual accessions of normally nodulating plants may not nodulate with specific rhizobia. An example is the pea cultivar Afghanistan, which nodulates normally with strains of *R. leguminosarum bv viciae* from the Middle East, but which carries a single recessive gene preventing nodulation with most European pea rhizobia. A parallel gene overcoming resistance to nodulation has been identified in *R. leguminosarum* strain TOM. Similarly, improved soybean varieties developed in the United States do not nodulate with indigenous bradyrhizobia from African soils, though unimproved soybean lines from Malaysia and Indonesia can.

Each rhizobium has the ability to nodulate some, but not all legumes. Cross-inoculation groups that contain legumes nodulated by the same bacteria can be further subdivided because some host–rhizobial combinations form nodules but are ineffective in N_2 fixation, as shown in Table II. As a consequence, legumes introduced into new areas of production are unlikely to find suitable indigenous rhizobia and will need inoculation to ensure good nodulation and N_2 fixation (see Fig. 6). More than 100 inoculants are needed to satisfy the requirements of the major crop, tree, and pasture legumes.

Fig. 6. Response to inoculation of soybean in newly cultivated areas of Puerto Rico. Photo courtesy of R. Stewart Smith, with permission.

TABLE II
Cross-Inoculation Groups and Effectiveness Subgroups

Cross-inoculation group	Effectiveness subgroup[a,b]	Common leguminous species
Bean		*Phaseolus vulgaris, P. coccineus*
Chickpea		*Cicer arietinum*
Clover	A	*Trifolium alexandrinum, T. incarnatum, T. resupinatum*
	B	*Trifolium canescans, T. dubium, T. hybridum, T. pratense, T. procumbens, T. repens*
	C	*T. affine, T. vesiculosum*
	D	*T. cherleri, T. hirtum, T. subterraneum*
	Specific	*T. ambiguum, T. fragiferum, T. glomeratum, T. rueppelianum, T. semipilosum, T. tembense, T. usambarense*
Cowpea	A	*Acacia albida, Indigofera hirsuta, Lespedeza cuneata, L. stipulacea, L. striata, Mucuna deeringiana, Phaseolus lunatus, Vigna angularis, V. mungo, V. radiata, V. unguiculata*
	B	*Arachis hypogaea, A. pintoi*
	C	*Cajanus cajan, Lablab purpureus, Macrotyloma uniflorum, M. axillare*
	D	*Macroptilium purpureum, M. lathyroides, Pueraria phaseoloides, Calopogonium caeruleum*
	E	*Centrosema spp.*
	F	*Desmodium spp*
	G	*Stylosanthes guyanensis, S. humilis*
	Specific	*Calopogonium caeruleum, Stylosanthes hamata*
Crownvetch	A	*Coronilla varia*
	B	*Onobrychis viciaefolia*
	C	*Dalea purpurea, D. candida*
Lotus	A	*Anthyllis vulneraria, Biserrula pelecinus, Lotus corniculatus, L. pedunculatus, L. tenuis*
	Specific	*Lotus uliginosus*
Lupin	A	*Lupinus spp.*
	B	*Ornithopus spp.*
Medic	A	*Medicago sativa, M. falcata, Melilotus alba, M. indica, M. officinalis, M. sulcata*
	B	*Medicago arabica, M. hispida, M. lupulina, M. minima, M. orbicularis, M. polymorpha, M. rigidula, M. rotata, M. rugosa, M. scutellata, M. truncatula, M. tuberculata, Trigonella foenum-graecum*
	Specific	*Medicago arborea, M. laciniata*
Pea	A	*Lathyrus hirsutus, Pisum sativum*
	C	*Lens esculenta Vicia faba, V. monantha, V. pannonica, V. villosa*
	Specific	*Lathyrus sativus, Vicia ervilla, V. sativa*
Soybean		*Glycine max, G. soja*
Others not grouped		*Astragalus, Hedysarum, Lotononis, Sesbania*

[a] Letters used to distinguish groups of legumes refer only to this table, not to inoculants made by specific manufacturers.

[b] Effectiveness subgroups determine the range of different inoculants needed for legumes. However, the recommendation of individual manufacturers may vary from the divisions shown. For example, some inoculant manufacturers provide different inoculants for peas/vetch and faba beans/lentils, while others use a single inoculant for all four species.

VI. ENVIRONMENTAL AND SOIL FACTORS AFFECTING NODULATION

A. Soil-Fertility Status, Including Level of Soil N

Well-nourished plants nodulate and fix N_2 better than those that are nutrient-limited, but several elements including P, Fe, and Ca have specific functions in nodulation. Combined forms of N can also influence nodulation.

The high P requirement of nodulated legumes reflects the energy cost of nodule formation and maintenance. Nodules are an important P sink, and plants that are deficient in this element will usually be poorly nodulated and have a low nodule mass. P also is important in attachment and signal transduction. The three-way symbiosis between legumes, mycorrhiza, and rhizobia can be critical to the N and P supply to the plant, with nodulation and N_2 fixation markedly improved by mycorrhizal colonization.

Iron is critical for symbiosis. It forms part of the active site of the nitrogenase enzymes and leghemoglobin, occurs at high levels in legume root hairs, and has an unspecified function in nodulation. Plants deficient in Fe produce nodule initials, but these may not develop until they receive additional fertilization with Fe. Both host plant and rhizobia can differ in iron-use efficiency, making for significant interaction under conditions of low Fe concentrations in soil.

Calcium chelators such as EGTA ([ethylene-bis(oxyethylenenitrilo)]tetra acetic acid) can inhibit the nodulation of alfalfa at concentrations as low as 0.4 mM, with the Ca requirement for nodulation of alfalfa at pH 4.8, six times greater than at pH 5.6. Because low Ca concentration limits attachment, the Ca requirement is often assumed to be for binding between rhicadhesin and rhizobia. However, Ca flux in root hairs during infection, high Ca levels in the infection thread, and Ca function in outer-membrane stability in *Rhizobium* must also be considered as contributing to the need for Ca in nodulation.

Combined N is needed for early plant growth, and small "starter" doses of N fertilizer will often enhance early nodulation and N_2 fixation. However in cases in which soil N and fertilizer N application exceeds 50 kg ha^{-1}, progressive inhibition of nodule formation can be expected. Suggested mechanisms include modified carbohydrate distribution to the root, inactivation of leghemoglobin, and degradation of hormones involved in infection.

B. Environmental Factors

Environmental factors including pH, temperature, and water availability can limit rhizobial survival and number in the soil, affect plant growth and development, and impact the nodulation process per se.

Soils with a pH less than 4.8 occupy more than 800×10^6 ha in Latin America alone, and are common throughout Africa, Asia, and the eastern United States. Their use in agriculture is often beset by problems of Al or Mn toxicity, or micronutrient deficiency. The reduced survival of rhizobia in acid soils is common. One study reported an average of 87,000 cells g^{-1} of the acid-sensitive *S. meliloti* in soils of near neutral pH; and only 37 cells g^{-1} in soils of less than pH 6.0. Similar data exist for *R. leguminosarum* and *R. etli*. Acid-sensitive steps in nodulation have also been identified, and are related to attachment and root-hair curling. Approaches used to resolve this problem include lime-pelleting seed and inoculant to provide microenvironments of higher pH that protect rhizobia and nodulation sites until after infection, and the identification of host cultivars and rhizobia more tolerant to acid soil conditions. Acid-tolerant germplasm has been important in extending soybean production in the cerrado of Brazil, and in the improved growth of annual medic species in acid-soil areas of Australia.

Rhizobia are mesophilic organisms, with an optimum temperature for growth and nodulation in the range of 20–30°C. Exceptions include rhizobia associated with arctic or high-altitude plant species, or from high-temperature environments such as the hot dry Sahel savannah of Africa. Rhizobia from *Gliricidia*, *Lonchocarpus*, and *Leucaena* that are able to nodulate and fix significant N_2 when exposed to 40°C for 8 hr day^{-1} have also been identified. High temperatures during inoculant shipment and storage can lead to loss of the symbiotic plasmid and to a reduction in the number of cells, limiting subsequent nodulation.

See Also the Following Articles

NITROGEN FIXATION • PLANT DISEASE RESISTANCE • SOIL MICROBIOLOGY

Bibliography

Bhuvaneswari, T. V., Bhagwhat, A. A., and Bauer, W. D. (1981). Transient susceptibility of root cells in four legumes to nodulation by rhizobia. *Plant Physiol.* **68,** 1144–1149.

Boivin, C., Ndoye, I., Molouba, F., Lajudie, P., Dupuy, N., and Dreyfus, B. (1997). Stem nodulation in legumes: Diversity, mechanisms, and unusual characteristics. *Crit. Rev. Plant Sci.* **16,** 1–30.

Boogerd, F. C., and van Rossum, D. (1997). Nodulation of groundnut by *Bradyrhizobium*—a simple infection process by crack entry. *FEMS Microbiol. Rev.* **21,** 5–27.

Bryan, J. A., Berlyn, G. P., and Gordon, J. C. (1996). Toward a new concept of the evolution of symbiotic nitrogen fixation in the leguminosae. *Plant Soil* **186,** 151–159.

Doyle, J. J., and Doyle, J. L. (1997). Phylogenetic perspectives on the origins and evolution of nodulation in the legumes and allies. *In* "Biological Fixation of Nitrogen for Ecology and Sustainable Agriculture" (A. Legocki *et al.,* eds.), Springer, Berlin. pp. 307–312.

Fahraeus, G. (1957). The infection of clover root hairs by nodule bacteria studied by a simple glass slide technique. *J. Gen. Microbiol.* **16,** 374–381.

Hirsch, A. M. (1992). Developmental biology of legume nodulation. *New Phytol.* **122,** 211–237.

John, M., Rohrig, H., Schmidt, J., Walden, R., and Schell, J. (1997). Cell signalling by oligosaccharides. *Trends Plant Sci.* **2,** 111–115.

Long, S. R. (1996). *Rhizobium* symbiosis: Nod factors in perspective. *Plant Cell* **8,** 1885–1898.

Miller, D. D., de Deruijter, N. C. A., and Emons, A. M. C. (1997). From signal to form—aspects of the cytoskeleton plasma membrane cell wall continuum in root hair tips. *J. Exp. Bot.* **48,** 1881–1896.

Nutman, P. S. (1967). Varietal differences in the nodulation of subterranean clover. *Aust. J. Agric. Res.* **18,** 381–425.

Pawlowski, K. (1997). Nodule-specific gene expression. *Physiol. Plant.* **99,** 617–631.

Rhizobium Web site: *http://www.rhizobium.umn.edu*

Sahlman, K., and Fahraeus, G. (1963). An electron microscope study of root-hair infection by rhizobium. *J. Gen. Microbiol.* **33,** 425–427.

Schultze, M., and Kondorosi, A. (1996). The role of lipochitooligosaccharides in root nodule organogenesis and plant cell growth. *Curr. Opin. Genet. Dev.* **6,** 631–638.

Tate, R., Patriaca, E. J., Riccio, A., Defez, R., and Iaccarino, M. (1994). Development of *Phaseolus vulgaris* root nodules. *MPMI* **7,** 582–589.

Nucleotide Metabolism

Per Nygaard

University of Copenhagen

Hans Henrik Saxild

Technical University of Denmark

GLOSSARY

analogs Naturally occurring or synthetic compounds with structural similarity to natural nucleobases or nucleosides.

nucleosides Substances composed of a heterocyclic nitrogenous base (a nucleobase) joined by an *N*-glycosidic bond to ribose (ribonucleosides) or 2'-deoxyribose (deoxyribonucleosides).

nucleotides Substances composed of a nucleoside with one or more phosphoroyl groups joined by phosphodiester bonds to the pentose moiety.

nucleotide biosynthesis The process that creates nucleotides *de novo* from small precursor molecules such as carbon dioxide, formate, amino acids and 5-phosphoribosyl-α-1-pyrophosphate (PRPP), or from PRPP and bases and from nucleosides through the salvage pathways.

NUCLEOTIDES IN CELLS are the building blocks of RNA and DNA. Free nucleotides are involved in all major aspects of metabolism and the importance of this is reflected in the careful regulation of their intracellular levels. ATP is the major substance used for the transfer of energy from energy-yielding reactions to energy-requiring processes, and GTP is of special importance for protein synthesis. Among other nucleotides are activated intermediates, such as nucleoside diphosphate sugar, involved in the synthesis of carbohydrates and lipids, and they are components of major coenzymes and of histidine. Nucleotides act as metabolic signal molecules, either as nucleoside 5'-phosphates or as modified nucleotides such as 3',5'-cyclic AMP and guanosine tetraphosphate (ppGpp). Nucleotides are synthesized *de novo* from small precursor molecules and from preformed bases and nucleosides by salvage pathways. The focus here is on the nucleotide metabolism of *Bacillus subtilis*. This gram-positive bacterium shows properties that resemble well-studied bacteria such as *Escherichia coli* and lower eukaryotes.

I. NUCLEOTIDES, AN OVERVIEW

A. Occurrence

The natural habitat of microorganisms contains highly variable levels of nucleic acid components arising as degradation products from decaying organisms or as products excreted by other microorganisms. To use the available degradation products, nucleotides, nucleosides, and nucleobases (see Fig. 1). Most microorganisms have developed transport and enzyme systems for their metabolism. Intracellularly, nucleosides and bases are continuously formed from the turnover of nucleic acids, nucleotides, and other metabolites.

B. Physiological Aspects

1. Enzymes and Regulation of Enzyme Activity

The enzymes catalyzing the metabolism of bases and nucleosides serve two major functions; this first is anabolic and ensures the reuse (salvaging) of nu-

Nucleobases

Purines

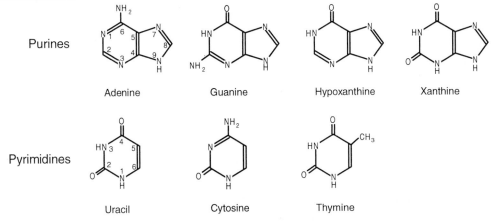

Adenine   Guanine   Hypoxanthine   Xanthine

Pyrimidines

Uracil   Cytosine   Thymine

Pentoses

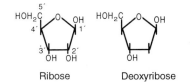

Ribose   Deoxyribose

Nucleosides

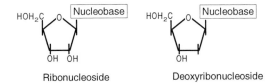

Ribonucleoside   Deoxyribonucleoside

Nucleotides

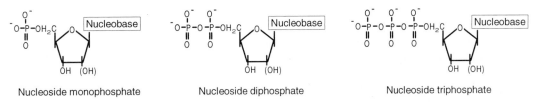

Nucleoside monophosphate   Nucleoside diphosphate   Nucleoside triphosphate

Fig. 1. Structure of purine and pyrimidine bases, pentoses, nucleosides, and nucleotides.

cleosides and bases for nucleotide synthesis, and the second is catabolic and makes bases and nucleosides available as nitrogen and carbon sources. A number of microorganisms can degrade purine and pyrimidine compounds to ammonia and carbon dioxide, whereas others can only partially degrade these compounds. The activity of the anabolic enzymes often are subjected to feedback control, and the synthesis of the catabolic enzymes is induced when bases and nucleosides are available. When purine and pyrimidine compounds are present in the growth medium, the contribution from the biosynthetic pathway of nucleotide synthesis *de novo* is greatly reduced. This suppression is accomplished by increased feedback inhibition of key enzymes of the pathway and by repression of the synthesis of the enzymes catalyzing the reactions of the synthesis *de novo* pathway. Some microorganisms depend on an exogenous purine or pyrimidine source for growth. All of the salvage and biosynthetic enzymes and some of the catabolic enzymes have been purified and characterized in detail from several microorganisms, and their genes have in many cases been cloned and the nucleotide sequences determined.

2. *Starvation*

In microorganisms starved for nutrients, the metabolism of purines and pyrimidines is changed. A substantial degradation of RNA may occur and this may provide a basis for the survival of the organism. Some microorganisms not only enter a resting stage, they start to differentiate and form spores. Differentiation in gram-positive bacteria and fungi seems to be initiated under conditions that result in the lowering of the level of GTP. At an early step in sporulation synthesis of nucleotides depends on the turnover of breakdown products of preexisting nucleic acids. In some bacteria, it has been observed that the enzymes of the purine and pyrimidine biosynthetic pathways are degraded. Very few if any nucleotides are synthesized in the dormant spore. In the germinating spore, no net nucleotide synthesis occurs at the early germination stage, simply because the biosynthetic enzymes are not present. Nucleosides and bases from the surroundings or from an endogenous storage pool may provide a source of nucleotides and may stimulate spore germination.

II. BIOSYNTHESIS *DE NOVO* OF NUCLEOTIDES

The first complete purine nucleotide formed is inosine monophosphate (IMP). IMP is then converted to adenosine monophosphate (AMP) and guanosine monophosphate (GMP) (Fig. 2). These pathways are also important for the use of purine bases and nucleosides. In the pyrimidine biosynthetic pathway, uridine monophosphate (UMP) is the precursor for all pyrimidine nucleotides (Fig. 3). Whether synthesized *de novo* or via the salvage pathways, AMP, GMP, UMP, and CMP are converted to the triphosphate level in two steps. The phosphorylation of ADP occurs through oxidative phosphorylation or substrate-level phosphorylation, whereas the other nucleoside diphosphates are phosphorylated by the nonspecific nucleoside diphosphokinase. The synthesis of deoxyribonucleotides is catalyzed by a single enzyme, which converts ribonucleoside diphosphates or triphosphates to their corresponding 2'-deoxyribonucleotide. The synthesis of thymidine triphosphate (dTTP) proceeds via deoxyuridine monophosphate (dUMP) Fig. 4.

A. Biosynthesis *De Novo* of Purine Nucleotides

1. *Synthesis of AMP and GMP*

The purine biosynthetic pathway consists of the enzyme-catalyzed reactions shown in Fig. 2. Basically, the same pathway is found in all organisms capable of purine biosynthesis *de novo*. The initial substrate is the highly phosphorylated sugar PRPP. An amino group is transferred from glutamine to PRPP, catalyzed by the enzyme glutamine PRPP amidotransferase. After that, it takes four reactions to complete the imidazole ring and an additional five reactions to synthesize the pyrimidine moiety and thereby to complete the purine molecule. In some bacteria (e.g., *B. subtilis* and *E. coli*), the third step is also catalyzed by a second enzyme (*purT*, see Fig. 2) that uses formate instead of a tetrahydrofolic acid derivative as formyl donor. The sixth step has been found to be a two-step reaction, which in some organisms is catalyzed by a single enzyme. The end product of the purine biosynthetic *de novo* pathway is inosine monophosphate (IMP). After the formation of IMP, purine synthesis splits into two separate pathways, leading to the synthesis of AMP and GMP.

2. *Regulation*

IMP synthesis is controlled both enzymatically and genetically. The control point for feedback inhibition by purine nucleotides and activation by PRPP is on the activity of the first enzyme of the pathway glutamine PRPP amidotransferase. The branching from IMP (Fig. 2) is regulated by ATP and GTP to ensure a proper balance between ATP and GTP. The regulation of the level of purine biosynthetic enzymes is controlled by the expression of the purine genes (*pur* genes, which encode the individual enzymes of the biosynthetic pathway). In *B. subtilis* (gram-positive) and *E. coli* (gram-negative), *pur* gene expression is controlled by specific repressor proteins that use PRPP or purine bases as low-molecular-weight effector molecules.

3. *Gene Organization*

The availability of the genome sequence of several bacterial species and of yeast allows us to obtain an overview of the *pur* gene organization in a diverse group of microorganisms. Table I lists the data from

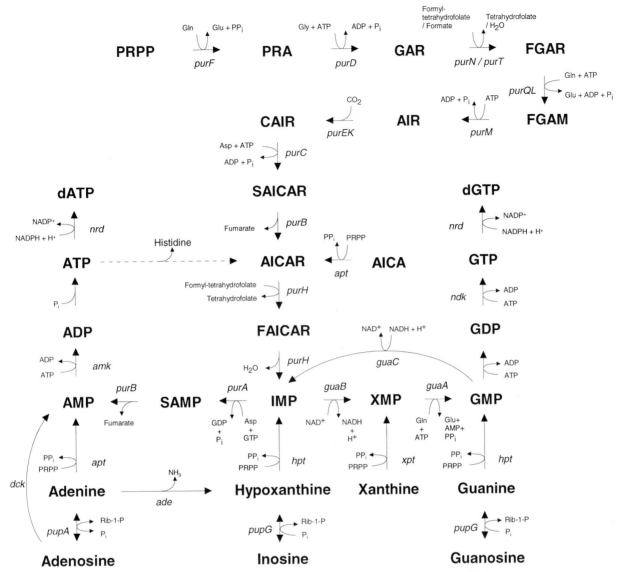

Fig. 2. Purine biosynthesis, salvage, and interconversion in *B. subtilis*. (The chemical structures of the various compounds can be looked up in any textbook of biochemistry). *ade,* adenine deaminase; AICA, aminoimidazolecarboxamide; AICAR, phosphoribosyl-aminoimidazole-carboxamide; AIR, phosphoribosylaminoimidazole; ADP, adenosine diphosphate; AMP, adenosine monophosphate; *apt,* adenine phosphoribosyltransferase; Asp, aspartic acid; ATP, adenosine triphosphate; CAIR, phosphoribosyl-carboxyaminoimidazole; dATP, deoxyadenosine triphosphate; *dck,* deoxycytidine-adenosine kinase; dGTP, deoxyguanosine triphosphate; FAICAR, phosphoribosyl-formamidoimidazole-carboxamide; FGAM, phosphoribosyl-formylglycinamidine, FGAR, phosphoribosyl-formylglycinamide; GAR, phosphoribosyl-glycinamide; GDP, guanosine diphosphate; Gln, glutamine; Glu, glutamate; Gly, glycine; GMP, guanosine monophosphate; GTP, guanosine triphosphate; *guaA,* GMP synthetase; *guaB,* IMP dehydrogenase; *guaC,* GMP reductase; *hpt,* hypoxanthine-guanine phosphoribosyltransferase; IMP, inosine monophosphate; *nrd,* nucleoside triphosphate reductase; PPi, pyrophosphate; PRA, phosphoribosylamine; PRPP, 5-phosphoribosyl-α-1-pyrophosphate; *pupA,* adenosine/deoxyadenosine phosphorylase; *pupG,* guanosine/deoxyguanosine-inosine/deoxyinosine phosphorylase; *purA,* adenylosuccinate synthetase; *purB,* adenylosuccinate lyase; *purC,* phosphoribosyl-aminoimidazole-succinocarboxamide synthetase; *purD,* phosphoribosyl-glycinamide synthetase; *purEK,* phosphoribosyl-aminoimidazole carboxylase I and II; *purF,* glutamine phosphoribosyl-pyrophosphate amidotransferase; *purH,* phosphoribosyl-aminoimidazole-carboxamide formyltransferase and IMP cyclohydrolase; *purM,* phosphoribosyl-aminoimidazole synthetase; *purN,* formyltetrahydrofolate-dependent phosphoribosyl-glycinamide transformylases; *purQL,* phosphoribosyl-formylglycinamidine synthetase I and II; *purT,* formate-dependent phosphoribosyl-glycinamide transformylase; Rib-1-P, ribose-1-phosphate; SAICAR, phosphoribosyl-aminoimidazole-succinocarboxamide; SAMP, adenylosuccinate; XMP, xanthosine monophosphate; *xpt,* xanthine phosphoribosyltransferase. Dashed lines indicate multiple enzyme-catalyzed steps.

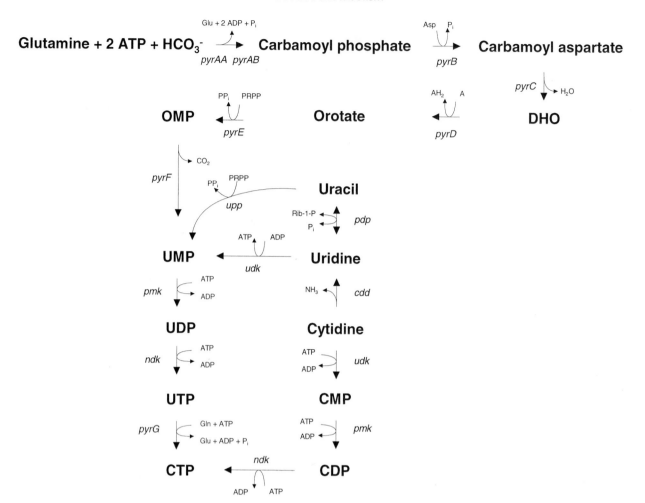

Fig. 3. Pyrimidine biosynthesis and salvage in *B. subtilis*. (The chemical structure of the various compounds can be looked up in any textbook of biochemistry.) A, oxidized electron acceptor; AH2, reduced electron acceptor; *cdd*, cytidine/deoxycytidine deaminase; CDP, cytidine diphosphate; CMP cytidine monophosphate; CTP, cytidine triphosphate; DHO, dihydroorotic acid; *ndk*, ribonucleotide diphosphate kinase; OMP, orotidine monophosphate; Pi, inorganic phosphate; *pdp*, pyrimidine nucleoside phosphorylase; *pmk*, pyrimidine ribonucleoside monophosphate kinase; *pyrAA* and *pyrAB*, carbamoyl phosphate synthetase subunit I and II; *pyrB*, aspartate carbamoyltransferase; *pyrC*, dihydroorotase; *pyrD*, dihydroorotate dehydrogenase; *pyrE*, orotate phosphoribosyltransferase; *pyrF*, orotidine-5′-phosphate decarboxylase; *pyrG*, CTP synthetase; *udk*, uridine-cytidine kinase; UDP, uridine diphosphate; UMP, uridine monophosphate; *upp*, uracil phosphoribosyltransferase; UTP, uridine triphosphate. For other abbreviations, see Fig. 2.

13 species and except for *B. subtilis*, which has all the genes involved in the synthesis of IMP organized in one large operon, *pur* genes in general are organized as single genes or in small operons. Genes encoding the enzymes involved in purine biosynthesis *de novo* reside on the chromosome. However, in *Borrelia* the *guaA* and *guaB* genes have been located on a plasmid.

B. Biosynthesis *De Novo* of Pyrimidine Nucleotides

1. Synthesis of UMP and CTP

UMP biosynthesis *de novo* consists of a six-step enzyme-catalyzed pathway common for all organisms capable of pyrimidine biosynthesis (see Fig. 3). The enzyme carbamoyl phosphate synthetase combines an amino group from glutamine with carbonate and ATP to form carbamoyl phosphate. Three additional steps are required for the final formation of the pyrimidine ring. The phosphoribosyl moiety is donated by PRPP to form orotate monophosphate, which is subsequently decarboxylated to UMP. Some bacteria contain two different dihydroorotate dehydrogenase enzymes. One is used for biosynthesis and the other is believed to function in the oxidative degradation of orotate. In bacteria the biosynthetic steps are catalyzed by monofunctional enzymes, whereas in yeast the formation of carbamoyl aspartate, which includes the first two reactions, is catalyzed by a bifunctional enzyme encoded by the URA2 gene. UMP is phosphorylated to UTP in two steps. CTP is formed by CTP synthetase by transferring an amino group from glutamine to UTP (Fig. 3).

2. Regulation

Pyrimidine biosynthesis is regulated by feedback inhibition of the first enzyme, carbamoyl phosphate synthetase, by the end products of the pathway, UMP, UDP, and UTP. Furthermore, carbamoyl phosphate synthetase is allosterically activated by PRPP and GTP. Because carbamoyl phosphate is also the substrate for arginine biosynthesis, some bacteria (e.g., *B. subtilis*) contain two carbamoyl phosphate synthetases. One is inhibited by arginine and the other by pyrimidine nucleotides. In bacteria with only one carbamoyl phosphate synthetase (e.g., *E. coli*) the enzyme activity is inhibited by pyrimidine nucleotides and stimulated by intermediates of the arginine biosynthetic pathway. In *E. coli*, in which aspartate carbamoyltransferase is the first enzyme specific for pyrimidine synthesis, this enzyme is in-

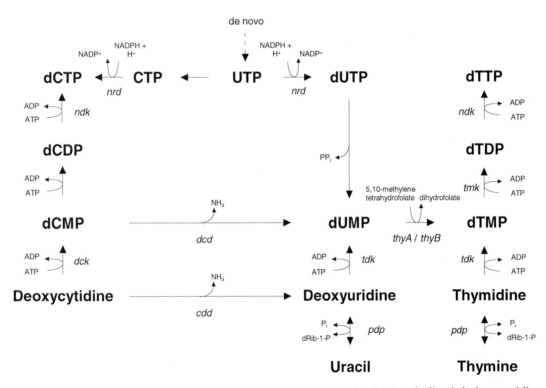

Fig. 4. Pyrimidine deoxyribonucleotide synthesis and interconversion in *B. subtilis*. *dcd*, deoxycytidine monophosphate deaminase; dCDP, deoxycytidine diphosphate; dCMP, deoxycytidine monophosphate; *dck*, deoxycytidine kinase; dCTP, deoxycytidine triphosphate; dTDP, thymidine diphosphate; dTMP, thymidine monophosphate; dTTP, thymidine triphosphate; dUTP, deoxyuridine triphosphate; dUMP, deoxyuridine monophosphate; *tdk*, thymidine kinase. For other symbols and abbreviations, see Figs. 2–3.

TABLE I

Organization of Purine and Pyrimidine Biosynthetic Genes in Genome Sequenced Microorganisms

	Gene organization[a]	
Organism	*Purine biosynthetic genes*	*Pyrimidine biosynthetic genes*
Eukarya		
Saccharomyces cerevisiae	Single genes on seven chromosomes	Single genes on six chromosomes
Archaea		
Methanococcus jannaschii	One small operon and single genes	Single genes
Methanobacterium thermoautotrophicum	Two small operons and single genes	Single genes
Bacteria		
Bacillus subtilis	One large operon and single genes	One large operon
Escherichia coli	Three small operons and single genes	One small operon and single genes
Synechocystis	Single genes	Single genes
Archaeoglobus fulgidus	Two small operons and single genes	Two small operons and single genes
Treponema pallidum	One gene identified	One gene identified
Borrelia burgdorferi	Two genes identified	One gene identified
Haemophilus influenzae Rd	Four small operons and single genes	Single genes
Helicobacter pylori	Single genes	Single genes
Mycobacterium tuberculosis	Three small operons and single genes	Two operons and single genes
Aquifex aeolicus	Single genes	Single genes

[a] Small operons consist of two or three genes.

hibited by CTP. Control of CTP synthesis is mediated by the inhibition of CTP synthetase (Fig. 3) by its product CTP and by activation by GTP. Also, the expression of the genes encoding the pyrimidine biosynthetic enzymes is regulated on the transcriptional level. In *B. subtilis* the control of *pyr* gene expression is mediated by a repressor protein and UMP, whereas in *E. coli* the expression of the genes is regulated in a noncoordinated manner by multiple control mechanisms, with some genes responding to fluctuations in the nucleoside triphosphate pools.

3. Gene Organization

Table I lists the genomic organization of the pyrimidine biosynthetic genes (*pyr* genes) of 13 microorganisms. It appears that the *pyr* genes tend to be distributed as single genes, except for *B. subtilis*, in which all *pyr* genes are organized in one large operon, and *Lactococcus lactis*, in which the genes are organized in small operons.

III. BIOSYNTHESIS OF DEOXYRIBONUCLEOTIDES

A. Ribonucleotide Reductase

Three of the DNA precursors, dATP, dGTP, and dCTP, are derived directly from their corresponding ribonucleotides. A single enzyme catalyzes the reduction of all four nucleoside triphosphates or diphosphates (Fig. 4). The substrate specificity is determined by allosteric effectors, which are the products of the enzymatic reaction and ATP. Some microorganisms may contain more than one enzyme, with different cofactor requirements. Typically, one type of enzyme is synthesized under aerobic growth conditions whereas another is synthesized under anaerobic conditions.

B. Synthesis of dTTP

1. Synthesis De Novo

The synthesis of dTTP proceeds via the intermediate formation of dUMP (see Fig. 4). Thymidylate synthase catalyzes the synthesis of dTMP from dUMP, which may be generated through phosphorolysis of dUTP formed from UTP or from the deamination of dCTP. An alternative route involves the deamination of dCMP. Some organisms such as *B. subtilis* and *Lactobacillus acidophilus* contain dCMP deaminase activity but not dCTP deaminase activity, whereas the reverse situation is found in *E. coli* and *S. typhimurium*.

2. Synthesis from Thymine or Thymidine

To be salvaged, thymine must be converted to thymidine (see Fig. 4). The reaction requires deoxyribose-1-phosphate, which normally is not present in the cell. Thymidine is converted to dTMP by thymidine kinase. This enzyme is abundant, and its level of synthesis is often increased when cells synthesize DNA. However, certain yeast types and *Neurospora crassa* do not contain this enzyme. Instead the *N. crassa* converts thymidine to uridine.

C. Alternative Routes for the Synthesis of Deoxyribonucleotides

Some microorganisms possess a kinase that phosphorylates, deoxyadenosine, deoxyguanosine, and deoxycytidine to their corresponding deoxyribonucleotide. Two enzymes are present in *B. subtilis* and in other gram-positive bacteria, whereas *E. coli* do not possess these activities. Nucleoside deoxyribosyltransferases are found in *Lactobacilli* and *Streptococci* and provide a route by which all four deoxyribonucleosides can be synthesized from a single deoxyribonucleoside and the appropriate base, that is, thymine + deoxyadenosine ⇌ adenine + thymidine. Infection of microorganisms with certain phages leads to the degradation of one of the normal deoxyribonucleotides. In *B. subtilis*, some bacteriophages contain uracil or 5-hydroxymethyluracil instead of thymine in their DNA, whereas some *E. coli* bacteriophages contain no cytosine but instead have hydroxymethylcytosine. Infection with these phages results in the synthesis of new enzymes that affect the normal deoxyribonucleotide content, resulting in the synthesis of the unusual deoxyribonucleotide.

IV. PURINE AND PYRIMIDINE SALVAGE

A. Purines and Pyrimidines from the Environment

Several microorganisms are known to require a purine or a pyrimidine compound for growth. Examples are *Lactobacilli* species, which require a deoxyribonucleoside or thymine and a purine or pyrimidine compound; *Mycoplasma* and *Acholesplasma* species, which have specific requirement for guanine, uracil, and thymine; and the free-living protozoa *Tetrahymena*, which requires both exogenous purine and pyrimidine bases for growth. Protozoan parasites such as *Trypanosoma*, *Leishmania*, *Chrithidia*, and *Plasmodia*, requires only a purine source from the host, whereas *Giardia* relies on the salvage of purines, pyrimidines, and thymidine. Certain intracellular parasites, such as *Chlamydia* and *Rickettsia*, requires all four ribonucleoside triphosphates from the host cell. For other microorganisms, the use of exogenous nucleotides requires a dephosphorylation catalyzed by nucleotidases located on the surface of the organism. The pathways by which purine and pyrimidine bases and nucleosides are metabolized in *B. subtilis* are shown in Figs. 2–3. The uptake is mediated by transport systems, either by facilitated diffusion or by an energy-requiring process. The specificity of the transport systems range from those that react specifically with a single compound to systems with broader specificities, which transport both purines and pyrimidines. In most situations, the uptake of bases is strictly coupled to the intracellular formation of nucleotide from the transported base. Different species, however, show significant variations in their abilities to metabolize purines and pyrimidines. Few reports have dealt with purine and pyrimidine salvage in archaea. A preliminary conclusion is that archaea possess the same pathway as do bacteria; however they have fewer salvage enzymes than *B. subtilis*. The use of radiolabeled bases and nucleosides for *in vivo* studies of nucleotide and nucleic acid synthesis has helped to establish the salvage and interconversion pathways.

B. Purine Salvage

1. Bases

The four naturally occurring purine bases, adenine, guanine, hypoxanthine, xanthine, and AICA can be salvaged and serve as the sole purine source in many microorganisms. A single enzyme catalyzes the phosphoribosylation of adenine to AMP. Alternatively, adenine is converted to IMP in several steps (Fig. 2). Mutants defective in the *purA* or *purB* gene cannot synthesize AMP and require adenine for growth. One or two enzymes are involved in the phosphoribosylation of hypoxanthine, guanine, and xanthine (Fig. 2). Hypoxanthine is converted to IMP and then to AMP and GMP, whereas guanine and xanthine (via XMP) are converted to GMP. A *guaA* mutant requires guanine for growth, whereas a *guaB* mutant can grow on either xanthine or guanine.

2. Ribonucleosides

Purine ribonucleosides are converted to their corresponding ribonucleotide in two steps in *B. subtilis*. The first is the phosphorolytical cleavage to the free base and ribose-1-phosphate, and the second is the phosphoribosylation to the nucleoside monophosphate. *Lactobacilli* species contain a ribonucleoside hydrolase that hydrolyzes ribonucleosides to the base and ribose. In addition, some microorganisms contain nucleoside kinases that catalyze the direct phosphorylation of the nucleoside to the nucleoside monophosphate. The uptake of nucleosides in some organisms parallels their incorporation into nucleotides. However, in *E. coli* and some other bacteria, the catabolism of the pentose moiety exceeds that of the purine moiety which is excreted and can be used

by other microorganisms. Rapid catabolism is a result of induced synthesis of enzymes and transporter proteins of nucleoside catabolism.

3. Interconversion Reactions

Adenine and guanine compounds are interconvertible through the common precursor IMP. This conversion cannot take place by the reversal of the biosynthetic reactions (Fig. 2). When adenine compounds serve as a purine source, the branching from IMP to AMP becomes dispensable, and when guanine compounds serve as purine source, the branching from IMP to GMP is no longer required. Adenine compounds can be converted to GMP by two routes in *B. subtilis* and in yeast, either via the histidine biosynthetic pathway in which AICAR, an intermediate in the purine biosynthetic pathway, is formed from ATP as a by-product, or by a more direct route from adenine to hypoxanthine and then to IMP (Fig. 2). In other organism such as *B. cereus, E. coli*, and *S. typhimurium*, the conversion of adenine compounds proceeds via adenosine to inosine and then further on to hypoxanthine. An alternative route is the deamination of AMP to IMP, catalyzed by AMP deaminase, an enzyme that has only been identified in few microorganisms. A single enzyme is essential for the conversion of guanine compounds to AMP, namely GMP reductase, which catalyzes the reductive deamination of GMP to IMP. Several microorganisms do not possess this enzyme and as a result cannot use guanine compounds as a general purine source.

4. Toxicity

The inhibition of growth by naturally occurring purine bases and nucleosides can be exerted in a wide variety of microbial systems. The type of biochemical effects and apparent mechanisms of action associated with these biological effects also are quite diverse. An example is the addition of purine compounds to the growth medium of most microorganisms, which stops purine synthesis *de novo* and may inhibit growth. When purine synthesis is stopped, AIR (Fig. 2) is no longer synthesized, and this also stops thiamine synthesis and hence growth because AIR is a precursor of thiamine. Growth will be restored upon addition of thiamine.

C. Pyrimidine Salvage

1. Bases

A key enzyme in salvaging pyrimidine bases and nucleosides is uracil phosphoribosyltransferase, which seems to be present in all the microorganisms studied. Uracil is phosphoribosylated to UMP. The first step in the metabolism of cytosine is the deamination to uracil (Fig. 3). Many fungi and bacteria, including *B. subtilis* and several *Lactobacillus* species, *Lactococcus lactis,* and some coryneform bacteria, lack the ability to deaminate and therefore cannot metabolize cytosine.

2. Ribonucleosides

In *B. subtilis,* uridine is converted to UMP, either directly via phosphorylation or via a phosphorolytical cleavage to uracil. A single enzyme catalyzes the phosphorolysis of uridine, deoxyuridine, and thymidine in *B. subtilis* (Figs. 3, 4), whereas other organisms, such as *E. coli* and *Lactobacillus casei* possess a uridine and a thymidine phosphorylase. Certain *Lactobacilli* and yeasts can degrade uridine hydrolytically to uracil and ribose. Cytidine and deoxycytidine can be phosphorylated to CMP and dCMP, respectively. Neither cytidine nor deoxycytidine can be phosphorolytically cleaved. However, both compounds can be deaminated, to uridine and deoxyuridine, respectively. Pyrimidine nucleosides, as are purine nucleosides, are rapidly catabolized in some bacteria.

V. CATABOLISM OF NUCLEOTIDES, NUCLEOSIDES, AND BASES

Microorganisms, whether free living in the nature or in association with a host, are responsible for the major part of the turnover of degradation products of nucleic acids. Intracellularily there is constant breakdown of ribonucleotides, resulting in nucleosides and bases that can be salvaged. However, under stress conditions the nucleosides and bases are further degraded. Deoxyribonucleotides are not normally degraded inside bacterial cells.

A. Nucleosides

Ribonucleosides are degraded to the free base and ribose-1-phosphate from which ribose-5-phosphate is formed. This reaction is catalyzed by phosphopentomutase in *B. subtilis,* an enzyme encoded by the *drm* gene. Ribose-5-phosphate then enters pentose metabolism. The *drm* gene and the gene encoding guanosine phosphorylase are localized in an operon in both *B. subtilis* and in *E. coli* and are induced by ribonucleosides. Catabolism of deoxyribonucleosides results in the free base and deoxyribose-1-phosphate, which is converted to deoxyribose-5-phosphate by phosphopentomutase. Deoxyribose-5-phosphate is then cleaved by deoxyriboaldolase to acetaldehyde and glyceraldehyde-3-phosphate. The *pdp* gene, encoding pyrimidine nucleoside phosphorylase, and the *dra* gene, encoding deoxyriboaldolase, are located in *B. subtilis* in an operon that includes the *nupC* gene encoding a pyrimidine nucleoside transporter. This operon plus the *drm–pupG* operon is induced when deoxyribonucleosides are present in the growth medium. *E. coli* and *S. typhimurium* have also evolved a highly regulated set of operons and single genes that are involved in the transport and catabolism of ribonucleosides and deoxyribonucleosides.

B. Catabolism of Nucleobases

Many microorganisms can meet their demand for carbon and nitrogen by degrading purine and pyrimidine compounds. The synthesis of the degradative enzymes is induced only when substrates or intermediates of the pathways are present and when a good nitrogen source is not available. Some bacteria express the catabolic pathways constitutively and often require purines or pyrimidines as nitrogen and energy sources.

1. Purines

In order to enter the purine degradative pathway, adenine, hypoxanthine, and guanine must be converted to xanthine. Adenine is converted to hypoxanthine by adenine deaminase, hypoxanthine is oxidized to xanthine by xanthine dehydrogenase, and guanine is deaminated to xanthine by guanine deaminase. In the aerobic degradative pathway (Fig. 5), xanthine is oxidized in three steps to allantoic acid and allantoic acid can be cleaved to yield glyoxylate and urea by two alternative routes. Urea is degraded to ammonia and carbon dioxide by the action of

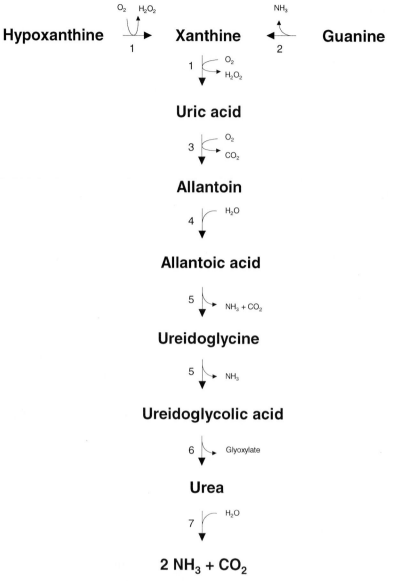

Fig. 5. Aerobic degradation of purine bases in *B. subtilis.* 1, xanthine dehydrogenase; 2, guanine deaminase; 3, uricase; 4, allantoinase; 5, allantoic acid amidohydrolase; 6, ureidoglycolase; 7, urease.

urease. Glyoxylate is either excreted or enters the tricarboxylic acid cycle. Certain bacteria such as *Bacillus fastidiosus,* specifically require allantoin or urate as nitrogen and carbon sources. In the anaerobic pathway, xanthine is degraded in five steps to formiminoglycine, which is further degraded to organic acids and glycine.

2. Pyrimidines

In some microorganisms, cytosine and orotate must be converted to uracil before they enter the pyrimidine catabolic pathways. In the reductive pathway, uracil is converted in three steps to β-alanine, ammonia, and carbon dioxide. Thymine is degraded to γ-aminoisobutyrate, ammonia, and carbon dioxide

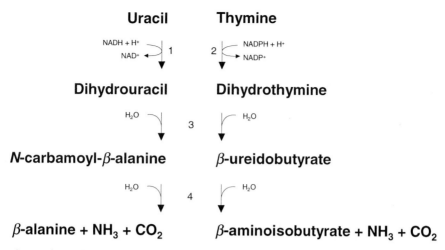

Fig. 6. The reductive degradation of pyrimidine bases in fungi. 1, dihydrouracil dehydrogenase (NAD^+); 2, dihydrouracil dehydrogenase ($NADP^+$); 3, dihydropyrimidinase; 4, ureidopropionase.

by the same pathway (Fig. 6). An oxidative pathway by which uracil is converted to barbituric acid and further on to urea also exists.

VI. SELECTION OF MUTANTS AND USE OF ANALOGS

A. Selection of Mutants

There are several distinct types of inhibitors of nucleotide biosynthesis. Many of these have been important for the elucidation of nucleotide metabolism, and some are used in the treatment of microbial infections. The isolation and characterization of mutants defective in purine and pyrimidine metabolism has provided a powerful tool in the understanding of the purine and pyrimidine pathways. Mutagenesis and penicillin counterselection has been used to isolate purine and pyrimidine auxotrophs and salvage mutants. Another approach involves the use of analogs that inhibit nucleotide metabolism, either directly or after conversion to toxic compounds inside the cell. It is possible to select mutants resistant to an analog because most analogs react with the same enzymes as the natural compounds. Resistant mutants make either a defective protein or an altered protein that no longer reacts with the analog. Other

types of mutants exhibit an altered expression of purine or pyrimidine genes, for example, the increased synthesis of purine biosynthetic enzymes.

B. Industrial Use of Mutants

Strains with multiple mutations in the purine pathways are used in the production of purine compounds in the food industry. IMP and GMP are used as flavor enhancers in food and are produced by bacteria that excrete these compounds or their nucleoside derivatives into the growth medium. Nucleosides, nucleotides, and their analogs are important materials in the synthesis of pharmaceuticals and compounds used in research, such as $2',3'$ dideoxynucleoside triphosphates. Interesting developments are the use of bacteria to synthesize nucleoside analogs. The *Lactobacillus helveticus* nucleoside deoxyribosyltransferases have been employed for the synthesis of nucleoside analogs used in the treatment of hairy-cell leukemia. A number of microorganisms produce purine and pyrimidine-related antibiotics, which are excreted. For example, decoyinine is produced by *Streptomyces* species. Decoyinine specifically inhibits guanine nucleotide synthesis in gram-positive bacteria and is used to stimulate sporulation.

C. 5-Fluorouracil and 5-Fluoroorotate

The synthetic pyrimidine analog 5-fluorouracil is toxic to most bacteria. To become toxic it must be converted to 5-fluoroUMP or 5-fluorodeoxyUMP both of which interfere with pyrimidine metabolism. 5-fluorodeoxyUMP is a potent inhibitor of thymidylate synthase. Some mutants resistant to 5-fluorouracil become defective in either uracil phosphoribosyltransferase or uracil transport. Other mutants isolated in *E. coli* and *S. typhimurium* have reduced UMP kinase activity, exhibit derepressed levels of the pyrimidine biosynthetic enzymes, accumulate UMP, and excrete uracil. In *upp* mutants, 5-fluorouracil can become toxic again if the cells are grown in the presence of a deoxyribonucleoside, which can be degraded to deoxyribose-1-phosphate by a nucleoside phosphorylase. 5-Fluorouracil can then react with deoxyribose-1-phosphate to form 5-fluorodeoxyuridine, which can be converted to 5-fluorodeoxyUMP by thymidine kinase. A mutation in one of the genes catalyzing one of these reactions, therefore, will lead to resistance. Among antifungal agents, one finds 5-fluorocytosine. The compound by itself is not toxic, but after deamination it gives rise to the toxic 5-fluorouracil. The safety of this drug in treatment is due to the lack of cytosine deaminase in humans. Based on development of resistence to 5-fluoroorotate, a widely used vector system for the cloning of DNA in yeast has been developed. In the first step, strains resistant to 5-fluoroorotate in the presence of uracil are isolated. Among them, strains defective in one of the last two enzymes of UMP biosynthesis are picked. The vector containing the wild-type UMP gene plus the gene of interest now can be introduced by selecting for growth in the absence of uracil. If it is desired, the cells can be cured for the plasmid by plating the cells on 5-fluoroorotate and uracil.

D. Inhibition of Folate Biosynthesis

Synthetic antifolate drugs such as trimethoprim are toxic to many bacteria and are used to treat bacterial infections. The analogs are potent inhibitors of dihydrofolate reductase, the enzyme that catalyzes the reduction of dihydrofolate to tetrahydrofolate. This reaction is the only one by which dihydrofolate, formed in the synthesis of thymidylate (Fig. 4), can be converted to tetrahydrofolate. Microorganisms that are resistant to trimethoprim overproduce thymidylate synthase or dihydrofolate reductase. A combination of a sulfonamide (e.g., sulfamethoxazole), a structural analog of p-aminobenzoic acid, and trimethoprim or related analogs becomes a potent inhibitor of bacterial growth and is used in treatment of infections with bacteria such as *E. coli* and *Salmonella*. Together, these compounds act by depleting the intracellular level of tetrahydrofolate derivatives, resulting in a shortage of nucleotides.

See Also the Following Articles

Bacillus Subtilis, Genetics • Carbohydrate Synthesis and Metabolism • Starvation, Bacterial

Bibliography

Mathews, C. K., and van Holde, K. E. (1996). "Biochemistry." Benjamin/Cummings, Menlo Park, California.

Neuhard, J., and Kelln R. A. (1996). Biosynthesis and conversion of pyrimidines. In "Escherichia coli and Salmonella typhimurium Cellular and Molecular Biology" (F. C. Neidhardt et al., eds.), pp. 580–599. American Society for Microbiology, Washington, DC.

Nygaard, P. (1993). Purine and pyrimidine salvage pathways. In "Bacillus subtilis and Other Gram-Positive Bacteria" (A. L. Sonenshein et al., eds.), pp. 359–378. American Society for Microbiology, Washington, DC.

Switzer, R. L., and Quinn, C. L. (1993). De novo pyrimidine nucleotide synthesis. In "Bacillus subtilis and Other Gram-Positive Bacteria" (A. L. Sonenshein et al., eds.), pp. 343–358. American Society for Microbiology, Washington DC.

Suhadolnik, R. J. (1979). "Nucleosides as Biological Probes." John Wiley & Sons, New York.

Zalkin, H. (1993). De Novo purine nucleotide synthesis. In "Bacillus subtilis and Other Gram-Positive Bacteria" (A. L. Sonenshein et al., eds.), pp. 335–341. American Society for Microbiology, Washington, DC.

Zalkin, H., and Nygaard, P. (1996). Biosynthesis of purine nucleotides. In "Escherichia coli and Salmonella typhimurium Cellular and Molecular Biology" (F. C. Neidhardt, et al., eds.), pp. 561–579. American Society for Microbiology, Washington, DC.

Nutrition of Microorganisms

Thomas Egli

Swiss Federal Institute for Environmental Science and Technology

GLOSSARY

anabolism   The process of synthesis of cell components from a metabolic pool of precursor compounds.

assimilation   The incorporation of a compound into biomass.

catabolism   The breakdown of nutrients to precursor compounds for anabolism or for dissimilation.

chemoautotrophy   The use of reduced inorganic compounds and CO_2 as the primary sources of energy and carbon for biosynthesis.

chemoheterotrophy   The process in which organisms are using organic compounds as the primary sources of carbon and energy for biosynthesis.

dissimilation   The oxidation of a reduced (in)organic compounds to provide energy for biosynthesis and cell maintenance.

growth medium   An aqueous solution containing all the nutrients necessary for microbial growth.

limitation of growth   The restriction on microbial growth by the availability of the nutrient that is first to be consumed to completion. This growth-limiting nutrient determines the maximum amount of biomass that can be formed in this system; at low concentrations in batch culture and in the chemostat, it also determines the rate (kinetics) of growth.

nutritional categories of microorganisms   Categories based on the principal carbon (CO_2 or reduced organic compounds) and energy sources (light or reduced (in)organic compounds) of microorganisms; there are four nutritional categories: photoautotrophs, photoheterotrophs, chemoautotrophs, and chemoheterotrophs.

nutrient   An organic or inorganic compound that is used by microorganisms as a building block for the synthesis of new cell material. In a wider sense, also a compound not incorporated into the microorganism, but serving as a source of energy or as terminal electron acceptor. Nutrients are grouped into classes depending on the physiological purpose they serve, the quantity required, and whether or not they are essential for growth.

photoautotrophy   The use of light and CO_2 as the primary sources of energy and carbon for biosynthesis.

photoheterotrophy   The use of light and reduced organic compounds as the primary sources of energy and carbon for biosynthesis.

TO GROW AND DIVIDE, microbial cells take up precursors and building blocks (nutrients) from the environment. In a wider sense, nutrients are also compounds that are not directly incorporated into cell material but are used by microbes to obtain the energy necessary to drive this synthesis and maintain cell integrity. Different nutritional types of microorganisms exist using different forms of carbon (CO_2 or reduced organic compounds) and energy (light or chemical energy) as the primary sources for biosynthesis. Nevertheless, the cellular composition of all microbial cells with respect to bulk components and the elemental composition is rather similar. Because of this, it is possible to estimate the general requirement of different nutrients for growth and to design and analyze microbial growth media. In well-designed growth media a particular, identified nutrient is growth-limiting and determines the amount of biomass that can be formed, whereas all other nutrients are present in excess (Liebig's principle). Cell metabolism and performance are strongly influenced by the nature of the growth-limiting nutrient. Therefore, many industrial fermentation processes are based on restricting the availability of a particular nutrient.

I. CLASSIFICATION OF MICROORGANISMS AND NUTRIENTS

Growth and production of offspring is the ultimate goal of each microbial cell and to achieve this, cells take up nutrients from the environment for two purposes, either to serve as a source of building blocks or precursors for the synthesis of new cellular constituents or to generate energy to drive biosynthesis. Individual members of the microbial world are extremely diverse and often unique with respect to their nutritional requirements and abilities. Hence, only the main patterns of the nutritional requirements and behavior of microorganisms will be delineated here.

Two approaches are traditionally taken to describe the nutritional behavior and requirements of living cells. The two approaches do not contrast, but rather complement each other. One is to categorize organisms on the basis of the principal sources of the carbon and of the energy they are able to use for growth; the other is to categorize them on the basis of quantitative and elemental aspects of the nutrients used for growth.

A. Nutritional Categories of Organisms

Based on their principal carbon and energy sources, (micro)organisms are classified into four nutritional categories (Table I). Most microorganisms using light as their principal source of energy are photoautotrophs (sometimes also referred to as photolithoautotrophs), whereas photoheterotrophs are a small group of specialists (certain purple and green bacteria). The ability to grow chemoautotrophically (i.e., in the dark in a medium containing only inorganic nutrients, including a reduced inorganic compound as a source of energy) is specific for bacteria and is lacking in eukaryotic microorganisms. In these three types of nutrition, the sources

TABLE I

Nutritional Types of Organisms Based on the Sources of Carbon and Energy Used for Growth

Energy source	Carbon source	Nutrition type
Light	CO_2	Photoautotroph (photolithoautotroph)
Light	Reduced organic compounds	Photoheterotroph (photolithoheterotroph)
Reduced inorganic compounds	CO_2	Chemoautotroph (chemolithoautotroph)
Reduced organic compounds	Reduced organic compounds	Chemoheterotroph (chemorganoheterotroph)

for carbon and for energy are clearly separated. This clear-cut distinction between carbon source and energy source is not valid for the big group of chemoheterotrophic organisms that obtain their energy from the oxidation of reduced organic compounds and at the same time use them as a source of building blocks. The terms "litho-" and "organotroph" are sometimes used to indicate in addition the source of hydrogen and electrons.

Some microbial strains are nutritionally rather flexible and could be placed in different nutritional categories. For example, the nutritional versatility of some photoautotrophic microalgae is such that they can employ equally well a chemoheterotrophic lifestyle, growing in the dark at the expense of organic carbon sources. Also it should be mentioned that, when given the chance, most autotrophs can take up and assimilate considerable amounts of reduced organic compounds (not only growth factors, as described later) and use them to feed their anabolism. The nutritional category of such microorganisms is usually based on the simplest nutritional requirements, in which phototrophy and autotrophy precede chemotrophy and heterotrophy, respectively. The degree of nutritional flexibility, in addition, is indicated by describing organisms as either obligate or facultative photo(chemo)autotrophs.

In any of the four nutritional categories there are auxotrophic strains that require low amounts of specific organic compounds, the growth factors. Unlike prototrophic strains, auxotrophs are unable to synthesize these growth factors from the principal source of carbon supplied in the medium.

B. Classes of Nutrients

In everyday use, the term "nutrient" is restricted to compounds either fully or at least partly incorporated into cell material. However, biosynthesis requires energy in addition to building blocks. Frequently, the compounds involved in the generation of energy are not incorporated into biomass, but only take part in redox processes. Hence, based on their physiological purpose, compounds essential for microbial growth can be divided into two major groups (Fig. 1).

1. Compounds that are either fully or partly incorporated into components of the biomass ("nutrients"), and
2. Compounds that are not incorporated into biomass but are essential for the generation of energy (electron donors or acceptors).

This distinction cannot always be made in such a clear way because there are nutritional categories of organisms in which some compounds can fulfil both functions at the same time. For example, reduced carbon sources are employed by chemoheterotrophs to obtain carbon precursors for biosynthesis as well as to generate energy, or ammonia can be used by particular chemolithoautotrophs as a source of both energy and nitrogen.

The chemical elements contained in the nutrients consumed and incorporated into new cell material can be divided into five different classes. The division is mainly based on the quantities of these elements required for growth and their occurrence in dry biomass (Table II). Not considered in this table is water, which is a main constituent of all cells, making up approximately 75% of the fresh cell weight. Table III indicates that typically some 95% of the dry biomass are made up of the eight elements C, N, O, H, P, S, K, and Mg. These elements are indispensable for microbial growth. Class two elements are required in significant amounts, whereas those in class three and four are usually referred to as trace elements. For elements categorized in classes two and three, it can be demonstrated experimentally that they are essential, whereas it is difficult to prove that elements of class four are essential for growth. These elements are required in such low amounts that they are usually introduced into media as impurities of the bulk salts. Finally, a special class of nutrients are the growth factors required by auxothrophic strains. This includes a diverse group of organic compounds. The physiological role of the main nutrients is discussed in more detail in Section III.

II. ELEMENTAL COMPOSITION OF BIOMASS

The composition of a microbial cell is highly dependent on the cultivation conditions. The type of cellular constituents present (e.g., ribosomes, particular enzymes, membrane and cell wall components, and compounds in the metabolic pool) and their amount can vary enormously. Herbert (1961) has emphasized this point saying that it is useless to give the cellular composition of a microbial cell without at the same time specifying both the exact growth conditions under which this cell has been cultivated and its growth history.

Despite this diversity and variability with respect to cell constituents, the elemental composition of microbial biomass—including cell material from archaea, eubacteria, and eukaryotes—varies in a sur-

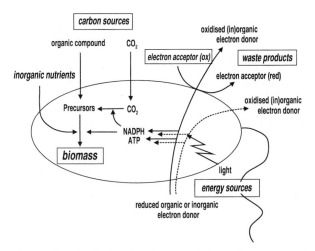

Fig. 1. Simplified sketch of the physiological function of nutrients for the growth of microorganisms.

TABLE II

Classes of Nutrients Used for Microbial Growth Based on Their Incorporation and Occurrence in Dry Cell Mass[a]

Class 1	Always essential	Major elements: C, H, O, N
		Minor elements: P, S, K(Rb), Mg
Class 2	Mostly essential	Fe, Ca, Mn, Co, Cu, Mo, Zn
Class 3	Essential in special cases	B, Na, Al, Si, Cl, V, Cr, Ni, As, Se, Sn, I
Class 4	Very rarely essential (difficult to prove)	Be, F, Sc, Ti, Ga, Ge, Br, Zr, W
Class 5	Growth factors, essential for special strains	Aminoacids, purines and pyrimidines, vitamins, hormones, etc.

[a] Based on Pirt (1975).

TABLE III

Elemental Composition of Microbial Biomass[a]

| Element | % of dry weight[b] | | Typical sources used for microbial growth in the environment |
	Average	Range	
Carbon	50	45[c]–58[d]	CO_2, organic compounds
Oxygen	21	18[e]–31[f]	H_2O, O_2, organic compounds
Nitrogen	12	5[g]–17[h]	NH_3, NO_3^-, organically bound N
Hydrogen	8	6[g]–8[h]	H_2O, organic compounds
Phosphorus	3	1.2[i]–10[j]	PO_4^{3-}, organically bound P
Sulphur	1	0.3–1.3	SO_4^{2-}, H_2S, organically bound S
Potassium	1	0.2[k]––5[l]	K^+ (can often be replaced by Rb^+)
Magnesium	0.5	0.1[m]–1.1	Mg^{2+}
Calcium	1	0.02–2.0	Ca^{2+}
Chlorine	0.5		Cl^-
Iron	0.5	0.01–0.5	Fe^{3+}, Fe^{2+}, organic iron complexes
Sodium	1		Na^+
Other elements	0.5		Taken up as inorganic ions
Mo, Ni, Co, Mn, Zn, W, Se, etc.			

[a] Data from Tempest (1969), Pirt (1975), Herbert (1976), and from results obtained in author's laboratory.

[b] Cells consist on average to 70% of their weight of water and 30% of dry matter. Average is gram-negative cells growing with excess of all nutrients at μ_{max} in batch culture.

[c] Carbon-limited cells containing no reserve materials.

[d] Nitrogen-limited cells storing PHA or glycogen in the presence of excess C-source.

[e] Cells grown N-limited accumulating neutral lipids.

[f] Cells grown N-limited accumulating glycogen.

[g] Nitrogen-limited cells storing PHA or glycogen in the presence of excess C-source.

[h] Cells growing at high μ containing high levels of rRNA.

[i] Cells grown P-limited.

[j] Cells accumulating the reserve material polyphosphate.

[k] Gram-positive *Bacillus* spores.

[l] Gram-positive bacilli.

[m] Magnesium-limited cells at low growth rates.

TABLE IV
Major Polymeric Constituents Found in Microbial Cells and Their Average Elemental Composition[a]

Constituent	% of dry weight Average[b]	% of dry weight Range	%C	%H	%O	%N	%S	%P
Protein	55	15[c]–75	53	7	23	16	1	—
RNA[d]	21	5[c]–30[e]	36	4	34	17	—	10
DNA[d]	3	1[c]–5[f]	36	4	34	17	—	10
Peptidoglycan	3	0[g]–20[h]	47	6	40	7	—	—
Phospholipids	9	0[i]–15	67	7	19	2	—	5
Lipopolysaccharides	3	0[h]–4[j]	55	10	30	2	—	3
Neutral lipids	—	0–45[k]	77	12	11	—	—	—
Teichoic acid[d,h]	—	0[l]––5[d]	28	5	52	—	—	15
Glycogen	3	0–50[k]	45	6	49	—	—	—
PHB	—	0–80[k]	56	7	37	—	—	—
PHA (C8)[m]	—	0–60[k]	68	9	23	—	—	—
Polyphosphate[d]	—	0–20[n]	—	—	61	—	—	39
Cyanophycin[o]	—	0–10	42	15	25	27	—	—

[a] Adapted from Herbert (1976) and extended. The figures given for the range have been collected from different organisms and, therefore, may not be applicable for particular strains.

[b] Average composition of an exponentially-growing gram-negative cell (*E. coli*), from Neidhard *et al.* (1990).

[c] Cells storing carbonaceous reserve materials.

[d] Inclusion of the highly negatively charged polymers such as RNA, DNA, polyphosphate, or cell wall components is paralleled by the presence of appropriate amounts of counter-cations, Mg^{2+}, Ca^{2+}, or K^+.

[e] At high growth rates.

[f] Cells growing slowly.

[g] Parasitic cell-wall less species.

[h] Gram-positive bacteria.

[i] Strains replacing phospholipids under P-limited growth conditions with P-free analog.

[j] Gram-negative bacteria.

[k] Cells grown N-limited.

[l] Grown P-limited.

[m] PHA consisting of 3-hydroxyoctanoic acid.

[n] Some yeasts and bacteria.

[o] Some cyanobacteria contain the nitrogen storage material cyanophycin [(asp-arg)$_n$].

prisingly narrow range. This is documented in the overview of the average elemental composition of microbial biomass and its variability in Table III. The relative constant composition of microbial biomass with respect to the major elements results from the fact that most of the dry biomass (typically some 95%) is made up of a limited number of organic macromolecules and only a small fraction consists of monomers (metabolites and inorganic ions). Because protein, RNA, and phospholipids are the dominating components, massive changes in the content of a particular cell component are required before the overall elemental composition

of the biomass is significantly affected (Table IV). For example, a significant increase in the carbon content of dry biomass is observed only when cells store high amounts of poly(3-hydroxyalkauoate) (PHAs) or neutral lipids, whereas it is the cellular oxygen content is primarily affected when cells accumulate glycogen. Note that an extensive incorporation of carbonaceous reserve materials results in a dilution and, hence, in a reduction of the relative content of other elements in dry biomass. A typical example is the reduced cellular nitrogen content found in cells accumulating PHA or glycogen.

III. REQUIREMENTS AND PHYSIOLOGICAL FUNCTIONS OF PRINCIPAL ELEMENTS

A. Carbon

Dried microbial biomass consists of roughly 50% carbon, virtually all of it present as one of the many reduced organic cell constituents. Hence, as discussed (Table I), the most obvious physiological function of carbon is as a source of building material for organic biomolecules. When its most oxidized form, CO_2, is used as the sole source of carbon for autotrophs, reduction to the level of organic cell material and the formation of carbon–carbon bonds is required. This process requires significant amounts of reducing equivalents (NADPH) and energy (ATP). (See Fig. 1.) CO_2 is also employed as a terminal electron acceptor by methanogens and acetogens.

In contrast, heterotrophs use reduced carbon compounds to build their cell material and in most cases (an exception are the photoheterotrophs) the carbon compound fulfills a dual function, namely it acts as both a carbon and an energy source. In some fermenting organisms, reduced carbon compounds can act as terminal electron acceptors. Typically, heterotrophic cells utilize the same carbon source for both purposes, oxidizing part of it to CO_2 (a process called dissimilation) and using the energy derived from this oxidation to synthesize cell material from the other part (assimilation). The ratio of dissimilated to assimilated carbon is essentially dependent on the degree of reduction of the carbon substrate used. The more oxidized the carbon compound, the more of it has to be dissimilated in order to provide the necessary energy to drive the synthesis processes, and the less of it can be assimilated. This is reflected in the maximum growth yield observed for various carbon sources when plotted as a function of their energy content (i.e., their degree of reduction, or heat of combustion), as shown in Fig. 2. Most extreme is the case of chemoheterotrophs growing at the expense of oxalate (HOOC — COOH). To generate energy, this compound is initially oxidized to CO_2, which is then assimilated in an autotrophic manner.

Heterotrophic microorganisms are extremely diverse with respect to the spectrum of carbon sources that they can use for growth. Whereas some are

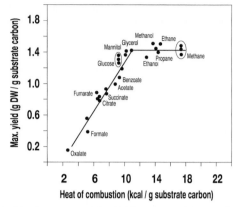

Fig. 2. Maximum growth yields reported for various carbon substrates observed for heterotrophic organisms, as a function of the energy content of the carbon substrate (adapted from Linton and Stephenson, 1973).

restricted to only a few carbon compounds (e.g., some methanotrophs appear to use only methane and methanol), others are able to metabolize and assimilate more than a hundred carbon compounds for growth. It should be added here that all heterotrophic organisms also assimilate a substantial amount of their cell carbon (typically 5–10%) from CO_2 (mainly for replenishing the tricarboxylic acid cycle when it is used as a source of building blocks for biosynthesis). Normally, this requirement for CO_2 is masked because CO_2 is produced in large amounts intracellularly from the catabolism of organic growth substrates. However, especially in freshly inoculated dilute cultures, its absence can slow down or even prevent growth on organic substrates (Pirt, 1975), and some heterotrophic mircoorganisms even require elevated concentrations of CO_2 in the culture medium.

In the case of energy excess, carbon compounds can be stored intracellularly as reserve materials in the form of polyphydroxyalkanoates, glycogen, or neutral lipids. In the case of carbon starvation, these internal carbon and energy sources are broken down to support cellular rearrangement and adaptation to the new conditions and to ensure survival.

B. Hydrogen

In cells, hydrogen is present in the form of water and as an element of all organic cell constituents. The main source of hydrogen for biosynthetic purposes is NADPH. The need for hydrogen is particularly evident for the reduction of CO_2 in autotrophs. In photo- and chemoautotrophs hydrogen equivalents used for CO_2 reduction originate from water, from the oxidation of reduced inorganic compounds. Chemoheterotrophs obtain their reducing equivalents from the oxidation of their primary carbon substrate.

C. Oxygen

As with carbon and hydrogen, oxygen is omnipresent in cells. It occurs in most of the organic components of cell material. The main sources of oxygen for the biosynthesis of particular cell components are water, molecular oxygen (but not in obligate anaerobes, where oxygen is frequently toxic) and, less obviously, CO_2. In aerobes, molecular oxygen is introduced into organic molecules with the help of mono- and dioxygenases. In addition to its function as a cell constituent, O_2 also serves as a terminal electron acceptor in aerobes.

D. Nitrogen

The cellular requirement for nitrogen is significant because it is a constituent of all major macromolecules (Tables III and IV). In cell components, nitrogen is mainly found in the reduced form (i.e., as primary, secondary, or tertiary amino groups). Oxidized forms (nitro- and nitroso-groups) are rarely found. Organic and inorganic forms of nitrogen in all states of oxidation, from NH_4^+ to N_2 to NO_3^-, can be used by microorganisms as sources of cell nitrogen (although some are unable to reduce oxidized forms). Note that the microbiological fixation of molecular nitrogen is of special interest to agriculture because of its availability in the air. Frequently, microbial cells exhibit nitrogen requirements in the form of special amino acids (L-forms for incorporation into proteins, or D-forms for the synthesis of cell-wall components) or peptides. Intracellularly, the assimilation of nitrogen occurs at the level of ammonia. Therefore, all more oxidized forms have to be reduced to this level before they can be used as a source of nitrogen.

Nitrogen compounds also play a major role in energy metabolism. Reduced forms (e.g., ammonia and nitrite) are used as sources of energy by nitrifying bacteria, whereas oxidized inorganic nitrogen compounds (e.g., nitrate and nitrite) are employed as terminal electron acceptors by denitrifying microbes.

E. Phosphorus

Inorganic phosphate is typically supplied in growth media as the only source of phosphorus. However, many organisms can also derived phosphorus from organic phosphates, such as glycerophosphate (organic P-sources can be used to avoid the precipitation of inorganic phosphate salts in the medium at basic pH values). Phosphate is primarily incorporated into nucleic acids, phospholipids, and cell-wall constituents. Some organisms may also store it as polymetaphosphate, which can be reused as a source of internal phosphorus or for the generation of ATP. Intracellularly, the main fraction of phosphorus is contained in ribosomal RNA, whereas ATP and other nucleic acids make up only a minor fraction of the total cellular phosphorus.

F. Sulfur

The bulk of intracellular sulfur is found in proteins (cysteine and methionine). An important function of cysteine is its involvement in the folding of proteins by the formation of disulfide bridges. Frequently, these amino acids are also found in reactive centers of enzymes (e.g., in the coordination of reactive iron centers). The sulfur-containing co-enzymes and vitamins (e.g., CoA, biotin, thiamine, glutathione, and lipoic acid) are small in quantity but physiologically very important. Intracellularly, sulfur is present in a reduced form ($-SH$), but it is usually supplied in growth media as sulfate salt. Some organisms are not able to catalyze this reduction and therefore must be supplied with a reduced form of sulfur, for instance, cysteine or H_2S.

Many inorganic sulfur compounds are also involved in the generation of energy. Whereas reduced sulfur compounds are used as electron donors (H_2S, thiosulphate, and S^0), oxidized forms are employed as terminal electron acceptors (SO_4^{2-}, S^0).

G. Major Cations

1. *Magnesium*

Magnesium is one of the major cations in cell material. Its intracellular concentration is proportional to that of RNA, which suggests that it is partly counterbalancing the negative charges of the phosphate groups in nucleic acids. Hence, its cellular concentration increases with growth rate. It is required for stabilizing the structure of ribosomes. Many enzymes are activated by or even dependent on the presence of Mg^{2+}; some important examples are enzymes catalyzing reactions dependent on ATP or chlorophylls. Magnesium is also found bound to the cell wall and the membrane, where it seems to be responsible for stabilizing the structure together with other cations.

Interestingly, the molecular ratio of $Mg:K:RNA$ nucleotide$:PO_4$ in gram-negative bacteria is always approximately $1:4:5:8$, independent of growth rate, temperature, or growth-limiting nutrient (Tempest, 1969). In gram-positive organisms, this ratio is $1:13:5:13$, except under phosphate-limited growth conditions in continuous culture, where it is also $1:4:5:8$. The higher K and PO_4 content of gram-positive bacteria is due to the presence of phosphate-containing cell-wall polymers (teichoic acids), which are replaced under phosphate-limited growth by nonphosphate-containing analogs (teichuronic acids).

2. *Potassium*

Potassium makes up a large part of the inorganic cations in biomass (Table III). Only a small fraction of K^+ present in cells seems to be associated with binding sites of high affinity and specificity because it can be rapidly exchanged with other monovalent cations. A large fraction of K^+ is bound to RNA, for which it seems to have a stabilizing function. Therefore, as with magnesium, the requirement for it increases with growth rate. Significant amounts are also found associated with the cell wall. K^+ activates a number of different enzymes, both nonspecifically (contributing to the ionic strength) or specifically (e.g., peptidyltransferase). Ions of similar size such as Rb^+ or NH_4^+ can frequently take over the function of K^+ (in contrast to magnesium, which cannot be replaced by other cations). The growth rates of many organisms are reduced when cultivated in media that is low in potassium. A variety of growth conditions affect the intracellular concentration of potassium, including the osmolarity of the medium, temperature, pH, or sodium concentration. Therefore, this cation should always be added to growth media in significant excess.

3. *Iron*

Frequently considered a trace element, iron is used in significant amounts by virtually all organisms, not only by obligate aerobes (lactobacilli seem to be the only bacteria that do not need iron for growth). Iron is the catalytic center of a number of enzymes, especially those involved in redox reactions. Most essential are the various iron-containing cytochromes in the respiratory chain, flavoproteins, or the enzymes essential for the detoxification of reactive oxygen species such as catalase or superoxide dismutase. Many of the mono- and dioxygenases initiating the breakdown of pollutants are as also iron enzymes. Most bacteria require concentrations of free iron exceeding 10^{-8} M for growth. Iron(III), which is the species that prevails in aerobic environments, easily forms insoluble hydroxides and other complexes. Therefore, the acquisition of iron is a major problem for growing organisms. Many organisms react to iron limitation by excreting iron-complexing organic compounds with a high affinity to iron, the siderophores. In mineral media, iron is therefore frequently supplied complexed with an organic ligand.

A number of anaerobic bacteria (in particular nitrate-reducing strains) can use Fe^{3+} (or Mn^{4+}) as a terminal electron acceptor, reducing it to Fe^{2+} (or Mn^{2+}). On the other hand, some specialist bacteria can use Fe^{2+} (or sometimes also Mn^{2+}) as a source of energy by oxidizing it to Fe^{3+}.

4. *Calcium*

In most organisms, Ca^{2+} is present intracellularly in significantly lower amounts than Mg^{2+}, which has similar properties. The role of Ca^{2+} is not always clear; however, it seems to have important functions in stabilizing the cell wall and controlling membrane permeability. Changes in cell morphology and cell-surface properties have been reported for several microorganisms in the absence of Ca^{2+}. Ca^{2+} activates many exoenzymes such as amylase. Furthermore, a number of uptake processes are stimulated by the presence of Ca^{2+}; an example is the uptake of exogenous DNA. It appears that calcium plays extracellularly the role that magnesium plays in the cytoplasm. Often Mg^{2+} cannot replace Ca^{2+} in these extracellular functions, but strontium can. In growth media attributing a clear function to calcium is difficult because of the presence of competing divalent cations that are essential for growth, such as Mg^{2+} or Mn^{2+}.

H. Trace Elements

1. Sodium

Microorganisms isolated from freshwater do not usually require sodium. For such organisms, it is difficult to demonstrate that this cation is essential for growth because its requirements are low and sodium is present in all bulk salts as an impurity. However, some extremely halophilic microorganisms require high concentrations of NaCl for growth. For example, in order not to disintegrate, *Halobacterium* needs more than 2.5 M NaCl in the growth medium. Sodium is also essential for certain photosynthetic bacteria and cannot be substituted for by other monovalent cations. In some marine bacteria, energy generation is even linked to the use of a Na^+-gradient, rather than an H^+-gradient. Furthermore, in many microorganisms this ion is involved in the regulation of intracellular pH using a Na^+–H^+ antiport system.

2. Manganese

As a substitute for the iron-containing catalase, lactobacilli produce a manganese-containing pseudocatalase for protection against molecular oxygen. A high requirement for manganese is typical for lactic acid bacteria. Many of the lignolytic peroxidases contain manganese.

3. Cobalt

Co^{2+}-containing coenzymes and cofactors are widespread, the best known being the coenzyme B_{12}. Cobalamines (cobalt-containing biomacrocyclic compounds) are found in bacteria as well as in humans,

but the highest levels are usually found in methanogenic bacteria.

4. Nickel

Methanogens require unusually high amounts of Ni^{2+} for growth. It was found that it is a component of the coenzyme F430 in these organisms. Furthermore, this divalent cation is a constituent of virtually all hydrogenases in both aerobic and anaerobic microorganisms that either use or produce molecular hydrogen.

5. Copper

Copper is the key metal in the active center of many redox-reaction catalyzing enzymes. It is present in many terminal oxidases of the respiratory chain. A number of other enzymes, such as peptidases, laccases, some nitrite reductases, or methane monooxygenase contain Cu^{2+}.

6. Molybdenum

A whole family of enzymes, the molybdoenzymes (also referred to as molybdenum hydroxylases) contain Mo^{2+}. This family includes the central enzyme in the reduction of nitrate to nitrite, nitrate reductase. Furthermore, the nitrogen-fixing nitrogenase contains a MoFe cofactor, which is clearly different from the molybdenum cofactor shared by other Mo-containing enzymes.

7. Zinc

Many bacterial (extracellular) metalloproteases contain Zn^{2+} (e.g., elastase). Many of these proteases are produced by pathogenic strains and play an important role in the pathogenesis. Other zinc-containing enzymes are alkaline phosphatase and the long-chain alcohol dehydrogenases.

IV. FEAST AND FAMINE: UNRESTRICTED VERSUS NUTRIENT-LIMITED GROWTH

In a typical laboratory shake flask culture, all the nutrients supplied in a well-designed growth medium are initially present in excess and the cells grow exponentially at the highest rate possible under these conditions. However, in every environmental and technical system, microbial growth cannot proceed unrestricted for a long time. A simple calculation makes this obvious. After 2 days of exponential growth, a single microbial cell doubling every 20 min will have produced roughly 2×10^{43} cells. Assuming an average cell weight of 10^{-12} g, this amounts to 2×10^{31} g of the biomass, or approximately 4000 times the weight of Earth. Hence, in every compartment, growth is always soon limited by the exhaustion of one or several nutrients.

A. The Concept of the Limiting Nutrient

The term "limiting nutrient" is used with meanings, which, unfortunately, are frequently mixed up. The availability of nutrients can restrict the growth of microbial cultures in two distinct ways, namely stoichiometrically and kinetically. The stoichiometric limitation is defined by the maximum amount of biomass that can be produced from the limiting nutrient in this system ("Liebig's principle," from Justus von Liebig's agricultural fertilization studies around 1840, in which he found that the amount of a particular nutrient determined the crop on a field as long as all other nutrients were present in excess; Eq. 1). The kinetic limitation arises at low nutrient concentrations (typically in the low milligram to microgram per liter range), at which the limiting nutrient also controls the specific rate of growth of cells (μ). This kinetic control of growth rate usually follows a saturation kinetics and the Monod equation (Eq. 2) is typically used to describe the relationship between the concentration of the growth-rate-controlling nutrient and μ.

$$X = X_0 + (S_0 - s) \cdot Y_{X/S} \qquad (1)$$

$$\mu = \mu_{max} \frac{s}{K_s + s}, \qquad (2)$$

where S_0 is the initial and s the actual concentration of the limiting nutrient S; X (X_0) is the (initial) biomass concentration; $Y_{x/s}$ is the growth yield for nutrient S, μ (μ_{max}) is the (maximum) specific growth rate, and K_s is the Monod apparent substrate affinity constant. This is visualized in Fig. 3 for growth in a closed

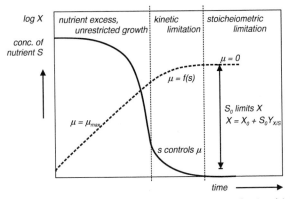

Fig. 3. Kinetic and stoichiometric limitation of microbial growth in a batch culture by the concentration of the limiting nutrient (substrate)S. S_0, initial concentration of S; s, actual concentration of S; X, biomass concentration; X_0, initial biomass concentration; $Y_{x/s}$, growth yield for nutrient S.

batch culture system, in which the cells initially grow unrestricted until the consumption of the limiting nutrient leads to growth at a reduced rate, and then to growth stoppage. This determines the final concentration of biomass that can be reached. In flow-through systems, such as a continuous culture in which fresh medium is continuously added and surplus culture is removed, the rate of addition of limiting nutrient (thought to be a single compound) controls simultaneously μ and the concentration of biomass obtained in the culture (Pirt, 1975; Kovárová and Egli, 1998).

In laboratory cultures, it is possible to cultivate cells under well-defined conditions in which the growth-limiting nutrient is known. Quantitative and practical aspects of the supply of nutrients to microbial growth media for controlled cultivation will be discussed in the Section V. For the cultivation of heterotrophic microorganisms for research purposes and the production of biomass, media are commonly designed with limiting carbon and energy sources, all other nutrients being supplied in excess. However, for biotechnological processes, limitation by nutrients other than carbon is frequently employed to manipulate the physiological state and metabolic performance of microbial cultures. The restriction (limitation) of specific nutrients induces or enhances the formation of many microbial metabolites and enzymes. Examples are the increased productivity in antibiotics fermentation by growth in phosphate-limited media, the production of citric acid under Fe-, Mn, or Zn-limited batch-culture conditions, the synthesis of NAD under Zn–Mn limitation, and the accumulation of the intracellular reserve materials PHB or PHA ("bioplastic") by limiting the supply of nitrogen.

In contrast to cultivation on the laboratory and industrial scales, it is difficult to assess the nutritional regimes that govern the growth of microbial cells in environmental systems (especially the identification of the kinetic control). There are indications that microbial growth is frequently controlled not by a single nutrient, but by combinations of two or more nutrients simultaneously (Kovárová and Egli, 1998).

V. DESIGN AND ANALYSIS OF DEFINED MINIMAL GROWTH MEDIA

To grow and synthesize their own cell material, organisms must obtain all the required building blocks (or their precursors) and the necessary energy from their environment. Consequently, to cultivate microbial cells in the laboratory these nutrients must be supplied in a culture medium in adequate amounts and in a form accessible to the organism.

As a result of the physiological diversity of the microbial world, a myriad of media of different compositions have been published, for either selective enrichment or cultivation of particular microorganisms (consult, for example, LaPage et al., 1970; Balows et al., 1992; Atlas, 1997). All these media contain components whose nutritional function is obvious, particularly when considering their elemental or energetic function. Nevertheless, most nutritional studies made have been qualitative rather than quantitative and different nutrients have been added in more or less arbitrary amounts. Also, many of the media contain components whose reason for inclusion cannot be clearly identified because their inclusion is based more on experience or tradition than on a clear purpose.

The identification of the nutritional requirements of microbial cells usually calls for the use of defined synthetic media. The design of defined culture media is based on quantitative aspects of cell composition and it allows the influence of the growth of a microbial culture at three major levels. First, the choice of which nutrient is to limit the growth of the culture stoichiometrically and kinetically is made. Second, for nutritionally flexible microbial strains, the choice of the type of metabolism that the organism is to perform is made by the selection of the compounds that are supplied to fulfill a particular nutritional requirement, including electron donors and acceptors. Third, and often linked with the second point, the choice of the maximum specific growth rate to be achieved during unrestricted growth in batch culture is set.

A. Setting μ_{max} during Unrestricted Growth

In addition to physicochemical parameters such as temperature or pH, the maximum specific growth rate of a microorganism is influenced by the composition of the nutrients supplied in the medium. This has been elegantly illustrated for the growth of *Salmonella typhimurium* by Schaechter *et al.* (1958), who used 22 media of different compositions to obtain growth of the culture at differing rates under nutrient excess conditions (a selection is given in Table V). Although the four media supporting the highest specific growth rates are undefined, the other media consist of a minimal salt medium to which different carbon sources or amino acid mixtures are added. Hence, the selection of the quality of precursors supplied in the mineral medium allowed the adjustment of the growth rate of the culture in a defined and reproducible way.

B. Medium Design and Experimental Verification of the Limiting Nutrient

1. Designing a Growth Medium

In the design of a defined growth medium, the initial decisions to be made are the choice of the maximum concentration of biomass the medium should allow to produce (X_{max}), and the definition of the growth-limiting nutrient (according to Liebig's principle). Typically, defined growth media for

TABLE V

Composition of a Selection of Media Used to Set the Maximum Specific Growth Rate of *Salmonella tymphimurium* in Batch Culture[a]

No.	Medium	Comments[b]	μ_{max} (hr^{-1})
1	Brain + heart infusion	Full strength	1.94
5	Nutrient broth	Diluted 1:2 with medium No. 14	1.80
6	Nutrient broth	Diluted 1:5 with medium No. 14	1.66
7	Casamino acids	1.5% + 0.01% tryptophan in medium No. 14	1.39
9	20 amino acids	20 natural amino acids + mineral salt solution	1.27
10	8 amino acids	8 natural amino acids + mineral salts solution	1.01
14	Glucose salt	0.2% glucose + mineral salts solution	0.83
15	Succinate salt	0.2% succinate + mineral salts solution	0.66
19	Methionine salt	0.06% methionine + mineral salts solution	0.56
22	Lysine salt	0.014% lysine + mineral salts solution	0.43

[a] Adapted from Schaechter *et al.* (1958).

[b] Mineral salts solution contained $MgSO_4$, Na_2HPO_4, $Na(NH_4)HPO_4$, KCl, and citric acid as chelating agent. It supported no visible growth without the addition of a carbon source.

heterotrophic microbes are designed with a single carbon–energy source restricting the amount of biomass that can be produced, whereas all other nutrients (each of them usually added in the form of a single compound) are supplied in excess. Having set X_{max}, it is possible to calculate the minimum concentration of the different elements necessary in the culture medium to produce X_{max} using the individual average elemental growth yields ($Y_{X/E}$). To ensure an excess of all the nonlimiting nutrients in the medium, their concentrations are multiplied by an excess factor (F_E). In this way, the concentrations of the nutrients required in the growth medium (E_{req}) are present in a theoretically x-fold excess with respect to the carbon source.

$$E_{req} = \frac{X_{max}}{Y_{X/E}} \cdot F_E. \qquad (3)$$

An example of the design of a carbon-limited medium supporting the production of 10 g/liter of dry biomass is given in Table VI. Note that in this medium the ingredients are chosen in such a way that it is possible to change the concentration of each of the elemental nutrients individually (for example, by including $MgCl_2$ plus $NaHSO_4$ instead of $MgSO_4$). In addition, this medium is only weakly buffered, hence, it might be necessary to control the pH during growth.

This approach works well for the design of media for the cultivation of aerobic microorganisms at low to medium biomass concentrations. More problematic is the design of media for anaerobic cultures in which many of the medium components precipitate easily at the required redox potential, or for high-cell-density cultures in which solubility or toxicity problems of some of the medium ingredients are encountered (see also Section V.B.2).

An estimate for most of the elemental growth yield factors $Y_{X/E}$ can be obtained from an elemental analysis of dry biomass cultivated under unrestricted growth conditions in batch culture (compare Table III). For carbon, oxygen, and hydrogen, $Y_{X/E}$ cannot be calculated directly from the elemental composition of cells because these elements are not only incorporated into the biomass, but also serve other metabolic functions. For example, carbon is not only assimilated by heterotrophs but is also oxidized to CO_2 to supply energy (see also Fig. 2). Also, not included in this table is the amount of electron acceptor that has to be supplied to ensure growth. Table VII shows the yield coefficients for oxygen, for some of the other common electron acceptors, and for some electron donors that support chemolithoautotrophic growth.

Two points influence the choice of excess factors. First, for elements whose cellular content does not

TABLE VI
Design of a Carbon-Limited Minimal Medium Allowing the Production of 10 g/liter of Dry Biomass[a,b]

Medium constituent	Source of, function	Growth yield assumed (g dry biomass/g element)	Excess factor assumed with respect to carbon	Mass of element (g/liter)	Mass of constituent (g/liter)
Glucose	C, energy	1	1	10	25.0
NH_4Cl	N	8	3	3.75	14.33
NaH_2PO_4	P	33	5	1.52	5.88
KCl	K	100	5	0.5	0.95
$NaHSO_4$	S	100	5	0.5	1.87
$MgCl_2$	Mg	200	5	0.25	0.98
$CaCl_2$	Ca	100	10	1.0	2.77
$FeCl_2$	Fe	200	10	0.5	1.13
$MnCl_2$	Mn	10^4	20	0.02	0.046
$ZnCl_2$	Zn	10^4	20	0.02	0.042
$CuCl_2$	Cu	10^5	20	0.002	0.0042
$CoCl_2$	Co	10^5	20	0.002	0.0044

[a] Based on elemental growth yields obtained from the composition of dry biomass (see Table I).

[b] Based on Pirt (1975) and Egli and Fiechter (1981). Elemental growth yields for C and the trace elements Zn, Cu, Mo, and Mn were taken from Pirt (1975). Excess factors were chosen taking into account their variation observed in dry biomass.

TABLE VII
Some Growth Yield Factors for Electron Donors and Electron Acceptors

Electron donors	
Molecular hydrogen	$Y_{X/H_2} \approx 12$ g/mol
Thiosulfate	$Y_{X/S_2O_3} \approx 4$ g/mol
Fe^{2+}	$Y_{X/Fe^{2+}} \approx 0.35$ g/mol
NH_4^+ to NO_2^-	$Y_{X/NH_4^+} \approx 1.3–2.6$ g/mol
NO_2^- to NO_3^-	$Y_{X/NO_2^-} \approx 0.9–1.8$ g/mol
Electron acceptors	
Molecular oxygen	$Y_{X/O_2} \approx 10^a–42^b$ g/mol
NO_3^- to N_2	$Y_{X/NO_3} \approx 27$ g/mol[c]
NO_2^- to N_2	$Y_{X/NO_2} \approx 17$ g/mol[c]
N_2O to N_2	$Y_{X/N_2O} \approx 9$ g/mol[c]

[a] For growth with reduced substrates such as methane or n-alkanes.

[b] For growth with more oxidized substrates such as glucose.

[c] For growth of *Paracoccus denitrificans* with glutamate as carbon substrate.

vary considerably as a function of cultivation conditions, excess factors can be set low (N, P, and S), whereas for elements that are known to vary considerably (e.g., with growth rate), they are set higher. Second, the chemical behavior of the medium component in the growth medium has to be also taken into account when choosing F_E. For example, most of the trace elements easily precipitate in growth media at neutral and basic pH and as a result their biological availability is reduced (and difficult to assess). Therefore, they are added in a 10- to 20-fold excess despite the fact that a metal complexing agent is usually added to the medium to keep them in solution (see Bridson and Brecker, 1970).

For biotechnological purposes, for which batch and fed-batch processes are primarily used, it would be advantageous to design media that contain all the nutrients in exactly the amount required, so that all nutrients would be consumed to completion at the end of growth. This, however, is difficult to achieve due to the variability of the yield factors for the individual elements and their dependence on the cultivation conditions. Nevertheless, one of the important points in medium optimization in biotechnology is certainly to optimize the consumption of nutrients and to minimize their loss.

2. Some Practical Comments on the Preparation of Media

It is appropriate to add some comments on some of the most important precautions to be taken when preparing a growth medium.

Many sugars easily deteriorate during sterilization at basic pH (especially in the presence of phosphates and peptones). This leads to a browning of the medium. The products that are formed can be inhibitory for growth. This can be avoided sterilizing the medium at a slightly acidic pH, or by sterilizing the sugars separately from the medium.

It is well known that all trace metals easily form highly insoluble phosphate salts and precipitate in growth media. This can be avoided by the addition of metal-chelating agents such as EDTA, NTA, or sometimes also carboxylic acids such as citrate or tartrate. The addition of chelating agents has a twofold effect. On one hand, it prevents the precipitation of trace metals; on the other hand, it acts as a sink for these metals, and in this way reduces their toxicity by lowering their free (for the microbes, accessible) concentration.

At medium pH > 7, the alkaline earth metals calcium and magnesium (as the trace metals) easily precipitate in the presence of phosphate (or in the presence of carbonate ions when using a bicarbonate buffered medium, or if only hard water is available) to form highly insoluble phosphate salts. These precipitates are sometimes difficult to see with the eye, especially in shake flasks, due to the small volume of the medium. To avoid this, the medium can be sterilized either at a slightly acidic pH (which requires the possibility that the pH will be adjusted later) or the phosphate salts are sterilized separately from the rest of the medium and combined after cooling.

For an elaborate treatment of this subject and more detailed information, especially on some of the established media, see the review by Bridson and Brecker (1970).

3. Experimental Identification of Growth-Limiting Nutrient

The variability of yield factors for the various nutrients, depending on the organism used and the compounds included in a medium, requires that the nature of the growth-limiting nutrient be experimentally verified for each case. For this, the maximum concentration of biomass (X) that can be produced in such a medium is determined as a function of the initial concentration of the medium component (S_0) that is supposedly growth-limiting, with the concentration of all other medium components kept constant. This experiment can be done in either batch or continuous culture.

The typical (theoretical) relationship obtained in such an experiment is visualized in Fig. 4. Ideally, the relationship between S_0 of a growth-limiting nutrient and X is initially a straight line that passes through the origin; that is, when no growth-limiting nutrient is added to the medium no biomass can be produced. When the S_0 exceeds a certain concentration, a deviation from the linear relationship will be observed. It is at this concentration that another nutrient becomes growth-limiting. Deviations from this relationship can be observed when one of the bulk salts used for medium preparation contains low amounts of the limiting nutrient as an impurity, or

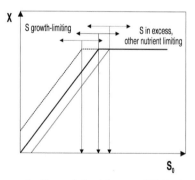

Fig. 4. Concentration of dry biomass (X) that can be produced in a medium, as a function of the initial concentration of the growth-limiting substrate (S_0). (-----), X in a case where low amounts of the limiting substrate are introduced with an other medium component; (......), X in a case where part of the limiting substrate is not available for the cells, for instance, due to precipitation with another medium component.

when a certain amount of the limiting nutrient becomes inaccessible in the medium, (e.g., due to precipitation with another medium component). (Note that variations in pH due to increasing concentrations of biomass or excreted toxic products can affect cultivation conditions and influence biomass yield also).

In practice, the linear relationship between X and S_0 is often observed, although interpretation of the data is frequently not as straightforward as suggested by Fig. 4. This is demonstrated in Fig. 5 for *Pseudomonas oleovorans* growing in continuous culture at a fixed dilution rate with a mineral medium in which either carbon or nitrogen was limiting growth, depending on the ratio of the two nutrients. By keeping

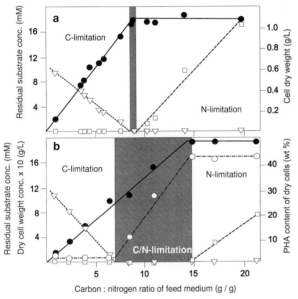

Fig. 5. Growth of *Pseudomonas oleovorans* with either citrate (a) or octanoate (b) as the sole source of carbon in continuous culture at a fixed dilution rate of 0.20/hr[1] as a function of the C:N ratio of the feed medium. The C:N ratio of the feed medium was varied by keeping the concentration of the nitrogen source (NH_4^+) constant and changing the concentration of the carbon source. (Adapted from Durner R. (1998). Ph.D. thesis No. 12591, Swiss Federal Institute for Environmental Science and Technology, Zürich, Switzerland). ● indicates cell dry weight; ▽, ammonium-N concentration; □, concentration of citric acid (a) or octanoic acid (b) in the culture; ○, polyhydroxyalkanoate (PHA) content of cells.

the concentration of ammonia constant and increasing the concentration of the carbon source this bacterium was cultivated at different C:N ratios. In the resulting biomass, ammonia and carbon-source concentrations were measured at steady state. When *P. oleovorans* was cultivated with citric acid as the sole source of carbon–energy the steady-state biomass concentration in the culture initially increased linearly with the concentration of the carbon source (Fig. 5a). Accordingly, the residual concentration of excess nitrogen decreased in the culture broth with increasing feed C:N ratios. At a C:N ratio of ~8.5 g/g, ammonia was consumed to completion. A further increase of the carbon concentration in the feed medium led to no further increase of the biomass produced. Instead, excess citric acid accumulated in the culture. Thus, one growth regime clearly limited by carbon with nitrogen in excess, and one limited by nitrogen with carbon in excess, can be recognized. When the same experiment was performed with octanoic acid as the sole carbon source, the biomass concentration also increased initially when growth was carbon-limited with excess nitrogen present in the culture (Fig. 5b). At a feed C:N ratio of 7.0, the residual concentration of nitrogen in the culture became undetectable. Despite this, the concentration of the biomass in the culture continued to increase linearly, with all the carbon consumed to completion when the concentration of octanoate was further increased in the feed medium. Only when the C:N ratio in the feed exceeded 14.5 did unutilized octanoate became detectable in the culture liquid. Thus, limitation based on the pattern of the biomass concentration growth became nitrogen-limited above a C:N ratio of 18.3. The residual concentration of the nitrogen source in the culture indicated the limitation of the culture by nitrogen already at C:N feed ratios higher than 7.0. The analysis of the cells showed that the effect observed was a result of channelling the surplus carbon into the formation of the reserve material polyhydroxyalkanoate (in other organisms this may be polyhydroxybutyrate, glycogen, or lipids). This dual-nutrient-(carbon–nitrogen)-limited growth regime is always observed when the organism has the ability to store carbonaceous reserve materials. Here, the extension of this zone between the two single-nutrient-limited growth regimes depends on the storage capacity of the organism for reserve material and the growth rate (Egli, 1991; for effects observed in batch culture see Wanner and Egli, 1990). Such multiple-nutrient-limited zones are not only observed for the interaction of carbon with nitrogen but also for other combinations of nutrients, such as C–P, C–Mg, or N–P. Furthermore, the extension of this zone will be determined by the limits a microorganism exhibits with respect to its elemental composition under differently limited growth conditions.

Thus, even when a linear relationship between the biomass produced is obtained, care has to be taken in the interpretation. In such a case, it is of advantage to know which of the nutrients is the second limiting component in the medium.

C. Assessing the Quality of Media and Some Notes of Caution

This approach cannot only be used to design growth media limited by a specifically selected nutrient, but it can also be used to assess the quality of various media. (Never assume that growth media reported in the literature are perfect. Many times they are not, nor are they necessarily employed for the purpose they were designed!) Such an analysis gives usually a good understanding of the capacity of a growth medium with respect to the maximum biomass that it can support, the nature of the limiting nutrient, and the degree of excess of other nutrients (Egli and Fiechter, 1981).

Note that most of the classical media used before the 1960s did not include trace elements. Their addition was usually not necessary because trace elements were contained as impurities in the bulk minerals used for the preparation of the medium. Modern media are frequently prepared with "ultrapure" salts, and, not surprisingly, they fail to produce good growth unless they are supplemented with a trace-element solution. A typical example is the classic synthetic medium M9 that is used widely to grow *E. coli* in genetic studies. This medium in its original composition does not support the growth of *E. coli* for more than a few generations, after which growth slows down and finally comes to a halt.

Acknowledgment

The author is indebted to Peter Adriaens for his careful reading of the manuscript and for valuable comments.

See Also the Following Articles

AUTOTROPHIC CO_2 METABOLISM • CONTINUOUS CULTURE • GROWTH KINETICS, BACTERIAL • LOW-NUTRIENT ENVIRONMENTS • ORE LEACHING BY MICROBES

Bibliography

Atlas, R. M. (1997). "Handbook of Microbiological Media," 2nd ed. CRC Press, Boca Raton, FL.

Balows, A., *et al.* (1992). "The Prokaryotes: A Handbook of the Biology of Bacteria," 2nd ed. Springer Verlag, New York.

Bridson, E. Y., and Brecker, A. (1970). Design and formulation of microbial culture media. *In* "Methods in Microbiology" (J. R. Norris and D. W. Ribbons, eds.), Vol. 3A, pp. 229–295. Academic Press, New York.

Egli, T. (1991). On multiple-nutrient-limited growth of microorganisms, with special reference to dual limitation by carbon and nitrogen substrates. *Antonie van Leeuwenhoek* 60, 225–234.

Egli, T., and Fiechter, A. (1981). Theoretical analysis of media used for the growth of yeasts on methanol. *J. Gen. Microbiol.* 123, 365–369.

Herbert, D. (1961). The chemical composition of microorganisms as a function of their environment. *Symp. Soc. Gen. Microbiol.* 11, 391–416.

Herbert, D. (1976). Stoicheiometric aspects of microbial growth. *In* "Continuous Culture 6: Application and New Fields" (A. C. R. Dean *et al.*, eds.), pp. 1–30. Ellis Horwood, Chichester, UK.

Kovárová, K., and Egli, T. (1998). Growth kinetics of suspended microbial cells: From single-substrate-controlled growth to mixed-substrate kinetics. *Microbiol. Molec. Biol. Rev.* 62, 646–666.

LaPage, S. P., Shelton, J. E., and Mitchell, T. G. (1970). Media for the maintenance and preservation of bacteria. *In* "Methods in Microbiology" (J. R. Norris and D. W. Ribbons, eds.), Vol. 3A, pp. 1–133. Academic Press, New York.

Linton, J. D., and Stephenson, R. J. (1978). A preliminary study on growth yields in relation to the carbon and energy content of various organic growth substrates. *FEMS Microbiol. Lett.* 3, 95–98.

Neidhardt, F. C., Ingraham, J. L., and Schaechter, M. (1990). "Physiology of the Bacterial Cell: A Molecular Approach." Sinauer, Sunderland, MS.

Pirt, S. J. (1979). "Principles of Microbe and Cell Cultivation." Blackwell Scientific, Oxford, UK.

Schaechter, M., Maaløe, O., and Kjeldgaard, N. O. (1958).

Dependency on medium and temperature of cell size and chemical composition during balanced growth of *Salmonella typhimurium. J. Gen. Microbiol.* **19**, 592–606.

Tempest, D. W. (1969). Quantitative relationship between inorganic cations and anionic polymers in growing bacteria. *Symp. Soc. Gen. Microbiol.* **19**, 87–111.

Thauer, R. K., Jungermann, K., and Decker, K. (1977). Energy conservation in chemotrophic anaerobic bacteria. *Bacteriol. Rev.* **41**, 100–180.

Wanner, U., and Egli, T. (1990). Dynamics of microbial growth and cell composition in batch culture. *FEMS Microbiol. Rev.* **75**, 19–44.

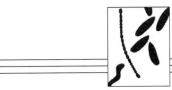

Oil Pollution

Joseph P. Salanitro
Shell Oil Company

GLOSSARY

biodegradation The complete or partial breakdown of a substance, usually by microorganisms present in soils, sediments, biosludges, marine water, and freshwater.

bioremediation (microbial remediation) A treatment process in which nutrients (e.g., ammonia-nitrogen, NH_4^+ and phosphate-phosphorus, PO_4^{3-}) are added to wastes, soils, or waters to enhance the degradation (oxidation or reduction) of organic or inorganic compounds by indigenous or applied microorganisms (bacteria, fungi) under aerobic (oxygen-supplied) or anaerobic (oxygen-deficient) conditions.

crude oil A natural liquid petroleum deposit containing hundreds of compounds derived from the high-temperature and -pressure aging of buried plants, animals, and microbes deep in the ground; it is refined into oil products. A barrel of petroleum oil is 42 gallons and is less dense than water.

hydrocarbons Organic compounds present in crude oils, natural gases, and fuel-oil products composed primarily of carbon and hydrogen atoms. Major classes of compounds include alkanes, alkenes, branched alkanes (isoalkanes), naphthenes (cyclic saturated rings), and mono- and polyaromatic structures.

oil products Fuel products such as gasoline, diesel, jet fuel, heating oil, and marine (bunker) ship fuel that are derived from refining (using distillation, catalytic cracking, and reforming processes) crude oils.

OIL POLLUTION is the discharge of crude oil or oil products (fuels) into the marine, freshwater, and land environments. This spilled hydrocarbon can originate from natural deep subsurface seepage of oil deposits and the accidental releases from oil-producing platforms, tanker ships, and above-ground or below-ground storage tanks and pipelines. Land and water ecosystems have been known to contain naturally occurring microbes that can degrade several types of hydrocarbon compounds present in crude oil and fuel.

In the 1940s, Zo Bell described some of the first laboratory studies on the assimilation of straight-chain and branched-chain hydrocarbons by marine microorganisms. Since that time, oil-degrading species (bacteria, fungi, and yeasts) have been shown to be biochemically diverse and ubiquitous in many bodies of water and in the terrestrial environment, including marine water and sediments and freshwater and sediments, surface and subsurface soils, groundwater, and municipal or industrial biosludges. These organisms can biodegrade a variety of hydrocarbon classes (alkanes, isoalkanes, and aromatics) present in crude oils and refined oil products (fuels).

The management of microbial oil-degrading activity, however, has been more commonly applied to treat oily wastes and oil-contaminated soils and hydrocarbons in groundwater in which the oil pollution source and concentration are controllable (i.e., biotreaters and land treatment). There have also been attempts to enhance oil biodegradation in open-water pollution incidences due to marine spills, but these appear to be difficult to assess. Here, sources of oil pollution and the potential and limitations of microbial oil-degrading activity to remediate oil pollution in water and land environments will be discussed.

I. OIL-POLLUTION SOURCES

Since the 1920s, oil and gas exploration and production have increased to enormous proportions in developed and developing countries (United States, Europe, Middle East, Far East, Africa, and South America). The petroleum industry has provided products for the development of the vast transportation, manufacturing and chemical sectors of the world societies. World crude oil production is about 60 million barrels/day (b/d). In the United States, daily oil consumption is 16–18 million b/d or 25–30% of the world's output. About 70% of each barrel of crude oil is converted in petroleum refineries to oil products such as automobile gasoline and truck diesel and jet fuels. The remaining 30% of the barrel is made into heating and lubricating oils, bunker fuel, and asphalt. In plants containing catalytic and reformer processes, a significant portion of the barrel can also be converted into feedstocks for the chemical industry.

Oil pollution by crude oil and fuels is a consequence of oil and gas exploration and production, its storage and transportation, and its manufacture into useful products. In addition, the combustion of oil and fuels for transportation and for use in domestic and industrial heating and electricity production are sources of air, water, and land pollution. The origins of major oil pollution sources in open-water bodies and land environments are given in Table I. It is difficult to determine precisely the amounts of oil and fuel products discharged to land and sea through accidents, human error, and process and equipment failures. However, in 1990 it was estimated that 16.4 million barrels or 0.07% of the total world oil production (21.9 billion barrels/year) may have been spilled from all sources, including natural seepage, production, storage, transportation, and manufacturing of oil and oil products. Natural seeps of oil from buried deposits of petroleum under high temperature and pressure from inside the Earth continuously enter ocean-floor sediments and Earth's

TABLE I
Sources of Pollution

Environment	Possible causes	Spill
Open water bodies (oceans, seas, rivers)	Natural seeps	Crude oil
	Offshore oil well platform (blowout, drilling muds, produced water)	Crude oil, fuel
	Subsea pipeline leak	Crude oil
	Supertanker accident	Crude oil, fuel
	Tank barge accident	Crude oil, fuel
	Marine storage tank leak	Crude oil, fuel
	Coastal refinery	Crude oil, fuel
Land (surface soil and water and subsoils and groundwater)	Natural seeps	Crude oil
	Onshore oil wells (blowout, drilling muds, produced water)	Crude oil, fuel
	Bulk storage terminal tank (leak)	Crude oil, fuel
	Pipeline leak	Crude oil, fuel
	Refinery process explosion	Crude oil, fuel
	Refinery tank leak	Crude oil, fuel
	Truck/rail accident	Fuel
	Retail service station release	Fuel

surface at a very low rate ($< 10\%$ of all ocean oil spills). Examples of such seeps are those of the seabed off the coast of Santa Barbara, CA, the Brea tar pits of Los Angeles, CA, and the asphalt lakes of Trinidad. Spills of oil at offshore platforms from incidences of oil well blowouts, discharges of spent drilling muds onto the seabed, and the discharge of oil in produced water represent about 7% of the total oil spilled to the oceans (1.15 million barrels). The magnitude of rare well blowouts has varied from 100,000 barrels (in the North Sea, 1977) to the largest, over 2 million barrels (in Ixtoc, Gulf of Mexico, 1979). Most large sources of oil pollution have occurred, however, with the use of supertanker shipping, tank barging in seaports and rivers, pipeline transmission, and surface transportation (trucking) of crude oil and fuels. Tankers carry oil from oilfields and refined products from terminals across the oceans to refinery depots with the potential of fouling open waters, nearshore and coastal beach, and marsh environments. Records of spillage of crude oil from oil tanker accidents have varied from 260,000 barrels (*Exxon Valdez,* Prince William Sound, Alaska, 1989) to 1.6 million barrels (*Amoco Cadiz,* Brittany Coast, 1978). Marine terminals, harbors, and inland rivers involved in the loading, storing oil and fuels, and processing of oily wastes (ship bilge water) are also potential oil-spill sources and are responsible for the contamination of shorelines and sediments. Coastal refineries may discharge small amounts of oil due to unloading from barges to tanks. Catastrophic oil and fuel releases from the destruction of oil wells and oil or fuel tank storage terminals due to sabotage and war should also be mentioned. For example, in the 1991 Persian Gulf War, 6–8 million barrels of oil were discharged along the Gulf area into inlet waterways, beaches, and sediments, and was the single largest oil spill in history.

Estimates of the amount of oil and fuel spilled on surface and subsurface soils are not known, but major releases can occur from pipeline leaks. The United States has thousands of miles of surface and subsurface 12- to 40-inch pipelines carrying large volumes of oils and fuels under high pressure between refineries and storage terminals. Several thousand gallons of the product can be lost from a pipeline. In 1997, pipeline leaks represented 70% of the oil spilled

(40,000 barrels) on land. Most of the oil or fuel is usually contained in subsoils, but hydrocarbons may be transported to surface bodies of water (streams) or enter aquifer subsoils and create groundwater plumes of contaminant. In this respect, fuel leaking from underground storage tanks and from service station gasoline pumps represents another major source of hydrocarbon pollution in groundwater. Storage-tank leaks (due to corrosion or valve failure) in land-based facilities (e.g., refineries and bulk terminals) also represent a source of oil pollution to surface soils, subsoils, and groundwater. In 1997, storage-tank failures represented 20% of oil spilled in the United States. The fate and transport of hydrocarbons to the subsurface in these incidences parallel those from pipeline releases.

II. MICROBIAL DEGRADATION OF HYDROCARBONS IN CRUDE OIL AND OIL PRODUCTS

Although indigenous microbes in bodies of water and soils represent a major means of breaking down spilled oil in the environment, not all hydrocarbons in crude oils and fuels are readily biodegraded or transformed to CO_2 or nontoxic compounds. Table II is a summary of hydrocarbon classes present in oil and fuels. Varying amounts of normal alkanes (e.g., octane), isoalkanes (branched alkanes, e.g., isooctane), cycloalkanes (naphthenes, e.g., cyclohexane), monoaromatics (e.g., benzene), polyaromatics (e.g., naphthalene and phenanthrene), and heterocyclics (thiophenes) are present in crude oils and fuels. Crude oils contain several hundred compounds and isomers and can vary in composition with light and medium crudes (from the Middle East, Texas, Michigan, Alaska, and the Gulf of Mexico) containing carbon numbers C_2–C_{40} and heavy crudes (from California, Mexico and South America) containing C_{10}–C_{60+} hydrocarbons. Heavy crude oils also contain high-molecular-weight compounds, cycloalkano-aromatics, and hydrocarbons with S, N, and O atoms (resins and asphaltenes). Refined oil products (fuels) vary in hydrocarbon class distribution, and with the exception of bunker oil, contain alkane, isoalkane, cycloalkane, and aromatic compounds of C_5–C_{24}.

TABLE II
Composition and Degradability of Major Classes of Hydrocarbons in Crude Oil and Refined Oil Products[a]

Substance (carbon no. range)	Alkanes[b]	Isoalkanes	Cycloalkanes[c]	Monoaromatics	Polyaromatics	Heterocyclics[d]
Crude oil[e] (C_2–C_{60}+)	20	4	4	4	3	0–2
Gasoline (C_5–C_{11})	22	28	3	28	6	—
Kerosene fuel (C_7–C_{21})	30	5	ND	3	2	—
Jet fuels (C_7–C_{24})	16–28	2–22	3–39	3–26	1–15	—
Heating oil (C_8–C_{24})	20	27	20–39	19–26	15	—
Bunker C (C_9–C_{60}+)	2	11	15	2	34	36
Motor/lubricating oil (C_{10}–C_{24})	24	20	35	4	10	≤0.1
Biodegradability[f]	+	+/−	−	+	+/−	+/−

[a] For identified hydrocarbon compounds. Values in wt%.
[b] Includes some alkenes.
[c] Includes some cycloalkane aromatics. ND indicates no data.
[d] Contains polar hydrocarbons with N, S, and/or O atoms.
[e] Medium gravity oil (API 30-35).
[f] + indicates many; +/−, some (low molecular weight); and −, few compounds degraded by microbes.

The literature on the microbial metabolism of hydrocarbons by pure and mixed cultures from seawater, freshwater, sediments, soils, and groundwater is extensive. In general, compounds with carbon range C_1–C_{30} have been shown to be degraded in laboratory cultures and environmental media. Alkanes up to C_{30} are degraded but branched-chain alkanes, especially those with several methyl groups, closely aligned alkyl groups in a carbon chain, or tertiary carbon structures, are refractory, partially oxidized, or slow to biodegrade. Representative compounds in this group are 3,3-dimethylpentane and pristane (2,6,10,14-tetramethylpentadecanoic acid). Only a few cycloalkanes (e.g., cyclohexane and methylcyclohexane) have been shown to be metabolized by microbes. Highly branched cycloalkanes or dicyclic alkanes (e.g., tetralin) have not been shown to be biodegraded. In this respect, the tetracyclic alkanes represented by the class of C_{30} hopane molecules are recalcitrant and have been used as stable biomarkers to follow the biodegradation of other hydrocarbons in oily wastes and spills in sediments and soil. Monoaromatics and up to three-ring polyaromatics have been shown to be readily metabolized by a variety of microbes under aerobic and anaerobic (nitrate-, iron-, and sulfate-reducing and methanogenic) conditions. Finally, there are some data on the transformation of several heterocyclic hydrocarbons (e.g., thiophenes), but not cyclic alkane carboxylic acids (e.g., naphthenic acids), in soil and sewage sludge.

III. BIOREMEDIATION OF OIL SPILLS IN OPEN WATERS

The fate of crude oil has been studied in laboratory-simulated oil slicks and actual spills to the marine environment. Evaporation of hydrocarbons from spreading oil slicks is rapid and can account for 15–30% of the oil loss, depending on the crude-oil type, wind and water-current velocities, and air and water temperature. Dissolution of hydrocarbons into the water column represents a small fraction of the oil because hydrocarbon solubilities in water are low (20–60 mg/liter). UV sunlight oxidation usually accounts for only 1% of the spilled oil. The tendency of crude oils to form emulsions in water (30–50% water) is common in marine spills and depends on the viscosity and presence of surface-active compounds (acids and heterocyclics), asphaltenes and wax content of the spilled oil, and tidal action of the water body. The emulsified oil as a semisolid mass develops with continued weathering into the consistency of "chocolate mousse," which may sink to sediments or get washed ashore (as much as 30–50% of spilled oil) as oil–mineral aggregates or tarballs.

The actual fraction of spilled oil degraded by indigenous microbes present in the water column has been difficult to assess. Major factors that limit significant natural bioremediation of oil slicks include the presence of very low levels of nutrients (0.1–1 mg/liter N and 0.1 mg/liter P) in seawater, the limited oxygen-transfer rate to the water column and oil emulsion to stimulate aerobic degradation, the inaccessibility of hydrocarbons sequestered in oil emulsions and tarballs, and the low environmental temperature (e.g., Alaska) of some areas. Most oil hydrocarbons that are readily degraded (i.e., alkanes and aromatics) require 2–3 mg oxygen/mg hydrocarbon for significant removal to be observed. Also, the biodegradation of hydrocarbons at low oxygen levels (e.g., in sediments) or under anaerobic conditions is much slower. Early reports from the 1960s of the degradation of oil in seawater ranged from 35–350 mg/liter/year. However, in simulated seawater microcosms, dodecane was degraded at a rate of only 0.01–1 mg/liter/day. Estimates of the contribution of microbial degradation of hydrocarbons in the water column, ocean sediments, and onshore beaches are difficult to assess because of the rapid removal processes of evaporation, current and tidal movement of the oil slick, and the changing properties of the spilled oil. In the *Amoco Cadiz* oil spill it has been estimated that about 4.5% of the hydrocarbons may have been degraded by microbes in water, sediments, and beaches. In the case of the *Exxon Valdez* spill, about 10% of the oil may have been biodegraded. However, extensive field experiments (1989–1990) on enhancing the bioremediation of oil washed onto beaches of the Prince William Sound were carried out. The application of a water-soluble N and P fertilizer, a slow-release N–P agent, and an oleophilic nutrient (N–P, surfactant, fatty acid, and Inipol EAP-22) indicated that hydrocarbon degradation was stimulated by at least three- to fivefold over untreated plots. These positive results were based on the reduction of oil-hydrocarbon concentrations and increases in the populations of oil-degrading bacteria in beach material. The large tidal fluxes in the spill area, however, made sampling and statistical analyses difficult. Beaches treated with the nutrient fertilizer were also "visually cleaner" than control areas.

The application of microbial cultures (seeding) onto oil slicks and emulsions has been attempted after two tanker spills in 1990 in the Gulf of Mexico (*Mega Borg* and *Shinoussa*). Preparations of bacteria and nutrients were applied to oil slicks and in marshes impacted by oil. The efficacy of microbial seeding in these applications was difficult to interpret because of the tidal influence on the disruption of the slicks. Furthermore, because of the hydrophilicity of the applied cultures and fertilizer nutrients, much of the applied material could readily disperse and dilute into the water column. Little degradative activity, therefore, would be associated with the oil slick or emulsion for the previously mentioned reasons of the inaccessibility of the hydrocarbons for microbial attack. The efficacy of seeding of oil slicks in open-water bodies with nutrients or microbes appears to lack good scientific support regarding the extent of oil slick-emulsion deterioration by microbial oxidation. Also, the addition of nonindigenous microbes and nutrients to seawater may have adverse consequences such as toxicity for invertebrates and fish, a sudden increase in planktonic organisms, and changes in species survival and diversity.

IV. BIOREMEDIATION OF OIL SPILLS ON LAND

The use of indigenous microbes in surface and subsurface soils to biodegrade spilled crude oil and fuel oils has been widely practiced in the petroleum and remediation industries since the 1960s. Various types of techniques of land treating, soil composting (aerated oily-soil piles) and soil slurry-phase reactors (10–30% soil/water) have been used with the application of fertilizer nutrients (NH_4^+-N or NO_3^--N and PO_4^{-3}-P), mixing, moisture adjustment, and aeration to the waste soil. Many refined products (gasoline, diesel, and aviation fuel) contain a significant fraction ($<C_{12}$) of volatile hydrocarbons, and treatment of fuel–contaminated soils is a combination of evaporation and enhanced biodegradation. Fuel spills can, therefore, be remediated to <90% of the spilled hydrocarbon. Land treating of oil- and fuel-containing soils has shown that the extent of bioremediation varies with the crude-oil type, hydrocarbon-structure composition, soil type, and the initial concentration of oil and nutrients added. In oily soils, the influence of nutrient amendment (N and P) has been extensively studied, but it is not clear why hydrocarbon degradation is similar for various levels of nutrients. Laboratory and field soil-treatment experiments carried out over months and years have shown that 10–85% of oil hydrocarbons can apparently be degraded by naturally occurring microbes. Many soil experiments support the pure-culture laboratory findings that alkanes and monoaromatics are degraded more readily than polyaromatics, cycloalkanes, and polar compounds. Also, low-viscosity crudes (light and medium oils) degrade more extensively in soils than heavy oils. Our understanding of soil-phase bioremediation indicates that the availability of hydrocarbons from sorption to soil organic matter and clays, and from slow desorption and dissolution, as well as their sequestration into soil pores, significantly limits the microbial oxidation of even readily metabolizable compounds.

In subsoils and aquifers contaminated by fuel-oil releases from leaking underground storage tanks, the bioremediation of hydrocarbons has also been extensively investigated. Releases of gasoline from retail service stations from tanks, flow lines, or dispensers can be transported to shallow aquifers. The fuel spill migrates on the top of the water table and water-soluble gasoline hydrocarbons (mainly the monoaromatics, benzene, toluene, ethylbenzene, and xylenes; together designated BTEX) are transported with the groundwater flow field. These aromatic compounds move in the direction of the groundwater, creating a plume of varying concentrations. Laboratory microcosms and field studies reported in the literature have shown that aerobic BTEX-degrading organisms present in all soils are a major factor for the intrinsic bioremediation (bioattenuation) of these regulated substances from impacting drinking water wells. Dissolved oxygen levels in groundwater and the extent of natural reaeration of the aquifer (e.g., air moving from soil zone above the water table and water-table fluctuations), govern, to a large extent, whether these hydrocarbon plumes are stable or "shrink" in size with biodegradation. Anaerobic aquifer soil studies, however, have also shown that BTEX may be degraded under nitrate-, iron-, and sulfate-reducing and methanogenic conditions, but the contribution of such mechanisms in the field is unclear where aquifer reaeration may be occurring and strict anaerobic conditions may be limited.

There is clear evidence with the use of sufficient monitoring wells, hydrogeological description, and computer modeling of a site that aromatic-hydrocarbon plumes are degraded by microbial processes. This type of bioattenuation for readily biodegradable

compounds is an effective tool for managing fuel pollution in the subsurface. Current research is also underway, however, to determine whether gasoline-vapor plumes, created by the residual spilled fuel, in the unsaturated zone of soil above the water table are similarly biodegraded by soil bacteria.

V. SUMMARY

Oil-pollution prevention and control will continue to be a challenge with the quest for new sources of petroleum energy, its recovery, transportation, and use as fuels. Federal, state, and international legislative bodies and cooperative agreements between government and industry must ensure the reduction of oil spills and the availability of appropriate remediation technologies in the protection of natural resources, ecosystems, and human health.

Indigenous oil-degrading microbes in the environment will play a part, to some extent, in the destruction of petroleum spilled into open-water bodies, sediments, beaches, marshes, surface and subsurface soils, and groundwater. Our understanding of hydrocarbon biodegradation in these media, however, indicates that there are limitations to the microbial breakdown of oils and fuels. For example, oil contains a variety of hydrocarbon classes with carbon numbers of C_2–C_{60}+. The isomers associated with a carbon number increase exponentially. A C_{10} compound may have 75 different possible isomeric structures, but a C_{20} and C_{30} compound could have 366,000 and 4 billion different structures, respectively. It is not possible, therefore, to analyze all the compounds in any crude oil by the most advanced instrumentation, nor is it feasible to show that microorganisms in a water column, sediment, or soil can completely degrade all the various isomers in a spilled sample. Many compounds in an oil or fuel may be recalcitrant or only partially metabolized because some structures sterically hinder attack by microbial oxidative enzymes. These carbon compounds do not support microbial growth very well and the ability to sustain adequate populations in environmental media will be minimal, even in the presence of adequate nutrients.

The use of microbial remediation techniques (fertilizer and nutrient application, and inoculation) for crude-oil releases in open waters will be limited because of the inability to control the contact of oil slicks with microbes and nutrients. The transformation of spilled oil into emulsions and tarballs also makes biodegradation in sediments and beaches slow because of the inaccessibility of hydrocarbons to microbial attack.

Although the application of bioremediation to releases of oils and fuels to soils, however, may represent better controlled processes (contamination concentration and nutrient and aeration mixing), work indicates that even readily degradable hydrocarbons are not completely destroyed by microbiota because a significant portion of the oil is sequestered in the soil matrix. In this respect, risk-assessment evaluations involving the nonbioavailability and reduced toxicity of compounds to ecological and human receptors may be considered in cost-effective restoration of impacted sites. Finally, fuel-oil spills from underground storage tanks to groundwater often represent a threat to drinking water supplies. However, the bioattenuation of aquifer plumes has been readily applied in the field because most of the water-soluble compounds (monoaromatics) from fuel spills are readily degraded and support the growth and survival of natural hydrocarbon-degraders.

See Also the Following Articles

BIODEGRADATION • BIOREMEDIATION • INDUSTRIAL EFFLUENTS: SOURCES, PROPERTIES, AND TREATMENTS

Bibliography

Atlas, R. (1984). "Petroleum Microbiology." McGraw-Hill, New York.

Atlas, R. M., and Bartha, R. (1992). Hydrocarbon biodegradation and oil spill bioremediation. *Adv. Microbial Ecol.* **12**, 287–338.

Clark, R. B. (1992). "Marine Pollution," 3rd ed. Clarendon Press, Oxford.

Geizer, R. A. (1980). "Marine Environmental Pollution, 1 (Hydrocarbons)" Elsevier, New York.

Gundlach, E. R., Boehm, P. D., Marchand, M., Atlas, R. M., Ward, D. M., and Wolfe, D. A. (1983). The fate of Amoco Cadiz Oil. *Science* **221**, 122–129.

Leahy, J. G., and Colwell, R. R. (1990). Microbial degradation of hydrocarbons in the environment. *Microbiol. Rev.* **54**, 305–315.

Mackay, D., and McAuliffe, C. D. (1988). Fate of hydrocarbons discharged at sea. *Oil and Chemical Pollution* **5**, 1–20.

Morgan, P., and Watkinson, R. J. (1994). Biodegradation of components of petroleum. *In* "Biochemistry of Microbial Degradation" (C. Ratledge, ed.), pp. 1–31. Kluwer Academic, Boston, MA.

National Research Council. (1985). "Oil in the Sea—Inputs, Fates and Effects." Steering Committee for the Petroleum in the Marine Environment Update. National Academy Press, Washington, DC.

National Research Council. (1993). "In Situ Bioremediation: When Does it Work?" National Academy Press: Washington, DC.

Norris, R. D., Hinchee, R. E., Brown, R., McCarty, P. L., Semprini, L., Wilson, J. T., Kampbell, D. H., Reinhard, M., Bouwer, E. J., Borden, R. C., Vogel, T. M., Thomas, J. M., and Ward, C. H. (1994). "Handbook of Bioremediation." Lewis, Boca Raton, FL.

Potter, T. L., and Simmons, K. E. (1998). "Total Petroleum Hydrocarbon Criteria Working Group Series, Vol. 2, Composition of Petroleum Mixtures." Amherst Scientific Publishers, Amherst, MA.

Riser-Roberts, E. (1998). "Remediation of Petroleum Contaminated Soils" Lewis, Boca Raton, FL.

Salanitro, J. P. (1993). The role of bioattenuation in the management of aromatic hydrocarbon plumes in aquifers. *Ground Water Monit. Remed.* **13**, 150–161.

Söhngen, N. L. (1913). Benzin, Petroleum Paraffinol und Paraffin als Kohlenstoff und Energiequelle für Mikroben. *Centralbl. Bakt., Abt. II* **37**, 595–609.

Swannell, R. P. J., Lee, K., and McDonagh, M. (1996). Field evaluations of marine and spill bioremediation. *Microbiol. Rev.* **60**, 342–365.

U.S. Congress, Office of Technology Assessment. (1991). "Bioremediation for Marine Oil Spills." Background Paper, OTA-BP-0-70. U.S. Government Printing Office, Washington, DC.

Wolfe, D. A., Hameedi, M. J., Galt, J. A., Watabayashi, G., Short, J., O'Claire C, Rice, S., Michel, J., Payne, J. R., Braddock, J., Hanna, S., and Sale, D. (1994). The fate of the oil spilled from the Exxon Valdez. *Environ. Sci. Technol.* **28**, 561A–568A.

Zo Bell, C. E. (1946). Action of microorganisms on hydrocarbons. *Bact. Rev.* **10**, 1–49.

Oncogenic Viruses

Anh Ngoc Dang Do, Linda Farrell, Kitai Kim, Marie Lockstein Nguyen, and Paul F. Lambert

University of Wisconsin Medical School

GLOSSARY

apoptosis A form of cell death in which the cell is argued to actively commit suicide. Apoptosis is characterized by the nucleolytic digestion of the cellular chromatin at points between nucleosomal units, the breakdown of the nucleus, and blebbing at the cell surface.

immortalization The capacity of a cell in tissue culture to grow ad infinitum. Normally somatic cells have a finite lifespan, at the end of which the cells senesce. Immortalization requires that the cell override the mechanisms that normally cause senescence.

oncogene A gene that encodes a factor which can confer a transformed phenotype onto cells in tissue culture or a tumorigenic phenotype *in vivo*.

provirus The double-stranded DNA replicative intermediate of a retroviral genome found integrated into the host chromosome.

transformation Morphological alteration or altered growth properties of a cell in tissue culture that are thought to reflect oncogenic properties; i.e., contact-uninhibited growth, anchorage-independent growth, and immortalization.

tumor suppressor gene A gene whose loss of function is associated with the tumorigenic phenotype. Commonly, such genes when reintroduced into a tumor cell inhibit its growth or tumorigenic phenotype.

ONCOGENESIS is believed to be a multifactorial process in which cells must acquire multiple aberrant properties that permit them to grow in an unregulated manner, to invade tissue, and to metastasize—that is, to establish residence in new locations within the organism. These properties are thought to be conferred mostly through inheritable changes to a cell which occur from both genetic and epigenetic events. The contributions to cancer of tumorigenic viruses mostly reflect genetic alterations to the cell. These alterations include the alteration of cellular genes (e.g., the insertionally mutagenic retroviruses) or the introduction of viral genes that directly contribute to oncogenesis (e.g., the human papillomaviruses that cause cervical cancer). In some cases, it is not apparent that a tumorigenic virus causes any genetic or epigenetic change to a cell that directly contributes to oncogenesis. Here it is possible that the host response to the infection is the contributing cause of virally-induced cancer (e.g., human hepatitis B virus). In this article, we describe the current understanding of how various viruses contribute to cancer. Emphasis is placed on describing the role of viruses in human cancers. We also describe the important insights gained from the study of animal viruses, which have served historically as important models for the study of tumor viruses.

I. GENERAL PRINCIPLES OF VIRALLY INDUCED CANCERS

Infectious agents, including viruses, bacteria, and parasites, are thought to be the etiologic agents in approximately 16% of human cancers. Human oncogenic viruses include hepatitis B and hepatitis C viruses (associated with hepatocellular carcinoma), Epstein–Barr virus (associated with B cell lymphomas), human papillomaviruses (associated with cervical carcinoma), human T cell lymphotropic virus I (HTLV-1; associated with T cell lymphomas), and human herpes-virus type 8 (associated with Kaposi's sarcoma). The initial recognition that viruses cause cancer arose from studies of animal viruses. In the early twentieth century, transmittable agents were demonstrated to cause tumors in chickens and in rabbits. The respective agents were later identified to be Rous sarcoma virus (RSV), the first-studied RNA tumor virus, and Shope papillomavirus, the first-studied DNA tumor virus. Much of our understanding of viral oncogenesis derived initially from the study of such animal viruses. In recent years, much attention has been focused on the study of human tumor viruses. The study of virally induced cancers has provided many basic insights into cancer, the most important of which is the identification and functional characterization of many oncogenes and tumor suppressor genes. From these collective studies several generalizable principles of virally induced cancers can be proffered. First, the onset of virally associated cancers is characterized by long latent periods following initial infection. This suggests that the viruses alone are not sufficient to cause cancer. Also, given this long latency, oncogenic viruses must be able to persist in the host for long periods of time. Second, the cancers induced by viruses do not constitute a natural part of the viral life cycle. Rather, virally associated cancers appear to be dead-end streets for viruses—by-products of the natural infection that provide no advantage to the virus evidenced by the fact that it is uncommon for progeny virus to be produced in the associated tumors. The third principle is that viral properties that contribute to the induction of tumors are ones that commonly play a critical role in the life cycle of the virus. In the following sections, we describe the basic properties of RNA and DNA tumor viruses and the major insights gained regarding how they contribute to cancer.

II. RNA TUMOR VIRUSES

The field of tumor virology initially arose from the study of retroviruses that cause tumors in birds and mammals other than humans. The oncogenic retro-viruses can be divided into three classes: (i) transducing retroviruses, (ii) insertionally mutagenic retroviruses, and (iii) true oncoretroviruses.

Transducing retroviruses are ones that acquire genetic information from the host genome. RSV is one example. Most other examples also were isolated from avian hosts, although a few have been isolated from feline, murine, and simian hosts. The cellular genetic information being transduced is a cellular proto-oncogene which, as present in and expressed from the transducing retrovirus, has oncogenic activity, i.e., it contributes to tumor formation *in vivo* and cellular transformation in tissue culture. The study of transducing retroviruses led to the discovery of more than a dozen oncogenes, including *src, ras,* and *myc.* It was not recognized until the 1970s that these viral oncogenes actually had been picked up by the virus from the host organism through gene transduction. This led to the recognition that animals encode oncogenes. The cellular copies of these genes were initially referred to as proto-oncogenes because it was recognized that the viral copies of these genes had undergone genetic changes that led to the increased expression and/or activity of the encoded gene product. These changes led to the oncogenic nature of the gene. In contrast, the cellular homologs do not contribute to cancer in their wild-type state. For example, the virally encoded *src* and *ras* genes have amino acid substitutions at positions in these proteins that lead to their constitutive activity in signal transduction pathways, and the virally encoded *myc* gene is missing regions of the cellular gene that normally lead to a tight regulation of its expression at the transcriptional and post-transcriptional levels. No examples of transducing retroviruses have been found for human tumors; rather, most such viruses were identified from the evaluation of laboratory stocks of retroviruses that infect laboratory animal species. Generally, these transducing, oncogenic retroviruses are replication defective because of the disruption of one or more essential viral gene by the insertion of the cellular proto-oncogene. The rare exception is RSV, which is a replication-competent virus in which the transduced cellular *src* gene did not disrupt function of any viral gene. How a transducing oncogenic retrovirus arises remains unclear. It may result from a provirus that integrates close to a cellular proto-oncogene. This could result in the formation of a chimeric RNA of viral and cellular origin which, when packaged into progeny retrovirus and on reverse transcription, undergoes rearrangement to produce the actual transducing viral genome.

Insertionally mutagenic retroviruses are ones that, as a consequence of insertion of the proviral genome into the host chromosomes, alter the structure and/or expression of a cellular proto-oncogene or tumor suppressor gene. These retroviruses are also known as the weakly oncogenic retroviruses since they only rarely cause cancers in laboratory animals, reflective of the low likelihood that the provirus insertion occurs at or near a cellular proto-oncogene. There are several mechanisms by which insertion of provirus can lead to activation of a cellular proto-oncogene. First, the virus may integrate upstream of the cellular gene in the same "sense" orientation such that read-through transcription from the viral promoter leads to the generation of chimeric virus : cell mRNAs that can encode the cellular proto-oncogene. Second, the viral transcriptional enhancer can upregulate the activity of the cellular gene's own transcriptional promoter. Finally, the virus can disrupt the integrity of the cellular mRNA leading to the loss of negative regulatory elements such as mRNA instability elements in the mRNA encoding the cellular proto-oncogene. Examples of each of these three cases exist in the literature. Many of the insertionally mutagenic retroviruses cause lymphomas/leukemia in their natural host. In these tumors, the provirus is found nearby to the c-*myc* locus, leading to increased expression of the c-*myc* gene product.

True oncoretroviruses are ones that, through virally encoded activities, alter the host cell or the host cell environment, thereby making cells more susceptible to cancer development. There are but a few cases of true oncoretroviruses. One is HTLV-1, which causes T cell lymphomas in humans. The role of HTLV-1 in human tumors is addressed in detail later.

III. DNA TUMOR VIRUSES

Whereas the study of RNA tumor viruses was largely responsible for the discovery of oncogenes, the study of DNA tumor viruses led to the discovery and elucidation of the function of cellular tumor suppressor genes, most notably *p53* and *pRb*. Much of the initial knowledge about DNA tumor viruses came from the study of SV40, a simian virus, and human adenoviruses. Neither of these are thought to cause tumors in their natural host; however, both can transform cells in tissue culture. Recently, SV40 has been found in the human population and in certain human cancers; however, an etiologic role of SV40 in human cancer is not yet established.

SV40 efficiently transforms cells and causes tumors in hamsters. Several viral gene products were found expressed in these transformed/tumorigenic cells and were identified because hamsters made antibodies against them. Most notable of these was the SV40 large tumor antigen (T Ag). This protein is one of the most potent oncogenic factors known. Its oncogenic activity is multifactorial, arising from its ability to bind and inactivate the cellular tumor suppressor proteins p53 and pRb. p53 is now recognized to be functionally disrupted in most human cancers. The protein was originally identified by virtue of its association with SV40 T Ag.

Initial cloning and functional analysis of *p53* led to its designation not as a tumor suppressor gene but an oncogene. This was due to the fact that the initial *p53* clones were isolated from established cell lines. In these cell lines the *p53* gene had undergone mutations that cause the encoded gene product to possess dominant-negative activity; that is, it could inactivate the normal p53 protein when put back into cells. It was not until the 1980s that the wild-type *p53* gene was recognized to be a tumor suppressor gene. *p53* is thought to play an important role in orchestrating the cellular responses to certain forms of stress, including insult from DNA-damaging agents. Activation of *p53* leads either to growth arrest, which is believed to allow cells time to repair damaged DNA, or apoptosis (programmed cell death). Without functional *p53* cells have a greater propensity for undergoing genetic alterations, some of which can contribute to tumorigenesis. It is now understood that when SV40 T Ag binds p53 protein, it inactivates the latter's function. SV40 T Ag also can bind the gene product of the retinoblastoma tumor susceptibility locus (Rb). Rb and related proteins, p107 and p130, are critical regulators of the cell cycle. In the G_1 stage of the cell cycle pRb binds to and inactivates a family of transcription factors, E2Fs, that regulate the activity of cellular genes involved in DNA synthesis and cell cycle control. Normally, pRb's activity is regulated by cyclin-associated kinases which phosphorylate pRb, leading to its dissociation from E2Fs. T Ag upon binding to pRb or its relatives p107 and p130 can also cause this dissociation and thereby lead to cell cycle deregulation. SV40 T Ag's inactivation of pRb and pRb-like proteins may also result from its ability to induce the degradation of these cellular factors.

Adenoviruses cause respiratory diseases in humans but are not associated with any human cancers. However, these viruses can efficiently transforms cells in tissue culture. The transforming activities of adenoviruses were mapped to a region of the viral genome that encodes sets of gene products, referred to as E1A and E1B, normally expressed early after infection. The E1A gene products were demonstrated to bind a set of cellular factors with different molecular weights. The ability for E1A to transform cells correlated with its ability to bind a subset of these associated factors. E1A was also known to activate transcription of viral promoters, including that for the viral E2 gene. E1A's regulation of the E2 promoter mapped to a promoter proximal DNA element that bound a cellular factor, appropriately labeled E2F. E1A's regulation of the E2 promoter through E2F correlated with its ability to bind a subset of cellular factors including ones of molecular weight p105, p107, and p130. One of these factors was later recognized to be pRb (p105) and the others were demonstrated to be pRb-like proteins. The E2F factor is the same as the aforementioned E2F that regulates expression of cellular genes involved in DNA synthesis and cell cycle regulation. Thus, it was the analysis of E1A that led to the initial understanding of the function of pRb as a regulator of transcription factors, the appreciation that multiple proteins (p107 and p130) share functions with pRb, and the discovery of the E2F family of transcription factors.

The E1B region, which also contributes to transformation by adenoviruses, encodes a factor (E1B55kd) that binds and inactivates p53, like SV40 T Ag. Another E1B gene product (E1B19kd) can directly in-

hibit apoptosis, a process that is induced by *p53*. Thus, both SV40 and adenoviruses have evolved the ability to regulate the *pRb* and *p53* tumor suppressor genes. The importance of these capacities in the context of the viral life cycle is not completely understood but is thought to relate to the need for the virus to induce a DNA synthesis-competent environment in the infected host cell, thereby allowing for production of progeny virus. Inactivation of pRb is likely to be critical for this process. However, why inactivate p53? Normally, it is not active in cells and only becomes active under conditions of stress. One theory is that the aberrant induction of DNA synthesis by these viruses leads to the induction of p53 which, unless overcome, would inhibit viral DNA replication. Recently, multiple links in the pathways that regulate pRb and p53 activity have been made, providing a potential understanding for why inactivation of pRb might trigger activation of p53. The insights gained from the study of SV40 and adenovirus have shed light on how human papillomaviruses cause cervical cancer.

IV. HUMAN TUMOR VIRUSES

In the following sections, we provided brief descriptions of four human tumor viruses, the evidence for their role in human cancers, and the current understanding of their mechanism of action in carcinogenesis.

A. Hepatitis B Virus

Infection by hepatitis B virus (HBV) is strongly associated with the development of hepatocellular carcinoma (HCC), one of the 10 most common carcinomas in the world. This oncogenic association categorizes HBV as a tumor virus. HBV is a hepatotropic, circular, partially duplex DNA virus of 3.2 kb and replicates via an RNA intermediate and reverse transcription. Infection by HBV can lead to liver damage and such pathological diseases as chronic hepatic insufficiency and cirrhosis, leading to HCC. HCC is a difficult cancer to treat because symptoms occur late in the development of the carcinoma, often too late for surgical treatment. Unfortunately, the survival rate for individuals diagnosed with HCC is low. The association of HCC and HBV was determined from worldwide epidemiological studies. In those areas of the world where a higher incidence of HCC was observed (e.g., sub-Saharan Africa and Southeast Asia), there was a higher rate of HBV. Though HCC is considered rare in America and other more developed countries where HBV infection is low, it is a leading cause of cancer in those areas where HBV infection rates are high. This association was demonstrated in a study performed on men in Taiwan in which 191 of 194 of those studied who died of HCC tested positive for the HBsAg surface antigen of the virus. Studies in animal systems have also demonstrated a correlation between infection with HBV and the occurrence of HCC. For example, 100% of woodchucks infected with woodchuck hepatitis virus (WHV) develop HCC. There is a long latency period of at least 30 years between infection by HBV and the development of HCC in humans. During this time, the virus integrates into the host genome and viral protein expression is ablated.

Infection by HBV can lead to either acute or chronic hepatitis. Acute infection involves symptoms such as fever, fatigue, anorexia, and nausea, which are soon followed by jaundice. In adults, the disease is usually cleared by an immune system response to viral surface antigen. In 1–8% of infected adults, however, infection persists and the virus undergoes active replication. Chronic infection rates are much higher (approximately 90%) in the case of perinatal and early childhood infections. It is these chronically or persistently infected patients that are at highest risk for developing HBV-associated HCC. Persistently infected individuals may be asymptomatic for years, although the HBsAg surface antigen can be detected during this time. In other patients, chronic hepatitis B occurs, with symptoms including fatigue, anxiety, and anorexia, associated with hepatocellular necrosis and inflammation. A fraction of those patients with chronic hepatitis will develop cirrhosis of the liver. Cirrhosis can then lead to HCC. How HBV causes HCC is a matter of debate, although there are two main pathways by which HBV might be acting: directly through a viral protein or integration or indirectly, as an effect of liver damage due to viral infection. HBV may directly induce HCC either in *trans*, by an activity of one of its gene products early on in cancer development, or by a *cis*-acting result of viral integration. For HBV to cause oncogenesis through a *trans*-acting mechanism, it must be actively undergoing transcription. Since most advanced HCCs display no viral gene expression, this putative pathway must act solely at the initial stages of cancer cell growth while viral genes are still being expressed. There are multiple viral gene products which have been argued to play a role in promoting uncontrolled cell growth.

It is common to find HBV genomes integrated into the host chromosome in HCC. These integrations commonly lead to the retention of a portion of the viral genome, including the X and preS/S regions. The X gene product of HBV has transcriptional activation activity. The activity of X in the context of the *in vivo* viral life cycle is unclear, although X has been shown to activate many promoters in reporter gene assays, including the *herpes simplex virus* (HSV) *tk* promoter, the HIV long-terminal repeat (LTR), and the HBV *EnI*. Cellular promoters activated by X include those for α-globulin, MHC class I, c-*myc*, and RNA polymerase III. It is possible that X is activating cellular oncogenes, leading to uncontrolled cell growth. X has also been shown to inhibit nucleotide excision repair, leading to accumulated mutations. This could cause an increased mutation rate induced by chemical carcinogens such as aflatoxin B1. X has also been found to bind to the p53 tumor suppressor protein and may therefore interfere with the latter's role in DNA damage response. Some mice transgenic for X driven by the X promoter develop HCC. However, this effect is not seen using heterologous promoters or subgenomic or genomic HBV DNAs containing the X region. Further support that X plays an active role in tumor development stems from the fact that avian hepatitis viruses which lack X do not cause HCC. The HBV truncated major S protein may also play a direct role in HCC formation for some of the same reasons as X does because it too possesses transcriptional activation activity. The pre-S protein product of HBV has also been shown to activate transcription of TGF-α, an effect which can lead to uncontrolled cell growth. Integration of the HBV genome may, in rare circumstances, cause the formation of a fusion protein with oncogenic potential.

One such example of this was observed in a particular HCC in which the HBV pre-S region was fused to the retinoic acid receptor locus.

HBV may lead to cancer via a *cis*-acting mechanism, whereby the integration of the virus into the host genome can cause interruption of a tumor suppressor gene or activation of an oncogene. An example of insertional activation is observed in WHV-induced HCC. Using Southern blotting analysis, WHV has been found to be integrated near two N-*myc* loci in 40% of HCCs. Using the more broad-scanning technique of pulse-field gel analysis, an even higher percentage of tumors have been found to be integrated near N-*myc*. However, insertional activation by HBV has not been regularly demonstrated in human HCCs using either technique. However, this does not mean that this phenomenon is not occurring because current techniques allow detection only within a few kilobases of a gene of interest.

Another major school of thought proposes that HBV can lead to HCC via an indirect pathway. Infection of the liver by HBV can cause liver damage, which leads to a chronic proliferative response in the liver to replace lost cells. The resulting increase in DNA synthesis increases the probability of mutations occurring, leading to uncontrolled cell growth. This theory is supported by the fact that most diseases of the liver leading to liver damage, such as alcoholic cirrhosis and Wilson's disease, are associated with an increased risk for the development of HCC. The Large S (L) protein product of HBV, when expressed, is toxic to cells and can lead to liver damage. Transgenic mice engineered to overexpress the HBV L protein display hepatocellular necrosis and regeneration, which eventually leads to HCC. The damage can also be due to immune system response to the L protein. Although this result is important, it should be interpreted carefully because the levels of L protein in these mice are much higher than those in an HBV infection.

Although the association of HBV infection with the development of HCC is high, the mechanism by which the virus causes cancer is debatable. It is possible that a combination of pathways are contributing to the oncogenic activity of HBV. Further clarification of this effect will pave the way for more effective treatments of HCC associated with HBV. Meanwhile, efforts are underway to eradicate HBV infections through mass immunizations.

B. Epstein–Barr Virus

In the 1950s, while working as a missionary doctor in Africa, Denis Burkitt described the childhood tumor Burkitt's lymphoma (endemic BL) and postulated that an infectious agent could possibly act as the etiologic agent of BL. In 1961, a collaboration among Denis Burkitt, Tony Epstein, and Yvonne Barr resulted in the establishment of BL-derived cell lines in culture. Electron microscopy examinations of the cell lines showed herpesvirus-like particles that are biologically and antigenically dissimilar to other human herpesviruses. Subsequent seroepidemiological studies indicated that the candidate human tumor virus named Epstein–Barr virus (EBV), recently renamed human herpes-virus 4, presents ubiquitously throughout the human populations in two different types and is persistently infectious. The strong association between the virus and BL, its transforming ability for human B lymphocytes into permanent lymphoblastoid cells, and a predominance of a single glycoprotein in the viral outer surface further differentiate EBV from other herpes members. All three subtypes of BLs—endemic, sporadic, and AIDS-related—are thought to be associated with EBV infection. Early-onset endemic BL, the most common childhood cancer in equatorial Africa, occurs with an unusually high incidence of 5–10 cases per 100,000 individuals per year. Virtually all of the malignant cells of every analyzed endemic BL tumor are EBV genome-positive. The sporadic BL, more commonly found in the United States and other Western countries, has a 50- to 100-fold lower incidence in comparison to the endemic BL with about 15–25% EBV genome-positive tumors. Also, the sporadic BL tumor has a slightly later age peak, a different pattern of presentation, and frequently involves the abdomen and bone marrow instead of the jaw, as in the cases of the endemic BL. The AIDS-BL that develops in approximately 10% of AIDS patients in Western societies has both early- and late-onset B-cell malignancies. However, even with such a high incidence of AIDS-BL tumors, only 30–40% of these are EBV genome-positive. Remarkably, all known forms of BL have the t(8:14), t(2:8), and t(8:22) chromosomal translocations that are likely deleterious to B-lymphocyte development. In addition to BL, EBV-associated malignancies also include immunoblastic B-cell lymphomas in immunosuppressed individuals, nasopharyngeal carcinoma, Hodgkin's disease, and EBV-associated T-cell lymphomas.

Acute infection of EBV causes mononucleosis. How EBV initially infects an individual is not clear, but it may do so through infection of epithelial cells lining the oral cavity. Usually, EBV infections become persistently latent, in which the viruses induce B-lymphocyte proliferation concurrently with viral DNA replication. Stable retention of EBV episomes is mandatory for the continual growth of EBV-positive BL cells. Viral protein EBNA1 is critical for the maintenance of the EBV replicon in infected cells. Although it had long been suspected that EBNA1 contributes directly to the replication of the viral genome, recent studies suggest its role is primarily in ensuring the efficient inheritance of the viral replicon during cell division.

How EBV contributes to human cancer is not well understood. In tissue culture, EBV can efficiently induce the immortalization of B lymphocytes, a property that is likely to be related to its oncogenic properties *in vivo*. The EBV genome is a 172-kbp linear, double-stranded DNA. In infected cells the EBV genome is maintained as a circular episome. Although the EBV genome can encode many genes, the fact that in EBV-immortalized B cells only a handful of viral genes are expressed helped researchers identify the genes relevant for immortalization. Currently, one viral gene, latent membrane protein 1 (LMP1), has been demonstrated to be critical for EBV's immortalization of B lymphocytes and in transgenic mice is sufficient to induce lymphomas. As its name implies, LMP1 is an integral membrane protein found in the cytoplasmic membrane of latently infected cells. LMP1 has characteristics of a cell surface receptor; it displays a rapid turnover that is associated with endocytosis and it signals to NF-κB through association with TRAFFs and TRADDs. This signaling capacity of EBV appears to be required for its capacity to induce B lymphocyte proliferation. In this regard, LMP1 is functionally analogous to CD40, a cellular receptor normally expressed in B lymphocytes which can induce B lymphocyte proliferation upon stimulation by its ligand. Unlike CD40, LMP1 is not known to have an associated ligand; consistent with this, LMP1 lacks a prominent extracellular domain to which an extracellular ligand could bind. Rather, LMP1's signal transduction activity appears to be constitutively active. In the past, LMP1 has been argued to cause the immortalization of B lymphocytes in part by inducing expression of the cellular anti-apoptotic factor Bcl-2; however, it is unclear whether the increased levels of Bcl-2 found in some EBV-infected cells are a direct consequence of LMP1 action or an acquired phenotype that correlates with improved survival of those cells in tissue culture. Another EBV-associated gene, EBNA2, has been shown to be critical for EBV's immortalization of B lymphocytes; however, its role is likely indirect because it is known to activate the transcription of LMP1.

With the recent advances in understanding the role of EBNA1 in the persistence of the EBV genome in human cells and the role of LMP1 in EBV's immortalization of B lymphocytes, scientists now have two relevant targets for developing novel antiviral therapies that could reduce the incidence of EBV-associated malignancy in the future.

C. Human Papillomaviruses

Papillomaviruses are the causative agents of warts. A subset of human papillomaviruses (HPVs) that are sexually transmitted cause warts of the anogenital tract. After decades of latency, a small fraction of these HPV-induced anogenital lesions progress to carcinomas. The cancers most highly associated with infection by these so-called "high-risk" HPVs is cervical cancer. Interestingly, cervical cancer was previously thought to be casually associated with infection by another sexually transmitted virus, HSV; however, this association was dismissed after it became clear that HSV was not present in the cancers. In contrast, more than 95% of all cervical cancers contain papillomavirus DNA. The presence of HPV DNA in cervical cancers and the strong epidemiological association of HPV infection with cervical cancer have led the National Cancer Institute to conclude that HPV is the main etiologic agent of cervical cancer in humans.

Papillomaviruses are non-enveloped icosohedral viruses which contain a double-stranded DNA genome of about 8000 bp that encodes early and late genes. These viruses infect the poorly differentiated basal cells within the epidermis of the skin and express the early viral genes. Progeny virus are only produced when the infected basal cell differentiates; this triggers amplification of the viral genome and expression of the viral late genes that encode the capsid proteins.

In the majority of cervical cancers, the viral DNA is found integrated into the cellular genome and no progeny virus is produced. The viral integration event almost always leads to a disruption in the E1 and E2 open reading frames. These open reading frames encode proteins involved in viral transcription and replication and their disruption may lead to derepression of viral transcriptional promoters as well as loss of viral episomal replication activity. The E6 and E7 open reading frames, on the other hand, are consistently intact in cancers. Integration disrupts an mRNA instability element which allows increased expression of E6 and E7. E6 and E7 are directly implicated in oncogenesis. E7 can cooperate with E6 or the activated *ras* oncogene to transform fibroblasts in tissue culture, and E6 and E7 alone or together can immortalize epithelial cells in tissue culture. When the expression of E6 or E7 is directed to the skin of transgenic mice, the mice exhibit thickening of the skin and spontaneously develop skin tumors.

As is the case with SV40 and adenoviruses, papillomaviruses have the ability to inactivate tumor suppressors. E7 binds to and induces the degradation of the tumor suppressor, pRb. The inactivation of pRb allows E2F to be constitutively active, which leads to increased proliferation in E7-expressing cells. The regions of E7 required for its inactivation of pRb have been mapped to conserved regions (CR)1 and 2, so named because of their similarity to pRb binding regions in the adenovirus oncoprotein E1A. The CR1 and CR2 domains of E7 are both necessary for E7's induction of tumors in animals and transformation of cells. An exception is that when the CR1 and CR2 domains of E7 are mutated in the context of

the whole papillomavirus genome, cells transfected with the genome still undergo transformation. This suggests that E7 performs functions other than pRb inactivation. This is further supported by recent findings that E7 can bind to and inactivate p21 and p27, inhibitors of cylin-associated kinases that play critical roles in the control of the cell cycle.

The discovery of the ability of E6 to inactivate the tumor suppressor, p53, was prompted by the oncogenic similarities between papillomaviruses and SV40 and adenovirus. E6 was found to form a tertiary complex with p53 and a cellular ubiquitin ligase, E6-AP. This leads to the ubiquitination of p53 which is then targeted for degradation. The p53 tumor suppressor is involved in the G_1 to S phase transition of the cell cycle and is induced in response to cellular stresses such as DNA damage. Although the importance of *p53* in other cancers suggests that it is important in papillomavirus-associated cancers, inactivation of *p53* cannot fully substitute for the activities of E6. For example, E6 was found to abrogate differentiation of skin epithelial cells in tissue culture. However, when the p53 in these cells was inactivated by overexpressing a dominant-negative form of p53, the cells could still differentiate. Also, *p53*-null mice differ from E6 transgenic animals in that they do not display thickened skin or develop skin tumors. E6 has recently been shown to bind a variety of protein partners, including E6-BP, paxillin, E6-TP, bak, and the human homolog of *Drosophila* disc large tumor suppressor gene. In addition, E6 is known to activate telomerase, the enzyme which adds DNA to the ends of telomeres. Mutational analysis of E6 has revealed that these novel E6 activities can be separated from E6's ability to degrade p53. Thus, these aforementioned proteins may be involved in unknown E6 pathways. Ongoing research of E6 and E7 is aimed at further elucidating their roles in carcinogenesis.

Human papillomavirus infection is the most prevalent sexually transmitted disease. Due to the papillomaviruses' intimate association with cervical cancer, much effort has been put forth to prevent these infections. The strategy that seems most promising is immunization using virus-like particles (VLPs). VLPs are ordered structures that resemble the natural papillomavirus capsid. They can be formed when the papillomavirus capsid proteins L1 and/or L2 are expressed at high concentrations. Immunization of animals with VLP's has proven successful in protecting against papillomavirus infection and tumor formation. Currently, clinical studies are underway to test the efficacy of similar vaccines in humans. During the past half century, use of Pap smears has reduced the incidence of frank cervical cancer through early detection of preinvasive lesions.

D. Human T Cell Leukotropic Virus Type 1

HTLV-1 is an oncogenic retrovirus associated with a variety of human diseases, including adult T cell leukemia/lymphoma (ATL), myelopathy, uveitis, and arthropathy. The epidemiology of ATL suggests that cumulative genetic defects may be responsible for the acute T cell malignancy in a given T cell clone. In HTLV-1 infection, the time-dependent emergence of infected T cell clones is well studied, and it has been shown that the uncontrolled growth of a single clone causes ATL. *In vitro*, T cell immortalization by HTLV-1 occurs within a few months of culture. HTLV-1 transforms human T cells *in vivo* and in cell culture. The development of T cell transformation can be divided into two stages. In the early stage, the virus induces interleukin-2 (IL-2)-dependent T-cell proliferation, which mimics the action of antigens. In the normal T cells, antigen-stimulated T cells cease growth after a few weeks. However, HTLV-1-infected T cells show infinite proliferation (considered immortalization). The uncontrolled cell proliferation is believed to facilitate secondary genetic changes. Therefore, the immortalized T cells can progress to the second stage of transformation, which has a typical property—IL-2-independent growth. Recent evidence suggests that HTLV-1 not only induces the proliferation of host T cells but also protects the infected cells from undergoing apoptosis.

The HTLV-1 provirus genome is 9032 bp long and contains *gag*, *pol*, and *env* genes that encode the viral matrix (capsid and nucleocapsid proteins), enzymes (reverse transcriptase, integrase, and protease) and envelope protein (surface glycoprotein and transmembrane protein), respectively. The HTLV-1 genome contains unique regulatory genes, the *rex* and *tax* genes (analogous to the HIV *rev* and *tat* genes), at the 3′ end of the genome which are not common in other retroviruses. Rex is a 27-kDa nuclear phosphoprotein that regulates viral RNA processing. Rex enhances the expression of single-spliced mRNA and unspliced viral genomic RNA encodes the *gag-pol* and *env* gene products but reduces the expression of double spliced *tax/rex* mRNA. At the early stage of viral gene expression in the viral life cycle, double-spliced mRNAs for Tax and Rex proteins are mainly produced. Newly synthesized Tax protein transactivates the transcription of HTLV-1. However, accumulated Rex protein enhances the expression of unspliced and single-spliced viral RNA for viral genomic RNA (*gag-pop* and *env* genes) and suppresses the expression of *tax/rex* mRNA. Therefore, by the function of Tax and Rev regulatory proteins, viral expression is regulated both positively and negatively, and it helps the transient expression of HTLV-1. Transient expression of the viral genes may be one of the mechanisms to escape from immune surveillance of the host. Tax protein is a 40-kDa nuclear phosphoprotein and is considered an oncogenic viral protein that transactivates the transcription of HTLV-1 and also binds to cellular transcription factors or other cytoplasmic cellular molecules involved in the fundamental cell function such as IL-2, IL-2R, and c-*fos* and the parathyroid hormone-related peptide. HTLV-1 has no preferential site of integration, which could explain why a transformation process is initiated. The oncogenic property of Tax protein is well documented in the development of ATL. Indirect evidence showed that the 3′ end of the HTLV-1 genome (*tax* gene) is necessary and sufficient for cell immortalization. In addition, cotransfection of Tax with Ras can also induce transformation of primary rat embryo fibroblasts. Finally, Tax transgenic mice develop several pathologies, including leukemia, mesenchymal tumor, and neurofibromas.

V. SUMMARY

It is clear that tumorigenic viruses contribute to cancer through multiple distinct mechanisms. In cases in which viruses are thought to encode genes that directly contribute to cancer, why is this so? In these cases (EBV, HPV, and HTLV-1) it is not obvious that the cancers contribute positively to the viral infections; rather, the cancers represent dead-end streets. For example, in HPV-induced cervical cancers the virus is commonly integrated, with the integration event causing disruption of the viral genome, and the cells, being poorly differentiated, cannot support progeny virus production. Recent evidence suggests that papillomaviruses use their oncogenes during the normal viral life cycle within a wart to reprogram differentiating epithelial cells to support the late stages of the viral life cycle in which the viral genome is amplified. This occurs within a compartment of the epithelium in which normally no DNA synthesis is thought to occur. The E7 oncogene is thought to trigger the infected, differentiating epithelial cell to express cellular genes that are necessary to provide a DNA synthesis-competent environment. Another example is provided by EBV in which LMP1 causes B cells to remain in a proliferating state, thereby providing a means by which the infected cell population is expanded and the viral episomal genome is allowed to persist within the organism. In the case of other viruses, the carcinogenic event appears not to be due to the virus encoding an oncogene but rather to the virus altering expression of cellular genes that contribute to cancer or protect cells from becoming tumorigenic. The most clear examples of this are the insertionally mutagenic retroviruses in which the provirus integration event can cause increased expression of a cellular proto-oncogene. The most enigmatic tumor viruses are those in which no direct mechanism of tumorigenesis has been elucidated. The prime example is human hepatitis B virus in which, while regions of the viral genome often are retained, the viral genes are not expressed commonly in the cancers and the viral genome is not thought to cause deregulation of any cellular gene through the integration events. In this case, it may well be the host's own immune response to the infection that causes the virally-induced cancers to occur.

The study of tumor viruses during the past century has provided many insights about cancer. Their study led to the discovery of oncogenes and tumor suppressor genes. Current studies continue to shed new insight into the process of carcinogenesis. For example, recent studies on adenoviruses and human papillomaviruses have led to the identification of novel cellular targets, the deregulation of which likely contributes to carcinogenesis. These include the human homolog for the *Drosophila* disc large tumor suppressor gene, a target for the adenovirus E4ORF1 and human papillomavirus E6 gene products, which is thought to alter β-catenin-mediated signal transduction through its interaction with the gene product of the adenomatous polyposis coli gene, APC, a tumor suppressor important in colon cancer. Another example is XRCC1, a putative target for the human papillomavirus E6 oncogene which functions in DNA repair. It is reasonable to predict that the continued study of tumor viruses will shed new insights into the carcinogenic process.

See Also the Following Articles

HEPATITIS VIRUSES • RETROVIRUSES • SEXUALLY TRANSMITTED DISEASES • TRANSFORMATION, GENETIC • TRANSGENIC ANIMAL TECHNOLOGY

Bibliography

Ganem, D. (1990). Of marmots and men. *Nature* 347, 230–232.

Nevins, J., and Vogt, P. (1996). Cell transformation by viruses. *In:* "Fields Virology" (B Fields, D. Knipe, and P. Howley, Eds.), 3rd ed., pp. 301–344. Lippincott-Raven, Philadelphia.

Tooze, J. (1981). "DNA Tumor Viruses," 2nd ed., pp. 943–1054. Cold Spring Harbor Laboratory Press, Cold Spring Harbor, NY.

zur Hausen, H. (1994). Molecular pathogenesis of cancer and its causation by specific human papillomaviruses. *In:* "Human Pathogenic Papillomaviruses" (H. zur Hausen, Ed.), pp. 131–156. Springer-Verlag, Heidelberg.

Oral Microbiology

Ian R. Hamilton and George H. Bowden
University of Manitoba

GLOSSARY

acquired pellicle The organic film on the surface of the tooth, consisting mainly of protein and glycoproteins from saliva.

alveolar bone Bone of the jaws in which the teeth are embedded.

approximal Adjacent, referring to the surfaces of teeth.

biofilm Microorganisms immobilized at a surface and frequently embedded in an organic layer (acquired pellicle) composed of polymers of host and microbial origin.

clone A population of cells descended from a single parental cell.

community A collection of bacterial populations growing together in a defined habitat.

dental plaque The microbial biofilms on teeth.

genomic species A group of organisms that share a high (~70%) DNA–DNA similarity value.

gingival margin The soft tissue (gums) next to the tooth.

indigenous flora Bacteria that make up the normal or resident flora of the ecosystem.

interproximal Between, referring to adjacent teeth.

occlusal Used for chewing, referring to the surface of a tooth.

periodontal pocket Pocket created adjacent to the tooth below the gingival margin.

population Bacteria of the same species, serovar, phage type, or DNA fingerprint found in a bacterial community in a given habitat.

subgingival plaque Bacteria growing below the gingival margin between the teeth and gingival tissue.

supragingival plaque Bacteria growing on the tooth surface above the gingival margin.

ORAL MICROBIOLOGY is the study of the bacteria that are the natural inhabitants of the oral cavity. Although most of the information on these bacteria relates to those genera and species that can be found in the human mouth, some information is available on the bacteria present in the mouths of a variety of animal species. Most attention has been paid to bacteria that can be found on human tooth surfaces in microbial communities known as dental plaque because the bacteria in plaque are etiological agents of dental caries. In addition, specific bacteria growing at the margin of the tooth and the soft tissue or gingiva (gums) are responsible for the condition known as gingivitis, which results in inflammation and bleeding of the gums. Failure to remove the bacteria at the gingival margin can lead to more complex pathological conditions, collectively known as periodontal disease, which is generally characterized by the destruction of the tissue adjacent to the tooth creating "periodontal pockets." The ultimate form of periodontal disease is destructive periodontitis, a chronic condition in which pockets of up to 8–10 mm can be observed containing a wide variety of anaerobic bacteria. Associated with tissue destruction and pocket formation is the concomitant loss of the alveolar bone that anchors the teeth in the jaws and, as a consequence, advanced periodontal disease often leads to tooth loss.

I. ORAL MICROBIAL ECOLOGY

The characteristics and properties of the microflora at various sites in the mouth, as are those of any microbial ecosystem, are regulated by the nature of the habitat and the associated environment.

In this respect, some discussion of the microbial ecology of the mouth is important, particularly because the two major diseases of dental caries and periodontal disease arise through alterations of the normal, or indigenous, flora rather than the invasion of foreign or alien infectious agents from outside the mouth.

A. General Characteristics of the Oral Environment

The mouth has two major types of surface that can be colonized by bacteria, the shedding mucosal surface of the soft tissues and the nonshedding mineralized surface of the teeth. As a consequence of the unique properties of these two types of surface, both the concentration and characteristics of the flora colonizing these areas are different. In addition, each type of surface harbors many distinct microbial ecosystems that have arisen under the various microenvironments in the mouth. In general, the flora is regulated by the nutrient supply, the pH of the environment, the oxygen level, antimicrobial agents, and microbial interactions, as well as the mechanical forces of mastication (chewing) and saliva flow. Most of the bacteria in the developed ecosystems in the mouth are either facultative or obligate anaerobes. Aerobic bacteria are found in significant proportions in the early stages of microbial biofilm development, but are replaced later by the anaerobic flora and may be restricted to the surface layers of the community. Saliva plays a significant role in the development and maintenance of the oral microflora by providing nutrients and antimicrobial factors, as well as buffers and electrolytes. Adequate salivary clearance of sugar substrates and acid end products resulting from carbohydrate metabolism by plaque bacteria is recognized as being essential for limiting the acid demineralization of the tooth enamel associated with dental caries.

1. Mineralized Tissues

The surfaces of teeth support a variety of complex microbial ecosystems known collectively as dental plaque, which includes the plaque microflora, extracellular matrix, and associated fluid phase (plaque fluid). Depending on an individual's oral hygiene, supragingival plaque can contain approximately 10^{11} organisms per gram wet weight, with the microbial communities comprising more than 31 genera and perhaps 300–400 taxa. Early studies of dental plaque viewed the oral microflora as being similar at all sites on the tooth surface, and the properties and characteristics of plaque were studied in pooled samples. However, research on the microflora of plaque since the 1960s has involved more sophisticated microbiological methods, particularly with regard to the cultivation and identification of anaerobic bacteria, and with this has come the concept that each small site on the tooth surface has a specific and unique microflora.

The development of the plaque biofilms on a clean tooth surface in the mouth begins with the deposition of an organic layer, the acquired pellicle, that is made up almost entirely of salivary proteins. The subsequent disruption of established dental plaque, however, can introduce to this organic layer proteins, polysaccharides, and lipoteichoic acid of bacterial origin. The acquired pellicle serves as a nutrient source, as a homeostatic barrier to retard the loss of mineral from the tooth surface, and as a promotor plaque formation. A significant amount of research has been directed at the specific and non-specific factors involved in the adherence of oral bacteria to tooth enamel (hydroxyapatite) and to salivary-coated hydroxyapatite. This research has indicated that selected salivary proteins and glycoproteins have affinity for the hydroxyapatite surface, whereas certain oral bacteria, such as *Streptococcus sanguis, S. mitis, S. oralis,* and *Actinomyces* species, have higher affinity for this surface than other members of the oral microflora. If left undisturbed, highly complex and diverse microbial communities can be established over the mineralized tooth surface and adjacent soft tissue within 3–5 days. This development involves the specific coaggregation among a wide variety of bacteria, often involving the interaction of specific cell-surface components, as well as salivary macromolecules.

Studies of the changes in oral bacterial communities during their accumulation with time have been limited to enamel surfaces (i.e., dental plaque). Early studies on the development of bacterial communities *in vivo* considered bacteria that accumulated after 24 hours and the subsequent increase in biomass complexity over periods of up to 14 days. These studies did not survey the total flora; however, the bacterial species that could be isolated most readily from the enamel at 24 hours were *S. sanguis* and *Neisseria* species, with some species of *Rothia, Veillonella,* and *Actinomyces*. The predominantly coccal plaque persisted for 72 hours, after which it gradually became populated with filamentous bacteria (*Actinomyces, Leptotrichia,* and *Eubacterium*). Concomitant with this increase in diversity was a reduction in the oxidation–reduction potential that allowed the flora to become more anaerobic, and anaerobic bacteria began to compete effectively with strict aerobes, which decreased with time. Importantly, species within the facultative genera *Streptococcus* and *Actinomyces* maintained themselves in high numbers in plaque, competing effectively with other members of the community.

Dental plaque can be divided into five major types based on the location of the habitats involved—supragingival, fissure, carious lesions, gingival margin, and subgingival. Supragingival plaque is a general term that includes bacteria in the many small ecosystems that reside on the tooth surface above the gingival margin. For example, unique microbial communities can be observed on smooth tooth surfaces, in the interproximal areas between the teeth, and in small pits that can range over the tooth surface. Bacteria in supragingival plaque receive their nutrients both from dietary constituents and from saliva, and a significant fraction of the bacteria in these communities use carbohydrates as their principal energy source. Large fissures, particularly those on the biting (occlusal) surfaces of the teeth, constitute a second major and unique type of microbial ecosystem, whereas open (overt) carious lesions are a third type that also support unique bacterial communities.

The organisms at the gingival margin between the gum line and the tooth form a fourth general type of bacterial community. Bacteria in this region colonize a transition zone between supragingival and subgingival habitats, with the latter habitat supporting a complex community below the gingiva in subgingival plaque. In normal healthy individuals, a small pocket (sulcus) 1–2 mm deep exists below the gingival margin and the bacteria in this region receive small amounts of a protein-rich liquid (crevicular fluid) that has its origin in plasma. Thus, the bacteria in the communities at the gingival margin have three nutrient sources, diet, saliva, and crevicular fluid, the last of which contains plasma proteins and certain factors such as hemin that can be used by the more fastidious oral bacteria. Crevicular fluid also contains significant amounts of the immunoglobulins IgG and IgM.

Failure to remove the bacteria at the gingival margin can lead to inflammation (gingivitis) and eventual destruction of the soft tissues next to the teeth, creating a pocket containing a largely gram-negative anaerobic flora with a considerable capacity for proteolysis. Individuals with such pockets have periodontal disease and the ultimate stage of the disease is known as destructive periodontitis. This condition is characterized by the presence of large pockets 8–10 mm deep created by the destruction of the adjacent soft tissue and the periodontal ligament that anchors the tooth to the alveolar bone of the jaw. Alveolar bone destruction also occurs in advanced cases, leading to the destabilization of teeth, which can often lead to tooth loss.

2. Soft Tissue

The number of bacteria on the cheek and palatal surfaces is on the order of 5–25/epithelial cell. The tongue, on the other hand, is a unique structure possessing numerous crypts and papillae, and epithelial cells from the dorsum of the tongue can carry more than 100 bacteria each per cell. Because the surface irregularities of the tongue allow the development of relatively high densities of bacteria, the tongue is believed to be the major source of the bacteria in saliva. Normally, the number of bacteria on oral mucosal surfaces is limited because of the continuous loss of surface epithelial cells (desquamation) and by the low growth rate of the bacteria on these surfaces; the biomass of the bacteria on oral mucosal surfaces is estimated to double only two to three times per day. Thus, epithelial cell desquamation in the mouth is an important host defense mechanism that limits the number of bacteria present on the surface at any one time. Because each surface epithelial cell has a finite life span, the growth rate of the attached bacteria is not a significant factor in determining the proportion of bacteria in the surface population. The colonization of epithelial cell surfaces appears to be determined largely by the capacity of the organism to attach and by the concentration of bacteria in the fluid phase near the surface.

3. Saliva

Saliva secreted into the oral cavity from the salivary glands is sterile, but soon becomes contaminated by the microflora in the existing saliva and on the various surfaces in the mouth. The bacterial concentration in saliva can approach 10^9 organisms/ml and fluctuates during the day, with a high number being found following periods of sleep when the saliva flow in the mouth is significantly reduced. In addition, higher numbers are present during mastication, due in part to the partial removal of bacteria from the surfaces of the mouth. The concentration of some bacteria in saliva is directly proportional to the number of the same bacteria in dental plaque. This observation has been used as the basis for a variety of assays of saliva to test for the presence of certain bacteria, such as species of *Streptococcus* and *Lactobacillus*, that are associated with dental caries.

B. Factors Influencing Oral Microbial Ecosystems

The oral microflora, like any natural bacterial community, is influenced by its environment. Changes in the community can be brought about by variation in the characteristics of the environment. These may be nutritional (e.g., a change in the balance of nutrients available, such as the proportion of carbohydrate in the diet), physical (e.g., the loss of a surface for colonization, or tooth loss), physiological (e.g., a change in the oxidation–reduction potential of the environment), or competitive (e.g., the effective competition of certain bacterial populations with other populations, resulting in the exclusion or reduction of the latter bacteria). Thus, if the microflora is left undisturbed, there will be a modification or a succession of the bacterial populations in the mouth until a degree of stability (climax community) is reached. Normally, stable climax plaque communities will be observed 5–7 days after the start of plaque development. The environmental factors influencing the oral flora are of two types, those determined by the host and those of an external nature.

1. Host-Associated Factors

A variety of physical–chemical factors influence the types and concentrations of the bacteria in the mouth. These include the influence of saliva and other body fluids, the nature of the site or habitat, the age of the host and the microbial community, and microbial interactions.

a. Body Fluids

Saliva interacts with all bacteria in the mouth except those found in deep fissures and carious lesions of the tooth and in periodontal pockets. The constituents of saliva include a variety of proteins, glycoproteins, lipids, carbohydrates, and antimicrobial factors, as well as buffers and electrolytes (Table I). The various polymers in saliva include enzymes, immunoglobulins, nonimmune agglutinins, glycoproteins

TABLE I
Components of Human Saliva[a]

Enzymes	*Complex carbohydrates*
Amylase	Glycoproteins
Lysozyme	Blood-group substances
Salivary peroxidase	*Immune system components*
Phosphatases	sgA, IgG, IgM
Esterases	Secretory component
β-Glucuronidase	Complement factors
Transaminases	*Small molecules*
Ribonucleases	Amino acids
Dehydrogenases	Glucose
Kallikreins	Lactate, citrate
Lipase	Ammonia, urea
Other macromolecules	Cholesterol
Albumin	Uric acid
Lactoferrin	Cretinine
Lipoproteins	*Cations/Anions*
Ceruloplasmin	Na^+, K^+
Proline-rich proteins	Ca^{2+}, Mg^{2+}
Statherin	Cl^-
Cystatins	Phosphates
Histatins	Carbonates
Orosomucoid	Thiocyanates
	F^-, Br^-, I^- (trace)

[a] Adapted from Mandel, I. D. (1974). *J. Dent. Res.* **53**, 246–266.

that are involved in adherence of bacteria to surfaces and in the binding of iron (lactoferrin), and other molecules. Although the flow rate of saliva can vary widely depending on the degree of stimulation of the salivary glands, saliva provides a ready nutrient source for bacteria on the various surfaces of the mouth. This has been shown in studies with subjects who obtained their nutrition by stomach tube. Bacteria developed on the teeth even though the subjects were not ingesting food via the mouth. Such plaque is composed of fewer cells than plaque found in individuals eating normally and contained fewer acid-producing (acidogenic) bacteria capable of metabolizing carbohydrate. Saliva has recently been shown to support the growth of a variety of bacteria that have the capacity to degrade salivary proteins and glycoproteins to their constituent amino acids and sugars, which are then used by bacteria for growth.

The flow rate and buffering of saliva are powerful forces for the clearance of toxic components, such as acids, from the tooth surface and for the prevention of acid demineralization of the enamel. One of the best demonstrations of the significance of the effects of saliva on dental plaque is in subjects with little or no salivary flow. Some patients with oral tumors undergo radiation therapy and, as a result, the output of the salivary glands is often seriously impaired, with salivary flow reduced by up to 95%. Absence of saliva is called xerostomia and patients with this condition develop caries at a rapid rate because of increases in the proportion of acidogenic and acid-tolerant (aciduric) bacteria in the plaque microflora following the reduction in saliva flow. The control of the caries can be achieved by the use of fluoride mouth rinses, good oral hygiene, and reduction of the carbohydrate content of the diet. Saliva also contains antimicrobial agents, such as secretory IgA, salivary peroxidase and thiocyanate, lactoferrin, and lysozyme. Because the resident or indigenous flora are resistant to these agents, these factors promote the homeostasis of the plaque microbial ecosystems by the elimination of invading microorganisms.

Another fluid of importance to bacteria near the gingival margin is gingival crevicular fluid that seeps from the sulcus at the tooth–tissue junction. This fluid contains plasma components and is, therefore, rich in proteins, vitamins, and factors that promote the growth of the more sensitive anaerobic oral bacteria in the gingival margin and sulcus area.

The microflora in the mouth is exposed to antibodies from two major sources, saliva and crevicular fluid. Salivary immunoglobulins are a component of the mucosal immune system and as such consist of dimeric IgA with a secretory piece (SIgA) that is relatively resistant to proteases. However, some oral bacteria produce an SIgA protease that cleaves IgA1 at the hinge region, and it is thought that this enzyme could play a role in facilitating the colonization by SIgA protease-producing bacteria. Serum immunoglobulins, including IgM and IgG, enter the mouth in crevicular fluid and consequently one might expect the diversity and specificity of antibodies of these classes to reflect those in the host's serum. In addition, there is local production of immunoglobulin by plasma cells in the gingiva.

Generally, it is thought that oral antibodies are most likely to exert their effect by influencing oral colonization by bacteria. However, the examination of the specificities of either SIgA or IgG and IgM antibody in saliva shows that they react with resident oral bacteria. Despite this interaction, the presence of a specific antibody does not appear to influence the survival of the oral flora. The reasons why this is so are not well understood, but several mechanisms have been proposed: (1) that once the bacteria are established in the plaque biofilm, the antibody is not effective; (2) that the antigens recognized are not significant in adhesion and subsequent colonization; (3) that bacteria among the resident flora are coated with host molecules from saliva and hidden from antibody; (4) that oral bacteria may express antigens closely similar to those of the host (molecular mimicry); (5) that the antibodies recognizing oral bacteria are natural, generated against common antigens to the host and of low avidity; and (6) that host tolerance is generated to oral bacteria during the initial colonization of infants. It seems likely that all of these mechanisms, and others that we do not understand, could influence the response of the host to the oral resident flora and also the activity of the immune system against oral organisms. There is some evidence that tolerance to *Actinomyces* occurs during initial colonization of infants. Host responses to the oral flora are of obvious significance when immune prophylaxis or control of oral diseases, such as caries or periodontal disease, is contemplated. In caries, active and passive immunity to *Streptococcus mutans* in animals and, in some cases, in humans, influences the extent of oral colonization by this species.

b. Habitat

Bacteria that grow successfully on the tooth surface do not compete with those on mucosal surfaces because the two sites represent distinctly different environments. Although numerous individually unique microbial communities are found on the surface of a single tooth, there will be some bacteria common to all sites; however, the species and their ratios will vary from site to site even in adjacent areas on a tooth surface. This is particularly true of the microflora in approximal sites on adjacent teeth and the approximal plaque from identical sites on the left and right sides of the same mouth. Teeth can also possess pits and fissures, and the communities present in these sites are less complex than those from smooth or approximal surfaces, primarily because the environment of a fissure is usually acidic and aciduric bacteria will survive in the deeper layers of this habitat. For this reason, the composition of fissure plaque is relatively stable and resembles in many ways the flora of a caries lesion.

c. Age of the Host

The age of the host has an influence on the complexity and distribution of the oral flora. At birth, the infant's mouth is sterile, but it is colonized rapidly by bacteria from the mother. Maternal transfer has been confirmed by bacteriocin and genetic typing of isolates from mother and child. Streptococci are common among the early colonizing bacteria with *S. mitis* biovar 1, *S. oralis*, and *S. salivarius* predominating, whereas *S. mitis* biovar 2, *S. sanguis*, *S. anginosus*, and *S. gordonii* are isolated less frequently and in lower numbers. Strains of *Neisseria* and *Veillonella* may also be present during the early days of life. Some organisms isolated from the infant's mouth are transients. These strains can originate from the environment and from other habitats associated with the mother. Typical of transient bacteria in the mouths of infants immediately following birth are maternal fecal lactobacilli. These bacteria fail to survive in the mouth, presumably because of the differences between the fecal and oral environments. The eruption of the primary dentition causes significant increases in the complexity (diversity) of the oral flora. Tooth surfaces provide hard nonshedding surfaces for the formation of complex biofilms and juxtaposition of the gingiva and tooth enamel provide new habitats. After tooth eruption, *Actinomyces* become established in the mouth, and *Streptococcus* spp. in the plaque biofilm show different proportions to those on soft tissues. The diversity of the oral flora increases over time and within 5 years, many of the bacteria commonly isolated from adults are resident in the child's mouth.

d. Microbial Interactions

In the absence of dental hygiene procedures, the accumulation of bacteria on the nonshedding surfaces of the teeth results in the formation of many complex and highly diverse plaque ecological systems. Microbial interactions are known to stimulate or inhibit microbial growth in these habitats. As plaque develops following cleaning, the initial predominantly gram-positive flora gives way to a more gram-negative and anaerobic flora. The reduction in the redox potential that permits the emergence of the anaerobic bacteria arises through the metabolic action of the facultative populations, which use oxygen as a terminal electron acceptor. This reduces the partial pressure of oxygen in the deeper layers of plaque and allows the facultative and obligate anaerobes to proliferate. The commensalistic relationship between facultative and obligate anaerobes was one of the first observations made in the field of oral microbial ecology in the early 1960s. Subsequent studies have demonstrated other beneficial elements, such as food chains, the generation of growth factors and extracellular energy polymers, competition for nutrients, and the formation of growth-inhibiting products, including hydrogen peroxide, acids, and bacteriocins.

2. External Factors

Diet and the use of antimicrobial agents are two factors having significant influence on the oral flora.

The complex nature of the microflora and the natural antimicrobial agents of the host preclude the invasion of this environment by alien bacteria unless the host has been severely compromised by disease or by the application of drugs.

a. Diet

As mentioned previously, the consumption of food is not essential for the development of bacterial communities in the mouth; however, many studies have established a direct relationship between the consumption of refined sugar (sucrose) and dental caries. Although little is known of the changes in plaque on small localized areas of the tooth surface, it is known that significant changes in the biochemical properties and, to some extent, the proportions of various bacterial populations occur when the sugar content of the diet increases. In general, the flora becomes more acidogenic with lactic, acetic, and formic acids being the predominant acid end products of sugar metabolism. Increased sugar intake by the host also results in an increase in the rate of sugar consumption and acid formation by the saccharolytic oral flora as these bacteria adapt to increases in this energy source. Because high-carbohydrate diets result in the formation of a more acidic plaque environment, aciduric bacteria such as *Lactobacillus* species and *Streptococcus mutans* tend to flourish and dominate the plaque ecosystems when high sugar intake is maintained for a period of time. Some studies have also reported concomitant increases in the lactic acid-using *Veillonella*, which are probably associated with the increased levels of lactic acid in the plaque matrix. The detection of high levels of *Lactobacillus* species and *S. mutans* strains in human saliva is being used in some countries as a measure of caries activity. The conclusion to be drawn from studies on dental plaque from a variety of groups (Japanese, western European, Russian, New Guinea, and Australian Bushmen) is that, in general, plaque composition is relatively independent of diet. It seems likely that the bacterial communities can adapt to a wide range of diets and only a change in the ratios of the populations may occur. These changes in population ratios in the microbial communities may be most easily detected after the consumption of high carbohydrate diets.

b. Antimicrobial Agents

Firm evidence exists for the alteration and the partial elimination of the oral flora by applications directly to the mouth of high concentrations of chemical agents, such as fluoride, iodine, chlorhexidine, and other compounds. More controversial is the potential antimicrobial effect of low levels of water-borne fluoride (1 μg/ml). Such levels are too low to influence the growth of oral bacteria, but result in the partial inhibition of carbohydrate metabolism by the saccharolytic flora, thereby, reducing the rate of acid formation on the tooth surface. The administration of antibiotics to the host can have profound effects on large communities of bacteria, often leading to an imbalance in the microbial ecological equilibrium in the mouth. An example of this would be the overgrowth of the oral cavity by *Candida ablicans* and related species, producing a condition known as candidiasis. These yeastlike fungi are normal inhabitants of the mouth and the elimination of the bacterial flora permits the transformation of these organisms from a commensal to a pathogenic state.

II. ORAL MICROFLORA

A. Normal Oral Flora

Any description of the resident flora of the oral cavity must take into consideration the variety of small ecosystems in the various oral habitats. The normal flora of such ecosystems describes those bacteria that are regularly present in all healthy members of the host population. The qualitative assessment of oral bacterial communities indicates that they consist of a relatively stable collection of species, and that strains can be found in similar habitats in the mouths of different members of the same species of animal. As mentioned previously, variations occur in the number and ratio of the organisms in a given mouth and can be related to diet, physiology, oral hygiene, disease, and the impact of agents such as fluoride, which are in general use by the human population.

Comparisons made among various species of animals show that although some genera are common to several mammals, different animal species show variations in the complexity of their flora and harbor different species of a given microbial genus. Among the best-studied animals, apart from humans, are those that naturally develop caries or periodontal disease, and those in which these diseases can be produced experimentally. The oral flora of the nonhuman primates are similar to those of humans, and dogs and cats are also known to have complex microbial communities. The flora of other animals, such as hamsters, rats and mice, are less complex. Few studies of the oral flora in herbivores, insectivores, and carnivores, other than cats and dogs, have been made and nothing is known of the microflora that are resident in the oral cavities of reptiles and fishes. Even in humans and primates, few studies have been directed at defining the number of microbial species and the diversity of oral bacterial communities. Examinations have been limited to detecting and enumerating strains at the genus level, with more detailed classification reserved for the specific bacteria associated with caries and periodontal diseases. An overview of the flora associated with the oral cavities of humans and animals is shown in Table II. This includes some examples of genera present in the oral

TABLE II
Bacteria Common to Humans and Animals[a]

Bacteria	Human	Monkey	Cattle	Giraffe	Tiger	Cat	Dog	Rat	Mouse
Streptococcus	+	+	+	+	+	+	+	+	+
S. sanguis	+	+	−	ND	ND	ND	+	−	−
S. mitis/mitor[b]	+	+	−	+	ND	ND	+	−	−
S. salivarius	+	+	−	ND	ND	ND	ND	−	−
Actinomyces	+	+	+	+	+	+	+	+	+
A. viscosus[c]		+	(+)	(+)	(+)	(+)	(+)	(+)	(+)
A. naeslundii	+	+	(+)	(+)	(+)	(+)	(+)	(+)	(+)
A. howellii	ND	ND	+	−	−	−	−	−	−
A. denticolens	ND	ND	+	−	−	−	−	−	−
A. slackii	ND	ND	+	−	−	−	−	−	−
Bacteroides[d]	+	+	+	+	+	+	+	+	−
Pigmented proteolytic	+	+	−	+	+	+	+	−	−
Pigmented saccharolytic	+	+	−	+	+	+	+	−	−
Nonpigmented	+	+	−	+	+	+	+	−	−
Porphyromonas	+	+	−	−	−	+	+	−	−
P. gingivalis	+	(+)	−	−	−	(+)	(+)	−	−
Porphyromonas spp.[e]	+	+	−	−	+	+	+	−	−
Prevotella	+	+	(+)	(+)	(+)	(+)	(+)	−	−
P. intermedia	+	+	−	−	−	−	−	−	−
P. melaninogenica	+	+	−	−	−	−	−	−	−
Prevotella spp.[f]	+	+	(+)	(+)	(+)	(+)	(+)	−	−
Fusobacterium	+	+	−	ND	+	+	+	−	−
Veillonella	+	+	−	ND	ND	−	+	−	−
Neisseria	+	+	+	+	+	+	+	−	−
Lactobacillus	+	+	−	−	−	−	−	−	+

[a] − denotes no information; ND, not detected; (+), presence of bacteria that resemble the species.

[b] Strains designated *Streptococcus mitior* have been reclassified as *Streptococcus mitis*.

[c] *Actinomyces viscosus* is now to be limited to strains in animals other than humans. Human strains are now included in *A. naeslundii* genomic species 2.

[d] Several oral *Bacteroides* have been reclassified into *Porphyromonas* and *Prevotella*.

[e] *Porphyromonas* isolated from animals (see Table IV).

[f] *Prevotella* includes oral organisms previously classified as saccharolytic *Bacteroides*.

cavities of animals, along with a listing of a few important species to indicate the variation among animals. The animals examined were all in captivity and little is known of the oral microflora of animals in their natural habitats.

It is likely that given similar environments in different animal species, similar microbial genera will be present. For example, spirochetes and *Peptostreptococcus* spp. are present in dogs with periodontal disease. However, *Staphylococcus,* which is unusual in human mouths, can be isolated from the mouths of mice, reflecting a different habitat. Different species having similar niches may also be present in the mouths of different animals. One example of this is the presence of *Actinomyces denticolens, A. slackii,* and *A. howellii* in cattle. These three species are physiologically similar to *A. naeslundii* genomic species 1 and 2 isolated from humans and probably occupy the same niches in the oral communities in cattle as the latter group of organisms do in humans.

Some 31 genera have been identified in the human oral flora (Table III). The approaches to defining the taxonomy of oral bacteria do not differ from those used for organisms associated with other surface habitats of humans; however, knowledge of the diversity within genera at the species level and below is limited. This is particularly true for species and strains that have not been associated with oral disease and that might be considered members of the resident or natural flora. However, as the studies of the microflora of diseased sites have become more detailed, more precision has been used in the classification of isolates in keeping with the realization that several different species or strains of oral bacteria can be involved in oral disease. The increased precision and discrimination provided by more sophisticated tests used in analysis of larger numbers of isolates has resulted in the recognition of more species diversity within some oral genera and also in the revision of older classifications to define new genera and species

TABLE III
Bacterial and Fungal Genera of the Human Oral Flora

Bacteria			
Gram-positive rods and filaments	*Gram-negative rods and filaments*	*Gram-positive cocci*	*Gram-negative cocci*
Facultative/aerobic			
Actinomyces	Actinobacillus	Streptococcus	Neisseria
Corynebacterium	Capnocytophaga	Stomatococcus	Branhamella
Lactobacillus	Eikenella	Enterococcus	
Rothia	Hemophilus		
	Simonsiella		
Anaerobic			
Actinomyces	Bacteroides	Peptostreptococcus	Viellonella
Bifidobacterium	Campylobacter		
Eubacterium	Centipeda		
Propionibacterium	Fusobacterium		
	Leptotrichia		
	Mitsuokella		
	Porphyromonas		
	Prevotella		
	Selenonomas		
	Treponema		
	Wolinella		
Fungi			
Candida			

TABLE IV
Oral Species within Selected Bacterial Genera Isolated from Humans

Genus	Species
Streptococcus	*sanguis* (biovars 1–4), *parasanguis*, *gordonii* (biovars 1–3), *mitis* (biovars 1, 2), *oralis*, *salivarius*, *vestibularis*, *salivarius*, *intermedius*, *anginosus*, *constellatus*
	Mutans group: *mutans, sobrinus, cricetus, rattus,* (*ferus, macacae, downei,* isolated from the oral cavity of animals)
Actinomyces	*israelii, gerencseriae, naeslundii* (genomic species 1, 2), *odontolyticus, meyeri, georgiae,* (*bovis, viscosus, denticolens, howellii, slackii, hordeovulneris, hyovaginalis, pyogenes,* isolated from animals)
Prevotella	*melaninogenica, intermedia, nigrescens, loeschii, denticola, buccae, buccalis, heparinolytica, oralis, oris, veroralis, zoogleoformans, oulora*
Porphyromonas	*gingivalis, endontalis, assacharolytica* (*macacae, levii, cangingivalis, catoniae, crevioricans, cansulci,* similar organisms from animals)
Fusobacterium	*alocis, sulci, peridonticum, nucleatum* (*simiae,* from monkeys)
Treponema	*denticola, macrodentium, oralis, socranskii, vincentii, scoliodontium*

(Table IV). *Streptococcus* is a good example of the former, whereas the revision of the classification of the oral Bacteroides to define *Prevotella* and *Porphyromonas* typifies the latter case. Also, studies of representatives of oral genera isolated from sites other than the mouth has contributed to revealing the potential diversity of species of oral genera. In particular, the identification of gram-positive rods from infections in humans and animals has extended our knowledge of *Actinomyces* with new species (such as, *A. bernardiae, A. neuii, A. radingae, A. turicensis,* and *A. hyovaginalis*) being described. The origin of these organisms in infections is not known, but they may be resident in mouth, as are the other species of this genus. It has been suggested that some of these newly described *Actinomyces* should be placed in *Arcanobacterium.*

In common with bacteria from other habitats, clonal diversity has been demonstrated among strains of species of oral bacteria. It is generally accepted that often one or relatively few clones of a pathogenic bacterial species may be responsible for causing disease. In contrast, resident or commensal species will be represented in a given habitat by several clones. The term "clone" describes a strain descended from a single parent and theoretically identical to its parent. However, parental DNA may not be preserved intact in daughter cells due to its modification during division through intragenic mutations and exchange of DNA between cells. Given this proviso, clones (genetic variants) among oral bacterial species have been demonstrated by a variety of techniques (e.g., multilocus enzyme electrophoresis, restriction-fragment-length polymorphism, ribotyping, and abitrary primed polymerase chain reaction) with different degrees of discrimination. Examples include 14–61 types of *S. mutans*, 24–93 types of *S. mitis*, 19–114 types of *A. naeslundii*, 8–38 types of *Actinobacillus actinomycetemcomitans*, and 10–78 types of *P. gingivalis*. Some oral species are represented by several clones in an individual, whereas others exist as a single predominant clone. For example, one clone of *P. gingivalis* predominates in a person, whereas up to 20 clones of *A. naeslundii* genomic species 2 can be isolated. The biological significance, if any, of extensive clonal diversity within oral resident species is not known, although diversity in an individual may enhance the survival of a species. Multilocus enzyme electrophoresis has revealed 78 electromorphic types among 100 strains of *P. gingivalis* from periodontal disease samples, suggesting that each has a similar association with the disease and, therefore, that this species is best described as an opportunist pathogen. Genetic-typing methods have also proved useful to demonstrate the transfer of strains of species from mothers to infants and among family members.

1. Flora of Soft Tissues

Studies on the oral flora of humans and animals have tended to concentrate on the organisms in dental plaque, whereas few studies have been made on the microflora of soft tissues. Some data are available on the flora of the human tongue, which include species of *Actinomyces* (*A. naeslundii* and *A. odontolyticus*), *Bacteroides melaninogenicus* (*Prevotella melaninogenica*), *Neisseria*, *Streptococcus* (*S. salivarius*, *S. oralis*, and *S. mitis*), and *Stomatococcus mucilagenosus*, with the tongue probably representing the natural habitat of this last organism. Although the same genera are found on other surfaces in the mouth, different species are found on soft and hard tissue surfaces. For example, *S. salivarius* regularly colonizes the soft tissues, but is relatively uncommon in dental plaque, whereas *S. sanguis* and *S. mutans* are found more commonly on tooth surfaces. Similarly, *A. naeslundii* genomic species 1 is thought to occur more commonly on soft tissues than *A. naeslundii* genomic species 2.

2. *Flora of Dental Plaque*

Dental plaque can be separated into two types, supragingival and subgingival, based on the site of its accumulation. Supragingival plaque is a significant reservoir for members of the resident flora and is consistently present in health or disease. In contrast, although a small gingival sulcus is present at the junction of teeth and gingiva in health, subgingival plaque accumulates in the pockets between the gingivae and the tooth surface formed during periodontal disease.

The most diverse supragingival plaque communities develop in the protected interproximal areas between adjacent teeth (Table V), whereas more exposed areas, such as occlusal surfaces, support a less complex community. After a period of undisturbed development in a protected area, dental plaque may reach equilibrium with the local environment and be considered a climax community (i.e., one in balance with its environment). This community will be the most diverse and include a wide range of bacteria; however, as plaque is normally disrupted frequently during normal mastication and oral hygiene procedures (e.g., tooth brushing and flossing), it is unlikely that climax communities develop in all habitats.

The subgingival habitat in periodontal disease provides a distinctly different environment and, consequently, supports a significantly different flora (Table V). Relatively few of the genera found in large numbers in supragingival plaque are isolated subgingivally, and the flora are dominated by gram-negative and anaerobic bacteria. It should be noted that although a pocket may still exist after tissue healing, the subgingival flora changes with increases in gram-positive species and a return to a community similar to that of supragingival plaque.

B. Oral Flora in Disease

1. *Oral Microbial Pathogenesis*

There is little doubt that all of the bacterial diseases that cause destruction of oral tissues involve more than one type of organism and should be considered mixed infections. Although individual organisms from such infections may be virulent in pure culture in experimental animals, they seldom, if ever, cause pure-culture infections *in situ* in the mouth. It can be proposed that various members of the infecting mixture enhance the virulence and allow the establishment of potential pathogens. Studies of mixed infections in experimental animals have shown that certain combinations of organisms are more virulent than others and that nutritional factors and immunity to various host defense mechanisms are factors in their pathogenesis. For example, succinate produced by *Actinomyces* is a growth factor for *Prevotella*, and the proteases of *Porphyromonas gingivalis* will degrade complement factors and immunoglobulins (antibodies), thereby, protecting bacteria against the host's immune system.

A significant feature of caries, and possibly of periodontal diseases, is that changes in the environment of the tissue habitat can result in changes to the local oral flora, encouraging the development of a potentially or overtly pathogenic bacterial community. For example, in the later stages of the succession of bacteria in dental plaque, a single species, serovar, or strain may be dominant in the community. These organisms are often those that are sufficiently virulent to produce tissue destruction in experimental animals and it can be assumed that they are playing a significant role in the natural disease process. Typical of these organisms are two of the "mutans streptococci," *S. mutans* and *S. sobrinus*, which increase in numbers in supragingival plaque in association with the decalcification of tooth enamel leading to dental

TABLE V
Selected Species among Supragingival and Subgingival Dental-Plaque Communities

Genus	Interproximal supragingival plaque	Subgingival plaque in periodontal disease
Streptococcus	S. sanguis biovars 1–4, S. gordonii biovars 1–3, S. mutans, S. sobrinus, S. anginosus, S. intermedius, S. oralis, S. mitis	S. intermedius, S. oralis
Actinomyces	A. naeslundii genomic species 1 and 2, A. israelii, A. gerencseriae, A. odontolyticus	A. naeslundii, A. israelii, A. gerencseriae, A. meyeri, A. georgiae, A. odontolyticus
Lactobacillus	L. casei, L. acidophilus, L. fermentum, L. plantarum, L. brevis, L. salivarius	L. uli
Propionibacterium	P. proprionicus, P. acnes	P. proprionicus, P. acnes
Corynebacterium	C. matruchotti	
Bifidobacterium	B. dentium	B. dentium
Leptotrichia	L. buccalis	
Eubacterium	E. alactolyticum, E. saburreum	E. alactolyticum, E. saburreum, E. timidum, E. nodatum, E. brachy
Peptostreptococcus		P. anaerobius, P. micros
Veillonella	V. parvula, V. dispar, V. atypica	V. parvula
Actinobacillus	A. actinomycetemcomitans	A. actinomycetemcomitans
Haemophilus	H. segnis, H. paraphrophilus, H. aphrophilus	H. segnis
Campylobacter		C. concisus, C. curvus, C. rectus
Prevotella	P. melaninogenica, P. nigrescens, P. buccae, P. oris, P. denticola, P. loescheii	P. melaninogenica, P. intermedia, P. nigrescens, P. denticola, P. oris, P. oralis, P. buccae, P. veroralis, P. tannerae
Porphyromonas		P. gingivalis, P. endontalis
Bacteroides		"B. forsythus" (closely related to Porphyromonas)
Fusobacterium	F. nucleatum ssp. polymorphum	F. nucleatum ssp. nucleatum, ssp. polymorphum, ssp. vincentii, ssp. fusiforme, F. periodontium, F. alocis
Centipeda		C. periodontii
Selenomonas		S. sputigena, S. noxia, S. infelix, S. dianae
Treponema		T. socranskii, T. vincentii, T. denticola, T. pectinovorum
Wolinella		W. recta

caries. However, the importance of nonbacterial factors of the local environment in the development of disease must be emphasized, as high concentrations of *S. mutans*, and possibly *S. sobrinus*, also occur in plaque in the absence of caries development.

2. Dental Caries
a. Enamel Caries

Dental caries results from the destruction of teeth by acids produced by the bacteria in dental plaque metabolizing dietary carbohydrate. The consumption of sugar results in a typical plaque pH-versus-time curve (Stephan curve), which is characterized by the immediate reduction in plaque pH to a minimum value followed by a return to more neutral pH values. The extent and duration of the pH minimum, which can reach values lower than pH 4.0, is determined by the concentration of sugar in the diet, the period of food intake, the numbers of acidogenic bacteria in plaque, and the flow rate and buffering capacity of saliva at the time of eating. The reduction in plaque pH indicates that the rate of acid formation by plaque bacteria at any one time has exceeded the buffering and clearing properties of saliva. Generally, enamel demineralization occurs in a region near pH 5.5, the critical pH region, whereas remineralization will occur above this pH because saliva is supersaturated with calcium and phosphate, the main components of enamel. Thus, small areas of enamel below dental plaque can be subjected to cycles of demineralization and remineralization during the normal course of food consumption each day. The greater the amount of time that the enamel surface at a site is subjected to an acidic environment, the greater the opportunity for caries development at that location.

The plaque bacteria associated with caries generally have the capacity to degrade carbohydrate substrates rapidly to acid end products, and also to grow and metabolize in acidic environments. Thus, these organisms are both acidogenic (acid-producing) and aciduric (acid-tolerant), properties that enable them to maintain a low environmental pH in association with the surfaces of teeth. Most significant among these cariogenic bacteria are those included among the "mutans streptococci" (Table IV). In humans, the species most closely associated with caries are *S. mutans* and *S. sobrinus*. A second genus specifically associated with caries in *Lactobacillus*, which colonizes the tooth surface later than *S. mutans,* probably after the destruction of the surface enamel. Lactobacilli are found in high numbers in the carious enamel lesions, which are habitats with characteristics different from those of nondecalcified enamel. Thus, caries represents a mixed infection in which bacterial succession and the emergence of opportunistic pathogens are associated with a changing environment. As such, caries can be used as a model for other mixed infections in the mouth. Most recently, it has been shown that early decalcification of enamel (i.e., within 4 days of plaque development in a protected area) can be associated with organisms other than *S. mutans* and *S. sobrinus*. This suggests that there are other organisms in the mouth that can initiate enamel decalcification. A potential role for other organisms in caries is suggested by studies with experimental animals in which caries can be produced by different species of *Streptococcus, Actinomyces,* and *Lactobacillus.*

b. Caries of Root Surfaces

As a person ages, the gingiva may recede so that the root of the tooth becomes exposed to the oral cavity and, because the mineral of the tooth root is less acid-resistant than the enamel, caries can develop readily on these surfaces. Recent studies of samples from well-defined areas of root caries lesions in extracted teeth have demonstrated the complexity of the flora of lesions, which includes gram-positive and gram-negative anaerobes, and also provided evidence for the association of different communities within lesions during their progression. The flora of some lesions is dominated by *A. naeslundii*, whereas in others *A. gerencseriae* and *A. israelii* form a high proportion of the gram-positive flora. In contrast, the flora of other lesions are dominated by mutans streptococci and *Lactobacillus* spp. Given the possible variation in local environments and the dynamic nature of lesion formation, root caries can be described as a polymicrobic infection typified by the variety of different communities associated with carious tooth roots.

3. Periodontal Diseases

The term "periodontal disease" is used to describe a collection of diseases of the gingiva ranging from inflammation to extensive tissue destruction. Some of the diseases are well defined clinically, whereas in others the signs and symptoms may vary and an accurate diagnosis can be difficult. Thus, the inflammation of the gingiva (gingivitis), which can be reversed by oral hygiene procedures, can readily be diagnosed and described, as can juvenile periodontitis. Some of the other periodontal diseases may represent more than one disease and are less well defined. These include adult periodontitis, rapidly advancing periodontitis, and refractory periodontitis, with the recognition of the last being based on its lack of response to normal methods of treatment.

The flora of periodontal pockets (Table V) are complex with a high diversity of bacteria. Over 500 taxa have been isolated from periodontal pockets and one study of adult periodontitis identified 47 species as being most numerous in samples. Despite the complexity of the flora and difficulties with accurate diagnosis, it has been possible to associate specific bacteria with lesions of certain periodontal diseases. In particular, although opinions have differed, *Porphyromonas gingivalis* and *Actinobacillus actinomycetemcommitans* are regarded by most oral microbiologists as significant periodontal pathogens.

a. Inflammatory

Gingivitis is unique among diseases affecting the gingiva in that it can be reversed by the use of rigorous dental-hygiene procedures. Experimental gingivitis has been studied in humans and the early microbiology of gingivitis is similar to that of supragingival plaque at the gingival margin. However, as inflammation becomes more severe and bleeding of the gingiva occurs, certain species become common in the plaque. These species include *Actinomyces israelli, Fusobacterium nucleatum, Streptococcus intermedius, Campylobacter sputagena, Prevotella nigrescens, Peptostreptococcus micros,* and *Campylobacter ochracea. Bacteroides intermedius,* which has been associated with pregnancy gingivitis, now includes *Prevotella intermedia* and *Prevotella nigrescens.*

b. Destructive

The site associated with this type of disease is the pocket created when there is a loss of attachment of the gingiva to the tooth. These mixed anaerobic infections can cause extensive tissue destruction, including the loss of connective tissue and the resorption of alveolar bone, which can result in tooth loss. Perhaps the best example of a relationship between a form of periodontal disease and a specific bacterium is that of juvenile periodontitis and *Actinobacillus actinomycetemcommitans.* This organism is often dominant in the microflora associated with the lesions of this disease and strains producing leucotoxin are particularly virulent. Certain other bacteria are also associated with areas of gingival and bone destruction, the most significant being *Porphyromonas gingivalis, Prevotella melaninogenica, Prevotella intermedia, Fusobacterium nucleatum, Capnocytophaga, Selenomonas,* and *Treponema.* The most virulent of these is *Porphyromonas gingivalis* (formerly known as *Bacteroides gingivalis*). This organism can cause fatal infections in mice and produces potent proteases that degrade collagen, complement, and immu-

noglobulins. The other organisms that may be significant in periodontal disease are *Bacteroides forsythus, Prevotella buccae, Fusobacterium* spp., *Peptostreptococcus* spp., and *Wolinella recta.* Although various species have been identified as potential pathogens in periodontal disease, a specific role for them in the disease process has yet to be established. Several unclassified bacteria have also been found in high numbers in individuals with periodontal disease, and allocating specific roles for the range of bacteria in the subgingival flora is a major task for research.

4. Other Infections

The biofilms on teeth, tongue, and the soft tissues of the mouth are reservoirs for resident oral bacteria that are significant opportunist pathogens. Infection is generally accepted to be the result of the entry of bacteria into the bloodstream via the gingival crevice during eating, tooth brushing, and other simple procedures that manipulate oral tissues. These bacteremias are usually transient and a normal occurrence in healthy hosts. However, in immunosuppressed subjects and those with some tissue damage or indwelling protheses, these transients can cause life-threatening infections.

Oral streptococci (*S. sanguis, S. gordonii,* and *S. mutans*) are among the most common organisms colonizing biofilms on prostheses and damaged tissues and in particular are responsible for subacute bacterial endocarditis. *S. oralis, S. intermedius, S. anginosus,* and *S. constellatus* cause liver and brain abscesses among the general population, and also serious blood-borne infections in immunosuppressed patients. Actinomycosis is a chronic significant mixed infection that can occur at any site in the body and it is generally accepted that the organisms associated with this condition originate from the mouth. The pathogen, which gives its name to the disease, is usually *A. israelii,* although *A. gerencseriae* and *Propionibacterium proprionicus* produce a similar disease. The infection is almost invariably mixed and includes oral bacteria. *Actinobacillus actinomycetemcomitans* is commonly found in association with *Actinomyces* and *Propionibacterium* in actinomycosis.

C. Control and Prevention of Oral Diseases of Bacterial Etiology

The control of bacterial infections in the mouth other than caries and periodontal diseases follows the general principles of treatment for infections elsewhere in the body and is based primarily on antibiotic therapy. Caries and periodontal diseases are approached differently, although antibiotics are used as adjuncts to the therapy for some types of periodontal disease. Emphasis is placed on the prevention of caries and periodontal disease through an improvement of oral hygiene. Oral hygiene procedures, such as brushing and flossing, cause not only disruption of dental plaque, but serve to reduce its mass and its potential to develop into a pathogenic community.

Apart from physical disruption, other methods aimed more directly at the control of bacteria are used to prevent caries and periodontal diseases. Perhaps the most effective and useful of these methods is the application of fluoride to prevent dental caries. Fluoride reduces caries by several mechanisms, including the enhancement of remineralization of the carious lesion and the conversion of the normal hydroxyapatite mineral structure of tooth enamel to the less acid-sensitive fluorapatite. Fluoride also has significant inhibitory effects on bacterial metabolism, and these are enhanced at low pH. The major antimicrobial effect of fluoride is the inhibition of carbohydrate metabolism by bacteria, which naturally reduces the rate and extent of acid end-product formation, the direct cause of caries. This reduces the ecological advantage afforded aciduric bacteria by making the environment in dental plaque less acidic. Consequently, fluoridation remains one of the best examples of the public health application of a preventive agent for control of a disease of bacterial etiology.

For the prevention of periodontal disease, there is no compound as effective as fluoride is in caries prevention. However, short-term use of mouth rinses can be helpful for the reduction of dental plaque and for the elimination of bacteria. The effectiveness of many mouth rinses as antibacterials is questionable, although chlorhexidine (Hibitane) is recognized by most workers to be effective in reducing oral flora. This agent is available as a mouth rinse, varying in concentration from 0.1–0.2%, and is also available as a tooth gel in some countries. Chlorhexidine is used clinically for patients with extensive periodontal disease to restore the level of oral hygiene before dental treatment, for elderly and handicapped people, and to reduce the numbers of bacteria in the mouth of immunocompromised patients prior to surgery.

A considerable body of evidence now exists establishing the association of S. mutans with dental caries in humans and, as a result, immunization has been proposed as a preventive measure. Considerable research has centered on isolating protective antigens and defining the optimum time and conditions for immunization. The antigens that have been tested in experimental animals include those on the surfaces of whole cells of S. mutans, cell-surface proteins, and enzymes, such as glucosyltransferase, which are considered significant in the etiology of caries. Two main routes of immunization, systemic and oral, have been explored, as has passive immunity in rat pups from immunized mothers. Recently, passive immunity by monoclonal antibodies in humans has been shown to affect the colonization of their mouths by S. mutans. Also, humans rinsing their mouths with milk from cows immunized with S. mutans had reduced numbers of this organism relative to controls and the strains isolated from subjects taking milk antibodies exhibited colonial variation. There is now little doubt that it is possible to influence the numbers and colonization of S. mutans in the human mouth through a variety of immunization procedures. The problems of cross-reactivity of whole cells of S. mutans to heart tissue have been overcome by the use of pure preparations of cell-surface proteins or enzymes, and recombinant proteins. Although human volunteers have been immunized with S. mutans by the oral route, there have been no clinical studies of the effectiveness of any vaccine in controlling caries in humans. Immune prevention and control of periodontal disease has also been proposed, based on the generation of active immunity against putative periodontal pathogens. Generally the antigens are cells or specific components (fimbriae) of P. gingivalis. Data from experimental animal infections indicate that immunization can influence oral colonization by P. gingivalis and bone loss in infections.

See Also the Following Articles

BIOFILMS AND BIOFOULING • ECOLOGY, MICROBIAL • GASTROINTESTINAL MICROBIOLOGY • SKIN MICROBIOLOGY

Bibliography

Bowden, G. H. (1991). Which bacteria are cariogenic in humans? *In* "Risk Markers for Oral Disease, Vol. 1, Dental Caries" (N. W. Johnson, ed.), pp. 266–286. Cambridge University Press, Cambridge, UK.

Bowden, G. H., Ellwood, D. C., and Hamilton, I. R. (1979). Microbial ecology of the oral, cavity. *In* "Advances in Microbial Ecology" (M. Alexander, ed.), Vol. 3, pp. 135–217. Plenum, New York.

Bowden, G. H., and Hamilton, I. R. (1998). *Crit. Rev. Oral Biol. Med.* **9**, 54–85.

Brailsford, S. R., Lynch, E., and Beighton, D. (1998). The isolation of *Actinomyces naeslundii* from sound and root surfaces and root caries lesions. *Caries Res.* **32**, 100–106.

Genco, R. *et al.* (eds.) (1994). "Molecular Pathogenesis of Periodontal Disease." ASM Press, Washington, DC.

Hamilton, I. R. (1990). Biochemical effects of fluoride on oral bacteria. *J. Dent. Res.* **69**, 660–667.

Kolenbrander, P. E., and London, J. (1992). Ecological significance of coaggregation among oral bacteria. *In* "Advances in Microbial Ecology" (K. C. Marshall, ed.), Vol. 12, pp. 183–217. Plenum, New York.

Liljemark, W. F., and Bloomquist C. (1996). Human oral microbial ecology and dental caries and periodontal disease. *Crit. Rev. Oral Biol. Med.* **7**, 180–198.

Lang, N. P., Holtz, P. P., Gusberti, F. A., and Joss, A. (1987). Longitudinal clinical and microbiological study on the relationship between infection with *Streptococcus mutans* and the development of caries in humans. *Oral Microbiol. Immunol.* **2**, 39–47.

Marcotte, H., and Lavoie, M. C. (1998). Oral microbial ecology and the role of salivary immunoglobulin A. *Microbiol. Mol. Biol. Rev.* **62**, 71–109.

Moore, W. E. C., and Moore, L. V. H. (1994). The bacteria of periodontal diseases. *Perio. 2000* **5**, 66–77.

Novak, M. J. (ed.) (1997). "Biofilms on Oral Surfaces: Implications for Health and Disease." 14th International Conference on Oral Biology. *Adv. Dent. Res.* **11**.

Scannapieco, F. A. (1994). Saliva-bacterium interactions in oral microbial ecology. *Crit. Rev. Oral Biol. Med.* **5**, 203–248.

Schüpbach, P., Osterwalder, V., and Guggenheim, B. (1996). Human root caries: Microbiota of a limited number of root caries lesions. *Caries Res.* **30**, 52–64.

Sevensater, G., Larsson, U-B., Grief, E. C. G., Cvitkovitch, D. G., Hamilton, I. R. (1997). Acid tolerance and survival by oral bacteria. *Oral Microbiol. Immunol.* **12**, 266–273.

Ore Leaching by Microbes

James A. Brierley
Newmont Metallurgical Services

GLOSSARY

bioleaching Solubilization of metal values from an ore by microbially mediated oxidation of the minerals.

biooxidation pretreatment Microbial oxidation of minerals in order to unlock a metal value contained in the mineral.

chemolithotrophic Describing or referring to a type of metabolism in which inorganic substrates are oxidized for energy and carbon dioxide serves as the carbon source for biosynthesis.

flotation concentrate High-value product of separation and concentration of specific minerals from surrounding rock material by use of reagents that promote the flotation of specific minerals.

heterotrophic Describing or referring to a type of metabolism in which the microorganism obtains energy from the oxidation of reduced carbon compounds and use of organic carbon compounds as the source of carbon for biosynthesis.

refractory Referring to locked precious metals, usually in sulfide minerals; their recovery by conventional processes is prevented (metallurgical term).

MICROORGANISMS are capable of solubilization of metals from a diversity of mineral substrates. This phenomenon is a consequence of the chemolithotrophic metabolism of the bacteria, whereby energy from mineral oxidation is harnessed for growth and survival. Mineral oxidation and associated metal solubilization have environmental and technological significance. Acid rock drainage results from microbial oxidation of sulfide minerals causing metal contamination and acidification of receiving waters. This is a natural process, enhanced by the exposure of sulfide minerals to the atmosphere. Such sulfide mineral exposure can occur through mining activities. The same bacterial process that causes acid rock drainage can be harnessed for the beneficial economic recovery of metals. This beneficial process, called bioleaching, is the bacterially mediated process for solubilization of metals such as copper, nickel, uranium, and zinc from sulfide minerals. These solubilized metals are subsequently recovered for commercial use. Precious metal ores in which gold, silver, and platinum group metals are occluded sulfide minerals (refractory precious metal ores) can also be oxidized by bacteria to liberate the precious metals. The liberated precious metal values are then extracted and recovered by conventional metallurgical processes. Biooxidation of refractory precious metal ores is not bioleaching in the true sense because the precious metals are not directly solubilized in the process of bacterial oxidation. Rather, this process is termed mineral oxidation or refractory ore pretreatment.

I. BASIC PRINCIPLES OF BIOLEACHING AND BIOOXIDATION OF MINERALS

The group of diverse bacteria capable of sulfide-mineral oxidation is characterized by the need for aerobic conditions and ability to grow in acidic conditions. Growth can occur in systems at less than pH 1, preferably in a range of pH 1.3–3.5. However, the acidity must be due to sulfuric acid, which is also a product of the bacterial oxidation of sulfur and sulfide minerals. These bacteria do not grow well in hydrochloric acid systems because of toxicity, attributed to higher concentrations of chloride ion. Another common characteristic of these bacteria is the ability to use CO_2 as a carbon source. This is important for practical use of bioleaching, as no expensive organic compounds are required for the process. Mineral-oxidizing bacteria have representation in diverse temperature groups. There are mesophilic types growing in the 10–40°C range, moderate thermophiles growing from 30–55°C, and thermophiles, which oxidize minerals at 50 to over 70°C. A number of sulfide minerals can be biooxidized, including but not limited to pyrite (FeS_2), arsenopyrite (FeAsS), bornite (Cu_5FeS_4), chalcocite (Cu_2S), chalcopyrite ($CuFeS_2$), covellite (CuS), digenite, (Cu_9S_5), millerite (NiS), and sphalerite (ZnS). The biooxidation of pyrite and arsenopyrite has commercial importance, because these minerals can encapsulate precious metals and their oxidation is required for the economic recovery of the precious metals. The biooxidation of copper, nickel, and zinc sulfide minerals is an inexpensive process for the extraction of these base metals and transforms otherwise uneconomic ores and mine wastes to metal resources.

An important reaction associated with mineral biooxidation is the bacterial oxidation of ferrous iron (1).

$$4FeSO_4 + O_2 + 2H_2SO_4 \rightarrow 2Fe_2(SO_4)_3 + 2H_2O \quad (1)$$

Bacteria, such as *Thiobacillus ferrooxidans*, oxidize ferrous iron for energy, using a membrane-bound glycoprotein for the transport of ferrous iron into the cell membrane for initial oxidation by a Fe(II) oxidase (Rawlings, 1997). The membrane-associated electron-transport system involves a cytochrome c_{552}, rusticyanin, and a cytochrome c oxidase of the aa_3-type for the reduction of oxygen.

The end product of ferrous iron oxidation, ferric iron, is soluble in the acidic conditions required for its biooxidation. Ferric iron is also an oxidant of sulfide minerals. The process whereby the metabolic end product of biooxidation, ferric iron, facilitates the solubilization of metals from sulfide minerals is termed "indirect bioleaching." The indirect bioleach-

ing of copper from chalcocite and chalcopyrite is illustrated by reactions (2) and (3).

$$Cu_2S + 2Fe_2(SO_4)_3 \rightarrow 4FeSO_4 + 2CuSO_4 + S° \quad (2)$$

$$CuFeS_2 + 2Fe_2(SO_4)_3 \rightarrow 5FeSO_4 \\ + CuSO_4 + 2S° \quad (3)$$

The ferric iron becomes reduced to ferrous iron in the oxidation of the copper sulfides, and the copper is solubilized as a sulfate. The ferrous iron is then reoxidized by the bacteria (reaction 1) to ferric iron, creating a cyclic process of oxidation and reduction of the iron and the solubilization of the copper from the minerals.

A second hypothesized mechanism for bioleaching of metals is called "direct bioleaching." In this process, the sulfide minerals are oxidized by the bacteria without the apparent use of the intermediary ferrous-ferric oxidation–reduction process (reactions 4 and 5).

$$2Cu_2S + 5O_2 + 2H_2SO_4 \rightarrow 4CuSO_4 + 2H_2O \quad (4)$$

$$2CuFeS_2 + 8.5O_2 + H_2SO_4 \rightarrow 2CuSO_4 \\ + Fe_2(SO_4)_3 + H_2O \quad (5)$$

Direct contact between the bacteria and the mineral is required. In commercial practice, it is difficult to discern the relative importance of indirect and direct bioleaching processes, as iron is always present to facilitate the ferric leach of the indirect bioleaching process.

As in indirect bioleaching, the direct bioleaching process may also involve the cyclic oxidation and reduction of iron. Figure 1 shows a model in which bioleaching of a mineral is mediated by the ferric iron-containing exopolymers of the bacterial cell. The exopolymer layer of the bacterium is rich in complexed iron. The cell attaches to the mineral surface, and the ferric iron oxidizes the sulfide mineral (per reactions 2 and 3). The ferrous iron and partially oxidized sulfur species are biooxidized to ferric iron and sulfate, respectively, and the metal values solubilized.

Soluble iron is produced from biooxidation of pyrite (6).

$$4FeS_2 + 15O_2 + 2H_2O \rightarrow 2Fe_2(SO_4)_3 + 2H_2SO_4 \quad (6)$$

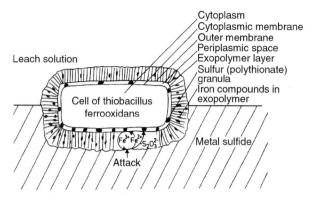

Fig. 1. Schematic model for biooxidation of sulfide minerals by attached bacteria. The exopolymeric layer is exaggerated to show the role of iron present in the layer in oxidation of the mineral. Reprinted from Sand, W., *et al.* (1995). Sulfur chemistry, biofilm, and the (in)direct attack mechanism—A critical evaluation of bacterial leaching. *Appl. Microbiol. Biotechnol.* **43**, 961–966, by courtesy of Springer-Verlag, New York.

Biooxidation of pyrite also becomes an indirect bioleaching process, with ferric iron serving as an oxidant of pyrite.

$$FeS_2 + 7Fe_2(SO_4)_3 + 8H_2O \rightarrow 15FeSO_4 + 8H_2SO_4 \quad (7)$$

The biooxidation of pyrite, illustrated in Fig 2, results in the extensive corrosion of the mineral.

The biooxidation of arsenopyrite can be illustrated as either an indirect bioleaching process (8) or a direct bioleaching process (9).

$$2FeAsS + 6O_2 + Fe_2(SO_4)_3 + 4H_2O \rightarrow 2H_3AsO_4 + 4FeSO_4 + H_2SO_4 \quad (8)$$

$$2FeAsS + 7O_2 + H_2SO_4 + 2H_2O \rightarrow 2H_3AsO_4 + Fe_2(SO_4)_3 \quad (9)$$

These reactions have commercial and environmental significance. Gold is often associated with arsenic in ores, and the biooxidation of precious metal ores requires the solubilization of arsenopyrite to release occluded gold. To prevent a potential water pollution problem due to the transport of soluble arsenic, the solution pH is increased to greater than 4, precipitating an environmentally stable ferric arsenate compound.

The ability of certain bacteria to oxidize sulfur is also an important bioleaching process.

$$2S + 3O_2 + 2H_2O \rightarrow 2H_2SO_4 \quad (10)$$

Sulfur can form as a product of oxidation of mineral sulfides (reactions 2 and 3). The sulfur can potentially impede further the oxidation of minerals by forming a passivating layer on the mineral surface. Bacteria, such as *Thiobacillus thiooxidans*, oxidize elemental sulfur and facilitate leaching by removing the sulfur formed on mineral surfaces. Furthermore, the oxidation of the sulfur to sulfuric acid promotes the bioleaching process by maintaining the acidic conditions required for the growth of the bacteria and promoting the solubility of the leached metals.

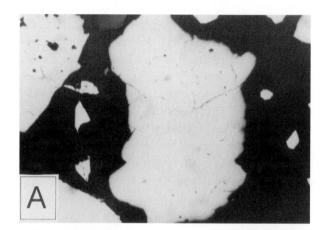

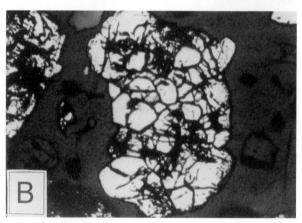

Fig. 2. The corrosive effect of sulfide mineral-oxidizing bacteria resulting from bioleaching pyrite. (A) Polished surface of pyrite before exposure to *T. ferrooxidans*. (B) The same surface after bioleaching in the presence of *T. ferrooxidans*. Magnification 130×.

II. MICROORGANISMS

Early research on bacteria associated with mineral oxidation focused on *Thiobacillus ferrooxidans.* This acidophilic, iron- and sulfur-oxidizing bacterium was initially found in association with acid rock drainage and later showed the capability of bioleaching sulfide minerals. The microorganism is chemolithotrophic, requiring no organic nutrient supplementation. It can function in acid conditions, pH 1–3, required for maintaining the solubility of leached metals. It is also tolerant of high concentrations of dissolved metals, over 20–30 g/liter Cu or Zn, up to 15 g/liter As, and over 50 g/liter Fe. However, *T. ferrooxidans* is not the only microbe with the characteristics required for bioleaching, nor is it the only microbe inhabiting the complex bioleaching environment. Table I lists some of the bacteria that are capable of bioleaching sulfide minerals and their temperature requirements for growth. Temperature is an important parameter for bioleaching processes. The oxidation of pyrite, arsenopyrite, and other mineral sulfides generates heat. Unless the bioleaching process plant is cooled, the mesophilic bacteria can be killed by the heat. The *T. ferrooxidans* and *L. ferrooxidans* often share the bioleaching environment, with the *L. ferrooxidans* population predominating when acidity increases, and particularly at pH less than 1.3. *L. ferrooxidans,* unlike *T. ferrooxidans,* is not able to oxidize sulfur. Because iron alone serves as an oxidizable energy source for *L. ferrooxidans,* this microbe conducts the indirect bioleaching process. An association of *L. ferrooxidans* and *T. thiooxidans* facilitates

bioleaching, even though the *T. thiooxidans* cannot leach metal sulfides. *L. ferrooxidans* catalyzes the ferric oxidation of the mineral sulfides (reactions 2, 3, 7, and 8), and *T. thiooxidans* oxidizes sulfur (reaction 10), which can form during sulfide mineral oxidation (reactions 2 and 3) and potentially passivate the leaching process by forming a coating on the mineral surface.

The moderately thermophilic *Sulfobacillus* species differ significantly from the *Thiobacillus* and *Leptospirillum* groups and function at higher temperatures. The *Sulfobacillus organisms* are gram-positive bacteria capable of forming endospores, a survival mechanism for unfavorable growth conditions. These bacteria are relatively large, 5 μm and greater in length. Microscopically, *Sulfobacillus* can be differentiated from smaller *T. ferrooxidans* (1–2 μm long) and the vibrio morphology of *L. ferrooxidans. Sulfobacillus* species have been found that can oxidize either sulfur or iron, thereby promoting the bioleaching process at temperatures of 50°C.

In the Archaea genera, *Acidianus, Metallosphaera,* and *Sulfolobus,* are effective for bioleaching at thermophilic conditions above 60°C. This group has other characteristics favoring their use for bioleaching. The copper sulfide chalcopyrite is poorly leached at mesophilic temperatures, but does bioleach well at thermophilic conditions using the Archaea. Furthermore, *Acidianus* has also been shown to bioleach molybdenite (MoS_2), which is highly refractory and has as its product soluble molybdenum, which is toxic to the mesophilic bacteria. These bacteria are all capable of bioleaching sulfide minerals, and all of them can oxidize iron, promoting the indirect bioleaching process.

There is also a diversity of microorganism types found in bioleaching environments. Fungi, yeast, algae, protozoa, and heterotrophic bacteria exist in the acidic-high-soluble-metal leaching systems. *Acidiphilium* species are also found in the bioleaching processes, and their contribution is believed to be the consumption of trace concentrations of organic compounds that could otherwise inhibit the iron-oxidizing chemolithotrophic bacteria. Heterotrophic *Acidimicrobium ferrooxidans* can also oxidize ferrous iron and facilitate the bioleaching of metals from mineral sulfides. The complex association of diverse

TABLE I
Bacteria and Archaea Capable of Bioleaching Mineral Sulfides

Genus and species	Temperature range for growth (°C)
Thiobacillus ferroxidans	10–40
Leptospirillum ferrooxidans	10–40
Sulfobacillus thermosulfidooxidans	30–55
Sulfolobus acidocalcarius	50–70
Metallosphaera sedula	55–80
Acidianus brierleyi	50–70

microorganisms in the bioleaching environment is not fully understood or well defined.

III. COMMERCIAL APPLICATIONS OF MICROBIAL MINERAL OXIDATION

Bioleaching has been commercially practiced for the extraction of two metals, copper and uranium, and for the pretreatment of refractory gold ores and concentrates to enhance precious-metal recovery. The recovery of copper from sulfide minerals may, however, data back at least 2000 years, when the Chinese unknowingly practiced bioleaching. However, the modern use of bioleaching first occurred in the 1950s for the recovery of copper from run-of-mine material that was too low in grade for smelter processing (Fig. 3). This commercial application dates to when the role of the bacteria in the process was established. In the 1980s, the first commercially designed bioleaching process for uranium recovery was operated in an underground mine. The biooxidation pretreatment of gold locked in sulfide minerals began on a commercial scale in 1986. This plant processed high-value flotation concentrates prepared from refractory ores. The pretreatment process used stirred-tank vessels for the mineral biooxidation pro-

cess. In 1995, a commercial-scale biooxidation heap process for pretreatment of refractory low-grade ores was conducted.

A. Copper Bioleaching

Bioleaching has been estimated to account for 15–25% of the world's copper production. At 15%, this amounts to about 1.1×10^6 tons copper per year. The bioleaching process has been particularly applicable for use with lower-grade copper sulfide ores containing 0.3–0.5% copper. Without the bioleaching process, these ores would be considered mine waste and the copper would be lost to production. This same material, if handled as a waste rather than as a resource, could produce problematic metalliferous acidic rock drainage, if not properly handled.

The first bioleaching mine dumps processed a low-grade run of mine material. The rock was not crushed and the process was not designed to promote bacterial activity. An acidic solution is percolated through the mine dump, placed on impervious pads to prevent solution losses, and the copper is solubilized as indicated by reaction (2), with the ferrous iron biooxidized to the reactive ferric species (reaction 1). The soluble copper can be recovered from the bioleaching solution either by precipitation on scrap

Fig. 3. Aerial photograph of mine dumps for the bioleaching of copper from mine waste material.

iron or solvent extraction-electrowinning (SX-EW). In SX-EW, the copper is separated from the acidic bioleaching solution using selective solvents and deposited on cathodes in electrolysis cells.

In the 1990s, bioleaching has been commercially applied to extract copper from higher grade (>1% copper) secondary copper minerals, principally chalcocite, digenite, and covellite. The process for these ores is carried out in simple inexpensive heap bioreactors. Copper ores are crushed to increase the reactive surface area and stacked on lined, impervious pads with underlying aeration pipes. The recycling of solution from the copper recovery circuit to the bioleaching heap provides the bacterial inoculum. Bioleaching is conducted with forced aeration to supply the microbes with oxygen and carbon dioxide. These systems are termed bioleach heaps. The solubilized copper is most commonly recovered from the bioleach heap solutions by SX-EW technology.

B. Uranium Bioleaching

The mineral uraninite (UO_2) contains uranium in an insoluble tetravalent state. The uranium can be leached by oxidation to the soluble hexavalent state. Bioleaching provides the mechanism for extraction by maintaining iron in an oxidized ferric form, which oxidizes and solubilizes the uranium as uranyl sulfate.

$$UO_2 + Fe_2(SO_4)_3 \rightarrow UO_2SO_4 + 2FeSO_4 \quad (11)$$

A commercial uranium bioleaching process was developed and operated in an underground mine in Canada. The process involved blasting low-grade ore contained in isolated cells within the mine. Each cell was flooded with acidic mine drainage containing bioleaching bacteria and ferric iron. Following the flooding, the cells were drained and aerated to facilitate the bacterial oxidation of iron and pyrite to generate additional acid and maintain iron in an oxidized state to facilitate leaching. The bioleached uranium was recovered by washing the contained rock and extracting the uranium from solution with an ion-exchange resin. The application of the bioleaching of uranium has been performed only on a limited scale because of the low demand and price for uranium.

C. Refractory Gold Biooxidation Pretreatment

The first commercial application of biooxidation pretreatment of refractory gold ores evolved from studies demonstrating the ability of *T. ferrooxidans* to oxidize arsenopyrite (reactions 8 and 9). Arsenopyrite is commonly associated with refractory gold ores, and it is a mineral that encapsulates gold.

Initially, biooxidation pretreatment was commercialized to treat flotation concentrates of pyrite and arsenopyrite minerals containing gold. Aerated stirred-tank bioreactors are used, most often several tanks operating in series to provide sufficient retention time to biooxidize the sulfide minerals. The first plant, operating in South Africa, treats 35 metric tonnes of concentrate per day. This plant requires a cooling system to remove the excessive heat generated during sulfide oxidation and maintain the bacteria–concentrate suspension at 40°C, a suitable temperature for the activity of *T. ferrooxidans*, *L. ferrooxidans*, and *T. thiooxidans*. One commercial system employs moderately-thermophilic mineral-oxidizing bacteria, such as the *Sulfobacillus* type, capable of operating at temperatures up to 55°C (Fig. 4). Following biooxidation, the concentrates are washed, neutralized, and leached, using cyanide with activated carbon to recover the gold. Because of high capital and operating costs, stirred-tank bioreactor systems are used for high-grade, high-value concentrates for which metal recoveries exceed 95%. These reactors are not suitable for the biooxidation pretreatment of low-grade, low-value whole ores.

Biooxidation heap reactors offer an economic alternative to the stirred-tank reactor systems for the commercial pretreatment of lower-grade refractory ores. One approach to promoting the biooxidation of the sulfide minerals is to inoculate the crushed ore as the heap is formed. This is accomplished with either a ferrous iron-grown culture or the recycling of solution percolating through previously formed heaps sprayed on the ore as it is stacked. This process assures the distribution of an active population of bacteria throughout the entire mass of the heap and promotes the rapid intiation of the biooxidation process. The addition of the acidic bacterial culture to the ore also benefits the process by facilitating a

Fig. 4. Three of six 500 m³ bioreactors treating 40,000 metric tonnes per year of flotation concentrate and producting about 60,000 oz gold at a sulfide refractory gold biooxidation plant at the Youanmi Mine, western Australia (courtesy of BacTech Metallurgical Solutions, Ltd.).

degree of agglomeration of the ore fine particles to reduce plugging. The inoculation process can be performed with a mixture of bacteria capable of sulfide mineral oxidation. The *Thiobacillus–Leptospirillum* group can be used in conjunction with moderate thermophiles (*Sulfobacillus*) and thermophiles (*Acidianus, Metallosphaera,* and *Sulfolobus*). The thermophiles are used because sufficient heat can be generated in the mineral oxidation process to increase the temperature a heap to over 60°C. Biooxidation pretreatment can require up to 270–360 days, depending on the amount of sulfide to be oxidized and the rate of oxidation. The biooxidized ore must be adjusted to an alkaline pH prior to leaching by the addition of lime as the ore is relocated to a permanent leach pad. Precious-metal recovery is achieved using conventional metallurgical gold leaching by cyanide, ammonium thiosulfate, or other gold-solubilizing reagents.

IV. CONCLUSION

Ore leaching by microbes, a process of biomining, is a technological and commercial application of spe-cific bacteria's ability to oxidize metal sulfide minerals. There is a diverse group of bacteria that directly participate in the bioleaching of metals and the biooxidation pretreatment of refractory precious metal ores. These bacteria, including species of *Thiobacillus, Leptospirillum, Sulfobacillus, Acidianus, Sulfolobus,* and *Metallosphaera,* are able to oxidize iron, which facilitates the oxidation of metal sulfide minerals.

The biooxidation of metal sulfide minerals is applied on a commercial scale. Low-grade copper ores and waste materials are processed using bioleaching in order to provide an economical recovery process. The bioleaching of copper is conducted using mine dumps and heaps as inexpensive bioreactors. Similarly, uranium has also been extracted using bioleaching.

Refractory gold ores and mineral concentrates, in which the precious metal is encapsulated within a mineral sulfide matrix, are biooxidized to expose the gold to conventional metallurgical processing. High-value refractory gold concentrates are bacterially pretreated in large stirred-tank reactor systems. Lower-value whole ores can be bacterially pretreated in heap bioreactors.

Ore leaching by microbes has potential for use in the extraction of other metals, for example zinc, cobalt, and nickel. Novel reactor systems will be developed to increase the efficiency of bioleaching in terms of cost and kinetics. These innovations are expected to extend the scope of bioleaching applications.

See Also the Following Articles

ARSENIC • HEAVY METAL POLLUTANTS: ENVIRONMENTAL AND BIO-TECHNOLOGICAL ASPECTS • METAL EXTRACTION AND ORE DISCOVERY • SULFIDE-CONTAINING ENVIRONMENTS

Bibliography

Ehrlich, H. L., and Brierley, C. L. (eds.) (1990). "Microbial Mineral Recovery." McGraw-Hill, New York.
"International Biohydrometallurgy Symposium IBS97 Biomine 97." (1997). Australian Mineral Foundation, Glenside, Australia.
Rawlings, D. E. (ed.) (1997). "Biomining: Theory, Microbes and Industrial Processes." Springer-Verlag, Berlin.
Vargas, T., Jerez, C. A., Wiertz, J. V., and Toledo, H. (eds.) (1995). "Biohydrometallurgical Processing," Vols 1–2. University of Chile, Santiago.

Origin of Life

William F. Loomis

University of California, San Diego

GLOSSARY

α-helix A polypeptide structure in which amino acids spiral around a central axis. There are 3.6 amino acids per turn.

β-strand A polypeptide structure in which the backbone zigzags and the side groups of the individual amino acids stick out on opposite sides.

exon The portion of an RNA molecule that is retained in final product and usually codes for a protein.

intron The portion of an RNA molecule that is cut out so that exons can be ligated together.

LIFE arose on Earth about 4 billion years ago and has flourished ever since. Over a hundred years ago, Charles Darwin pointed out that natural selection can lead to the origin of species, but he was unable to conceive of processes that could lead to the original appearance of life. More recently, A. I. Oparin, J. B. S. Haldane, S. Miller, M. Eigen and others have considered sets of chemical reactions that might lead to self-perpetuating systems. Their suggestions are based more on plausibility than on any new data from the dawn of life, but stay rigorously within the bounds set by the laws of physical chemistry. The surface of our planet is now teeming with life, but when it first condensed, about 4 billion years ago, there were only rocks, sand, and water in an atmosphere rich in nitrogen, methane, ammonia, and carbon dioxide, and devoid of oxygen. For a few hundred million years, Earth was bombarded by meteorites as it swept up the rocks in its orbit around the sun. Then the surface cooled down and life evolved. There are fossil traces of bacteria in rocks that are 3.5 billion years old. The conditions on Earth seem to have been favorable for the rapid evolution of life. We do not know exactly how the first microorganisms formed on the young planet long ago, but we know what it takes for an organism to live under conditions that prevailed billions of years ago. The early steps took place so long ago that there is little hope that they left any traces. However, the chemical processes that may have generated the building blocks for life can be carried out in test tubes to determine which ones are plausible and which ones are not. Biologists are beginning to understand some of the steps that may have led to life as we know it. We can link these together into a series that leads to the origin of life.

I. THE NATURE OF LIFE

We all agree that rocks are not alive. They do not grow, divide, make more rocks, or evolve into different-shaped rocks. Some semicrystalline clays of aluminum silicate can dissolve in water and reform when the water evaporates. The form of the original material determines the form of subsequent deposits to some extent, but the microscopic structures are ephemeral and subject to many outside forces. Clays cannot control their forms sufficiently to be considered alive. Living things must have the ability to grow and reproduce themselves accurately in the proper environment. All living organisms are constructed of cells in which the components necessary for survival and reproduction are surrounded by a protective layer of lipids and proteins. The information necessary for making all the vital components and arranging them in their proper places is carried inside in the form of long polymers of nucleic acids.

A. Cells

Cells come in many sizes and shapes, from bacteria that are less than 1 μm long to meter-wide fungal cells with thousands of nuclei. Most cells are no bigger than 10 μm across. Mammals and trees are just aggregates of a huge number of tiny cells. The earliest cells in the fossil record appear to be filamentous bacteria about 5 μm long. Larger cells and organisms do not appear in the fossil record until several billion years later. It seems that for most of the history of life of Earth, bacteria and other microorganisms have been the whole story. When considering the origin of life, we have the goal of seeing how the first bacterium might have arisen. The subsequent evolution of large organisms depended on the bacteria that preceded them.

Bacteria are defined as cells without internal compartments. The first cells that we would consider alive probably had a single membrane made of lipids and some proteins surrounding them. Later, many of these cells gave rise to bacteria with strong cell walls surrounding their outer membrane, but this structure does not seem to be absolutely essential because some bacteria survive today without a cell wall. Putting all the components in a membrane-enclosed space keeps them together and lets them evolve as an integrated system.

B. Powering the Process

The cellular membrane is also used to convert available energy sources into the chemical form that is most commonly used by all living cells, ATP. When a cell finds itself in an environment with more protons outside than inside, it can let a few protons in at a time by association with a protein embedded in the membrane. If this protein is coupled to an enzyme on the inside of the cell that has the ability to convert ADP to ATP, then the cell can generate ATP in this way. In fact, the coupling of proton-motive force to phosphorylation is the major source of biological energy in every organism today. It is the powerhouse of photosynthesis.

A cell cannot rely on encountering an environment with proton potential whenever it needs it, and so most cells have evolved a mechanism that lowers the internal concentration of protons by pumping them out. They will then have a lower concentration of protons inside than outside. However, pumping against a gradient takes energy and something has to be coupled to the proton pump for it to work. Cells have evolved a wide variety of ways to drive proton pumps, but the most prevalent uses the energy of sunlight. Pigments inside cells absorb light energy and use it to drive unfavorable reactions. Pigments such as chlorophyll are embedded in membranes and are coupled to proteins that drive protons against the concentration gradient. The cell can then let some protons back in through the proteins that are coupled to ADP phosphorylation and thereby generate ATP.

ATP is a very unstable molecule that can easily be activated to donate a phosphate to another molecule. Many proteins function as enzymes that catalyze such reactions. The catalytic site of an enzyme of this type is made from six repeats of about 40 amino acids that are strung together in a particular sequence such that they spontaneously take up the specific shape that has been called the mononucleotide binding fold. One-half of the chain of amino acids in the active site of the enzyme forms α-helices and the other half forms β-sheets. These secondary structures are the most stable states that the particular amino acids sequence of these enzymes can take, and so they always form spontaneously. Together the α-helices and the β-sheets make a pocket in which ATP is held in just the right orientation for it to react with another molecule held close by. The reaction then goes much faster than in the absence of the enzyme.

Charged molecules such as sugar phosphates cannot cross the lipid membrane. This makes it hard for a cell to take them up from the environment, but it is useful for concentrating them inside a cell once they are made in an enzyme-catalyzed reaction. Sugar phosphates can be used later as a source of energy. Cells can take up uncharged molecules such as glucose from the environment or synthesize them from simpler molecules that may be available. To keep glucose in the cell, it is often necessary to add a phosphate from ATP and thereby convert it to glucose-phosphate, which will stay in the cell. When more ATP is needed, sugar phosphates and other such molecules can be metabolized in reactions catalyzed by specific enzymes into other molecules that will be able to donate the phosphate to ADP and make new ATP. These reactions go on, back and forth, in every living thing all the time.

C. Growing

Growth is one of the attributes we look for when we are trying to decide if something is alive. Even before there were living cells, biological molecules were growing and dividing. Proteins and nucleic acids make up most of the living material of cells and these have been synthesized since the earliest times. Proteins are long chains of amino acids of defined sequences. The order of amino acids, which can have any one of the 20 distinct amino acids at each position, is rigorously prescribed as the protein is synthesized. This process is carried out on the surface of ribosomes under the direction of specific messenger RNAs (mRNAs); each mRNA molecule directs the synthesis of only one of the thousands of distinct proteins found in living cells. The messenger RNA is itself a long chain of nucleic acid bases in which the order of the four bases, adenine (A), guanine (G), uracil (U), or cytidine (C), is also prescribed. When a new protein is synthesized, the code carried in the sequence of bases of the mRNA is translated by transfer RNA (tRNA) molecules that bind to three bases in the mRNA at a time. There are at least 20 tRNAs, each one specific for a certain sequence of bases in the mRNA (such as GGC) and each one carrying a specific amino acid that can be added to the growing peptide. In this way, a mRNA molecule that is 300 bases long directs the synthesis of a new molecule of a protein that is 100 amino acids long. Cells grow by accumulating proteins and other macromolecules.

The synthesis of the nucleic acids, RNA and DNA, is simpler than protein synthesis. The sequence of

bases in the new polymer is directly determined in a one-to-one fashion by the sequence in another polymer as a result of the complementarity of the bases, themselves. For instance, whenever there is a G at a specific position in the template molecule, a C is added at that position to the growing molecule; likewise, C specifies G, A specifies U, and U specifies A. The chemistry of the bases is such that G-C and A-U pairs are favored over all others. Therefore, nucleic acids can direct the synthesis of complementary copies without any participation of proteins, although enzymes usually participate to speed up the reaction. The complementary copy can then direct the synthesis of copies of the original sequence.

D. Reproducing

When a cell has doubled in size, it can divide in two to generate copies of its former self. Cell division is a complex process in bacteria and other microorganisms that carefully positions equal amounts of all the essential components in each daughter cell. However, it should be kept in mind that bacteria have been selected over billions of years to reproduce at optimal rates, as well as to survive the competition of other microorganisms. When life was first evolving, competition was less fierce and any reproduction was a major success. The chief concern was to retain enough of the interacting molecules that sometimes both daughters were able to continue growing.

A cell with 100 RNA molecules coding for a certain enzyme can divide roughly in half and generate two cells each with about 50 molecules of the RNA. Sometimes one might get 70 and the other 30 molecules, but because RNA can replicate and direct the synthesis of new copies, any imbalance can be corrected. Modern-day replication is complex because there are thousands of highly selected genes in even the simplest organism and it is necessary to link a single copy of each into one long molecule of DNA. This chromosome is replicated from one end to the other before a copy is given to each daughter cell. DNA-based reproduction appears to have evolved to replace RNA-based reproduction at some time early in the history of life. The initial genes were probably exclusively RNA molecules, and reproduction was not quite as accurate as it is now.

Random errors generate the variants from which improvements are selected. Evolution depends on mutations that occur during reproduction, but success depends on accurately passing on the improvements that might have occurred. A balance has to be struck in which the rate of mutation is high enough to generate new stratagies for survival and yet low enough that the lucky few can become the lucky many. The replication of DNA in modern bacteria is highly accurate, making only one mistake every 10^8 bases. This level of accuracy is achieved by careful proofreading enzymes that correct most errors. Such meticulousness is essential in highly adapted organisms, but was probably unnecessary in the simpler cells that first arose. A mistake every 10^4 bases could have been easily tolerated.

II. THE PREBIOTIC SOUP

A. The Spark Flask

In 1953, when he was a graduate student, Stanley Miller carried out a simple experiment at the suggestion of his adviser, Harold Urey. For several days he let a spark discharge every few seconds in an atmosphere of methane, ammonia, hydrogen, carbon monoxide, and water. This is a reducing environment like the one that is thought to have covered this planet during the first few billion years. At the end of the experiment, Miller analyzed the contents of the flask. The warm solution contained high concentrations of more than 10 distinct amino acids, as well as many other interesting compounds that could form biologically important molecules. The concentration of many of the compounds was greater than $1 \mu M$. Under slightly different conditions, the purine and pyrimidine bases that go into making nucleic acids were made from equally simple starting materials. Likewise, sugars and lipids will form in reactions catalyzed by metalic ions. These results make it clear that the prebiotic conditions favored the formation of the subunits of proteins and nucleic acids. At times, these ingredients could be concentrated when puddles or droplets dried up or when the surrounding water froze. In some locales the subunits coated the surface of clays and minerals, where they reacted with each other to form even more complex

molecules. One of the molecules with the most potential may have been the precursor to RNA. The nucleic acid bases could condense with glycerol phosphate and then be strung together in random order. Such acyclic nucleic acids have been made in the laboratory and shown to direct the synthesis of complementary copies.

B. The RNA World

At some point before the appearance of life, nucleic acid bases linked to the cyclic sugar D-ribose were polymerized into short chains. The sequence of bases in these oligomers may have been partially determined by their hybridization to complementary sequences in preexisting acyclic nucleic acids and thereby derived some hereditary information from the sequences that preceded them. There are viruses alive today in which the genes are carried in the form of RNA rather than DNA.

Individual nucleic acid bases attached to riboses can spontaneously add to the end of an RNA chain, thereby elongating it. The choice of which of the four bases is attached at any given point is determined by the complementarity rules (G pairs with C; A pairs with U). Therefore, when a short RNA molecule is bound to a slightly longer one, the shorter one will grow by adding the complementary base.

```
UCGUG^G            UCGUGG             UCGUGG^U
||||||    goes to  |||||||   goes to  |||||||    etc.
AGCACCAA           AGCACCAA           AGCACCAA
```

This reaction is driven by the energy inherent in the complementation of bases with each other.

So far we have not invoked any catalysts at all. However, it turns out that RNA itself has the ability to catalyze the making and breaking of the bonds between monomers and polymers by acting as an intermediate. Such RNA molecules have been called ribozymes and several have been found to still be functioning in present-day cells. Ribozymes have been shown to lead to the polymerization of about 30 bases per hour. The polymers that are made are all linked correctly and have the expected sequences 99% of the time. Thus, RNA can make RNA.

RNA can also catalyze the transfer of phosphates between unrelated molecules and so could have been the central player in a set of simple but interactive reactions that occurred before peptides joined the story. This RNA world would have had considerable stability as well as a limited ability to evolve greater complexity. However, when RNA functions together with single amino acids or short peptides, the level of complexity can go up rapidly and greater specificity can be given to the system as a whole.

Fig. 1. An autocatalytic cycle. If a peptide happens to favor its own synthesis, more and more of the peptide will be made. Randomly generated nucleic acid sequences partially specified certain peptides that entered into such cycles.

C. Coding for Peptides

Peptides cannot replicate themselves and so must depend on sequences in RNA to determine their amino acid sequences. Translating base sequences in RNA into amino acid sequences in peptides is facilitated by tRNAs. These short polymers take up unique three-dimensional structures that distinguish them so that only a single type of amino acid is added to a specific tRNA. A given tRNA molecule then binds to three adjacent bases in the coding RNA using the complementarity rules and presents its amino acid for addition to a growing peptide. When the system was just getting started, the function of tRNAs might have been played by RNA polymers less than 15 bases long in which the sequence was self-complementary. That is, they could bend back on themselves and form the base pairs that stabilize their three-dimensional structure. The end of the hairpin could accept amino acids, whereas the bases in the loop at the top could bind to the template RNA molecule as they decoded it.

One of the biggest problems in understanding how translation could have occurred before there was life on Earth is to account for the specific attachment of only one kind of amino acid on each kind of tRNA. The structure of tRNA itself is not sufficient for this specificity and so it is likely that randomly synthesized peptides were involved from the start and that by chance some of the systems coded for these very peptides. When this occurred, the first positive feedback system was established and the system could rapidly expand.

When amino acids dry out on rocks and clays, they will sometimes polymerize into random polymers. There may have been other environments as well in which short polymers of 10 to 12 amino acids were formed. Some of these peptides probably had the ability to bind certain amino acids, as well as a specific hairpin RNA that could serve as a tRNA. Those that increased the rate of aminoacyl tRNA formation stood a chance of entering an autocatalytic cycle (Fig. 1). However, they had to be in the same microenvironment with an RNA that happened to have a sequence coding for a similar peptide. Because millions of sequences were being made and broken down all the time, it probably did not take long for this chance to come along. The sequence did not have to code for exactly the same sequence of amino acids in the peptide because many of the 20 amino acids have very similar chemical properties. For instance, glycine, valine, and alanine are all uncharged hydrophobic amino acids whereas both asparatate and glutamate are negatively charged hydrophilic amino acids. As long as one of the hydrophobic amino acids was encoded, it probably did not matter much which one it was. Similarly, for the hydrophilic amino acids, the system only had to distinguish from among about five kinds of amino acids, and if it was correct only 10% of the time, it could still expand. Before the appearance of life, there were no enzymes that would break down peptides or nucleic acids, and so even very inefficient, primitive systems had the time to carry out each step. The biggest problem was keeping the system stable. Error catastophies in which one weak link leads to the next undoubtedly destroyed many promising early systems.

D. Hypercycles

The solution to finding greater stability came from linking cycles together into hypercycles. If some peptides increased the selectivity of the addition of amino acids to tRNAs, whereas others increased the rate of replication of RNAs and peptides, they would all help to make more of the others. The increased fidelity of translation would help make the peptides that made more peptides and RNAs, as well as the peptides responsible for the increased selectivity. Similarly, the peptide that made more RNA would help make more copies of the RNA encoding itself, as well as the tRNAs and the RNAs that encoded the peptides for adding the proper type of amino acid (Fig. 2). This kind of bootstrapping could lead to relatively rapid increases in both stability and specificity.

The original hairpin RNA molecules of a dozen bases grew longer and were able to take up more specific shapes. Modern tRNAs are about 60 nucleotides long and have specific sequences that result in cloverleaf shapes because the bases along the stems are complementary. Each tRNA is charged by a specific amino acid and uses three bases in its anticodon to bind to three complementary bases in mRNA. Hairpin RNAs in early hypercycles may have been specific to only one of the bases in a triplet codon, but still provided some sequence specificity to the newly made peptides. There has always been a reward for greater accuracy, and when the tRNAs of one hypercycle happened to use two or even three bases for translational specificity, they all benefitted. The present-day code by which the sequence of bases in RNA is translated into proteins gives some clues as to how it happened to arise. The first base in the codons for glycine, valine, alanine, aspartate, and glutamate is G in all cases. It certainly looks as if this system evolved from one in which any tRNA that recognized that G would add one of these amino acids as long as the second base in the codon was not A. Codons with the sequence GAN (where N is any of the four bases) would be recognized by tRNAs charged with either aspartate or glutamate. Later on, tRNAs evolved with specificity to the other bases in the second position and allowed glycine, valine, and alanine to be distinguished. At a later time, when the system was already greatly improved, tRNAs could evolve with specificity to the third base in the codons such that aspartate and glutamate could be distinguished. Such stepwise increases in the specificity of coding could have gradually brought each of the 64 codons into use. The complete genetic code did not have to arise all at once.

The same is true of the activating enzymes that add specific amino acids to tRNAs. They could have gradually improved on what was initially a very inaccurate system. The examination of modern-day activating enzymes shows that they evolved from more primitive ones that worked on several different amino acids. There may have been only a few of these primitive catalysts that were then duplicated and diverged to gave rise to the highly specific enzymes we have today.

When hypercycles started to have more than a few dozen interacting components, stability was more limited by the environment than by the system itself. Mixing could disperse the needed components or bring in ones less adapted to the particular system. Even before this crisis was reached, lipid droplets enclosed small volumes. Some of these contained all of the interacting components of a successful hypercycle and went on to develop new properties.

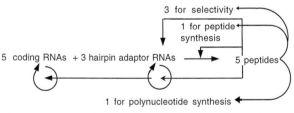

Fig. 2. An integrated set of autocatalytic cycles. Eight randomly generated RNAs could be integrated into a hypercycle by the function of the peptides they happened to encode. If the products of three RNAs increased the accuracy of translation of the codons used by all the five coding RNAs and the product of another increased the replication of them all, then the system would benefit from using three hairpin RNAs as tRNAs.

III. COUPLED SYSTEMS

A. Primitive Metabolism

Droplets that encompassed semistable hypercycles generating peptides and nucleic acids could grow and evolve if they were able to manipulate their environment. Organic reactions that had been catalyzed by simple metal ions or prebiotically generated co-enzymes such as nicotinamide (vitamin B3) or pyridoxal phosphate (vitamin B6) could be improved by peptides that bound these cofactors. RNA molecules that happened to code for such peptides might have been trapped by the droplets or generated by the error-prone replication system. They did not have to be long sequences; 25 to 30 bases would do because the peptide asparagine, phenylalanine, asparagine, proline, histidine, lysine, tryptophan (or NFNPHKW in the standard one-letter abbrevations) is sufficient to bind pyridoxal phosphate and is actually found in such enzymes of both vertebrates and invertebrates. It is likely that similar peptides linked to only a few other amino acids would bring pyridoxal phosphate together with substrates for a variety of reactions. Other short peptides might bind zinc or iron atoms and catalyze certain reactions. The most useful would be those that provided the substrates for protein and nucleic acid synthesis. Equally important would be peptides that could derive energy from available sources.

A peptide that bound magnesium ions and ADP could convert ADP to ATP by coupling the reaction to the dephosphorylation of diphosphoglycerate. The product, phosphoglycerate, could be further metabolized by distinct but related peptides to generate another molecule of ATP and pyruvate. All that it takes to bind ADP is a mononucleotide-binding fold consisting of two dozen amino acids in a sequence that will spontaneously form a β-strand connected to an α-helix followed by another β-strand. If the connecting sequence has three glycines and a properly positioned lysine, a hydrophobic pocket will be made. The aggregation of six such molecules would form a barrel into which ADP would fit tightly. Today, these six peptides are all linked in a longer protein, but originally they may have associated only by the interaction of their hydrophobic surfaces.

Coupling the oxidation or reduction of a wide variety of compounds to the phosphorylation of ADP would generate ATP, which could be used in biosynthetic reactions even before the energy of sunlight was harnessed. However, fairly early on a droplet that could derive energy directly from trapping the energy in sunlight had a big advantage because it would be more independent of the environment. The first photophosphorylations may have coupled an ADP-binding protein to a heme-binding protein that was activated by sunlight. Whether these coupled peptides had to be embedded in the lipid layer that surrounded the droplet is not clear, but that might not be improbable in any case. With a steady source of energy, peptide and nucleic acid synthesis could proceed rapidly. Competition between droplets for raw materials then became the decisive factor.

B. Protocells

Simple lipid droplets containing a few peptides that increased the rates of a few reactions cannot be considered alive. The early peptides were nowhere near as effective catalysts as are modern enzymes and did not have the exquisite specificity we have come to appreciate. Moreover, the replication system was so error prone that no two peptides were exactly alike. Nevertheless, such droplets could increase in mass, replicate some of the essential components, and divide them among smaller droplets when they broke up. Each droplet then became a partially independent unit that could respond to selective pressures. The more stable, more accurate systems lasted longer and multiplied. Controlling the properties of the droplets themselves may have been the next major step toward the origin of life.

There would be strong selection for membrane proteins that could affect the physical properties of the droplets. Hydrophobic peptides are spontaneously embedded in lipid bilayers and change the characteristics of the membrane. For instance, some membrane proteins facilitate the transport of ions, such as potassium or protons, whereas others can pump glucose or amino acids into cells. Peptides with similar properties may have been early arrivals on the scene. A droplet able to scavenge the environment had a strong advantage. Of course, if it were to pass on this ability to subsequent droplets, it would have to have an RNA encoding the peptide that gave it this advantage.

Membrane proteins could also serve as anvils on which to hammer out new peptides. Template RNA molecules could be held to the inner surface of the droplet while tRNAs bound to them and donated their amino acids to growing peptides. This function is now carried out on ribosomes within cells, but ribosomes are large complex organelles containing several RNA molecules and over 50 complex proteins. There had to be something simpler before there

TABLE I
Fifty Vital Sequences

1	Nucleic acid polymerase
2	Purine synthetases
2	Pyrimidine synthetases
1	Phosphotransferase
2	Peptide elongation factors
3	Ribosomal proteins
2	rRNAs
10	Amino actyl tRNA synthetases
10	tRNAs
4	Amino acid metabolic enzymes
2	Oxidation–reduction enzymes
5	Fermentation catalysts
4	Membrane proteins
2	Lipid metabolic enzymes

were ribosomes. The inner face of the droplets might have served this function when specialized peptides were embedded in the lipid layer.

As each advance was made, the system became more stable, more resilient, and better able to multiply than less-favored systems. A minimum of 50 vital sequences seems to have been necessary for a droplet to have had much of a hope of surviving (Table I). It would have been a complicated system, but still far from alive.

C. Survival by Replication

A droplet with 50 vital sequences would have to have hundreds of copies of each one to ensure that it would give at least one copy of each sequence to smaller droplets when it broke up. Further expansion of the genetic repertoire required a more orderly manner of distributing the information. All living cells use DNA rather than RNA for their genetic material. The genes are strung together in a few long molecules that can be easily distributed to daughter cells. Moreover, DNA is almost always double-stranded, so the information is carried redundantly in the two complementary strands. If one strand is harmed, it can be repaired with no loss of information. There are several other advantages of DNA over RNA, and the early protocells soon shifted over to using it. Besides using $2'$ deoxyribose-linked nucleic

acid bases, DNA uses thymine rather than uridine. The enzyme that converts uridylate to thymidylate evolved long before bacteria started to diverge and has been conserved ever since. There are only eight differences in the active site found in the bacteria *Lactobacillus casei* and *Escherichia coli,* and not many more in the active site found in humans. The active site is about 30 amino acids long and uses a hydrogen bond between cysteine and arginine to catalyze the reaction. A similar short peptide probably carried out this same reaction before the dawn of cells.

Once the sequences of RNA were copied into long molecules of DNA, both the number and the length of genes could expand. Each gene could be several hundred bases long and code for proteins of 100 or so amino acids that had greater stability and specificity. The rate of growth and division could increase as several hundred genes were strung together and distributed equally to daughter cells.

When more than half of the progeny cells were as functional as the original cells, the population would expand. At this point, the seas rapidly became filled with cells all competing with each other. Those that were able to replicate faster than others had a distinct advantage. When weaker cells broke, other cells scavenged parts from the wreckage. It is hard to conceive of the huge number of cells that can live on this planet. There are over 10^{20} ml in just the top 10 m of the oceans and there can be millions of cells in each milliliter. With 10^{26} cells growing and dividing every day, events as rare as 1 in 10^{30} would occur every century. When a rare cell acquired a strong advantage, its progeny could take over the world in a few hundred years, and all subsequent improvements would take place in the background of the new improved model. Without much question, these cells were alive.

IV. THE EARLIEST CELLS

A. Common Descent

It is impossible to be certain of the steps that led to the first cells, but it is possible to know something about what the first cells were like because all living things are descended from then. There is ample evidence that one line came to predominate and gave rise to all subsequent organisms. All cells use the same triplet code for the translation of nucleic acid sequences into protein sequences as well as the same punctuation marks, and all use the same basic metabolic processes. Even the sequences of amino acids in the active sites of some enzymes have been conserved since earliest times. For instance, the following sequences are found in triose phosphate isomerase, the enzyme that makes 3-phosphoglycerate:

Escherichia coli	QGAAAFEGAV	IAYEPVWAIG	TGKSATPAQ
Bacillus stearothermophilus	QLTPQEVKII	LAYEPLWAIG	TGKSSTPAQ
Rabbit	DNVKDWSKVV	LAYEPVWAIG	TGKTATPQQ
Fish	DDVKDWSKVV	LAYEPVWAIG	TGKTASPQQ
Corn	EKIKDWSNVV	VATEPVWAIG	TGKVATPAQ

Although there have been changes in the amino acid sequence of this enzyme over the last few billion years, a run of 12 amino acids (AYEPVWAIGTGK), as well as other components of the active site, have remained practically the same. The chance occurrence of 12 amino acids in a row being identical is 20^{12}, because there are 20 different amino acids that could be at any one of the positions. Together with the other similarities in the primary sequence of triose phosphate isomerase in these bacteria, vertebrates, and plant, there is no question that they are descended from a common sequence that was functioning even before bacteria had evolved. The only other possible route to such sequences would be convergence from independently evolved sequences. Convergence of more than a dozen amino acids would be so rare that it can be safely ignored. Selective forces appear to have kept this and other portions of triose phosphate isomerase almost unchanged for billions of generations. Humans that have a single amino acid change in the active site of triose phosphate isomerase suffer from hemolytic anemia and neuromuscular disorders. Their chance of passing on the varient triose phosphate isomerase is exceedingly low.

There are many other enzymes in which the active site has remained almost the same since it arose in the earliest cells. These include pyruvate kinase, thymidylate synthetase, several transcarbamylases and dehydrogenases, the CO_2-fixing enzyme ribulose-bisphosphate carboxylase, the nucleic acid poly-

merases, and proteins involved in electron transfer such as the ferridoxins. We can be fairly sure that genes encoding very similar proteins were functioning in the earliest cells. Copies of these genes have duplicated in the genomes of various organisms and given rise to distinct but related proteins. Thus, the whole genetic repertoire of bacteria and larger organisms is derived from the first few hundred genes of the cell line that first populated Earth.

Most improvements in early genes came from adding on pieces of other genes, rather than from changing the core sequence. This process of accretion can be seen in the mosaic structure of many ancient enzymes. They often have a common sequence flanked by sequences found in unrelated enzymes as well as pieces that have diverged so much that their origins can no longer be recognized. This mix-and-match approach is the consequence of rare errors in replication of DNA that result in fusion of one sequence to another. Now and then, such a translocation resulted in an advantage to the cell and the mutation was fixed in the population. The eminant microbiologist Francois Jacob has described evolution working as a tinkerer. Natural selection can only work with the random mutations that are present and so it is impossible to aim for a set goal. Mutations that seriously change a vital sequence are lethal, but those that just fiddle with what is working and happen to make it work better can be saved for use by posterity.

B. Introns

The fusion of sequences to generate complex genes is a powerful mechanism that can lead to improved products without compromising existing advantages. But it has serious side effect. The fusions seldom start or end exactly where coding regions do, and so there will be intervening sequences separating the two portions. If there are signals for the termination of translation in the intervening sequence, the mRNA will not be fully translated. However, certain RNA molecules have the ability to splice out portions of their own sequences and fuse two or more coding portions or exons. Fused genes that happen to make a self-splicing RNA that removes introns can function as well as the far rarer fusion genes, in which the

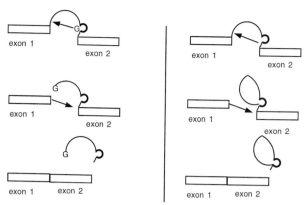

Fig. 3. Self-splicing RNAs. There are two related mechanisms by which RNAs catalyze the removal of their own introns. In Group I RNAs (left), a free guanine (G) is bound and used to cleave the 5'-end. In Group II RNAs (right), an internal adenosine attacks the exon–intron junction and then forms a lariat of the intron. The cleavage of the 3'-junction and ligation of the exons forms the final functional product. Most mRNAs belong to Group II.

joint was perfectly in phase. Although introns are prevalent in eukaryotic organisms, they are rare in bacteria. Until recently, this was taken as evidence that exon splicing only evolved at the time of appearance of eukaryotic cells about 1 billion years ago. However, during the late 1990s, a number of bacterial genes have been found to carry introns that are removed by self-splicing (Fig. 3). It is becoming clear that many genes may have initially carried introns, but that there was a strong selective pressure within the last billion years for the replacement of these genes by copies derived from their processed (spliced) mRNAs. A few remain in anaerobic bacteria and in the viruses of aerobic bacteria.

C. Regulation

A system without controls has little chance of survival. The integration of coupled processes requires that the individual components be able to regulate each other to arrive at the most selectively advantageous mix. Early cells undoubtedly had mechanisms for the regulation of both the transcription of mRNA and translation of proteins. We can see the kinds of proteins that may have mediated these controls by comparing transcriptional and translational regulators found in both bacteria and eukaryotic cells.

One of the motifs found in DNA binding proteins is called a zinc finger. These proteins have a section about 20 amino acids long in which four cysteines or histidines are so positioned that they can all bond to a central zinc ion. This forms a loop or "finger" that binds to three bases in the major groove of DNA and holds the protein near those genes that have this particular sequence of three bases. There are usually several zinc fingers in each DNA binding protein, so that 12–15 bases are used to provide specificity. A sequence of 15 bases will only occur randomly once in a billion bases ($4^{15} \approx 10^9$). This combinatorial use of short sequences is found in almost every case in which a protein associates with a specific region of DNA.

Another motif found in DNA binding proteins is the helix-turn-helix. In these proteins, a sequence of a dozen amino acids forms an α-helix that can fit in the major groove of DNA and associate with four bases. The second helix is positioned at right angles to the first one and holds it in place. These proteins usually function as dimers recogizing palindromic sequences of eight bases. When such a protein is bound to DNA, it can either facilitate or inhibit transcription from nearby genes, depending on the the nature of the rest of the DNA-binding protein. Because a given eight-base sequence occurs every 10,000 bases or so along a DNA chain, there are thousands of binding sites in any genome. The binding of a helix-turn-helix protein to one of these is stabilized by interactions with other DNA-binding proteins that have recognition sites nearby. Transcriptional initiation is dependent on the formation of a complex consisting of several independent proteins, all of which have to have binding sites within several hundred bases of the gene in question.

Proteins that bind RNA are less well understood and common structural motifs have not been recognized. They appear to recognize specific RNAs on the basis of secondary structures that the RNA take up by internal complementarity. If the secondary structure is disrupted by interactions with other RNA-binding proteins or with ribosomes during translation, then the recognition site is removed. Thus, the regulation of translation also works in a combinatorial fashion that can pick out one mRNA sequence from the thousands of different RNA sequences in a cell.

Feedback loops form when a macromolecular or metabolic process interacts with a specific regulatory mechanism controlling another gene. The link is usually made through a small molecule effector such as cAMP or cGMP. The activity of DNA-binding proteins is controlled by interactions with such effectors at independent sites in the regulatory protein. This is usually the first step in a hierarchical control mechanism. Regulation can come in at a higher level as well, by controlling the genes for the regulatory proteins themselves. In this way, the function of thousands of genes can be carefully orchestrated to result in a harmonious response to the state the cell finds itself in at any given time.

V. MICROBIAL DIVERSIFICATION

The complete genomes of a dozen microorganisms have been sequenced such that all their genes can now be inspected. The comparison of the most closely related genes in a variety of bacteria confirmed the ancient divergence of the eubacteria and archaebacteria about 3 billion years ago. For the first billion years of life, genes may have been passed among different bacteria so frequently that the concept of separate species would not apply. Species that arose from this pool of shared genes gradually put up barriers to lateral transfer and proceeded to evolve for the most part independently. As bacteria further specialized to exploit different environments, their genomes diverged more and more. About 2 billion years ago, it appears that endosymbiosis of an archaebacterium and a eubacterium gave rise to the eukaryotes and their organelles, respectively. Many genes were transfered to the nucleus, where they were organized into chromosomes.

The fossil record gives little to go on when all life was microbial, and so the timing of events that led to the deepest branching that separates the archaebacteria from the eubacteria and the more recent formation of the eukaryotes and major bacterial

groups has had to be extrapolated from comparisons of the amino acid sequences in orthologous genes found in present-day members of these clades. Nevertheless, data are accumulating that gram-positive, gram-negative, and cyanobacteria diverged about the same time that an α-proteobacterium became established within an Archaea-like bacterium to give rise to eukaryotes. Some of the newly formed eukaryotes subsequently lost their mitochondria but retained many of the α-proteobacterial genes. As oxygen began to build up in the atmosphere, others diversified to give rise to protists, plants, fungi, and animals. Starting about 600 million years ago, the history of life on this planet can be traced in rocks where the variety of living things can be clearly seen in well-preserved fossils.

VI. CONCLUSION

Once an autocatalytic process was set in motion over 3.5 billion years ago, there was very little that could stop life from expanding, diversifying, and filling every conceivable niche on the surface of this planet. All life descends from the first cells that arose spontaneously from the prebiological soup. Those early cells were more like bacteria than they were like eukaryotes, but it is hard to say if they were more like archebacteria or eubacteria. There is no question that they were anaerobes because oxygen had not yet built up in the atmosphere. Photosynthetic bacteria released oxygen for several billion years before the capacity of Earth's surface to reduce it to iron oxides and other minerals was exhausted. Thereafter, aerobic bacteria and eukaryotes evolved from the existing anaerobes.

The wide distribution and impressive extent of stromatolites from the some of the oldest rocks on Earth attest to the ability of early bacteria to proliferate. Stromatolites are laminated rocks several meters in height that form when sand is trapped in the filaments of bacterial colonies. They have been found in formations over 3.5 billion years old in Australia, South Africa, the Sahara, Greenland, and Siberia. It is clear that life originated on this planet a few hundred million years after the oceans cooled and the meteoric bombardment was over. For several billion years, life remained confined to single-cell organisms that resembled modern-day bacteria in many ways. During this long period, natural selection could work to fine-tune each gene for optimal adaption to an enormous number of specialized environments. Bacteria lived in seawater, freshwater, and brines. They lived in thermal hot springs at 100°C and in polar seas at 0°C.

When the atmosphere started to fill with oxygen, the bacteria that were not killed could capitalize on the chemical potential of the air and evolve aerobic metabolism. That allowed multicellular eukaryotic organisms such as humans to evolve. Most of our genetic heritage is derived from the bacteria of a billion years ago and theirs, in turn, was inherited from the earliest cells that had learned to survive with only a few hundred simple genes. When life first appeared, it was a fragile thing, but it had the power to evolve into almost anything.

See Also the Following Articles

DIVERSITY, MICROBIAL • DNA REPLICATION • EVOLUTION, THEORY AND EXPERIMENTS • RNA SPLICING, BACTERIAL

Bibliography

DeDuve, C. (1991). "Blueprint for a Cell." Patterson, Burlington, NC.

Feng, D-F., Cho, G., and Doolittle, R. F. (1997). Determining divergence times with a protein clock: Update and reevaluation. *Proc. Natl. Acad. Sci. U.S.A.* **94**, 13028–13033.

Huynen, M., and Bork, P. (1998). Measuring genome evolution. *Proc. Natl. Acad. Sci. U.S.A.* **95**, 5849–5856.

Joyce, G. (1998). Nucleic acid enzymes: Playing with a fuller deck. *Proc. Natl. Acad. Sci. U.S.A.* **95**, 5845–5847.

Loomis, W. F. (1988). "Four Billion Years." Sinauer, Sunderland, MA.

Miller, S., and Orgel, L. (1974). "Origins of Life of Earth." Prentice-Hall, Englewood Cliffs, NJ.

Rivera, M., Jain, R., Moore, J., and Lake, J. (1998). Genomic evidence for two functionally distinct gene classes. *Proc. Natl. Acad. Sci. U.S.A.* **95**, 6239–6244.

Osmotic Stress

Douglas H. Bartlett

University of California, San Diego

Mary F. Roberts

Boston College

GLOSSARY

compatible solutes Low-molecular-weight organic compounds that enhance the ability of organisms to survive in environments in which the extracellular solute concentration exceeds that of the cell cytoplasm.

halophile An organism that requires NaCl for growth; salt-lover.

hyperosmotic Pertaining to an increase in osmotic pressure or osmolality.

osmolality Osmotic pressure expressed in terms of osmols or milliosmols per kilogram of water.

osmophile An organism that requires high osmotic pressure; osmatic-pressure-lover.

turgor pressure The pressure exerted by the contents of a cell against the cell membrane or cell wall.

water activity The ratio of the vapor pressure of water in equilibrium with an aqueous solution divided by the vapor pressure at the same temperature of pure water. Values of a_w vary between 0 and 1.

WATER IS THE SOLVENT OF LIFE, and the degree to which it is available exerts a profound influence on all living systems. The movement of water from regions of higher to lower concentration (higher-solute concentration) across a semipermeable membrane is called osmosis. In biology, osmosis frequently involves the movement of water through a membrane phospholipid bilayer. The osmotic state of a system can be quantitated by one of several measurements. Osmotic pressure, for example, is the external pressure that is just sufficient to prevent osmosis between a solution and pure water. It is often expressed in terms of osmolality. An osmole is a gram molecular weight of osmotically active solutes in 1 kg of water and can be determined by one of several colligative properties (such as freezing-point depression). Osmotic stress results when the process of osmosis generates a cytoplasm of osmolality that is either less than or greater than optimal for a particular organism.

I. GENERAL CONSIDERATIONS

When a microbial cell is in a medium that is hyperosmotic relative to its cytoplasm, the cell tends to lose water by osmosis. Conversely, if the cell is in a medium that is hypoosmotic by comparison to its own fluids, it will accumulate water. In the absence of an appropriate response there may be growth inhibitory if not lethal consequences. The control of internal osmolality is related to other aspects of cell physiology. At low osmolality, the need to preserve internal pH, ionic strength, metabolite concentration, and the concentration of specific ions a certain range is an important aspect of the adaptational response. For example, K$^+$ is the most abundant cation in the cell, and it is kept the range of 100–150 mM in the case of *Escherichia coli*. In contrast, at high osmolality, the need to supplement the cellular milieu with solutes that are least inhibitory to cellular processes will predominate. Adaptation to osmolalities near the upper limit of tolerance typically also confers enhanced resistance to other stressful conditions, such as heat or cold shock.

Prokaryotes and other osmoconforming organisms such as plants and invertebrates are generally considered to be close to being isoosmotic with their environment. However, in fact, because of the increased osmolality of the cytoplasm relative to the environment and the resulting inward flow of water, prokaryotic and eukaryotic microorganisms and the cells of many other osmoconforming organisms retain a positive hydrostatic pressure, termed turgor pressure, which pushes out against their cell membranes and cell walls. In response to turgor pressure, the prokaryotic cell wall expands elastically until its tension exerts an opposing force equal to the hydrostatic pressure. In this way the cell wall acts like a pressure vessel. The pressure inside gram-negative bacterial cells is on the order of 2 atmospheres (30 pounds per square inch) and can reach up to about 20 atmospheres in the case of some gram-positive bacteria. Measurements of bacterial turgor pressure indicate that it does not vary much during cell division or during the various phases of growth. It has been proposed that turgor pressure is maintained within narrow bounds in order to produce the necessary mechanical stress for cell-wall expansion. This then could be one of the reasons why it is important for cells to accumulate osmolytes as they grow and divide.

In describing the osmotic adaptations of microorganisms, another important term is water activity, abbreviated a_w. This is the ratio of the vapor pressure of water in equilibrium with an aqueous solution divided by the vapor pressure at the same temperature of pure water. The values of a_w vary between 0 and 1. Microbial life has evolved widely different preferences for a_w and the amount of change in a_w that can be tolerated. Thus, osmotic stress is different to different prokaryotes. *E. coli* typically contains 70% of its cytoplasm as water, although this value may vary by close to twofold, depending on the osmolality of its environment. Bacteria such as certain species of *Caulobacter* and *Spirillum* can grow at a_w close to 1.000 (pure water); marine bacteria are adapted to the a_w of seawater, 0.980; some members of the Archaea can grow at a_w down to 0.750; and

certain fungi can still grow at a_w as low as 0.700. Similarly, endolithic lichen and bacteria of deserts must be adapted to extremely low a_w values. The bacterial species *Salinivibrio costicola* as well as bacteria of the family *Halomonadaceae* are particularly fascinating because of the tremendous range of a_w they will tolerate. They will grow over a range of water activities between 0.98 (close to freshwater) to 0.86 (close to saturated NaCl). Decreasing water activity by freeze-drying or salting is one of the principal means of preventing food spoilage by microorganisms. Water potential, a related energy term that is expressed in units of pressure, is also used to describe the concentration of water in some scientific fields.

Organisms that can grow in media of high and average solute concentrations are called osmotolerant, and those that actually prefer elevated osmolalities are termed osmophiles. In the environment, osmotic adaptations are usually correlated with the presence of particular solutes. Because of this it is important to distinguish general osmotic requirements from those regarding specific chemicals. Marine microorganisms, for example, typically have a specific need for a certain amount of sodium ions in addition to a moderately increased osmolality over their terrestrial microbial counterparts. They are thus moderate halophiles. Certain members of the Archaea domain, such as *Haloferax* and *Halobium* species, have still higher requirements for sodium ions and are described as extreme halophiles. In terms of the salt concentrations enabling growth, microorganisms may grow from close to 0 M to close to saturating concentrations at 5.2 M sodium chloride. *Salinivibrio costicola* and bacteria of the family *Halomonadaceae* are described as haloversatile.

Dessication results in extreme osmotic stress and low a_w values. Some microbes die almost instantly in air, whereas the spores of algae, fungi, and bacteria and the cysts of protozoa are very resistant to drying. The vegetative cells of many gram-positive cells and bacteria of the family *Deinococcaceae* are also quite resistant to dessication. In the former case, this is believed to stem in part from their thick cell walls, which may help protect membrane integrity. Protection of damage to chromosomal DNA during drying is also critical for cell survival. The extremely efficient DNA-repair capabilities of the *Deinococcaceae* have been found to be required for their dessication resistance.

The focus here is on the physiological processes by which microorganisms adapt to osmotic stress. However, note that one of the most important responses of motile microorganisms to changes in osmolality is to swim away into a new more favorable osmotic environment. This process is known as osmotaxis. Bacteria are repelled by pure water, attracted to their particular optimal osmolality, and repelled again by supraoptimal osmolality. By moving back into a region of optimal osmolality, the need for substantial physiological modification can be greatly diminished. How microorganisms sense osmolality and modulate their motility is largely unknown. However, it appears to be mediated by a direct effect on the rotation switch at the base of the flagellum. Chemoreceptor proteins in the cell membrane also help cells adjust their motility in response to particular solutes such as NaCl.

II. EFFECTS OF INCREASED OSMOLALITY ON PROTEINS

During exposure to hyperosmotic stress conditions, solutes will increase in concentration inside cells. Many of the deleterious effects of hyperosmotic stress can be interpreted in the context of solute effects on macromolecules. Increased osmolality tends to promote the dehydration of proteins. In a sense, proteins can be considered as behaving similar to a semipermeable membrane. The exposure of a protein to an environment in which the a_w is lower than that of the water bound up to the protein will result in the release of the bound water to the surrounding solution.

One line of evidence indicating the profound influence of osmolality on protein structure has come from the isolation of osmoremedial mutants. These microorganisms encode a mutant protein that is nonfunctional when cells are grown in media of low ionic strength, but that regains at least partial function when the cells are grown in media of higher osmolality. Osmoremedial mutations have been identified in genes encoding both soluble and membrane proteins, of monomeric or multimeric con-

figuration. Most if not all temperature-conditional mutants are also osmoremedial.

Different solutes can have different effects on proteins and their structure and activity. K^+ and glutamate are important intracellular osmolytes at low concentration, but, as do other ions, will inhibit many enzymes when present at high concentrations. Solutes that carry a net electric charge are generally more damaging to protein stability than nonpolar or zwitterionic solutes. The charged solutes can be further subdivided according to the Hofmeister series (also called lyotropic series) according to their effects on proteins. Stabilizing ions do not bind strongly to proteins or interfere with the water molecules coating the protein surface. These ions tend to preserve the native protein structure. Thus, stabilizing solutes distribute themselves away from the protein, which is preferentially hydrated near its surface. Because the exclusion of stabilizing solute molecules from the water around the protein reduces the entropy of the system, it is accompanied by an energetic cost. This cost can be reduced by minimizing the area on the surface of the protein that is exposed to solvent, thus resulting in a more compact protein structure. Thus, even in the presence of stabilizing solutes, in the absence of protein modification, increased ionic strength will alter protein structure.

The Hofmesiter series does not apply to all proteins. An exception are the proteins of some of the extremely halophilic Archaea. Proteins isolated from these organisms are stable only under conditions of high ionic strength. Indeed the interior of Archaea of the family Halobacteriaceae may accumulate more than 5 M KCl. Their proteins have an excess of acidic over basic residues when compared to their corresponding nonhalophilic proteins. Because of this acidic character, these proteins are able to fold correctly only in the presence of sufficient counterions, such as K^+. High potassium levels may also balance the amount of water bound to proteins, and the burial of weakly hydrophobic residues such as alanine and glycine.

III. K+ UPTAKE

Instead of passive cell-volume regulation, prokaryotes respond to osmotic increases by selectively increasing the concentration of certain solutes, thus reducing water activity but returning cell volume and turgor pressure to approximately prestress levels. The exposure of *E. coli* cells to hyperosmotic stress results in a sequential series of adaptive steps. The first line of defense is a large increase in the rate of potassium (K^+) uptake. This appears to be necessary for the resumption of growth of the plasmolyzed cells. Subsequently, the synthesis of glutamate is induced. This charged amino acid serves as the principal counterion for K^+. In some organisms (e.g., *Salmonella typhimurium*), the accumulation of glutamate is necessary for optimal growth in media of high osmolality. However, because glutamate is a substrate for a variety of cellular enzymes (specifically, a key intermediate in nitrogen metabolism and a component of proteins), its concentration in cells may be regulated.

With the uptake or synthesis of compatible solutes (see later), K^+ levels in *E. coli* decrease as a result of efflux and glutamate levels decrease as a result of turnover. The synthesis of trehalose results in this dissaccharide eventually replacing K^+ (and glutamate) as the major osmolyte inside the cell. Mutants impaired in K^+ uptake produce trehalose more rapidly and mutants unable to synthesize trehalose exhibit increased intracellular K^+ concentrations. Thus, *E. coli* and presumably many other bacteria have evolved a signaling process for communicating and controlling the levels of the osmolytes K^+ and trehalose together inside the cell. K^+ is believed to act as a second messenger that influences not just trehalose levels, but many of the later steps of the cellular response to hyperosmotic stress. Consistent with this hypothesis, there is a dependency on external K^+ for osmoadaptation in the absence of external compatible solutes.

E. coli has three K^+ uptake systems, the predominant Trk, the inducible Kdp, and the minor Kup systems. Although all of these transport systems respond to osmotic upshift by increasing K^+ influx, an interesting aspect of the Kdp system is the transcriptional regulation of the genes encoding this transporter in response to osmolarity. Two proteins, the integral membrane protein KdpD and the soluble KdpE, control Kdp-transporter gene expression as members of the two-component sensor kinase–response regulator protein family in which one protein, in this case KdpD, acts as a sensor of some signal, and the other protein, in this case KdpE, is the response regulator that controls the cell's response to the perceived signal. Kdp gene expression is principally controlled by K^+ levels, but is also influenced by osmolality because increases in medium osmolality also increase the threshold K^+ concentration for induction. Because osmotic upshock transiently increases Kdp expression it was once thought that the decrease in cell turgor pressure following such a stress is the signal governing osmolality control of Kdp gene expression. Later experiments with KdpD mutants insensitive to K^+ indicated that KdpD can also control steady-state Kdp gene expression as an effective osmosensor. In order to do so, it may respond to osmotically induced changes in membrane structure.

IV. COMPATIBLE SOLUTES

A. Introduction

The response of the vast majority of organisms in all three domains of life to hyperosmotic stress is the eventual intracellular accumulation either by synthesis or transport of organic molecules (charged or neutral) for osmotic balance. The exception to this rule are all members of the family Halobacteriaceae within the Archaea domain and some groups of anaerobes within the Bacteria domain (e.g., *Halobacteroides* and *Haloanaerobium*), which have high intracellular inorganic cation concentrations. The most halophilic Archaea can accumulate very high (up to ~7 M) concentrations of inorganic ions, chiefly K^+ and Na^+. The intracellular concentration of these cations varies with external NaCl; hence they can be considered osmolytes.

However, in most microorganisms, the accumulation of inorganic cations such as K^+ is only a transient response that is followed by the accumulation of zwitterionic organic solutes such as proline or glycine betaine. An organic molecule whose intracellular concentration can vary widely (reaching molar levels in halophiles) without affecting the activity of enzymes or the integrity of cellular structures is

TABLE I
Sugars and Polyols Used by Cells for Osmotic Balance

Solute	Structure	Organisms
Neutral		
glycerol		Algae (*Dunaliella* sp.), cyanobacteria, fungi, yeast
arabitol		Fungi (*Dendryphiella* sp.), sugar-tolerant yeasts
mannitol		Algae (*Dunaliella* sp.), cyanobacteria, fungi, vascular plants
sorbitol		Algae, gram-negative bacteria (e.g., *Zymomonas mobilis*), animals (e.g., renal medulla cells of mammals), vascular plants, yeasts (*Hansenula anomala*)
myo-inositol		Plants, mammalian brain
fructose		Plants
glucose		Plants
glucosylglycerol		Marine cyanobacteria, freshwater cyanobacteria (*Synechocystis* sp. and *Microcystic firma*), phototrophic eubacterium (*Rhodobacter sulfidophilus*), *Pseudomonas mendocina*, *P. pseudoalcaligenes*
sucrose		Cyanobacteria (*Anabaena* sp., *Phormidium autumnale*, and *Chroococcidiopsis* sp.), plants and animals

continues

Continued

Solute	Structure	Organisms
trehalose		Cyanobacteria (*Anabaena* sp., *Phormidium autumnale*, *Chroococcidiopsis* sp.), plants and animals
β-fructofuranosyl-α-mannopyranoside		Soil bacterium (*Agrobacterium tumefaciens*)
Anionic		
glucosylglycerate		*Methanohalophilus portucalensis*, marine heterotroph
β-mannosylglycerate		*Archeoglobus fulgidus, Thermus thermophilus* (α-Isomer) *Rhodothermus marinus*
sulfotrehalose		*Natronobacterium* sp., *Natronococcus occultus*
di-*myo*-inositol-1,1′-phosphate (DIP)		Hyperthermophilic archaea: *Methanococcus igneus, Pyrococcus furiosus, Pyrococcus woesei, Thermatoga maritima*
diglycerol phosphate		*Archeoglobus fulgidus*

termed a compatible solute. The distribution of commonly occurring organic osmolytes found in Bacteria, Eukarya, and Archaea, shown in Tables I–IV, is limited to only a few classes of organic compounds including sugars and polyhydric alcohols, free α- and β-amino acids and their derivatives, methylamines (with glycine betaine as perhaps the most ubiquitous osmolyte), and other simple net neutral compounds (e.g., β-dimethylsulfoniopropionate, taurine, and urea) and anionic molecules (e.g., diglycerol phosphate and cyclic 2,3-diphosphoglcyerate, cDPG). In most cases the accumulation of the organic solutes

TABLE II
Amino Acids and Derivatives Used as Osmolytes in Cells

Solute	Structure	Organisms
α-Amino acids		
glycine		*Listeria monocytogenes*
proline		*E. coli, Bacillus subtilis, Listeria monocytogenes, Salmonella typhimurium, Staphylococcus aureus, Streptomyces clavuligerus,* vascular plants
hydroxyproline		*Listeria monocytogenes*
ectoine		*Ectothiorhopospira, E. coli, Brevibacterium linens, Halomonas* SPC1, *Marinococcus halophilus*
hydroxyectoine		Actinomycete A5-1 *Nocardiopsis* sp., *Halomonadaceae*
glutamate		*E. coli, Enterobacter aerogenes, Halomonas elongata, Methanococcus* sp., *Methanobacterium thermoautotrophicum, Natronococcus occultus, Pseudomonas fluorescens, Salmonella typhimurium*
β-Amino acids		
β-glutamate		*Methanococcus thermolithotrophicus, M. igneus, M. jannaschii*
β-glutamine		*Methanohalophilus portucalensis*
Nε-acetyl-β-lysine		*Methanosarcina thermophila, Methangenium cariaci, Methanphalophilus* sp., *Methanococcus thermolithotrophicus*
Peptides		
N-acetylglutaminyl-glutamine amide (NAGGN)		*Pseudomonas fluorescens, Rhizobium meliloti*

TABLE III
Methylamines Used as Osmolytes in Cells

Solute	Structure	Organisms
Glycine betaine		Enterobacteria, plants, fish, humans, cyanobacteria, *Bacillus subtilis, Ecto-thiorhodospira halo-chloris, Listeria monocyto-genes, Methanohalophi-lus portucalensis, Staphy-lococcus aureus*
Sarcosine		Marine cartilaginous fish
Choline		*Staphylococcus aureus*
Glycerophosphorylcholine		Renal medullary cells
Trimethylamine *N*-oxide (TMAO)		Cartilaginous fish, marine invertebrates, coelacanth

by transport from the culture medium is energetically preferred over biosynthesis of osmolytes. There is also a general sense that the chemical nature of the solute is important in determining the degree of osmotolerance of the organism.

B. Sugars and Polyhydric Alcohols

Sugars and polyols, such as glycerol, glucosylglycerol, sucrose, and trehalose, are commonly used for osmotic balance among diverse organisms, including prokaryotes, yeast, plants, algae, and mammalian cells (Table I). Sugars and polyols are often accumulated by organisms that must deal with dessication. Sorbitol is a good example that can either be internalized from the medium (if it is present) or synthesized *de novo*. High concentrations of this polyol have been shown to protect proteins during dehydration by

osmotic or thermal stress and to preserve proteins during storage. Sorbitol has been shown to function as an osmolyte in sugar-tolerant yeasts (*Hansenula anomala*), as well as in the gram-negative, strictly fermentative and ethanologenic bacterium *Zymomonas mobilis*. Other polyols, notably glycerol, mannitol, and arabitol, are accumulated by several species of yeast and algae. As specific examples, the marine algae *Dunaliella* sp. accumulates glycerol, whereas the fungi *Dendryphiella* sp. uses arabitol and mannitol as osmolytes.

Soluble carbohydrates, including glucose and fructose, are relatively common in the osmotic adaptation of plants. Disaccharides (e.g., sucrose and trehalose) are also frequently accumulated as osmolytes by a wide range of organisms, particularly if they are available in the growth medium. *Escherichia coli* synthesizes trehalose in response to hyperosmotic stress,

TABLE IV
Miscellaneous Osmolytes Used by Cells

Solute	Structure	Organisms
β-Dimethyl-sulfoniopropionate		Algae (e.g., *Enteromorpha intestinalis*)
Arsenobetaine		Crustaceans and mollusks, brine shrimp, Mono Lake gram-negative bacteria
Cyclic-2,3-diphosphoglycerate (cDPG)		*Methanobacterium thermoautotrophicum*
Taurine		*Staphylococcus aureus*
Urea		Marine cartilaginous fishes, coelacanth, crab-eating frogs, mammalian kidney, estivating amphibians, lungfishes, and snails

and mutants defective in trehalose synthesis display increased osmosensitivity (as well as heat sensitivity). Trehalose has been shown to stabilize membrane structure and lower the temperature for phase transition from gel to liquid crystalline state, which can be important during the rehydration of dessicated cells.

A more unusual disaccharide with a documented osmotic response in cells is β-fructofuranosyl-α-mannopyranoside; this solute is found in the soil bacterium *Agrobacterium tumefaciens*. Glucosylglycerol is the organic osmolyte that accumulates in cyanobacteria in response to increased salinity. Its accumulation has been documented primarily in cyanobacteria from marine habitats. This solute has also been reported in two nonphotosynthetic salt-tolerant (1.2 M NaCl) microorganisms (*P. mendocina* and *P. pseudoalcaligenes*).

Although most of the carbohydrates that function as osmolytes are neutral, there are several anionic sugars (modified with carboxylate, sulfate, or phosphate groups) that are used for osmotic balance in Archaea. The very high intracellular K$^+$ present in some of these organisms requires appropriate counterions. Glucosylglycerate, a negatively charged

structural analog of glucosylglycerol, is synthesized and accumulated in *Methanohalophilus portucalensis*, an extreme halophile, grown under nitrogen-limiting conditions. Although trehalose is a common solute accumulated in Bacteria, a charged variant of this, sulfotrehalose (the 1 → 1 α-linked glucose disaccharide with a sulfate group attached to one of the glucose moieties at C-2 position) functions as an osmolyte in halophilic, alkaliphilic Archaea such as *Natronococcus* and *Natronobacterium* spp. Concentrations of the sulfotrehalose are balanced by the intracellular K$^+$ ions. Studies of *Natronococcus occultus* showed that these cells could transport and accumulate exogenous neutral disaccharides (sucrose, trehalose, and maltose) for osmotic balance. The accumulation of solutes from the medium suppressed sulfotrehalose biosynthesis.

There are two other types of charged polyols detected in Archaea that are symmetric phosphodiesters. Diglycerol phosphate is the major organic solute detected in *Archaeoglobus fulgidus*. Di-*myo*-inositol-1,1'-phosphate (DIP) has been found to have a role as an osmolyte in hyperthermophilic Archaea including *Methanococcus igneus*, several *Pyrococcus* species, and

Thermotoga maritima. In both *M. igneus* and *P. furiosus,* DIP was synthesized and accumulated to high intracellular concentrations only at supraoptimal growth temperatures, suggesting the possibility that DIP has a role as a thermoprotectant in addition to its use as an osmolyte. Derivatives of DIP, notably di-2-O-β-mannosyl-DIP, are also observed in members of the order *Thermotogales.*

C. α-Amino Acids and Derivatives

Free amino acids, such as glutamate and proline, are also used for osmotic balance in a wide range of organisms (Table II). Often high-affinity uptake systems exist to internalize these solutes. The neutral solute proline is accumulated from the external medium by an osmotically induced proline transport system, ProP and ProU, in enteric bacteria. The importance of proline to osmotic adaptation of *E. coli* is exemplified by the fact that proline-overproducing mutants are more resistant to osmotic stress. In *Staphylococcus aureus,* the most halotolerant, gram-positive nonhalophile bacterium, proline acts as an osmoprotectant along with choline, glycine betaine, and taurine in high-osmotic strength medium. Proline- (and glycine-) containing peptides, as well as amino acid pools, resulting from exogenous peptone in the environment, also contribute to maintaining turgor and stimulating cell growth at high osmolarity in the gram-positive intracellular parasite *Listeria monocytogenes.*

Ectoine, a neutral cyclic amino acid derivative that is also a tetrahydropyrimidine (for structure see Table II), is synthesized by a number of halotolerant and halotrophic bacteria. This unusual solute was first detected in the extremely halophilic phototrophic bacteria of the genus *Ectothiorhodospira* and is assumed to have a protective function similar to that of proline and glycine betaine. Hydroxyectoine, differing from ectoine by a hydroxy group in position 3 of the pyrimidine ring, has also been detected in some salt-stressed halotolerant bacteria.

In addition to being a precursor and a nitrogen donor for the biosynthesis of several amino acids, glutamic acid also serves as a compatible solute in many species. Accumulation of this amino acid can occur either by transport from the medium or from *de novo* synthesis. Because glutamate is an anionic solute, it is usually accompanied by a monovalent cation to maintain charge balance in the cell. Bacteria with lower salt tolerances, such as *E. coli, Enterobacter aerogenes,* and *Pseudomonas fluorescens,* accumulate glutamate in response to changes in salinity. In more halophilic Bacteria such as *Halomonas elongata,* moderate concentrations of glutamate are accumulated, but they are relatively insensitive to extracellular NaCl. However, in the most halophilic Archaea, other organic anions are accumulated.

Peptides may also be used by organisms to maintain osmotic balance. Interestingly, the constituent amino acids include those with documented roles as osmolytes (i.e., glutamate, proline, hydroxyproline, and glycine). A specific dipeptide, N-acetylglutaminylglutamine amide (NAGGN), was found to accumulate in *Rhizobium meliloti,* the root nodule symbiont of alfalfa, as well as *Pseudomonas fluorescens.* The more general situation is for cells to internalize peptides from complex media enriched with peptone in response to osmotic stress.

D. β-Amino Acids and Derivatives

For use as osmolytes, amino acids have the advantage of high solubility and ease of synthesis, but the disadvantage of they being coupled to protein (and other metabolite) biosynthesis. β-Amino acids are relatively rare in nature and they are not incorporated in proteins. The transfer of the amino group from C-2 to C-3 produces solutes with ideal properties for use as compatible solutes. The *de novo* synthesis of β-amino acids and derivatives, such as β-glutamate (3-aminoglutaric acid), β-glutamine, and Nε-acetyl-β-lysine, as osmolytes is an unusual feature found in many methanogenic Archaea. These β-amino acid (for structures see Table II) pools in methanogens exhibit very slow turnover rates (e.g., β-glutamate turns over much more slowly than L-α-glutamate)—a characteristic of a true compatible solute. β-Glutamate is negatively charged and serves, along with L-α-glutamate, as a counterion to intracellular K^+. In many methanogens, it is the major solute synthesized and accumulated in response to increased NaCl. However, in some methanogens (e.g., *M. thermolithotrophicus*), there appears to be a maximum intracellular concentration of K^+ (and by charge balance β-glutamate and L-α-glutamate) in response to external NaCl. As the external NaCl is increased above this threshold, the cells adapt by synthesizing and accumulating Nε-acetyl-β-lysine. This zwitterionic molecule is used as the major compatible solute in response to elevated external NaCl levels in both marine and nonmarine methanogens. Transformation of anionic β-glutamate to zwitterionic β-glutamine by glutamine synthetase yields the third β-amino acid osmolyte. β-Glutamine has only been documented in the most halophilic methanogens.

E. Methylamines and Analogs

Glycine betaine (a trimethylated derivative of the amino acid glycine) is found widely in nature and is accumulated in species as distantly related as Archaea, enterobacteria, plants, and humans. Glycine betaine has been adopted by a wide variety of organisms as an efficient osmoprotectant in high-osmolarity environments (Table III). It has also been shown to confer cryotolerance in the gram-positive food-borne pathogen *Listeria monocytogenes*. Many organisms lack the ability to synthesize glycine betaine *de novo*. However, glycine betaine can often be synthesized from exogenous choline in various organisms, including nonhalophilic gram-positive eubacterium *Staphylococcus aureus,* gram-positive soil bacterium *Bacillus subtilis*, methanogenic Archaea (in particular *Methanobacterium thermoautotrophicum* strain Marburg), and *Escherichia coli*. Related methylamines (e.g., sarcosine and trimethylamine N-oxide) can also serve as osmolytes in higher organisms such as cartilaginous fish.

β-Dimethylsulfoniopropionate is a methylated sulfur analog of glycine betaine that has been reported as an osmolyte in some organisms (Table IV). Along with sulfotrehalose, it is the only other documented sulfur compound with a role in osmotic balance. In some algae (e.g., *Entermomorpha intestinalis*), β-dimethylsulfoniopropionate, the major osmolyte at normal salinity, is held constant during conditions of hyperosmolarity, favoring the accumulation of other types of solutes (i.e., sucrose and proline) that are compatible at high concentrations. A similar compound, arsenobetaine, has been detected in marine crustaceans and mollusks and can be internalized and used as an osmolyte by bacteria isolated from Mono Lake. Mono Lake is an alkaline, moderately hypersaline environment with relatively high levels of arsenic and it has been proposed that the algae in the lake convert the arsenic to arsenosugars, which are then converted to arsenobetaine by brine shrimp. Once released to the lake, this methylated betaine analog can be used for osmotic balance.

V. OSMOTIC DOWNSHIFT

Because the cytoplasmic membrane is highly permeable to water, maintaining appropriate internal osmolality as well as turgor pressure in dilute environments necessitates a series of adaptive steps that in some instances are the converse of those undertaken during high osmolality conditions and that in some cases are unique. Most of what is known about adaptation to osmotic downshifts has come from studies of bacteria. After exposure to a low osmolality environment (one that is hypoosmotic to the cytoplasm) water will enter the bacterial cell and turgor pressure will increase until a pressure of about 6 atm is achieved. The peptidoglycan layer of the cell envelope is critical for resisting cell damage at this pressure. Alterations in the cell wall, produced enzymatically or chemically, greatly increases the susceptibility of cells to swelling and lysis when exposed to dilute conditions. The periplasmic space of gram-negative bacterial cells provides an osmotic buffer between the environment and the cytoplasm. During low-osmolality conditions cells synthesize large amounts, up to 50 mM, of periplasmic glucans, also called membrane-derived oligosaccharides. The accumulation of these acidic compounds together with the cations that bind to them makes a significant contribution to the maintenance of a higher osmolality in the periplasmic space than that present outside the cells. This reduces the turgor inside the cytoplasm and allows for a higher osmolality in the cytoplasm than would otherwise be possible. In bacteria such as *Rhizobium meliloti* and *Agrobacterium tumefaciens,* mutants blocked in the production of

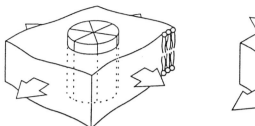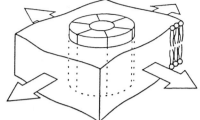

Fig. 1. Model of a mechanosensitive channel. Diagram indicating a channel that is closed at low mechanical stretch force (left) and open at high mechanical stretch force (right). Taken from Sukharev *et al.* (1997) with permission.

these compounds are more sensitive to hypoosmotic stress.

In response to gradual decreases in osmolality, bacteria employ specific efflux systems to release turgor pressure. K$^+$, glutamate, proline, betaine, and trehalose can all be specifically released from cells. Sudden increases in turgor require more dramatic measures and can lead to loss of many osmolytes including even ATP, through less-specific channels. Stretch-activated channels may be activated under these conditions. Mechanosensitive ion channels have been found in all three domains of life including the extremely halophilic archaeon *Haloferax volcanii*. These channels are believed to be important in the efflux of osmolytes in response to a sudden shift to lower osmolality. These proteins exhibit gating by mechanical force on the membrane alone. MscL is one such identified protein. A figure portraying a channel, such as MscL, opening in response to membrane stretch is shown in Fig. 1. Such channels may also operate in the outer membrane of gram-negative bacterial cells, but have not been discovered in this membrane environment yet.

Another consequence of bacterial exposure to sudden hypoosmotic stress may be the need to expel water as well as osmolytes to relieve excess turgor pressure. Many bacterial cells have been found to possess a specific water-channel protein, designated aquaporin, the product of the *aqpZ* gene in the case of *E. coli*. Genome sequencing has suggested the presence of water channels in *Haemophilus influenzae*, *Mycoplasma genitalium* and *Synechocystis* PCC6803. The transport of water across cell membranes is essential to life and similar water channels have been detected in plants and animals. In prokary-

otes, such channels may be less essential under isoosmotic growth conditions because the uptake of water needed for bacterial volume increase and cell growth and division can be met by simple diffusion, taking into account the water permeability of simple lipid bilayers and the high surface-to-volume ratio of microorganism. However, AqpZ may be needed for the efflux of water under conditions of hypoosmotic stress. Consistent with this hypothesis *aqpZ* gene transcription is highest at low osmolality and a *aaqpZ* knockout mutant is inhibited in growth at low osmolality.

VI. ADDITIONAL OSMOSENSORS

Osmoregulation of virulence gene expression and of outer-membrane proteins has been observed in many bacteria. In these cases the regulation does not appear to be required for osmotic adaptation per se, but rather to conditions that correlate with a given osmolality, such as the presence of a possible host organism, a high- or low-nutrient environment, or the presence of toxic chemicals. Such osmoregulation in enteric bacteria is mediated by the EnvZ and OmpR proteins. This regulatory system is one of the premier paradigms of prokaryotic signal transduction. These two proteins control the expression of the outer-membrane protein genes *ompC* and *ompF* in response to medium osmolality. OmpR is a DNA-binding protein that can be phosphorylated by the inner-membrane-spanning protein EnvZ. These proteins, like the KdpD and KdpE proteins (Section III), are members of the two-component sensor kinase–response regulator protein family. In this case EnvZ acts as the sensor and OmpR is the response regulator. The induction of *ompC* expression and repression of *ompF* expression results from increased phosphorylation of OmpR at high osmolality.

Several facts indicate that EnvZ does not respond specifically to osmolality, but to one or more related factors. For example, the high osmolality signal is most effective in the case of solutes that cannot traverse the outer membrane. Also, supplementation of media with betaine inhibits high-osmolality signaling by EnvZ. Because betaine supplementation would be predicted to affect cytoplasmic osmolyte levels, the betaine effect is most easlily interpreted as altering a cytoplasmic signal. However, because the periplas-

mic domain of EnvZ is known to be important for osmosensing, a periplasmic signal must also be necessary. EnvZ may be sensing osmolytes on both sides of the cytoplasmic membrane.

Another widespread osmosensor, designated ToxR, is found bacteria of the family Vibrionaceae. Many comparisons can be drawn between this protein and EnvZ and OmpR. The ToxR protein of the human pathogen *Vibrio cholerae* coordinates the activation of virulence genes, as well as the inverse expression of a pair of outer-membrane protein-encoding genes in a fashion similar to that seen in the EnvZ–OmpR system. Also, like EnvZ, ToxR appears to function as a membrane-spanning oligomer and its

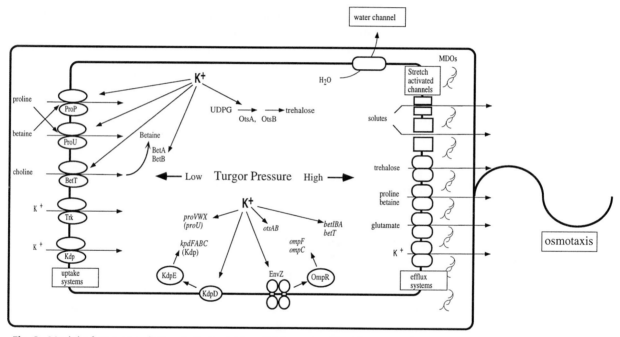

Fig. 2. Model of osmoregulation in *Escherichia coli*. The outer box represents the outer membrane and the inner box the inner membrane. The lower left portion of the schematic indicates the K⁺ uptake systems, which are activated by decreased turgor pressure resulting from exposure to increased medium osmolarity. K⁺ uptake may be responsive to the effect of a change in turgor on membrane structure. Uptake of proline, betaine, and choline is dependent on K⁺ uptake. The right portion of the model indicates the efflux systems, which may be activated at increased turgor pressure resulting from exposure to decreased medium osmolarity. This includes the transport out of the cell of specific osmolytes, such as K⁺, glutamate, betaine, and trehalose, as well as the less specific release of osmolytes by stretch-activated channels and the release of water through the aquaporin water channel protein AqpZ during sudden increases in turgor. Membrane-derived oligosaccharides (MDOs) preferentially accumulate in the periplasmic space during osmotic downshifts. K⁺ uptake also regulates the expression of many osmoregulated genes, as indicated in the lower portion of the figure, and enzyme activity, as indicated in the upper portion of the figure. Osmotaxis will be used to bias cell movement toward favorable osmotic environments and away from unfavorable osmotic environments. Modified from Csonka and Epstein (1996) with permission.

periplasmic domain is also critical for osmosensing. Temperature influences the activity of both the EnvZ–OmpR and the ToxR systems. Unlike the EnvZ—OmpR system, ToxR lacks the cytoplasmic portion of the EnvZ sensor and the amino-terminal domain of the OmpR regulator. As a result, osmolality signals perceived by ToxR may be directly transmitted to its DNA-binding domain the cytoplasm without the need for a two-component signaling system. Furthermore, unlike EnvZ, ToxR function is influenced by a second membrane protein, the ToxS protein, which may enhance ToxR oligomerization. To what extent ToxR is truly responsive to osmolality and not ionic effects remains to be determined. The fact that ToxR exhibits a biphasic response to medium salt concentration suggests that, like EnvZ, ToxR does not represent a true osmosensor.

VII. OSMOREGULATION MODEL

A model for bacterial osmoregulation based in large part on that presented by Csonka and Epstein (1996) for *E. coli* and *S. typhimurium* is shown in Fig. 2. The initial signal for a change in osmolality is proposed to be a change in turgor pressure or a consequence of a change in turgor, such as membrane stretch or some other structural change in the membrane or cell wall. During hyperosmotic conditions, the decrease in cytoplasmic turgor pressure results in the net accumulation of K^+ (reflecting the sum of its uptake and efflux). This in turn will induce the synthesis or transport of the principal counter ion for K^+, glutamate. At osmolalities above 250 mosM, solutes less perturbing than K^+ are transported or synthesized. So, the increased transport of K^+ leads to increased transport and synthesis of compatible solutes such as betaine, proline, and trehalose. The accumulation of these compounds increases turgor above prestress levels, which results in the efflux of K^+ and glutamate. During osmotic downshift, the increase in turgor will stimulate specific efflux systems and the synthesis of periplasmic glucans. If the magnitude of the osmotic change is great enough, stretch-activated ion channels and possibly the water channels will provide a release valve mechanism for

the rapid removal of both solutes and water. Osmotaxis is another important aspect of the osmotic response, as it provides a mechanism for cell movement into more favorable osmotic environments.

This model is largly based on information gathered from studies of enteric bacteria. Much more work is needed on microorganisms possessing adaptations for extremes of the levels and ranges of a_w that they will tolerate so that a more universal understanding of osmotic stress adaptation can be unveiled. The growing list of genetic methodologies available for use in extreme halophiles and haloversatile bacteria indicate a great opportunity for advanced studies of osmotic stress adaptation in these organisms.

See Also the Following Articles

Amino Acid Function and Synthesis • Extremophiles • Freshwater Microbiology

Bibliography

Calamita, G., Kempf, B., Bonhivers, M., Bishai, W. R., Bremer, E., and Agre, P. (1998). Regulation of the *Escherichia coli* water channel gene aqpZ. *Proc. Natl. Acad. Sci. U.S.A.* **95**, 3627–3631.

Csonka, L. N. (1989). Physiological and genetic responses of bacteria to osmotic stress. *Microbiol. Rev.* **53**, 121–147.

Csonka, L. N., and Epstein, W. (1996). Osmoregulation. *In* "*Escherichia coli* and *Salmonella typhimurium* Cellular and Molecular Biology" (F. C. Neidhardt *et al.*, eds.), pp. 1210–1223. American Society for Microbiology Press, Washington DC.

DasSarma, S. (1995). Halophilic Archaea: An overview. *In* "Archaea, a Laboratory Manual" (F. T. Robb *et al.*, eds.), pp. 3–11. Cold Spring Harbor Laboratory Press, Cold Spring Harbor.

Desmarais, Jablonski, D., P., Fedarko, N. S., and Roberts, M. F. (1997). 2-Sulfotrehalose, a novel osmolyte in halo-alkaliphilic Archaea. *J. Bacteriol.* **179**, 3146–3153.

Grant, W. D., Gemmell, R. T., and McGenity, T. J. (1998). Halophiles. *In* "Extremophiles: Microbial Life in Extreme Environments" (K. Horikoshi and W. D. Grant, eds.), pp. 93–132. John Wiley & Sons, New York.

Imhoff, J. F. (1986) Osmoregulation and compatible solutes in eubacteria. *FEMS Microbiol. Rev.* **39**, 57–66.

Le Rudulier, D., Strom, A. R., Dandekar, A. M., Smith, L. T., and Valentine, R. C. (1984). Molecular biology of osmoregulation. *Science* **224**, 1064–1068.

Martins, L. O., Huber, R., Stetter, K. O., Da Costa, M. S., and

Santos, H. (1997). Organic solutes in hyperthermophilic Archaea. *Appl. Environ. Microbiol.* **63,** 896–902.

Robertson, D. E., and Roberts, M. F. (1991). Organic osmolytes in methanogenic archaebacteria. *BioFactors* **3,** 1–9.

Severin, J., Wohlfarth, A., and Galinski, E. A. (1992). The predominant role of recently discovered tetrahydropyrimidines for the osmoregulation of halophilic eubacteria. *J. Gen. Microbiol.* **138,** 1629–1638.

Somero, G. N., Osmond, C. B., and Bolis, C. L., (1992). "Water and Life: Comparative Analysis of Water Relationships at the Organismic, Cellular, and Molecular Levels." Springer-Verlag, New York.

Sukharev, S. I., Blount, P., Martinac, B., and Kung, C. (1997). Mechanosensitive channels of *Escherichia coli*: The MscL gene, protein, and activities. *Annu. Rev. Physiol.* **59,** 633–657.

Ventosa, A., Nieto, J. J., and Oren, A. (1998). Biology of moderatley halophilic aerobic bacteria. *Microbiol. Molec. Biol. Rev.* **62,** 504–544.

Vreeland, R. H. (1987). Mechanisms of halotolerance in microorganisms. *CRC Crit. Rev. Microbiol.* **14,** 311–356.

Outer Membrane, Gram-Negative Bacteria

Mary J. Osborn

University of Connecticut Health Center

GLOSSARY

β-barrel A protein-folding motif in which several amino acid sequences forming β-sheets fold in such a way as to form a barrel-like structure.

EDTA Ethylenediaminetetracetic acid, a chelator of divalent cations.

GlcNac N-acetyl-D-glucosamine.

glycosyl transferase An enzyme catalyzing the transfer of a sugar residue, generally from its activated nucleotide sugar derivative to an acceptor molecule.

Kdo 2-keto-3-deoxy-D-manno-octulosonate.

lipid bilayer The fundamental structural basis of biological membranes, in which two layers of lipids are apposed, with hydrophilic groups of each facing the aqueous medium and hydrophobic fatty acyl chains forming the interior.

lipopolysaccharide A complex polymer exposed at the cell surface and responsible for major antigenic specifities and endotoxic properties.

periplasmic space The region between the cytoplasmic and outer membranes, containing the cell wall, hydrolytic enzymes, components of active transport systems, and proteins required for maturation of newly synthesized periplasmic proteins.

plasmolyzed cells Cells incubated in hypertonic medium to retract the inner membrane and expand the periplasmic space.

SDS-PAGE Polyacrylamide gel electrophoresis, incorporating the denaturant sodium dodecyl sulfate in sample and gel buffers.

THE CELL ENVELOPE OF GRAM-NEGATIVE BACTERIA consists of an inner cytoplasmic membrane, the peptidoglycan (murein) cell wall, and an outer membrane that bounds the cells. The murein cell wall underlies the outer membrane and is generally anchored to it covalently through the murein lipoprotein and by noncovalent interactions with the outer-membrane protein, OmpA. The outer membrane is the interface between the cell and its external environment. It acts as a protective barrier, mediates interactions between bacteria and between bacteria and the animal or plant host cells, and anchors externally disposed organelles such as pili. The composition and structure of the outer membrane is highly specialized, reflecting its specialized functions. The enteric bacteria, specifically *Escherichia coli* and *Salmonella typhimurium,* provide the paradigm for outer-membrane structure and function and will form the major basis of this review. However, it should be recognized that the specifics of structure and function may differ significantly as one moves away from the enteric family.

I. OUTER-MEMBRANE COMPOSITION AND STRUCTURE

A. Isolation of Outer Membrane

Two general methods have been employed to separate inner and outer membranes. The first and more rigorous takes advantage of the difference in buoyant densities between the two membranes. Cells are broken by French press or by lysozyme plus osmotic shock and membranes are fractionated by one of several protocols for isopycnic sucrose gradient centrifugation. The outer membrane bands at an apparent buoyant density between 1.22 and 1.27 g/cc, whereas the inner membrane band is recovered in the range of 1.12 to 1.15 g/cc. Unseparated material bands between the two fractions, as do several minor specialized membrane fractions that can be further purified by additional sedimentation and flotation gradient centrifugations (Ishidate *et al.*, 1986).

The second method of outer-membrane isolation takes advantage of the differential resistance of outer membrane to detergents or alkali. Detergent treatment (Triton X-100, in the absence of EDTA or Sarkosyl) is used to dissolve inner membrane, leaving the outer membrane presumably intact. Alternatively, major outer-membrane protein species and lipopolysaccharide (LPS) remain insoluble following treatment with alkali, whereas inner membrane is solubilized. These methods are far quicker and more convenient than isopycnic centrifugation, but risk the solubilization of the peripheral outer-membrane-associated proteins. The assumption that all inner-membrane proteins are solubilized is also difficult to verify.

B. Outer-Membrane Proteins

In contrast to the inner cytoplasmic membrane, the outer membrane is essentially metabolically inert and contains a limited number of major protein species. These include the porins, which are responsible for solute permeation across the membrane and proteins (murein lipoprotein, OmpA) that interact covalently or noncovalently with the underlying murein. In addition, a variety of less abundant species mediate interactions with the external environment (e.g., adherence), assembly of surface structures such as pili and flagella, conjugation, and secretion of proteins to the external medium.

C. Outer-Membrane Lipids

1. Lipopolysaccharide Structure

Lipopolysaccharide is a unique and major constituent of the outer membrane and is responsible for the endotoxic properties of gram-negative bacteria. The LPS of enteric bacteria consists of three structural regions—Kdo-lipid A, a unique glycolipid anchoring the molecule into the outer membrane; a core oligosaccharide region (subdivided into backbone and outer core); and peripheral O-antigen

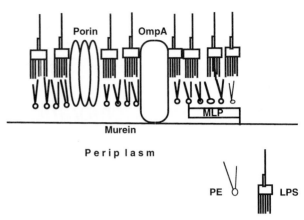

Fig. 1. Schematic structure of the outer membrane. Mlp, murein lipoprotein; PE, phosphatidylethanolamine.

chains, which are responsible for major immunological specificities of the intact organism. (See Fig. 1.)

Kdo-Lipid A is highly conserved among gram-negative bacteria. It consists of a fatty acylated phosphorylated glucosamine disaccharide characterized by N-linked and O-linked 3-hydroxymyristoyl residues (or shorter chain 3-hyroxy fatty acids in some nonenteric genera), as well as saturated fatty acyl substitutions on the 3-hydroxy fatty acids. Other decorations (phosphoryl ethanolamine, 4-aminoarabinose) are variably present. A Kdo disaccharide is linked to the 4′ position of the disaccharide and provides the attachment site for the core saccharide unit.

The inner-core backbone region contains two or three L-glycero-D-mannoheptose residues, and again may have additional phosphate and pyrophosphorylethanolamine decorations. The outer-core oligosaccharide is glycosidically linked to the nonreducing heptose residue, and in the enterics typically contains galactose, glucose, and N-acetylglucosamine. The structures are similar, but not identical, among the enteric genera. Genera such as *Hemophilus* and *Neisseria* produce only lipoligosaccharides (LOS), which are analogous to the core LPS of enterics.

O-antigen consists of long polysaccharides, generally composed of oligosaccharide repeating units. The structure is characterized by the frequent presence of unusual deoxy, dideoxy, and amino sugars; the composition, structure, and immunospecificity vary widely within and between genera, and form the basis of serological typing. Wild-type bacteria with the complete O-antigenic LPS are referred to as smooth, whereas mutants lacking O-antigen are called rough. In certain rough mutants of enteric bacteria, a second polysaccharide, enterobacterial common antigen (ECA), takes the place of the O-chains. A form of ECA that is widely distributed in the enterics is anchored in the outer membrane by attachment to a phosphoglyceride. ECA structure is conserved among the enterics.

Both the O-antigenic LPS of enteric bacteria and the LOS of *Neisseria* and *Hemophilus* are subject to changes in structure and antigenicity due to altered patterns of gene expression. In *Salmonella,* for example, a number of temperate converting phages carry genes that encode glycosyl transferases or change the expression of host glycosyl transferases or the pattern LPS acetylation. *Neisseria* are capable of altering LOS glycosyl transferase expression by a process of antigenic variation.

2. Enterobacterial Common Antigen

Enterobacteria share a widely distributed surface-exposed antigenic specificity called enterobacterial common antigen (ECA). ECA is a heteropolysaccharide composed of the unusual amino sugars, *N,O*-diacetyl glucosamine, *N*-acetylaminomannuronic acid, and *N*-acetylfucosamine. It can exist in two forms in the outer membrane. A form anchored to the external leaflet of the outer membrane by linkage of the reducing terminus to a phosphatidic acid residue is generally present. The second form, in which ECA replaces the O-chains of LPS, is found in some LPS mutants that lack O-antigen.

3. Phospholipid Composition

In addition to lipopolysaccharides, outer membranes contain conventional phospholipids. However, the phospholipid composition is typically very simple, with phosphatidylethanolamine accounting for well over 90% of the total in the enteric family.

D. Outer-Membrane Structure

1. The Enteric Paradigm

The lipid bilayer of the outer membrane is highly asymmetric, with LPS confined exclusively to the external leaflet of the membrane, with phosphatidyl ethanolamine restricted, probably exclusively, to the inner periplasmic leaflet. The total number of acyl chains in the two types of lipid are calculated to be roughly similar. The asymmetric distribution of lipids in the membrane is disrupted physiologically by deep rough mutations in the LPS backbone, which give a phenotype deficient in major outer-membrane proteins and result in the appearance of substantial amounts phospholipid in the outer membrane. (See Fig. 1.)

The major species of proteins (porins, OmpA) are integral transmembrane membrane proteins; however, certain proteins, (e.g., murein lipoprotein) are anchored specifically to the inner leaflet of the membrane, and others to the surface-exposed outer leaflet. The outer membrane is linked covalently to the underlying murein cell wall by the murein lipoprotein and noncovalently through interaction with the periplasmic domain of OmpA. Experimental treatments that disrupt the attachment of outer membrane to peptidoglycan can result in randomization of lipopolysaccharide across the bilayer, and perhaps other alterations in the normal membrane topology.

2. Zones of Adhesion

Using light and electron microscopy of plasmolyzed cells, sites are observed at which the inner membrane fails to retract from outer membrane and the two membranes appear to be more or less firmly associated in some way. Such zones of adhesion were early postulated to function as bridges in the translo-

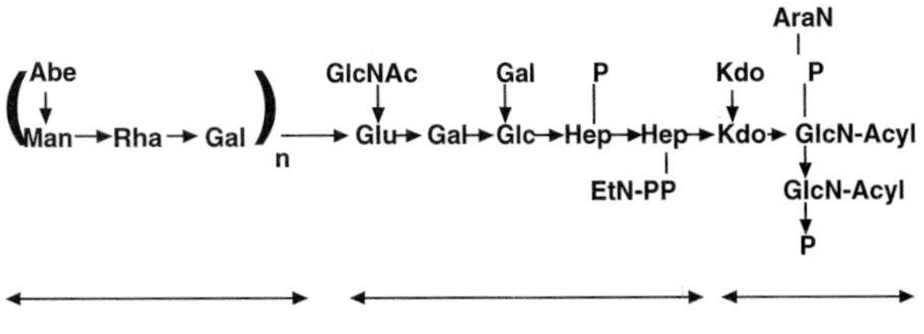

Fig. 2. Structure of *S. typhimurium* LPS. Abe, abequose; AraN, 4-aminoarabinose; EtN, ethanolamine; Gal, galactose; Glc, glucose; GlcN-acyl, *N,O*-di-3-hydroxymyristoyl glucosamine and *N,O* di-3-hydroxy(3-O-lauroyl or myristoyl)myristoyl glucosamine; GlcNAc, N-acetylglucosamine; Hep, L-glycero-D-mannoheptose; Kdo, 2-keto-3-deoxyoctulosonate; Man, mannose; Rha, rhamnose.

cation of proteins, polysaccharides, and lipids to the outer membrane, but their biochemistry remains largely obscure. It seems likely that the contact sites observed represent several different kinds of structures. For example, annuli traversing the cell circumference, the periseptal annuli, are thought to play a role in septum formation and appear to be relatively stable. Type III secretion systems and the biogenesis of Type IV pili both include a number of periplasmically oriented proteins that are thought to function in secretion of virulence proteins or pilus subunits across the outer membrane. Other proteins, such as TonB, which is required for a number of specific transport functions, are also postulated to bridge between the inner and outer membranes.

II. OUTER-MEMBRANE PROTEINS

A. Porins

1. *General Pores*

In order to reach active transport systems in the inner membrane, nutrient solutes must first be able to diffuse across the outer membrane to the periplasmic space. A family of ubiquitous major 34- to 37-kDa outer-membrane proteins, including OmpC, OmpF, and PhoE of *E. coli* and other enteric bacteria, provide fixed nonspecific diffusion pores for permeation of small hydrophilic solutes. The proteins typically exist in the outer membrane as homotrimers. Their amino acid sequences show no extended hydrophobic regions and no extensive predicted α-helical structure. Rather, the crystal structure of the OmpF porin of *E. coli* has revealed an amphipathic β-barrel structure, with the hydrophobic faces lying among the three subunits and at the external surface of the trimer. Each of the monomer units contains a transmembrane hydrophilic pore approximately 7×11 Å in size. By diffusion measurements, the OmpF channel is slightly larger than that of OmpC. The fixed pores of OmpF, OmpC, and PhoE allow diffusion-limited permeation of small hydrophilic solutes up to a limit of about 600 Da, with the rate of permeation highly dependent on the diameter of the hydrated solute. OmpC and OmpF are nonspecific, but favor neutral or cationic solutes over anions. PhoE, though part of the phosphate regulon, is not specific for phosphate and is considered a general channel with a preference for anions over cations.

The expression of OmpC and OmpF is subject to complex reciprocal regulation, at both transcriptional and posttranscriptional levels. OmpF is repressed by high temperature (37°C), high osmolarity, and probably other factors of the host environment, and is therefore repressed in the animal host. Presumably, the repression of the porin with the larger channel in favor of OmpC, whose channel is smaller, provides additional protection against bile salts and other toxic molecules involved in host defenses. Transcriptional regulation is mediated by the EnvZ–OmpR two-component system, which responds to osmotic stress and results in the down-regulation of OmpF and activation of OmpC. In addition, a small antisense RNA, micF, is divergently transcribed from the OmpC operator and is thought to inhibit the translation of OmpF mRNA in response to environmental cues.

2. *Specific Diffusion Channels*

The group of nonspecific porins will not accomodate solutes larger than 600 Da, and specialized diffusion channels are required for the uptake of solutes such as maltodextrins, vitamin B12, and iron–siderophore complexes. LamB, a component of the inducible maltose regulon of *E. coli*, has a trimeric structure similar to that of OmpF, and indeed the LamB pore allows the permeation of a variety of small molecules in addition to maltose and maltodextrins. However, the LamB channel is distinguished by the presence of a binding site for maltose oligosaccharides, which confers a higher degree of specificity for this family saccharides. Diffusion of iron–siderophore complexes is mediated by a group of at least five outer-membrane channel proteins in *E. coli*, whereas B12 uptake is dependent on the Btu channel. In these cases, the channel protein exhibits high-affinity binding sites for its substrate and permeation is substrate-specific. It is of interest that most of these outer-membrane channel proteins also act as receptors for specific bacteriphages or colicins. LamB, as the name suggests, was originally identified as the receptor for phage lambda.

B. OmpA

OmpA is a highly abundant outer-membrane protein of approximately 35 kDa. The protein is monomeric in the membrane and consists of two domains; the N-terminal domain has a transmembrane β-barrel structure, whereas the globular C-terminal region is exposed at the periplasmic face of the membrane. The protein is only partially denatured in SDS at temperatures below 100°C, the basis for its initial identification as a "heat modifiable" protein in SDS-PAGE. Two physiological functions have been identified. OmpA is required for *E. coli* to act as an efficient recipient in conjugation, a function of its surface-exposed N-terminal domain, and anchors the outer membrane noncovalently to the underlying murein cell wall through its periplasmically oriented C-terminal domain. Mutants lacking OmpA tend to form spherical cells with unstable outer membranes. It is of interest that OmpA, although monomeric, can act as a weak nonspecific porin in reconstituted systems. This is probably not physiologically significant in *E. coli,* but in *Pseudomonas* species, which lack OmpC/F homologs, the OmpA homolog, OprF, is the major nonspecific porin.

C. Outer-Membrane Lipoproteins

1. Murein Lipoprotein

Murein lipoprotein is a small (7.2 kDa), highly abundant protein that provides covalent linkage between the outer membrane and murein. The protein is hydrophilic and predominantly α-helical. The protein lies at the periplasmic face of the outer membrane and is integrated into the bilayer by its N-terminal lipid modification. The C-terminal cysteine residue is modified by thioether linkage to a diglyceride residue, and, in addition, the α-amino group is fatty acylated. About one-third of the murein lipoprotein, the bound form, is covalently attached to the underlying murein by isopeptide linkage of the carboxyl group of murein diaminopimelic acid (DAP) to the ε-amino group of the C-terminal lysine residue. Murein lipoprotein mutants unable to form the attachment to murein have unstable outer membranes that leak periplasmic proteins and release outer-membrane blebs into the medium.

2. *Other Outer-Membrane Lipoproteins*

Over 25 lipoprotein genes have been identified in *E. coli*, encoded on chromosomal, phage, or plasmid genomes. The majority have been localized to the outer membrane, although a significant number are found in inner membrane. A variety of functions have been assigned to members of the outer-membrane class, including involvement in the efflux pump for acriflavine, osmoregulation, surface exclusion, serum (complement) resistance, and release of colicins.

Lipoproteins are not limited to gram-negative bacteria. For example, in *Bacillus* species, some binding proteins required for the active transport of various substrates, as well as certain secreted proteins, are anchored to the external face of the cytoplasmic membrane as lipoproteins. Spirochetes offer a particularly dramatic example; lipoproteins apear to make up the majority of proteins in the outer membranes of treponemes and *Borrelia*. Indeed, close to 100 lipoprotein genes and pseudogenes have been identified in the *Borrelia* genome (Fraser *et al.*, 1997). Their functions are largely unknown, but presumably mediate host–pathogen interactions. Derived amino-terminal lipopeptides are highly immunogenic and induce strong inflammatory responses.

III. FUNCTIONS OF OUTER MEMBRANE

A. Barrier Function

In comparison with the cytoplasmic membrane and other biological membranes, the outer membrane exhibits strikingly reduced permeability to a wide variety of lipophilic compounds, including bile salts and other detergents, lipophilic antibiotics, and dyes. Lipopolysaccharide is primarily responsible for the barrier properties of the membrane. Permeation of lipophilic compounds across membranes requires the initial intercalation of the compound into the hydrophobic interior of the lipid bilayer. Penetration into the outer membrane is hindered in the presence of LPS for two reasons. Lipid A fatty acids are saturated; therefore, they are more highly ordered and of lower mobility than those of a conventional phospholipid leaflet. In addition, each LPS molecule contains 6–7 fatty acyl residues, allowing an increased number of intermolecular hydrophobic interactions than phosphoglycerides having only two fatty acids.

The barrier function is disrupted under conditions that compromise the molecular organization of the membrane. Treatment with EDTA removes divalent cations that cross-bridge LPS phosphate groups and enhance LPS–LPS association. Mutations affecting the biosynthesis of the heptose-containing backbone region of LPS result in a deep rough phenotype, characterized by the reduced abundance of outer-membrane porins and appearance of substantial amounts of phosphatidylethanolamine in the external leaflet of the outer membrane. Polymyxin and other polycations also disrupt LPS organization and the barrier function by electrostatic interaction with the polyanionic LPS. Mutants of *Salmonella typhimurium* resistant to polymyxin have markedly increased amounts of the aminosugar 4-aminoarabinose in their lipid A, thus reducing LPS acidity and binding to added polycations.

B. Host–Pathogen Interactions

The outer membrane plays a crucial role in many aspects of pathogenesis, including the evasion of host defenses and the adherence of invading bacteria to host surfaces, which is important for the initial establishment of infection as well as intracellular invasion and survival. LPS O-antigen is antiphagocytic and both outer-membrane proteins and LPS have been implicated in resistance to serum complement. Outer-membrane proteins can themselves act as adhesins or invasins, mediating specific receptor–ligand interactions with host-cell surfaces (e.g., the PII (opa) protein of *N. gonorrheae* and the invasin of *Yersinia* species). Alternatively, adherence may be mediated by pili or fimbriae (e.g., the P pilus of uropathogenic *E. coli*) whose assembly requires outer-membrane usher proteins. Type III secretion systems required for intracellular invasion by many gram-negative bacteria include outer-membrane components that form multimeric structures mediating the direct injection of virulence proteins into host-cell cytosol. Other outer-membrane proteins specifically facilitate the secretion of protein toxins into the medium.

IV. ASSEMBLY OF OUTER MEMBRANE

A. Export of Outer-Membrane Proteins

Outer-membrane proteins are synthesized as precursor proteins with classic signal sequences and are exported to the periplasm by the Sec system for polypeptide secretion. The signal sequence is removed by the leader peptidase (or, in the case of lipoproteins, by a specific prolipoprotein signal peptidase), located at the periplasmic face of the inner membrane. The interaction of soluble porin monomers in the periplasm with (presumably nascent) lipopolysaccharide is thought to promote a conformation switch allowing its assembly into mature trimers. Oligomerization is required for integration into the outer membrane. The role of LPS is consistent with the fact that deep rough LPS mutants are deficient in outer-membrane porins. Similarly, a newly synthesized OmpA monomer is found in the periplasm in an open conformation that is much more sensitive to proteolysis than is the mature protein. Interaction with nascent LPS is not required for the maturation of the periplasmic intermediate or for integration into the outer membrane, but interaction with LPS within the outer membrane is necessary for OmpA function. A periplasmic protein, Skp/OmpH, has been shown to bind both porins and OmpA, and is required for effective assembly into outer membrane. Presumably, Skp/OmpH acts as a chaperone, facilitating the maturation of the nascent periplasmic proteins.

The targeting of outer-membrane proteins from periplasm to the outer membrane appears to depend on tertiary or quaternary structures of the proteins, rather than on any obvious conserved targeting sequence. Amphipathic β-barrel structures have been identified or predicted for the major transmembrane porins and for OmpA. This structure is necessary for integration of these proteins into the membrane, and may be sufficient.

B. Biogenesis of Lipopolysaccharide

1. Biosynthetic Pathways

The modular structure of LPS—Kdo-lipid A, core, and O-antigen—is reflected in its genetics and biosynthetic pathways. In enteric organisms, most genes for biosynthesis of core and O-antigen are clustered in the *rfa* and *rfb* regions, respectively, although the genes required for synthesis of the Kdo-lipid A moiety are scattered in several locations. Core LPS and O-antigen are synthesized by two independent pathways, involving different mechanisms of polysaccharide chain elongation. The core pathway uses the classic mechanism in which single sugar residues are transferred from their respective nucleotide sugars to the nonreducing terminus of the growing chain, whereas the O-antigen pathway uses the membrane isoprenoid coenzyme bactoprenyl-P (undecaprenyl-P) for the polymerization of oligosaccharide repeating units.

The synthesis of the Kdo-Lipid A portion (Fig. 3) begins in the cytosol with 3-hydroxymyristoylation of UDP-GlcNAc, followed by deacetylation of the amino sugar and addition of a second, N-linked 3-hydroxymyristoyl residue. A portion of the diacyl nucleotide sugar is then hydrolyzed to diacyl glucosamine-1-P (lipid X), which acts as acceptor for transfer of the diacylglucosamine residue from the UDP derivative. An additional phosphate residue is then transferred from ATP to the 4' position of the resulting tetraacyl disaccharide-P, followed by transfer of two Kdo residues from CMP-Kdo to the 6' position. In the final steps, the two 3-hydroxymyristoyl residues of the nonreducing glucosamine are esterified with saturated acyl residues (laurate and myristate in *E. coli*). Lipid A also contains variable amounts of phosphorylethanolamine and 4-aminoarabinose, but little information is available on the enzymology of their addition. The addition of Kdo is unusual in that the two residues are added in concerted fashion

by a single bifunctional Kdo transferase. The number of Kdo residues varies among genera. *Chlamydia* LPS contains three, added by a trifunctional transferase, whereas in *Hemophilus influenzae,* only a single Kdo and a monofunctional transferase are present.

It is of interest that UDP-GlcNAc is the starting substrate for the synthesis of both lipidA and murein. It is clearly important for the cell to regulate entry into the two pathways in such a manner as not to compromise either because both products are essential for viability. Regulation occurs at the level of the second enzyme of the lipid A pathway, deacetylation of UDP-*O*-myristoyl-*N*-acetylglucosamine.

The addition of the heptosyl residues of the backbone and the sugars of the outer core is catalyzed by a series of peripheral membrane-associated glycosyl transferases with transfer single-sugar residues successively from the nucleotide sugars to the nonreducing terminus of the growing chain. The transferases, as well as the enzymes required for synthesis of ADP-L-glycero-D-mannoheptose, are encoded by genes of the *rfa* locus. The expression of both *rfa* and *rfb* genes is regulated by the transcriptional activator, RfaH, which is also required for expression of the hemolysin of enteric bacteria and surface functions encoded by the F-factor.

O-antigen chains are typically composed of oligosaccharide repeating units, whose synthesis and polymerization are determined by the *rfb* operon. The oligosaccharide unit is assembled on the membrane-bound coenzyme bactoprenyl (undecaprenyl)-phosphate (Fig. 4), and all the steps of the pathway are catalyzed by membrane-associated enzymes. The pathway begins with transfer of the reducing terminal sugar-1-P from its nucleotide sugar to the coenzyme to form bactoprenyl-pyrophosphoryl-monosaccharide. The remaining sugars are then transferred sequentially to the nonreducing terminus of the growing oligosaccharide. It should be noted that murein biosynthesis also requires bactoprenyl-P, and in certain mutants in which incomplete lipid-linked intermediates in the O-antigen pathway accumulate, the coenzyme is unavailable for murein synthesis and cell lysis ensues.

Polymerization, catalyzed by O-polymerase, takes place by a mechanism in which new oligosaccharide units are introduced at the reducing end of the grow-

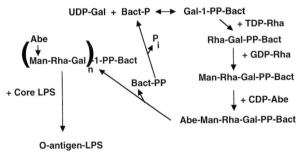

Fig. 3. Lipid A biosynthetic pathway. ACP, acyl carrier protein; other abbreviations as in Fig. 2.

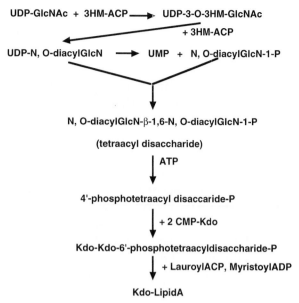

Fig. 4. O-antigen biosynthetic pathway. Bact, bactoprenol; other abbreviations as in Fig. 1.

ing polymer by transfer of polymer to the incoming oligosaccharide unit. The process is analogous to the mechanisms of elongation of polypeptides and fatty acids. In the final step of LPS assembly, polymeric O-antigen chains are added to the independently synthesized core LPS by O-ligase. O-antigen-containing LPS forms a ladder in SDS-PAGE, indicative of a high degree of heterogeneity in O-chain length. The distribution of polysaccharide chain lengths is presumably determined in part by the kinetics of competing O-polymerase and O-ligase, but is also under genetic control by the the *wzz* (*rol*) gene.

2. Biosynthesis of Enterobacterial Common Antigen

In mutants lacking O-antigen, ECA can be attached to core LPS as though it were an *O*-polysaccharide. The ECA polymer is composed of oligosaccharide repeating units, such as O-Antigen, and biosynthesis parallels that of O-antigen, using a similar bactoprenyl-P pathway. Presumably, the phosphoglyceride form of the ECA polysaccharide is formed by the transfer of the completed polymer chain to a phosphatidic acid acceptor.

3. Topology of LPS Biogenesis and Export to Outer Membrane

Both the core LPS and oligosaccharide intermediates of O-antigen synthesis are assembled at the cytoplasmic face of the inner membrane, yet the attachment of O-antigen to the core takes place at the periplasmic face of the inner membrane. Thus, the flip-flop of both across the inner membrane must be invoked. Similar considerations hold for the assembly of murein from its comparable lipid-linked intermediates. The mechanisms of these transmembrane translocation steps are not well understood. The presence of Kdo residues is necessary for the efficient translocation of lipid A to the outer membrane. In addition, evidence (Zhou *et al.*, 1998; Chu *et al.*, 1995) strongly suggests that the *msbA* gene of *E. coli* and the *abcA* gene of *Aeromonas salmonicida*, both members of the ATP-dependent ATP binding cassette (ABC) superfamily of transporters, are required for the translocation of cytosolically oriented core LPS or O-antigen intermediates to the periplasmic face of the inner membrane. Other ABC systems are known to facilitate the secretion of capsular polysaccharides, whose synthesis takes place in cytosol. The mechanism of LPS translocation to the outer membrane also remains unclear. The process is independent of ongoing protein synthesis, and directly or indirectly requires both ATP and a membrane potential across the inner membrane. Immunoelectron microscopy shows LPS newly incorporated into the outer membrane preferentially localized over zones of adhesion between the inner and outer membranes, but the molecular nature of such contact sites is unknown.

C. Export of Phospholipid to Outer Membrane

The problem of the mechanism of translocation of phospholipid to outer membrane is somewhat analogous to that for LPS. Phosphatidylethanolamine and other phospholipids are believed to be synthesized in the cytoplasmic leaflet of the inner membrane and rapidly transposed to the periplasmic leaflet. The translocation of phosphatidylethanolamine to the periplasmic leaflet of the outer membrane is energy-dependent, rapid, and apparently reversible. Although phospholipid-exchange protein has been identified in the periplasm of purple bacteria, efforts to detect such activity in enteric bacteria have been unsuccessful. It has been postulated that translocation occurs at zones of adhesion, but direct evidence is lacking.

See Also the Following Articles

ADHESION, BACTERIAL • ENTEROPATHOGENIC BACTERIA • LIPOPOLYSACCHARIDES

Bibliography

Chu, S., Noonan, B., Cavaignac, S., and Trust, T. J. (1995). Endogenous mutagenesis by and insertion sequence element identifies Aeromonas salmonicida AbcA as an ATP-binding cassette transport protein required for biogenesis of smooth lipopolysaccharide. *Proc. Natl. Acad. Sci. U.S.A.* **92**, 5754–5758.

Fraser, C. M., Casjens, S., Huang, W. M, *et al.* (1997). Genomic sequence of a Lyme disease spirochaete, *Borrelia burgorferi. Nature* **390**, 580–586.

Ishidate, K., Creeger, E. S., Zrike, J., *et al.* (1986). Isolation of differentiated membrane domains from Escherichia coli and Salmonella typhimurium, including a fraction containing attachment sites between the inner and outer membranes and the murein skeleton of the cell envelope. *J. Biol. Chem.* **261**, 428–443.

Nikaido, H. (1996). Outer membrane. *In* "Escherichia coli and Salmonella" (F. Neidhardt, ed.), 2nd ed., pp. 29–47. ASM Press, Washington, DC.

Raetz, C. R. H. (1996). Bacterial lipopolysaccharides: A remarkable family of bioactive macroAmphiphiles. *In* "Escherichia coli and Salmonella" (F. Neidhardt, ed.), 2nd ed., pp. 1035–1063. ASM Press, Washington, DC.

Rick, P. D., and Silver, R. P. (1996). Enterobacterial common antigen and capsular polysaccharides. *In* "Escherichia coli and Salmonella" (F. Neidhardt, ed.), 2nd ed., pp. 104–122. ASM Press, Washington, DC.

Wu, H. C. (1996). Biosynthesis of lipoproteins. *In* "Escherichia coli and Salmonella" (F. Neidhardt, ed.), 2nd ed., pp. 1005–1014. ASM Press, Washington, DC.

Zhou, Z., White, K. A., Polissi, A., Georgopolous, C., and Raetz, C. R. (1998) Function of Escherichia coli MsbA, an essential ABC family transporter, in lipid A and phospholipid biosynthesis. *J. Biol. Chem.* **273**, 12466–12475.

Oxidative Stress

Pablo J. Pomposiello and Bruce Demple

Harvard School of Public Health

GLOSSARY

oxidative stress The excess production or insufficient disposal of intracellular oxidants.
reactive oxygen species (ROS) Partially reduced oxygen derivatives, such as superoxide ($O_2^{\cdot-}$) and hydrogen peroxide (H_2O_2).

AEROBIC METABOLISM requires the exposure of cells to oxygen, which sometimes reacts non-enzymatically with cellular components to generate free radicals and other reactive molecules. These reactive by-products can damage all biological macromolecules and thus interrupt growth and cause mutations. This damage is limited by small molecules such as glutathione, which neutralizes some free radicals, and by enzymes such as superoxide dismutase (SOD) and catalase, which eliminate specific reactive species. Under some circumstances, the rate of free-radical production increases or cellular-defense activities are diminished, which results in oxidative stress. Oxidative stress can arise in many ways—through metabolic changes (e.g., inactivation of some components of the electron transport chain); through exposure to environmental agents that divert electron flow (e.g., the herbicide paraquat); and through immune responses to bacterial infection (e.g., superoxide and nitric oxide generated by activated macrophages). Aerobic organisms have evolved inducible defense mechanisms against various types of oxidative stress. From single proteins to complex self-regulating genetic networks, these defenses scavenge reactive oxygen species and mediate the repair of cellular damage. The knowledge of these antioxidant mechanisms is more advanced in enteric bacteria than in any other group of organisms, although progress is being made in gram-positive bacteria and yeast. It is clear that genes that originally aided microorganisms in colonizing an oxidizing atmosphere were in many cases recruited into different regulatory networks, helping the cell to fine-tune its metabolism according to a wide variety of metabolic and environmental conditions.

I. SOURCES OF OXIDATIVE STRESS

A. Aerobic Metabolism

Various metabolic pathways involving the exchange of electrons between biochemical intermediates have the potential to generate oxidative damage by the anomalous transfer of single electrons. In this fashion, aerobic metabolism, photosynthesis, and denitrification are metabolic functions whose oxidant by-products have to be kept at concentrations that are compatible with cellular integrity. In a sense, the reactive by-products may be considered to be unavoidable "leaks" in these natural processes (Fig. 1). Oxygen competes with respiratory components to oxidize some elements of the electron-transport chain and thus yields superoxide, which is readily converted to H_2O_2 by the enzyme SOD. H_2O_2 can be safely disposed of by catalases, or it can react further, especially with reduced metals such as

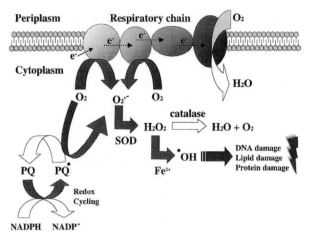

Fig. 1. Reactive oxygen species, oxidative stress, and cellular damage. The majority of the oxygen that enters the cell is reduced to water by the respiratory chain, a reaction that consumes four electrons. However, a small proportion of the oxygen molecules can be reduced in a series of one-electron reactions. Molecular oxygen forms superoxide (O_2^-) by reaction with reduced components of the respiratory chain. Superoxide can also be formed by reaction with redox-cycling drugs such as paraquat (PQ), which is enzymatically re-reduced at the expense of NADPH. Superoxide is eliminated by superoxide dismutase (SOD) to form hydrogen peroxide (H_2O_2). Hydrogen peroxide can either be detoxified by conversion into water and oxygen by catalase, or react with reduced transition metals such as iron and copper to form a hydroxyl radical ($^\cdot$OH). The hydroxyl radical is a highly reactive molecule than can damage virtually all the fundamental cellular components. Solid arrows indicate reactions that yield oxidants. Open arrows indicate reactions that yield innocuous products. See color insert.

Fe^{2+}, to generate still more unstable products, notably the highly reactive hydroxyl radical ($^\cdot$OH; see Fig. 1). DNA is a critical target for oxidative damage, and the DNA lesions caused by oxidants can disrupt replication and lead to mutations. The metal centers of enzymes constitute another group of targets for oxidative damage, with critical sensitive activities such as ribonucleotide reductase, essential for producing DNA precursors, and aconitase, a pivotal component of the citric acid cycle. Unsaturated lipid components of the membrane react to form lipid peroxides in chain reactions with many products derived from a single free radical. The ultimate breakdown of lipid peroxides yields still another reactive

compound, malondialdehyde, which can form mutagenic DNA damage. Thus, the action of ROS can exert widespread effects in the cell, both directly and indirectly.

B. Chemical and Physical Agents

In addition to these normal metabolic sources, environmental compounds can divert single electrons to generate oxygen radicals. These redox-cycling compounds undergo enzymatic reduction and are then reoxidized by O_2, a cyclic process that generates a flux of superoxide catalytically (Fig. 1). The variety of these superoxide-generating compounds is large and includes many types of quinones, naphthoquinones, and nitroquinolones, which can act as efficient sources of oxidative stress. Several physical agents can impose oxidative stress; ionizing radiation produces ROS by radiolysis of water, whereas ultraviolet light produces H_2O_2 through photochemical reactions involving various chromophores, including the amino acid tryptophan.

C. Photosynthesis

Aerobic metabolism is not the only source of oxidative stress. If photosynthetic cells are exposed to light in excess of their synthetic capabilities, light-harvesting antennae can transfer excitation energy to ground-state oxygen and yield singlet oxygen, another highly reactive species (though not actually a free radical). In addition, light-driven electron transport systems may divert electrons to oxygen instead of $NADP^+$, resulting in the same reactive derivatives as found for aerobic respiration. This light-dependent production of oxygen derivatives is termed photooxidative stress.

D. Nitric Oxide

During bacterial denitrification, the aerobic conversion of nitrate to N_2, nitric oxide ($NO^\cdot$) is formed as a product of nitrite reduction. The accumulation of this toxic intermediate is minimized by the catalytic conversion of $NO^\cdot$ into nitrous oxide (N_2O) by the enzyme nitric oxide reductase. $NO^\cdot$ is also a key cytotoxic weapon of the mammalian (and other) immune systems, and it is produced by many nonimmune cell types during inflammatory responses. Moreover, NO· reacts very rapidly with superoxide to generate another unstable and even more reactive compound, peroxynitrite (ONOO⁻). Thus, the ultimate effects of one reactive species may be entwined with the effects of another.

The bacteriostatic and bactericidal effects of ROS have been exploited by macrophages and other phagocytic cells to attack microbial infections. After endocytosis of bacteria, macrophages and neutrophils are activated by the bacterial lipopolysaccharide coat to produce copious amounts of superoxide and H_2O_2. Macrophages and many other cell types produce NO· during inflammatory responses.

II. PHYSIOLOGY OF OXIDATIVE STRESS

The diverse sources of oxidative stress have probably modeled the evolution of adaptations to aerobic life. Consistent with this idea, aerobic organisms display both constitutive and inducible defenses against ROS. Several types of small molecules in the cytoplasm have clear roles in scavenging ROS and aiding the repair of potential damage.

A. Glutathione

Glutathione (γ-L-glutamyl-L-cysteinylglycine; GSH) is present in bacteria, fungi, animals and plants as the major low-molecular-weight thiol, typically at millimolar concentrations. GSH acts as a chemical scavenger of radicals such as ·OH, and also as a H-atom donor to restore macromolecules that have been attacked by free radicals. GSH in eukaryotic cells, including yeast, is also a cofactor for the H_2O_2-destroying enzyme GSH peroxidase and can be enzymatically conjugated to oxidative products to mark them for disposal from the cell.

B. NAD(P)H

Reduced GSH is maintained by the enzyme GSH reductase, which reduces the oxidized form (GSSG). In organisms where GSH is not present, usually a related peptide thiol is present, together with the respective thiol reductase. These critical reactions depend on the reducing power of NADPH, which is equilibrated in the cell with NADH (nicotinamide adenine dinucleotide). NADPH also supports other antioxidant enzymes, such as alkyl hydroperoxide reductase. Ironically, NAD(P)H is involved in redox-cycling reactions that generate high fluxes of superoxide in the cell (Fig. 1).

C. Thioredoxins and Glutaredoxins

A family of small proteins, the thioredoxins and glutaredoxins, act as efficient thiol donors to many cellular proteins. Thioredoxins and glutaredoxins have a pair of conserved cysteine residues that become oxidized to a cystine disulfide as other proteins are reduced. Reduced thioredoxin is in turn regenerated by a reductase using NADPH; reduced glutaredoxin is regenerated using GSH as a reducing donor.

D. Superoxide Dismutase

The conversion of ROS into less dangerous products is catalyzed by several enzymes that are almost universal among aerobic organisms, which reflects the universality of the biochemistry of oxidative stress. Superoxide dismutase catalyzes the dismutation of superoxide into hydrogen peroxide and oxygen and plays a central role in the protection of aerobic organisms against ROS. All aerobic organisms have at least one form of SOD. *E. coli* has three isozymes, encoded by the *sodA*, *sodB*, and *sodC* genes. The products of the *sodA* and *sodB* genes are cytoplasmic, whereas the *sodC* product is periplasmic. The three SODs of *E. coli* differ in the metals at their active sites; the SodA protein contains manganese, SodB iron, and SodC copper and zinc. Bacterial strains lacking both of the cytoplasmic SOD enzymes suffer DNA damage during aerobic growth, which results in an elevated mutation rate. The aerobic growth of SOD-deficient *E. coli* in rich media is only slightly impaired, but in minimal media growth it is abolished unless amino acids, particularly the branched-chain types, are provided. This conditional multiple auxotrophy, which is not observed during anaerobic growth, is probably due to oxidative inactivation of biosynthetic enzymes that contain iron–sulfur centers.

E. Catalases and Peroxidases

Catalases and peroxidases are heme-containing enzymes that eliminate H_2O_2 by related mechanisms. Catalases redistribute electrons among H_2O_2 molecules by alternate two-electron oxidation and reduction, which generates oxygen and water. Peroxidases oxidize an organic compound while generating H_2O. *E. coli* has two catalases—HPI, encoded by the *katG* gene and predominant during exponential growth; and HPII, encoded by *katE* and present mainly in stationary phase.

III. BIOCHEMICAL BASIS OF THE RESPONSES

Many organisms exhibit adaptive responses to oxidative stress; that is, the exposure of the cells to a sublethal level of oxidative stress enhances the resistance to subsequent, higher levels of oxidative stress. These adaptive responses depend on protein synthesis, and different sets of proteins may be synthesized upon exposure to different types of oxidative agents. The induction of some key proteins in response to oxidative stress has been known for some time. For example, SOD activity increases in *E. coli* that is grown in high levels of oxygen or exposed to the redox-cycling agent paraquat. Paraquat also induces glucose-6-phosphate dehydrogenase, evidently to replenish NADPH used up in antioxidant reactions (e.g., by GSH reductase). Exposure to H_2O_2 induces catalase activity in many organisms and increases GSH reductase levels in *E. coli* and *S. typhimurium*. We now know that these inductions reflect the activation of coregulated groups of genes affecting the expression of many additional proteins.

The number of oxidative stress-inducible proteins described ranges from one in *Mycobacterium bovis*, to at least 80 in *E. coli*. The number, identity, and degree of activation of the proteins induced by oxidative stress has been studied by two-dimensional gel electrophoresis. In this technique, radioactively labeled proteins are resolved in two dimensions, first by isoelectrofocusing and then by denaturing gel electrophoresis. Two-dimensional gel analysis allows the comparison of the overall pattern of proteins synthesized during various growth conditions, and in various genetic backgrounds. Thus, this type of analysis has been applied to mutant strains that either overexpress or fail to induce sets of oxidative stress proteins, which has led to the isolation of genes coding for global regulators. Note, however, that some inductions can be overlooked by focusing solely on two-dimensional gel analysis, and the number established by this method should be taken only as a lower limit. The complexity of some responses is truly daunting; for example, the number of oxidative stress proteins in *E. coli* is probably more than 100 if evidence from other methods is combined with the two-dimensional gel approach.

IV. GENETIC BASIS OF THE RESPONSES

Bacterial gene function is regulated mainly at the transcriptional level. Cells have evolved transcriptional modulators that sense oxidative stress and activate genes whose products avert or repair the damage caused by ROS. These transcriptional modulators usually activate multiple and unlinked promoters, which as coregulated groups constitute regulons. In most cases, one or more activated genes decrease the activating stimulus in a type of negative feedback, which often makes the responses self-regulating (for reviews, see Hidalgo and Demple, 1996; Jamieson and Storz, 1998).

A. OxyR

OxyR is a redox-sensitive protein and a member of the LysR family of DNA-binding transcriptional modulators. OxyR is a homotetramer in solution and it activates as many as 10 genes in response to an increase in the intracellular concentration of H_2O_2 (Fig. 2). The protein exists in two forms, reduced and oxidized, and exhibits redox-regulated DNA binding. Only oxidized OxyR binds tightly to the promoters of its target genes and activates transcription, evidently through contacts with the α-subunit of RNA polymerase. Reduced OxyR binds only to the *oxyR* promoter itself, and thereby limits the expression of this regulatory protein. Active oxidized OxyR has an intramolecular disulfide bond between cysteines 199

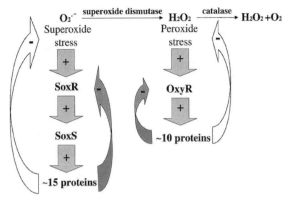

Fig. 2. Regulation and homeostasis of the responses to oxidative stress in *Escherichia coli*. A rise in the intracellular concentration of superoxide ($O_2^{.-}$) activates the redox-sensing protein SoxR, which controls the induction of the protein SoxS. SoxS enhances the transcription rate of at least 15 genes, including *sodA,* and this concerted response alleviates the superoxide stress through proteins that scavenge superoxide and repair oxidative damage. Superoxide is reduced to hydrogen peroxide (H_2O_2) either spontaneously or enzymatically by SOD. A rise in the concentration of hydrogen peroxide activates the protein OxyR, which in turn enhances the transcription rate of as many as 10 genes, including *katG*, and this response alleviates the hydrogen peroxide stress. Both the SoxRS and the OxyR responses are homeostatic; the induction of the effector proteins results in a decrease of the stress signal, and therefore a reduction of the response (open arrows). The induced proteins may directly switch off the regulators (solid arrows), as glutaredoxin and GSH do for OxyR.

and 208, which is evidently formed by direct reaction with H_2O_2 or another oxidant. Reduced inactive OxyR is maintained by thiol-disulfide exchange proteins, such as glutaredoxin, which is dependent on GSH. OxyR activates the transcription of *katG* (HPI catalase), *ahpFC* (alkyl hydroperoxide reductase), *dps* (general DNA-binding protein that may exclude Fe), *gorA* (GSH), *grxA* (glutaredoxin-A), and *oxyS* (small untranslated RNA that may regulate other genes posttranscriptionally). The induction of *gor* and *grxA* may ultimately down-regulate the response by regenerating reduced OxyR.

B. SoxR and SoxS

The SoxR protein is a redox-sensing transcriptional activator that belongs to the MerR family of DNA-binding proteins. SoxR is a homodimer of 17-kDa subunits, each one containing a redox-active [2Fe-2S] center. The DNA-binding activity of SoxR does not depend on the iron–sulfur clusters, but only SoxR with oxidized [2Fe-2S]centers activates transcription. The activation of SoxR through the one-electron oxidation of its [2Fe-2S] centers therefore corresponds to an allosteric transition in the promoter DNA–SoxR complex. This transition does not increase binding by RNA polymerase, but rather stimulates the formation of the "open" complex essential for initiating transcription. It is not known what cellular activities operate to maintain the [2Fe-2S] centers of SoxR in the reduced state. After the exposure of bacteria to superoxide-generating agents, such as the redox-cycling compound PQ (Fig. 1), or to NO (either directly or following phagocytosis by murine macrophages), oxidized SoxR activates the transcription of *soxS*, a gene encoding a second transcriptional modulator (Fig. 2). The SoxS protein, a 13-kDa monomer homologous to the C-terminal portion of AraC protein, binds to specific sequences in 15 or more promoters and recruits RNA polymerase to stimulate their transcription. The SoxS-regulated genes in *E. coli* include *sodA* (Mn-containing SOD), *zwf* (glucose-6-phosphate dehydrogenase), *micF* (an antisense RNA that inhibits translation of the porin OmpF), *nfo* (endonuclease IV, a repair enzyme for oxidative DNA damage), *fpr* (ferredoxin reductase), *acrAB* (cellular efflux pumps), and *fumC* (redox-resistant fumarase). However, SoxS in other organisms may induce different genes: activation of *S. typhimurium soxRS* induces Mn-SOD and endonuclease IV, but not fumarase-C or glucose-6-phosphate dehydrogenase. Like many other regulatory proteins, both SoxS and SoxR limit their own expression by binding the promoters of their own structural genes. So far, SoxR and SoxS have been identified only in Enterobacteria.

C. Sigma S

The alternative sigma factor, sigma S (σ^s), encoded by the *rpoS* gene, is expressed during starvation or in the stationary phase in a variety of bacterial species. RNA polymerase containing σ^s activates the expression of several genes that counteract oxidative stress, and bacteria that are starved or in stationary phase are more resistant to oxidative stress than are cells growing exponentially. The σ^s-activated genes include *katE* (catalase-hydroperoxidase HPII), *dps* (protective DNA-binding protein), *xthA* (exonuclease III, another DNA-repair enzyme for oxidative damage), and *gorA* (glutathione reductase). The regulation of σ^s activity is poorly understood and seems to be under complex transcriptional, translational, and proteolytic control.

D. Fur

Antioxidant defense genes are also controlled by proteins of the Fur family. *E. coli* Fur protein is a repressor governing a system of genes involved in iron uptake, and the system is switched on when bacteria are grown in media with limiting Fe concentrations. It is unclear whether Fur senses iron levels directly or through some other signal. Fur in *E. coli* also regulates the *sodA* gene and thus Mn–SOD activity. More complex regulation involving Fur-related proteins has been described in *Bacillus subtilis,* in which the PerP protein regulates genes encoding a catalase, an alkyl hydroperoxide reductase, and a Dps-like protein. Although this collection of gene products is reminiscent of the *E. coli oxyR* system, the regulatory mechanism is quite different. PerP is a repressor, possibly with a metal corepressor, so that gene induction depends on lowered DNA binding by the regulator (i.e., derepression rather than positive control). Mutant strains lacking PerP have increased resistance to H_2O_2.

V. BIOLOGICAL IMPLICATIONS

Genetic responses to oxidative stress are widespread in microbes, although the molecular mechanisms of stress signal transduction are known in only a few cases. These responses mobilize diverse biochemical activities that operate to exclude environmental toxins, repair oxidative damage, eliminate ROS, and diminish radical production. These actions can collectively alleviate the stress, even if elevated radical production continues, and are thus homeostatic. The continuous variation of OxyR activation during the normal aerobic growth of *E. coli* provides an especially clear example.

The sensors of oxidative stress take advantage of reactions that are deleterious for most other proteins. The formation of a disulfide bond in a cytoplasmic protein is often inactivating, but this same modification is the signal that OxyR has evolved to use for its activation. Similarly, some iron–sulfur centers are damaged by oxidation, but SoxR exploits this chemistry as a signal transduction device connecting oxidation to gene activation.

For both the *oxyR* and the *soxRS* regulons, several regulon genes are involved in multiple regulatory pathways. The *sodA* gene is an excellent example, as it is controlled by at least five different regulatory systems in addition to *soxRS*. The *katG* and *dps* genes are both regulated jointly by *oxyR* and *rpoS*. Whether this theme will reemerge in other species awaits additional experimentation.

Microbial responses to oxidative stress are by no means limited to bacteria. The yeast *Saccharomyces cerevisiae* has received particular attention. Physiological and genetic analysis shows that distinct regulatory systems govern adaptive resistance to H_2O_2 compared to redox-cycling (superoxide-generating compounds). The response to H_2O_2 involves a large number of inducible proteins, including some of the same defense functions mentioned previously: catalase, SOD, glucose-6-phosphate dehydrogenase, and GSH reductase (Godon *et al.,* 1998). The modulated expression of metabolic enzymes in H_2O_2-treated yeast is proposed to remodel metabolism to increase the regeneration of NADPH at the expense of glycolysis. The expression of a portion of the H_2O_2-inducible proteins in yeast (e.g., GSH) is controlled by YAP1, a yeast homolog of the c-Jun—c-Fos transcription activators of mammalian cells. Although YAP1 itself does not seem to be the redox-sensing component of this system, the signaling mechanism that activates this transcription factor in response to oxidative stress is unknown.

What of the evolutionary pressures that have shaped the oxidative-stress responses in bacteria? Clearly, the colonization of the aerobic world required antioxidant defenses, and SOD and catalase appear to have ancient origins. One could suppose that the ability to regulate these defenses was especially useful for organisms exposed to changing levels of oxygen, as in the facultative aerobic lifestyle of *E. coli* and *S. typhimurium*. The chemical warfare conducted among organisms may also have selected for inducible genetic systems, and for the inclusion of certain other genes within the coregulated groups. The driving force for such inducible resistance is not necessarily restricted to the mammalian immune system; plants also employ H_2O_2 and NO to mediate systemic immunity (Dangl, 1998). The diversity of microbial interactions might well have elicited diverse systems for coping with oxidative stress.

See Also the Following Articles

AEROBIC RESPIRATION • OSMOTIC STRESS • pH STRESS

Bibliography

Dangl, J. (1998). Plants just say NO to pathogens. *Nature* 394, 525–526.

Godon, C. *et al.* (1998). The H_2O_2 Stimulon in *Saccharomyces cerevisiae. J. Biol. Chem.* 273, 22480–22489.

Hidalgo, E., and Demple, B. (1996). Adaptive responses to oxidative stress: The *soxRS* and *oxyR* regulons. *In* "Regulation of Gene Expression in *Escherichia coli*" (E. C. Lin and A. Simon Lynch, eds.), pp. 435–452. R. G. Landes, Austin, TX.

Jamieson D. J., and Storz, G. (1997). Transcriptional regulators of oxidative stress responses. *In* "Oxidative Stress and the Molecular Biology of Antioxidant Defenses" (John G. Scandalios, ed.) pp. 91–116. Cold Spring Harbor Laboratory Press, New York.

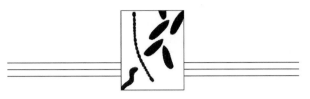

Paramyxoviruses

Suxiang Tong, Qizhi Yao, and Richard W. Compans
Emory University

I. Taxonomy
II. Structure of the Virion
III. Organization of the Genome
IV. Proteins
V. Life Cycle
VI. Diseases
VII. Control

GLOSSARY

hemagglutinin A transmembrane glycoprotein on the virus surface that binds to sialic acid residues on target cells. It also promotes the agglutination of red blood cells by cross-linking the red cells.

negative-stranded RNA virus A virus with an RNA genome that is complementary to the mRNA (positive) strand that directs the synthesis of virus-encoded proteins.

neuraminidase An enzymatic activity of the hemagglutinin protein that cleaves sialic acid residues from cell and virus surfaces.

parainfluenza virus (PIV) An important respiratory-tract pathogen of young children, also responsible for many acute respiratory tract infections in older children and adults.

respiratory syncytial virus (RSV) The most important cause of viral lower-respiratory-tract infections in infants and children worldwide. Related viruses are found in sheep and cattle.

rinderpest virus A morbillivirus that causes severe gastroenteritis in cattle, with inflammation and necrosis.

sendai virus (SN) A parainfluenza virus that is indigenous to mice.

simian virus 5 (SV5) A virus related to human parainfluenza viruses, which was originally isolated as a contaminant of monkey kidney cell cultures. It is a natural cause of respiratory disease in dogs.

THE FAMILY PARAMYXOVIRIDAE consists of a number of enveloped negative-stranded RNA viruses. They are responsible for a number of serious diseases, including mumps, measles, croup, and pneumonia in humans; and Newcastle disease, distemper, and rinderpest in animals. The paramyxoviruses are closely related to two other families of negative-strand RNA viruses, the *Orthomyxoviridae* (whose envelope glycoproteins have similar biological functions) and the *Rhabdoviridae* (whose nonsegmented genome has a similar organization). These viruses have been identified only in vertebrates, and almost exclusively in mammals and birds. Infection is often cytolytic, inducing cell fusion, but noncytopathic and persistent infections also are common. The transmission of these viruses appears to be mainly through the aerosol route, and no vectors are known. All members of the family contain two major surface glycoproteins, a receptor-binding protein (hemagglutinin–neuraminidase, HN; hemagglutinin, H; or G protein, G), which facilitates attachment to cells, and a fusion protein (F), which enables the penetration of the viral genome into the host cell. The genomes are contained in helical nucleocapsids that also have associated proteins that possess RNA transcriptase activity.

I. TAXONOMY

The best-studied members of the family *Paramyxoviridae* are listed in Table I. The family contains two subfamilies, the *Paramyxovirinae* and the *Pneumovirinae*. The distinguishing features of members of the *Paramyxovirinae* are that their genome encodes 6–7 transcriptional elements, in contrast to the 10 tran-

Encyclopedia of Microbiology, Volume 3
SECOND EDITION

533

TABLE I
Taxonomic Structure of the Paramyxoviridae

Subfamily	Genus	Members
Paramyxovirinae	Respirovirus	Sendai virus
		Human parainfluenza virus types 1 and 3
		Bovine parainfluenza virus type 3
	Morbillivirus	Measles virus
		Canine distemper virus
		Dolphin distemper virus
		Rinderpest virus
	Rubulavirus	Newcastle disease virus (avian paramyxovirus 1)
		Mumps
		Simian virus 5
		Human parainfluenza virus types 2, 4a, and 4b
Pneumovirinae	Pneumovirus	Human respiratory syncytial virus
		Pneumonia virus of mice
		Bovine respiratory syncytial virus

scriptional elements in the member species of the *Pneumovirinae*. Based on morphology, the organization of the genome, sequence homology, and physical or chemical properties of their encoded proteins, the *Paramyxovirinae* have been divided into three genera, *Respirovirus, Morbillivirus,* and *Rubulavirus*. The genus *Respirovirus* is fairly homogenous and comprises the bovine and human strains of parainfluenza virus type 3 (PIV 3), Sendai virus (SN), and human parainfluenza virus type 1; it is more closely related to the *Morbillivirus* genus, which contains measles virus, rinderpest, dolphin and equine morbillivirus, and canine distemper virus. The genus *Rubulavirus* is more heterogeneous and includes Newcastle disease virus (avian paramyxovirus 1) (NOV), simian virus 5 (SV5), mumps virus, and human parainfluenza types 2 and 4. All species of the genus *Respirovirus* have hemagglutinin and neuraminidase activities, have six transcriptional elements, and encode a C protein in the unedited P mRNA. Viruses of the genus *Morbillivirus* share many features with viruses in the genus *Respirovirus*, such as an identical gene order, number of transcriptional elements, size of intergenic sequences, nucleotide-sequence similarities of the corresponding proteins, and genome structure. However, morbilliviruses lack a neuraminidase activity and some species do not have hemagglutinin activity. Viruses of the third genus, *Rubulavirus*, also have

hemagglutinin and neuraminidase activities. They exhibit low levels of sequence relatedness of their corresponding proteins. Some members contain an extra gene (SH) and none of them have a C protein open reading frame (ORF).

The subfamily *Pneumovirinae* contains only the single genus *Pneumovirus*. The pneumoviruses are more distant relatives than other paramyxoviruses and are particularly noteworthy for the possession of 10 nonoverlapping genes, smaller average gene size, a narrower nucleocapsid, one additional nonglycosylated membrane-associated protein (M2), and extensive O-linked glycosylation of the attachment (G) protein. The G protein is unrelated in sequence to the HN or H proteins of the other genera. Pneumoviruses also lack neuraminidase, and hemagglutinin is only present in pneumonia virus of mice (PVM), but is absent in bovine or human respiratory syncytial viruses.

II. STRUCTURE OF THE VIRION

The virion structure is illustrated schematically in Fig. 1. Virions are often spherical, although filamentous forms also occur and marked pleomorphism is observed after negative staining. The spherical virions have a diameter of 150–300 nm, and filamentous

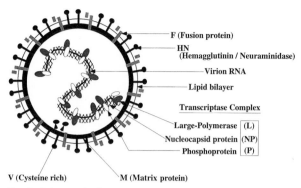

F (Fusion protein)
HN (Hemagglutinin / Neuraminidase)
Virion RNA
Lipid bilayer
Transcriptase Complex
Large-Polymerase (L)
Nucleocapsid protein (NP)
Phosphoprotein (P)
V (Cysteine rich)
M (Matrix protein)

Fig. 1. Schematic diagram of the arrangement of the structural components of a paramyxovirus.

particles are approximately 100–120 nm in diameter and up to 10 μm in length. The purified virions have a composition of 70–80% protein, 15–20% lipid, 5–7% carbohydrate, and 1% RNA. A coiled helical nucleocapsid composed of protein and the single-stranded RNA genome is enclosed within the envelope. The envelope contains a lipid bilayer derived from the plasma membrane of the infected host cell, and a number of spike-like projections, which are approximately 8–12 nm (*Respirovirus* and *Morbillivirus* genera) or 11–20 nm (*Pneumovirus* genus) long and are spaced at intervals of 6–10 nm. These are composed of two transmembrane glycoproteins. One of these spike proteins (HN, H, or G) is important for virus attachment and the other spike protein (F) consists of two disulfide-linked subunits (F1 and F2) and mediates membrane fusion and hemolytic properties of the virus. The envelope of respiratory syncytial virus (RSV) (*Pneumovirus*) also contains a 22-kDa protein designated M2. The inner layer of the envelope of all paramyxoviruses is formed by a nonglycosylated protein, the M or matrix protein. Evidence has indicated that the M protein binds to both the nucleocapsid and viral envelope, thereby linking the two structures within viral particles, and it is believed to be responsible for maintaining structural integrity and to be involved in virus maturation. Because the viral envelope has a lipid bilayer, organic solvents or detergents rapidly inactivate the virions by disrupting their envelopes.

Within the envelope, the RNA is encapsidated by 2200–2600 nucleoprotein (NP) protein molecules. Together, these two components form a coiled flexi-

ble left-handed helical nucleocapsid, with a length of around 1 μm. The diameter of the nucleocapsid in respiroviruses and morbilliviruses is about 18 nm, whereas that of RSV and PVM in the *Pneumovirus* genus is smaller (12–15 nm), with a central hollow core of 4 nm and a helical pitch of approximately 5 nm. The RNA is of a negative polarity, is single-stranded, and consists of approximately 15×10^3 nucleotides with a sedimentation value of around 50S. The phosphoprotein (P) and large (L) protein, which have RNA-dependent RNA polymerase activity, are also closely associated with the nucleocapsid core and are distributed in discrete clusters.

III. ORGANIZATION OF THE GENOME

Figure 2 illustrates diagrammatically the genome structures in the 4 genera of the family *Paramyxoviridae*. The genome consists of a single strand of negative-sense RNA and contains a 3′ extracistronic leader sequence 50–55 nt long, and a 5′ extracistronic trailer sequence 40–55 nt long. By analogy with other nonsegmented negative-stranded RNA viruses, these two regions presumably contain the promoter sequences for transcription and replication. Between the leader and trailer regions are the coding sequences for the structural proteins. The respiroviruses and morbilliviruses exhibit the same basic pattern, possessing six genes. The rubulaviruses possess seven genes, and the pneumoviruses possess ten genes. The genes are linked by intergenic sequences that vary in size and sequence among individual viruses.

There are at least six major genes and proteins common to all *Paramyxoviridae*. These are the nucleocapsid protein gene (NP), phosphoprotein gene (P), matrix protein gene (M), fusion protein gene (F), attachment protein gene (HN, H, or G) and large protein gene (L). Most genes are uniquely monocistronic, using a single start codon and producing a single primary translation product, with exception of the phosphoprotein (P) gene mRNAs, which are polycistronic and encode multiple proteins (P, V, and C proteins) by using multiple start codons. In respiroviruses and morbilliviruses, the P gene encodes P, V, and C, three distinct proteins in overlap-

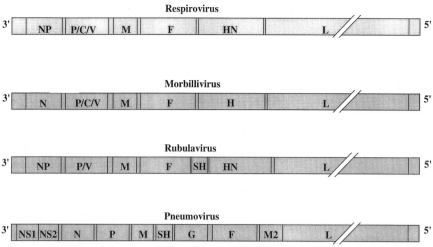

Fig. 2. Genome organization of the four genera of the family *Paramyxoviridae*. Each box represents an individual transcribed mRNA. The lengths of each box and intervening spaces are not to scale. Each letter identifies an ORF and multiple ORFs within a single mRNA are indicated by slashes. The L gene is interrupted by two slashes.

ping reading frames. Rubulaviruses lack a C protein but their P genes encode P and V proteins in a discontinuous reading frame. Some rubulaviruses have an additional gene coding for a small hydrophobic (SH) integral membrane protein inserted between the F and HN genes. The genes have been ordered in the genome, based upon nucleotide sequence analysis, as 3' N-P(C/V)-M-F-(SH)-HN-L 5'.

Pneumoviruses are unique among the family in having two additional membrane-associated proteins (M2 and SH), as well as two small nonstructural proteins of unknown function, NS1 and NS2, which precede the normally 3'-terminal N-protein gene. The attachment (G) protein and fusion (F) protein genes are inverted in the gene order relative to other members of the family. Further, there is an overlap between the M2 and L transcriptional elements in some pneumoviruses.

IV. PROTEINS

A. NP Protein

The NP protein is the major nucleocapsid protein, ranging from 489–553 amino acids in length (Mr 53,167–57,896). It binds tightly along the genome and antigenome RNAs in the replicative cycle. During viral replication, it encapsidates the genome RNA into an RNAse-resistant nucleocapsid and associates with the P-L proteins to form the transcriptase complex. Additional soluble NP proteins, which bind to the nascent RNA, are required in RNA replication. NPs may also interact with the M protein during virus assembly.

B. P Gene-Encoded Proteins

The P protein is a phosphorylated nucleocapsid-associated protein. For respiroviruses and morbilliviruses, P proteins are 507–603 amino acids; and for rubulaviruses, 245–397 amino acids. The smallest are 241 amino acids for pneumoviruses. A complex of P and L is the viral transcriptase, and P-NP⁰ (soluble NP) is probably the active form in RNA encapsidation. The carboxy terminal domain of the P protein contains sites for binding to the L protein and nucleocapsids and is essential for transcription.

The C proteins are expressed from an ORF that overlaps the N-terminal portion of the P gene. They are present in the respirovirus and morbillivirus genera but not in rubulaviruses. Recent studies with Sendai virus indicate that this protein appears to be required for multiple cycles of replication in mice and may affect viral genome replication.

The V proteins are products of mRNA produced by RNA editing from the P gene, and occur downstream of the C-protein ORF. The V and P proteins are amino coterminal for 164 residues in the SV5 virus. The carboxyl terminus of the V protein contains seven cysteine residues and resembles a zinc finger. This protein may function during virus replication.

C. L Protein

The L protein is a large nucleocapsid-associated protein, approximately 2200 amino acids in length. It is the least abundant protein in the virion, and forms a complex with P that possesses polymerase activity with NP:RNA templates. The amino-terminal half of the L protein is responsible for binding with P protein. The L subunit is probably responsible for cotranscriptional modifications such as capping, methylation, and polyadenylation.

D. M Protein

The M protein is a nonglycosylated matrix protein located between the nucleocapsid and the envelope and is the most abundant protein in the virion. It contains 341–375 amino acid residues (Mr approximately 38,500 to 41,500). The M protein interacts with nucleocapsids as well as the F and HN cytoplasmic tails, and thus it is considered to be the central organizer of viral morphogenesis.

E. Attachment Protein

1. HN Protein

The HN protein is the viral attachment protein in the *Respirovirus* and *Rubulavirus* genera. It is a type II integral membrane protein with an amino terminus located inside the virus and a carboxy terminal ectodomain. It has 565–582 amino acids and a molecular weight of 69–72 kDa. There are four to six predicted N-linked glycosylation sites. The structurally important glycine residues, cysteine residues, and hydrophobic regions tend to be highly conserved. A conserved sequence, NRKSCS, has been implicated in neuraminidase activity. Many paramyxovirus HN proteins form disulfide-linked oligomers, such as a tetramers formed by SN HN proteins. The structure of the ectodomain of the HN spike has not been solved, but electron micrographs suggest that it consists of a large terminal globular domain, or "head," supported by a stalk extending out from the membrane. A schematic diagram of HN is shown in Fig. 3a.

HN is the dominant antigen of respiroviruses and rubulaviruses. It is responsible for binding to sialic acid residues, which serve as viral receptors on the target cells, and for agglutination of erythrocytes. It also possesses a neuraminidase, which is involved in cleavage of sialic acid from the surfaces of virions and infected cells, and is thought to be important in the facilitation of the release of viral particles during and after budding at the cell surface. Recent studies revealed that HN also has fusion-promotion activity, and that a type-specific interaction of the two surface glycoproteins, HN and F, is required for cell fusion induced by most paramyxovirus.

2. H Protein

The attachment protein of morbilliviruses is also a type II protein and shares some sequence homology with the HN of other paramyxoviruses. It contains 617 amino acids, with a molecular mass of about 78 kDa. This protein interacts with the measles virus cellular receptor CD46 and is a hemagglutinating protein, but it does not exhibit neuraminidase activity. H contains 13 highly conserved cysteines and forms a disulfide-liked homodimer on the surface of the virion the infected cell. A schematic diagram of H is shown in Fig. 3b.

3. G Protein

The *Pneumovirus* attachment protein is quite different from those of other *Paramyxoviridae*, but it is also a type II protein. The RSV G protein has neither hemagglutinating nor neuraminidase activity. The cellular receptor for this virus is unknown. In RSV, the nucleotide sequence of the G protein is only about 298 amino acids, which is about half the size of HN or H, and the predicted molecular weight is only about 33 kDa. Another major difference is the extensive carbohydrate modification of this protein.

a. HN Protein.

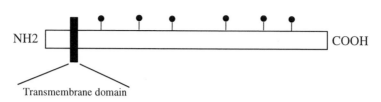

Transmembrane domain

b. H Protein.

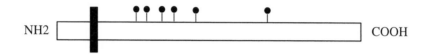

c. G Protein.

d. F Protein.

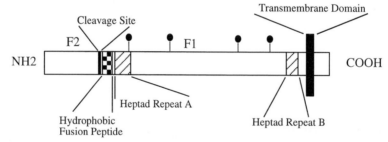

Fig. 3. Schematic diagram of paramyxovirus glycoproteins. (a) HN protein, (b) H protein, (c) G protein, and (d) F protein. ⸸ represents N-linked glycosylation sites, which are found in various positions on glycoproteins of different paramyxoviruses. | represents O-linked glycosylation sites found in pneumoviruses.

It not only has N-linked sugars (about four potential sites), but also extensive O-linked glycosylation (over 70 potential acceptor serine or threonine residues). The positions at which the O-linked sugars are attached have not been mapped. Because of the extensive glycosylation of the G protein, its molecular weight is 84 to 90 kDa. The G protein also has a high proline content. The structural properties of the G protein are somewhat similar to the mucins, which are highly O-glycosylated proteins produced by cells of mucosal epithelia. A schematic diagram of G is shown in Fig. 3c.

F. F Protein

The F protein is a type I glycoprotein with a cleaved N-terminal signal peptide. It contains 540–580 amino acid residues and has an estimated molecular weight of 60 kDa. There are three to six potential N-linked glycosylation sites, and cysteine residues are found to be conserved in all F_1 sequences. Studies of SV5, NDV, and PIV3 F proteins have shown that paramyxovirus F proteins exist as a trimer. A schematic diagram of F is shown in Fig. 3d.

The F protein is synthesized as an inactive precursor (F_0), which is cleaved by a protease to form biologically active F_1–F_2 subunits, which are linked by disulfide bonds. The cleavage results in the exposure of a new hydrophobic N-terminus on the F_1 subunit, which is believed to participate directly in the fusion process and has therefore been termed the "fusion peptide." The cleavage of F_0 is a key determinant of the viral host range, tissue tropism, and pathogenicity. Sequences upstream of the cleavage activation site contain a stretch of basic amino acid residues that varies from one virus to another, from one residue in the F protein of SN or PIV-1 to five in SV5. For F proteins with multibasic residues at the cleavage site, cleavage occurs intracellularly; in contrast, F protein having a single basic residue at the cleavage site are cleaved by an exogenous protease. The localization and intensity of paramyxovirus infections in the respiratory tract are at least partially determined by the susceptibility of the F protein to cleavage by host-cell proteases. Although the fusion peptide in the F protein is considered to be directly involved in the fusion process, other regions, including heptad repeats, may also play important roles in membrane fusion. There are two heptad repeats, A, which is adjacent to the fusion peptide; and B, which is adjacent to the transmembrane domain, which together form a triple-stranded coil of three α-helices.

G. Nonstructural Proteins, NS1 and NS2

NS1 and NS2 are only found in the *Pneumoviruses*. They are encoded by separate mRNAs, unlike the C or V proteins in the *Respiroviruses*. Only small amounts of NS1 and NS2 can be detected in purified virions. The functions of these proteins are unknown. They may be involved in regulating RNA synthesis or morphogenesis of the virion, or possibly they have some functions in interacting with the host immune system.

H. SH Protein

The SH gene is located between F and HN in rubulaviruses (SV5 and mumps virus) and between M and G in pneumoviruses (RSV), respectively. SH is a type II integral membrane protein whose function is unknown.

V. LIFE CYCLE

The viral infectious cycle begins with attachment to receptors, followed by the viral nucleocapsid entering cells by virus–cell membrane fusion, and then by mRNA transcription and translation, RNA replication, and virus-particle assembly and release from the infected cells. The life cycles do not appear to differ significantly for each paramyxovirus, but the kinetics of replication and the lengths of the phases are highly variable. For example, the eclipse period is 3–5 hr for parainfluenza viruses, 8–10 hr for Newcastle disease virus, 16–18 hr for mumps virus, and 9–12 hr for measles viruses. Other viruses such as SV5 have a prolonged growth cycle and cause virtually no cytopathic effects in some cells. The replication process takes place entirely in the cytoplasm. Fig. 4 shows a schematic diagram of the events in the replication cycle of a paramyxovirus.

A. Adsorption

The first step in the life cycle is the adsorption of the virus particle to the surface of a susceptible cell. It is begun by a random collision between the virus particle and the cell surface, and then involves an interaction between specific sites on the viral surface and specific cellular receptors. These specific sites reside on the attachment glycoproteins on the surface of the virion. Sialic acid-containing molecules (glycoprotein or lipid) found on cell surface mainly func-

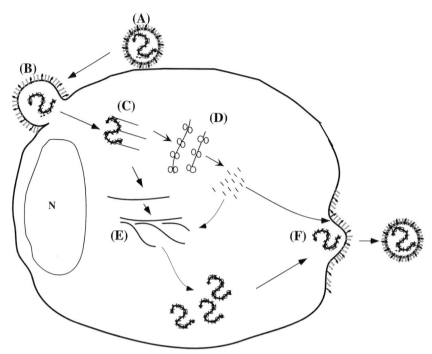

Fig. 4. Diagram of the life cycle of a paramyxovirus. A, adsorption to receptors; B, fusion of the virus to the plasma membrane and uncoating; C, transcription; D, translation; E, genome replication; F, virion assembly and release. All of the intracellular steps take place in the cytoplasm. N, nucleus.

tion as receptors for respiroviruses and rubulaviruses. Another identified receptor is the cell-surface protein CD46, which serves as a receptor for measles virus. Cells lacking CD46 receptors are resistant to measles virus infection, but become susceptible after transfection with the gene encoding CD46. The cellular receptor for the *Pneumoviruses* is not known.

B. Penetration and Uncoating

After the binding of the virus to one or more cellular receptors on the cell surface, the next step is penetration and uncoating. This is the result of the fusion of the virus envelope with a cellular membrane with entry of only the nucleocapsid into cytoplasm of the cell. Fusion with the cellular plasma membrane by paramyxoviruses is known to occur in a pH-independent manner (usually at neutral pH), unlike influenza viruses, which undergo fusion at low pH. A specific viral envelope protein (fusion protein, F) is involved in the induction of membrane fusion. Although under certain conditions F alone of some

viruses can cause cell fusion, most paramyxoviruses require the coexpression of both F and HN (or H or G) for fusion to occur. The F protein is synthesized as an uncleaved precursor F_0, and for membrane fusion and penetration to occur, it must be cleaved by a cellular proteolytic enzyme. The cleaved F protein (consisting of F_1 and F_2 subunits) interacts with its homotypic HN and act to fuse the viral membrane with the host cell plasma membrane. The details of the molecular interactions between F and HN are being actively investigated. As a consequence of such fusion, the nucleocapsid is delivered directly into the cytoplasm where it begins to direct RNA and protein synthesis.

C. Transcription

Following virus entry, primary transcription of the negative-stranded genome into the complementary positive-stranded mRNAs is the first event in virus replication. Most knowledge of primary transcription has resulted from *in vitro* systems. The viral nucleocapsids contain an RNA polymerase that is required for early mRNA synthesis and processing. All viruses in the family possess a single promoter site adjacent to the 3′-terminus and the RNA polymerase transcribes the genome in a linear, sequential, stop–start manner into 6–10 separate, 5′ capped, 3′ polyadenylated subgenomic, viral-complementary mRNAs. Between the gene boundaries are intergenic sequences that contain regulatory signals for the stop and start of transcription, and are not copied in the mRNAs. These positive strands function as messenger RNA species that direct the synthesis of the individual viral polypeptides. The viral genome encodes virion proteins involved in mRNA transcription and RNA replication, as well as components of virus particles, and these are assembled into progeny nucleocapsids, which amplify the level of transcription during infection. Evidence shows a progressive attenuation of transcription with increasing distance from the promoter, so that the 3′-proximal NP, N, or NS1 and NS2 mRNA is the most abundant species of viral RNA in the infected cell, and the 5′-proximal L-protein mRNA is the least abundant. Two other mechanisms of regulation of gene expression at the transcriptional level are also used by paramyxoviruses. The first involves reading through some junctions at a higher frequency, producing bicistronic mRNAs, from which only the upstream ORF is translated. For example, measles uses this mechanism to down-regulate M gene expression during long-term persistent infections of the human brain. The second mechanism is by genes encoding overlapping mRNAs, such as in the RSV L gene, for which transcripts begin within the preceding M2 gene, which ends within the L gene. In this case a majority of transcripts for the L mRNA are terminated.

The mRNAs produced are generally monocistronic, except for the transcripts of the P gene. The majority of the nonstructural proteins (such as C and V proteins) of paramyxoviruses are derived from the viral P gene. A feature of most paramyxoviruses is the use of alternative reading frames and RNA editing within their P genes, resulting in the increased coding capacity of the genome. C proteins found in Sendai virus, HPIV3, measles virus, and canine distemper virus are the products of ORFs that are initiated in the +1 reading frame relative to the start of the P ORF. Another unique method for generation of nonstructural proteins is the cotranscriptional editing of P gene mRNA transcripts by inserting a single nontemplated G residues at a specific site within the P gene (such as in Sendai virus and measles virus) or by inserting two non-templated G residues within the P gene (such as in SV5 and mumps virus.) The consequence of this RNA editing is the use of alternative reading frames during translation. The resulting different proteins are amino coterminal, but divergent at the C termini. In the case of Sendai and measles viruses, the edited (+1) transcript encodes the V protein, whereas for SV5 and mumps virus the unaltered mRNA codes for the V protein, and the edited (+2) mRNA encodes the P protein. It is believed that insertion of nontemplated nucleotides during RNA editing occurs by the polymerase complex "stuttering" on a region of template and reiteratively reading a template base.

D. Translation

All of the mRNAs are translated into viral polypeptides in the cytoplasm of the infected cell. The nonglycosylated proteins are synthesized on free cytoplasmic polyribosomes. The viral glycoproteins are translocated into the lumen of the rough endoplasmic reticulum where they are glycosylated by the normal cellular machinery, and are transported through the Golgi complex where further processing of the oligosaccharides occurs. They are then transported to the cell surface by a vesicular transport process.

E. Genome Replication

The first stage of RNA replication requires that the negative strand be copied into a full-length positive strand; and then RNA replication proceeds through the use of these positive-strand replicative intermediates as templates. Although transcription does not require protein synthesis *de novo,* RNA replication is dependent on continued protein synthesis and thus the expression of viral genes is a prerequisite condition for continued replication. Genome RNA replication is carried out via a semiconservative mechanism. During the replication process, it appears that accumulation of the N or NP protein is required in the cytoplasm, which then binds to the leader sequence of the nascent positive-stranded RNA and allows the RNA polymerase to proceed along the template strand. Newly synthesized positive- and negative-stranded RNAs immediately associate with N or NP proteins as well as P and L to form RNase-resistant nucleocapsids. The RNA genome synthesis and encapsidation appear to occur concomitantly and this may enable the RNA polymerase to ignore the intergenic signals, which result in the termination of transcription.

F. Virion Assembly and Release

Paramyxovirus particles are assembled and released simultaneously by budding at the plasma membrane. Virion assembly can be divided into two parts, the assembly of the nucleocapsid, and the envelopment of the nucleocapsid. Nucleocapsid formation occurs in the cytoplasm and is thought to be a spontaneously occurring self-assembly process. The nucleocapsids appear to be assembled in two steps. First free NP subunits associate with the RNA to form the helical RNP structure. Then, the P-L protein complex associates. It appears that the nucleic acid–protein and protein–protein interaction are concerned in nucleocapsid formation. By analogy with rhabdoviruses, it is believed that the initial interaction between NP and viral RNA in the formation of helical nucleocapsids occurs at a site near the 5′-terminus of the viral RNA. The P protein interacts with the L protein, forming the viral transcriptase complex. The assembly of viruses that contain membranes requires the modification of a cell membrane by the insertion of virus-specific transmembrane glycoproteins. The nucleocapsids interact with membranes that contain the glycoproteins, which is believed to be mediated by the M Protein lining these regions of cell membranes. The binding of nucleocapsids may provide the stimulus for the budding of progeny virions from these differentiated cell membranes where the viral envelope glycoproteins are targeted. The virus detaches from the membrane by a process that can be considered the reverse of viral entry. Viruses that bind to sialic acid residues also possess a neuraminidase activity in their HN protein,

which serves to cleave sialic acid from the viral glycoproteins as well as from the cell surface, and is thus likely to be involved in the release of the virus from the cell. Activation of the fusion protein by proteolytic cleavage is also required to generate infectious virus particles. Also, the activated F protein can cause fusion with nearby cell surfaces, leading to the formation of large multinucleated syncytia.

VI. DISEASES

All *Paramyxoviridae* are highly contagious. Very efficient infection occurs through aerosolized virus. The virus enters the body by infecting cells on one of the mucosal surfaces, usually nasal or conjunctival, and after a relatively short incubation period local symptoms are induced through cellular destruction. Some viruses (parainfluenza and RSV) cause localized respiratory infections, whereas mumps and measles viruses are able to traverse the mucosal surface and cause systemic infections. The following sections briefly describe some of the most commonly known diseases caused by viruses of this family.

A. Parainfluenza Virus

Parainfluenza viruses primarily cause respiratory disease in infants and young children, ranging from mid and upper respiratory infections to croup or pneumonia. Parainfluenza viruses are believed to account for 4–22% of the respiratory illnesses in children, and are second only to respiratory syncytial virus in this regard. Type 1 and type 2 viruses are the most common causes of croup in young children. Type 3 appears to be the most virulent, frequently causing bronchiolitis, laryngitis, and pneumonia in children, especially during the first month of life; and 67% of children are infected with PIV type 3 during the second year of life. Type 4, which consists of two subgroups (4A and 4B), induces the least severe symptoms and is a less common etiologic agent. The viruses enter by the respiratory route, and in most adults they multiply and cause inflammation only in the upper segments of the tract, although in infants and young children bronchi, bronchioles, and lungs are occasionally involved. Immunity probably requires antibody responses in mucosal secretions, and is therefore of short duration. Thus, reinfections with these viruses are common and the viruses also cause a sizable proportion of respiratory illness in healthy older children and adults. Parainfluenza viruses, as do other paramyxoviruses, can readily establish a persistent infection *in vitro;* however, no conclusive data exist indicating that a parainfluenza virus is the cause of any chronic disease. In addition, parainfluenza viruses can also cause respiratory disease in cattle, sheep, monkeys, and dogs.

B. Respiratory Syncytial Virus

Respiratory syncytial virus (RSV) is a major cause of lower respiratory tract disease during infancy and early childhood throughout the world. Almost 30% of infants in the first year of life have a medically attended illness, which often is diagnosed as bronchiolitis or pneumonia due to RSV infection. Initially, RSV infection and inflammation involve the mucous membranes of the nose and throat. In most cases the inflammatory process is extended into the trachea, bronchioles, and the parenchyma of the lung, leading to bronchiolitis and pneumonitis, especially in young infants. In the presence of normal immunity, productive infection is probably limited to the respiratory epithelium. In cases of fatal RSV infection, virus spread was observed outside the respiratory tract and into other organs, including the kidney, liver, and myocardium. It is estimated that the incubation period from exposure to virus to development of fever and signs of RSV lower respiratory disease was approximately 4.5 days and illness lasted an average of 5.5 days. Like parainfluenza viruses, protective immunity appears to be of short duration and reinfections with these viruses are common.

C. Mumps

Mumps is an acute infectious disease of humans characterized by the swelling of the parotid and submaxillary salivary glands. The virus is highly contagious, usually entering the mouth and nose by droplet infection, and a primary round of viral replication takes place in epithelial cells of the nasal mucus or upper respiratory tract, followed by spread of the virus to regional lymph nodes. A viremia then occurs and the virus spreads to other parts of the body. The salivary glands are the primary target organs. The mean incubation period from exposure to the appearance of clinical symptoms is 18 days. The disease lasts only a short time and usually is benign. Mumps may, however, be accompanied by complications, such as meningitis or meningoencephalitis, gastroenteritis, oophoritis, or orchitis. Severe cases of mumps orchitis can result in atrophy of the germinal epithelium with scarring and fibrosis. Over 50% of mumps cases occur in children between 5 and 9 years of age, but the virus rarely attacks infants under one year old, who may be protected by maternal antibody. Although humans are the only known natural host for mumps virus, mumps infection can be experimentally induced in a variety of mammalian species and the virus can be propagated in many primate cell lines.

D. Measles

Measles is one of the most contagious infections of humans. The virus is transmitted via secretions of the respiratory tract and eye, and in urine, and enters the upper respiratory tract, or perhaps the eye, and multiplies in the epithelium and regional lymphatic tissues. Viremia occurs toward the end of the incubation period (which is about 11–14 days), permitting further widespread dissemination of the virus to the lymphoid tissue and skin, which causes the symptoms of coryza, conjunctivitis, dry cough, sore throat, headache, fever, Koplik's spot, and later the typical red, maculopapular rash over the whole body. Some measles cases (up to 10%) are associated with complications, the commonest being otitis, pneumonia, diarrhea, or encephalitis. A rare delayed complication that occurs an average of 6 years after the initial measles infection is subacute sclerosing panencephalitis, which is characterized by the onset of cerebral dysfunction and progresses to coma and death within 2 years. Measles' acute-case fatality is currently limited to a very low rate in developed countries because of disease control through the use of measles vaccine, whereas a high death rate still occurs in developing countries.

E. Newcastle Disease

Newcastle disease virus (NDV) is primarily a respiratory tract pathogen of birds, particularly chickens. It is characterized by being very lethal to chickens and giving rise to severe hemorrhage of the intestine. The virus attacks the epithelium of the respiratory tract, causing the destruction of the ciliated cells and also endothelial cells in the interstitial tissues of the lung. Newcastle disease can also occur in humans, but is almost exclusively confined to poultry workers or laboratory personnel. Infection occurs mainly by two routes, via the conjunctiva or by inhalation. Conjunctivitis can be acute and is accompanied by pain and lachrymation, and the local preauricular lymph gland is often swollen. Infection by inhalation can lead to pharyngitis, accompanied by influenza-type symptoms including headache, malaise, and disturbance of the upper respiratory tract.

VII. CONTROL

Reducing the attack rate of respiratory diseases is an important social and economic goal. The following approaches have been taken toward preventing and controlling diseases caused by paramyxoviruses.

A. Killed and Attenuated Live-Virus Vaccines

Mumps and measles viruses infection can be prevented by immunization. Two types of vaccine have been designed for controlling these viral diseases. One is a vaccine killed by formalin or ultraviolet light, and the other is an attenuated live-virus vaccine. Formalin-inactivated mumps and measles vaccines have been shown to induce antibody responses and to be clinically effective, but only confer temporary protection. The administration of live attenuated mumps or measles vaccines results in the production of antibodies that remain for at least 10 years and probably also induces cell-mediated immune responses. These vaccines have afforded long-term protection against the disease. The live attenuated measles and mumps vaccines have been safely and effectively used when given at an early age (9–15

months) and have reduced the incidence of infection and lowered the morbidity and mortality to a small fraction of the preimmunization level in many countries. For developing countries, however, the use of measles vaccine was not introduced until 1985. Since then, the global number of reported measles cases has fallen, but thousands of measles cases continue to occur every year in many developing countries, especially in Africa, and measles is still a major cause of death. About 1 million children die from measles each year. Because vaccination in the Americas has resulted in the virtual disappearance of measles from that hemisphere, it is believed that measles can be controlled by vaccination. Because the disease is limited to humans, and there is no significant antigenic variation or natural reservoir for the virus, the goal of global eradication of measles should be feasible.

Early attempts to develop safe and effective RSV or parainfluenza viruses vaccines have not met with success. Formalin-inactivated RSV vaccines were found to enhance the disease severity in vaccinated infants who were subsequently infected. These observations have resulted in great caution in subsequent efforts to develop effective vaccines for these viruses. Recent experimental approaches to develop live vaccines have included the evaluation of the closely related bovine PIV-3, as well as the production of live attenuated mutants of PIV-3 and RSV by cold adaptation.

B. Subunit Vaccines

The use of purified glycoproteins as experimental subunit vaccines has been actively investigated in animal models. The attachment (G or HN) and fusion (F) proteins of RSV and human parainfluenza virus are of primary importance in inducing virus-neutralizing antibodies and immunity against reinfection in experimental animals. The development of subunit vaccine has focused on immunization with purified HN or G and F proteins from native PIV and RSV viruses and producing the protein by recombinant expression systems. It was shown that immunization with a mixture of both HN and F glycoproteins from human parainfluenza virus induced significant resistance to subsequent challenge infection, whereas immunization with HN or F alone was much less effective. A trivalent prototype PIV-1,2,3 subunit vaccine containing the HN and F glycoproteins from each parainfluenza has been prepared, and induced significant levels of PIV-1,2,3, neutralizing antibodies. The immunization of cotton rats with affinity-purified attachment protein G or the fusion protein (F) of RSV was shown to provide complete pulmonary resistance and partial nasal resistance against challenge infection with live virus. In measles virus, both surface glycoproteins also appear to be needed for effective subunit vaccines. Clinical trails with some purified glycoprotein vaccines such as RSV F glycoprotein and PIV3-HN and F glycoproteins are underway.

C. Genetically Engineered Viruses

Recently, the technology of reverse genetics to recover infectious viruses from a cDNA clone of single-stranded, negative-sense RNA viruses, including RSV, SN, HPIV3, and measles virus, has been developed. This provides the first available method for direct genetic manipulation of the viral genome by introducing designed mutations separately or in various combination into the full-length cDNA clones. Using these experimental procedures, RSV and PIV3 attenuated vaccines are being developed.

D. Antiviral Agents

For RSV and PIV infections, because no vaccines are available, chemotherapeutic approaches are of importance. Ribavirin, a synthetic guanosine nucleoside that interferes with the viral replication process, has broad-spectrum antiviral efficacy and is the only approved therapy for lower respiratory tract disease due to RSV. There is some controversy concerning the overall benefits of ribavirin aerosol therapy, and the 1996 American Academy of Pediatrics guidelines recommended that ribavirin treatment may be considered for patients hospitalized with RSV lower respiratory tract disease, specifically, infants at high risk of severe or complicated RSV infection. Ribavirin has also shown some antiviral effects against PIV in cell cultures, and has been used in certain recipients with upper and lower respiratory tract infection with PIV, but it is not licensed for the treatment of PIV infection.

More recently, passive intravenous immunization with human polyclonal immune globulin or humanized monoclonal antibody directed against the conserved fusion protein of RSV has been shown to prevent severe RSV disease. Also, the humanized RSV monoclonal antibody is licensed for use in premature infants and infants with bronchopulmonary dysplasia, and is being evaluated in infants with cardiac disease.

See Also the Following Articles

ANTIVIRAL AGENTS • VIRUS INFECTION • VACCINES, VIRAL

Bibliography

Collins, P. L., Chanock, R. M., and McIntosh, K. (1996). Parainfluenza viruses. *In* "Virology" (B. N. Fields *et al.*, eds.), 3rd ed., pp. 1205–1241. Lippincott-Raven, Philadelphia.

Collins, P. L., McIntosh, K., and Chanock, R. M. (1996). Respiratory syncytial virus. *In* "Virology" (B. N. Fields *et al.*, eds.), 3rd ed., pp. 1313–1351. Lippincott-Raven, Philadelphia.

Griffin, D. E., and Bellini, W. J. (1996). Measles virus. *In* "Virology" (B. N. Fields *et al.*, eds.), 3rd ed., pp. 1267–1312. Lippincott-Raven, Philadelphia.

Lamb, R. A., and Kolakofsky, D. (1996). Paramyxoviridae: The viruses and their replication. *In* "Virology" (B. N. Fields *et al.*, eds.), 3rd ed., pp. 1177–1204. Lippincott-Raven, Philadelphia.

Murphy, B. R., and Collins, P. L. (1997). Current status of respiratory syncytial virus (RSV) and parainfluenza virus type 3 (PIV3) vaccine development: Memorandum from a joint WHO/NIAID meeting. *Bull. WHO* 75, 307–313.

Ray, R., and Compans, R. W. (1990). Immunochemistry of paramyxoviruses. *In* "Immunochemistry of Viruses II" (A. R. Neurath, ed.), pp. 217–236. Elsevier Biomedical, Amsterdam, The Netherlands.

Rima, B., Alexander, D. J., Billeter, M. A., Collins, P. L., Kingsbury, D. W., Lipkind, M. A., Nagai, Y., Orvell, C., Pringle, C. R., and ter Meulen, V. (1995). Paramyxoviridae. *In* "Virus Taxonomy" (F. A. Murphy *et al.*, eds.), 6th report, pp. 268–274. Springer-Verlag, New York.

Wolinsky, J. S. (1996). Mumps virus. *In* "Virology" (B. N. Fields *et al.*, eds.), 3rd ed., pp. 1243–1265. Lippincott-Raven, Philadelphia.

The Patenting of Living Organisms and Natural Products

S. Leslie Misrock, Adriane M. Antler, and Anne M. Schneiderman

Pennie & Edmonds LLP

I. Introduction
II. Biotechnological Inventions as Patentable Subject Matter
III. Utility and Enablement
IV. Novelty and Nonobviousness Requirements
V. Written Description and Enablement Requirements for Patentability and Special Problems Relating to Living Organisms
VI. Uses of Discovered DNAs
VII. Conclusion

GLOSSARY

Board of Patent Appeals and Interferences The branch of the U.S. Patent and Trademark Office that reviews inventors' appeals from negative decisions by the office regarding the patentability of their inventions. The board also decides who is entitled to a patent when two or more inventors patent or attempt to patent the same invention.

Court of Appeals for the Federal Circuit The federal court that handles appeals from decisions made by the U.S. Patent and Trademark Office regarding the patentability of an invention and the U.S. district courts regarding infringement and/or validity of a patent.

enablement, enabling disclosure A statutory requirement for patentability mandating that the written description of an invention that is included in a patent specification enable a person "skilled in the art" to make and use the invention. In the case of biotechnological inventions, in certain circumstances the enablement requirement may be satisfied for a genetically engineered organism by depositing the organism with a depository recognized under the Budapest Treaty on the International Recogni-

tion of the Deposit of Microorganisms for the Purposes of Patent Procedure, pursuant to the provisions of the treaty.

interference contest An administrative proceeding conducted by the U.S. Patent and Trademark Office to determine who is entitled to a patent on an invention when more than one inventor (or group of inventors) claim the same invention.

Patent and Trademark Office, United States The branch of the U.S. Department of Commerce that issues patents. The office handles patent applications, determines the patentability of inventions, and issues patents.

patent examiner A Patent and Trademark Office employee who reviews patent applications, determines the patentability of inventions, and, if the invention is patentable, determines the scope of patent protection to be granted. Examiners have technical backgrounds in the fields of the patent applications that they examine.

patent prosecution The process carried out by the Patent and Trademark Office in determining the patentability of inventions and in issuing patents. Patent prosecution includes the examination of a patent application by a patent examiner and the interactions that ensue between the examiner and the inventor (or the inventor's representatives) in determining the patentability and scope of the invention.

prior art The previous developments in the field of an invention against which the novelty and nonobviousness of an invention are assessed, including developments that have been described, published, or disclosed publicly.

reduction to practice The inventor's act of physically carrying out his or her invention after initially conceiving of it (termed "actual reduction to practice") or the inventor's act of filing a patent application (either a provisional or a regular application) on the invention that satisfies certain statutory requirements (termed "constructive reduction to practice").

I. INTRODUCTION

Biotechnological inventions frequently involve naturally occurring substances or living organisms as their subject matter. A unique set of issues has arisen with respect to the patentability of biotechnological inventions and, in particular, the patentability of genetically engineered organisms and genes and their encoded proteins. These issues range from legal and ethical considerations of whether living organisms should be properly classified as patentable subject matter to more technical considerations of the novelty of naturally occurring substances and the complexity of living organisms. Living organisms often defy written description and thus pose patentability problems if they are not adequately "enabled" or described by a written disclosure in a patent application that allows one to carry out the invention. This article focuses on issues that have arisen with respect to (i) the novelty requirements of the United States patent laws as applied to biotechnological inventions, (ii) the classification of certain biotechnological inventions as patentable subject matter, and (iii) the enablement requirement for patentability as it is applied to biotechnological inventions that are not especially amenable to written description.

A. What Is a Patent?

A patent is a document that consists of a written description or specification of the invention, followed by a legal section, the patent claims. The specification is a technical disclosure of how to make and use the invention. The claims, listed at the end of the patent, are what define the patented invention. The claims thus describe and define the scope of the protection granted by the patent. A United States patent entitles an inventor to a limited monopoly on his or her invention in exchange for the inventor's full disclosure of his or her discovery to the public. The United States patent system, therefore, rewards the inventor for his or her discovery by securing for the inventor the exclusive rights to his or her invention for a limited time and, most important, benefits the public by promoting the full disclosure of such inventions. A patent allows the owner of the patent to exclude others from making, using, selling, importing, and offering for sale what is covered by the claims of the patent. After the expiration of the limited monopoly on the invention afforded by the patent laws (generally, 20 years after filing the patent application), the patent may no longer be used to exclude others from making, using, selling, importing, or offering for sale the invention.

An inventor's right to the limited monopoly defined by the claims of his or her patent is derived from a right granted in the United States Constitution. Article I, Section 8 of the Constitution provides that "The Congress shall have Power . . . To Promote the Progress of Science and useful Arts, by securing for limited Times to Authors and Inventors the exclusive Right to their respective Writings and Discoveries."

It is from this constitutional grant that both the patent and the copyright laws of the United States derive. In accordance with the congressional power conferred by Article I, Section 8 of the Constitution, Congress enacted the "Patent Statute," Title 35 of the United States Code. The Patent Statute sets forth the statutory criteria for securing a patent. For the purposes of this article, it is important to stress that the same general rules and tests for patentability apply to all inventions, whether they are biotechnological or not.

B. Patentable Subject Matter

Section 101 of the Patent Statute, entitled "Inventions Patentable," defines four classes of "new and useful" subject matter that may be patentable:

Whoever invents or discover any *new and useful* process, machine, manufacture, or *composition of matter,* or any new and useful improvement thereof, may obtain a patent therefore, subject to the conditions and requirements of this title. [italics added]
—35 U.S.C. §101

Thus, Section 101 establishes that an invention involving subject matter that can be classified as a process, a machine, an article of manufacture, or a composition of matter (or an "improvement" on any of these four classes) may be patented provided it is both new and useful and also that the invention meets the other important statutory requirements for patentability: inter alia, it must be novel (35 U.S.C §102), it must be nonobvious (35 U.S.C. §103), and the patent application must contain a written description of the invention that enables (i.e., teaches) someone "skilled in the art" (i.e., skilled in the field of the invention) to make and use the invention (35 U.S.C. §112).[1] These important statutory requirements will be discussed in detail with respect to biotechnological inventions.

Although the enumerated list in Section 101 covers a broad range of subject matters that could be patented, its scope is not unlimited. In particular, discoveries of nature, laws of nature, physical phenomena, and abstract ideas remain outside the scope of patentable subject matter. Thus, new minerals found in the earth, new plants found in the wild, and Einstein's theory of relativity are subject matters that cannot be properly patented.

Typically, the subject matter of a biotechnological invention involves the discovery, identification, synthesis, purification, or alteration of naturally occurring substances, molecules, or organisms. Because these substances and organisms may be found in nature, many interesting questions arise when the patentability of such inventions is considered. First, are naturally occurring substances and/or living organisms found in nature really "new" and, therefore, can they be considered patentable subject matter? Second, did Congress intend the terms "article" and "composition of matter" to encompass living organisms as patentable subject matter when it enacted the Patent Statute? The answers are relevant to the patentability of subject matter such as DNA, protein, and transgenic and "knockout" animals.

A body of case law has developed from the decisions of federal and state courts that further defines and explains what is meant by the terms "new" and compositions of matter as they are used in Section 101 of the Patent Statute. The case law, discussed in more detail later, establishes two main legal principles. First, the case law supports the proposition

that naturally occurring substances that have been purified, altered, or engineered in some way so that they are different in kind and/or form from their naturally occurring counterparts are new compositions of matter that may be patentable provided that the other statutory requirements for patentability are met (*inter alia,* utility, nonobviousness, and the Section 112 requirements of written description, enablement, and best mode). Second, the case law supports the proposition that living organisms, including both unicellular microorganisms and multicellular organisms such as plants and animals, are compositions of matter that may be patented provided that they are novel, useful, nonobvious, and meet the Section 112 requirements.

II. BIOTECHNOLOGICAL INVENTIONS AS PATENTABLE SUBJECT MATTER

A. The Meaning of "New and Useful"

The Patent Statute requires that a composition of matter must be both new and useful (with legal emphasis on the conjunctive term "and") in order to be considered patentable. Many biotechnological products have industrial, diagnostic, or therapeutic uses. Disclosing such a utility in the patent specification should be done to satisfy the law's requirement of having at least one utility to support patentability of the claimed composition. Many biotechnological inventions such as genes and proteins have naturally occurring counterparts. However, in order to evaluate the patentability of a biotechnological product, the Patent and Trademark Office (or a court if an inventor appeals an unfavorable decision of the Patent and Trademark Office) must determine whether such products, which have naturally occurring counterparts, are new within the meaning of the patent law.

The United States Supreme Court enunciated the general rule that "discover[ies] of the phenomena of nature" are not patentable in *Funk Bros. Seed Co. v. Kalo Inoculant Co.,* 333 U.S. 127,130 (1948). The subject matter whose patentability was at issue in *Funk* were species of nitrogen-fixing bacteria (genus *Rhizobium*) that occur naturally and infect the roots of leguminous plants (e.g., beans, clover, and alfalfa).

---

1. Section 112 of the Patent Statute sets forth the stated written description and enablement requirements; this section also includes the best mode requirement, which mandates that the patent application disclose the best mode known to the inventor for practicing his or her invention.

Through a complex symbiotic relationship, the bacteria fix nitrogen from the air in a form that can be used by the plant for growth. The patentee (i.e., the inventor to whom the patent was granted) had discovered that certain known species of these nitrogen-fixing bacteria were noninhibitory to each other when mixed together. Prior to this discovery, mixed cultures of nitrogen-fixing bacteria had proved unsatisfactory because the different species had inhibitory effects on each other's ability to fix nitrogen. The product claimed in the patent was a composite mixture of the noninhibitory strains of bacteria packaged for use to inoculate leguminous plants. The Supreme Court held that claims to the mixture of bacteria, as a product, were not patentable because each of the bacteria existed in nature prior to the inventor's discovery (*Funk*, 333 U.S. at 131). Each species of bacteria performed in the same way and had the same effect on nitrogen fixation that it had had prior to the patentee's discovery. The Court held that discovering the noninhibitory nature of the combination of bacteria was a mere discovery of a principle of nature and not the product of invention (*Funk*, 333 U.S. at 131-32). Thereafter, the *Funk* case was regarded as standing for the general rule that products found in nature are not patentable.

As is the case with most legal "generalities," exceptions were soon carved out of the "general rule" of the *Funk* case. A series of cases involving natural products such as vitamin B_{12} [*Merck & Co. v. Olin Mathieson Chemical Corp.*, 253 F.2d 156 (4th Cir. 1958); *Merck & Co. v. Chase Chemical Co.*, 273 F. Supp. 68 (D.N.J. 1967)] and prostaglandins [*In re Bergstrom*, 427 F.2d 1394 (C.C.P.A. 1970)], and culminating in the renowned "strawberry case" [*In re Kratz*, 592 F.2d 1169 (C.C.P.A. 1979)] established the principle that the status of a product as "naturally occurring" does not preclude patentability provided that the product claimed is purified or altered in some way so as to comprise a "new and useful" composition of matter.

The inventors in the *Merck* vitamin B_{12} cases were the first to separate and purify vitamin B_{12} from fermentates of a vitamin B_{12} activity-producing strain of fungus. The patents at issue in the *Merck* cases claimed the purified vitamin B_{12} product of a particular range of specific activities. The challengers of the patents contended that the claims covered a product

of nature, and therefore the claimed purified vitamin B_{12} was not patentable. In upholding the patentability of the purified vitamin B_{12}, the court pointed out that it was a product of such purity that it differed not only in degree but also in kind from the naturally occurring substance, and that it was of great medicinal and commercial value in the treatment of pernicious anemia (*Merck v. Olin Mathieson*, 253 F.2d at 164). Previously, the active substance for treating pernicious anemia had been unidentified and unknown. There was nothing comparable in the "prior art," i.e., the previous developments in the field. By contrast, the fermentates containing the naturally occurring vitamin B_{12} were not known to contain the desired activity and had no therapeutic or commercial utility until converted into the patented compositions. The court concluded that purified vitamin B_{12} was not the same as vitamin B_{12} found in nature but was a new and useful composition of matter entitled to patent protection (*Merck v. Olin Mathieson*, 253 F.2d at 164).

The inventor in the *Bergstrom* case had purified prostaglandins PGE_2 and PGE_3 in crystalline form from a natural source—sheep prostate glands. The patent application, which claimed the purified prostaglandins as products that could be used to stimulate smooth muscle and lower blood pressure, was rejected by the Patent and Trademark Office and the decision upheld by the Patent Office Board of Appeals. In reversing the decision of the board of appeals and upholding the patentability of the purified prostaglandins, the Court of Customs and Patent Appeals (the "C.C.P.A.," the predecessor court to the current Court of Appeals for the Federal Circuit) pointed out that the existence of a compound as an ingredient of another substance does not negate the novelty of a claim to the pure compound. It did not matter that the claimed prostaglandins existed elsewhere (i.e., were "inherent") in nature in less pure form. The pure forms were not naturally occurring, and therefore they were patentable (*Bergstrom*, 427 F.2d at 1401-02). The *Bergstrom* court also pointed out that the criteria for determining whether a claimed subject matter is new within the meaning of Section 101 of the Patent Statute are no different than the "novelty" requirements of a subsequent section, Section 102 (see Section IV). The question of "newness" or "novelty" turns on whether the claimed pure materials are novel compared with the less pure materials of the prior art (*Bergstrom,* 427 F.2d at 1401).

This line of reasoning culminated in the famous *Kratz* "strawberry case," another decision by the Court of Customs and Patent Appeals. *Kratz* addressed the interpretation of the term new composition of matter as it applied to naturally occurring substances. The inventor had purified a chemical (2-methyl-2-pentenoic acid) from strawberries and determined that it was responsible for the flavor and fragrance of strawberries. The patent application, which claimed a composition containing a specified amount of substantially pure 2-methyl-2-pentenoic acid, was rejected by the Patent and Trademark Office as claiming a substance that was not new. The Patent and Trademark Office cited the natural constituent of strawberries as prior art against the claims. The court reversed the Patent and Trademark Office decision and upheld the patentability of the composition as claimed, i.e., substantially pure 2-methyl-2-pentenoic acid. In upholding patentability, the court pointed out that the chemical did not exist in a substantially pure form in nature; thus, the substantially pure product was new (*Kratz,* 592 F.2d at 1173-75).

The *Funk, Merck, Bergstrom,* and *Kratz* cases clearly stand for the proposition that natural substances that are changed in some way by man's intervention so that they are new when compared with the natural product or with compositions made and reported by others could be patented provided they meet the further requirements of patentability as provided by the Patent Statute. What are the implications of these court decisions for the patentability of genes and proteins? In practice, patent attorneys draft patent applications so that "isolated" or "purified" genes and proteins are claimed. Such claimed genes and proteins are therefore new when compared with their naturally existing cellular counterparts.

B. The Meaning of "Composition of Matter"

With respect to recombinant cells, transgenic organisms, and "gene knockout" organisms, additional questions have arisen regarding their patentability.

The impact of biotechnology has led to an intense examination of the law regarding the patentability of living organisms. However, it was not until 1980, in the landmark decision *Diamond v. Chakrabarty,* 447 U.S. 303 (1980), that the Supreme Court squarely faced the question of whether a living organism qualifies as a composition of matter which may be patentable. The Supreme Court's answer to this question was most definitely in the affirmative.

The *Chakrabarty* invention involved a bacterium from the genus *Pseudomonas* that had been genetically engineered in the laboratory to contain new genes that enabled in to degrade components of crude oil. Such genetically engineered bacteria were useful in the treatment of oil spills since the bacteria could break down multiple components of crude oil in the spill. The Patent and Trademark Office allowed (i.e., agreed to grant a patent on) the patent application's process claims for a method of producing the genetically engineered bacteria and its product claims for an inoculum composed of a carrier material and the new bacteria. The Patent and Trademark Office rejected, however, the product claims for the bacteria themselves. The inventor appealed to the Patent Office Board of Appeals, which subsequently upheld the Patent and Trademark Office's rejection on the grounds that the scope of subject matter enumerated in the Patent Statute (i.e., a process, machine, article of manufacture, or composition of matter) was never intended to cover living things such as laboratory-created microorganisms.

The C.C.P.A. reversed the board of appeals' decision on the authority of the court's prior decision in *In re Bergy,* 563 F.2d 1031 (C.C.P.A. 1977), which held that "'the fact that microorganisms . . . are alive . . . [is] without legal significance' for the purposes of the patent law" (*Chakrabarty,* 447 U.S. at 306).

The Supreme Court affirmed the C.C.P.A. decision reversing the board's rejection, and upheld the patentability of Chakrabarty's microorganisms. In its decision, the Supreme Court distinguished this case from *Funk,* nothing that Chakrabarty had "produced a new bacterium with markedly different characteristics from any found in nature and one having the potential for significant utility. His discovery is not nature's handiwork, but his own; accordingly it is patentable subject matter under [Section] 101" (*Cha-*

krabarty, 447 U.S. at 310). The Court also enunciated the rule that the enumeration of a process, machine, manufacture, or composition of matter in Section 101 of the Patent Statute included "anything under the sun that is made by man" (*Chakrabarty*, 447 U.S. at 309). (Chakrabarty *et al.*, U.S. Patent No. 4,535,061, entitled "Bacteria capable of dissimilation of environmentally persistent chemical compounds," was subsequently issued on August 13, 1985.)

In summary, the Supreme Court decision in *Chakrabarty* stands for the proposition that living things are not excluded from the patentable subject matter enumerated in the Patent Statute. Such patentable subject matter, the Court held, was meant to include anything under the sun that is made by man except for abstract ideas and phenomena of nature. Thus, microorganisms are patentable, provided that they are not as found in nature, and that, *inter alia* they are novel, nonobvious, and enabled by a written description in the patent specification. With the *Chakrabarty* decision, the stage was set for future legal battles regarding the patentability of higher life forms.

C. Multicellular Organisms as Patentable Subject Matter: Plants and Animals

The *Chakrabarty* decision provided a broad definition of patentable subject matter. When limited narrowly to its facts, however, *Chakrabarty* could be interpreted to apply only to microorganisms and not to multicellular or higher life forms. The Patent and Trademark Office Board of Appeals and Interferences ("board of appeals") first faced the question of the patentability of multicellular organisms in *Ex Parte Hibberd*, 227 U.S.P.Q. 443 (Bd. Pat. App. & Int. 1985), in which the board of appeals decided that plants, seeds, and plant tissues are patentable subject matter.

The subject matter of the invention in *Hibberd* involved plant technologies, including seeds, plants, and plant tissue cultures, which were engineered to contain or give rise to plants with increased levels of tryptophan, an amino acid essential to human nutrition. In denying patentability, the Patent and Trademark Office acknowledged that in view of *Cha-*

krabarty, it appeared that man-made life forms including plant life could be patentable subject matter (*Hibberd*, 227 U.S.P.Q. at 444). However, the Patent and Trademark Office explained that by Congress' enactment of subsequent legislation, the Plant Patent Act and the Plant Variety Protection Act, which occurred after the enactment of Section 101 of the Patent Statute and which was more specific than Section 101, "Congress has specifically set forth how and under what conditions plant life covered by these Acts should be protected" (*Hibberd*, 227 U.S.P.Q. at 444). The Patent and Trademark Office's position was that these plant-specific acts were the exclusive forms of protection for the plant life covered by the acts, and thus such plant life could not be protected under the Patent Statute. The board of appeals disagreed.

In examining the legislative history of the plant-specific acts, the board of appeals determined that no specific language or explicit support existed in the acts that restricted the scope of patentable subject matter under the Patent Statute. Noting the "cardinal rule" of statutory construction set forth by the Supreme Court in *United States v. Borden Co.*, 308 U.S. 188 (1939), the board pointed out that the Patent and Trademark Office's attempt to "implicitly" narrow the scope of subject matter protectable under the Patent Statute was improper:

The overwhelming weight of authority is to the effect that repeals by implication are not favored and that when there are two acts on the same subject. . . [footnote omitted] . . . the rule is to give effect to *both* unless there is such a "positive repugnancy" or "irreconcilable conflict" that the statutes cannot co-exist. [italics added]
—*Hibberd*, 227 U.S.P.Q. at 445

The board applied the cardinal rule and determined that despite the overlap of subject matter protectable under each act, no such "positive repugnancy" or "irreconcilable conflict" existed between the Patent Statute and the plant-specific acts (*Hibberd*, 227 U.S.P.Q. at 445-46). Thus, the *Hibberd* decision upheld the patentability of genetically engineered plants.

In 1987, the board of appeals established precedent with respect to the patentability of animals in its decision in *Ex Parte Allen*, 2 U.S.P.Q. 2d 1425 (Bd.

Pat. App. & Int. 1987). The subject matter of the claimed invention involved oysters that were genetically altered to be polyploid so that their growth would be increased. The Patent and Trademark Office rejected the claims on the grounds that multicellular animals are not patentable subject matter under Section 101 of the Patent Statute. In rejecting the application, the Patent and Trademark Office cited *In re Bergy*, 563 F.2d 1031 (C.C.P.A. 1977) and explained that while living microorganisms are covered by the terms "manufacture" and "composition of matter" in Section 101, such microorganisms are "more akin to inanimate chemical compositions such as reactants, reagents, and catalysts than they are to horses and honeybess or raspberries and roses" (*Allen*, 2 U.S.P.Q. 2d at 1426).

However, the board of appeals refused to sustain the Patent and Trademark Office's rejection under Section 101. In reversing the Patent and Trademark Office's rejection under Section 101, the board cited *Chakrabarty* and *Hibberd* as controlling authority that Congress intended patentable subject matter to include anything under the sun that is made by man whether living or not (*Allen*, 2 U.S.P.Q. 2d at 1426). However, the Patent and Trademark Office's rejection of the claimed oysters as obvious over a prior reference was sustained (*Allen*, 2 U.S.P.Q. 2d at 1428-29). Thus, the man-made oysters were ultimately found to be obvious and, therefore, unpatentable.

Shortly after the *Allen* decision, the Commissioner of Patents and Trademarks issued a notice advising that nonnaturally occurring nonhuman multicellular organisms, including animals, are patentable subject matter under Section 101 of the Patent Statute and that the Patent and Trademark Office would not reject such claims directed to nonhuman organisms on the grounds of nonstatutory subject matter. [See Commissioner of Patents and Trademarks, *Policy Statement on Patentability of Animals,* 1077 Off. Gaz. Pat. Office 24 (April 7, 1987) reprinted in Donald S. Chisum, 9 *Chisum on Patents* app. 24-1 (1998).] This notice was met by strong adverse reaction from various animal rights groups which petitioned the Patent and Trademark Office to reverse its policy. These opponents contended that such policy raised moral and ethical issues of such magnitude and im-

portance that it could not simply be left for the Patent and Trademark Office to decide. The subject was litigated in favor of the Patent and Trademark Office [*Animal Legal Defense Fund v. Quigg,* 710 F.Supp. 728 (N.D. Cal. 1989); *aff'd.* 932 F.2d 920 (Fed. Cir. 1991)].

In 1988, the Patent and Trademark Office granted the first United States product patent covering a genetically engineered nonhuman animal: a "cancer-prone" transgenic nonhuman animal engineered to contain an activated oncogene (Leder *et al.,* United States Patent No. 4,736,866, entitled "Transgenic non-human mammals"). After a $3\frac{1}{2}$-year hiatus, three additional animal patents were issued in December 1992 covering transgenic mice: (i) transgenic mice that continuously produce a sufficient level of βinterferon to provide antiviral activity (Wagner *et al.,* United States Patent No. 5,175,385, entitled "Virus-resistant transgenic mice"); (ii) transgenic mice engineered so that they are incapable of developing T cells (Krimpenfort *et al.,* United States Patent No. 5,175,384, entitled "Transgenic mice depleted in mature T-cells and methods for making transgenic mice"); and (iii) male transgenic mice engineered to develop benign prostate hyperplasia or hypertrophy (Leder *et al.,* United States Patent No. 5,175,383, entitled "Animal model for benign prostatic disease").

Subsequent to the Patent and Trademark Office ruling and the grant of the first animal patent in 1988, the issue of patenting animals became the subject of a public policy debate in the legislature. Legislation had been introduced in sessions of Congress to place a moratorium on patenting life forms to allow congressional consideration of the broader implications of such patents. A proposal introduced in February 1993 (S387) would impose a 2-year moratorium on granting animal patents, makes a number of "findings" justifying congressional action in this area (including an evaluation of the Human Genome Project and the National Institutes of Health proposal to patent more than 2000 DNA sequences), and would add a new Section 106 to Title 35 that would exclude human beings, human organs (genetically engineered or otherwise), and genetically engineered animals as patentable subject matter. Precedent exists for congressional action. Congress has twice before prohibited the patenting of an entire class of otherwise patentable subject matter: in 1946, a total ban on inventions in atomic energy, and in 1954 a ban on inventions useful "solely in the utilization of special nuclear material or atomic energy in an atomic weapon" [42 U.S.C. §2181(a)]. However, none of the introduced legislation limiting the patenting of animals was passed by Congress.

Nevertheless, it appears settled that human beings are not patentable subject matter. The Patent and Trademark Office routinely rejects patent claims if they include human beings within their scope. The Patent and Trademark Office has stated that "[a] claim directed to or including within its scope a human being will not be considered to be patentable subject matter under 35 U.S.C. 101. The grant of a limited, but exclusive property right in a human being is prohibited by the Constitution" (Commissioner of Patents and Trademarks, in 9 *Chisum on Patents* at app, 24-1). Under current law, the Patent and Trademark Office could reject patent claims directed to human beings under the Thirteenth Amendment, which prohibits slavery, or the Fourth Amendment, which includes the right to be protected against unreasonable search and seizure (Commissioner of Patents and Trademarks, in 9 *Chisum on Patents* at app. 24-3). To avoid this rejection, the Patent and Trademark Office has suggested that claims directed to categories of nonplant multicellular organisms that could include human beings be worded to include the limitation "nonhuman" (9 *Chisum on Patents* at app. 24-3).

The Patent and Trademark Office could also apply a rarely invoked, nineteenth-century legal standard, the standard of moral utility: "All that the law requires is that the invention should not be frivolous or injurious to the well-being, good policy, or sound morals of society" [*Lowell v. Lewis,* 15 F. Cas. 1018, 1019 (D. Mass. 1817) (No. 8568)]. Under this standard, the Patent and Trademark Office could decide that claims directed to human beings are immoral or injurious to the well-being of society and therefore are not patentable [U.S. Patent and Trademark Office, Media Advisory, April 1, 1998; Facts on Patenting Life Forms Having a Relationship to Humans (visited Jun. 24, 1999) (*http://www.uspto.gov/web/offices/com/ speeches/98-06.htm*); Thomas A. Magnani, *The Pat-*

entability of Human-Animal Chimeras, 14 Berkeley Tech. L. J. 443, 444].

In summary, man-made nonhuman living organisms, including recombinant cells and transgenic organisms, are statutory subject matter. These products are all eligible for patent protection provided that the other statutory requirements for patentability are satisfied.

III. UTILITY AND ENABLEMENT

Under Section 101 of the Patent Statute, in order to be patentable, an invention must have utility. In *Brenner v. Manson,* 383 U.S. 519 (1966), the Supreme Court enunciated the standard for patentable utility. The Court stated that to have utility, an invention must have a beneficial use in society, i.e., where a specific benefit exists in currently available form (*Brenner,* 383 U.S. at 534-35). There is no patentable utility where the sole "utility" of a chemical compound consists of its potential role as an object of use testing. In other words, a patent is not a "hunting license" (*Manson,* 383 U.S. at 536).

The utility requirement is viewed by the courts as requiring a "practical utility," one that attributes a "real-world" value to the claimed subject matter, i.e., a discovery that can be used in a manner that provides some immediate benefit to the public [*Nelson v. Bowler,* 626 F.2d 853, 856 (C.C.P.A. 1980)]. In the early 1990s, the Patent and Trademark Office implementation of this standard, particularly with respect to claimed therapeutic methods, was problematic. The Patent and Trademark Office had been requiring statistically significant and reproducible human clinical data or such data from an art-accepted animal model for the particular therapeutic indication. This was problematic for newly cloned genes and their encoded proteins whose sole apparent utility was as potential therapeutics.

In 1995, new "Utility Examination Guidelines for Patent Examiners" were promulgated to bring the United States Patent and Trademark Office practice in line with the existing case law precedent (60 Federal Register 36263; effective date July 14, 1995). Under the relevant case law and the "Examination Guidelines," a therapeutic or pharmacological invention has patentable utility if an asserted utility for the claimed invention would be considered credible by a person of ordinary skill in the art in view of all the evidence of record. Significantly, demonstration of efficacy and/or safety in humans is not required. If reasonably correlated to the asserted utility, data generated from the results of *in vitro* assays and/or in animal models are sufficient to meet the utility requirement. Moreover, the animal model used for demonstration need not be an industry-accepted model for the disease or condition. If one of ordinary skill in the art would accept the animal tests as being reasonably predictive of utility in humans, evidence from those tests should be considered sufficient to support the credibility of the asserted utility. Proof beyond a reasonable doubt is not required. Instead, the evidence will be sufficient if, considered as a whole, it leads a person of ordinary skill in the art to conclude that the asserted utility is, more likely than not, true.

The revised guidelines are applicable to all utilities, not just to therapeutic ones. In practice, however, it is usually easier to obtain data that indicate or prove a diagnostic or practical research utility for a gene than it is to obtain data that indicate or prove a therapeutic utility. Thus, strategically it is advantageous to disclose in the patent specification as many utilities as are reasonable, that can be enabled, and among which the inventors may be able to provide proof of one utility, if necessary, during the course of patent prosecution. One need prove only one patentable utility for a product claim in order to satisfy the utility requirement.

With respect to utility, even a trivial utility is generally acceptable as a patentable utility. A question currently debated concerns the utility of expressed sequence tags (ESTs), which are partial cDNA sequences synthesized from mRNA transcripts of unknown, uncharacterized genes, and single nucleotide polymorphisms (SNPs). In the past few years, a number of parties have published and deposited in public databases the sequences of ESTs and SNPs. The question raised is whether one may patent ESTs or SNPs in view of the utility requirement for patentability. The answer depends on whether the specification discloses and enables any real-world use. Even when an EST or SNP lacks direct utility in identifying a gene of known function, it may have other utilities. Possible utilities for ESTs that have been proposed include identification of an expressed gene, use as a site marker to locate a gene on a physical map, individual or forensic identification, tracing ancestry and parentage, tissue typing, production of antibodies where the antibody has a practical use, construction of DNA "fingerprints" for correlation with disease, and diagnostic use. Utilities can also arise from identification of an open reading frame (ORF) in the EST with homology to a protein of known function or identification of a domain associated with a function or biological activity. Therefore, if the specification of a patent application discloses and enables a utility, then claims covering ESTs and/or SNPs may be patentable.

The Patent and Trademark Office indicated in a February 14, 1997 announcement by the Patent and Trademark Office acting deputy commissioner that it will allow claims to ESTs based on their utility as probes (presumably only if sufficiently described and enabled by the patent specification), and it has indicated that ESTs and SNPs may be patentable (Doll, 1998). In the fall of 1998, the first patent containing claims to ESTs was issued (Au-Young *et al.*, U.S. Patent No. 5,817,479, entitled "Human Kinase Homologs," issued October 6, 1998).

IV. NOVELTY AND NONOBVIOUSNESS REQUIREMENTS

To be patentable, an invention must be novel and nonobvious. Section 102 of the Patent Statute specifies that a person shall be entitled to a patent on an invention unless the invention is not novel. Section 102 describes several conditions that destroy novelty and thus render the invention unpatentable. These conditions include, *inter alia*, where the invention was known or used by others in the United States, or patented or described in a printed publication in the United States or a foreign country, before the invention of it by the applicant for the patent; where the invention was publicly used, offered for sale or sold, or described in a printed publication more than 1 year prior to a patent application being filed on it; and where the invention was not invented by the individual filing the application, or the invention was invented by another before the individual filing the application invented it (35 U.S.C. §102). Thus, for example, claims of a patent application covering a purified protein that was published or publicly used prior to the patent applicant's invention of the protein would not be patentable.

Section 103 of the Patent Statute specifies that a person shall be entitled to a patent on an invention unless the invention is obvious, such that "the differences between the subject matter sought to be patented and the prior art are such that the subject matter as a whole would have been *obvious* at the time the invention was made to a person having ordinary skill in the art to which said subject matter pertains" (35 U.S.C. §103) [italics added].

The objective standard for obviousness under Section 103 was set forth clearly by the Supreme Court in *Graham v. John Deere Co.*, 383 U.S. 1 (1966). The objective standard for obviousness requires that the examiner ascertain (i) the scope and content of the prior art, (ii) the level of ordinary skill in the art, and (iii) the differences between the claimed subject matter and the prior art (*Graham*, 383 U.S. at 17). The obviousness or nonobviousness of the claimed subject matter must be determined in light of these inquiries. Moreover, the *Graham* court also explained that secondary considerations, such as commercial success, long-felt but unsolved needs, and failure of others, might be utilized in determining the obviousness or nonobviousness of the invention (*Graham*, 383 U.S. at 17-18).

Following *Graham*, the C.C.P.A. and its successor, the Court of Appeals for the Federal Circuit (C.A.F.C.), have held the following considerations to be objective evidence of nonobviousness: long-felt need, commercial success, failure of others, copying, and unexpected results [See *Avia Group Int'l, Inc. v. L. A. Gear California. Inc.*, 853 F.2d 1557 (Fed. Cir. 1988); *In re Sernaker*, 702 F.2d 989 (Fed. Cir. 1983).] In fact, the C.A.F.C. has consistently made clear that when evidence of such secondary considerations is present, it must be considered by the examiner or a court in determining a question of obviousness. [See *Hybritech, Inc. v. Monoclonal Antibodies, Inc.*, 802 F.2d 1367, 1379-80 (Fed. Cir. 1986), *cert. denied*, 480 U.S. 947 (1987); *Stratoflex, Inc. v. Aeroquip Corp.*, 713 F.2d 1530, 1538-39 (Fed. Cir. 1983).]

The test for obviousness has also been formulated as requiring in the prior art both a suggestion of the invention and a reasonable expectation of success. The relevant inquiry is whether the prior art disclosure would have both suggested the invention and provided one of ordinary skill in the art with a reasonable expectation of successfully achieving the invention [*In re O'Farrell*, 853 F.2d 894, 903-04 (Fed. Cir. 1988); *In re Vaeck*, 947 F.2d 488, 495 (Fed. Cir. 1991)].

Three prominent federal circuit decisions have dealt with the obviousness/nonobviousness of claimed nucleic acid molecules: *Amgen, Inc. v. Chugai Pharmaceutical Co., Ltd.*, 927 F.2d 1200 (Fed. Cir. 1991), cert. denied, 502 U.S. 856; *In re Bell*, 991 F.2d 781 (Fed. Cir. 1993); and *In re Deuel*, 51 F.3d 1552 (Fed. Cir. 1995). In *Amgen*, the defendants alleged that the claim to an isolated DNA encoding human erythropoietin (EPO) was obvious since it was obvious to use the known monkey EPO gene, which has about 90% homology to the human gene, as a probe. The court disagreed, reasoning that the invention was nonobvious because neither the DNA nucleotide sequence of the human EPO gene nor its exact degree of homology with the monkey EPO gene were known at the time, and the degenerate probing method of cloning was speculative at the time (*Amgen*, 927 F.2d at 1208-09. The court concluded that although it may have been obvious to try to probe a human genomic DNA library with a monkey cDNA probe, there was no reasonable expectation of success (*Amgen*, 927 F.2d at 1208-09). (Note, however, that the court's conclusion of "no reasonable expectation of success" was based on the existing state of molecular biological science in the early 1980s, and that a court today might reach a different conclusion.)

In *Bell*, nucleic acid molecules containing the human coding sequences for insulin-like growth factors (IGF)-I and IGF-II were claimed. The C.A.F.C. found that the prior art disclosure of the complete amino acid sequence of IGF-I and IGF-II, in combination with a general cloning method (that emphasized the importance of unique codons), did not render obvious the human IGF gene sequences claimed, owing to the high degree of degeneracy of the genetic code for the disclosed amino acid sequence (*Bell*, 991 F.2d at 784). The court determined that more than 10^{36}

DNA sequences could be derived from the prior art disclosure, and that the prior art did not suggest which of the 10^{36} possibilities corresponded to the human gene. However, in reaching its decision of nonobviousness, the court emphasized that the claims did not cover all the 10^{36} DNA sequences that might potentially code for the gene product; rather, the claims were limited in scope to coding sequences of the human gene (*Bell*, 991 F.2d at 784-85).

In *Deuel*, isolated DNAs encoding human and bovine heparin-binding growth factors (HBGFs) were claimed. The clones were isolated from human placental and bovine uterine cDNA libraries. The C.A.F.C. found that the prior art disclosure of a partial amino acid sequence of heparin-binding brain mitogens (identical to the N-terminal sequence of human and bovine HBGF) in combination with the general cloning methods described in a well-known laboratory reference manual of cloning techniques did not render obvious the particular human or bovine cDNA sequences claimed or broader claims covering degenerate sequences encoding the human and bovine HBGF proteins (because only a partial and not the complete amino acid sequence was disclosed) (*Deuel*, 51 F.3d at 1556, 1560).

In reaching its decision, the court did question whether the broader claims were adequately enabled and supported by the specification in question, which failed to describe how to obtain any DNA except for the disclosed cDNA molecules. The court noted, however, that the claims covered compounds, and not methods. To be rendered obvious, the court reasoned, there must have been some suggestion of the structure, i.e., the sequence of the cDNAs claimed. In *Deuel*, the prior art did not disclose any structurally similar compounds (i.e., relevant DNAs). Instead, the prior art disclosed proteins, including a partial amino acid sequence. The court found that the knowledge of a protein does not give one conception of a particular DNA encoding it and stated that until the claimed molecules were actually isolated, it was highly unlikely for one of ordinary skill to contemplate what was ultimately obtained and "[w]hat cannot be contemplated or conceived cannot be obvious" (*Deuel*, 51 F.3d at 1558). The court acknowledged that the case might have been decided differently

had the protein been sufficiently small, simple, and lacking in degeneracy so that each possible DNA would have been obvious over the protein. The court reaffirmed the principle stated in *Bell*—that the existence of a general method of isolating cDNA or DNA molecules is essentially irrelevant to the question of whether the specific molecules would have been obvious, in the absence of other prior art that suggests the claimed DNAs, i.e., structurally similar DNAs, or the amino acid sequence of a small, simple protein with little degeneracy.

In summary, under current case law, a known amino acid sequence, without more, will not render obvious claims to the particular nucleotide sequence encoding the amino acid sequence. However, in rendering decisions of patentability, the Patent and Trademark Office has generally tried to limit the precedent set by *Deuel* to its facts. For example, in *Ex Parte Goldgaber*, 41 U.S.P.Q. 2d 1172 (Bd. Pat. App. & Int. 1995), the claimed subject matter was DNA that encodes the β-amyloid polypeptide of Alzheimer's disease. A prior art patent disclosed the full-length amino acid sequence of β-amyloid polypeptide and taught (i.e., enabled or described) two sets of degenerate probes that could be used to pull out the clone. Based on the facts of the case, the board of appeals distinguished its decision from *Deuel* and *Bell*, and held that the invention was obvious.

In view of the foregoing case law (*Amgen, Bell,* and *Deuel*) it would follow that, in appropriate instances, a partial gene sequence (EST) should not render the full-length gene obvious. This question has not been directly addressed by any published decision of which we are aware. Nonetheless, in determining the patentability of a claim to a nucleotide sequence encoding a gene of interest, one must consider the obviousness of the gene sequence in view of the publication of the EST sequence. The pertinent questions should be whether the prior publication of the partial (EST) sequence suggests the structure of the claimed gene or cDNA and whether one of ordinary skill in the art would reasonably expect success using the EST sequence to identify and clone the entire gene. In such a case, the determination of patentability should be factually driven. In our judgment, factors that should thus impact on the analysis of obviousness might include whether the prior art suggests a purified mRNA, a useful cellular source for the mRNA, or source for the mRNA from a cDNA or gDNA library from which the gene or coding sequence could be obtained; evidence of the failure of others; and evidence of unexpected results and any other factors affecting the expectation of success in cloning the gene, such as evidence of cross-hybridization, whether the mRNA is rare or expressed in low abundance, whether the mRNA is extremely large, and whether the secondary structure of the RNA was problematic. For example, if the published EST cannot be used to obtain the gene for technical reasons (e.g., low abundance of the mRNA or secondary structure of the mRNA that precludes full-length cDNA synthesis), then in our opinion a good case for nonobviousness could be made.

V. WRITTEN DESCRIPTION AND ENABLEMENT REQUIREMENTS FOR PATENTABILITY AND SPECIAL PROBLEMS RELATING TO LIVING ORGANISMS

Section 112 of the Patent Statute, entitled "Specification," states

> The specification shall contain a *written description* of the invention, and of the manner and process of making and using it, in such full, clear, concise, and exact terms as to *enable any person skilled in the art* to which it pertains, or with which it is most nearly connected, to make and use the same, and shall set forth the *best mode* contemplated by the inventor of carrying out his invention. [italics added]
> —35 U.S.C. §112

Section 112 further states that the specification shall conclude with claims "particularly pointing out and distinctly claiming the subject matter which the applicant regards as his invention."

As previously explained, once it is established that the subject matter of the invention is within the scope of Section 101 of the Patent Statute, several additional requirements must be met before a patent is awarded: The invention must be novel, useful,

nonobvious, and enabled by a written description in the patent specification that sets forth the best mode (i.e., best way) for making and using the invention. The requirement that there be a written description of the invention (the written description requirement) and that the description enable one skilled in the art to practice the invention (without "undue" experimentation) (the enablement requirement), as set forth in Section 112, can pose special problems for biotechnological inventions. Novelty and nonobviousness may be decided on a case-by-case basis, measuring the invention against the prior art. By contrast, meeting the written description and enablement requirements with respect to living organisms that often defy written description (e.g., cell lines such as hybridomas, mutant cells, and mutant organisms) imposes another complication in obtaining patent protection for living organisms.

Many cases have set out certain "ground rules" with respect to enabling others to make and use microorganisms that are the subject of a patent. These cases provide that microorganisms that are the subject matter of an invention may be deposited in order to satisfy the enablement and written description requirements for such microorganisms. The deposit is made pursuant to certain conditions with an organization that maintains such cultures of microorganisms much the way a library maintains its books—in other words, in an international depository such as the American Type Culture Collection that is recognized under the relevant treaty (there exists an international treaty to which the United States is a signatory—the Budapest Treaty on the International Recognition of the Deposit of Microorganisms for the Purposes of Patent Procedure).

Two cases, *In re Argoudelis*, 434 F.2d 1390 (C.C.P.A. 1970) and *In re Lundak*, 773 F.2d 1216 (Fed. Cir. 1985) discuss acceptable methods that can be used to deposit such microorganisms to meet the written description and enablement requirements of Section 112 for patenting microorganisms. In *Argoudelis*, the patent applicant's prior deposit of the claimed microorganism with an internationally accepted microorganism depository that makes deposited strains freely available to the public upon issuance of a patent was held to satisfy the requirement for an enabling written disclosure (*Argoudelis*, 434 F.2d 1390 at 1396). The *Lundak* decision adds a gloss to the deposit requirement in that the microorganism need not be deposited prior to filing the patent application, provided, however, that (i) the Patent and Trademark Office has access to the microorganism during the pendency of the application, and (ii) the public has access after the grant of a patent (*Lundak*, 773 F.2d at 1222). However, most foreign countries require deposits to be made on or before filing of the patent application.

The Patent and Trademark Office also has issued guidelines concerning such deposits of microorganisms. The guidelines are in accord with more recent decisions that specify that biological material need not be deposited if it is known and readily available to the public or can be made or isolated without undue experimentation [see 37 C.F.R. §1.802(b) (1998); *Amgen*, 927 F.2d 1200; *In re Wands*, 858 F.2d 731 (Fed. Cir. 1988)].

Thus, with respect to microorganisms, it is clear that the enablement and written description requirements can be met by depositing the subject strains with an appropriate depository. This logic for satisfying the enablement requirement was followed in the *Hibberd* case, in which the seeds that formed the subject matter of the patent were deposited with an organization that accepts plant seeds. With respect to genetically engineered animals, deposit of the DNA construct used to engineer the animal, or of germline cells or frozen embryos in a "germplasm bank," could possibly be utilized to fulfill the enablement requirement.

Another important issue is what is necessary for an adequate written description of a gene (an isolated nucleic acid). The significant cases in this area—*Amgen; Fiers v. Revel*, 984 F.2d 1164 (Fed. Cir. 1993); and *Regents of the University of California v. Eli Lilly & Co.*, 119 F.3d 1559 (Fed. Cir. 1997), *cert. denied*, 118 S. Ct. 1548 (1998)—establish that an inventor must have a "conception" of the invention in order to have an adequate written description of a claimed nucleic acid. Conception is the mental act of invention which in combination with an "actual" or "constructive" reduction to practice completes the making of an invention.

What, then, are the requirements for conception of a cloned nucleic acid? Conception of an invention, according to the *Amgen* court, is the

> formation in the mind of the inventor of a definite and permanent idea of the complete and operative invention, as it is hereafter to be applied in practice [citation omitted]. Conception requires *both* the idea of the invention's structure *and* possession of an operative method for making [the invention]. [Italics added]
> —*Amgen*, 927 F.2d at 1206

Reduction to practice is either actual (i.e., the actual practice of the invention) or constructive (i.e., the filing of a patent application that satisfies the Section 112 requirements and thus describes the invention in sufficient detail to enable the practice of the invention). In some instances, an inventor may be unable to establish a date of conception until he or she has actually reduced the invention to practice, resulting in simultaneous conception and reduction to practice. Under current case law, such as *Amgen*, conception of a gene has been found to occur simultaneously with its reduction to practice, i.e., the date on which the gene is in fact isolated and its sequence known. The case law also requires that to have a conception of the gene, the inventor must be able to envision characteristics of the gene sufficient to distinguish it from other genes.

Amgen involved recombinant DNA encoding EPO. Amgen owned a patent with claims directed to isolated DNA encoding human EPO and recombinant host cells. Amgen sued the defendants for producing recombinant EPO and thereby infringing the patent. The defendants cited Section 102 of the Patent Statute, alleging prior invention of human EPO by another who had conceived a successful strategy for isolating the human gene and was diligent in reducing it to practice. The court of appeals for the federal circuit held "that when an inventor is unable to envision the detailed constitution of a gene so as to distinguish it from other materials, as well as a method for obtaining it, conception has not been achieved until reduction to practice has occurred, i.e., until after the gene has been isolated" (*Amgen*, 927 F.2d at 1206). Thus, conceiving of the strategy for isolating the gene was not deemed a conception of the isolated gene itself.

Another case defining conception of a nucleic acid, *Fiers*, involved an interference contest between three parties over claims to DNA encoding human fibroblast β-interferon. In this case, the party who was first to file an application disclosing the entire gene sequence prevailed. To have an adequate written description, the specification must reasonably convey to the artisan that the inventor had possession of the claimed subject matter. Citing *Amgen*, the court held that conception of a DNA requires definition other than by its functional utility: It requires conception of the DNA's structure—definitive chemical or physical properties (*Fiers,* 984 F.2d at 1168-69). Conception did not occur upon conception of a method for obtaining the DNA. The *Fiers* court then addressed the written description requirement and held that "one cannot describe what one has not conceived" (*Fiers,* 984 F.2d at 1171).

The court found, moreover, that one losing party's written description of mRNAs and their proposed cloning methods for obtaining the gene was insufficient because the losing party's application did not disclose the nucleotide sequence "or an intact complete gene" (*Fiers,* 984 F.2d at 1170). In this regard, the court explained that

[A]n adequate written description of a DNA requires more than a mere statement that it is part of the invention and reference to a potential method for isolating it; what is required is a description of the DNA itself. . . . Claiming all DNAs that achieve a result without defining what means will do so is not in compliance with the description requirement; it is an attempt to preempt the future before it has arrived.
—*Fiers,* 984 F.2d at 1170-71

In *University of California,* the C.A.F.C. held that the nucleotide sequence of a rat insulin cDNA is not a sufficient written description of human, vertebrate, or mammalian insulin sequences. What was needed, the court reasoned, was a disclosure in the patent specification of a number of representative members of the genus or of structural characteristics common to the genus. The statement that the human gene was part of the invention, coupled with a description of the rat cDNA and proposed general methods for isolating the human cDNA, was deemed not adequate to fulfill the statutory written description requirement, and thus claims directed to "human," "vertebrate," and "mammalian" cDNAs were held invalid (*University of California,* 119 F.3d at 1568).

VI. USES OF DISCOVERED DNAs

Under current law, the new uses of genes and gene products for diagnostics, drug screening, and/ or therapy are potentially patentable. First, claimed uses of patentable genes and/or gene products for diagnosis, therapy, and drug screening should be patentable in view of the Biotechnological Process Patents Legislation enacted on November 1, 1995 (Public Law PL 104-41). This act amends Section 103 of the Patent Statute by adding a new Subsection 103(b), which provides that biotechnological processes using or resulting in a composition of matter that is novel under Section 102, and nonobvious under Section 103, shall be considered nonobvious provided that certain filing and ownership requirements are met. However, the legislation only addressed the issue of obviousness; the ultimate patentability of such biotechnological processes also depends on meeting the written description, enablement and best mode requirements of Section 112, and the utility requirement of Section 101 (as well as the novelty requirement).

The enactment of this legislation resolved an apparent contradiction in the previous case law and resolved a long-standing controversy that denied patentability of biotechnological processes based on the principles of *In re Durden,* 763 F.2d 1406 (Fed. Cir. 1985). *Durden* involved a chemical (rather than a biotechnological) invention. Both the starting compound and the final product were found to be novel, nonobvious, and patentable, but the claimed process was admittedly obvious and found to be unpatentable. The court held that an otherwise obvious process is not made patentable by the use of a new, nonobvious starting material or by the production of a new product (*Durden,* 763 F.2d at 1411). In contrast, under similar sets of facts in two other chemical cases, *In re Pleuddemann,* 910 F.2d 823 (Fed. Cir. 1990) and *In re Ochiai,* 71 F.3d 1565 (Fed. Cir. 1995), claims covering methods of using a novel, nonobvious starting material to make a novel, nonobvious product were held to be patentable.

For purposes of the Biotechnological Process Patents Legislation, biotechnological processes include processes for genetically engineering single or multicellular organisms (to express exogenous genes or to decrease or increase expression of endogenous genes), cell fusion processes for hybridoma production, and methods using products produced by such processes. Thus, processes that utilize novel and nonobvious genes, host cells, gene products, and/or antibodies (e.g., to diagnose or treat a disease and screen for drugs) should be patentable provided that such processes are supported by the disclosure of the application under Section 112 of the Patent Statute and are useful within the meaning of Section 101.

Quite apart from the amendment of Section 103, new and nonobvious uses discovered for known, published gene sequences should likewise be patentable, assuming the Section 112 requirements are met, since Section 100(b), in combination with Section 101 of the Patent Statute, provides that potentially patentable processes include new uses of known compositions of matter, materials, or processes. For example, a known sequence may be identified as a new target for designing diagnostics or therapeutics for a particular disease under investigation. Methods that utilize the known sequence in new diagnostic and therapeutic approaches may be patentable, provided that the prior art does not suggest and provide a reasonable expectation of success for such approaches (e.g., perhaps by disclosing the correlation of a particular gene with a particular disease).

VII. CONCLUSION

As the law currently stands, naturally occurring substances that have been purified, altered, or engineered in some way so that they are different in kind and/or form from their naturally occurring counterparts are new compositions of matter that may be patentable provided that, *inter alia*, they are also nonobvious and their manufacture and use are enabled by the patent specification. Patentable subject matter may be anything under the sun made by man except for abstract ideas and phenomena of nature. Man-made, nonhuman living organisms, including recombinant cells and transgenic organisms, are eligible for patent protection, provided that they meet the statutory requirements (e.g., *inter alia*, they are novel and nonobvious, and their manufacture and use are enabled by a written description in the

patent specification). With respect to microorganisms, it is clear that the enablement and written description requirements for such microorganisms can be met by deposits of the subject microorganisms with appropriate depositories under the Budapest Treaty on the International Recognition of the Deposit of Microorganisms for the Purposes of Patent Procedure.

Under current case law, a known amino acid sequence, without more, will not render obvious claims to the particular cDNA sequence encoding the amino acid sequence. The current case law also requires for an appropriate conception, and thus written description of a gene, that the inventor be able to envision characteristics of the gene that are sufficient to distinguish it from other genes. With respect to a gene, conception has been found to occur simultaneously with its reduction to practice, i.e., the date on which the gene is in fact isolated and its sequence known. Finally, processes that utilize novel and nonobvious genes, host cells, gene products, and/or antibodies (e.g., to diagnose or treat a disease and screen for drugs) should also be nonobvious and thus patentable, provided that such processes meet the other statutory requirements in that they have utility and are supported by an enabling written description in the patent specification that sets forth the best mode known to the inventor practicing the processes.

See Also the Following Articles

GENETICALLY MODIFIED ORGANISMS: GUIDELINES AND REGULATIONS FOR RESEARCH • INTERNATIONAL LAW AND INFECTIOUS DISEASE

Cases Cited

1. *Allen, Ex Parte*, 2 U.S.P.Q. 2d 1425 (Bd. Pat. App. & Int. 1987).
2. *Amgen, Inc. v. Chugai Pharmaceutical Co., Ltd.*, 927 F.2d 1200 (Fed. Cir. 1991).
3. *Animal Legal Defense Fund v. Quigg*, 710 F. Supp. 728 (N.D. Cal. 1989); *aff'd*, 932 F.2d 920 (Fed. Cir. 1991).
4. *Argoudelis, In re*, 434 F.2d 1390 (C.C.P.A. 1970).
5. *Avia Group Int'l, Inc. v. L. A. Gear California, Inc.*, 853 F.2d 1557 (Fed. Cir. 1988).
6. *Bell, In re*, 991 F.2d 781 (Fed. Cir. 1993).
7. *Bergstrom, In re*, 427 F.2d 1394 (C.C.P.A. 1970).
8. *Bergy, In re*, 563 F.2d 1031 (C.C.P.A. 1977).
9. *Borden Co. v. United States*, 308 U.S. 188 (1939).

10. *Brenner v. Manson* 383 U.S. 519 (1966).

11. *Chakrabarty v. Diamond,* 447 U.S. 303 (1980).

12. *Deuel, In re,* 51 F.3d 1552 (Fed. Cir. 1995).

13. *Durden, In re,* 763 F.2d 1406 (Fed. Cir. 1985).

14. *Fiers v. Revel,* 984 F.2d 1164 (Fed. Cir. 1993).

15. *Funk Bros. Seed Co. v. Kalo Inoculant Co.,* 333 U.S. 127 (1948).

16. *Goldgaber. Ex Parte,* 41 U.S.P.Q. 2d 1172 (Bd. Pat. App. & Int. 1995).

17. *Graham v. John Deere Co.,* 383 U.S. 1 (1966).

18. *Hibberd, Ex Parte,* 227 U.S.P.Q. 443 (Bd. Pat. App. & Int. 1985).

19. *Hybritech, Inc. v. Monoclonal Antibodies, Inc.,* 802 F.2d 1367 (Fed. Cir. 1986), *cert. denied,* 480 U.S. 947 (1987).

20. *Kratz, In re,* 592 F.2d 1169 (C.C.P.A. 1979).

21. *Lundak, In re,* 773 F.2d 1216 (Fed. Cir. 1985).

22. *Merck & Co. v. Chase Chemical Co.,* 273 F. Supp. 68 (D.N.J. 1967).

23. *Merck & Co. v. Olin Mathieson Chemical Corp.,* 253 F.2d 156 (4th Cir. 1958).

24. *Nelson v. Bowler,* 626 F.2d 853 (C.C.P.A. 1980).

25. *Ochiai, In re,* 71 F.3d 1565 (Fed. Cir. 1995).

26. *O'Farrell, In re,* 853 F.2d 894 (Fed. Cir. 1988).

27. *Pleuddemann, In re,* 910 F.2d 823 (Fed. Cir. 1990).

28. *Sernaker, In re,* 702 F.2d 989 (Fed. Cir. 1983).

29. *Stratoflex, Inc. v. Aeroquip Corp.,* 713 F.2d 1530 (Fed. Cir. 1983).

30. *University of California v. Eli Lilly & Co., Regents of the,* 119 F.3d 1559 (Fed. Cir. 1997), *cert. denied,* 118 S. Ct. 1548 (1998).

31. *Vaeck, In re,* 947 F.2d 488 (Fed. Cir. 1991).

32. *Wands, In re,* 858 F.2d 731 (Fed. Cir. 1988).

Suggested Readings/Web Sites

Barton, J. H. (1991, March). Patenting life. *Sci. Am.* **264**(3), 40–46 (March 1991).

Berkeley Technology Law Journal: *http://www.law.berkeley. edu/journals/btlj*

Burchfiel, K. J. (1995). "Biotechnology and the Federal Circuit." Patent Resources Group, Charlottesville, VA.

Burchfiel, K. J. (1997). "Biotechnology and the Federal Circuit: 1997 Supplement." Patent Resources Group, Charlottesville, VA.

Cooper, I. P. (1982). "Biotechnology and the Law." Clark Boardman Callaghan, Deerfield, IL.

Conlin, D. G., *et al.* (1995). "Biotechnology Patent Practice." Patent Resources Group, Charlottesville, VA.

Doll, J. J. (1998). The patenting of DNA. *Science* **280**, 689–690.

Harvard Journal of Law & Technology: *http://jolt.law.harvard.edu*

Intellectual Property Magazine: *http://www.ipmag.com*

Kevles, D. J. (1994). Ananda Chakrabarty wins a patent: Biotechnology, law, and society, 1972–1980. *Historical Stud. Phys. Biol. Sci.* **25**, 111–135.

Stanford Technology Law Review: *http://stlr.stanford.edu*

United States Patent and Trademark Office: *http://www. uspto.gov*

Wegner, H. C. (1992). "Patent Law in Biotechnology, Chemicals & Pharmaceuticals," Stockton Press, New York.

Pectinases

Fred Stutzenberger

Clemson University

GLOSSARY

endopolygalacturonase Poly(1,4 - α - D - galacturonide) glycanhydrolyase that randomly cleaves polygalacturonates to yield a series of intermediate oligomers.

exopolygalacturonase Poly(1,4-α-D-galacturonide) galacturonohydrolase that cleaves polygalacturonates from the nonreducing ends to yield either galacturonate or digalacturonate.

oligogalacturonate lyases A group of diverse, often cytoplasmic, enzymes which catalyze the eliminative removal of unsaturated terminal residues from D-galacturonate oligosaccharides.

pectate lyases Poly(1,4-α-D-galacturonide) lyases, acting in either an endo- or exo- manner, that cleave polygalacturonates via a transelimination mechanism to yield the double-bonded 4-deoxy-L-threo-5-hexoseulose uronic acid.

pectin Heterogeneous plant polysaccharide consisting primarily of methylated α-D-galacturonate residues linked (1-4) with small amounts of L-rhamnose-linked β (1-2) and β (1-4) to the D-galacturonate.

pectin methylesterase Pectin demethoxylase that hydrolyzes the methoxyl groups from the 6-carboxyl groups of the galacturonate residues in the pectin chain.

protopectinase Polygalacturonase that hydrolyzes protopectin (the water-insoluble, highly cross-linked parent compound of the pectic substances found in the middle lamella of plants) to a water-soluble form with varying degrees of polymerization.

PECTIN occurs in varying abundance in plant biomass. It contributes only approximately 4% to the dry weight of woody tissue but >30% in beet pulp and citrus fruit skins. This heterogeneous family of polysaccharides is important to the physical structure of the plant.

Pectins are found most abundantly in the middle lamella (the tissue that lies between adjacent cells) and act as an adhesive, plastifying gel. In the primary cell wall, pectin forms extensive polysaccharide networks (Fig. 1) similar to those composed of cellulose and hemicellulose.

In addition to its importance in plant structure, pectin is the source of pectic fragments which act as chemical messengers in the development, growth, senescense, and biochemical protection of plants. In this last role, pectin also contributes to the texture and appearance of commercial fruits (the shine on "an apple for the teacher" can be attributed to its pectin-rich surface). Pectin is also an important component in the diet of man and other animals; foods high in soluble fiber lower blood cholesterol and reduce glucose intolerance in diabetic animals. Pectin also alters digestive microflora by stimulating selected populations of pectinophilic bacteria such as *Lachnospira pectinoschiza*, a novel species of anaerobic, motile bacterium in the colonic contents of pigs maintained on a high pectin diet; pectin, polygalacturonate, and gluconate are the only substrates which support growth of these obligate pectinophiles. This article describes the various forms of pectic substances, the complex array of enzymes that depolymerize them, the cultural conditions influencing enzyme production by the pectolytic bacteria and fungi,

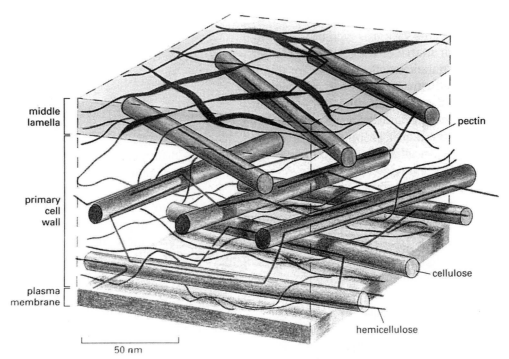

middle
lamella

pectin

primary
cell
wall

cellulose

plasma
membrane

hemicellulose

50 nm

Fig. 1 Scale model of a portion of a plant primary cell wall showing the major polysaccharide networks. The orthogonally arranged layers of cellulose microfibrils are cross-linked into a network by hydrogen-bonded hemicellulose, which in turn is coextensive with the pectic polysaccharide network. The cellulose/hemicellulose network provides tensile strength while the pectin network resists compression. The three classes of polysaccharides are approximately in equal proportions in the primary cell wall. In addition to compression resistance, the pectin in the middle lamella cements adjacent cells together (reproduced with permission from Alberts *et al.*, 1994, p. 1002).

and some biotechnological applications of pectinase activity.

I. PECTIC SUBSTANCES

Primary pectin chain composition consists largely of partially esterified D-galacturonic acid residues linked α (1-4) with small amounts (2–4%) of L-rhamnose-linked β (1-2) and β (1-4) to the D-galacturonate units (Fig. 2). In addition to the primary chain structure, various side chains are composed of arabinose, glactose, and xylose. Apple pectin, which has been well characterized, consists of "hairy" regions (backbones of rhamnogalacturonan carrying arabinogalactan and xylogalacturonan side chains) and "smooth" regions consisting of homogalacturonans with a 70–80% residue methylation in an apparently random manner. Modified hairy region

polymers released by the synergistic action of pectinases, hemicellulases, and cellulases on apple juice have been characterized as highly branched rhamnogalacturonans, with arabinose the most abundant sugar (55 mol%).

Fig. 2 Chemical structure of a segment of pectin polymer showing an extended stretch of D-galacturonate units in pyranose ring form. Some of the C6 carboxyl groups are methylated, whereas others are in the -COOH or -COO⁻ form, depending on the pH of the solution. Pectic substances are most stable at pH 3 or 4; at higher pH, spontaneous demethylation occurs, whereas at lower pH the chain undergoes hydrolysis (reproduced with permission from Whitaker, 1990, p. 136).

Protopectin, the parent compound of pectic substances, exists as huge, insoluble polymers of α-D-galacturonic acid residues; the majority of the -COOH groups of the residues are esterified with methyl groups to form -COOCH$_3$. The polymers are cross-linked at unesterified groups with divalent cations such as Ca^{2+} and Mg^{2+}. Water-soluble pectic substances (which may be derived from the parent compound by enzymatic action) are as follows: pectin (in which at least 75% of the galacturonate residues are esterified with methanol), pectic acid (in which all the methoxyl groups are removed from the galacturonate residues), and pectinic acid (which is 1–75% methylated). These pectic substances may be hydrolyzed to form oligogalacturonates and oligomethylgalacturonates, each having two or more residues. These oligosaccharides have been particularly useful in determining the size requirements of pectinase catalytic sites.

Pectin can be analyzed in a variety of ways, depending on the specific pectic substance to be measured. Galacturonic acid is measured by its reaction with *m*-hydroxyl biphenyl to form a colored product in heated acidic media. Base titration before and after saponification is useful for the simultaneous determination of degree of methylation and galacturonate content. The degree of methylation can also be determined by saponifying methyl ester groups and analyzing for free methanol by gas chromatography. Carboxyl and methyl ester groups can be quantitated against known standards in deuterated water by infrared spectroscopy. Neutral sugars can be volatilized for separation on gas chromatography after hydrolysis and derivatization.

II. PECTOLYTIC ENZYMES

Most of the studies on pectolytic enzymes have been prompted by the importance of their roles in plant pathogenicity of both bacteria and fungi. The phytopathogenicity and epidemiology of these microbes is a multifaceted process which may include one or several of the following: transmission vectors (such as insects and airborne pollen), binding of microbial cells at receptor sites on host plant cells (such as the binding of the β-glucans of the *Agrobac-*

terium cell wall lipopolysaccharide to pectin complexes on the plant cell wall), secretion of invasive enzymes which degrade plant cell walls (particularly prominent in the soft-rot diseases that are characterized by a watery disintegration of the parenchyma tissues), and elaboration of host-specific toxins (low-molecular-weight metabolites such as the polypeptide, *victorin,* which causes rapid dissociation of the plant plasmalemma, and *T-toxin,* a polyketol which inhibits ATP synthesis in plant mitochondria).

In addition to these factors, some plant-pathogenic fungi employ elaborate structures to mechanically penetrate host plant tissue at the infection site. The appressorium is one such structure; it secretes a sticky substance which serves as an attachment adhesive and produces a penetration peg. This peg penetrates deep into the plant tissue via both mechanical pressure and a range of depolymerizing enzymes. The pectolytic enzymes are particularly important in this penetrating process because they degrade the middle lamella; this degradation kills the surrounding host cells and releases soluble sugars for growth of the fungus.

Plant pathogens have not only caused a continued and extensive economic loss in agriculture but also triggered widespread ecological and demographic changes. For example, the canker fungus, *Cryphonectria parasitica,* which produces a potent endopolygalacturonase as part of its invasive pectolytic armamentarium, destroyed the chestnut (*Castanea dentata*) tree population in the eastern forests of North America from Maine to Georgia. The result from the loss of this single species was devastating to the viability of rural communities because chestnuts constituted approximately 25% of the total Appalachian tree population. The chestnut trees were valued for their rot-resistant lumber, for their nuts which were a nutritious staple for man and animals, and for their tannins which were widely used in the leather processing industry. By the 1920s, the fungus had invaded more than 80% of the chestnut range, causing tremendous economic and ecological devastation (which eventually spread to Europe through the export of diseased chestnut wood). The tendency of the chestnut to send up secondary shoots from the base of the dying tree led the poet Robert Frost to hold out hope for the eventual survival of the species in this poem "Evil Tendencies Cancel":

"Will the blight end the chestnut?
The farmers rather guess not.
It keeps smoldering at the roots
And sending up new shoots
Til another parasite
Shall come to end the blight."

Today, a few chestnut trees have survived, mostly as secondary shoots, in isolated areas.

Not all microbial invasions of plant tissue have such disastrous consequences as the chestnut blight. One of the most interesting and beneficial plant–microbe interactions is the symbiosis between legumes of agricultural importance (soybeans, lentils, clover, alfalfa, and peas) and heterotrophic soil bacteria of the genera *Rhizobium*, *Bradyrhizobium*, and *Azorhizobium*. Infection of the roots by these bacteria leads to the formation of root or stem nodules which can convert gaseous nitrogen to combined nitrogen. The initial infection in the root hairs of leguminous

plants by the bacteria is facilitated by an array of plant cell wall-decomposing enzymes (cellulase, hemicellulase, and pectinase). After the bacteria gain access to the tissue of the main root, they form an endosymbiotic relationship with the plant via the formation of bacteroids (swollen, pleiomorphic forms which are surrounded by portions of the plant cell membrane). The symbiosis continues until seed formation, at which time the nodules disintegrate and release the nitrogen-fixing rhizobia back into the soil to infect new root hairs.

The array of pectolytic enzymes which facilitate plant pathogenesis or endosymbiosis is even more complex than the types of plant molecules they depolymerize. The major classes of pectolytic enzymes are listed in Table I.

A. Protopectinases

Of the recognized pectolytic enzymes, protopectinases have been the least studied; the extracellular enzymes of *Kluvuerornyces fragilis* (IFO 0288), *Ga-*

TABLE I
Classification of the Pectolytic Enzymes[a]

Name	EC No.	Primary substrate	Products	Mechanism
Esterase				
Pectin methylesterases (pectinesterases)	3.1.1.11	Pectin	Pectic acid + methanol	Hydrolysis
Polygalacturonases				
Protopectinases		Protopectin	Pectin	Hydrolysis
Endopolygalacturonases	3.2.1.15	Pectic acid	Oligogalacturonates	Hydrolysis
Exopolygalacturonases	3.2.1.82	Pectic acid	Monogalacturonate	Hydrolysis
Oligogalacturonate hydrolases		Trigalacturonate	Monogalacturonate	Hydrolysis
$\Delta 4:5$ Unsaturated oligogalacturonate hydrolases		$\Delta 4:5$ (Galacturonate)$_n$	Unsaturated monogalacturonate and saturated $(n-1)$	Hydrolysis
Endopolymethylgalacturonases		Pectin	Methyl oligogalacturonates	Hydrolysis
Lyases				
Endopolygalacturonate lyases (endopectate lyases)	4.2.2.2	Pectic acid	Unsaturated oligogalacturonates	Transelimination
Exopolygalacturonate lyases (exopectate lyases)	4.2.2.9	Pectic acid	Unsaturated digalacturonate	Transelimination
Oligogalacturonate lyases	4.2.2.6	Unsaturated digalacturonate	Unsaturated monogalacturonate	Transelimination
Endopolymethylgalacturonate lyases (endopectin lyases)	4.2.2.10	Pectin	Unsaturated methyl oligogalacturonates	Transelimination

[a] From Whitaker (1990).

lactomyces reesii (University of Tokyo Collection, No. JAM 1247), and *Trichosporon penicillatum* (ATCC 42397) have provided most of the currently available information in this regard. When aerobically grown in a glucose–peptone–yeast extract medium at pH 5.0 and 30°C, these fungi synthesize protopectinase, reaching a maximum at 15–25 hr. Purification procedures (precipitation with ammonium sulfate, ion-exchange chromatography on carboxymethyl Sephadex, and size-exclusion chromatography on Sephadex G-75) have yielded crystalline preparations that are electrophoretically homogeneous. Some properties of the pure protopectinases are listed in Table II. One unit of protopectinase activity is defined as the liberation of pectic substances equivalent to 1.0 μmol D-galacturonic acid/ml of reaction mixture at 37°C in 30 min.

The mechanism by which these enzymes release pectin from protopectin has been postulated from studies of their action on galacturonic acid oligomers. Based on reaction products from oligomers with degrees of polymerization from 3 to 5, the enzymes apparently cleave the glycosidic bonds of the parent compound at sites having at least three unmethoxylated galacturonate residues. Both the K_m and the V_{max} values for the enzymes are influenced greatly by chain length (for a summary of their action, see Table III). Because most of the galacturonate residues in the protopectin molecules are methoxylated at random, protopectinase action can occur only at a limited number of sites. Therefore, the product retains a high degree of polymerization. Although protopectinase activity involves the cleaving of polygalacturonate chains, other polygalacturonases have relatively low affinity for protopectin and probably do not contribute significantly to protopectin solubilization in nature.

B. Pectin Methylesterases

This enzyme is a carboxylic acid esterase that has been referred to in the literature as pectinesterase, pectase, pectin methoxylase, pectin demethoxylase, and pectolipase. The products of its action on pectin include deesterified pectin (which may eventually yield pectic acid), methanol, and a H^+ from the ionization of the newly formed carboxyl group. One unit of activity is usually defined as either the amount of enzyme causing a decrease of 0.1 in the pH of the reaction mixture in 30 min or the liberation of 1 mg CH_3OH in 30 min.

Pectin methylesterase is produced by a variety of microbes, including many fungal genera (*Aspergillus, Botrytis, Fusarium, Helminthosporium, Heterobasidian, Lachnospira, Myrothecium, Penicillium, Phytophthora, Saccharomyces,* and *Tubercularia*) and a variety of bacteria (*Bacillus, Clostridium, Corynebacterium, Pseudomonas,* and *Thermomonospora* species). Fungal pectin methylesterases are produced by aerobic growth at 20–30°C in chemically defined medium. As with the propectinases, the addition of pectin to such media generally does not enhance

TABLE II
Characteristics of Fungal Protopectinases

Characteristic	*Kluyveromyces fragilis*	*Galactomyces reesei*	*Trichosporon penicillatum*
Molecular weight	35	33	33
Sugar content (%)	5.9	5.1	1.7
Isoelectric point	5.0	8.5	7.7
Optimal pH	5.0	5.0	5.0
Thermostability (°C)	<40	<50	<55
K_m (mg/ml) for protopectin	90	50	30
Yield (%) of total pectin from			
Orange peel	63	97	88
Radish	99	100	82
Carrot	73	95	82

TABLE III
Action Mode of Protopectinases on Galacturonic Acid Oligomers[a]

Enzyme	Substrates	Reaction products	K_m (mM)	V_{max} (μM/U/min)
Kluyveromyces fragilis protopectinase			$>10^2$	$<10^{-10}$
			1.82	1.90×10^{-2}
			5.95×10^{-1}	2.95×10^{-1}
Galactomyces reessii protopectinase			3.98	1.96×10^{-3}
			2.77	2.37×10^{-2}
Trichosporon penicillatum protopectinase			7.09×10^{-1}	9.79×10^{-2}
			4.26	4.79×10^{-4}
			2.20	5.20×10^{-2}
			8.69×10^{-1}	2.85×10^{-1}

[a] Reactions were done under the optimum conditions for each enzyme. Hexagons express D-galacturonic acid molecules. (Adapted from Sakai, 1988, p. 348.)

methylesterase production, although the presence of orange peel pulp or similar sources of complex plant material will often result in increased enzyme levels. Characteristics of these enzymes from a variety of fungal species are relatively similar in several aspects: (i) their M_r values are in the range of 26–45, (ii) their isoelectric points generally fall in the pH range of 6 to 7, (iii) their pH optima fall in two ranges (3.5–4.8 and 6.5–7.5), and (iv) their activity is stimulated by NaCl concentrations of 0.05–0.25 M.

In contrast to the constitutive nature of the fungal pectin methylesterases, these enzymes are inducible in bacteria. Furthermore, in *Corynebacterium* species isolated from yam soft rot, pectin methylesterase is induced during growth in plant tissue but not in media containing purified pectin. This suggests that additional factors other than substrate are necessary for induction. In *Clostridium multifermentans,* polygalacturonic acid is a better inducer of pectin methylesterse than is pectin because esterase and lyase activities reside in the same protein. It is interesting in this dual-function protein that esterase activity can be eliminated by heating for 30 min at pH 7 without loss of lyase activity. The molecular characteristics of only a few bacterial pectin methylesterases have been reported to date. Their M_r ranges from 41 in *Pseudomonas solanacearum* to 400 in *C. multifermentans*. Their pH optima range from 7 to 9, and their stability is markedly pH dependent (a characteristic responsible for large losses in activity during purification).

Pectin methylesterases have almost exclusive specificity for the ester linkages of polymethylgalacturonides; substrate characteristics that influence their affinities and reaction rates are the degree of polymerization of the pectin chain and the distribution of methoxyl groups along its length. A free carboxyl group adjacent to a methylated group is required at the site of hydrolysis; therefore, chains of fully methylated residues are relatively resistant to enzymatic attack.

The action of pectin methylesterase is obviously very important to the action of other pectolytic enzymes in nature. Because the polygalacturonases and polygalacturonate lyases cannot act on highly methylated chains, the inability of a microbe to synthesize pectin methylesterase severely limits its ability to use pectin as a carbon and energy source. However, some phytopathogens may circumvent their deficiencies in pectolytic enzyme production by synergy with host enzyme systems (e.g., *Bacillus subtilis* polygalacturonate lyase acts in conjunction with carrot polygalacturonase and pectin methylesterase to macerate host tissue). *In vitro*, traces of pectin methylesterase activity contaminating purified preparations of polygalacturonases and polygalacturonate lyases introduce drastic errors in kinetic characterizations. For example, in a study on the polygalacturonate lyase from *Thermomonospora curvata*, the K_m for 60% esterified pectin was 625 mg/ml when determined using a highly purified preparation lacking pectin methylesterase activity; in the presence of the esterase, this apparent value was reduced almost 2000-fold.

A sufficient amount of pectin methylesterase also appears to be the key in the application of pectolytic enzymes in vegetable pulp saccharification. Lack of the esterase in *Trichoderma reesei* and *Talaromyces emersonii* severely limits the ability of their pectolytic enzymes to solubilize agricultural residues. When their enzyme systems are supplemented with pectin methylesterase from another source, such plant materials are readily esterified.

C. Endopolygalacturonases

Endopolygalacturonases are relatively common in the microbial world. They are secreted by a variety of fungi, including the following genera: *Aspergillus, Botytris, Corticium, Fusarium, Kluyveromyces, Penicillium, Rhizoctonia, Rhizopus, Talaromyces, Trichoderma, Trichosporon,* and *Verticillium*. Cultures are generally grown at 20–30°C (45°C for *Talaromyces*) in aerated, glucose-free mineral salts–pectin media or in macerated plant tissue such as beet or carrot pulp at pH 4–6. Polygalacturonase activity (either endo- or exo-) is generally measured by the increase in reducing groups; a unit of activity is defined as the release of 1 μmol of reducing groups (as D-galacturonic acid) per minute at 30°C. Measurement of viscosity changes using an Ostwald viscometer is less convenient for routine measurement but is useful in distinguishing between endo- and exoaction. Endopolygalacturonases can be visually detected in zymograms using a coupled substrate (D-galacturonan DP10 with Ostazin Brilliant Red S-5B dye) in slab gels after electrophoresis or isoelectric focusing.

The properties of fungal endopolygalacturonases are, with few exceptions, remarkably uniform from species to species. The M_rs of these enzymes mostly fall in the range of 30–60 (*Aspergillus* and *Botrytis* species produce endopolygalacturonases of 85 and 69, respectively). Their isoelectric points generally range from pH 6 to 8 (3.8–4.5 for *Aspergillus* and *Penicillium*). A notable exception to this generalization is the nonglycosylated 34-kDa endopolygalacturonase produced by the barley pathogen, *Cochliobolus sativus*; it has an isoelectric point at pH 9.3. The pH optima of the fungal endopolygalacturonases are in the range of 4–6, except the enzyme from *Corticium rolfsii* has maximal activity at pH 2.5. The K_m values for pectin are within a 10-fold range (0.1–1.0 mg/ml), although when comparing these values it must be remembered that traces of pectin methylesterase activity greatly influence the apparent values. Although most of these widely studied fungi are mesophiles (with the exception being *Talaromyces*), the endopolygalacturonases that they secrete are often quite thermostable. Much of this thermostability is apparently conferred by stabilizing factors secreted into culture fluid. For example, the partially purified endopolygalacturonase from *Penicillium capsulatum* had a half-life of 3 hr at 60°C but only 3.8 min when purified to homogeneity.

Endopolygalacturonase is an important phytopathogenic factor in the fungi. Damage to plant cell walls by endopolygalacturonase is accompanied by the release of pectic fragments and their accumulation in intercellular spaces and in the aggregated cytoplasm. In studies on the anthracnose fungus, *Colletotrichum lindemuthianum,* the localization of the enzyme can be traced in bean leaf cells via a polygalacturonate-binding agglutinin.

Bacteria (including *Bacillus, Erwinia,* and *Pseudomonas* species) also produce endopolygalacturonases. For highest production, they are generally cultured for 24–48 hr at 28–30°C in vigorously aerated (antifoaming agent usually required) peptone or yeast extract mineral salts media containing pectin or polygalacturonate. In *Erwinia* cultures, the addi-

tion of 50 mM ethylenediaminetetraacetate (EDTA) stimulates polygalacturonase production.

The bacterial endopolygalacturonases have not been extensively characterized. *Erwinia carotovora* secretes a galacturonase that has a pH optimum of 5.3 and that is stimulated by Na^+, K^+, and NH_4^+. Its major products from polygalacturonic acid are mono- and digalacturonic acids. Hexameric galacturonic acid is cleaved to trimers; digalacturonic acid is not cleaved. *Pseudomonas solanacearum* produces two forms of the enzyme, with pH optima at 5.5 and isoelectric points at pH 5.0 and 7.5, respectively.

D. Exopolygalacturonases

Little information is available on the microbial exopolygalacturonases compared to the corresponding enzymes in plants; a diverse set of exopolygalacturonases are expressed during plant pollination (there are 12 known exopolygalacturonase genes in maize). Several fungal species (*Aspergillus, Coniothyrium,* and *Rhizopus*) produce the enzyme, which is distinguished from the endopolygalacturonases by differential rates of reducing sugar production and viscosity reduction in polygalacturonic acid solutions. The usual product of the exo-acting galacturonases is monogalacturonic acid, although *Erwinia* and *Pseudomonas* species produce poly(1,-4α-D-galacturonide) digalacturonhydrolases, which cleave diagalacturonic acid units from the nonreducing ends of polygalacturonates. Enzymes that remove mono- or digalacturonic acid units liberate much reducing sugar but decrease viscosity very little compared to the endo-acting galacturonases.

The exopolygalacturonase from *Erwinia chrysanthemi* has a M_r of 67, an isolectric point at pH 8.3, a pH optimum of 6.0, and a K_m of 0.05 mM for galacturonate as substrate. An interesting aspect of exopolygalacturonase production in *Erwinia* is its regulatory function exerted on the biosynthesis of polygalacturonate lyase isozymes. The exopolygalacturonase releases digalacturonic acid, which is an inducer of the lyase. This activity may increase the sensitivity of catabolite repression toward changes in pectate lyase activity by continuously sampling the nonreducing ends of the pectic polymer population. Its product exerts less catabolite repression than does the 4,5-unsaturated digalacturonic acid product from the lyase. The ratio of the two products may constitute a feedback mechanism keyed to the changes in the activity of the lyase in the environment. Moreover, the expoplygalacturonase enables *Erwinia* to utilize polygalacturonic acid as a carbon and energy source under conditions in which the lyase is inactive due to the unavailability of sufficient Ca^{2+}.

Among the few exopolygalacturonases to have been highly purified and characterized, the mercury-activated intracellular enzyme produced by *Aspergillus niger* is perhaps the most unusual. The hydrolysis rate is highest on di- and trigalacturonic acids at pH 4.5 and is increased 10-fold by the presence of 1 mM Hg^{2+} (K_a for Hg^{2+} was $6 \times 10^{-8} M$). Sulfhydryl compounds such as 2,3-dimercaptopropanol or chelating agents such as EDTA cause almost complete loss of activity. Other divalent cations (Ca, Zn, Cu, Co, Ni, Ba, Sr, Fe, Cd, Pb, and Mn) had no effect on exopolygalacturonase activity. The role of this novel enzyme in the metabolism of the fungus is intriguing because one would not expect cytoplasmic levels of Hg^{2+} to be sufficient for activation.

E. Oligogalacturonate Hydrolases

These enzymes are identical in their action to the polygalacturonases and are measured by the same reducing group methods (viscometry is not useful in their assay). However, they differ in regard to substrate affinity insofar as they have higher specific activities on oligogalacturonate chains with low degrees of polymerization. For example, for the oligogalacturonate hydrolases produced by *Bacillus* species, the order of preference is trimer > tetramer > pentamer > dimer > acid-soluble pectic acid (degree of polymerization, 15–20 galacturonate residues). The other major difference is in enzyme location. In contrast to the endopolygalacturonases, the oligogalacturonases are cell-associated and are released only in old cultures by autolysis. Extraction strategies usually employ cakes of frozen cells subjected to mechanical disruption or to organic solvents such as toluene or acetone.

The oligogalacturonate hydrolases can be divided into two categories: One type acts on saturated oligo-

galacturonides, whereas the other acts on $\Delta 4 : 5$ unsaturated oligogalacturonides. The *Bacillus* enzyme is highly specific; it is completely inactive against the unsaturated chains and cleaves the terminal glycosidic bond at the nonreducing ends of saturated oligogalacturonides. The *Aspergillus* enzyme is unique in this regard, attacking both substrates (although the hydrolysis rate on unsaturated digalacturonate is only approximately 2% of that on the saturated digalacturonate).

F. Endopolymethylgalacturonases

The distribution and even the existence of this group of endopectin hydrolases in the microbial world are currently under dispute. The reaction is presumed to occur through the random hydrolysis of polymethylgalacturonates to oligomethylgalacturonates. A unit of activity is defined as production of 1 μmol of reducing groups per minute at 30°C.

A variety of fungi (*Alternaria*, *Aspergilus*, *Botrytis*, *Penicillium*, and *Sclerotinium*) and bacteria (*Erwinia* and *Xanthomonas*) have been reported to synthesize endopolymethylgalacturonases. However, this reaction has never been conclusively demonstrated in reaction mixtures containing 100% methylated polygalacturonate as substrate and enzyme preparations devoid of pectin methylesterase and endopolymethylgalacturonate lyase activities. Further work using highly purified enzyme preparations with well-characterized substrates will be required to substantiate claims for the existence of this category of pectinase. Such conclusive proof might not be so easily obtained because microbes purported to produce endopolymethylgalacturonases also produce the other enzymes previously described.

G. Endopolygalacturonate Lyases

The endopolygalacturonate lyases, although ubiquitous in the microbial world, have not been demonstrated in higher life forms. In addition to the unique mechanism of glycosidic bond cleavage (transelimination rather than hydrolysis), they also differ from the hydrolases in that their pH optima (8–10.5) are some of the highest known. These enzymes have an absolute requirement for Ca^{2+} (although Mg^{2+} and

Mn^{2+} can partially replace Ca^{2+} in some). In the soft-rotting pseudomonads (such as *Pseudomonas fluorescens*), in which endopolygalacturonate lyase is the sole enzyme responsible for maceration of plant tissue, Ca^{2+} induces higher levels of enzyme than polygalacturonate. Moreover, Ca^{2+} not only induces the enzyme but also controls its localization; in Ca^{2+}-deficient media, most of the endopolygalacturonate lyase remains cell associated, whereas with adequate Ca^{2+} most of the enzyme is found free in the medium. When deprived of Ca^{2+} by the presence of chelating agents such as EDTA, attack by pseudomonads on plant tissue is considerably diminished.

A unit of endopolygalacturonate lyase activity is most often defined as the generation of 1 μmol of $\Delta 4:5$ unsaturated galacturonic acid per minute. Rates are usually measured spectrophotometrically at 235 nm (the extinction coefficient of the unsaturated product is 4600 M^{-1} cm^{-1}). Activity can also be measured by reducing group generation or viscosity reduction. The unsaturated oligosaccharides (degree of polymerization of 2–11 residues) which are generated from polygalacturonate lyases can be separated by ion-pair reverse-phase high-performance liquid chromatography (HPLC) to determine whether the enzyme has exo- or endo- activity.

Although many fungi produce endopolymethylgalacturonate lyases, only a few genera (*Cephalosporium* and *Hypomyces*) have been shown to produce endopolygalacturonate lyase; the enzyme is more common among the bacteria (*Arthrobacter, Bacillus, Enterobacter, Erwinia, Klebsiella, Micrococcus, Pseudomonas, Streptomyces, Thermomonospora,* and *Xanthomonas*). Highest production is usually achieved in mineral salts media containing 0.5–1.0% polygalacturonate as sole carbon source (although pectin is often a much better substrate due to its slower rate of degradation and hence lower tendency to cause catabolite repression).

As might be expected from the broad range of sources, the molecular characteristics of these enzymes are relatively varied compared to the other pectinases. The M_r ranges from 18 to 70, and the isoelectric points from pH 4.6 to 10.3. Pectates are generally good substrates for the endopolygalacturonate lyases. The K_m values for pectate range from 6 μg/ml in *Bacillus polymyxa* to 5 mg/ml in *Erwinia*

rubifaciens. These values may be misleading when used for comparative purposes if identical substrates are not used for all enzymes since the degree of pectin esterification has widely varied effects on K_m determinations, depending on the source of the enzyme. In *Arthrobacter* and *Bacillus,* pectinic acids with degrees of esterification ranging from 21 to 44% were the best substrates for lyase activity, whereas in *Thermomonospora* an increase in the degree of esterification from 0 to 44% increased the K_m by approximately 16,000-fold. The influence that the degree of esterification has on pectate lyases raises some question as to whether or not some of the enzymes previously included in this group should be reclassified as low methoxyl pectin lyases.

Reaction rates also depend on substrate chain length, with the hexamer or pentamer supporting the highest activity and the trimer or tetramer representing the lowest degree of polymerization allowing detectable activity. The essential role for Ca^{2+} is apparent because calcium tetragalacturonate is cleaved, whereas the tetragalacturonate anion in the presence of EDTA is not.

H. Exopolygalacturonate Lyases

In contrast to the widespread occurrence of endopolygalacturonate lyases in the bacteria, exopolygalacturonate lyases are far less common. Principle producing genera include *Clostridium, Erwinia,* and *Streptomyces.* The preferred substrate is polygalacturonate, from which $\Delta 4:5$ digalacuronates (or trigalacturonates in the case of *Streptomyces*) are split from the nonreducing chain ends by trans-elimination. The method for measurement of exopolygalacturonate lyase activity is identical to that for the other lyases, i.e., accumulation of the transelimination product determined spectrophotometrically at 235 nm. Other than substrate specificity and mode of action, the characteristics of the exolyases are similar to those of the endolyases; they have generally high pH optima and a requirement for Ca^{2+}. It is interesting that the Ca^{2+} requirement for the exopolygalacturonate lyase from *Streptomyces nitrosporeus* varies with substrate chain length; reaction with oligogalacturonic acids (trimer to hexamer) is markedly promoted by high Ca^{2+} levels, whereas its activity on polygalacturonate is inhibited by them. The endo- and exopolygalacturonate lyases (like the polygalacturonases) may be distinguished by their ratios of rates of viscosity reduction versus production of the transelimination product. The exolyases convert approximately 20–25% of the polygalacturonate substrate to the unsaturated dimer before a 50% reduction in viscosity is detected, whereas the endolyases achieve the same viscosity reduction with only 1 or 2% cleavage of substrate.

I. Oligogalacturonate Lyases

Some of the plant pathogens, notably *Erwinia* and *Pseudomonas,* produce oligogalacturonate lyases that attack their substrates from the reducing end (unlike the oligogalacturonate hydrolases) to yield $\Delta 4:5$ unsaturated monomers from oligogalacturonates with degrees of polymerization of 2–20. The *Erwinia* enzymes have a preference for the unsaturated dimer over its saturated counterpart by a $3:1$ margin, whereas the *Pseudomonas* enzyme prefers tetragalacturonate. These enzymes, which have pH optima in the range of 6.5–7.5, are only slightly stimulated by Ca^{2+}, the cofactor required by other lyases for activity. Only a few oligogalacturonate lyases have been studied in detail. Their role in pectin metabolism should be better defined because they might be essential to the utilization of pectins as carbon and energy sources in some microbial species such as *Clostridium lochheadii,* which is part of the bovine rumen microflora. The pectate lyases from various bacteria present an interesting array with a commonality of mechanism, but they are difficult to group in well-defined classes due to their divergent and overlapping characteristics.

J. Endopolymethylgalacturonate Lyases

In contrast to the pectate lyases, which are produced almost exclusively by bacteria, the endopolymethylgalacturonate lyases, with few exceptions, are restricted to the fungi. *Alternaria, Aspergillus, Dothidea,* and *Phoma* species produce these enzymes, often in multiple forms, but with close conformity with regard to molecular weight (M_r in the range of 28–

35), K_m for pectin (1–5 mg/ml), and stimulation by (but not absolute requirement for) Ca^{2+}. Such conformity is unusual in view of the marked differences in amino acid composition (isoelectric points from pH 3 to 9). Their pH optima range from 5 to 9, making them suitable for a wide range of applications in the food industry, particularly for substances having large concentrations of highly esterified pectin.

Of this group of enzymes, the *A. niger* pectin lyase B is one of the most highly characterized. Crystallographic studies at a resolution of 1.7 Å revealed a structure with 359 amino acids and 339 water molecules. The polypeptide folds into a right-handed parallel β helix, from which loops and folds (the largest of which is 53 amino acids) protrude from the central structure and confer function. The substrate binding site consists of an extensive network of highly conserved tryptophane and histidine residues and is therefore less hydrophilic than that of the *Erwinia chrysanthemi* pectate lyase C (which has a novel type of domain structure in which side chain atoms form linear stacks that include ladders of asparagine, serine, aliphatics, and ringed residues). The pectin lyase B structure provides an explanation for the lack of a catalytic requirement for Ca^{2+} in the pectin lyase family compared to the pectate lyases; the positive charge in the catalytic site is provided by a conserved arginine instead of Ca^{2+}.

Like the other lyases, the activity of this enzyme is most readily measured spectrophotometrically by an increase in A_{235} during generation of the $\Delta 4:5$ double bonds produced at the nonreducing ends of the unsaturated products. Activity can also be measured by reducing group formation. The two methods used together give valuable information with regard to the purity and mechanism of the enzyme preparation because the molar ratios of double bonds and reducing groups should be equal when produced by a homogeneous enzyme preparation.

The bacteria have been studied relatively little for their ability to produce endopolymethylgalacturonate lyases. *Erwinia aroideae* produces an endo-acting lyase (M_r 30, pH optimum 8.1), and *E. carotovora* produces an exo-acting lyase, which releases both unsaturated monomer and dimer from highly methylated pectin. This lyase, produced in continuous culture, is inducible only in media containing highly methylated pectin supplied at a very narrow range of dilution rates.

K. Arabinofuranosidases and Arabanases

The quantities of L-arabinose and L-arabinan in plant tissue are small, but arabinose is found in association with pectic substances in the middle lamella. Arabinans are among the substances that cause the formation of "haze" in the juices of apples and other fruits. During enzyme treatment of juices, the arabinofuranosidases commonly present in commercial pectinase preparations reduce the degree of branching in the arabinans and increase the chance for chain-to-chain association, resulting in haze formation. Two alternatives are available to avoid this undesirable event: purification of enzyme preparations to eliminate arabinofuranosidase activity completely or addition of an arabinase which degrades the branched arabinans completely.

The arabinofuranosidases (EC 3.2.1.55) hydrolyze terminal nonreducing arabinofuranosyl groups from α-L-arabinofuranosides, arabinans, arabinoxylans, and arabinoglactans. The substrates most widely used for their assay are p-nitrophenyl α-L-arabinofuranoside (for which its A_{400nm} is measured), phenyl α-L-arabinofuranoside, and beet L-arabinan (measurement of released L-arabinose by either HPLC or reducing sugar assays). The enzymes from most sources will act on all three substrates, but some show preferences for either the low-molecular-weight or high-molecular-weight substrates. The endoarabinases randomly hydrolyze arabinofuranosidic linkages in (1,5) arabinans. The preferred substrate is beet L-arabinan. A unit of activity is defined as the release of 1 μmol of L-arabinose per minute under standard assay conditions.

A variety of microbes can release L-arabinose from natural polysaccharides such as pectic and hemicellulosic substances; sources of these enzymes include the fungi *Aspergillus*, *Botrytis*, *Corticium*, *Myrothecium*, and *Sclerotinia* and the bacteria in the genera *Bacillus*, *Clostridium*, and *Streptomyces*. The enzymes are inducible, and their production requires 0.5–1.0% arabinan or arabinose in the culture medium.

Because these enzymes are extracellular, culture fluid must be clarified of cells and the soluble proteins concentrated by ultrafiltration or precipitation with ammonium sulfate and then purified by conventional chromatographic and electrophoretic techniques. The arabinofuranosidases from *C. rolfsii* and *Streptomyces purpurescens* and the arabinase from *B. subtilis* have been purified 60- to 160-fold to homogeneity by this general strategy. The range of M_r for these enzymes is 30–500, and their isoelectric points range from pH 3 to 10. The pH optima for the fungal enzymes are 2.5–4.5 and 5.0–6.5 for their bacterial counterparts. The arabinosidases in general are quite stable at low pH, a property that may increase their usefulness as anti-hazing agents in apple juice (pH 3.0–3.5) and wines (pH 2.9–4.0).

III. REGULATION OF PECTINASE BIOSYNTHESIS

The importance of pectolytic enzymes in plant pathogenesis, the potential for their application in food technology, and the possibility that pectinase activity may facilitate microbe–plant symbiosis have stimulated interest in their regulation at the gene level. Classically defined, if the rate of enzyme biosynthesis is constant regardless of carbon source, the enzyme is considered to be constitutive. If enzyme production is substantially increased (often by orders of magnitude) by addition of its substrate to the medium, the enzyme is inducible. These criteria are not simple to apply in pectinase production. Some pectolytic enzymes, including several fungal protopectinases and pectinesterases, clearly appear to be constitutive. Others, such as the *Thermomonospora* endopolygalacturonate lyase, are clearly inducible (induced : constitutive ratio of about 1700). However, pectinases such as the endopolygalacturonate lyase of *Bacillus* species RK9 present a third category in which pectic substances reduce enzyme production 5- to 10-fold compared to growth on sucrose or glucose. *Alternaria brassicola* produces endo- and exopolygalacturonases, pectin methylgalacturonase, pectin lyase, and pectin methylesterase during growth on 20% glucose but produces only pectate lyase during growth on pectin or pectate. Exopolygalacturonase production is repressed by pectin but stimulated by pectate. Both pectic substances repress pectin methylgalacturonase but stimulate pectin lyase synthesis.

The regulation of pectinases produced by *A. flavus* and *A. parasiticus* has attracted recent attention since these fungi cause diseases of cotton and other crops. A DNA probe generated by PCR of the *A. niger pg* all gene was used to isolate, clone, and characterize a novel polygalacturonase-encoding gene (*pec*A) from *A. parasiticus*. The PecA sequence had significant base and amino acid homology with other polygalacturonases of fungal origin. In *Aspergillus nidulans*, the pectate lyase gene (*pel*A) contains two short introns and encodes a peptide of 326 amino acids. It is more closely related in its nucleic acid sequence to pectate lyase genes of plants than to those of other fungi. Pectate lyase is induced by polygalacturonic acid and repressed in the presence of preferred carbon sources such as glucose. Gel mobility shift analysis suggests that production is regulated via a fusion protein (composed of the CREA catabolite repressor and glutathione-*S*-transferase) which binds to a *Pst*I–*Sph*I fragment from the *pel*A promoter region.

Helminthosporium sacchari, a pathogen which causes leaf blight in maize and oats, produces pectate lyase on glucose or pectate and polygalacturonase and pectin methylesterase on pectin. Galactose represses pectate lyase formation but induces polygalacturonase. With fructose, sorbose, xylose, and cellobiose as carbon sources, pectin lyase production is stimulated, whereas during growth on lactose as a carbon source, pectate lyase production is stimulated. Our understanding of pectinase regulation is further complicated by requirements in some microbes for undefined factors. For example, in defined mineral salts–pectin medium, *Bacillus*, *Corynebacterium*, and *Vibrio* isolates produce only hydrolytic depolymerizing enzymes (no lyases or esterases). However, in cultures supplemented with plant extracts such as orange peel pulp, hydrolase production is decreased and lyase and esterase are induced. In *Thermomonospora*, the amount of lyase in culture fluid during growth on plant material is regulated by an extracellular protease that is only induced in the presence of cellulose or its degradation products.

Of all the pectinase-producing microbes, the soft-rot erwinias (*E. chrysanthemi*, *E. carotovora*, *E. atroseptica*, *E. betavasculorum*, *E. cypripedii*, and *E. rhapontici*) have probably received the most attention at the gene level, with *E. chrysanthemi* wild-type strain 3937 emerging as the model system. This pectolytic system is known to include an extracellular pectin methylesterase (encoded in the *pem*A gene), a membrane-bound pectin methylesterase (encoded in *pem*B), an exopolygalacturonase (encoded in the *peh*X gene) which generates unsaturated digalacturonate as its sole product, and five major isozymes of pectate lyase (encoded in the *pel*A, *pel*B, *pel*C, *pel*D, and *pel*E genes which are arranged in two widely separated clusters on the bacterial chromosome). In addition, there are at least four secondary pectate lyases (for which only one gene, *pel*L, has been cloned) that are produced at extremely low levels when grown in synthetic medium.

The two sets of plant-inducible pectate lyase isozymes are independently regulated. The first set is coinduced with exopolygalacturonosidase. The other set is composed of enzymes with diverse characteristics and isoelectric points ranging from highly acidic to highly alkaline. This second set of enzymes is still capable of macerating plant tissue when the genes for the first set are inactivated by directional deletion or insertions. The contribution of each pectate lyase isoenzyme to *Erwinia* phytopathogenicity varies according to the plant infected. Only *pel*A [the acidic isoenzyme which produces a series of oligomers (dimer to dodecamer) from polygalacturonic acid through a random endolytic polymerization] appears to be essential for soft rot in pea plants, whereas in chicory leaf all the pel genes contribute to maceration except *pel*E, which encodes a basic isoenzyme that catalyzes a nonrandom endolytic cleavage yielding dimer as the principle product. Moreover, pectate lyase inducibility is plant dependent; although the basic pectate lyase isozymes are produced by *Erwinia* strains during growth on both melons and cucumbers, the acidic isozymes are produced only on melon.

Regulation of *pel* genes in *Erwinia* is exerted by at least five mechanisms: (i) catabolite repression by glucose or other easily metabolized carbon sources not related to pectin degradation or "self-catabolite" repression by products of polygalacturonate breakdown (both reversible by exogenous cyclic adenosine monophosphate); (ii) negative control by the *gpi*R gene, which is in a chromosomal location different than that of the *pel* genes (mutations in *gpi*R derepress *pel* transcription, suggesting that the gene encodes a pectate lyase repressor protein); (iii) positive control by the *pec*A gene, in which mutations result in the reduction of expression of all pectolytic enzymes; (iv) negative control by the *pec*R gene specifically for *pel*A, one of the five genes encoding pectate lyases (*pel*A shows little sensitivity to the action of *gpi*R or *kgd*R); and (v) selective regulation by the self-regulating *pec*T gene which produces a 35-kDa protein that represses the expression of *pel*C, *pel*D, *pel*E, *pel*L, and *kdg*C, activates *pel*B, and has no effect on the expression of *pel*A or the pectin methylesterase genes *pem*A and *pem*B. In addition, there is a *kdg*R regulon, the product of which is a 68-kDa dimeric protein that negatively regulates all the pectinase genes rather than the *pel* genes specifically. In this regard, the synthesis of pectin lyase in *Erwinia* is regulated very differently from that of pectate lyase. Pectin lyase appears to be expressed as part of the SOS response (a mechanism involving nine or more genes that come into play only following irradiation with large doses of ultraviolet light). The enzyme is induced after exposure to DNA-damaging agents produced in several plant species. Therefore, pectin lyase induction may play an initial role in overcoming host plant defenses. In *Erwinia*, there may also be a global regulator, similar to the sigma factors in *Escherichia coli*, which controls a large number of genes expressed in stationary phase or under various stress conditions. This factor, termed RpoS, regulates extracellular levels of pectate lyase, polygalacturonase, and cellulase via positive control of the *rsm*A gene. The effect of this regulation on phytopathogenicity is shown by the fact that RpoS⁻ mutants are more effective macerators of plant tissue than their parent strains.

Pectate lyase inducers not only include recognized pectin degradation products (galacturonate and saturated and unsaturated digalacturonate) but also undefined glucidic factors in crude plant extracts. Compounds in plant extracts which induce pectate lyase in *Erwinia* are thermostable, low molecular weight (350–1000 Da), and very hydrophilic. The transcription of pectinase genes is also subject to a variety of environmental factors (temperature, oxygen levels, iron limitation, osmolarity, and nitrogen starvation).

The action of this array of pectolytic enzymes produces saturated and unsaturated oliogalacturonates which are degraded to monomers by an oligogalacturonate lyase (encoded by the *ogl* gene). The monomers D-galacturonate and 5-keto-4-deoxyuronate are transported into the cell via active transport systems encoded in the *exu*T gene and the *kdg*T gene, respectively. These carbon sources are then metabolized by the products of the genes *kdu*I, *kdu*D, *kdg*K, and *kdg*A. A summary of pectin depolymerization and metabolism is shown in Fig. 3.

The phytopathogenic invasiveness of the bacterium is augmented by the cellulolytic activity of the extracellular endoglucanase Z (encoded by the *cel*Z gene). The secretion of the majority of these enzymes is facilitated via signal sequences in a type II secretion mechanism dependent on the products of 15 genes organized into five transcriptional units (*out*S, *out*B, *out*T, *out*CDEFGHIJKLM, and *out*O). Much remains to be learned about the regulation of pectolytic enzymes in *Erwinia* at the transcriptional level.

IV. COMMERCIAL APPLICATIONS OF PECTOLYTIC ENZYMES

All fruits and berries of nutritional significance contain large amounts of pectin, often up to one-third of the total dry weight. In the unripened fruit, most of the pectic substances are in the form of protopectin; as the fruit ripens, conversion to more soluble forms occurs and softening of the flesh results. Because of its solubility in ripened fruit, the pectin passes into the juice. Fruit juices are usually obtained from the pulps by mechanical separation (pressing, sieving, and centrifuging) of fluid from cell wall fragments. The high pectin content in fluids of ripened fruits makes them viscous and difficult to separate from the flesh. The result is a relatively low yield of both juice and coloring or flavoring

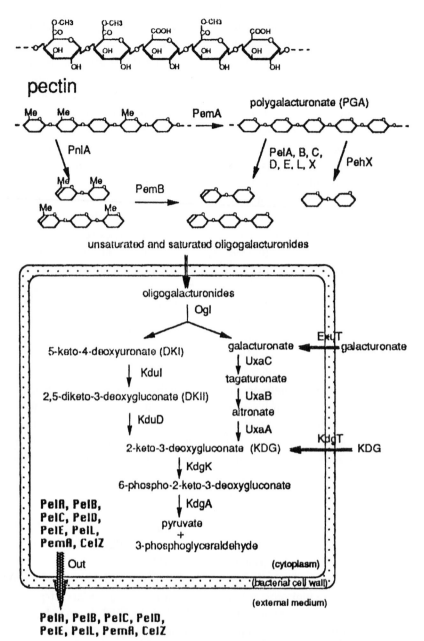

Fig. 3 Pectin catabolism in *Erwinia chrysanthemi*. (Top) The detailed structure of pectin. (Bottom) The diagram outlines the action of the different pectinases outside the bacterial cell (but note that PelX and PemB are a periplasmic and a membrane protein, respectively). The Out proteins (large arrow) are involved in the secretion of most of the pectinases and of the cellulase Celz. Double arrows indicate sugar transport systems. The catabolism of oligogalacturonides occurs in the cytoplasm (reproduced from Hugouvieux-Cotte-Pattat *et al.,* 1996, with permission from the *Annual Review of Microbiology,* Volume 50, © 1996, by Annual Reviews).

compounds. Furthermore, the extracted juice is difficult to filter and retains a high turbidity. These difficulties may be overcome by fermenting the fruit pulp in some cases before juice extraction or by pectinase treatment. In both cases, depolymerization of pectin to highly soluble oligomers allows higher yields of juice, faster filtration rates, better color, and increased clarity. Pectolytic treatment of citrus peel juice results in a viscosity decrease of ~90%, which is much more than can be obtained by fermentation with baker's yeast. Treatment of orange peel with dilute (0.06–0.5%) sulfuric acid at 100°C increases the efficiency of the process by releasing pectin from its association with hemicellulose to allow greater enzyme accessibility.

The efficacy of pectinase treatment is often very application specific; when the extracellular pectin lyase from *Penicillium italicum* was employed in a continuous-flow ultrafiltration membrane reactor to reduce the viscosity of grape, peach, and melon juices, reductions of up to 60% were achieved, but the enzyme had no detectable effect on the viscosity of apple or pear juices. When grapes or other berries are extracted for either juice or wine, pectinase is added most effectively between pulping and pressing. Precipitation of turbidity-forming polymers occurs, allowing most of them to settle and the remainder to be removed by filtration or centrifugation.

Commercial pectinase preparations for use in the food industry are usually obtained from *Aspergillus* or *Rhizopus* species. They are currently marketed by several companies under various descriptive trade names such as Macerase. Some are produced for the processing of apples, pears, grapes, and berries and for use in the citrus industry. Others increase apple juice extraction and clarity of the extracted product. Companies also produce a combined enzyme preparation (cellulase, hemicellulase, and pectinase) for use in processing underripe fruits and vegetables, when extensive maceration is the primary goal, or for fruit juice and wine clarification. Currently, pectinases account for approximately 10% of industrial enzyme sales.

Mixtures of pectolytic enzymes have also found some use as food additives to improve the digestibility of forage grasses and grains in both livestock and poultry. The principle underlying this practice is that the pectinases act on the plant cell wall outer parenchyma (where pectin accounts for approximately 70% of the acid-insoluble carbohydrate). The depolymerization of pectin in this area subsequently makes other cell wall components (cellulose and hemicellulose) more vulnerable to rapid enzymatic attack by the indigenous microflora in the animals' digestive tracts. The same principle applies to the use of pectinase in combination with other depolymerizing enzymes in the protoplasting of plant cells for various research purposes.

Limitations to the further expansion of the use of pectinases include the economics of production, elimination of undesired activities or toxicities from enzyme preparations, stability under usage conditions, and potential for recycling. The economics may be improved by the use of thermostable pectinases from thermophilic sources (*Bacillus stearothermophilus* or *Thermomonospora fusca*) or by the immobilization of pectinases from mesophilic sources such as *Rhizopus* or *Aspergillus* species. When a pectinase purified from *A. niger* was linked via carbodiimide coupling (which activates both carboxyl and amino groups of proteins) to alginate–magnetite latex beads, the level of activity remained stable for 3 hr at 60°C compared to a half-life of 9 min in the free state; immobilization had no detectable effect on the kinetic properties of the pectinase. Alternate immobilization systems which show promise for pectinase application include binding of the enzyme to macroporous titania membranes (which allow passage of viscous preparations containing particulate matter) or binding to some polymer which would allow retention of the pectinase by even a very large pore filter system.

Introduction of genetically engineered plants with increased resistance to phytopathogenicity may inadvertently create another factor which may restrict the expansion of commercial pectinase application. Fruits and vegetables that contain inhibitors of polygalacturonases or which synthesize pectins more resistant to enzymatic degradation are obviously not ideal candidates for pectinase treatment. Pectinase inhibitors (some of which have been characterized as heat-stable proteins in the M_r range of 15–50) may be induced in plants as a response to microbial pectinase exposure during cultivation. Selected plant strains (such as potato hybrids, which have twice the pectin esterification in stem cell walls, making them more resistant to infection by the soft-rot *Erwinia* species) will also be more recalcitrant to commercial pectinase treatment. Environmental factors during the early stage of growth also play a role in plant cell wall susceptibility to pectolytic action; in barley seedlings, a brief heat treatment (30–60 s at 50°C) altered their cell walls to make them more resistant to pectinase attack. Protein engineering, by which a cloned gene can be overexpressed to produce large enzyme quantities, an enzyme can be selectively modified to produce a novel form, or an entirely new enzyme can be synthesized by fusing of gene fragments encoding selected domains, should eventually provide tailor-made pectinases that are more thermostable and inhibitor resistant for specific industrial applications.

In the application of pectinases (as in the application of other microbial enzymes such as cellulases and proteases) to large-scale industrial processes, consideration must be given to their potential as hypersensitizing agents in workers who routinely come into contact with dusts or aerosols. The pectate lyases produced by some of the phytopathogenic *Erwinia* species have extensive amino acid sequence homology with the well-characterized *Cry* JI allergen from Japanese cedar (*Cryptomeria japonica*) and with pollen allergens from short ragweed (*Ambrosia artemisiifoila*). Repeated inhalations of airborne allergens can cause acute or chronic bronchial or interstitial lung disease symptoms collectively known as either occupational asthma (type I IgE-mediated hypersensitivity which causes wheezing, dyspnea, and elevated pulse and blood pressure within minutes after exposure) or hypersensitivity pneumonitis (type III IgG-mediated hypersensitivity causing fever, chills, dry cough, malaise, and profound dyspnea within 8 hr after exposure). The major mechanism in occupational asthma is mast cell degranulation with release of histamine and other mediators, whereas hypersensitivity pneumonitis is an immune complex reaction dependent on the presence of high levels of complement-fixing IgG antibody in the sera of affected workers.

If workers are to be protected from respiratory damage during industrial-scale use of pectinases or other microbial enzymes, the following practices must be implemented:

1. Isolation of the various stages of the process
2. Use of automation so as to allow utilization of a minimal number of employees
3. Rotation of work assignments on a regular basis
4. Regular medical testing for development of hypersensitivities
5. Respiratory protection during routine equipment cleanup or repair
6. Specific instructions for emergency measures in the event of process spillage or aerosolization.

See Also the Following Articles

Cellulases • Enzymes in Biotechnology • *Erwinia*: Genetics of Pathogenicity Factors • Nodule Formation in Legumes • Plant Disease Resistance

Bibliography

Alberts, B., Bray, D., Lewis, J., Raff, M., Roberts, K., and Watson, J. D. (1994). "Molecular Biology of the Cell," 3rd ed. Garland, New York.

Collmer, A. (1986). The molecular biology of pectic enzyme production and bacterial soft rot pathogenesis. *In* "Biology and Molecular Biology of Plant Pathogen Interactions" (J. A. Barley, Ed.), pp. 277–289. Springer, New York.

Coughlan, M. P., Mehra, R. K., Considine, P. J., O'Rorke, A., and Puls, J. (1986). Saccharification of agricultural residues by combined cellulolytic and pectinolytic enzyme systems. *Biotechnol. Bioeng. Symp.* **15**, 447–458.

Fishman, M. L. (1992). Pectic substances. *In* "Encyclopedia of Food Science and Technology" (Y. H. Hui, Ed.) Vol. 3, pp. 2039–2043. Wiley, New York.

Forster, H. (1988). Pectinesterases from *Phytophthora infestans. Methods Enzymol.* **161**, 355–361.

Gilliespie, A. M., Keane, D., Griffin, T. O., Tuohy, M. G., Donaghy, J., Haylock, R. W., and Coughlan, M. P. (1990). The application of fungal enzymes in flax retting and the properties of an extracellular polygalacturonase from *Penicillium capsulatum. In* "Biotechnology in Pulp and Paper Manufacture: Applications and Fundamental Investigations" (T. K. Kirk and H. M. Chang, Eds.), Proceedings of the 4th International Congress on Biotechnology in the Pulp and Paper Industry, pp. 211–219. Butterworth-Heinemann, Boston.

Hugouvieux-Cotte-Pattat, N., Condemine, G., Nasser, W.,

and Reverchon, S. (1996). Regulation of pectinolysis in *Erwinia chrysanthemi. Annu. Rev. Microbiol.* **50**, 213–257.

Lim, D. (1998). "Microbiology," 2nd ed., pp. 418–423. WCB/McGraw-Hill, Boston.

Lonsane, B. K., Ghildyal, N. P., Ramakrishna, M., and Stutzenberger, F. (1993). Health and safety aspects of working with solid substrate processes. *In* "Solid Substrate Cultivation" (H. W. Doelle, D. A. Mitchell, and C. E. Rolz, Eds.), pp 441–474. Elsevier, London.

Madigan, M. T., Martinko, J. M., and Parker, J. (1997). "Biology of Microorganisms," 8th ed., pp. 595–603. Prentice Hall, Upper Saddle River, NJ.

Mukherjee, A., Cui, Y., Ma, W, Liu, Y., Ishihama, A., Eisenstart, A., and Chatterjee, A. K. (1998). RpoS (Sigma-S) controls expression of *rsmA*, a global regulator of secondary metabolites, Harpin, and extracellular proteins in *Erwinia carotovora. J. Bacteriol.* **180**, 3629–3634.

Pitt, D. (1988). Pectin lyase from *Phoma medicaginis* var. *pinodella. Methods Enzymol.* **161**, 350–354.

Sakai, I. (1988). Protopectinase from yeasts and a yeastlike fungus. *Methods Enzymol.* **161**, 335–350.

Schumann, G. L. (1991). "Plant Diseases: Their Biology and Social Impact," pp. 79–80, 267–271. APS Press, St. Paul, MN.

Solis, S., Flores, M. E., and Huitron, C. (1997). Improvement of pectinase production by interspecific hybrids of *Aspergillus* strains *Lett. Appl. Microbiol.* **24**, 77–81.

Sylvia, D. M., Fuhrmann, J. J., Hartel, P. G., and Zuberer, D. A. (1998). "Principles and Applications of Soil Microbiology," pp. 84–89. Prentice Hall, Upper Saddle River, NJ.

Tagawa, K., and Kaji, A. (1988). Polygalacturonase from *Corticum rolfsii. Methods Enzymol.* **161**, 361–365.

Visser, J., and Voragen, A. G. J. (1996). Pectins and pectinases. *In* "Progress in Biotechnology." Elsevier, Amsterdam.

Voragen, A. G. J., and Pilnik, W. (1989). Pectin-degrading enzymes in fruit and vegetable processing. *In* "Biocatalysis in Agricultural Biotechnology" (I. R. Whitaker and P. E. Sonnet, Eds.), American Chemical Society (ACS) Symposium Series No. 389, pp. 93–115. ACS, Washington, DC.

Voragen, A. G. J., Schols, H. A., Siliha, H. A. L., and Pilnik, W. (1986). Enzymic lysis of pectic substances in cell walls: Some implications for fruit juice technology. *In* "Chemistry and Function of Pectins" (M. L. Fishman and J. J. Jens, Eds.), American Chemical Society (ACS) Symposium Series No. 310, pp. 230–247. ACS, Washington, DC.

Whitaker, J. R. (1990). Microbial pectolytic enzymes. *In* "Microbial Enzymes and Biotechnology" (W. M. Fogarty and C. T. Kelly, Eds.), 2nd ed., pp. 133–176. Elsevier, London.

PEP: Carbohydrate Phosphotransferase Systems

Pieter Postma

University of Amsterdam

GLOSSARY

catabolite repression Inhibition of the expression and/or protein activity which is caused by the presence of a rapidly metabolizable carbon source in the growth medium; sometimes incorrectly referred to as glucose repression or diauxic growth.

chemotaxis Movement of the bacterial cell toward certain compounds.

crr carbohydrate repression resistant gene.

PTS Phosphoenolpyruvate: carbohydrate phosphotransferase system, a major bacterial transport system that was discovered in 1964 and that catalyzes the transport and phosphorylation of many carbohydrates in gram-negative and gram-positive bacteria.

THE PHOSPHOENOLPYRUVATE : CARBOHYDRATE PHOSPHOTRANSFERASE SYSTEM (PTS) is a major bacterial transport system which was discovered in 1964 by Roseman and co-workers in *Escherichia coli*. It catalyzes the transport of many carbohydrates in numerous gram-negative and gram-positive bacteria. It is distinguished from the other bacterial transport systems in that it phosphorylates its substrates concomitantly with their transport, i.e., carbohydrates are accumulated in the cell as their respective phosphate esters. The phosphoryl group is derived from phosphoenolpyruvate and is transferred to the carbohydrate substrates via reversible phosphorylation of many proteins, collectively called PTS proteins. In addition to transport, the PTS is involved in other important cellular processes. Components of the PTS function in chemotaxis toward PTS carbohydrates. Furthermore, depending on the bacterial species, the phosphorylation state of particular PTS proteins plays a central role in the regulation of the metabolism of many non-PTS compounds by regulating the activity of enzymes and/or the expression of metabolic operons. In this article, we will give an overview of the essential properties of the PTS. Since more than 1500 articles have been published on the PTS since its discovery 35 years ago and several hundred PTS proteins are known, only the most important facets of this complex system can be dealt with here. More detailed information can be found in the reviews listed in the Bibliography.

I. OVERVIEW OF THE PHOSPHOTRANSFERASE SYSTEM

The phosphoenolpyruvate: carbohydrate phosphotransferase system (PTS) is involved in the uptake and phosphorylation of numerous carbohydrates and catalyzes the following reaction:

$$\text{carbohydrate}_{out} + \text{PEP}_{in} \rightarrow \text{carbohydrate-phosphate}_{in} + \text{pyruvate}_{in}$$

The PTS has been found in many eubacteria but not in archaea and eukaryotes. The basic composition of the PTS in both gram-negative and gram-positive bacteria is remarkably similar. It consists (with a few exceptions mentioned later) of two types of proteins:

Encyclopedia of Microbiology, Volume 3
SECOND EDITION

580

(i) the general PTS proteins which are required for the transport and phosphorylation of all PTS carbohydrates, enzyme I (EI), and HPr (*heat-stable protein* or *histidine-containing protein*) (strains lacking EI or HPr cannot phosphorylate any of the PTS carbohydrates), and (ii) the carbohydrate-specific enzyme II (EII) complexes which are involved in the recognition of the various carbohydrates in the extracellular medium, the actual transport across the cytoplasmic membrane, and phosphorylation. It is important to stress that a carbohydrate can be a PTS substrate in one organism but be transported via a non-PT system in another organism. An example is lactose, which is accumulated in *Escherichia coli* by the lactose permease, an H^+-symport system, but is a PTS carbohydrate in several gram-positive bacteria. The specificity of EII complexes can vary, with some recognizing only one or at most a few carbohydrates and others having broad specificity. EIIs will be designated according to the most preferred carbohydrate.

The phosphoryl group is sequentially transferred from phosphoenolpyruvate (PEP) to EI, HPr, and the various EII complexes (Fig. 1). In most cases, a particular histidine residue is phosphorylated (see Section II). In general, cells contain one type of EI and one type of HPr. As discussed later, in a number of bacteria EI-like and HPr-like proteins have been detected but in many cases their role has not been elucidated.

Although some EII complexes consist of a single polypeptide (e.g., *E. coli* EIICBA^Mtl for mannitol), other EII complexes may consist of two to four different polypeptides (e.g., IICB^Glc and IIA^Glc for glucose and IIAB^Man, IIC^Man, and IID^Man for mannose). In Section II, I discuss in more detail the structure of the various EII complexes and the domains which they contain.

Although the composition of the PTS is quite conserved throughout the various bacterial species, the detailed study of the PTS in certain bacteria has revealed variation. In addition, the recent genome projects have uncovered many genes that encode proteins homologous to either the general PTS proteins, EI or HPr, or a member of one of the EII families. The physiological function of only a few of these new proteins has been elucidated. However, in a number of cases, the role of alternative PTS proteins is known.

Although it was stated previously that the general PTS protein HPr is involved in the phosphorylation of all PTS carbohydrates, there are exceptions. For instance, the fructose PTS of *E. coli* and *Salmonella typhimurium* has its own HPr-like molecule, FPr. Mutants lacking HPr still grow on fructose and FPr can functionally replace HPr as a general phospho-carrier protein. Even more surprising, *Rhodobacter capsulatus* contains a polypeptide composed of a fructose-specific EII domain, an HPr domain, and an EI domain.

In the following sections, the structure and properties of the various PTS proteins, the genes encoding these proteins, and the roles of particular PTS proteins in the regulation of overall cell metabolism will be discussed.

II. STRUCTURE AND PROPERTIES OF PTS PROTEINS

A. Structure of PTS Proteins

1. Enzyme I and HPr

A number of EIs have been purified and several *ptsI* genes encoding EI have been sequenced. The molecular weights of the *E. coli* and *S. typhimurium* EI are 63,412 and 63,489, respectively, and those of other EIs are in the same range. EI undergoes a monomer/dimer transition. In the presence of PEP and Mg^{2+}, the dimer, but not the monomer, autophosphorylates. After dissociation, the phosphorylated monomer transfers its phosphoryl group to HPr. Using various techniques, it has been concluded that EI consists of two domains. The C-terminal domain (EI-C) is important for autophosphorylation and dimer formation, whereas the N-terminal domain (EI-N), which contains the His residue, can accept the phosphoryl group from P~HPr. The EI-N domain can be complemented by the EI-C domain for phosphoryl transfer both *in vitro* and *in vivo*. The structure of the complete EI molecule is not known, but the structure of the N-terminal domain has been determined by x-ray diffraction and nuclear magnetic resonance (NMR). By using the structures of EI-N and HPr, the interaction between both proteins has been modelled. Most likely, interaction does not require major conformational changes in EI.

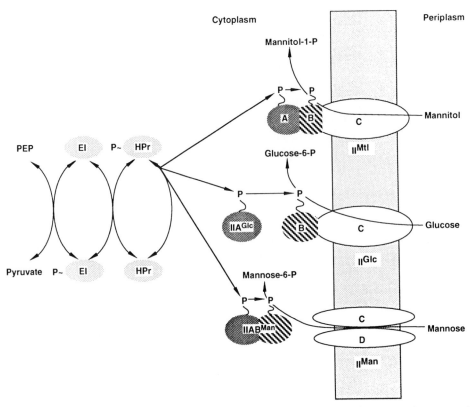

Fig. 1. The phosphoenolpyruvate: carbohydrate phosphotransferase system. The general PTS proteins are enzyme I (EI) and HPr. Of the many carbohydrate-specific EII complexes, three are shown. IICBA^{Mtl} is specific for mannitol (Mtl), the IICB^{Glc}–IIA^{Glc} complex for glucose (Glc), and the IIAB^{Man}–IIC^{Man}–IID^{Man} for mannose (Man). P~EI and P~HPr are the phosphorylated forms of EI and HPr, respectively. The phosphorylated forms of the A and B domains/proteins of the various EII complexes are indicated (reproduced with permission from Postma *et al.*, 1996).

HPr molecules of several bacteria have been purified. The molecular weight of most HPrs is approximately 9000. The structures of several HPr molecules from both gram-positive and gram-negative bacteria have been determined by x-ray crystallography and NMR, and the overall structures are quite similar. In addition, the structure of some phosphorylated HPrs has been elucidated. Comparison of the structure of *E. coli* HPr and P~HPr shows that changes on phosphorylation are limited to the region around the active site His-15.

2. Enzyme II Complexes

Biochemical and genetical studies have led to the realization that although EII complexes may at first seem to be quite different, they are in fact built from a number of modules that function in similar ways. For example, consider *E. coli* II^{Mtl} (Fig. 1). The enzyme consists of three domains: IIA, IIB, and IIC. In this and other EIIs, the IIA and IIB domains are hydrophilic, whereas the IIC domains are to a large extent hydrophobic. They contain six to eight transmembrane-spanning segments and a large hydrophilic loop which together form the transmembrane channel through which the carbohydrates enter. Comparison of all EIIs analyzed to date allows the following generalizations. First, most EIIs consist of three domains or proteins: A (first phosphorylation site), B (second phosphorylation site), and C (transmembrane domain). In most cases, the three domains/proteins of EIIs consist of approximately 650 residues and the domains, when present in a single polypeptide, are connected via flexible linkers. Second, if two or more domains are present in one polypeptide chain, the order of the domains can vary. Third, EIIs can be classified into four families (the glucose, mannitol, lactose, and mannose families) in which the similarity between members of one family is more than that between members of different families. Finally, in the mannose family, including the *E. coli* mannose PTS and the *Bacillus subtilis* fructose PTS, the transmembrane channel consists of two different polypeptides, IIC and IID.

A comparison of all EIIs suggests that they have arisen through evolution by fusion and splicing of domains, duplication, and shuffling. The latter property is illustrated by IICBA^{Nag} (specific for *N*-acetylglucosamine) and IIBCA^{Bgl} (specific for β-glucosides) in which the order of the B and C domains in the polypeptide chain is reversed. Splicing is illustrated by IICBA^{Nag} and IICB^{Glc} + IIA^{Glc}. The II^{Nag}, II^{Bgl}, and II^{Glc} complexes belong to the same glucose family.

The tertiary structure of a number of PTS proteins has been determined by using both x-ray diffraction and 3-D-heteronuclear NMR. In addition, the structures of several IIA and IIB proteins/domains have been elucidated, including the *E. coli* IIA^{Glc} protein and the IIB^{Glc}, IIA^{Mtl}, and IIA^{Man} domains; the *B. subtilis* IIA^{Glc} domain and IIB^{Lev} protein (fructose PTS); and the *Lactococcus lactis* IIA^{Lac} protein (lactose PTS). Thus, the structures of IIA proteins/domains representing each of the four families have been determined. All four structures are completely different from each other, displaying different folds. None of the more hydrophobic IIC domains have been crystallized. The secondary structures of those domains have been determined in a number of cases, and on the basis of *lacZ* and *phoA* protein fusions six to eight transmembrane helices have been proposed.

B. Phosphoryl Transfer Reactions

The fact that the PTS catalyses the covalent modification of its substrates has greatly facilitated the study of this transport system. The general proteins EI and HPr of various bacteria have been purified, as have several EII complexes, and the phosphoryl transfer reactions between the various purified proteins have been studied. In a number of cases the residues in the PTS proteins that are covalently phosphorylated have been identified.

In most cases, a histidine residue in the PTS proteins is phosphorylated. The phosphoryl group, which originates from PEP, is bound first to the N3 of a histidine residue of EI. EI is a homodimer consisting of two identical monomers. Two phosphoryl groups are bound per dimer, and the dimer is the active form. Subsequently, the phosphoryl group of EI is transferred to the N1 of the His15 residue in HPr. HPr molecules of both gram-positive and gram-negative bacteria exhibit a striking similarity near the His15 residue. The phosphoryl group is subsequently transferred from P~HPr to the EII complexes. Each EII complex is phosphorylated at two sites: a histidine residue in the A domain and a cysteine or histidine residue in the B domain. Finally, the phosphoryl group is transferred to the carbohydrate for which the EII complex is specific.

Only in a limited number of cases (e.g., the *E. coli* II^{Glc}, the II^{Mtl}, and the II^{Man} complexes) has it been shown that the EII complex is phosphorylated on two distinct residues using various techniques to identify the residues that are phosphorylated by various means. Based on sequence similarities and mutation studies, it is assumed that in all cases two phosphorylation steps exist in each EII complex. Including the phosphorylation sites in EI and HPr and the final phosphoryl transfer to the carbohydrate, it appears that five phosphorylation steps are involved in each phosphotransferase system.

Studies with the *E. coli* II^{Mtl} complex revealed that the second phosphorylation site in the IIB^{Mtl} domain is not a histidine but a cysteine residue. P~HPr donates its phosphoryl group to His554, located in the C-terminal domain of EII^{Mtl}. Subsequently, the ~P group is transferred to the Cys384 residue. Phosphocysteine has been identified as a catalytic intermediate with [^{31}P]phospho-NMR. In II^{Glc} a cysteine residue has also been identified as the second phosphorylation site, and it is assumed to be the phosphorylation site in other EIIs belonging to the glucose, mannitol, and lactose families. In the case of *E. coli* mannose PTS, no cysteine seems to be involved. Two phosphorylated histidine residues have been identified in the IIAB^{Man} polypeptide and none in the other two EII^{Man} polypeptides that make up the mannose PTS—IIC^{Man} and IID^{Man}.

The phosphoryl transfer between the various PTS proteins is a reversible reaction. This has been studied in detail with *E. coli* EI, HPr, and IIA^{Glc}. The phosphate group transfer potentials of P~EI, P~HPr, and P~IIA^{Glc} are close to that of PEP, and the equilibrium constant K_{eq} for the reaction PEP + IIA^{Glc} ↔ pyruvate + P~IIA^{Glc} is 10–15. This is important for PTS-mediated regulation (see Section V).

To date, *E. coli* IICBA^{Mtl}, made up of 637 amino acid residues, is probably the best studied EII and is discussed in detail as an example of how an EII may work. The N-terminal, membrane-bound IIC domain is hydrophobic and contains the binding site for mannitol, whereas the C-terminal, cytoplasmic part contains the hydrophilic IIA and IIB domains which each harbor a phosphorylation site. By expressing the three domains separately or by introducing mutations in specific residues, including the phosphorylation sites, interactions between those domains could be studied as could the role of each domain in the overall transport and phosphoryl transfer. The active form of EII^{Mtl} is a dimer and its formation is dependent on the IIC domain. The IIBA domain can be phosphorylated by P~HPr in the absence of the IIC domain. However, transfer of the phosphoryl group to mannitol requires the IIC domain.

Does phosphoryl transfer in an EII dimer occur within one subunit or can the His residue in one subunit phosphorylate a Cys residue in the other subunit? Studies with IICBA^{Mtl}, IICB^{Glc}/IIA^{Glc}, and IIAB^{Man} suggest that intersubunit phosphotransfer can occur. It has also been shown that heterodimer formation occurs because IICBA^{Nag} (specific for N-acetylglucosamine) can transfer the phosphoryl group from its A domain to the B domain of IICB^{Glc}, substituting for IIA^{Glc}. However, in recent studies it was proposed that phosphoryl transfer between domains in the same subunit of II^{Mtl}, resulted in maximal rates.

The enzymes II can catalyze a partial reaction, transphosphorylation, that involves transfer of a phosphoryl group from a carbohydrate phosphate to another carbohydrate. For example, EII^{Mtl} catalyzes transphosphorylation between mannitol 1-phosphate and mannitol. Whereas P~HPr-dependent mannitol phosphorylation requires a complete II^{Mtl}, containing the two phosphorylation sites, transphosphorylation can be catalyzed by the N-terminal domain containing only the Cys384 residue but lacking the His554 residue.

III. TRANSPORT AND PHOSPHORYLATION OF PTS CARBOHYDRATES

An important feature of the PTS is phosphorylation of its substrates concomitant with transport. The free carbohydrates are not detected in the cell until after some time due to the action of phosphatases. The PTS can be described as a translocator coupled to a trapping kinase. However, in the absence of PTS-mediated phosphorylation, the movement of the free carbohydrate across the membrane via a nonphosphorylated EII complex is very slow compared to PTS-dependent transport activity. Thus, facilitated diffusion of PTS carbohydrates via the nonphosphorylated EIIs seems to be absent or at least very slow. In *S. typhimurium ptsH,I* deletion strains, which lack EI and HPr and in which the EIIs are nonphosphorylated, the rate of uptake of 2-deoxyglucose or methyl α-glucoside (non-metabolizable substrates of the mannose and glucose PTS, respectively) is approximately 0.1% of the rate of uptake in the *ptsH$^+$, I$^+$* parent strain. A detailed kinetic study of II^{Mtl} in inside-out membrane vesicles has shown that the phosphorylated enzyme (phospho-II^{Mtl}) can catalyze mannitol translocation without phosphorylation at a rate two or three orders of magnitude higher than that of non-phosphorylated EII^{Mtl}. Finally, reconstitution of several EII complexes (e.g., II^{Mtl}, II^{Nag}, and II^{Man}) in liposomes shows that the rate of substrate uptake in the absence of enclosed PEP, EI, and HPr is very low compared to that in the presence of the general PTS proteins. The following steps in mannitol transport and phosphorylation have been proposed. Mannitol binds to the IIC domain of P~II^{Mtl} at the periplasmic side of the membrane and is translocated to the cytoplasmic side. Part of the mannitol molecules are phosphorylated via phospho-Cys384 and released as mannitol 1-phosphate in the cytoplasm. The remainder of the molecules can be released as the free carbohydrate, bind again to P~II^{Mtl} at the cytoplasmic side, and can be phosphorylated.

Mechanistically, there is no obligatory coupling between translocation and phosphorylation. Phosphorylation of an EII lowers the energy barrier for translocation. However, in the physiological sense, 100% of all PTS carbohydrate is phosphorylated. From this proposal, it follows that PTS carbohydrates, generated intracellularly from other carbohydrates, can be phosphorylated via EII complexes at the cytoplasmic side of the membrane. This has been observed in a few cases, and *E. coli* mutants have been isolated which are unable to transport glucose and its analogs via II^{Glc} but can still phosphorylate glucose when provided intracellularly. The model also provides an explanation for the observation that PTS carbohydrates can efflux from the cell in a nonphosphorylated form at rates comparable to uptake rates. The coupling between transport and phosphorylation is at the level of the translocation of the substrate.

From *S. typhimurium ptsH,I* deletion strains which are unable to generate any phospho-EII, mutants have been isolated that can catalyze facilitated diffusion of glucose, for example. The mutations are localized in the *ptsG* gene, encoding IICB^{Glc}, and allow the diffusion of glucose. Subsequent phosphorylation is catalyzed by an ATP-dependent glucokinase. The rate of uptake is not very different from that via the intact PTS, but the K_m for glucose has been increased from approximately 10 μM in the *pts*$^+$ parent to 0.6–12 mM in these so-called "uncoupled" mutants. Most substitutions are in residues in two of the transmembrane helices of the IIC^{Glc} domain. It has been proposed that EII^{Glc} is a closed pore in the parent unless it is phosphorylated, whereas the mutated EII^{Glc} is always in the "open" conformation.

A surprising aspect of transport via an EII should be mentioned. It was noted that infection of *E. coli* with phage λ required the mannose PTS, and it was shown that both IIC^{Man} and IID^{Man}, but not IIAB^{Man}, are essential in λ DNA penetration. The related EIIs of the *B. subtilis* fructose PTS can replace the *E. coli* proteins.

IV. GENES ENCODING PTS PROTEINS

Before the advent of genome projects, most genes encoding PTS proteins were identified by classical genetical and complementation studies in homologous or heterologous systems. This resulted in information on the *ptsI* and *ptsH* genes, encoding EI and HPr, respectively, from many different bacteria as well as genes encoding many EII complexes. Genome projects allow us to identify many genes in various bacteria that may encode uncharacterized PTS functions. For instance, *E. coli* contains at least 50 genes encoding known and putative PTS proteins, whereas the estimate for *B. subtilis* is more than 30.

The *ptsH,I* genes have been studied extensively, and in most bacteria the two genes are linked. In enteric bacteria and in *B. subtilis* the two genes form an operon. Growth of enteric bacteria on a PTS substrate, including glucose, generally increases the level of HPr and EI three- to fivefold. However, expression of the *ptsHI* operon is also dependent on the CRP/cyclic AMP (cAMP) signal. Transcription of the *pts* genes is complex and depends on two promoters, P0 and P1, upstream of *ptsH*. Depending on the strength of catabolite repression, transcription initiates from different start sites and utilizes preferentially one or both promoters. A mechanism of promoter switching has been proposed.

The genes for many EII complexes have been identified, cloned, and sequenced. In enteric bacteria, most genes encoding a particular EII complex are contained in an operon or regulon together with the enzyme involved in the first step of the metabolism of the particular carbohydrate phosphate. One notable exception is glucose. The *crr* (carbohydrate repression resistant), gene, encoding IIA^{Glc}, is located next to the *ptsHI* operon and separated by approximately one-fourth of the *E. coli* chromosome from *ptsG*, encoding IICB^{Glc}. Although the *ptsH,I* and *crr* genes constitute an operon in *E. coli*, more than 85% of *crr* transcription originates from promoter elements at the end of the *ptsI* gene.

Most PT systems are induced by growing cells on the particular PTS carbohydrate. In some cases an appreciable basal level is present in the cell, irrespective of the source of carbon during growth (e.g., in the case of the glucose and mannose PTS in enteric bacteria). The nature of the inducer and its site of action in many cases have not been determined. Although in some cases it has been shown that a carbohydrate phosphate binds to a specific repressor and promotes its release from DNA (e.g., fructose 1-phosphate and FruR, the repressor of the fructose

PTS, and *N*-acetylglucosamine 6-phosphate and NagC, the repressor of the *N*-acetylglucosamine PTS), it is not always clear whether the inducer is an intracellular carbohydrate phosphate or an extracellular carbohydrate. A case in point is glucose, which acts as an inducer of II^{Glc} and II^{Man} in *E. coli* compared to growth on glycerol. Although synthesis of both II^{Man} and II^{Glc} is dependent on CRP/cAMP, the level of II^{Glc} is considerably higher after growth on glucose compared to glycerol. A Mlc protein has been identified which can bind to regulatory regions in front of the *ptsG, manXYZ* (encoding the II^{Man} complex), and *malT* genes/operons and acts like a repressor. Glucose metabolism is supposed to result in an unidentified compound that should bind to and release Mlc. Several sugars that generate glucose intracellularly (i. e., lactose, maltose, and melibiose) do not act as inducers.

An interesting case of regulation by the PTS at the transcriptional level is provided by the *E. coli bgl* operon, encoding the β-glucoside PTS, and the *B. subtilis* sac operons, encoding the sucrose PTS. Expression of the *bgl* and *sac* operons is under the control of transcription antitermination. In the absence of external β-glucosides, expression of the *bgl* operon is not required. The small amount of II^{Bgl} always present is in the phosphorylated form since no drain of the phosphoryl group by external β-glucosides can occur. Under these conditions, BglG, the antiterminator, becomes phosphorylated as a result of the transfer of a phosphoryl group from the phospho-IIB^{Bgl} domain to a specific His residue in BglG. Phospho-BglG is a monomer and has no activity. A terminator can occur in the *bgl* operon and transcription is terminated. When β-glucosides are detected, the phosphoryl group of phospho-II^{Bgl} will be transferred to the carbohydrate and phospho-II^{Bgl} will be dephosphorylated. As a consequence, phospho-BglG will be dephosphorylated as well, become a dimer, and can then act as a transcription antiterminator since the nonphosphorylated form is an RNA-binding protein and prevents formation of the terminator. As a result, the *bgl* operon is expressed, as required for utilization of β-glucosides. Thus, II^{Bgl}, the *bglF* gene product, acts both as a kinase and as a phosphatase.

Several other *B. subtilis* operons are under the control of regulators of which the activity is dependent on the phosphorylation state. In several cases, these regulators are under the control of multiple phosphorylation on different residues. The phosphoryl group can originate from the associated P-EII, as in the case of *E. coli* BglG or *B. subtilis* SacY, or directly from P~HPr. Examples of the latter class include SacT, LicT, and LevR, which are involved in the regulation of the high-affinity sucrose, the β-glucoside, and the fructose system, respectively. Whereas the EII-mediated phosphorylation in general results in an inactive regulator, HPr-mediated phosphorylation results in positive regulation. The simultaneous activation and inhibition may allow precise tuning of enzyme activity.

V. PTS-MEDIATED REGULATION IN ENTERIC BACTERIA

In classic growth studies with *E. coli* and *B. subtilis*, Monod noted that certain carbohydrates are utilized in preference to others when a mixture of two carbohydrates was presented. A well-known example is the growth of *E. coli* cells on a mixture of glucose and lactose. First, all glucose is consumed but no lactose. When glucose is exhausted, growth stops, but after a short period the cells resume growth by utilizing the second substrate, lactose. The enzymes specifically required for lactose metabolism, the lactose permease and β-galactosidase, are not synthesized as long as glucose is present in the growth medium. The presence of glucose represses the expression of the *lac* operon, encoding these enzymes. This phenomenon was called diauxic growth, but the whole process or various aspects of it were later also referred to as glucose repression or catabolite repression. Diauxic growth, and catabolite repression, has often been interpreted as being caused by changes in the cAMP level, with cAMP being an alarmone in enteric bacteria and required for the expression of many catabolic genes and operons. I use the term catabolite repression in a very general sense of inhibition of the expression and/or protein activity caused by the presence of a rapidly metabolizable carbon source in the growth medium and discuss the role of the PTS in certain forms of catabolite expression.

It was previously noted that *E. coli* and *S. typhimurium ptsH,I* mutants display a pleiotropic phenotype. They are unable to grow on all PTS carbohydrates, as expected. In addition, these mutants cannot grow on many non-PTS carbon sources, including lactose, maltose, melibiose, and glycerol (referred to as class I compounds), as well as xylose, rhamnose, some other sugars, and Krebs cycle intermediates (class II compounds). It should be stressed that this behavior is found only in the Enterobacteriaceae. Although *B. subtilis* displays a related pleiotropic phenotype, a completely different mechanism plays a role (see Section VI). The remainder of this section applies only to *E. coli* and *S. typhimurium* and related enteric bacteria.

Why is the absence of HPr and/or EI deleterious to the cell? Also, is there a connection between the *ptsH,I* phenotype and Monod's observations and later extensions, described previously, that certain carbohydrates such as glucose can prevent the utilization of class I and class II compounds? Growth of *ptsHI* mutants on class I or class II compounds can be restored in either of two ways. First, mutations restoring growth of *ptsH,I* mutants on class I compounds were found in the *crr* gene. Second, addition of 3′,5′-cyclic AMP (cAMP) restores growth on both class I and class II compounds. Thus, it has been concluded that *ptsH,I* mutants are defective in cAMP synthesis.

A. Inducer Exclusion

The *crr* gene was found to be the structural gene of IIA^{Glc} and is localized next to the *ptsHI* operon. A clue to the role of IIA^{Glc} was found in transport studies with *ptsI* mutants which contain a small amount (0.5%) of EI activity. It was found that in such leaky *ptsI* mutants, but not in a *ptsI crr* double mutant, transport of class I substrates is strongly inhibited by PTS carbohydrates, even by nonmetabolizable analogs such as methyl α-glucoside (a glucose analog). This phenomenon is called PTS-mediated inducer exclusion and does not occur in *crr* mutants lacking IIA^{Glc}. It may explain why *ptsH, I* deletion mutants do not grow on class I substrates: Transport is impaired.

What is the mechanism by which IIA^{Glc} is involved in inducer exclusion? Figure 2 shows how the PTS is thought to regulate the entry and metabolism of numerous non-PTS carbon sources. Central to PTS-mediated regulation is the phosphorylation state of IIA^{Glc} which is determined by the rate of phosphorylation via EI/HPr and the rate of dephosphorylation via EII^{Glc} in the presence of its substrate, glucose. In *ptsH, I* mutants or in *pts⁺* strains in the presence of glucose, one expects IIA^{Glc} to be mainly or exclusively in the dephosphorylated state. This has been confirmed experimentally. It has been proposed that nonphosphorylated IIA^{Glc} inhibits a number of non-PTS uptake systems by binding directly to the target proteins (Fig. 2). The *crr* mutation eliminates the inhibitor IIA^{Glc} and thus restores growth of *ptsH, I* mutants on class I compounds. PTS carbohydrates other than glucose can bring about dephosphorylation of phospho-IIA^{Glc} via dephosphorylation of phospho-HPr. It should be remembered (see Section II) that the phosphoryl transfer between the PTS proteins is a reversible reaction.

B. Regulation of Adenylate Cyclase Activity

Figure 2 also shows the other mechanism by which IIA^{Glc} can regulate metabolism. The phosphorylated form of IIA^{Glc}, P∼IIA^{Glc}, is essential for activation of adenylate cyclase. In *crr* mutants, lacking both forms of IIA^{Glc}, adenylate cyclase activity is 10–20% of that in the parent *crr⁺* cells. It was also shown that the addition of PTS substrates to toluenized *E. coli* cells (i.e., conditions in which PTS proteins are dephosphorylated) inhibited adenylate cyclase activity. Since cAMP is required for the expression of many catabolic operons, a low rate of cAMP synthesis will lower or even prevent gene expression, depending on the actual concentration of cAMP in the cell and the requirement of each promoter for cAMP.

Previously, the glucose–lactose diauxie phenomenon, which has often been presented as the paradigm for cAMP-dependent regulation of gene expression, was discussed. Although glucose certainly regulates cAMP synthesis and cAMP is certainly important for expression of, for example, the *lac* operon, recent experiments show that the major determinant in at

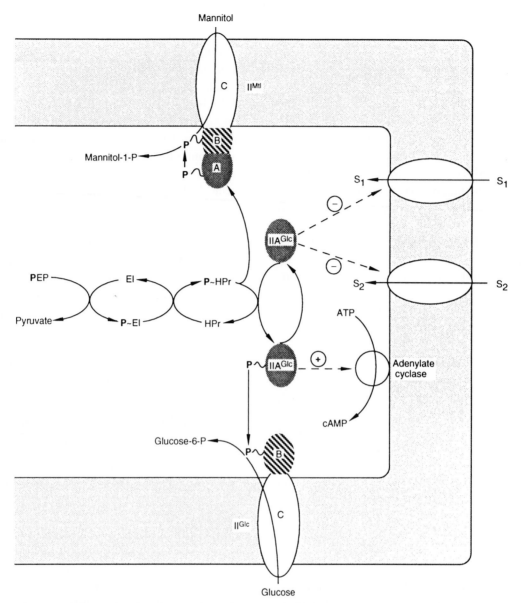

Fig. 2. Regulation by the PTS in enteric bacteria. IIA^{Glc} can exist in the phosphorylated form (P~IIA^{Glc}) and in the non-phosphorylated form, depending on the rate of phosphorylation via phospho-HPr (P~HPr) and the rate of dephosphorylation via II^{Glc} or HPr. P~IIA^{Glc} is involved in activation (+) of adenylate cyclase. Nonphosphorylated IIA^{Glc} is an inhibitor (−) of a number of non-PTS uptake systems, denoted by S$_1$ and S$_2$, which represent lactose, maltose, melibiose, and glycerol. Other symbols are as defined in the legend to Fig. 1 (reproduced with permission from Postma *et al.*, 1996).

least glucose–lactose diauxie (and presumably in many other combinations) is inducer exclusion and not a decrease of the intracellular cAMP concentration. In fact, it was shown that the cAMP level was the same in the glucose and the lactose phases, with a higher concentration only in the lag period when the cell is switching between glucose and lactose consumption. It was also shown that the level of CRP, the cAMP-binding protein, decreased in glucose-grown cells rather than the cAMP level.

C. Interaction between PTS and Non-PTS Proteins

What is the biochemical evidence for these regulatory roles of IIA^{Glc}? It was shown that IIA^{Glc}, but not phospho-IIA^{Glc}, binds to the lactose carrier (LacY) and glycerol kinase (GlpK) and inhibits their enzymatic activity, as predicted by the model. Binding to MalK, a component of the maltose transport system, has also been demonstrated. Interestingly, binding occurs in all cases only when the substrate of the target protein, a β-galactoside, glycerol, or maltose is present. It suggests that the target protein has to undergo a conformational change before it can bind IIA^{Glc}. In a physiological context this means that in cells only those systems are inhibited for which a substrate is available in the medium. The cell does not "waste" its IIA^{Glc}, of which it has only a fixed amount.

Has the interaction of P~IIA^{Glc} with adenylate cyclase and subsequent activation of the enzyme been demonstrated? In this case, the biochemical evidence is much less convincing. A small activation of adenylate cyclase activity has been observed in a cell-free *E. coli* extract after addition of purified phospho-IIA^{Glc}. However, this activation is much too small to explain the large variations in the rate of cAMP synthesis. Possibly, other factors are required in addition to IIA^{Glc}. Studies with truncated adenylate cyclase show that the C-terminal domain is required for activation by IIA^{Glc}.

IIA^{Glc} is able to specifically recognize many non-PTS proteins that are quite different from each other in their primary structure. Apart from interacting with various PTS proteins, IIA^{Glc} binds to LacY, GlpK, the MalK protein (the ATP binding subunit of the maltose transport system), and (most likely) the melibiose carrier. Are the same IIA^{Glc} residues involved in binding to different target proteins? Since defined mutations in IIA^{Glc} result in loss of interaction with all four target systems, most likely the same residues of IIA^{Glc} interact with the various target proteins. Is there a common binding motive in the target proteins? Mutations have been isolated in all four target proteins which render those proteins insensitive to regulation by IIA^{Glc}. In the MalK protein the mutations are distributed across two-thirds of the C-terminal part. In LacY, mutations are localized both in the N-terminus and in the middle part of the protein, whereas inhibition of the melibiose carrier by IIA^{Glc} is influenced by mutations in the C-terminal part of the molecule. No clear domain or conserved amino acid sequence can be recognized. In the case of glycerol kinase, the 3-D structure of the complex between GlpK and IIA^{Glc} has been determined by x-ray crystallography. It can be concluded that a IIA^{Glc} molecule binds to each of the four monomers of the GlpK tetramer, mostly by hydrophobic interaction between a limited number of residues. At least in the case of the IIA^{Glc}–GlpK interaction, it was found that Zn^{2+} ions decreased the apparent inhibition constant from 16.6 μM in the absence of Zn to 0.28 μM in the presence of 0.1 mM Zn. The Zn^{2+} ion is coordinated by the two important His residues in IIA^{Glc} (His75 and His90), a Glu418 residue in GlpK, and a water molecule. Phosphorylation of IIA^{Glc} on His90 prevents binding to GlpK by introducing a negative charge which protrudes out of the hydrophobic ring of residues which form the IIA^{Glc} binding site. It has been shown that no large conformational changes occur upon IIA^{Glc} phosphorylation.

Can IIA^{Glc} still be active in transport and phosphorylation while its regulatory functions are abolished or vice versa? Replacing His75 with a glutamic acid prevents phosphoryl transfer but does not affect regulation. The reverse has also been observed. Some IIA^{Glc} mutants can still catalyze phosphoryl transfer and activate adenylate cyclase but are inactive in mediating inducer exclusion. Clearly, the various functions of IIA^{Glc} can be separated and may be confined to separate domains.

D. Physiological Consequences

PTS-mediated regulation via IIA^{Glc} also has important quantitative aspects. Since IIA^{Glc} and its target proteins form stoichiometric complexes, it can be predicted that cells can escape from inducer exclusion when the number of target proteins is larger than the number of IIA^{Glc} molecules (which is constant in an *E. coli* cell). This has been verified experimentally by lowering the amount of IIA^{Glc} in the cell (using plasmids that produce defined amounts of IIA^{Glc}), by increasing the amount of one of the target proteins, by inducing two of these target proteins at the same time, or by using membrane vesicles with various amounts of included IIA^{Glc}. In the case of GlpK, it could be demonstrated that glycerol metabolism was completely inhibited when cells contained four times as many IIA^{Glc} molecules as GlpK (tetramer) molecules, in agreement with the data on the structure of the complex. In each of these examples, uptake of class I compounds becomes insensitive to inhibition by PTS carbohydrates in these cells or vesicles when the level of IIA^{Glc} becomes too low compared to the level of the target protein studied.

The formation of complexes between IIA^{Glc} and various target proteins has another consequence. Not only do the target protein become inactive but also the number of free and active IIA^{Glc} molecules decrease. Thus, the class I systems not only are regulated by the PTS but also can control the activity of the glucose PTS under conditions in which the amount of IIA^{Glc} controls the rate of glucose uptake and metabolism.

Regulation of the so-called class I transport systems, as described previously, occurs at two levels: directly by inhibition of the uptake of substrates (short-term regulation) and indirectly by a lowered expression of the respective operons (long-term regulation). Utilization of class II compounds, such as xylose, rhamnose, and Krebs cycle intermediates, is only regulated at the level of gene expression (i.e., the level of cAMP synthesis). Their respective uptake systems are not subject to inducer exclusion via IIA^{Glc}.

Recently, IIA^{Glc}-dependent inducer exclusion was extended to non-PTS carbon sources. Although carbon sources such as glucose 6-phosphate cannot directly result in dephosphorylation of P∼IIA^{Glc}, they nevertheless can result in inducer exclusion of lactose, for example. It has been shown that the metabolism of carbon sources such as glucose 6-phosphate results in low intracellular PEP:pyruvate ratios. As a consequence of this low ratio, the flux of phosphoryl groups through the PTS is reversed, generating PEP from pyruvate, and P∼IIA^{Glc} is dephosphorylated. This dephosphorylation of IIA^{Glc} has been measured directly in intact cells. It has been concluded that the metabolism of any carbon source, which results in a low PEP:pyruvate ratio, will result in an increased level of dephosphorylated IIA^{Glc} and thus in increased inducer exclusion and lowered intracellular cAMP levels.

It was previously mentioned that several EIIs of both gram-positive and gram-negative bacteria contain C-terminal domains that show considerable identity with IIA^{Glc} from enteric bacteria. Can such EIIs substitute for IIA^{Glc} in *Enterobacteriaceae* and can PTS-mediated regulation occur at the level of inducer exclusion and adenylate cyclase in other, especially gram-positive, bacteria? In *E. coli crr* mutants lacking IIA^{Glc}, EIIs such as II^{Nag} and II^{Bgl} from *E. coli* can restore the uptake of glucose or its nonmetabolizable analog methyl α-glucoside via IICB^{Glc}. At least II^{Nag} can substitute for IIA^{Glc} in inducer exclusion as well (implying that a membrane-bound IIA^{Glc}-like domain can also interact with the various non-PTS target proteins). Currently, there is no evidence that this type of PTS-mediated regulation involving IIA^{Glc} occurs outside *Enterobacteriaceae*, although IIA^{Glc}-like domains occur in gram-positive bacteria. For instance, *B. subtilis ptsH,I* mutants do not grow on non-PTS compounds such as maltose and glycerol, similar to *E. coli*. However, unlike *E. coli ptsH,I crr* mutants, *B. subtilis* mutants which in addition lack the IIA^{Glc} domain are unable to grow on maltose and glycerol. Since *B. subtilis* contains no adenylate cyclase, PTS-mediated regulation is likely to occur by a different mechanism.

VI. PTS-MEDIATED REGULATION IN GRAM-POSITIVE BACTERIA

In previous sections, phosphoryl transfer reactions from PEP to the various PTS proteins which involved in most cases histidine residues (the exception being cysteine) were discussed. It was found that HPr of gram-positive bacteria can be phosphorylated on two residues. Phosphorylated EI transfers its phosphoryl group on a His residue (His15 in *B. subtilis*) resulting in P-His-HPr, whereas a conserved serine residue (Ser46 in *B. subtilis*) can be phosphorylated by an ATP-dependent protein kinase. The resulting P-Ser-HPr can be dephosphorylated by a phosphoprotein phosphatase. Both activities have been purified and the gene encoding the kinase (*ptsK* or *hprK*) has been cloned recently from *B. subtilis*. Although the serine residue is also conserved in *E. coli* HPr, no phosphorylation of this serine residue by gram-positive kinases has been observed. *Escherichia coli* lacks kinase and phosphatase activity. However, from genome sequence programs it is clear that HPr kinase homologs are present not only in gram-positive bacteria but also in a few gram-negative bacteria.

Phosphorylation of the serine residue has been implicated in regulation of expression of many genes and operons as well as in solute transport. It has been shown that P-Ser-HPr is much less readily phosphorylated by P~EI. Thus, serine phosphorylation may lower the flux through the PTS pathway. However, this slower EI-dependent phosphorylation is increased when IIA domains or proteins are present, at least in *in vitro* studies. On the other hand, P-His-HPr is a poor substrate for the ATP-dependent kinase. HPr kinase is activated by fructose 1,6-diphosphate and possibly other glycolytic intermediates and inhibited by P_i, whereas the phosphatase activity is stimulated by P_i. As a consequence, the rate of glycolysis and the steady state concentration of the glycolytic intermediates will affect the phosphorylation state of HPr and thus its activity. It is not possible to predict the phosphorylation state of HPr, which can range from HPr to doubly phosphorylated HPr. Attempts have been made to determine the various forms in intact cells using antibodies raised against HPr and non-denaturing gelelectrophoresis.

Phosphorylation of HPr on a serine residue has been linked to the process of inducer expulsion. In this process, certain carbohydrate phosphates such as lactose phosphate, generated by the lactose PTS, are actively expelled from the cell in the presence of, for instance, glucose, another PTS carbohydrate. Efflux is energy requiring and II^{Lac} (in the case of lactose) is involved. Lactose phosphate (or its analog TMG phosphate) is dephosphorylated by a phosphatase and expelled as the free carbohydrate. This process of inducer expulsion coincides with phosphorylation of the serine residue in HPr. Although it has been suggested that P-Ser-HPr is essential in this expulsion process, its role remains obscure.

Involvement of the PTS in global regulation of gene expression in gram-positive bacteria is not via the phosphorylation state of the IIA^{Glc} protein or domain but rather via HPr. The CcpA protein (catabolite control protein), which is a DNA-binding protein, was first discovered in *B. subtilis* and CcpA-like proteins were subsequently detected in several other gram-positive bacteria. Many genes and operons contain a conserved CRE (catabolite-responsive element) site to which CcpA can bind. The CRE site can be localized near the promoter region or in the structural gene and transcription is prevented upon binding of CcpA to the CRE site. Studies with the purified CcpA protein and several CRE binding sites suggest that CcpA binds only tightly in the presence of P-Ser-HPr. Thus, the phosphorylation state of HPr determines the level of expression of CcpA-regulated genes. Catabolite repression is released in a *B. subtilis* mutant in which the Ser46 residue of HPr is replaced by Ala, but release is only partial for a number of genes. A second HPr-like protein, Crh, has been found that contains Ser46 but in which His15 is replaced by Glu. In HPr/Crh double mutants, release of catabolite repression is complete.

It is instructive to compare this global regulatory mechanism with that in enteric bacteria in which CRP (the cAMP-binding protein) and cAMP regulate transcription of many catabolic operons. In gram-positive bacteria, the metabolism of glucose or other PTS carbohydrates may result in increased phosphorylation of HPr on its serine residue. Binding of P-Ser-HPr [and possibly P-Ser-(P-His-)HPr] to CcpA results in inhibition of transcription of catabolic operons. In enteric bacteria, glucose or other PTS carbohydrates lower the phosphorylation state of P-IIA^{Glc}, resulting in decreased adenylate cyclase activity and lower intracellular cAMP levels. Under these conditions, less of the CRP–cAMP complex is available and transcription of catabolic operons is not at all or less activated. Thus, although CcpA acts as an inhibitor and CRP as an activator, in both systems the flux of PTS carbohydrates and most likely other carbohydrates controls gene expression by controlling the phosphorylation state of particular PTS proteins.

Other differences exist between PTS-mediated regulation in gram-negative and gram-positive bacteria. Remember that glycerol kinase was inhibited by nonphosphorylated IIA^{Glc} in enteric bacteria. In the gram-positive bacterium *Enterococcus faecalis*, glyc-erol kinase is activated by phosphorylation of a histi-dine residue by P-His-HPr.

VII. OTHER INTERACTIONS BETWEEN PT AND NON-PT SYSTEMS

Previously, various mechanisms by which the PTS influences non-PT systems in both gram-negative and gram-positive bacteria were described. Are there other interactions between the PTS and the remain-der of cellular metabolism?

A possible link may exist between the PTS and oxidative metabolism in *E. coli* based on the finding that *in vitro* phosphoryl groups can be transferred between the phosphorylated form of purified acetate kinase and the active site His residue of EI. In princi-ple, it would allow the PTS to operate at low PEP levels since EI would be phosphorylated by ATP via acetate kinase. Whether this reaction is operative under physiological conditions in the intact cell is unknown. Another ATP-dependent kinase has been detected in *E. coli* which is able to phosphorylate EI at the active site His residue. Its physiological function, if any, also remains unknown.

The connection between the PTS and the global CRP/cAMP regulatory system was previously dis-cussed. A second example involves a protein that was first identified as the repressor of the fructose PTS, FruR, which regulates the expression of the *fruFKA* operon, encoding FPr (the HPr-like protein), fructose 1-phosphate kinase, and II^{Fru}. In addition, *fruR* mutants are unable to grow on gluconeogenic substrates such as pyruvate and lactate. *fruR* mutants lack the enzyme PEP synthase which converts ATP and pyruvate into PEP, AMP, and P$_i$ and which is obligatory for growth of *E. coli* and *S. typhimurium* on pyruvate and lactate. The activity of PEP carboxy-kinase and many other enzymes is also lowered in *fruR* mutants, although the effect is strongest for PEP synthase and PEP carboxykinase. It was shown that FruR is an activator of *ppsA* transcription (encoding PEP synthase) as well as a number of other enzymes, mostly involved in gluconeogenesis. Expression of other enzymes is repressed, however. Genes to be activated by FruR have a conserved FruR binding site upstream of the promoter, whereas a FruR bind-ing site downstream of the promoter results in inhibi-tion. Since *in vitro* studies have shown that low con-centrations of fructose 1-phosphate can displace FruR from its binding site, carbon sources that pro-duce fructose 1-phosphate will inhibit expression of FruR-activated genes while they stimulate expression of FruR-inhibited genes. FruR has been renamed Cra (*c*atabolite *r*epressor/*a*ctivator).

A third non-PTS system which shows interaction via PTS proteins involves chemotaxis proteins (Che pro-teins). The movement towards chemicals, chemotaxis, has been studied in great detail in enteric bacteria as well as some others. The CheA and CheY proteins be-long to the family of two-component systems and change their rate of phosphorylation in response to many external signals which are sensed via receptor proteins. The phosphorylation state of CheY, the re-sponse regulator, signals to the flagellar motor whether it should turn clockwise or counterclockwise.

PTS carbohydrates are also chemoattractants and feed their signal into the Che pathway, but they utilize the various EIIs as chemoreceptors. Mutants lacking EI or HPr are defective in chemotaxis toward all PTS carbohydrates. The PTS pathway feeds early into the general signal transduction pathway and it has been shown that phosphorylated EI inhibits the autophosphorylation of CheA by protein–protein in-teraction. No evidence has been found for transfer of a phosphoryl group from a PTS protein to one or more of the Che proteins.

See Also the Following Articles

ABC Transport • Bacillus subtilis, Genetics • Chemotaxis • Escherichia coli and Salmonella, Genetics

Bibliography

Hueck, C. J., and Hillen, W. (1995). Catabolite repression in *Bacillus subtilis*: a global regulatory mechanism for the gram-positive bacteria. *Mol. Microbiol.* **15**, 395–401.

Lengeler, J. W., and Jahreis, K. (1996). Phosphotransferase systems or PTSs as carbohydrate transport and signal trans-duction systems. *In* "Handbook of Biological Physics" (W. N. Konings, H. R. Kaback, and J. S. Lolkema, Eds.), pp. 573–598. Elsevier, Amsterdam.

Lengeler, J. W., Jahreis, K., and Wehmeier, U. F. (1994). Enzymes II of the phosphoenolpyruvate-dependent phos-photransferase systems: their structure and function in carbohydrate transport. *Biochim. Biophys. Acta* **1188**, 1–28.

Meadow, N. D., Fox, D. K., and Roseman, S. (1990). The bacterial phosphoenolpyruvate : glycose phosphotransferase system. *Annu. Rev. Biochem.* **59**, 497–542.

Postma, P. W., Lengeler, J. W., and Jacobson, G. R. (1993). Phosphoenolpyruvate : carbohydrate phosphotransferase systems of bacteria. *Microbiol. Rev.* **57**, 543–594.

Postma, P. W., Lengeler, J. W., and Jacobson, G. R. (1996). Phosphoenolpyruvate : carbohydrate phosphotransferase systems. *In* "*Escherichia coli* and *Salmonella.* Cellular and Molecular Biology" (F. C. Neidhardt, R. Curtiss, III, J. L. Ingraham, E. C. C. Lin, K. B. Low, B. Magasanik, W. S. Reznikoff, M. Riley, M. Schaechter, and H. E. Umbarger, Eds.), pp. 1149–1174. ASM Press, Washington, DC.

Reizer, J., and Reizer, A. (1996). A voyage along the bases: Novel phosphotransferase genes revealed by in silico analy-ses of the *Escherichia coli* genome. *Res. Microbiol.* **147**, 458–471.

Reizer, J., and Saier, M. H. (1997). Modular multidomain phosphoryl transfer proteins of bacteria. *Curr. Opin. Struct. Biol.* **7**, 407–415.

Rutberg, B. (1997). Antitermination of transcription of catabolic operons. *Mol. Microbiol.* **23**, 413–421.

Saier, M. H., Chauvaux, S., Cook, G. M., Deutscher, J., Paulsen, I. T., Reizer, J., and Ye, J.-J. (1996a). Catabolite repression and inducer exclusion in gram-positive bacteria. *Microbiology* **142**, 217–230.

Saier, M. H., Ramseier, T. M., and Reizer, J. (1996b). Regulation of carbon utilization. *In* "*Escherichia coli* and *Salmonella.* Cellular and Molecular Biology" (F. C. Neidhardt, R. Curtiss, III, J. L. Ingraham, E. C. C. Lin, K. B. Low, B. Magasanik, W. S. Reznikoff, M. Riley, M. Schaechter, and H. E. Umbarger, Eds.), pp. 1325–1343. ASM Press, Washington, DC.

Pesticide Biodegradation

Li-Tse Ou

University of Florida

GLOSSARY

aged residue Pesticide that over time diffuses into the internal soil matrix and is no longer available for microbial attack.

bioavailability Property of a chemical that is in the solution phase and available for microbial attack.

bioremediation Use of microorganisms or microbial processes to control and detoxify environmental contaminants.

bound residue Chemical species originating from a pesticide that are covalently bound to soil organic matter.

cometabolism Biodegradation of an organic chemical (pesticide) by a microorganism that cannot utilize the chemical for energy and growth.

insertion sequence element Specific nucleotide sequence that is involved in the transfer and integration of pieces of DNA to various sites in plasmid and chromosome.

plasmid Extrachromosomal DNA with a characteristic copy number that replicates autonomously.

transposon Mobile genetic element that is flanked by an insertion sequence element at each end.

xenobiotics Man-made chemicals that do not occur naturally.

BIODEGRADATION OF PESTICIDES is defined in this article as the transformation of pesticides by microorganisms isolated from soils and water or in soils and water where microorganisms are responsible for the degradation. Bacteria and fungi are the two main classes of microorganisms responsible for the degradation of pesticides in the environment. Some pesticides can be utilized by microorganisms as a sole source of carbon and energy, and are converted to innocuous end products, CO_2, H_2O, and inorganic ions. Others are degraded cometabolically and are likely to be more persistent, resulting in accumulation of organic metabolites. Degradation of pesticides in soils is a complex process. Pesticides can be mineralized to CO_2, H_2O, and inorganic ions; assimilated into microbial mass; transformed to toxic and nontoxic organic metabolites; and converted to bound residues associated with organic matter. Repeated field applications of some pesticides such as carbamate compounds may induce enhanced degradation of these chemicals, which may result in reduced crop production due to failure in controlling target pests.

I. USE OF PESTICIDES

Pesticides have been an integral part of agriculture and public health since World War II for controlling pests and disease carriers, respectively. Approximately 1.1 billion pounds are utilized in the United States annually, and the total amounts applied and number of pesticides in the market have increased steadily since the 1960s, when the first reliable records of production were established. Herbicides, insecticides, and fungicides are considered to be the major classes of pesticides. Minor classes include nematicides, fumigants, miticides, molluscides, and rodenticides. Herbicides and insecticides are the first and second most-used pesticides in industrialized

Encyclopedia of Microbiology, Volume 3
SECOND EDITION

countries respectively, whereas in developing countries, especially in tropical regions, the order is reversed. The majority of pesticides are synthetic organic chemicals. They encompass many different classes of compounds, from highly volatile gaseous forms to nonvolatile solids and from simple chemicals such as methyl bromide to complex chemicals such as DDT. The vast majority of the pesticides belong to several classes of organic chemicals, including organochlorines, carbamates, phenoxy compounds, organophosphates, triazines, urea compounds, pyrethroids, nitroanilines, anilides, amines, and amides.

Use of pesticides has undoubtedly resulted in increased crop production and quality, reduction of labor costs, and control of vectors for public health hazards. However, public concerns about potential adverse effects on the environment, and on human and animal health, have also increased.

II. MICROORGANISMS INVOLVED IN PESTICIDE DEGRADATION

Bacteria and fungi are the two major groups of microorganisms responsible for extensive degradation of pesticides in soil and water. A large number of different bacteria that degrade pesticides have been isolated from soils and water. They belong to various genera of gram negative and gram positive aerobic and anaerobic organisms. The most important aerobic gram negative bacteria are *Pseudomonas*, *Sphingomonas*, *Burkholderia* (both *Sphingomonas* and *Burkholderia* previously classified as *Pseudomonas*), *Alcaligenes*, *Acinetobacter*, *Flavobacterium*, methane-oxidizing bacteria and nitrifying bacteria; gram-positive bacteria include *Arthrobacter*, *Nocardia*, *Rhodococcus*, and *Bacillus*. Some denitrifying bacteria, sulfate-reducing bacteria, and methanogenic bacteria are involved in the anaerobic degradation of pesticides. Key fungi that are involved in aerobic pesticide degradation include *Phanerochaete* (white rot fungi), *Penicillium*, *Aspergillus*, *Trichoderma*, and *Fusarium*. Some strains of *Phanerochaete* sp. also have the capacity to degrade structurally complex pesticides such as DDT.

III. BIOCHEMISTRY OF PESTICIDE DEGRADATION

A. Initial Biochemical Reactions

The majority of pesticides are composed of one or two benzene rings and short-chained aliphatic hydrocarbons linked with ether, carbamate, phosphate, amine, amide, anilide, ester, or thioether groups. In addition, the benzene rings may be halogenated, especially with chlorine. Consequently, initial biochemical reactions of pesticides are mainly via hydrolysis, oxidation, and reduction. Generally, pesticidal activity will be lost after hydrolysis, whereas the corresponding oxidation or reduction products often retain pesticidal activity. Figure 1 illustrates the initial biodegradation of some widely used pesticides. The initial degradation of the herbicide atrazine is either through oxidation to deethylatrazine or via hydrolysis to hydroxyatrazine. It should be noted that deethylatrazine has herbicidal activity similar to that of atrazine, whereas hydroxyatrazine does not have such activity. The insecticide carbofuran also undergoes hydrolysis to carbofuran phenol or oxidation to 3-hydroxycarbofuran. Similar to atrazine, 3-hydroxycarbofuran retains insecticidal activity, but carbofuran phenol does not. On the other hand, the insecticide parathion is initially hydrolyzed to *p*-nitrophenol or reduced to aminoparathion. Even the simple fumigant methyl bromide is initially degraded via oxidation to formaldehyde.

Other initial biochemical reactions include methylation, acetylation, and conjugation. Products of methylation may be volatile, resulting in losses to the atmosphere. Products of hydrolysis, oxidation and reduction also may undergo dimerization or polymerization. 3,4-Dichloroaniline, a hydrolysis product of the herbicide propanil, can be dimerized to form 3,4,-3′4′-tetrachloroazobenzene. The oxidation product of the herbicide 2,4-D—2,4-dichlorophenol—also can be dimerized and even polymerized.

B. Degradation Pathways

Degradation pathways of aliphatic hydrocarbons and simple aromatic hydrocarbons (benzene, phenol and benzoic acid) are well documented. For aliphatic

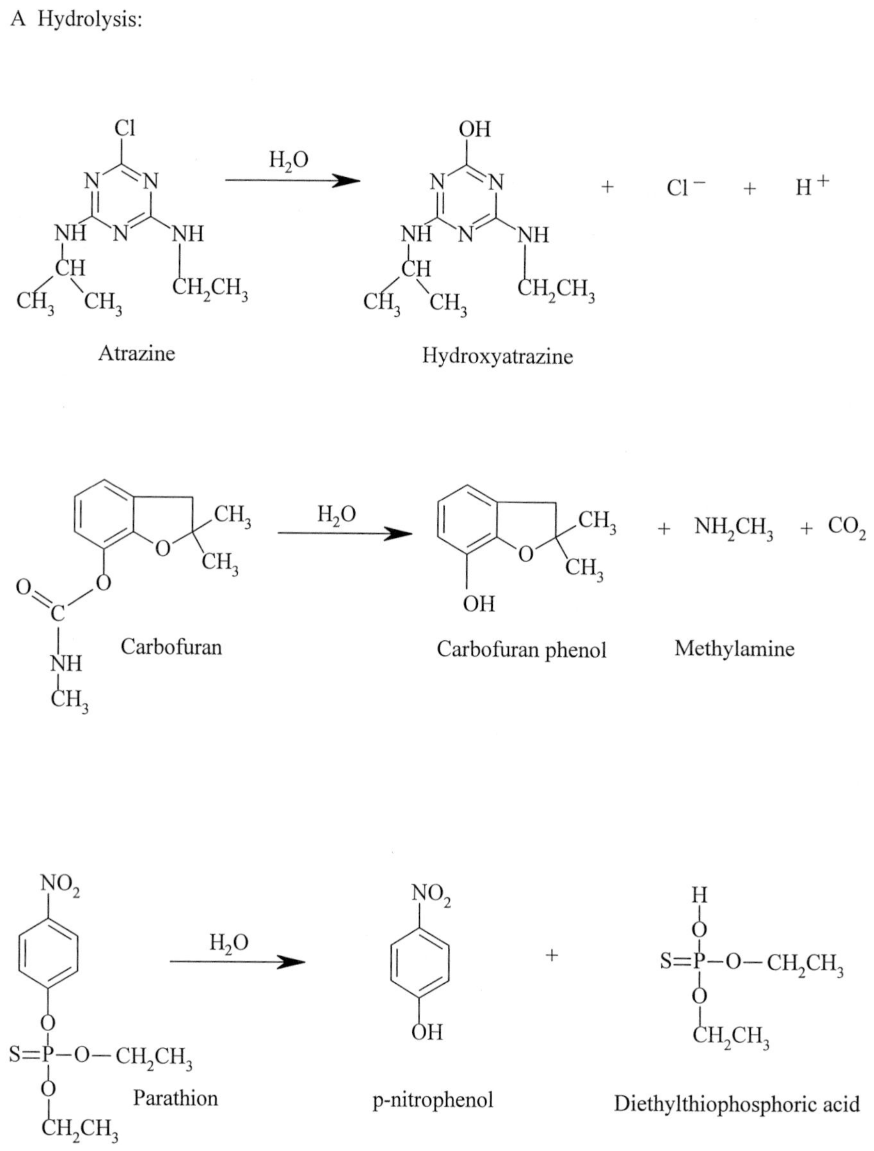

A Hydrolysis:

Fig. 1. Initial biochemical reactions for selected pesticides.

hydrocarbons to be completely degraded to the final oxidation products CO_2 and H_2O, these compounds must undergo a series of enzymatic oxidations to alcohols, aldehydes, and eventually to alkanoic acids. Alkanoic acids are then subject to β-oxidation to acetic acid and finally to CO_2 and H_2O (Fig. 2). Simple aromatic compounds, such as benzene, phenol, and benzoic acid are initially oxidized by hydroxylation by monooxygenase or dioxygenase to catechol (Fig. 2), which is then subject to ring fission via ortho or meta cleavage to *cis, cis*-muconic acid or

2-hydroxymuconic semialdehyde, respectively. The two intermediates are subsequently degraded to 3-ketoadipic acid or to pyruvic acid and acetaldehyde, respectively, which are common metabolic intermediates and rapidly channeled into central biochemical pathways before eventually yielding CO_2 and H_2O.

As mentioned previously, aromatic compounds comprise the major group of pesticides, with degradation resulting in the formation of phenolic compounds, benzoic acids, and amino aromatics. These products are then oxidized to catechols, which are

B Oxidation:

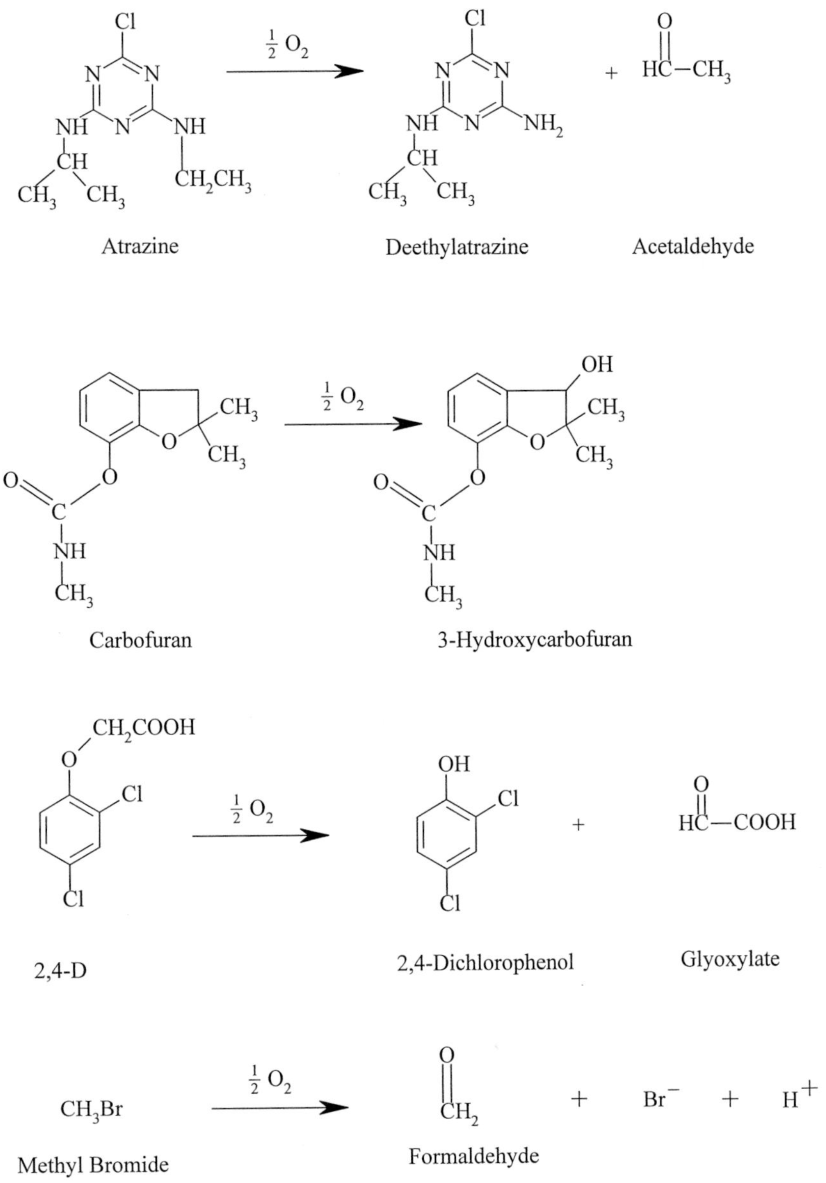

Atrazine → Deethylatrazine + Acetaldehyde

Carbofuran → 3-Hydroxycarbofuran

2,4-D → 2,4-Dichlorophenol + Glyoxylate

Methyl Bromide → Formaldehyde + Br^- + H^+

Fig. 1. (*continued*)

subsequently metabolized as described previously. Many pesticides are halogenated compounds and predominately chlorinated compounds. If benzene rings are halogenated, extensive dehalogenation will occur after the ring structures are broken, resulting in the release of the halogen ion. Complete transformation of organic chemicals to inorganic forms is termed "mineralization."

The most studied pesticide is 2,4-D. The biochemistry and genetics of 2,4-D degradation, and the ecology of 2,4-D degrading bacteria in soils, have been well characterized. A soil bacterium which was isolated from an Australian soil, *Alcaligenes eutrophus* JMP134, has been extensively studied. The first step of 2,4-D degradation by this organism is oxidation to 2,4-dichlorophenol by 2,4-D-dioxygenase, fol-

C Reduction

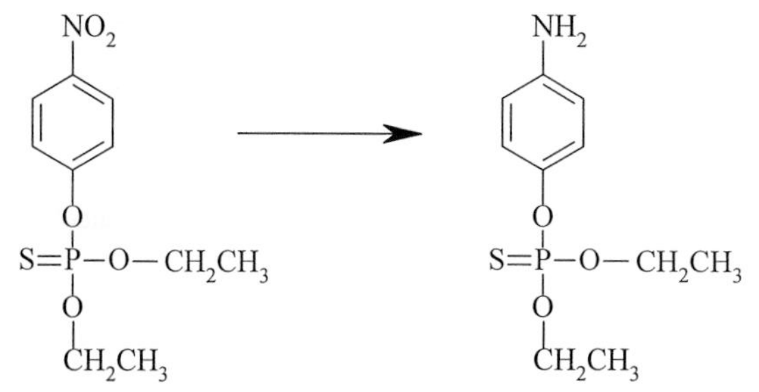

Parathion   Aminoparathion

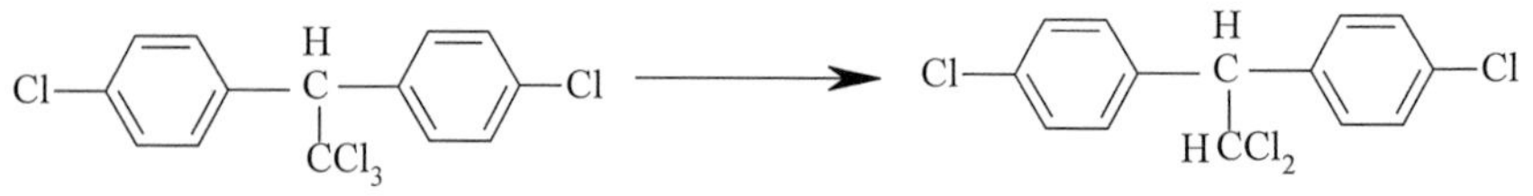

DDT   DDD

Fig. 1. *(continued)*

lowed by hydroxylation by 2,4-D-dichlorophenol hydroxylase to 3,5-dichlorocatechol (Fig. 3). The ring structure is then opened up through ortho cleavage by chlorocatechol 1,2-dioxygenase to yield 2,4-dichloromuconic acid. During the conversion of this compound to *trans*-2-chlorodiene lactone, a chlorine is detached to form a chloride ion. Eventually, the degradation process leads to the formation of succinic acid, and the last chlorine is dechlorinated. Succinic acid is a common biochemical intermediate, which is rapidly channeled into central metabolic pathways to either yield CO_2 and H_2O or to be incorporated into microbial biomass. Soil bacteria that have been found capable of degrading 2,4-D include strains of *Alcaligenes, Arthrobacter, Pseudomonas, Flavobacterium, Burkholderia, Rhodoferax, Acinetobacter, Corynebacterium, Rhodopesudomonas,* and *Sphingomonas.*

Many of these degrade 2,4-D in a similar manner to *A. eutrophus* JMP134.

Degradation pathways for most pesticides, even for widely used chemicals, are still not fully characterized. Atrazine is the most widely used pesticide in the United States, and many soil bacteria including *Pseudomonas, Rhizobium, Alcaligenes, Agrobacterium,* and *Rhodococcus* spp. have the capacity to degrade this herbicide. *Pseudomonas* strain ADP sequentially hydrolyzes chlorine, *N*-ethyl, and *N*-isopropyl moieties to cyanuric acid (Fig. 4), but the subsequent degradation pathway has not been elucidated. Since this organism completely degrades atrazine to CO_2, H_2O, and ammonia, it is suggested that this organism has the capacity to cleave the ring structure, but the mechanism of ring cleavage is not clear. Similarly, biodegradation of carbofuran has been studied for

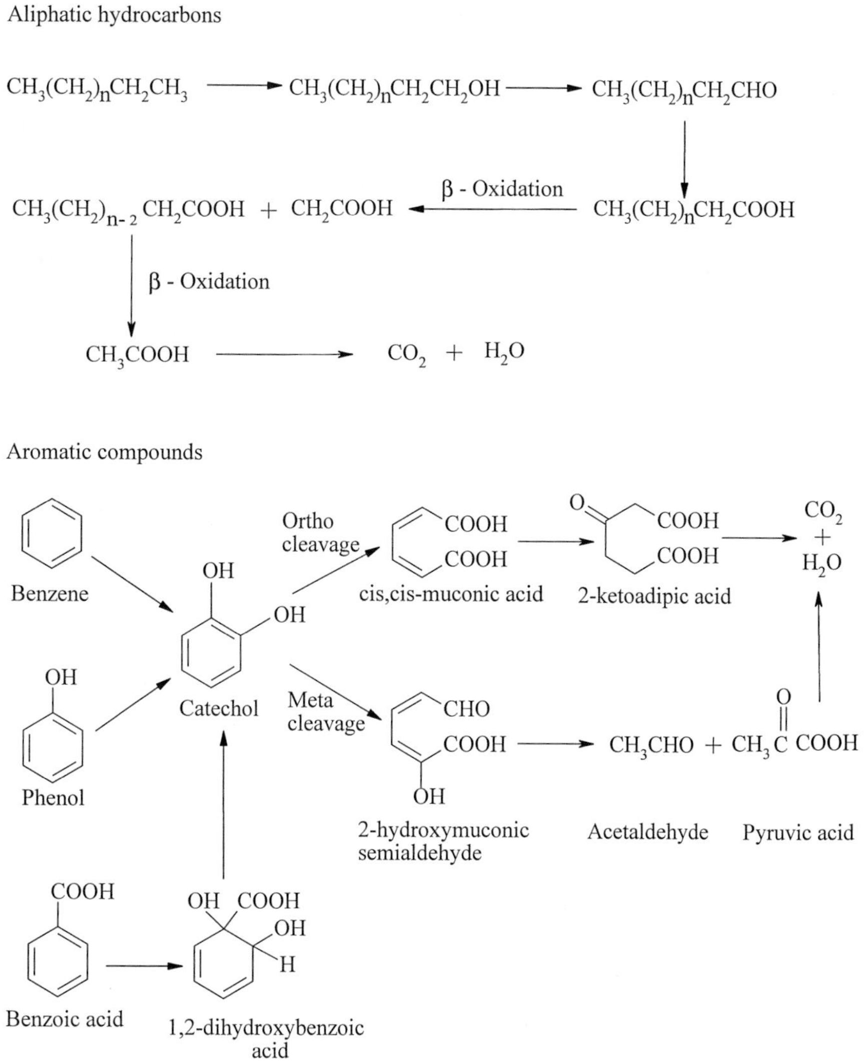

Aliphatic hydrocarbons

Aromatic compounds

Fig. 2. Degradation pathways for aliphatic hydrocarbons and for three aromatic hydrocarbons: benzene, phenol, and benzoic acid.

more than 30 years, but only the initial steps of degradation are known (Fig. 5), mainly via hydrolysis to phenolic metabolites and oxidation to hydroxycarbofuran and ketocarbofuran. The oxidation products retain insecticidal activity.

C. Mineralization and Cometabolism

When organic forms of C, N, P, and S are converted to inorganic forms, energy in the form of ATP is often generated and, at the same time, some of the atoms may be incorporated into biomass. Microorganisms (bacteria and fungi) can utilize pesticides, such as some of the phenoxyacids, carbamates, triazines, and organophosphates, as sole sources of C, N, P, and/or S for growth and/or energy and concomitantly convert organic forms of C, H, O, N, P, S, and Cl to CO_2, H_2O, and inorganic ions. As mentioned previously, many soil bacteria have the capacity to utilize 2,4-D as a sole source of C for energy and growth, resulting in the formation of CO_2, H_2O, and Cl^-. In comparison, atrazine and the insecticide methyl parathion are mineralized to CO_2, H_2O, and ammonium (or nitrate), or to CO_2, H_2O, and inorganic phosphate; respectively.

Partial or complete degradation of a pesticide to

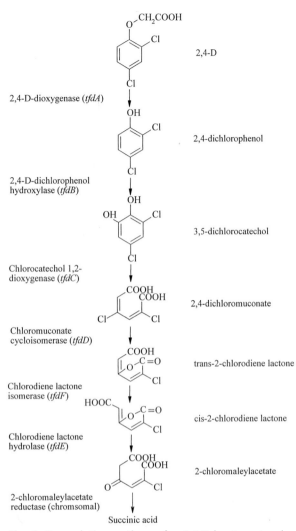

Fig. 3. Degradation pathway for 2,4-D by *A. eutrophus* JMP134 (redrawn with permission from Top *et al.*, 1995).

its final oxidation products (H_2O and CO_2) may require a concerted effort by a bacterial consortium rather than by a single organism. In this case, a second organic substrate that can serve as a carbon source for growth of the consortium is needed. The second chemical may or may not be structurally similar to the pesticide. A pesticide alternatively may be partially degraded by an axenic bacterial culture when grown on a second organic substrate, resulting in accumulation of organic metabolite(s). This phenomenon is termed "cometabolism." A strain of *Pseudomonas diminuta*, when grown using a readily degradable carbon source such as yeast extract,

peptone, or tryptone, hydrolyzes the insecticide parathion to *p*-nitrophenol. An axenic bacterial culture, a strain of *Bacillus* sp. capable of utilizing methyl parathion for growth, was isolated from a soil, but no axenic culture could be isolated from a different soil. When a bacterial consortium was enriched using glucose for growth, however, this mixed culture mineralized the insecticide. Only the first step of the degradation, hydrolysis to *p*-nitrophenol, is a cometabolic process. Because a bacterium associated with the mixed culture utilizes *p*-nitrophenol as a sole

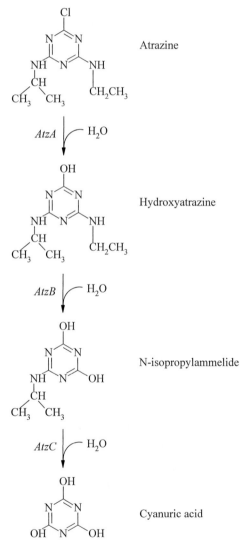

Fig. 4. Pathway for atrazine degradation to cyanuric acid by *Pseudomonas* sp. strain ADP (redrawn with permission from de Souza *et al.*, 1998).

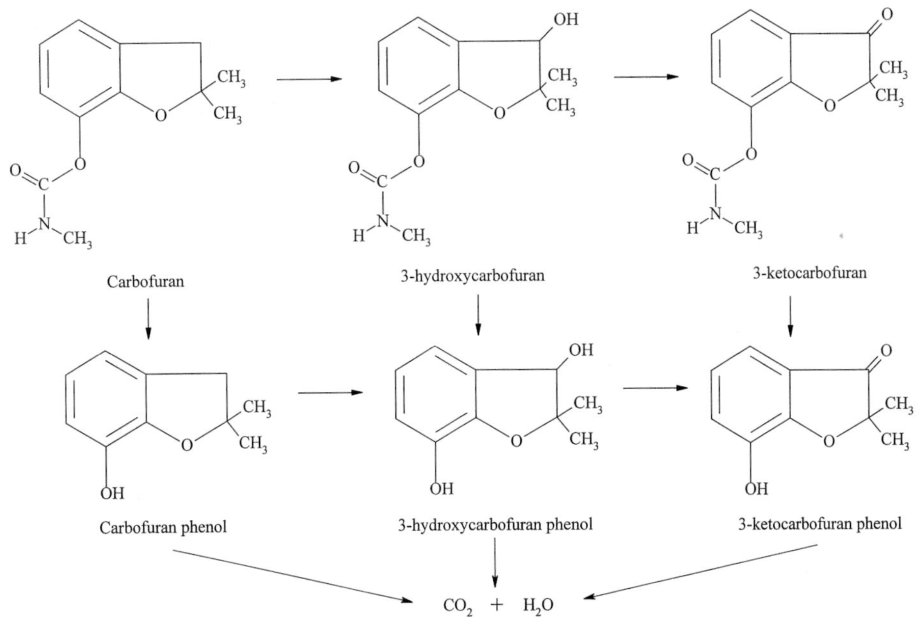

Fig. 5. Degradation pathway for carbofuran (redrawn with permission from Ou *et al.*, 1982).

source of carbon for growth, the nature of interactions among members of the consortium that leads to pesticide degradation is poorly understood. Cometabolic degradation of pesticides in soils and water appears to be fairly common.

IV. MOLECULAR BIOLOGY OF PESTICIDE DEGRADATION

Because pesticides are principally xenobiotic chemicals, exposure of these chemicals to microorganisms is relatively recent (<50 years for most pesticides). Metabolic pathways for natural chemicals have evolved after long times. However, microorganisms exposed to pesticides frequently respond by producing enzymes that degrade these chemicals. Genes are responsible for encoding the enzymes specific for pesticide degradation. Most of the current understanding of the genes and corresponding enzymes involved in pesticide degradation is derived from studies of 2,4-D degrading bacteria, especially of *A. eutrophus* JMP134.

Plasmids are often involved in the pesticide degradation that has been studied. Such pesticides include

2,4-D, parathion, methyl parathion, atrazine, carbofuran, and the herbicides 2,4,5-T and EPTC. Plasmids are extrachromosomal DNA elements with characteristic copy numbers within the host, and they replicate autonomously. It should be pointed out that genes responsible for pesticide degradation may also be located on chromosomal DNA. Investigations of 2,4-D-degrading bacteria isolated from soils provided the first evidence of the involvement of plasmids in the degradation of pesticides. The most studied 2,4-D-degradative plasmid is pJP4 from *A. eutrophus* JMP134. This broad host-range plasmid is an 80-kilo-base pairs, self-transmissible plasmid. It consists of six structural genes (*tfdA, tfdB, tfdC, tfdD, tfdE,* and *tfdF*) and two regulatory genes (*tfdR* and *tfdS*). The six structural genes encode six enzymes that are responsible for degrading 2,4-D to 2-chloromaleylacetate (Fig. 3). These genes are organized into three operons: *tfdA, tfdB,* and *tfdCDEF*. Genes responsible for subsequent degradation are chromosome-borne.

The current body of knowledge on the genetic diversity of pesticide-degrading microorganisms in soils is limited. Nevertheless, diversity exists among genes responsible for some pesticide degradation by soil bacterial populations. Notable examples are the soil bacteria capable of degrading 2,4-D. Many 2,4-D degrading bacteria have been isolated from various soils, principally from the United States, Australia, and Canada. The majority of the isolates exhibit homology to either one of the six genes of the plasmid pJP4 of *A. eutrophus* JMP134, especially the *tfdA* gene, or none at all. Moreover, most degraders possess combinations of *tfdA*-, *tfdB*-, and *tfdC*-like elements that show various levels of homology using gene probes of the three genes. Some degraders that are classified as *Alcaligenes, Burkholderia,* and *Rhodoferax* carry 60% or more sequence homology to *tfdA* and *tfdC*. Some strains of *Sphingomonas* that degrade 2,4-D hybridize only to *tfdB* at low stringency. It appears that extensive interspecies transfer of a variety of homologous degradative genes has been involved in the evolution of 2,4-D-degrading bacteria. Genetic diversity of bacteria that degrade other pesticides are likely to exist as well. For example, many bacteria that degrade carbofuran have been isolated from soils. The majority of them do not hybridize with the plasmid-borne *mcd* gene that encodes for carbofuran hydrolase in *Achromobacter* sp. strain WM111. Apparently, most carbofuran-degrading bacteria from soils carry a carbofuran hydrolase gene not related to the *mcd* gene. The carbofuran hydrolase catalyzes the hydrolysis of carbofuran to carbofuran phenol.

Evolution of bacteria capable of degrading xenobiotic chemicals such as synthetic pesticides in soils and water appears to be remarkable. Most synthetic pesticides have been on the market for less than 50 years and many pesticides are found to be biologically degraded in such systems. Bacteria and fungi capable of complete degradation of some widely used pesticides are often isolated from these sources. Initial degradation of some pesticides may result in the formation of metabolites that are naturally occurring chemicals. For example, the natural chemical 2,4-dichlorophenol is the initial product of 2,4-D degradation by soil bacteria such as *A. eutrophus* JMP134. In essence, 2,4-dichlorophenol-degrading bacteria may gain the capacity to degrade 2,4-D by acquiring just one gene, the 2,4-D-dioxygenase gene. Similarly, methyl parathion and parathion are xenobiotic chemicals, but hydrolysis during initial degradation

of the two chemicals results in the formation of *p*-nitrophenol. Large bacterial populations that degrade *p*-nitrophenol exist in soils and surface water. Thus, these organisms need to recruit only one gene—the methyl parathion hydrolase gene or the parathion hydrolase gene—to completely degrade methyl parathion or parathion. The question remains as to where the genes involved in the initial degradation come from. Soil organic matter is rich in methoxy and ethoxy groups. Monooxygenases and dioxygenases are responsible for the breakdown of these groups from the main chemical structures. Recruitment of one of these genes in conjunction with mutation or other means may allow 2,4-dichlorophenol degraders to completely degrade 2,4-D. Also, *p*-nitrophenol-degrading bacteria only need to acquire one gene—the methyl parathion hydrolase gene or the parathion hydrolase gene—to have the capacity to degrade this chemical completely.

Products of the metabolism of many pesticides are not natural chemicals, however, but xenobiotics. In this case, bacteria have to acquire the genes responsible for the entire degradation that leads to the formation of common biochemical intermediates, and they likely evolved this capacity more slowly than the bacteria that needed to gain only one gene. Bacteria that mineralize the ring structure of 2,4-D were isolated from soils in the late 1960s. Although atrazine has been used in agriculture since 1959, only recently (since 1993) have bacteria capable of complete degradation of the atrazine ring been isolated from soils. Degradation products of atrazine including hydroxyatrazine, *N*-isopropylammelide, deethylatrazine, deisopropylatrazine, and deethyldeisopropylatrazine are all xenobiotics.

Plasmids that harbor pesticide degradative genes are often conjugative (self-transmissible) plasmids. Genetic transfer among bacteria in soils is mainly by means of conjugation. The DNA regions where pesticide degradative genes are located are often rich in insertion sequence (IS) elements or transposons. Thus, pesticide degradative genes can be disseminated from one plasmid to another or to a chromosome through IS elements or transposons. Interspecies (horizontal) transfer can occur through conjugative plasmids.

V. DEGRADATION OF PESTICIDES IN SOILS

Degradation of pesticides in soils is a complex phenomenon, with no single factor contributing solely to the degradation. Biological, chemical, and photochemical degradation all may be involved in such degradation. Photochemical reactions occur only on top of the soil surface, and are not an important factor contributing to pesticide degradation in soils. Soil surface- and pH-induced reactions and abiotic hydrolysis can all be involved in chemical degradation in soils. Biological degradation is by far the most important factor in the removal of pesticides (except for highly volatile compounds) from soils and often leads to complete degradation to innocuous end products, CO_2, H_2O, and inorganic ions (mineralization). The extent of chemical and photochemical reactions is generally limited and these reactions also generally lead to accumulation of organic products.

A. Degradation Rates

As mentioned previously, pesticides consist of various classes of chemicals, and most of them are xenobiotics. As a result, some are very persistent in soils, whereas others are not. Degradation rates are one of the key factors determining the fate of pesticides in the environment. Table I shows average half-life values in soils for 17 widely used pesticides that range from 5 to 2000 days. DDT is a highly persistent chemical with a half-life of 2000 days; whereas methyl parathion and EPTC are relatively nonpersistent with half-lives of 5 and 6 days, respectively. Degradation rates are typically expressed in one of three forms: mineralization rate, parent chemical disappearance rate, or total toxic residue (TTR) disappearance rate. Mineralization rate represents the rate of complete degradation of a pesticide to inorganic forms, CO_2, H_2O, and inorganic ions. Generally, CO_2 evolution during pesticide degradation is measured to represent the mineralization rate of the chemical in soil. To differentiate CO_2 evolution associated with the pesticide from soil respiration, the pesticide typically is labeled with radioactive carbon-14 so that $^{14}CO_2$ evolution represents the mineralization rate

TABLE I
Average Half-Life Values in Soils for Some Commonly Used Pesticides[a]

Pesticide	Use[b]	Half-life (days)
Alachlor	H	15
Aldicarb	I,N	30
Atrazine	H	60
Carbaryl	I	10
Carbofuran	I,N	50
Chlorothalonil	F	30
Chlorpyrifos	I	30
2,4-D	H	10
DDT	I	2,000
EPTC	H	6
Fenamiphos	I,N	50
Glyphosate	H	47
Methyl parathion	I	5
Metolachlor	H	90
Parathion	I	14
2,4,5-T	H	30
Trifluralin	H	60

[a] Values from Hornsby, A. G., Wauchope, R. D., and Herner, A. E. (1995). "Pesticide Properties in the Environment." Springer-Verlag, New York.
[b] F, fungicide; H, herbicide; I, insecticide; and N, nematicide.

of the pesticide, thus avoiding confusion with CO_2 evolution associated with soil organic matter breakdown. Parent chemical disappearance rate represents the disappearance of the pesticide from the soil over time. However, for some pesticides, metabolites that have a similar pesticidal activity as that of the parent compound are formed and can persist in soils. In this case, it is better to measure TTR disappearance rate, which includes toxic metabolites as well as the pesticide, rather than simply the parent chemical alone.

Not all organic carbon associated with pesticides in soils is mineralized to CO_2. In addition to CO_2 evolution during mineralization, some of the organic carbon will be incorporated into microbial biomass (if the degradation is not a cometabolic process), some will be accumulated in the form of organic metabolites, and some will be converted to bound residues associated with soil particles. As a result, the mineralization rate of the pesticide in soil is less than the parent chemical disappearance rate. Due to the inclusion of toxic metabolites that have similar pesticidal activity, the TTR disappearance rate should be less than the parent chemical disappearance rate but more than the mineralization rate. Table II illustrates the differences among the three rates in terms of half-life values for carbofuran and the insecticides–nematicides fenamiphos and aldicarb. It should be pointed out that more toxic metabolites are formed from fenamiphos and aldicarb, especially from fenamiphos, than for carbofuran. As a result, half-life values of TTR for fenamiphos and aldicarb are much larger than the corresponding values of parent chemical disappearance. For carbofuran, in contrast, the two values are very similar.

Soil is a complex heterogeneous medium. Even though enzymes are responsible for biodegradation of pesticides in soils, the overall process is fairly complex but can often be described by first-order rate kinetics. First-order kinetic equation can be expressed as $-dC/dt = k_1 C_1$, where C is pesticide concentration (mg/g soil), t is time (days), and k_1 is a first-order rate constant (1/days). Based on first-order kinetics, the half-life of a pesticide in soil can be estimated using the following equation: $t_{1/2} = 0.693/k_1$ where $t_{1/2}$ is the half-life of the pesticide (days). The half-life values shown in Tables I and II were obtained assuming first-order kinetics.

B. Enhanced Degradation

The phenomenon of enhanced degradation was first observed in field studies reported between 1979 and 1981. After field treatments with two carbamate pesticides (EPTC and carbofuran) in New Zealand and the United States for many years, it was observed that the pesticides failed to control target pests ade-quately. This resulted in crop failure. Failure to control the target pests was due to rapid degradation of the chemicals, which was faster than that in soils from nearby sites with no prior history of exposure to the chemicals. Since then, many pesticides have been found to exhibit enhanced degradation (or accelerated degradation) following repeated field applications of the chemicals. These include compounds related to the carbamates, phenoxy acids, organophosphates, amides, and chlorinated short-chain aliphatic hydrocarbons. Microorganisms are responsible for enhanced degradation in soils. Pesticides that are structurally related may exhibit cross reactions. In other words, a soil with enhanced degradation of a pesticide such as EPTC may also exhibit enhanced degradation of structurally similar chemicals such as the carbamate herbicides butylate and vernolate.

One field application of some pesticides such as carbofuran and fenamiphos may suffice to induce enhanced degradation. As the number of field applications increases, the extent of enhanced degradation for a given pesticide typically increases as well. Enhancement for some pesticides such as 2,4-D after application of the chemical is due to an increase in microbial population (or biomass) capable of utilizing the chemical as a sole source of carbon and energy for growth. Thus, the degrading population will progressively increase as the number of pesticide applications increases, resulting in a progressive increase in degradation rate. On the other hand, for pesticides such as carbofuran, repeated field applications of the chemical do not result in any increase in population capable of degrading the chemical. Rather, the result occurs from an increase in enzyme activity per cell.

Knowledge of the duration of enhanced degradation of a pesticide at a given site after field application

TABLE II
Estimated Half-Life Values Based on the Three Types of Degradation Rates of Aldicarb, Carbofuran, and Fenamiphos

	Half-life (days)		
Pesticide	Mineralization	Parent chemical disappearance	TTR disappearance
Aldicarb	130	2	42
Carbofuran	293	24	25
Fenamiphos	274	2	130

ceases may prevent crop failure due to enhanced degradation. The duration of enhanced degradation is influenced by soil type (soil physical, chemical, and biological characteristics) and climate. Durations of enhanced degradation in temperate-zone soils for pesticides studied to date typically last more than 3 years, although periods of enhanced degradation are likely shorter in tropical soils. One of the tactics to prevent crop failure due to enhanced degradation is to use alternative pesticides that are not structurally related to the pesticide until the degradation rate of the latter pesticide returns to a level similar to that for untreated soil. Enhanced degradation may reduce the chances of pesticides leaching to groundwater.

C. Bound Residues

When a ^{14}C-labeled pesticide is used to study biodegradation in soils, a part of the ^{14}C activity originally from ^{14}C-pesticide—ranging from <1% to >50% of the applied ^{14}C activity, depending on pesticide and soil type—is not associated with the parent chemical, metabolites, and evolved ^{14}CO$_2$ after exhaustive extraction with polar and nonpolar solvents. The nonextractable ^{14}C activity is associated with soil organic matter, is mainly covalently bound to humic acid, fulvic acid, and humin, and is termed "bound residues." Both biological and chemical mechanisms may be involved in the formation of bound residues. Pesticides for which the initial degradation products are phenolic compounds, amino compounds, or anilines may form large amounts of bound residues in soils, and these include carbofuran, parathion, 2,4-D, and propanil. Bound residues in soils appear to be fairly stable and are very slowly degraded. In addition, they appear to be harmless to target and nontarget organisms.

D. Aging and Bioavailability

Because soil is a complex heterogeneous medium, interactions of pesticides with soil particles are complex and not fully understood. Nevertheless, the interactions include sorption to soil surfaces, dissolution into soil solution, evaporation into soil pore space, diffusion into internal soil matrices via macro- and micro-pores, and remaining as a solid on soil surfaces. Microorganisms degrade pesticides mainly in the solution phase. Thus, sorbed pesticides may not be available for biodegradation until the pesticides have been desorbed into the solution phase. This phenomenon is termed "bioavailability." For hydrophobic pesticides, the degradation rate may depend on the rate and extent of desorption from soil surfaces or to a lesser extent, on dissolution from solid or gaseous phase to the solution phase. The rate of diffusion of a pesticide into the internal soil matrix, and especially into micropores, is likely dependent on the volatility and solubility of the chemical. This process is called "aging." Pesticide residues that remain in the internal soil matrix are called "aged residues" and may not be accessible for microbial attack. Aged pesticide residues cannot be extracted with organic solvents at ambient temperature but are extractable at elevated temperatures. Aged pesticide residues may constitute a source of eventual groundwater contamination by the pesticide. Contamination of groundwater by the fumigant EDB was found to have originated from soil in Connecticut that contained aged EDB residues, even though the soil had not been treated with EDB for more than 10 years. EDB residues could not be extracted with methanol at room temperature, although the residues could be extracted at an elevated temperature of 75°C.

VI. USES OF BIODEGRADATION

Biodegradation of pesticides has been studied for more than 50 years, especially in soils. Many principles generated from the study of the biodegradation of pesticides can be extended to biodegradation of industrial organic chemicals. Similar to pesticides, some industrial organic chemicals may serve as the sole source of carbon and energy for growth of microorganisms (e.g., phthalic acid), whereas others are degraded cometabolically (e.g., trichloroethylene). Degradation of industrial chemicals in soils and water often exhibits similar patterns to pesticide degradation.

An important extension of our knowledge concerning biodegradation of pesticides is for the bioremediation of pesticide-contaminated soils, water, and wastes. Knowledge of biological and nonbiological factors that influence the biodegradation rate of

APPENDIX
Common Names and Chemical Names of Pesticides Given in Text

Common name	Use[a]	Chemical name
Alachlor	H	2-Chloro-N-(2,6-diethylphenyl)-N-(methoxylmethyl)acetamide
Aldicarb	I,N	2-Methyl-2-(methylthio)propionaldehyde O-(methylcarbamoyl) oxime
Atrazine	H	6-Chloro-N-ethyl-N'-(1-methylethyl)-1,3,5-triazine-2,4-diamine
Butylate	H	S-ethyl bis(2-methylpropyl)thiocarbamate
Carbaryl	I	1-naphthyl-N-methylcarbamate
Carbofuran	I,N	2,3-Dihydro-2,2-dimethyl-7-benzofuranyl-N-methylcarbamate
Chlorothalonil	F	2,4,5,6-Tetrachloro-1,3-benzenedicarbonitrile
2,4-D	H	(2,4-Dichlorophenoxy)acetic acid
DDT	I	1,1,1-Trichloro-2,2-bis(4-chlorophenyl)ethane
EDB	Fu	1,2-Dibromoethane
EPTC	H	5-Ethyl dipropylthiocarbamate
Fenamiphos	I,N	O-ethyl[3-methyl-4-(methylthio)phenyl](1-methylethyl phosphoramidate)
Glyphosate	H	N-(phosphonomethyl)glycine
Methyl bromide	Fu	Bromomethane
Methyl parathion	I	O,O-dimethyl O-(4-nitrophenyl)phosphrothioate
Metolachlor	H	2-Chloro-N-(2-ethyl-6-methylphenyl)-N-(2-methoxyl-1-methylethyl) acetamide
Parathion	I	O,O-diethyl O-(4-nitrophenyl)phosphorothioate
Propanil	H	N-(3,4-dichlorophenyl)propanamide
2,4,5-T	H	(2,4,5-Trichlorophenoxy)acetic acid
Trifluralin	H	2,6-Dinitro-N,N-dipropyl-4-(trifluoromethyl)benzeneamine
Vernolate	H	S-propyl dipropylthiocarbamate

[a] F, fungicide; Fu, fumigant; H, herbicide; I, insecticide; and N, nematicide.

a pesticides in soil or water can be used to assess the potential for bioremediation (natural or engineered bioremediation). Bacteria or fungi isolated from soil or water that degrade a given pesticide can be used as inoculum to accelerate removal of the pesticide residues and toxic metabolites from contaminated soil or water. These organisms also can be introduced into bioreactors or land treatment units for decontamination of pesticide wastes. Furthermore, information based on the mechanisms of enhanced degradation of pesticides in soils may be used to accelerate the removal of pesticide residues from contaminated sites.

See Also the Following Articles

BIOREMEDIATION • PLASMIDS, CATABOLIC • SOIL DYNAMICS AND ORGANIC MATTER, DECOMPOSITION

Bibliography

Alexander, M. (1994). "Biodegradation and Bioremediation." Academic Press, New York.

de Souza, M. L., Wackett, L. P., and Sadowsky, M. J. (1998). The atzABC genes encoding atrazine catabolism are located on a self-transmissible plasmid in *Pseudomonas* sp. strain ADP. *Appl. Environ. Microbiol.,* **64,** 2323–2326.

Hornsby, A. G., Wauchope, R. D., and Herner, A. E. (1995). "Pesticide Properties in the Environment." Springer-Verlag, New York.

MacRae, I. C. (1989). Microbial metabolism of pesticides and structurally related compounds. *Rev. Environ. Contam. Toxicol.* **109,** 1–87.

National Academy of Sciences. (1972). "Degradation of Synthetic Organic Chemicals in the Biosphere." National Academy Press, Washington, DC.

Ou, L.-T., Gancarz, D. H., Wheeler, W. B., Rao, P. S. C., and Davidson, J. M. (1982). Influence of soil temperature and soil moisture on degradation and metabolism of carbofuran in soils. *J. Environ. Qual.* **11,** 293–298.

Ou, L.-T., Edvardsson, K. S., and Rao, P. S. C. (1985). Aerobic and anaerobic degradation of aldicarb in soils. *J. Agric. Food Chem.* **33,** 72–78.

Top, E. M., Holben, W. E., and Forney, L. J. (1995). Characterization of diverse 2,4-dichlorophenoxyacetic acid-degradative plasmids isolated from soil by complementation. *Appl. Environ. Microbiol.* **61,** 1691–1698.

Phloem-Limited Bacteria

Michael J. Davis
University of Florida

I. Microflora of the Phloem
II. Association with Plant Diseases
III. Plant Phloem Habitat
IV. Transmission of Phloem-Limited Bacteria

GLOSSARY

disease association A circumstance that exists when the pathogenicity of a microorganism has not been established by rigorous standards that eliminate all other possible causes of disease, but the possibility that the microorganism causes the disease is strongly supported by circumstantial evidence, such as a constant association with diseased but not healthy plants both in nature and following experimental transmission of the disease, and by remission of symptoms following treatment of diseased plants with antibiotics.

latex A colorless or milky sap that exudes from wounds of some plants, such as milkweed.

laticifers Cells or a series of cells forming ducts containing latex and occurring in some vascular plants. Elements of the laticiferous system are found in the phloem of some plant species.

phloem The food-conducting tissue of vascular plants, consisting of sieve tubes, fibers, parenchyma, sclereids, and sometimes laticifers. Sieve tubes and laticifers provide throughways for the spread of phloem-limited bacteria within plants.

sieve tube A series of cells, called sieve cells or elements, joined end to end, forming a tube through which nutrients are conducted.

vascular system A network in vascular plants composed of two tissues, the phloem and the xylem, involved in the transport of water and nutrients.

vector An organism, such as an insect or nematode, that carries disease-causing microorganisms from one host to another. Vectors are known to transmit some phloem-limited bacteria from diseased plants to healthy plants. However, vectors have not been identified for many diseases associated with phloem-limited bacteria. Vectors may also serve as hosts for the bacteria.

xylem A tissue that conducts water and provides structural strength to the aerial portion of plants.

PHLOEM-LIMITED BACTERIA comprise a relatively recently discovered diverse group of microbes that are associated with plant diseases.

That bacteria can cause plant diseases has been known for over a century. Nevertheless, the concept that noncultivable or extremely fastidious prokaryotes might cause plant disease has developed only since 1967 when it was discovered that wall-free bacteria, now known as phytoplasmas, were found constantly associated with some diseases. This discovery inspired renewed investigations into the causes of numerous, poorly understood, infectious plant diseases, many of which had been assumed to be caused by viruses. Consequently, not only were more phytoplasmas found associated with plant disease but also several new groups of fastidious prokaryotes were discovered. These organisms include both helical, wall-free bacteria called spiroplasmas and bacteria with cell walls. Both phytoplasmas and spiroplasmas inhabit the phloem of plants. The walled bacteria were usually found in either the xylem or the phloem and thus became known as either xylem-limited or phloem-limited bacteria. These fastidious prokaryotes are usually transmitted from plant to plant by insect vectors that also serve as hosts. At first, the walled bacteria were frequently referred to as "rickettsia-like" because their fastidious nature, morphology, ultrastructure, and insect transmission resembled those of human pathogens in the Rickettsiaceae, but this resemblance now appears to be only superficial in most cases. All of the xylem-limited bacteria known to science have been isolated in pure culture and taxonomically classified; most are members of the genera *Xylella* or *Clavibacter*. On the other hand, the phloem-limited bacteria, cautiously called "bacteria-like organisms" by some, with few possible exceptions have not been isolated in pure culture and their phylogenetic relationship to one another and to other bacteria remains largely unknown.

I. MICROFLORA OF THE PHLOEM

Phloem-limited bacteria for the most part are obligate, intracellular parasites that can live only within their plant and insect hosts. Electron microscopy was first used to reveal their existence within diseased plants (Figs. 1 and 2). Although there have been a few reports of the *in vitro* cultivation of the phloem-limited bacteria associated with certain plant diseases, these reports have not been independently confirmed. They include the cultivation in chick embryos of a bacterium from yellows-diseased grapevine and the cultivation in pure culture of the bacterium associated with citrus greening disease. The inability to routinely cultivate phloem-limited bacteria outside of their natural hosts makes it difficult to study their biology and interactions with their hosts and to demonstrate unequivocally their role as plant pathogens.

In general, phloem-limited bacteria are small, elongate organisms (0.1–0.5 × 1.0–3.0 μm). Their dimensions are at the lower limit of the size range for bacteria. They are so small that they are visible by phase-contrast microscopy but not by conventional brightfield microscopy. Occasionally, small spherical bodies are found associated with the rod-shaped cells of some phloem-limited bacteria. Phloem-limited bacteria appear to divide by binary fission. Granular structures resembling ribosomes and fibrilar strands resembling DNA have frequently been observed by transmission electron microscopy within the cytoplasm of the bacteria. The cell wall ultrastructure of

these bacteria resembles that of typical gram-negative bacteria (Fig. 1B). The cell wall is thin (20–30 nm in thickness) and consists, in part, of outer and inner three-layered, unit membranes (each ca. 5–8 nm wide) surrounding a periplasmic space varying in width and in which an apparent peptidoglycan or R-layer has sometimes been observed. The outer membrane usually has a wavy outline without regularity in longitudinal section; however, some of the bacteria have a rippled outline. Both ultrastructure and the susceptibility of the organisms to penicillin derivatives that inhibit bacterial cell wall biosynthesis suggest the presence of typical bacterial cell wall constit-

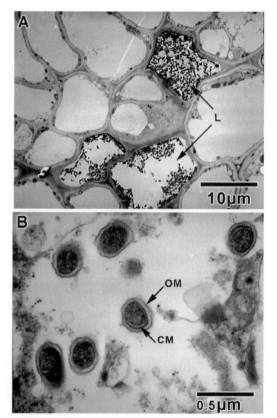

Fig. 1. Transmission electron micrographs showing the bacterium associated with papaya bunchy top disease in transverse sections of infected papaya petioles. (A) Laticifers colonized by the bacteria. (B) Cross section of bacteria showing a cell wall ultrastructure that resembles that of gram-negative bacteria. Outer membranes are seen separated from cytoplasmic membranes by a periplasmic space. L, laticifer; OM, outer membrane; CM, cytoplasmic membrane.

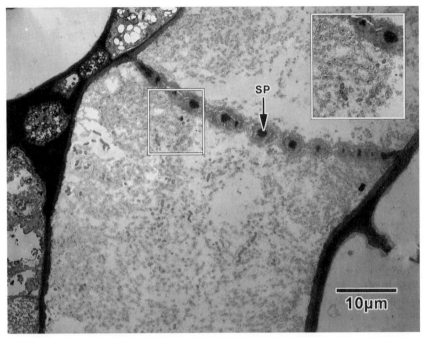

Fig. 2. Transmission electron micrograph showing a longitudinal section of a sieve tube element in cantaloupe colonized by the bacterium associated with yellow vine disease of cucurbits. Portions of two sieve cells separated by a sieve plate are shown. Bacteria are dispersed within the cytoplasm of both sieve cells. An inset enlarged 1.5× shows the bacteria near the sieve plate (courtesy of Benny D. Bruton).

uents. Under phase-contrast microscopy, the bacterium associated with clover club leaf disease exhibited an undulatory motility in 30% glycerol that ceased upon the addition of mercuric chloride or potassium penicillin G. However, motility has not been reported for other phloem-limited bacteria.

Identification of the different phloem-limited bacteria has usually been based on symptoms of the associated plant disease, plant host range, insect vector specificity, and electron microscopy. Polyclonal antibodies for the clover club leaf bacterium and grapevine yellows bacterium and monoclonal antibodies for the citrus greening bacteria have been produced using bacteria extracted from plants. The antibodies were incorporated into various serological tests for the specific detection of the bacteria in their plant and insect hosts. However, there is very little information about the antigenic relationships, if any, among different phloem-limited bacteria. DNA hybridization was developed for detection of the citrus greening organism, and polymerase chain reaction

(PCR) assays were developed for detection of the bacteria associated with papaya bunchy top disease and yellow vine disease of cucurbits. PCR also provides a means to isolate specific DNA fragments for studies based on DNA sequence comparisons to determine whether various species share a common ancestry and how closely they are related. This approach is beginning to produce a more comprehensive understanding of the relationships among the different phloem-limited bacteria. In this way, the bacteria associated with greening disease of citrus in Africa and Asia were found to be members of the α-subdivision of the *Proteobacteria* on the basis of 16S ribosomal DNA sequence homologies. The citrus greening bacteria appear to be members of a new lineage within the α-subdivision, their closest relatives being members of the α-2 subgroup. By using this approach, the bacteria associated both with yellow vine disease of cucurbits and with marginal chlorosis of strawberries were found to be γ-3 proteobacteria. The bacterium associated with papaya bunchy top disease was determined to be related to members of the α-1 subgroup of the *Proteobacteria* in the genus *Rickettsia* on the basis of the DNA sequence homology of several different genes, including the 16S ribosomal RNA gene. Thus, at least one of the fastidious prokaryotes associated with plant disease is truly rickettsia-like. These phylogenetic studies on phloem-limited bacteria in citrus, papaya, cucurbits, and strawberry indicate that they are a diverse assemblage belonging to several distinctly different taxa.

II. ASSOCIATION WITH PLANT DISEASES

The association of phloem-limited bacteria with certain plant diseases has been established by detecting them within tissues or extracts from diseased but not healthy plants. Transmission electron microscopy has been the most widely used means of detection used to establish the associations initially. Once the associations have been established, other visual detection methods have been found to be more convenient for routine detection. Phase-contrast microscopy has been used to detect the phloem-limited bacteria in extracts from diseased plants, and epifluorescence microscopy has been used to detect phloem-limited bacteria both in extracts and within tissues of diseased plants after staining with DNA-binding fluorescent stains. However, visual detection methods are not sufficiently sensitive and specific. For greater accuracy, the amplification of diagnostic DNA fragments by PCR provides a better method of detection, but this methodology still has not been developed for most phloem-limited bacteria. PCR assays have been developed for the bacteria associated with papaya bunchy top and cucurbit yellow vine diseases. PCR has also been used to detect the papaya bunchy top bacterium in its leafhopper vector, *Empoasca papayae,* thus providing further evidence of the association of this bacterium with the disease.

Phloem-limited bacteria are associated with diseases of a wide range of plants (Table I). The host range of these bacteria is not well established. Disease symptoms have been observed in species of five plant families following transmission of the clover club leaf

pathogen by its leafhopper vector, *Agaliopsis novella.* The citrus greening bacterium has been experimentally transmitted to numerous *Citrus* spp. and other rutaceous plants by graft or insect transmission. However, attempts to transmit the bacterium to several non-rutaceous plant species either by dodder (*Cuscuta* spp.), a plant parasitic plant, or its insect vectors have been unsuccessful, except for transmission to periwinkle (*Catharanthus roseus*) by dodder.

Diseases associated with phloem-limited bacteria have sometimes been described as yellows or yellows-like because of similarities to diseases such as aster yellows caused by phytoplasmas. Both phytoplasmas and walled bacteria have been observed within different plants of the same species with similar symptoms; for example, a phytoplasma and a phloem-limited bacteria were observed in sugar beet with rosette disease and latent rosette disease, respectively, and similarly in strawberries with yellows diseases. Furthermore, both types of organisms have been found in carrot and tomato plants with yellows symptoms and in proliferated *Wissadula periplocifolia.* Symptoms of diseases associated with phloem-limited bacteria can vary considerably even when incited by the same bacterium in closely related plant cultivars or species. Entire plants often exhibit stunting, dieback, degeneration, or death. Shoots may become proliferated with a resulting witches' broom appearance. Roots may be necrotic and stunted. The phloem of plants may be necrotic or degenerate. Leaves may be deformed or reduced in size, and they may also be chlorotic, exhibiting general chlorosis, marginal chlorosis, or intervienal chlorosis. Flowers may be dwarfed, proliferated, or exhibit viresensce or the greening of petals. Symptoms are sometimes mild, and spontaneous recovery may be a significant feature. Some symptoms, such as the proliferation of shoots or floral virescences, are more characteristic of diseases caused by phytoplasmas than those caused by phloem-limited bacteria.

Remission of disease symptoms after treatment of plants with antibiotics has been used as further evidence that disease-associated, phloem-limited bacteria are the likely causal agent of individual diseases. Plants infected by these bacteria often exhibit a remission of disease symptoms when treated with ei-

TABLE I
Plant Diseases Associated with Phloem-Limited Bacteria

Disease association	Plant host	Geographic location
Pollen sterility of garlic	*Allium sativum*	Germany
Sugar beet latent rosette	*Beta vulgaris*	Europe
Papaya bunchy top	*Carica papaya*	Caribbean islands, Central America
Yellow vine disease of cucurbits	*Citrullus lanatus, Cucumis melo, Cucurbita* spp.	United States
Citrus greening	*Citrus* spp., *Poncirus trifoliata,* and other rutaceous plants	Asia, Africa
Coconut palm decline	*Cocos mucifera*	Tanzania
Dodder malformation	*Cuscuta subinclusa*	France
Carrot yellows	*Daucus carota* subsp. *sativus*	USSR
Strawberries yellows	*Fragaria X ananassa*	Australia
Strawberry marginal chlorosis	*Fragaria X ananassa*	France, Spain
Brown blast of rubber trees	*Hevea brasiliensis*	China
Hop crinkle	*Humulus lupulus*	Eastern Europe
Larch witches' broom	*Larix decidua*	Germany
Tomato stolbur-like	*Lycopersicon esculentum*	Eastern Europe
Proliferation and stunting	*Melaleuca armilaris*	Israel
Potato leaflet stunt	*Solanum tuberosum*	Israel
Little leaf	*Sida cordifolia*	Puerto Rico
Spinach witches' broom	*Spinacia oleracea*	Italy
Rugose leaf curl of clover	*Trifolium* spp.	Australia
Yellows of clover	*Trifolium repens*	Canada
Clover club leaf	*Trifolium repens*	United States, England
Wheat yellow leaf curl	*Triticum* spp.	China
Yellow disease of grapevine	*Vitis vinifera*	Germany, Greece
Infectious necrosis of grapevine	*Vitis vinifera*	Czechoslovakia
Shoot proliferation	*Wissadula periplocifolia*	Jamaica

ther penicillin or tetracycline. Permanent cures are not obtained, and cessation of treatment results in the reoccurrence of the disease. Since penicillin acts by inhibiting peptidoglycan synthesis in bacterial cell walls, especially those of gram-negative bacteria, symptom remission due to penicillin treatment has been cited as additional evidence that a walled bacterium, and not a wall-free prokaryote such as a phytoplasma or spiroplasma, may be responsible for the corresponding disease.

III. PLANT PHLOEM HABITAT

Phloem-limited bacteria are most frequently found within sieve tubes in plants (Fig. 2). The phloem of plants consists, in part, of sieve tubes that comprise sieve cells (or sieve tube elements) that form a long, pressurized conduit. Pores in sieve plates between sieve cells provide continuity of the fluids in the tissue. Although the pores may be <1 μm in diameter, they may be as wide as 14 μm and average more than 2 μm. Thus, prokaryotic organisms can generally pass unrestricted from sieve cell to sieve cell. Therefore, electron micrographs of microorganisms jammed into sieve pores probably are artifacts resulting from the release of pressure upon sampling of the tissue. Phloem-limited bacteria reportedly have been observed not only in sieve tube elements but also in other phloem cell types. Such observations also have been made for phytoplasmas. The passage through the plasmadesmata of the plant cell walls would be necessary for organisms to gain access to cells adjoining the sieve tube elements without disruption of the walls. Also, cells of developing phloem tissue are difficult to identify. Therefore, some or all the reports of phytoplasmas in cells other than sieve tube elements can be attributed to misidentification, either of sieve tube elements as parenchyma cells or of membrane-bound vesicles of host origin as phytoplasmas. Certainly, the difficulties facing phytoplasmas in such plasmadesmatal passages would be even more insurmountable for the walled (and therefore less plastic) bacteria unless some possess the ability to actively breach barriers within plants. Such movement might help to explain reports of fastidious bacteria simultaneously inhabiting the phloem and xylem tissue and occasionally the parenchyma and meristematic cells.

In rubber trees with brown blast disease, the bacteria were observed in both sieve tubes and laticifers, and in papaya with bunchy top they appeared to be limited to laticifers (Fig. 1A). Laticifers are cells or a series of fused cells that contain a fluid called latex. Laticifers are found in many different plant species. Presumably, the latex is involved in the protection of plants against insect pests. Latex composition varies among plant species, and the latex from some plant species is harvested for commercial use. For example, the latex from rubber trees is used to produce rubber, and that from papaya contains the proteolytic enzyme papain used for various purposes such as meat tenderizers. Laticifers may form a network throughout plants, and elements of this laticiferous system may be found in the phloem. In papaya with bunchy top disease, one characteristic symptom of the disease is the cessation of the flow of latex from wounds in affected parts of the plants.

In the phloem of susceptible plants, the microorganisms take advantage of intracellular fluids that are rich in many potential nutrients for bacteria, especially since the phloem transports plant nutrients. These include inorganic cations and anions, organic acids, amino acids, proteins, and carbohydrates. The composition of sieve cell sap can be favorably compared with that of insect hemolymph. Together, the two fluids comprise an ecological niche that has been filled by several kinds of prokaryotes. The ecological niche provided by laticifers, however,

would seem to be even more unique considering the specialized nature of these cells. Other than the papaya bunchy top and rubber brown blast-associated bacteria, the only other microbes frequently associated with laticifers are flagellated protozoans.

IV. TRANSMISSION OF PHLOEM-LIMITED BACTERIA

The means of transmission of phloem-limited bacteria from plant to plant, if known, is usually by insects. However, nematode transmission is suspected for the bacterium associated with yellows disease of grapevine, and soil-borne transmission was reported for the bacterium associated with stunted and witches' broom diseased larch trees. The plant host range of their vectors significantly restricts the habitat breadth of phloem-limited bacteria. Colonization of the phloem by microorganisms usually involves insects that have become specialized in using this tissue as a source of nutrition. These insect vectors are often leafhoppers (Homoptera; Cicadellidae); however, psyllids (Homoptera; Psyllidae) and a piesmid (Heteroptera; Piesmidae) have been found to be the vectors of individual bacterial species. Vector specificity undoubtedly plays an important role in the epidemiology of diseases associated with these bacteria. The leafhopper vector, *Agaliopsis novella,* readily transmits the clover club leaf bacterium, but related leafhoppers were unable to transmit the pathogen. The phloem-limited bacterium associated with rugose leaf curl of clover, but not that associated with a yellows-type disease of strawberries, is transmitted by the leafhopper, *Austrogallia torrida.*

Isolates of phloem-limited bacteria have been maintained in host plants by plant to plant transmission or propagation of cuttings from diseased plants. The bacteria are not seed-borne and can not be transmitted by mechanical means. Periwinkle has been found to be a convenient experimental host for the clover club leaf and citrus greening agents where they are maintained by graft transmission. In some situations, transmission by dodder has been found useful, especially when other vectors are unknown or otherwise not available. The citrus greening bacterium was initially transmitted to periwinkle by dodder, and the phloem-limited bacterium associated with rossette disease of sugar beet has been transmitted between sugar beet plants by dodder. Dodder, however, appears to be more adversely affected by infection with some phloem-limited bacteria than by others, thus restricting its use for transmission. For example, numerous attempts to establish the bacteria associated both with marginal chlorosis of strawberries and with yellow vine disease of cucurbits in periwinkle by dodder transmission have been unsuccessful. However, whether or not different dodder species respond in a similar manner is unknown.

In most respects, knowledge of insects as habitats for phloem-specialized bacteria is scanty. Most of these bacteria are known only as suspected plant pathogens; although they are assumed to be insect borne, in most cases the vector has not been identified. With the leafhoppers, *A. novella* and *A. torrida*, which are vectors of the phloem-limited bacteria associated with different clover diseases, transmission of the bacteria from adult females to their progeny through eggs occurs at rates as high as 99%. The demonstration that the clover club leaf pathogen was transmitted vertically through multiple generations was the first evidence that a microorganism could multiply in both animal and plant reservoirs in the course of a complex biological cycle.

See Also the Following Articles

Outer Membrane, Gram-Negative Bacteria • Phytoplasma • Polymerase Chain Reaction

Bibliography

Avila, F. J., Bruton, B. D., Fletcher, J., Sherwood, J. L., Pair, S. D., and Melcher, U. (1998). Polymerase chain reaction detection and phylogenetic characterization, of an agent associated with yellow vine disease of cucurbits. *Phytopathology* **88**, 428–436.

da Graca, J. V. (1991). Citrus qreening disease. *Annu. Rev. Phytopathol.* **29**, 109–136.

Davis, M. J. (1991). Fastidious bacteria of plant vascular tissue and their invertebrate vectors. *In* "The Prokaryotes, A Handbook on the Biology of Bacteria, Ecophysiology, Isolation, Identification, Applications" (A. Balows, H. G. Truper, M. Dworkin, W. Harder, and K. H. Schleifer, Eds.), Vol. 4, pp. 4030–4049. Springer-Verlage, New York.

Davis, M. J., Ying, Z., Brunner, B. R., Pantojas, A., and Ferwerda, F. H. (1998). Rickettsia relative associated with papapaya bunchy top disease. *Curr. Microbiol.* **36**, 80–84.

Jagoueix, S., Bove, J. M., and Garnier, M. (1994). The phloem-limited bacterium of greening disease of citrus is a member of the α subdivision of the *Proteobacteria*. *Int. J. Systematic Bacteriol.* **44**, 379–386.

Nienhaus, F., and Sikora, R. A. (1979). Mycoplasmas, spiroplasmas, and rickettsia-like organisms as plant pathogens. *Annu. Rev. Phytopathol.* **17**, 37–58.

Zreik, L., Bove, J. M., and Garnier, M. (1998). Phylogenetic characterization of the bacterium-like organism associated with marginal chlorosis of strawberry and proposition of a Candidatus for the organism, "Candidatus *Phlomobacter fragariae*." *Int. J. Systematic Bacteriol.* **48**, 257–261.

Phosphorus Cycle

Ronald D. Jones

Florida International University

GLOSSARY

biologically available phosphorus Phosphorus that is readily available to all biota including microorganisms either directly or indirectly; consists of mainly biologically available organic and inorganic phosphates and the phosphorus contained in living and dead organisms.

essential chemical element An element required by organisms to complete their normal life cycle.

mineralization The conversion of an element from an organic form to an inorganic state as a result of microbial activity.

mycorrhizae Mutualistic relationship between plant roots and fungi that aids the plants in obtaining phosphorus and other elements.

phosphate A compound chemically identified as PO_4^{3-} that in its soluble reactive forms can be readily utilized by organisms; the principal form of phosphorus found in the earth's crust.

redox potential An indication of the oxidation/reduction state of the system; measured in millivolts, where a positive Eh represents environmental conditions favorable for oxidation reactions and a negative Eh represents environmental conditions favorable for reduction reactions.

total phosphorus The total quantity of phosphorus contained in a sample; consists of both biologically available and biologically refractory forms of organic and inorganic phosphorus and includes mineral phosphates, living and dead organisms, and soluble reactive phosphates.

PHOSPHORUS is an essential chemical element required in varying amounts by all living beings for the normal completion of their life cycles. Given the importance of phosphorus in cellular growth and metabolism, it is interesting to note that most texts relegate the microbial cycling of phosphorus to the broad category of "other cycles."

I. INTRODUCTION

Phosphorus is a nonmetallic element with an atomic weight of 30.973 and an atomic number of 15. Elemental phosphorus occurs in four forms: white, red, violet, and black. It was discovered circa 1669 by a German alchemist, Henning Brand, in his attempts to convert silver to gold. Freshly prepared phosphorus is white, but it turns light yellow when exposed to light. It is highly poisonous in its elemental form, and it ignites spontaneously in air. It is stored under water and is insoluble in water and sparingly soluble in some organic solvents.

Phosphorus is the 11th most abundant element in the earth's crust (estimated at 10^{15} kg P) and occurs almost exclusively as phosphate (PO_4^{3-}). Despite this relatively high abundance in the earth's crust, phosphorus is not an abundant component in the biosphere and is often a growth-limiting nutrient. The majority of phosphorus is found in apatite mineral forms with PO_4^{3-} being bound largely to calcium and to a lesser extent to Mg, Al, and Fe. The phosphorus in mineral forms is sparingly avilable to biota.

II. PHOSPHORUS CYCLE IN THE ENVIRONMENT

The phosphorus cycle, unlike those of carbon, nitrogen, and sulfur, does not involve the alteration of the oxidation state because phosphorus remains in its pentavalent state as phosphate molecules (PO_4^{3-}). Phosphorus also does not have significant biologically induced gaseous fluxes to and from the atmosphere as do these other elements. Some scientists believe that some sediment microorganisms may have the ability to utilize phosphate as a terminal electron acceptor in the absence of oxygen, nitrate, and sulfate. This results in the formation of gaseous phosphine (PH_3). This process is of little significance and as such (PO_4^{3-}) is key to the biological and chemical cycling of phosphorus.

Cycling of phosphorus is thus essentially the conversion of phosphate from the organic to the inorganic forms and the reverse. Microorganisms play a key role in these transformations. There are two pools of phosphorus that must be considered: a large, slowly cycled pool of mineral phosphate contained in the earth's crust and sediments, and a much smaller pool of actively cycled phosphate (approximately 0.1% of global total P). This article deals almost exclusively with the small but actively cycled pool of phosphorus. A simplified diagram of the phosphorus cycle is provided in Fig. 1.

III. DIRECT MICROBIAL CONTRIBUTIONS TO THE PHOSPHORUS CYCLE

Direct microbial contributions to the phosphorus cycle can be found in four areas. Primarily, microorganisms are key in the mineralization of organic phosphorus released during the death and decay of other organisms. As the microorganisms utilize carbon compounds for cellular metabolism and catabolism, they contribute to the release of soluble inorganic phosphorus. Second, microorganisms produce a series of extracellular enzymes called phosphatases which break down both soluble and particulate organic phosphorus and release soluble inorganic phosphates. The most widely studied enzymes are alkaline and acid phosphatases, named for the pH at which they are most effective at bond cleaving. These

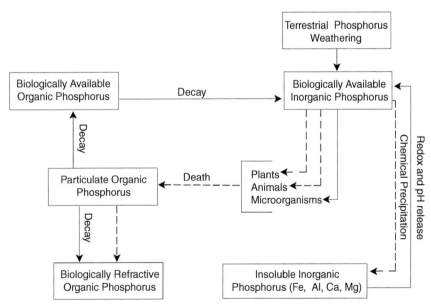

Fig. 1. Phosphorus cycle in the biosphere. Solid lines represent microbially mediated processes.

enzymes function by hydrolyzing organic phosphate esters as well as inorganic pyrophosphates and several other inorganic phosphates. Without phosphatases organic phosphorus would act as a terminal sink for phosphorus, and all biological growth would be dependent on continental weathering of minerals as the only source of soluble, inorganic phosphate.

Third, microorganisms also incorporate phosphate into cellular material and thus convert inorganic phosphates to organic phosphates. Microorganisms have a high affinity for phosphate uptake and often outcompete and therefore regulate the rate at which higher organisms acquire phosphate necessary for their growth and metabolism.

Finally, some fungi enter into a mutualistic relationship with plant roots called mycorrhizae. In this case, the fungi actually are integrated into the physical structure of the roots. The fungus derives nutritional benefits and the plants receive enhanced uptake of water and mineral nutrients, particularly phosphorus and nitrogen. This enhancement of nutrient uptake is facilitated by both direct microbial mineralization of phosphates and the increased surface area of plant roots.

IV. INDIRECT MICROBIAL CONTRIBUTIONS TO THE PHOSPHORUS CYCLE

In addition to the direct role that microorganisms play in the mineralization and release of phosphorus, they also play three indirect roles in phosphorus cycling. First, phosphates have a tendency to bind with calcium and magnesium to form relatively insoluble mineral phosphates and are thus less available. Many bacteria and fungi produce organic and inorganic acids which mediate the release of dissolved inorganic phosphate from calcium and magnesium phosphates.

Second, phosphates combine with high affinity to ferric iron to form $FePO_4$, one of the least soluble and least avilable forms of phosphate. Not only do microorganisms mediate the conversion of iron from

its soluble ferrous (2+) to its insoluble ferric (3+) forms and reverse, but also the redox conditions in most environments are controlled by microbial activities. The formation of iron phosphates occurs in aerobic environments. Under anaerobic conditions, Fe^{3+} is microbially reduced (solubilized) to Fe^{2+} with subsequent release of bound phosphate. In addition, sulfate-reducing bacteria form hydrogen sulfide, which has a higher affinity for iron than phosphate and thus forms FeS and results in the release of soluble phosphate. Given that much of the phosphorus in lakes and oceans is buried in the sediments in the form of $FePO_4$, these microbial processes are very important in returning the insoluble inorganic phosphates to the soluble and biologically available form.

Finally, all microorganisms contain approximately 3% phosphorus (on a dry weight basis) and others have the ability to store polyphosphates in their cells. These organisms then act as a temporary sink for the phosphates that would be avilable to other organisms or that would be complexed into long-term fixation in minerals. On the other hand, bacteria and fungi serve as a food source for many organisms and therefore can serve as a direct source of phosphorus to other organisms. Phosphorus that is not directly incorporated by the consumers is excreted in either its organic or inorganic forms and thus returned to the cycle.

V. CONCLUSIONS

Despite the fact that there are no significant oxidations or reductions of phosphorus, or biologically induced gaseous states, microorganisms still play the key role in the cycling of phosphorus. Phosphorus is likely to be the nutrient limiting primary production in most environments. Therefore, the rate at which microorganisms recycle the biologically available phosphorus is key to ecosystem productivity. Thus, despite the fact that "phosphorus cycling" is given little direct attention in the literature because of the simplicity of the phosphorus cycle, microbial cycling of phosphorus is of paramount importance for ecosystem function.

See Also the Following Articles

ECOLOGY, MICROBIAL • ENZYMES, EXTRACELLULAR • MYCORRHIZAE

Bibliography

Atlas, *R. M. S.,* and Barth, R. (1998). "Microbial Ecology: Fundamentals and Applications," 3rd Benjamin/Cummings, Menlo Park, CA.

Boers, P. C. M., Cappenberg, Th. E., and Raaphorst, W. van (Eds.) (1993). "Developments in Hydrobiology: Proceedings of the Third International Workshop on Phosphorus in Sediments." Kluwer, Dordrecht.

Fenchel, T., and Blackburn, T. H. (1979). "Bacteria and Mineral Cycling." Academic Press, London.

Torriani-Gorini, A., Rothman, F. G., Silver, S., Wright, A., and Yagil, E. (Eds.) (1987). "Phosphate Metabolism and Cellular Regulation in Microorganisms." ASM Press, Washington, DC.

Photosensory Behavior

Brenda G. Rushing, Ze-Yu Jiang, Howard Gest, and Carl E. Bauer

Indiana University

I. Phototaxis
II. Eubacterial Scotophobic Response
III. Archaebacterial Scotophobic Response

GLOSSARY

action spectrum Wavelengths of light that cause a photoresponse.

photophobia Literally, "fear of light"; a photophobic response is a change in cell motion that occurs when encountering a field of repellent light, resulting in movement away from that light.

photoreceptor A protein with an attached photopigment molecule that can absorb light energy. Types of photopigments include bacteriochlorophyll, carotenoid, flavin, retinal, and phycobilin.

photoresponse A general term for the reaction of a cell to changes in light direction and quality, or intensity of light.

photosynthesis A series of processes in which electromagnetic energy is converted to chemical energy that is used for biosynthesis of organic cell materials.

phototaxis Oriented movement of a motile organism with respect to the direction of light. Phototaxis can either be toward (positive) or away (negative) from light.

reaction center A membrane-spanning complex composed of pigments and proteins. Electromagnetic energy is converted to chemical energy within the reaction center.

scotophobia Literally, "fear of darkness"; a scotophobic response consists of a reversal or stoppage of flagellar rotation when a motile cell that is illuminated with attractant light begins to cross a light–dark boundary or is subjected to a decrease in light intensity.

MOVEMENT OF ALL MOTILE PHOTOSYNTHETIC ORGANISMS is influenced by the quantity and qual-
ity of light. Motile algal and cyanobacterial cells exhibit directed (oriented) movement toward or away from a light source using a process that is termed phototaxis. Anoxygenic (nonoxygen evolving) photosynthetic bacteria and archaebacteria display a scotophobic response which typically results in cells accumulating in visible or infrared light as well as a photophobic response to ultraviolet light. In addition, colonies of one species of purple photosynthetic bacteria exhibit phototactic behavior. This article discusses these photosensory behaviors.

I. PHOTOTAXIS

Phototaxis is defined as the oriented movement of an organism with respect to the direction of light. In contrast to a biased, random walk exhibited by anoxygenic bacteria, phototactic organisms can orient a specific axis of the cell relative to the direction of light and subsequently move toward (positive) or away (negative) from the light source. An outstanding feature of phototaxis, recognized quite early, is that the response is to the direction of light rather than to light intensity. Thus, it could be demonstrated with certain eukaryotic organisms (*Chlamydomonas, Euglena,* and *Volvox*) that phototactic cells will move toward a light source even when the path of motion occurs via a gradient of decreasing light intensity. Phototaxis has been observed and studied for more than a century, and the early history of research on the subject is reviewed by Mast (1911).

It is commonly believed that, in organisms with photosynthetic capability, phototaxis aids cells in moving to locations where the physical and chemical conditions are more optimal for light-dependent metabolism and growth. However, several kinds of evidence indicate that the mechanism of phototaxis per se can operate independently of photosynthesis. In fact, phototaxis has been observed in various non-photosynthetic organisms, such as fungal cells (e.g., zoospores of the fungus *Allomyces reticulatus*).

A. Eukaryotic Green Algae

Phototaxis of *Chlamydomonas* has been extensively studied and illustrates major aspects of the phenomenon in single-celled eukaryotic algae. In this organism, an "eyespot" within the chloroplast contains layers of pigmented droplets. The photoreceptor for phototaxis is believed to be located in the plasma membrane just above the eyespot. The photoreceptor consists of a retinal-containing protein, as do photoreceptor proteins used in animal vision such as human rod rhodopsin. *Chlamydomonas* swims helically through the medium and, depending on the orientation of the cell with respect to light, the photoreceptor is shaded by the eyespot for varying periods of time (Fig. 1). The cell rotation allows the "antenna" (eyespot plus associated structures) to scan light in the environment periodically. Foster and Smyth (1980) note, "The cell processes this periodic signal to determine when to make a corrective response that will realign the path to the light. The response of the flagella, by changing the orientation of the cell, changes the shape of the signal during the next scan cycle."

Euglena gracilis is a unicellular organism with a single anterior flagellum. At the base of the flagellum, a swelling called the paraflagellar body contains the photoreceptor for phototaxis. The eyespot of *Euglena* consists of 20–50 orange-red lipid droplets that contain carotenoids. It is located so that, in most orientations of the cell, it must modulate the light that reaches the paraflagellar body as the cell rotates. Although the eyespot contains carotenoids, the action spectrum for phototaxis in this organism suggests that the photoreceptor is a flavoprotein, not a carotenoprotein.

Volvox is a multicellular, colonial green alga consisting of a spherical array of 500–60,000 biflagellate cells. The mechanism of light tracking by this organism involves a so-called "stop response" of flagella in the anterior portion of the colony (the portion oriented toward the light source). As in *Chlamydomonas*, individual cells of *Volvox* contain eyespots, and the photoreceptor pigment appears to be a carotenoprotein. Absorption of light by anterior cells of *Volvox* causes their flagella to stop beating, whereas flagella of lateral cells continue to beat. *Volvox* colonies rotate, and if the direction in which the colony is moving deviates from the light direction, continued fla-

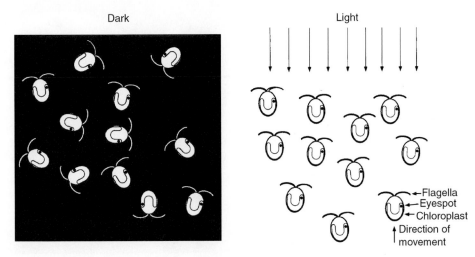

Fig. 1. Oriented motility demonstrated by phototactic microorganisms as exemplified by *Chlamydomonas*. (Left) Random orientation of motile cells in darkness. (Right) Oriented movement of cells toward a light source.

gellar motion in more shaded lateral cells results in a torque that turns the colony back toward the direction of light.

B. Filamentous Cyanobacteria

Cyanobacteria do not have flagella or other obvious motor organelles. In Oscillatoriaceae, such as *Phormidium*, filaments rotate around the long axis during gliding and exhibit positive but not negative phototaxis. In general, such photomovements are quite slow. The action spectrum of phototaxis corresponds to absorption maxima of carotenoids and phycobilins. In Nostocaceae, exemplified by *Anabaena*, there seems to be a true steering mechanism. The filament does not rotate during movement, and the action spectra of positive and negative phototaxes are significantly different; chlorophyll *a* and phycocyanins function as photoreceptors.

C. Anoxygenic Photosynthetic Bacteria

In photosynthetic bacteria, phototactic behavior has thus far been observed in only one species, the nonsulfur purple bacterium *Rhodospirillum centenum*. This thermotolerant bacterium has an assortment of unique properties, including the ability to differentiate. When grown in liquid media, the cells bear a single polar flagellum. Cells grown on agar media, however, are peritrichously flagellated and form "swarm colonies" that exhibit authentic phototactic behavior; that is, movement of the colony toward or away from light sources, depending on the spectral characteristics of the light source.

Unilateral illumination induces colony migration rates as high as 45 mm/hr. Positive phototaxis is associated with the spectral region that overlaps the *in vivo* infrared absorbance bands of bacteriochlorophyll. In contrast, negative phototaxis is induced by visible light that includes the 590-nm absorption band of bacteriochlorophyll and the absorbance maxima of carotenoid pigments (475–550 nm). The unique colony phototaxis of *R. centenum* identifies this bacterium as an attractive experimental organism for applying molecular biological and genetic approaches to identify, isolate, and characterize

genes governing phototaxis and other photosensory phenomena.

II. EUBACTERIAL SCOTOPHOBIC RESPONSE

In contrast to *R. centenum* swarm colonies, true phototactic motility has not been observed in anoxygenic photosynthetic bacteria growing in liquid environments. Rather, when these organisms begin to cross a light–dark boundary, or experience a sudden decrease in light intensity (a step-down), flagellar rotation either reverses direction or stops; this process is known as a scotophobic (fear of darkness) response (Fig. 2). The tumbling motion caused by a scotophobic response results in a random reorientation of cells that is followed by resumption of smooth swimming in a new direction. A decrease of as little as 2% of light intensity is sufficient to elicit a scotophobic response. When going up a light gradient, the cells suppress tumbling, whereas when going down a gradient of intensity the cells increase the frequency of tumbling. The net result is that the cells move toward light via a biased random walk as seen in bacterial chemotaxis. The scotophobic photosensory behavior is often referred to in the literature as

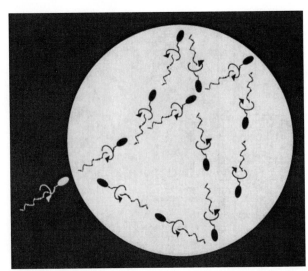

Fig. 2. Scotophobic response exhibited by anoxygenic microorganisms. The figure shows how a scotophobic bacterium would become trapped in a light field.

"phototaxis" by authors who use this term in a more general sense to include a broad range of light intensity-dependent changes in motility. However, we prefer to use a more restricted definition of phototaxis that reserves its use for light direction-dependent responses, and we use the terms scotophobic and photophobic for responses that involve tumbling in response to changes in light intensity.

A. Photoreception

Work by Manten and Clayton on *Rhodospirillum rubrum* in the late 1940s to mid-1950s showed that the bacterial scotophobic response involved the same wavelengths of light that are utilized for photosynthesis. This discovery led to the erroneous proposal that bacteriochlorophyll and carotenoids are directly used as photoreceptor pigments for photoresponses. Instead, several groups have shown that a scotophobic response by purple photosynthetic bacteria involves sensing whether or not the process of photosynthesis is occurring. For example, research with *R. rubrum* by Harayama and Iino in the 1970s suggested that alterations in either the membrane potential of the cells or the transfer of electrons through the photosynthetic apparatus generate the primary signal that causes a scotophobic response (Fig. 3). The use of antimycin A to inhibit photocyclic electron transport in these experiments dramatically reduced the accumulation of bacteria in the light; however, this compound also causes concomitant changes in the membrane potential. It was unclear, therefore, whether the signal was a change in the proton gradient that results from electron transport or if the redox state of an electron transfer component was being sensed. The addition of valinomycin and potassium, which depolarize the membrane while leaving electron transport intact, also causes inhibition of the photoresponse, indicating that changes in membrane potential may be sensed. However, the conclusion that membrane potential is the primary signal should be taken with caution since the effect observed with valinomycin and potassium may also be mediated by a change in the redox state of an intermediate in the electron transfer system.

Research by Armitage and co-workers further supports the hypothesis that the primary signal for the scotophobic response involves the redox state of the cell. Specifically, mutant strains of *R. rubrum, Rhodobacter sphaeroides,* and *Rhodobacter capsulatus* that do not contain a functional photosystem (i.e., have no photosynthetic electron flow but contain the photosynthetic pigments) also do not exhibit a scotophobic response. Studies of *Rb. sphaeroides,* using electron transport inhibitors, demonstrated that there was no measurable photoresponse when the photosynthetic electron transport was blocked. It has also been shown that *Rb. sphaeroides* displays an even greater scotophobic response in the absence of oxygen, suggesting that photosynthetic and respiratory electron transfer are competing processes (Fig. 3). When oxygen is present, respiration reduces the amount of photosynthetic electron flow, leading to a decrease in the signal required for a photoresponse. The Armitage group also reported that the addition of the proton ionophore FCCP (carbonyl cyanide 4-trifluoromethoxyphenylhydrazone), which disrupts membrane potential while maintaining electron transfer, did not affect the photoresponse. Thus, it appears that the primary signal for the scotophobic response in *Rb. sphaeroides* involves sensing a change in electron transport, possibly related to the redox state of one of the electron transfer intermediates.

B. *Rhodospirillum centenum*

Although it is generally believed that anoxygenic photosynthetic bacteria do not exhibit a true phototactic response, there is one exception, as noted previously. In the early 1990s, Ragatz *et al.* demonstrated that *R. centenum* is capable of forming swarm colonies that rapidly move toward or away from light. Colony motility passes a key phototactic test, similar to studies in algae, in that a colony moves toward an infrared light source irrespective of whether the colony is going up or down a light gradient. In contrast to colony motility, individual *R. centenum* cells in liquid culture are known to exhibit a scotophobic response, resulting in accumulation of cells in a zone of light. It is unclear why *R. centenum* exhibits different types of photosensory responses during individual cell versus colony motility.

Involvement of electron transport in *R. centenum* colony taxis is suggested by the research of Zannoni

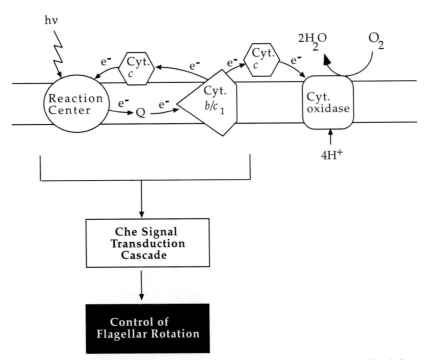

Fig. 3. Photosynthetic membrane from a typical anoxygenic photosynthetic bacterium showing the overlap between photosystem and the respiratory electron transport components. Cyt, cytochrome; Che, chemotaxis.

and associates using chemical inhibitors. Myxothiazol, which blocks photosynthetic electron transfer, strongly repressed colony phototaxis, whereas the cytochrome c oxidase inhibitor sodium azide stimulated phototaxis. Again, removal of the competition for electrons by respiration increases the signal for phototaxis (Fig. 3).

Extensive mutational analysis of *R. centenum* colony phototaxis was undertaken by Jiang and coworkers (1998), who isolated several categories of mutants that show defects in the phototactic responses. Category I mutants have defects in photoperception. Many of these mutants exhibited disruptions in the photosystem such as defects in the reaction center. Other mutants were defective in electron transport components, such as the cytochrome bc_1 complex. Category II mutants contain sensory defects affecting a photoreceptor or transduction of a signal from a photoreceptor to the flagellar motor switch complex. Category III mutants are defective in motility as a result of a mutation in one of the approximately 40 genes that are involved in synthesis of flagella.

A particularly interesting finding by Jiang *et al.* (1997) is that the phototactic response of *R. centenum* shares the same signal transduction components that the cells use for chemotaxis. In other words, the bacteria use the same *che* gene products for relaying signals from a photoreceptor or chemoreceptor to the flagellar motor. The photoreceptor also shares considerable sequence similarity with chemoreceptor membrane proteins. Integration of phototaxis with chemotaxis may be advantageous because a photosynthetic bacterium such as *R. centenum* must coordinate both nutrient and light signals when searching for a niche that is optimal for growth.

C. Archaebacterial Scotophobic Response

Archaebacteria do not synthesize a tetrapyrrole-based photosynthetic apparatus that converts solar energy into useful cellular energy. However, the Archaeon *Halobacterium salinarum* does synthesize a retinal-based photoreceptor that generates a signal

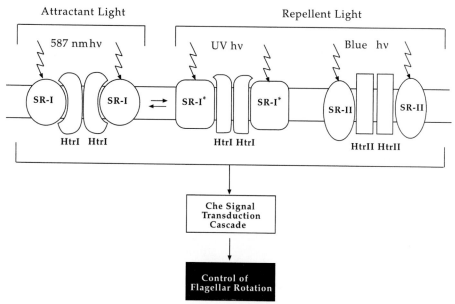

Fig. 4. Photosensory transduction components present in the membrane of *H. salinarum.* Details of sensory transduction are given in the text. Arrows depict a photon of light (hν).

in response to excitation by photons. When motile *H. salinarum* cells exit a field created by 587-nm light, they display a scotophobic response that is very similar to that observed with anoxygenic photosynthetic bacteria. Additionally, these cells exhibit a photophobic response which involves a reversal of motility upon entering a light field generated by ultraviolet (UV; 373 nm) or blue (490 nm) light.

Genetic studies have identified two retinal-containing photoreceptors—sensory rhodopsin I (SR-I) and sensory rhodopsin II (SR-II)—that are involved in mediating the *H. salinarum* photo-responses (Fig. 4). Both SR-I and SR-II undergo a photocycle in which the retinal undergoes a transformation upon excitation. In its dark state, SR-I absorbs light at 587 nm, which acts as an attractant by suppressing reversals of cell motility. Excitation of SR-I generates a long-lived intermediate (SR-I*) that absorbs UV light at 373 nm, thus functioning as a repellent receptor. The absorption of the UV light causes the return of SR-I to the dark state. Thus, the ratio of SR-I to SR-I* controls the attraction/repulsion response to these wavelengths. Excitation of SR-II is caused by blue light (490 nm), generating an intermediate that is responsible for move-

ment away from this region of the light spectrum. The integration of these responses allows the cells to maintain themselves in light without DNA damage caused by wavelengths in the blue or UV range.

Signals from SR-I and SR-II are transmitted to the chemotaxis phosphorylation transduction cascade by interactions with membrane-spanning proteins HtrI and HtrII, respectively. HtrI and HtrII have features that are similar to methyl chemoreceptors, such as a CheW docking domain and CheB/CheR methylation/demethylation domains. This suggests that HtrI and HtrII function to transmit a signal from SR-I and SR-II to the flagellar motor by affecting the Che phosphorylation cascade. As in anoxygenic photosynthetic bacteria, light perception in archaebacteria functions in a manner analogous to that used in chemoreception.

See Also the Following Articles

CHEMOTAXIS • ENERGY TRANSDUCTION PROCESSES • FLAGELLA

Bibliography

Armitage, J. P. (1997). Behavioural responses of bacteria to light and oxygen. *Arch. Microbiol.* **168**, 249–261.
Bickel-Sandkotter, S., and Gartner, W. (1996). Conversion of

energy in halobacteria: ATP synthesis and phototaxis. *Arch. Microbiol.* **166**, 1–11.

Foster, K. W., and Smyth, R. D. (1980). Light antennas in phototactic algae. *Microbiol. Rev.* **44**, 572–630.

Gest, H. (1995). Phototaxis and other photosensory phenomena in purple photosynthetic bacteria. *FEMS Microbiol. Rev.* **16**, 287–294.

Häder, D.-P. (1987). Photosensory behavior in procaryotes. *Microbiol. Rev.* **51**, 1–21.

Hoff, W. D., Jung, K.-H., and Spudich, J. L. (1997). Molecular mechanisms of photosignaling by Archaeal sensory rhodopsins. *Annu. Rev. Biophys. Biomol. Struct.* **26**, 223–258.

Jiang, Z.-Y., Gest, H., and Bauer, C. E. (1997). Chemosensory and photosensory perception in purple photosynthetic bacteria utilize common signal transduction components. *J. Bacteriol.* **179**, 5720–5727.

Jiang, Z.-Y., Rushing, B. G., Bai, Y., and Bauer, C. E. (1998). Isolation of *Rhodospirillum centenum* mutants defective in phototactic colony motility by transposon mutagenesis. *J. Bacteriol.* **180**, 1248–1255.

Mast, S. O. (1911). "Light and the Behavior of Organisms." Wiley, New York.

Nickens, D., Fry, C. J., Ragatz, L., Bauer, C. E., and Gest, H. (1996). Biotype of the purple nonsulfur photosynthetic bacterium, *Rhodospirillum centenum*. *Arch. Microbiol.* **165**, 91–96.

Nultsch, W., and Häder, D.-P. (1988). Photomovement in motile microorganisms—II. *Photochem. Photobiol.* **47**, 837–869.

Ragatz, L., Jiang, Z.-Y., Bauer, C. E., and Gest, H. (1995). Macroscopic phototactic behavior of the purple photosynthetic bacterium, *Rhodospirillum centenum*. *Arch. Microbiol.* **163**, 1–6.

Sistrom, W. R. (1978). Phototaxis and chemotaxis. *In* "The Photosynthetic Bacteria" (R. Clayton and W. Sistrom, Eds.), pp. 899–905. Plenum, New York.

pH Stress

Joan L. Slonczewski

Kenyon College

GLOSSARY

acid resistance, acid survival, or acid tolerance The ability of a microbial strain to exist for an extended period at pH values too acidic for growth, retaining the ability to be cultured after the acidity is neutralized.

acid shock Sudden decrease of pH of the growth medium to a level of acid at or below the limit for growth.

acidophile A species whose optimal external pH for growth is in the acidic range, generally lower than pH 6. Hyperacidophiles may grow below pH 3.

alkaliphile A species whose optimal external pH for growth is in the alkaline range, generally higher than pH 8.

base resistance The ability of a microbial strain to exist for an extended period at pH values too alkaline for growth, retaining the ability to be cultured after the alkali is neutralized.

membrane-permeant weak acid (permeant acid) A weak acid, usually an organic acid, which can permeate the cell membrane in the hydrophobic protonated form and then dissociate in the cytoplasm, producing hydronium ions and depressing intracellular pH.

membrane-permeant weak base (permeant base) A weak base, usually an organic base, which can permeate the cell membrane in the hydrophobic unprotonated form and then become protonated in the cytoplasm, producing hydroxyl ions and increasing intracellular pH.

neutrophile A microbial species which grows best in the neutral range of external pH, generally pH 6–8.

pH homeostasis The maintenance of intracellular pH within a narrow range during growth over a wider range of extracellular pH.

protonmotive force The proton potential across the cell membrane (Δp or $\Delta \mu H$), composed of the chemical gradient of protons ($Z\Delta pH$) minus the transmembrane electrical potential ($\Delta \psi$).

thermoacidophile (extreme thermophile; hyperthermophile) One of several archaean species, usually sulfur oxidizers, that grow optimally in extreme heat and extreme acid.

transmembrane pH difference (ΔpH) The difference in pH between the cytosol and the external medium. Usually the pH difference is maintained across the cellular membrane or the inner membrane of gram-negative organisms. The chemical gradient, $Z\Delta pH$, is a component of the proton potential.

THE BALANCE BETWEEN HYDRONIUM AND HYDROXYL ion concentrations in aqueous solution is most commonly represented by pH, the negative logarithm of hydronium concentration. pH affects microbial growth in numerous ways. In any given environment, from geothermal springs to human tissues, pH determines which species survive. Extracellular pH can be a signal for microbial behavior, whereas intracellular pH affects enzyme activity and reaction rates, protein stability, and structure of nucleic acids and many other biological molecules. In theory, every macromolecule is a "pH sensor."

Historically, pH has played many key roles in the development of microbiology. Since ancient times, fermentation has produced storable food products containing inhibitory acids (dairy products and vinegar) or bases (Japanese natto from soybeans; African dadawa from locust beans). In mining, leaching by acidophilic lithotrophs contributes to the recovery of valuable minerals; unfortunately, acidophiles also contribute to the decay of monuments. Changes in pH caused by growth on indicator media are used to identify microbial species—for example, fermentation of sugars in MacConkey media or the respiration of TCA cycle components in Simmons agar. The function of many modern therapeutic agents, including food preservatives, antibiotics, and dental fluoride, requires a transmembrane pH gradient for concentration within the cell.

I. RANGE OF pH FOR GROWTH OF MICROORGANISMS

All microbes have evolved to grow within a particular range of external pH. Acidophiles are defined approximately as growing optimally within the range pH 0.5–5, neutrophiles within pH 5–9, and alkaliphiles within pH 9–12. For example, the neutrophile *Escherichia coli* in rich broth can grow at pH 4.4 –9.2, although its rate of growth is greatly decreased at the extremes. Some species have growth ranges more narrow than those previously mentioned, whereas others overlap two ranges; for example, yeast grows in the acidic range as well as the neutral range.

Among eubacteria, the acidophiles, such as *Thiobacillus ferrooxidans*, tend to be iron or sulfur metabolizers which produce H_2SO_4. Among archeans, many sulfur oxidizers live in extreme heat and acid (the "thermoacidophiles"). These species are isolated from geothermal springs as well as deep-sea hydrothermal vents. Their cells are known for unusual membrane phospholipids and glycolipids of exceptional strength, derived from isopranoid diether or tetraether. They can maintain a transmembrane pH difference (ΔpH) of several units, which is in part compensated by an inverted electrical potential (inside positive).

Neutrophiles include the majority of organisms which grow in association with human bodies. The enterobacteriaceae are neutrophiles, including *E. coli*, in which the effects of pH stress have been studied extensively. An interesting property of neutrophiles is the ability to grow either with an outwardly directed ΔpH (at low external pH) or with an inverted ΔpH (at high external pH), requiring the cell to "spend" much of its electrical potential to maintain the inverted gradient (Fig. 1).

The best studied alkaliphiles are the *Bacillus* species, often found in soda lakes. In some alkaliphilic *Bacillus* species, cell walls rich in negetively charged amino acids may assist the exclusion of hydroxyl ions. Krulwich has studied the ability of species such as *B. firmus* to grow faster at pH 10.5 than at pH 7.5, maintaining oxidative phosphorylation despite an inverted ΔpH which takes up most of the proton potential. Models involving sequestration of protons by the respiratory chain complex or the proton-translocating ATPase have been proposed.

During the study of pH and other stresses, a factor often neglected is the change of pH caused by growth of the culture. Microbes possess enormous capacity to change the pH of their environment, as exemplified by their well-known role in food production. Nevertheless, investigators commonly attempt to study microbial growth in media lacking adequate pH control. In any experiment, pH must be controlled either by a chemostat or by growth in buffers

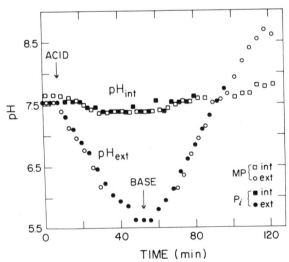

Fig. 1. pH homeostasis in *Escherichia coli*. As acid or base or base are added over time, a suspension of *E. coli* cells maintain internal pH within a relatively narrow range. Both internal and external pH were measured independently by P^{31} NMR resonances of inorganic phosphate (P_i) and of methylphosphonate (MP) (reprinted with permission from Slonczewski *et al.*, 1981).

TABLE I
Buffers for Microbial Growth Media[a]

Buffer abbreviation	Chemical name	pk_a at 37°C	Useful pH range
HOMOPIPES	Homopiperazine-N,N'-bis-ethanesulfonic acid	4.55	3.9–5.1
MES	2-(N-morpholino)ethanesulfonic acid	5.96	5.5–6.7
PIPES	Piperazine-N,N'-bis(2-ethanesulfonic acid)	6.66	6.1–7.5
MOPS	3-(N-morpholino)propanesulfonic acid	7.01	6.5–7.9
TAPS	3-[N-tris-(hydroxymethyl)methylamino]-propanesulfonic acid	8.11	7.7–9.1
CAPSO	3-(Cyclohexylamino)-2-hydroxy-1-propanesulfonic acid	9.43	8.9–10.3
CAPS	3-(Cyclohexylamino)-1-propanesulfonic acid	10.08	9.7–11.1

[a] Examples of organic sulfonate buffers used to maintain external pH during microbial growth (data from Research Organics 1998/1999 "Catalog of Biochemicals").

of appropriate pk_a for the pH range of interest (Table I). The concentration of buffer typically needs to be as high as 100 mM, and the pH of the media must be measured both before and after microbial growth. In poorly buffered media, some genes whose expression control was initially ascribed to anaerobic regulation turn out in fact to be regulated by pH.

II. MEASUREMENT OF INTERNAL pH

Most microbial species maintain some degree of pH homeostasis; that is, the internal or cytoplasmic pH is maintained at a value different from that of the external medium, usually closer to neutrality. Study of pH stress generally requires measurement of intracellular pH; in the case of eukaryotes, measurement of mitochondrial and other organellar compartments is also required. The measurement of intracellular pH remains fraught with difficulties. Bacterial cells are too small to permit introduction of microelectrodes, but several alternative methods have been used, including equilibration of radiolabeled permeant acids and bases, nuclear magnetic resonance (NMR) of pH-titratable phosphates, and fluorescence microscopy of pH-titratable dyes.

A. Radiolabeled Permeant Acids

Radiolabeled permeant acids can equilibrate across the cell membrane primarily by permeation of the hydrophobic protonated form, HA. If the dissociation constant of the acid is well below the external pH,

then the transmembrane concentration gradient of the acid (primarily in the deprotonated form) will approximately equal the ΔpH. This measurement, together with an independent determination of the intracellular volume, can be used to calculate the internal pH (pH_{in}), according to the formula:

$$pH_{in} = \log[(A_{in}/A_{ex})\,(10^{pk} + 10^{pHex}) - 10^{pk}]$$

where A_{in} is the total internal concentration of radiolabeled acid, A_{ex} is the external concentration, and pH_{ex} is the external pH. Padan and others have applied this method to observe ΔpH in many bacterial species.

Considerations for use of radiolabeled weak acids are reviewed by Kashket (1985). The advantage of the use of radiolabeled acids is the relative ease of measurement of large numbers of samples. Disadvantages include the large number of potential sources of error and the loss of sensitivity that occurs at small values of ΔpH. Where internal pH is lower than external pH, sensitivity falls off rapidly, and a permeant base must be used. Some organic acids and bases are actively transported by the cell; this effect, however, is usually small compared to the equilibration time of the acid or base.

B. P[31] NMR of Titratable Phosphates

A method of measurement permitting independent observation of internal and external pH is that of NMR observation of pH-titratable P[31] shifts in phosphates. This method enables highly accurate and reproducible pH measurement on time scales as short at 10 s. Both positive and inverted pH gradients can be observed in the same experiment. Figure 1 shows the results of P^{31} NMR of inorganic phosphate (P_i) and methylphosphonate (MP) used to demonstrate internal pH homeostasis in *E. coli* during gradual addition of acid (HCl) or base (NaOH). A drawback of the NMR method, however, is the requirement for highly concentrated suspensions of cells so that the intracellular volume fills 5–10% of the total suspension in order for sufficient intracellular signal to be obtained.

For eukaryotic microbes such as yeast, fluorescent indicators have been developed which can distinguish the pH of different compartments. Use of this method is likely to increase, although it remains impractical for most bacteria.

III. pH HOMEOSTASIS

Most microbes are capable of protecting their cytoplasm from rapid pH change by maintaining internal pH within a range more narrow than that of the external environment. The degree or tightness of regulation differs for different organisms. *Escherichia coli* regulates internal pH very closely, even sacrificing proton potential to maintain an inverted ΔpH at high external pH (Fig. 1). Other organism, such as *Enterococcus faecalis,* can maintain only a positive ΔpH. Still others appear to sacrifice pH homeostasis in order to avoid a large ΔpH which can drive the uptake of toxic organic acids.

Maintenance of internal pH can be vital for cell survival even under conditions in which growth does not occur. As shown for many enteric bacteria, cells can survive exposure to extreme pH for several hours, followed by regeneration of colonies at neutral pH. During survival in extreme acid, internal pH decreases below the level found during growth; but it must remain higher than a critical level in order for viability to be maintained.

A. Mechanisms of pH Homeostasis During Growth and Survival

The full mechanisms of bacterial pH homeostasis remain unknown, even in *E. coli,* in which the prob-

lem has long been studied. Different species have evolved different mechanisms of homeostasis. In *E. faecalis,* the proton-translocating ATPase is induced at low pH to pump out more protons, thus maintaining internal pH slightly above 7. In *E. coli,* however, ATPase mutants maintain pH normally. At low external pH, internal pH is maintained by flux of potassium ion through several transporters, as shown by Epstein and colleagues. At high alkalinity, sodium ion regulates pH through the base-inducible sodium–proton antiporter (Padan and Schuldiner, 1987). In alkaliphilic *Bacillus* species, Krulwich has shown that sodium–proton antiporters maintain an inverted Δp of up to H 2.3 units during growth at pH 11.

In dormant cells, forced to survive outside the pH range for growth, internal pH deviates from the optimum but may still be maintained at a difference from external pH. In *E. coli,* there appears to be a critical value of internal pH (approximately pH 5), below which cells lose the ability to recover and grow in neutralized media. Many different factors contributing to survival are being studied, including the σ^s (stationary-phase sigma factor) regulon and the glutamate decarboxylase system. It is interesting that two different "steps" of pH maintenance can be defined: one for growth (approximately pH 7. 5) and one for survival (approximately pH 5).

B. Avoiding Uptake of Permeant Acids

In some species, particularly the lactococci and the streptococci, bacteria may forego pH homeostasis in order to avoid uptake of permeant acids. As shown by Russell and colleagues, the buildup of fermentation acids such as acetate eventually slows and halts the growth of all microbes, but some organisms are far more resistant to fermentation acids than others. Fermentation acid-resistant bacteria such as *Lactococcus lactis* and *Streptococcus bovis* appear to allow their intracellular pH to decrease below pH 6 in order to diminish the uptake of acids by the transmembrane pH gradient. This question remains controversial because of the difficulties of measuring the transmembrane pH difference accurately using radiolabeled permeant-acid probes.

C. Microbial Modification of External pH

In addition to metabolic adjustment of cytoplasmic pH, another strategy of bacteria is to modulate the external pH so as to lessen pH stress. Amino acid decarboxylases and deaminases can produce organic bases or acids, respectively, to neutralize external acid or base. In pathogens such as *Helicobacter pylori* and *Proteus mirabilis,* the enzyme urease neutralizes their host environment, playing a critical role in virulence. Microbial effects on pH contribute significantly to the stability of aquatic ecosystems.

IV. pH STRESS AND ORGANIC ACID STRESS

pH stress has been defined differently by different researchers. Generally, the operational definitions of pH stress fit one of two categories: (i) adaptation to growth at the extreme acidic or alkaline end of the pH range and (ii) survival outside the pH range of growth, with recovery of colony-forming capacity after the pH is brought within the growth range. With respect to extreme acid, the latter category is variously termed acid survival, acid tolerance, or acid resistance. The definition of an acid-resistant strain depends on the degree of acidity designated "extreme," the time of exposure, and the death rate or percentage of recoverable colonies designated as "resistant." One commonly used definition of acid resistance in *E. coli* is the recovery of 10% or more colonies following incubation of a culture for 2 hr at pH 2.5. It is increasingly recognized, however, that even smaller levels of resistance could still provide a competitive edge to a pathogen.

The definition of pH stress is complicated by the presence of membrane-permeant organic acids or bases, whose uptake into the cell may be concentrated by the transmembrane pH gradient (ΔpH). Even small traces of such organic acids can halt the growth of lithotrophic acidophiles. The fermentative production of acids or bases contributes enormously to food production.

The potential effects of pH stress are manifold. Protein stability is affected; pH stress can cause un-

predictable auxotrophies as the result of a single stressed enzyme in a biochemical pathway. For this reason, organisms generally show a broader pH range of growth in complex media than in minimal media. At high pH, the protonmotive force is "spent" in order to maintain an inverted ΔpH; also, DNA can be destabilized, resulting in induction of SOS response in *E. coli*. At low pH, cell membranes may become destabilized, and the cell then becomes hypersensitive to organic acids and uncouplers. It is not surprising that cells have evolved many genetic responses to alleviate pH stress and even to take advantage of it, particularly in pathogenesis.

V. GENETIC RESPONSES TO pH STRESS

Bacterial responses to pH stress include many kinds of genetic regulation. Gene products induced by acid or base may function to neutralize the internal and/or external pH, to transport protons, to avoid production of acids or bases, or to turn on a cascade of virulence factors. Genes regulated by pH are most commonly coregulated by other environmental factors, such as cations, anaerobiosis, and enzyme substrates. Certain genes respond specifically to permeant acids at low external pH; their regulation may actually respond to internal acidification. Gene products conferring acid resistance may also offer cross-protection to other environmental stresses such as starvation or osmotic shock.

A. Mechanisms to Neutralize Acid or Base

An example of a pH response cycle in *E. coli*, dissected by Olson and by Bennett, is that of lysine decarboxylase, the *cadBA* operon (Fig. 2). Lysine decarboxylase (CadA) removes carbon dioxide from lysine, thus producing cadaverine, an amine which protonates to generate OH−. When both lysine and acidity (pH 5 or 6) occur externally, they interact with membrane-bound receptors LysP and CadC. CadC contains a pH sensor domain as well as a DNA-binding domain, which binds the promoter of *cadBA*, inducing expression of the lysine/cadaverine antiporter CadB as well as CadA. CadB brings in lysine

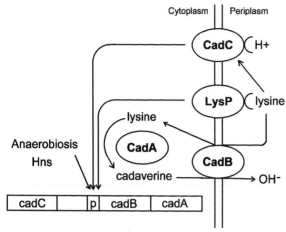

Fig. 2. Acid regulation of lysine decarboxylase and cadaverine transport: A mechanism to reverse external acidification.

and then exports the product cadaverine, which protonates in acid, increasing the external pH. Expression of *cadBA* is also enhanced by anaerobiosis; this is logical because anaerobic growth generally produces acid.

An example of a cation transport system regulated by pH is the sodium–proton antiporter, NhaA, whose expression is regulated by NhaR. As shown by Padan, NhaA is coinduced by sodium ion at high pH and functions to protect *E. coli* from both stresses. At high pH, Na^+ is exported in exchange for H^+, which helps reverse internal alkalinization. In alkaliphilic *Bacillus firmus,* Krulwich has shown that a more elaborate system of antiporters maintains internal pH, and pH homeostasis at high pH requires Na^+.

B. Virulence Regulators

For many virulence regulators, pH provides a signal to initiate pathogenesis. In *Vibrio cholerae,* the regulator ToxR, which induces production of cholera toxin, is activated by acidification. Presumably, acid within the human stomach is the signal. In the plant pathogen *Agrobacterium tumefaciens,* the virulence factor VirG shows similar activation by acid. Reversal of pH stress can enhance pathogenesis; for example, *Proteus mirabillis* produces urease to alkalinize the urinary tract, resulting in calcification and increased colonization by the pathogen.

The multiple drug-resistance regulon *mar* is induced by permeant acids, especially benzoate derivatives, at low external pH. Thus, acidic pH stress may increase antibiotic resistance of enteric bacteria in the host environment.

VI. FOOD MICROBIOLOGY AND OTHER APPLICATIONS OF pH STRESS

Historically, the production of growth-limiting organic acids or bases has been essential to production of diverse foods in various human societies. The acids or bases produced function to prevent spoilage, improve digestability, and add interesting flavors.

A. Acidic Fermentation

In western European traditions, the best known food-making processes involve fermentation acids such as lactate or acetate. Dairy fermentations involve conversion of lactose to lactic acid, initially by streptococci (lactococci), succeeded by the more acid-resistant lactobacilli. Final pH reached in yogurt or cheese is in the range of pH 4 or 5, in combination with hundred-millimolar concentrations of lactate. The addition of lactose to sausage facilitates a similar fermentation process. Fermentation of vegetables, such as cabbage to produce sauerkraut, usually involves *Leuconostoc* species, reaching pH 3.5. Fruit juices are usually fermented by yeast to produce ethanol, but the ethanol can be fermented to acetate (final pH 3. 5) to make vinegar.

The role of fermentation acids in inhibiting growth of gastrointestinal pathogens deserves further study. The concentration of fermentation acids in the human colon is comparable to concentrations which retard growth of pathogens *in vitro,* particularly in infants whose colonic pH is approximately 5. Microbial tolerance to fermentation acids may prove to be an important virulence factor; for example, *E. coli* O157:H7 is more resistant to acetate than are nonpathogenic strains of *E. coli.*

B. Alkaline Fermentation

In Southeast Asian and African societies, an important group of foods are produced by alkaline fermentations. Most commonly, alkali-tolerant *Bacillus subtilis* and related species conduct fermentation with production of ammonia. In Japan, soybeans are fermented to produce natto; the breakdown of proteins to amino acids significantly enhances digestibility of the food. In West and Central Africa, the inedible locust beans are fermented to an edible product, dadawa. The fermentation process decreases the levels of toxic substances in the bean. Similarly, melon seed is fermented to ogiri, and leguminous oil beans are fermented to ugba. All of the alkaline fermentations produced provide critical sources of protein as well as culturally significant sources of flavor. Nutritional value is enhanced by hydrolysis of proteins and by increase of vitamin levels. The products have a long shelf life and often serve to make inedible foods edible.

VII. ENVIRONMENTAL ROLE OF pH STRESS

In aquatic and soil environments, pH can be the crucial factor determining which microbial species grow and what forms of metabolism are sustained. Acid stress in lakes can reduce the rate of bacterial decomposition, but bacteria can also help reverse aquatic acidity. Diatoms are so sensitive to pH that the species distribution of diatoms in sediment (diatom stratigraphy) is used to study historical rates of aquatic acidification. Figure 3 is a calibration curve in which a calculation based on populations of acidophilic, neutrophilic, and alkaliphilic diatoms accurately predicts aquatic pH.

In soil, bacteria generally grow best in the neutral range, whereas fungal species grow better at pH 5

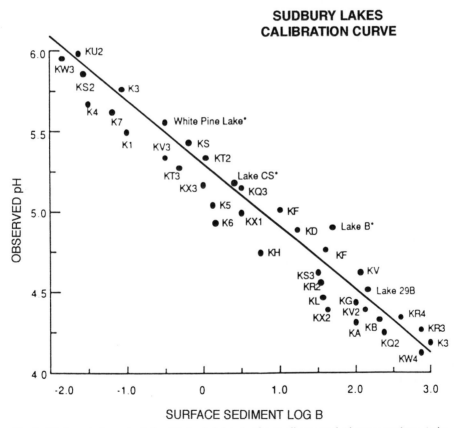

Fig. 3. Diatom-inferred pH for 38 study lakes in the Sudbury and Algoma regions. Lake pH was plotted versus the diatom species index log B of Renberg and Hellberg (1982) (reprinted with permission from Dickman and Rao, 1989).

or 6. Increased acidification can affect bacteria either directly, by interfering with pH homeostasis, or indirectly, by increasing the concentration of toxic metal ions.

Acidity in soil and fresh water is often increased by "acid rain," the deposition of sulfuric and nitric acids in part caused by human-made chemical pollution. The effects of acid rain on microbial processes such as decomposition are generally seen at pH levels a unit lower (pH 5) than those which interfere with aquatic invertebrates or fish (pH 6). In soil, acidification to below pH 3.5 (or liming to increase pH to 7) can inhibit mycorrhizal fungi; the significance of this inhibition for agriculture is unclear.

As in food, microbes in soil or water can dramatically alter the pH of their own environment by metabolic production or consumption of acids or bases. Consumption or production of CO_2 can shift water pH by up to 3 units before atmospheric reequilibration occurs; for example, heterotrophic metabolism beneath ice cover can decrease pH, whereas photosynthesis by algal blooms can increase pH. Other processes increasing acidity include ammonium assimilation and oxidation of HS or Fe(II); processes increasing alkalinity include nitrate assimilation, ammonification, sulfate assimilation, and Fe(III) oxidation.

Overall, the interplay between microbes and pH is of profound significance from the standpoint of microbial physiology and genetics, virulence, ecology, and industrial applications. Stress from extreme pH has diverse effects on growth of microorganisms, which in many cases respond by altering the pH of their environment. Nonetheless, although the effects of pH stress are well documented, the mechanisms of microbial response and adaptation continue to pose many important questions for research.

See Also the Following Articles

AGROBACTERIUM • CHOLERA • EXTREMOPHILES • FOOD SPOILAGE AND PRESERVATION

Bibliography

Booth, I. R. (1985). Regulation of cytoplasmic pH in bacteria. *Microbiol. Rev.* **49**, 359–378.

Chadwick, D. J., and Cardew, G. (Eds.) (1999). "Bacterial Responses to pH," Novartis Foundation Symposium No. 221. Wiley, Chichester, UK.

Dickman, M., and Rao, S. S. (1989). *In* "Acid Stress and Aquatic Microbial Interactions," p. 123. CRC Press, Boca Raton, FL.

Hall, H. K., Karem, K. L., and Foster, J. W. (1995). Molecular responses of microbes to environmental pH stress. *Adv. Microbiol. Physiol.* **37**, 229–272.

Hofmann, A. F., and Mysels, K. J. (1992). Bile acid solubility and precipitation in vitro and in vivo: The role of conjugation, pH, and Ca^{2+} ion. *J. Lipid Res.* **33**, 617–626.

Kashket, E. R. (1985). The proton motive force in bacteria: A critical assessment of methods. *Annu. Rev. Microbiol.* **39**, 219–242.

Krulwich, T. A., Ito, M., Gilmour, R., Sturr, M. G., Guffanti, A. A., and Hicks, D. B. (1996). Energetic problems of extremely alkaliphilic aerobes. *Biochim. Biophys. Acta* **1275**, 21–26.

Myrold, D. D., and Nason, G. E. (1992). Effect of acid rain on soil microbial processes. *In* "Environmental Microbiology" (R. Mitchell, Ed.), pp. 59–81. Wiley–Liss, New York.

Olson, E. R. (1993). Influence of pH on bacterial gene expression (MicroReview). *Molec. Microbiol.* **8**, 5–14.

Padan, E., and Schuldiner, S. (1987). Intracellular pH and membrane potential as regulators in the prokaryote cell. *J. Membrane Biol.* **95**, 189–198.

Rao, S. S. (Ed.) (1989). "Acid Stress and Aquatic Microbial Interactions." CRC Press. Boca Raton, FL.

Russell, J. B., and Diez-Gonzeles, F. (1998). The effects of fermentation acids on bacterial growth. *Adv. Microbial Physiol.* **39**, 205–234.

Slonczewski, J. L, and Foster, J. W. (1996). pH-regulated genes and survival at extreme pH. *In* "*Escherichia coli* and *Salmonella typhimurium*: Cellular and Molecular Biology" (F. C. Neidhardt, R. I. Curtiss, C. A. Gross, J. L. Ingraham, and M. Riley, Eds.), 2nd ed., pp. 1539–1552. ASM Press, Washington DC.

Slonczewski, J. L., Rosen, B. P., Alger, J. R., and Macnab, R. M. (1981). *Proc. Natl. Acad. Sci. USA* **78**, 6272.

Wang, J., and Fung, D. Y. C. (1996). Alkaline-fermented foods: A review with emphasis on pidan fermentation. *Crit. Rev. Microbiol.* **22**, 101–138.

Phytophthora infestans

William E. Fry

Cornell University

GLOSSARY

gene-for-gene interaction   The relationship characterizing the interaction of many specialized plant pathogens with their hosts. Incompatibility (lack of pathogen growth) is conditioned by recognition of a specific pathogen genotype by a specific host genotype.

late blight   The disease of potatoes caused by *Phytophthora infestans* that led to the Irish potato famine in the mid-nineteenth century. The same name applies to infections of *P. infestans* on tomatoes.

mating type   The characteristic of an oomycete that determines its sexual response to another organism. In *P. infestans*, the trait is governed by a single locus, and there are two alleles (A1 and A2). In the presence of an A2 thallus, an A1 will produce sexual structures (antheridia and oogonia). The converse is also true.

near-obligate parasite   An organism that survives mainly as a parasite in its natural habitat.

oomycetes   The group of filamentous organisms, closely related to some algae, that may appear superficially like fungi. Cells of these organisms have many diploid nuclei in the vegetative stage.

oospore   The spore that results from fertilization of an oogonium by an antheridium. This spore form in *P. infestans* is thick-walled and is the stage that is best able to survive adverse conditions.

R gene   The plant gene that in response to a message from a specific pathogen attribute initiates a signal cascade that results typically in death of the host cell and cessation of further pathogen growth. Most R genes are dominant.

sporangia   Asexual reproductive structures formed by oomycetes. Some can be easily removed from the filamentous cells bearing them (sporangiophores) and dispersed via wind or splashed water. Sporangia of *P. infestans* can germinate either to produce a filamentous multinucleate germ tube or to produce several uninucleate motile zoospores.

zoospore   Uninucleate, wall-less swimming spores produced in sporangia. For *P. infestans* these spores are biflagellated with an anterior-directed tinsel flagellum and a posterior-directed whiplash flagellum.

PHYTOPHTHORA INFESTANS   is the filamentous plant pathogen that causes the late blight disease of potatoes (Fig. 1) and tomatoes. It is the devastation caused by this disease, beginning in the fall of 1845, which led to the Irish potato famine. This plant disease remains a major concern in potato production, with a recent resurgence in importance. *Phytophthora infestans* is the type species of the genus *Phytophthora*, which contains many important plant pathogens, most of which are near-obligate parasites. Although *P. infestans* appears superficially to resemble true fungi, it is phylogenetically unrelated to the true fungi (kingdom Fungi). Despite the phylogenic distinction, the ecology of *P. infestans* is similar to that of many true fungi.

I. TAXONOMY

The genus *Phytophthora* occurs within the phylum Oomycota (oomycetes), class Peronosporomycetidae in the order Pythiales and in the family Pythiaceae. There are two genera in the family with many impor-

Fig. 1. Potato late blight. Shown is a potato shoot tip that is infected by *Phytophthora infestans.* The darkened area is infected, and the white hairiness is sporulation of *P. infestans.* In addition to stem/petiole infections, leaflets can also be infected. Infected tissue is often killed within 6–10 days after the initiation of infection.

tant plant pathogens: *Pythium* and *Phytophthora.* Species of *Pythium* tend to be less specialized than species of *Phytophthora.* Other important plant pathogenic organisms in the class are the downy mildews (e.g., *Bremia, Sclerospora, Peronospora,* and *Plasmopara*). All plant pathogenic oomycetes are filamentous and microscopic. Vegetative growth occurs by extension and branching of the filaments (mycelium) with the rare cross walls, producing a multinucleate (coenocytic) mycelium.

It has recently become clear that oomycetes belong to a kingdom distinct from those of fungi, plants, animals, and bacteria. Although the composition and name of the kingdom are still being determined, it is generally accepted that oomycetes are closely related to some algae (brown and golden brown). Names such as Chromista and Protista have been suggested for the appropriate kingdom. Characteristics common to these organisms are cell walls of β-1-3 glucans and gametic meiosis with a dominant vegetative diploid stage. Sexual structures for *P. infestans* are oogonia and antheridia, with the fertilized oogonium developing into an oospore (Fig. 2a) which can survive adverse conditions.

II. LIFE HISTORY/ECOLOGY

In nature, *P. infestans* can exist with asexual stages only or with both sexual and asexual stages. However, epidemics of the late blight disease result from repeated asexual cycles. In some habitats, survival is associated with asexual stages in dormant plant parts, but in other habitats survival may occur through dormant spores produced in the sexual process.

Asexual reproduction is via sporangia produced on sporangiophores (Fig. 2b). Sporangia are lemon shaped and typically about 22×30–35 μm size. Each has an apical pore through which zoospores are released (Fig. 2d) or through which the germ tube develops. Sporangia are borne on sporangiophores and each sporangiophore may produce many sporangia sequentially. The sporangium is borne at the tip of the sporangiophore, but as it elongates the sporangiophore swells slightly and turns the attached sporangium to the side. Thus, the sporangiophore has periodic swellings at points where sporangia were produced. When sporangia are mature they are readily dislodged from sporangiophores and can then be disseminated via wind or water.

a

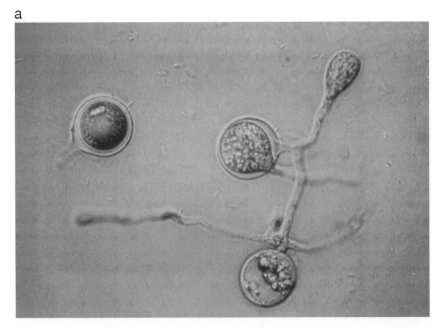

b

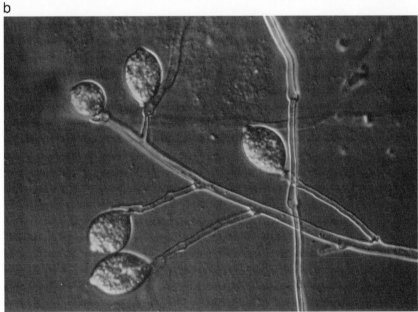

Fig. 2. Developmental stages of *Phytophthora infestans.* (a) Germinating sexual spores (oospores) (typically ca. 35 μm in length). (b) Sporangia (asexual spores) on a sporangiophore. Sporangia are typically 30–35 μm in length. (continues)

Germination of sporangia requires very moist conditions, such that in the absence of free moisture germination may be inhibited. There are two modes of germination—direct and indirect. Direct germination is via a germ tube, whereas indirect germination occurs when zoospores produced within the sporangium are released. Sporangia are apparently "programmed" to germinate indirectly when they are produced, but this method of germination is influenced by temperatures. Higher temperatures (20–26°C) suppress indirect germination and allow direct germination; however, direct germination is much less

c

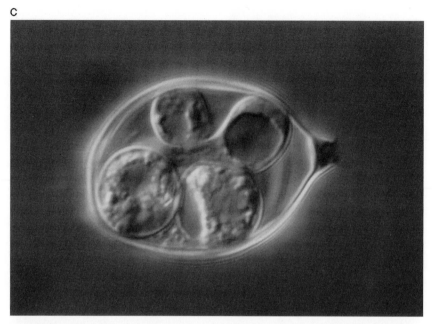

d

Fig. 2. (*continued*) (c) Sporangium with zoospores delimited within. (d) Zoospore emerging from a sporangium. (Figs. 2a and 2b by R. C. Shattock; Figs. 2c and 2d by B. G. Turgeon).

certain than indirect germination, and usually fewer than half of the sporangia will germinate. The developing germ tube is also multinucleate. Indirect germination (release of uninucleate zoospores) is favored by lower temperatures (<15°C). Zoospores are biflagellate with a forward-directed tinsel flagellum and a rear-facing whiplash flagellum. After a short period of motility (minutes or perhaps more than 1 hr), zoospores lose their flagella, encyst, and can then germinate.

The germ tube from either a sporangium or the encysted zoospore either penetrates directly into a host epidermal cell or gains ingress to host tissue via stomates to initiate pathogenesis/parasitism.

Growth of the subsequent mycelium is both intracellular and intercellular. In susceptible potato tissue, macroscopic symptoms may be visible within 3–5 days as very small areas of necrosis. *Phytophthora infestans* is semi-biotrophic: The pathogen is able to establish a parasitic relationship with the host in which host cells initially remain viable and apparently functional; however, after several days the host cells first affected die. Thus, a lesion on a leaf typically has a necrotic center surrounded by tissue that is still green and in which the pathogen is actively growing (Fig. 1).

Temperature and moisture have important governing roles in growth. The temperature optimum varies with the process, but it is also influenced by the genotype. Mycelial growth in culture and in host tissue is usually optimal at about 18–22°C. However, indirect germination is usually optimal at temperatures lower than 15°C, and direct germination is stimulated at temperatures higher than 18°C. In natural or agricultural habitats, environmental moisture also has a governing role. In addition to the requirement for free moisture for germination of sporangia, zoospores require free moisture for continued motility, and sporulation of *P. infestans* external to infected tissue requires very high moisture in the environment. Moisture is also important in the dissemination of sporangia. As the relative humidity decreases, sporangiophores twist and mature sporangia dehisce and are carried away by wind currents. Rainfall will also remove mature sporangia from a sporangiophore so that splash dispersal can be a factor. The capability for both splash dispersal in rain and wind dispersal of sporangia enables both rapid short-range splash dispersal (centimeters to meters) and long-range wind dispersal over several kilometers.

Sexual reproduction requires two mating types, designated A1 and A2. Each individual produces both antheridia and oogonia, but different strains may have different degrees of maleness and femaleness. Sexual structures are apparently stimulated by hormones from the opposite mating type. The result of fertilization is a thick-walled spore (oospore) which is dormant and can survive adverse conditions that would inactivate sporangia or zoospores. For example, whereas sporangia can survive in soil for days or weeks, oospores survive months and possibly years.

In nature, *P. infestans* is essentially an obligate parasite. It has no known significant saprophytic stage. Although it can be cultured on artificial media, it generally grows slower and has more fastidious requirements for culturing than many other oomycetes and true fungi. In ecosystems with a constant supply of living hosts, survival of *P. infestans* occurs through repeated asexual cycles. However, in ecosystems with seasonal availability of hosts, survival depends either on oospores or on infections in dormant host tissue. In potato agroecosystems. *P. infestans* survives commonly in infected potato tubers. Infected tubers can survive in soil and, if they are not frozen, during the winter, but infected tubers can also survive the between cropping periods in storage. Thus, *P. infestans* is particularly well suited to a potato agroecosystem.

III. POPULATION GENETICS

The center of genetic diversity and the probable center of origin of *P. infestans* is in the central highlands of Mexico. Prior to knowledge about the diversity of *P. infestans* in central Mexico, this species was considered to consist essentially of a single mating type (A1), and sexual reproduction was considered to be absent or extremely rare. However, at the end of the 1940s, it was discovered that A2 mating types as well as A1 mating types existed in the highlands of central Mexico. The population of *P. infestans* there has the features of a sexually reproducing population with significant genetic and genotypic diversity. Both selectively neutral and ecologically important markers have indicated this diversity. In contrast to the situation in the highlands of central Mexico, analyses of populations in other locations worldwide indicated that prior to the 1970s these were all very closely related and were all dominated by the same clonal lineage.

A. Initial Migrations

Reports of a new disease that we now know to be late blight were recorded in the early 1840s in the northeastern United States and in Europe in 1845, where it destroyed potatoes for the many years and initiated the Irish potato famine. It seems likely that these were the first occurrences of late blight on the domesticated potato, *Solanum tuberosum*. This is possible because potatoes evolved in the South American Andes, whereas *P. infestans* evolved with other *Solanum* spp., in the highlands of central Mexico. Genetic analyses of isolates from old collections in the United States, Europe, and other locations as well as the chronology of reports are all consistent with the following scenario. Somehow *P. infestans* was transported to the northeastern United States from central Mexico sometime shortly before 1843. From there it was probably transported to Europe. This hypothesis includes a drastic decrease in the genetic diversity in the migrating population (genetic bottleneck), first from Mexico to the United States and again in subsequent migrations to Europe and worldwide. Alternative hypotheses are (i) that first the pathogen went from Mexico to South America and then to the United States or Europe and (ii) that *P. infestans* originated in the South American Andes with subsequent loss of genetic diversity in that location and migration from there to other worldwide locations.

B. Twentieth Century Migrations

During the twentieth century, there have been a series of migrations of strains of *P. infestans* from Mexico to other parts of the world. Most "migrations" are associated with commerce in potatoes or tomatoes. Infected potato tubers or tomato fruits that are transported can be effective vehicles for introducing genotypes into a new area. For example, a shipment of 25,000 metric tons of potatoes from Mexico to Europe in 1976 and 1977 apparently introduced a diverse population of *P. infestans* to Europe, where previously there had been a very homogeneous population dominated by a single clonal lineage. Because *P. infestans* can be transported very efficiently in infected tubers used subsequently as "seed," subsets of the European population were subsequently transported to other world-wide locations, including several in South America and the Middle East. The population of *P. infestans* introduced in the 1970s and 1980s displaced the previous population in Europe and in several other locations worldwide.

Other migrations have also been important in changing the structures of populations in other locations. An apparently independent migration introduced a new strain to Japan and Korea in the late 1980s. New strains have been detected in the United States and Canada during the 1980s and 1990s. It seems likely that a series of migrations have occurred. The introduced strains were more aggressive and were usually insensitive to the fungicide metalaxyl.

Unfortunately, the introduction of exotic strains to worldwide locations has been accompanied by a global worsening of late blight. New strains have been demonstrated to be more aggressive (causing more rapid disease), more virulent (able to cause disease in a larger number of potato cultivars), and to have fungicide resistance. All of these factors explain why late blight has recently increased in importance world-wide. The increased aggressiveness of introduced strains relative to indigenous strains may be an example of mutational load accumulating in asexual lineages. Presumably, sexual recombination provides a mechanism to generate new genotypes freed of the deleterious mutations that accumulate in lineages of asexual lines. The phenomenon of increased aggressiveness in recently introduced strains has surprised growers and scientists during the latter quarter of the twentieth century. It has had disastrous consequences in many locations.

IV. PATHOGENICITY

Phytophthora infestans is a near-obligate parasite. It can be cultured on artificial media, but in nature it probably requires a living host for growth. It is a hemi-biotroph in that it can grow in host tissues for a few days without causing necroses. However, tissues that have been affected for more than 2 or 3 days are typically necrotic. Thus, a lesion on a leaf typically has a necrotic center surrounded by host tissue invaded by the pathogen, which is not necrotic.

In some interactions, the host cells respond rapidly to invasion by the pathogen, and cell death rapidly follows invasion of tissues. In these situations, the pathogen does not establish a successful (= compatible) interaction with the host and the pathogen does not survive. The genetic control of these responses appears to be simple, governed both by specific genes in the host and specific genes in the pathogen in a typical "gene-for-gene" interaction.

Gene-for-gene interactions occur in many pathogen–plant systems. Typically, a single gene in the host (usually a dominant allele) allows the plant to recognize the pathogen and often results in a signalling process that results in rapid cell death. Alternatively, specific genes in the pathogen (typically recessive alleles) enable the pathogen to avoid recognition and to then establish a successful interaction with the host plant. Whereas there is high specificity in individual recognition events, the signalling process and cellular reaction seem common to the "incompatible" interaction. There are currently many efforts to clone the specificity genes from the host (R genes for "resistance genes") and complementary efforts to clone the specificity genes from the pathogen (termed "avirulence genes"). In some interactions, it appears that elicitins (glycoproteins secreted by *Phytophthora* spp.) may function as "avirulence genes" in some interactions.

V. DISEASE MANAGEMENT

Suppression of potato and tomato late blight is a constant concern in most locations where these plants are grown commercially. Integration of several tactics is now necessary to suppress the disease adequately. An initial tactic is to eliminate the pathogen from the potato/tomato agro-ecosystem. Although complete elimination is very difficult, it is possible to decrease the initial pathogen population and thus lower the probability that disease will occur. Factors that help decrease the initial pathogen population include (i) planting only seed tubers that are free of *P. infestans* and (ii) destroying potatoes in the agro-ecosystem that might be infected—those left in the ground after harvest and those that were discarded during grading.

Despite efforts to eliminate the pathogen, disease may still occur, therefore, other factors that reduce the pathogen growth rate during the season are necessary. Some varieties of potatoes are less susceptible than others and these retard pathogen growth rates. Resistant plants should be used where possible. Investigations are currently under way to learn how to synthetically enhance normal resistance responses.

In many locations, the most visible efforts to suppress disease include the use of fungicides or antibiotics. There are location-specific guidelines (forecasts) for using these efficiently. Some fungicides have no curative effect and must be applied to plants before infection is initiated; others enter plant tissues and can suppress pathogen growth even after infection. In many production systems, repeated application of disease-suppressive compounds is required during the growing season.

A major focus of current research is to identify novel approaches to achieve plant resistance so that planting of resistant varieties will become an increasingly available option. Approaches include transfer of genes for resistance through classical plant breeding, use of tissue culture techniques to transfer genes that were intractable through classical plant breeding, and use of biotechnology to transfer genes. Genes for resistance are sought in wild potatoes, in unrelated plants, and in other organisms. Additionally, synthetic (engineered) resistances are being investigated.

See Also the Following Articles

PLANT DISEASE RESISTANCE • PLANT PATHOGENS • RUST FUNGI

Bibliography

Dowley, L. J., Bannon, E., Cooke, L. R., Keane, T., and O'Sullivan, E. (Eds.) (1995). "*Phytophthora infestans* 150." Boole Press, Dublin.

Erwin, D. C., and Ribeiro, O. K. (1996). "*Phytophthora* Diseases Worldwide." American Phytopathological Society Press, St. Paul, MN.

Fry, W. E., and Goodwin, S. B. (1997 a). "Re-emergence of potato and tomato late blight in the United States." *Plant Dis.* **81** (12) 1349–1357.

Fry, W. E., and Goodwin, S. B. (1997b). "Resurgence of the Irish potato famine fungus." *Bioscience* **47**, 363–371.

Goodwin, S. B., Cohen, B. A., and Fry, W. E. (1994). "Panglobal distribution of a single clonal lineage of the Irish potato famine fungus." *Proc. Natl. Acad. Sci. USA* **91**, 11591–11595.

Hardham, A. R., and Hyde, G. J. (1997). Asexual sporulation in the oomycetes. *Adv. Bot. Res.* **24**, 353–398.

Lucas, J. A., Shattock, R. C., Shaw, D. S., and Cooke, L. R. (1991). "*Phytophthora*" British Mycological Society/Cambridge Univ. Press, Cambridge, UK.

Phytoplasma

Robert E. Davis and Ing-Ming Lee

U.S. Department of Agriculture, Agricultural Research Service

GLOSSARY

mycoplasmalike organisms Nonhelical, cell wall-less, prokaryotic microbes that resemble mycoplasmas (members of class Mollicutes) in morphology and ultrastructure and are found in phloem tissues of infected plants and in the bodies of insects, principally leafhoppers, that transmit them from plant to plant.

nested polymerase chain reactions Sequential polymerase chain reactions in which each successive reaction is primed by an oligonucleotide pair whose annealing positions on template DNA are located (nested) within the annealing positions of oligonucleotides used to prime the preceding reaction.

PHYTOPLASMAS (previously called mycoplasmalike organisms; MLOs) are minute wall-less bacteria that are found in the phloem tissues of plants and in the bodies of insect vectors that transmit them from plant to plant. The history of research on phytoplasmas began with investigations of what were believed to be virus diseases.

A group of plant diseases known as "yellows diseases" affects several hundred plant species worldwide. After decades during which yellows diseases of plants were presumed to be caused by viruses, Japanese scientists reported a startling discovery. They had observed minute, cell wall-less microorganisms in the phloem of diseased plants and found that both disease and the microorganisms were sensitive to treatment with tetracycline antibiotics. They recognized that these microorganisms resembled mycoplasmas known in humans and animals, and the term "mycoplasmalike organism" was adopted for these plant-inhabiting microbes. The research that followed resulted in the recognition of cell wall-less bacteria in association with numerous plant diseases and led to the discovery of other types of plant pathogens.

The wall-less bacteria associated with plant diseases are of two distinctly different types, the nonhelical MLOs and the characteristically helical spiroplasmas. In recent years, the term "phytoplasma" has been adopted in place of the term mycoplasmalike organism.

I. BACKGROUND

Prior to 1967, numerous plant diseases known as yellows diseases were presumed to be caused by viruses. These diseases are characterized by the following symptoms: yellowing of leaves, general stunting of plants, excessive development of shoots to give growths having a "witches' broom" appearance, abnormally small leaves, virescence (loss of normal flower color and development of green flowers), and phyllody (development of leafy structures in place of normal floral parts). The diseases are also characterized by transmission of the disease-causing agents by leafhopper insects, by grafting, and by the parasitic plant dodder (*Cuscuta* spp.).

In 1967, a new hypothesis of yellows disease etiology emerged with reports of mycoplasmalike organisms (MLOs) in the phloem of yellows diseased plants in Japan. Numerous reports have confirmed the association of such organisms with plant yellows diseases, and it is widely accepted that organisms that morphologically and ultrastructurally resemble mycoplasmas are the cause of plant yellows diseases. During the past decade, studies of conserved gene sequences have revealed that the plant-inhabiting MLOs are related to but distinct from other members of the class Mollicutes, and the term phytoplasma has been accepted for reference to these organisms. Phytoplasmas are now believed to be the cause of diseases in more than 300 plant species worldwide.

II. DISEASES

Because it has not been possible to isolate and culture any phytoplasma in artificial nutrient medium, it has not been possible to subject phytoplasmas to rigorous tests to prove their pathogenicity. Observations that favor the conclusion that phytoplasmas cause diseases in plants include (i) their constant association with diseased but not healthy plants, (ii) pathogenic reactions in cells invaded by phytoplasmas and in surrounding cells, (iii) disease development following transmission of phytoplasmas to previously healthy plants, (iv) remission of disease symptoms and disappearance of phytoplasmas from plants after treatment with tetracycline antibiotics, and (v) the therapeutic effect of thermal treatments that eliminate or inhibit the development of phytoplasmas.

Phytoplasmas cause reduced plant growth and productivity, poor quality of crop produced, sometimes plant death, and economic losses. Diseases attributed to phytoplasmas are among the most damaging diseases of plants. For example, coconut lethal yellowing disease has been cited as the cause of the death of millions of coconut trees in the Caribbean basin. Peach yellows disease has been cited as a major cause for abandonment of peach production in large acreages in mid-Atlantic states of the United States. Paulownia witches' broom disease has greatly reduced populations of the highly useful Paulownia tree in regions of Asia. Sandal spike disease has had

a major negative impact on sandal-wood and sandal oil production in India. Grapevine yellows diseases are among the most serious diseases of cultivated grapevine in Europe. These and several other diseases attributed to phytoplasmas are listed in Table I.

Modern molecular methods for the detection of phytoplasmas in plant and insect tissues are greatly aiding the diagnosis of phytoplasmal diseases of plants. The methods are highly useful in searches for insect vectors that transmit specific phytoplasmas, and the data gathered from studies are being used to design more effective measures to control disease spread. Despite this progress, it is still difficult to attribute disease in an individual plant to a specific phytoplasma in some cases because more than one phytoplasma species may naturally infect a single plant in the field.

III. DETECTION AND CLASSIFICATION

A. Detection

Serological methods, the use of stains such as Dienes' stain and DAPI, and DNA-based methods have been used for the detection of phytoplasmas in infected plant and insect hosts. Some methods combine a serological procedure for the capture and concentration of phytoplasma cells and a polymerase chain reaction (PCR) procedure for the amplification of phytoplasma DNA. DNA-based methods have been the most widely applied. Their application has made possible the rapid, sensitive detection of phytoplasmas in infected hosts.

The use of phytoplasma-specific cloned DNA and monoclonal antibodies as molecular probes has made it possible to detect phytoplasmas on the basis of DNA–DNA homology and serological reactivity. Dot hybridizations using cloned phytoplasma DNA probes have enabled the detection and differentiation of distinct phytoplasmas. However, the development of PCR assays, using phytoplasma-specific primers, was necessary to achieve the level of sensitivity required for detection of phytoplasmas that occurred in relatively low concentration in infected hosts. For detection of some phytoplasmas, in woody plant hosts and in plants that are simultaneously infected

TABLE I
Classification of Phytoplasmas on the Basis of Analysis of 16S rRNA Gene Sequences[a]

16S rRNA group (subgroup)	Plant disease (phytoplasma strain)	Source location[b]
16SrI (aster yellow group)		
I(A)	Tomato big bud (BB)	Arkansas
I(C)	Clover phyllody (CPh)	Canada
I(D)	Paulownia witches' broom (PaWB)	Taiwan
I(K)	Strawberry multiplier (STRAWB 2)	Florida
16SrII (peanut witches' broom group)		
II(A)	Peanut witches' broom (PnWB)	Taiwan
II(B)	Lime witches' broom (WBDL) (*Candidatus* Phytoplasma aurantifolia)	Oman
16SrIII (X-disease group)		
III(A)	X-disease (CX)	Canada
III(B)	Clover yellow edge (CYE)	Canada
III(H)	Poinsettia branch-inducing (PoiBI)	United States
III(I)	Virginia grapevine yellows (VGYIII)	Virginia
16SrIV (palm lethal yellows group)	Coconut lethal yellowing (LY)	Florida
16SrV (elm yellows group)	American elm yellows (EY1)	United States
16SrVI (clover proliferation group)		
VI(A)	Clover proliferation (CP)	Canada
VI(B)	"Multicipita" phytoplasma	Canada
16SrVII (ash yellows group)	Ash yellows (AshY1)	United States
16SrVIII (loofah witches' broom group)	Loofah witches' broom (LfWB)	Taiwan
16SrIX (pigeon pea witches' broom group)	Pigeon pea witches' broom (PPWB)	Florida
16SrX (apple proliferation group)	Apple proliferation (AP)	Germany
16SrXI (rice yellow dwarf group)	Rice yellow dwarf (RYD)	Japan
16SrXII (stolbur group)		
XII(A)	stolbur (STOL)	Serbia
XII(B)	Australian grapevine yellows (AUSGY) (*Candidatus* Phytoplasma australiense)	Australia
16SrXIII (Mexican periwinkle virescence group)	Mexican periwinkle virescence (MPV)	Mexico
16SrXIV (Bermuda grass white leaf group)	Bermuda grass white leaf (BGWL)	Thailand

[a] Phytoplasmas were classified on the basis of RFLP analysis of 16S rRNA gene sequences. Each group comprises at least one species of phytoplasma. For illustration, most subgroups are omitted.

[b] Geographical location of the source of the representative phytoplasma strain is given.

by two or more phytoplasma species, it has been advantageous to use nested PCRs consisting of two sequential PCRs in which template DNA for the second reaction consists of DNA amplified in the first reaction.

Oligonucleotides that have been used to prime PCRs for the amplification and detection of phytoplasmal DNA have been based on the nucleotide sequences of randomly cloned phytoplasma DNA fragments of unknown function and on the sequences of phytoplasmal known genes. PCR using phytoplasma group-specific or universal primers derived from conserved 16S ribosomal (r)RNA gene sequences has provided the most widely used and most sensitive means for detection of a broad array of phytoplasmas from infected plants or insect vectors. The strains listed in Table I indicate the breadth of diversity among phytoplasmas detected to date.

B. Classification

For many years, characteristics of transmission by specific insect vectors, similarities and differences in plant host ranges, and the type of symptom(s) induced in infected plants were the only criteria available for estimating the extent of relatedness among different phytoplasmas. Although phytoplasmas resembled mycoplasmas (members of the class Mollicutes) in morphology and ultrastructure, their relationship to mycoplasmas remained uncertain. During decades of studies of plant yellows diseases, it was not possible to determine how many distinct species were represented by the phytoplasmas that were present in the numerous diseases with which they were associated. In recent years the situation has changed dramatically with the use of molecular-based techniques. In particular, the use of DNA-based methods has made it possible to identify unknown phytoplasmas associated with many plant diseases and to develop comprehensive systems for the classification of phytoplasmas.

Taxonomically, phytoplasmas are classified in the class Mollicutes (Latin *mollis*, "soft"; *cutis,* "skin"). Also classified in the class Mollicutes is a group of plant pathogenic, wall-less bacteria known as spiroplasmas. All phytoplasmas and some spiroplasma species are pathogenic in plants. However, phytoplasmas and spiroplasmas comprise two mutually exclusive clades in the class Mollicutes, and both are distinct from other microbes in the class.

Phytoplasmas differ from spiroplasmas in several important features. Phytoplasmas vary in cell shape from rounded bodies to filamentous; spiroplasmas are characteristically helical in cell shape and exhibit motility. Plant pathogenic phytoplasmas and spiroplasmas are found as parasites in the phloem tissue of infected plants and in the bodies of insects that transmit them from plant to plant. Whereas phytoplasmas are known only from plants and insect vectors, spiroplasmas have been reported in diseased plants, in insect vectors of plant pathogenic spiroplasma species, in the bodies of other insects, in ticks, and on the surfaces of flowers. Spiroplasmas can be isolated in artificial culture in the laboratory, but scientists have been unable to isolate and cultivate any phytoplasma in artificial medium. No phytoplasma is known to be pathogenic in humans or animals, and although certain species of spiroplasma are pathogenic in certain newborn animals when experimentally inoculated, no spiroplasma has been demonstrated to be pathogenic in humans or animals under natural conditions.

In early attempts to identify and classify phytoplasmas, the determination of biological properties was often time-consuming, laborious, and sometimes unreliable. The less laborious molecular-based analyses introduced in the past decade have proved to be more accurate and reliable. The use of molecular probes, such as phytoplasma-specific cloned DNA and monoclonal antibodies, made it possible to classify phytoplasmas on the basis of DNA–DNA homology and serological data.

The cloning of random DNA fragments from phytoplasmas was begun principally as an approach to phytoplasma detection. This approach provided for the development of labeled DNA probes for use in dot hybridizations and restriction fragment length polymorphism (RFLP) analyses of chromosomal and extrachromosomal DNAs. Results from the use of such probes revealed genomic cluster relationships among various phytoplasmas, and these relationships gave rise to a system of classification of phytoplasmas based on genomic clusters, with each cluster containing closely related phytoplasmas.

Current systems of phytoplasma classification have as their basis the genetics of phytoplasmas reflected in chromosomal DNA homologies and nucleotide sequences of ribosomal genes. Investigation of ribosomal (r) gene sequences yielded a phylogenetic classification that coincided with RFLP groupings (genomic clusters) that were recognized through the use of DNA hybridization probes. The principal ribosomal gene sequence that has been investigated in the development and use of phytoplasma classification schemes is the 16S rRNA gene. Phytoplasma DNA that has been detected by the use of PCR is analyzed by RFLP analysis and/or by nucleotide sequencing. The grouping of phytoplasmas based on RFLP analysis of 16S rRNA genes is consistent with the genomic strain cluster grouping based on DNA–DNA homology and serological data and with classification based on nucleotide sequence analyses.

The major phytoplasma groups, and examples of

subgroups, that have been established through the use of RFLP analyses of 16S rRNA gene sequences (DNA) are listed in Table I. Each group in Table I carries the designation 16Sr and Roman numeral (to indicate the identity of the 16S rDNA RFLP group) and the name of the phytoplasma tentatively designated as representative of the group. Classification by RFLP analysis of PCR-amplified phytoplasmal 16S rDNA has provided a simple and rapid method that can be used to differentiate and identify a large number of unclassified phytoplasmas in a relatively short period of time. However, when few restriction enzymes are used in RFLP analyses, the identification and classification of a phytoplasma is not always consistent with that based on phylogenetic analysis of 16S rRNA gene sequences.

In practice, the 16S rRNA gene sequence, or most of the gene, is amplified in the laboratory by the use of a PCR that is primed by a pair of phytoplasma-universal oligonucleotides that are designed to yield specific amplification of DNA from phytoplasmas. Several primer pairs have also been designed for the group-specific amplification of phytoplasma DNA to facilitate the identification of unknown strains. Amplified DNA is then subjected to enzymatic RFLP analysis.

Other conserved gene sequences in addition to that of 16 rRNA have been used in the development of phytoplasma classification schemes, including the 16S–23S rRNA gene spacer region, ribosomal protein gene operon sequences, and the elongation gene EF-TU (tuf). All these classification schemes are mutually consistent, although, as expected, the analysis of less conserved gene sequences, such as that of the ribosomal protein gene operon, tend to yield finer distinctions among closely related phytoplasmas than do analyses of more highly conserved genes.

IV. PHYLOGENY AND TAXONOMY

A. Phylogeny

Comparisons of the nucleotide sequences of the 16S rRNA gene from phytoplasmas with the 16S rRNA gene from *Acholeplasma laidlawii* and other Mollicutes revealed that phytoplasmas are members of the class Mollicutes and are more closely related to acholeplasma than to the animal mycoplasmas. Results from additional phylogenetic analyses of 16S rRNA and ribosomal protein (rp) gene sequences placed phytoplasmas definitively among members of the class Mollicutes and revealed that phytoplasmas form a large discrete monophyletic clade within the expanded Anaeroplasma clade. The phytoplasma clade is paraphyletic to *Acholeplasma* species. Within the phytoplasma clade, at least 20 distinct subclades (monophyletic groups or taxa) have been recognized. The phylogenetic interrelationships among representative phytoplasmas validated phytoplasma classification schemes that were based on RFLP analysis of PCR-amplified 16S rDNA.

B. Taxonomy

Characterization of pure cultures is required for the naming and description of species in the class Mollicutes. Because it has not been possible to isolate phytoplasmas in pure culture, the convention of "*Candidatus* Phytoplasma" species has been adopted to make it possible to refer to distinct lineages of phytoplasmas. Three *Candidatus* phytoplasma species have been named to date: *Candidatus* Phytoplasma aurantifolia (lime witches' broom phytoplasma), *Candidatus* Phytoplasma australiense (Australian grapevine yellows phytoplasma), and *Candidatus* Phytoplasma australasia (papaya yellow crinkle phytoplasma).

V. GENOME

Phytoplasmas have the smallest genomes known among cellular plant parasites and pathogens. Estimates indicate that genome sizes vary among diverse phytoplasmas from approximately 600 to 1200 kbp. This size range is similar to that of genomes in the genus *Mycoplasma*, and it is smaller than that of the genomes of their close relatives, members of the genus *Acholeplasma*, both of which genera are classified, along with phytoplasmas, in the class Mollicutes.

The phytoplasma chromosome has been physically mapped for several phytoplasmas, including those associated with diseases known as western X-disease of stone fruits, sweet potato little leaf, Australian tomato big bud, and apple proliferation. The phytoplasma genome contains two ribosomal RNA gene operons. Interoperon sequence heterogeneity has been reported in several phytoplasma species.

In addition to a large circular double-stranded DNA chromosome, phytoplasmas also contain extrachromosomal DNA molecules. Some such DNAs may derive from viruses that infect phytoplasmas or from phytoplasma plasmids.

The G + C contents of phytoplasma chromosomal DNA are in the range of 23–29 mol%. Genes encoding 16S rRNA, 23S rRNA, ribosomal proteins, and elongation factors G and EF-TU have been characterized in various phytoplasmas. The 16S rRNA gene has been sequenced from more than 60 distinct phytoplasmas. 16S rRNA oligonucleotide sequences (signatures) unique to phytoplasmas distinguish them from other members of the class Mollicutes.

VI. EPIDEMIOLOGY

Phytoplasmas are spread (transmitted) in nature by insects that feed in the phloem tissues of plants. Phytoplasmas are spread primarily by insect vectors belonging to the families Cicadelloidea (leafhoppers) and Fulgoridea (planthoppers). The phytoplasma cells multiply in the bodies of the insect vectors and in the plant phloem.

In nature, the range of insect species capable of spreading a phytoplasma varies with the phytoplasma, as does the range of plant species that become infected by the phytoplasma. Some phytoplasmas have a low insect vector specificity (i.e., are transmitted by many insect species), whereas others have a high vector specificity (i.e., are transmitted by one or few insect species). Examples of phytoplasmas having low insect vector specificity are California aster yellows phytoplasma, which is transmitted by 24 species of leafhoppers, and peach X-disease phytoplasma, which is transmitted by at least 15 leafhopper species. Phytoplasmas with high insect vector specificity include beet leafhopper-transmitted virescence phytoplasma and American elm yellows phytoplasma; each of these phytoplasmas has only one

or few known species of insect vector. Some insect species are capable of transmitting more than one phytoplasma species.

Although some phytoplasma species have a wide plant host range in nature, others have a narrow natural plant host range. Some plant species can serve as hosts of two or more distinct phytoplasma species in nature, and plants may be simultaneously infected by two or more distinct phytoplasmas.

Laboratory studies have shown that some species of plants, for example, *Catharanthus roseus* (periwinkle), can become infected by many different species of phytoplasma. However, in nature a plant species is generally found to be infected by one or few phytoplasma species. Insect vectors appear to play an important role in determining the species of plants that become infected by a particular phytoplasma in a given geographical region.

Natural insect vectors remain unknown for phytoplasmas that are associated with many plant diseases. Examples include phytoplasmas causing pecan bunch and ash yellows in the United States and apple proliferation in Europe. Vectors of these phytoplasmas have not been found among insects that have been collected from the respective plants species infected by the phytoplasmas. Such phytoplasmas may be spread largely by insect vector species that feed on these plants only occasionally. That is, it is possible that the plants are not preferred hosts of the respective insect vectors.

A possibly parallel example of transmission of a Mollicutes pathogen to a crop plant is the case of citrus subborn disease. The pathogen, *Spiroplasma citri*, is transmitted by the beet leafhopper, *Circulifer tenellus*. This leafhopper does not survive on citrus, although it feeds and completes its life cycle on numerous other plants. Thus, the vector of the citrus stubborn disease pathogen visits and feeds on citrus only occasionally and is not normally found among insects collected from plants of citrus.

The geographic distributions of various insect vectors and the preferential plant host(s) of each vector are major factors that determine whether a given plant species will be infected by a particular phytoplasma and whether the plant species may become infected by single or multiple phytoplasmas.

Many species of agricultural crop plants and weed

plants are known to harbor phytoplasmas. Weeds can be an important source of phytoplasmas that can be carried by insect vectors to susceptible crop plants. Because phytoplasmas systemically infect their host plants, phytoplasmas may be introduced into new geographic regions through their presence in vegetative plant parts, such as cuttings, bulbs, corms, and tubers. Since various phytoplasmas can be seriously damaging to the crop plants that they infect, national and international quarantine measures are in force throughout the world to prevent the entry of specific phytoplasmas into regions where they do not occur.

See Also the Following Articles

Diagnostic Microbiology • Phloem-Limited Bacteria • Plant Pathogens

Bibliography

Davis, R. E., and Lee, I.-M. (1982). Pathogenicity of spiroplasmas, mycoplasmalike organisms, and vascular-limited fastidious walled bacteria. *In* "Phytopathogenic Prokaryotes, Vol. 1" (M. S. Mount and G. Lacy, Eds.), pp. 491–513. Academic Press, New York.

Kirkpatrick, B. C., and Smart, C. D. (1995). Phytoplasmas: Can phylogeny provide the means to understand pathogenicity? *Adv. Bot. Res.* **21,** 188–212.

Lee, I.-M., and Davis, R. E. (1986). Prospects for *in vitro* culture of plant pathogenic mycoplasmalike organisms. *Annu. Rev. Phytopathol.* **24,** 339–354.

Razin, S., Yogev, D., and Naot, Y. (1998). Molecular biology and pathogenicity of mycoplasmas. *Microbiol. Mol. Biol. Rev.* **62,** 1094–1156.

Schreiber, L. R. (1992). Dutch elm disease and elm yellows. *Encycloped. Microbiol.* **2,** 23–28.

Tully, J. G. (1992). Mollicutes (Mycoplasmas). *Encycloped. Microbiol.* **3,** 181–191.

Pigments, Microbially Produced

Eric A. Johnson

University of Wisconsin

GLOSSARY

astaxanthin A reddish carotenoid pigment found in salmon, crustacean shells, and bird feathers (e.g., pink flamingos).

bacteriochlorophyll A form of chlorophyll present in photosynthetic bacteria.

carotenoids A group of yellow, orange, or red oil-soluble polyene pigments produced in plants and certain microorganisms that are divided into carotenes (hydrocarbons) and xanthophylls (oxygenated carotenes).

chlorophyll The generic name of any of several oil-soluble tetrapyrrole pigments that function as photoreceptors of light energy in photosynthesis.

color The sensation resulting from stimulation of the retina of the eye by light waves of certain lengths.

light The form of electromagnetic radiation that acts on the retina of the eye.

melanins High-molecular-weight polymeric indole quinone pigments having brown or black colors occurring predominately in vertebrates and insects and occasionally in microorganisms.

Monascus pigments Water-soluble red pigments produced by the fungus *Monascus*.

phenazines Compounds based on the trycyclid phenzaine ring system, some or which are colored, and are secreted by certain species of bacteria.

pigment Any coloring matter in microbial, plant, or animal cells.

pyrrole A compound containing the pyrrole ring, classified into mono-, di-, tri-, and tetrapyrroles.

riboflavin (vitamin B$_2$) A water-soluble, yellow-orange fluorescent flavin-based pigment that is essential to human nutrition.

vitamin B$_{12}$ A group of polypyrrole pigments containing the metal cobalt.

ALL LIFE ON EARTH ultimately depends on light energy from the sun, which can be converted to other forms of energy or to biomass by plants and photosynthetic microorganisms. The electromagnetic spectrum (Fig. 1) includes the wavelength range humans recognize as visible light (380–750 nm). When pigments are illuminated by visible light, part of the energy may be absorbed, and the emitted light will appear as color.

Pigments and dyes contain a chromophore ("color-bearing") group, which directly produces color by light absorption or transfer energy to auxochromes ("color increasers"), giving characteristic colors. Various chemical classes of pigments occur in nature, of which the greatest number is produced by microorganisms. In this article, the primary pigments are described with an emphasis on pigments of keen biological and industrial importance.

Some microbial pigments have important biological functions for the producing organisms, such as the production of energy and fixation of matter by bacteriochlorophylls and phycobilins in microbial photosynthesis, whereas in some organisms the only known attribute of pigments is to give color. Some microbial pigments such as the phenazines are secondary metabolites, and their natural functions are not well understood. Certain classes of pigments have economic importance, for example, as food,

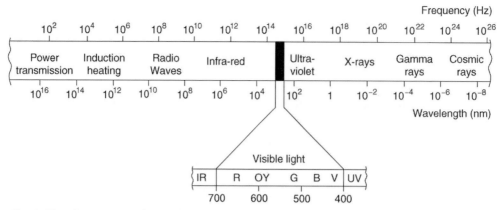

Fig. 1. The electromagnetic spectrum; 1 Hz is one cycle per second (reproduced with permission from Nassau, 1983).

feed and drug colorants, or as essential vitamins in nutrition. Animal pigments including human hemoglobin have been cloned and expressed in microorganisms and are being evaluated as pharmaceuticals. Production of pigments in heterologous hosts could contribute to the industry of biotechnology, and microbially produced pigments could add value to the food, feed, and pharmaceutical industries.

I. QUANTITATION OF MICROBIAL PIGMENTS

The color of natural pigments is described by the wavelength (λ) of maximum absorption (λ_{max}) in the visible portion of the electromagnetic spectrum (380–750 nm). The concentration of a solution of pigment can be calculated using Beer's and Lambert's laws by determining its maximum absorption (A_{max}) at λ_{max} in a spectrophotometer. The quantity of light absorbed by a pigment in solution is proportional to the concentration of the absorbing biomolecule (Beer's law) and to the length of the pathway through which the light passes (Lambert's law). The concentration is calculated using the molar absorptivity or extinction coefficient (ε) of a 1 M solution of a compound in a 1-cm light path. The 1% extinction coefficient or absorbance of a 1% solution at A_{max} is often used in older literature. The concentration of pigment (mol liter^{-1}) = $(A_{max})/[(\varepsilon)\,(1\text{ cm})]$.

II. CLASSES OF MICROBIAL PIGMENTS

Microbial pigments can be classified into relatively few major biosynthetic and structural classes: the pyrroles and tetrapyrroles, tetraterpenoids or carotenoids, N- or O-heterocyclic pigments, and metalloproteins. Some microbial pigments do not share obvious structures with these groups, and some other pigments such as hemoglobin are rare in the microbial world but they have been expressed by cloning the genes in microbial hosts.

A. Pyrroles and Tetrapyrroles

The pyrroles, particularly the tetrapyrroles, are natural pigments that exhibit a wide range of colors. They are one of the most abundant and important groups of pigments globally. The N-heterocyclic compound pyrrole is the basic precursor to this pigment class, which includes the cyclic tetrapyrroles and porphyrins, highlighted by chlorophylls and heme (Fig. 2). These tetrapyrroles do not function free in cells but are attached as the prosthetic group to proteins in which they function.

Chlorophyll is the fundamental pigment of life because it converts part of the sun's energy to chemical energy and biomass. Green plants, algae, and certain bacteria all contain chlorophylls and contribute to global photosynthesis. Cyanobacteria and some algae and protists possess the tetrapyrrole class of pigments referred to as phycobilins, including the blue-green compound phycoyanin. Phycobilins and related pigments such as phycoerythrin are present as prosthetic groups in phycobiliproteins and function as accessory pigments in energy acquisition from light. Related tetrapyrroles, including the red compound heme, are important in mammals as the prosthetic group of hemoglobin in blood and in cytochromes, which are important in energy metabolism as well as in biosynthesis of sterols and other important cellular components. Heme is also present in a large class of enzymes termed cytochrome P450s, which metabolize xenobiotics, carcinogens, and pollutants. The biotechnology industry is developing technologies to produce cytochrome P450s in high yields for industrial applications.

Vitamin B_{12} (cyanocobalamin) consists of a modified tetrapyrrole ring and contains cobalt instead of magnesium (chlorophyll) or iron (heme). It has a pink color in solution and crystallizes as red needles. Vitamin B_{12} is synthesized in nature only by microorganisms and is an essential constituent of human and most animal diets. Some animals appear to have intestinal microbes that can provide Vitamin B_{12} for nutrition. Vitamin B_{12} attaches to proteins in which it is able to perform its functional roles in biosynthesis. Since the chemical synthesis of Vitamin B_{12} involves approximately 70 chemical steps, it is a very costly synthesis and is produced more economically by microbial industrial fermentation. Various microorganisms, including specialized strains of *Bacillus, Streptomyces, Brevibacterium, Propionibacterium,* and *Pseudomonas,* are used for industrial production of vitamin B_{12}. The worldwide production is approximately 3 metric tons per year.

Other natural microbial pigments are derived from pyrrole. Prodiogisin, produced by some bacteria including *Serratia,* consists of a linear tripyrrole. Linear tetrapyrroles termed bilins are formed by algae and present in the energy-related protein phycobilin. The fungus *Wallemia* produces a pigment in its fruiting bodies in which a monopyrrole is conjugated to a polyene.

B. N- and O-Heterocyclic Pigments

The N- and O-heterocyclic pigments comprise a large group of pigments that all contain one or more N- or O-substituted aromatic rings. Among the more common members are the pale-yellow flavonoids and their derivatives, the anthocyanins, that give plant flowers many of their brilliant colors. The flavonoids, anthocyanins, and phenolic compounds are mainly present in plants and not in microbial cells. Pterins are pigments mainly displayed in insects, fishes, reptiles, and birds. They are derived from tetrahydrofolic acid, which is synthesized by many microorganisms.

Riboflavin (Vitamin B_2) is an orange-yellow pigment with a strong yellowish-green fluorescence that has been produced on an industrial scale by fermentation processes utilizing microorganisms. The chemical structure of riboflavin is 6,7-dimethyl-9-(D-1′-ribtyl)-isoalloxazine (Fig. 2). Riboflavin is of considerable biological importance as the functional component of various enzymes and coenzymes. Riboflavin occurs universally in all living organisms but cannot be synthesized by animals and is an essential constituent of their diet.

A nonpolymeric N-heterocyclic pigment is produced by the yeasts *Saccharomyces cerevisiae* and *Kluyveromyces lactis* that contain mutations in the adenine biosynthesis pathway. These mutant yeasts accumulate intense red pigmentation in their vacuoles when starved for adenine. The pigments have not been completely characterized but appear to be composed of a peptide and purine. The pigmentation is used extensively as a nonselective marker for genetic analyses of these yeasts and for complementation of gene mutations in purine biosynthesis from other organisms including humans. Several microorganisms also deaminate guanine to xanthine, which absorbs strongly in the ultraviolet and can be seen by insects and some other animals.

The phenazines comprise a family of approximately 60 organic pigments, often with striking color and representing every basic color of the visible electromagnetic spectrum. Naturally occurring phenazines are produced by bacteria in the genera *Pseudomonas, Streptomyces, Brevibacterium,* and *Nocardia* and certain other genera. They are formed as secondary metabolites in the shikimate pathway and possess characteristic nitrogenous aromatic rings. Apart from their color, some phenazine pigments also have antibiotic activity.

A

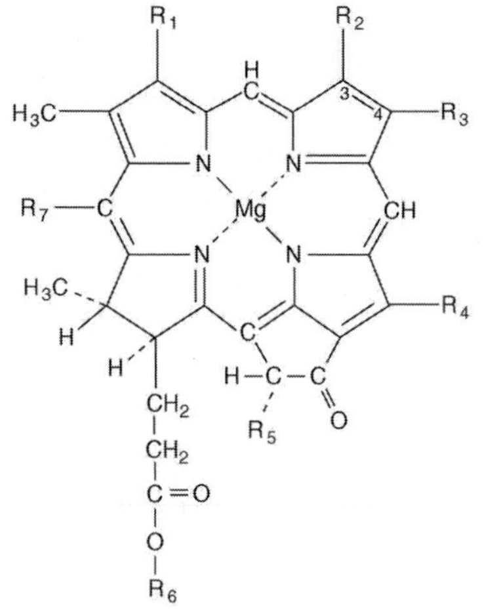

B

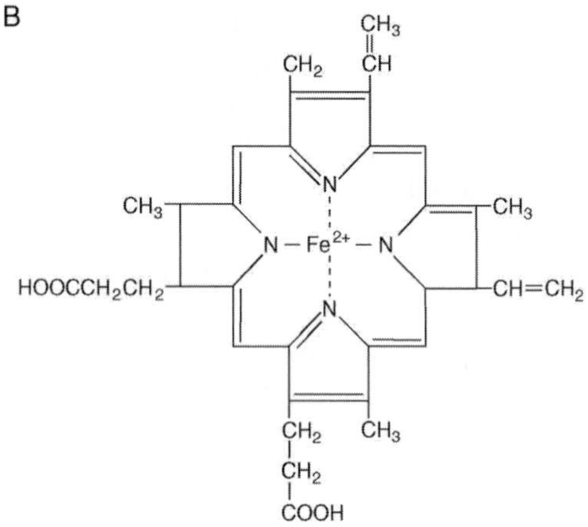

C

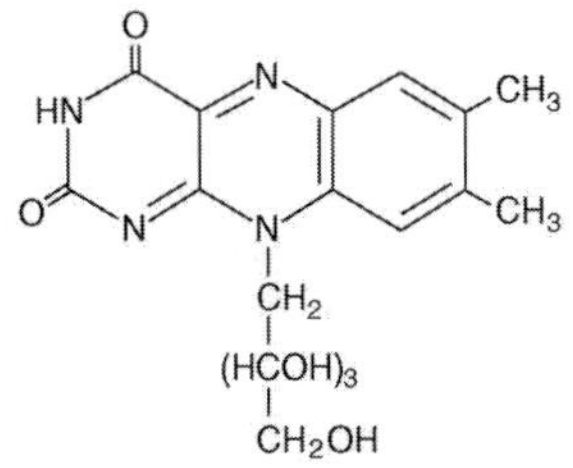

D

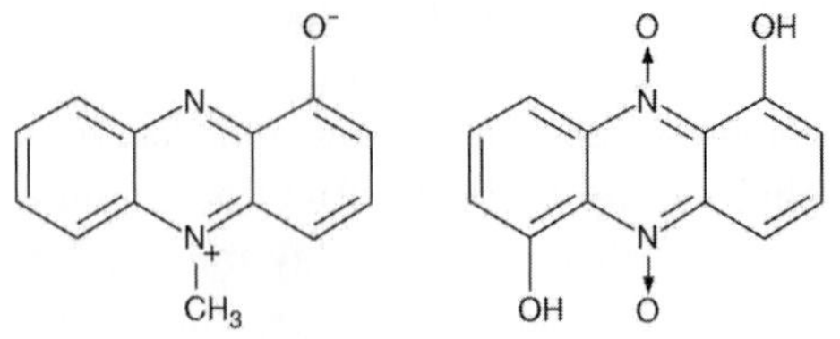

E

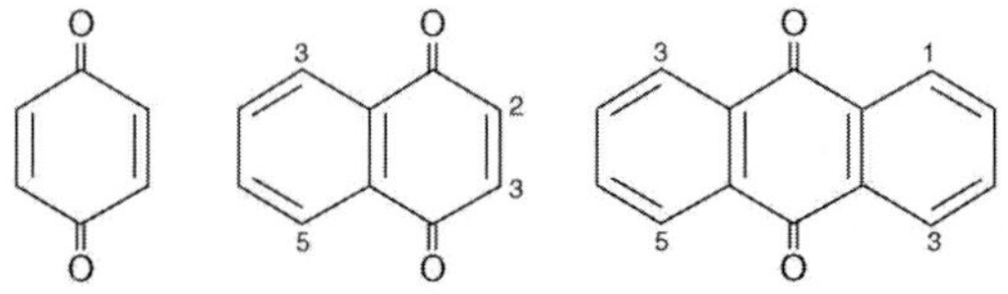

F

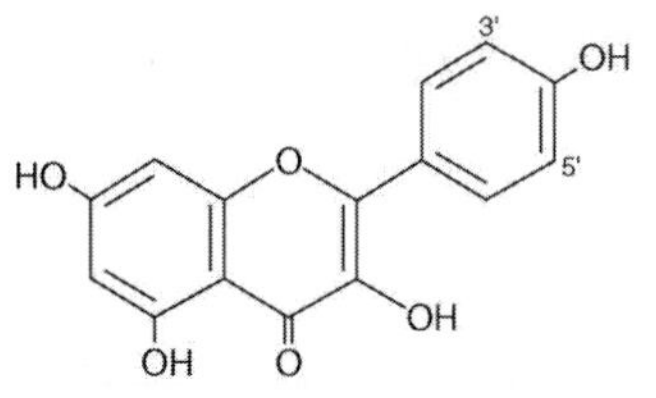

G

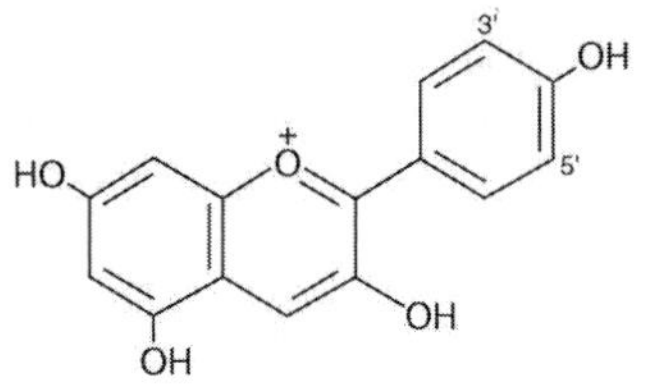

H

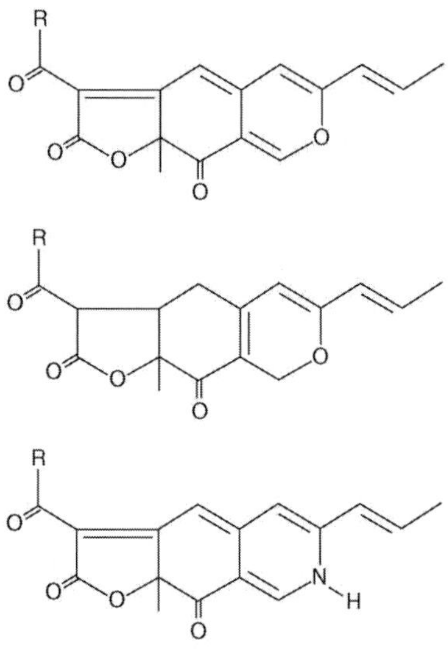

I

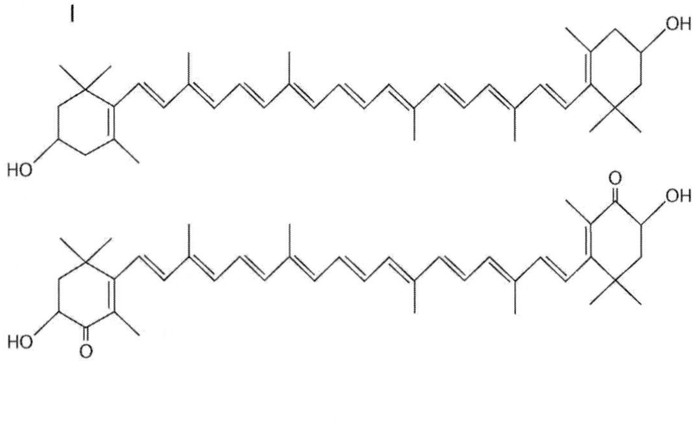

Intensely colored N-heterocyclic pigments of miscellaneous structures are produced by certain bacteria. The indole derivative violacein is produced by *Chromobacterium violaceum* and the pigment has been demonstrated to have antibiotic activity and also to be involved in quorum sensing in which a diffusible signal molecule couples gene expression with cell population density.

The deep-red-colored betalains are most commonly found in plants, but they have also occasionally been isolated from fungi. A violet betacyanin and several yellow betaxanthins have been detected in the deathcap mushroom, *Amanita muscaria*. Although it is unlikely that pigments could be used for food uses from such sources, it is likely that other fungi produce betalains as well.

The quinones are a large group of pigments having colors ranging from yellow to orange, red, purple, brown, and black. The most commonly occurring natural quinones are benzo-, naptho-, and anthraquinones. Microbial quinones are produced by many yeasts and fungi as well as by certain bacteria. Besides pigmenting microbes, quinones also impart the vivid colors of some invertebrates such as sea urchins and starfish. The quinones consist chemically of conjugated cyclic diketones rather than aromatic systems. The polyketide pathway is most commonly used for synthesis of quinones in microorganisms. Benzoquinone isoprenoid derivatives such as ubiquinones have important roles in energy metabolism. The terphenylquinones such as pulvinic acid produce intense colors, including a characteristic deep blue on injury of bolete fungi. Certain quinone pigments were used in ancient times as dyes, and others such as yellow-brown henna are used today for dyeing and for cosmetic purposes. Recently, quinones have been investigated for antiviral and antimicrobial activities.

Indigo is one of the oldest dyes used by humans and traditionally has been obtained by chemical modification of precursors from plants. *Pseudomonas indigofera* produces indigoidine or "bacterial indigo" but this is not used commercially. Although indigo is currently produced by chemical synthesis, the microbial synthesis has been accomplished by introduction of naphthalene oxidation genes in *Escherichia coli*. These studies illustrate the potential of recombinant DNA technology for the microbial production of pigments commonly found only in plants or animals.

Melanins have traditionally been known as dark or black pigments produced by a variety of organisms, including bacteria, fungi, plants, and animals. They also include natural brown, red, and yellow colors, particularly in feathers and hair and skin of animals and humans. Melanins are insoluble high-molecular-weight polymers produced from phenolic substrates. Relatively little is known regarding the biosynthesis and functions of this pigment group, largely because of technical difficulties in studying them. In addition to providing pigmentation, melanins are associated with virulence of certain microbial pathogens such as *Cryptococcus neoformans*, *Wangiella dermatidis*, and *Mycobacterium leprae*. Like certain other pigments, the presence of conjugated double bonds appears to render organisms less sensitive to killing by free radicals.

The fungus *Monascus* produces a group of yellow, orange, and red pigments of polyketide origins. This fungus has been utilized in East Asia for centuries for the production of red koji starter and red rice (ang-kak), and the water-soluble red pigments have also been used in coloring of various foods. Interest has increased in recent years for using the pigments as natural food colorants to replace synthetic food dyes. Several *Monascus* pigments have

---

Fig. 2. Representative structures of microbial pigments. Pigment groups: (A) chlorophyll—the R groups represent modifications in different bacteriochlorphylls; (B) heme; (C) riboflavin; (D) phenazine pigments—pyocyanine (left) and iodinin (right); (E) quinone pigments—benzoquinone, napthoquinone, and anthraquinone (left to right); (F) flavonoid, represented by flavonol; (G) anthocyanin, represented by anthocyanidin; (H) *Monascus* pigments—monascin, rubropunctatin, and rubropunctamine (top to bottom) (diversity is increased by substitution of R groups); (I) carotenoids—zeaxanthin (top) and astaxanthin (bottom).

been characterized chemically and for toxicological activity.

C. Carotenoids

Among the microbial pigments, carotenoids are probably the most widely distributed in nature, in which they serve important biological functions. They occur universally in photosynthetic organisms but sporadically in nonphotosynthetic bacteria and eukaryotic microorganisms. Although carotenoids are not synthesized by animals, some can be metabolized or converted to essential nutrients, such as vitamin A and retinoids. Since many animals of agricultural importance, including salmonids and poultry, require carotenoids for pigmentation of their flesh, plumage, and eggs, these pigments are industrially important as components in animal feeds. They are also important colorants and sources of vitamin A in human foods. In addition to providing necessary nutrients, carotenoids scavenge reactive oxygen species (ROS) and provide protection against damaging consequences of light (photoprotection). Due to their ability to scavenge potentially harmful by-products of metabolism including ROS, carotenoids such as α-carotene, β-carotene, lycopene, and astaxanthin are currently being clinically evaluated as pharmaceuticals to prevent or delay onset of diseases such as cancer, arteriosclerosis, and chronic diseases associated with aging. In addition to these attributes, carotenoids bestow brilliant pigmentation and aesthetic value to various organisms, including crustacea, amphibia, birds and their ova, and higher animals. Traditionally, carotenoids for agriculture and food uses were obtained by extraction of plant sources (e.g., marigold flowers and corn). Chemical syntheses were developed after 1950 for many economically important carotenoids and the industry has been dominated by synthetic sources. Currently, there is interest in developing microbial and enzymatic processes for production of industrially important carotenoids.

Carotenoids are isoprenoids containing a characteristic polyene chain of conjugated double bonds. Most carotenoids are tetraterpenoids composed of eight isoprenoid precursors. Two general groups of carotenoids are recognized: the hydrocarbons (carotenes) and the oxygenated derivatives (xanthophylls). More than 650 natural carotenoids have been isolated and characterized.

The chromophore of carotenoids responsible for light absorption and color is the polyene chain consisting of conjugated double bonds. The wavelength of λ_{max} is directly dependent on the number of conjugated double bonds. Normally the polyene chain has a characteristic absorbance spectrum with three peaks referred to as persistence. Introduction of certain chemical groups such as carbonyls results in loss of persistence and gives a rounded symmetrical peak of absorbance and different λ_{max}. Most carotenoids impart yellow, orange, or red hues to the producer organisms or foods, but some give a blue color when conjugated with proteins such as the pigment crustacyanin composed of astaxanthin and protein in the American lobster *Homarus americanus*.

Although approximately 650 carotenoids have been isolated from nature, only approximately 12 are used industrially as food colors, primarily in animal feeding, as human nutrients, and as coloring agents for pharmaceuticals and cosmetics. The primary natural carotenoids used industrially are α- and β-carotene, lycopene, canthaxanthin, lutein, zeaxanthin, and astaxanthin. The current world market for carotenoids exceeds $500 million, and this is expected to increase significantly as farming practices such as aquaculture becomes more common and when carotenoids are used as human food supplements as agents to prevent diseases. Recently, the genes for biosynthesis for industrially important carotenoids, such as β-carotene, lycopene, and astaxanthin, have been cloned and expressed in noncarotenogenic bacteria and yeasts such as *E. coli* and the food and feed yeasts *S. cerevisiae* and *Candida utilis*. Expression of carotenoids in these organisms illustrates the potential for production in generally recognized as safe host microorganisms.

Several fungi are prolific producers of terpenoids and it is not surprising that certain genera have been found to accumulate specific sesquiterpenoids at levels that impart red, orange, green, and purple pigmentation to the fruiting body. It should be possible to genetically modify fungi to produce terpenoid pigments that were previously only found in high concentrations in plants.

D. Metalloproteins

Metalloproteins have characteristic colors in which the color is derived from association of metals such as iron or copper, or when metal-containing prosthetic groups such as porphyrins are conjugated in the protein. Several copper proteins are blue or blue-green, the protein hemovanadin containing V is apple green, and protoporphyrin containing Zn is pink-brown. Some fungal iron enzymes have brown or reddish colors. Pigmented metalloproteins have been studied mostly in animals, but as cloning and expression systems are developed metalloproteins will be produced to a greater extent in microbial systems.

III. PERSPECTIVES

There are several advantages of microorganisms for the study of biosynthesis and function of pigments. Bacteria and fungi offer a tremendous resource in that they produce hundreds to thousands of various pigments, and it has been estimated that only 5–7% of fungi and 10–12% of bacteria have been isolated from nature. It is likely that many more natural pigments will be isolated from microorganisms. The fungi are prolific producers of relatively few classes of pigments, whereas the bacteria rival plants and animals in the diversity of pigments produced. Microorganisms offer advantages over plants and animals for pigment production. Microorganisms have an exceedingly small size compared to animal and plant cells and consequently a tremendous surface-to-volume ratio, which facilitates high rates of nutrient transport and an intensive rate of metabolism and high growth rates. Microorganisms can perform a variety of metabolic reactions and can adapt to adverse environments. Microbial cells provide an excellent host environment for the expression of foreign genes. Certain mammalian and plant pigments, such as hemoglobin and nonmicrobial carotenoids, have been expressed in microorganisms, and it is certain that microbes will be more extensively utilized in the future for basic and applied studies of pigments.

See Also the Following Articles

Secondary Metabolites • Tetrapyrrole Biosynthesis in Bacteria • Vitamins and Related Biofactors, Microbial Production

Bibliography

Britton, G. (1983). "The Biochemistry of Natural Pigments." Cambridge Univ. Press, Cambridge, U.K.

Demain, A. L. (1983). Industrial microbiology. *Science* **214**, 987–995.

Ensley, B. D., Ratzkin, B. J., Osslund, T. D., Simon, M. J., Wackett, L. P., and Gibson, D. T. (1983). Expression of naphthalene oxidation genes in *Escherichia coli* results in the biosynthesis of indigo. *Science* **222**, 167–169.

Fox, D. L. (1979). "Biochromy: Natural Coloration of Living Things." Univ. of California Press, Berkeley.

Harborne, J. B. (1993). "Introduction to Ecological Biochemistry," 4th ed. Academic Press, London.

Hendry, G. A. F., and Houghton, J. D. (Eds.) (1996). "Natural Food Colorants," 2nd ed. Blackie Academic, London.

Isler, O. (Ed.) (1971). "Carotenoids." Birkhäuser-Verlag, Basel.

Johnson, E. A., and Schroeder, W. A. (1995). Microbial carotenoids. *Adv. Biochem. Eng.* **53**, 119–178.

Luckner, M. (1990). "Secondary Metabolism in Microorganisms, Plants, and Animals," 3rd ed. Springer-Verlag, Berlin.

Margalith, P. Z. (1992), "Pigment Microbiology." Chapman & Hall, London.

Misawa, N., and Shimada, H. (1998). Metabolic engineering for the production of carotenoids in noncarotenogenic bacteria and yeasts. *J. Biotechnol.* **59**, 169–181.

Nassau, K. (1983). "The Physics and Chemistry of Color: The Fifteen Causes of Color." Wiley, New York.

Packer, L. (Ed.) (1992). Carotenoids. *Methods Enzymol.* **213/214**.

Turner, J. M., and Messenger, A. J. (1986). Occurrence, biochemistry and physiology of phenazine pigment formation. *Adv. Microbiol. Physiol.* **27**, 211–275.

Vandamme, E. J. (1989). "Biotechnology of Vitamins, Pigments, and Growth Factors." Elsevier, New York.

Plague

Elisabeth Carniel

Institut Pasteur

GLOSSARY

enzyme-linked immunosorbent assay Technique using an antigen fixed to a plastic well to detect serum antibodies specific for the fixed antigen. The antigen–antibody reaction is revealed with a second antibody conjugated to an enzyme that allows colorimetric detection of the complex.

pathogenicity island (PAI) Large chromosomal DNA region that carries virulence functions and mobile elements. PAIs are often bordered by a tRNA locus, their G+C% is different from that of the rest of the chromosome, they are sometimes capable of spontaneous excision, and they carry phage signatures, suggesting that they have been acquired by horizontal transfer from bacteriophages. PAIs are found in numerous bacterial species.

plasmid Extrachromosomal self-replicating DNA. A self-transferable plasmid possesses the machinery required for its transfer from one bacterial cell to another bacterial cell in close contact.

polymerase chain reaction Amplification of a target DNA sequence by a heat-stable polymerase which uses small DNA segments specific to each extremity of the target as primers.

ribotype Specific profile obtained after hybridization of digested bacterial DNA with a ribosomal RNA or DNA probe.

siderophore Small nonprotein molecule having a high affinity for iron and secreted by different microorganisms to scavenge ferric iron bound to eukaryotic proteins. Siderophores are separated into different biochemical groups.

Yersinia pestis Causative agent of bubonic and pneumonic plague.

PLAGUE has been one of the most devastating infectious diseases in human history and has profoundly altered the structure of human societies.

Cases of plague were described as long ago as the pre-Christian era but their numerical and historical importance remain unknown. During the Christian era, three well-documented plague pandemics occurred. The first pandemic, known as the "Justinian's plague," probably started from central Africa. It reached lower Egypt and then the Mediterranean countries during the sixth century, causing an estimated population loss of 50–60% (actually, not entirely attributable to plague) from 541 to 700 AD. The second pandemic, subsequently called "Black Death," originated in central Asia, reached the Crimean ports during the fourteenth century, and invaded all of Europe and the north of Africa. This pandemic killed one-quarter to one-third of the European population and continued late into the seventeenth century in the form of intermittent outbreaks. The third pandemic began in the war-torn Yunnan province of China and reached Hong Kong in 1894. The network of steamships and railways favored a rapid spread of the disease over all continents and the colonization of previously unscathed areas. Major advances in the knowledge of plague were made during the early years of the last pandemic. In 1894, Alexandre Yersin identified the etiologic agent

(*Yersinia pestis*) and showed that rats were the reservoir of the disease. Four years later, Paul-Louis Simond demonstrated that plague was transmitted by fleas. As a result of the advent of effective public health measures, mass vaccination in 1934, and antibiotic therapy in 1938, the number of victims reported during the third pandemic was not comparable to those of the two previous pandemics.

I. PLAGUE: CURRENT SITUATION

Despite the considerable reduction in mortality and morbidity of plague during the twentieth century, the disease has never been eradicated. Endemic foci persist in Africa, Asia, and North and South America. During the past 20 years, 27,520 cases of human plague and 2488 deaths have been reported to the World Health Organization (WHO) by 24 countries (Table I). The most affected countries are Madagascar and Tanzania in Africa, Vietnam in Asia, and Peru in the Americas. No plague cases have been reported recently in Europe or Oceania. The worldwide numbers are most likely underestimated because of unrecognized cases and the failure of some countries to report their plague cases.

Since the beginning of the 1990s, a sharp increase in the number of human plague cases has been observed (Fig. 1). This increase is partly attributable to the reappearance of epidemics in countries or regions where human plague had disappeared for several decades. This was the case in the sea port of Mahajanga (Madagascar) in 1991, in Malawi, Zimbabwe, Mozambique, and India in 1994, and in Zambia and in some provinces of China in 1996. These features have led WHO to categorize plague as a reemerging disease. The reasons for the recent reemergence of this disease remain mostly unknown.

II. THE PATHOGEN

The agent of plague, *Y. pestis*, is a small gram-negative bacillus belonging to the genus *Yersinia*. This genus was created in 1944 by the division of the genus *Pasteurella* into three distinct genera: *Pasteurella sensu stricto*, *Francisella*, and *Yersinia*. The genus *Yersinia* is part of the family Enterobacteria-

TABLE I
Geographic Distribution of Human Plague Cases Reported to WHO during a 20-Year Period (1977–1996)

Africa		Asia		America	
Country	*Number of human cases*	*Country*	*Number of human cases*	*Country*	*Number of human cases*
Angola	27	China	366	Bolivia	322
Botswana	173	India	876	Brazil	722
Congo	2,829	Kazakhstan	10	Ecuador	83
Kenya	442	Laos	10	Peru	1848
Libya	19	Mongolia	64	United States	286
Madagascar	4,272	Myanmar	1995		
Malawi	9	Vietnam	5040		
Mozambique	325				
Uganda	660				
Tanzania	6,744				
Zambia	1				
Zimbabwe	397				
Total	15,898		8361		3261

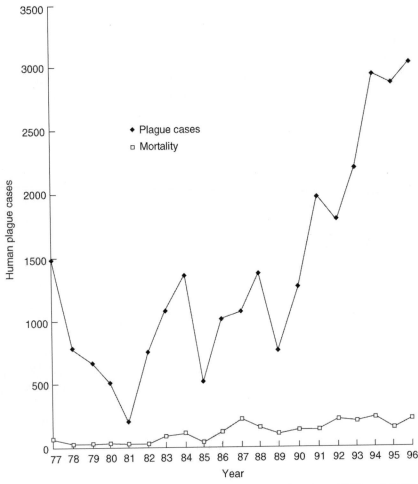

Fig. 1. Annual human plague cases reported to WHO between 1977 and 1996.

ceae but displays the following characteristics that differentiate it from the other members of this family: (i) a G+C content of 46–50%, (ii) an optimal growth temperature of 28°C, (iii) slow *in vitro* growth (colonies are clearly visible on agar plates after 48 hr), (iv) the ability to multiply at 4°C, and (v) its motility at 25°C but not at 37°C (except for *Y. pestis*, which is always nonmotile). The genus *Yersinia* is composed of 11 species: *Y. pestis, Y. pseudotuberculosis, Y. enterocolitica, Y. intermedia, Y. kristensenii, Y. frederiksenii, Y. mollaretii, Y. bercovieri, Y. aldovae, Y. rohdei,* and *Y. ruckeri*; only the former 3 are human pathogens. *Yersinia pseudotuberculosis* and *Y. enterocolitica* differ epidemiologically and clinically from *Y. pestis* because they are enteropathogenic bacteria

transmitted by the oral route. Nonetheless, *Y. pestis* and *Y. pseudotuberculosis* are genetically almost identical (>90% and almost 100% chromosomal and plasmid DNA relatedness, respectively) and it has been proposed that they be grouped into one single species, *Y. pseudotuberculosis,* composed of two pathovars: *Y. pestis* and *Y. pseudotuberculosis.* However, this proposal has not been adopted because of the potential hazard this would pose for clinical laboratories.

Yersinia pestis possesses the Kunin enterobacterial antigen and has a typical cell wall, but its lipopolysaccharide lacks the O-group side chain and is characterized as rough. *In vivo,* the bacterium is surrounded by a capsular-like structure called F1 antigen which is also synthesized *in vitro* upon incubation at 37°C. This facultative anaerobe displays auxotrophies for several amino acids and is inactive in most routine biochemical tests. *Yersinia pestis* is a phenotypically homogeneous species with only one serotype, one phage type, and three biotypes: Antiqua [ferment glycerol (G^+) and reduces nitrate to nitrite (N^+)], Medievalis (G^+, N^-), and Orientalis (G^-, N^+). Genetically, the species displays a higher degree of polymorphism and is divided into more than 20 ribotypes that correlate with the three biotypes. Most of the genetic tools used in *Escherichia coli* are suitable for *Y. pestis*.

III. EPIDEMIOLOGY

Plague is a zoonosis that primarily affects rodents, but several other animal species including cats, rabbits, camels, and humans can also be infected. More than 200 rodent species in 73 genera worldwide are susceptible to various degrees to the disease. Sporadically or periodically, explosive outbreaks occur among populations of susceptible rodents. The animal species constituting the natural plague reservoir vary greatly depending on the geographical location of the foci.

Plague is classically transmitted via the bites of fleas. Maintenance of plague in nature is absolutely dependent on cyclic transmission between fleas and mammals. More than 80 different species of fleas are implicated in maintaining the plague cycle. While feeding on an infected host at a bacteremic/septicemic stage, the ectoparasite ingests the bacilli present in the bloodstream. The bacteria multiply in the midgut of the insect, eventually forming a solid mass that blocks its proventriculus (blocked fleas). During repeated attempts to feed on a new host, the hungry blocked flea is unable to pump blood and subsequently regurgitates the bacteria into the bite wound.

Rodent-to-human plague transmission most commonly occurs by infected fleas associated with peridomestic animals (rats and cats) or wild rodents. Although rare, infection may also result from direct exposure (through conjunctive or cutaneous excoriations) to contaminated fluids or by inhalation of infected aerosols, for instance, during the manipulation of animal furs. Human-to-human transmission occurs either by means of the so-called human flea *Pulex irritans* or by inhalation of infected respiratory droplets spread by a patient suffering from pneumonic plague.

IV. CLINICAL DISEASE

A. General Symptoms

The two major clinical forms of plague are bubonic and pneumonic plague. In both forms, the incubation period usually varies from a few hours to 5 or 6 days, although longer times have been reported. In the majority of plague attacks, the onset of the disease is sudden and severe. It is characterized by a rapid temperature increase which reaches 39.5 to 40°C (103 or 104°F) in a few hours accompanied by an incessant quickening of the pulse and a decrease in blood pressure, indicative of progressive heart failure. Symptoms of alteration of the nervous system are very common and show individual variations characterized by delirium and restless agitation in some patients and apathy and stupor in others. Unless adequate specific treatment is administered, the condition of the patient deteriorates incessantly and death most commonly occurs within a period of 3–5 days.

B. Bubonic Plague

Bubonic plague, the most common form of the disease, is acquired after the bite of an infected flea. At the site of the bite, the bacteria multiply locally and sometimes cause a small vesicle that develops into a painful, dark, and necrotic carbuncle. Although inconstant, the presence of this skin lesion is a valuable orientation for the clinical diagnosis of plague. The bacilli then disseminate via the lymphatic vessels and reach the proximal draining lymph node where they multiply, causing an extremely tender tumefaction known as the bubo. In the best cases, the bubo spontaneously suppurates (ruptures and drains) and the patient recovers from the infection. Frequently, the bacterium disseminates via the lymphatic and blood vessels to the spleen, liver, and sometimes the lungs, causing a rapidly fatal septicemia. In the absence of adequate treatment, the mortality rate of bubonic plague ranges from 40 to 70%

C. Pneumonic Plague

Although less frequent, pneumonic plague is an even more severe form of the disease. Individuals are infected upon inhalation of infected droplets spread by a patient who had developed a lung infection either as a secondary complication of a bubonic form or after a primary pulmonary contamination. In addition to the general symptoms described previously, the patients exhibit signs of lung involvement such as cough, muco-purulent and bloody sputum, dyspnea, and pain in the chest. Without treatment, the outcome of pneumonic plague is invariably fatal, usually in less than 3 days.

D. Other Clinical Forms of Plague

Although relatively infrequent, septicemic, meningeal, hemorrhagic, gastro-intestinal, benign, or fulminant forms of plague can also be observed and may lead to a misdiagnosis of the disease.

V. TREATMENT AND PREVENTION

A. Vaccines

The first mass vaccination campaigns against plague were carried out simultaneously in 1934 by Girard and Robic (strain EV) in Madagascar and by Otten (strain Tjiwidej) in Java with live attenuated plague bacilli. These campaigns resulted in a significant decrease in human morbidity and mortality, but the adverse reactions and the brief protection conferred by these vaccines led to a gradual decline in their use. Formalin-killed plague organisms have also been used to induce immunity in humans, but the relatively low degree and short duration of protection along with the adverse effects conferred by these vaccines considerably limit their use. Today,

several acellular or cellular recombinant vaccines are under study, but they have not been tested in the natural conditions of endemic and epidemic plague foci.

B. Prophylactic Therapy

Prophylactic treatments are administered to close contacts of pneumonic plague patients or to relatives of bubonic plague patients living in the same house. The treatment consists in the administration of sulfonamide (in a single dose or repeated doses) or tetracycline.

C. Antibiotic Therapy

The use of antibiotics to treat plague started in 1938 with sulfonamide and in 1946 with streptomycin and led to a dramatic decrease in human mortality. With the current utilization of effective antibiotic therapy, the worldwide fatality rate attributable to plague has decreased to ca. 10%. Today, the three main antibiotics used to treat this disease are streptomycin, tetracycline, and chloramphenicol. The former is the drug of choice, whereas the latter is often restricted to meningeal forms. Administration of mono-, bi-, or tritherapies depends on the form of plague infection and on the severity of the symptoms. The chances of patients' survival are directly linked to the precocity of the antiobiotic therapy, especially in the more severe pneumonic and septicemic forms of the disease.

Until recently, the plague bacillus was considered to be universally susceptible to all antibiotics used to treat gram-negative infection. In 1995, the first multiresistant strain of *Y. pestis* was isolated in Madagascar from a patient with symptoms of bubonic plague. This resistance involved all the antibiotics recommended for therapy (chloramphenicol, streptomycin, and tetracycline) and prophylaxis (sulfonamide and tetracycline) of plague and also included antibiotics that may represent alternatives to classical therapy, such as ampicillin, kanamycin, spectinomycin, and minocycline. The resistance determinants were carried by a self-transferable plasmid that was probably acquired from a member of the Enterobacteriaceae family.

VI. DIAGNOSIS

A. Bacteriological Tests

Bacteriological identification of *Y. pestis* is carried out on biological samples such as bubo aspirates, blood, sputum, cerebrospinal fluids, and organs (liver, spleen, and lungs) following postmortem examination. Rapid presumption of plague infection can be obtained by Wayson staining or by a fluorescent antibody test, but confirmation of the diagnosis requires the isolation of the organism. *Yersinia pestis* is a bacterium sensitive to heat and drought and is easily overgrown by contaminant microorganisms. Therefore, if bacteriological analysis of the sample cannot be performed rapidly, storage of the sample at 4°C in the transport medium of Carry–Blair is recommended. *Yersinia pestis* grows well on most routine laboratory culture media incubated at 25–28°C for 2 days. Biochemical characters are studied on strains grown at 28°C whereas enzymatic tests are performed at 37°C. *Yersinia pestis* has few of the usual biochemical marker activities and can be mistaken for *Y. pseudotuberculosis*. Differential diagnosis between these two species relies on a few characters which are negative in *Y. pestis* and positive in *Y. pseudotuberculosis*: rhamnose fermentation after 24 hr, urease production, and motility at 25°C. However, some *Y. pestis* strains able to produce urease upon primary isolation have been reported. Lysis by a *Y. pestis* phage is widely used for rapid bacteriological orientation in the field, although this phage is not specific for *Y. pestis* and is lytic in some strains of *Y. pseudotuberculosis* and *E. coli*.

B. Animal Inoculation

The direct recovery of *Y. pestis* from suspected biological samples is often difficult, and the inoculation of infected materials into experimental animals such as guinea pigs or mice is a valuable complementary test to increase the chances of bacterial isolation.

Animals are infected subcutaneously and are observed until they become moribund. *Yersinia pestis* are recovered in high numbers from their spleen and/or blood.

C. Serological Tests

Several tests, all based on the F1 antigen, are used for the serodiagnosis of plague. The passive hemagglutination test is the simplest and cheapest technique, but it lacks sensitivity and specificity. Radioimmunoassay, although more sensitive, is not readily applicable in most routine laboratories. Enzyme-linked immunosorbent assay appears to be the simplest and most reliable serological technique to use for plague. Although isolation of the causative agent remains the only confirmatory test, serological techniques may be useful for retrospective diagnosis when a treatment has been administered to the patient without taking biological specimen or when no plague bacilli have been isolated from the specimen.

D. Other Diagnostic Techniques

Several other techniques of antigen detection by DNA hybridization, polymerase chain reaction, or antigen capture have been recently developed for diagnosing plague, but they are not widely and routinely used by clinical and field laboratories. They may prove useful when analyzing large numbers of samples or when only dead bacteria are present in the samples.

VII. GENETICS OF VIRULENCE

Based on the natural fatality rate of human plague and on the extremely low lethal doses during experimental infection of mice ($LD_{50} \approx 1$ bacterium by intravenous injection and ≤ 10 bacteria by subcutaneous injection), *Y. pestis* may be regarded as one of the most pathogenic bacterial species. However, the reasons for such a high degree of pathogenicity remain unclear and very few virulence factors have been identified in the plague bacillus until now. The genome of *Y. pestis* is composed of three natural plasmids and one chromosome.

A. Plasmids

Of the three plasmids harbored by *Y. pestis*, two (pFra and pPla) are specific for this species, whereas the third (pYV) is common to the three pathogenic species of *Yersinia* (*Y. pestis*, *Y. pseudotuberculosis*, and *Y. enterocolitica*). These three plasmids have recently been entirely sequenced.

1. pFra

pFra (101 kb) carries approximately 80 putative genes of which the best characterized are those coding for F1 antigen (F1 Ag) and murine toxin. F1 Ag, widely used in diagnostic tests, forms a capsule-like structure surrounding the bacteria grown *in vivo* or *in vitro* at 37°C. The locus coding for this structure (*caf*) is composed of four genes: *caf1* (capsular subunits), *caf1M* (chaperone), *caf1A* (capsule anchoring), and *caf1R* (regulation). Although the presence of F1 Ag has been associated with resistance to phagocytosis by monocytes, mutational inactivation of its structural gene does not affect the virulence of the mutant in different animal models, suggesting that F1 Ag is not a virulence factor.

The second best characterized locus, *ymt*, codes for a so-called murine toxin which may act as a β-adrenergic antagonist but whose role in human infection has not been clearly established. Recently, it has been shown that this factor promotes the colonization of the flea gut, thus playing a role in the plague cycle by enhancing flea transmission of the bacterial agent.

pFra also harbors several insertion sequences which may contribute to its spontaneous integration into and excision from the chromosome. Some reports showing that a *Y. pestis* deprived of the entire pFra keeps an intact virulence potential for mice suggest that this plasmid may be required for vector-borne plague transmission but not for virulence.

2. pPla

pPla (9.6 kb) carries the genes encoding a plasminogen activator (*pla*), the bacteriocin pesticin (*pst*), the pesticin immunity protein (*pim*), and one copy of the IS*100* insertion sequence. Pesticin does not play any role in virulence. The Pla protease cleaves the C3 component of the complement and degrades several secreted proteins (Yops) encoded by the pYV. However, its function *in vivo* remains unclear. Mutagenesis of *pla* or curing of pPla has a deleterious effect on the mouse virulence of only some isolates, suggesting that pPla may not be systematically required for virulence of *Y. pestis*.

3. pYV

pYV (70.5 kb) has been extensively studied in the three pathogenic species of *Yersinia*. This plasmid carries an integrated antihost system. A major part of pYV is occupied by a type III secretion/translocation apparatus involved in the delivery of effector proteins (Yops) into the cytosol of eukaryotic cells upon contact of the bacteria with the cells. At least six effector Yops are delivered into the eukaryotic cytosol: the YopE and YopT cytotoxins, the YopH tyrosine phosphatase, the YpkA (YopO) serine/threonine kinase, YopM (whose function is not completely resolved), and YopJ (YopP: which induces apoptosis of the host cell). The entire system is positively regulated at 37°C by the VirF activator and is directly or indirectly inhibited by the LcrQ (YscM) component of the secretion apparatus. Two genes encoding outer membrane proteins (the adhesin YadA and the lipoprotein YlpA) are interrupted in *Y. pestis* but are functional in *Y. pseudotuberculosis* and *Y. enterocolitica*. The role of pYV in virulence is critical because the absence of this plasmid results in a complete loss of virulence in the three pathogenic *Yersinia* species.

B. Chromosome

Sequencing of the entire chromosome of *Y. pestis* is under way at the Sanger center in Cambridge (UK) and should provide highly valuable tools for the understanding of *Y. pestis* virulence. Currently, the very few characterized *Y. pestis* chromosomal virulence loci (i.e., loci whose mutation reduces the pathogenicity of the mutant strain in an animal model) are the *psa* locus that codes for fimbriae and the high pathogenicity island. *hms* is another key chromosomal locus which is not essential for pathogenesis but is important for plague transmission.

1. Fimbriae

The *psa* locus encodes a surface structure originally known as the pH6 antigen because its synthesis was observed at acidic pH and at 37°C. The pH6 antigen is now known to correspond to fimbriae that have homologs in the two other pathogenic *Yersinia* species and in *E. coli*. The *psa* locus is composed of at least five genes (*psaEFABC*) that act in concert to assemble PsaA subunits of 15 kDa and form a fibrillar structure at the surface of the bacterium. This structure is synthesized *in vivo* in the liver and spleen of mice and plays a role in the virulence of *Y. pestis*. The exact function of the Psa fimbriae during the plague infectious process has not been elucidated.

2. High Pathogenicity Island

A pathogenicity island designated "high pathogenicity island" or HPI is present in the three pathogenic species of *Yersinia*. This 35- to 45-kb DNA sequence carries a locus involved in siderophore-mediated iron acquisition. The siderophore, called yersiniabactin, chelates the iron molecules bound to eukaryotic proteins (transferrin, lactoferrin, etc.) and transports them back into the bacteria, thus allowing their *in vivo* growth and dissemination. This locus comprises genes encoding (i) a large cytosolic protein complex involved in the nonribosomal biosynthesis of yersiniabactin (*ybt*), (ii) an outer membrane receptor (*psn*) and a transporter (*ybtP* and *ybtQ*) of the siderophore, and (iii) an activator (*ybtA*) of the system. In addition to the yersiniabactin locus, the HPI carries genes encoding one or several insertion elements and phage-like genes which are probable remnants of the acquisition of the island by horizontal transfer from a bacteriophage. The HPI is inserted in an asparagine tRNA locus (which contains a phage P4-like attachment site) and retains the ability to excise from the host chromosome (at least in *Y. pseudotuberculosis*), probably by site-specific excision mediated by a P4-like integrase encoded on the HPI.

3. Hemin Storage

The hemin storage locus (*hms*) is in the vicinity of the HPI on the *Y. pestis* chromosome. Both loci are located on an unstable 102-kb DNA segment (the *pgm* locus) that can spontaneously delete upon homologous recombination between two flanking IS*100* elements. The *hms* locus is composed of at least four genes (*hmsHFRS*) that enable *Y. pestis* to form red colonies at 25°C on Congo red agar plates *in vitro*. *In vivo,* this locus is not involved in virulence for mammals but it plays a key role in plague transmission by participating in the flea proventriculus blockage.

See Also the Following Articles

PLASMIDS, BACTERIAL • SURVEILLANCE OF INFECTIOUS DISEASES • ZOONOSES

Bibliography

Mollaret, H. H. (1972). La peste. *Encyclopéd. Médico-Chirurgicale* **8035**(10), 1–16.

Perry, R. D., and Fetherston, J. D. (1997). *Yersinia pestis*—Etiologic agent of plague. *Clin. Microbiol. Rev.* **10**, 35–66.

Pollitzer, R. (1954). "Plague," WHO Monograph Series No. 22. World Health Organization, Geneva.

World Health Organization (1998). Human plague in 1996. *Weekly Epidemiol. Records* **73**, 366–369.

Plant Disease Resistance: Natural Mechanisms and Engineered Resistance

Karl Maramorosch and Bradley I. Hillman

Rutgers University

I. Natural Resistance
II. Engineered Resistance

GLOSSARY

antisense Region of DNA or RNA sequence (usually short) that is complementary, is of opposite polarity, and hybridizes to a naturally functional (sense) target sequence, resulting in decreased function.

avirulence gene A gene in a plant pathogen that encodes a protein for which there is a corresponding dominant resistance gene in a plant whose expression results in an active defense response.

coat protein-mediated resistance Phenomenon whereby expression of a viral coat protein gene confers resistance to plant infection by the same or a closely related virus (with >60% amino acid homology in the coat protein).

cross-protection Conferring plant protection to severe virus infection by the use of a related mild strain.

epistatic Differential genetic effect on phenotype in which a gene suppresses the activity of another, nonallelic gene.

horizontal resistance Resistance conferred by the combined action of several genes (synonym: field resistance); can vary from low to high susceptibility.

hypersensitivity Condition in which plant cells are so highly susceptible that they are killed very soon upon infection so that the attacking pathogen is prevented from causing further damage.

hypovirulent An isolate of a pathogen with low virulence; the term is typically used to describe fungi that are reduced in virulence because of infection with a transmissible virus and may therefore be used for biological control of the pathogen.

immunity The ability to prevent infection, which results in no development of disease and no detectable level of pathogen multiplication, i.e., total resistance.

interference Process whereby the presence or expression of part or all of a pathogen's genome competes with or inhibits the normal replication cycle of an incoming pathogen.

pathogen-derived resistance Resistance in a plant that results from the natural presence or transgenic expression of all or part of the genome of a plant pathogen, for example, cross-protection or coat protein-mediated resistance.

plasmodesmata Cytoplasmic connections between plant cells.

resistance The ability to counteract the development of disease, following infection; can vary from low resistance (high susceptibility), when disease development is very extensive, to high resistance (low susceptibility), when disease development is very limited.

ribozyme Catalytic, self-cleaving RNA sequence, first described in Tetrahymena ribosomal RNA and in the satellite RNA of tobacco ringspot virus, which forms a "hammerhead" configuration and cleaves after GUC~ in *cis* or in *trans* without the need for any proteinaceous enzyme.

somaclonal variation Clones produced from single cells grown in tissue culture, called somaclones, and differing in resistance or other characteristics from parental plants.

tolerance Condition in which a host plant is tolerant when its response to infection by a pathogen is negligible in depressing the yield.

transgenic plant Stably transformed plant expressing a foreign gene of interest from an inserted DNA cassette containing a plant-active transcriptional promoter and terminator.

vertical resistance Resistance conferred by the action of a single gene exerting a major effect expressed as immunity or hypersensitivity.

viroids Smallest known plant pathogens, consisting of co-valently closed circular RNA molecules of low molecular weight that do not encode proteins but replicate using host plant enzymes.

PLANT DISEASE RESISTANCE refers to a plant's ability to counteract the development of disease, following infection, and can be the result of either natural or artificial, engineered mechanisms. The induction of resistance to diverse pathogens, vectors, and pests represents the most efficient and most desirable way to increase the productivity and quality of food plants that are essential to human survival. This article deals with resistance of plants to microorganisms (bacteria, fungi, spiroplasmas, and phytoplasmas), viruses, and viroids as well as resistance to vectors and pests.

I. NATURAL RESISTANCE

The high costs of chemical and biological pathogen and pest control of plant diseases can be reduced by better techniques through the use of resistant crop varieties. The development of resistant cultivars has been important in modern agricultural and horticultural practices.

Plant breeders and farmers have long recognized that disease and pest control achieved through host resistance is very desirable. It avoids undesirable ecological side effects of chemical applications and the cost involved with chemical control.

Resistance is the opposite of susceptibility. A resistant plant can be naturally extremely resistant, moderately resistant, or slightly resistant. In many instances, the protective mechanism is not well understood. The aim of plant breeders is the development of high-yielding crop varieties that are pest and disease resistant. Cooperative plant breeding programs seek the development of new plant varieties that are suitable for a given area with regard to yield, quality, adaptability, vigor, and pest and disease resistance. Natural and artificial selection over many centuries has resulted in highly heterogeneous plant populations with some degree of resistance in local environments. Although nonhost resistance is usually permanent, resistance of cultivated plants is seldom, if ever, permanent, because new strains of pathogens and pests evolve that threaten the newly developed varieties. Genetic diversity of the world's major crops is narrow due to inbreeding. Also, monoculture over large areas increases epidemic risks.

A. Resistance Mechanisms

Resistance involves different mechanisms that are not well understood but that can be classified as follows: (i) Resistance at the species level, or nonhost immunity, where all individuals of a species remain unaffected by a pathogen. Most plants are nonhosts for most pathogens: (ii) resistance at the cultivar level, where cultivars contain one or more genes that confer resistance against a pathogen to which cultivars of the species are susceptible: and (iii) resistance at the individual plant level, where an individual plant of a susceptible cultivar becomes resistant due to specific chemical, physical, or environmental treatment. If this resistance is acquired by genetic engineering, it becomes heritable.

B. Pathogenicity

Pathogenicity refers to the ability of a pathogen to cause disease, whereas virulence defines the relative capacity of a pathogen to cause disease.

Disease development in plants is a dynamic relationship that results from the interaction between resistance of the host and pathogenicity of the pathogen. There are two types of pathogenicity. The first type is conferred by the action of single genes, exerting a major effect. This type of pathogenicity is qualitative and is expressed as virulence. The second type of pathogenicity is conferred by the additive action of several genes that singly exert a small effect. It is quantitative in effect and expressed as aggressiveness of fitness.

Resistance begins when avirulence genes are matched with resistance genes. Regardless of the aggressiveness of a pathogen, single-gene resistance is epistatic.

C. Vertical Resistance

Vertical resistance protects a host plant against only specific strains of a pathogen. It is based on a single gene. Plant breeders describe vertical resistance based on major gene resistance as race-specific. This type of resistance is usually chosen by plant breeders because a single gene with a major effect is relatively easy to identify and manipulate and, if introduced into a plant, it can be expected to confer complete resistance. Major genes offer only short-term control in the case of epidemic plant disease, resulting in a recurring battle between the plant breeder and the pathogen. Various isogenic or single gene lines can be combined in so-called blended or synthetic varieties that differ in their resistance to a specific pathogen.

The "gene-for-gene" hypothesis is an oversimplification because it does not consider possible effects of modifier genes or additive effects of different resistance genes at different loci. It makes simplistic assumptions about the genetics of virulence because the gene's function is not solely the control of virulence. However, the customary gene-for-gene analysis helps to predict reactions of different strains of a pathogen or different plant hosts.

D. Horizontal Resistance

Horizontal resistance is defined as resistance for which there is no specific interaction with genetic variants of a pathogen. Horizontal resistance protects host plants to a greater or lesser degree against all genetic variants of a pathogen.

Horizontal, or multigene, resistance has been explained by physical or chemical barriers to pathogen entry and multiplication, inhibitors, or the lack of specific receptors. Cultivar resistance comprises very effective and less effective mechanisms. In virus diseases, acquired resistance can be attributed to a particular plant gene [e.g., the N gene against tobacco mosaic virus (TMV) in tobacco] or can be induced by prior infection with a related but mild virus strain (= cross-protection) or, in transgenic plants, by the coat protein (CP) genes of viruses. Probably the primary mechanism of resistance in both instances is inhibition of the early stages of uncoating of the viral

RNA by the virus CP present in the cells. Many plants contain proteinaceous toxins or inhibitors that affect the initial infection by viruses and other pathogens. Other chemical substances in plants may act as resistance inducers.

The mechanism of horizontal resistance is not well understood. To produce more stable resistance, breeders rely on the incorporation of several major genes into a plant variety. A variety with incorporated multiple major genes for resistance will not become susceptible to the pathogen as quickly as one with a single major gene because the probability that several mutations or recombinations occur simultaneously is remote. In practice, the incorporation of several major genes is only useful if these genes are at different loci.

The development of horizontal resistance is more difficult and time-consuming than the development of vertical resistance. Horizontal resistance theoretically provides a fairly stable resistance in preventing epidemics of plant diseases. Because the resistance is not absolute, it does not exert an extreme selection pressure on the pathogen population for the evolution of new races. Because it is polygenic, it is not subject to sudden breakdown. Success in preventing late blight of potato by horizontal resistance has been demonstrated during nearly three decades of field trials in Toluca, Mexico, where epiphytotics of late blight occur annually and, as a result of the presence of sexual stages of the pathogen, physiological races are prevalent. Horizontal resistance in potatoes in Mexico, where potatoes originated, has been found within existing cultivars. The total number of genes involved in horizontal resistance can be very large.

Horizontal resistance breeding methods have proven to be cost-effective and beneficial to both third-world and industrial countries. In recent years, it has been shown that breeding for horizontal resistance is not necessarily different from conventional breeding for resistance, except for the selection methodology, the choice of the challenging pathogen genotypes, and the evaluation of the results obtained. Breeding for durable horizontal resistance has been carried out successfully with potatoes resistant to late blight, coffee berry disease caused by *Glomerella cingulata,* coffee leaf rust caused by *Hemileia vastatrix,* and wheat diseases caused by *Helminthosporium sativum, Fusarium, Colletotrichum graminicola, Septoria nodorum,* powdery mildew, and barley yellow dwarf virus. Other plants in which horizontal resistance has been developed include chickpeas and broad beans. In the breeding programs in which horizontal resistance is to be developed, the breeder starts with susceptible parents and eliminates vertical resistance (i.e., one must ensure vertical susceptibility). Contrary to vertical resistance, horizontal resistance provides a suitable control measure for many important crops.

E. Hypersensitivity

A different type of single-gene (vertical) resistance is caused by the hypersensitive reaction, common in certain viral, fungal, and bacterial infections. The phenotype of a hypersensitive response may be minute, observable as the death of only one or a few cells, or it may be a larger localized necrotic lesion at the initial point of pathogen attack. Resistance mechanisms that prevent systemic pathogen invasion are usually gene dosage-dependent, incompletely dominant, and determined by quantitative interactions between host- and pathogen-specified functions. Only rarely does completely recessive resistance occur. Virulent virus isolates often overcome the resistance genes, and extreme durability of resistance is rare.

Resistance mechanisms that result in localization usually involve an inducible, positive inhibitor of pathogen spread or multiplication. Such inhibitors tend to be responsive to environmental cues such as temperature and humidity. Most positive interactions of pathogen avirulence (Avr) genes and plant R genes result in a hypersensitive response. Hypersensitivity and corresponding cell death may be the result of a rapid production and release of toxic compounds, such as reactive oxygen species, pathogenesis-related (PR) proteins, and phytoalexins, or due to induction of the normal cascade that results in programmed cell death (apoptosis).

F. Systemic Acquired Resistance

One of the consequences of a plant's active localized defense against pathogen attack is induction of the systemic acquired resistance (SAR) response. When induced, SAR results in resistance at distal portions of a plant that is somewhat pathogen nonspecific. Infection by many fungal, viral, and bacterial pathogens is reduced by SAR, but there is little or no protection against other members of the same pathogen groups. The mobile signal molecule that is thought to be critical to induction of SAR within a plant is salicylic acid. There is evidence that a derivative, methyl salicylate, may serve as a volatile signal capable of inducing SAR in nearby plants. The ecological significance of methyl salicylate as an airborne signal is unknown, but it could have implications for the study of disease in plant populations.

G. Qualitative (Vertical) and Quantitative (Horizontal) Resistance

Plant resistance to viruses has been characterized by qualitative and quantitative traits. Qualitative virus resistance is complete and conferred by hypersensitivity or extreme resistance to the spread of a virus. Quantitative resistance is partial and usually polygenically controlled. This type of resistance occurs more commonly than qualitative resistance, and it is more stable and thus preferred by plant breeders.

Quantitative resistance may be expressed as extended incubation periods of viruses, milder signs of disease, decreased accumulation of virus, few infected plants, nonsystemic virus invasion, and reduced economic losses compared to sensitive genotypes. Quantitative resistance to viruses may be modified gradually by the virulence of virus strains, dose of infecting virus, host resistance, time of infection, and environmental conditions, especially temperature and light intensity. Qualitative durability of resistance is important to plant breeders and farmers. Resistance may be effective for many years or it may be rendered ineffective within a short period. The loss of resistance occurs when a cultivar no longer possesses the capability to resist a new variant of the pathogen. Certain fungi (e.g., yellow rust caused by *Puccinia striiformis* and brown rust caused by *Puccinia recondita*) rapidly produce new races that overcome host resistance. Resistance to viruses is usually more durable, and resistance for periods exceeding

50 years has been known to occur in potatoes, sugar beet, tobacco, and other plants. The reason that resistance to viruses is so durable is not known because viruses mutate readily and new strains are commonly produced in nature. The durability is probably caused by combined polygenic and major gene-linked resistance. Breeders are trying to combine different types of resistance so as to obtain durable resistance.

H. Sources of Resistance

The first step in breeding for resistance is the identification of a source for resistance. Sometimes the genetic structure of an existing commercial cultivar is already known. If the resistance is controlled by a major gene(s), the development of a new cultivar is usually very rapid. Some resistance genes may protect a cultivar from several pathogens (e.g., from a fungal and viral pathogen). For example, in dwarf beans (*Phaseolus vulgaris*), the dominant gene I provides resistance to halo blight caused by the bacterium *Pseudomonas syringae* pv. *phaseolicola,* to anthracnose caused by the fungus *Colletotrichum lindemuthianum,* and to common bean mosaic virus; in lettuce, it provides resistance to mildew and lettuce mosaic virus due to the "Gallega" gene.

When no known source of resistance exists, breeding lines, primitive cultivars, commercial cultivars, and germ plasm banks must be tested or screened; if resistance is found and other characteristics are acceptable, the resistant cultivar can be introduced or used for breeding. Such resistant cultivars can be used as parents in a breeding program to produce a large number of commercially acceptable cultivars. If no resistant plants are found in a given species, wild relatives must be tested and used in the breeding program. This approach involves prolonged and extensive breeding, and problems of fertility of crosses between distantly related species of plants may be encountered. Tissue culture production of F1 plants from abnormal embryos may overcome these difficulties. Extensive backcrossing is usually required to obtain commercially desirable cultivars of the resistant plants.

Control of plant diseases through the use of resistant cultivars is the cheapest and most effective method of plant protection. Environmental pollution caused by the use of chemicals is avoided, as is the development of insect vector resistance to chemical insecticides. Several resistance genes have been successfully incorporated from certain cross-species (e.g., *Nicotiana genes*). However, producing a resistant variety may take a long time and the cost is sometimes high.

Germ plasm banks play an important role in current breeding programs for resistance. Such banks have been established in various countries to collect and preserve germ plasm of genetically diverse species planted in traditional agricultural systems. *In vitro* methods have been developed to provide a germ plasm and protect it against pest and disease attack. *In vitro* shipment of germ plasm to breeders reduces the risk of disseminating diseases and permits rapid and efficient distribution.

I. Plant Immunization

Plant resistance by immunization has not been used as widely as has immunization against animal and human diseases. The main reason for this is that, unlike in animals, "immunization" of plants as generally defined is not dependent on induction of an internal immune response with long-term memory for specific pathogens. Rather, it depends either on pathogen-derived resistance (PDR; either natural or transgenic) or on SAR.

The most common form of natural PDR involves cross-protection of plants with mild strains of a virus that may otherwise be a serious pathogen. The mechanism of natural cross-protection remains poorly understood, but studies of engineered virus PDR and defective plant virus RNAs are helping to elucidate these mechanisms. Widespread release of attenuated "mild" virus strains has been opposed for fear of back-mutations to more virulent forms of the virus. In the past few years, however, these objections have been muted in some countries. In The Netherlands, cross-protection has been used to control TMV losses in tomatoes, and a mild TMV strain has been commercially produced for several years. Cross-protection works when plants are "immunized" with the mild virus hours or a few days before exposure to the more severe strain. Large-scale cross-protection of papaya with mild strains of papaya ringspot virus and of citrus with citrus tristeza virus has also been applied successfully.

SAR is an immunity response in the sense that it can be induced by a pathogen that is not virulent on the specific plant in question, and the response has memory. In fact, SAR was originally termed "acquired physiological immunity" when described in 1963. SAR is a general rather than a specific immune response, however, and can also be induced by abiotic factors and several chemicals. Indeed, potential widespread field use of SAR as a general plant immunizing procedure in nontransgenic plants probably depends on its induction with chemicals. However, in China, a satellite RNA of cucumber mosaic virus (see Section II.B.1c) designated as the biological control agent S52 has been used successfully to induce resistance to certain fungal diseases in several crops. In tomatoes, strong resistance to *Phytophthora infestans* and to *Cladosporium fulvum* was induced. In cucumbers, S52 induced resistance to *Alternaria alternata*.

J. Somaclones

The enhanced genetic variability in plants regenerated from tissue culture comprises genetic changes affecting biochemical and morphological characteristics, some of which might confer resistance. Tissue culture of callus, single cells or protoplasts derived from a genetically stable parent plant may produce whole plants that are genetically different from the parent plant. Such clones are called somaclones, and their variation is called somaclonal variation. In several instances, somaclones proved resistant to pathogens to which the parent plant was susceptible. Resistance of potato somaclones to early blight of potato, caused by *Alternaria solani,* and to late blight, caused by *P. infestans,* has been achieved. Sugar-cane somaclones, produced from single-cell callus cultures, have been found to be resistant to mosaic virus and to Fiji disease virus. The mechanism responsible for somaclonal variation and the resulting resistance is not well understood, but it is possibly caused by chromosome changes induced in individual cells during tissue culture.

K. Tolerance

Tolerance is sometimes used incorrectly to describe plant responses that result from low levels of pathogen multiplication due to resistance. Tolerance means that when the plant is infected by a pathogen the disease symptoms are mild or at least commercially acceptable. Tolerance to certain viruses is heritable and used in breeding programs to control losses caused by citrus tristeza, barley yellow dwarf, cacao swollen shoot, curly top virus in beans, and other diseases. However, propagation of such tolerant plant varieties creates potential reservoirs of viruses that may present a threat to crops of a susceptible variety grown nearby.

L. Plant Resistance to Insect Vectors

The following terms are used to describe two basic types of plant virus transmission by vectors: (i) persistent transmission, which is characterized by the retention of the ability to transmit for prolonged periods, sometimes for the remainder of the vector's life, and (ii) nonpersistent transmission, which is characterized by immediate transmission from an infected to a healthy plant and rapid loss of the transmitting ability.

Three types of insect vector resistance have been recognized: nonpreference for a plant host, antibiosis, and tolerance. Nonpreference occurs when a vector lands on a plant, attempts to feed, and quickly departs. Plants possessing nonpreference traits have an advantage in controlling the spread of viruses carried in a persistent (permanent and replicative) manner by insect vectors, but they are at a disadvantage insofar as nonpersistently transmitted viruses are concerned. The latter are actually spread to more plants that are carrying the nonpreference trait because probing is more frequent on numerous plants. Antibiosis is defined as a condition in which the growth and multiplication of a vector are inhibited. By reducing the vector population, plants possessing the antibiosis trait diminish the spread of vector-borne viruses. Tolerance implies the ability of a cultivar to withstand the attack of an insect. This trait is important in controlling losses caused by insect pests, but it does not control virus spread.

Several aphid-transmitted viruses have been controlled successfully with insect vector resistance: peanut rosette transmitted by *Aphis craccivora,* cucumber mosaic virus (CMV) transmitted to melons by *Aphis gossypii,* and several viruses transmitted to raspberries by *Amphorophora rubi.*

Nonpreference has also reduced losses due to curly top virus transmitted by leafhoppers. Considerable progress in breeding rice plants resistant to leafhoppers has been achieved in recent years at the International Rice Research Institute in the Philippines. Nonpreference resistance introduced into rice plants resulted in dual benefits, preventing losses caused by leafhopper pests and losses caused by the rice tungro virus infection.

Plant resistance to pathogens can be combined with resistance to insect pests and vectors. Breeding plants resistant to pests and vectors involves not only knowledge of plant breeding techniques but also knowledge of insect physiology and behavior and insect genetics. Both vertical and horizontal resistance to pests and vectors have been developed by plant breeders. Field resistance to pathogens, vectors, or pests shown under natural field conditions by a host plant is of considerable practical importance.

Until recently, most conventional breeding targeted resistance to pathogens and, to a much lesser degree, resistance to vectors. With the advent of biotechnology and genetic engineering, plant resistance to pathogens and to vectors has been developed and exploited.

II. ENGINEERED RESISTANCE

Genetically engineered disease resistance involves specific introduction of foreign genetic material into the plant genome such that its expression, via RNA and/or protein, interferes with the normal infection and replication cycle of one or more plant pathogenic agents. Native plant genes associated with disease resistance, part(s) of the genetic material of the pathogen, or genes encoding specific antipathogen molecules have been used for these studies. To date, constitutive expression throughout the transformed plant has taken precedence; however, tissue-specific

or signal-induced promoters of gene expression have been used and may offer better protection of the plant. Furthermore, targeted gene expression may facilitate product licensing/marketing through reduced gene expression in the edible or processed parts of the plant. Considerable insight into the mechanism(s) of action of some genetically engineered crop protectants has been gained in the past few years, although others remain obscure. Their utility, however, continues to increase in the search to reduce the use of environmentally toxic agrochemicals.

It has been more than 10 years since coat protein-mediated TMV resistance was reported as the first successful example of engineered resistance to a plant pathogen. Since then, many other examples using different strategies and genes have been reported. In some cases, results have followed predicted or preliminary results as expected. Other cases have not been so straightforward. Some genes that were not predicted to have an effect on disease resistance did so, and other genes that were predicted to confer resistance did not.

A. Engineering Natural Resistance Genes

1. *Pathogenesis-Related Proteins*

The HR (horizontal resistance) acts against a wide variety of fungal, bacterial, and viral pathogens. It is associated with the rapid appearance of necrotic local lesions, restricted spread of the pathogen, and induction of many newly synthesized PR proteins, oligosaccharide cell wall fragments (elicitors), and antimicrobial compounds called phytoalexins. PR proteins have been classified into five groups based on size, sequence homology, biochemical characteristics, and enzymatic activity (Table I). Subgroupings have also been made on the basis of charge (acidic or basic forms). Induction of the HR (and PR proteins) results from specific recognition between the plant and the pathogen, often via single dominant genetic elements in both—an embodiment of the so-called gene-for-gene hypothesis.

Genes for most of the tobacco PR proteins have been cloned and sequenced, and several have been expressed constitutively in transgenic plants. The

TABLE I
Pathogenesis-Related Proteins in Tobacco

Group	Acidic Xanthi-nc	Acidic Samsun-NN	Basic Samsun-NN (kDa)	Function
1	1 a,b,c	1 a,b,c	19	Unknown
	b_1 b_2 b_3	16 kDa		
	16 kDa			
2	2, N, O	2, N, O	33	β-1,3-Glucanase
	b_4 b_5 b_{6b}	2 a,b,c		
	31–35 kDa	31–35, 40 kDa		
3	P,Q	P,Q	34	Chitinase
	b_{7b} b_{8b}	3 a,b		
	21–24 kDa	28 kDa		
4	R′	R	—	Unknown
	b_{9b} b_{9c}	4		
	15 kDa	15 kDa		
5	R	S	—	Protease inhibitor, thaumatin-like
	b_{9a}	5		
	23 kDa	23 kDa		

importance of PR proteins in the plant defense response is highlighted by the synergistic increase in resistance in plants co-expressing two PR proteins compared to plants expressing either PR protein alone. Salicylic acid, an endogenous plant hormone implicated as the signal molecule for induction of SAR, has also been implicated in PR protein synthesis and the HR, which are integral parts of SAR.

2. Single Resistance Genes

Numerous dominant plant genes that conform to the classical gene-for-gene hypothesis and confer resistance to a particular virus, bacterium, or fungus have been described genetically. Examples of natural plant genes conferring resistance to fungi include the *Rp1* complex (resistant to maize rust, *Puccinia sorghi*), *Pc-2* (resistance to crown rust in oats), and *mlo* (powdery mildew resistance in barley). Genes conferring resistance to races of *Pseudomonas* or *Xanthomonas* bacterial species, particularly important in the early characterization of Avr genes, have been identified. The *N* gene of tobacco, identified as conferring resistance to TMV in 1929, has practical application for TMV resistance in the field and has been of considerable importance in studies of pathogen interaction with plants. Some selected examples of plant-derived, pathogen-derived, and other genes that have been used to confer engineered resistance to plants are summarized in Table II.

In 1992, the first major gene for resistance to a plant pathogen was cloned and characterized. This was *Hm1*, which encodes the HC-toxin reductase responsible for detoxifying the *Helminthosporium* (*Cochliobolus*) *carbonum* pathogenicity factor, HC-toxin, and rendering it a non-pathogen. Such a gene, however, is not thought of as a classical resistance gene. The first classical R gene that conferred race-specific resistance to pathogens expressing particular Avr genes was the *Pto* gene of tomato, which confers resistance to *Pseudomonas syringae* pv *tomato*. *Pto* was cloned and characterized in 1993, and at least 12 other classical R genes have since been characterized, including genes that confer resistance to fungal, bacterial, and viral diseases. Putative functional domains of R genes are categorized into at least four classes: serine/threonine kinase domains, protein-binding domains denoted by leucine-rich repeats, nucleotide-binding domains, and leucine zipper domains, thought to be important in the role of transcription regulation.

TABLE II
Selected Examples of Single Genes Associated with Natural or Engineered Plant Disease Resistance

Gene	Source	Wild-type function	Protected against
Pathogen-derived genes			
TMV coat protein	TMV	Coat protein	TMV virus, many strains
PVX coat protein	PVX	Coat protein	PVX virus and RNA
TMV 54 K	TMV	Viral replicase	TMV, strain specific
TMV 30 K	TMV	Cell-to-cell movement	Several viruses
CMV satellite	CMV	Replicating satellite RNA	CMV, TAV
BYMV antisense	BYMV	CP and 3′ noncoding	BYMV
PSTVd ribozyme	PSTVd	Modified PSTVd	PSTVd
Ustilago maydis virus	*Ustilago maydis*	Killer toxin	*Ustilago maydis*
Plant-derived genes			
Hml	Maize	Toxin reductase	*Cochliobolus carbonum*
Pto	Tomato	Protein kinase	*Pseudomonas syringae* pv. *tomato*
Cf-9	Tomato	Unknown, leucine-rich repeat	*Cladosporium fulvum*
N	Tobacco	Unknown, leucine-rich repeat, nucleotide binding motif	TMV
PR proteins	Tobacco	Chitinase, glucanase	Several fungi
Other genes			
AMCV antibody	H + L chain	Recombinant anti-CP MAb	AMCV
Opsin proton pump	*Halobacterium halobium*	Proton pump	Several viruses, bacteria fungi
Pokeweed antiviral protein	Pokeweed	Unknown, ribosome inactivating protein	Several viruses, fungi
Barnase	*Bacillus amyloliquefasciens*	RNase	*Phytophthora infestans*

Note. Abbreviations used: TMV, tobacco mosaic virus; PVX, potato virus X; CMV, cucumber mosaic virus; TAV, tomato aspermy virus; BYMV, bean yellow mosaic virus; PSTVd, potato spindle tuber viroid; AMCV, artichoke mottled crinkle virus.

Resistance genes have typically been cloned by transposon tagging or by map-based cloning. Recently, polymerase chain reaction with degenerate oligonucleotide primers was used to identify candidate resistance genes from lettuce. If candidate genes identified by this method prove to be functional resistance genes, the speed with which resistance genes could be cloned from diverse plant species in the future could be increased considerably. Formal demonstration that a putative R gene is responsible for initiating the resistance response in a plant requires transgenic complementation in an otherwise susceptible plant that is capable of mounting a resistance response when induced.

Most of the previously discussed examples provide resistance to pathogen attack by allowing the plant to initiate a hypersensitive resistance response, with attendant induction of PR proteins and SAR. In addition to these strategies involving natural R genes, numerous other molecules have been examined for their potential in disease control. Many of these are pathogen-derived nucleic acids, which may or may not require translation to functional proteins in order to provide resistance in transgenic plants; others are nucleic acids foreign to a particular host–pathogen system.

B. Engineering Interfering Molecules

1. Nucleic Acids

a. Defective Interfering RNA or DNA
Natural or synthetically produced subfragments, reassortants, or chimeras of the genomes of plant

pathogens can interfere with the normal replication cycle in an otherwise susceptible host plant. This approach has shown the most potential for applicability among plant viruses and viroids, which rely entirely on the metabolism of the host cell for replication of their relatively simple genomes.

More than three-fourths of the nearly 700 known plant viruses contain positive-sense RNA genomes. For several of these, a "sense" approach to block virus replication *in trans* has been found to work both *in vitro* and *in vivo*. Thus, competitor RNA fragments of virus sense from the 3' end of the genome, or mutant viral components altered in the 3' sequence required for correct replication, can be shown to reduce overall virus multiplication. The most notable example of the sense approach to depression of RNA virus multiplication came from efforts to determine the mechanism of resistance in transgenic plants that were transformed with selected potyvirus coat protein constructs. It was found that even transformed DNA constructs expressing untranslatable RNA resulted in plant protection; indeed, untranslatable constructs led to much higher levels of plant protection.

During replication of some plant viruses, notably the RNA-containing tombusviruses and carmoviruses and the DNA-containing geminiviruses, defective interfering (DI) nucleic acids molecules arise spontaneously by complex deletion events during replication. As their name suggests, such nucleic acid species could form the basis for crop protection by interfering with virulent field isolates of related viruses. The potential for such protection has been demonstrated in the laboratory but not in the field.

Transgenic plants expressing defective beet curly top geminivirus DNA may be resistant to infection by closely related strains of the same virus. Whether resistance is the result of viral DNA, transcribed RNA, or protein has not been determined unambiguously. The strain specificity of geminiviral defective DNAs renders this strategy of questionable value on an agricultural scale. Similarly, chimeric DI RNAs of tombusviruses may be rescued during infection with the parent virus RNA. Unfortunately, protection against virus infection is dependent not only on viral strain relatedness but also on plant host. Several plants of agronomic importance that are susceptible to tombusvirus attack are not protected by the presence of DI RNA, and many others have yet to be examined.

b. Antisense/Ribozyme

By engineering a plant to produce the complementary (antisense) RNA strand of a particular gene native to the plant or an essential gene of a pathogen (such as a virus), the cytoplasmic level of sense RNA expression can be reduced dramatically. As a positive step toward down-regulation of gene expression, the antisense approach has proven both useful and attractive in plants. Thus, tomatoes with delayed ripening characteristics, or with increased solids content in the pulp, have been created by antisense interference with the endogenous messenger RNA (mRNA) for polygalacturonase. Nuclear gene inhibition may result from double-stranded RNA formation in the nucleus and the consequent effects on mRNA processing and nucleocytoplasmic transport. The main problems that must be overcome in targeting antisense RNA against a viral RNA are (i) that most plant viruses have no nuclear component to their replication cycle and, thus, the antisense molecule must be efficiently exported in a stable form to the cytoplasm and (ii) that pathogenic (sense) RNA molecules in the cytoplasm are in high copy number with most RNA viruses, so there is a need for a huge stoichiometric excess of the antisense species. Therefore, antisense-mediated transgenic resistance against some cytoplasmic RNA viruses, including TMV and tomato spotted wilt virus, has been reported, but many failures using this strategy also have been noted. These problems may be circumvented to some extent by selecting a low-copy-number nuclear target. Thus, the relatively rare replicase protein mRNA of tomato yellow leaf curl geminivirus was targeted, and resulting transgenic plants were resistant to superinfection.

To create a more inhibitory interfering molecule, a catalytic "ribozyme" RNA cleavage sequence can be sandwiched between two antisense RNA arms. The *cis*-autolytic RNA sequence from the plus strand of satellite tobacco ringspot virus RNA has been shown to function *in trans,* provided that the active part of the RNA structure (the ribozme) can be positioned precisely next to a susceptible GUC sequence in the RNA strand that is targeted for cleavage (Fig. 1). Natural satellite RNA replication involves a rolling circle template RNA strand, producing a repetitive head-to-tail linear satellite RNA structure. The latter has the remarkable ability to self-cleave spontaneously into precise monomeric units between 330 and 360 nucleotides without requiring any proteinaceous, biological catalyst (i.e., an enzyme). Only magnesium ions are required for this autolytic processing, which is thought to be one of the earliest reactions of nucleic acids that occurred before cellular life evolved. To cleave itself, the satellite RNA normally forms an unusually folded structure, described in two dimensions as a "hammerhead," to position the RNA cleavage site (substrate) next to the catalytic part (ribozyme). The minimal sequences required to achieve this activity are shown in Fig. 1. Provided the antisense RNA arms (zones C in Fig. 1) are not too long, the lower catalytic molecule (ribozyme) can melt from the upper (substrate) strand after cleavage.

Ribozymes have been used successfully to target virus or viroid sequences in several transgenic plant species, including CMV and potato spindle tuber

viroid (PSTVd). Results with PSTVd were particularly promising for effective disease control. Transgenic plants expressing ribozymes targeted to the minus sense strand of the viroid were highly resistant to superinfection with PSTVd, and resistance was stably inherited. As with the antisense strategy, targeting a subcellular pathogen with a nuclear component to its replication cycle may be the most promising approach for ribozyme deployment.

c. Satellite RNA

In addition to their useful self-cleaving properties, several plant viral satellite RNAs have been found to ameliorate symptom severity caused by infection with the helper virus. The mechanism whereby this occurs is not clear. It cannot be due simply to reduced viral replication through template competition, although this occurs in some virus satellite combinations. Unfortunately, not all satellite RNAs reduce viral severity; some exacerbate it. Like the DI RNAs discussed previously, symptomatology also depends on the specific plant–virus combination. This can lead to potential problems when the protection strategy involves large-scale propagation of a potentially harmful subpathogen. Consequently, the widespread use of satellite RNAs has not proved popular in many countries but has been used effectively in some. In China, protection of pepper plants against CMV symptoms has relied on the intentional release of satellite-attenuated strains of the virus. Transgenic tobacco plants expressing this same satellite RNA were found to resist CMV infection as well as the related tomato aspermy virus (TAV). In transgenic plants, the low constitutive level of expression of CMV satellite RNA was amplified dramatically through cytoplasmic RNA replication *in vivo* after CMV or TAV inoculation.

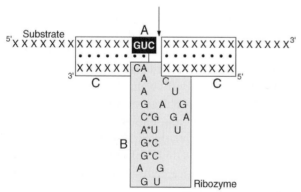

Fig. 1. Generic model for a *trans*-active ribozyme, or "gene shears," based on satellite tobacco ringspot viral sequences. Three structural domains (A–C) are boxed. A, minimum conserved sequence for the RNA substrate next to the cleavage site (arrow); B, highly conserved sequence for the catalytic module ("ribozyme"); C, flanking antisense complementary RNA sequences, forming base-paired double helices to juxtapose domains A and B (reprinted with permission from Haseloff and Gerbach, 1988).

2. Proteins
a. Coat Protein
Although classical cross-protection of crops against a severe viral pathogen by preinfection with a mild strain of a closely related virus could be due to interference at all possible levels of virus replication and spread in the plant, in many cases the dominant feature appeared to be the interaction between

the homologous CPs of the protecting and challenge viruses. Thus, the idea was born that expression of only the viral CP in a transgenic plant would protect that plant against challenge inoculation with a whole virus.

In 1986, the first report of successful CP-mediated protection appeared using the tobacco–TMV system. Since then, numerous RNA virus–CP transgenic plant systems have been shown to be protected. In most cases, challenge inoculation with unencapsidated viral RNA overcomes some protection, and the higher the level of CP expressed *in planta,* the higher the level of protection (as measured by delay in symptom expression or reduced replication or spread of virus). Typically, CP expressed at 0.05–0.1% (w/w) of the total soluble leaf protein confers protection. The spectrum of viruses to which a particular CP+ plant is resistant is relatively narrow, and virus concentrations >1 μg/ml begin to overcome the protection. As long as the challenge virus CP is >60% homologous at the amino acid sequence level, measurable protection generally occurs. The more unrelated the virus (CP), the less resistance is seen. Demonstration of effective protection may be done by mechanical inoculation with whole virus or RNA or vector delivery of virus. Transgenic CP mediated resistance has successfully been applied to more than 40 plant–virus combinations, many of which have been field tested extensively.

Less success has been achieved with CP-mediated resistance against DNA-containing plant viruses. Although there is a report of resistance against TYLCV (tomato yellow leaf curl virus), other attempts to protect against single or double-stranded DNA viruses have been unsuccessful.

As with many other genes, expression of a viral CP gene *in planta* apparently does not adversely affect the agronomically beneficial traits of the crop (e.g., fruit yield) and, thus, represents a valuable positive addition of a single new resistance gene without lengthy breeding programs. Simultaneous expression of two or more viral CP genes has been shown to provide multiple virus protection in several crop species.

Extensive studies on the mechanism(s) of CP-mediated resistance with RNA viruses have shown that challenge virus disassembly is the first step to be inhibited substantially but that later steps in virus replication and long-distance spread are also affected. There may also be a sense-suppression mechanism functioning at the RNA level to protect plants, as discussed previously.

b. Nonstructural Proteins

Transgenic expression of plant viral proteins other than CPs (i.e., nonstructural proteins) often provides protection against subsequent virus inoculation. Replicase genes or portions of those genes have been used successfully to protect transgenic plants against many RNA viruses. Perhaps the most celebrated of these was the identification of strain-specific protection of tobacco against TMV infection by transgenic expression of the C-terminal 54 kDa of the 183 kDa replicase gene. Several mechanisms may account for replicase-mediated resistance, but in most cases resistance is RNA mediated and homology dependent.

c. Movement Proteins

In many instances, transgenic plants that express viral movement proteins (MPs) can complement movement-deficient mutants defective in cell-to-cell spread. Plant viral MPs line the lumen of the plasmodesmata and relax the so-called gating limit (permeability) of the connecting bridges between plant cells known as plasmodesmata (singular, plasmodesma), thereby facilitating intercellular communication. MPs also act as single-stranded nucleic acid-binding proteins that transport viral RNA intracellularly and from cell to cell. There have been numerous reports of interviral complementation and spread attributed to the MP of a helper virus that functions in a particular host plant. The strategy of engineering into a crop a defective viral MP gene that saturates the plasmodesmal receptor sites and thereby competes for and inhibits viral RNA spread has been used successfully. The fact that MPs may be similar among plant viruses whose genomes are otherwise only distantly related led to the prediction that such MP-mediated resistance may be relatively broad spectrum against viruses with related MPs. This has turned out to be the case, with transgenic plants expressing dysfunctional copies of the TMV MP resistant against infection by cucumoviruses, tobraviruses, and potexviruses in addition to tobamoviruses.

3. Whole Viral/Viroid Genomes

Transgenic plants expressing a complete viral or viroid RNA genome can provide good protection against severe isolates of the same pathogen. This approach clearly requires a cloned, mild or symptomless strain of the virus or viroid and the interference could operate at any of several levels, in much the same way that classical cross-protection operates. This transgenic approach is less favored than those described earlier due to the obvious risk of mutations increasing the severity of the "protecting" RNA and because it increases the total quantity of pathogen in the crop.

C. Novel Strategies

1. Bacillus thuringensis Toxin

Transgenic tomato, tobacco, and cotton plants expressing (at 0.05–0.1% w/w of the total soluble protein) an insecticidal toxin gene from *Bacillus thuringensis* (*Bt*) have been shown to be protected selectively against herbivorous larvae of lepidopteran, coleopteran, or dipteran pests. Different *Bt* toxins have different spectra of activity. The protein disrupts the intestinal brush border membranes in susceptible insect larvae, leading to death. It is harmless to animals, and crystals of the toxin have been applied as sprays to crops for many years. The toxin has a short lifetime in the soil, being degraded by other bacteria. In field tests, the level of insect control has been excellent. In some transgenic crop plants, however, the specter of *Bt*-resistant insect populations has become apparent.

2. Mammalian Antibodies

In 1989, the first report of successful transformation of tobacco plants with complementary DNAs (cDNAs) to mouse immunoglobulin heavy (H) or light (L) chains appeared. More remarkable was the demonstration that by cross-pollinating these transgenic plants, some of the progeny expressed both H and L chains, assembled functional mouse monoclonal antibody, and exported it from the plant cell in the same way that mouse immune system cells handle these proteins.

Since then, it has become relatively straightforward to create transgenic plants that express a functional mouse monoclonal antibody to any specified epitope on a pathogen simply by taking H- and L-mRNAs from a particular hybridoma cell culture, using polymerase chain reaction primers to create cDNAs, and transforming plants with these constructs. Regenerated plants that express antibodies to plant viral CP epitopes have been created by several groups. The most success achieved using this strategy was with transgenic plants expressing antibodies directed against artichoke mottled crinkle tombusvirus. Recently, production of plants resistant to stolbur phytoplasma with transgenic plants expressing antibodies directed against a membrane protein of the phytoplasma was reported. The transgenic antibody approach is relatively labor-intensive and the chance of success seems relatively low, considering the number of systems on which the method has been attempted. Still, it may be a useful approach when applied to pathogens that are otherwise recalcitrant, such as phytoplasmas.

3. Fungal Killer Systems

Some haploid isolates of *Ustilago maydis,* which causes boil smut of maize, produce low-molecular-weight polypeptide toxins that inhibit the growth of sensitive strains of *U. maydis* and other cereal smuts. Three different killer toxins have been described. All are encoded by the double-stranded RNA genomes of a group of multicomponent spherical viruses found in most, if not all, field isolates of *U. maydis*. Resistance to the toxins is also encoded by the dsRNA genomes. It has been demonstrated that transgenically expressed killer toxin is successfully exported by the plant cell and is functional, providing resistance against the smut fungus. The utility of this approach has yet to be shown in field trials.

4. Ribosome Inactivating Proteins

Pokeweed antiviral protein (PAP) is one of several plant-derived proteins determined to have antiviral properties. These antiviral proteins are ribosome inactivating proteins, removing a single exposed adenine from the large rRNA of eukaryotic and prokaryotic ribosomes and thus rendering them nonfunctional. Although their unregulated expression may be toxic to plants, the toxic region of these proteins has been separated from the region that confers antiviral activity, allowing for production of transgenic plants that expressed antiviral activity. Surprisingly, these transgenic plants expressing modified PAP were resistant to infection by many fungi. Antifungal activity appears to be due to a mechanism that is different from that of the antiviral activity. Whereas antiviral activity could be demonstrated in *in vitro* assays, antifungal activity could not, requiring expression in whole plants. Thus, the antifungal effect may be primarily a consequence of SAR induction in the plant.

5. Antifungal/Antibacterial Enzymes and Peptide Antibiotics

Chitinase has the capacity to degrade fungal cell walls selectively. As mentioned earlier, some PR proteins have chitinase activity, and transgenic plants expressing high levels of this wall-degrading enzyme have shown some fungus resistance.

Other naturally occurring antimicrobial peptides, including those from amphibian skin (the magainins, 23 amino acids; xenopsin, 8 amino acids; and caerulein, 10 residues), from honey-bee venom (mellitin, 26 amino acids), and from insects (the cecropins, attacins, and lysozyme) have been or could be engineered into plants to provide a protective function against bacterial infections, preferably in ornamental or nonedible parts of the plant. Several synthetic antifungal peptide antibiotics have been engineered into plants as well. Because of potential toxicity problems, these approaches are more promising for ornamentals than for crop plants.

6. Proteinase Inhibitors

Inhibitors of serine proteinases or endopeptidases (serpins) are classified into 13 unique gene families, although all act as competitive inhibitors. Plant tissues contain comparatively high levels of serpins in their seeds and tubers (1–10% of total protein) and even 50% in some fruits. Serine proteinase inhibitors protect against foreign proteolytic enzymes such as those of herbivorous insects, fungi, and some viruses.

See Also the Following Articles

Agrobacterium and Plant Cell Transformation • Antisense RNAs • Genetically Modified Organisms: Guidelines and Regulations for Research • Plant Virology, Overview

Bibliography

Baulcombe, D. C. (1996). Mechanisms of pathogen-derived resistance to viruses in transgenic plants. *Plant Cell* 8, 1833–1844.

Beachy, R. N., Loesch-Fries, S., and Tumer, N. E. (1990). Coat protein-mediated resistance against virus infection. *Annu. Rev. Phytopathol.* 28, 451–474.

Bent, A. F. (1996). Plant disease resistance genes: Function meets structure. *Plant Cell* 8, 1757–1771.

Hammond-Kosack, K. E., and Jones, J. D. G. (1996). Resistance gene-dependent plant defense responses. *Plant Cell* 8, 1773–1791.

Haseloff, J., and Gerlach, W. L. (1988). Simple RNA enzymes and highly specific ribonuclease activities. *Nature* 334, 585–591.

Lomonossoff, G. P. (1995). Pathogen-derived resistance to plant viruses. *Annu. Rev. Phytopathol.* 33, 323–343.

Maramorosch, K. (1993). Plant resistance to mycoplasma diseases. *In* "Management of Plant Diseases Caused by Fastidious Prokaryotes" (S. P. Raychaudhuri and D. S. Teakle, Eds.), pp. 69–72. Associated Publishing, New Delhi.

Maramorosch, K., and Wilson, T. M. A. (1992). Plant disease resistance: natural mechanisms and engineered resistance. *In* "Encyclopedia of Microbiology" (J. Lederberg, Ed.), Vol. 3, pp. 417–429. Academic Press, New York.

Palukaitis, P., and Zaitlin, M. (1997). Replicase mediated resistance to plant virus disease. *Adv. Virus Res.* 48, 349–377.

Saxena, R. C. (1989). Durable resistance to insect pests in irrigated rice. *In* "Proceedings of the International Rice Research Conference," Hangzhou, China.

Tien, P., and Wu, G. (1991). Satellite RNA for the biocontrol of plant disease. *Adv. Virus Res.* 39, 321–339.

Tricoli, D. M., Carney, K. J., Russel, P. F., McMaster, R. J., Groff, D. W., Hadden, K. C., Himmel, P. T., Hubbard, J. P., Boeshore, M. L., and Quemada, H. D. (1995). Field evaluation of transgenic squash containing single or multiple virus coat protein gene constructs for resistance to cucumber mosaic virus, watermelon mosaic virus 2, and zucchini yellow mosaic virus. *Bio/Technology* 13, 1458–1465.

Tumer, N., Rong, D., Hudak, K., Wang, P., Coetzer, C., and Zoubenka, O. (1998). Pokeweed antiviral protein and its applications. *In* "Current Topics in Microbiology and Biochemistry. Plant Biotechnology: New Products and Applications." Springer-Verlag, New York.

Plant Pathogens

George N. Agrios

University of Florida

GLOSSARY

disease cycle The chain of events involved in disease development, including the stages of development of the pathogen and the effects of the disease on the host.

facultative Having the ability to act in a certain way.

haustoria Feeding organs of some fungi and parasitic higher plants that absorb water and nutrients from host cells.

life cycle The stage or successive stages in the growth and development of an organism that occur between the appearance and reappearance of the same stage (e.g., spore) of the organism.

parenchyma Plant tissue composed of thin-walled cells which synthesize or store foodstuffs and usually leave intercellular spaces between them.

pathogen A living organism that can cause disease on certain other living organisms (hosts).

phloem Tube-like cells of the plant conductive system that carry sugars and other organic substances from leaves to other parts of the plant.

vector A specific insect, mite, nematode, or fungus that can acquire and transmit a specific pathogen from an infected to a healthy host.

virulence Relative ability of a pathogen to cause disease on a given host plant.

xylem Tube-like cells of the plant conductive tissue that carry water and minerals from the roots to other parts of the plant.

PLANTS, like humans and animals, are affected by diseases (Fig. 1). In all organisms, diseases are caused by internal or external abiotic factors, such as nutritional or environmental conditions; by quasi-biotic factors, such as genetic abnormalities; or by pathogens. Pathogens are living entities, mostly microorganisms, such as certain fungi, prokaryotes such as bacteria and mollicutes, viruses, protozoa, nematodes, etc. that can attack other organisms and cause disease. In addition to these pathogens, diseases in plants can also be caused by several plants that parasitize other plants and by some green algae. With the exception of the few diseases caused by parasitic plants and green algae, however, the vast majority of diseases in plants are caused by the same groups of pathogenic microorganisms as those that cause diseases in animals and humans. Plant pathogens vary considerably in size (Fig. 2), shape (Figs. 2–4), and multiplication (Fig. 5). Like all pathogens, those affecting plants vary considerably in their host specificity; some are able to infect all or most plants belonging to one species, a genus, or several plant families, whereas others can infect only one or a few varieties within one plant species.

Plant pathogens cause disease in plants by disturbing the metabolism of plant cells and tissues through enzymes, toxins, growth regulators, and other substances they secrete and by absorbing foodstuffs from the host cells for their own use. Some pathogens may also cause disease by growing and multiplying in the xylem and phloem vessels of plants, thereby blocking the upward transport of water and minerals and the downward translocation of sugars. Infected plants develop a variety of symptoms (Fig. 1) that may vary in severity, ranging from insignificant to death of the entire plant. It is estimated that plant pathogens are responsible for a 16% loss (approximately $190 billion) of the attainable annual world crop production estimated at $1.2–1.3 trillion.

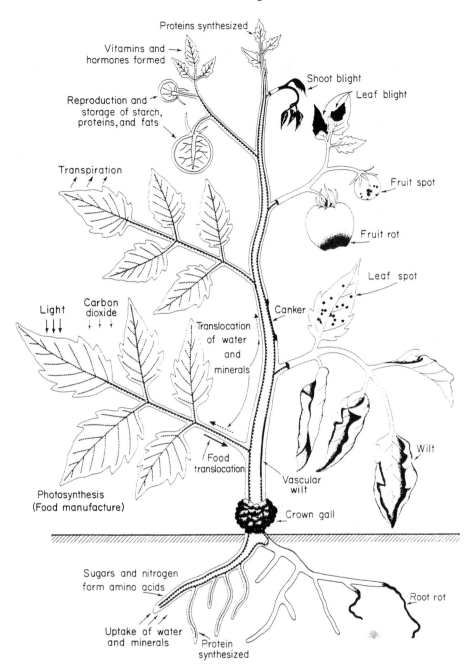

Fig. 1. Schematic representation of the basic functions in a plant (left) and the interference in these functions (right) caused by some common types of plant diseases (from Agrios, 1997, p. 5).

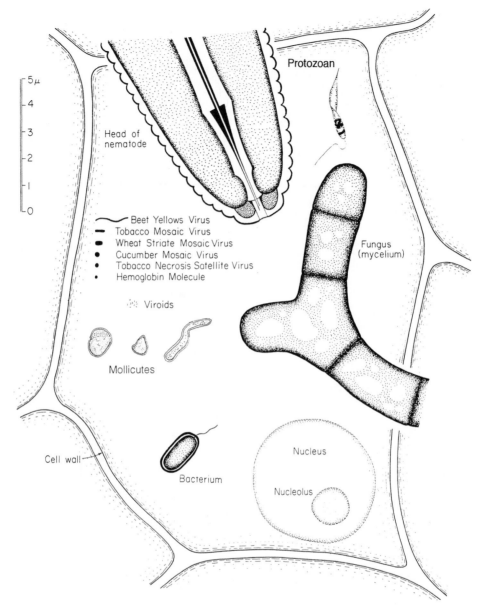

Fig. 2. Schematic diagram of the shapes and sizes of certain plant pathogens in relation to a plant cell (from Agrios, 1997, p. 6).

I. TYPES AND GENERAL CHARACTERISTICS OF PLANT PATHOGENS

A. Fungi

Most pathogens of plants are fungi. They cause the majority (approximately 70%) of all plant diseases.

More than 10,000 species of fungi can cause disease in plants. Some of the plant pathogenic fungi are obligate parasites (biotrophs) because they can grow and multiply only by remaining in constant association with their living host plants. Others are nonobligate parasites and either require a living host plant for part of their life cycles but are able to complete their cycles on dead organic matter or they are able

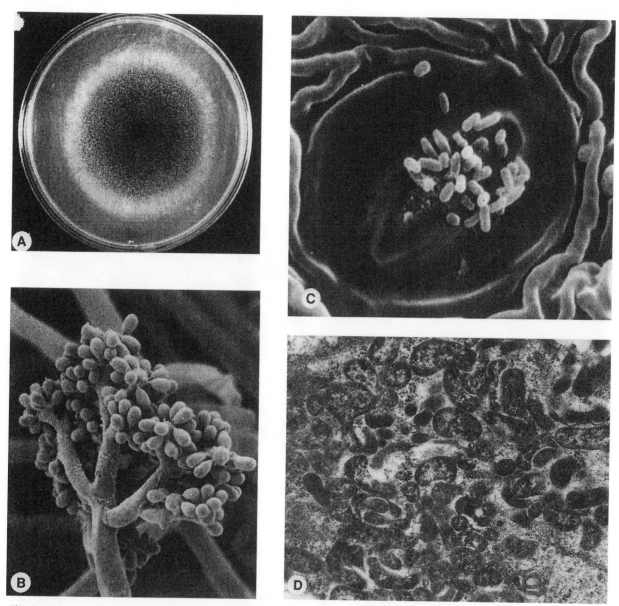

Fig. 3. Three types of pathogenic microorganisms that cause plant diseases. (A) Fungus growing out of a piece of infected plant tissue placed in the center of a culture plate containing nutrient medium. (B) Mycelium and spores of a plant pathogenic fungus (*Botrytis* sp.) (600×) (C) Bacteria at a stoma of a plant leaf (2500×); (D) Mollicutes in a phloem cell of a plant (5000×); (photo B courtesy M. F. Brown and H. G. Brotzman; photo C courtesy L. Mansvelt, I. M. M. Roos, and M. J. Hattingh; photo D courtesy J. W. Worley) (from Agrios, 1997, p. 8).

to grow and multiply on dead organic matter and on living plants.

1. Morphology

Most fungi have a branching filamentous body called a mycelium (Figs. 2, 3, and 5). Mycelium produces numerous branches that grow outward in a radial fashion and produce a colony (Fig. 3). Each branch of the mycelium is called a hypha. Hyphae are tubular and generally of uniform thickness (1–5 μm in diameter). In some fungi the mycelium is a more or less continuous tube containing many nuclei; in

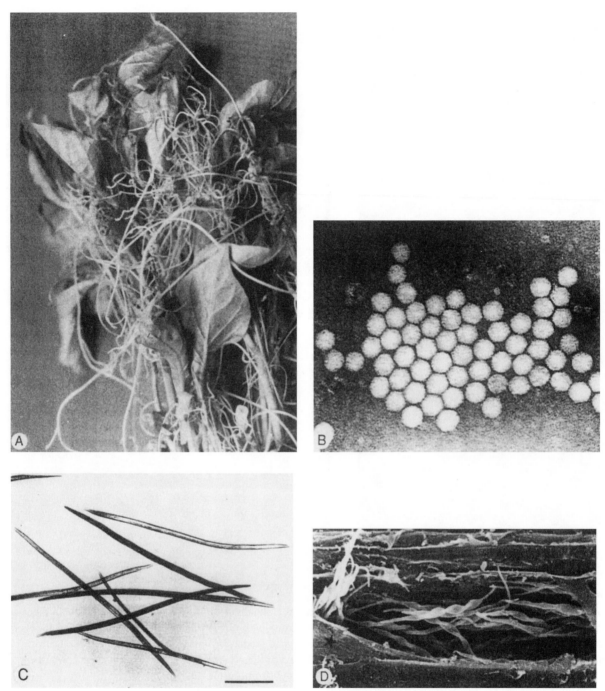

Fig. 4. Four types of pathogens that cause plant diseases. (A) The thread-like parasitic higher plant dodder (*Cuscuta* sp.) parasitizing pepper seedlings. (B) Tobacco ringspot virus isolated from infected tobacco plants (200,000×). (C) Plant parasitic nematodes (*Ditylenchus* sp.) isolated from infected onion bulbs (80×). (D) Protozoa (*Phytomonas* sp.) in a phloem cell of a palm root (4000×) (photo A courtesy G. W. Simone; photo C by N. Greco, supplied courtesy R. Inserra; photo D courtesy W. de Sousa) (from Agrios, 1997, p.9).

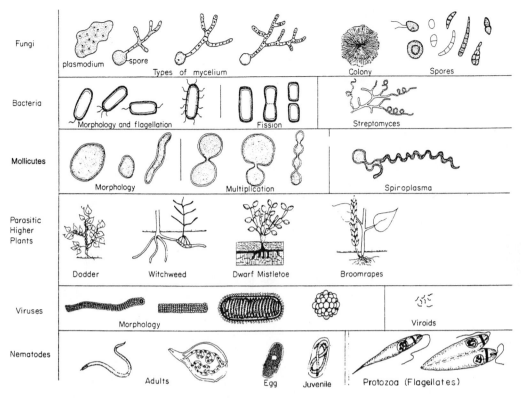

Fig. 5. Morphology and multiplication of some of the groups of plant pathogens (from Agrios, 1997, p. 7).

others, the mycelium is partitioned into cells by cross-walls, with each cell containing one or two nuclei. The length of the mycelium in some fungi is only a few millimeters, but in others it may be several to many centimeters long. In some lower fungi, now thought to be protozoa, the body consists of a mass of wall-less protoplasm containing numerous nuclei, and it is called a plasmodium.

2. Reproduction

Fungi reproduce primarily by means of spores (Fig. 6) which may consist of one or a few cells. Spores may be formed asexually, like buds on a twig, or sexually as the result of a sexual fertilization. Asexual spores in some fungi are produced inside sacs called sporangia and are referred to as sporangiospores. In some of these fungi, the sporangiospores have flagella with which they can swim and are therefore called zoosporangiospores or zoospores. In most fungi the asexual spores are called conidia and are produced by the cutting off of terminal or lateral cells from special hyphae called conidiophores. Conidiophores and conidia in some fungi are produced naked on the mycelium, whereas in others they are produced inside thick-walled containers called pycnidia. Sexual reproduction occurs in most but apparently not all fungi. Generally, two cells of similar or dissimilar size and appearance fuse to produce a zygote that, depending on the group of fungi, is produced and undergoes meiosis inside a germinating zygospore (in zygomycetes), oospore (in oomycetes), ascus (in ascomycetes), or basidium (in basidiomycetes). In some fungi, the deuteromycetes or imperfect fungi, no sexual spores have been found. Some fungi seem to never produce any kind of spores and are known as sterile fungi. Generalized life cycles of the groups of fungal plant pathogens are shown in Fig. 7.

3. Classification

The "fungal" pathogens of plants include some microorganisms, referred to as fungal-like organisms

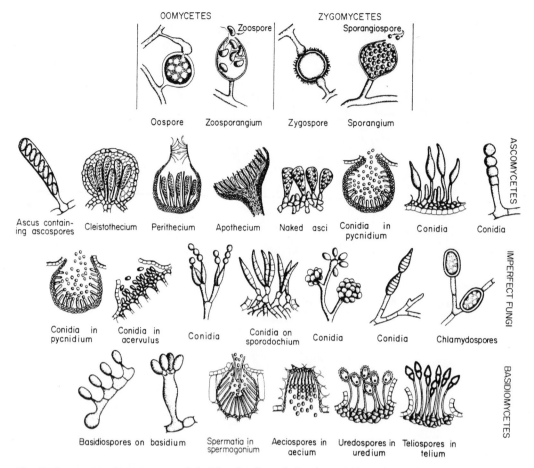

Fig. 6. Representative spores and fruiting bodies of the fungal-like oomycetes and of the main groups of fungi (from Agrios, 1997, p. 247).

or pseudofungi, that were recently shown to belong to the kingdoms of Protozoa or Chromista rather than that of Fungi. Thus, the following is a sketchy classification of fungal pathogens of plants:

Fungal-like Organisms or Pseudofungi
  Kingdom: Protozoa
    Phylum: Myxomycota—produce a plasmo-dium; surface slime molds
    Phylum: Plasmodiophoromycota—endoparasitic slime mold diseases
  Kingdom: Chromista
    Phylum: Oomycota—produce mycelium that has no cross-walls; produce oospores and zoospores; cause many root rots, seedling diseases, foliar blights, and the downy mildews

The True Fungi
  Kingdom: Fungi
    Phylum: Chytridiomycota—have round or elongated mycelium; zoospores
    Phylum: Zygomycota—No zoospores; produce conidia in sporangia; cause bread molds and fruit rots in storage; some are mycorrhizal fungi
    Phylum: Ascomycota—have mycelium that has cross-walls; produce sexual spores (ascospores) within sacs (asci) that are either naked or contained in fruiting structures of different shapes (cleistothecia, perithecia, and apothecia); produce asexual spores (conidia) on naked hyphae or in containers (pycnidia) or other structures; cause most plant diseases (leaf, stem, and fruit spots

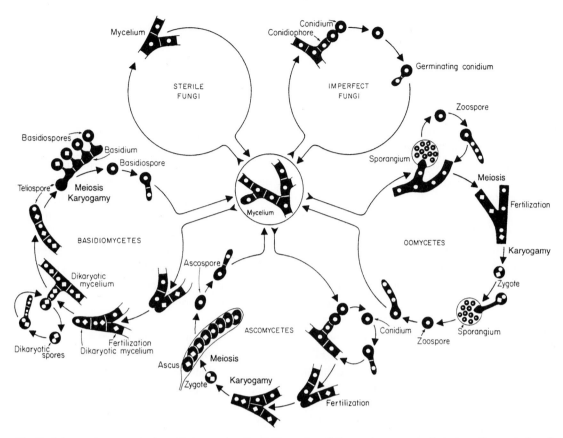

Fig. 7. Schematic presentation of the generalized life cycles of the main groups of phytopathogenic fungi (from Agrios, 1997, p. 259).

and blights, root rots, fruit rots, cankers, vascular wilts, seed rots, etc.)

This group of fungi also includes the imperfect fungi or deuteromycetes, which are similar to Ascomycota in all respects except that they either do not produce sexual spores (ascospores) or such spores are rarely produced and have not been found in these fungi.

Phylum: Basidiomycota—have mycelium, often with binucleate cells; sexual spores (basidiospores) produced externally on a club-like structure called a basidium; some of them produce several types of spores and spore-bearing structures; cause the serious rust diseases of grain and other crops and trees, the smuts of grain crops, and the

wood and root rots and decays of trees and timber.

4. Ecology

Plant pathogenic fungi spend all or part of their lives on their host plants and part in the soil or in plant debris on the soil. During the parasitic phase, some fungi, such as the powdery mildews, grow outside the plant surface but obtain their food from their host plants by sending feeding organs, called haustoria, into epidermal plant cells. Other fungi enter plants through wounds, natural openings such as stomata, or by direct penetration of the cuticle and epidermal cells. Once inside, fungi invade smaller or larger areas of the plant by producing mycelium primarily in various parenchyma tissues either intercellularly (between cells) or intracellularly (through the cells). Some fungi, as soon as they enter the plant, find and grow into the xylem vessels of plants, resulting in wilting of the infected plants, and remain in the vessels until the plants die, at which point they invade other tissues of the plant and sporulate at its surface.

5. Dissemination

Spores of fungal plant pathogens produced on a plant are disseminated to other plants by wind, light air currents, water, wind-blown rain, insects and other animals, etc.

6. Infection

Once in contact with a plant, and provided that there is enough moisture and the temperature is favorable, spores germinate and secrete a number of enzymes and other compounds with which they attack and enter the host. In the process of penetration and invasion, germinating spores and the resulting mycelium may secrete cutinases, pectinases, cellulases, ligninases, and other enzymes. These break down the plant cuticle and cell wall substances into smaller molecules which not only lead to the disintegration of the plant tissue but also can be used by the fungus as food. Inside the plant, fungal pathogens produce proteinases, amylases, lipases, and other enzymes that break down plant proteins into amino acids, starch into sugars, fats and lipids into fatty acids, etc., all of which can be used by fungi as food. Fungal enzymes usually result in the death of cells and disintegration (rotting) of plant tissues, especially of fleshy fruits and vegetables. Several fungal pathogens also secrete into the host plant toxins that may be host specific, or they may affect a wide range of plants. Toxins kill cells without causing them to break down and may affect cells at some distance from where the fungus produces them. Some plant pathogenic fungi also produce growth regulators, such as auxins, gibberellins, cytokinins, and ethylene, which, instead of killing or disintegrating plant cells, stimulate cells to enlarge and divide, thereby producing swellings or galls on plant parts.

As the fungal pathogen comes in contact with a host plant, some of the molecules secreted by the fungus, or released from host plant tissues by enzymes of the pathogen, react with receptor molecules in host cells. The plant receptor molecules recognize such fungal or fungus-induced plant molecules (elicitors) as harbingers of a pathogen attack and quickly trigger a cascade of defense reactions in the attacked and the surrounding cells. If the defense reactions are quick and effective, the plant stops the progress of infection and such a plant is considered resistant to this pathogen. If no defense reactions occur, or if they are ineffective or too slow, the plant becomes infected and such a plant is considered susceptible to this pathogen. Generally, resistance in commercial crop plants can be, and usually is, the result of one or more resistance genes. In almost all wild plants and in many cultivated plants, however, resistance in a plant to infection by one or more pathogens is the result of expression of many plant genes controlling various defensive structural and physiological characteristics of the plant.

7. Genetics of Disease

Pathogenic fungi have genes for pathogenicity, that is, genes that allow pathogens to produce enzymes, toxins, growth regulators, etc. that enable the pathogens to cause disease on their host plants. A particular pathogenic (race of a) fungus, however, may be virulent (can cause disease) on a particular (susceptible) host plant variety or it may be avirulent to a particular (resistant) host plant variety. The resulting reaction depends on the genes for virulence or avirulence present on the particular pathogen race and the genes for resistance or susceptibility present in the particular host plant variety. In many host plant–pathogen interactions there is a gene-for-gene relationship; that is, for each gene for virulence/avirulence in the pathogen there is a corresponding gene for susceptibility/resistance in the host plant. In a host–pathogen interaction, if the pathogen has a gene for avirulance, it then produces an elicitor molecule. If the host has a gene for resistance corresponding to the pathogen's gene for avirulence, the host plant produces a receptor molecule that reacts with and thereby recognizes the elicitor produced by the pathogen.

The recognition of the pathogen elicitor by the plant receptor molecule triggers many defensive structural and chemical reactions in the plant, including production of pathogenicity-related proteins toxic to the pathogen, a burst of oxidative reactions and the appearance of active oxygen radicals, lipoxygenation and breakdown of cellular membranes, loss of cellular compartmentalization, loss of electrolytes, and activation of phenoloxidase enzymes and production and oxidation of phenolic compounds. Eventually, the attacked cells and possibly a few surrounding cells die (hypersensitive response or apoptosis). The hypersensitive response also causes the death of the pathogen and stops the further development of infection. In such a host–pathogen combination, the specific pathogen is avirulent, the specific host plant is resistant, and the plant remains resistant. If the specific pathogen lacks an elicitor, or the specific host lacks genes for resistance, no defense reactions are triggered or develop and the plant becomes infected (it is susceptible).

8. Epidemiology

Plant pathogenic fungi cause disease in plants when a virulent pathogen lands on a susceptible host plant and the environmental conditions, primarily temperature and moisture, are favorable for the pathogen to germinate and grow. Once the pathogen establishes infection, temperature continues to be important for pathogen growth and invasion of the host plant as well as for sporulation and dissemination of the fungal spores. With some fungi, moisture is also essential for release of the spores from the infection area. The newly produced spores are then disseminated to other parts of the plant, other plants, or other fields where they cause new infections, and the cycle is repeated. The duration of a disease cycle depends on the prevailing temperature and the periodic availability of moisture. With some plant–fungus combinations and favorable temperature and moisture conditions, a disease cycle may last as little as 4 days; with others, especially under less favorable conditions, it may take several more days to several weeks. The number of disease cycles per growth season, therefore, may vary from 1 or 2 to more than 20. At the end of the growing season, plants may be absent or it may be too cold for new infections, and these require that fungal pathogens overwinter as dormant spores or mycelium in plant debris on the soil or in a perennial host. The following spring the overwintering spores germinate and begin a new series of infection cycles.

9. Management or Control

Plant pathogenic fungi are sensitive to numerous chemicals (fungicides). Fungicides applied on the surface of plant foliage, blossoms, stems, and fruits protect these organs from infection by inhibiting spore germination and growth or by killing the spore. Fungicides are also effective against surface or soil fungi when applied on the surface of seeds, tubers, roots of transplants, etc. Some fungicides are absorbed by the plant foliage, roots, seeds, etc. and are distributed systemically through the plant, stopping fungal infections wherever they may be initiated. Plant pathogenic fungi are also inhibited in their growth and ability to cause disease when exposed to unfavorable temperatures or to low moisture and relative humidity; when they land on nonhost or resistant plants; and when certain antagonistic or parasitic fungi, bacteria, or other microorganisms are present in the vicinity of plant pathogenic fungi or actually parasitize the fungal pathogens (biological control).

B. Prokaryotes: Bacteria and Mollicutes

Approximately 100 species of bacteria and an unknown number of mollicutes cause diseases in plants. Most plant pathogenic bacteria are facultative saprophytes and can be grown on synthetic nutrient media but survive and multiply best in contact with their host plant. Some fastidious vascular bacteria, however, survive in nature only inside their living hosts, are difficult to grow in culture, and some have yet to even be grown in culture. Plant pathogenic mollicutes also survive and multiply in nature only inside their living hosts. Mollicutes of only one genus, *Spiroplasma,* can be grown in culture; all others, usually known as phytoplasmas, can be maintained but do not multiply on nutrient media.

1. Morphology

Most plant pathogenic bacteria are rod shaped, ranging from 0.5–1.0 μm in diameter to 0.6–3.5 μm in length. In some bacteria, and in older cultures, the bacteria may be longer and they may be branched. Bacteria of the genus *Streptomyces* are filamentous. Most species of bacteria have one or more flagella which can be found at the polar ends or over the entire surface of the bacteria. Bacteria have a firm cell wall surrounding a cytoplasmic membrane. The cell wall of most bacterial species is enveloped by a more or less viscous material, which appears as a slime layer or capsule. Mollicutes have a cell membrane but lack cell wall, and as a result they have no regular shapes but are polymorphic, ranging from spherical to tubular (Figs. 2, 3, and 5).

2. *Reproduction*

Plant pathogenic bacteria and mollicutes reproduce in ways similar to those of other bacteria and mollicutes, i.e., by fission (Fig. 5). A few bacteria (gram-positive) also produce endospores, whereas *Streptomyces* bacteria produce conidia-like spores.

3. *Classification*

The prokaryotic pathogens of plants can be classified approximately as shown below. The shapes of the bacteria and the types of plant symptoms caused by the most important genera of bacteria are shown in Fig. 8.

Kingdom: Prokaryotae
  Bacteria—Have cell membrane and cell wall
    Gracilicutes—gram-negative bacteria
      Proteobacteria—Mostly single-celled bacteria
      Enterobacteriaceae—*Erwinia;* causing wilts, blights, soft rots
      Pseudomonadaceae—for example, *Pseudomonas, Ralstonia,* and *Xanthomonas;* causing leaf spots, blights, wilts, soft rots
      Rhizobiaceae—for example, *Agrobacterium, Rhizobium;* causing crown gall, hairy root, root nodules
      Still unnamed—*Xylella;* causing leaf scorch and dieback of trees

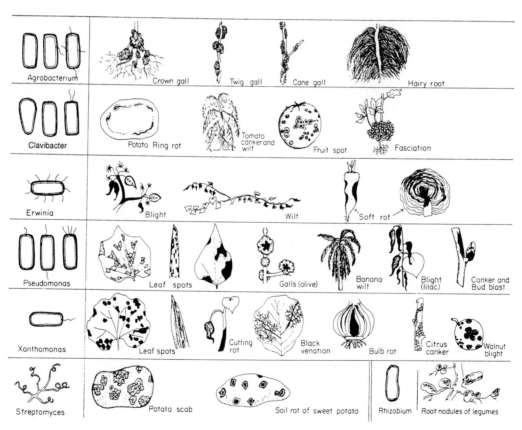

Fig. 8. The most important genera of plant pathogenic bacteria and the kinds of symptoms they cause (from Agrios, 1997, p. 411).

Fermicutes—gram-positive bacteria
> Fermibacteria—mostly single-celled bacteria
>> *Bacillus;* causing rot of tubers, seeds, seedings
>> *Clostridium;* causing rot of stored tubers and leaves, wetwood of trees
> Thallobacteria—branching bacteria; for example, *Arthrobacter, Clavibacter, Curtobacterium, Rhodococcus, Streptomyces;*causing blights, wilts, malformations, scabs

Mollicutes—mycoplasma-like organisms having cell membrane but no cell wall
> Spiroplasmataceae—single celled, spiral bodies
>> *Spiroplasma;* causing corn stunt, citrus stubborn disease, etc.
> Still unnamed—polymorphic bodies
>> *Phytoplasma;* causing numerous yellows, proliferation, and decline diseases in trees and some annuals

4. Ecology

Most plant pathogenic bacteria develop mostly inside the host plant as parasites, on the plant surface (especially buds) as saprophytes, or in the soil as saprophytes. The preferred environments of different bacterial species vary greatly. In the host plant, bacteria often grow and move intercellularly in parenchyma tissues until the host cells break down. Some bacteria, however, move into and multiply in the xylem vessels, causing wilting and death of the host or causing leaf scorch and dieback of the host. Mollicutes multiply and move in the phloem cells of their host plants, causing stunting, yellowing, proliferation, and decline of the plant. The populations of most bacteria released into the soil are reduced drastically within a few days, whereas those of others (e.g., *Agrobacterium*) may survive in the soil for more than 1 year.

5. Dissemination

Plant pathogenic bacteria spread from plant to plant primarily by wind-driven rain, rain splashes and run-off, insects and other animals, tools and cultural practices, and occasionally wind. Over long distances, bacteria are spread in contaminated or infected seeds, other propagative organs, and transplants.

6. Infection

Once in contact with the plant, in the presence of surface moisture bacteria enter the plant through wounds or through natural openings, such as stomata, lenticells, hydathodes, and nectarthodes. Some bacteria and all mollicutes are introduced into either the xylem vessels (xylem-inhabiting bacteria) or into phloem vessels (mollicutes) by specialized insect vectors. Inside the plant tissues most typical bacteria produce a variety of proteolytic, pectinolytic, and other enzymes that break down the corresponding plant substances. This results in plant cell death and breakdown, the appearance of spots or maceration of tissues and organs, and development of the various disease symptoms. Some bacteria, in addition to enzymes, also secrete toxins that kill cells without breaking them down; others also secrete growth regulators, such as auxins and cytokinins, which cause abnormal cell growth and division and result in galls, excessive branching of shoots or roots, or malformed organs. Most bacteria also secrete in their environment a mixture of polysaccharides. Some of these are toxic, but all may play a role in clogging vascular tissues and interfering with the movement of water and nutrients within the plant. Bacteria infecting parenchymatous tissues multiply and move intercellularly and through broken-down plant cells until they reach the surface of their host, from which they spread to other organs or plants. Bacteria invading the xylem and mollicutes invading the phloem usually multiply and attain their highest populations in the younger leaf veins.

7. Genetics of Disease

Plant pathogenic bacteria have genes for pathogenicity, which make them pathogens, and genes for virulence or avirulence corresponding to genes of susceptibility or resistance in the host, which make them pathogenic on some varieties of the host plant but not pathogenic on other varieties of the same host. The avirulence genes are thought to code for molecules that, when recognized by specific receptor molecule produced by resistance genes in the plant, act as elicitors of the host resistance response which stops further development of infection and disease, as occurs in plant–fungus interactions. In addition to the virulence/avirulence genes, several other types of genes seem to play important roles in the development and expression of disease. The genetics of disease in mollicute-induced plant diseases are still unknown.

8. Epidemiology

Bacteria overwinter as bacteria in host plants, seeds, and other propagative organs and transplants, in plant debris, and in soil. In the spring, bacteria ooze out of infected hosts and are disseminated to other plants. Following infection of the plant and multiplication and invasion of the bacteria, the latter ooze out of the plant through wounds or natural openings and are disseminated to additional plants and fields. Disease cycles of bacterial diseases of plants also depend on favorable temperature and moisture and may last from a few days to several weeks. Bacterial plant diseases are more common and much more severe in warm, humid areas that have frequent rains. The number of disease cycles per growth season may vary from a few to more than 30. Mollicutes and the xylem-limited bacteria live continually in the phloem and xylem vessels, respectively, and never ooze out of the plant but are disseminated by specialized insect vectors. Both of these pathogens grow and invade their hosts slowly and therefore their disease cycles last longer than those of typical bacteria.

9. Management or Control

Plant diseases caused by bacteria and mollicutes are difficult to manage and control. Usually a combination of control measures is required to avoid severe losses. Such measures include planting bacteria-free seed, transplants, etc. in a bacteria-free soil; planting disease-resistant varieties when available; reducing fertilizing and watering; and when necessary, spraying with bactericidal chemicals, such as certain copper compounds, which help but are not very effective in controlling bacterial diseases of plants. Some antibiotics, such as streptomycin and oxytetracycline, have been effective in some cases but are avoided because they lead to the early appearance of bacterial strains resistant to these antibiotics. Against some

bacterial diseases such as crown gall, treating plants before planting with antagonistic bacteria provides biological control of the disease.

C. Viruses and Viroids

More than 600 viruses causing disease in plants have been identified and their properties have been studied. Viruses, of course, consist of one or a few molecules of nucleic acid (RNA or DNA) each enveloped in a protein coat consisting of one or a few types of protein molecules. Plants are also affected by more than 20 viroids—that is, small, low-molecular-weight RNAs that can infect plants, replicate themselves, and cause disease. All viruses and viroids are intracellular obligate parasites and only a few of them survive, but do not multiply, in plant debris or outside cells.

Viruses and viroids, because they are too small to be seen with the compound microscope, are detected and identified by several indirect techniques, including determining the presence of plant symptoms characteristic of viral infections, such as oak leaf or ring patterns on leaves, stems, or fruit; searching for virus-specific inclusion bodies in infected young plant cells; transmission of the symptoms to healthy plants by sap, grafting, and specific insect and other vectors; and using serological tests or nucleic acid probe tests against known viruses. Viruses in plant sap can sometimes be seen with the electron microscope and, once purified, viruses and viroids can be seen with the electron microscope.

1. Morphology

Nearly half of the known plant viruses are elongate (rigid rods about 15×300 nm or flexuous threads $11 \times 480–2000$ nm), and almost as many are spherical (isometric or polyhedral about $17–100$ nm in diameter) (Figs. 4 and 5), with the rest being rhabdoviruses—that is, cylindrical bacillus-like rods ($52–75 \times 300–380$ nm) (Fig. 9). Many viruses have split genomes, consisting of two or more distinct nucleic acid strands each encapsidated in similar- or different-sized particles made of the same protein molecules (subunits). The surface of viruses consists of a definite number of protein subunits, which are spirally arranged in the elongated viruses and packed

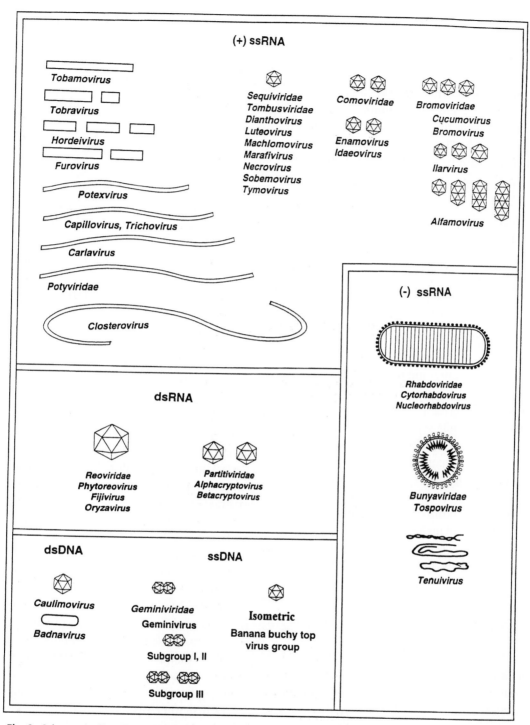

Fig. 9. Schematic diagram of families and genera of viruses that infect plants (modified from Murphy *et al.*, (1995) with permission: from Agrios, 1997, p. 501).

on the sides of the polyhedral viruses. In cross section, the elongated viruses appear as hollow tubes of spirally arranged protein subunits, with the nucleic acid strand embedded between successive spirals of protein subunits. In spherical viruses the visible shell consists of protein subunits, whereas the nucleic acid is inside the shell but it is not known how it is arranged. The rhabdoviruses and a few spherical viruses also have an outer membrane composed of lipoprotein.

2. Reproduction (Replication)

Plant viruses and viroids replicate only within living plant cells or protoplasts. The mechanisms of replication are quite complex and the complexity varies depending on whether the nucleic acid of the virus is RNA or DNA, single or double stranded, and positive or negative strand. Basically, in the replication of viruses and viroids, the infected plant cell provides all the structural materials (nucleic acids and proteins), the energy, and the machinery needed to make more virus, whereas the viruses (or viroid) supply only the blueprint in the form of their nucleic acid. Virus (and viroid) replication is analogous to making copies of an original page (the virus) with the help of a modern copier (the plant cell). Once new viral nucleic acid and viral protein are produced in the cell, they are assembled into virus particles. The virus particles, in turn, either accumulate in the cell singly or as parts of inclusion bodies or they move to other adjacent plant cells and, possibly through the phloem sieve elements, throughout the plant.

3. Classification

The classification of viruses is unlike that of any other microorganism. The main characteristics considered in the classification of viruses are the nature and number of their nucleic acids and the shape and size of their particle(s). A very sketchy classification of plant viruses is as follows (Fig. 9):

Kingdom: Viruses
  RNA viruses
    Single-stranded positive RNA
      Rod-shaped particles—e.g., *Tobamovirus* (tobacco mosaic virus)
      Filamentous particles—e.g., *Potyvirus* (potato virus Y)
      Isometric particles—e.g., *Luteovirus* (barley yellow dwarf virus)
    Single-stranded negative RNA
      Rhabdoviridae—e.g., potato yellow dwarf virus
      Bunyaviridae—e.g., *Tospovirus* (tomato spotted wilt virus)
    Double-stranded RNA
      Isometric viruses—e.g., Reoviridae, *Phytoreovirus* (wound tumor v.)
  DNA viruses
    Double-stranded DNA
      Isometric—e.g., *Caulimovirus* (cauliflower mosaic virus)
      Non-enveloped bacilliform—e.g., *Badnavirus* (rice tungro bacilliform v.)
    Single-stranded DNA
      Geminate (twin) particles—*Geminivirus* (bean golden mosaic virus)
      Single isometric particles—*Nanovirus* (e.g., banana bunchy top virus)

Ecology. *Plant* viruses and viroids, in general, survive in nature only in living plant cells. Very few of them *survive* in nature in dead plant debris or dry sap. Some viruses also survive in their insect, nematode, or fungal vectors.

4. Dissemination

In nature, a few plant viruses are disseminated from plant to plant by the rubbing of infected leaves with healthy ones during strong winds that cause minute to larger injuries to leaves or by cultural practices and tools that carry sap among plants. A few are also disseminated when pollen from infected plants fertilizes flowers of healthy plants or when natural root grafts develop between infected and healthy plants. Many viruses are disseminated from infected mother plants to offspring plants through propagative organs, such as buds and grafts, tubers, and corms, and through seeds. In nature, however, most virus transmission occurs by specific vectors, such as insects, mites, nematodes, and fungi. A particular virus is transmitted by one specific or a few related vectors and vice versa. The most common and most effective insect vectors are aphids, followed in importance by leafhoppers, whiteflies, thrips, and a few others. Some of the insect vectors, such as aphids, acquire the virus after feeding for only a few seconds on an infected plant and can transmit it after feeding for a similarly short time on a healthy plant. Of the other insect vectors and the few mite vectors, some must feed for several hours or days before they acquire enough virus, some must wait several days or weeks after feeding before they can transmit the virus, and most must then feed on a plant for several hours or days before they can transmit the virus to the healthy plant. Nematodes, which are tiny worm-like animals feeding on roots of plants, and a few soil fungi transmit viruses by first feeding on roots of infected plants and then moving on and feeding on roots of healthy plants.

5. Infection

Once a virus (or viroid) enters a plant cell, the virus sheds its protein coat and the naked nucleic acid (RNA or DNA) associates with appropriate cell organelles, membranes, enzymes, etc. which replicate the nucleic acid into hundreds of thousands or millions of copies. Soon, translation of part of the nucleic acid strand results in the production of protein molecules that are the subunits of the viral coat protein, and these assemble with the nucleic acid strands and form the complete virus particles.

In the process of virus replication and accumulation, the metabolism of the host is affected. The infected plant's defense mechanisms are activated, and if the defenses are effective (if the plant is resistant) virus replication and movement may be stopped, the infected and a few surrounding cells may die, and no further infection develops beyond a tiny visible or invisible local lesion. If, however, the host defenses are ineffective or inadequate, the virus moves from cell to cell, replicating in each of them, and finally reaches the phloem sieve elements that make up part of the plant veins. Once in the phloem, the virus is distributed throughout the plant, and from the phloem cells it moves into adjacent parenchyma cells and to the other living cells in which it again replicates. Depending on the age and size of the plant, viruses can spread throughout the plant within a few days to a few weeks or months.

Infected plants are often stunted; their leaves, shoots, and fruits become smaller, discolored, or malformed; yields are reduced; and with some virus–host combinations some plant parts or whole plants may be killed.

6. Epidemiology

Viruses overseason in perennial plants, in vegetative propagative organs such as bulbs and tubers, in seeds, and in their insect, nematode, or fungal vectors. Very few viruses overwinter in plant debris or on surfaces smeared with infected plant sap. In the spring, vectors transmit the viruses to appropriate annual and perennial host plants of the viruses, and field-workers and implements may also spread some of the viruses to their host plants. When the virus infects and spreads throughout a newly infected plant, the transmission of virus by vectors or workers may be repeated until potentially all the plants in a field may become infected. Vector-transmitted viruses may complete from a few disease cycles per season (2–5 for nematode-transmitted ones) to many (10–20 or more for aphid-transmitted viruses). The early presence of a virus in a field, whether in weeds, in volunteer infected plants from the previous season, in transplants, or in infected seed or vegetative organ propagated plants, leads to tremendous virus buildup and spread of the virus if the latter is also transmitted by a vector.

7. Management and Control

The use of virus-free seed, tubers, transplants, etc. is the single most important measure for avoiding or reducing virus diseases of many crops, especially those that do not have insect vectors. There are no effective viricides for management or control of plant viruses and viroids once the plant is infected. An effective way to avoid losses from plant viruses is to keep viruses out of an area through a system of inspections that lead to certification of propagative materials as virus free and through quarantine of plants carrying the virus. Virus-infected propagative materials can sometimes be freed of the virus by exposure to hot air or hot water for varying periods of time. Control of virus vectors is desirable, but in only some cases is it sufficiently effective to justify the cost and effort. In some cases, inoculating plants with mild strains of a virus protects them from infection by severe strains of the same virus. In the past few years, successful control of plant viruses has been achieved by genetically engineering plants to carry and express the coat protein and other genes of the virus, which makes the plants resistant to subsequent infection by that virus.

D. Nematodes

Nematodes are usually microscopic animals that are worm-like in appearance but quite distinct taxonomically from true worms. Most nematode species live freely in fresh or salt water, but numerous species attack and cause disease on humans and animals and several hundred species feed primarily on plant roots and cause diseases in plants. Symptoms of infected plants include root lesions or root galls, devitalized root tips, excessive root stunting or branching, distortion of above-ground plat organs, general yellowing and stunting of the plant, and death of the plant. The annual worldwide losses caused by nematodes on the various crops are estimated to be $80 billion.

1. Morphology

Plant parasitic nematodes are small and eel-shaped, but only 300– 1000 μm, with some up to 4 mm long $\times$ 15–35 μm wide (Figs. 2, 4, 5, and 10). The females of some of these nematodes become swollen at maturity and have pear-shaped or spherical bodies. Nematode bodies are transparent, unsegmented, and have no legs or other appendages. Plant parasitic nematodes have a hollow stylet or spear that is used to puncture holes in plant cells and through which they withdraw nutrients from the plants.

2. Reproduction

Nematodes have well-developed reproductive systems that distinguish them as female and male nematodes. The females lay eggs usually after fertilization by males but in some cases without fertilization. Many species lack males. Nematode eggs hatch into juveniles that resemble the adult nematodes but are smaller. Juveniles grow in size and each juvenile stage is terminated by a molt. All nematodes have four juvenile stages, with the first molt usually occurring in the egg. After the final molt the nematodes

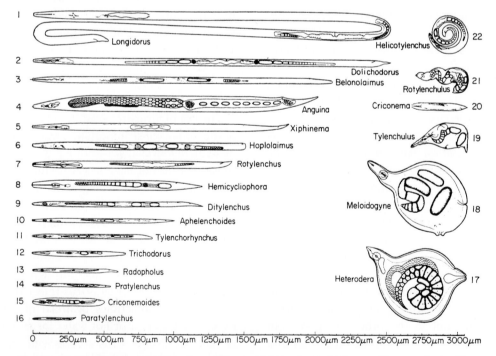

Fig. 10. Morphology and relative sizes of the most important plant parasitic nematodes (from Agrios, 1997, p. 568).

differentiate into males and females and the new females can produce fertile eggs in the presence or absence of males. One life cycle from egg to egg may be completed within 2–4 weeks in favorable weather, longer in cooler temperatures. In some nematodes only the second-stage juvenile can infect a host plant, whereas in others all but the first juvenile and adult stages can infect. When the infective stages are produced, they must feed on a susceptible host or they will starve to death. In some species, however, some juveniles may dry up and remain quiescent, or the eggs may remain dormant in the soil for years.

3. Classification

The plant pathogenic nematodes can be classified as follows:

Kingdom: Animalia
  Order: Tylenchida
    Anguinidae—*Anguina* (seed-gall nematode), *Ditylenchus* (bulb and stem n.)
    Belonolaimidae—*Belonolaimus* (sting n.), *Tylenchorhynchus* (stunt n.)
    Pratylenchidae—*Pratylenchus* (lesion n.), *Radopholus* (burrowing n.)
    Hoplolaimidae—*Hoplolaimus* (lance n.), *Rotylenchus* (spiral n.)
    Heteroderidae—*Globodera* (round cyst n.), *Heterodera* (cyst n.), *Meloidogyne* (root knot n.)
    Criconematidae—*Criconemella* (ring n.), *Hemicycliophora* (sheath n.)
    Paratylenchidae—*Paratylenchus* (pin n.)
    Tylenchulidae—*Tylenchulus* (citrus n.)
    Aphelenchoidea—*Aphelenchoides* (foliar n.), *Bursaphelenchus* (pine wilt n.)
  Order: Dorilaimida
    Longidoridae—*Longidorus* (needle n.), *Xiphinema* (dagger n.)
    Trichodoridae—*Paratrichodorus* and *Trichodorus* (stubby root nematodes)

4. Ecology

Almost all plant pathogenic nematodes live part of their lives in the soil, especially in the top 15–30 cm, where most of the plant roots are located. Many nematodes live freely in the soil, feeding superficially on roots and underground stems. Others spend much of their lives inside their hosts, but often their eggs, preparasitic juvenile stages, and the males spend some of their lives in the soil. Soil temperature, moisture, and aeration affect survival and movement of nematodes in the soil. A few nematodes survive in the tissues of the plant they infect and in their insect vector, and they seldom, if ever, enter the soil.

5. Dissemination

Nematodes spread through the soil slowly under their own power, probably no more than 1 to possibly a few meters per growth season. Nematodes, however, can be easily spread by anything that moves soil particles, such as irrigation or runoff water, farm equipment, animal feet, and dust storms. A few plant pathogenic nematodes are spread from plant to plant by their specific insect vector.

6. Infection

When an infective nematode juvenile or adult reaches the surface of a plant, it places its mouthparts (six lips) in contact with the plant surface while the stylet moves back and forth, exerting pressure as it tries to pierce the outer epidermal cell wall. Once that occurs, the nematode, through the stylet, secretes saliva into the cell, the contents of which are liquefied and absorbed through the stylet. Some nematodes are rapid feeders and feed only on epidermal cells, moving their stylet from one cell to the next without ever entering the plant (ectoparasitic nematodes). Others (endoparasitic nematodes) enter the plant parts and feed slower while inside the plant. Some of the latter (sedentary nematodes) become attached to one area of the plant and do not move. Some of the nematodes (migratory) enter the plant, feed internally for varying lengths of time, and then exit the plant and move about freely. Depending on where the female of a particular nematode species is feeding, she lays her eggs inside or outside the plant. When the eggs hatch, the new infective juveniles either cause further infections inside the plant or infect new plants. The mechanical injury caused by nematodes in the infected area, as well as the removal of nutrients from plants by nematodes, certainly has a detrimental effect on the plant. It is thought, however, that much greater damage by nematodes on plants is caused by the enzymes, growth regulators, and toxic compounds contained in the secretions of the nematode into the plant and by the ports of entry for other pathogens (fungi and bacteria) created by nematodes on plant roots.

7. Epidemiology

Plant pathogenic nematodes overseason as eggs in the soil or plant debris, as juveniles or adults in roots or stems of perennial plants, in vegetative propagative organs and some infected seeds, and as dehydrated juveniles or adults in contact with plant tissues. At the beginning of the growth season, the eggs, often stimulated by hatching factors excreted by host plants, hatch and produce juveniles which are attracted to host plants and cause infection. Dormant adults and juveniles are also activated and may cause new infections. Before or following infection, the nematode goes through its remaining juvenile stages, if any, develops into an adult, and subsequently lays eggs that can hatch; the juveniles or adults then infect and repeat the disease cycle. Nematodes are favored in their survival and spread by moderate temperatures and moisture. Too high or too low temperatures kill the nematodes, as do dry soils and flooding.

8. Management and Control

As with other plant pathogens, nematodes are best controlled by keeping them out of a field through quarantines and use of nematode-free propagative materials, seeds, and transplants. Infected propagative materials are freed of nematodes by treating them with hot water. When nematodes are present in a field, they can be managed primarily by planting resistant plant varieties and by applying pre-plant or post-plant nematicides. Several cultural practices, such as plowing the field repeatedly before planting to expose more nematodes to the sun and dry air or flooding the field to kill nematodes by suffocation, are helpful but difficult to carry out. Planting trap crops that attract nematodes but do not allow them to reach the egg-laying stage or destroying the trap plants before the nematodes reproduce are also possible but difficult. Recently, several antagonistic bacteria and fungi have been found that provide experi-

mental control of plant pathogenic nematodes, but these are not available commercially.

E. Protozoa

To date, only a small number of trypanosomatid flagellate protozoa have been found to parasitize and to cause disease in plants. Some flagellate protozoa parasitize the latex-bearing cells (the laticifers) of laticiferous plants, and it is not clear whether these protozoa cause disease to their hosts. Some have been isolated from fruits such as tomato, on which they cause considerable damage. Several flagellate protozoa, however, parasitize the phloem of plants. These parasites are obligate parasites that live and reproduce in the phloem sieve elements of their host plants and in their insect vector. It has been difficult to isolate, grow in pure culture, and inoculate these parasites into healthy plants to reproduce the disease; therefore, their role as causes of plant diseases was uncertain. Nevertheless, protozoa have been found associated and are now considered as the causes of a disease of coffee trees, several diseases of palm trees, and possibly a disease of cassava plants. Plant pathogenic protozoa have been found in the tropic and semi-tropic areas of the Americas.

1. Morphology

The protozoa found associated with plant diseases to date are spindle-shaped single cells that have a typical nucleus and one or more long, slender flagella at some or all stages of their life cycle (Fig. 2, 4, and 5). The size of flagellate protozoa in the phloem of infected plants varies, even for the same pathogen, depending on the stage of life cycle at which the pathogen is observed. The sizes observed are $12–18 \times 1.0–2.5$, $4–14 \times 0.3–1.0$, and 3 or $4 \times 0.1–0.2$ μm.

2. Reproduction

Plant pathogenic flagellate protozoa reproduce by longitudinal fission. At different stages of infection, reproduction of the flagellates produces the forms and sizes listed previously.

3. Classification

The following is a classification of plant pathogenic protozoa:

Kingdom: Protozoa
Phylum: Euglenozoa
Order: Kinetoplastida
Family: Trypanosomatidae
Genus: *Phytomonas*

Most plant pathogenic protozoa seem to belong to one genus (*Phytomonas*), but some, especially those affecting fruits, probably belong to one or more genera other than *Phytomonas*.

4. Ecology

Plant pathogenic protozoa survive and multiply in the phloem elements of their host plants and in their insect vectors. Some of them, of debatable pathogenicity, survive and multiply in the latex-bearing cells (laticifers) of laticiferous plants or in tissues of fruits they colonize.

5. Dissemination

Plant pathogenic protozoa are disseminated from plant to plant primarily by insect vectors belonging to the families Pentatomidae, Lygaeidae, and Coreidae. Slower dissemination occurs through natural root grafts between infected and healthy trees.

6. Management or Control

There are no effective controls against plant diseases caused by protozoa. Management practices such as planting protozoa-free nursery plants and planting them away from infected ones may help reduce the disease, as may efforts to control the insect vectors of these pathogens.

F. Parasitic Higher Plants

More than 2500 species of higher plants live parasitically on other higher plants. Relatively few of these plants, however, affect and cause disease on cultivated plants or on forest trees of commercial significance. The parasitic plants produce flowers and seeds like all plants (Figs. 4 and 5). They belong to widely separated botanical families and vary greatly in dependence on their host plants. Some parasitic plants (e.g., mistletoes), have chlorophyll but no roots; therefore they depend on their hosts only for water and nutrients. Others (e.g., dodder) have little or no chlorophyll and no true roots, therefore, in addition to water and inorganic nutrients, they also depend on their hosts for photosynthetic products. Parasitic higher plants obtain water and nutrients from their host plants by producing and sinking into the vascular system of their host stems or roots food-absorbing organs, called haustoria. The following are the most important parasitic higher plants and the botanical families to which they belong:

Cuscutaceae—*Cuscuta* sp., the dodders
Viscaceae—*Arceuthobium,* the dwarf mistletoes of conifers
Phoradendron—the American true mistletoe of broad-leaved trees
Viscum—the European true mistletoes
Orobanchaceae—*Orobanche,* the broomrapes of tobacco
Scrophulariaceae—*Striga,* the witchweeds of many monocotyledonous plants

Parasitic higher plants vary in size from a few millimeters to several centimeters in diameter and from 1 cm tall to upright green plants more than 1 m tall. Some, however, are orange or yellow leafless vine strands that may grow to several meters in length and entwine around the stems of many adjacent host plants (Fig. 4). Parasitic higher plants reproduce by seeds. Seeds are disseminated to where host plants grow by wind, run-off water, birds, and cultivating equipment. Seeds of some parasitic plants are forcibly expelled to significant distances (10 m or more). Parasitic higher plants overseason on perennial hosts or as seeds on the host or on the ground. In the spring, the seeds germinate and the seedling infects a new host plant. Control or management of parasitic plants depends on removing infected plants carrying the parasites and avoiding bringing seeds of parasitic plants into new areas.

G. Parasitic Green Algae

Although green algae are primarily free-living organisms, many genera of green algae live as endophytes of many hydrophytes and seem to cause little or no damage to their host plants. The green alga genus *Rhodochytrium* of the family Chlorococcaceae and the genus *Phyllosiphon* of the family Phyllosiphonaceae infect numerous weeds and occasionally a few minor cultivated plants. Green algae of the genus *Cephaleuros* of the family Trentepohliaceae, however, are true parasites of many wild and cultivated terrestrial plants and cause diseases of economic significance.

Cephaleuros green algae, particularly *Cephaleuros virescens,* occurs commonly as a leaf spot and stem pathogen on more than 200 plant species growing primarily in the tropics between latitudes 32°N and 32°S. Fruit lesions are less frequent but occur on many host plants. Some of the economically most important plants affected by green algae are tea, coffee, black pepper, cacao, citrus, and mango. *Cephaleuros* green algae consist of a disc-like vegetative thallus composed of symmetrically arranged cells. Algal filaments are produced that grow mostly between the cuticle and the epidermis of host leaves and, under some conditions, between the palisade and mesophyll cells of leaves. These green algae reproduce by means of zoospores in zoosporangia, which can be disseminated by wind, rain splashes, and wind-driven rain and can infect new leaves, shoots, and fruit of plants. Most infections occur at the end of the rainy season. Upon infection, plant cells next to the invading thallus turn yellow, whereas adjacent cells enlarge and divide. In stressed plants the infecting thallus expands, whereas cells in earlier invaded tissues die and form a lesion. Lesions on leaves and shoots may be so numerous that they nearly cover the entire surface. Control of parasitic green algae, when needed, can be obtained by spraying plants likely to be affected with appropriate fungicides at the time most infections occur.

II. CONCLUSIONS

Plant pathogens belong to the same groups of organisms that include the pathogens that cause diseases in humans and animals. Most plant diseases are caused by fungi, followed by viruses, prokaryotes (bacteria and phytoplasmas), nematodes, and protozoa, respectively. The species of these organisms that infect plants do not infect humans or animals—with the possible exception of a few bacteria that may affect humans and a few storage plant organs. The nature, replication, disease-causing mechanism, etc., of plant pathogens are identical or very similar to those of their counterparts that infect humans and animals. They differ from them, however, not only in their host range but also in their ecology, dissemination, epidemiology, and control. In addition to the previously mentioned common pathogens, plants are also affected by viroids, parasitic higher plants, and parasitic green algae, none of which has been shown to affect humans or animals.

Bibliography

Agrios, G. N. (1997). "Plant Pathology," 4th ed. Academic Press, San Diego.

Alexopoulos, C. J., Mims, C. W., and Blackwell, M. (1996). "Introductory Mycology," 4th ed. Wiley, New York.

Dollet, M. (1984). Plant diseases caused by flagellated protozoa (*Phytomonas*). *Annu. Rev. Phytopathol.* **22,** 115–132.

Goto, M. (1992). "Fundamentals of Bacterial Plant Pathology." Academic Press, San Diego.

Joubert, J. J., and Rijkenberg, F. H. J. (1971). Parasitic green algae. *Annu. Rev. Phytopathol.* **9,** 45–64.

Kuijt, J. (1969). "The Biology of Parasitic Flowering Plants." Univ. of California Press, Berkeley.

Matthews, R. E. F. (1991). "Plant Virology," 3rd ed. Academic Press, New York.

Nickle, W. E. (Ed.) (1991). "Manual of Agricultural Nematology." Dekker, New York.

Plant Virology, Overview

Roger Hull

John Innes Centre

GLOSSARY

cistron The basic unit of genetic function. It usually refers to a gene or coding region for a protein.

hyperplasia Increase in size of organ or tissue due to an increase in the numbers of component cells.

icosahedral symmetry A type of cubic symmetry found in isometric viruses. The basic form is 12 pentameric subunits forming the vertices of the particle and giving twofold, threefold and fivefold rotational symmetry axes. For larger icosahedra the number of subunits is increased in a regular manner by the addition of hexameric subunits in defined numbers and places.

local lesion Symptom sometimes found in a plant leaf inoculated with a virus caused by the death of, or changes in, cells around the original point of entry.

open reading frame A set of codons for amino acids uninterrupted by stop codons.

subliminal infection Infection in which the virus multiplies in the originally infected cell but is unable to spread to adjacent cells.

triangulation number A description of the triangular face of an icosahedron indicating the number of triangles into which the face is divided.

THERE ARE APPROXIMATELY 960 VIRUSES which infect plants, several of which are of major agricultural importance. Although there are no full collations of the amounts of crop losses due to viruses, some data assembled by Hull (1994b) give an indication of the significant damage that viruses can cause. It should be noted that these data do not account for the costs of insecticides and other control measures used.

I. INTRODUCTION

The viral aetiology of plant diseases was first recognized in the late nineteenth century in the work of Mayer, Iwanowski, and Beijerinck on what is now known as tobacco mosaic tobamovirus (TMV). They demonstrated that viruses infecting any organism differed from bacteria, and from the term "contagium vivum fluidum" is derived the term virus. The study of plant viruses has pioneered the understanding of viruses in many aspects, including the recognition that they are nucleoproteins and, that the nucleic acid is the infectious entity, and the concepts of virus particle shape and structure.

Plant viruses are usually named after the host in which they were first described, the symptoms they usually cause, and the virus genus to which they belong. This naming does not necessarily reflect the importance of a virus; for instance, arabis mosaic nepovirus was first described from an ornamental garden plant but causes a significant problem in soft fruit crops such as strawberries and raspberries. The classification of plant viruses is based primarily on a variety of characteristics, such as particle morphology, composition and organization of the genome, and biological properties. Plant viruses are currently

TABLE I
Classification of Plant Viruses

Genome[a]	Family	Genus	Type member	No. spp.	No. tentative spp.
dsDNA (RT)	Caulimoviridae	Badnavirus	Commelina yellow mottle virus	10	4
		Caulimovirus	Cauliflower mosaic virus	6	6
		"Soybean chlorotic mottle virus-like"	Soybean chlorotic mottle virus	2	0
		"Cassava vein mottle virus-like"	Cassava vein mottle virus	1	0
		"Petunia vein clearing virus-like"	Petunia vein clearing virus	1	0
		"Rice tungro bacilliform virus-like"	Rice tungro bacilliform virus	1	0
ssDNA	Geminiviridae	Mastrevirus	Maize streak virus	12	2
		Curtovirus	Beet curly top virus	4	1
		Begomovirus	Bean golden mosaic virus	127	15
	No family	Nanovirus	Subterranean clover stunt virus	4	1
dsRNA	Reoviridae	Fijivirus	Fiji disease virus	5	0
		Oryzavirus	Rice ragged stunt virus	2	0
		Phytoreovirus	Wound tumor virus	3	0
	Partitiviridae	Alphacryptovirus	White clover cryptic virus 1	16	10
		Betacryptovirus	White clover cryptic virus 2	4	1
	No family	Varicosavirus	Lettuce big-vein virus	2	2
ssRNA (−)	Rhabdoviridae	Cytorhabdovirus	Lettuce necrotic yellows virus	8	0
		Nucleorhabdovirus	Potato yellow dwarf virus	6	0
		Unassigned		61	0
	Bunyaviridae	Tospovirus	Tomato spotted wilt virus	9	2
	No family	Tenuivirus	Rice stripe virus	6	4
		Ophiovirus	Citrus psorosis virus	1	2
ssRNA (+)	Bromoviridae	Bromovirus	Brome mosaic virus	6	0
		Alfamovirus	Alfalfa mosaic virus	1	0
		Cucumovirus	Cucumber mosaic virus	3	0
		Ilarvirus	Tobacco streak virus	16	0
		Oleavirus	Olive latent virus 2	1	0
	Comoviridae	Comovirus	Cowpea mosaic virus	15	0
		Fabavirus	Broad bean wilt virus	4	0
		Nepovirus	Tobacco ringspot virus	31	10
	Potyviridae	Potyvirus	Potato virus Y	91	97
		Macluravirus	Maclura mosaic virus	2	0
		Ipomovirus	Sweet potato mild mottle virus	1	0
		Tritimovirus	Wheat streak mosaic virus	2	0
		Bymovirus	Barley yellow mosaic virus	6	0
		Rymovirus	Ryegrass mosaic virus	5	2
	Tombusviridae	Tombusvirus	Tomato bushy stunt virus	13	0
		Carmovirus	Carnation mottle virus	12	8
		Necrovirus	Tobacco necrosis virus A	5	2
		Dianthovirus	Carnation ringspot virus	3	1
		Machlomovirus	Maize chlorotic mottle virus	1	0
		Avenavirus	Oat chlorotic stunt virus	1	0
		Aureusvirus	Pothos latent virus	2	0
		Panicovirus	Panicum mosaic virus	2	0

continues

Continued

Genome[a]	Family	Genus	Type member	No. spp.	No. tentative spp.
	Sequiviridae	Sequivirus	Parsnip yellow fleck virus	2	0
		Waikavirus	Rice tungro spherical virus	3	0
	Closteroviridae	Closterovirus	Beet yellow virus	4	21
		Crinivirus	Lettuce infectious yellows	1	6
	Luteoviridae	Luteovirus	Barley yellow dwarf virus (MAV strain)	3	0
		Polerovirus	Potato leaf roll virus	6	0
		Enamovirus	Pea enation mosaic virus-1	1	0
		Unassigned		7	0
	No family	Benyvirus	Beet necrotic yellow vein virus	2	0
		Carlavirus	Carnation latent virus	29	32
		Capillovirus	Apple stem grooving virus	5	1
		Furovirus	Soil-borne wheat mosaic virus	5	5
		Hordeivirus	Barley stripe mosaic virus	4	0
		Idaeovirus	Raspberry bushy dwarf virus	1	0
		Marafivirus	Maize rayado fino virus	3	0
		Ourmiavirus	Ourmia melon virus	2	1
		Pecluvirus	Peanut clump virus	2	0
		Pomovirus	Potato mop top virus	3	0
		Potexvirus	Potato virus X	21	21
		Allexivirus	Shallot virus X	6	2
		Foveavirus	Apple stem pitting virus	2	1
		Sobemovirus	Southern bean mosaic virus	9	5
		Tobamovirus	Tobacco mosaic virus	13	5
		Tobravirus	Tobacco rattle virus	3	0
		Trichovirus	Apple chlorotic leafspot virus	4	3
		Tymovirus	Turnip yellow mosaic virus	20	1
		Umbravirus	Carrot mottle virus	7	4
		Vitivirus	Grapevine virus A	4	1

[a] Genome type: dsDNA (RT), double-stranded DNA (replicating by reverse transcription); ssDNA, single-stranded DNA; dsRNA, double-stranded RNA; ssRNA ($-$), ($-$) sense RNA; ssRNA ($+$), ($+$) sense RNA (mRNA). Virus species and genome type: dsDNA, 3.2%; ssDNA, 17.2%; dsRNA, 4.7%; ssRNA ($-$), 10.2%; ssRNA ($+$), 64.7%. Data from Murphy *et al.* (1995), Pringle (1998a, b), and the Association of Applied Biologists CD-ROM on descriptions of plant viruses (1998).

grouped by the International Committee on the Taxonomy of Viruses into 70 genera and two unassigned groups making up 12 families (Table I). However, classification of these viruses is ongoing and further families and genera are being suggested (see Web sites *http://biology.anu.edu.au/research-groups/MES/vide/* and *http://www.ncbi.nlm.nih.gov/ictv*).

Most of the virus families are unique to plants, although, in some cases, there are some with features common to viruses of animals. These virus families (e.g., Reoviridae, Rhabdoviridae, and Bunyaviridae) have members which also infect invertebrates, and it is thought that members of these families were originally viruses of insects which have become adapted to plants.

II. HOST RANGE AND SYMPTOMS

Viruses can infect plants at different levels. At one end of this range, termed subliminal infection, the

TABLE II
Properties of Plant Viruses[a]

Family	Genus	Host range[b]	Vector[c]	Particle shape[d]	No. genome segments
Caulimoviridae	Badnavirus	n	m	b	1
	Caulimovirus	n	as	i	1
	"Soybean chlorotic mottle virus-like"	n	u	i	1
	"Cassava vein mottle virus-like"	n	u	i	1
	"Petunia vein clearing virus-like"	n	u	i	1
	"Rice tungro bacilliform virus-like"	n	lh	b	1
Geminiviridae	Mastrevirus	n	lc	g	1
	Curtovirus	n–w	lc	g	1
	Begomovirus	n	w	g	1 or 2
No family	Nanovirus	n	ac	i	6 to 7
Reoviridae	Fijivirus	n	pp	ic	12
	Oryzavirus	n	pp	ic	10
	Phytoreovirus	w–n	lp	ic	12
Partitiviridae	Alphacryptovirus	n	s	i	2
	Betacryptovirus	n	s	i	2
No family	Varicavirus	n–m	f	r	–
Rhabdoviridae	Cytorhabdovirus	n	lp, ap	bc	1
	Nucleorhabdovirus	n	lp, ap	bc	1
	Unassigned	n	lp, ap, mtp	bc	1
Bunyaviridae	Tospovirus	w	tp	ic	3
No family	Tenuivirus	n	pp	fl*	4
	Ophiovirus	n	u	fl	3
Bromoviridae	Bromovirus	n	u	i	3
	Alfalfamovirus	w	an	b	3 (4)
	Cucumovirus	n–w	an	i	3
	Ilarvirus	w	s	(i)	3
	Oleavirus	n	u	b	3
Comoviridae	Comovirus	n	b	i	2
	Fabavirus	w	an	i	2
	Nepovirus	n–w	n	i	2
Potyviridae	Potyvirus	n–m	an, s	f	1
	Macluravirus	n	an	f	1
	Ipomovirus	w	w	f	1
	Tritimovirus	n	mt	f	1
	Bymovirus	n	f	f	1
	Rymovirus	n	mt	f	1
Tombusviridae	Tombusvirus	n	u, s, f	i	1
	Carmovirus	n	u, s, f	i	1
	Necrovirus	w	f	i	1
	Dianthovirus	m	u	i	1
	Machlomovirus	n	s	i	1
	Avenavirus	n	u	i	1
	Aureusvirus	n	u	i	1
	Panicovirus	n	u	i	1

continues

Continued

Family	Genus	Host range[b]	Vector[c]	Particle shape[d]	No. genome segments
Sequiviridae	Sequivirus	n	as	i	1
	Waikavirus	n	ls	l	1
Closteroviridae	Clostervirus	n	as, w, m	f	1
	Crinivirus	w	w	f	2
Luteoviridae	Luteovirus	n	ac	i	1
	Polerovirus	n	ac	i	1
	Enamovirus	n	ac	i	1
	Unassigned	n	ac	i	1
No family	Benyvirus	n	f	r	4
	Carlavirus	n–m	as, w, s	f	1
	Capillovirus	n	s	f	1
	Furovirus	n	f	r	2
	Hordeivirus	n	c, s	r	3
No family	Idaeovirus	n–w	s	i	2
	Marafivirus	n	lp	i	1
	Ourmiavirus	m	u	b	3
	Pecluvirus	n	f, s	r	2
	Pomovirus	n	f	r	2
	Potexvirus	n	c	f	1
	Allexivirus	n	mt	f	1
	Foveavirus	n	u	f	1
	Sobemovirus	n	b, s	i	1
	Tobamovirus	m	c	r	1
	Tobravirus	w	n	r	2
	Trichovirus	n	s, m, as	f	1
	Tymovirus	n	b	i	1
	Umbravirus	n	h	ih	1
	Vitivirus	n–m	as	f	1

[a] Data from Murphy *et al.* (1995) and Pringle (1998a,b).

[b] Host range: n, narrow (1–5 natural hosts); m, medium (6–20 natural hosts); w, wide (more than 20 natural hosts).

[c] Vector: ac, aphid, circulative; an, aphid, nonpersistent; ap, aphid, propagative; as, aphid, semipersistent; b, beetle; c, contact; f, fungus; h, requires a helper virus, usually a luteovirus; lc, leafhopper circulative; lh, leafhopper but requiring helper virus; lp, leafhopper propagative; ls, leafhopper, semipersistent; m, mealybug; mt, mite; mtp, mite, propagative; n, nematode; pp, planthopper propagative; s, seed; tp, thrip, propagative; u, unknown; w, whitefly.

[d] b, bacilliform; bc, complex bacilliform; f, filamentous rod; fl*, flexuous particle but may be a complex particle; g, geminate; i, isometric; (i), possibly isometric; ic, complex isometric; ih, encapsidated by helper virus giving isometric particle; r, rigid rod.

virus has the ability to replicate in the initially infected cell but then it cannot spread to adjacent cells. Studies using protoplasts show that, for several viruses, the range of subliminally infectable species is often much wider than that of systemically infectable species, which could indicate that cell-to-cell movement also involves host factors. Some viruses are restricted, usually by host defenses, to the initially inoculated leaf, giving local infection. The most apparent manifestation of viral infection occurs when it spreads systemically to usually most parts of the plant.

Viruses differ in the number of plant species which they can infect systemically (Table II). Although obviously not all plant species–virus combinations have been tested, most viruses are reported to infect a few to a few tens of species. Some viruses have very wide host ranges, as demonstrated by cucumber mosaic cucumovirus (CMV) which infects more than 800 species in 85 families. Others have very restricted host ranges infecting one or a very few species (e.g., apple stem grooving capillovirus). Some viruses, although they spread systemically, are restricted to certain organs within the plant. For instance, luteoviruses are restricted to the vascular tissue and necroviruses are usually restricted to roots.

The breadth of the natural host range of a virus may differ from that of the experimental host range and in many cases is much narrower, as indicated in Table II. Among the factors which control this difference is the specificity of the biological vector in determining to which plant species a virus can be naturally transmitted.

Virus infection of plants results in a wide range of symptoms ranging from no or very mild symptoms (latent or cryptic infection) to death of the plant. Because it is usually to the "advantage" of the virus to maintain its host, it is generally considered that a virus causing minimal effect is more adapted to that plant species than one that kills the host.

The symptom type is usually specific to the virus or even to the virus group and can also vary with different environmental conditions (e.g., temperature and light). In some virus–host combinations there can be a local reaction around the point of virus entry into the leaf giving chlorotic or necrotic lesions. On mechanical inoculation the numbers of lesions can be related to virus concentration and local lesion assays are frequently used for virus quantification. Certain plant species (e.g., *Chenopodium* spp.) are local lesion hosts for many viruses. Necrotic, and sometimes chlorotic, local lesions often limit the virus to the point of entry. In other virus–host combinations local infection can be expressed as ringspots, with a necrotic (or chlorotic) ring surrounding a green or chlorotic area. The rings frequently spread and coalesce to give an "oak-leaf pattern."

The most common systemic symptom, reflected in the frequency of its use in virus names, is a mottle or mosaic pattern of light and dark green and sometimes yellow areas on the leaf. The distribution of the different colors is usually delimited by the veins on the leaf giving an irregular pattern in dicotyledonous species and striping in monocotyledons. Mosaic-like patterns can be found in other parts of the plant usually giving striping on petioles and stems and striping or color break in flowers. Other systemic color symptoms include general chlorosis or reddening of leaves, usually associated with phloem-limited viruses such as beet yellows closterovirus (BYV), chlorotic bands alongside the veins (chlorotic vein banding), vein clearing (often an early symptom), chlorosis of the veins, and chlorotic ringspots which can resemble those found on inoculated leaves and which can become necrotic.

Viruses can cause growth abnormalities in plants. The most common of these is stunting. Other less common growth abnormalities include reduction of the leaf lamina leading to strap-like leaves (e.g., infection of tomato with CMV and/or TMV), enations or outgrowths of the veins on the underside of the leaf (e.g., pea enation mosaic enamovirus), swelling of the stem (e.g., cocoa swollen shoot badnavirus), and tumor-like growths at wound sites (e.g., wound tumor phytoreovirus).

The symptoms induced by viruses often reflect histological and cytological changes within the plant. There are three main types of histological changes: necrosis, hypoplasia, and hyperplasia. Necrosis (discussed previously) results in local lesions or systemic necrotic reactions which can lead to plant death. Leaves with mosaic symptoms frequently show hypoplasia in the yellow areas with mesophyll cells being less differentiated, having fewer chloroplasts and fewer or no intercellular spaces. Among the hyperplastic reactions are vein clearing symptoms often associated with enlargement of cells around the veins and abnormal division and expansion of cells resulting in enations or stem swelling.

The cytological effects of virus infection on plants are twofold: effects on the plant cellular structure and organelles and manifestation of the virus. Most obvious effects are seen on chloroplasts which reflect the different color regions of mosaic or chlorotic symptoms. Some effects are also seen on mitochondria and nuclei. There is an increasing appreciation that viral replication usually involves membrane surfaces, and organellar membranes are implicated in some cases. For instance, tymovirus infection is associated with "flask-like" structures on the chloroplast membranes, tobravirus infections appear to associate with mitochondrial membranes, and nucleorhabdoviruses appear to involve a nuclear membrane. Characteristic inclusion bodies are formed from the accumulated virus particles or gene products of members of several virus groups and these are sometimes used for virus identification. For instance, crystalline inclusions are formed from the rod-shaped particles of viruses, such as TMV, BYV, and potato virus X potexvirus (PVX), and the isometric particles of viruses, such as tobacco necrosis necrovirus, broad bean wilt fabavirus, and turnip yellow mosaic tymovirus. Other gene products can form inclusion bodies, such as the nuclear and cytoplasmic inclusions of potyviruses, the X bodies of TMV, and the caulimovirus inclusions (transactivator protein).

III. SYMPTOM PRODUCTION

Obviously, symptoms are the expression of the interactions of the viral genome with that of the plant host. The application of molecular techniques to isolate host genes and to manipulate and mutate cloned infectious viral genomes is providing an increased understanding of virus–host interactions. For instance, the N gene, which controls necrotic local lesion production of TMV in *Nicotiana* spp., has been isolated and shown to have a similar structure to those of genes giving hypersensitive reactions to fungi and bacteria in plants. Even small changes in the viral sequence can give different symptoms. Thus, single nucleotide changes in the RNA genome of TMV or in the DNA genome of maize streak geminivirus results in different symptom expression.

Similarly, the application of techniques such as microscopy coupled with serology and hybridization is increasing the understanding of the changes occurring at the cellular level. For instance, in some potyviral infections host gene expression is altered, either inhibited or enhanced, in the cells in which the virus is replicating.

IV. PLANT-TO-PLANT SPREAD

To infect a plant, a virus particle has to cross the cuticle, cellulose cell wall, and the plasmamembrane.

The crossing of at least the first two barriers requires mechanical damage and all plant-to-plant virus spread involves mechanisms which inflict this sort of damage.

The experimental transmission of many viruses which are not phloem restricted is by mechanical inoculation, which usually involves rubbing sap from the infected plant or purified virus, together with an abrasive, onto a leaf of the healthy plant. In some cases, mechanical inoculation can be effected by pin pricking or injection.

Natural transmission of viruses occurs through a wide range of damage-inducing mechanisms (Table II) which, in most cases, are virus-specific. Some viruses, such as TMV and PVX, are naturally mechanically transmitted through, for example, broken hairs of infected and healthy plants rubbing together or through root contact. However, the vast majority of viruses require a biological vector, including arthropods which feed by piercing and sucking (e.g., aphids, leafhoppers, and mites), nematodes, fungi (Plasmodiophoromycetes or Chytridiomycetes), and pollen; some viruses are transmitted vertically by seed. Transmission requires specific interactions between the virus and its vector (Table II). Thus, if a virus is transmitted by an insect it will not be nematode or fungus transmitted. Furthermore, it will only be transmitted by one type of insect, for example, aphid and not leafhopper, and in many cases by only one or a few aphid species or even aphid biotypes. However, there may be transmission by different vector types within a viral genus (Table II).

Virus–vector interactions are classified into two basic forms: externally borne, in which the virus does not enter the vector, and internally borne, in which the virus enters the vector's body or cell. Most details are known about interactions between viruses and insect vectors. In the externally borne interaction there is usually rapid transmission and short persistence of the virus association with its vector. These interactions have been subdivided into non-persistent and semi-persistent (Table II) depending on the length of time that the insect is able to transmit the virus. Such viruses interact with a specific region(s) of the vector's anterior food canal: the non-persistent with the ends of the stylets and the semi-persistent with regions further down the food canal but anterior to the first valve. Non- or semi-persistent transmission involves interactions directly between the viral coat protein and a specific site in the vector's mouthparts (e.g., CMV) or a virus-encoded non-structural protein is required. This protein is termed a helper or transmission protein (component) and is found, for example, in potyviruses (helper component) and caulimoviruses (aphid transmission factor or P18). There are several theories regarding how transmission proteins function; the most widely accepted is that it links between specific sites on the virus particle and specific sites in the vector's mouthparts.

Viruses internally borne in arthropods are also termed persistent or circulative (Table II) because they enter the vector's body and circulate therein. In some cases they multiply in the insect and thus are called propagative. Internally borne viruses require longer feeds (hours to days) than do externally borne viruses for acquisition; there is a latent period before the vector becomes infective but the vector remains able to transmit for long periods, even the whole of its life in propagative viruses. All circulative viruses have to cross several barriers, usually the gut wall to enter the haemocoel and the salivary gland basal lamina to gain entry to the salivary glands, to be transmitted. Luteoviruses cross these barriers in their aphid vectors by endocytosis and movement through various vesicles. Little is known about the interactions between the virus and receptor sites at these barriers; these interactions presumably control vector specificity.

Some viruses require the presence of other viruses of effect their transmission. For instance, the genomes of umbraviruses are encapsidated in the coat protein of their helper virus usually a luteovirus, and are transmitted in the same manner as their helper virus. The leafhopper transmission of rice tungro bacilliform virus is thought to be dependent on a transmission protein provided by its helper virus, rice tungro spherical waikavirus (Table II).

V. MOVEMENT THROUGH THE PLANT

The initial infection of a plant is usually to one or a few cells and, to spread to the rest of the plant, the virus has to pass through other cells within the leaf (cell-to-cell movement) to reach the vascular tissue, where it spreads to more distal parts of the plant (long-distance spread). For cell-to-cell spread the virus has to cross the cellulose cell wall bounding each cell. Plant cells communicate with one another via cytoplasmic connections through the cell wall termed plasmodesmata. The effective diameter of plasmodesmata (the size exclusion limit, approximately, 1 kDa) is too small to enable virus particles to pass directly through them, but viruses have found ways of overcoming this problem. Two basic mechanisms are currently recognized. In the first, which is exemplified by TMV, the virus encodes a protein (P30), termed the movement protein, which is thought to temporally increase the exclusion limit of the plasmodesmata to approximately 10 kDa. P30 also binds to the TMV RNA, forming a long, narrow nucleoprotein structure which can pass through plasmodesmata. Details of the mechanisms are still being determined and recent observations indicate that plant cytoskeleton structures, microtubules and actin filaments, are involved. After it has been used for cell-to-cell movement, P30 is phosphorylated and "stored" near plasmodesmata, where it can be observed by immuno-electron microscopy.

In the second mechanism, which is demonstrated by cowpea mosaic comovirus (CPMV) and cauliflower mosaic caulimovirus (CaMV), one or more virus gene products (P58/48 of CPMV and gene 1 product of CaMV) are involved in forming tubules which pass from cell to cell, most likely through plasmodesmata. Virus particles which are thought to be moving from cell to cell have been observed in these tubules.

The processes determined to date allow the movement of the virus through mesophyll tissues to the vascular bundle sheath. It is indicated that in many plant species a different mechanism(s) of cell-to-cell movement operates within the bundle sheath.

Long-distance movement of viruses is usually via the phloem sieve elements, although the sobemoviruses also use xylem elements. Virus movement in the phloem follows the source-sink movement of photoassimilates. Little is known about how viruses enter and leave the phloem. Persistently-transmitted viruses, such as luteoviruses, are probably introduced directly into the phloem by their vectors and thus must be able to exit as complete virus particles. For many other viruses the viral coat protein is required for long-distance movement.

VI. VIRUS PARTICLE STRUCTURE

The viral genome of most plant viruses has a relatively simple structure that is surrounded by a capsid comprising one or more virus-encoded coat protein species. The capsid protects the genomic nucleic acid from degradation by nucleases and is also the interface between the virus particle and the external environment. It is involved in interactions such as those in the movement of the virus particles from plant to plant, and it is often involved in long-distance movement within the plant. There are two basic simple particle structures: rod-shaped and isometric (Table II).

The shapes of rod-shaped particles of plant viruses range from short, rigid rods (e.g., TMV, 300 $\times$ 18 nm) to long, flexuous rods (e.g., citrus tristeza closterovirus, 2000 $\times$ 12 nm). In these particles the coat protein subunits form a helix in which the RNA genome is embedded; the structure is stabilized by both protein:protein (including in TMV unusual carboxyl:carboxylate bonds) and protein:RNA links. The structure and assembly of TMV particles have been studied in detail. These particles are composed of approximately 2100 subunits of a single coat protein species arranged in a right-hand helix with a pitch of 2.3 nm. Each turn of the helix is made up of 16 1/3 subunits and each subunit associates with three nucleotides of the genomic RNA, which is set in a groove at approximately 4-nm radius. One end of the rod, that covering the 5' end of the RNA, is concave, and the other is convex. The particles have a central canal with a radius of 2 nm which is involved in particle assembly. Particle assembly is initiated by the binding of a double-disk of 17 subunits of coat protein per disk to a sequence-specific stem loop structure in the RNA termed the origin of assembly and which is about 1100 to 900 nucleotides from the 3' end of the RNA. This interaction results in the double-disk changing to a helical (lock washer) form with both ends of the RNA pro-

truding from the same side of the disk. A second double-disk adds to the first on the side away from the protruding RNA tails and interacts with the RNA 5' to the origin of assembly. Thus, the helical rod grows in the 5' direction with the addition of further double-disks pulling the RNA through the axial hole until all the RNA 5' of the origin of assembly is encapsidated. Assembly in the 3' direction occurs through the addition of small aggregates of coat protein to the helical rod and is much slower than assembly in the 5' direction.

Much less is known about the structure and assembly of other rod-shaped viruses. Most of the viruses with short, rigid particles have a helical repeat similar to that of TMV, but those with longer, flexuous particles have a more relaxed structure with a helical repeat at about 3.5 nm. Some closteroviruses have a short section of a second virus-encoded protein, similar to the main coat protein, at one end of their particles; it is thought that this protein is associated with insect transmission of these viruses.

Isometric virus particles have icosahedral symmetry with a quasi-equivalent arrangement of coat protein subunits according to strict surface lattice criteria. The structure of many plant virus isometric particles, those with diameters of approximately 25–35 nm, is that of triangulation number $T = 3$ (180 coat protein subunits). Technically, comoviruses which have particles in this range of diameters have $T = 1$ particles (60 subunits) but each "subunit" comprises two protein species, one with two structural domains. Some viruses with a diameter in the range of 15–20 nm have true $T = 1$ particles (e.g., satellite tobacco necrosis virus), and caulimoviruses have $T = 7$ symmetry (420 subunits). Geminivirus particles comprise two $T = 1$ icosahedra fused together. Isometric particles are stabilized by interactions which range from electrostatic protein:RNA bonds (e.g., CMV) to strong protein:protein bonds (e.g., comoviruses and nepoviruses). In the latter case, virus preparations contain a proportion of "empty" particles which do not contain nucleic acid. In some viruses (e.g., southern bean mosaic sobemovirus) divalent cations such as Ca^{2+} are involved in particle stabilization as well as protein:RNA and unusual carboxyl:carboxylate bonds similar to those found in TMV.

The bacilliform particle structure found in several viruses (e.g., alfalfa mosaic alfamovirus) is considered to be a modification of the isometric structure, the particles being wholly or mainly stabilized by protein:nucleic acid bonds. This bonding determines the number of coat protein subunits that have to interact with the viral genome and the bacilliform structure avoids the strict structural requirements of icosahedral symmetry. Particles of ilarviruses are thought to have a loose quasi-icosahedral structure which also overcomes the constraints of icosahedral symmetry.

Some viruses have more complex structures. For instance, the particles of members of the Reoviridae are composed of two shells with icosahedral symmetry together with some other proteins, resulting in surface projections or spikes. Those of members of the Rhabdoviridae have a nucleoprotein structure (comprising the viral RNA associated with virus-encoded proteins) encapsidated in a bacilliform-type structure which is then encased in a membrane through which virus-encoded glycoprotein spikes protrude. The glycoprotein-containing membrane is required for cell-to-cell spread in the animal (insect) vector in which these viruses also multiply. Similarly, tospoviruses have a nucleoprotein structure in a quasi-isometric particle surrounded by a glycoprotein-containing membrane. The nucleoprotein filamentous rods of tenuiviruses are usually found in plants without any surrounding membrane, but it is thought that more organized membranous structures might be found in their insect vectors.

VII. VIRAL GENOMES

Approximately 80% of plant viruses have RNA genomes (Table I), and approximately 65% have a genome of single-stranded (+)-sense RNA. The double-stranded DNA viruses are pararetroviruses which replicate by reverse transcription: there are no plant viruses with double-stranded DNA genomes replicated by direct DNA–DNA mechanisms.

The genome organizations of representatives of most of the 70 genera listed in Table I have been determined by sequencing. Basically, each virus encodes products which facilitate their replication, encapsidation, and in many cases their movement

within and between plants. Some also encode products such as proteases and translational control elements which are involved in the full expression of the genome.

A. Single-Stranded (+)-Sense RNA Viruses

As noted previously, this is the most common type of genome found in plant viruses. Genome sizes range from approximately 4 kb in many single component isometric viruses to more than 20 kb in some of the closteroviruses. The genome required for infection can be either one molecule (monopartite) or divided into two, three, or four molecules (multicomponent) (Table II). Multicomponent genomes are often encapsidated into several different particles (components), e.g., comoviruses into two isometric components and, hordeiviruses into three rod-shaped components.

The RNA of these viruses acts directly as an mRNA on entry into the cell. A range of structures are found at the ends of the viral genome. Many viruses have a methylated blocked 5′ terminal group $(m7G^{5'}ppX^{(m)}Y^{(m)}p. . .$, where $X^{(m)}$ and $Y^{(m)}$ are methylated bases), and others have a virus-encoded protein, the VPg, at the 5′ end. At the 3′ end, structures include polyadenylate sequences and tRNA-like sequences which can accept amino acids; these are thought to be involved in replication.

B. Single-Stranded (−)-Sense RNA Viruses

Two of the three families which have (−)-sense RNA genomes (Rhabdoviridae and Bunyaviridae) resemble, and have many genome organizational features in common with, viruses which infect animals; they also replicate in their insect vectors. For instance, Sonchus yellow net nucleorhabdovirus has a genome organization identical to those of animal rhabdoviruses except that it has an extra gene product most likely involved in cell-to-cell movement in plants.

Tospoviruses have three genome segments, the largest of which is (−)-sense monocistronic. The other two each encode two proteins, one from the virion sense RNA and the other from the complementary RNA; this is termed ambisense expression.

The genomes of tenuiviruses comprise four RNA segments. The largest is (−)-sense and encodes the RNA-dependent RNA polymerase. The other three have an ambisense arrangement.

C. Double-Stranded RNA Viruses

The genome organization of phytoreoviruses resembles those of animal-infecting reoviruses, with the genome being divided between 10 or 12 segments, depending on genus (Table II). These RNAs range from approximately 2.5 to 0.8 kb and, in most cases, are monocistronic. They encode various proteins which make up the virion structure and also non-structural proteins, including the RNA-dependent RNA polymerase. The cryptic viruses (Partitiviridae) have genomes comprising two monocistronic segments—one encoding the RNA-dependent RNA polymerase and the other the coat protein.

D. Single-Stranded DNA Viruses

The three genera of geminiviruses have circular single-stranded DNA genomes of 2.7–3.0 kb. Two of the genera, the Mastreviruses and Curtoviruses, have monopartite genomes, whereas the genomes of many members of the Begomovirus genus are divided between two molecules (Table II). In all cases the genome is transcribed bilaterally from a characteristic region (the common region) with the transcripts terminating on the opposite side of the genome, one transcript being from the virus-sense DNA and the other from the complementary DNA. Each of these transcripts expresses one or more proteins using mechanisms such as different transcript origins and splicing.

The genomes of the recently recognized Nanovirus genus comprises several (up to six or seven) small, circular, single-stranded DNA molecules of approximately 0.8–1 kb. Each DNA species is thought to be monocistronic.

E. Double-Stranded DNA Viruses

The genomes of plant double-stranded DNA viruses are circular molecules of approximately 7.2–8.0 kbp with one discontinuity in one strand and one or more in the other strand; these discontinuities represent the priming sites for DNA synthesis. These viruses have between three and six open reading frames depending on the genus, which mostly encode single proteins. However, one open reading frame of members of the badnavirus and "rice tungro bacilliform virus" genera encodes a polyprotein which is processed by a virus-encoded protease to give the functional proteins. These viral DNAs are transcribed to provide a more-than-genome-length RNA with terminal redundancy (often termed the 35S RNA) which is the template for the reverse transcription phase of replication and is the mRNA for several of the gene products.

VIII. VIRAL GENE EXPRESSION

As noted previously, plant viruses express several genes from their genomic nucleic acid, which in many cases is polycistronic. Because expression is in eukaryotic cells, viruses have to overcome the problem of eukaryotic ribosomes only translating the 5′ cistron on an mRNA; usually open reading frames downstream of the 5′ one are effectively translationally closed. There are two basic mechanisms by which plant (and other eukaryotic) viruses overcome this problem: either by dividing the RNA into several monocistronic units or translating the genetic information into a single protein (a polyprotein) which is then cleaved into functional units.

The mechanism of division of the RNA genome into several mRNAs is found in various forms in many plant viruses. Viruses such as TMV and PVX have a single genomic RNA from which 3′ co-terminal subgenomic mRNAs which encode the downstream genes at their 5′ ends are produced during virus replication. Thus, TMV expresses the 5′ replicase genes from the genomic RNA and the downstream movement and coat proteins from subgenomic mRNAs. In the multicomponent viruses, most, if not all, genome segments are effectively monocistronic. All of the genome segments of members of the Reoviridae are monocistronic. The larger two RNAs of members of the Bromoviridae are monocistronic; the smallest RNA is bicistronic, with the 3′

gene being expressed from a subgenomic RNA produced during replication.

The polyprotein mechanism is found in viruses such as potyviruses, comoviruses, and sequiviruses. In these viruses, the polyprotein is processed to the functional units by virus-encoded proteinases through complex pathways.

Various strategies maximize the capacity of the viral genome to express many products. There are some examples of bicistronic translation in which both the 5' and the 3' cistrons are translated (e.g., BYV); in these cases there are few, if any, extra translation initiation codons (AUG) in the 5' cistron. Two adjacent open reading frames can be expressed in a controlled manner from the same mRNA either by frameshift, in which the ribosome changes reading frame between two overlapping open reading frames (e.g., the 5' two cistrons of luteoviruses), or by readthrough of a weak stop codon (e.g., the 5' two cistrons of TMV). Both these mechanisms yield two proteins, the larger of which has the same N-terminal sequence as the smaller and a unique C-terminal sequence. Frameshift and readthrough occur at a low frequency, which controls the relative production of the two proteins. Other mechanisms include splicing (found in caulimoviruses, rice tungro bacilliform virus, and geminiviruses) and complex ribosome shunts and translational transactivation found in many of the double-stranded DNA viruses.

IX. VIRAL GENOME REPLICATION

Replication of plant viruses, as with all viruses, comprises four stages: uncoating, translation (which is often divided into two phases—early and late), replication of the viral genome, and encapsidation of the viral genome. Details of these stages differ depending on the types of genomes of the viruses.

A. Single-Stranded (+)-Sense RNA Viruses

Many of the (+)-strand RNA viruses are uncoated by a process termed cotranslational disassembly. In this process, the particle structure becomes relaxed, for example, by chelation of stabilizing divalent cat-ions or by protonation of the unusual carboxyl: carboxylate bonds, and ribosomes locate the 5' end of the viral genome. The ribosomes then translocate along the RNA translating the 5' cistron and at the same time displacing viral coat protein subunits. For some viruses (e.g., sobemoviruses and, alfamovirus) the whole genome is probably uncoated by this process. The genome of TMV is uncoated from the 5' end to the origin of assembly by ribosomes and then is thought to be uncoated from the 3' end to the origin of assembly by the viral replicase that is translated from the 5' end of the RNA.

All RNA viruses encode an RNA-dependent RNA polymerase (RdRP) which synthesizes both (−) and (+) strands. RdRPs do not have proof reading ability and thus there is a high error rate in newly synthesized RNA. The enzyme has several functional domains including the catalytic, capping, methylase, and often helicase domains, and the replication complex also contains various host factors. Replication usually occurs on membranes where the complex is assembled and starts with priming at the 3' end of the input RNA proceeding towards the 5' end and giving a (−) strand. This (−)-strand is then the template for synthesis of genomic (+) strands and, if required, subgenomic RNAs.

As noted previously, initial translation often occurs during particle disassembly and results in products (e.g., RdRP) required early in virus replication. Translation of products required late in replication (e.g., coat protein) is thought to be mainly from newly synthesized (+)-strand RNA. Coat protein subunits associate with the newly synthesized (+)-strand RNA molecules to assemble new virus particles.

B. Single-Stranded (−)-Sense RNA and Double-Stranded RNA Viruses

The basics of replication of these two groups of viruses are similar to those of (+)-strand viruses except in the early stages. Because the encapsidated genome cannot be translated directly, uncoating cannot occur by cotranslational disassembly; the mechanism of uncoating is unknown. Virus particles contain RdRP, which effects the transcription of (+) strands from which the rest of the replication cycle can then be undertaken.

C. Single-Stranded DNA Viruses

Viruses with single-stranded DNA genomes replicate by a rolling circle mechanism of DNA–DNA synthesis using both viral gene products and host proteins. It is thought that viral gene products interact with the host cell cycle and host products to overcome the limitation of one round of DNA replication per cell division.

D. Double-Stranded DNA Viruses

As noted earlier, plant viruses with double-stranded DNA genomes are pararetroviruses. The viral DNA is transcribed by the host DNA-dependent RNA polymerase II in the nucleus to give a more-than-genomic-length RNA which is the template for reverse transcription in the cytoplasm. The enzymes for the reverse transcription phase (reverse transcriptase and RNaseH) are virus coded.

X. VIRUS CONTROL

There are three basic approaches to controlling plant viruses. For perennial crops the use of clean planting material is essential since viruses will pass to vegetatively propagated progeny. It is possible to free certain plant species from certain viruses by using techniques such as tissue culture and heat treatment, but this is only economically viable for high-return crops such as ornamentals and fruit. For long-term perennial plantation crops, such as cacao and citrus, roguing infected plants is used, but this requires considerable labor in virus diagnostics and can lead to loss of crop production while the trees are being replaced.

The most common approach to controlling viruses of annual (and some perennial) crops is to control the vector. In the case of insects, the application of insecticides can be effective unless resistance to the chemical occurss. The efficiency of control also depends on the virus–vector interaction and is less for externally-borne viruses which are acquired and transmitted rapidly. Other cultural techniques, such as planting to avoid the vector and removal of sources of infection, can be effective.

The third approach is to breed resistance to the virus. However, there are relatively few resistance genes available in sexually compatible species for the most prevalent viruses of major crops. Recent advances in gene technology are leading to the ability to transform plants with transgenes which confer resistance. Many of these transgenes are derived from viral sequences which will interfere with or block the normal infection cycle of the virus. This transgenic approach is very promising, but because it is very recent the durability of the transgenes has yet to be fully determined.

Bibliography

Bisaro, D. M. (1996). *In* "DNA Replication in Eukaryotic Cells," pp. 833–854. Cold Spring Harbor Laboratory Press, Cold Spring Harbor, NY.

Bos, L. (1963). "Symptoms of Virus Diseases of Plants." Centre for Agricultural Publications and Documentation, Wageningen.

Caspar, D. L. D., and Klug, A. (1962). Physical principles in the construction of regular viruses. *Cold Spring Harbor Symp. Quant. Biol.* **27**, 1–24.

Dinesh-Kumar, S. P., Whitham, S., Choi, D., Hehl, R., Corr, C., and Baker, B. (1995). Transposon tagging of tobacco mosaic virus resistance gene N: Its possible role in the TMV-N-mediated signal transduction pathway. *Proc. Natl. Acad. Sci. USA.* **92**, 4175–4180.

Gildow, F. E. (1987). Virus–membrane interactions involved in circulative transmission of luteoviruses by aphids. *Curr. Topics Vector Res.* **4**, 93–120.

Granoff, A., and Webster, R. G. (Eds.) (1999). "Encyclopedia of Virology." 2nd Edition, Vols. 1–4. Academic Press, San Diego.

Hull, R. (1994a). Molecular biology of plant virus–vector interactions. *Adv. Dis. Vector Res.* **10**, 361–386.

Hull, R. (1994b). Resistance to plant viruses: Obtaining genes by non-conventional approaches. *Euphytica* **75**, 195–205.

Mason, W. S., Taylor, J. M., and Hull, R. (1987). Retroid virus genome replication. *Adv. Virus Res.* **32**, 35–95.

Matthews, R. E. F. (1991). "Plant Virology," 3rd ed. Academic Press, San Diego.

Mezitt, L. A., and Lucas, W. J. (1996). Plasmodesmatal cell-to-cell transport of proteins and nucleic acids. *Plant Mol. Biol.* **32**, 251–273.

Murphy, F. A., Fauquet, C. M., Bishop, D. H. L., Ghabrial, S. A., Jarvis, A. W., Martelli, G. P., Mayo, M. A., and Summers, M. D. (1995). "Virus Taxonomy: Classification and Nomenclature of Viruses." Springer-Verlag, Wien.

Pirone, T. P., and Blanc, S. (1996). Helper-dependent vector transmission of plant viruses. *Annu. Rev. Phytopathol.* **34**, 227–247.

Pringle, C. (1998a). The universal system of virus taxonomy of the International Committee on Virus Taxonomy (ICTV) including new proposals ratified since publication of the Sixth ICTV Report in 1995. *Arch. Virol.* **143**, 203–210.

Pringle, C. (1998b). Virus taxonomy—San Diego 1998. *Arch. Virol.* **143**, 1449–1459.

Rothnie, H. M., Chapdelaine, Y., and Hohn, T. (1994). Pararetroviruses and retroviruses: A comparative review of viral structure and gene expression strategies. *Adv. Virus Res.* **44**, 1–67.

Séron, K., and Haenni, A.-L. (1966). Vascular movement of plant viruses. *Mol. Plant Microbiol. Int.* **9**, 435–442.

Plasmids, Bacterial

Christopher M. Thomas
The University of Birmingham

GLOSSARY

bacteriophage or phage   Virus infecting a bacterium.

conjugative transfer   Transfer of plasmid DNA from one bacterium to another by a process that involves physical contact between donor and recipient bacteria.

phenotype   Property conferred on the host organism by a gene or group of genes.

plasmid   Extranuclear genetic element that can reproduce autonomously.

transduction   Carriage of nonphage DNA between bacteria protected by a phage coat and injected by the normal phage injection process.

transformation   Uptake of naked DNA by a bacterium followed by recombination, normally resulting in a change in the properties of the bacterium; for some bacteria, complex transformation processes result in conversion of DNA to single-stranded linear form so that only plasmid molecules which are dimeric, containing two tandem copies, can be converted back to circular form.

transposable elements   Defined segments of DNA which can move from one site to another in plasmids, chromosomes, or viruses, in some cases leaving a gap at the orginal sites but in other cases replicating the element so that one copy remains at the original site; transposition depends on recombination enzymes encoded within the element, although defective elements can sometimes be activated by another element in the same cell.

THE MICROBIOLOGY OF BACTERIAL PLASMIDS encompasses the study of the constituent parts of these genetic elements, the phenotypes they confer on their hosts, and the genetic processes they promote within the bacteria carrying them.

The existence of a sexual system promoting genetic exchange and recombination between mutant strains of *Escherichia coli* was discovered in the late 1940s and the role of a genetic factor called F (fertility) was established a few years later. The name plasmid was coined by Lederberg in 1952 to describe all extranuclear genetic elements capable of autonomous replication, but he also included temperate bacteriophage, such as lambda, under this definition. The role of plasmids in capturing antibiotic resistance genes and promoting the spread of multiple resistance determinants as single units was recognized in the late 1950s. The physical nature of plasmids as circles of DNA was demonstrated in the late 1960s and by the mid-1970s their potential as vehicles for recombinant DNA technology employing restriction enzymes and DNA ligase was already being exploited. At the same time, plasmids carrying biodegradation determinants were identified and there were the first hints that plasmid DNA could be transferred to plant cells, resulting in "plant cancers." Molecular biology has provided a good understanding of how plasmids multiply and spread, although many questions remain. Comparison of the DNA sequences and predicted replication proteins of many plasmids has identified the main groups to which plasmids belong. Linear plasmids have been discovered in many species as a result of new isolation and electrophoresis procedures. A challenge for the future is to assess the importance of plasmids as the basis of a pool of genes which are available to many species, thereby promoting both diversity and adaptability in bacteria.

I. BACTERIAL PLASMIDS: DEFINITIONS

A plasmid is a genetic element that can replicate in a controlled way physically separate from the chromosome. Although plasmids are known in a few eukaryotic systems, this article is limited to bacterial plasmids. Bacterial plasmids can be minimal, consisting only of the functions needed for controlled replication, or they can be almost indistinguishable from bacterial chromosomes in their size and number of genes carried. However, a variety of properties are used to distinguish plasmids from chromosomes. The most important criterion is that the plasmid does not carry genes that are essential for the structure and growth of the bacterium. In other words it should be possible to displace the plasmid from the bacterium without loss of bacterial viability as long as the environment is rich in nutrients (containing many alternative carbon, nitrogen, and energy sources) and does not contain specific harmful compounds, such as antibiotics or toxic ions. Recent DNA sequencing of some genetic elements in the size range 100–500 kb which could be plasmids has shown them to carry many genes of primary metabolism, typical of the bacterial chromosome in *E. coli*, and they have therefore been classified as chromosomes rather than plasmids. A second important criteria is that the replication of the plasmid is not coupled to the chromosome replication cycle. Although there has been dispute regarding whether plasmids such as the sex factor F replicate in step with the chromosome, the general pattern seems to be that plasmids control replication so that it occurs at a random time during the cell cycle, whereas all chromosomal replication origins initiate at a specific cell mass and thus maintain a 1:1 ratio with each other. Although the first plasmids identified were double-stranded DNA molecules, some plasmids have been found to have a double-stranded linear form. The first linear plasmids were found to have proteins attached to their ends like linear phage genomes, whereas other classes of linear plasmids have telomeres that contain a hairpin joining the two strands, making the ends resistant to exonucleases. Circular plasmids are generally covalently closed on both strands and are often negatively supercoiled, which means that they are stressed by having fewer turns around the helical axis than DNA has when allowed to adopt its most stable conformation. This stress gives the DNA a tendency to melt more easily. To compensate for this stress the molecule twists up on itself and forms a more compact form which increases its resistance to shear forces, an important principle exploited in plasmid isolation.

Plasmids can range in size from approximately 1 kb to hundreds of kilobases. Their copy number can vary from 1 per chromosome to 50 or more per chromosome. They may confer no identifiable phenotype (cryptic) or they may carry multiple phenotypic determinants. Any gene or genes can be carried by a plasmid, as illustrated by the range of phenotypes listed in Table I, but those that are found commonly on plasmids seem to be those that can function without the need for additional genes. For example, antibiotic resistance in many cases is conferred by a single enzyme which degrades or exports a specific antibiotic or modifies the target for the antibiotic. Other phenotypes such as heavy metal resistance or degradative pathways often involve multiple proteins, but on plasmids all necessary genes are generally found as a block, although not necessarily as a single operon. The other factor that may be important in selecting genes which are favored by a plasmid location is the selection of those which confer a phenotype that can be strengthened by an increase in gene dosage. For example, the level of resistance to penicillin conferred by β-lactamases is strongly dose dependent.

II. OVERVIEW OF PLASMID EVOLUTION

Some plasmids may be as ancient as the chromosomes of the bacteria that carry them. Others may have evolved recently. A plasmid comes into existence when a segment of DNA gains the ability to replicate autonomously. This involves the creation of a replicator region by mutation. Such a unit of replication is called a "replicon." The term originally referred to the unit of replication—that is, the segment of DNA that is replicated—but it currently is

TABLE I
Examples of Phenotypes Conferred by Bacterial Plasmids

Phenotype	Element	Microbe
Antibiotic production	SCP1	*Streptomyces coelicolor*
Antibiotic resistance	RP4 (IncP)	*Pseudomonas aeruginosa*
Bacteriophage resistance	pNP40	*Lactococcus lactis*
Bacterocin	p9B4–6	*Lactococcus lactis*
Biphenyl/4-chlorobiphenyl degradation	Tn4371	*Ralstonia eutropha*
Capsule production	pXO2	*Bacillus anthracis*
Chemotaxis/chemosensor	pNod	*Rhizobium leguminosarum*
Colicin immunity	ColE2–P9$^+$	*Escherichia coli*
Colonization antigens	pK88	*Escherichia coli*
Conjugative transfer	F	*Escherichia coli*
Crystal protein (insecticide)	pHD2	*Bacillus thuringiensis*
Ecological competence in soil	pRtrW14-2c	*Rhizobium leguminosarum*
Electron transport proteins	pTF5	*Thiobacillus ferroxidans*
Enterotoxins	pTP224	*Escherichia coli*
Exopolysaccharide production (galactoglucan)	pRmeSU47b	*Rhizobium meliloti*
Galactose epimerase	pSa (IncW)	*Escherichia coli*
Gas vacuole formation	pHH1	*Halobacterium*
H$_2$S production	pNH223	*Escherichia coli*
Hemolysin production	pJH1	*Streptococcus faecalis*
Herbicide degradation (2,4-D)	pJP4	*Alcaligenes eutrophus*
High rate of spontaneous mutation	pMEA300	*Amycolatopsis methanolica*
Hydrogen uptake	pIJ1008	*Rhizobium leguminosarum*
Insecticide degradation (carbofuran)	pCF01	*Spingomonas* spp.
Iron uptake	pJM1	*Vibrio angularum*
Lactose fermentation	pLM3601	*Streptococcus cremoris*
Lysine decarboxylation	pGC1070	*Proteus morgani*
Melanin production	pNod	*Rhizobium leguminosarum*
Metal resistance	pMERPH (IncJ)	*Pseudomonas*
Nitrogen fixation	pIJ1007	*Rhizobium leguminosarum*
Nodulation functions (Sym plasmid)	pPN1	*Rhizobium trifoli*
Oncogenic suppression of Ti plasmid	pSa	*Shigella*
Pigmentation	pPL376	*Erwinia herbicola*
Plant alkaloid degradation	pRme41a	*Rhizobium meliloti*
Plant tumors	Ti plasmid	*Agrobacterium*
Protease production	pLM3001	*Streptococcus lactis*
Restriction/modification	pR1eVF39b	*Rhizobium leguminosarum*
Reverse transcriptase (mitochondrial plasmid)	pFOXC1	*Fusarium oxysporum*
Sex pheromones	pAD1	*Streptococcus faecalis*
Siderophore production	pDEP10 (IncF1me)	*Escherichia coli*
Sucrose utilization	CTnscr94	*Salmonella senftenberg*
Sulfur oxidation (dibenzothiophene)	pSOX	*Alcalignes eutrophus*
Tolerance to acidity	pRtrANU1173b	*Rhizobium leguminosarum*
Tolerance to NaCl	pRtrW14-2b	*Rhizobium leguminosarum*
Toluene degradation	Tol plasmids	*Pseudomonas putida*
UV protection	R46 (IncN)	*S. typhimurium*
UV sensitization	R391 (IncJ)	*Escherichia coli*
Virulence	pXO1	*Bacillus anthracis*

generally used to define the genetic elements needed for autonomous replication. Many strategies for replication are employed by plasmids (see Section V). The strategy which is easiest to imagine evolving is that employed by ColE1 and similar plasmids, and it involves a transcriptional unit mutating so that the transcript it produces remains associated with the DNA template so that it is able to prime replication. Mutant replicons which replicate more efficiently are likely to proliferate up to the point that they start depressing host growth significantly because of metabolic burden or other deleterious effects. Mutants which acquire a way of autogenously regulating their replication so that initial replication is rapid but reaches a limit will then be selected. Regulation by production of antisense transcripts is one of the simplest ways in which such an autogenous circuit could appear: Transcription of the opposite strand to the pre-primer RNA creates an inhibitor that builds up as the plasmid copy number increases and thus eventually switches off replication when a critical threshold is reached.

Starting from the autonomous replicon, gene acquisition by recombination and transposition could lead to possession of auxiliary maintenance functions and transfer functions, which provide an advantage for the element, and phenotypic markers, which may aid the plasmid by giving its host a selective advantage. The first acquisition of extra DNA would occur in a location which does not inactivate the replicon. Subsequent insertions would tend to occur in the same region because this region, which can accommodate insertions, is now enlarged. There will therefore tend to be apparent clustering of such insertion events, and thus the segments that are acquired in this way will also appear clustered.

The plasmids we currently see have probably been through many rounds of recombination and transposition and deletion and insertion. Events which are deleterious would not survive, whereas events which give a fitter plasmid will displace the parent. Rearrangements which result in functionally complementary genes becoming clustered may be selected because such clustering reduces the chances of these functions becoming separated. Once such compact groups of complementary functions exist, they will tend to be inherited as modules—recombination

events within the module will be disfavored compared to the ones that preserve the module. On the other hand, homologous recombination will lead to reassortment of allelic differences within a modular unit. Modules built up in this manner may be gained by a different genetic system and exploited in a different way. For example, the complex units that provide plasmids with a key part of their transfer apparatus can be shown by sequence alignments to derive from the same genes as those which drive the export of toxins from the cell. The logic is that both processes involve transport of proteins to the surface of the bacterial cell. Similar gene order over a 10 or 11 gene cluster indicates that the genes were inherited as assembled gene blocks.

Plasmids are thus mosaics of genes, each of which may have its own history. However, to understand plasmids one must consider their component parts and then examine the groupings in which they are found. Some combinations of genetic fragments may be longer lived than others, as illustrated by the backbone of the IncP-1 plasmids (Fig. 1). However, given the diversity of plasmids, it is only now, with DNA sequencing and computer databases, that this approach is feasible. Previously, other approaches were used to provide order from the bewildering diversity.

III. INCOMPATIBILITY AND PLASMID CLASSIFICATION

The number of plasmids now known is very large. Catalogs describing some of these plasmids are available from public culture collections, but many plasmid descriptions are buried in the literature. Methods for grouping plasmids are essential for the cataloging process. The most universally accepted way is to start by grouping plasmids on the basis of their replication and maintenance systems. A method of determining whether two plasmids use the same system was devised in the 1960s—the incompatibility test. Plasmids with different selectable markers were introduced into the same cell line, selecting for both plasmids. Selection for one or both of the plasmids was then removed and a culture grown. After approximately 20 generations of growth, the bacteria were

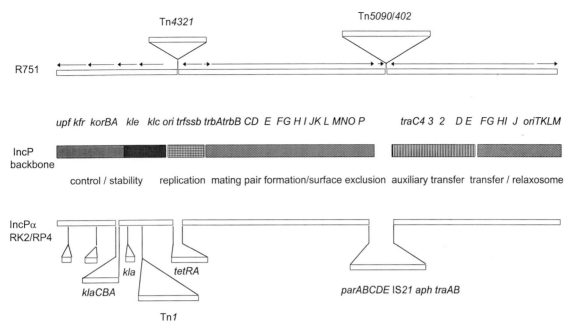

Fig. 1. Organization of two related plasmids belonging to the same incompatibility group (see Section III). The IncP-1 plasmids were originally isolated in *Pseudomonas aeruginosa* but they can transfer efficiently between and maintain themselves stably in all gram-negative bacterial species. Their transfer system can also promote transfer of DNA to gram-positive bacteria and yeast. The complete sequences of both plasmids are known. The core of replication, stable inheritance, and transfer genes can be aligned as shown. For simplicity, they are shown linearized at an arbitrary point, although these are circular plasmids. Arrows indicate the transcriptional organization of these genes. Although the two plasmids carry different phenotypic markers, the insertion of DNA carrying these genes, often but not always on transposable elements, has occurred repeatedly in the same places in the backbone. These plasmids, isolated in the late 1960s and early 1970s, confer antibiotic resistance, but related plasmids found in the environment carry biodegradation capability.

spread on selective medium to determine whether both plasmids were still present in all bacteria or whether segregation had occurred to give bacteria with one plasmid or the other but not both. Plasmids that caused each other to segregate were called "incompatible" and intuitively it was concluded that this was because they depended on the same replication and maintenance functions. Therefore, incompatible plasmids were categorized into the same group, whereas plasmids that were compatible were categorized into different groups. Since these groupings depended on the use of the same host, different incompatibility groupings were established for enteric bacteria (particularly *E. coli*), *Pseudomonas* species (since most *Pseudomonas* plasmids seemed able to replicate in more than one *Pseudomonas* species), and *Staphylococcus aureus*, in which much early plasmid work was done because of the prevalence of

resistance to commonly used antibiotics such as penicillin. Some plasmids were never classified for a variety of reasons. Other plasmids showed anomalous behavior, displacing but not being displaced or showing apparently more than one type of incompatibility. Many of these properties are the result of the plasmids being cointegrates of plasmids of two groups so that they can use alternative replication systems. Thus, if one of the two systems is competing with that on another plasmid, the second replication system in the hybrid can continue to maintain the plasmid. This competition will therefore add a selective pressure for plasmids which are not hybrids to evolve new specificities rapidly, leading to related plasmids which have novel initiation and control specificity so that they are not subject to direct competition. Because of the complexity of real incompatibility data and because DNA sequencing is becoming easier, comparison of plasmid sequences is now a much simpler and more reliable indicator of plasmid groups. However, it does not provide information about incompatibility, which can only be determined empirically. In some cases, the real incompatibility data are very important for assessing plasmid behavior in a microcosm because they define those that coexist and those that compete.

IV. THE HORIZONTAL GENE POOL

A bacterium can carry many plasmids. These can be completely unrelated to each other or may be derived from a common ancestor whose replication/replication control systems have diverged sufficiently so that they do not compete with each other. Low-copy-number plasmids can be hundreds of kilobases in size and may carry approximately 10% of the bacterial genome, although as noted previously an element of this size is generally referred to as a minichromosome if it carries a standard density of housekeeping genes, as has been established for certain *Rhodobacter* and *Pseudomonas* species. The presence of two or more large plasmids (as in *Rhizobium* species) may result in 10–20% of the cellular DNA being plasmid encoded. Two strains may differ simply in the plasmid(s) they carry. These plasmids may be lost and replaced by others. These plasmids may come from a closely related strain or from an unrelated species which carries broad host range plasmids. The potential number of genes that these populations of plasmids can carry into a bacterium is very extensive. This pool is called the horizontal gene pool since it is not confined to a single species. Any gene that moves onto a plasmid enters this pool as long as the plasmid is not confined to its host. Current knowledge of the bacterial kingdom is limited to a relatively small sector. Sampling of plasmids from new bacteria from uncharacterized environmental niches suggests the existence of many plasmids unrelated to plasmids that are already known. However, the sequencing studies performed to date have provided a base to which it is increasingly possible to find matches for new plasmids when some or all of their DNA sequence is determined. A major research priority for bacterial genetics is to sample

the horizontal gene pool of all culturable bacteria and, if possible, the nonculturable bacteria in order to establish the diversity of this pool and provide data on the extent of exchange occurring within this pool.

V. REPLICATION STRATEGIES

A general characteristic of plasmids is their feedback control loop, which responds to bacterial growth to allow replication to occur randomly throughout the cell cycle. Plasmid molecules appear to be randomly selected for replication. Recent evidence suggests that DNA PolIII may be located primarily at the predivision site (PDS) at the cell midpoint. For some groups of plasmids there is evidence that the Rep protein functions at a membrane site. Selection of a plasmid DNA molecule for replication depends on both initiation and elongation competence. One of the problems for all plasmids is that they must move away from the PDS prior to cell division and migrate to the new zone of PolIII replication potential. This will particularly be a problem for low-copy-number plasmids. For plasmids such as F, P1, and RK2, this may explain why partitioning appears to have evolved to take their DNA to the one-quarter and three-quarter positions where the next round of replication should occur. Many mechanisms have evolved to allow initiation. These have been studied in detail for circular plasmids and are described later, followed by discussion of the mechanisms of replication control. For linear plasmids, recent studies have shown that replication initiates internally and proceeds towards the ends. How the ends are replicated is the subject of research.

A. Rolling Circle Replication

One of the most ubiquitous replication strategies involves the introduction of a nick into one strand of the plasmid DNA by a Rep protein encoded by the plasmid (Fig. 2). This double-strand replication origin (*dso*) generally consists of at least two inverted sequence repetitions (IVR) in the DNA. One IVR is the specific binding site for the Rep protein dimer, whereas the other is the target for the nicking activity of the Rep protein, where it becomes attached to the

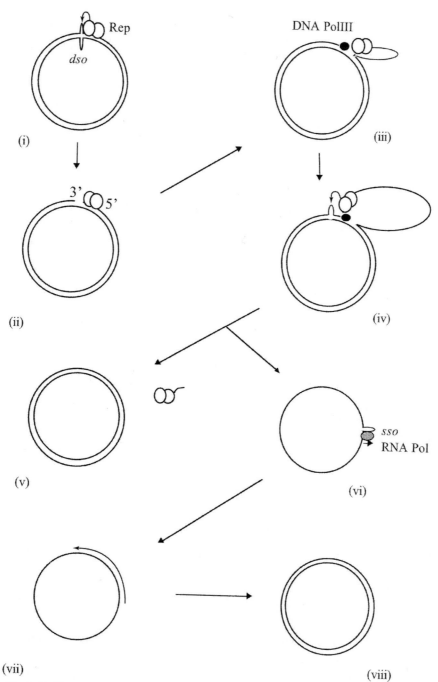

Fig. 2. Rolling circle replication. Dimeric Rep protein binds to its target in the double-strand origin (*dso*) next to an inverted sequence repetition which may be extruded as hairpins in supercoiled DNA (i). Nicking of one strand occurs and Rep remains covalently attached to the 5′ end (ii). The 3′ end is extended by DNA PolIII, displacing the paired strand (iii). When replication goes past *dso* again, Rep attacks it (iv), reforming the circle (v), displacing a ssDNA circle (vi), and leaving a short ssDNA tail. The ssDNA circle is converted back to dsDNA after RNA polymerase has created a primer on a hairpin stem formed at the single-strand origin (*sso*) (vi–viii).

5′ end of the nick, leaving the 3′ end to prime leading strand synthesis. Replication from this 3′ end copies the intact DNA circle, displacing the nicked strand. In phage replication this strategy can generate linear molecules consisting of many copies of the circular genome by repeated replication around the circle—hence the name rolling circle replication (RCR). There is controversy whether the nicked *ori* IVR is extruded as a hairpin from the supercoiled plasmid DNA molecule: Current data suggest that extrusion is not essential. For some groups of RCR plasmids the Rep protein has been identified in the membrane fraction, and this may help to bring the plasmid to PolIII which will copy the leading strand and this will also hold the 5′ end close to the replication fork—a vital feature to allow termination which occurs when PolIII replicates past the origin again. This allows Rep to nick the *dso* again, transferring the 5′ end which was attached to Rep to the 3′ end and thus reforming the DNA circle. The extra piece of DNA past the origin is transferred to Rep, where it remains, resulting in inactivation of Rep after one round of replication. Lagging strand synthesis occurs as a separate process starting from an origin (*sso* or single-strand origin) which forms a hairpin in the single-stranded lagging strand DNA which has been displaced by the replication fork. The hairpin is recognized by RNA polymerase in a highly species-specific way to create an RNA primer which is first elongated by DNA PolI and then by PolIII. How termination of lagging strand synthesis occurs is not known. Although most common in gram-positive bacteria, such plasmids are also found increasingly in gram-negative bacteria. Some such plasmids (e.g., the well-characterized plasmid pLS1 from *Streptococcus*) can replicate efficiently in both gram-positive and gram-negative species. The most limiting factor seems to be the efficiency of *sso* function.

B. DNA PolI-Dependent Replicons

One of the conceptually simplest forms of replicon is illustrated by plasmids such as ColE1, whose relatives have been exploited in gram-negative bacteria as cloning vectors of the common pUC series. These plasmids will continue to replicate in bacteria after protein synthesis is inhibited but will not replicate

if transcription is inhibited or if DNA PolI is lacking, as in a PolA⁻ strain. Studies during the 1970s and 1980s showed that replication depends on a transcript produced by RNA polymerase from a promoter approximately 500 nucleotides upstream of the *ori* region (Fig. 3). This preprimer is folded into a complex secondary structure which allows it to remain associated with the DNA of the *ori* region, where it is processed by RNaseH to generate a primer which is used by DNA PolI to initiate leading strand synthesis. After leading strand extension for approximately 100 nucleotides, a primosome assembly site is reached which allows lagging strand synthesis to be initiated, and at this stage DNA PolIII takes over leading strand synthesis.

Many other plasmids seem to use a similar initiation process based on the processing of a transcript created by RNA polymerase. In the case of the pAMβ1 family of plasmids from the gram-positive bacteria, it seems clear that the role of the Rep protein is to promote processing of the transcript at the replication origin. A similar situation may apply to the *repFIC* replicon found in the well-studied plasmid R1, although this plasmid does not appear to use DNA PolI to initiate leading strand synthesis.

C. Iteron-Activated Replicons

The replication origins of many plasmids contain tandemly repeated sequences adjacent to an AT-rich region, where directed melting of the DNA occurs and leading strand synthesis is initiated. The repeats termed iterons (because they are iterated), each approximately 21 bp in length, are the binding sites for a plasmid-encoded Rep protein. This arrangement is similar to that in all known chromosomal replication origins (*oriCs*), in which the iterons bind DnaA protein, as well as in the origins of such bacteriophage as lambda, in which the iterons bind the O protein. Many plasmid replication origins of this sort also contain binding sites for DnaA, and these plasmids display a varying degree of dependence on the activity of this host protein. Thus, in *E. coli* mutants that are temperature sensitive for replication due to a mutation in *dnaA*, the F plasmid can suppress the temperature sensitivity if it integrates into the chromosome so that chromosome replication can be initi-

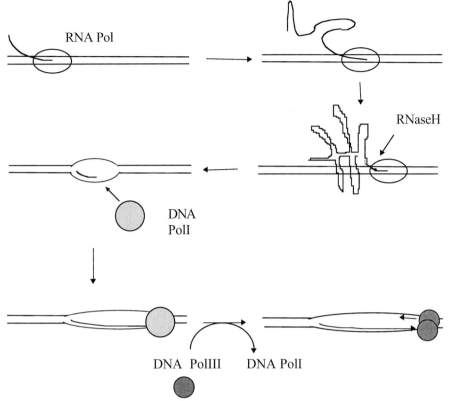

Fig. 3. DNA polymerase I-dependent replication as established for ColE1. RNA polymerase generates a transcript that folds into a three-dimensional structure that promotes association of the RNA with the DNA at the replication origin. Here, it is processed by RNaseH to generate a primer which is used by DNA PolI to initiate leading strand synthesis. After approximately 100 nt the replication fork reaches a primosome assembly site which initiates lagging strand synthesis. DNA PolIII takes over from DNA PolI. In the IncFII plasmids a similar creation of an RNA primer may occur by Rep-mediated processing.

ated from within the F plasmid ("integrative suppression"). The broad host range IncP-1 plasmids have four DnaA binding sites upstream of a group of five iterons in the replication origin, and these are essential for replication in most bacterial species except *Pseudomonas aeruginosa,* in which their removal causes only a partial defect in replication resulting in a lowered copy number. The role of DnaA in these plasmids may be to lead in other host proteins such as DnaBC, whose helicase activity is needed to unwind the replication origin so that DnaG primase can enter and generate the primer for leading strand synthesis. However, clearly DnaA is not essential in all circumstances; therefore, some Rep proteins can

perform the recruitment process. Many Rep proteins are identified largely as dimers, but in general it appears that the active form of Rep is the monomer. Also, various host chaperone proteins (proteins that bind to and act catalytically to promote the normal folding and assembly of the active form of other proteins) are needed to convert dimers to monomers. Dimers can actually be inhibitory as described in Section V.D. The IncQ plasmids are unusual in coding for three Rep proteins. One of these proteins is an iteron-binding protein (like Rep of other plasmids), whereas the other two proteins are a helicase and a primase, making the plasmid largely independent of the host replication machinery, apart from DNA polymerase. This correlates with an extremely broad host range.

D. Replication Control

An essential feature of plasmids appears to be the ability to sense their own copy number and to switch off their replication as the copy number increases to higher than the norm. It is generally accepted that copy number control systems conform in some way to the "inhibitor dilution model," in which the plasmid produces an inhibitor either constitutively so that its concentration is proportional to the copy number or in a burst after replication so that further replication is prevented until the bacterium has grown enough to dilute the inhibitor to below its critical inhibitory level. This principle was confirmed by the demonstration that in a hybrid between a low-copy-number and a high-copy-number plasmid, the high-copy-number replicon dominates, silencing the low-copy-number replicon, unless the high-copy-number replicon is inactivated. The production of an unstable inhibitor that senses copy number is illustrated best through the high-copy-number plasmids ColE1 (from *E. coli*) and pT181 (from *S. aureus*) and the low-copy-number plasmid R1 (from *E. coli*). In each case, a highly structured antisense RNA attacks an RNA target that is essential for replication. In ColE1 RNAI binds to preprimer RNA and prevents it from folding to the form that can be processed into primer essential for replication initiation. In pT181, Cop RNA induces transcription termination on the mRNA that codes for the Rep protein. In R1, Cop prevents translation of a leader peptide needed to allow translation of the rep mRNA. In each of these cases, the antisense RNAs are highly reactive, binding to their targets through single-stranded loops, and it is the speed of these reactions that is essential for their activity.

Iteron-containing replicons seem to use a different control strategy that is not fully understood. In these plasmids, it is generally found that overproduction of the Rep protein does not result in runaway replication as might be expected, and in some cases overproduction of Rep is actually inhibitory. Mutations affecting copy number show changes in the ability of Rep to form multimers and to interact with iterons in the replication origin. Since the active initiator is a monomer, it is suggested that dimers link replicated molecules together (handcuffing) and block further replication. The mechanism by which handcuffed molecules are pulled apart is not clear, although it may normally be connected to an active partitioning cycle. Some Rep proteins control their own synthesis by binding to an inverted repeat of the iteron binding sequence, suggesting that there are different multimeric forms of the Rep protein able to alternatively bind an inverted repeat or two direct repeats on different molecules. Often, there are also additional groups of iterons, adjacent to the replication origin, which are not essential for replication but which modulate initiation frequency, possibly by titration of Rep or by formation of inhibitory complexes.

VI. STABLE INHERITANCE

Plasmid copy number is controlled as described in Section V.D. Control systems have generally evolved to maintain large plasmids (>50 kb) at copy numbers similar to that of the bacterial chromosome, whereas small plasmids tend to be present at much higher levels (15–20 per chromosome or more). At such high copy numbers, even in slowly growing bacteria there would be 30–40 copies of the plasmid in pre-divisional cells. In faster growing bacteria there will be proportionately more chromosomal equivalents per cell (because the speed at which an individual chromosome can be replicate does not increase as growth rate increases so that increased replication rate is achieved by more frequent initiation of the replication cycle), and there will generally be an increase in the number of plasmids, although not necessarily in direct proportion to the chromosome. The probability of producing a plasmid-free cell is $P = 2^{-(2n-1)}$, where n is the copy number per baby cell, assuming that each copy of the plasmid is distributed randomly. For example, a plasmid that is present at 1 copy per baby cell will, on average, be present at 2 copies prior to cell division, and in half of divisions both plasmid molecules will be transmitted to only one of the two daughter cells. Thus, one-fourth of cells in each round of cell division will have lost the plasmid. For a plasmid that is present at 10 copies per baby cell, there will be 20 copies in predivisional cells and the probability of a plasmid-free cell occurring is $0.5^{19} = 1.9 \times 10^{-6}$; thus, approximately 1 in a million bacteria would be expected to lose the plasmid at each round of cell division. Accordingly, high-copy-number plasmids have generally acquired those stable inheritance functions related to ensuring that individual plasmid genome copies do segregate randomly. On the other hand, low-copy-number plasmids have also acquired genes which allow them to segregate in a nonrandom way and to combat plasmid loss in other ways.

A. Multimer Resolution

All plasmids have the potential to form multimers by homologous recombination between identical DNA molecules in the bacterial cell. This should lead to great segregational instability because dimers and higher multimers tend to take over if unchecked. Also, by reducing the number of physically separate DNA molecules, they reduce the efficiency of both random and active partitioning for a given number of copies of the plasmid genome. A dimer has two origins and is therefore twice as likely to be chosen for replication as a monomer. However, once initiated it will produce two plasmid genome units and thus this effectively counts as a double-replication event. If the plasmid controls replication by producing a repressor in proportion to the number of plasmid copies present, replication of a dimer doubly reduces the potential for monomers to replicate and the cell enters a runaway situation in which dimers take over ("dimer catastrophe"). This scenario should not apply to RCR plasmids since the replication mechanism will terminate when it reaches the second origin in the dimer and will generate a monomer.

Many multimer resolution systems have been identified, the best studied of which is the *cer* system of the high-copy-number *E. coli* plasmid ColE1, whose replicon is related to the replicon found in many of the high-copy-number cloning vectors. The *cer* locus encodes a target for four host proteins: ArgR, PepA, XerC, and XerD. ArgR is the repressor that coordinates expression of operons encoding arginine metabolism genes, whereas PepA is a peptidase involved

in polypeptide processing. XerC and -D are recombinases also involved in chromosomal site-specific recombination. Thus, ColE1 has recruited many host proteins, including two with apparently unrelated function, to promote its stable inheritance. Both ArgR and PepA are hexamers composed of two layers of trimers. They have the potential to stack on top of each other in alternating layers. They form the core around which the plasmid DNA wraps and holds XerC and XerD on two sides of this trimeric structure. Potentially, this recombinosome could promote recombination between sites either in the same molecule (resolution) or in different molecules (dimerization). Mutant sites can be obtained which catalyze both of these reactions, but normally the resolution pathway is favored to the exclusion of the dimerization pathway. It is thought that this is due to the need for the protein complexes at the *cer* site to be held next to each other in a supercoiled dimer for long enough to allow the slow strand exchange reaction to occur. This would not be possible in noncovalently joined molecules. The *cer* region also contains an additional stability determinant—a promoter that directs the transcription of a short region called *rcd* which causes retardation of cell division. The promoter appears not to function in monomeric plasmids but may be activated in the nucleoprotein complex at the paired *cer* sites. The transcript folds into two hairpins which occupy almost the whole length of the RNA. Although the target for this molecule is not known, it seems to serve the purpose of slowing growth of bacteria with dimers or completely halting division of bacteria which have been taken over by multimers. It therefore seems to preempt the need for postsegregational killing (*psk*) systems.

B. Killing of Plasmid-Free Segregants

psk systems were first discovered when it was found that the *ccd* locus on plasmid F, which appeared to block cell division in bacteria with a plasmid that could not replicate, was responsible for generating two populations of the bacteria as the plasmid was lost: viable bacteria with the plasmid and nonviable bacteria which had lost the plasmid. It was subsequently shown that the CcdB protein is a poison of DNA gyrase when it is not associated with an antidote protein, CcdA. Both CcdA and CcdB are produced from the plasmid, but CcdA decays rapidly when the plasmid is lost and the action of CcdB kills these plasmid-free segregants. Plasmid loss can be mimicked by addition of rifampicin, which blocks transcription, and many of the *psk* systems have been discovered or confirmed by this test.

The best studied *psk* system is the *hok/sok* system of plasmid R1 (Fig. 4). Host killing is achieved by the 52-amino acid polypeptide Hok, which causes

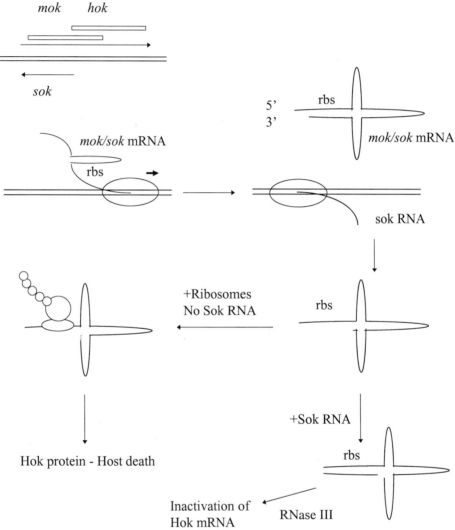

Fig. 4. Postsegregational killing by the *hok/sok* system of the *E. coli* IncFII plasmid R1. The locus consists of cotranscribed *mok* and *hok* and countertranscribed *sok*. *hok* is cotranslationally coupled to *mok*. During transcription the *mok/hok* transcript folds so that translational signals are buried. The final mRNA refolds so that the 3′ tail now takes over the inhibitory role, but slow degradation from the 3′ end eventually uncovers the translational signals (rbs, ribosome binding site). If the plasmid is still present, then the Sok RNA it produces binds to the *mok/hok* RNA and blocks translation: however, if the plasmid has been lost and the Sok RNA degraded then the mRNA is activated, Hok is produced, and the bacteria die.

membrane depolarization. Translation of *hok* depends on activation by translation of the preceding and overlapping gene *mok*. The need for *mok* thus helps to silence *hok* when it is repressed. As it is transcribed, the primary transcript of the *mok/sok* region forms a metastable structure which buries the *mok* and *hok* translational signals in an inaccessible form. When the transcript is complete the 3′ tail interacts with the 5′ end, causing refolding and favoring an alternative structure in which the translational signals for *mok* and *sok* are now blocked by pairing with the 3′ tail. Degradation of the mRNA occurs from the 3′ end so that as removal of nucleotides proceeds the *mok* and *hok* translational signals become accessible again. In cells with plasmid, an unstable countertranscript (Sok) is constantly being produced and blocks translation of the activated *hok* transcript. However, if the plasmid has been lost, then by the time the *hok* mRNA has been activated the Sok RNA has disappeared and Hok is produced, resulting in cell death. The action of this *hok/sok* locus is sometimes referred to as programmed cell death. It certainly is one of the most interesting examples of a timed sequence of events in bacteria and shows how different cellular responses can be triggered depending on how the system is set by the state of the cell.

C. Active Partitioning

Many low- or medium-copy-number plasmids appear to be lost at a lower frequency than one would expect on the basis of their copy number. Genetic regions required for this activity have been mapped. In some cases (e. g., the plasmid pSC101), the defect can be suppressed by mutations in host genes such as topoisomerase, suggesting that the partitioning conferred by the region involves control of plasmid DNA conformation. The discovery of DNA gyrase binding sites in the *par* region confirms this hypothesis. However, in other cases there is increasing evidence that the increased stability conferred by *par* genes is due to a cycle of pairing and directed separation, akin to the eukaryotic mitotic cycle. The best studied of these systems belong to a family first discovered on the plasmid F and the prophage (dormant or lysogenic phase of a bacteriophage) P1, which

exists as a plasmid. Another interesting system is found on the low-copy-number plasmid R1. All these systems encode two *trans*-acting proteins and a *cis*-acting sequence which is termed the centromere-like sequence because, like that region in eukaryotic chromosome, it appears to provide the handle which allows directed movement of the plasmid. Of the two *trans*-acting proteins, the B protein, encoded second in the operon, binds to the centromere-like sequence, whereas the A protein is an ATPase that may provide energy for a vital step in the partitioning process. The A and B proteins of F and P1 belong to large protein families, encoded by many plasmids as well as the majority of bacterial chromosomes, and it is thought that this may reflect a common partitioning process for both plasmids and chromosomes—perhaps completing a stage of pairing followed by directed separation.

The behavior of these systems has been visualized by fluorescence microscopy. This is achieved by fixing and permeabilizing the bacteria followed either by binding of fluorescent antibodies to the plasmid proteins or by annealing fluorescent DNA probes specific for the plasmid. Alternatively, the plasmid-encoded Par proteins can be tagged with the green fluorescent protein (GFP) from *Aequoria victoria* or the plasmid can be tagged by insertion of the *lac* operator so that it will bind the Lac-repressor-GFP fusion protein. In this way, the movement of plasmid or proteins can be followed in real time in living cells. Such visualization shows a general pattern in which plasmid replicates at the center of the bacterial cell and then moves away symmetrically to ¼ and ¾ positions, where the next round of replication occurs. The nature of the machinery which drives this symmetrical, directed movement is not known. For a variety of reasons, it is generally thought that prior to separation the plasmids should pair. *In vitro* paired DNA molecules have been seen for plasmid DNA with the centromere-like sequence of plasmid R1 in the presence of Par proteins ParM and ParR. In the IncP plasmid partitioning system, which is related to the F and P1 systems, pairs can be seen *in vivo* when the A protein is inactivated. However, it is not clear whether the cycle of replication and partitioning involves pairing for most plasmids.

VII. PLASMID TRANSFER

Plasmid transfer can occur in a variety of ways: transformation, transduction, and conjugation. The most common means probably varies with host and environment.

A. Transformation

Binding to the surface of clay particles may help DNA to remain in a nuclease-resistant but biologically availabe form in some environments. In environments such as tropical seawater, the level of DNA can be high enough that there is a reasonable chance that bacteria will encounter naked DNA so that transformation may occur if the bacteria are in a competent state. Such a competent state occurs naturally for species such as *Bacillus subtilis, Haemophilus influenzae,* and *Neisseria meningitidis.* However, the transformation process, at least for *B. subtilis,* is more suited to taking up fragments of DNA in a form that allows recombination with homologous endogenous DNA rather than taking up circular plasmids *de novo.* Circular plasmids can only establish themselves if they are presented to such bacteria as dimers: On entry and conversion back to a double-stranded form, there is terminal sequence redundancy which allows circularization by recombination. Uptake of plasmid DNA by bacteria which do not have sophisticated transformation apparatuses has been shown to be surprisingly common under conditions which are likely to occur in nature: Bacteria and DNA mix in the presence of divalent cations such as calcium which promote binding of DNA to the bacterial surface. Conditions such as electrical discharges are also expected to cause bacteria to become transiently permeable to DNA so that bound DNA enters the cell. Microcosm experiments are being undertaken to assess the importance of transformation under natural conditions. In bacteria growing closely together, transfer of plasmids that are too small to be self-transmissible by conjugation can occur in a contact-dependent way. This may be due to release and uptake of DNA rather than to cell fusion.

B. Transduction

Another general way in which plasmids may transfer is by transduction: Bacteriophage infecting a bacterial strain carrying a plasmid may result in plasmid DNA being packaged into the phage particle. A generalized version of this is provided by the 90-kb P1 element, which is generally referred to as a temperate bacteriophage because of its ability to coexist with its bacterial host and to switch into a lytic cycle when stress or damage to its host result in repressor inactivation. In its quiescent state this element is a plasmid, multiplying at low copy number in the bacterial cytoplasm via an iteron-containing replicon (see Section V.C) , using an active partitioning system of the *parAB* type, killing plasmid-free replicons via a protein killer system called *phd/doc,* and resolving multimers via a site-specific recombinase. In its lytic state this plasmid multiplies exponentially, produces bacteriophage heads and tails, and is packaged and released so that it can be transported to another host. Other plasmids can be transported by transduction in the particles of completely unrelated phage. For example, the small plasmids of *S. aureus* are very efficiently transported by bacteriophage of this bacterium. An intriguing question is how many plasmid molecules are packaged together in such particles because these plasmids are very small and bacteriophage particles normally assemble by a "head full" mechanism. Again, the role of phage in transferring plasmids in natural environments is currently under investigation.

C. Conjugative Transfer

The mechanism of gene spread that is traditionally plasmid associated is conjugative transfer, a process that depends on direct contact between donor and recipient bacteria. In recent years, gene systems promoting conjugative transfer have been found as part of elements that are not plasmids, emphasizing the mosaic nature of genetic elements found in bacteria as referred to previously. For example, certain transposons can promote their own spread via conjugation, and other chromosomal segments such as some recently identified pathogenicity islands (chromosomal segments encoding virulence determinants that, based on sequence characteristics such as G + C composition and codon usage, appear to be of heterologous origin, resulting from horizontal transfer) also appear to be able to promote conjugative transfer. Conjugative transfer systems can vary in complexity. Plasmids of *Streptomyces* species do not encode a complex surface to promote fusion between donor and recipient bacteria. *Streptomyces* grow as mycelia, and when two such colonies grow into the same space fusion between different organisms may occur at reasonable frequency, allowing plasmids to move from one cytoplasm to another. The major protein needed to promote this process is an ATPase which may bind to the DNA at the pore between the two cytoplasms and help to translocate the DNA past the membrane junction. One or more additional proteins are needed to help the plasmid spread throughout the mycelium of the new host.

A more sophisticated transfer mechanism found in other gram-positive bacteria is the pheromone-induced transfer system found in Enterococci. Adhesion between donor and recipient bacteria is promoted by specific exported proteins encoded by the plasmid, although no surface appendage such as a pilus is made. Normally, these transfer genes are switched off and only switched on in the presence of potential recipients which are sensed because they produce pheromones (short hydrophobic peptides) that accumulate in the medium. Donors import the pheromone by an oligopeptide uptake system and this causes the induction of the transfer genes. A range of pheromones are produced naturally by Enterococci. Each plasmid responds to just one pheromone. On entry into a new host the plasmid inhibits production of its cognate pheromone so that it can no longer act as a recipient. The plasmid also produces an antagonistic peptide that ensures that low levels of the pheromone still inside the cell do not switch the plasmid into transfer mode. These plasmids are therefore highly specialized to transfer between strains of Enterococci.

The best studied conjugative plasmids are those from gram-negative bacteria, belonging either to the IncF group or the IncP-1 group (Fig. 5). The IncF plasmids include the sex factor F from *E. coli*, which was probably noticed for its ability to promote genetic exchange between different mutant strains because its transfer genes are switched on continuously as a result of a mutation that knocks out a key regulator of these genes. Approximately 40 genes are needed for conjugative transfer, and a large proportion of these are involved in producing the apparatus in the bacterial envelope that promotes formation of

pairs between donor and recipient bacterium. A key part of this apparatus is a long, thin, flexible protein tube called the pilus. This makes the initial contact with a suitable recipient and then appears to disassemble from the base, bringing the two bacteria into a close contact that is sufficiently stable to maintain the pairing in unshaken liquid culture. This seems particularly suited to promote transfer in a fluid environment such as the animal gut. In contrast, the IncP-1 plasmids make a short, rigid pilus that is more prevalent in detached form in the liquid surrounding the bacteria than attached to the bacterial surface. Conjugative transfer of plasmids encoding these sort of pili is very inefficient in liquid but occurs at high frequency on solid surfaces—agar in the laboratory and probably in biofilms on soil particles or on stone surfaces in lakes or rivers.

The role of pili is not resolved, although the consensus of opinion is in favor of the pilus facilitating contact between bacteria, which then leads to localized membrane fusion, degradation of a small area of the cell wall, and formation of a pore through which the DNA can move by a process that is still not understood properly. That the apparatus is focused on assembly of this pore rather than simply with assembly of a pilus is suggested by the fact that many proteins encoded as part of plasmid mating pair apparatuses show similarity to protein export systems as well as to more obviously analogous systems, such as those that promote DNA uptake in naturally competent bacteria. Nevertheless, some researchers are convinced that the pilus is a protein tube down which DNA can travel from one bacterium to another, protected from the environment. Recently, new plastic materials have been bored with uniform diameter holes just big enough to accommodate a pilus, and these have been used to separate donor and recipient bacteria in tightly sealed containers. Although recombinant bacteria appeared to be obtained, these experiments have failed to sway the general view that the pilus is normally just a means for selecting a mate and moving close to it. To prevent donors from trying to mate with bacteria already in possession of a related plasmid, most plasmids produce one or more proteins that are localized to the bacterial surface and which prevent the surface contacts from becoming more than a temporary association (surface exclusion).

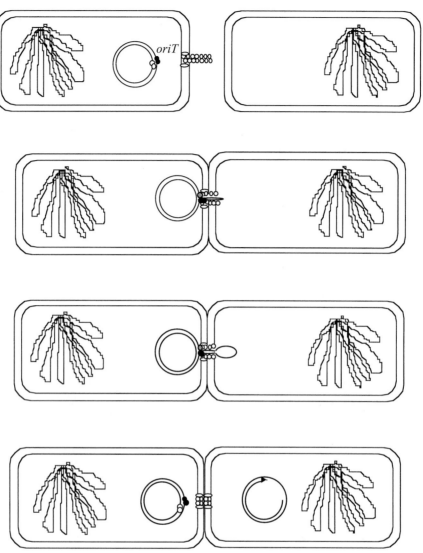

Fig. 5. The transfer cycle established for conjugative plasmids of *E. coli*. A plasmid-encoded pilus (shown here consisting of two types of protein components) on the donor bacterium makes contact with the surface of the recipient and promotes contact which leads to local fusion and pore formation. Plasmid DNA carrying a relaxasome (a protein complex at *oriT*; small open and closed circles) associates with the pore, and rolling circle replication (see Fig. 2) is initiated, extruding a single-stranded loop into the recipient. When the replication process reaches *oriT* again, the donor molecule is reformed and a ssDNA circle is released into the recipient. Primase initiates the conversion back to dsDNA. The small circles and elipses do not imply the actual structure of any one system but rather are used to indicate a possible structure of any system.

The other essential component of these conjugative transfer apparatuses is the relaxasome, a nucleic acid–protein complex at a specific site on the plasmid which introduces a nick into one strand of the plasmid DNA and inititates the rolling circle replication process that results in the transfer of a copy of the plasmid into the recipient bacterium. This complex is called a relaxasome because plasmid DNA normally exists in a supercoiled and thus stressed state. When the complex at *oriT* introduces a nick, it dissipates this stress because if the proteins are stripped off then the strands become free to rotate about each other and the plasmid becomes relaxed. Relaxasomes on different plasmids vary in the number of proteins associated, but three is average. Probably the most important of these proteins is the relaxase, which introduces the nick into *oriT* and becomes covalently joined to the 5′ side of the gap so that the 3′ OH can serve as the primer for elongation and that the relaxase is in a position to reform a circular plasmid molecule when the replication process passes the *oriT* sequence again. In many respects this process is very similar to the replication of RCR plasmids described previously; indeed, sequence motifs in both the *oriT* region and in the relaxase protein are conserved in *dso* and Rep sequences, respectively, of RCR plasmids. There are also features in common with the transfer of DNA from bacteria to plant cells promoted by the Ti plasmid of *Agrobacterium tumefaciens*. The relaxasome is linked to the Mpf apparatus by an adapter protein (TraG in the case of IncP-1 plasmids) that determines the specificity of the system and may relay to the relaxasome the signal that the bacterium has docked with a potential recipient.

Some plasmids are unable to transfer on their own but can do so in the presence of a specific conjugative plasmid. They are said to be mobilizable. An essential requirement is that they have an *oriT* which can attract and function with the Tra proteins of the conjugative plasmid or that they have their own Mob functions and these can link to the transfer apparatus of the conjugative plasmid by the TraG-like adapter protein. The discovery of the role of the specific adapter protein has led to much greater understanding of the interactions between different transfer systems and mobilizable plasmids.

Expression of transfer genes is controlled, but in different systems this has different consequences for transfer frequency. Transfer system related to F (but not F itself) are controlled very tightly so that they are normally switched off. A key element of this negative control is an antisense RNA (FinP), which provides the regulatory specificity, and an accessory protein FinO, which promotes the binding of the antisense RNA to its target. These control circuits respond to a variety of physiological stimuli and at low frequency generate transfer-proficient bacteria. After transfer there is a delay before FinO and FinP accumulate to repressing levels, and so new transconjugants are transfer proficient. As long as there are potential recipients available, this cycle will continue, creating a wave of transfer through the plasmid-negative population. Eventually, when there are no more potential recipients the transfer genes will get switched off. Ti and related plasmids of *Agrobacterium* species, as well as symbiosis plasmids of *Rhizobium* species, control their transfer genes in response to donor cell density. Plasmid-positive bacteria manufacture a homoserine lactone at low levels; however, when this accumulates because there are many bacteria, feedback occurs and the transfer genes are switched on. For Ti plasmids a second stimulus is needed—the presence of the complex amino acids that are manufactured by the plant tumors generated by the plasmids. The transfer of T-DNA to plant cells is controlled by a chemical signal (acetosyringone) which is synthesized in response to wounding, a necessary prerequisite for invasion by the bacterium.

An alternative general strategy adopted by some plasmids is to have one or more autogenous circuits which allow the transfer genes to be expressed continuously but kept at a low level once the transfer apparatus has been assembled so that the plasmid is as little a burden on its host as possible. In this way, the plasmid can always transfer at high frequency but the host growth rate is not burdened. The IncP-1 plasmids are the best studied example of this type of strategy.

D. Plasmid Establishment in a New Host

Many factors may limit the efficiency with which a plasmid establishes itself after transfer. The most obvious of these is degradation of the incoming DNA by restriction endonucleases if the DNA is not appropriately modified. Although it has been suggested that transfer by conjugation through a single-stranded DNA intermediate somehow allows a plasmid to avoid restriction systems, until recently there was no indication regarding how this might occur. Comparison of the frequency of transfer into strains with and without restriction shows very clearly that restriction systems do reduce transfer. When additional restriction sites are introduced into such a plasmid, the difference is even more dramatic. Plasmids have evolved two strategies to cope with this problem. One strategy is illustrated by the broad host range plasmids of the IncP-1 group which can transfer between almost all gram-negative bacterial species and in so doing must clearly encounter diverse restriction barriers. Examination of the complete nucleotide sequence of these plasmids showed that they are depleted for restriction targets for those restriction systems found in the *Pseudomonas* species, which are thought to be their natural hosts. This depletion could easily occur by point mutation because a base substitution at any one of six positions should abolish cutting of the DNA by the cognate restriction enzyme.

A second strategy is to allow transfer replication to deliver multiple copies of the plasmid to the recipient so that the restriction system is effectively saturated. A third, more complicated strategy exhibited by some plasmids is to encode a protein that interferes with the restriction system in the recipient. The *ardA* gene encodes such a protein which blocks the type I restriction systems encoded by enteric bacteria. Clearly, it would be important for such a protein to work before the plasmid DNA has been converted into a double-stranded form that would be sensitive to restriction. Unless the plasmid carries the protein across with it, this poses a problem because gene expression does not normally occur on single-stranded DNA. The solution is similar to that found for priming DNA replication on the lagging strand of rolling circle replication plasmids: Extended, imperfect inverted repeats in the single-stranded DNA form double-stranded hairpins that are recognized by RNA polymerase and this initiates transcription to express the genes in the leading region of the plasmid—the segment that enters the recipient bacterium first. Therefore, the protection afforded by *ardA* can only be effective if the plasmid goes through a single-stranded phase—that is, it protects during conjugative transfer but not after simple uptake of double-stranded DNA during transformation or transduction.

VIII. CHROMOSOME MOBILIZATION AND GENE ACQUISITION

One of the most important properties of plasmids is their ability to move genes from one bacterium to another without the need for recombination to allow the genes to be established in their new host. Continuous acquisition of new genes by plasmids is important to provide the variation in the plasmid population on which selection can act, as bacterial populations experience changing conditions. Transposable elements provide much of this gene movement and at the same time can promote interactions between different genomes. Transposition of an element from the chromosome to a plasmid can lead to the formation of a cointegrate in which the plasmid becomes temporarily a part of the chromosome and as such can promote conjugative movement of chromosomal DNA from one bacterium to another if the conditions are favorable for plasmid transfer. Alternatively, the presence of homologous transposable elements in the chromosome and in the plasmid can allow recombination between these repeated elements to achieve the same effect (cointegrate formation). The F plasmid contains multiple copies of various common insertion sequences which allow it to integrate into the chromosome in this way. (By chance one of these elements is inserted into the *finO* gene that is needed for repression of the transfer genes.) This results in a high frequency of transfer, and if the plasmid integrates into the chromosome then one also observes a high frequency of chromosome mobilizing ability (Cma). This property led to the discovery of F, the first plasmid identified. Although Cma of plasmids may be an unselected consequence of the conjugative transfer apparatus that has evolved for advantage of the plasmid, Cma illustrates the importance of plasmids as elements which promote the adaptability and diversity of bacteria.

See Also the Following Articles

Bacteriophages • Conjugation, Bacterial • DNA Replication • Fimbriae, Pili • Transduction: Host DNA Transfer by Bacteriophages

Bibliography

Clewell, D. B. (ed.) (1993). "Bacterial Conjugation." Plenum, New York.

Summers, D. (1996). "Bacterial Plasmids." Blackwell, Oxford.

Thomas, C. M. (ed.) (1999). "The Horizontal Gene Pool: Bacterial Plasmids and Gene Spread." Harwood Academic. Reading, UK.

Plasmids, Catabolic

Anthony G. Hay, Steven Ripp, and Gary S. Sayler

University of Tennessee

GLOSSARY

biodegradation   The process of biochemical catabolism of a pollutant as opposed to chemical or physical degradation.

cointegrate   Structure produced via the fusion of two replicons, one of which contains a transposon; the cointegrate will possess a copy of the transposon at both junctions of the replicon.

constitutive expression   Gene from which RNA synthesis occurs continuously without the requirement of an inducing molecule or complex.

degradative pathway   Specific enzyme catalyzed chemical reactions for the metabolism of a particular substrate.

enzymes   Protein molecules which catalyze specific chemical reactions by reducing the energy of activation.

gene   Specific segment of DNA in which the nucleotide base sequence determines the synthesis of an rRNA, tRNA, or mRNA.

genetic regulation   Physiological controls by which expression of genes is turned on (induced) and/or off (repressed).

inducible expression   Increased transcription of specific genes in the presence of an inducing molecule or complex.

operon   A unit of gene expression that is coordinately transcribed.

plasmid   Autonomous self-replicating extrachromosomal DNA molecule.

replicon   Region of DNA that contains the origin for initiation of replication.

transposon   Nonreplicating segment of DNA that is capable of inserting itself into other DNA molecules via an event called transposition.

THE WORD CATABOLIC is derived from two Greek words, *cata* and *bolē*. *Cata* means down, away, through, or backwards, and *bolē* means to throw. Thus, catabolism can be defined as the biochemical process whereby a compound is broken down or in some way transformed to a lower or less energetic state.

I. INTRODUCTION

Although all heterotrophic organisms rely on catabolic pathways to convert primary productivity into the chemical building blocks required for growth, the term catabolic pathway has taken on a more specific meaning in recent years. This is largely due to research done concerning the ability of microorganisms to degrade environmental pollutants. These compounds, generated primarily from industrial and agricultural sources, have become pervasive in many of our natural ecosystems. Restoration of contaminated environments usually requires the removal and long-term storage or incineration of hazardous pollutants. Alternatively, biodegradation, utilizing microbes to degrade pollutants, has been proposed for the remediation of some contaminated environments. At the most basic level, a microorganism's capability to transform toxic compounds is dependent on the genetic information encoded within the cell. The enzymes needed to convert compounds from one form to another via defined chemical reactions are encoded by specific genes. These genes may be located on the chromosome or on extrachromosomal, self-replicating DNA referred to as plasmids. Plasmids that have been found to harbor genes encoding the transformation of environmental pollutants have come to be known as catabolic plasmids. Organisms that maintain catabolic plasmids have been isolated from almost every conceivable environment, including soil, rivers, sewage treatment plants, termites, plant roots, industrial facilities, and marine ecosystems. The size of these plasmids and the substrate specificity of the enzymes they encode are also very diverse. Although in most cases these enzymes catalyze transformations that allow the host organism to obtain energy or growth factors, such as carbon, nitrogen, phosphorus, or sulfur, from the degradation of pollutants, this is not always the case. In some cases, the genes associated with the growth on one class of pollutant, encode enzymes that are also able to degrade other, more recalcitrant pollutants. This process is referred to as cometabolism. Although cometabolism of recalcitrant pollutants does not usually provide the degrader organism with either energy or growth factors, this is an important mechanism for the catabolism of compounds such as trichloroethylene and polychlorinated biphenyls. Although not exhaustive, Table I provides a list of representative compounds which are utilized for growth or degraded cometabolically by the enzymes of plasmid-encoded, catabolic pathways. This article provides a review of the biochemistry of some of the best-studied plasmid-encoded catabolic pathways. The prevalence of regulation via LysR-type proteins will also be discussed, as will the high frequency of association found between well-studied plasmids and catabolic transposons.

II. BIOCHEMISTRY

The catabolic pathways for the transformation of many of the compounds in Table I have been elucidated and some of the enzymes which catalyze these transformations have been studied in detail. From these studies it is apparent that there is wide variation in both the number and type of catabolic steps encoded by plasmid-borne genes. This section will briefly describe the transformation of nine well-studied organic compounds: *n*-alkanes, atrazine, bromoxynil, dichloroethane, 2,4-dichlorophenoxy acetate, naphthalene, nylon, parathion, and toluene.

Linear or *n*-alkanes are a significant component of naturally occurring hydrocarbon deposits. The genes which encode for the degradation of medium-length *n*-alkanes (C6–C12) have been isolated from many microorganisms. In *Pseudomonas oleovorans* these genes are located on the OCT plasmid and appear to be divided between two distinct operons. The *alkBFGHJKL* operon encodes for seven proteins and is transcribed as a single mRNA molecule 7.3 kb in length. The *alkST* operon encodes both a structural (AlkT) and a regulatory protein (AlkS) and is separated from the other operon by more than 40 kb. Initially, alkanes are hydroxylated at one end by alkane hydroxylase (Fig. 1), a multicomponent monooxygenase encoded by *alkB* (a membrane-bound hydrolase), *alkG* (a rubredoxin), and *alkT* (a rubredoxin reductase). The latter two proteins serve as a short electron transport chain, shuttling reducing equivalents from NADH to the hydrolase. Molecular analysis of *alkF* shows that it encodes a truncated, nonfunctional form of the rubredoxin encoded by *alkG*. The proteins encoded by *alkJ* and *K* catalyze sequential oxidations of the alcohol, first to the aldehyde and then to the acid. The acidic hydroxyl group is then covalently bound to coenzyme A (CoA) by the gene product of *alkL*. The resultant acyl-CoA molecule then undergoes β-oxidation multiple times, yielding acetyl CoA, a tricarboxylic acid (TCA) cycle intermediate. β-Oxidation, the normal process whereby fatty acids are degraded, is encoded by genes found on the chromosome, as are additional copies of genes encoding the last three enzymes *alkj, -K,* and *-L*.

Atrazine is a herbicide that has been widely used for weed control during the past 30 years. Although plasmids bearing the catabolic genes for atrazine degradation have been found in several strains, the plasmid pADP-1 of *Pseudomonas* sp. ADP is the only plasmid reported to harbor all the genetic information required for the production of enzymes capable of transforming atrazine to cyanuric acid. Although

TABLE I
Compounds Degraded by the Enzymes of Plasmid-Encoded Catabolic Pathways

Substrate	Plasmid	Size (Kb)	Host
α-Methylstyrene	pBS253	130	*Pseudomonas* sp.
1,2-Dichloroethane	pXAU1	200	*Xanthobacter autotrophicus*
2,2-Dichloropropionate	pFL40	nd	*Alcaligenes xylosidans*
2,4-Dichlorophenoxy acetate	pJP4	78	*Rolstonia eutropha*
2,4-Dinitrotoluene	pJS1	180	*Pseudomonas* sp.
2-Aminobenzoate	pKB740	8	*Pseudomonas* sp.
2-Chloroacetate	pU01	65	*Morexella* sp.
2-Chlorobenzoate	pRC10	45	*Pseudomonas aeruginosa*
2-Fluoroacetate	pU01	65	*Morexella* sp.
3-Chlorobenzoate	pAC25	117	*P. putida*
3-Chloropropionic acid	pU202	230	*Pseudomonas* sp.
4-chlorobenzoic acid	pSS50	53	*R. eutropha*
4-chlorobiphenyl	pSSD50	117	*R. eutropha*
6-Aminohexanoic acid dimer	pOAD2	44	*Flavobacterium* sp.
Alkylbenzenesulfonates	ASL	91	*P. testosteroni*
Aminobenzenesulfonate	pSAH	nd	*Alcaligenes* sp.
Anthracene	pKA1	101	*P. fluorescens*
Atrazine	pADP-1	97	*Pseudomonas* sp.
Benzoate (anaerobic)	pCBI	17	*A. xylosidans*
Camphor	CAM	>200	*Pseudomonas* sp.
Carbofuran	pDL11	120	*Achromobacter* sp.
Carboxydiphenylether	pPOB	40	*P. pseudoalcaligenes*
Cinnamic acid	pCINNS	75	*P. stutzeri*
cis-1,3-Dichloropropene	nd	50	*P. cichorii*
Di- and trichlorobenzene	pP51	110	*Pseudomonas* sp.
Di- and trichlorobenzoate	pPH111	120	*P. putida*
Dibenzofuran	NAH7	83	*P. putida*
Dibenzothiophene	pSB1	34	*Nocardioides* sp.
Dihydroxybiphenyl	pTA431	560	*Rhodoococcus erythropolis*
E-aminocaproic acid	pBS271	500	*Pseudomonas* sp.
E-caprolactam	pBS271	500	*Pseudomonas* sp.
Ethylbenzene	OCT	>200	*P. oleovorans*
Hydroxy and methylphthalate	pMOP	225	*P. cepacia*
Isopropylbenzene	pBD2	210	*R. erythropolis*
Methylcarbamate	nd	77	*Rhodococcus* sp.
Morpholine	pMOR2	28	*Mycobacterium cheloni*
Naphthalene	NAH7	83	*P. putida*
Nicotine	AO1	160	*Arthrobacter oxidans*
Nitrotoluene	pWWO	117	*P. putida*
Octane	OCT	>200	*P. olevorans*
p- and m-xylene	pWWO	117	*P. putida*
Parathion	pPDL2	43	*Flavobacterium* sp.

continues

Continued

Substrate	Plasmid	Size (Kb)	Host
p-Cresol	pND50	nd	*P. putida*
Phenanthrene	pKA2	31	*Beijerinckia* sp.
Phenol	pPGH1	200	*P. putida*
Phenyl acidic acid	pWW17	270	*Pseudomonas* sp.
Polyhydroxybuturate	pRmeSU47b	1400	*Rhizobium melaloti*
Quinolone	nd	225	*P. aeruginosa*
Salicylate	Sal1	84	*P. putida*
s-Ethyl-*N*,*N*-dipropylthiocarbamate	nd	50	*Arthrobacter* sp.
Styrene	pEG	37	*P. fluorescens*
Toluene	pWWO	117	*P. putida*
Toluene sulfonate	pT2T	50	*Comomonas testosteroni*
Trichloroethylene	pJP4	78	*R. eutropha*
Trimethylbenzene	pGB	85	*P. putida*

Note. nd, not determined.

the genes encoding the enzymes for the degradation of cyanuric acid are not found on pADP-1, chromosomally encoded genes allow strain ADP to degrade cyanuric acid and thus utilize atrazine as a sole source of nitrogen.

The first step in the degradation of atrazine is the removal of the chlorine from the triazine ring via hydrolytic dehalogenation (Fig. 2). This is a reaction in which the hydroxyl group of a water molecule displaces the chlorine. The reaction is catalyzed by atrazine chlorohydrolase, resulting in the release of hydrochloric acid. Atrazine chlorohydrolase, encoded by *atzA*, is constitutively expressed. After dehalogenation the N-ethyl side chain of hydroxyatrazine is attacked by an aminohydrolase encoded by *atzB*. The action of AtzB results in the hydrolytic removal of ethylamine to yield N-isopropylammelide. This compound is further degraded via another hydrolytic deamination step in which an aminohydrolase, encoded by *atzC*, catalyzes the removal of isopropylamine, yielding cyanuric acid.

Bromoxynil is a dibrominated hydroxybenzonitrile herbicide that is rapidly metabolized in soil. *Klebsiella ozaenae*, a soil isolate, maintains the plasmid pBrxn1 which encodes a single enzyme (*bxn*) for the transformation of this compound. Bxn, a constitutively expressed nitrilase, catalyzes the addition of

two water molecules across the triple bond, reducing the nitrogen moiety to ammonia and oxidizing the aromatic moiety to dibromohydroxybenzoate (Fig. 3). Ammonia is subsequently used for growth, thereby allowing bromoxynil to serve as a sole source of nitrogen for this microorganism.

Industrially, dichloroethane (DCE) has been widely used as a chemical building block for synthetic reactions and as a solvent in a variety of processes. Improper disposal of this compound has resulted in significant contamination of both soil and groundwater. Although many organisms have been reported to fortuitously degrade chlorinated alkanes via cometabolism, few organisms have been reported to use them as sole sources of carbon and energy. *Xanthobacter autotrophicus* GJ10 is able to use DCE as a sole source of carbon and energy because it harbors pXAU1, a plasmid that encodes enzymes for the degradation of DCE and several other short-chained haloalkanes, including bromoethane and dichloropropane. The first step in the catabolism of DCE is hydrolytic dehalogenation mediated by a constitutively expressed haloalkane dehalogenase encoded by the gene *dhlA* (Fig. 4). This protein transforms DCE to 2-chloroethanol, which is subsequently converted to 2-chloroacetaldehyde by a chromosomally encoded alcohol dehydrogenase. The *ald*

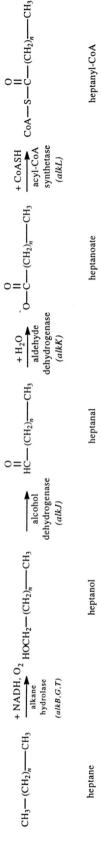

CH$_3$—(CH$_2$)$_n$—CH$_3$ $\xrightarrow[\text{(alkB,G,T)}]{\substack{+ \text{ NADH, O}_2 \\ \text{alkane} \\ \text{hydrolase}}}$ HOCH$_2$—(CH$_2$)$_n$—CH$_3$ $\xrightarrow[\text{(alkJ)}]{\substack{\text{alcohol} \\ \text{dehydrogenase}}}$ $\overset{\overset{\text{O}}{\|}}{\text{HC}}$—(CH$_2$)$_n$—CH$_3$ $\xrightarrow[\text{(alkK)}]{\substack{+ \text{ H}_2\text{O} \\ \text{aldehyde} \\ \text{dehydrogenase}}}$ $^-$O—$\overset{\overset{\text{O}}{\|}}{\text{C}}$—(CH$_2$)$_n$—CH$_3$ $\xrightarrow[\text{(alkL)}]{\substack{+ \text{ CoASH} \\ \text{acyl-CoA} \\ \text{synthetase}}}$ CoA—S—$\overset{\overset{\text{O}}{\|}}{\text{C}}$—(CH$_2$)$_n$—CH$_3$

heptane   heptanol   heptanal   heptanoate   heptanyl-CoA

Fig. 1. Pathway for the degradation of heptane encoded by genes found on the OCT plasmid. The number of methylene groups (*n*) is five; however, 4 ≦ *n* ≦ 10.

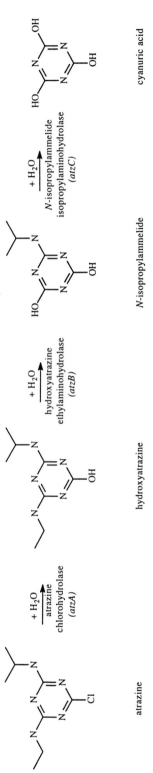

atrazine   hydroxyatrazine   *N*-isopropylammelide   cyanuric acid

Fig. 2. Pathway for the degradation of atrazine encoded by genes found on pADP-1.

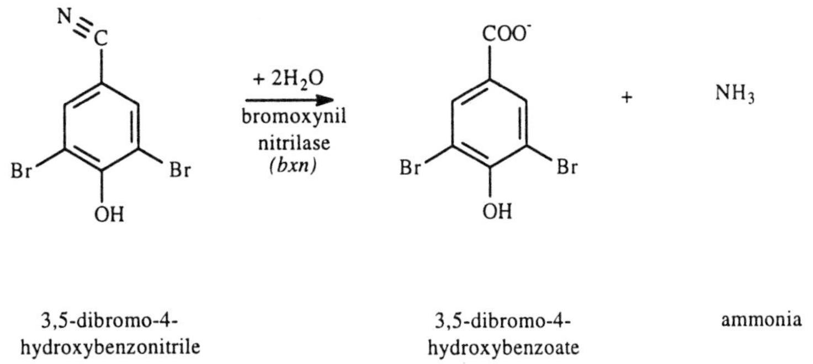

Fig. 3. Pathway for the degradation of bromoxynil encoded by a gene found on pBrxn1.

gene, maintained on both the chromosome and the plasmid, encodes for an inducible aldehyde dehydrogenase that oxidizes chloroacetaldehyde to chloroacetate. Chloroacetate is then hydrolytically dehalogenated by a chromosomally encoded haloalkanote dehalogenase, yielding glyoxylate.

The herbicide 2,4-dichlorophenoxy acetate (2,4-D) remains one of the most widely used weed-control agents in U.S. agriculture. Microorganisms capable of degrading this compound are readily isolated from most environments with a history of 2,4-D application. *Ralstonia eutropha* JMP134 is the archetypal 2,4-D degrader and harbors the plasmid pJP4 that encodes the inducible enzymes responsible for 2,4-D degradation in three main operons. In an uninduced state, low levels of the enzymes involved in 2,4-D degradation are produced constitutively. The first step in 2,4-D degradation is the cleavage of the ether linkage between the acetate and dichlorophenyl moieties (Fig. 5). This step is mediated by a 2-ketoglutarate-dependent dioxygenase encoded by *tfdA*. This reaction yields glyoxylate and 2,4-dichlorophenol. Dichlorophenol hydroxylase (*tfdB*) transforms this

latter product to 3,5-dichlorocatechol. Insertion of molecular oxygen between the vicinal hydroxyl groups by chlorocatechol-1,2-dioxygenase (*tfdC*) cleaves the aromatic ring, yielding *cis,cis*-2,4-dichloromuconate. Although a downstream metabolite, 2,4-dichloromuconate is the effector molecule which helps regulate transcription of the genes involved in 2,4-D degradation. Degradation of this effector molecule is catalyzed by chloromuconate cycloisomerase (*tfdD*). This reaction results in the spontaneous loss of chloride yielding *cis*-2-chlorodiene lactone. This lactone is cleaved by *cis*-2-chlorodiene lactone hydrolase (*tfdE*) to form chloromaleylacetate, which is subsequently reduced by chloromaleylacetate reductase (*tfdF*), eliminating the remaining chloride and generating 3-oxoadipate. This last product is a substrate for the chromosomally encoded adipate pathway, eventually yielding succinate and acetate, which are TCA cycle intermediates.

Naphthalene is a minor component of naturally occurring hydrocarbon deposits. However, synthesis and refining processes, such as those utilized at gas manufacturing plants in the early part of this century,

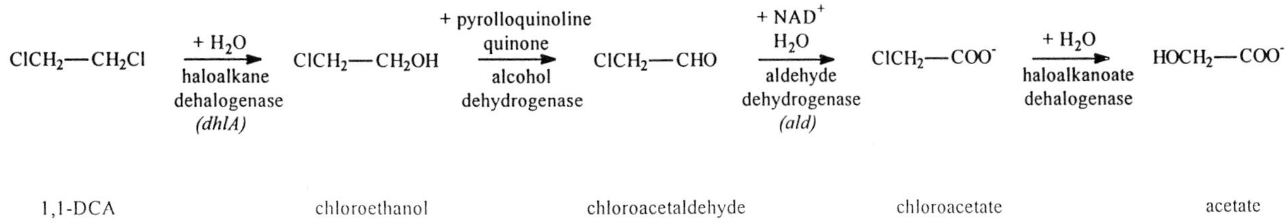

Fig. 4. Pathway for the degradation of 1,1-dichloroethane encoded by genes found on pXAU1.

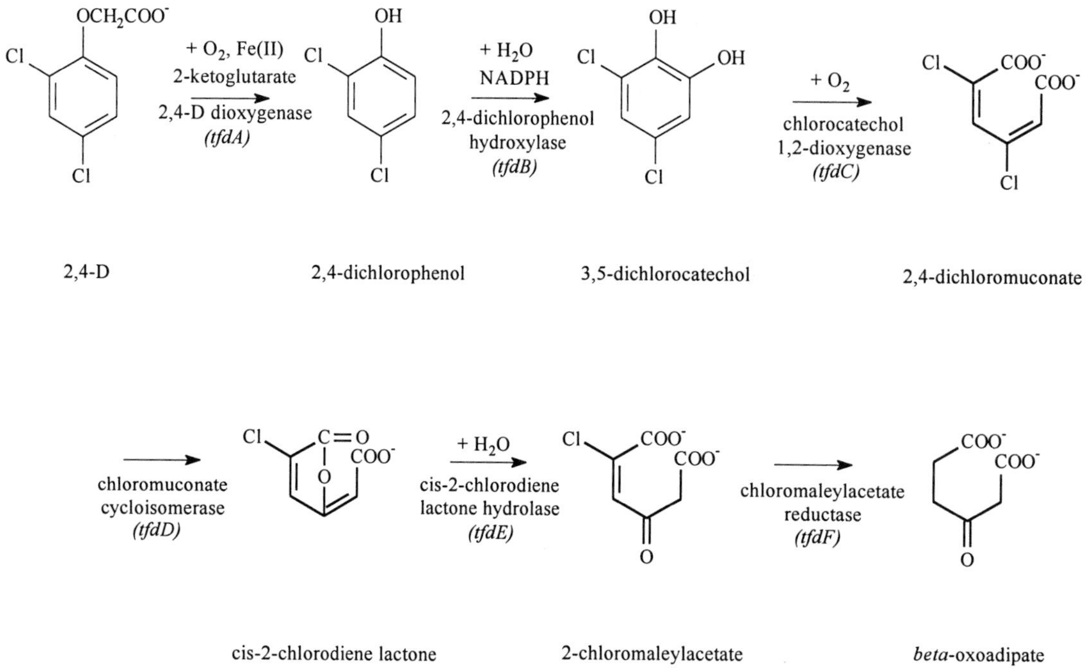

Fig. 5. Pathway for the degradation of 2,4-dichlorophenoxy acetate encoded by genes found on pJP4.

resulted in high levels of localized naphthalene contamination. Microorganisms are readily isolated from these environments and typically transform naphthalene in a manner similar to the pathway encoded by the NAH7 plasmid of *Pseudomonas putida* PpG7. The catabolic genes found on this plasmid are organized into two operons. The upper operon encodes the enzymes required to convert naphthalene to salicylate, whereas the lower operon encodes the enzymes required to convert salicylate to TCA cycle intermediates. A multicomponent enzyme, naphthalene dioxygenase, which inserts molecular oxygen across the double bond in the 1,2 position, catalyzes the first degradative step of the upper pathway (Fig. 6). Naphthalene dioxygenase is composed of four subunits encoded by the *nahA* gene cluster—specifically *nahA*$_a$, *-A*$_b$,*-A*$_c$, and *-A*$_d$—and yields *cis*-naphthalene dihydrodiol. The hydroxylated ring of this compound is rearomatized by a dihydrodiol dehydrogenase (*nahB*) resulting in the elimination of two protons. The enzyme 1,2-dihydroxynaphthalene dioxygenase (*nahC*) then adds molecular oxygen across the double bond cleaving the hydroxylated ring. This ring cleavage product spontaneously

isomerizes to 2-hydroxychromene-2-carboxylate. A second isomerization, catalyzed by 2-hydroxychromene-2-carboxylate isomerase (*nahD*), yields *trans*-2- hydroxybenzyli-denepyruvate. The aliphatic side chain of this molecule is cleaved and then oxidized by the same enzyme that has both hydratase and aldolase functions (*nahE*), yielding pyruvate and salicylate.

The first step in the lower pathway is the monooxygenation of salicylate by salicylate hydroxylase (*nahG*). This results in the formation of catechol, which is degraded via *meta*-fission. Although the genes encoding the *meta*-fission pathway are part of the salicylate operon on the NAH7 plasmid, the metabolism of catechol is common to many microorganisms that degrade aromatic compounds. The genes encoding catechol degradation have been localized to the chromosome in some organisms and to plasmids in others, whereas some organisms maintain copies on both the chromosome and plasmids.

The first step in the oxidation of catechol is ring cleavage. Catechol-2,3-dioxygenase, encoded by *nahH*, catalyzes this step, adding molecular oxygen across the double bond adjacent to one of the hy-

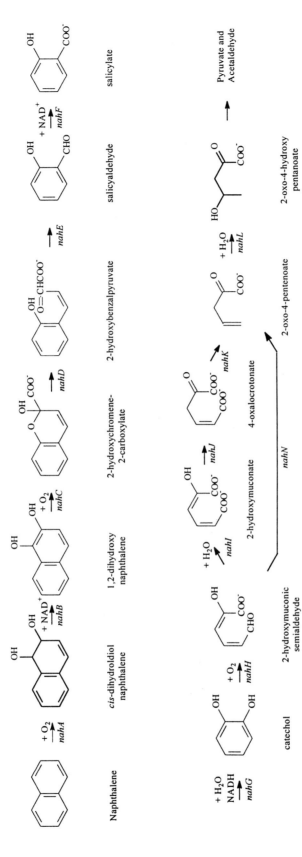

Fig. 6. Pathway for the degradation of naphthalene encoded by genes found on the NAH7 plasmid.

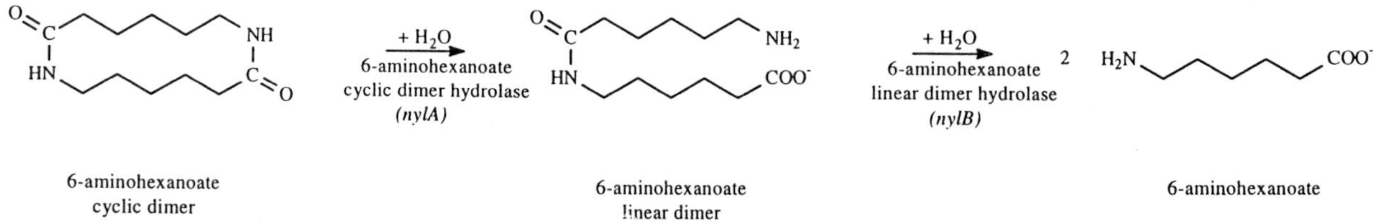

Fig. 7. Pathway for the degradation of nylon oligomers encoded by genes found on pOAD2.

droxyl groups. This *meta*-cleavage differs from the *ortho*-cleavage observed during the degradation of 2,4-D metabolites when molecular oxygen was added between adjacent hydroxyl groups. The resultant 2-hydroxymuconic semialdehyde can be further metabolized by hydrolytic removal of the aldehyde group (*nahN*). Alternatively, the aldehyde is first oxidized to 2-hydroxymuconate (*nahI*) and then tautomerized to 4-oxalocrotonate (*nahJ*) before being decarboxylated by oxalocrotonate decarboxylase (*nahK*). Both variations yield 2-oxo-4-pentenoate, which is hydroxylated by 2-oxo-4-pentenoate hydratase (*nahL*) to give 2-oxo-4-hydroxypentanoate. Cleavage of this molecule by an aldolase yields pyruvate and acetaldehyde, both of which can feed into the TCA cycle.

6-Aminohexanoate is the basic building block of nylon, a widely used polymer in the textile industry. *Flavobacterium* sp. strain KI1225 is capable of using this compound as a sole source of carbon and nitrogen. One of the three plasmids maintained by this strain, pOAD2, harbors genes which encode three enzymes able to catalyze the depolymerization of short nylon oligomers (2–20 covalently joined molecules) to 6-aminohexanoate acid. Although each of these enzymes is capable of adding water across the amide bond, they differ in type of polymer attacked. The gene product of *nylC*, endo-type 6-aminohexanoate oligomer hydrolase, is capable of linearizing and depolymerizing *N*-carbobenoxy-6-aminohexanoate trimers. Cyclic dimers, however, require separate enzymes for decyclization and depolymerization. *nylA* encodes 6-aminohexanote cyclic dimer hydrolase which cleaves one of the amide bonds, linearizing the dimer (Fig. 7). Another hydrolase encoded by *nylB* is then able to hydrolyze the remaining amide bond yielding two molecules of 6-aminohexanoate.

Parathion is one of a broad class of organophosphorus neurotoxins commonly used for pest control in agriculture. Interest in the breakdown of parathion and related compounds is high because large quantities of related organophosphorus compounds have been stockpiled in the United States' now defunct chemical weapons arsenal. *Pseudomonas diminuta* MG maintains the plasmid pMCS1, which harbors the gene *opd*, encoding a broad-spectrum organophosphorus anhydrase capable of cleaving one of the phosphotriester bonds of parathion. This reaction is a result of the nucleophilic addition of a water molecule to the anhydride bond, yielding diethyl thiophosphate and *p*-nitrophenol (Fig. 8). The enzyme

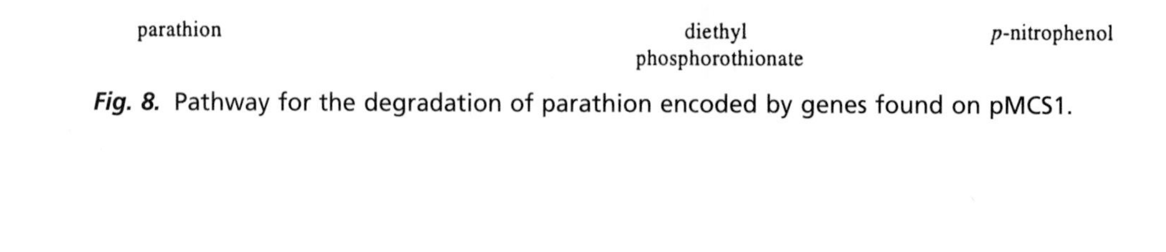

Fig. 8. Pathway for the degradation of parathion encoded by genes found on pMCS1.

encoded by *opd* is constitutively expressed. However, neither of parathion's breakdown products can be used by strain MG to support growth.

Toluene, also found in naturally occurring hydrocarbon deposits, is a significant component of widely used fuels and is a common industrial chemical. Leakage from fuel storage tanks is one of the ways whereby groundwater in many areas has become contaminated with toluene and related fuel hydrocarbons. Five different pathways have been reported for the degradation of this compound, two of which are plasmid borne. The best studied pathway for the degradation of toluene is encoded on the TOL plasmid, pWWO, originally isolated from *Pseudomonas putida* mt-2. The initial step in the degradation of toluene is catalyzed by a multicomponent monooxygenase (*xylA* and *M*) yielding benzyl alcohol (Fig. 9). Successive oxidations carried out by benzyl alcohol dehydrogenase (*xylB*) and benzaldehyde dehydrogenase (*xylC*) transform benzyl alcohol first to benzaldehyde and then to benzoate. The addition of molecular oxygen across the double bond by a multicomponent toluate (benzoate) dioxygenase (*xylX, Y, Z*) yields 1,2-dihydroxycyclohexa-3,4-diene carboxylate which is decarboxylated and rearomatized by 1,2-dihydroxycyclohexa-3,4-diene carboxylate dehydrogenase (*xylL*). This results in the production of catechol which is metabolized via enzymes of the *meta*-fission pathway which are also encoded on pWWO.

These examples illustrate the diverse nature of substrates whose transformation is encoded on catabolic plasmids. From simple single-step reactions to complex multistep pathways, plasmids appear to be a versatile means of enabling microorganisms to gain metabolic capacities for the exploitation of otherwise unavailable resources. Inoculation of plasmid-bearing strains into contaminated environments has demonstrated that the genetic information contained on plasmids is so useful that these genetic elements are often transferred to and maintained by indigenous microorganisms, even when the introduced host is unable to survive. This type of genetic exchange, termed horizontal transfer, may account for the high degree of sequence homology found in catabolic genes that allow distantly related microorganisms to transform the same substrates. Although homology between many catabolic genes is noticeable, significant divergence has also been noted in both DNA sequence and regulatory control of some more complex pathways.

This is evidenced by a comparison of the *clc, tcb,* and *tfd* operons. All three of these operons encode a regulatory protein and enzymes for the degradation of chlorocatechol which share a high degree of homology. Limited sequence divergence, however, has changed the specificity of the regulatory proteins: The regulatory proteins of the *clc* and *tcb* pathways are interchangeable, as are the *clc-* and *tfd*-encoded proteins. However, the *tcb-* and *tfd*-encoded proteins are unable to substitute for one another. Thus, although these genes may have had a common origin, changes have occurred, possibly due to the accumulation of mutations over time or some other selective pressure.

III. TRANSCRIPTIONAL REGULATION

Many of the complex catabolic operons studied to date are regulated by proteins belonging to the LysR family, which represents a diverse group of evolutionarily related DNA-binding proteins responsible for the regulation of prokaryotic transcription. LysR-type transcriptional regulators are usually transcribed divergently from a promoter which overlaps the promoter region of one or more structural genes. Genes encoding these proteins can be recognized through sequence comparisons that typically identify similar DNA-binding domains consisting of a helix-turn-helix motif near the amino terminus, an effector recognition site, and a site required for DNA binding and effector response. The presence of an effector causes the LysR-type protein to change conformation. This change alters the affinity with which the protein binds DNA. The result is operon specific because some LysR protein–effector complexes appear to activate transcription by stabilizing RNA polymerase, whereas others negatively regulate transcription by apparently competing with RNA polymerases for promoter binding sites. The genetic orga-

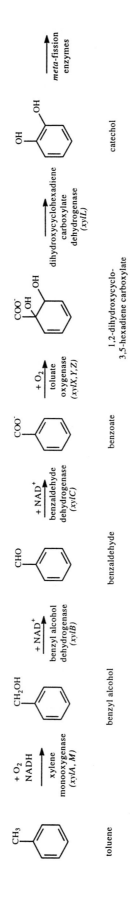

Fig. 9. Upper pathway for the degradation of toluene encoded by genes found on the TOL plasmid pWWO.

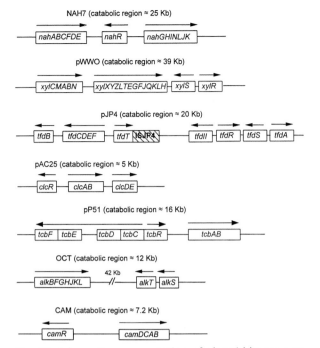

Fig. 10. Schematic representation of plasmid-borne catabolic operons (not drawn to scale). Arrows represent proposed direction of transcription.

nization of several LysR-type operons is shown in Fig. 10.

In the NAH7 plasmid of *P. putida* PpG7 the *nahR* region encodes the transcript for the 36-kDa LysR-type regulatory protein NahR. This protein is required for upregulating transcription of the genes encoding the catabolic enzymes for naphthalene degradation. The upper pathway is responsible for the conversion of naphthalene to salicylate and these genes are constitutively expressed at low levels. The lower pathway converts salicylate to 2-oxo-4-hydroxypentanoate. NahR binds to the promoter regions of the upper (*nahABCFDE*) and lower (*nahGHINLJK*) pathway operons. The binding of NahR to the upper and lower pathway promoters occurs in both the presence and the absence of the effector, salicylate. However, when salicylate is present, the conformation of NahR changes such that it binds more tightly to portions of the promoter. Associations between DNA-bound NahR and the RNA polymerase increase the frequency with which successful transcription is initiated. Thus, salicylate acts as a

positive regulator for the transcription of the naphthalene genes. Unlike some LysR-type regulons, transcription of the genes encoding naphthalene degradation does not appear to be repressed by compounds more readily utilized by microorganisms, such as succinate, glucose, or complex-rich media. Overall, regulation of naphthalene degradation represents an effective means of control over a metabolic pathway at the molecular level. Such efficiency is important considering that bacterial populations subsist in environments that are highly limited for metabolic resources. Evolution has therefore provided tightly regulated metabolic systems that can be rapidly and efficiently adjusted to allow bacterial cells to utilize compounds that may be present at extremely low concentrations.

Numerous other LysR-type transcriptional regulators have been identified and characterized in catabolic plasmids. The TOL operon from the pWWO plasmid consists of an upper degradative pathway (*xylCMABN*) that converts toluene or *m*- and *p*-xylenes to aromatic carboxylic acids and a lower pathway (the *meta* operon) that converts aromatic carboxylic acids to pyruvate and acetyl CoA, both TCA cycle intermediates. This process is regulated by two LysR-type regulatory proteins, XylR and XylS, which are the products of *xylR* and *xylS* respectively. The upper pathway is positively regulated by XylR. In the presence of toluene, XylR binds to the promoter, upregulating the normally low level of constitutive transcription of both the upper operon and the *xylS* gene. Because XylS is the regulatory protein of the lower pathway, regulation of both operons is maintained by XylR in slightly different ways. XylS normally binds weakly to the promoter of the lower operon. Metabolism of toluene by the upper pathway enzymes results in the accumulation of effector molecules, such as benzoate. In the presence of the effector, XylS activates transcription of the lower pathway. If benzoate is supplied directly as a carbon source, XylS can also function independently of XylR, upregulating transcription of the lower operon. The lower pathway may also be activated by a chromosomally encoded LysR-type regulatory protein that is responsible for transcription of a chromosomally encoded benzoate dioxygenase gene. The presence of glucose or succinate has been shown to repress toluene degradation.

The genes for 2,4-D degradation by *Ralstonia eutropha* strain JMP134 are maintained on plasmid pJP4 and are encoded for in four main operons: *tfdAS, tfdRD*$_\parallel$*, tfdB,* and *tfdCDEF*. The LysR regulatory protein in this system, designated TfdS, is encoded for by two identical genes, *tfdR* and *tfdS*. This protein is responsible for positively regulating transcription of the two operons *tfdA* and *tfdB*, which encode for enzymes that degrade 2,4-D to 3,5-dichlorocatechol, and also for the regulation of the *tfdCDEF* operon and *tfdD*$_\parallel$. Although the promoter sequences for *tfdAS* and *tfdCDEF* vary slightly, TfdS binds to both these regions. Binding is enhanced in the presence of dichloromuconate, a 2,4-D metabolite produced by TfdD, and results in the upregulation of transcription. Although the reason for duplication is not clearly understood, it is interesting to note that an additional copy of *tfdD*, named *tfdD*$_\parallel$, is divergently transcribed from the *tfdR* promoter, providing for greater production of TfdD. Another gene, *tfdT*, originally identified as a LysR-type regulatory protein, is a remnant from an insertion event and yields a nonfunctional, truncated protein.

Another chlorocatechol degradative pathway exists in *P. putida* strains harboring the catabolic plasmid pAC25. The *clcABDE* operon is responsible for the conversion of 3-chlorobenzoate to maleylacetate. The LysR-type regulatory protein is ClcR, which likely binds upstream of the *clcA* gene. The complete regulatory mechanism of ClcR has not been determined. A similar chlorocatechol degradative pathway can be found on the *Pseudomonas* sp. strain B13 catabolic plasmid pWR1 (also referred to as pB13), which has been shown to be nearly identical to plasmid pAC25. A chlorocatechol pathway can also be found on the *tcbCDEF* operon on plasmid pP51 which is able to promote growth on chlorobenzenes. The *tcbCDEF* operon encodes for chlorocatechol ring-fission enzymes while a second operon, *tcbAB*, encodes for a chlorobenzene dioxygenase and dehydrogenase. The *tcbR* gene product serves as the LysR-type regulatory protein. As mentioned previously, many of these isofunctional proteins are also interchangeable.

Pseudomonas oleovorans is able to utilize linear alkanes as sole sources of carbon and energy due to a degradative pathway located on the catabolic OCT plasmid. The specific genes responsible are *alkBFGHJKL* and *alkST,* which encode for the enzymes required for converting alkanes to acyl-CoA derivatives that can be used in the preliminary β-oxidation pathway leading to the TCA cycle. The *alkST* operon is necessary for activation of expression of the *alkBFGHJKL* operon and for recognition of inducer. The AlkS protein exhibits characteristics of a LysR-type protein but its specific regulatory mechanism is unknown.

The catabolic CAM plasmid is responsible for the conversion of camphor to isobutyrate through enzymes encoded by the *camDCAB* operon. Regulation is mediated by the LysR-type protein CamR, which negatively regulates expression of itself (autoregulation) as well as expression of *camDCAB*. Binding of CamR occurs in a single overlapping promoter region located between the *camDCAB* operon and the *camR* gene. In the presence of the effector camphor, CamR undergoes a conformational change, allowing *camDCAB* and *camR* to be expressed.

IV. ROLE OF TRANSPOSONS

Studies concerned with elucidating the previously mentioned pathways have revealed evidence of genetic rearrangements and gene duplications involving certain catabolic operons. It has since been shown that some of these catabolic pathways are located on transposable elements, which promote genetic rearrangements through an event referred to as transposition. Catabolic operons located on transposons can be transferred to the plasmid or chromosomal DNA of recipient strains, even when the parent plasmid is unable to replicate in the recipient. Such mobility increases the potential for horizontal gene transfer beyond that obtained with plasmids alone. Despite findings that homologous catabolic genes have been found in isolates that are unrelated either taxonomically or geographically, relatively few catabolic transposons have been identified.

Transposon Tn4651 harbors the toluene degradation (TOL) pathway from plasmid pWWO. Although Tn4651 is situated within a second transposon (Tn4653), the full toluene degradative pathway is located on Tn4651. Tn4651 contains three identified genes that participate in transposition events: *tnpA*, *tnpS*, and *tnpT*. The *tnpA* gene product mediates the formation of cointegrates whereas *tnpS* and *tnpT* encode enzymes responsible for resolving cointegrates. A *tnpR* gene, located on Tn4653, also resolves cointegrates but is dependent on Tn4651 for full activity. A *res* region, signifying the site of resolution, is located upstream of the *tnpR* gene. Tn4655 is the naphthalene catabolic transposon located on the NAH7 plasmid. Tn4655 is only capable of transposition in the presence of a complementing transposase. It does, however, contain its own *tnpR* and *res* sites. Both Tn4651 and Tn4655 are members of the type II family of transposons. Members of this family utilize conserved inverted repeats sequences and rely on transposases and resolvases which are closely related at the amino acid level. They also share similar transposition mechanisms.

Alcaligenes sp. strain BR60 contains the plasmid pBRC60 that carries a transposon-related (Tn5271) catabolic pathway for benzoate degradation. The operon responsible is *cbaABC*. A *tnpA* gene is located at either end of Tn5271 within inverted repeat sections designated IS1071. No *tnpR* or *res* sites have been identified. Recently, DNA regions flanking *atzA*, the plasmid-encoded gene for atrazine chlorohydrolase in *Pseudomonas* sp. ADP, were found to be 95% identical to IS1071. It is assumed that pADP-1 therefore either carries or has carried a catabolic transposon.

Tn5280 represents the catabolic transposon on the pP51 plasmid harboring the chlorocatechol (*tcb*) degradative pathway. Two operons, *tcbAB* and *tcbCDEF*, allow for growth on chlorobenzene. However, only the *tcbAB* operon is located on Tn5280.

Inverted repeats with varying degrees of homology to known insertion sequences have also been found associated with the *alk* genes on the OCT plasmid, the *dhl* gene on plasmid pXAU, and the *nyl* genes on plasmid pOAD2. IS6100 is found five times on this last plasmid. The same sequence is also found on plasmid pNAD2, the only other plasmid reported

to harbor genes encoding enzymes for nylon oligomer degradation. The inverted repeats are not found in the chromosome of either host nor on other plasmids maintained by these organism; thus, these repeats appear to have originated from some ancestral plasmid or transposable element. That this sequence plays a role in the evolution and stability of the plasmids in which it resides is evidenced by a deletion which occurred in pOAD2. Homologous recombination between two identical sites resulted in the loss of *nylA* and led to the discovery of *nylC*.

The 2,4-D (*tfd*) catabolic pathway from plasmid pJP4 was recently shown to contain an insertion sequence designated ISJP4. ISJP4 harbors the *tfdT* gene, which was shown to produce a protein homologous to other LysR-type transcriptional regulators. The *tfdT* gene product, however, proved to have no discernable function and failed to activate the *tfdCDEF* operon. The presence of ISJP4 has since been shown to disrupt the *tfdT* open reading frame, leading to the production of a truncated, nonfunctional TfdT protein.

Tn4371 is a transposon that encodes enzymes for the degradation of biphenyl and some polychlorinated biphenyls. The gene cluster consists of the operons *bphEGF*, *bphA1A2A3*, *bphBCD*, and *bphA4*, all of which are located on Tn4371. The Tn4371 transposon frequently associates with broad-host-range IncP1 plasmids and often is found integrated into host chromosomal DNA.

Although few catabolic plasmids have been well characterized, the previously mentioned examples demonstrate that associations between transposons and catabolic plasmids are relatively common. Unfortunately, the genetic analysis often stops when the catabolic genes of interest have been located. The evidence available suggests that if sequencing efforts are extended beyond the catabolic genes, many of the poorly characterized plasmids reported in the scientific literature may be found to have had their evolution affected by transposons.

See Also the Following Articles

Biological Control of Weeds • Oil Pollution • Pesticide Biodegradation • Transcriptional Regulation in Prokaryotes

Bibliography

Horikoshi, K., Fukua, M., and Kudo, T. (Eds.) (1997). "Microbial Diversity and Genetics of Biodegradation." Japan Scientific Societies Press, Tokyo.

Sayler, G. S., Hooper, S. W., Layton, A. C., and King, J. M. H. (1990). Catabolic plasmids of environmental and ecological significance. *Microbial Ecol.* **19**, 1–20.

Schell, M. A. (1993). Molecular biology of the LysR family of transcriptional regulators. *Annu. Rev. Microbiol.* **47**, 597–626.

Wyndham, R. C., Cashore, A. E., Nakatsu, C. H., and Peel, M. C. (1994). Catabolic transposons. *Biodegradation* **5**, 323–342.

Plasmodium

John E. Hyde

University of Manchester Institute of Science and Technology

GLOSSARY

adjuvant One of a heterogeneous group of substances that enhances the immune response to an antigen in a nonspecific manner, and therefore is an important component of vaccines; several mechanisms may be involved, but prolonging exposure of the antigen to immunocompetent cells is thought to be a major factor.

apicoplast A multi-membraned organelle apparently common to the apicomplexan parasites which resembles plastids found in plants but lacks genes involved in photosynthesis; analysis of the circular extrachromosomal DNA that constitutes the genome of this organelle suggests that it was acquired by a common ancestor of the Apicomplexa by secondary endosymbiosis, possibly from an alga or dinoflagellate.

cytokine A member of a group of low-molecular-weight proteins secreted from cells of the immune system that help regulate immune responses by exerting a range of effects on other cells of the system; the term includes the monokines, lymphokines, and interleukins.

gametocyte An erythrocytic form of the malaria parasite that has differentiated into either a male (microgametocyte) or a female (macrogametocyte) cell, appearing after several rounds of blood-stage infection; upon ingestion by the mosquito, the gametocytes mature into male and female sex cells (gametes) ready for fusion.

merozoite Invasive form of the malaria parasite; first released from liver cells after maturation of the hepatic schizonts and then from erythrocytes in the cycles of blood-stage infection.

oocyst A mosquito stage of the malaria parasite in which, after embedding in the stomach wall as the ookinete, the parasite rapidly grows into a spherical cyst enclosed by an elastic membrane.

ookinete A mosquito stage of the malaria parasite which forms a few hours after fertilization; the zygote resulting from fusion becomes a mobile, invasive ookinete which seeks out the stomach wall, passing through the layer of epithelial cells and coming to rest at the basement membrane.

parasitophorous vacuole The vacuole within which the malaria parasite resides after invasion of the red blood cell; the vacuole is formed initially by the internal membrane of the erythrocyte and grows as the parasite expands to fill the cell; the parasitophorous vacuole membrane is actively involved in the transport of nutrients to, and waste products from, the parasite.

rosetting Adhesion of uninfected erythrocytes to an infected erythrocyte, particularly common in *Plasmodium falciparum* infections, though apparently strain dependent; the resulting clustering of cells is thought to exacerbate the blockage of blood capillaries in vital organs, particularly the brain.

schizont Multinucleate forms of the malaria parasite; hepatic schizonts develop in the liver and contain many thousands of maturing merozoites, which are released into the bloodstream when the host cell ruptures; erythrocytic schizonts develop in the red blood cells and, depending on the species of *Plasmodium*, each gives rise to 8–24 merozoites.

sporozoite The worm-like invasive form of the malaria parasite that develops in thousands in the oocyst and migrates to the salivary glands of the mosquito after the oocyst bursts. When the female mosquito takes a blood meal from the vertebrate host, the sporozoites are injected along with the saliva and rapidly invade the liver cells.

telomere The specialized end region of a linear chromosome that carries multiple copies of a short DNA sequence and which protects the chromosome from enzymic degradation and inappropriate joining to other chromosomes.

trophozoite   The feeding forms of the malaria parasite; hepatic trophozoites absorb nutrients from the liver cell, grow rapidly, and then divide internally to give multinucleate hepatic schizonts; erythrocytic trophozoites ingest hemoglobin from the red cell and break it down to provide free amino acids, as well as importing nutrients from the vertebrate plasma, before differentiating into erythrocytic schizonts.

PLASMODIUM   is the genus to which the malaria parasites belong. These protozoan parasites infect a wide range of vertebrates, including lizards, birds, rodents, and primates including humans. Globally, malaria has maintained its position as one of the most important and devastating diseases of *Homo sapiens* from ancient times to the present day. The advent of modern molecular techniques, especially recombinant DNA technology, has led to an enormous effort during the past 20 years to learn more about the biology, biochemistry, and genetics of the causative organisms, particularly in the case of the lethal agent *Plasmodium falciparum,* and thus to determine possible new routes to combating the disease. Much of this effort has been directed toward the development of vaccines against malaria and of new drugs that will permit at least a temporary solution to the increasing problem of ever-spreading resistance to the current clinically effective inhibitors.

I. THE PARASITE AND THE DISEASE

A. Taxonomy and Distribution of *Plasmodium*

Unambiguous taxonomic relationships among protozoa are difficult to establish and still not fully determined. Although there is some variation among authorities, *Plasmodium* can be placed in the domain Eukarya, kingdom Protista, phylum Apicomplexa, class Sporozoea, subclass Coccidia, order Eucoccidi-ida, suborder Hemosporina, family Plasmodiidae, of which *Plasmodium* is the only genus. These organisms are obligate intracellular parasites for most of their life cycle and have two hosts: the vertebrate,

in which two phases of asexual reproduction occur, and a blood-sucking insect, in which fertilization between male and female forms of the parasite occurs followed by another phase of asexual reproduction. The phylum to which *Plasmodium* belongs takes its name from the apical complexes found in these parasites and their near relatives, which are intracellular organelles utilized when the parasite invades host cells. The suborder is named from the fact that all of its members parasitize the blood of the vertebrate host at some stage of their life cycle.

Although about 170 species of parasite have been placed in the genus *Plasmodium,* only 4 of these are specific to humans: *P. falciparum, P. vivax, P. malariae,* and *P. ovale.* The last three are placed in the subgenus *Plasmodium* along with many other species that infect primates, whereas *P. falciparum* is placed in the subgenus *Laverania* which it shares only with *P. reichenowi,* a parasite of the chimpanzee. *Plasmodium falciparum* is different from its three cousins in several ways, and it is the most dangerous of the human malaria parasites, giving rise to the fatal form of the disease. Of the four species, *P. vivax* and *P. falciparum* account approximately equally for the great majority of human infections (>90%) when considered globally. *Plasmodium vivax* causes severe and debilitating attacks but is not generally a lethal organism, although fatal rupture of an enlarged spleen can occasionally occur. It has a broad geographical distribution in both temperate and tropical areas, predominating in Central America, southern Asia, and the Indian subcontinent, whereas *P. falciparum* is largely confined to tropical and subtropical areas, particularly in Africa and Southeast Asia. *Plasmodium malariae* is found over a similar range as that of *P. falciparum,* but it is much rarer, and *P. ovale* is mainly confined to tropical areas of West Africa. Historically, the disease can occur essentially anywhere between the latitudes of about 64°N and 32°S if temperatures are in the range of about 16–33°C and altitudes are less than about 2000 m. Outside of these parameters, the parasite cannot complete its life cycle successfully in the mosquito.

The host specificity of *Plasmodium* parasites means that studying the human disease in animal models is extremely difficult. Only *P. malariae* is transmissible to animals (chimpanzees) naturally, and experimental infections with *P. falciparum, P. vivax,* and *P. malariae* are mainly confined to Colombian owl (or night) monkeys (*Aotus trivirgatus*), squirrel monkeys (*Saimiri sciureus*), or splenectomized chimpanzees (*Pan satyrus*), all of which are highly limited in accessibility for experimental studies. Mention should therefore be made of the most important animal parasites because they have been studied extensively as laboratory models for human malaria, despite often quite distant relationships to the human parasites both phylogenetically and in the details of their pathogenesis. Of the primate parasites, *P. cynomolgi* (which resembles *P. vivax*) and *P. knowlesi* are the best studied, but the rodent parasites have provided by far the most accessible *in vivo* models and have been extremely important in the elucidation of the detailed biology of the parasite. The most widely used species are *P. berghei* and *P. yoelii,* which preferentially invade immature erythrocytes (reticulocytes), together with *P. chabaudi* and *P. vinckei,* which invade mature erythrocytes. All of the rodent parasites are placed in the subgenus *Vinckeia,* along with many other species that infect nonprimate animals.

B. Historical Perspective

Malaria is one of the world's oldest recorded diseases, with obvious references appearing from around 3000 BC in Sumerian writings, in a Chinese medical treatise from the same era, in Vedic scripts from about 1600 BC in India, and in Egyptian hieroglyphic temple and papyrus inscriptions. The parasites are thought to have spread to the New World from the sixteenth century onwards, mainly from Europe and Africa, coincident with the increase in exploration and development of the slave trade. The influence of malaria on the course of world history from ancient times to the present has been profound. The rapid decline of Alexander the Great's vast empire followed his premature death at the age of 32 which is believed to have been from malaria. A pivotal event in the transition of power from classical Greece to Rome was the defeat of Athens by Sparta

after the Athenian forces were decimated by malaria during the siege of Syracuse in 413 BC. The area around Rome was one of the most malarious in the world from classical times until the 1930s, and the disease also played a part in the collapse of the Roman empire. The Crusaders of the Middle Ages were constantly hampered by it in their various campaigns in the Holy Land from the eleventh to the thirteenth centuries. The outcomes of many battles of the American war of independence, the American civil war, the Spanish civil war, the two world wars, and the Vietnam war were also heavily influenced by malaria casualties.

Currently, malaria parasites retain their position as the world's most important agents of parasitic disease and are a threat to more than one-third of humanity, with an estimated number of infections per year of 300–500 million in a total of about 100 countries; however, the worst affected are the nations of sub-Saharan Africa, India, Sri Lanka, Brazil, Colombia, and Vietnam. Mortality is thought to be between 1.5 and 2.5 million per year, mostly among children under 5 years old in Africa. In addition to the burdens of mortality and morbidity, the economic impact of malaria is enormous, particularly in poorer countries.

The ability to culture *P. falciparum* continuously *in vitro* was established in 1976 when it was realized that, to better simulate *in vivo* conditions, the parasites should be exposed to a reduced level of oxygen and increased carbon dioxide, relative to the normal atmosphere. Successful *in vitro* culture of *P. vivax* has been reported only recently, and long-term maintenance is complicated by the need for regular addition of human reticulocytes. Thus, the majority of research on human malaria has concentrated on *P. falciparum* during the past two decades, which is appropriate given its position as the most lethal species. Unless otherwise stated, the following discussion refers to data obtained with *P. falciparum*.

II. THE PARASITE LIFE CYCLE

Human malaria parasites are transmitted by female mosquitoes of the genus *Anopheles* when feeding on

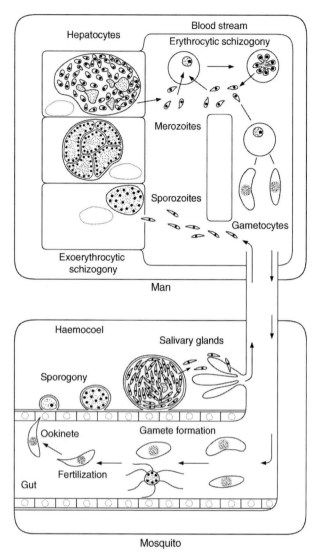

Fig. 1. Life cycle of *Plasmodium falciparum* (reprinted with permission from Wilson, 1979).

falciparum malaria in Africa. In areas of high ende-micity with intense transmission, unprotected indi-viduals are likely to experience multiple infective bites every night.

The form of the parasite carried in the salivary glands is the infective sporozoite, a long, thin cell about 11 μm in length and 1 μm in diameter, which enters the bloodstream as the mosquito feeds (Fig. 2). An infected mosquito will normally inject several tens to several hundreds of sporozoites during a sin-gle bite. In less than 1 hr all surviving sporozoites have entered hepatocytes, the major type of cell in the liver. Within these cells, the parasites transform into a rounded form known as the preerythrocytic schizont. These grow and undergo many nuclear divisions, initially without division of the cytoplasm. Just before the swollen hepatocyte bursts, the multi-nucleate schizonts mature into individual invasive merozoites, 10,000–30,000 of which are released from a single cell depending on the species. The erythrocytic phase now begins; this is responsible for all the clinical symptoms of malaria. The pear-shaped merozoites, measuring about 1 × 1.5 μm, rapidly adhere to red blood cells, reorienting their apical ends to make a specific interaction with the erythrocyte membrane, releasing molecules from the organelles (rhoptries and micronemes) of the apical complex that induce the invagination of the erythro-cyte surface (Fig. 3). Within 1 min, the parasite

blood, a process essential to the development of the eggs (Fig. 1). As a genus, these insects are common in most tropical and temperate regions and are highly adaptable organisms with enormous breeding capa-bility. About 70 species of *Anopheles* naturally trans-mit the human disease, although in most regions only a small number predominate. For example, in sub-Saharan Africa, just 3 species constitute the ma-jor vectors, the most notorious of which is *Anopheles gambiae,* which has a marked preference for feeding on human blood and is responsible for much of the

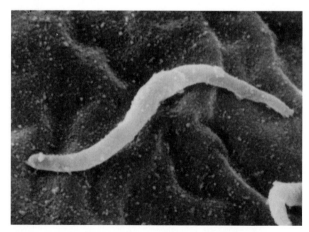

Fig. 2. Sporozoite of *Plasmodium yoelii* (×12,500) (cour-tesy of Prof. R. E. Sinden, Imperial College, London).

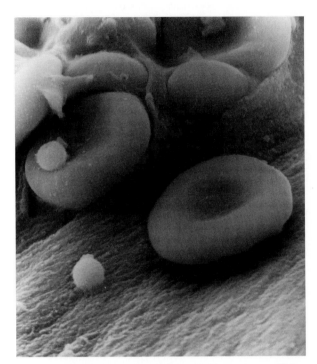

Fig. 3. Merozoites of *Plasmodium falciparum*, one still free and the other attached to a red blood cell at the start of invasion (×6000) (courtesy of Dr. L. A. Bannister, King's College, London).

occupies a cavity known as the parasitophorous vacuole, which is initially lined with a membrane derived from the surface of the erythrocyte. The young ring form (so-called because of its resemblance to a signet ring) now feeds on the cytoplasm of the host erythrocyte, breaking down the hemoglobin to obtain amino acids. As it passes through a mature trophozoite stage to the schizont stage, the parasite fills almost all of the red cell and contains between 8 and 24 nuclei after nuclear division. These again mature into invasive merozoites, released upon rupture of the red cell. This erythrocytic replication cycle repeats approximately every 48 hr in the infections of *P. falciparum, P. vivax,* and *P. ovale* and every 72 hr for *P. malariae.* In this way, the number of parasites in the human host can build up to more than 10^{12}. Parasitemias (the fraction of red cells infected) will normally stabilize at around 0.2–2%, although in *P. falciparum* infections much higher levels (up to 50%) are sometimes observed.

The synchronized production of a large number of merozoite progeny that is seen in both the liver and the blood stages is thought to be a defining feature of the Apicomplexa, in which nuclear divisions are complete before the progeny begin to bud from the parent cell. However, within the erythrocytes, some merozoites differentiate into another morphologically distinct type called the gametocyte, a presexual stage with male and female forms. These forms are key to successful continuation of the life cycle in the mosquito. When taken up in a blood meal by a feeding mosquito, the male gametocyte starts to divide within 10 min of reaching the insect's gut, giving rise after three mitotic divisions to eight highly motile, thread-like microgametes (Fig. 4) that swim toward the female macrogamete, which has lost the protection of the erythrocyte membrane and is ready for fertilization. After nuclear fusion, the resulting zygote rapidly develops into an oval-shaped ookinete, another motile invasive form that escapes from the blood meal and embeds itself in the epithelial cell layer of the mosquito midgut. Further differentiation results in an oocyst, which enlarges considerably during the next 10–12 days (Fig. 5). Numerous nuclei are generated, which partition into newly formed sporozoites. These are released from the oocyst (Fig. 6) and migrate to the salivary glands of the mosquito, where they appear to be viable indefinitely. The cycle is thus completed.

Following the release of merozoites from the liver stages of *P. falciparum* and *P. malariae,* the liver is clear of parasites until another infective bite from a mosquito; exoerythrocytic growth and multiplication is thus single phased. Typical infections with these parasites last a few months. However, for *P. vivax* and *P. ovale,* an additional form of the parasite, termed a hypnozoite, remains dormant in the liver and can cause renewed blood infections with full clinical symptoms months or years later without further infection by a mosquito. This type of multiple-phased disease is called relapsing malaria, and it is not known what triggers activation of the hypnozoite. *Plasmodium malariae* infections can exceptionally persist for decades, but the parasite maintains itself at a very low density in the bloodstream and not by virtue of a hypnozoite form in the liver.

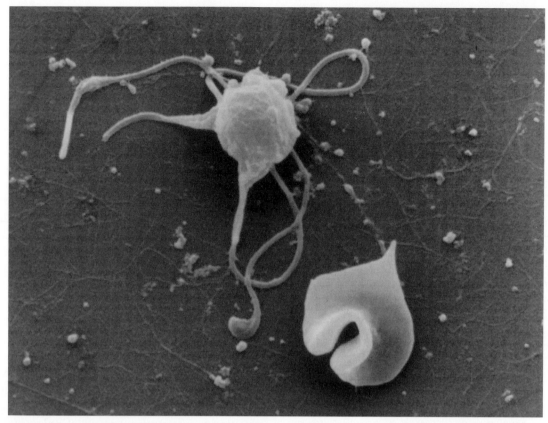

Fig. 4. The flagellated form of the microgamete of *Plasmodium yoelii* just prior to exflagellation and fertilization of the macrogamete (×12,000) (courtesy of Prof. R. E. Sinden, Imperial College, London).

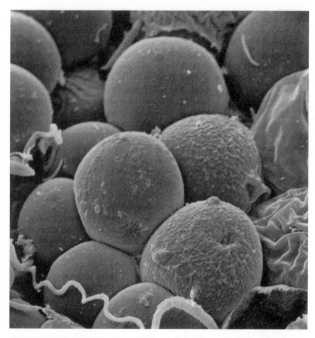

Fig. 5. Oocysts of *Plasmodium falciparum* embedded in the midgut of the mosquito (×1200) (reprinted with permission from Sinden and Strong, 1978).

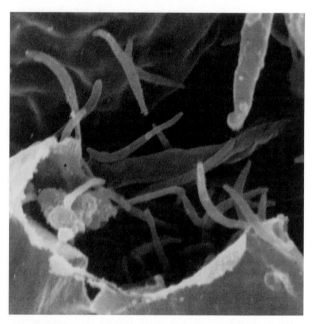

Fig. 6. Bursting oocyst of *Plasmodium falciparum* and emerging sporozoites (×8000) (reprinted with permission from Sinden and Strong, 1978).

III. PATHOLOGY OF MALARIA

A. Clinical Responses

An infection with *P. falciparum* gives rise within 2 weeks to bouts of chills, fever, and sweating, with the soaring body temperature (up to 41°C) reflecting multiplication within, and reinvasion of, the red blood cells. The periodicity of symptoms results from the 48-hr erythrocytic cycle but is less regular and pronounced in *P. falciparum* infections than with the other human parasites, and fever can be more or less continuous. In *P. vivax* and *P. ovale* infections, fevers peak every 48 hr, whereas for *P. malariae,* the cycle is 72 hr. The spleen swells as defensive white blood cells and parasitized red cells accumulate there. This organ plays a crucial role in clearing parasites and protecting the host. Lysis of infected cells leads to anemia, which is possibly exacerbated by a reduced level of red cell production from the stem cells in the bone marrow. The great danger of *P. falciparum* infections results from the expression of parasite proteins on the erythrocyte surface that cause infected cells to adhere to endothelial cells in the microcapillaries of vital organs. Thus, only younger parasites (<24 hr postinvasion) are seen in the peripheral blood, with the older forms being sequestered in the deep vascular beds of many organs. This appears to be a ploy by the parasites to reduce their flow and destruction via the spleen, which can remove erythrocytes infected with the older but not the early forms. Death of the host can result from aggregation of parasitized erythrocytes in the kidneys, the lungs, and most critically, the brain. Some strains also give rise to a phenomenon known as "rosetting" in which up to 10 or more uninfected erythrocytes clump together around an infected cell, presumably contributing further to dangerous blockages and possibly enhancing the efficiency of reinvasion. Individual symptoms in the patient can vary widely and depend, among other things, on age, genetic makeup, previous exposure, and geographical location. Organ failure is much more common in Asian and South American adults than in African children, whereas the latter are particularly susceptible to fatal cerebral malaria. In areas of very high transmission (several hundred infected bites per year), anemia is the more common cause of death, whereas in areas where rates are lower, less stable, or highly seasonal, cerebral malaria is more typical.

Electron microscopy of *P. falciparum*-infected cells reveals numerous knob-like structures that are absent from the smooth surfaces of normal red cells. The importance of these knobs was demonstrated by disruption of the gene encoding the "knob-associated histidine-rich protein." This protein lies under the surface of the knobs, and its loss resulted in "knob-less" cells that no longer adhered to the receptors found on endothelial cells under conditions simulating the flow of erythrocytes through blood vessels. Although inhibitors of the knob protein would not prevent the disease, they could ameliorate the most dangerous complications of falciparum malaria.

During a malaria attack, cytokines are released from the cells of the host immune system, especially tumor necrosis factor and interleukins 1 and 6. These are inflammatory molecules, whose raised levels correlate with the severity of the clinical symptoms; however, it is unclear whether these molecules are directly responsible for causing the pathology of severe malaria or whether their marked elevation is also a result of other events. The fever response is thought to represent another of the host's mechanisms for combating the parasite because there is evidence that schizonts are damaged by the high blood temperatures that result. However, in some individuals such temperatures can lead to persistent neurological sequelae, whereas the very high levels of cytokines induced may contribute toward a fatal outcome.

B. Immunity to Parasite Infection

In endemic areas, if the young child survives beyond the age of about 5 years, naturally acquired immunity to malaria builds up, though only slowly, presumably due to the high degree of antigenic diversity exhibited by the parasites. The severity of attacks subsides as the individual progresses to adulthood. An immune mother passes on factors (mainly antibodies) to her newborn, such that severe disease is rare for the first few months of life. In addition, fetal hemoglobin (HbF) in the new baby's erythrocytes is not conducive to parasite development. The period of greatest danger is from about 6 months to 3 years. However, an effective degree of immunity requires constant challenge and it is thought that complete ("sterile") immunity to parasite invasion never occurs. Thus, symptomless adults in endemic areas are nevertheless usually parasitemic, and even they can be subject to further attacks if they leave their local endemic area, although this is unlikely to involve life-threatening disease. These considerations have major strategic implications for the possible development of vaccines against malaria (see Section VI,C).

Although host genetic factors are known to influence the course and outcome of malaria infection at different levels, they are in general poorly understood. The best characterized phenomenon, in which a gene type has clearly been selected for protection against malaria, is the sickle-cell trait in which a mutant hemoglobin gene (encoding hemoglobin S; HbS) is found at high levels in susceptible populations. Thus, there are regions of Africa where more than 20% of the population are heterozygous for HbS and HbA (where HbA is the normal form), despite the sickle-cell anemia resulting from HbS homozygosity being almost always fatal. Although AS heterozygotes appear to be as commonly infected as AA homozygotes, severe clinical effects are much rarer; estimates indicate that this trait provides approximately 90% protection against cerebral malaria and severe anemia. The basis for this is not fully understood, but it seems that the increased tendency of AS cells to sickle (become abnormally shaped) results in a lower degree of cytoadherance and rosetting as well as more efficient clearance of infected cells by the spleen. In contrast, SS homozygotes, whose spleens are compromised, are much more liable to malaria fatality, indicating that there is no major impediment to parasite development in HbS-containing cells.

IV. MOLECULAR BIOLOGY OF *P. FALCIPARUM*

A. DNA and RNA

The malaria parasites are haploid organisms (i.e., they contain only a single copy of their genetic information) for the great majority of their life cycle.

DNA replication occurs during the liver stages, blood stages, maturation of the gametocytes, in the zygote and the oocysts. The rate of replication in the male gametocytes is particularly impressive, with three rounds of DNA replication to the octoploid stage occurring within 10 min (possibly the fastest known example), before exflagellation leads to the eight mature male gametes. Upon fertilization, the newly formed diploid zygote doubles its DNA content and then rapidly undergoes two meiotic divisions, reverting to haploidy. Often, self-fertilization occurs, but if the zygote has resulted from the fusion of different strains of the parasite, the random segregation of chromosomes, as well as crossover events between homologous chromosomes, will generate new genotypes.

The nuclear DNA content of *P. falciparum* has been estimated from DNA extraction yields and chromosomal analysis to be 25–30 Mb, approximately six times that of the gut bacterium *Escherichia coli* (4.67 Mb) and more than twice that of the brewer's yeast, *Saccharomyces cerevisiae* (12.07 Mb). This DNA is distributed among 14 chromosomes in *P. falciparum*, whose sizes range from about 0.7 to 3.5 Mb as estimated on pulsed-field gel electrophoresis systems and recently by DNA sequencing; however, there can be considerable size variations for a given chromosome among different strains (Fig. 7). The telomeres resemble those of other eukaryotes, although internal to these are characteristic subtelomeric regions consisting of complex and simple repeat sequences interspersed with a nonrepetitive region. In addition to this nuclear DNA, the parasite possesses a 35-kb circular DNA molecule found in a multi-layered membrane-bound organelle whose sequence characteristics strikingly resemble those of plant chloroplasts. This plastid is thought to have originated from engulfment of a single-celled alga or dinoflagellate (which carried the remnants of a light-harvesting bacterium) by remote ancestors of the malaria parasites. The *P. falciparum* 35-kb circle (termed the "apicoplast" because it is found in other protozoa of the Apicomplexa phylum) is maternally inherited, the smallest known vestigial plastid genome of its type, and carries genes for ribosomal RNA (rRNA), ribosomal proteins, subunits of a prokaryotic-type RNA polymerase, and 25 different transfer RNAs (tRNAs), among others.

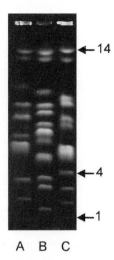

Fig. 7. Pulsed-field electrophoresis gel of chromosomes extracted from three different strains of *Plasmodium falciparum* (A and B are from different areas of Africa and C is from Thailand), showing the marked degree of size polymorphism among the 14 chromosomes. Selected chromosome numbers are indicated for strain C; not all of the chromosomes can be resolved in a single gel run.

Although the functions of the apicoplast are not understood, it appears to be essential for viability of the parasite. There is also a mitochondrial genome of 6 kb, found as multiple copies in linear concatamers, which is the smallest mitochondrial genome characterized to date. It appears to be transcribed polycistronically as a complete unit and encodes just three mitochondrial proteins (cytochrome *b* and two subunits of cytochrome oxidase) as well as highly fragmented large and small subunit rRNA sequences.

The nuclear DNA of *P. falciparum* has an unusual base composition, representing one of the most AT-rich genomes yet discovered, with an overall content of 81% AT. This is unevenly distributed between coding regions that average about 70% and noncoding regions of about 86%. A substantial minority of characterized genes contain introns (approximately 30–40%), which are usually few in number and relatively short, averaging about 200 bp in length. The splice junction sequences that delineate the boundaries of introns closely resemble those of other eukaryotes. The AT-rich nature of the genome has hindered characterization of *P. falciparum* genes because the DNA is less stable and less easy to manipulate in bacterial cloning systems than DNA of a more

balanced composition. However, a significant number of important genes have been analyzed, and chromosome mapping and large-scale sequencing promise to reveal the entire genetic makeup of *P. falciparum* before 2002. The accumulating data are accessible at several web sites, including *http://www.nibi.nlm.nih.gov/malaria/* and *http://www. sanger.ac.uk/projects/P_falciparum/*.

Another major breakthrough has been the recent development of techniques to genetically transform the parasite, a process now considered routine for many organisms such as yeast or *E. coli* but which remains a considerable challenge in *Plasmodium*. Protocols are now established for several species, including *P. falciparum* and *P. berghei*. This has provided many possibilities to increase our understanding of parasite gene functions and has already been exploited to study the influence of mutations in critical genes involved in antimalarial drug resistance.

The messenger RNA transcripts from protein-encoding genes of the parasite are polyadenylated and appear to follow the standard eukaryotic pattern, although many have quite long 5′ and 3′ untranslated regions compared to those of other organisms, often accounting for one-third to one-half of the transcript length. Little is known of the details of ribosome biogenesis, but *P. falciparum* and other species of malaria parasites are unusual in possessing very few nuclear rRNA genes: only eight copies in *P. falciparum*, four copies in *P. berghei*, and six to eight copies in *P. lophurae* (an avian parasite), which are unlinked and show marked sequence variation. This can be compared to other eukaryotes in which the nuclear rRNA genes are usually highly repeated (hundreds to thousands of identical copies) in one or several tandem arrays. In addition, *Plasmodium* is unique to date as the only known genus in which individual rRNA genes are differentially transcribed at various stages of the life cycle. Each species appears to possess either two or three distinct types of rRNA gene. For instance, in the asexual blood stages of *P. vivax*, mature rRNA is derived from an A-type gene to be replaced by transcripts from an O-type during development in the mosquito of the oocyst from the zygote, which in turn yields to an S-type after differentiation to the sporozoite stage. A similar process has also been demonstrated in other *Plasmodium* species, including *P. falciparum*. These curious phenomena set *Plasmodium* apart from other organisms, including the vertebrate host.

B. Evolutionary Relationships

As with many other organisms, the phylogenetic relationships of *Plasmodium* species have been explored by gathering sequences of the genes that encode the highly conserved ribosomal RNA molecules, particularly the small subunit ("18S") RNA. Trees have also been constructed on the basis of an extensive collection of circumsporozoite protein (CSP) genes. These studies have indicated that *P. simium*, a parasite of New World primates, and *P. vivax* are so closely related that they may even be the same species; this is also the case for *P. brasilianum* (also from New World primates) and *P. malariae*. A recently identified *P. vivax*-like species that can infect humans has also been tentatively identified as identical to *P. simiovale*, a parasite of Old World macaques. *Plasmodium falciparum*, on the other hand, and its close relative *P. reichenowi*, which parasitizes chimpanzees, do not group with other primate parasites but appear to be more closely related to avian parasites such as *P. gallinaceum* and *P. lophurae*. This may be reflected in the distinctive characteristics of malaria caused by *P. falciparum* compared to the other human parasites.

V. IMPORTANT BIOCHEMICAL PATHWAYS

A. Reorganizing the Red Cell

Understanding the detailed biochemistry of malaria parasites has been limited by the technical difficulties of growing the parasites in the laboratory and isolating sufficient quantities of active material. Even though they are single-celled parasitic organisms, they are complex eukaryotes that must pass through a wide diversity of environments within the vertebrate and insect hosts and are thus endowed with a plethora of metabolic pathways. Only those of special significance will be considered here. The parasite obviously has much more complex metabolic needs

than the largely inactive erythrocyte host cell, and transport of the relevant molecules involves passage across the erythrocyte plasma membrane, the parasite vacuole membrane, and the parasite plasma membrane. To meet some of its requirements, the parasite induces profound changes in the permeability of the red cell membrane to a wide range of low-molecular-weight substrates upon infection, including certain cations and anions, amino acids, sugars, choline, purine bases, and nucleosides. Some of these appear to be actively transported, whereas others pass by diffusion. The parasite also possesses a feeding organelle, the cytostome, which can directly ingest host-cell cytoplasm, leading to the formation of a transport vesicle that fuses to the food (or digestive) vacuole. There is also evidence for a mechanism by which the parasite can access the extracellular environment directly, bypassing the cytoplasm of the erythrocyte, although the details of the structures and molecules involved are controversial. The parasite elaborates distinct tubular membrane structures that appear to begin at the vacuolar membrane surrounding the parasite and extend outwards toward the inner surface of the erythrocyte. This tubovesicular membrane network may provide ducts through which certain molecules may directly pass to the parasite.

B. Energy Production

The malaria parasite, in common with other parasitic organisms, undergoes phases of rapid multiplication for which it requires considerable amounts of energy. Unlike higher organisms and some other parasitic protozoa, which deposit carbohydrate as glycogen stores (polymerized glucose), *Plasmodium* relies, at least in the vertebrate bloodstream, on importing large quantities of glucose from the plasma and processes it through the glycolytic pathway to generate ATP. Thus, an infected erythrocyte consumes approximately 100 times as much glucose as a normal cell. This represents the major energy-producing pathway and contributes to the host hypoglycemia characteristic of severe malaria. The glucose is not completely oxidized because the pyruvate formed is reduced by NADH to lactate and excreted by the parasite rather than undergoing further oxidation in a Krebs' (tricarboxylic acid) cycle. Although the parasite does possess a single mitochondrion, it appears, at least in the case of *P. falciparum,* to lack the well-defined extensive folding and large surface area of the inner mitochondrial membranes (christae) typical of other eukaryotes and is deficient in the machinery for oxidative phosphorylation. However, a functional mitochondrion is essential for growth of blood-stage parasites and is thought to mainly serve as an electron sink for dihydroorotate dehydrogenase catalysis in the *de novo* biosynthesis of pyrimidines (see Section V,C).

In common with most other parasites, oxidation of stored lipid is not utilized as a contribution to energy levels. The lipids that are required by the parasites, for membrane synthesis and other purposes, must be obtained externally because they are unable to synthesize long-chain fatty acids from simple precursors. In culture, lipids are traditionally supplied in the serum or plasma added to the medium, although these sources can be replaced by more reproducible lipid-rich preparations of bovine serum albumin, which can be important for certain experiments. Such lipids are crucial for the many membrane-bound organelles in the parasite, such as the food vacuole, rhoptries, micronemes, and mitochondrion, as well as for the membrane forming the parasitophorous vacuole, which must expand as the parasite grows in the red cell.

C. DNA Production

To make the DNA needed for rapid multiplication, the parasite must ensure supplies of the component D-ribose sugar, pyrimidine, and purine bases. Enzymes of the hexose monophosphate shunt pathway, starting from glucose-6-phosphate, have been identified, but it is thought that this contributes to only a minor degree to the ribose pool and that the majority derives from reactions using fructose-6-phosphate and glyceraldehyde-3-phosphate or inosine. The parasite possesses a *de novo* pathway for pyrimidine synthesis but not for purines. Pyrimidine salvage appears to be possible but of little importance under natural conditions, whereas purine salvage is essential. Starting from carbamyl phosphate and aspartate, a series of enzymes lead to the production of deoxy-

uridine monophosphate (dUMP). The subsequent synthesis of the thymidylate required for DNA production (as dTMP) involves addition of a $-CH_2$ group to dUMP. Like all organisms, malaria parasites require a source of reduced folates to act as cofactors in many essential 1-carbon transfer reactions, including the dUMP to dTMP conversion. The folate biosynthetic pathway is thus of particular importance to the parasite and has long been the target of the important class of antimalarials, the antifolates. These include the drugs pyrimethamine and cycloguanil, which inhibit dihydrofolate reductase (DHFR), and the sulfonamide/sulfone class, which inhibits dihydropteroate synthetase (DHPS) (see Section VI,B). The pathway starts at GTP, which is converted to 7,8-dihydrofolate (DHF) in seven steps, incorporating *p*-aminobenzoate and glutamate along the way. This part of the pathway is absent from vertebrates, which must salvage preformed folate from the diet, and it thus represents an attractive target area for inhibitors, exploited to date in antimalarial chemotherapy only in the inhibition of DHPS by the sulfur-based drugs. Subsequently, the DHF is reduced to tetrahydrofolate (THF) by DHFR, and a modified version of this donates the $-CH_2$ group to dUMP, with the regeneration of DHF. *Plasmodium falciparum* has also been shown to be capable of utilizing preformed folates in culture, but the relative contributions of folate synthesis and salvage *in vivo* are not known.

Plasmodium species, like virtually all known parasites, are incapable of synthesizing the purine doubling-ring system and must therefore resort to salvage pathways for which they appear to possess enzymes allowing uptake and interconversion of purine-containing compounds provided by the host. Cells in culture will take up high levels of adenosine, inosine, and hypoxanthine (which is the most favored). *In vivo*, this base probably derives mainly from dephosphorylation of ATP to AMP, deamination to IMP, and then conversion to hypoxanthine via inosine.

D. Amino Acid Metabolism and Proteases

The parasite derives the amino acids it requires for protein synthesis from several sources. It is capable of synthesizing a limited number of them, such as alanine, aspartic acid, glutamic acid, cysteine, and methionine. Others are obtained preformed by uptake from the host plasma (such as isoleucine, which is absent from human hemoglobin) or the erythrocyte itself, but these supplies (which overlap in their constituents) appear to be insufficient and the parasite is obliged to furnish much of its requirement by the degradation of the hemoglobin in the red cell to heme and free amino acids. Up to approximately 75% of the hemoglobin in an infected erythrocyte will be destroyed during the parasite's occupation. The main site of degradation is the acidic digestive or food vacuole, in which several proteases are utilized. Two closely related aspartic proteases, plasmepsins I and II, appear to have essential roles in initiating hemoglobin degradation, and the crystal structure of plasmepsin II (reported in 1996) was the first to be determined for a malarial protein. Plasmepsin I is active in the ring form soon after erythrocyte invasion, whereas plasmepsin II becomes active later in the cycle (in the mature trophozoite stage). A cysteine protease known as falcipain appears to act on denatured hemoglobin after initial cleavages made by the plasmepsins. These enzymes degrade the protein down to peptides averaging about eight amino acids in length, which are thought to be transported from the vacuole to the cytoplasm for exopeptidase conversion to single amino acids. The toxic heme moiety remaining after hemoglobin digestion is inactivated by polymerization and sequestration as hemozoin, a dark pigment that becomes clearly visible as granules as the parasite develops within the red cell.

In addition to their role in processing hemoglobin, proteases are important in both invasion of and parasite release from host cells. Various serine and cysteine proteases have been reported to be involved in these events and some of these enzymes have been partly or fully characterized and compared to the human homologs, where they exist. Several of their properties differ significantly, for instance, in substrate affinities, sensitivities to inhibitors, and pH optima, suggesting that antimalarial drugs based on differential inhibition of proteases could in principle be developed to interfere with these vital functions.

VI. CONTROL OF *PLASMODIUM* INFECTIONS

A. Vector Control

Although this article is primarily concerned with the parasite, an important aspect of malaria control involves interference with the anopheline mosquito host and its transmission of the parasite. This is primarily achieved by destruction of breeding habitats and application of insecticides to reduce vector numbers and the use of bednets to reduce contact with humans. Recently, the two latter strategies have been combined in the use of bednets impregnated with a pyrethroid insecticide (usually permethrin), and large-scale evaluations have been performed. Although deaths among African children were reduced by up to one-third in these trials, there is still controversy regarding the long-term desirability of such a strategy because it has implications for the development of natural immunity. Some workers believe that a reduced exposure in infancy may lead to greater susceptibility and more severe symptoms in later life, possibly resulting eventually in a greater number of deaths, particularly in areas of intense transmission. Although this is unlikely to be rigorously testable in practice, a counterargument is that even if the use of bednets causes some increase in deaths from cerebral malaria, this increase will be more than offset by the reduction in deaths brought about by malaria in conjunction with other pathogens in infancy.

The ease of eliminating or minimizing the habitats available for the mosquito to breed depends greatly on geographical location and the anopheline species involved. Drainage of the notorious marshes in the Campagna and Pontina areas to the west of Rome in the 1930s succeeded in eliminating transmission in one of the most malarious areas in the world. In Africa, the highly successful *Anopheles gambiae* can make use of small, temporary pools of water, such as those formed by foot- or hoofprints or tire marks, clearly posing almost insurmountable logistical problems. In other areas, species such as *A. stephensi* favor breeding in wells or cisterns, which are obviously more amenable targets.

Just as chemotherapy against the malaria parasite is threatened by spreading resistance to the available drugs (see Section VI,B), *Anopheles* mosquitoes have become increasingly resistant to insecticides such as DDT and its successors such as HCH, dieldrin, the organophosphates, and organochlorine compounds. This has largely occurred from agricultural use rather than malaria control, but it has become a major factor in the upsurge of malaria in many tropical countries. In at least one of the bednet studies mentioned previously, significant increases in resistance to permethrin were observed after just 1 year's practice. In many areas, such as Madagascar, parts of Southeast Asia, and Brazil, extensive deforestation has led to increased human contact with the vectors and encouraged conditions for breeding, leading to significant resurgences of the disease. In an attempt to find ways to circumvent these problems, research has been directed into the possible genetic manipulation of *Anopheles* as a novel but almost certainly long-term approach to disease control. Originally, this revolved around the idea of producing sterile mosquitoes, but current thinking favors the development of mosquitoes that are refractory to infection by *Plasmodium*. *Anopheles* lines resistant to the parasite have been identified in which the invader is either encapsulated or lysed, and attempts are being made to map and clone the genes responsible for these phenotypes. A complementary approach is to produce transgenic mosquitoes which carry foreign genes that encode products capable of attacking the insect-stage parasite and blocking development to the infective sporozoite. Clearly, many problems will have to be solved before such an approach can reach the stage of practical application, including the efficient dispersal of such genes through wild mosquito populations and the likely resistance to the release of genetically engineered organisms of this type. With regard to the former, it is hoped that a transposable element can be identified, akin to the P-element found in *Drosophila* genomes, that would behave like an infective agent and spread rapidly through the population, carrying with it the gene(s) encoding the agent inhibitory to parasite development.

B. Chemotherapy

Chemotherapy against malaria, in the West at least, can be regarded as starting with the use of prepara-

tions derived from the bark of the *Cinchona* tree that was exported by Jesuit missionaries ("Jesuit powder") in the early seventeenth century. The tree is native to the Andean region of South America, particularly Peru, but was transplanted to Java by the Dutch and to India by the British in the nineteenth century, when its importance as a source of purified quinine was realized. Until shortly before World War II, quinine represented the sole antimalarial drug. However, with the introduction of synthetic drugs such as chloroquine (1934), proguanil (1944), primaquine (1952), and pyrimethamine (1952), the repertoire quickly expanded, although quinine is still one of the most powerful drugs and continues to play an important role in treating life-threatening disease caused by parasites that are multiply resistant to the more modern synthetic inhibitors. By far the most successful antimalarial drug, for both prophylaxis and therapy, has been chloroquine, a member of the quinoline group of compounds which are structurally related to quinine. However, it is much less toxic than quinine, and it is very cheap. Chloroquine is selectively and rapidly taken up by infected erythrocytes, where the parasite is feeding, to give concentrations that are several hundred fold higher than in normal cells. Its mode of action appears to involve a blockage of the mechanism by which the parasite sequesters the toxic heme resulting from breakdown of the hemoglobin; thus, the parasite is essentially poisoned by its own waste. Although it still retains its position as the first-line drug in some areas of West Africa, the Middle East, and Central America, resistance to chloroquine has spread so extensively, from initial foci in Thailand and Colombia at the end of the 1950s, that its utility as monotherapy in combating falciparum malaria appears to be near an end. However, there is only limited resistance of *P. vivax* to this drug, which started to spread from Papua New Guinea at the end of the 1980s. The potential loss of chloroquine is a major setback because other available drugs are significantly more expensive. The only other class of drugs of a comparable cost are the antifolates, which target enzymes in the folate biosynthetic pathway (see Section V,C). The most common formulation from this class is the synergistic combination of pyrimethamine with sulfadoxine (Fansidar). Although currently being used as the first-line response to chloroquine-resistant parasites in many parts of Africa, resistance is spreading to levels that threaten to take these drugs out of the armory within a few years.

Among the small number of modern drugs that have been licensed for clinical use are mefloquine (1975), halofantrine (1989), and derivatives of artemisinin (early 1980s). Mefloquine, as a quinoline compound, is also related to quinine and has been used extensively to treat multiple drug-resistant parasites in Southeast Asia and as prophylaxis for travelers to areas of chloroquine resistance. It is a potent, long-acting drug that has the advantage of effecting cure after a single dose or requiring only weekly doses as a prophylactic. However, it appears to be unsuitable for people with a history of epilepsy or psychiatric disorders, and the antibiotic doxycycline is often prescribed as a prophylactic in such cases or for areas where mefloquine resistance is becoming a problem. Halofantrine is a phenanthrene methanol also used against multiple drug-resistant parasites. Artemisinin is the active component of qinghao (sweet wormwood, *Artemisia annua*), a Chinese remedy for malaria and other parasitic diseases that dates back approximately 2000 years. This and its derivatives, such as artemether and sodium artesunate, are sesquiterpene lactones with a mode of action quite different from those of the quinoline family, which is an important consideration in minimizing the likelihood of parasites becoming cross-resistant to the different drugs. These drugs concentrate on parasite membranes, interfere with protein synthesis, and cause changes in organelle morphology. They carry an epoxide bridge across the seven-membered component of their triple-ring system, whose oxidizing capability is triggered by contact with the iron complexed in the parasite. This is thought to be the cause of the observed inhibitory effects, which can rapidly clear parasites resistant to the other classes of antimalarials.

Another aspect of contemporary antimalarial chemotherapy is the use of inhibitor combinations to restore or prolong the utility of available drugs and to slow the spread of resistant parasites. For example, pyrimethamine could no longer be used alone only a short time after its introduction because resistance to it spread too quickly, but the powerful synergy it displays with the sulfur-based group of drugs means that it is still of great importance today. Similarly, the use of proguanil has recently (1997) been revitalized by combination with atovaquone, a naphthoquinone that inhibits the electron transport chain in the mitochondrion. Also, a combination of tetracycline with quinine has proved useful in areas of Southeast Asia where quinine alone is now less than 100% effective. Current practice is moving toward combinations of short and long half-life drugs, such as artesunate and mefloquine, in which exposure of parasites to sublethal levels of one drug (the ideal condition for selection of resistant mutants) is eliminated or minimized by the presence of the second drug acting on a different target over a different time scale. This combination, for example, has been used to good effect in Thailand and Cambodia, where resistance to mefloquine alone has become established.

C. Development of Malaria Vaccines

Vaccines that have made a major impact on the health of humankind have been directed against viral or bacterial pathogens that have relatively simple lifestyles and show little or no strain variation. Perhaps the prime example is the smallpox virus, which was officially declared to have been eradicated as a clinical pathogen in 1979. Protozoan eukaryotes are considerably more complex, and much of the evolution of parasites such as the agents of malaria has presumably been in the direction of evasion and suppression of the host immune responses. To date, there is no vaccine against a eukaryotic parasite in clinical use. However, considerable resources and ingenuity have been directed toward the development of malaria vaccines during the past 30 years because success would be expected to represent a highly cost-effective method of controlling the disease.

Although successful immunization has been achieved with irradiated sporozoite forms of the parasite in both animal and human experiments, it is impossible to scale up production of these cells (which cannot even be cultured *in vitro*) to a useful level. The approach has thus been to identify genes encoding antigenic proteins thought to play important roles in the development of protective immunity in endemic areas. After cloning of such genes, or relevant parts thereof, the protein product is recovered from a heterologous expression system, such as *E. coli* or yeast. Alternatively, peptides covering regions of interest can be synthesized chemically.

Three classes of vaccine have been produced for experimental evaluation, and several individual formulations have reached clinical trials. Preerythrocytic vaccines have been principally based on the major coat protein of the sporozoite stage, the circumsporozoite protein (CSP). CSP-based vaccines are intended to promote antibody attack on the sporozoite before it reaches the liver and T cell responses against parasites that successfully invade hepatocytes. If successful, such a vaccine would theoretically prevent all clinical symptoms by eradicating the parasites before blood-stage forms could be released from the liver. The second class of vaccines comprises candidates based on various antigens expressed during the blood stages that are on the surface of the merozoite or infected erythrocyte or become exposed during merozoite invasion. Important candidates include merozoite surface proteins 1 and 2, the ring infected erythrocyte surface antigen, and the apical membrane antigen-1. The rationale is to induce antibodies that would not only promote clearance of infected erythrocytes by triggering effector mechanisms, such as complement lysis and enhanced phagocytosis after opsonization, but also block the invasion process as well as bind to toxins that cause the disease symptoms. Finally, there is the so-called altruistic or transmission-blocking type of vaccine in which antibodies induced by the vaccine against sexual stage antigens are taken up, along with the gametocytes, when the mosquito feeds. As the parasite enters the sexual phase, these antibodies would block development or invasion of the mosquito midgut. Such a vaccine would have no direct benefit for the recipient, but it could contribute greatly to reducing disease transmission and sexual recombination, which contribute to the spread of important genetic traits such as drug resistance. Candidate vaccines of this type are based on surface antigens of gametes, zygotes, or ookinetes, principally the Pfs25 antigen found on the latter two stages.

An advantage of this type of vaccine compared to the other two types is that when the cognate antigen is expressed the parasite is not undergoing selective pressure from the immune system of the human host because this type of antigen only appears in the mosquito stages.

As both B cell and T cell epitopes on human-stage antigens have gradually been identified, it has become clear that they often encompass regions of the proteins that are polymorphic in the *P. falciparum* population. The degree of polymorphism is higher on surface-expressed molecules of the invasive sporozoite and merozoite forms than it is on sexual (insect) stage or internally expressed proteins. Such polymorphism is thus presumably an evolutionary response by the parasite to the polymorphism that is found in the human immune system, at the level of the MHC gene products that are crucial in eliciting T cell (and hence B cell) responses. This type of consideration has intensified the search for vaccine candidates based on protein regions that are apparently invariable or to identify the most frequent or most dangerous variants and attempt to incorporate all the corresponding protein sequences into the vaccine.

A successful vaccine is likely to be constituted from components corresponding to all three of the previously mentioned classes of vaccine, and already some multistage, multicomponent vaccine candidates have been produced. Currently, approximately 16 test vaccines have reached or are close to phase I or phase II human trials. Although there has been little problem with safety or side effects, results to date in terms of protection have been disappointing. Two areas that may contribute to improvements are DNA-based vaccines and the development of more powerful adjuvants suitable for human use. DNA vaccines are essentially plasmids that carry one or more malarial genes (or part genes), which are delivered to the body and internally express the protein product of their DNA inserts. This has been shown in experimental models to provide a more intense stimulation of the immune system than is achieved when using the protein alone in a conventional vaccine, and clinical trials of the first vaccines of this type have recently begun. Similarly, although the aluminum salts currently licensed for human vaccine use are safe, they are weak adjuvants. Much enhanced responses are obtained if the antigen is linked to synthetic polymer microparticles or mixed with plant saponins or other lipid-based complexes, and some of these more powerful adjuvants are currently undergoing safety tests for human use.

D. Antigenic Variation

As implied previously, a common survival strategy of pathogenic microorganisms is to vary the surface molecules that are exposed to the host immune system, thus ensuring that chronic infection can be maintained. An important recent development has been the discovery in *P. falciparum* of a large and diverse multigene family (the *var* genes) whose protein products of 200–350 kDa are transported to the surface of the infected erythrocyte. There appear to be 50–150 members of this gene family, distributed on all of the chromosomes and mainly associated with the subtelomeric regions; however, it is not known whether all these genes can be expressed as functional proteins. Their importance is derived from the fact that they encode the PfEMP-1 protein. The switch in PfEMP-1 type results in antigenic variation and occurs as a result of their possession of structurally variable extracellular domains. These molecules are localized to the knobs on the surface of the erythrocyte (see Section III,A) and are primarily responsible for the adhesion of the infected cells to the microcapillaries in the major organs, which can occur via interaction with a variety of host endothelial cell receptors. These discoveries have opened up the possibility of using chemotherapy that can block adhesion of infected erythrocytes to alleviate the symptoms and potential fatality of cerebral malaria.

E. Parasite Diagnostics

Tests for human (or animal) malarial infection are easily carried out by examination of a blood smear under the light microscope after the parasites have been highlighted with a stain such as Giemsa or Field. So-called thin and thick films are usually made from the same blood sample, with the latter having an erythrocyte density approximately 20–30 times greater than that of the thin film. Spotting parasites in the thick film is straightforward, but identification is easier in a thin film, in which each parasite in its red cell is clearly separated from its neighbor. Using a thin film, a trained microscopist can easily tell the difference between the four species of *Plasmodium* that infect humans from characteristics such as pigment color, morphology, and the ratios of the various stages. Such an examination is extremely important given the wide variety of clinical symptoms that can present as a result of plasmodial infection.

In recent years, several molecular-based diagnostic techniques have been developed with specialist purposes rather than for general diagnosis. As indicated in Section VI,C, analysis of the range and degree of polymorphism in parasite antigens exposed to the immune system is critical to understanding how the parasite evades immune clearance and in the rational formulation of vaccine strategies. Originally, parasites collected from different areas were assayed using banks of monoclonal antibodies that specifically recognized particular versions of a given antigen. Recently, assays based on the polymerase chain reaction (PCR) have been developed that can detect the changes in the coding sequences at the DNA level that underlie antigen polymorphism. Such tests have proved of great utility in analyzing the parasites carried in an infected mosquito in which the material available is extremely limited and microscopic examination requires highly skilled dissection. PCR using gene probes specific to *Plasmodium* sequences can be carried out on mosquito squashes because the host DNA or RNA does not interfere with the signal obtained. Similarly, with the demonstration that resistance to both the anti-DHFR and anti-DHPS components of the antifolate antimalarials involves point mutations in the genes encoding these enzymes, PCR assays have been developed that can differentiate between strains carrying the wild-type or the various mutant forms of these genes (Fig. 8). Such tests should assist with the management of clinical treatment and contribute to epidemiological surveys of the spread of drug resistance. The molecular basis of resistance to other antimalarials, such as chloroquine and mefloquine, appears to be more complex, and much work is currently in progress to identify the key DNA sequence alterations that could be exploited

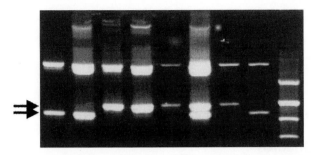

Fig. 8. Diagnostic PCR assays of mutations in the *dhps* gene of different patient samples of *Plasmodium falciparum*. The presence of the wild-type allele is indicated by a smaller band (lower arrow), whereas the larger band (upper arrow) corresponds to the mutant allele conferring elevated resistance to sulfur-based drugs. The sixth lane from the left shows a mixed infection; the ninth lane contains DNA size markers.

to give accurate predictions of the resistance status for these drugs in PCR reactions.

Our knowledge of *Plasmodium* has expanded enormously in the century since Ronald Ross proved that malaria is transmitted by the bite of a mosquito—particularly during the past 20 years with the advent of sophisticated molecular techniques. However, the prevalence and continued increase in malaria deaths and illness, essentially little different from when records began approximately 5 millennia ago, are a measure of the complexity and guile of these highly successful and fascinating microorganisms.

See Also the Following Articles

ANTIGENIC VARIATION • INTESTINAL PROTOZOAN INFECTIONS IN HUMANS • MALARIA

Bibliography

Borst, P., Bitter, W., McCulloch, R., Vanleeuwen, F., and Rudenko, G. (1995). Antigenic variation in malaria. *Cell* 82, 1–4.

Carlson, J., Olson, K., Higgs, S., and Beaty, B. (1995). Molecular genetic manipulation of mosquito vectors. *Annu. Rev. Entomol.* 40, 359–388.

Collins, F. H., and Paskewitz, S. M. (1995). Malaria: Current and future prospects for control. *Annu. Rev. Entomol.* 40, 195–219.

Gilles, H. M., and Warrell, D. A. (1993). "Bruce–Chwatt's Essential Malariology," 3rd ed. Arnold, London.

Good, M. F., Kaslow, D. C., and Miller, L. H. (1998). Pathways and strategies for developing a malaria blood-stage vaccine. *Annu. Rev. Immunol.* 16, 57–87.

Holder, A. A. (1999). Malaria vaccines. *Proc. Natl. Acad. Sci. USA* 96, 1167–1169.

Hyde, J. E. (1990). "Molecular Parasitology." Wiley, Chichester, UK.

Makler, M. T., Palmer, C. J., and Ager, A. L. (1998). A review of practical techniques for the diagnosis of malaria. *Ann. Trop. Med. Parasitol.* 92, 419–433.

Miller, L. H., and Hoffman, S. L. (1998). Research toward vaccines against malaria. *Nature Med.* 4, 520–524.

Reeder, J. C., and Brown, G. V. (1996). Antigenic variation and immune evasion in *Plasmodium falciparum* malaria. *Immunol. Cell Biol.* 74, 546–554.

Rosenthal, P. J. (1998). Proteases of malaria parasites: New targets for chemotherapy. *Emerg. Infect. Dis.* 4, 49–57.

Sinden, R. E., and Strong, K. (1978). An ultrastructural study of the sporogonic development of *Plasmodium falciparum* in *Anopheles gambiae*. *Trans. Roy. Soc. Trop. Med. Hyg.* 72, 477–491.

White, N. J., and Olliaro, P. L. (1996). Strategies for the prevention of antimalarial drug resistance: Rationale for combination chemotherapy for malaria. *Parasitol. Today* 12, 399–401.

Wilson, R. A. (1979). "An Introduction to Parasitology," 2nd ed. Cambridge University Press, Cambridge, UK.

Wu, Y. M., Kirkman, L. A., and Wellems, T. E. (1996). Transformation of *Plasmodium falciparum* malaria parasites by homologous integration of plasmids that confer resistance to pyrimethamine. *Proc. Natl. Acad. Sci. USA* 93, 1130–1134.

Polio

Ciro A. de Quadros

Pan American Health Organization

GLOSSARY

disease eradication Total absence of a disease and/or its infection agent from a given geographic area. When global eradication is achieved control measures can be terminated.

disease surveillance Actions that provide information on the determinants and conditions that affect individual or collective health and that generate appropriate responses for the prevention and control of such conditions.

flaccid paralysis Loss of muscle tone, usually due to the absence of motor neuron function.

immunity Protection from an antigen usually measured by the presence of serum antibodies.

inapparent infection The presence of infection in a host without recognizable clinical signs or symptoms. It can only be identified through laboratory tests.

pathogenicity The property of an infectious agent that determines the extent to which the disease occurs in an infected population.

serotype Classification of an infectious agent according to its antigenic constitution.

susceptibility Opposite of immunity; the absence of protection against an antigen.

POLIO is a viral infection recognized by its characteristic onset of flaccid paralysis. It is also known as infantile paralysis and poliomyelitis. The disease was known in ancient Egypt and was first recognized as a viral infection in the beginning of the twentieth century. By 1960 two highly effective vaccines in protecting against the disease had been developed. By 1991 the disease was eradicated from the Western Hemisphere and, following the initial successes of the eradication strategies employed in the Americas, the World Health Assembly targeted polio to be globally eradicated by the end of year 2000, a goal now on the verge of being achieved.

I. HISTORICAL NOTES

Polio is a disease that has been known to mankind since Ancient Egypt, having been depicted on steles during the Eighteenth Dynasty (1580–1350 BC). It is also believed that epidemics of "clubfoot" described by Hippocrates and Galen may have been due to poliomyelitis because they referred to both the congenital clubfoot and the clubfoot that occurs in early infancy.

Although it was an English pediatrician, Michael Underwood, who first described the relation of fever with the onset of the paralysis in the second edition of his *Treatise on Diseases of Children* in 1789, the first modern clinical description of the disease was made by the Italian physician Giovanni Battista Monteggia in 1813. This description was followed by references made by London surgeon John Shaw, who described the clinical aspects of the disease and mentioned that individuals also acquired the disease outside England. Subsequently, the German physician Jacob von Heine published a detailed description of the disease in 1840. Among its main characteristics he mentioned the age of the patients (between 6 and 36 months), the good general health preceding the first symptoms, the febrile manifestation that immediately preceded the paralysis, and the pain and in most cases the affliction of both lower extremities. Karl Oskar Medin, a Swedish pediatrician, had the opportunity to study a major epidemic in Scandinavia. He described several characteristics of the disease, including the milder, atypical form without paralytic illness, which is nonetheless important in the chain of transmission of the disease. This latter form was described by Ivar Wickman, a disciple of Medin, who proposed the name Heine–Medin for the disease.

The first publication suggesting that the disease had occurred in the United States (Ohio) was written by A. G. Walters, a surgeon from Pittsburgh. Coincidentally, his paper appeared in 1840, the year the disease was described by Heine. The name of the disease has evolved from the eighteenth-century "debility of the lower extremities" to "morning paralysis," "essential paralysis of infants," "Heine–Medin Disease," and finally to the twentieth-century "infantile paralysis" and "poliomyelitis," abbreviated to "polio."

The search for the causative agent of the disease, which spanned more than 50 years, was completed in 1908 when Karl Landsteiner, an Austrian immunologist, and his assistant, E. Popper, reproduced the disease in monkeys (*Cynocephalus hamadrias* and *Macaca mullata*), which showed lesions in their spinal cord exactly like the ones observed in human poliomyelitis. By 1909, the entire scientific community was certain of the viral etiology of the disease, and by 1912 it was generally agreed that poliomyelitis was an infectious disease with epidemic characteristics. The work of Landsteiner gave origin to the idea that a vaccine could prevent the disease. He was awarded the Nobel prize for his work on immunology.

In the United State, initial efforts to fight polio dealt mainly with the rehabilitation of its crippling aspects after Franklin D. Roosevelt was paralyzed by the disease in the early 1920s. In 1927, the Georgia Warm Springs Foundation was established which offered rehabilitation services to patients in a proper environment while also providing the medical profession with useful information on the disease and its rehabilitation through physical therapy. Although the foundation was primarily involved with the physical rehabilitation of patients, investigation on the

acute form of the disease soon followed. Funds for the foundation increased with the creation of the President's Birthday Ball Commission (PBBC), established after Roosevelt was elected president.

The search for a vaccine was intensified when the National Foundation for Infantile Paralysis (NFIP) was formed in 1938 and lead by Basil O'Connor, a former law partner of President Roosevelt. The NFIP was an offshoot of the PBBC, and its main focus was to raise funds to fight poliomyelitis, of which more than 10% was targeted at research. In the late 1940s, the newly created World Health Organization (WHO) established an Expert Committee on Poliomyelitis, which further outlined critical areas for research.

A major breakthrough came when John F. Enders, Thomas H. Weller, and Frederick C. Robbins cultivated strains of the three serotypes of poliovirus in a variety of extraneural human tissues. Their work earned them the Nobel prize and opened the way for the development of vaccines against poliomyelitis: first the killed (inactivated) injectable poliomyelitis vaccine (IPV) developed by Jonas Salk and tested successfully in a field trial conducted by Thomas Francis in 1954 in which more than 200,000 children (the "polio pioneers") received the vaccine without any serious reactions of accidents, and subsequently the live, attenuated, oral poliomyelitis vaccine (OPV) developed by Albert Sabin. Between 1958 and 1960, two meetings sponsored by the Pan American Health Organization (PAHO) addressed the issue of strains to be used for the attenuated vaccines and analyzed data from 20 field trials that had been carried out in several countries. In 1959, the WHO appointed Dorothy M. Horstmann, a Yale epidemiologist, to evaluate trials that had been conducted in the former Soviet Union involving millions of children. In 1960, after a meeting was convened by PAHO, the data on the vaccine were reviewed and OPV started to be employed widely by health services throughout the world.

In 1985, PAHO launched an initiative to eradicate poliomyelitis from the Western Hemisphere by 1990. In August 1991 the last case of poliomyelitis in the Western Hemisphere was detected in Peru and an international certification commission, headed by Frederick C. Robbins, in September 1994 declared that indigenous transmission of wild poliovirus was interrupted in this region. The initiative in the Americas led by PAHO resulted in the initiative for the global eradication of poliomyelitis launched by the World Health Assembly in May 1988.

II. THE DISEASE

A. Clinical Features

Many persons who are infected with the wild poliovirus exhibit minor illnesses, but these cannot be distinguished clinically from many other illnesses. Symptoms associated with these minor illnesses include mild fever, muscle pains, headache, nausea, vomiting, stiffness of the neck and back, and, less frequently, signs of aseptic (nonbacterial) meningitis. Inapparent (subclinical) infections with no symptoms are common: Depending on the strain of the polio virus, the estimated ratios of inapparent to apparent infections range between 100 : 1 to 1000 : 1.

Susceptible older children and adults run a greater risk of developing paralytic illness. The case fatality rate varies between 2 and 20% among persons who do develop the paralytic form of the disease. However, if there is bulbar involvement affecting the muscles of respiration, the case fatality rate may be as high as 40%. Most deaths occur within the first week following the onset of paralysis (Table I).

1. Differential Diagnosis

Every case of acute flaccid paralysis in persons younger than 15 years old that is clearly not due to severe trauma should be investigated. If there is strong suspicion of polio in persons older than 15 years of age, these cases should also be thoroughly investigated.

It is difficult to confirm paralytic poliomyelitis in the acute phase based on clinical signs and symptoms alone because many other diseases and conditions may cause similar symptoms. Laboratory confirmation is therefore critical to the final diagnosis. The two diseases most frequently confused with polio are Guillain–Barré syndrome (GBS) and transverse myelitis (Table II).

TABLE I
When Paralysis Due to Polio Occurs

It is **typically flaccid** (the muscles are not stiff or spastic).

Patients usually have **problems with standing and walking.**

It is commonly preceded by **symptoms of minor illness** such as sore throat, headache, backache, fever, and vomiting.

Paralysis develops rapidly, usually within 4 days.

Fever is usually present at onset of paralysis.

Most patients have **little or no sensory loss** (e.g., patients will feel a needle stick). This sign may be difficult to determine in children.

The legs are more commonly involved than arms, and the large muscle groups of the hand are at greater risk than the small ones. Proximal muscles of the extremities tend to be more involved than distal ones.

It is **usually asymmetric** (not affecting both sides equally). Although any combination of limbs may be paralyzed, the most common pattern is involvement of one leg only followed by one arm only. Less common is to have both legs or both arms affected. Quadriplegia is rarely observed in infants.

Sequelae is usually present at 60 days after onset.

Other conditions which may present similar to paralytic poliomyelitis include traumatic neuritis and tumors, and, less frequently, meningitis/encephalitis and illnesses produced by a variety of toxins. The most prominent difference between poliomyelitis and other causes of acute flaccid paralysis (AFP) is that for polio the paralytic sequelae are generally severe and permanent, whereas for many other causes of AFP paralysis tends to resolve or improve within 60 days of onset.

B. Etiology

The causative viral agent is an enterovirus of the family Picornaviridae. The genome is a single linear molecule of single-stranded RNA, appearing smooth and round in outline by electron microscopy (Fig. 1).

The enteroviruses, which are resistant to the acid in the stomach, infect via the alimentary tract. There are three antigenic types (serotypes) of polioviruses called types 1–3. *All three can cause paralysis,* although type 1 causes paralysis most often, type 3

TABLE II
Criteria for the Differential Diagnosis of Polio

	Polio	*GBS*	*Traumatic neuritis*	*Transverse myelitis*
Time for paralysis to develop	Usually 2 or 3 days	From hours to 10 days	From hours to 4 days	From hours to 4 days
Fever	Fever with onset of paralysis, disappearing within 3 or 4 days.	Not common	Usually present before, during, and after paralysis	Rarely present
Flaccid paralysis	Acute, asymmetrical, principally proximal (upper part of arms and legs)	Generally acute, symmetrical, and distal (lower part of arms and legs)	Asymmetrical, acute, and usually affecting only one limb	Acute, lower limbs affected symmetrically
Muscle tone	Reduced or absent in the affected limb	Reduced or absent	Reduced or absent in the affected limb	Reduced in lower limbs
Deep tendon reflexes	Decreased to absent	Absent	Decreased to absent	Absent in lower limbs
Sensation pain	Sensation usually normal, severe myalgia, backache	Cramps, tingling, reduced sensation of palms and soles	Pain in buttocks, feeling of reduced temperature	Anesthesia of lower limbs with sensory perception
Cranial nerve involvement	Only when bulbar involvement is present	Often present, low and high: Miller–Fisher variant	Absent	Absent
Respiratory insufficiency	Only when bulbar involvement is present	In severe cases, bacterial pneumonia	Absent	Often thoracic, with sensory perception
Autonomic signs & symptoms	Motor in coordination, weakening of the reflexes	Rare	Blood pressure alterations, sweating, temperature fluctuations	Hypothermia in affected limb
Cerebrospinal fluid	Inflammatory	High protein with relatively low cells	Normal	Normal or mild elevation in cells
Nerve conduction velocity at 3 weeks	Abnormal: anterior horn cell disease (normal during the first 2 weeks)	Abnormal: demyelinization	Abnormal: axonal damage	Normal or abnormal, no diagnostic value
Sequelae at 3 months up to 1 year	Severe, asymmetrical atrophy, skeletal deformities develop later	Mild	Symmetrical atrophy of peroneal muscles	Moderate atrophy in affected lower limb

Fig. 1. The causative viral agent of polio—enterovirus of Picornaviridae.

less frequently, and type 2 rarely does so. Most epidemics are due to type 1. Vaccine-associated cases are usually due to types 2 and 3.

C. Occurrence

Polio exists worldwide. It is seasonal, occurring more commonly in summer and early autumn in temperate climates. In tropical countries seasonality is less clearly defined. However, some areas experience increases in infections during the rainy season. In developing countries with low immunization coverage, poliomyelitis produces a significant amount of illness, death, and disability. Where poliomyelitis is common, 3–10 of every 1000 young children will develop paralytic disease. In 1991, more than 16,000 cases of poliomyelitis were reported worldwide to WHO. However, lameness surveys have previously shown that official reporting in developing countries usually identifies <10% of the actual number of paralytic cases. Thus, it is likely that the true global incidence was at least 160,000 cases annually.

D. Reservoir

Man is the only reservoir, and infection is spread from person to person. Given the large number of inapparent infections, it is difficult to find the source of a case. Long-term carriers are not known to occur.

E. Incubation

On average, the incubation period from exposure to the virus to the onset of first symptoms is 7–10 days (range, 4–30 days), followed by a few days relatively free of symptoms before the onset of paralysis.

F. Immunity

All nonimmunized persons are susceptible to polio. Epidemiological evidence shows that infants born to mothers with antibodies are protected naturally against paralytic disease for a few weeks. Immunity is obtained from infection with the wild virus and from immunization. Immunity following natural (including inapparent and mild infections) immunization or a completed series of immunizations with live OPV results in both humoral and local intestinal cellular responses. Such immunity is thought to be lifelong and can serve as a block to infection with subsequent wild viruses and therefore helps in breaking chains of transmission. Vaccination with IPV confers humoral immunity but relatively less intestinal immunity; thus, vaccination with IPV does not provide resistance to carriage and spread of wild virus in the community. There is thought to be little, if any, cross-immunity between poliovirus serotypes.

G. Pathogenesis, Pathology, and Transmission

The mouth is the usual site of entry and the virus first multiplies at the site of implantation in the lymph nodes in the pharynx and gastrointestinal tract. The virus is usually present in the pharynx and in the stools before the onset of paralytic illness. Once the virus has entered the body, it invades local lymphoid tissue, enters the bloodstream, and then may invade certain types of nerve cells. In the process of intracellular multiplication the virus may damage or completely destroy these nerve cells.

Fecal–oral transmission is most common in developing countries in which sanitation is poor, whereas oral–pharyngeal transmission is likely to be more common in industrialized countries and during outbreaks. One week after onset, little virus remains in the throat, but it continues to be excreted in stools for 6–8 weeks. Cases are probably most infectious during the first few days before and after the onset of symptoms.

H. Laboratory Diagnosis

Stool specimens collected for virus culture from both cases of AFP and their contacts provide the most sensitive and effective way to rule out wild poliovirus transmission.

1. Types of Specimen

- Stool: Virus usually can be found in the feces from 72 hr to 6 weeks after infection, with the highest probability in the first 2 weeks.
- Cerebrospinal fluid: Not likely to yield virus and therefore its collection is not recommended.
- Throat: Not likely to yield virus and therefore *not recommended* for collection.
- Blood: Not likely to yield virus, and current serologic tests cannot differentiate between wild and vaccine virus strains. Experience has shown that for polio, interpreting serological data can often be misleading. Collection of blood specimens is therefore not recommended.

In the event of death, a definite diagnosis of polio can be made or rejected by examining the spinal cord. It is important that a qualified and experienced pathologist do the examination, and that a specimen be sent directly to a reference laboratory so that efforts can be made to grow poliovirus.

Rectal swabs are not recommended, although rectal tubes may be used in special studies. Stool samples must be kept cold if they are to be adequate for reliable testing when they arrive at the laboratory. It is important that the status and results of the stool samples are conveyed to the individuals requesting such testing as soon as possible.

2. Virus Isolation

Failure to isolate a poliovirus from a fecal specimen does not exclude the diagnosis of poliomyelitis since many factors can influence isolation results, including intermittent excretion of the virus in the stool, insufficient material collected, collection too late in the course of the illness, inadequate storage and shipping of specimens, and laboratory technique. The proportion of specimens for which other kinds of enteroviruses are isolated should be reported because this serves as an indirect indicator of the quality of the specimens. In tropical areas at least 15–20% of specimens should have these other enteroviruses isolated.

3. Characterization of Poliovirus

All polioviruses isolated from the stools of patients with acute flaccid paralysis or from contacts are characterized by hybridization with strain-specific nucleic acid probes. This characterization determines whether the virus is "wild" or "vaccine-like." The initial identifications are confirmed by polymerase chain reaction (PCR) analyses using primer sets specific to each vaccine strain and to the predominant wild polioviruses indigenous to the region. Wild viruses identified by these procedures are further characterized by partial nucleotide sequencing of the virus genomes, which reveals the genetic relationships among virus isolates. Given that poliovirus genomes evolve rapidly during replication in humans, the proximity of epidemiologic links among cases may be estimated by the extent of nucleotide sequence relationships among isolate genomes. Sequence information is also used to aid systematic design of nucleic acid probes and PCR primers.

I. Molecular Epidemiology

The molecular epidemiology (i.e., strain identification, based on genetic polymorphisms determined by genomic sequencing) of wild poliovirus has recently proved to be useful in helping to identify whether different virus isolates originate from a common ancestral source of infection. With this information, geographic foci or reservoirs of transmission can be defined and sources of outbreaks can be traced across large geographical areas. For example, the eight culture-positive cases reported from Colombia in 1991 were evaluated by genomic sequencing; results indicated that all were indigenous to Colombia. Similarly, genomic sequencing results of the 14 wild-type 1 viruses isolated in 1990 indicated that all were indigenous Andean genotypes and not importations from other areas in Latin America or from other endemic countries throughout the world.

J. Polio Vaccines

There are currently two effective polio vaccines: IPV, which first became available in 1955, and live, attenuated trivalent OPV, first used in mass vaccination campaigns in 1959. In developing countries, OPV is the vaccine of choice not only because of ease of application but also because it simulates natural infection, induces both circulating antibody and intestinal resistance, and protects against secondary spread to susceptible contacts. Success of eradication depends on the manner in which OPV is used.

There are many key characteristics which favor the choice of OPV over IPV for use in an eradication program. These include the development of intestinal immunity and the ability to reduce intestinal spread of wild virus, duration of immunity, ease of administration in both routine and mass campaigns, and cost. Probably the most critical issue relates to the effect of the vaccine on wild poliovirus transmission. It has been well documented that the use of OPV can successfully interrupt wild poliovirus transmission in both developed and developing countries. IPV protects against clinical disease and suppresses pharyngeal excretion of the virus, but it has less effect on intestinal excretion. Vaccinating children with IPV would reduce the number of paralytic cases due to the vaccine but would have little effect on the transmission of the wild poliovirus, which in developing countries is primarily by the fecal–oral route. The overall risk in the United States for paralytic polio associated with OPV in vaccine recipients is one case per 5.2 million doses distributed. The risk of vaccine-associated paralytic polio in vaccine recipients for first dose is one case per 1.3 million doses. Results in the Americas have been similar. The overall risk in vaccine recipients in Latin America is one case per 4.2 million doses administered, whereas the risk for recipients of first doses is one per 1.5 million doses distributed.

To assess the risks and benefits of each vaccine, the Centers for Disease Prevention and Control (CDC) in the United States developed a mathematical modeling analysis to estimate risks and benefits over a 30-year period for two cohorts of 3.5 million children each: one cohort would have received OPV and the other IPV. The model assumed periodic importations of wild poliovirus, a coverage rate of 95%, and an efficacy of 98% for both vaccines. The model predicted seven times as many cases of paralytic disease if IPV, rather than OPV, were used.

In the Americas, the current OPV provides the most likely means to stop transmission if used properly. Polio can be eradicated by using mass campaigns to supplement routine vaccine delivery and by placing additional emphasis on reducing missed opportunities to a minimum. Mass vaccination campaigns will interrupt transmission in areas where routine immunization programs have failed.

1. Vaccine Efficacy

Under ideal conditions in temperate countries a primary series of three doses of OPV produces seroconversion to all three virus types in more than 95% of vaccine recipients and is thought to have a clinical efficacy of nearly 100%. Three properly spaced doses of OPV should confer lifelong immunity. In developing tropical countries, the serologic response to OPV may be only 85%. This may be due to suboptimal vaccine storage and transporation, interference with the vaccine's ability to produce intestinal infection because of the presence of other enteroviruses, the presence of diarrhea that causes excretion of the virus before it can attach to the mucosal cell, and other factors.

2. Schedule, Contraindications, and Adverse Events
a. Recommended Schedule

Although the schedule may vary in some countries, for routine services it is recommended to apply three doses of trivalent OPV at 4- to 8-week intervals, beginning at 6 weeks of age. (It may be given to children and adolescents up to 18 years of age.) An additional dose at birth is highly recommended in endemic areas, although it is not considered part of the primary series and is referred to as "OPV zero." Longer intervals between doses do not require restarting of the schedule. Polio vaccine may be given simultaneously with any other childhood immunization.

b. Contraindications

For purposes of the polio eradication program there are virtually no contraindications to vaccination with OPV. Although diarrhea is not a contraindication, a dose administered to a child with diarrhea should not be considered part of the series. Another dose should be given at the first opportunity 4 weeks later (if diarrhea is no longer present). In countries in which human immunodeficiency virus (HIV) infection is a problem, individuals should be immunized with the appropriate antigens according to standard schedules. This also applies to individuals with asymptomatic HIV infection. Nonimmunized individuals with clinical (symptomatic) AIDS in countries in which polio remains a serious risk should receive OPV vaccine according to established schedules.

c. Adverse Events

Other than the rare instances in which OPV is associated with paralysis in vaccine recipients or their contacts there is virtually no other adverse reaction to the vaccine.

3. Dosage, Administration, and Formulation

OPV should be administered orally, that is, directly into the mouth. Each single dose consists of two or three drops (approximately 0.1 ml) of live oral polio virus vaccine or the dosage recommended by the manufacturer.

PAHO/WHO recommends that the OPV formulation to be 1,000,000, 100,000, and 600,000 infectious particles for types 1, 2, and 3, respectively. This formulation was adopted after a randomized trial of alternative formulations of OPV in the Americas revealed that children who received a single dose of this vaccine had higher seroconversion rates than children who received the previously formulated vaccine of 1,000,000, 100,000, 300,000.

The importance of careful monitoring of vaccine potency is well illustrated by the fact that a substantial number of the cases that occurred in the Americas between 1990 and 1991 could be attributed to the use of a substandard vaccine formulation.

4. Storage and Supply

Polio vaccine (OPV) is one of the most heat-sensitive vaccines in common use. The vaccine can be stored for up to 1 year and should be kept frozen at a temperature below 8°C at all times. Unopened vials of polio vaccine may be kept in the refrigerator for up to 6 months at temperatures between 0 and 8°C and may be thawed and refrozen without damage.

III. THE GLOBAL ERADICATION OF POLIO

A. Program Strategy

The strategies used in the polio eradication program are based on knowledge about the disease, the vaccine, and effective methods for its control:

1. *The natural history of the disease:* The ratio of inapparent infections of poliomyelitis to paralytic cases is high—between 100 and 1000 to 1. Despite the large number of subclinical cases of poliomyelitis, if the wild virus continues to circulate in a community a paralytic case is likely to appear. Surveillance techniques are therefore useful to identify high-risk areas.

2. *Properties of the vaccine:* The OPV contains live poliovirus whose virulence has been reduced or attenuated. Given that the vaccine is live and is administered orally, mimicking the natural route of infection, it also can be transmitted from a vaccinated person to close contacts who have not been immunized. Its circulation interrupts transmission of the wild virus by displacing it. This effect is enhanced if the vaccine is applied to entire communities by means of national vaccination days.

3. *Effective control methods:* After mass vaccination campaigns with OPV, developed and developing countries have experienced a dramatic decrease in the number of cases of poliomyelitis. Many approached or achieved zero incidence within a few years of initiating mass campaigns. In many countries success has been achieved despite large gaps in coverage in the younger than 5 years age group. This success is due primarily to the emphasis on mass vaccination for community protection rather than on individual immunity alone.

With this background, the major components of the strategy used to interrupt transmission of wild poliovirus in the Americas were the achievement and maintenance of high levels of vaccination coverage in children younger than 5 years of age, the development of a surveillance system for the timely detection and investigation of persons with AFP ("probable" poliomyelitis cases), and the creation of a surveillance system for detection of wild poliovirus supported by a regional network of diagnostic virology laboratories.

To achieve and maintain high levels of vaccination coverage in the population at risk, routine vaccination services have been supplemented by annual national immunization days (NIDs) in all countries of the region where wild poliovirus was circulating in 1985. Mass vaccination campaigns have been conducted twice a year over a short period of time (usually 1 or 2 days) in which one dose of OPV is administered to all children in the target age group (generally children 0–4 years of age), regardless of prior vaccination history. The annual NIDs are conducted with an interval of 4–6 weeks between campaigns.

Following completion of the NIDs, vaccine coverage and surveillance data identified areas which had low coverage and/or in which wild poliovirus transmission persisted. These areas were targeted for "mop-up" activities in the form of "house-to-house" vaccination of all children younger than 5 years of age, regardless of previous vaccination status.

To monitor the adequacy of regional AFP surveillance, the following surveillance indicators were established:

- At least 80% of reporting units should report weekly on the presence or absence of AFP cases.

- The annual AFP rate per 100,000 population less than 15 years of age should be ≥1.
- At least 80% of reported AFP cases should be investigated within 48 hr of being reported.
- At least 80% of reported AFP cases should have two adequate stool specimens collected within 14 days of paralysis onset.

The minimum annual rate of at least one case of AFP per 100,000 population less than 15 years of age was based on the fact that in the absence of wild poliovirus transmission, cases of AFP due to other causes (e.g., GBS and transverse myelitis) will continue to occur. Therefore, a sensitive AFP surveillance system would be expected to detect these "background" cases even when poliovirus is not circulating in a country.

B. Impact

In 1994, vaccination coverage through routine services among children 1 year of age with at least three doses of oral poliovirus vaccine was 86% in the Americas. These coverage rates have been supplemented by the NIDs carried out by several Latin American countries.

By mid-1991, more than 20,000 sites (health facilities) were reporting regularly on a weekly basis on the presence or absence of persons with AFP in their jurisdiction. During the period 1988–1994, approximately 2000 AFP cases were reported annually. Each year the regional AFP notification rate exceeded 1 case/100,000 population younger than 15 years of age.

The number of cases investigated within 48 hr of notification has been >80% since 1992. The number of reported cases with adequate stool specimens collected from patients and their contacts has increased each year during the 7-year period. Although the number of stool specimens collected and tested increased each year, the number of wild polio isolates decreased. The last wild poliovirus isolate from an AFP patient was obtained in 1991.

Of the 453 confirmed polio cases reported in the period 1988–1991, 79 (17.4%) had wild poliovirus isolated from a stool specimen. Of these, 64 (83.5%) occurred in persons younger than 6 years of age;

moreover, almost 94% of the patients with wild poliovirus isolation in the years 1989–1991 were persons younger than 6 years of age.

Type 2 wild poliovirus was the first subtype to disappear from the Americas; the last reported type 2 poliovirus case occurred in Colombia in 1988. The last reported type 3 poliovirus case occurred in Mexico in 1990, and the last case associated with type 1 poliovirus isolation had paralysis onset in August 1991 in Peru (Fig. 2).

As of May 1999, more than 7 years had elapsed since a 2-year-old boy living in Junin, Peru, suffered from poliomyelitis; this was the last case of poliomyelitis associated with wild poliovirus isolation (type 1) in the Americas. Since this last confirmed case

TABLE III
Highlights in Global Polio Eradication

1908: Karl Landsteiner and Edwin Popper discover the causative agent of poliomyelitis in Vienna

1938: March of Dimes launched by President Roosevelt

1949: Enders, Weller, and Robbins cultivate the poliovirus in tissue culture

1954: Killed inactived poliomyelitis vaccine developed by Jonas Salk

1958: Attenuated oral poliomyelitis vaccine developed by Albert Sabin

1985: The Pan American Health Organization (PAHO) launches an initiative to eradicate polio from the Americas by 1990

1988: The World Health Assembly approves a resolution calling for the global eradication of polio by the Year 2000, adopting the strategies developed by PAHO

1991: Last case of poliomyelitis in the Americas is detected in a 3-year-old boy in Peru

1994: The International Certification Commission declares that transmission of wild poliovirus has been interrupted in the Western Hemisphere

1995: 300 million children immunized against polio during national immunization days conducted in endemic countries (about half of the world's children younger than 5 years old)

1996: Less than 5000 cases of polio reported worldwide—the lowest number on record.

1997: Last case of polio reported in the western Pacific region (in Cambodia)

1999: WHO director general intensifies campaign by declaring "war on polio"

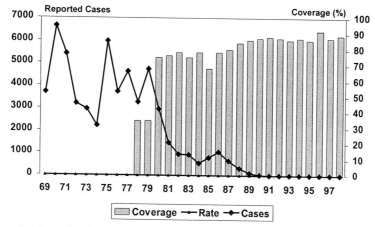

Fig. 2. OPV3 vaccination coverage and incidence of paralytic poliomyelitis—region of the Americas, 1969–1998 (coverage data are provisional) (source: HVP/PAHO).

was reported, more than 7000 AFP cases have been thoroughly investigated and none has been confirmed as paralytic poliomyelitis due to wild poliovirus. Furthermore, more than 36,000 stool specimens obtained from these cases and their contacts were tested and found to be negative for wild poliovirus.

After reviewing detailed national surveillance data in early 1994, all National Certification Commissions recommended that their countries be certified as being polio free. On September 29, 1994, during the quadrennial meeting of the Pan American Sanitary Conference, the president of the International Commission for the Certification of Poliomyelitis Eradication (ICCPE), Frederick C. Robbins, stated that, based on recommendations of the National Certification Commissions and after thoroughly reviewing regional polio vaccination coverage, AFP surveillance, and wild poliovirus surveillance laboratory data, the ICCPE had concluded that wild poliovirus transmission had been interrupted in the Americas.

The certification of the interruption of wild poliovirus transmission in the Americas is a milestone in the global effort to eradicate the poliovirus (Table III). Several factors made it possible for the region of the Americas to interrupt poliovirus transmission. First was the complete implementation of an effective strategy based on sound epidemiologic data that was continually adapted and refined with information provided by a solid surveillance

system. Furthermore, the implementation of this strategy was supported by a very high level of political commitment of the member governments, a high degree of community participation, a strong collaboration of various agencies and organizations through interagency coordinating committees, and the availability of well-managed resources.

C. Global Situation

The reported incidence of poliomyelitis globally in 1998 was 5298 confirmed cases compared with 35,000 cases reported in 1998 (Fig. 3). It has been almost 2 years since the last polio case was identified in the western Pacific region. The region is preparing to be certified in the Year 2000. In the European region, wild poliovirus was identified in 1998 only in southeastern Turkey. Poliovirus is now largely confined to south central Asia, west Africa, central Africa, and the Horn of Africa. In Asia, Afghanistan, Bangladesh, India, Nepal, and Pakistan constitute a major poliovirus reservoir. Together, these five countries reported more than 90% of all 1998 polio cases. However, it should be noted that significant improvements in surveillance have occurred in these countries so that this percentage is artificially high.

On the African continent, no virologically confirmed polio cases are being reported from North Africa and the countries of southern Africa. However, high-level surveillance has demonstrated the contin-

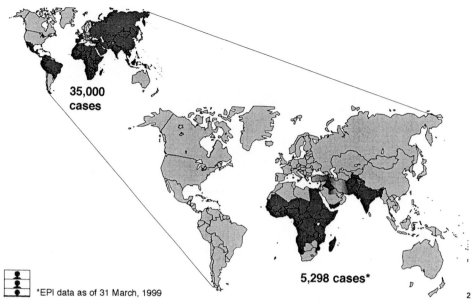

Fig. 3. Progress toward polio eradication, 1988–1998 (*Expanded Program on Immunization data as of March 31, 1999).

ued circulation of wild poliovirus in Egypt, in which at least 12 cases have been confirmed. Additional NID rounds are planned for the spring of 1999, with mopping-up in the autumn. Although poliovirus is not being found in east African countries, surveillance has not been developed sufficiently so that it can be stated with confidence that poliovirus has been eradicated there. Virological surveillance demonstrates major, separate reservoirs of poliovirus in West Africa, Nigeria, and the Democratic Republic of Congo. Surveillance is not sufficiently developed in many countries to give a clear picture of the circulation of polioviruses on the continent.

Excellent progress is being made toward the global polio eradication goal, and it is expected that the last case of polio in the world will occur in the Year 2000.

See Also the Following Articles

ENTEROVIRUSES • SURVEILLANCE OF INFECTIOUS DISEASES • VACCINES, VIRAL

Bibliography

Daniel, T. M., and Robbins, F. C. (1997). "Polio." Univ. of Rochester Press, Rochester, NY.

de Quadros, C. A. (1997). Global eradication of poliomyelitis. *Int. J. Infect. Dis.* **1**(3), 127–29.

Long, E. R. (1928). "A History of Pathology." Williams & Wilkins, Baltimore.

Pan American Health Organization PAHO (1971). International Conference on the Application of Vaccines Against Viral, Rickettsial, and Bacterial Diseases of Man, Scientific Publ. No. 226. PAHO.

Pan American Health Organization (PAHO) (1988). Polio eradication field guide, Technical Paper No. 6. PAHO.

Pan American Sanitary Bureau (1959). Live poliovirus vaccines, Scientific Publ. No. 44. Pan American Sanitary Bureau.

Paul, John, R. (1971). "A History of Poliomyelitis." Yale Univ. Press, New Haven, CT.

White, D. O., and Fenner, F. J. (1994). "Medical Virology," 14th ed. Academic Press, San Diego.

World Health Organization (WHO) (1997). "Polio: The Beginning of the End." WHO, Geneva.

Polyketide Antibiotics

Annaliesa S. Anderson, Zhiqiang An, and William R. Strohl

Merck Research Laboratories

I. History
II. Classification
III. Structure and Activity
IV. Biosynthesis
V. Evolution
VI. Future Perspectives

GLOSSARY

broad-spectrum antibacterial A compound with bacteriostatic or bacteriocidal activity against gram-negative and gram-positive bacteria.

domain A discrete amino acid sequence in a protein that can be equated with a particular function.

fatty acids Straight-chain monocarboxylic acids, especially those naturally occurring in fats. They can be saturated or unsaturated, and unsaturated fatty acids can be polyunsaturated or monosaturated.

fungi (pl. of fungus) A kingdom of life forms that are eukaryotic, mycelial or yeast-like, heterotrophic, achlorophyllous, are characterized by sexual and/or asexual reproduction, and are mostly aerobic.

immunosuppressant An agent capable of suppressing immune responses.

motilide Compounds (several of which are macrolides) possessing gastrointestinal motor-stimulating activity.

natural products Naturally occurring organic compounds that are structural materials or biologically active molecules.

phylogenetic Describing or referring to any form of analysis that groups characters according to evolutionary relationships.

prokaryote Cellular organisms that lack a true nucleus and have ribosomes of the 70S type; also synonymous with bacteria.

semisynthetic Describing or referring to compounds modified from natural products by chemical modifications.

POLYKETIDES are natural products, produced by representatives of the eukarya, prokarya, and planta, which are partly composed of a carbocyclic or alicyclic carbon chain derived from the condensation of carboxylic acids. The polyketides are one of the most diverse classes of biologically active molecules found in nature. This article will focus on biologically active polyketides produced by eubacteria and fungi. Polyketides exhibit a broad spectrum of biological activities, including anti-infective (e.g., erythromycin), anthelmintic (e.g., avermectin), immunosuppressive (e.g., rapamycin), antitumor (e.g., doxorubicin), and carcinogenic (e.g., aflatoxin) activities.

I. HISTORY

The antibiotic potential of polyketides was first discovered in 1950 with the discovery of oxytetracycline, an antibacterial compound produced by *Streptomyces rimosus*. Additional antibacterial polyketides discovered during this period include pikromycin (1950), erythromycin (1952), tetracycline (1953), and rifamycin (1959). Nystatin, a polyene and the first polyketide shown to possess antifungal activity, was discovered in 1950. The first members of the anthracycline class of polyketides, which possess cytotoxicity and antitumor activity, also were first discovered in the 1950s. During the 1960s, the potent polyketide antitumor drugs, daunorubicin and doxorubicin, were discovered, as were the animal feed additives, tylosin and monensin. The 1970s marked the discovery of avermectin, an extremely potent antiparasitic agent. The antimetabolic potential of polyketides was extended during the 1980s to the

TABLE I
Characterized Polyketide-Producing Organisms

Kingdom	Division	Subdivision/Order	Example
Eubacteria	Proteobacteria	Delta subdivision	*Sorangium* sp.
		Gamma subdivision	*Pseudomonas* sp., *Yersinia* sp.
	Firmicutes	Actinobacteria	*Streptomyces* sp., *Micromonospora* sp., *Actinomadura* sp.
	Cyanobacteria	Nostocales	*Nostoc* sp.
Eukaryotae	Fungi	Ascomycota	*Aspergillus* sp., *Colletotrichum* sp., *Penicillium* sp., *Cochliobolus* sp.
	Alveolates	Dinoflagellates	*Prorocentrum* sp., *Dinophysis* sp.
	Viridiplantae	Charophyta/Embryophyta	*Daucus carota*
	Metazoa	Mollusca	*Siphonaria* sp.
		Porifera	*Theonella* sp., *Geodia* sp.
		Arthropoda	*Rhytidoponera* sp.

areas of immunosuppression (tacrolimus), cholesterol-lowering activity (lovastatin), and motilides (erythromycin and analogs). In recent years, polyketides of the macrolide class also have demonstrated potential activity in a wide range of therapeutic areas, including nerve regeneration activity (neuroimmunophilin agonists), prevention of hypertrophic cardiomyopathy, and asthma treatment (Table 1).

In 1998, polyketides represented approximately 3% of the total world sales for prescription pharmaceuticals, with an estimated value of $8.5 billion. Of this value, approximately $5 billion was due to sales in the cholesterol-lowering field (e.g., lovastatin, simvastatin, and pravastatin), $1.5 billion from antibacterials (e.g., erythromycin and derivatives such as clarithromycin and azithromycin), $1 billion from anticancer therapeutics (e.g., doxorubicin, epirubicin, and aclarubicin), $250 million from the tetracyclines, and $300 million from the anthelmintics (e.g., avermectin, ivermectin, and doramectin). The world market for the imunosuppressants rapamycin and tacrolimus (formerly FK506) is currently at least $200 million and increasing rapidly.

II. CLASSIFICATION

Early classification systems, which were based on chemical structure, loosely categorized polyketides as either aromatic or nonaromatic (Fig. 1). Aromatic polyketides have a core structure composed of flat six-membered rings. The fungal aromatic polyketide 6-methylsalicylic acid is composed of a single aromatic ring with hydroxyl and methyl groups substituted on C1, C5, and C6, respectively. Tetracyclines are based on the naphthacene skeleton: four aromatic rings substituted with amino, methyl, keto, and hydroxyl groups. The anthracyclines have a tetrahydronaphthacenequinone chromophore substituted

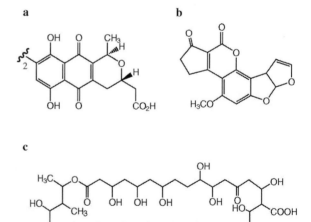

Fig. 1. Examples of the different classes of polyketides. Actinorhodin (a) and aflatoxin B (b) are examples of prokaryotic and fungal aromatic polyketide antibiotics. Nystatin (c) is a prokaryotic nonaromatic polyketide.

with glycosyl moieties. The nonaromatic polyketides include the macrolides, polyenes, and polyethers. The nonaromatic polyketides also are classified as macrocycles; this classification encompasses and differentiates simple lactones (e.g., erythromycin), large macrolactones (e.g., rapamycin), dilactones (e.g., boromycin), polylactones (e.g., macrotetrolides), and macrolactones condensed with other rings (e.g., brefeldin A).

From extensive biochemical and genetic analysis, the biosynthesis of polyketides has been found to be analogous to fatty acid biosynthesis, in which the fatty acid synthases (FASs) are divided into two groups: (i) type I FAS, which are large, multi-functional enzymes, almost exclusively eukaryotic in nature, and in which the various enzymatic functions are incorporated as active domains or motifs into the long polypeptide chain, and (ii) type II FAS, which are bacterial enzyme complexes composed of several discrete enzymes, each possessing one or a few enzymatic functions.

The assembly of polyketide structures and the architecture of the biosynthetic machinery follow approximately the same pattern as that found with fatty acid biosynthesis, although there are some important subtleties. Type I polyketide synthases are large, multifunctional enzymes, as are type I fatty acid synthetases. The type I polyketide synthases, however, are subdivided into two discrete subgroupings: (i) the modular, processive type, thus far exclusively bacterial in origin and exemplified by the erythromycin polyketide synthases, and (ii) an iterative type, exemplified by the fungal 6-methylsalicylic acid syn-

thase but found to have representatives in both bacteria and fungi (Table II). The type I iterative enzymes differ from the processive type I polyketide synthase (PKS) enzymes by the repeated use of active sites during product synthesis, rather than each active site catalyzing a single reaction step. The type II polyketide synthases, similar to type II fatty acid synthases, are exclusively bacterial in origin and are composed of several individual enzymes that form a functional complex. As far as is currently known, these enzymes exclusively form aromatic products that possess quinones (e.g., tetracyclines, anthracyclines, angucyclines, anthraquinones, and benzoisochromanequinones) (Table II).

III. STRUCTURE AND ACTIVITY

A. Macrolides

Macrolides, which are produced by a wide variety of actinomycetes, are classified based on the structure of the lactone ring. The name of the lactone ring varies according to the antibiotic: for example, erythronolide for erythromycin and tylactone for tylosin. A lactone ring is substituted with hydroxyl, alkyl, and ketone groups (12- and 14-membered ring macrolides). In addition, an aldehyde group is present for macrolides with 16-membered rings. In many cases, one or more of the hydroxyl groups of the aglycone nucleus are substituted with sugars, either amino or neutral 6-deoxy-hexoses, typically containing oxy, methyl, or methoxy moieties.

TABLE II
Biogenic Classes of Economically Important Natural Product Polyketides

Class	Subclass	Example	Therapeutic activity
Type I modular, processive	Macrolides	Erythromycin, tylosin	Antibacterial
	Pentacyclic lactones	Avermectin, milbemycin	Insecticidal, acaricidal
	Ascomycins	Tacrolimus	Immunosuppressive
	Ansamycins	Rifamycin, ansatrienins	Tuberculostatic, antibacterial
	Polyenes	Nystatin, amphotericin B	Antifungal
	Polyethers	Monensin	Growth promotants
Type I iterative	Statins	Lovastatin, pravastatin	Cholesterol lowering
Type II iterative	Tetracyclines	Oxytetracycline	Antibacterial
	Anthracyclines	Doxorubicin	Cytotoxic antitumor

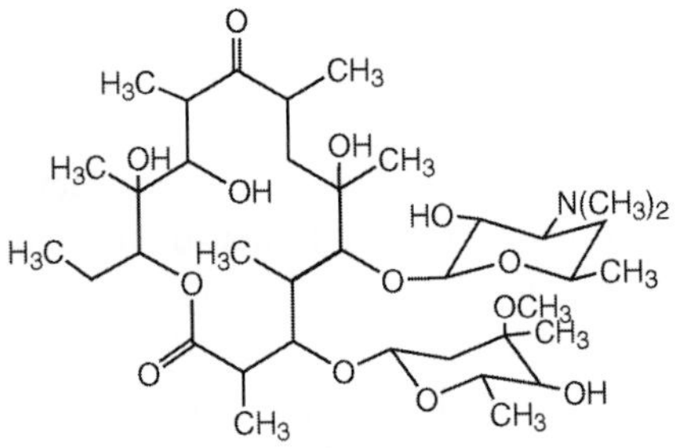

Fig. 2. Erythromycin A.

Fig. 3. Avermectin.

Macrolide antibiotics, which are bacteriostatic agents, inhibit bacterial protein synthesis by structure-based mechanisms: 14-membered macrolides such as erythromycin (Fig. 2) block translocation of peptidyl-tRNA, whereas 16-membered macrolides inhibit peptidyltransfer reactions. There are apparently additional interactions between macrolides of both groups with ribosomes, but after 30 years of research these are not fully understood. Rustmicin is a 14-membered macrolide possessing potent antifungal activity via specific interaction with the α-mannan of the cell wall by a potentially novel mode of action.

Although the ring sizes of macrocyclic compounds can range from 6-membered to more than 30-membered rings, most of the clinically relevant macrolides possess 14-membered or 16-membered rings. Current clinical antibacterial macrolides are mostly derived from erythromycin, although a wide variety of biologically active macrolides are used for several purposes across several therapeutic areas (Table III).

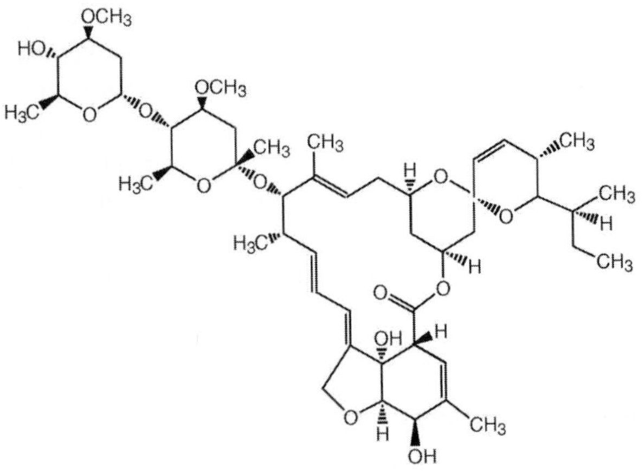

B. Pentacyclic Lactones

Avermectin (Fig. 3) and milbemycin are antiparasitic agents. Avermectins are composed of a 16-membered aglycone nucleus which possesses a spiroketal system between C17 and C25 and a disaccharide chain at C13. Avermectin is produced exclusively by *Streptomyces avermitilis*. Four major (A_{1a}, A_{2a}, B_{1a}, and B_{2a}) and four minor (A_{1b}, A_{2b}, B_{1b}, and B_{2b}) compounds are produced, of which the B series is the most biologically active. A semisynthetic derivative, "ivermectin", is composed of two derivatives, dihydro H_2B_{1a} and H_2B_{1b}, which differ by the presence of a methyl group on the side chain at position 25.

The avermectins act as γ-aminobutyric acid (GABA) agonists with extraordinary efficacy against a wide range of endoparasites and ectoparasites. They

TABLE III
Bioactive Areas in Which Type I Modular Polyketides Are Being Studied

Bioactive area	Example	Comments
Antibacterial	Erythromycin analogs	Traditional area of strength based on erythromycin
Antifungal	Rustmicin, galbonolides	New activities of interest; interaction with mannans
Antiparasitic	Avermectin, doramectin	Avermectin recently approved for human use in the United States
Anticancer	Epothilone	Tubulin stabilizer activity similar to taxol
Immunosuppressive	Tacrolimus, rapamycin	Substitutes for cyclosporin
Neuroimmunophilin	Rapamycin and analogs	Stimulation of neuronal growth
Motilin receptor	Motilide, ABT-229	Treatment of gastroesophageal reflux disorder
Respiratory	Roxithramycin	Antiasthmatic by multiple potential mechanisms
Agricultural	Tylosin	Antibacterial and growth-promoting feed additive
Phytotoxin	Coronatine	Produced by *Pseudomonas syringae* pv. *glycinea* PG4180

are used widely as antiparasitics and antifilarials in the agricultural/veterinary and companion animal veterinary areas, as well as to kill *Onchocerca volvulus,* the causative agent of river blindness in man.

C. Ascomycins

The ascomycins, produced by various actinomycetes, possess immunosuppressive activity analogous to, but significantly more potent than, that of cyclosporin A. Ascomycin (also called immunomycin or FK520), FK506 (tacrolimus), and FK523 are almost identical ascomycins (Fig. 4), differing only by a substituted side chain at C21. All three of these compounds are composed of a 23-membered ring structure with a terminal substituted cyclohexyl moiety. On the other hand, rapamycin is an ascomycin possessing a unique 31-membered ring. FK506 was approved in 1994 for use in liver transplants.

At the molecular level, FK506 and immunomycin bind to a protein known as FK506-binding protein FKBP-12, which belongs to a class of proteins called immunophilins, all of which possess peptidyl prolyl *cis-trans* isomerase (rotamase) activity. The protein ligand interactions are mediated from the last three extender units as well as the pipecolate and cyclohexanoyl residues (the east-northeast face of the molecules, as shown in Fig. 4). Whereas FK506 is a potent inhibitor of the rotamase activity of FKBP-12, neither the binding nor inhibition of rotamase activity are sufficient to effect immunosuppression. Instead, the immunosuppressant–immunophilin complex causes inactivation of the Ca^{2+}-dependent, serine–threonine phosphatase 2B, calcineurin, resulting in the inhibition of the signal transduction pathway that normally leads to the activation of T cells. Rapamycin, on the other hand, exerts its immunosuppressive activity via a different molecular mechanism. Although rapamycin also binds FKBP-12, the target for the rapamycin –FKBP-12 complex is not calcineurin but rather FRAP (FKBP-12–rapamycin-associated protein), a member of the phosphatidylinositide-3-kinase family. FRAP exhibits protein kinase activity in the signal transduction pathway which regulates protein synthesis required for the progression of IL-2-stimulated cells from G_1 to S phase. Inhibition of the protein kinase activity of

	R
Tacrolimus (FK506)	$CH_2CH=CH_2$
Immunomycin	C_2H_5
FK523	CH_3

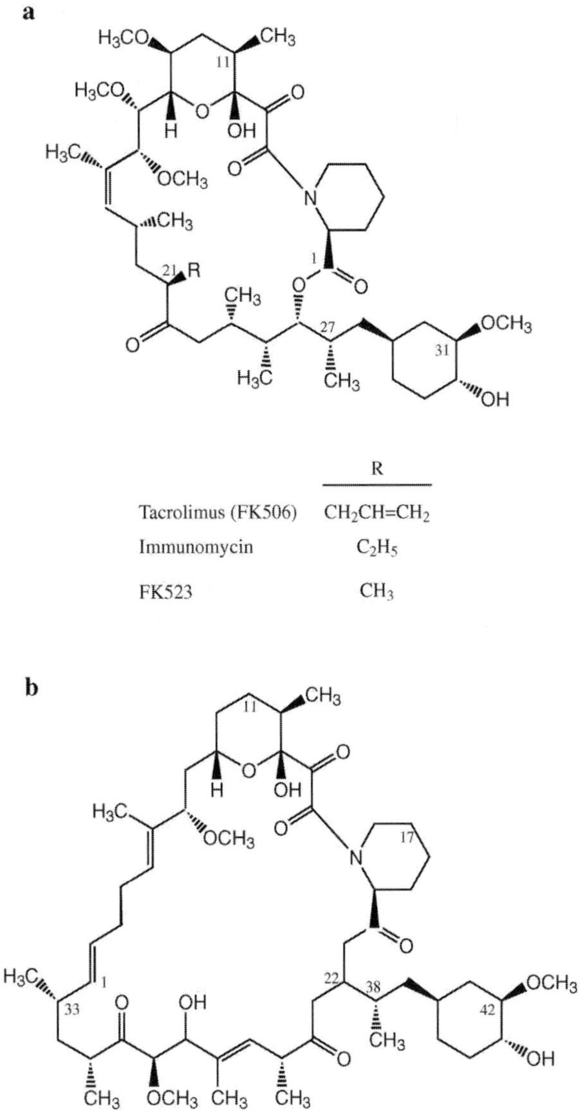

Fig. 4. Ascomycins: 23-membered lactone ring ascomycins (a) and rapamycin (b).

FRAP by the rapamycin–FKBP-12 complex therefore blocks IL-2-induced cell cycle progression, resulting in the suppression of T cell proliferation, though at a much later stages than in FK506 or cyclosporin.

Immunosuppression is not the only therapeutic activity associated with the ascomycins. Calcineurin has been implicated in causing hypertonic cardiomyopathy (HCM), an inherited heart disease that affects 1 in 500 individuals. In mice models FK506 inhibits the calcineurin, thus preventing HCM. The active site of FK506 also has nerve regenerative properties when bound to FKBP-12 that are independent of calcineurin inhibition, making FK506-like compounds effective candidates for the treatment of neural degenerative disorders such as Parkinson's disease. Bronchial inflammation in asthma is caused by an influx of eosinophils and CD4+ T lymphocytes into the bronchi. The bronchial infiltration by eosinophils, which is the most proinflammatory factor, is reduced by FK506. Though it is not feasible to treat these disorders with molecules possessing potent immunosuppressive activity, it does indicate potentially exciting new treatments for previously incurable disorders. Additionally, the ascomycin SDZ ASM 981 is effective for the topical treatment of atopic dermatitis.

D. Ansamycins and Macrolactams

The ansamycins are macrocyclic compounds produced by several actinomycetes which are structurally composed of an aromatic nucleus that is joined by an aliphatic bridge to form the lactone ring. Ansamycins include rifamycins, streptovaricins, halomycins, and benzoquinones. The rifamycins encompass a group of compounds that differ by the substituted group at C4 of the aromatic naphthalene nucleus (Fig. 5). They have antibacterial activity against gram-positive bacteria, including mycobacteria, staphylococci, and streptococci, and are front-line tuberculostats used in combination treatment of tuberculosis. The bacteriostatic activity of the rifamycins is mediated by the inhibition of transcription through binding to the RNA polymerase β subunit. The streptovaricins have a similar structure and

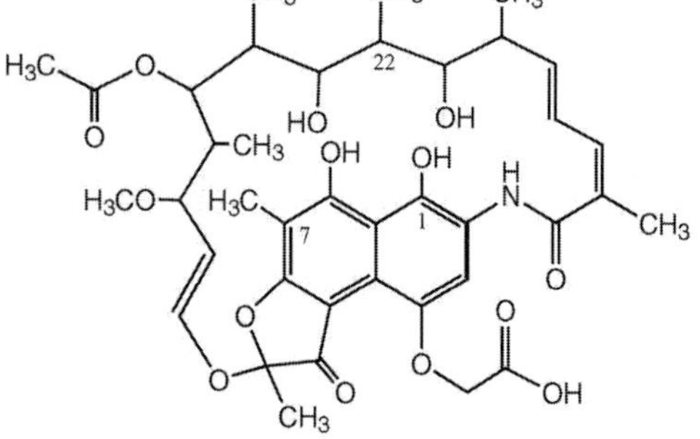

Fig. 5. Rifamycin B.

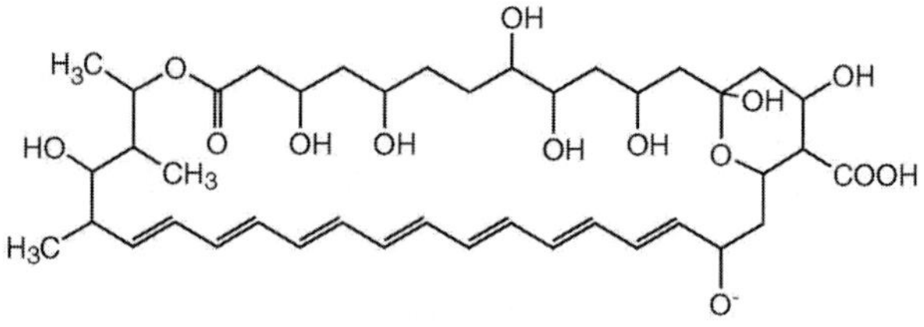

Fig. 6. Amphotericin B.

mode of action as those of the rifamycins, but they differ by the length of the aliphatic bridge.

Benzoquinone ansamycins (such as herbimycin and geldanamycin, which inhibit tyrosine kinases and heat shock protein 90, respectively) display potent anticancer activity. These compounds, produced by several *Streptomyces* spp., are still at the trial stage as they exhibit *in vivo* toxicity due to the presence of the quinone core.

E. Polyenes

The polyenes (Figs. 1 and 6), produced by several *Streptomyces* spp., are a group of 20- to 44-membered ring compounds that can be subdivided according to the number of double bonds that they posses (e.g., trienes, tetraenes, pentaenes, and hexaenes). The lactone ring may have additional substitutions including aminosugar, carboxyl, aliphatic, or aromatic groups. Members of this class include nystatin (tetraene), amphotericin B (heptaene), candicin (hepaene), and piramicin (tetraene).

Polyenes possess antifungal activity which is mediated by their effect on cell-membrane sterols resulting in the leakage of ions and small molecules. Polyene antibiotics have a greater affinity for ergosterol, the primary sterol of fungal cell membranes, than for cholesterol (mammalian cell membrane sterol), which indicates their specificity against fungi. In addition, amphotericin B is also active in experimental prion disease models and therefore may have a use in elucidating the pathogenic mechanisms involved in neurodegenerative prion disorders.

F. Polyethers

Polyethers, also known as open-chain macrotetralides, are composed of linked cyclic ethers. Poly-

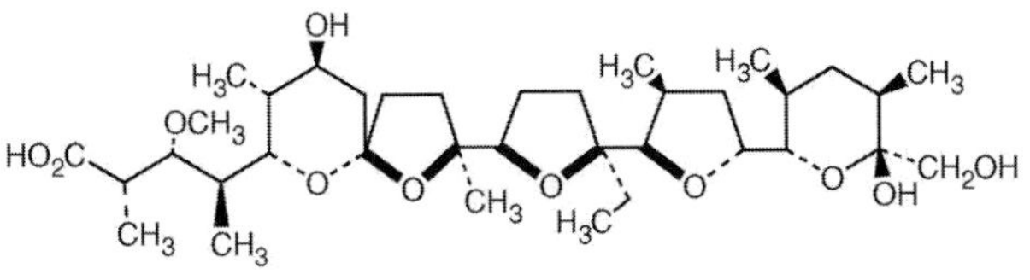

Fig. 7. Monensin A.

ethers, which are produced by various actinomycetes, typically possess coccidiostatic activity. They are used commercially as livestock growth promotants, as exemplified by the use of monensin A (Fig. 7) as a feed additive for ruminating livestock to increase the efficiency of food conversion by reducing acetic acid- and hydrogen-producing rumen bacteria, resulting in an increase in propionic acid producers and a suppression of methanogenesis which is energetically wasteful.

Polyethers act as ionophores which affect membrane function by structure-specific modes of action. For example, nigericin transports monovalent cations such as K^+ across membranes by losing a proton and binding to the ion, thereby generating a neutral complex that can then cross the membrane, resulting in an altered membrane pH gradient.

G. Statins

A class of fungal polyketides, generally known as "statins," was discovered to inhibit HMG-CoA reductase (3-hydroxy-3-methylglutaryl-CoA reductase). In the cholesterol biosynthetic pathway, HMG-CoA reductase catalyzes the rate-limiting step, i.e., the NADPH-dependent conversion of HMG-CoA to mevalonate, the first compound unique to sterol biosynthesis. Lovastain (ML236B) and pravastatin, the 6 β-hydroxy acid form of compactin, were the first two clinically important statins (Fig. 8) used to treat human hypercholesterolemia. Simvastatin, a 2,2-dimethylbutyrate derivative of lovastatin, and lipitor, a new synthetic compound developed from the lovastatin structural motif, are the major statins used to treat hypercholosterolemia today.

The biosynthesis of lovastatin by *Aspergillus terreus* involves two methylated polyketide chains, one of 18 carbons and another of 4 carbons, each synthesized from acetate units. The two methyl groups on the polyketide chains were derived from methionine. Pharmacologically, the active form of lovastatin is the β-hydroxy acid form; the lactone form is inactive. Removal of the side chain of lovastatin resulted in marked loss of potency.

The squalestatins are another class of fungal polyketides that exhibit inhibition of sterol biosynthesis. These compounds specifically inhibit squalene synthetase, an enzyme which catalyzes a committed step in the cholesterol biosynthetic pathway. Squalestatins and the structurally related zaragozic acids, also produced by filamentous fungi, possess unusual features of a highly functionalized bicyclic core with three carboxylic groups. Zaragozic acids differ from one another by varying side chains and long chain acyl groups. Zaragozic acids were shown recently to inhibit Ras farnesyltransferase, an enzyme which catalyzes the posttranslational farnesylation of all Ras proteins, and therefore they were suggested to have anticancer potential.

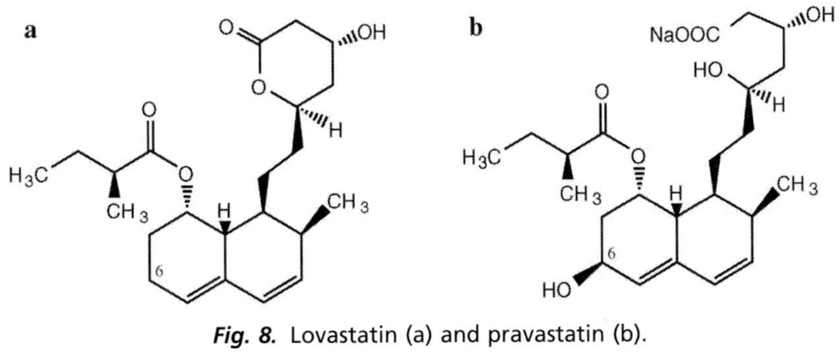

Fig. 8. Lovastatin (a) and pravastatin (b).

H. Tetracyclines

The tetracyclines (Fig. 9), which are produced by several *Streptomyces* spp., exhibit broad-spectrum antibacterial (bacteriostatic) activity by arresting the translational process, and they have broad human and veterinary use. For example, tetracyclines are used to treat stomach ulcers that are caused by *Helicobacter pylori* infections and as bulk chemical prophylactic antibiotics in fish hatcheries. Semisynthetic tetracyclines have also been developed which have anti-inflammatory action; for example, minocycline is used for the treatment of rheumatoid arthritis through its suppression of T cell proliferation.

I. Anthracyclines

Anthracyclines, produced by a variety of *Streptomyces* spp. and *Actinomadura* spp., were originally found in the 1950s to be potent antibiotics and cytotoxic agents. Daunorubicin, first described in 1964 as a natural product of *Streptomyces peucetius,* was the first anthracycline to possess reasonable efficacy to toxicity ratios for first-line therapy of several carcinomas. Doxorubicin (14-hydroxydaunorubicin; Fig. 10), discovered in 1969 as a product of an *S. peucetius* mutant, quickly became the cytotoxic drug of choice for chemotherapeutic treatment of a wide variety of cancers. Both daunorubicin and doxorubicin are "class I" anthracyclines which intercalate into the DNA and act primarily by trapping nicked DNA and topoisomerase II in what is known as the "cleavable

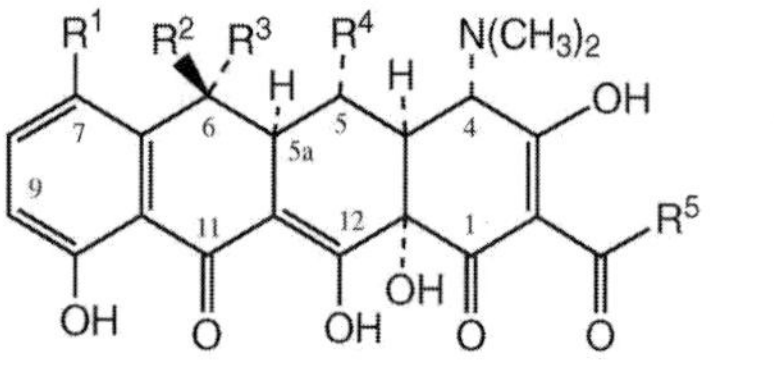

Fig. 10. The anthracyclines: doxorubicin and daunorubicin.

complex." Aclarubicin, a class II anthracycline discovered in 1975 and used as a chemotherapeutic agent in Japan, acts by inhibiting DNA–topoisomerase II interaction.

IV. BIOSYNTHESIS

Polyketides are synthesized in a manner similar to fatty acid biosynthesis in which thioesters of short-chain fatty acids are condensed onto an initial starter moiety. The resulting chains are then modified by other enzymes to yield diverse, complex molecules.

These enzymes have strong mechanistic and sequence similarities with the fatty acid synthases (FASs), which has resulted in the adoption of the FAS nomenclature (i.e., type I and type II) for PKS systems. The following are the major differences between FASs and PKSs: (i) FASs use primarily an acetyl moiety as the starter unit, whereas PKSs utilize a wide variety of starter units; (ii) FASs and type II PKSs use only the C2 unit derived from malonyl-SCoA as an extender, whereas type I PKSs utilize various extender units; (iii) FASs typically processively reduce and dehydrate the β-carbonyl of the growing chain to the saturated form, whereas PKSs may or may not completely reduce the β-carbonyl of the growing chain; and (iv) FAS products are linear, mostly reduced, fatty acids which are rarely modified postassembly, whereas PKS products typically possess either aromatic or lactonized ring structures which are extensively modified after assembly.

	R^1	R^2	R^3	R^4	R^5
tetracycline	H	OH	CH_3	H	NH_2
oxytetracycline	H	OH	CH_3	OH	NH_2
terramycin X	H	OH	CH_3	OH	CH_3

Fig. 9. Tetracycline.

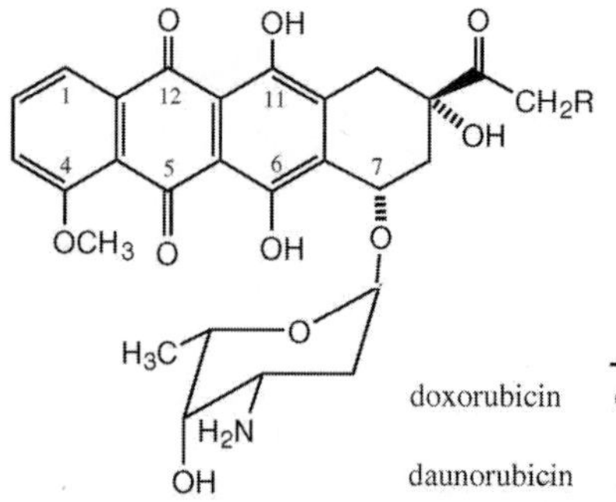

A. Products of Type I Processive Polyketide Synthases

Type I modular, processive polyketide synthases, exemplified by the erythromycin polyketide synthase (Fig. 11a), are large multifunctional enzymes which possess a separate, distinct, enzymatic active site for each biosynthetic step. For example, the core aglycone of the prototype macrocyclic polyketide, erythromycin, is synthesized by the huge multifunctional enzyme, 6-deoxyerythronolide B synthase (DEBS), which consists of three multidomain subunits, with each subunit having a molecular weight of approximately 300 kDa. The macrolide chain is synthesized by a set of enzymatic functional domains within the proteins in a linear, stepwise, or "assembly-line" fashion, with no enzymatic domain used more than once for the synthesis of a given molecule (Fig. 11a).

Interestingly, the enzymatic active sites are arranged in the polypeptides in the same order as their apparent function. A polyketide structure cannot necessarily be predicted by the motifs in the polypeptide chain, however, because it is difficult to determine from the amino acid sequence which extender units and what stereochemistry will be employed in the growing chain. Moreover, domains within the polypeptide chain may be inactive and apparently vestigial in nature. The decisions that are made in the assembly of polyketides by type I modular enzymes include type of starter unit, type and stereochemistry of extension, level of reduction at each extension step, number of extender units (dictated by the number of modules), and whether or not the fully assembled polyketide is cyclized. Most of the products of type I modular PKSs are modified postassembly by hydroxylation, glycosylation, or other modifications.

B. Products of Type I Iterative Polyketide Synthases

Type 1 iterative PKSs, found first in filamentous fungi but recently also discovered in *Pseudomonas* spp. and *Streptomyces* spp., are multifunctional polypeptides that have one set of catalytic domains (e.g., KS, AT, ACP, KR, DH, ER, and TE) used repeatedly during polyketide assembly. All type I iterative PKSs have KS, AT, and ACP domains, but not all have the complete set of the other domains. For example, the T-toxin PKS isolated from the filamentous fungus, *Cochliobolus heterostrophus,* contains six catalytic domains arranged in the following order. NH_2-KS-AT-DH-ER-KR-ACP-COOH. In contrast, the gene that encodes 6-methylsalicylic acid synthase (MSAS), cloned from *Penicillium patulum,* has a 5322-bp-long open reading frame encoding a protein of 1771 amino acids (molecular weight, 190,731 Da) with five catalytic domains sequentially arranged as follows: NH_2-KS-AT-DH-KR-ACP-COOH. Figure 11b illustrates the steps in 6-methylsalicylic acid biosynthesis by MSAS.

The design rules for products of type 1 iterative PKSs are completely uncharacterized. Different levels of modifications (i.e., reduction, dehydration, and enoyl-reduction) are dictated by the same enzymatic system at each iteration, and there is no information available regarding what dictates these differential reactions. The products of type I iterative PKSs are thought to encompass a very wide range of structures, including, but not exclusively, aromatic quinones, flavanoids, resorcylic acid lactones, pyrones, statins, aliphatic chains, and macrocyclic compounds.

C. Products of Type II Iterative PKSs

The type II PKSs, thus far found exclusively in bacteria, most closely resemble type II bacterial FASs. Type II PKSs typically contain the following discrete proteins encoded by genes in a tightly-linked cluster: (i) a ketosynthase (KSα), containing the classical cysteine active site; (ii) a second ketosynthase-like protein (KSβ) with moderate sequence similarity to the KSα but lacking the characteristic cysteine active site (this has sometimes been referred to as the "chain length factor"); (iii) a soluble, acidic acyl carrier protein (ACP); (iv) a polyketide reductase which acts to reduce a single carbonyl after the polyketide has been assembled; (v) one or more cyclase/aromatase types of enzymes; and (vi) in some systems, an oxygenase to complete the quinone is present. The order of type II PKS genes within the gene clusters varies, but typically the KSα, KSβ, and ACP genes are linked within a single operon. The enzymatic structure of the minimal PKS unit resulting in a polyketide has recently been shown to be $KS\alpha_2$, $KS\beta_2$, ACP_2. The

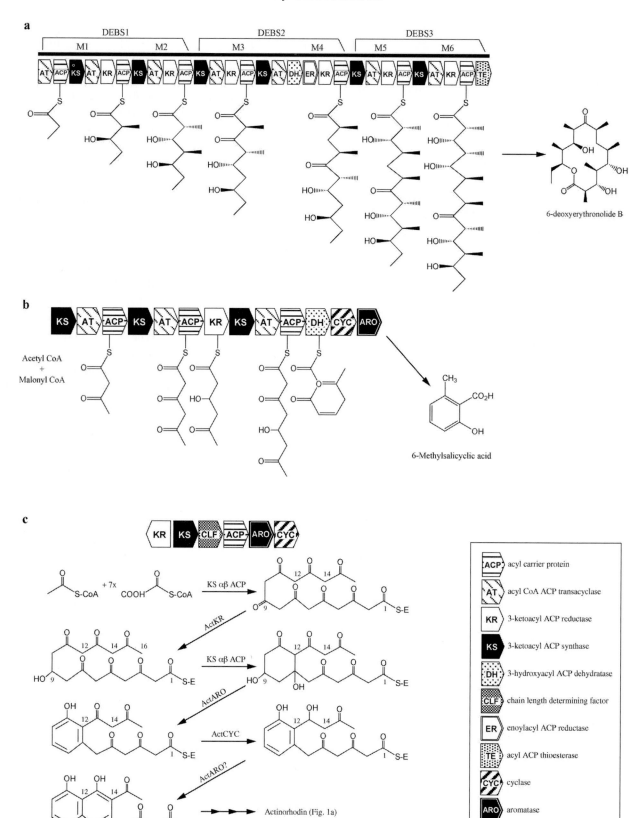

6-deoxyerythronolide B

6-Methylsalicyclic acid

Actinorhodin (Fig. 1a)

ACP   acyl carrier protein

AT   acyl CoA ACP transacylase

KR   3-ketoacyl ACP reductase

KS   3-ketoacyl ACP synthase

DH   3-hydroxyacyl ACP dehydratase

CLF   chain length determining factor

ER   enoylacyl ACP reductase

TE   acyl ACP thioesterase

CYC   cyclase

ARO   aromatase

hypothetical early biosynthetic reactions leading to actinorhodin are exemplified in Fig. 11c. In the biosynthesis of aromatic polyketides by type II PKSs, the choices made include starter unit (usually an acetyl moiety), number of extensions (chain length), ketoreduction (if present, usually at C9 from the carboxy terminus), and cyclization; a C2 unit derived from malonyl-ScoA is always used as the extender moiety, just as in fatty acid biosynthesis.

The products of type II PKSs are exclusively aromatic quinones, including tetracylines, anthracyclines, angucyclines, and benzoisochromane quinones such as actinorhodin. In nearly all cases, products of type II PKSs are significantly modified by various postassembly biosynthetic reactions, including hydroxylations, methylations, oxidations and reductions, glycosylations, and/or amino-acylations.

V. EVOLUTION

The production of polyketides and other secondary metabolites by microorganisms is not specie or genus specific; that is, similar products are found in distantly related taxa. For example, the immunosuppressant FK506 has been identified as a product of several *Streptomyces* spp. and an *Actinoplanes* sp. Likewise, more than eight different fungal genera produce the HGM-CoA reductase inhibitor lovastatin, a fungal polyketide metabolite that was first documented as being produced by *A. terreus*. Despite the structural and functional diversity found within the three polyketide classes, the biosynthetic pathways within each class are remarkably similar at both genome and protein levels. These similarities provide insight into the complex evolutionary history of these compounds. The currently accepted hypothesis is that polyketide synthetases have arisen from their primary progenitors and have been passed around by horizontal gene transfer between species. Evidence that supports this hypothesis is derived from the genetic analysis of the genes that encode the polyketide synthetases. Comparisons of the type II PKS KS and CLF domains by Revill and coworkers indicates that they originated by a duplication of an ancestral gene set and have since diverged. A similar hypothesis has been put forward by the Katz laboratory for the evolution of type I modular polyketide synthases. They argued that evolution of erythromycin resulted from a combination of horizontal and vertical transfer events because the erythromycin PKS cluster, which is composed of three open reading frames (ORFs) (Fig. 11a), evolved via an amplification event originating from module 4. This hypothesis can be further investigated by comparing the ketosynthase domains of type I modular PKS pathways (Fig. 12). This hypothesis appears to hold true for the rifamycin PKS KS domains, which form a tight cluster. It does not appear to be as straightforward for all type 1 PKS pathways; that is, typically the KS domains cluster according to the separate ORFs rather than according to the complete PKS. For example, the rapamycin PKS cluster is divided into three ORFs (Raps1–Raps3), and phylogenetic analysis of the KS domains clusters these into three disparate groups. The KS domains from Raps3 are distant from the Raps1 and Raps2 KS domains, indicating that the genes may have evolved separately. Similar clustering is observed for FK506, which has structural and functional parity to rapamycin. Further evidence for the horizontal transfer of PKS biosynthetic clusters is that KS domains from gram-negative (soraphen and pyrolutin) and gram-positive bacteria cluster together (tylosin and KS domains of unknown function from *Streptomyces coelicolor*) (Fig. 12). The KS domain for the polyketide component of coronatine, the phytotoxin produced by *Pseudomonas syringae*, does not cluster with the other gram-negative KS domains. This is interesting,

---

Fig. 11. Biosynthetic schemes for polykidide biosynthesis. (a) Biosynthesis of 6-deoxyerythronolide, the lactone core of erythromycin, is constructed via a type I modular PKS pathway; (b) type I iterative polyketide synthases are responsible for the biosynthesis of molecules such as 6-methylsalicylic acid: (c) aromatic bacterial polyketides are assembled by type II iterative PKS systems, for example the actinorhodin biosynthetic cluster.

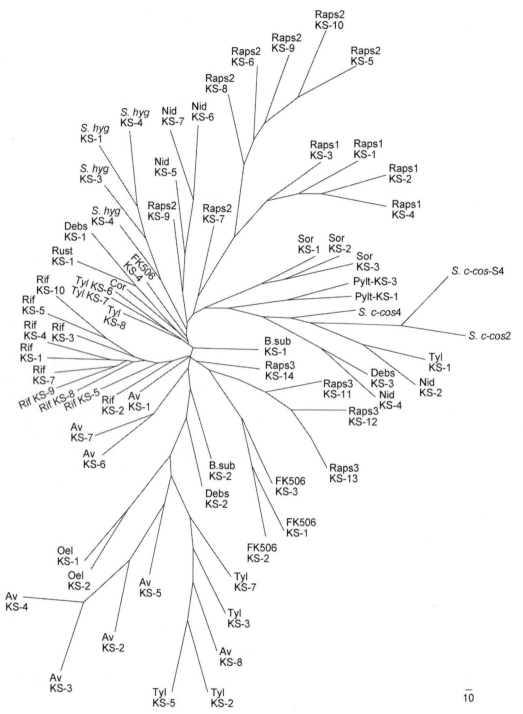

Fig. 12. Phylogenetic relationships of the ketosynthase domains (KS) from a selection of bacterial type 1 PKS pathways. DNA sequence data were analysed in PHYLIP using DNAdist and FITCH. The GenBank accession numbers are in parentheses. Av, *Streptomyces avermetilus* avermectin biosynthetic cluster (D. MacNeil, unpublished data); B. sub, *Bacillus subtilis,* unidentified function (Z14098); Cor, *Pseudomonas syringae* coronafacic acid biosynthetic cluster (U33313); Debs, *S. erythrea* erythromycin biosynthetic cluster (M63676); FK-506, *Streptomyces* sp. FK-506 biosynthetic cluster (Y10438); Nid,

because the coronafacic acid PKS cluster is located on a plasmid, suggesting a possible vehicle for horizontal transfer.

VI. FUTURE PERSPECTIVES

The structural diversity and broad spectrum of biological activities of polyketides have made this class of compounds a source of many successful medicines in the past. There is no doubt that many new polyketide structures and activities will be discovered by screening microorganisms using traditional fermentation (e.g., the continuing interest in finding new derivatives of fungal HMG-CoA reductase inhibitors) and by creating semisynthetic polyketides by modifying bioactive natural polyketides (biocombinatorial synthesis).

The first example of a "hybrid" polyketide, reported in 1985, was the formation of mederrhodin A by a medermycin-producing *Streptomyces* strain transformed with the *actVA* gene from the actinorhodin biosynthesis gene cluster. This approach of generating novel polyketides was largely empirical and required the activity of enzymes for late tailoring steps. The first example of the generation of novel polyketides through the genetic manipulation of PKS genes was reported in 1990 by the Strohl group. They demonstrated the biosynthesis of the anthraquinones aloesaponarin II and desoxyerythrolaccin (neither are bacterial products) through the cloning of actinorhodin biosynthesis genes in the aclarubicin-producing strain *Streptomyces galilaeus*. Shortly thereafter, a more combinatorial approach toward the generation of novel type II PKS products by mixing and matching various genes of the type II PKSs was devised and applied widely by the Hopwood and Khosla laboratories, thus generating dozens of novel compounds. This

approach has also led to the understanding of genetic programming of aromatic PKSs; consequently, the rational design of novel aromatic polyketides is also advancing. Moreover, a wide range of research efforts have shown that novel aromatic polyketides can be produced by manipulating not only the PKS-associated genes but also a wide variety of the postassembly tailoring reactions.

The era of rational design for novel antibiotic structures was ushered in by early successes in the synthesis of complex polyketides. It was demonstrated in the early 1990s that analogs of the erythromycin polyketide backbone could be generated by eliminating or substituting active sites within the PKS. By repositioning a chain-terminating cyclase domain from the C terminus of module 6 of DEBS3 to the C terminus of module 2 of DEBS1, it was possible to construct a multienzyme unit that catalyzed only the first two rounds of polyketide chain extension. The mutant produced a triketide lactone structure without any trace of erythromycin, the wild-type polyketide, indicating premature chain termination and cyclization. By expressing the entire DEBS gene cluster in a heterologous host, substantial quantities of 6-deoxyerythronolide B, the aglycone of the macrolide antibiotic erythromycin, were produced.

Similar principles and roles for genetically engineering bacterial polyketides could be applied to combinatorial synthesis of fungal polyketides, but specific protocols need to be developed for fungi since fungal gene regulation and structure are very different form those of bacteria. For example, fungal genes contain nontranslated DNA known as introns. The concept of making new metabolites by using chimeric genes is simple and obvious, but the challenge of constructing many chimeric gene clusters is great. This approach, though very promising, has yet to be proven practically.

S. cealestis nidamycin biosynthetic cluster (AF016585); Pylt, *P. fluorescens* pyoluteorin biosynthetic cluster (AF003370); Oel, *S. antibioticus* oeleandomycin biosynthetic cluster (L09654); Raps, *S. hygroscopicus* rapamycin biosynthetic cluster (X86780); Rif, *Amycolatopsis mediterrani* rifamycin biosynthetic cluster (AF040570); Rust, *Micromonospora chalcea* rustmicin biosynthetic cluster (unpublished data); Sor, *Sorangium cellulosum* soraphen biosynthetic cluster (U24241/188042/188045); S.c., *S. coelicolor* putative KS domains (AL021409/AL021530); S. hyg, *S. hygroscopicus* unidentified PKS biosynthetic cluster (AF007101); Tyl, *S. fradiae* tylosin biosynthetic cluster (SFU78289).

See Also the Following Articles

ANTIBIOTIC BIOSYNTHESIS • SECONDARY METABOLITES • *STREPTOMYCES*, GENETICS

Bibliography

Bryskier, A. J., Butzler, J. P., Neu, H. C., and Tulkens, P. M. (Eds.) (1993). "Macrolides: Chemistry, Pharmacology and Clinical Uses." Arnette Blackwell, Paris.

Hershberger, C. L., Queener, S. W., and Hegeman, G. (Eds.) (1989). "Genetics and Molecular Biology of Industrial Microorganisms." ASM Press, Washington, DC.

Hopwood, D. A. (1997). Genetic contributions to understanding polyketide synthases. *Chem. Rev.* **97**, 2465–2497.

Hopwood, D. A., and Khosla, C. (1992). Genes for polyketide secondary metabolic pathways in microorganisms and plants. *Ciba Found. Symp.* **171**, 88–106.

Hopwood, D. A., and Sherman, D. H. (1990). Molecular genetics of polyketides and its comparison to fatty acid biosynthesis. *Annu. Rev. Genet.* **24**, 37–66.

Hutchinson, C. R. (1995). Anthracyclines. *Biotechnol.* **28**, 331–357

Hutchinson, C. R., and Fujii, I. (1995). Polyketide synthase gene manipulation: A structure–function approach in engineering novel antibiotics. *Annu. Rev. Microbiol.* **49**, 201–238.

Katz, L., and Donadio, S. (1993). Polyketide synthesis: Prospects for hybrid antibiotics. *Annu. Rev. Microbiol.* **47**, 875–912.

Leadlay, P. F. (1997). Combinatorial approaches to polyketide biosynthesis. *Curr. Opin. Chem. Biol.* **1**, 162–168.

Mann, J., Davidson, R. S., Hobbs, J. B., Banthorpe, D. V., and Harborne, J. B. (Eds.) (1996). "Natural Products. Their Chemistry and Biological Significance." Addison–WelsleyLongman, Essex, UK.

Rawlings, B. J. (1997). Biosynthesis of polyketides. *Nat. Prod. Rep.* **14**, 523–556.

Strohl, W. R. (Ed.) (1997). "Biotechnology of Antibiotics." Dekker, New York.

Vining, L. C., and Stuttard, C. (Eds.) (1995). "Genetics and Biochemistry of Antibiotic Production." Butterworth-Heinemann, Boston.

Polymerase Chain Reaction (PCR)

Carol J. Palmer

Nova Southeastern University, Ft. Lauderdale, FL

Christine Paszko-Kolva

Accelerated Technology Laboratories, Inc., West End, NC

GLOSSARY

carryover In molecular biology, describing the amplification product of a previous PCR reaction that is carried to another PCR reaction, potentially leading to a false positive result.

DNA hybridization The process by which two complementary strands of DNA bind to each other in a mixture of DNA strands.

infectious agent Any bacterium, virus, protozoan, prion, or other agent that can cause infection or disease in a human, animal, or other host.

PCR (polymerase chain reaction) An *in vitro* method for the enzymatic synthesis of specific DNA sequences, using two oligonucleotide primers that hybridize to opposite strands and flank the region of interest in the target DNA. Following a series of repetitive cycles that involve template denaturation, primer annealing, and the extension of the annealed primers by *Taq* polymerase, there is an exponential accumulation of the specific target DNA.

RT-PCR (reverse transcriptase polymerase chain reaction) A technique similar to conventional PCR, except that the starting material is RNA rather than DNA. Because of this, a DNA copy must first be made from the RNA, utilizing an enzyme known as reverse transcriptase. Once the copy of the DNA is made, the PCR proceeds as usual.

Taq polymerase A thermostable DNA polymerase that was isolated from a bacteria *Thermus aquaticus*. This enzyme allowed PCR to be easily automated. Due to the heat stability of the enzyme, fresh enzyme no longer had to be added after each amplification cycle.

THE POLYMERASE CHAIN REACTION has proven to be a powerful new diagnostic tool for microbiologists. The technology is based on repeated cycles of enzymatic amplification of small quantities of specific DNA or RNA sequences in target organisms until a threshold signal for detection is obtained. This technology has been applied to microbiology via direct detection of microorganisms, detection of genes that code for virulence factors, identification of the presence of genes responsible for antimicrobial resistance, and typing of bacterial isolates in epidemiological investigations.

I. INTRODUCTION

There are various methods and approaches that can be taken to identify microorganisms. Bacteria are perhaps among the easiest to identify since they can be subjected to numerous detection methods, ranging from direct staining (a variety of differential stains are available), culture on artificial media, (agar or tissue culture), serological techniques, or a range of molecular methods (PCR, DNA hybridization, or direct sequencing). Many of the same methods used to identify bacteria can also be applied to the identification of protozoans and viruses with minor modifications. The focus of this article will be a review of the polymerase chain reaction (PCR) and the advances that have been made in microbial detection since the invention of the polymerase chain reaction.

Current detection methods date back to the beginnings of microbiology and utilize growth on artificial media, often impregnated with differential dyes, or various broths from pre-enrichment to selective pre-enrichment. These methods often require days to weeks to obtain visible growth and suspect colonies still need to undergo further confirmatory biochemical testing. Newer methods involve the use of both polyclonal and monoclonal antibodies, either alone or in conjunction with a separation matrix, such as magnetic beads, following a pre-enrichment culture step. Although these techniques often represent significant time savings over traditional cultural methods, they may lack sensitivity and specificity. Because of such shortcomings, there is a tremendous need for rapid, accurate, and sensitive pathogen detection methods. Molecular methods provide the answer in many cases. Quite often, molecular methods are used in conjunction with classical methods to obtain results even faster. The remainder of this section will discuss the application of polymerase chain reaction (PCR) technology to two divisions of microbiology: infectious disease and environmental analysis. In addition, a recent improvement in PCR, known as the 5′ nuclease assay, as it is applied to both clinical and environmental microbial identification, will be discussed.

II. PCR TECHNOLOGY

Since its discovery in the 1980s by Kary Mullis, the PCR has been modified and enhanced into a powerful tool for the genetic detection of agents of infectious disease, sequencing of entire genomes of microorganisms (as well as the ongoing sequencing of the human genome), environmental analysis, forensic science, detection of food-borne pathogens, and identification of specific DNA markers in individuals. New applications of this exciting technology will continue to appear in the literature as the imaginations of scientists around the world continue to exploit and modify the basic PCR technique.

The PCR requires two synthetic oligonucleotide primers that are complementary to regions on opposite strands of a target piece of DNA, a target sequence in a DNA sample that occurs between the pair of primers, a thermostable DNA polymerase and the four deoxyribonucleotides. The procedure, performed in a themocycler, usually consists of three basic cycles and is completed within minutes. The first step is denaturation, which increases the temperature within the sample vial to about 94°C, causing the double-stranded dDNA within the sample to separate into two pieces. The second step, renaturation, is completed by dropping the temperature within the thermocycler to about 55°C, which allows the primers to anneal with their complementary sequences in the source DNA. The final step is the synthesis portion of the reaction, wherein the temperature is raised to about 74°C, the optimum temperature for the catalytic functioning of Taq DNA polymerase. In this step, target DNA is extended, replicating to form additional copies of the target DNA. Within 30 cycles, over a million copies of the original target DNA can be reproduced.

III. DIAGNOSIS OF INFECTIOUS DISEASE

Initially, applications of PCR to infectious disease diagnostics focused on microorganisms that were impossible or slow to grow in culture, were difficult to cultivate, or posed significant health hazards using standard recovery techniques. Examples of these types of organisms include *Mycobacterium tuberculosis*, *Helicobacter pylori*, *Chlamydia trachomatis*, and HIV. The PCR has also played a key role in the identification of new or reemerging diseases, such as hantavirus and cyclospora, and has also been applied to retrospective analysis of samples from outbreaks or deaths caused by previously unidentified etiological agents.

Aother area in the application of PCR to infectious disease is related to the rapid global expansion of HIV and the need to monitor new drugs in its treatment. This has led to the advent of discoveries and automation both in tracking the disease and in therapy. Patients infected with HIV can now have automated analysis of precise HIV viral loads, which can inform the clinician as to which stage of the disease the patient is in and how they are responding to antiviral therapy. This has been particularly useful with the introduction of the new protease inhibitors. Clinicians can quickly assess the impact of these new antiviral drugs on the course of HIV disease by monitoring viral load.

IV. TROPICAL INFECTIOUS DISEASES

More recently, diagnosis of tropical diseases has received significant attention in relation to PCR technology. The PCR has been applied to identification of malaria species, dengue virus serotyping as well as rapid identification of infection by leishmania, chagas and leptospirosis. Since many of these diseases initially present with similar symptoms (fever, chills, headache), differential diagnosis is critical to proper patient treatment and follow-up. For example, antibiotic treatment for malaria caused by *Plasmodium vivax* and *Plasmodium falciparum* is different. Thus, proper identification of the infecting species is crucial to curing the patient of malaria and preventing recurrence of the disease. Importantly, PCR has also been applied to the identification of circulating drug-resistant strains of the malaria parasite. Since drug resistance is dramatically increasing in malaria endemic tropical areas, the PCR offers a precise methodology to monitor the spread of resistant parasites and to help identify early drug failures due to resistant strains.

Diagnosis of dengue fever is particularly problematic since the virus can only be detected in the peripheral blood during the first week of infection and must be cultivated in mosquito cell culture. Few laboratories in dengue endemic areas have the ability to perform this technique. Other dengue diagnostic techniques rely on IgM capture assays or hemagglutination inhibition assays, both of which require reagents that are difficult to obtain and require that paired serum samples be obtained at least a week apart to detect increases in antibody response. The PCR has provided a solution to this problem, since it does not depend on either growth of the virus or development of antibody. Thus, PCR provides a powerful tool for infectious disease diagnostics that

is less stressful to the patient, as serial serum samples are not required and results can be obtained in less than a day.

V. FUTURE ROLE OF PCR IN INFECTIOUS DISEASE IDENTIFICATION

Novel future applications of the PCR include the adaptation of this technology with high-density oligo arrays into microinstrumentation platforms. This advancement allows for hands-off amplification technology and can cut down or eliminate current problems in PCR, including cross-contamination of samples, user friendliness and throughput issues. Handheld field PCR devices are also under development which will allow for researchers to perform PCR without the need for clean-room laboratory space and extensive instrumentation. These handheld devices will be especially important in military operations that involve the use of biological warfare agents.

VI. ENVIRONMENTAL APPLICATIONS

Microbial contamination of water is a global public health problem. There are now numerous reports of the application of PCR to the detection of microorganisms in water. The technique is especially useful for detecting water-borne pathogens that are not easily cultured or when rapid detection and identification are critical. Many pathogenic bacteria have been identified in the environment using PCR, including toxigenic *E. coli*, and *Mycobacteria* species. Important water-borne protozoan parasites, including *Giardia* and *Cryptosporidium*, have also been detected using molecular techniques. The PCR has even been utilized to evaluate microbial contamination in laboratory eyewash stations and emergency showers.

Environmental molecular analysis has also integrated the use of reverse-transcriptase PCR (RT-PCR) to look for medically important viruses in the environment. In RT-PCR, RNA is first transcribed into DNA, then is followed by standard PCR methodology. Enteroviruses, hepatitis A virus, rotavirus, poliovirus, and Norwalk virus have all been detected in sewage, sludge, groundwater, river water, ocean water, and ice using this technique. Other studies have utilized RT-PCR to assess the potential for HIV to survive in water.

Other applications of PCR to environmental recovery of microorganisms are evidenced in the detection of pathogens in seafood. Shellfish, such as mussels, oysters, and clams, have frequently been associated with outbreaks of food poisoning. Microbial pathogens linked to shellfish-related food-borne outbreaks utilizing PCR have included bacterial pathogens, *Vibrio* species, *Salmonella* species, *Campylobacter*, *Listeria*, and several viral pathogens, including Norwalk, enterovirus, and hepatitis A virus. The PCR technology has also assisted in identifying pathogens affecting the health of shellfish. Several investigators have identified *Haplosporidium nelsoni*, a pathogen of the eastern oyster, *Crassostrea virginica*, and *Nocardia crassostreae* species, a pathogen of Pacific oysters, *Crassostrea gigas*, using newly developed PCR test methods.

VII. COMMERCIAL PCR TEST KITS FOR ENVIRONMENTAL ANALYSIS

The first commercially available PCR test for an environmental pathogen was developed in the early 1990s by Roche Molecular Systems. This kit allowed for the direct detection of environmental *Legionella* species and *Legionella pneumophila* and was marketed under the name EnviroAmp. The EnviroAmp kit represented the first attempt to make the PCR technique user-friendly, since the analyst had premixed reagents and the results were read by interpreting colored dots on a nylon strip. This kit offered semiquantitative results and represented a major advancement in the detection of *Legionella*, since culturing the organism could take several days to weeks to recover an isolate. The assay was based on the use of PCR to amplify specific sequences from a conserved region of Legionella's 5S rRNA and from the macrophage infectivity potentiator (*mip*) gene specific for the *L. pneumophila* species. The kit was well accepted by researchers; unfortunately, the market for this new test kit was often driven by water-borne outbreaks, which were not a regular occurrence, instead

of being integrated as a standard monitoring tool for hospitals and water utilities. Unfortunately, Roche Molecular Systems, the manufacturer of the EnviroAmp has discontinued the production of this kit.

Another commercialized environmental use of PCR was introduced several years later, when Hazard Analysis of Critical Control Points (HACCP) was promulgated in the United States. The DuPont Corporation (now Qualicon) introduced manual PCR assays under the trade name Bax to target the detection of *Salmonella* in food samles. Other tests are available for *E. coli* 0157:H7 and *Listeria* species. These tests are commercially available to track foodborne outbreaks caused by microbial contamination increases. In summary, molecular tests for environmental analysis typically are higher in cost per test and require additional instrumentation, however, the labor savings and a precise identification often compensates for the higher initial costs. The major benefits of molecular tests are rapid identification of contaminants and that they provide excellent screening tools, allowing scientists to concentrate their efforts on isolating cultures from those samples that tested positive.

VIII. NEW TECHNOLOGY

Although the PCR assay has represented a huge breakthrough that molecular biologists and microbiologists around the world were quickly learning and modifying, there were still too many steps in which errors could be introduced and erroneous results derived. As a result, researchers at Roche devised a brilliant improvement on the technique that allows automation and eliminates many of the original steps that could introduce error. The 5′ nuclease assay utilizes the 5′ → 3′ endonuclease activity of Taq DNA polymerase which is capable of digesting an internal probe labeled with a fluorescent reporter dye (6-carboxy-fluorescein [FAM] with an emission wavelength of 518 nm) and a fluorescent quencher dye (6-carboxytetramethyl-rhodamine [TAMRA] with an emission wavelength of 582 nm, see Fig. 1). In conventional PCR, the forward and reverse primers anneal to their appropriate target sequences; the "twist" in the 5′ nuclease assay is that the fluores-

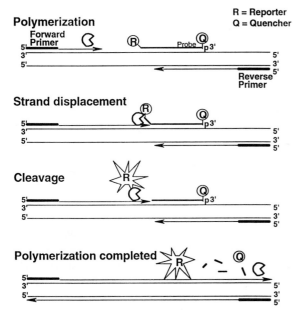

Fig. 1. A schematic diagram of the 5′ nuclease assay.

cently labeled probe anneals to the target sequence between the primers. The fluorescence of the intact probe is suppressed by the activity of the quencher dye. Digestion of the probe by Taq DNA polymerase results in separation of reporter and quencher dyes and a subsequent increase in the fluorescence of the reporter dye as a direct result of specific amplification. The assay uses a closed homogenous system that utilizes optical caps and tubes, which means that fluorescence readings can be taken directly through the caps. The major advantage of this detection scheme is a decrease in error resulting from pipetting and potential cross-contamination is also greatly reduced. A possible disadvantage is that nonviable cells are also detected by PCR and the 5′ nuclease assay, since it is the nucleic acid that is targeted. However, when combined with culture methods, rapid results can easily be obtained. The benefits of this technology make it deal for investigating food-borne outbreaks of organisms that are difficult or slow to culture. This technology has been evaluated for the detection of *E. coli* O157:H7 in ground beef and other food samples. This represents a major advance in diagnostic microbiology and, as,

in most cases, will undoubtedly be used for clinical applications first, before the technology is widely accepted for environmental applications.

The future of PCR remains bright and the various modifications will continue to lead to more user-friendly, integrated, high throughput pathogen detection systems that will translate into a better quality for life for us all. This quality improvement will come from a protected food/water supply and more rapid clinical diagnosis. This technology will be exploited in environmental and diagnostic microbiology, and numerous other fields.

See Also the Following Articles

DIAGNOSTIC MICROBIOLOGY • IDENTIFICATION OF BACTERIA, COMPUTERIZED • LEGIONELLA • WATER, DRINKING

Bibliography

Erlich, H. A. (ed.) (1989). PCR Technology: Principles and Applications for DNA Amplification. Stockton Press.

Ferre, F. (1994). Polymerase chain reaction and HIV. *Clin. Lab. Med.* **14**, 313–333.

Hill, W. E. (1996). The polymerase chain reaction: Applications for the detection of foodborne pathogens. *Crit. Rev. Food Sci. Nutr.* **36**, 123–173.

Hussong, D., Colwell, R. R., O'Brien, M., Weiss, A. D., Pearson, A. D., Weiner, R. M., and Burge, W. D. (1987). Viable *Legionella pneumophila* not detectable by culture on agar media. *Biotech* **5**, 947–950.

Innis, M. A., Gelfand, D. H., Sninsky, J. J., and White, T. J. (1990). "PCR Protocols: A Guide to Methods and Applications." Academic Press, San Diego.

Persing, D. H., Smith, T. F., Tenover, F. C., and White, T. J. (1993). Diagnostic Molecular Microbiology: Principles and Applications. American Society for Microbiology, Washington, DC.

Pillai, S. D. (1997). Rapid molecular detection of microbial pathogens: Breakthroughs and challenges. *Arch. Virol. Suppl.* 1367–1382.

Rabinow, P. (1996). Making PCR: A Story of Biotechnology. The University of Chicago Press, Chicago.

Stokes, N. A., and Burreson, E. M. (1995). A sensitive and specific DNA probe for the oyster pathogen *Haplosporidium nelsoni*. *J. Eukaryot. Microbiol.* **42**, 350–357.

Walliker, D. (1994). The role of molecular genetics in field studies on malaria parasites. *Int. J. Parasitol.* **24**, 799–808.

Potyviruses

John Hammond

United States Department of Agriculture–Agricultural Research Service, Beltsville, MD

GLOSSARY

cylindrical inclusion bodies Paracrystalline cytoplasmic arrays of ca. 5 nm periodicity, composed primarily of the virus-coded, nonstructural cylindrical inclusion (CI) protein. In cross section, the inclusions have curved arms radiating from a central axis, leading to the alternate name "pinwheel inclusions." Scrolls and laminated aggregates of CI protein are associated with the pinwheel inclusions of some potyviruses.

flower-breaking Irregular patterns of absence, reduction, or intensification of floral pigments, typically in feathery or mosaic patterns, caused by viral infection.

helper component A virus-coded accessory transmission factor necessary for vectored transmission. Thought to form a bridge between the virion and the mouthparts of the vector.

heteroencapsidation Also referred to as transcapsidation or genomic masking. The partial or complete encapsidation of the RNA of one virus in the coat protein of another isolate or distinct virus. The virion may have a mosaic of two different coat proteins.

immunodominant epitopes The major antigenic determinant of a particular antigen in its native state, stimulating the strongest antigen-specific immune response of the immunized animal.

insertional mutagenesis Introduction of an in-frame addition of one or more codons, resulting in altered local or global properties of the protein. Used to disrupt and, thereby, identify gene functions.

nonpersistent or stylet-borne transmission Vectored transmission which is relatively transitory and results from an association of virus particles, helper component, and the mouthparts of the vector (the stylets of an aphid). Virions dissociate from the vector mouthparts during the course of feeding, typically within a few minutes.

pathogen-derived resistance Resistance derived by expression of a gene or nucleic acid from the pathogen in a transgenic host, that interferes with the ability of the pathogen to complete its normal life cycle.

pathotype A pathogen phenotype distinguished by differential reactions on hosts carrying different resistance genes.

persistent transmission Vectored transmission by a more intimate association of virions with the vector. Virions may accumulate within salivary glands or other organs, from which they can later be passed back into plant tissue in saliva or regurgitant. Infectivity is retained for much longer periods than with nonpersistently transmitted viruses and, in some cases, for the life of the vector.

polyprotein A multicistronic protein translated from a single open reading frame. The polyprotein is subsequently cleaved into individual mature gene products, typically by one or more protease which is autocatalytically released from the polyprotein.

replicase The major virus-coded component of the viral replication complex; the RNA-dependent RNA polymerase. The replicase (NIb in the potyviruses) has motifs associated with nucleotide triphosphate binding and the core polymerase motif.

stylets The interlocking mandibles and maxillae of aphids; the mouthparts used to penetrate plant tissue and suck up sap.

VPg A virus-coded genome-linked protein, covalently attached at the 5′ end of the genomic RNA.

zoospores The nonsexual spores that attach to plant roots after release from the resting spores (dormant spores), especially of Plasmodiophoromycete fungi, such as *Polymyxa graminis*.

THE POTYVIRUSES, or *Potyviridae,* comprise the largest group of plant viruses. Over 100 definitive non-synonymous members, and about 90 tentative members, of the group are recognized. Some members have broad host ranges, infecting species in up to 30 families; most potyviruses infect a limited range of species. Most major crops are infected by one or more potyviruses, often with significant economic impact. All potyviruses form cytoplasmic "pinwheel" cylindrical inclusion (CI) bodies (Fig. 1) in infected cells, which can be a useful diagnostic feature, as no other viruses induce similar inclusions. Most potyviruses are readily transmitted by mechanical inoculation of susceptible hosts. Six genera are currently recognized. Most spe-

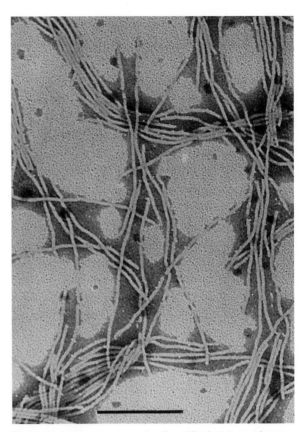

Fig. 2. Electron micrograph of purified virions of watermelon mosaic virus-II. Bar represents 325 nm. (Micrograph by M. Dienelt.)

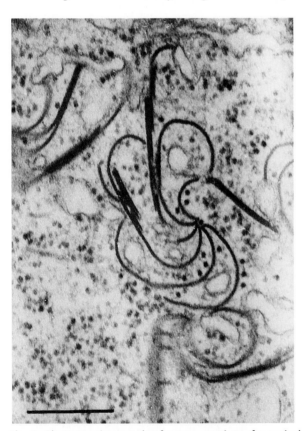

Fig. 1. Electron micrograph of a cross section of a typical "pinwheel" cylindrical inclusion body in sweet potato infected with sweet potato feathery mottle virus. Bar represents 225 nm. (Micrograph by S. Hurtt.)

cies are within the genus *Potyvirus* and are transmitted by various species of aphids. Other genera are differentiated, in part, by different vector specificities; genera are also differentiated by degree of genome similarity, serological properties, and genome division. All potyviruses have flexuous virions (Fig. 2) of 11–15 nm diameter and a helical pitch of about 3.4 nm, and 650–900 nm long for monopartite genera, composed of a single molecule of messenger-sense RNA encapsidated in about 2000 copies of a single species of coat protein. Most genera have monopartite genomes that are expressed as a single open reading frame, with individual gene products cleaved by virus-encoded endoproteinases. The bymoviruses have a bipartite genome that is otherwise organized in a similar fashion to the other genera. The genomic RNA is polyadenylated at the 3'-end and (at least, for the genus *Potyvirus*) has a protein (VPg of about 24 kDa), covalently linked to the 5'-terminal nucleotide.

I. TAXONOMY

Over the past few years, molecular data have been used to resolve many of the problems in potyvirus taxonomy. As a result, what was once thought to represent a continuum between isolates of a single virus and distinct species has been resolved to a large degree. Isolates previously considered as distinct viruses, because of differences in host range, have been recognized as members of the same species. Other isolates formerly described as strains of one virus have been determined to belong to distinct species. The basic characters of the genera are as follows:

Potyvirus. Type species potato virus Y, PVY. Particles 680–900 nm long, 11–13 nm wide (Fig. 2). Virions of some viruses are longer in the presence of divalent cations than in the presence of EDTA. The monopartite genomic RNA of ca.9.5–10 kb (Mr $3.0–3.6 \times 10^6$) forms about 5% of particle weight. The coat protein has between 251 and 332 amino acid residues and a molecular weight of ca.30–40 kDa. Viruses in the genus *Potyvirus* are generally transmitted by various species of aphids in a nonpersistent manner, as the result of interaction between the viral coat protein (CP) and the helper component (HC) domain of the viral-encoded, nonstructural gene product, helper component-proteinase (HC-Pro).

Rymovirus. Type virus ryegrass mosaic virus, RGMV. RNA of ca. 9–10 kb, CP of ca. 29 kDa (RGMV), particles 690–720 nm. The rymoviruses are more closely related to the genus *Potyvirus*, from which they differ primarily in the vector, having similar size CP. *Rymovirus* species have host ranges restricted to the Gramineae and are vectored by eriophyid mites of the genus *Abacarus*, primarily *A. hystrix*, which is also limited to gramineaceous hosts. Viruses in the genus *Tritimovirus* are vectored by eriophyid mites of a different genus. While there is some argument that the genus *Rymovirus* should be integrated into the genus *Potyvirus*, there is insufficient sequence information available at present to justify this change.

Tritimovirus. Type virus wheat streak mosaic virus, WSMV. RNA of ca. 9.5 kb, virions 690–700 nm. Tritimoviruses have proteolytic cleavage sites that have been less clearly defined than for other potyvirus genera; there may be additional cryptic cleavage sites that result in variable CP populations. The WSMV CP has ca. 298 amino acid residues, but various CP-related species with apparent MWs of ca. 30–45 kDa may be found in sap extracts. Tritimovirus CPs are clearly phylogenetically distinct from those of other potyvirus genera. Transmitted by the eriophyid mite *Aceria tulipae,* which has hosts in both the Gramineae and Liliaceae. The best characterized viruses infect wheat and grasses, but two putative tritimoviruses infect *Allium.*

Ipomovirus. Type virus sweet potato mild mottle virus, SPMMV. Virions ca. 800–950 nm, RNA ca. 10.8 kb, CP of 270–275 residues and ca. 38 kDa. Transmitted by the whitefly *Bemisia tabaci* in a nonpersistent manner. Motifs in HC-Pro and CP associated with aphid transmission of the genus *Potyvirus* are missing or altered and may account for the difference in vector specificity. Naturally occurring in sweet potato, but with experimental hosts in at least 45 species from 14 families.

Macluravirus. Type virus Maclura mosaic virus, MacMV. Transmitted by aphids, with particles ca. 650 nm long, ca. 13 nm wide, and CP of ca. 32–33 kDa. The genomic RNA is about 8 kb. There is limited CP amino acid sequence homology (14–23 %) with members of the genus *Potyvirus*, but significant homology of the replicase protein to other potyvirus genera. Genome organization is presumed similar to other members of the *Potyviridae*.

Bymovirus. Type virus barley yellow mosaic virus, BaYMV. The bymoviruses are distinguished from other potyvirus genera in two major features: the genome is divided between two RNAs encapsidated in separate particles, and these viruses are transmitted by the fungus *Polymyxa graminis*. RNA 1 is ca. 7.3–8 kb (Mr 2.6×10^6), and RNA 2 is 3.5–4 kb (Mr 1.5×10^6). Both RNAs are polyadenylated and have little homology apart from the 5′-untranslated region. Particles are 250–300 and 500–600 nm long and ca. 13 nm wide. The CP is ca. 29–33 kDa; BaYMV CP has 297 residues.

II. GENOME ORGANIZATION AND REGULATION

Viruses of the *Potyviridae* have genome organizations with similarities to other viruses in the picorna-like supergroup, including the *Picornaviridae, Comoviridae,* and *Hypoviridae.* Their genomes are single-stranded, positive sense RNA with a 5′-terminal genome-linked protein (VPg) and a polyadenylated 3′ terminus. The genomic RNA (or RNAs, in the case of divided genomes) are translated into a single polyprotein that is processed into at least nine individual mature gene products by virus-encoded proteases (see Fig. 3). Members of the genus *Potyvirus* are best studied in this regard and have three distinct proteases. The P1 protease autocatalytically cleaves itself from the genomic polyprotein at the C-terminus, as does the protease function of HC–Pro. The third protease, NIa–Pro (NI = nuclear inclusion), also cleaves itself from the viral polyprotein at the C-terminus and releases all of the remaining gene products by either intramolecular or intermolecular reactions.

The NIa–Pro cleavage site is a heptapeptide motif, with cleavage between residues 6 and 7. The recognition sequence differs both between potyviruses and between different sites within a single potyviral polyprotein, although some positions in the motif are highly conserved. The efficiency of cleavage at different sites is one means of regulation of gene expression, and incompletely cleaved products may fulfill different functions than do the final, completely processed gene products. This is an important feature for viruses that otherwise produce equimolar amounts of each gene product, despite requiring about 2000 copies of the CP but only one molecule of VPg per virion.

The formation of various inclusion bodies and differential stability of different gene products may also contribute to gene regulation. The cylindrical, or cytoplasmic inclusion body, is produced by all potyviruses and initially forms over plasmodesmata; as such, it has been proposed to have a role in intercellular transport of either virions or RNA. Later in infection, the cytoplasmic inclusions separate from the cell wall and accumulate in the cytoplasm. Some potyviruses also produce nuclear crystalline inclusions made up of equimolar amounts of the (incompletely processed) NIa and the NIb proteins, although not all isolates of a particular virus will form such inclusions. The HC–Pro protein of certain potyviruses aggregates into amorphous inclusions in the cytoplasm.

The organization of the bymoviruses is similar to that of the monopartite potyviruses, with gene products expressed from the larger RNA 1 being similar to the C-terminal 75% of other genera (see Fig. 3); an N-terminal product has similarity to P3, followed by cistrons equivalent to 6K1, CI, 6K2, NIa, NIb, and CP, in the same order as other potyviruses. The bicistronic RNA 2 has an N-terminal product with homology to the protease domain of HC–Pro. The P2 function of RNA 2 is likely to be involved in fungal transmission, as an isolate that has lost fungal transmissibility had a large deletion in the C-terminal domain of P2. P2 may, thus, have functional equivalence to the HC portion of HC–Pro as an accessory factor in vectored transmission. No Pro function is required in P2 because it is the C-terminal cistron

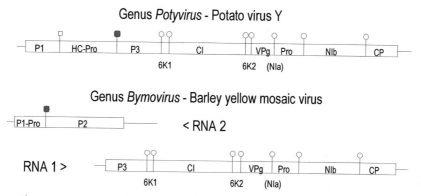

Fig. 3. Genome structure of *Potyvirus* potato virus Y and *Bymovirus* barley yellow mosaic virus, drawn approximately to scale. Symbols represent similar cleavage sites.

of the bymovirus RNA 2. The 5′ noncoding regions of RNA 1 and RNA 2 have extensive sequence homology, whereas the 3′ noncoding regions differ considerably in both length and sequence.

III. REPLICATION AND GENE FUNCTIONS

Replication is presumed to occur in the cytoplasm and involves multiple virus-encoded proteins, several of which also have other functions. The P1 protein is essential for genome amplification and must be self-cleaved from the viral polyprotein by its protease activity to fulfill this function. P1 also has similarities to cell-to-cell transport proteins of some other plant viruses. HC–Pro is also involved in genome amplification, as well as in long distance movement within the plant, aphid transmission, and self-cleavage from the polyprotein. The function of P3 is not clear, but insertional mutagenesis disrupted viral replication, indicating that the P3 product is also involved in replication. For plum pox virus, at least, a 6K protein (6K1; see Fig. 3) at the C-terminal end of P3 need not be processed to retain viability of the virus, and its function is unclear. The CI protein has both ATP-ase and helicase motifs and has been demonstrated to have catalytic activity *in vitro*; CI also has a role in cell-to-cell movement. The second 6K protein (6K2; see Fig. 3) is known to be membrane associated and involved in RNA replication. NIa–Pro is further processed to release the N-terminal domain, which functions as the VPg, and is associated with RNA replication, probably priming (-) strand RNA synthesis. The VPg has also been implicated in host range determination and long distance movement in specific host genotypes. The C-terminal proteinase domain of NIa functions to release most of the other gene products from the polyprotein by either inter- or intramolecular reactions. NIb has the NTP-binding site and hallmark motif of RNA-dependent RNA polymerases and is the major factor of the viral replicase complex. To date, no host-encoded factors of the complex have been identified. The viral CP is the C-terminal region of the polyprotein and has roles in both cell-to-cell and long distance move-

ment, and in aphid transmission, in addition to encapsidation of the genomic RNA.

IV. TRANSMISSION

Most members of the *Potyviridae* are readily transmitted by mechanical inoculation. Natural transmission is typically via appropriate vectors, or through vegetative propagation of infected plants, or, in some instances, through seed. Levels of seed transmission vary considerably between viruses and host species and may range from trace levels to close to 100%. Seed transmission may occur as a consequence of infection of the testa, and subsequent mechanical transmission to the germinating seedling, or by invasion of the embryo. Infection of seed by pollen transmission may also occur at lower levels. Embryonic invasion typically occurs in plants infected prior to flowering, but not in seed from plants infected later. Seed transmission is of great significance for plant germplasm movement and quarantine, as even very low levels of infection may result in introduction of viruses to new areas. Seed transmission of endemic viruses can result in significant losses, as foci of infection within a crop can lead to epidemic infections in the presence of appropriate vectors.

Aphid transmission of viruses in the genera *Potyvirus* and *Macluravirus* can lead to local spread from infection foci or to long-distance spread over large areas, under appropriate weather conditions. Transmission is nonpersistent and depends upon an interaction between the CP of encapsidated virus, HC–Pro from infected plants, and the aphids' stylets. Virus is most effectively acquired by aphids that have been starved prior to a short feeding period on the infected host, and transmission can occur during short duration probing of a second plant. There is relatively little virus/vector specificity, inasmuch as several aphid species are known to transmit many distinct potyviruses, and individual potyviruses may be transmitted by multiple species of aphid. However, aphids are the only arthropod vectors. Among the well-known vector species are *Myzus persicae*, *Macrosiphum euphorbiae*, and *Aphis gossypii*, each of which are polyphagous (capable of feeding on many different host species). It is not necessary for aphid vectors to colonize a plant for transmission to occur. In fact, much transmission may occur as the result of aphids alighting on a potential host, making a brief test probe to determine host palatability, and flying on to another plant. Virus may, thus, be spread to plant species by aphid species that rarely or never colonize that host. Under experimental conditions HC–Pro can be acquired by the vector first, followed by the virus; however, it is assumed that both are normally acquired at the same time. The HC–Pro of one virus can effectively interact with the CP of a distinct virus to effect transmission, but such interactions are not necessarily reciprocal. Another potential interaction between two potyviruses co-infecting a single host is heteroencapsidation (also called phenotypic mixing), in which the RNA of one virus is partially or wholly encapsidated in the CP of the second virus.

Natural occurrence of aphid nontransmissible isolates can be perpetuated either by vegetative propagation (as in the case of potato virus C, now considered to be an aphid nontransmissible isolate of PVY), or by interaction between the HC–Pro of one virus with virions of the other, or by heteroencapsidation. Mutations of specific amino acid motifs in the N-terminus of either HC–Pro or CP can result in loss of aphid transmissibility of a particular isolate, and both have been documented in naturally occurring aphid nontransmissible isolates.

Mite transmission of rymoviruses and tritimoviruses has not been as well studied as the aphid transmission of the genus *Potyvirus*. Mites can acquire the tritimovirus WSMV during the nymphal stages and transmit as adults; adults cannot acquire and then transmit the virus. Transmission is persistent and may last for the life of the vector. Mites can be distributed over long distances by wind.

Fungal transmission of the bymoviruses within the zoospores of the root-infecting obligate parasite *Polymyxa graminis* is also persistent. Virus can remain infectious in resting spores for periods of months or even years. While spores will normally be distributed only locally, and virus outbreaks often occur as small patches within fields, spores can also be spread in contaminated seed, on farm machinery, and over long distances as wind-blown particles.

The ipomovirus SPMMV is transmitted by the whitefly *Bemisia tabaci* in a nonpersistent fashion, but vegetative propagation from infected sweet potato may be of far greater significance to agriculture.

V. SYMPTOMS

Typical symptoms of potyvirus-infected plants are foliar mosaics, but there is much variability even between individual isolates of a single virus or between reactions of different genotypes of a particular host species. Other types of symptoms reported in specific virus–host combinations include mottle; necrotic etching, tip die-back, or systemic necrosis; streaking; stunting; generalized chlorosis; and ringspots. Rings or spots are characteristic symptoms of certain viruses on fruit or seeds, such as ringspots on papaya caused by papaya ringspot virus, and rings on the fruit and stones of apricots infected with plum pox virus. Viruses such as tulip breaking virus cause characteristic flower-breaking in some hosts; other potyviruses cause flower-break in gladiolus, bulbous iris, and crocus. Viral names often reflect the typical symptoms in the initially described host.

Potyviruses are often found in mixed infections, both with other potyviruses and also with viruses of other groups. In many cases, there are synergistic reactions (an increase in symptoms or titer of one or both viruses compared to a single infection) that are due to the potyvirus gene products P1 and HC–Pro acting as general transactivators of viruses of some other groups. For example, the combination of PVY with potato X potexvirus in tomato causes the severe disease "leaf drop streak," whereas each virus alone causes a relatively mild mosaic. Typically, the titer of the potyvirus is little affected compared to a single infection, but both the titer of the other virus and the severity of symptoms may be increased.

VI. DIAGNOSIS

Visible symptoms are rarely sufficient for identification of a specific virus. A variety of tests can be used to detect potyviruses, but the primary means of identifying specific potyviruses are host range, serology, and nucleic acid methods, such as dot-spot hybridization and combined reverse transcription/polymerase chain reaction. A series of staining techniques for light microscopy, using Luxol Brilliant Green and Calcomine Orange stains, can be used to rapidly visualize the "pinwheel" cytoplasmic inclusions that are diagnostic for the potyvirus group, as well as the crystalline nuclear inclusions and amorphous cytoplasmic inclusions that are produced by some members of the group. Additional detail can be gained by fixing and embedding tissue for electron microscopy.

Bioassay by mechanical inoculation to appropriate indicator plants is also useful. Symptoms will typically appear between four days and three weeks after inoculation, depending upon the virus and assay host. Differential indicator hosts can often be used to distinguish viruses in mixed infections with other potyviruses or nonrelated viruses.

Double antibody sandwich-type enzyme-linked immunosorbent assay (DAS-ELISA) or a variety of assays with virus-specific monoclonal antibodies can be used to identify specific viruses. Various forms of indirect ELISA, and especially antigen-coated plate ELISA with polyclonal antisera or crossreactive monoclonal antibodies, can be used to detect potyvirus infections without determining identity. Several monoclonal antibodies to epitopes in the highly conserved trypsin-resistant core of the potyvirus CP are extremely broadly reactive within the genus *Potyvirus* and are used commercially for disease diagnosis. The N-terminal domain of the CP typically contains immunodominant, virus-specific epitopes, which may also differ between isolates of a particular virus. Both the N terminus and C terminus of the CP are exposed on the virion surface; these domains can be removed from virions by digestion with trypsin under mild conditions.

Several sets of primers for nonspecific amplification of various portions of the potyvirus genome have proven useful for obtaining cDNA clones for virus-specific detection or sequence comparisons. Other primer sets for virus- or strain-specific amplification are useful for diagnosis and differentiation or can be combined with restriction mapping to differentiate isolates.

VII. ECONOMIC IMPORTANCE

Some potyviral infections are truly devastating, especially following introduction into regions where the virus was previously not known. Plum pox virus was initially discovered in Eastern Europe and has spread steadily westward throughout Europe and around the Mediterranean. Millions of stone fruit trees have been destroyed in efforts to contain the virus, as orchards can be devastated by infection. The raised "pox" spots on plums render the fruit unsalable even before the virus saps the vitality of the tree, and as much as 90% of the fruit may fall prematurely. No effective natural resistance is known.

The introduction of papaya ringspot virus (PRSV-P) almost eliminated the industry in Hawaii and severely limits the longevity and productivity of trees elsewhere in the world. Disease incidence may approach 100%, and yield reductions averaging 70% have been reported in South America. Pathotype PRSV-W infects cucurbits, but not papaya, and may also cause extremely high yield losses.

Lettuce mosaic virus is seed-borne in lettuce, and foci of seed-borne infection can result in total crop loss as a result of subsequent aphid transmission. The use of certified virus-tested seed lots is, therefore, necessary to minimize disease losses. Tolerance levels of zero in 1000 to zero in 30,000 seed have been established in different countries to control the disease.

Zucchini yellow mosaic virus was first reported in Italy in 1981, causing severe mosaic, yellowing, leaf and fruit distortion. It was soon after reported from countries as distant as Australia, Japan, and the United States. Several disease epidemics have occurred in countries bordering the Mediterranean, while disease losses reported in Australia have approached 100%. Isolates of the virus are extremely variable and have been grouped into at least three pathotypes, according to host responses.

VIII. CONTROL

The most effective control of potyvirus diseases is through genetic resistance in the crop. This is possible, to an extent, in some crops, such as beans and cucurbits, but there are other crop genera in which no effective resistance genes have been identified. Use of virus-free planting material and cultural practices designed to minimize infection rates are employed in many crops, but epidemics may still cause major losses. More recently several genetic engineering approaches are showing some promise.

Host resistance to certain potyviruses available in *Phaseolus vulgaris*, *Pisum sativum*, lettuce, and cucurbit germplasm is pathotype-specific, in some cases, and resistance to different pathotypes may be allelic. Shifts in the virus isolates prevalent in a particular area may negate the benefits of previously effective resistant crop genotypes. For example, new pathotypes of lettuce mosaic virus overcame the resistance genes in tolerant cultivars. Fortunately, testing of lettuce seed lots and rejection of lots not meeting tolerance levels for seed-borne infection is still an effective, although more expensive, means of disease control.

In the case of vegetatively propagated crops such as potato, virus-certified seed crops are frequently grown in areas that have climates and prevailing winds that are unfavorable for the aphid vectors, thus minimizing introduction of virus into the production crop. Vector control within fields is also important in maintaining plant health, although application of insecticides may encourage vector movement, resulting in additional transmission before the insect dies. Resistance of the crop plant to vectors through such adaptations as glandular trichomes that trap aphids is also valuable. Early planting, so the crop is more mature and more resistant to infection when aphid migrations occur, may be beneficial in some situations; in other situations, delaying planting until after vector movements take place may be more effective. In situations where the crop is the primary source of inoculum, enforcing a delay between the harvesting of one crop and planting of the next may be highly effective in limiting infection. In other cases, there are alternate hosts, including weed species, that serve as viral reservoirs, and these may also harbor vectors. In the case of perennial crops like stone fruit trees and plum pox virus, use of virus-free propagation material (both rootstocks and bud-

wood) is important, as is separation of new orchards from infected established plantings. If the virus is not well established within an area, aggressive eradication of infected trees may serve to protect the remaining orchards.

Cross-protection—the deliberate infection of a crop with a mild isolate to prevent the more damaging effects of infection with a severe isolate—has been most effectively applied to perennial crops in which infections by severe isolates are both likely to occur and economically significant. Under such circumstances, the yield loss due to the protective isolate (compared to healthy plants) is significantly less than losses due to severe infection. A mild isolate of PRSV-P obtained by nitrous acid mutagenesis was highly effective in protecting against severe infections in papaya.

Genetically engineered pathogen-derived resistance may confer significant benefits in crops where no effective natural resistance is known. The combination of genetically engineered resistance with natural resistance genes in other crops should result in superior and more durable resistance than either alone. Potyviral gene constructs shown to confer useful levels of resistance include the coat protein (CP), the replicase (NIb), the NIa protease, untranslatable CP RNA, and antisense RNA. The mechanisms of resistance are not clear in all instances, but, in many cases, appear to be RNA-mediated. In general, RNA-mediated mechanisms appear to be highly effective (with resistant lines typically not infected by target isolates), but to be quite strain- or virus-specific. In contrast, CP-mediated resistance appears to be effective against a broader spectrum of isolates and viruses but to confer only partial resistance in many cases (allowing some replication, with reduced symptom severity). Even incomplete resistance will result in a reduction of virus available for transmission to other crops. The HC–Pro genes have been shown to be involved in the synergistic interactions between potyviruses and various other virus groups, and the wild-type gene is, thus, not a good candidate for genetically engineered resistance.

The first virus-resistant transgenic crops to be commercially available were squash, transformed with the CP genes of both zucchini yellow mosaic and watermelon mosaic II potyviruses, and they were highly resistant to either mechanical or aphid transmission of both viruses.

See Also the Following Articles

BEET NECROTIC YELLOW VEIN VIRUS • LUTEOVIRIDAE • PLANT DISEASE RESISTANCE

Bibliography

Barnett, O. W. (ed.) (1992), "Potyvirus Taxonomy. Archives of Virology [Suppl. 5]". Springer-Verlag, Vienna.

Dougherty, W. G., and Carrington, J. C. (1988). Expression and function of potyviral gene products. *Annu. Rev. Phytopathol.* **26,** 123–143.

Edwardson, J. R., and Christie, R. G. (1991). "The Potyvirus Group". Monograph No. 16-I-16.IV. University of Florida, Agricultural Experiment Station, Gainesville.

Murphy, F. A., Fauquet, C. M., Bishop, D. H. L., Ghabrial, S. A., Jarvis, A. W., Martelli, G. P., Mayo, M. A., and Summers, M. D. (1995). "Virus taxonomy: Sixth report of the International Committee on Taxonomy of Viruses. Archives of Virology [Suppl. 10]". Springer-Verlag, Vienna.

Riechmann, J. L., Lain, S., and Garcia, J. A. (1992). Highlights and prospects of potyvirus molecular biology. *J. Gen. Virol.* **73,** 1–16.

Shukla, D. D., Ward, C. W., and Brunt, A. A. (1994). "The Potyviridae." CAB International, Wallingford.

Powdery Mildews

Alison A. Hall, Z. Zhang, and Sarah J. Gurr

University of Oxford

T. L. W. Carver

Institute of Environmental and Grassland Research, UK

GLOSSARY

appressorium A flattened, but swollen, hyphal structure from which the infection peg emerges to penetrate the host.

avirulent Pertaining to a pathogen race which cannot cause disease due to resistance of the potential host plant.

biotroph An organism that requires nutrients from a living host in order to grow and reproduce.

cleistothecium The closed, fruiting body of an Ascomycete fungus which contains sexually produced ascospores.

conidium An asexual fungal spore formed from a generative cell known as a conidiophore.

cuticle A thin layer, principally composed of wax and cutin, external to the wall of plant epidermal cells.

haustorium A structure formed by many obligate fungal pathogens upon penetration of the host cell wall. The haustorium forms within an invagination of the host plasma membrane and absorbs nutrients from the host.

hypersensitive response Localized plant cell death resulting from recognition of an avirulent pathogen race.

papilla Localized deposition of host material lying on the inner side of the plant cell wall subtending the site of fungal attack.

virulent Pertaining to a pathogen race capable of causing disease on a host plant.

POWDERY MILDEWS are among the most commonplace, widespread, and recongnizable of all plant diseases. They are aptly named, for infection produces a white lawn of fungal mycelium that covers the plant surface, while chains of aerial conidia give the characteristic powdery appearance.

Powdery mildews can infect a wide range of hosts, including over 9000 dicotyledonous and over 650 monocotyledonous plant species. The cereals, particularly wheat and barley, are among the most important agricultural crops that suffer from powdery mildew disease. Indeed, in temperate regions, barley powdery mildew can cause yield losses of some 5–20% and occasionally as much as 40%. Other plants that succumb to severe powdery mildew infections include cucurbits (cucumbers, squashes, and cantaloupe melons), legumes, strawberries, grapes, roses, apples, and oak trees. Curiously, powdery mildews do not infect maize, carrots, or celery. Taken collectively, powdery mildews cause greater losses in terms of crop yield than any other single type of plant disease.

The powdery mildew diseases are caused by many species of Ascomycete fungi, which are grouped into seven main genera. They are true obligate biotrophs, which means that growth and reproduction of these fungi depends on their parasitizing living host plants. The lack of methodologies for their axenic culture has, to date, precluded DNA-mediated transformation studies and mutational analysis. Nonetheless, significant progress has been made over the past decade toward understanding powdery mildew–host interactions at both the cellular and the molecular level.

As yet, the most intensively studied powdery mildew–host systems have been those which are important as crop diseases. *Erysiphe graminis* DC. [syn. *Blumeria graminis* (DC.) Speer] f.sp. *hordei* Marchal infects barley and is one such powdery mildew fungus. Indeed, *E. graminis* will be focused on here, reflecting the relatively detailed understanding of this system. However, more recently the interaction of the powdery mildew *Erysiphe cruciferarum* Opiz and Junell, with its host *Arabidopsis thaliana* L., has come to the forefront as a model system. While a vast amount of genetic information is now available for *A. thaliana* following the *Arabidopsis* genome sequencing initiative, far less is known about its fungal pathogen.

I. LIFE CYCLE

The powdery mildew fungi can reproduce both sexually and asexually. Sexual reproduction is common toward the end of the host's growing season or under unfavorable environmental conditions. However, by far the more significant mode of reproduction, in terms of epidemic disease, is the formation of prolific numbers of asexual spores or conidia. The conidia are ovoid and are produced in a process termed conidiogenesis, either singly or in long chains, from a specialized generative cell known as the conidiophore (Fig. 1). They are wind-dispersed

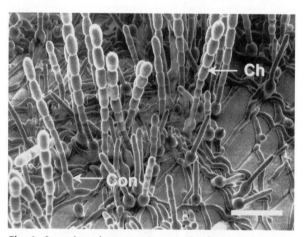

Fig. 1. Scanning electron micrograph of a mature, sporulating colony of *Erysiphe graminis* f.sp *hordei* (the barley powdery mildew fungus) growing on a susceptible barely leaf. Bulbous conidiophores (Con), specialized generative cells, protrude from the fungal mycelium that forms a lawn over the leaf surface. Chains (Ch) of asexual spores, conidia, formed by the conidiophores, give the colony its characteristic powdery appearance. White bar represents 50 μm.

and each conidium has the potential to initate a new fungal colony. Indeed, within 4–6 days of a conidium landing on a susceptible leaf surface, a fungal colony can mature and itself begin to sporulate. Thus, given the presence of an abundance of the host species, as is the case with many crop plants, the asexual mode of reproduction enables very rapid spread of infection.

The fungal mycelium is epiphytic in that it grows over the host plant surface. In most powdery mildews, specialized feeding bodies, or haustoria, are the only structures that are formed within the host. Haustoria are unique to biotrophic plant pathogens and, in the case of the powdery mildews, the haustoria range from being simple lobed structures to being intricately digitate (Fig, 2). They do not directly contact the host cytoplasm, but instead form within a sealed invagination of the host plasma membrane, which becomes modified during infection and is known as the haustorial sac or extrahaustorial membrane.

II. ESTABLISHMENT OF INFECTION

Following contact with a host plant, powdery mildew conidia undergo a series of developmental steps, which cluminate in establishment of biotrophy through the formation of a haustorium. Production of the haustorium is critical to successful infection as it is the means of siphoning nutrients from the host to support growth of the fungus.

A. Conidial Germination and Appressorial Differentiation

In *E. graminis* f. sp *hordei*, germination typically occurs within 2 hr and results in the production a short, aseptate germ tube, the primary germ tube (PGT). Soon afterwards, a second germ tube emerges. On the host, this second-formed germ tube elongates and then differentiates a terminal appressorium and is, therefore, known as the appressorial germ tube (AGT). This developmental sequence is illustrated in Fig. 3, A–B. The sequential production of the PGT and AGT is unique to *E. graminis*. All other species and genera produce a single germ tube, which itself

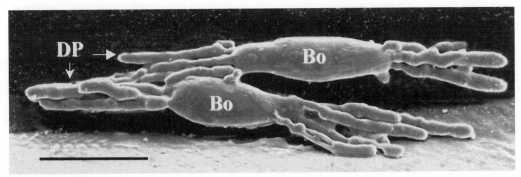

Fig. 2. Scanning electron micrograph of haustoria (feeding structures) of *Erysiphe graminis* f.sp *hordei* (the barley powdery mildew fungus) growing inside a barley leaf epidermal cell. The top of the cell has been cut away to reveal the two haustoria. Each haustorium has an ovoid central body (Bo) with several digitate processes (DP) projecting from each end. The connections (haustorial neck) to the surface mycelium have been lost during preparation. Black bar represents 30 μm.

elongates and forms the apical appressorium, as illustrated for the pea powdery mildew fungus, *Erysiphe pisi* DC. (Fig. 3, D–E). In *E. graminis*, elongation of the AGT (to ca. 40 μm) and differentiation of the swollen, hooked appressorium is completed by around 8–10 hr after spore deposition. In *E. pisi,* the single germ tube elongates and differentiates the clubbed appressorium by around 4–6 hr. The appressorium is separated from the body of the conidium by a septum. During the formation of the appressorium, the conidial nucleus divides and one of the daughter nuclei migrates with most of the cytoplasm into the expanding AGT, while the second nucleus remains in the body of the conidium surrounded by the residual cytoplasm.

As pointed out earlier, *E. graminis* differs from other powdery mildew fungi by producing two germ tubes, a PGT and an AGT. The evolutionary benefits to *E. graminis* of the PGT are unclear. There is evidence that the PGT may play important roles in adhesion, in signal perception, and in the uptake of water from the host. However, it is uncertain the extent to which these roles are unique to the PGT, as the single germ tube of other powdery mildew fungi may have similar functions. Indeed, it could be argued that the PGT is of some disadvantage to *E. graminis*, as this germ tube elicits response in the host epidermal cell to which it adheres. This early stimulation of the plant cell reponse (*ca.* 2 hr after inoculation) induces increased cellular resistance to subsequent attack by the appressorium (*ca.* 12 hr,

after inoculation). For this reason, the two germ tubes (PGT and AGT) usually emerge from different ends of the conidium, increasing the likelihood that the different germ tubes will contact different plant cells.

B. Host Penetration

In *E. graminis*, penetration occurs via an infection peg formed from the underside of the hooked tip of the appressorium, which is flattened against the host cuticle. Penetration is believed to be mediated, in part, by the build-up of turgor pressure within the AGT and, in part, by the localized action of enzymes, such as cutinase and cellulase, which degrade the host surface. The infection peg penetrates through the host cuticle and cell wall, but does not break through the host plasma membrane. Instead, the plasma membrane is invaginated, to form the extra-haustorial membrane that separates fungus from host cytoplasm. This may be crucial to development of a biotrophic relationship, allowing the plant cell to survive invasion and the parasite to feed on the living host cell. Following establishment of the haustorium, secondary hyphae ramify across the host surface, appressoria differentiate at intervals along the mycelium, further haustoria develop, and conidiophores are formed. Within its lifetime, a single barley mildew colony may produce up to 5000 haustoria and a similar number of conidiophores, with each conidiophore producing around 40 conidia. In this way, a

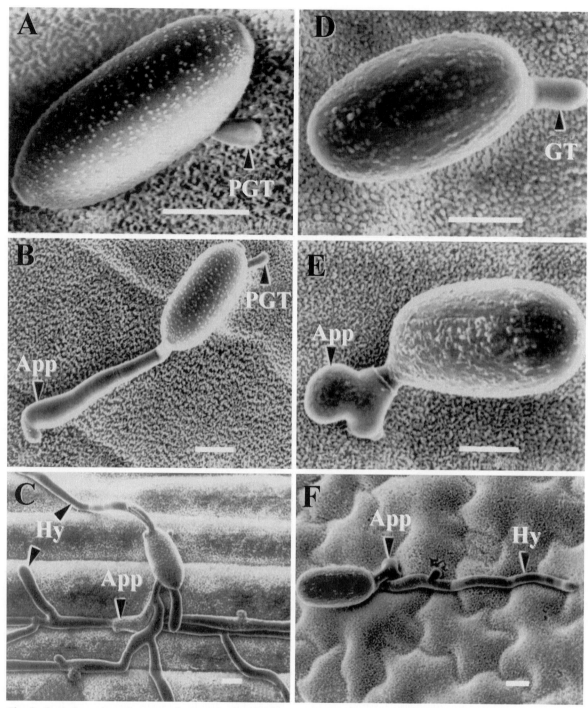

Fig. 3. A–C. Scanning electron micrograph showing the early development of *Erysiphe graminis* f.sp hordei (the barley powdery mildew fungus) and D–F, of *Erysiphe pisi* (the pea powdery mildew fungus) on barley and pea leaves, respectively. In all cases, white bar represents 10 μm. (A) *E. graminis* 0.5–2 hr after inoculation. The short, aseptate primary germ tube (PGT) has emerged from the conidium and made contact with the barley leaf surface, which is covered by a layer of crystalline wax plates. (B) *E. graminis* 9–10 hr after inoculation. The appressorial germ tube has emerged from the opposite end of the conidium, has elongated and swollen towards the tip, and a hooked, apical appressorium (App) with a single lobe has differentiated. (C) *E. graminis* 48 hr after inoculation.

single colony can produce in the region of 200,000 conidia, each capable of forming a new colony.

C. Regulation of Developmental Processes Following Germination

On host leaves, the germination and development of conidia is highly ordered and most viable conidia form functional appressoria. While germination can occur on many artificial surfaces, development is frequently impaired and, often, AGT fail to form or develop abnormally. This has led to the hypothesis that multiple leaf-derived signals are perceived by the conidia and regulate development. These signals are thought to include physical features of the leaf surface, such as the hydrophobicity of leaf waxes, as well as products of cuticle degradation, such as cutin and cellulose breakdown products. Recently, cAMP has been shown to play a role in appressorial development by *E. graminis* and, thus, is a potential intracellular mediator of the external leaf-derived signals.

III. THE HOST RESPONSE TO POWDERY MILDEW DEVELOPMENT

Powdery mildews are highly specialized parasites. At every step in their developmental program, they must successfully evade antifungal defenses. Success in this leads to the establishment of disease (compatibility), whereas failure results in host resistance to infection (incompatibility).

In general, plant pathogenic fungi, including the powdery mildews, are incapable of infecting most plant species, due to nonhost resistance. However, some plant genotypes that are potential hosts to a particular powdery mildew carry specific resistance genes that prevent infection by particular races of the fungus. In this case, the fungal races are said to be avirulent. Often, attack by an avirulent race causes hypersensitive death of the attacked plant epidermal cell. In common with other plant/pathogen interactions, for cereal powdery mildews, the expression of race-specific resistance due to interaction between plant resistance genes and fungal avirulence genes has been shown to follow the gene-for-gene hypothesis proposed by Flor to explain the flax/flax rust interaction. This monogenic, race-specific resistance tends to be ephemeral, as natural selection favors the rapid evolution of virulent races, especially where vast acreage of a crop carries a single resistance gene. An exception to this rule is provided by the *mlo* gene of barley. The stable resistance so far provided by alleles of this gene is discussed later. In contrast to the apparent immunity provided by many monogenic resistances, plants often possess a form of resistance that limits the severity of disease caused by virulent fungal races without totally preventing symptom development or precluding reproduction of the pathogen. This quantitative form of resistance is equally effective against all pathogen races. It is commonly controlled by many genes, each with minor effect, that act synergistically to confer the plant's defenses against attack. Such resistance is independent of fungal race (race nonspecific) and is durable.

A. Host Response to *E. graminis* in a Compatible Interaction

Some aspects of the plant's defense responses may be observed in the host even in a compatible interaction. For example, in the *E. graminis*-barley interac-

---

Penetration from the appressorium (App) has succeeded in forming a haustorium, which is hidden within the underlying leaf epidermal cell. Nutrients extracted by the haustorium from the leaf cell have supported growth of the mycelial hyphae (Hy), which have started to grow across the leaf surface. (D) *E. pisi* 1–2 hr after inoculation. The single germ tube (GT) has emerged from the conidium and made contact with the pea leaf surface. (E) *E. pisi* 6–8 hr after inoculation. The single germ tube has elongated and swollen to form a clubbed appressorium (App) with multiple lobes. This appressorium shows two lobes, but up to five can be formed. Penetration will be attempted from beneath only one of these lobes. (F) *Erysiphe pisi* 36 hr after inoculation. Penetration from the appressorium (App) has succeeded in forming a haustorium, which is hidden within the underlying leaf epidermal cell. Nutrients extracted by the haustorium from the leaf cell have supported growth of the mycelial hyphae (Hy), which have started to grow across the leaf surface. Reproduced from Carver, Ingerson, and Thomas (1996).

tion, the epidermal cells respond to the penetration attempts of both the PGT and AGT by the rapid formation of a papilla on the inner surface of the plant cell wall underlying the fungal germ tube contact site. The papilla is a localized reinforcement of the cell wall. It is a chemically complex structure, containing elemental constituents, autofluorogenic phenolic compounds and proteins, infused in a callose matrix. As early as 1900, Smith considered that papillae may provide a barrier to fungal penetration, but the contribution of different papilla constituents, and whether they provide a physical or chemical defense to penetration, remains unresolved. In a compatible interaction, the infection peg is able to breach the papilla. The means by which the fungus is able to either avoid recognition by its host and/or overcome the defense response triggered in a compatible interaction remains obscure.

B. The Host Response to *E. graminis* in an Incompatible Interaction

1. Race-Specific Resistance to E. graminis

Monogenic, race-specific resistance to *E. graminis* acts according to Flors' gene-for-gene hypothesis. Thus, race-specific resistance in barley is effective only against fungal isolates carrying complementary avirulence genes. Certain race-specific resistance genes, such as those at the *Mla, Mlc,* and *Mlh* loci in barley, appear to contribute to penetration resistance as well as conditioning hypersensitivity, while others, such as *Mlat, Mlk, MlLa,* and *Mlp,* seem solely to condition hypersensitivity. Hypersensitive cell death is often localized to the plant cell under attack or to cells immediately adjacent or subtending attacked cells. However, the time at which cell death occurs varies and it may take place subsequent to the apparent arrest of fungal development. Thus, it is uncertain in these cases whether cell death is the cause or consequence of arrested infection. The nature of the avirulence factors and the mode of their recognition by the host are, as yet, unknown.

2. mlo-Mediated Resistance in Barley

In barley recessive, loss-of-function alleles at the *Mlo* locus mediate resistance that is effective against most races of *E. graminis.* Barley lines that lack a functional *Mlo* protein respond to *E. graminis* infection by the generation of papillae that are both larger and more rapidly formed than those in nonmutant lines. Both the size and speed of papilla formation help prevent penetration, thereby halting infection. Recently, the *Mlo* gene has been isolated and found to encode a membrane-spanning protein. Expressed sequence tags with high homology to the barley *Mlo* gene have been identified in several plant species, such as rice, which is a nonhost for powdery mildew, as well as *Arabidopsis thaliana,* which is host to *E. cruciferarum.* However, their relevance to powdery mildew development in other host species has yet to be assessed.

C. Nonhost Resistance

Perhaps one of the most remarkable features of the powdery mildew–plant interaction is the narrow host range of many powdery mildew species. For example, while the Gramineae are infected only by *E. graminis,* specialized forms of the fungus (*formae speciales;* f.sp.) have evolved pathogenic capability that is restricted to different plant species. Thus, for example, *E. graminis f.sp. hordei* can infect only barley and *E. graminis* f.sp. *tritici* can infect only wheat. Not all species are so specialized: the less discriminate, polyphagous mildew species can infect several hosts, for example, *Erysiphe polygoni* has around 350 known host species. This poses an intriguing question: what are the determinants of host specificity? Is it determined by the fungus, the host, or both? It is clear that successful infection is dependent upon many factors. The fungus has to perceive the right signals for differentiation, overcome both preformed barriers and active defense responses, and be able to divert host metabolism to supply its own needs. This illustrates the close coevolution of host and parasite.

IV. INFLUENCE OF ENVIRONMENT ON POWDERY MILDEW INFECTION

Although powdery mildew diseases occur from the tropics to polar regions, they are found mainly in the temperate areas. Powdery mildew infection is affected by climatic conditions, temperature, moisture, and light, as well as by the nutritional status of the host plant, host plant age, and phenology. Both the direct and the interacting influences of these factors on disease development are complex. For example, while high humidity appears to favor mildew development, germination is inhibited by water. Moreover, environmental conditions also affect host susceptibility. For example, *mlo*-resistant barley lines show a marked increase in powdery mildew susceptibility following relief from drought. Thus, there appears to be a complex interplay between environmental factors that influence both sides of the host–mildew interaction and the physiological state of the host, and, probably, the pathogen.

V. PERENNATION

In the absence of living host tissue, powdery mildew fungi may enter a dormant phase, termed perennation. Frequently, this involves formation of fruiting structure, the cleistothecium, containing sexually produced ascospores. The cleistothecium has a characteristic spherical black appearance and the details of its structure are used in the taxonomic classification of powdery mildew fungi. In addition to its importance in perennation, the sexual phase also enables the generation of variation, including new virulence combinations. Some powdery mildews have been reported to overwinter as mycelium. Infection onto successive hosts can avoid the need for a sexual phase, particularly for mildew species with a wide host range or for species that infect crops that are planted in rapid succession.

VI. CONTROL

In addition to crop management practices, powdery mildews have traditionally been controlled by host breeding for specific resistance and the use of chemical control. However, both of these latter methods, which will be discussed in further detail, have generally shown limited durability, due to the relative ease by which they are overcome by the pathogen. Thus, the challenge is to identify more durable forms of host resistance. This has lead to the investigation of novel control methods, which include the use of genetically complex partial resistance, "immunization" of the plant to heighten its defense responses prior to attack, and the exploitation of natural parasites of powdery mildews as a form of biological control.

A. Host Resistance in Control

Monogenic race-specific resistance has been exploited to breed crop lines carrying specific resistance genes. Such resistance is useful and easy to exploit by plant breeding, but it is often short-term, being overcome by the emergence of new virulent races of the pathogen. More durable control may be afforded by the use of quantitative, polygenic resistance but, currently, the complexity of the genetic control of this form of resistance makes its exploitation problematic. In cereals, some attempt has been made to employ cultivar mixtures or multilines that differ in pathogen resistance specificities. This diversity provides a reduced density of susceptible plants, with resistant plants acting as a barrier between susceptible components of the crop. However, the effectiveness of this method for yield preservation is questionable.

In barley, *mlo*-mediated resistance has been used widely over the past decades to control powdery mildew infection, however, the durability of this form of resistance is uncertain. Fungal isolates that show increased virulence on *mlo*-lines have been identified through both natural and artificial selection. However, as yet pathogen virulence to *mlo*-barley has not become a problem in areas where this gene has been deployed in the field.

B. Chemical Control

Over the past century, chemical fungicides have been the major method of control of powdery mildew disease. Indeed, in 1885, grapevine powdery mildew was one of the first fungal plant diseases to be controlled by the copper-based "Bordeaux mixture". Subsequently, this century, sulfur was used as both a spray and a dust to control powdery mildews, along with chemical fungicide sprays, such as Dinocap and Benomyl. These compounds, although effective, had many disadvantages: principally, the rapid growth of new and unprotected plant tissue or frequent rainfall necessitated reapplication of these fungicides. The first major systemic fungicides to be used for control of powdery mildew included the 2-amino-pyrimidine, ethirimol, as a seed treatment, and tridimenol and trideomorph, as foliar sprays. However, although the systemic action of these fungicides meant they were more effective at controlling disease, they were also found to have a single target site in the pathogen. This resulted in the rapid emergence of resistant races. Recently, a new protectant fungicide, quinoxyfen, has been shown to be particularly potent against cereal mildews, grape, curbits, and tomato powdery mildew.

In addition to the development of new fungicidal compounds, there has been considerable interest over the past few years in the induction of host resistance using plant defense activators. Such defense activators have been found to induce the accumulation of defense-related gene transcripts *in planta* and, thus, enhance the plants' resistance to disease, rather than acting directly on the pathogen. Of these activators, benzo-(1,2,3)-thiadiazole-7-carbothioic acid S-methyl ester (BTH), marketed as Bion, and 2,6-dichloroisonicotinic acid (DCINA) are functional analogs of the naturally occurring compound salicyclic acid.

C. Concluding Remarks

The importance of powdery mildew as a disease of agricultural, horticultural, and ornamental plants ensures a continuing interest in both its consequences and its control. The rapidity with which novel pathotypes arise and are selected has overcome simply controlled genetic resistance in the host and has lead to insensitivity to many fungicides. Novel control methods, including exploitation of partial resistance, the use of chemical inducers of host defenses and biological control, offer the promise of improved durability. In addition, the advent of genetic engineering has created enormous scope for the introduction of novel disease resistance with both a greater degree of precision and greater speed than that available by conventional breeding. The adoption of transgenic methods, however, depends on the resolution of public concerns regarding their acceptability. It is clear improved control methods can arise solely from a greater understanding of the interactions of among the powdery mildew fungi, their hosts, and the environment.

See Also the Following Articles

Downy Mildews • Pesticide Biodegradation • Rust Fungi • Smuts, Bunts, and Ergot

Bibliography

Agrios, G. N. (1997). "Plant Pathology" (4th ed.). Academic Press, San Diego, CA.

Aist, J. R., and Bushnell, W. R. (1991). Invasion of plants by powdery mildew fungi and cellular mechanisms of resistance. *In* "The Fungal Spore and Disease Initiation in Plants and Animals" G. T. Cole and H. C. Hoch, (eds.), pp. 321–345. Plenum Press, New York.

Braun, U. (1996). "The Powdery Mildews (Erysiphales) of Europe." Verlag, Stuttgart & New York.

Carver, T. L. W., Ingerson, S. M., and Thomas, B. J. (1996). Influences of host surface features on development of *Erysiphe graminis* and *Erysiphe pisi. In* "Plant Cuticles" (G. Kerstein, ed.), pp. 255–266. BIOS Scientific Publishers Ltd., Oxford UK.

Prions

Christine Musahl and Adriano Aguzzi

University of Zurich

I. Molecular Biology of Prions
II. The "Protein Only" Hypothesis
III. Yeast Prions
IV. Prion Diseases
V. Peripheral Pathogenesis

GLOSSARY

amyloid plaques Characteristic aggregated fibrils with high beta-sheet content, stainable by a dye and pH indicator called Congo red (CR). CR intercalates with a universal structure shared by all amyloids, which consists of antiparallel beta sheet extensions arranged in a quasi-crystalline fashion.

knockout mice Genetically engineered mice lacking the ability to express a specific protein because the gene of interest was ablated.

PrP^C Cellular, normal host prion protein.

PrP^Sc Scrapie-associated, pathological, potentially "infectious" prion protein.

PrP protein; prion Cellular protein of unknown function, primarily expressed on cells of the central and peripheral nervous system, as well as on lymphocytes. PrP is expressed in at least two different isoforms, one of which is thought to be the infectious agent of transmissible spongiform encephalopathies (TSEs).

reactive gliosis Strong proliferation of reactive astrocytes as seen by staining with antibodies against glial fibrillary acidic protein (GFAP). While not specific for TSEs, gliosis is extremely prominent in CJD and scrapie.

spongiform changes Highly characteristic hallmark of most TSEs. Formation of vacuoles (microscopic holes within cells) in the gray matter of the brain, giving it a spongelike appearance.

transgenic mice Genetically modified mouse strains, expressing at least one foreign or altered gene.

transmissible spongiform encephalopathies (TSEs) Creutzfeldt–Jakob disease (CJD), Gerstmann–Sträussler–Scheinker syndrome (GSS), fatal familial insomnia (FFI), bovine spongiform encephalopathy (BSE), and scrapie are the most common transmissible neurodegenerative diseases.

THE PRION is defined as the infectious agent that causes transmissible spongiform encephalopathies (TSEs). It has been the subject of exciting discoveries and passionate controversies over the last 40 years. A large body of evidence indicates that prions do not contain informational nucleic acids. In 1996, S. B. Prusiner was awarded the Nobel Prize for Medicine for the hypothesis that prions consist of a modified form of the normal cellular protein called PrP^C.

Although the human forms of prion diseases are rare, the recent epidemic of bovine spongiform encephalopathy (BSE) in the UK has most dramatically raised the issue of transmissibility of these diseases from affected animals to humans.

I. MOLECULAR BIOLOGY OF PRIONS

A. The *Prnp* Gene

The gene encoding the cellular isoform of the prion protein (PrP^C, or simply PrP) is located on the short arm of chromosome 20 in humans and on chromosome 2 in the mouse. It was termed *Prnp* (in the mouse) or *PRNP* (in humans). PrP^C is encoded by a single open reading frame in all known mammalian and avian *Prnp* genes. The *Prnp* genes of mouse, sheep, and cattle contain 3 exons, while those of hamsters and humans span over 2 exons. PrP expression seems to be controlled by the SP1 transcription factor which binds to G+C rich nonamer-regions upstream of exon 1. *Prnp* is present in the genome as a single copy gene and does not show any homologies to other known cellular genes. *Prnp* mRNA is constitutively expressed in the adult organism and developmentally regulated during mouse embryogenesis.

B. The Cellular Prion Protein, PrP^C

The protein encoded by *Prnp* (Fig. 1) is anchored in the cellular membrane via a glycosyl–phosphatidyl–inositol residue (GPI-anchor) and is expressed at relatively high levels on the surface of neurons as well as of astrocytes, but also in kidney, lung, heart, muscle, spleen, and cells of the lymphoreticular system. PrP^C contains two N-linked complex-type oligosaccharides at positions 181 and 197. Accordingly, western blot analysis of PrP reveals three major bands, reflecting PrP that has two, one, or no glycosylation signals occupied. Upon translocation of PrP into the endoplasmic reticulum, an NH$_2$-

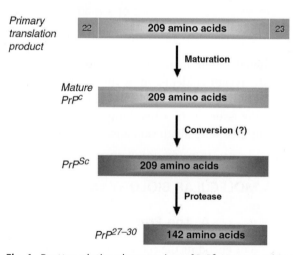

Fig. 1. Posttranslational processing of PrP^C. Upon reaching its destination on the cell surface, an amino-terminal secretory signal peptide of 22 amino acids is cleaved from the 254 amino acid PrP^C precursor protein. 23 carboxy-terminal residues are also processed during addition of the GPI-anchor to a serine residue at position 231. Upon completion of these modifications, mature PrP^C contains 209 amino acids.

terminal secretory signal peptide of 22 amino acids is cleaved from PrP, and 23 COOH-terminal residues are processed during addition of the GPI-anchor to the amino acid residue serine 231. Mature mouse PrP contains 209 amino acids (Fig. 1).

PrP^{Sc}, which accumulates in TSE-affected organisms is a modified form of PrP^C. PrP^C and PrP^{Sc} have an identical amino-acid sequence and share the same posttranslational modifications, as assessed by currently available methodology, but differ in their secondary and (presumably) tertiary structure.

The physiological isoform PrP^C is protease-sensitive and, thus, sometimes designated as PrP^{sen}, while the pathological isoform PrP^{Sc} is partially protease-resistant and, thus, also called PrP^{res}. The protease-resistant core of PrP^{Sc}, designated PrP 27–30, according to its apparent molecular weight on SDS-PAGE, provides a specific and reliable molecular marker for the presence of the infectious agent.

C. Hypotheses on the Function of PrP

The question of the physiological function of PrP has not been clarified so far. It is difficult to imagine that PrP^C, a protein highly conserved throughout evolution, whose gene is unrelated to any other known sequence, exists only to foster susceptibility to prion diseases. On the other hand, PrP^C-deficient mice suffer from very few defects: This observation speaks against an essential nonredundant function of PrP.

One intriguing thought is that the function of PrP^C might be related to its convertibility into PrP^{Sc}. Some structural equivalents of PrP^{Sc} may exist in liquid phase in equilibrium with PrP^C, in a noninfectious and nonneurotoxic form. Nongenetic propagation of structural information might be advantageous for a population, since it is much faster than transmission of genetic traits and can spread horizontally within the individuals belonging to one single generation. An analog function has been discussed for the so-called yeast prions (see Section III).

1. Prions Are Copper-Binding Proteins

All prion proteins investigated so far contain a stretch of 8 amino acids, which is repeated 4–5 times. This region is termed *octarepeat*. The presence of additional repeats is associated with hereditary prion diseases, like familial Creutzfeldt–Jakob disease (fCJD) and Gerstmann–Sträussler–Scheinker syndrome (GSS). Interestingly, the shortest known prion protein gene allele, which occurs in goat, has only three octapeptide repeats and is nonpathogenic.

Mammalian and chicken prion octarepeats bind copper tightly and specifically. Homology to known copper-binding proteins or domains may suggest a 3-D structure and enzymatic function for the octapeptide repeat region. Copper-binding proteins are often involved in oxidation, storage, and transport processes. A copper atom can be complexed with 4 (approximately planar) histidine residues. The highly conserved histidines and tetra-repeats that are found in all prion proteins suggest that binding of copper may be essential for the function of PrP^C.

D. Proteins Interacting with PrP

Another crucial area which has yielded little information so far is the search for proteins interacting with PrP^C and PrP^{Sc}. Research activity has concentrated on two aspects: cellular receptors which may mediate the normal function of PrP^C and chaperonelike molecules catalyzing the conversion of PrP^C into PrP^{Sc}. That PrP^{Sc} converts PrP^C directly into PrP^{Sc} *in vivo,* without any aid from ancillary molecules, is unlikely, since there is no demonstrable physical affinity between the two isoforms.

Genetic evidence clearly speaks in favor of a protein that may bind to the central region of both PrP^C and PrP^{Sc} and facilitate the conversion process. Studies on the transmission of human prions to transgenic mice suggest that another molecule, provisionally designated protein X, participates in the formation of nascent PrP^{Sc}. Substitution of a basic residue at positions 167, 171, or 218 prevent PrP^{Sc} formation: at a mechanistic level, these mutant PrP molecules appear to act as "dominant negatives" by binding protein X and rendering it unavailable for prion propagation. Many proteins, like the yeast chaperone Hsp 104 or molecular chaperones of the Hsp60 family, have been suggested to interact with PrP.

A technique known as complementary hydropathy was used to generate antibodies against a peptide that, theoretically, would mimic the part of the receptor responsible for prion binding. These antibodies recognize a hitherto uncharacterized 66 kDa protein on the surface of mouse neurons. In addition, the antibodies and the peptide itself can prevent a neurotoxic peptide derived from human PrP from killing neurons in culture.

Other researchers applied the yeast two-hybrid methodology to demonstrate that PrP^C interacts with a 37 kDa laminin receptor precursor. The amount of this protein correlates with abnormal PrP accumulation in the brain, spleen, and pancreas of scrapie-infected mice and hamsters. Further, Kurschner and Morgan reported in 1995 that a truncated fragment of PrP binds to the anti-apoptotic protein Bcl-2. Since Bcl-2 related molecules, such as Bax, may cap homodimers by interacting with their extramembraneous region, it is possible that PrP disrupts chains of Bcl-2 molecules at the homomeric association site in the transmembrane region.

Finally, the group of S. B. Prusiner created a probe by fusing PrP with alkaline phosphatase (AP). PrP–AP was used to screen a lambdagt 11 mouse brain cDNA library. By this method, fragments of the Nrf2 (NF–E2 related factor 2) transcription factor and Aplp1 (amyloid precursor-like protein 1) were identified as partners of PrP. The observation that PrP binds to a member of the APP (amyloid precursor protein) gene family is intriguing, in light of possible analogies of transmissible spongiform encephalopathies to Alzheimer's disease.

E. Structural Properties of the Prion Proteins

The pathological isoform PrP^{Sc} is invariably associated with transmissible spongiform encephalopathies (TSE). Extensive biochemical characterization failed to identify differences between the primary structures of PrP^C and PrP^{Sc}. By contrast, physical measurements have demonstrated dramatic conformational differences between the two PrP forms. Fourier transform infrared and circular dichroism spectroscopy indicate that the α-helical content of PrP^C is 40%, with little or no β sheet. In contrast, PrP 27–30 contains 50% β sheet and only 20% α helix. The solution structure of a bacterially expressed protein corresponding to the full-length mouse PrP^C has allowed a direct determination of its secondary and tertiary structure (Fig. 2). The structural differences between PrP^C and PrP^{Sc} are thought to be propagated by a sort of contagious interaction between the two isoforms of the protein. The high β sheet content of PrP^{Sc} is responsible for its amyloidogenic properties, resulting in the characteristic amyloid deposits in the brain, similar to the amyloids found in other neurodegenerative diseases, such as Alzheimer's disease.

II. THE "PROTEIN ONLY" HYPOTHESIS

Prions have properties which are fundamentally different from those of all known viruses or viroids. The so-called protein-only hypothesis still offers the best intellectual framework for accommodating many of the experimental findings on prions and prion diseases, despite its shortcomings in explaining certain phenomena, such as different strains of the infectious agent.

First formulated by Griffith in 1967, the "protein only" hypothesis states that the infectious agent causing transmissible spongiform encephalopathies (kuru, Creutzfeldt–Jakob disease, BSE, sheep scrapie, and other related ailments) consists of the modified form of a normal cellular protein (Fig. 3). In 1982, Stanley Prusiner proposed that prions would not contain nucleic acids and are identical with PrP^{Sc}, which is the main proteinaceous constituent of the amyloid plaques found in many TSEs. Enrichment for infectivity from scrapie-infected hamster brains by Prusiner and colleagues led to the identification of PrP 27–30 (the protease resistant core of PrP^{Sc}, named according to its molecular weight of 27–30 kDa) copurifying with infectivity. In a collaboration with Prusiner, Leroy Hood and Charles Weissmann were then able to determine parts of the sequence of the hamster prion protein in 1984. Oesch and colleagues, in 1985, cloned the cognate gene, which

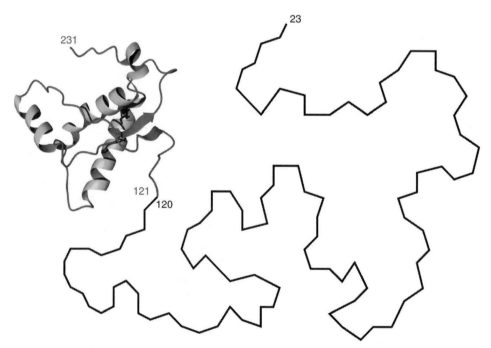

Fig. 2. Three-dimensional structure of recombinant prion protein produced in *E. coli,* which corresponds to the residues 121–231 of mouse PrP^C. The structure was obtained by nuclear magnetic resonance (NMR) analysis. Analysis of a full-length recombinant mouse PrP by the groups of R. Glockshuber and K. Wüthrich shows that the N terminus (amino acids 23–120, not shown) is extremely flexible, unstructured, and devoid of α-helix or β-sheet structure.

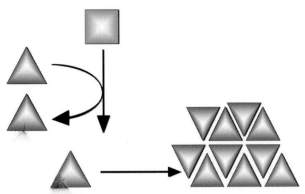

Fig. 3. The Protein-only Hypothesis. According to this theory, each PrP^C molecule (rectangle) is converted into the PrP^{Sc} conformation upon interaction with PrP^{Sc} (triangle). Newly converted PrP^{Sc} is then able to convert further molecules and tends to form insoluble aggregates.

turned out to be encoded by the host genome. The normal cellular protein encoded by the *Prnp* gene was termed PrP^C (c_cellular). These substantial discoveries allowed the prion hypothesis to be laid on firm scientific ground.

PrP^{Sc}, the partially protease-resistant isoform of PrP^C, can be detected by digesting protein extracts derived from infected brains with proteinase K and subjecting them to Western blot analysis. Following protease treatment, PrP^{Sc} looses 67 N-terminal amino acids, giving rise to PrP 27–30, which tends to aggregate into insoluble polymers. PrP^{Sc} was shown to accumulate intracellularly in cytoplasmic organelles and in the extracellular space in form of amyloid plaques. The isolation of fractions enriched for infectious materials from the brain of scrapie-sick animals leads to a enrichment of PrP^{Sc}. Conversely, the isolation of PrP^{Sc} using affinity chromatography leads to enrichment of the infectivity. A further landmark speaking in favor of the protein only hypothesis is the demonstration by Hsiao and colleagues of linkage between familial forms of the Gerstmann–Sträussler–Scheinker-disease (GSS) and mutations in the prion gene.

The molar ratio between the number of infectious particles and the number of PrP^{Sc} molecules in infectious brain extracts is usually in excess of $1:10^5$. This fact raises great analytical problems: it might be very difficult to purify the infectious agent if the latter consisted of subspecies of PrP^{Sc} or of a molecular modification of the PrP^C molecule which differs from what was operationally defined as PrP^{Sc}.

A. Mice Devoid of PrP

The protein only hypothesis proposes that PrP^{Sc}, when introduced into a normal host, causes the conversion of PrP^C into PrP^{Sc}. Therefore, an animal devoid of PrP^C should be resistant to prion diseases.

Homozygous *Prnp$^{o/o}$* ("PrP knockout") mice, first generated by the group of C. Weissmann in 1992, are viable and remain free of scrapie for at least 2 yrs after inoculation with prions. Also, there is no multiplication of prions in the *Prnp$^{o/o}$* animals. Accordingly, heterozygous *Prnp$^{o/+}$* mice, which express PrP^C at about half the normal level, show delayed development of disease when exposed to prions. Introduction of murine PrP transgenes into *Prnp$^{o/o}$* mice restores susceptibility to mouse scrapie.

A synaptic phenomenon called *long term potentiation* (LTP), which is presumed to be important for short-term memory and learning, is apparently impaired in homozygous *Prnp$^{o/o}$* mice. In addition, *Prnp$^{o/o}$* mice exhibit aberrant sleep patterns and degeneration of cerebellar neurons in age, but it is unclear whether these observations are causally related to PrP^C. A different strain of *Prnp* knockout-mice (generated independently) was reported to show ataxia and Purkinje cell degeneration developing after 70 weeks of age. However, since no such phenotype was observed in other PrP-knockout mice generated by Büeler *et al.*, it remains unclear whether this observation is related to a function of PrP or, rather, absence of neighboring regulatory elements within noncoding portions of the *Prnp* locus.

B. The Virus and Virino Hypotheses

The virino hypothesis has been proposed as an alternative to the protein-only hypothesis. It states that the infectious agent consists of viral nucleic acids complexed to host-derived PrP. This theory claims that PrP can be recruited by the viral nucleic acids as a sort of coat. The existence of many different strains of prions has been put forward as an argument in favor of this hypothesis. These strains retain their phenotypic characteristics even if they are propagated serially in one and the same inbred strain of susceptible experimental animals (mice, hamsters, or even minks).

On the other hand, no proof for the existence of a viral nucleic acid in infectious extracts was brought about yet. Finally, it has been convincingly demonstrated by Riesner and colleagues in 1991 that nucleic acid longer than 50–100 nucleotides cannot possibly be important for the infectivity of prion fractions. This finding is supported by the demonstration of the extraordinary resistance of the infectious agent toward treatments that damage or degrade nucleic acids, as shown by inactivation experiments with ultraviolet and ionizing radiation and the very low molar ratio between nucleic acids and infectious units in highly purified prion fractions.

C. Prion Strains and the Species Barrier

The existence of prion strains is a formidable intellectual obstacle for the protein only hypothesis, and no really satisfactory explanation has been found for this phenomenon to date.

Infectious material from different sources can produce distinct and reproducible patterns of incubation time, distribution of CNS involvement, and of proteolytic cleavage of PrP^{Sc}. These properties are retained even after several passages in isogenic mice. In contrast to pathogens with a nucleic acid genome, prions have to be able to encipher strain-specific properties in the tertiary structure of PrP^{Sc}. Indeed, there seem to be conformational differences in PrP^{Sc} that correlate with such behavior. For example, the TSE strains hyper (HY) and drowsy (DY) describe the behavior of the minks affected. They are associated with characteristic incubation time, locations of neuropathology, and characteristic degrees of proteinase K susceptibility of the respective PrP^{Sc}, including partial treatment producing different NH$_2$ termini.

An example of strain properties in human prion diseases shows that differences in the amino-acid sequence of PrP are not necessary to produce different strains: The human disease fatal familial insomnia (FFI), associated with the mutation D178N, shows a proteinase K-resistant PrP^{Sc} species of 19 kDa after deglycosylation. In contrast, both familial and sporadic Creutzfeldt–Jakob disease (CJD), associated with the same mutation, show a 21 kDa species. Inoculation of the respective human brain homogenates into PrP-deficient mice expressing a chimeric mouse–human mouse (MHuM) PrP transgene, produced disease associated with the respectively sized PrP^{Sc}. This suggests that the two distinct PrP^{Sc} species can act as template upon a single primary MHuM PrP structure and impart onto it their own respective conformation.

The different PrP conformations could represent either different tertiary structures or different quaternary assemblies of the same fold. Alternatively, it remains possible that modifications, such as N-linked glycosylation, confer the strain-specific properties. It has been discussed that molecules which are glycosylated in a certain fashion might interact with other glycosylated moieties in a *like-recruits-like* fashion (Fig. 4). Nevertheless, glycosylation is not necessary for acquisition of a proteinase K-resistant PrP^{Sc} in a cultured cell system. Strain-specific properties of incubation time and brain localization might also reflect the targeting of different forms of PrP^{Sc} to specific cell-types. These cells would then convert new molecules in the same way.

The primary sequence of PrP has been identified as one main determinant of the species barrier of prions. The disease develops very slowly, or does not develop at all, if prions are being transmitted from one animal species to another. If the disease is passaged serially in the new species, however, incubation time becomes dramatically shortened and stabilizes after a few passages.

Many important observations about the species barrier have been made possible by transgenic mice. This approach enables the expression of heterologous genes in a whole organism. If a hamster *Prnp* transgene is introduced into the germ line of a mouse, the mouse will express both mouse and hamster PrP^C. Such transgenic mice are susceptible to infection with hamster-derived prions. The brain extract prepared from those hamster-prion inoculated, scrapie-sick transgenic mice is highly infectious for hamsters, but contains only very little infectivity for wild-type mice. However,

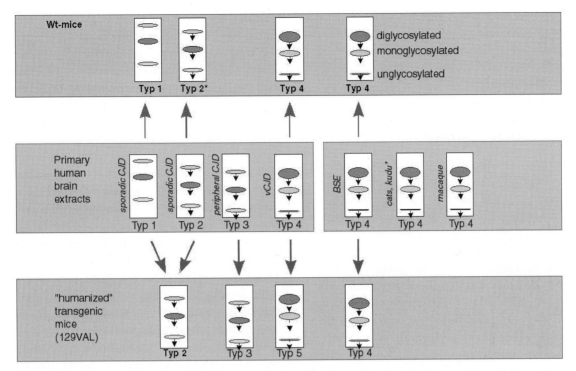

Fig. 4. Patterns of PrP glycosylation. Representation of the three glycosylated PrP^{Sc} moieties (non-, mono-, and di-glycosylated PrP^{Sc}) in immunoblots of brain extracts after digest with proteinase K. Different inocula result in specific mobilities of the three PrP bands as well as different predominance of certain bands (middle panel). These characteristic patterns can be retained, or they can change to other predictable patterns after passage in wild-type (upper panel) or humanized mice (PrP-deficient mice bearing a human PrP transgene, lower panel).

if one infects such transgenic mice with mouse prions, the brain extract derived from these mice is highly infectious for mice, but hardly for hamsters. Within the conceptual framework of the protein-only hypothesis, these remarkable findings can be interpreted as evidence that hamster PrP^{C}, but not mouse PrP^{C}, represents a suitable substrate for the conversion into hamster PrP^{Sc} as catalyzed by hamster prions. Conversely, mouse PrP^{C}, but not hamster PrP^{C}, is a good substrate for the conversion into mouse PrP^{Sc}. The conversion process, thus, seems to rely upon a homophilic interaction between the pathological and the normal isoforms of the PrP protein. Accordingly, introduction of murine PrP transgenes in *Prnp$^{o/o}$* mice renders them highly susceptible to mouse scrapie but not to hamster-prions, while the insertion of Syrian ham-

ster PrP transgenes leads to susceptibility to hamster prions.

Theoretically, there are two distinct events in transspecies transmission of TSE: the primary recruitment, where the infecting prion is directly involved in conversion of normal host PrP^{C}, and the secondary self-recruitment, where newly formed host prions convert further normal host PrP^{C}. Passaging in a constant and normal genetic background eventually stabilizes strain properties, since the infecting PrP^{Sc} is no longer contributing directly to the late stages of the conversion process. Nevertheless, its initial influence may determine the strain type.

The initial cross-species event differs from subsequent passages in third-generation animals. While, for example, bovine spongiform encephalopathy (BSE) crosses to new variant CJD (nvCJD) only with great difficulty, nvCJD might spread quite easily through blood transfusions in a human-to-human process. So any "species barrier" may quickly lose its comfort value as the situation changes to intraspecies recruitment.

D. PrP Glycosylation

Western blot analysis of PrP from extract of normal or prion-infected brains reveals three major bands, corresponding to PrP bearing two, one, or no polysaccharide chains. Elimination of these N-glycosylation sites, by replacing the threonine residues with alanine, abolishes transport to the cell surface and leads to intracellular PrP^C accumulation.

In 1995, Parchi and colleagues showed that protease-digested PrP^{Sc} (PrP 27–30) in sporadic CJD has different mobility patterns which, together with a methionine/valine polymorphism at codon 129 of *PRNP*, specify four clinical entities. Collinge *et al.* (1995) extended these studies to nvCJD and acquired CJD (Fig. 4). The relative intensities of the three bands that reflect glycosylation state in different prion strains define four types of patterns in human brain extracts derived from sporadic (type 1 or 2), acquired (type 3), or nvCJD (type 4). Transmission experiments with these human TSEs to wild-type mice or to mice that were transgenic for the human PrP genes indicate that the supposed conformational strain specificity is conserved after transfer between species. nvCJD shares a common pattern of di-, mono-, and nonglycosylated species with BSE-infected animals, distinct from that of sporadic or acquired CJD. The proteinase K-resistant diglycosylated species was particularly prominent, raising questions of whether this form of PrP^C is more susceptible to BSE-mediated conformational changes or whether a population of cells preferentially producing diglycosylated PrP may be more readily targeted by the BSE agent.

E. Proposed Conversion Mechanisms

To date, all attempts to achieve conversion of PrP^C to an infectious form *in vitro* have failed. Studies of the biosynthesis and processing of PrP^C and PrP^{Sc} in

cultured cells by Butler and Caughey and colleagues in the late 1980s indicate that normal PrP^C traverses the secretory pathway to reach the cell surface. PrP^C is attached to the plasma membrane via a GPI anchor. GPI-anchored proteins are known to localize at the cell surface in cholesterol-rich plasma membrane invaginations that are Triton X-100 insoluble, known as rafts or DIGS (detergent-insoluble glycosphingolipid-enriched membranes). It has to be mentioned that absence of the GPI anchor from a truncated PrP inhibits, but does not prevent, production of the proteinase K-resistant PrP^{Sc} species in cultured cells. On the other hand, Taraboulos and colleagues showed, in 1995, that treatment of the cultured cell system with lovastatin, an inhibitor of cholesterol biosynthesis, blocks the conversion process. It is unclear whether this effect is due to the disruption of the DIGS where conversion might take place or by lack of PrP^C on the cell surface.

In scrapie-infected cultured cells, PrP^C undergoes turnover with a $t_{1/2}$ of 6 hr. A small fraction (5%) of newly synthesized, metabolically labeled PrP^C ($t_{1/2} =$ 5–15 hr), undergoes conversion to a form that exhibits the same proteinase K resistance as PrP^{Sc}.

Further studies by Caughey and Raymond in 1991 showed that conversion to PrP^{Sc} can be blocked by addition of exogenous phosphatidyl–inositol-specific phospholipase C or by releasing PrP from the cell surface by proteases. Thus, PrP^C conversion seems to happen either at the cell surface or after internalization from the cell surface to the endocytic pathway. Accordingly, low temperature incubation at 18°C, which retards endocytosis, also blocks production of PrP^{Sc}.

The nature of the conversion of PrP^C into PrP^{Sc} is still completely obscure. It may consist of a chemical or of a conformational modification of PrP^C after it has reached the cell surface. While the hypothesis that one form of a protein can catalyze the refolding of native molecules into a distinct "misfolded" conformation might seem radical, such a process is, by no means, physically unreasonable. Two different models for the kinetics of conversion have been proposed during the last decade: template directed refolding vs nucleation (Fig. 5).

For the first model, Prusiner, in 1991, suggested that the PrP^{Sc} form is more stable than PrP^C, but

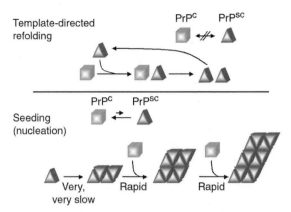

Fig. 5. Two models for conversion of PrP^C to PrP^{Sc}. In the template-directed model, kinetically inaccessible PrP^{Sc} (triangle) is more stable than PrP^C (rectangle). PrP^{Sc} facilitates conversion by catalyzing the rearrangement of PrP^C. According to the seeding hypothesis, conversion between PrP^C and PrP^{Sc} is reversible, but monomeric PrP^{Sc} is less stable than PrP^C. PrP^{Sc} aggregates, once formed, can grow by recruiting further monomeric PrP^{Sc} from the solution.

kinetically inaccessible. In this case, PrP^{Sc} could promote conversion by catalyzing the rearrangement of a molecule of PrP^C, or of any intermediate form, into the more stable PrP^{Sc} conformation. Thus, infectivity would rely on the ability of the PrP^{Sc} molecule to bind to and catalyze conversion of existing intermediate molecules. According to this template assistance model, the genetically inherited and spontaneous diseases result from mutations that increase the rate at which mutant PrP^C spontaneously converts to PrP^{Sc} and/or enhance the number of intermediate molecules.

In a second proposed mechanism, formation of PrP^{Sc} is a nucleation-dependent polymerization. The conversion between PrP^C and PrP^{Sc} is reversible in the absence of a preexisting aggregate, with a PrP^{Sc} monomer being less stable than PrP^C. PrP^{Sc} polymers, however, might bind to and stabilize the PrP^{Sc} conformation, thereby promoting the conversion of PrP^C. The initial nucleation process, thus, represents the barrier to a stable conversion, since formation of low-order aggregates is not favored until a minimum size nucleus is attained.

The requirement for the formation of a stable nucleus before conversion's being accomplished predicts certain characteristics of the aggregation process, such as kinetics displaying a *lag* phase and a critical protein concentration-dependence for the initial formation of aggregates. The *in vitro* conversion process, as shown by Caughey and colleagues in 1995, seems to show such features. The observation that fractions containing high-order PrP^{Sc} aggregates (> 300 kDa) can mediate the conversion to protease resistance in *in vitro* reactions, while smaller-sized fractions do not, can be explained by the relatively large size of the minimum stable nucleus, that would tend to make such a particle insoluble. Infection would, thus, circumvent the slow, thermodynamically unlikely step of nucleation by introducing a "seed" that initiates aggregation.

There are biological examples for both proposed mechanisms. In the case of the template-assisted conversion mechanism, PrP^C has a conformation that does not spontaneously form the more stable PrP^{Sc}. A number of proteins have been observed to be separated from their true free energy minima by a large barrier. These include influenza hemagglutinin and a number of proteases, including subtilisin and α-lytic protease. In the case of α-lytic protease, the conversion from a globulelike intermediate, I, to the native state, N, is extremely slow, with little or no conversion over a period of a month. But if the propeptide region (a naturally occurring polypeptide cleaved from the same translation product as α-lytic protease during its maturation) is bound in either cis or trans, conversion is dramatically accelerated.

There are also examples for the nucleation–polymerization. There is a resemblance, for example, to tubulin polymerization or bacterial flagellar polymerization. The soluble monomer flagellin is incorporated into the growing end of a flagellum. In liquid phase, flagellin units are unable to spontaneously polymerize even at nearly millimolar concentrations. If a seed of fragmented flagellum is placed into the mixture, polymerization rapidly takes place. The polymerizing monomers can even adopt the conformation of heterologous seed material, indicating a "templating" behavior.

It is important to mention that there could be a hybrid mechanism. The surface of an PrP^{Sc} aggregate, initially formed by a nucleation process, could catalyze the conformational change of PrP^C monomers.

III. YEAST PRIONS

Psi$^+$, a trait of the budding yeast, *Saccharomyces cerevisiae,* was first described by Brian Cox in 1965. The efficiency at which a certain suppressor-strain could misread UAA stop codons as sense is dependent upon the presence of a non-Mendelian factor, named [*PSI*$^+$]. Wickner suggested in 1994 that [*PSI*$^+$] is a prion form of the Sup35 protein (Sup35p). He proposed that, in [*psi*$^-$] cells, the conformation of Sup35p is fully functional (Sup35p^{psi-}) and promotes efficient termination at stop codons (we know now that Sup35p codes for the translational release factor eRF3). In [*PSI*$^+$] cells, some or all of the Sup35p is proposed to take on a new, biologically inactive conformation (Sup35p^{PSI+}), leading to less efficient termination and, thus, nonsense suppression. The SUP35 aggregates appear to act as a nucleus, which, similarly to the seed of a crystal, promotes the aggregation of newly synthesized SUP35 protein and allows the propagation of the PSI+ state in a manner analogous to the propagation of prions. There is now evidence that Sup35p exists in different structural states in [*PSI*$^+$] and [*psi*$^-$] cells. Sup35p in lysates of [*PSI*$^+$], but not [*psi*$^-$] strains, shows increased protease resistance and aggregation, two characteristics typical of vertebrate prions. These results can be interpreted in terms of the seeded nucleation model; however, the initial rate of unseeded fiber nucleation is not as dependent upon the concentration of soluble Sup35p monomers as predicted by the original polymerization model.

The chaperone protein Hsp 104 is known to facilitate the folding of proteins and is required for the propagation of [*PSI*$^+$]. Both deletion or overexpression of Hsp 104 result in disappearance of the SUP35 aggregates and loss of the PSI+ state. The fact that the impairment of Sup35p, either by Mendelian mutations in *SUP35* or by the presence of [*PSI*$^+$], causes similar phenotypes is consistent with the prion model. A multicopy plasmid carrying *SUP35* efficiently induces the *de novo* appearance of [*PSI*$^+$]. This can be interpreted as evidence for the prion model, since the *SUP35* overexpression increases the probability that a Sup35p^{psi-} molecule would take on a prion shape by chance. It has also been shown that [*PSI*$^+$] can reappear, arguing against the possibility that "curing" is due to the loss of a cytoplasmic nucleic acid with no nuclear master gene. Furthermore, a dominant mutation, which causes the loss of [*PSI*$^+$], PNM2 (Psi-No-More) is an allele of *SUP35* with a missense mutation.

Another cytoplasmically inherited genetic element in yeast, [Ure 3], appears to propagate by a similar mechanism, as shown by Wickner and Lindquist. There is reason to believe that the behavior of these unconventional genetic elements points to a molecular process that is broadly distributed and functions in a wide variety of biological contexts. Unfortunately, in the case of PrP conversion, no chaperone component (like Hsp 104) has been identified in the cellular components where conversion appears to occur. Therefore, it still remains to be verified whether the mammalian PrPSc can propagate itself through a mechanism analogous to those of the yeast prions.

IV. PRION DISEASES

A. A Historical Perspective

Scrapie is a naturally occurring disease which has been known to infect sheep in Great Britain for at least 250 years. In 1936, Cuille and Chelle showed that the disease could be transmitted via inoculation of a homogenate of spinal cord from sick animals into the brain of healthy sheep and goats. These findings have been confirmed by the experiments of Gordon in 1946: Inoculated animals developed a disease identical to that spontaneously occurring in sheep. In further experiments, atypical properties of the infectious agent were demonstrated, such as a very long incubation period, a remarkable resistance towards inactivation with high temperatures, treatment with alkalizing agents, aldehydes, and UV light irradiation. These properties were later on defined as some of the central features of prions.

In humans, a group of diseases with similar pathological characteristics was described to occur sporadically or in a hereditary fashion (Fig. 6). These diseases include Creutzfeldt–Jakob disease, Gerstmann–Sträussler–Scheinker syndrome (GSS), as well as an infectious disease called kuru, which

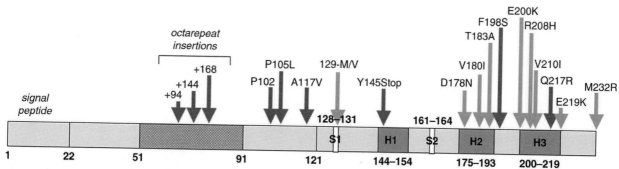

Fig. 6. Mutations and polymorphisms associated with the human hereditary prion diseases. The positions of the octarepeats, α-helices (H1–3) and β-sheet (S1, S2) conformations are indicated. Mutations correlated with plaque-type deposits in GSS, spastic paraparesis, and Alzheimer-like phenotype are indicated with dark arrows. Mutations correlated with synaptic deposits found in sporadic and familial CJD, as well as FFI, are shown with grey arrows. The position of the M/V polymorphism at codon 129 is indicated. It seems that amino-proximal mutations induce a GSS-like phenotype, while most mutations close to the carboxyl terminus result in a CJD-like phenotype. Based on drawings published by S. B. Prusiner and by J. Tateishi.

affected the Fore population in the northern provinces of New Guinea. Already 35 years ago, William Hadlow drew the attention of scientists to the fact that these diseases may represent a particular form of scrapie in humans. These speculations of Hadlow were later confirmed in inoculation experiments performed by D. Carleton Gajdusek and colleagues. In these seminal experiments, Gajdusek accomplished transmission of kuru, and also of CJD, to chimpanzees. For these experiments Gajdusek was awarded the Nobel Prize for Medicine in 1976.

In the last several years, a new disease was found to affect cows: bovine spongiform encephalopathy (BSE). It is very likely that feeding of cows with rendered offal derived from meat and bones of scrapie-infected sheep was responsible for the transmission of scrapie and for the outbreak of BSE.

B. Transmissible Spongiform Encephalopathies (TSE)

The neuropathological hallmarks of TSEs, or prion diseases, are spongiform changes, astrocytic gliosis, neuronal loss, and PrP-positive plaques (Fig. 7). Unfortunately, until now, diagnosis of TSE could be reliably made only postmortem. Although the human forms of these diseases are rare, the epidemic proportions of the bovine spongiform encephalopathy

(BSE) forced researchers worldwide to urgently reconsider the question of transmissibility to humans.

1. Animal Prion Diseases

Sheep scrapie is the prototype of the growing group of TSEs. The typical symptoms of scrapie-sick sheep include hyperexcitability, pruritus (a chronic itching of the skin, leading to the scraping behavior), and myoclonus (a brief, sudden, singular, shocklike muscle contraction). The disease is characterized by a rapid progression, which leads to tetraparesis (paralysis of all four legs) and, ultimately, to the death of the affected animal.

The clinical symptoms of BSE are insidious and consist of behavioral changes (including aggressive behavior, which is proverbially atypical in cows) and uncoordinated gait. Histologically, the brain exhibits lesions similar to those of TSEs in other species. It is not excluded that cases of BSE may have been seen in England as early as 1985, but probably not before that year, despite earlier anecdotal reports. The numbers of cases continued to increase, provoking a major epidemic, and peaked in summer 1992.

It has been suggested early on that one common exposure of cattle to prions came about through the use of a dietary protein supplement, meat and bone meal (MBM), that was regularly fed after weaning.

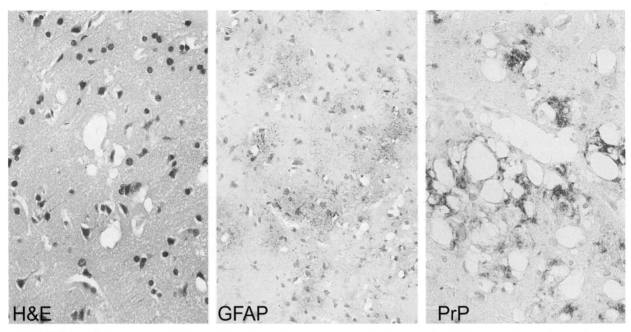

Fig. 7. Characteristic features of Creutzfeldt–Jakob disease. Hematoxylin-Eosin stain (left) shows the typical vacuoles in the brain of a CJD patient, which leads to the spongiform appearance. Proliferation of reactive astrocytes is visualized by staining with antibodies against glial fibrillary acidic protein (GFAP, middle). PrP protein deposits are shown with anti-PrP immunostaining. See color insert.

No credible alternative hypothesis has been put forward on the origin of the BSE epidemic, and the incidence of new cases has been precipitously declining some 4–5 years after a more-or-less effective ban of MBM was put in place.

2. Human Prion Diseases

Since the first description by A.M. Jakob and H.G. Creutzfeldt, five human diseases have been identified as TSEs. The disease bearing the authors' names, Creutzfeldt–Jakob disease, occurs sporadically, may be transmitted, and has a genetic basis in 10–15% of all cases. Creutzfeldt–Jakob disease generally presents as a progressive dementia.

Other genetic diseases are Gerstmann–Sträussler–Scheinker (GSS) syndrome and fatal familial insomnia (FFI). GSS is transmitted by autosomal dominant inheritance (Fig. 6). It is characterized by missense mutations of *PRNP*, associated with specific neuropathological lesions, and by multicentric amyloid plaques. The latter can be labeled by antibodies directed against the prion protein. This restrictive definition justifies retaining the name of Gerstmann–

Sträussler–Scheinker syndrome and excludes observations of hereditary prion diseases without multicentric amyloid plaques, as well as sporadic forms with multicentric plaques. It has been possible to transmit GSS to chimpanzees and to mice. An additional member of the group of spongiform encephalopathies, fatal familial insomnia (FFI), was also transmitted experimentally in at least three instances to mice. Kuru is a form of spongiform encephalopathies transmitted by ritual cannibalism, which affected the Fore population in the northern provinces of New Guinea.

The newest form of CJD in humans, new variant CJD (nvCJD), was first described in 1996 and has been considered evidence for a link between human TSEs and BSE. nvCJD has a distinct pathology characterized by abundant "florid plaques," decorated by a daisylike pattern of vacuolation. The age of onset in nvCJD is much lower than in sporadic CJD. The notion that nvCJD could be transmitted from cattle to primates is supported by several arguments, including the observation that intracerebral inoculation of BSE-infected brain extracts into macaque monkeys produced disease and pathology resembling that in the nvCJD patients.

C. Pathology of Prion Diseases

The questions revolving around the nature of the prion are important and fascinating. Less glittering, but certainly not less important, is the question of how prions induce brain damage. Very little is understood about the mechanisms of the latter phenomenon. The accumulation of some protein in brain, by itself, does not necessarily explain the dire consequences of TSE on the brain of its host. They include a highly characteristic vacuolation and, eventually, death of nerve cells, activation of astrocytes and microglial cells to an extent unparalleled by other pathological conditions, and, invariably, lethal impairment of the electrical functions of the brain.

Characteristic lesions under the light microscope consist of spongiform changes in the neuropil nerve cells and astrocytes with nerve cell degeneration and astrocytosis. These changes are often observed in the gray matter of the cerebrum and cerebellum. The distribution of the histopathologic lesions may vary according to the strain of the agent, as has been shown with inbred strains of mice. It may also vary with the site of inoculation. If the infection reaches the brain through the optic nerve, the spongy degeneration is clustered around the occipital lobe. Amyloid plaques containing PrP may be seen between cells in some TSEs (e.g., kuru, hamster scrapie) and stained with Congo red (Fig. 7). Electron microscopy shows twiglike structures 12–16 nm in width and 100–500 nm long, which are found only in TSE and are now called scrapie-associated fibrils (SAF).

Whether the damage is brought about by accumulation of the pathological prion protein or, rather, by the abrupt withdrawal of its normal isoform during the course of the disease is not clear to date. The fact that PrP-deficient mice are reasonably healthy only apparently refutes the latter hypothesis, since such mice may have had the time to adapt to a prionless life from early on. The crucial experiment to settle this question may be a "conditional knockout" allowing abrupt shutoff of PrP^C expression in adult life.

If PrP^{Sc} deposition was the cardinal event in pathogenesis, why is it that, in many instances of human and experimental TSE, extremely little PrP^{Sc} can be detected, even in terminal disease? And, if PrP^{Sc} deposition is an important event at all, it would seem that it can exert deleterious effect only through some PrP^C mediated processes, since PrP-deficient nerve cells are not affected, even after long-term exposure to PrP^{Sc}.

While the prime target of damage seems to be neuronal, profound neuronal loss is not always seen in TSEs. Instead, astrocytic activation occurs very early and in an extremely consistent fashion. It can be easily reproduced *in vitro* and leads to significant physiological effects, such as impairment of the blood–brain barrier. Since astrocytes belong to the few cell types identified that are capable of supporting prion replication, elucidation of the role of this cell type in TSE pathogenesis will be an exciting task for the years to come.

Growing evidence incriminates another cell type in brain damage, not only in TSE, but in diseases ranging from Alzheimer's to multiple sclerosis and even stroke: microglial cells. *In vitro* experiments seem to indicate that activation of microglia may be quite pivotal in effecting neuronal damage in TSE, and that this phenomenon is dependent on expression of PrP^C. The details of this pathway of cell death, however, still escape our understanding, and the proof that these phenomena occur in the brain during the course of the disease (and not only in petri dishes with explanted cells) is still missing.

V. PERIPHERAL PATHOGENESIS

A. Prions and the Immune System

It has long been observed that, even following intracerebral inoculation of mice with prions, there is early acquisition of infectivity in the spleen, long preceding any appearance of infectivity in the brain. Consistent with a primary replication step in the lymphoreticular system that favors neuroinvasion, SCID mice are relatively resistant to CNS disease following intraperitoneal inoculation.

Although prions are most effective when directly

administered to the brain of their hosts, this situation occurs mainly in experimental lab work and does not reflect the reality of the common routes of prion infections. In humans, most cases of iatrogenic CJD transmission were traced to intramuscular injection of growth hormone and, to a lesser extent, pituitary gonadotropins. The other example of massive, efficient human-to-human transmission is New Guinea's kuru, where oral uptake of infectivity was accomplished in the course of cannibalistic rituals.

Therefore, peripherally administered prions can reach the brain of their host. This neurotropism is remarkable, especially since no pathologies can be identified in organs other than the CNS. In addition, it may be important to identify "reservoirs" in which prions multiply silently during the incubation phase of the disease.

One such reservoir of PrP^{Sc} or infectivity is doubtlessly the immune system, and a wealth of early studies points to the importance of prion replication in lymphoid organs. The nature of the cells supporting prion replication within the LRS, however, is still uncertain. Inoculation of various genetically modified immunodeficient mice lacking different components of the immune system with scrapie prions revealed that the lack of B-cells renders mice unsusceptible to experimental scrapie. While defects affecting only T-lymphocytes had no apparent effect, all mutations affecting differentiation and responses of B-lymphocytes prevented development of clinical scrapie. Since absence of B-cells and of antibodies correlates with severe defects in follicular dendritic cells (FDCs), the lack of any of these three components may prevent clinical scrapie. Yet mice expressing immunoglobulins exclusively of the IgM subclass, without detectable specificity for PrP^{C}, developed scrapie after peripheral inoculation: Therefore, differentiated B-cells seem crucial for neuroinvasion of scrapie, regardless of B-cell receptor specificity. FDCs have been incriminated, because PrP^{Sc} accumulates in FDCs of wild-type and nude mice (which suffer from a selective T-cell defect).

Repopulation of immunodeficient mice with fetal liver cells (FLCs) from either PrP-expressing or PrP-deficient mice and from T-cell deficient mice, but not from B-cell deficient mice, is equally efficient in restoring neuroinvasion after *i.p.* inoculation of

scrapie prions. This suggests that cells whose maturation depends on B-cells or their products (such as FDCs) may enhance neuroinvasion. Alternatively, B-cells may transport prions to the nervous system by a PrP-independent mechanism.

We have recently learned that "Type 4"-PrP^{Sc}, one of the hallmarks of new variant CJD, accumulates in the lymphoid tissue of tonsil in such large amounts that it can easily be detected with antibodies on histological sections. Infectivity can accumulate also in intestinal Peyer's plaques, where it replicates almost immediately following oral administration of prions.

Immune cells are unlikely to transport the agent all the way from LRS to CNS, since prion replication occurs first in the CNS segments to which the sites of peripheral inoculation project. This implies that the agent spreads through the peripheral nervous system, analogously to rabies and herpes viruses.

B. Development of Drugs against Prions

While common neurodegenerative conditions, such as Alzheimer's disease, have attracted a very large mind share in the pharmaceutical industry, the same has not yet been the case for TSE. One unavoidable consideration is that the exceeding rarity of TSE in humans renders them an uneconomical target for therapeutical efforts. Sadly, it is not unlikely that the BSE epidemics may change the epidemiology of human TSE considerably.

It has been argued that the possibility that we may be witnessing an incipient nvCJD epidemic may not be attractive to pharmaceutical companies, since the former will be self-limited and will subside just as BSE is doing now, with a delay mainly determined by the (hitherto unknown) incubation time of the disease in humans. This viewpoint is not necessarily true: the time periods needed for the development cycle of new drugs are rapidly falling, as impressively demonstrated by the crop of antiretroviral agents which are now hitting the market. Given that the peak of the BSE epidemic in humans may be reached in 10–20 years, it would seem quite urgent and important to start development of therapeutic agents.

Unfortunately, by the time the first signs and symptoms of disease are recognized, significant CNS damage is consistently present. It, therefore, appears that, in the case of BSE transmission, postexposure prophylaxis may provide a more viable alternative, just as treatment of presymptomatic HIV-infected individuals is more effective than treatment of terminally ill, neurologically impaired, and often demented patients with large retroviral burdens in their CNS.

Perhaps at some point, it will be possible to treat patients even after prions have reached the brain and made themselves apparent. Two classes of substances have raised some (limited) hope to this end: amphotericins, sulfated polyanions, and anthracyclin antibiotics. While the mechanisms of action of the former is totally unknown, the latter is thought to intercalate into the highly ordered structures of prion amyloid and to disrupt it. As exciting as this approach may sound, the published data solely indicate that the apparent virulence of hamster prion preparation was only reduced after pre-incubation with iodo-deoxyrubicine, which procedure can hardly be described as "treatment." It seems a safe prediction that structural data on PrP^{Sc} and its aggregated forms will be extremely important for designing more effective intercalating agents.

See Also the Following Articles

PROTEIN SECRETION • TRANSGENIC ANIMAL TECHNOLOGY • VIRUSES, EMERGING • YEASTS

Bibliography

Aguzzi, A., and Weissmann, C. (1997). Prion research: The next frontiers. *Nature* **389**, 795–798.

Harris, D. A. (1999). "Prions: Molecular and Cellular Biology." Horizon Scientific Press.

Prusiner, S. B. (1996). Prions prions prions. *Curr. Top. Microbiol. Immun.* **207**. Springer Verlag.

Prusiner, S. B. (1998). Prions. *Proc. Natl. Acad. Sci. U.S.A.* **95**(23), 13363–13383.

Protein Biosynthesis

Rosemary Jagus and Bhavesh Joshi
University of Maryland Biotechnology Institute

GLOSSARY

aminoacyl tRNA Activated form of amino acid used in protein biosynthesis. Consists of amino acid linked through an acid–ester linkage to the hydroxyl group at the 3′ terminus of tRNA.

aminoacyl tRNA synthetase Enzyme that attaches amino acid to specific tRNA species.

anticodon Sequence of three nucleotides complementary to the three nucleotide codon on mRNA.

A-site Aminoacyl tRNA or entry site on the ribosome.

5′ cap 7-methyl guanosine residue linked to 5′ triphosphate at 5′ terminus of eukaryotic mRNA.

codon Sequence of three nucleotides in DNA or mRNA that codes for one amino acid.

elongation factor Protein factor required for the elongation step of protein biosynthesis.

E-site Ribosomal exit site for deacylated tRNA prior to its release from the ribosome.

initiation codon Start codon; determines correct reading frame. Usually AUG.

48 S initiation complex Initiation intermediate containing small ribosomal subunit, initiator tRNA, mRNA, and several initiation factors.

initiation factor Protein factor that promotes the correct association of ribosomes with mRNA; required for the initiation step of protein biosynthesis.

ORF Open reading frame, a nucleic acid sequence containing a series of codons which is uninterrupted by stop codons, and which potentially encodes a protein.

peptide bond Linkage between two amino acids.

peptidyl transferase Ribozyme that catalyzes the transfer of amino acid from the A-site to a growing polypeptide chain.

poly(A) tail Series of A residues attached posttranscriptionally to the 3′ end of eukaryotic mRNA.

P-site Peptidyl tRNA or donor site on the ribosome.

reading frame The phase in which nucleotides are read in sets of three (triplets) to encode a protein.

release factor Protein factor that promotes the termination of protein biosynthesis and release of the mRNA and newly synthesized polypeptide from the ribosome.

rRNA Ribosomal RNA; forms part of the structure of ribosome and participates in the synthesis of proteins. Comprises approximately 60% of ribosome.

ribosome Large complex of ribosomal RNA and ribosomal protein that brings components of protein biosynthesis together.

Shine–Dalgarno sequence Region in 5′UTR of eubacterial and archaeal mRNAs complementary to region of 16 S rRNA.

termination codon Stop codon; signals end of protein. Can be UAA, UGA, UAG.

translation The conversion of the nucleotide sequence of mRNA into protein.

translation factor Protein factor that promotes protein biosynthesis.

tRNA Adaptor molecule; translates information encoded in mRNA into appropriate amino acid sequence.

3′ UTR Untranslated nucleotide sequence 3′ of the termination codon of mRNA (also called trailer sequence).

5′ UTR Untranslated nucleotide sequence 5′ of initiation codon of mRNA (also called leader sequence).

PROTEIN BIOSYNTHESIS is the assembly of amino acids into polypeptides. Proteins are encoded by genes. The genetic information encoded by DNA is expressed by a two-stage process, transcription, which generates the single-stranded mRNA template from DNA, and translation, which converts the nucleotide sequence of mRNA into protein using a triplet code. The code is essentially the same in all living organisms, suggesting that all present-day cells have a common origin.

Proteins constitute more than half the dry mass of a cell, making protein synthesis central to cell maintenance, growth, and proliferation. In present-day organisms, an elaborate biochemical machinery directs protein biosynthesis via a complex multienzyme system, involving the participation of hundreds of macromolecules. Protein biosynthesis represents the translation of the mRNA template into the protein sequence. There are two aspects to the overall process. The first is the **decoding strategy**, the mechanism by which a base sequence in an mRNA molecule is translated into an amino acid sequence. The genetic code is represented by the mRNA and the code is translated by the tRNA molecules. The second is the **actual process of synthesis** of the protein, which takes place on the surface of ribosomes. Ribosomes are complex particles composed of ribosomal RNA (rRNA) and up to 50 different kinds of proteins. During protein biosynthesis, the ribosomes and tRNAs interact to link amino acids in a sequence determined by the base sequence of mRNA, read from the 5′ to the 3′ terminus. Proteins are synthesized from the amino- (NH_2) to the carboxy- (COOH) terminus. The decoding of the base sequence of mRNA on the ribosome, takes place in three distinct stages **initiation, elongation, and termination/release**, which denote the mechanisms by which the ribosome engages mRNA, the linking together of amino acids in the correct order, and the termination of translation with release of the finished polypeptide chain from the translational apparatus. Different sets of protein accessory factors, **translation factors**, assist the ribosome at each of the three stages. These are referred to as **initiation factors, elongation factors, and release factors**, to reflect the step at which they are involved. The central and universal features of translation, decoding and amino acid polymerization, utilize RNA-based chemistry that is aided by numerous structural and catalytic proteins, the ribo-

somal proteins and translation factors. Energy for the process is provided by ATP and GTP.

Although the ribosome is more complex in eukaryotes, it is fundamentally the same in cells from all three primary life domains (or kingdoms), the Bacteria (eubacteria), Archaea (archaebacteria), and Eucarya (eukaryotes). The appearance of the ribosome as the coordinator of the events of protein synthesis represented a crucial step in early evolution. Lineage-specific features in the protein biosynthetic apparatus suggest that many details of the translation process were separately perfected in each of the three life domains. Protein biosynthesis is considered to be an ancient property of life and probably existed prior to the development of cellular life. It therefore comes as no surprise that the decoding of the genetic code and amino acid polymerization, as well as the structure of the ribosome, are similar in all three primary life domains. However, the mechanisms of recognition of the correct reading frame differ, as do the mechanisms by which mRNA is recruited by the ribosome. In general, the Bacteria and Eucarya differ in both recognition of the open reading frame, as well as in mRNA recruitment, with concomitant differences in their translational initiation factors. The archaebacterial translational machinery includes both eubacterial and eukaryotic-like features, along with some unique characteristics. Recognition of the correct reading frame and mRNA recruitment in Archaea resembles that found in Bacteria. However, archaeal translational initiation factors most closely resemble those of eukaryotes, although they express homologs of the eubacterial initiation factors, in addition.

I. THE DECODING SYSTEM: tRNAs AND AMINOACYL tRNA SYNTHETASES

A. The Triplet Code

Each triplet of nucleotides, called a **codon**, specifies one amino acid. Since RNA is a polymer of four nucleotides, there are 64 unique codon triplets possible. Although only 20 amino acids are encoded, all 64 possible triplets carry some information. AUG is used as the initiation codon (or start codon) in all organisms, as well as being the codon for methionine. AUG is the usual, but not the only, initiation codon. GUG, which codes for valine, is also used as an initiation codon in the Bacteria and Archaea, although at a lower frequency. CUG and ACG, which code for leucine and threonine, respectively, can also be used as initiation codons, albeit very rarely. Three of the codons, UAA, UGA, and UAG, do not code for amino acids, but specify the termination sites of translation. These are referred to as stop codons, or termination codons. The remaining codons all correspond to amino acids. More than one codon can specify the same amino acid, making the code degenerate. For instance, serine is specified by UCU, UCC, UCA, and UCG, with redundancy at the third position. Like tRNA^{Ser}, several tRNA molecules recognize any of the four nucleotides in the third (wobble) position, making it possible to fit 20 amino acids to 61 codons with 31 tRNA molecules (eubacteria). Except for minor differences in mitochondria and ciliates, the same codon assignments are found in all organisms, i.e., the code is universal.

B. Reading Frames

The genetic code is read in nonoverlapping triplets from a fixed starting point. Successive codons are represented by successive trinucleotides. The use of a fixed starting point means that translation of a mRNA into a protein must start at one end of the mRNA template and work to the other. By reading a nucleotide sequence as nonoverlapping triplets, there are three possible ways of translating the sequence into protein. For instance, for the mRNA sequence

GACGCUACGAUGGCUACU

the three possible ways of reading the code are:

```
GAC GCU ACG AUG GCU ACU
Asp Ala Thr Met Ala Thr
    ACG CUA CGA UGG CUA CUG
    Thr Leu Arg Trp Leu Leu
        CGC UAC GAU GGC UAC UGA
        Arg Tyr Asp Gly Tyr STOP
```

These three possible ways of translating a nucleotide sequence give rise to three different amino acid sequences, which are referred to as reading frames.

For any mRNA, only one reading frame is normally used. There are a few examples from Bacteria and the Eucarya of the synthesis of proteins from two different reading frames. This can occur by the choice of an alternate initiation codon or by an unusual phenomenon called ribosomal frameshifting.

C. tRNAs

Amino acids do not by themselves recognize triplets of nucleotides. The decoding operation that converts the nucleotide sequence within an mRNA molecule to the amino acid sequence of a protein is accomplished by tRNAs and their corresponding aminoacyl tRNA synthetases. A tRNA and its corresponding synthetase are referred to as cognates. Specific tRNAs bind specific amino acids and their cognate aminoacyl tRNA synthetases recognize and interact with both tRNA and the amino acid. Figure 1A illustrates the two regions of a tRNA important for its decoding function. One of these is a sequence of 3 bases, referred to as the **anticodon**, that is complementary to the codon for a particular amino acid. The second region is the amino acid attachment site at the 3′ hydroxyl terminus. The amino acid that binds to this site is specified by the codon complementary to the anticodon region of the same tRNA species, although the amino acid is not recognized by the tRNA itself but by the cognate synthetase. tRNA recognition by its cognate synthetase spans the entire tRNA molecule rather than involving a localized site, with contacts along and around the inside of the L-shaped tertiary structure Fig. 1(B).

All sequenced tRNA molecules share a set of conserved nucleotides and range in size from 70 to 90 nucleotides. They all have the capacity to form a characteristic secondary structure, that of a cloverleaf, with four major arms, named for their function or structure. These are the **acceptor arm**, a base-paired stem that ends in an unpaired sequence, the 3′ terminus of which is the amino acid acceptor site; the **T-loop**, sometimes called the TΨC loop, named for the ribothymidine, usually as a TΨC triplet with another modified base, pseudouridine (Ψ); the **anticodon arm**, which contains the anticodon triplet in the center of a loop; and the **D arm**, named for its content of dihydrouridine (D). The most variable feature of tRNAs is the **extra arm**, which lies between

A

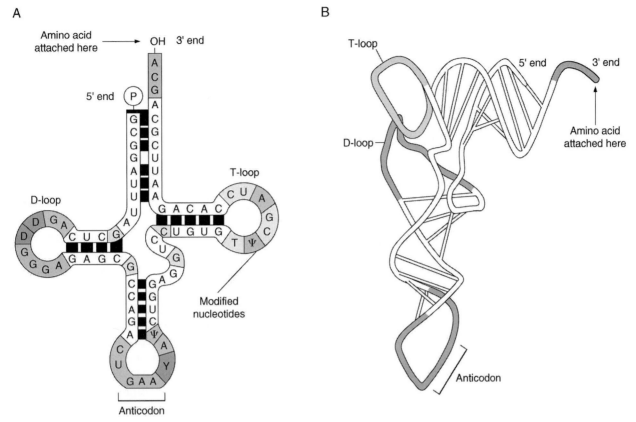

B

Fig. 1. Structure of phenylalanine tRNA. (A) Cloverleaf secondary structure showing base pairing (Ψ, pseudouridine). (B) 3-D shape, as revealed by X-ray diffraction analysis.

the T-loop and the anticodon loop. Although the function of the extra arm is not understood, it has been used to classify tRNAs into two classes: class 1, which represents 75% of all tRNAs containing a small extra arm between 3–5 nucleotides, and class 2, in which the extra arm is 13–21 nucleotides. The base pairing that maintains the secondary structure is virtually invariant. tRNA is unique among nucleic acids in its content of unusual nucleotides, mainly produced posttranscriptionally by modification of one of the four nucleotides. These nucleotide modifications confer on tRNA a much greater range of structural versatility through a wider variety of base pairing patterns facilitating the recognition of the appropriate mRNA codon by the tRNA molecule. The cloverleaf secondary structure folds to give a very compact L-shaped tertiary structure, as shown in Fig. 1B, that allows separation of the bound amino acid and the anticodon. This compact structure is consistent with need for tRNA to occupy the

small active sites on the ribosome and with the need for the aminoacyl group to be near the peptidyl transferase site on the large ribosomal subunit, while the anticodon pairs with a codon in the mRNA on the small ribosomal subunit.

D. Initiator tRNA

The first amino acid in all nascent peptides is methionine, which is decoded by a distinct initiator tRNA. In organisms from all three domains, the tRNA species that are used for the recognition of the initiation codon are different from those that provide methionine for internal codons. In eubacteria, the methionine attached to the initiator Met-tRNA is formylated to give *N*-formyl-Met-tRNA (often referred to as fMet-tRNA^{fmet} or fMet-tRNA$_{\text{i}}$). In eukaryotes, the initiator Met-tRNA is not formylated, but can be formylated *in vitro* by the eubacterial enzyme and so is often referred to as Met-tRNA$_{\text{f}}$, where f stands for "able to be formylated," or Met-tRNA$_i$, where i stands for "initiator." Initiator tRNAs in archaebacteria resemble those of eukaryotes.

E. Aminoacyl tRNA Synthetases

These enzymes make possible the conversion of the genetic information in nucleotide triplets to specific amino acids. Aminoacyl tRNA synthetases catalyze the ATP-dependent attachment of the appropriate amino acid to its corresponding tRNA by an amino acid-ester linkage. The aminoacylated tRNA is able to align the correct amino acid with a codon of the mRNA by interaction of the tRNA anticodon with the corresponding codon in mRNA. This alignment takes place on the ribosome. A second function of the amino acid attachment is to activate the amino acid by the generation of a high-energy linkage at the carboxyl end of the amino acid that promotes interaction with the adjacent amino acid in the protein sequence to form a **peptide bond**. Although each tRNA synthetase catalyzes the same general reaction, they are surprisingly heterogeneous with respect to sequence, subunit size, and quaternary structure. Despite this, a common theme is discernible in the structural organization in the arrangement of functional domains for aminoacylation and tRNA recognition. To these domains are added sequences that are dispensable for catalysis and which account for the size and quaternary structure polymorphism. These sequences serve other biological roles, such as regulation of expression or, in the eukaryotes, the assembly of multisynthetase complexes.

II. AMINO ACID POLYMERIZATION

The universal processes of amino acid polymerization, consisting of entry of aminoacyl tRNA, peptide bond formation, and translocation, take place at active sites on the ribosome under the direction of the protein synthesis elongation factors.

A. Ribosomes

Ribosomes are complex and dynamic ribonucleoprotein assemblies that provide the framework for amino acid polymerization in all organisms. Ribosomes are a major cellular component. In an actively growing bacterium, there can be 20,000 ribosomes per cell. Ribosomes contain approximately 10% of the total eubacterial protein and account for >80% of the total mass of cellular RNA. In eukaryotic cells, ribosomes also account for >80% of the cellular RNA. Although the proportion of total protein represented by ribosomes in eukaryotes is less, their absolute number per cell is greater. The reactions of protein synthesis take place on the ribosome, the function of which is to receive genetic information in the form of mRNA and translate that into the peptide chain. There are two aspects to this function: the ribosome must be able to process data and to catalyze peptide bond formation. Ribosomes from Bacteria to Eucarya have the same overall structure, reflecting the fact that aminoacyl tRNA binding, peptide bond formation, and translocation are virtually identical in all organisms. All ribosomes have two subunits consisting of ribosomal RNA (rRNA) and various ribosomal proteins. The large subunit binds tRNAs and mediates peptidyl transfer; the small subunit controls mRNA binding and decoding. In fact, these activities all occur at the interface between the two subunits. Ribosomes are basic and insoluble at physiological salt concentrations and pH. The molecular weights of ribosomal particles range from 2.5×10^6, which is typical of eubacterial ribosomes, to approximately 4.5×10^6 for typical eukaryotic ribosomes, with a ratio of RNA to protein of 60:40. Table I gives a summary of *E. coli* (eubacterial), *Sulfolobus solfataricus* (archaeal), and rat liver (eukaryotic) ribosome components. Because of historical precedent, the intact ribosomes and their subunits are named according to their sedimentation coefficients. The *E. coli* ribosome is termed a 70 S ribosome and the two subunits, which are unequal in size and composition, are the 30 S (small) and 50 S (large) subunits. The archaeal ribosome resembles the eubacterial ribosome in overall subunit size. Eukaryotic ribosomes are larger. The corresponding figures for rat liver ribosomes are 80 S, for the intact ribosome, and 40 S and 60 S, respectively, for the small and large ribosomal subunits. The small ribosomal subunit contains rRNA that has a sedimentation value of 16 S in *E. coli* and archaebacteria, compared with 18 S in plants and ani-

TABLE I
Composition of Ribosomes

Property	Bacteria (E. coli)	Eucarya (rat liver)	Archaea (Caldariella acidophila)
SMALL SUBUNIT			
S-value	30 S	40 S	30 S
Mr	0.9×10^6	1.44×10^6	1.15×10^6
RNA (S-value)	16 S	18 S	16 S
proteins (number)	21	33	28
proteins (range Mr)	$8.3–26 \times 10^6$	$11.2–41.5 \times 10^6$	
protein (% by weight)	40%	50%	52%
LARGE SUBUNIT			
S-value	50 S	60 S	50 S
Mr	1.6×10^6	2.8×10^6	1.8×10^6
RNA (S-value)	5 S	5 S	5 S
	23 S	5.8 S	23 S
		28 S	
proteins (number)	34	50	35
proteins (range Mr)	$5.3–24.6 \times 10^6$	$11.5–41.8 \times 10^6$	
protein (% by weight)	30%	35%	32%
COMPLETE RIBOSOME			
S-value	70 S	80 S	70 S
Mr	2.5×10^6	4.2×10^6	3×10^6
protein (% by weight)	34%	40%	39%

mals. The large ribosomal subunit contains rRNAs of 23 S in eubacteria and archeabacteria, compared with between 25 S and 28 S in plants and animals, respectively. In addition, the large subunit contains a 5 S rRNA species in all organisms. Eukaryotic large ribosomal subunits also have a 5.8 S rRNA, which is homologous to the 5′ end of eubacterial 23 S rRNA (28 S rRNA is homologous to the rest).

The protein components of eubacterial, archaeal, and eukaryotic ribosomes also differ. Ribosomal proteins have been defined operationally as those proteins that stay bound to the ribosome at 0.5 M salt and bind to specific regions of rRNA. The eubacterial small ribosomal subunit contains 21 different protein species, the eukaryotic 30 species. The eubacterial and eukaryotic large ribosomal subunits contain 32 and approximately 50 different proteins, respectively. Ribosomes from archaea, such as *Caldariella acidophila* and *Methanococcus jannaschi,* have a higher protein content than eubacterial ribosomes in terms of percentage of protein, number of protein

species, and average size of proteins. The majority of proteins associated with the ribosomal subunits (especially the small subunit) are present in both eubacteria and eukaryotes, although the protein-rich eukaryotic ribosomes contain additional proteins not found in the eubacterial ribosome. A smaller number of eubacterial-specific ribosomal proteins occur as well. The *M. jannaschi* genome contains all ribosomal proteins common to eukaryotes and eubacteria. It has no homologs to the eubacterial-specific ribosomal proteins, but does possess a number of the eukaryotic-specific ones. Ribosomal proteins are identified by an arbitrary numbering system, in which all the small ribosomal proteins have an S suffix, e.g., S6, S12, and all the large ribosomal subunits have an L suffix, e.g., L5. There is a separate (and arbitrary) numbering convention for bacteria, archaebacteria, yeast, plant, and animal cells. The most highly conserved ribosomal protein throughout the three life domains is L7 (eubacterial nomenclature) and its methylated counterpart, L12. Apart from L7/L12, which are present in four copies, there is only one copy of each ribosomal protein per subunit. Macromolecular structures that contain single copies of most components must be asymmetric, from which it follows that there are single sites for each process and only one polypeptide may be synthesized at a time.

While ribosomes from different kingdoms can be distinguished morphologically, they resemble each other closely, except that eukaryotic ribosomes are much bigger. The consensus structures, illustrated in Fig. 2, come from a combination of experimental approaches: cryo-electron microscopy, immune mapping, cross-linking, neutron diffraction, crystallography, and functional studies. The small subunit is divided into unequal parts, referred to as the **head**, **base**, and **platform**. The 5′ domain of 16 S rRNA forms the bulk of the body of the 30 S subunit. The middle domain of 16 S rRNA and a long terminal loop structure form the lateral platform. The 3′ domain forms the head of the subunit. The platform, which extends from the base and forms a **cleft** between it and the head, is the site of codon–anticodon recognition. Cryo-electron microscopy has revealed

a channel in the cleft region of the small ribosomal subunit, leading from the cytoplasmic side into the intersubunit space, which is thought to constitute a pathway for the mRNA. The incoming mRNA is envisioned as threading through the channel in the small subunit cleft, making a U-turn and exiting by a path circumscribed by the platform rim. The channel is located in a region thought to be traversed by a single-stranded segment of 16 S rRNA. Near one end of the channel, the end corresponding to the position of the 5′ terminus of mRNA, is the 3′ end of 16 S rRNA. The suggested locations of entry and exit of the mRNA restrict the possible sites of binding of the aminoacyl and peptidyl tRNAs (A- and P-sites) to a region in the cleft 200 nm wide. Ribosomal proteins in this region are involved in mRNA binding. The protein synthesis initiation factors bind near the cleft on the platform.

A consensus structure of the large ribosomal subunit is an irregular hemisphere, the flat face of which interacts with the small ribosomal subunit. It also includes a **central protuberance**, the **L7/L12 stalk**, the **L1 ridge**, and the **interface canyon**. A tunnel starts at the bottom of the interface canyon and runs through the middle of the subunit all the way to the back. The large subunit tunnel is envisioned as the conduit through which the nascent polypeptide exits the ribosome. The prime enzymatic activity of the large ribosomal subunit that catalyzes peptide bond formation is called peptidyl transferase, a ribozyme activity fulfilled by the central loop of domain V of 23 S rRNA. The stalk contains the two highly conserved ribosomal proteins L7/L12 (eubacterial nomenclature). These proteins mediate the GTP-dependent steps of protein synthesis (initiation and elongation). The central protuberance consists primarily of 5 S rRNA and its associated proteins. The two subunits enclose a space, the intersubunit space, that accommodates mRNA, the tRNAs, and associated factors. In a monomeric ribosome, the small ribosomal subunit is positioned on the large subunit, so that the partition between the head and base is approximately aligned with the notch of the large subunit. The platform containing the mRNA channel faces the central protuberance. This junction forms the translational domain. Once an active ribosome is formed, successive additions of amino acids are

A
Small ribosomal subunit

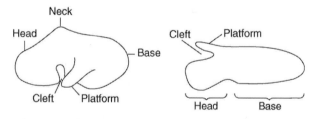

B
Large ribosomal subunit

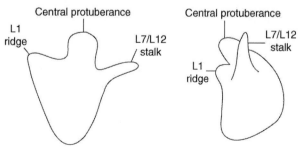

Fig. 2. Ribosomal subunits.

made to the growing peptide chain and the ribosome moves along the mRNA in a 5′ to 3′ direction. It is unusual to find only one ribosome traversing a mRNA. As one ribosome travels down, another ribosome joins. This cyclic process, initiation, ribosome movement, initiation of another ribosome, continues until the mRNA is covered with ribosomes. Such a translation unit is termed a polyribosome, or **polysome**, and is the usual form of translational unit in all cells.

B. Elongation

The fundamental and universal function of protein synthesis is the stepwise addition of amino acids to the growing peptide chain by formation peptide bond between the carboxyl group at the end of the growing peptide chain and a free amino group on the next amino acid. It consists of a series of interactions on the ribosome that is repeated as each codon is translated sequentially from the 5′ to the 3′ end of the mRNA, extending the nascent peptide from the NH$_2$ to the COOH terminus. Elongation is the most rapid step in protein synthesis. In *E. coli*, at 37°C, 20 amino acids are added to the growing polypeptide chain per second, so it takes about 20 s to synthesize a protein of 400 amino acids. In eukaryotes, the rate is lower; in rabbit reticulocytes, the elongation rate is 2 amino acids per second. In most cells, protein synthesis consumes more energy than any other biological process. At least four ATP molecules are split to make each new peptide bond; two of these are required to charge each tRNA and two more drive steps in the cycle of reactions occurring on the ribosome during peptide bond formation. Consistent with the antiquity of the polymerization of amino acids, the events and players involved in peptide bond formation are conserved in Bacteria, Archaea, and Eukarya.

1. Ribosomal Sites of Action

The ribosome must align tRNA accurately with mRNA for correct codon–anticodon base-pairing at the decoding site and for peptide bond formation. The classic model of elongation envisioned two ribosomal binding sites, one for the incoming aminoacyl tRNA (acceptor or A-site) and one for the peptidyl-

tRNA (P-site). More recently, a third site has been postulated, which represents an exit site for deacylated tRNA (E-site). Each site extends over both large and small ribosomal subunits. The A- and P-sites have been mapped to distinct nucleotides in both the eubacterial 16 S and 23 S rRNAs and the E-site to a set of nucleotides in 23 S rRNA. Although both ribosomal subunits contribute to the A- and P-sites, interaction of mRNA with the anticodon is a small ribosomal subunit function, whereas peptide transfer is primarily a large subunit function. The three basic reactions of elongation are: (1) occupation of the A-site by an aminoacyl tRNA according to the corresponding codon at the A-site; (2) peptide bond formation; (3) the translocation reaction, which moves the peptidyl tRNA to the P-site and the deacylated tRNA to the E-site. In the classic model of translation, tRNA movement on the large and small ribosomal subunit is coupled. Direct footprinting of tRNA-ribosome complexes, corresponding to different intermediate states of the elongation cycle, has led to the proposal of hybrid state. In the hybrid states model, binding to, and movement between the A- and P-sites of the small ribosomal subunit is uncoupled from its movement between the A-, P-, and E-sites of the large ribosomal subunit. The events of elongation outlined here will be presented as a three-site hybrid states model.

2. Entry of Aminoacyl tRNA into the A-Site

The entry of aminoacyl tRNA does not occur spontaneously, but is mediated by an elongation factor; EF1A (formerly, EF-Tu) in eubacteria, eEF1A and aEF1A (formerly eEF1α) in higher eukaryotes and Archaea. Once the aminoacyl tRNA is in place, the factor leaves the ribosome to sponsor entry of another aminoacyl tRNA. This cyclic association with, and dissociation from, the ribosome is characteristic of translation factors. The pathway for entry of aminoacyl tRNA to the A-site is illustrated in Fig. 3. EF1A binds to and is activated by GTP. The entry of the aminoacyl tRNA.EF1A.GTP complex occurs in multiple steps. Aminoacyl tRNA is delivered to the ribosome as a ternary complex of aminoacyl tRNA.EF1A.GTP. This ternary complex binds only to ribosomes in which the P-site is occupied by initiator tRNA or peptidyl tRNA. In this state, the anticodon

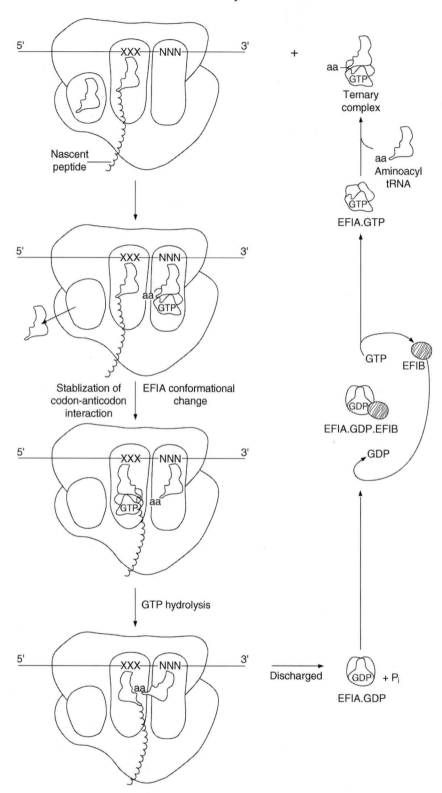

Fig. 3. Entry of aminoacyl tRNA into the ribosomal A-site.

to the A-site of the small ribosomal subunit, interacting with mRNA and 16 S rRNA. The acceptor arm (3′OH–CCA) is bound to EF1A, which, in turn, interacts with a loop in 23 S rRNA on the 50 S ribosomal subunit, locking the aminoacyl end of the tRNA in the A-site. In this intermediate conformation, the aminoacyl end of tRNA is shielded from the peptidyl transferase site. Once the aminoacyl tRNA.EF1A.GTP complex has bound to the ribosome.mRNA complex, the codon–anticodon interaction alters the conformation of EF1A, opening up a site with affinity for peptidyl tRNA. The two tRNA-binding sites of EF1A are topographically, as well as functionally, distinct. Using the second site, EF1A binds to peptidyl tRNA in the P-site, triggering hydrolysis of the bound GTP. EF1A releases the 3′ end of aminoacyl tRNA and leaves the ribosome. The significance of the second tRNA-binding site of EF1A is that GTP hydrolysis only occurs when proper codon–anticodon interaction has taken place at the A-site, which has the effect of increasing accuracy. After discharge of EF1A, the aminoacyl end of tRNA in the A-site is left free to move with respect to 23 S

rRNA loop region, allowing interaction of the aminoacyl end of tRNA with 23 S rRNA in the peptidyl transferase center.

EF1A is released from the ribosome as EF1A.GDP, in which conformation it is not able to interact with aminoacyl tRNA. Guanine nucleotide exchange is mediated by the elongation factor EF1B (formerly EF-Ts) in eubacteria and eEF1β in eukaryotes. EF1B displaces the EF1A-bound GDP and is subsequently replaced by GTP, reforming EF1A.GTP. This sequential activation/inactivation of EF1A by guanine nucleotide binding, hydrolysis, and exchange is a common motif in the repetitive events of translation and represents the mechanism whereby components of the translational machinery can form close associations at one stage of the process, but are able to disengage at a later step. Table II illustrates the properties of EF1A and EF1B, as well as the archaeal and eukaryotic equivalents.

3. Peptide Bond Formation

While decoding of the anticodon of tRNA takes place primarily on the small ribosomal subunit, pep-

TABLE II
Elongation Factors

| Name | Subunits | Mass (kDa) | | Function | Previous name(s) |
		SDS-PAGE	Deduced		
Bacteria (*E. coli*)					
EF1A		50	43.2	binds aminoacyl tRNA and GTP	EF-Tu
EF1B		74	74	binds EF1A by displacing GDP	EF-Ts
EF2		78	77.4	motifs for GTP binding, promotes translocation	EF-G
Eucarya (rabbit)					
eEF1A		51	50.1	binds aminoacyl tRNA and GTP	eEF-1α
eEF1B		114	106.1	promotes guanine nucleotide exchange on eEF1A	eEF-1β
	α-	30	24.8		eEF-1δ
	β-	36	31.2		eEF-1γ
	γ-	48	50.1		eEF-1β
eEF2		95	95.3	motifs for GTP binding, promotes translocation. ADP-ribosylated by diphtheria toxin	eEF-2
Archaea (*Sulfolobus solfataricus*)					
aEF1A		49		sequence like eEF1A, motifs for GTP and aminoacyl tRNA binding	
aEF1B		10		found as a dimer	
aEF2		78	81.7	sequence like eEF2, motifs for GTP binding. ADP-ribosylated by diphtheria toxin	

tide bond formation takes place primarily on the large ribosomal subunit. Peptide bond formation is the central function of the ribosome and is catalyzed by peptidyl transferase. Although current ideas view the enzymatic activity as a function of the central loop of domain V of 23 S rRNA, some role seems to be played by ribosomal protein L27 (eubacterial nomenclature). A peptide bond is formed between the C-terminal carbonyl group of the nascent polypeptide chain, attached to a P-site bound tRNA, and the α-amino group of the A-site-bound aminoacyl tRNA. The CCA terminus of tRNA interacts with peptidyl transferase in the large ribosomal subunit. Peptidyl transferase catalyzes peptide bond formation between the nascent polypeptide carried by the peptidyl tRNA in the P-site. Concomitant with this, the aminoacyl end of A-site-bound tRNA, now carrying the peptide chain, moves into the P-site, leaving the anticodon end in the A-site, to give an A/P hybrid state. This sequence of events of peptide transfer is illustrated in Fig. 4. The significance of the three-site model is that two adjacent codon–anticodon interactions are preserved simultaneously. This

seems to be important for fixing the mRNA and preserving the reading frame. The three-site model of the ribosome distinguishes two ribosomal conformations, referred to as pretranslocational and post-translocational conformations. In the pretranslocational conformation, the A- and P-sites are occupied by tRNAs, with the E-site in a low affinity state. Deacylated tRNA resides in the P-site and the acceptor arm (3′-end) moves to the E-site, although the anticodon remains in the P-site (P/E hybrid state). The peptidyl tRNA resides in the A-site with the peptidyl end towards the P-site (A/P hybrid state). In the posttranslocational conformation, the P- and E-sites are occupied with the A-site in a low affinity state. The significance of the hybrid state is that tRNAs are bound to the ribosome at one end or the other at all times, accounting for the processive character of elongation. In this model, the peptidyl moiety remains stationary; it is the tRNAs that move.

4. Translocation

The formation of the hybrid states described above indicates that translocation should be thought of as a multistep process, beginning during peptide transfer and ending with advancement of the ribosome. The cycle of sequential addition of amino acids is completed as shown in Fig. 5. Translocation is mediated by another protein factor, EF2 (formerly EF-G, eEF2 in animal cells). This is another factor activated by GTP, although it does not need a separate recycling factor. The three-dimensional structure of EF2 shows extensive similarities to that of the ternary complex of aminoacyl tRNA.EF1A.GTP, with some domains of EF2 mimicking the tRNA moiety in the ternary complex (Fig. 6). Consistent with this molecular mimicry, domain IV of EF2, which strikingly matches the structure of the tRNA anticodon in the ternary complex, is found in the same binding pocket on the small ribosomal subunit as that occupied by the ternary complex during entry of aminoacyl tRNA into the A-site. Once bound in the A-site, the ribosomal GTPase is activated and GTP is hydrolyzed to GDP, inducing a conformational change in EF2 and its release from the small ribosomal subunit. This, in turn, triggers a conformational transition in the ribosome that allows movement of the anticodon end of the A-site-bound tRNA to the P-site, which then

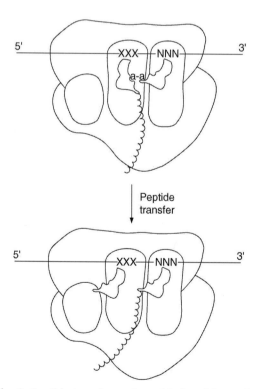

Fig. 4. Peptide transfer and peptide bond formation.

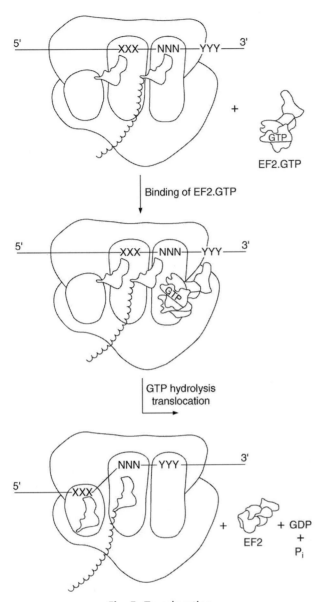

Fig. 5. Translocation.

with respect to the small subunit. Although it is usual to describe translocation in terms of tRNA movement from one ribosomal active site to another along the mRNA, this most probably occurs by the relative movement of (at least parts of) the small ribosomal subunit relative to the large subunit to provide a ratcheting effect. It is important to note that two tRNAs are present on the ribosome before and after translocation. Deacylated tRNA is only released from the E-site upon binding of a new aminoacyl tRNA to the A-site.

C. Termination and Release of the Polypeptide

The final stage of protein biosynthesis is the termination of amino acid polymerization, which occurs when the ribosome encounters a termination (stop) codon. This may be UAA, UAG, or UGA. Termination is accompanied by release of the ribosome and the newly synthesized polypeptide from the mRNA. The overall process resembles the elongation process, except that a stop codon is decoded at the A-site. Cytoplasmic proteins called **release factors** bind directly to any termination codon that reaches the A-site. Bacteria have three release factors, RF1, RF2, and RF3. RF1 and RF2 exhibit codon specificity, recognizing UAG/UAA and UGA/UAA, respectively. They each form a complex with RF3. RF3, a GTP-binding protein that has significant homology to EF2 and EF1A, binds to RF1 or RF2 and carries these factors to the ribosome. Translational termination in eukaryotes is governed by two release factors, eRF1 and eRF3, which are structurally conserved but distinct from the bacterial factors. Analogous to the elongation steps of translation, the termination step involves hydrolysis of GTP by RF3 (or eRF3). Release factor-binding alters the activity of the peptidyl transferase, causing it to catalyze the addition of water, instead of an amino acid, to the peptidyl tRNA. This reaction frees the carboxyl end of the growing polypeptide chain from its attachment to the tRNA molecule, thus releasing the completed polypeptide. As with EF1A and EF2, GTP stimulates the binding of RF-3 to the A-site and GDP dissociates the interaction. This mechanism of translational termination poses a problem in decoding, since it entails RNA-

holds the complete peptidyl tRNA. At the same time, the anticodon end of deacylated tRNA from the previous cycle is translocated from the P-site to the E-site. In the hybrid-states model of elongation, the peptidyl moiety remains stationary; it is the tRNAs that move. The ribosome advances three nucleotides along the mRNA and a new codon triplet is exposed in the A-site. Movement (translocation) of tRNAs takes place in two steps: both tRNAs move with respect to the large ribosomal subunit, after which both tRNAs, along with their bound mRNA, move

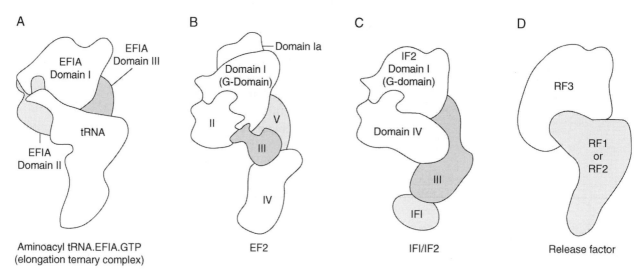

Fig. 6. Domain organization of translation factors illustrating structural similarities to the aminoacyl tRNA.EFIA.GTP complex.

protein recognition instead of codon–anticodon recognition. However, the release factors contain regions homologous to those domains in EF2 which confer its ability to mimic tRNA structure. The tRNA mimicry domain suggests that release factors have the ability to recognize the stop codon through an anticodon-mimicry element. The mRNA sequence immediately downstream of the stop codon is important for termination efficiency. In *E.coli*, a U as the first base downstream gives very efficient termination, whereas an A base gives inefficient termination. Thus, termination is ruled by a signal of four bases, instead of the three that constitute the stop codon itself.

The ribosome also releases the mRNA and goes on to dissociate into two subunits prior to beginning a new round of protein biosynthesis. This requires additional protein factors, the bacterial ribosome release factor, RRF, also called RF4, and the translocation factor EF2, known to be needed for the function of RRF in ribosome recycling. RF1 and RRF have mutually exclusive and, perhaps, overlapping binding sites on the ribosome. These data can be explained by a three-step model for termination, where the first step is RF1-promoted hydrolysis of peptidyl-tRNA. The second step is an intrinsically slow dissociation of RF1, which is accelerated by RF3. The third step, catalyzed by RRF and EF2, leads to mobility of the ribosome on mRNA, allowing it to enter a further round of translation.

III. mRNA RECRUITMENT AND RECOGNITION OF THE CORRECT OPEN READING FRAME

A. Structure of mRNAs

mRNA has a nucleotide sequence complementary to the antiparallel strand of the DNA, except that thymidine is replaced by uridine. All mRNAs contain two types of regions: **coding regions** and **noncoding regions**. Generalized structure for all mRNAs may be seen in Fig. 7. By convention, the base sequence of mRNA, like any other nucleotide sequence, are always described from 5′–3′. Translation of mRNA into protein occurs in the 5′–3′ direction, so the initiation codon is closer to the 5′ end and the termination codon is closer to the 3′ end. The coding sequence, which consists of a series of codons representing the amino-acid sequence, begins at the initiation codon and ends at the termination codon. However, most mRNAs are longer than the coding region; extra regions are present at both the 5′ and the 3′ end and are referred to as the noncoding regions (eubacteria) or **5′ untranslated region (5′ UTR)** and the **3′ untranslated region (3′ UTR)**, respectively (eukaryotes), sometimes called the leader and trailer regions.

A major difference between eubacterial and eukaryotic mRNAs is that eubacterial mRNAs are polycis-

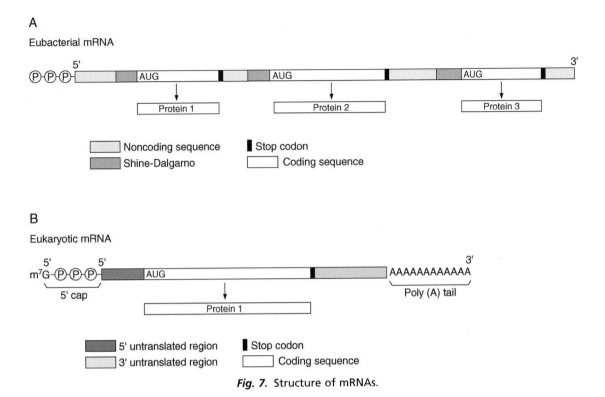

A

Eubacterial mRNA

Protein 1

Protein 2

Protein 3

☐ Noncoding sequence   ■ Stop codon
▨ Shine-Dalgarno   □ Coding sequence

B

Eukaryotic mRNA

5′ cap

Protein 1

Poly (A) tail

▨ 5′ untranslated region   ■ Stop codon
☐ 3′ untranslated region   □ Coding sequence

Fig. 7. Structure of mRNAs.

tronic, whereas eukaryotic mRNAs are not; compare **Fig. 7(A) and (B)**. This reflects the fact that eubacterial genes are almost always in clusters and that the eubacterial mRNA recruitment mechanism allows the polycistronic primary transcript to be translated. The Archaea also have polycistronic mRNAs. In polycistronic mRNAs, intercistronic regions lie between the coding regions. Like the 5′ and 3′ untranslated regions, these also vary greatly in size and may be up to 30 nucleotides, although they can be as short as 1 or 2 nucleotides. In certain instances, the last base of the termination codon is known to serve as the first base of the initiation codon of the next coding sequence.

An almost universal feature of eubacterial mRNAs is a purine-rich region, centered about 10 nucleotides upstream of the initiation codon and termed the Shine–Dalgarno sequence. In most eubacterial, and many archaeal mRNAs, this corresponds to part or all of the sequence **5′-AGGAGG-3′**. This purine tract is complementary to the 3′ terminal sequence of 16 S rRNA, **5′-PyACCUCCUUA-3′**, and represents part of the ribosome-binding site in eubacteria. Methano-

genic and sulfate-reducing Archaea possess Shine–Dalgarno sequences, although halophilic Archaea, in general, have a very short or no 5′ UTR for the first coding sequence in a polycistronic mRNA.

A generalized mRNA structure for eukaryotic mRNAs may also be seen in Fig. 7(B). The length of the 5′ UTR ranges from 3–>700 nucleotides, with most cellular mRNAs having sequences between 40 and 80 nucleotides. Another significant difference between eubacterial and eukaryotic mRNA is that eukaryotic mRNA undergoes important posttranscriptional modifications at both 5′ and 3′ ends. Eukaryotic mRNAs have a methylated cap structure at the 5′ terminus, which is added posttranscriptionally. The initial 5′ mRNA sequence may be represented as:

$$5′pppA/GpNpNpNp . . .$$

The first nucleotide is usually a purine (A or G). This first nucleotide retains its 5′ triphosphate group and makes the usual phosphodiester bond from its 3′ position to the 5′ position of the next nucleotide. After transcription, an additional base, G, is added to the 5′ end by a 5′–5′ triphosphate linkage, catalyzed by guanylyl transferase, a nuclear enzyme:

$$5′\text{Gppp} + 5′\text{pppA/GpNpNpNp} \rightarrow \rightarrow \rightarrow$$
$$5′\text{Gppp5′A/PpNpNpNp} \ldots$$
$$\text{guanylyl transferase}$$

This structure is termed the 5′ cap. It is a substrate for methylations. The first methylation occurs in all eukaryotes and takes place on the 7 position of the 5′ terminal guanine. The enzyme responsible for this is guanine-7-methyltransferase, a cytoplasmic enzyme. In viral, and most animal cell, mRNA, methylation also takes place on the ribose of the second nucleotide, catalyzed by 2′-O-methyltransferase, giving a structure with two methyl groups, termed cap 1. In some cases, methylation also occurs on the ribose of the third nucleotide, to give a structure with a total of three methyl groups, termed cap 2. The m^7G moiety is important for mRNA stability and functions as the recognition site for the eukaryotic ribosome. The functions of the additional methylations are still being determined.

The length of the 3′ UTR shows considerable variation in eukaryotes, averaging 50–100 nucleotides. Most, although not all, eukaryotic mRNAs also have a terminal stretch of A residues, commonly described as the poly(A) tail, downstream from the 3′ UTR. The poly(A) tail is also added posttranscriptionally, in the nucleus usually, and is catalyzed by poly(A) polymerase, which adds between 40–200 residues. Although the length of the poly(A) tract is variable, the factors that determine length are not understood. The poly(A) tail affects mRNA stability and translatability (and allows easy purification of eukaryotic mRNAs).

B. The Initiation Process and Establishment of the Reading Frame

As detailed in the previous sections, most of the components and basic processes of protein biosynthesis are universal. The exception is translational initiation, in which recognition of mRNA, location of the initiation codon, and the components are dissimilar in eubacteria and eukaryotes. The eubacterial initiation process is relatively simple, but the eukaryotic process is complex, involving many steps catalyzed by the initiation factors. Because of the complexity, many of the details of initiation are still uncertain. In the past, these two translational initiation processes were seen as the result of separate evolutionary paths in eubacteria and eukaryotes after the split from a "Universal Ancestor." The two systems have traditionally been seen as analogous, with the implication that translational initiation at the universal ancestor stage was either nonexistent or very different from the processes today. However, the knowledge of archaeal translational initiation components has brought to light a higher degree of similarity between the process of initiation in the three domains than had been previously recognized. Archaea present an interesting mosaic, having the mRNA recruitment signal of eubacteria, but using many of the initiation factors used in eukaryotes. These are among the considerations that led to the speculation that such "design commitments" evolved in the pre-organismic stage of life with three ill-defined types of organization, each distinctly different, which became the three cell types.

One universal aspect of translational initiation is that the small ribosomal subunit interacts with mRNA prior to the large ribosomal subunit. Another universal aspect of initiation is the use of a special initiator tRNA.

1. Initiation of Protein Biosynthesis in Bacteria

The series of reactions that make up the initiation of protein synthesis are finalized with the formation of a 70 S (eubacterial) or 80 S (mammalian) ribosome, containing both ribosomal subunits, initiator tRNA and mRNA. As shown in Fig. 8A prior to the binding of the large ribosomal subunit, the small ribosomal subunit interacts with the mRNA in such a way that the initiation codon lies within that part of the P-site carried by the small ribosomal subunit. In all organisms, this alignment is stabilized by binding of the initiator tRNA. In bacteria, initiator tRNA is able to enter directly into the partial P-site by recognition of the initiation codon in bound mRNA. In eukaryotes, the initiator tRNA steers the alignment of the mRNA. The large ribosomal subunit then interacts to generate a completely functioning ribosome, in which the initiator tRNA lies in a now-completed

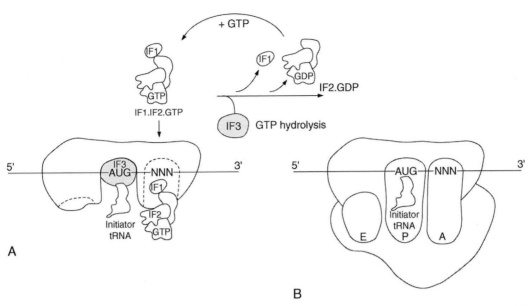

Fig. 8. Alignment of initiator tRNA (eubacterial scheme). (A) Alignment of initiator tRNA in partial P-site. (B) Alignment of initiator tRNA in completed P-site.

P-site and the A-site is available for entry of the aminoacyl tRNA complementary to the second codon Fig. 8(B).

In *E. coli,* the initiation process requires ribosomes, mRNA, fMet-tRNAi, and three initiation factors, IF1, IF2, and IF3. ATP and GTP are also needed. Characteristics of the eubacterial initiation factors are shown in Table III. During initiation, the initiation factors fulfill a role that is similar to that of EF2 during elongation. A key intermediate of initiation is the 30 S initiation complex, consisting of the 30 S ribosomal subunit, mRNA, fMet-tRNAi, and the initiation factors IF1, IF2.GTP, and IF3, as shown in Fig. 8(A).

During the formation of this intermediate, fMet-tRNAi and mRNA containing the appropriate initiations signals, bind to the 30 S ribosomal subunit in random order. IF1 is a small, monomeric, basic thermostable protein of 9 kDa. During initiation, IF1 binds to the ribosomal A-site at the same position as the anticodon domain of the aminoacyl tRNA during elongation, suggesting that the function of IF1 may be to prevent premature binding of elongation Met-tRNA during the initiation process. IF2, the largest eubacterial initiation factors, is a member of the GTP GDP-binding protein superfamily. This factor was believed to function as a carrier of fMet-tRNAi to the 30 S ribosomal P-site, in much the same way as EF1A functions as a carrier of aminoacyl tRNAs to the A-site of 70 S ribosomes. However, the finding that IF2 is universally conserved from eubacteria, archaea, and eukaryotes and is required for efficient initiation in all three life domains has called this function into question. In particular, the eukaryotes, as will be described, have a specific factor that binds the initiator tRNA, but still require IF2 for efficient initiation. These findings, coupled with the sequence comparisons of IF1 and IF2 with EF2, along with modeling from the three-dimensional structure of EF2, suggests that the theme of molecular mimicry can be

TABLE III
Characteristics of Eubacterial Initiation Factors

Name	Mass (kDa)	Function
IF1	9	Binds partial A-site, directs binding of fMet-tRNAi to partial P-site. Recycling of IF2
IF2	100	Binds GTP, interacts with IF1. Promotes fMet–tRNAi binding to partial P-site
IF3	23	mRNA binding and subunit dissociation

further extended to include the IF1/IF2 (see Fig. 6). IF2 promotes the association of the 50 S subunit with the 30 S initiation complex, at the same time activating the ribosomal GTPase activity. All the structural domains of EF2 have a counterpart in IF1 and IF2, with Domain III of IF2, together with IF1, filling the space occupied by the anticodon domain of the aminoacyl tRNA during elongation. In view of this, IF2 bound to IF1 is now viewed as binding to the A-site and directing fMet-tRNA$_i$ to the ribosomal P-site during initiation.

IF3 has a double function during the initiation of translation, those of subunit dissociation and tRNA discrimination. These functions are separable into two discrete domains of IF3. IF3 binds to 16 S rRNA, stimulating the dissociation of vacant 70 S ribosomal couples and, thereby, allowing recycling of the ribosomal subunits for a new initiation event. IF3 also interacts with the anticodon domain of fMet-tRNA$_i$. Filling of the partial P-site by fMet-tRNA$_i$ occurs preferentially when a translational initiation domain of mRNA is within the channel on the small ribosomal subunit. Binding of mRNA to the small ribosomal subunit through interaction with the Shine–Dalgarno sequence brings the initiation codon into the partial P-site, altering its confomation to favor tight binding of the initiator tRNA. Once mRNA and fMet-tRNA$_i$ are aligned in the partial P-site, the IF2-bound GTP is hydrolyzed. The conformational changes induced by GTP hydrolysis have two consequences, the discharge of IF1 and IF2 from the small ribosomal subunit and the completion of the ribosomal A- and P-sites by binding of the 50 S subunit. fMet-tRNA$_i$ occupies the completed P-site and the A-site is ready to accept aminoacylated tRNAs. IF1 plays an additional role at this stage, the recycling IF2.GDP to IF2.GTP.

2. Initiation of Protein Biosynthesis in Eukaryotes

The initiation of protein biosynthesis follows the same basic mechanisms in eukaryotes; the initiator tRNA is bound to the small ribosomal subunit in GTP-dependent reaction, and the large ribosomal subunit interacts only after both initiator tRNA and mRNA are in place on the small ribosomal subunit. However, there are many significant differences in the eukaryotic system, the main distinguishing features being: many more protein synthesis initiation factors and protein complexes are involved; additional components modulate the activities of the initiation factors; mRNA does not contain sequences complementary to the 3′ UTR of rRNA; the 5′ terminus is capped; the mRNAs are usually monocistronic; and the sequence of events is more complex. In eukaryotes, there is an abundance of controls at the level of initiation, which provides for a wealth of regulatory effects on gene expression during development, differentiation, and coordination of the metazoan state. Many of the initiation factors are regulated by reversible phosphorylation or interaction with proteins that modulate their activity. The cap-dependent mechanism of initiation accounts for initiation of most cellular mRNAs. The 3′-terminal poly(A) tail enhances the recruitment of capped mRNAs synergistically by the ribosome. There are also a collection of alternate initiation mechanisms that are responsible for initiation of a small number of mRNAs, such as those for certain cellular regulatory and viral proteins. Table IV lists the characteristics of the eukaryotic initiation factors. The eukaryotic factors are termed eIF1, eIF2, etc., with the number giving some indication of the stage at which the factors work. It is not entirely clear whether the factors 4 represent discrete entities or constitute subunits of a multisubunit factor.

a. Cap-Dependent Initiation

Figure 9 illustrates a consensus model for cap-dependent initiation. The first step is the generation of small ribosomal subunits competent to take part in initiation. The inactive ribosome couple dissociates and the 40 S ribosomal subunit is captured for initiation by binding to eIF1A (17 kDa) and eIF3 (~700 kDa). Under intracellular conditions, the equilibrium between free ribosomal subunits and ribosome couples is strongly weighted to the couples. eIF6 (25 kDa) promotes dissociation by binding to the large ribosomal subunit and decreasing the rate constant of the reaction. The factors eIF1A and eIF3 bind to newly dissociated 40 S subunits and delay reassociation with the 60 S ribosomal subunit long enough to permit their recruitment for initiation. The enormous size of eIF3 changes the sedimenta-

TABLE IV
Characteristics of Eukaryotic Initiation Factors

| *Name* | *Subunits* | *Mass (kDa)* | | *Function* | *Other names* |
		SDS-PAGE	*Deduced*		
IF2		175	139	binds partial A-site, directs binding of Met-tRNA$_i$ to partial P-site.	
eIF1		15	12.6	involved in scanning and recognition of initiator codon, binds eIF3p110.	eIF-1
eIF1A		17	16.5	delays reassociation of subunits, promotes scanning	eIF-4C
eIF2		**125**		binds initiator tRNA, recognition of initiator codon	eIF-2
	α-	36	36.1	regulatory subunit	
	β-	37	38.4	binds Met-tRNAi, interacts with eIF2B$_\varepsilon$ and eIF5	
	γ-	52	51.8	binds GTP, binds Met-tRNAi	
eIF2B		**270**		guanine nucleotide exchange factor for eIF2	GEF
	α-	30	33.7	regulatory subunit	
	β-	39	39	regulatory subunit	
	γ-	58	50.4	guanine nucleotide exchange	
	δ	66	57.1	regulatory subunit	
	ε	85	80.2	guanine nucleotide exchange	
eIF2C		94		stabilizes ternary complex	co-eIF2A
eIF3		**700**		dissociates ribosomes, binds to 18 S rRNA, recognition of initiation codon, interacts with eIF4G and eIF5	eIF-3
	p35	35			
	p36	36	36.5	assembly and maintenance of eIF3 complex	
	p40	40	39.9	Mov-34 family	
	p44	44	35.4	RNA binding protein, binds eIF5	
	p47	47	37.5	Mov-34 family	
	p48	48	52.2	regulates eIF3 activity	
	p66	66	63.9	binds RNA	
	p110	10	105.2	involved in Met-tRNAi binding, interacts with eIF5	
	p116	116	98.8	interacts with p170	
	p170	170–180	167	interacts with eIF4B	
eIF4A		45	44.4	ATPase, helicase, binds mRNA	eIF-4A
eIF4B		80	69.8	binds mRNA, interacts with p170 of eIF3, forms homo-dimer	eIF-4B
eIF4E		25	25.1	binds to m^{7}G cap	eIF-4α, CBP1
eIF4G		>220	171.6	binds eIF3, eIF4A, eIF4E, poly(A) binding protein	eIF-4γ, p220
eIF5		58	48.9	stimulates GTPase activity on eIF2, recognition of initiation codon, release of factors from 48 S initiation complex. Interacts with eIF3p110, eIF2β and eIF2Bε	eIF-5
eIF6		25		binds to 60 S subunits, promotes subunit dissociation	eIF-3A

tion value of the small ribosomal subunit to 43 S, which is referred to as the 43 S preinitiation complex. Biophysical and electron microscope analyses of eIF3 show that it has the shape of a flat triangular prism, which binds to the small ribosomal subunit in the region that interfaces with the large ribosomal subunit. eIF3 serves to organize the interaction of mRNA and initiation factors at the active site of initiation in the cleft of the small ribosomal subunit.

Following ribosomal subunit dissociation, the

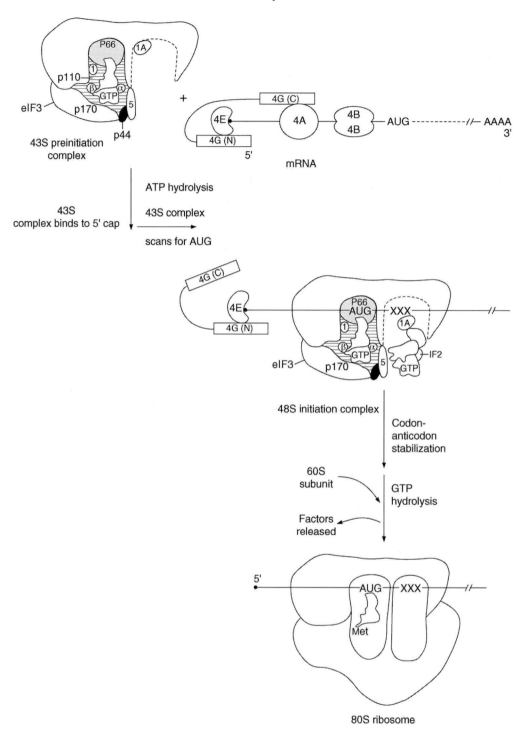

Fig. 9. Cap-dependent initiation.

competent small ribosomal subunit interacts with the initiator tRNA, Met-tRNA$_i$, to generate a 43 S preinitiation complex (initiation intermediate). Met-tRNA$_i$ is recruited to the competent small ribosomal subunit with the help of eIF3, as part of a ternary complex with the initiation factor eIF2 and GTP. GTP-bound eIF2 brings the initiator tRNA to the partial P-site on the small ribosomal subunit, in readiness for ribosome recognition of the initiation codon. Recent evidence suggests that the eukaryotic homolog of eubacterial IF2 also plays a role in this step, by binding to the partial A-site, preventing occupation by elongation aminoacyl tRNAs. eIF2 is a complex of three polypeptide chains, α (36 kDa), β (38 kDa), γ (52 kDa), which remain associated throughout the initiation cycle. The γ-subunit is homologous to eubacterial EF2. This subunit binds GTP and contains the three consensus GTP-binding elements, appropriately spaced, for a GTP-binding protein. The mammalian β-subunit (but not that from yeast or *Drosophila*) contains two of these GTP-binding elements; they do not seem to be required for eIF2 function. The β-subunit, which has a zinc finger motif, functions in recognition of the initiation codon, as well as in the interaction of eIF2 with the GTP recycling factor eIF2B. Both β- and γ-subunits cross-link to Met-tRNA$_i$.

The 43 S preinitiation complex recruits the mRNA in a two-phase ATP-dependent step: (i) interaction with the 5′ terminus of mRNA, under the influence of the eIF4 group of initiation factors and (ii) the migration (scanning) of the ribosome along the mRNA to the initiation codon. eIF4E is a factor of 25 kDa, which binds to the 5′ cap structure of mRNA. eIF4A (45 kDa) is an RNA-dependent ATPase that has ATP-dependent duplex unwinding activity. Both eIF4E and eIF4A interact with eIF4G ($>$220 kDa), which functions to bring together, in the correct orientation and in the close proximity to the cap, the components necessary to unwind secondary structure and place the small ribosomal subunit at the 5′ end of mRNA. eIF4G fulfills this function through interaction with eIF3. The enhancement of cap-mediated initiation by the 3′ poly(A) tail also involves a direct interaction between poly(A) binding protein, which is bound to the poly(A) tail, and eIF4G. In mammals, eIF4E, eIF4A, and eIF4G can

be isolated in a complex called eIF4F. In plants, *Drosophila,* and yeast, a binary complex of eIF4E and eIF4G (and iso4E and iso4G in plants) can be isolated. Another eIF4 group member is eIF4B (80 kDa), which functions as a dimer to promote the ATPase reaction of eIF4A. eIF4B has an RNA-recognition motif (RRM) near its amino-terminus and a central domain rich in aspartate, arginine tyrosine, and glycine. The eIF4B dimer can bind two molecules of RNA simultaneously and interacts directly with 18 S rRNA and the p170 subunit of eIF3. eIF4B is thought to promote ribosome binding to the 5′UTR by bridging it to 18 S rRNA. It is generally accepted that eIF4E recruits the mRNA by binding to the 5′ cap and that the function of eIF4G, through its interaction with eIF4E and the small ribosomal subunit-bound eIF3, is to bring the mRNA to the small ribosomal subunit. It is also generally accepted that the helicase activity of eIF4A, either alone, or as part of eIF4F, unwinds secondary structures from the 5′ UTR, which would otherwise impede the small ribosomal subunit. Once bound to mRNA, the resulting initiation intermediate in mammalian cells can be distinguished as a 48 S initiation complex. The small ribosomal subunit with its attached factors migrates along the 5′ UTR in an ATP-dependent process known as scanning, until it encounters an initiator AUG codon. For this, additional factors, eIF1 and eIF1A and ATP are needed. In keeping with current understanding of the role of IF2, it seems likely that a complex between eIF1A and IF2 functions as a tRNA/elongation factor mimic, to occupy the partial A-site during scanning. A dependence of scanning on GTP hydrolysis, therefore, seems likely but has not been demonstrated.

The scanning process continues until the small ribosomal subunit recognizes the initiation codon. The minimal feature of an mRNA essential for initiation codon recognition is an AUG in a favorable context, the consensus sequence for which is **GCCGCCA/GCCAUGG**. Genetic studies in the yeast show that eIF1, eIF1A, all three subunits of eIF2, eIF5 (50 kDa), and the smallest subunit of eIF3 are involved in the AUG recognition process, along with Met-tRNA$_i$. Once Met-tRNA$_i$, is lined up in the partial P-site with the initiation codon, the complex is recognized by eIF5, which is thought of as a fidelity

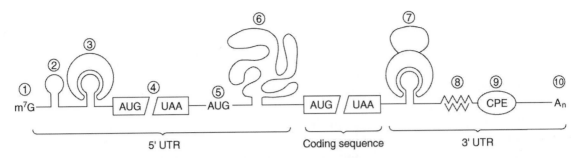

Fig. 10. Features of eukaryotic mRNA untranslated regions that confer translational regulation (after Gray and Wickens, 1998). Key: 1, modification of cap structure; 2, secondary structures; 3, 5′UTR/protein interaction; 4, upstream open reading frames; 5, upstream AUGs; 6, IRES elements; 7, 3′UTR/protein interaction; 8, 3′UTR/protein interaction; 9, cytoplasmic polyadenylation elements; 10, poly(A) tail length.

factor that assesses the correctness of the Met-tRNA$_i$-initiation codon interaction. Through a bipartite motif, the **AA-box**, rich in aromatic and acidic residues, eIF5 can interact with lysine-rich motifs, **K-boxes**, in the p110-subunit of eIF3, as well as the β-subunit of eIF2. It seem likely that eIF5 is recruited to the small ribosomal subunit, through its association with eIF3 p44 and p110 subunits, and is triggered to interact with eIF2β when initiation codon–anticodon recognition occurs. This results in activation of GTPase, hydrolysis of the eIF2-bound GTP, and the release of eIF2, eIF3, and probably other factors from the initiation complex. Although eIF5 is known to contain motifs typical of the GTPase superfamily, it can only promote GTP hydrolysis in the presence of the

small ribosomal subunit. After release of the initiation factors, the 60 S subunit binds and elongation begins.

After GTP hydrolysis, eIF2 is released from the ribosome as eIF2.GDP. For the next round of initiation, the GDP on eIF2 must be exchanged for GTP by the action of eIF2B, a guanine nucleotide exchange factor. eIF2B is a high molecular mass complex of five subunits, α (26 kDa), β (39 kDa), γ (58 kDa), δ (67 kDa), and ε (82 kDa). The eIF2Bγ- and ε-subunits are the catalytic subunits that promote guanine nucleotide exchange on eIF2γ. eIF2Bγ alone has no catalytic activity, but stabilizes the binding of eIF2 to eIF2Bε. As with eIF5, the AA-box of eIF2Bε interacts with the K-box of eIF2β. The other subunits are involved in the regulation of activity that occurs through phosphorylation of eIF2α.

b. Variations of Cap-Dependent Initation

An AUG in optimal context does not ensure efficient initiation. The frequency and efficiency of initiation is also determined by the accessibility of the 5′ terminus of the mRNA and subunit migration can be limited by structural features of the 5′ UTR, which can impede or derail the scanning process. The translational efficiency of an mRNA can also be influenced by sequences at or near the 3′ UTR. Figure 10 illustrates some of the forms of translational regulation by the 5′ and 3′ UTRs of the mRNA. In general, the first AUG encountered by a scanning ribosome is selected as the site of initiation. However, an AUG may be recognized

TABLE V
Initiation Factors Encoded by Genome of *M. jannaschii*

Gene ID Number	Proposed Homolog
0445	eIF1A
0117	eIF2α
0097	eIF2β
1261	eIF2γ
1574	eIF4A
1505	eIF4A
0669	eIF4A
1228	eIF5
0454	eIF2Bα (disputed)
0122	eIF2Bδ (disputed)
0264	IF2

inefficiently or ignored, if it is located close to the cap, or if it lies in poor context. Inefficient recognition of an initiation codon allows a percentage of small ribosomal subunits to continue to scan and initiate at a downstream site. This is called **leaky scanning.** Some mRNAs have short upstream open reading frames (**uORFs**), which serve to inhibit utilization of a mRNA except in special circumstances. The coding sequence can only be used if a percentage of the small ribosomal subunits are able to stay on the mRNA after passing a termination codon and reinitiate at the authentic initiation codon downstream. A new ternary complex and some factor 4 family members are required for this to take place. A small number of viral RNAs have been descrebed that contain 5 UTR structures unfavorable to the scanning process. For these unusual mRNAs, the small ribosomal subunit begins by initiating at the 5′ cap and scanning downstream, only to bypass the unfavorable structure by a process known as **hopping** (or shunting, jumping). The mechanism of hopping has not been elucidated, but involves special sequence elements in the 5′ UTR.

c. Cap-Independent Initiation

A small proportion of animal virus mRNAs and a smaller proportion of cellular mRNAs are not recruited through their 5′ cap structures. Instead, the small ribosomal subunit binds to a secondary structure within the 5′ UTR referred to as an internal ribosome entry site (**IRES**). IRESs can be functionally discriminated from other 5′ UTR secondary structures by their ability to mediate translation of a downstream ORF of a bicistronic reporter mRNA, independently of translation of the first ORF. Most of the general initiation factors described in the preceding section seem to be required, except for eIF4E. Translation from many IRESs require additional *trans*-acting factors, such as the polypyrimidine tract binding protein (**PTB**), the function of which may be to maintain the structure of the IRES in a favorable conformation. Internal initiation allows picornavirus mRNAs, which do not contain 5′ caps, to escape the shutdown of initiation that occurs during picornavirus infection. Recruitment of host mRNA is prevented by the cleavage of eIF4G. This cleavage sepa-

rates the eIF4E-binding domain of eIF4G from the domains that interact with eIF3 and eIF4A, thereby separating the cap-binding function of eIF4G from the helicase and ribosome binding activities of eIF4A and eIF3. However, the remaining fragment of eIF4G can substitute for intact eIF4G in IRES-mediated recruitment of mRNA.

3. Initiation of Protein Biosynthesis in Archaea

The genome-sequencing projects have provided insights into the relationship between the process of initiation in Archaea, compared to eubacteria and eukaryotes. In general, the Archaea seem to use a mixture of eubacterial and rudimentary eukaryotic features. However, little experimental verification of the conclusions based on sequence has yet been made. Translation initiation signals in archaeal mRNAs resemble those of eubacteria, in that Archaea frequently utilize Shine–Dalgarno sequences to identify initiation codons. Table V lists the proposed initiation factor homologs encoded by the genome of *M. jannaschii*. The archaeal genomes contain the three eubacterial initiation factors, IF1, IF2, and IF3, along with homologs to eukaryotic eIF1A, all subunits of eukaryotic eIF2, and an eIF4A-like helicase. Although homologs to two subunits of eIF2B have also been claimed, this homology has been called into question. Since eubacterial IF2 and eIF2 apparently perform the same function, accurate delivery of the initiator tRNA to the small ribosomal subunit, this seemed somewhat of a puzzle. However, subsequent to the archaeal discoveries, it has been found that yeast and humans also have IF2 (referred to as yIF2 and hIF2, repectively). Further examination has suggested that IF2 fulfills the same function in all three domains, binding to the partial A-site on the small ribosomal subunit, thus facilitating proper binding of the initiator tRNA to the partial P-site. eIF5, crucial to hydrolysis of eIF2-bound GTP, is present in Archaea although neither of the subunits of eIF2B involved in guanine nucleotide exchange are found. Since these are the AA-box-containing factors in eukaryotes, which interact with the K-box in eIF2β, it is of interest to note that eIF2β in Archaea does not contain the K-box domain crucial for interactions with eIF2B and eIF5. Archaea do not possess any of the factors 4 family involved in recruitment of capped mRNAs.

See Also the Following Articles

Bibliography

Brock, S., Szkaradkiewicz, K., and Sprinzl, M. (1998). Initiation factors of protein biosynthesis in bacteria and their structural relationship to elongation and termination factors. *Mol. Microbiol.* **29**, 409–417.

Buckingham, R. H., Grentzmann, G., and Kisselev, L. (1997). Polypeptide chain release factors. *Mol. Microbiol.* **24**, 449–456.

Czworkowski, J., and Moore, P. B. (1996). The elongation phase of protein synthesis. *Progr. Nucl. Acids Res.* **54**, 293–332.

Gray, N. K., and Wickens, M. (1998). Control of initiation in animals. *Annu. Rev. Cell Dev. Biol.* **14**, 399–458.

Green, R., and Noller, H. F. (1997). Ribosomes and translation. *Annu. Rev. Biochem.* **66**, 679–716.

Hershey, J. W. B., Mathews, M. B., and Sonenberg, N. (1996). "Translational Control." Cold Spring Harbor Laboratory Press, Cold Spring Harbor, NY.

Moore, P. B. (1998). The three-dimensional structure of the ribosome and its components. *Ann. Rev. Biophys. Biomol. Struct.* **27**, 35–58.

Pestova, T. V., and Hellen, C. U. T. (1999). Ribosome recruitment and scanning: What's new? *TIBS* **24**, 85–87.

Lee, J. H., Choi, S. K., Roll-Mecak, A., Burley, S. K., and Dever, T. E. (1999). Universal conservation in translation initiation revealed by human and archaeal homologues of bacterial translation initiation factor IF2. *Proc. Natl. Acad. Sci. USA* **96**, 4342–4347.

Protein Secretion

Donald Oliver

Wesleyan University

Jorge Galan

Yale University

GLOSSARY

general secretion pathway Pathway by which proteins containing signal peptides or membrane signal–anchor domains are integrated into or transported across the cytoplasmic membrane of bacteria.

protein export or secretion Transport of proteins to extracytoplasmic compartments, including the exterior of the cell or to other cells.

sec-machinery or translocon Components that promote protein export via the general secretion pathway.

signal peptide or secretion signal A region of a secreted protein that contains information for interaction with the secretion machinery.

PROTEIN SECRETION, or protein export, is the movement of proteins from one cellular compartment to another or to the exterior of the cell. In bacteria, this involves the movement of presecretory and membrane proteins from the cytoplasm, their initial site of synthesis, to a number of aqueous or membranous compartments, depending on the type of bacterium under study (see Fig. 1). In recent years, it has become evident that a variety of pathogenic bacteria are also capable of secreting their proteins into animal and plant cells.

Bacteria possess several different pathways for promoting protein secretion, and these are generally classified according to the secretion machinery uti-

lized for protein export (Table I). The most commonly utilized pathway is the Sec-dependent or general secretion pathway (GSP). The GSP promotes protein translocation across the cytoplasmic membrane. It is driven by a collection of cytoplasmic, peripheral, and integral membrane proteins that are collectively known as the Sec machinery, or translocon. The GSP can be thought of as the major housekeeping protein secretion pathway of the bacterial cell. It is homologous to the pathway promoting import of presecretory and membrane proteins into the endoplasmic reticulum of eukaryotic cells. Certain gram-negative bacteria possess terminal branches of the GSP for secretion of proteins across the outer membrane. Besides the GSP, there are several other specialized protein secretion pathways that are utilized by more limited sets of proteins. These include (i) the twin-arginine protein secretion pathway that promotes the export of redox-cofactor-containing enzymes, (ii) ABC protein secretion pathways, that are driven by ABC (ATP-binding cassette) exporters, and promote the secretion of toxins, proteases, lipases, and specific peptides, and (iii) the type III or contact-dependent protein secretion pathways that promote binding and import of proteins into animal and plant cells. These latter two pathways are typically associated with bacterial pathogens and specialized microbial niches.

I. GENERAL SECRETION PATHWAY

A. Targeting Signals on Preproteins

The hallmark of a GSP-dependent presecretory protein (preprotein) is the possession of a character-

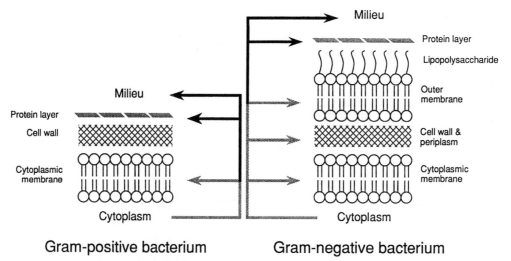

Fig. 1. Destinations of secreted proteins in bacteria. Black arrows indicate destinations of extracellular secreted proteins, while shaded arrows indicate destinations of intracellular secreted proteins. [From Pugsley, A. (1993). *Microbiol. Rev.* **57,** 51.]

istic signal peptide that is often located at its amino terminus (Fig. 2). Such signal peptides serve as recognition elements for the Sec-machinery, and they also delay the rate of protein folding, a potentially important feature, since protein export by the GSP must occur prior to the completion of protein folding. Secretory preproteins can interact with the Sec-machinery in either a cotranslational or posttranslational fashion. Sec-dependent signal peptides range in length from 18 to 30 amino acid residues (typically being longer in gram-positive bacteria). Although they possess little in the way of primary sequence homology, they do contain three characteristic regions: (i) a short, positively charged N-domain, (ii) a hydrophobic H-domain of 7 to 15 uncharged amino-acid residues, and (iii) a short, polar C-domain that contains a signal peptidase cleavage site. Two consensus cleavage motifs exist within the C-domain, one for general secretory proteins and one for lipoproteins. Lipoproteins need to be modified with diglyceride at a cysteine residue at the beginning of the mature region to allow signal peptide cleavage. Mutational studies show that the H-domain is the most critical for signal peptide function, and truncation or introduction of charged residues within the H-domain often prevents export of the protein. In contrast, the N-domain appears less critical, although its overall charge can affect the kinetics of protein

secretion. Alteration of the C-domain may prevent signal peptidase cleavage; while the resulting preprotein is still secreted, it remains tethered to the cytoplasmic membrane by its uncleaved signal peptide. The signal peptide is thought to function as a loop, allowing insertion of an amino-terminal hairpin of the preprotein into the cytoplasmic membrane. Whether such insertion occurs directly into the lipid bilayer, by analogy to model systems, or is promoted by prior interaction with the translocon is unknown.

Additional recognition elements are contained within preproteins since signal peptide deletion still allows a low level of protein secretion to occur. Residual export can be greatly augmented by certain mutations in the Sec machinery that generally suppress signal peptide defects (see Prl following). One type of recognition element for chaperones that targets preproteins to the translocon appears to be exposed hydrophobic protein surfaces. It is thought that delayed folding kinetics of preproteins, with resultant hydrophobic surface exposure, facilitates preprotein interaction with specific chaperones for maintenance of the export-competent state. However, the structural basis for such recognition has remained largely illusory.

The final sorting of secretory proteins to the periplasm, cell wall, and outer membrane or protein layer occurs subsequent to translocation across the

TABLE I

Classification of Bacterial Protein Secretion Pathways

Pathway	Alternative name	Distribution	Translocation pathway	Protein substrates	Location of protein secretion signals	Machinery components
General secretion pathway	Sec-dependent pathway	Ubiquitous Bacteria and Archaea	Cytoplasmic membrane traversal or integration	Most periplasmic, cell wall, and outer membrane proteins, and some cytoplasmic membrane proteins	N-terminal signal peptide or signal-anchor domain, hydrophobic patches on molten globule state of preproteins?	Sec machinery or translocon
Main terminal branch of the general secretion pathway	Type II pathway	Limited Gram-negative animal and plant pathogens	Outer membrane traversal	Restricted number of hydrolytic enzymes and toxins	Patch signal(s) on tertiary or quaternary structure?	Type II exporter
Twin-arginine protein secretion pathway		Nearly ubiquitous? Bacteria and Archaea	Cytoplasmic membrane traversal or integration	Redox enzymes	N-terminal twin arginine signal peptide, patch signal(s) on tertiary structure?	TAT exporter
ABC protein secretion pathway	Type I pathway	Limited Bacteria and Archaea	Single step traversal of cytoplasmic and outer membrane (if present)	Restricted number of hydrolytic enzymes and toxins	Folded C-terminal secretion signal	ABC exporter
Type III protein secretion pathway	Contact-dependent protein secretion pathway	Limited Gram-negative animal and plant pathogens	Single step traversal of both bacterial and target cell membranes	Restricted number of agonists and antagonists of eukaryotic target cell responses	N-terminal secretion and targeting signals	Type III exporter

A

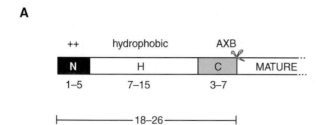

B

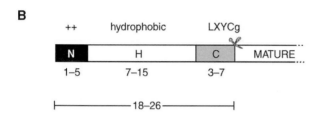

C

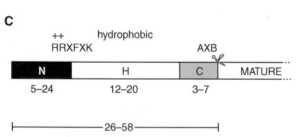

Fig. 2. Structure of signal peptides. N-, H-, and C-domains for each signal peptide are indicated, along with their approximate length in amino acids. ++ indicates a net positive charge. The scissors indicate the signal peptidase cleavage site. (A) Signal peptide of a typical preprotein utilizing the GSP. The consensus cleavage motif AXB is indicated, where A indicates small uncharged or larger aliphatic amino acid residues, X indicates any residue, and B indicates small uncharged residues. (B) Signal peptide of a lipoprotein utilizing the GSP. The consensus cleavage motif LXYCg is indicated, where L indicates leucine, X and Y tend to be small uncharged amino-acid residues, and Cg indicates glyceryl-modified cysteine. (C) Signal peptide of a preprotein utilizing the twin-arginine secretion pathway. The twin-arginine motif RRXFXK is indicated, utilizing single amino-acid code, except X indicates any amino acid. The AXB cleavage motif is the same as in (A). [Adapted from Fekkes, P., and Driessen, A. (1998). *Microb. Rev.*]

cytoplasmic membrane. The periplasm is probably a default pathway that requires no specific sorting signals to achieve proper targeting. By contrast, topogenic signals are required to promote association with the cell wall or outer membrane. The nature of these signals is only now being clarified

for specific systems, and it is unclear, at present, how general such signals are. For outer membrane protein targeting, the possession of β-structure or the ability to bind to lipopolysaccharide appears to be important. In *E. coli* lipoproteins, the identity of the second mature amino-acid residue is critical for distinguishing inner from outer membrane destinations. In certain gram-positive bacteria, an LPXTGX motif (X indicates any amino acid residue), followed by a carboxy-terminal hydrophobic domain and charged tail, promotes anchoring of cell wall proteins.

Certain integral membrane proteins have been shown to utilize the GSP for their biogenesis, particularly when they contain large periplasmic domains. In this case, the signal-anchor domain (transmembrane segment) fulfills a function analogous to the signal peptide, except that no cleavage takes place. For complex polytopic membrane proteins, odd-numbered transmembrane segments serve to initiate membrane protein insertion, while even-numbered segments serve as membrane anchor domains to arrest translocation (Fig. 3). Insertion of an amino-terminal portion of a membrane protein in an N_{out}–C_{in} orientation (so-called amino-terminal tail insertion) has been shown to be independent of the GSP and may require a novel pathway promoted by a homolog of Oxa1p, a protein responsible for amino-terminal tail insertion of mitochondrial membrane proteins.

B. Chaperones and Targeting Factors

As nascent or newly completed chains of presecretory and membrane proteins emerge from the ribosome, they encounter a variety of chaperones that maintain them in a unfolded state and target them to the translocon. While a number of chaperones have been shown to play at least minor roles in protein export (i.e., trigger factor, a ribosome-associated proline isomerase, and heat shock proteins GroE, DnaK, DnaJ, and GrpE), two major chaperones have been characterized in *E. coli*: SecB and signal recognition particle (SRP). SecB is present in gram-negative bacteria (its existence in gram-positive bacteria is uncertain), while SRP is conserved among all three domains.

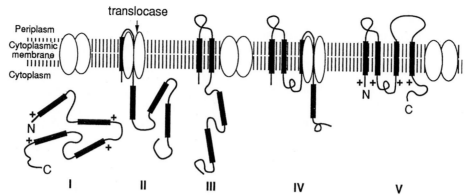

Fig. 3. Sec-dependent insertion of integral membrane proteins utilizing signal-anchor domains. Signal-anchor domains are indicated by black rectangles and + indicates positively charged regions that help promote the correct membrane topology. I, II, III, IV, and V indicate successive stages of membrane protein integration. [From Pugsley, A. (1993). *Microbiol. Rev.* **57**, 70.]

SecB is a small tetrameric protein that binds a particular group of preproteins, either cotranslationally or posttranslationally, and targets them to the translocase through its specific interaction with SecA, a subunit of translocase. SecB is nonessential for cell viability, although its absence results in substantial secretion defects for SecB-dependent proteins. Unlike the more complex heat shock chaperones, SecB is not an ATPase, and it can only prevent protein folding, not reverse it. The structural basis of SecB-preprotein recognition is unclear, although it requires a hydrophobic patch on SecB. Of interest, this region overlaps with acidic residues that are the SecA-binding site on SecB. These results have lead to a model (Fig. 4) where preproteins are passed from SecB to SecA by a "hand-off" mechanism.

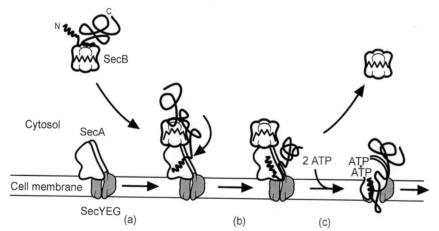

Fig. 4. "Hand-off" model for preprotein transfer from SecB to SecA. (a) SecB targets the preprotein (wavy black line) to membrane bound SecA, where signal peptide binding to SecA stimulates interaction between SecA and SecB. (b) SecA–SecB interaction causes the release of the mature portion of the preprotein from SecB. The zigzag symbol represents the region of SecB that associates with both SecA and preprotein and which changes conformation during the hand-off. (c) Upon binding ATP, SecA changes conformation, resulting in initiation of protein translocation and release of SecB. [From Driessen *et al.* (1998). *Curr. Opin. Microbiol.* **1**, 218.]

Bacteria possess an SRP particle and SRP receptor that is homologous to its eukaryotic counterparts and are essential for cell viability. The *E. coli* homologs of SRP54 and 7S RNA are Ffh or P48 and 4.5S RNA, respectively. FtsY is the *E. coli* homolog of eukaryotic SRP receptor α subunit, and no SRP receptor β subunit homolog has been identified yet. While there is no evidence for a protein translation arrest by bacterial SRP, the membrane targeting function of bacterial SRP appears to occur similarly to eukaryotic SRP. The interaction of Ffh with proteins must occur when they are still ribosome-associated, and it is dependent on the hydrophobicity of the signal peptide or signal-anchor domain. Of the three domains of Ffh (amino-terminal N domain, GTPase G domain, and methionine-rich M domain), the N and G domains appear to contain the signal peptide-binding region, while the M domain is involved in interaction with 4.5S RNA. Membrane-bound FtsY and GTP hydrolysis are required for signal peptide release from SRP. Upon release from SRP, presecretory or membrane proteins associate with the translocon. Although the details of this "hand-off" reaction are not yet known, it appears likely that the protein is targeted to SecA. While characterization of *E. coli* SRP–SRP receptor has proceeded well, considerable controversy surrounds the identity of the *in vivo* substrates of SRP. While SRP mutation or depletion affects somewhat the secretion of certain periplasmic and outer membrane proteins, recent data indicates that this system plays an important role in the biogenesis of particular integral membrane proteins. The effects of SRP depletion on periplasmic and outer membrane secretion may be indirect, since SecY protein requires SRP receptor (and presumably SRP) for its biogenesis.

C. The *E. coli* Translocon

The bacterial translocon consists of core and accessory subunits (Fig. 5). The core structure is composed of two essential proteins: cytoplasmic membrane protein SecYE and membrane-dissociable SecA ATPase. This catalytic core of translocase has been shown to be required for both *in vivo* and *in vitro* protein translocation. Three accessory proteins have also been characterized in some detail: integral membrane proteins SecD and SecF (which probably func-

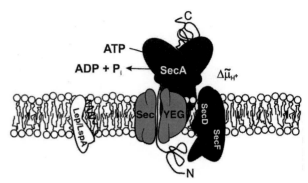

Fig. 5. Structure of the translocon of *E. coli*. Core subunits SecA and SecYE are indicated, along with accessory subunits SecG, which associates with SecYE, and SecDF. A partially translocated preprotein (wavy black line) is shown as well as the two energy sources, ATP and $\Delta\tilde{\mu}_{H^+}$. Lep/LspA indicate signal peptidase I and II, respectively. [From Fekkes, P., and Driessen, A. (1998). *Microb. Rev.*]

tion together and are contained in a single protein in *B. subtilis* and certain other bacteria) and SecG protein. These accessory proteins are essential for cell growth only at low temperatures, and they enhance *in vivo* and *in vitro* protein translocation under a variety of conditions.

SecYE, in addition to constituting the SecA receptor, has been proposed to form a channel for protein translocation. This would be analogous to the Sec61 channel present in the endoplasmic reticular membrane of eukaryotes. SecYE/Sec61$\alpha\gamma$ homologs are present in all three domains of life. Bacterial SecY consists of 10 transmembrane domains with small cytoplasmic and periplasmic loops flanking these domains. By contrast, SecE protein is much smaller, consisting of only three transmembrane domains in *E. coli* (only one of which is essential) and a single transmembrane domain in *B. subtilis*.

SecA is a large dimeric protein present only in the bacterial domain. Most bacterial SecA proteins probably contain at least five domains: two ATP-binding domains, a preprotein-binding domain, and domains for binding SecB and SecYE proteins. This organization allows SecA to serve as an assembly factor for binding preproteins or SecB-bound preproteins to translocase. SecA interacts with preproteins by recognition of both signal peptide and mature portions, and the basic N-domain and hydrophobic H-domain of the signal peptide are important for binding. The order of assembly of the translocation complex has not been defined, although it appears likely that SecA binding to the cytoplasmic membrane and SecYE precedes the preprotein binding step.

SecA also serves as a motor protein to drive protein translocation, utilizing its translocation ATPase activity. This activity appears to correspond to successive cycles of membrane insertion and retraction by SecA protein, and it has been proposed that a mobile domain of SecA acts like a molecular sewing machine needle to drive protein translocation in a stepwise manner (Fig. 6). Cross-linking studies suggest that the translocating preprotein is contained in a channel that is formed by SecA and SecY proteins, which is shielded from lipid. Secretory proteins must be driven through the translocon to the periplasmic side of the membrane, while Sec-dependent membrane proteins undergo limited translocation, until a signal-anchor domain arrests further translocation and allows their release into the lipid bilayer. Little is known about how the translocon allows such lateral release of membrane proteins.

SecG is a small protein with two transmembrane domains that undergoes a cycle of membrane topology inversion during protein translocation. This cycle is coupled to SecA membrane cycling and enhances it in some way. Recent work suggests that SecG promotes a conformational change in SecA that allows it to insert into the membrane at SecYE.

It has been suggested that SecDF is important for the stabilization of membrane-inserted SecA, thereby preventing backward slippage of preproteins during the translocation cycle. However, this suggestion seems at odds with the observation that archaea lack a *secA* homolog but possess *secD* and *secF* homologs. Since SecD and SecF each possess a large periplasmic domain, it has been speculated that they may play a role late in protein secretion, such as in the folding or release of translocated proteins, their presentation to leader peptidase, or the disassembly and recycling of the Sec machinery.

Genetic studies suggest that the translocon has a proofreading activity that recognizes signal peptides on proteins. Mutations allowing export of preproteins with defective signal peptides (designated *prl*

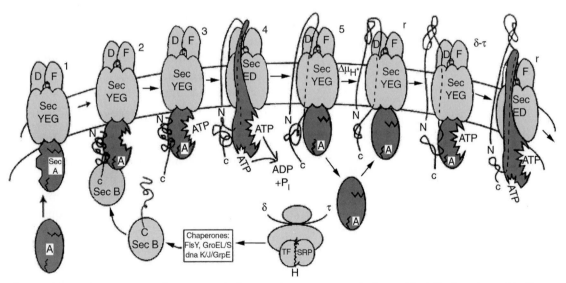

Fig. 6. Proposed protein translocation cycle for *E. coli* GSP. Preprotein (wavy black line) and ATP-binding to SecA promote a cycle of SecA membrane insertion and retraction that drives translocation of the first segment of the preprotein (steps 1–4). A later step (step 5) in protein translocation is driven by PMF ($\Delta\mu_H^+$). In the lower portion of the figure, the role of SecB and other chaperones in delivering preprotein to SecA is depicted, along with the ability of SecYEG-bound SecA to exchange with cytosolic SecA. [From Duong *et al.* (1997). *Cell* **91**, 568.]

for protein localization) are alleles of *secA* (*prlD*), *secE* (*prlG*), *secG* (*prlH*), and *secY* (*prlA*). *prlA* mutants are the strongest suppressors, and they can efficiently secrete proteins that lack a signal peptide (although not cytoplasmic proteins). Genetic analysis of this system suggests that certain regions of SecY and SecE are involved in signal peptide recognition, and this step is regulated by SecA ATPase activity, which is likely to be utilized as a molecular clock to define the kinetic window for preprotein recognition.

D. Energetics and Mechanism of Protein Translocation

Studies of protein translocation indicate that ATP hydrolysis is essential for *in vitro* protein translocation, whereas PMF stimulates the rate of translocation of many preproteins. Hydrolyzable ATP serves to drive SecA membrane cycling, which is required for insertion of the initial loop of the preprotein into the translocator. Translocation of the remainder of the preprotein can occur by either SecA-generated force or PMF. The exact nature of PMF-dependent translocation is uncertain, although it is clear that PMF stimulates the activity of the translocon in some manner.

The energetics and mechanism of protein translocation in archaea is not known. Their SecY/Sec61α and SecE/Sec61γ homologs are more eukaryotic-like, and they lack any detectable SecA homolog. This suggests that protein translocation may be more similar to eukaryotes, where translation drives protein translocation via a tight junction between the ribosome and the translocon. Alternatively, PMF or an as-yet-undiscovered ATPase may drive protein secretion in this domain.

E. Distal Steps during Protein Export

As preproteins emerge on the periplasmic side of the plasma membrane, they must undergo structural modifications that typically include signal peptide removal, protein folding, disulfide bond formation, and oligomerization. Studies reveal that there are a variety of protein catalysts that promote these events.

Signal peptidases have been characterized for general secretory proteins and lipoproteins. Signal or leader peptidase I enzymes cleave signal peptides from nonlipoproteins that utilize the GSP. These enzymes are membrane-anchored and contain a periplasmic catalytic domain that represents a novel type of serine protease, which appears to employ a serine–lysine dyad in the catalytic mechanism. Their substrate specificity is primarily determined by amino-acid residues at the -1 and -3 (relative to the cleavage site) positions of the preprotein, where small uncharged residues at -1, and small uncharged or larger aliphatic residues at -3, are required (see Figs. 2A and 2C). Preprotein recognition is also an important element in catalysis, as the catalytic rate differs by many orders of magnitude for preprotein versus simple peptide substrates. *In vivo,* signal peptide processing occurs relatively late in protein secretion, and little is known about the interplay between the translocon and processing enzymes. Bacterial species differ in whether they possess one essential (e.g., *E. coli*) or multiple, redundant, signal peptidase I enzymes. In *B. subtilis,* for example, there are five closely related signal peptidase I enzymes, two of which are coordinately induced at the onset of maximal protein secretion.

Signal peptidase II enzymes are specific for lipoproteins that utilize the GSP. They are membrane-embedded enzymes that recognize a sequence around the processing site, Leu-X-Y * gCys or Leu-X-Y-Z * gCys (where X, Y, and Z tend to be small uncharged amino-acid residues, gCys indicates glycerylcysteine and * is the cleavage site) (see Fig. 2B). The presence of the glyceride modification is a prerequisite for processing and, subsequent to cleavage, an additional fatty acid is added to the glycerylcysteine through an amide linkage. The antibiotic globomycin inhibits signal peptidase II enzymes. Subsequent to processing, lipoproteins may remain anchored in the cytoplasmic membrane, they may be transported to the outer membrane, or they may be released into the extracellular media, in certain instances. Transport to the outer membrane of gram-negative bacteria is accomplished by a transport pathway consisting of two components: LolA is a periplasmic chaperone that promotes lipoprotein release from the inner membrane, and LolB is an outer membrane receptor for Lo1A that binds the Lo1A–lipoprotein complex and promotes lipoprotein assembly into the outer membrane.

Secretory proteins and periplasmic domains of membrane proteins must undergo folding as they exit the translocon. Chaperones are thought to work not by direct promotion of protein folding per se, but rather, by preventing improper interactions (e.g., aggregation) that sidetrack a protein from its correct folding pathway. Bacterial periplasmic chaperones need to be dissimilar from ATP-dependent chaperones of Hsp60 and Hsp70 families present in many eukaryotic organelles or the bacterial cytoplasm, since the periplasm contains little, if any, nucleoside triphosphate.

Pilus- and fimbriae-specific chaperones represent a large and well-characterized group. Unlike cytoplasmic chaperones that recognize their substrate in an unfolded state, these periplasmic chaperones recognize their substrate in a more native state and function by capping interactive surfaces that function in oligomerization and assembly reactions. In one prototypic system from uropathogenic *E. coli*, the periplasmic chaperone–substrate complex interacts with an outer membrane "usher," which determines the order of assembly of the various pilus subunits in the outer membrane on the basis of its affinity for them. The genes for pilus- and fimbriae-specific chaperones are ordinarily located within operons that encode the structural components of these organelles, and they are typically contained on plasmids or chromosomal segments unique to pathogenic bacterial strains.

Proline isomerases (peptidyl–prolyl cis/trans isomerases or rotamases) catalyze the cis/trans isomerization of pepidyl–prolyl residues within proteins, normally a slow step. Such enzymes are found in both the bacterial cytoplasm and periplasm. *E. coli* contains at least three periplasmic proline isomerases, each belonging to a different family: PpiA (RotA) (cyclophilin family), FkpA (FKBP family), and SurA (PpiC family). *ppiA* is regulated by the Cpx two-component system that senses perturbations of the periplasmic milieu.

The bacterial periplasm contains an array of enzymes that promote the formation and correct distribution of disulfide bonds in secretory and membrane proteins. Enzymes that promote disulfide bond formation in proteins are termed disulfide oxido-reductases, while those that promote the rearrangement of disulfide bonds within proteins are termed disulfide isomerases. At the heart of the mechanism of such enzymes is a catalytic pair of cysteines, often in a Cys-X-X-Cys (where X represents a nonspecific amino-acid residue) configuration, which can undergo oxidation (to promote substrate reduction, in the case of an isomerase) or reduction (to promote substrate oxidation, in the case of an oxido-reductase). These enzymes must undergo the reverse redox reaction catalyzed by a recharging enzyme in order to be regenerated. Several different disulfide oxido-reductases/isomerases are known in *E. coli* and other bacteria (Fig. 7). DsbA appears to be the major enzyme catalyzing net disulfide bond formation of periplasmic and outer membrane proteins, and it is reoxidized by its partner, DsbB, a transmembrane protein. The reoxidation of DsbB itself appears to be dependent on the respiratory chain. DsbC appears to function primarily as an isomerase to correct aberrant disulfide bonds that form within proteins. Reducing potential for DsbC reduction comes ultimately from cytoplasmic thioredoxin and is transferred to the inner membrane protein DsbD before reaching DsbC. Additional Dsb proteins also exist. It is clear from these studies that a delicate redox balance is present in the periplasm that allows the correct array of disulfide bonds to be formed and maintained.

II. TERMINAL BRANCHES OF THE GENERAL SECRETION PATHWAY

Certain species of gram-negative bacteria possess pathways for transporting proteins across the outer membrane once they have reached the periplasm via the GSP. A number of such pathways are specific for a single protein, and they may require none (e.g., *N. gonorrhoeae* IgA protease) or only a single additional protein (e.g., *S. marcescens* hemolysin) to promote the export process. Space does not allow coverage of this iconoclastic group of proteins. The most important pathway for outer membrane traversal is the main terminal branch (MTB) of the GSP or type II protein secretion pathway. Hallmarks of the MTB

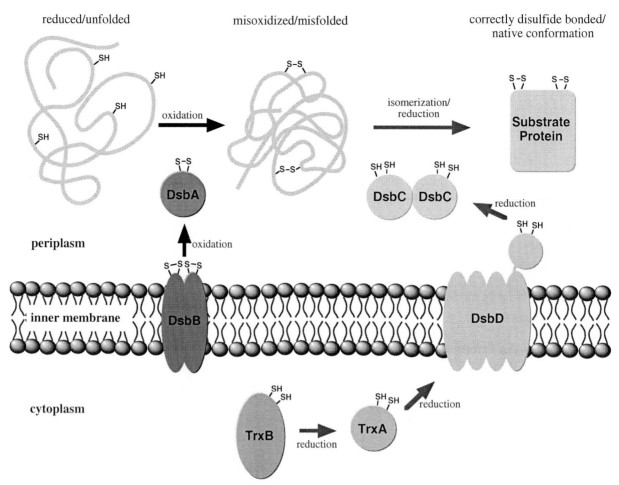

reduced/unfolded   misoxidized/misfolded   correctly disulfide bonded/ native conformation

Fig. 7. Model for disulfide bond formation in the *E. coli* periplasm. The roles of DsbA as an oxido-reductase and DsbC as an isomerize are depicted along with their regeneration utilizing DsbB and DspD, respectively. TrxA and TrxB indicate thioredoxin 1 and thioredoxin reductase, respectively. [From Rietsch, A., and Beckwith, J. (1998). *Annual Rev. of Genetics* **32**, 163–184, with permission from the *Annual Review of Genetics,* Vol. 32, ©1998, by Annual Reviews.]

include: (i) this pathway is present in a limited number of plant and animal pathogens (currently, *Klebsiella oxytoca, Erwinia chrysanthemi* and *carotovora, Xanthomonas campestris, Pseudomonos aeruginosa, Aeromonas hydrophila,* and *Vibrio cholerae*), (ii) it is responsible for the secretion of a limited number of proteins (often referred to as exoproteins) that include hydrolytic enzymes, such as cellulase and pectinase for plant pathogens, and proteases, lipases, and toxins for animal pathogens, (iii) secretion occurs on folded protein substrates that are present in the periplasm, (iv) a given exporter can secrete several structurally diverse exoproteins but not structurally

similar exoproteins from a heterologous exporter, (v) they are complex systems that require 12–14 secretion machinery components that are located in both the inner and outer membranes, and (vi) the genes encoding these components are tightly linked in a single regulon, while the genes encoding exoproteins are often unlinked.

Correctly folded protein substrates are required for secretion by the MTB. The starch-degrading enzyme pullulanase ordinarily contains disulfide bonds prior to its secretion, although the presence of disulfide bonds is not requisite for secretion. Secretion of cholera toxin requires prior oligomerization of the A monomer and the B pentamer within the periplasm. Even the presence of single amino-acid alterations is sufficient to block secretion of particular protein substrates (e.g., *A. hydrophila* aerolysin). These examples suggest that surface elements within the final tertiary structure of exoproteins are recognized by the MTB secretion machinery (i.e., a patch signal), although particular recognition motifs have not been defined further. The existence of multiple, redundant secretion signals may hamper their definition in certain cases.

Similar recognition motifs are probably present on structurally unrelated exoproteins, since a given MTB exporter may secrete several diverse exoproteins (e.g., *P. aeruginosa* Xcp secretes elastase, exotoxin A, alkaline phosphatase, and phospholipase C). In contrast, homologous and structurally similar exoproteins from related bacteria are generally not secreted from bacteria that possess a different MTB exporter. The basis for specific recognition by these systems is yet to be clarified.

MTB exporters are complex systems, often consisting of over one dozen proteins. The best characterized systems are the *K. oxytoca* Pul and the *E. chrysanthemi* Out systems (Fig. 8), which have been reconstituted in *E. coli*. DNA sequence analysis indicates that all MTB exporters contain a number of similar components (identities from 30–60%) (Fig. 9). Several of these proteins are similar to ones that are required for the assembly of type IV pili. In particular, four pilinlike proteins (pseudopilins) are present in the Pul, Out, and other systems (Pul/Out-GHIJ). Remarkably, pseudopilin signal peptides are cleaved by a specific peptidase (Pul/OutO) that is functionally related to prepilin peptidase. In addition, there are two other components (Pul/OutEF), that are also homologous to proteins required for pilin assembly. One of these proteins (Pul/OutE) contains a predicted ATP-binding site and is likely to function as an ATPase or kinase. However, rather than serving as an energy source for exoprotein secretion or regulating it by phosphorylation, it seems more likely that this protein functions in exporter assembly, in keeping with the role of its homolog in pilus assembly. The striking similarities between two seemingly different systems suggest that pseudopil-

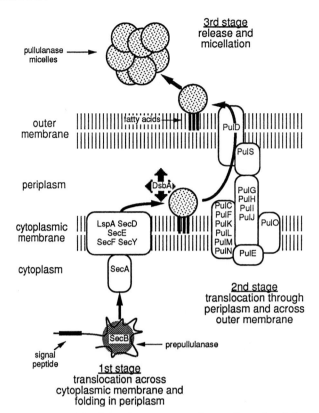

Fig. 8. Model of the three stages of secretion of pullulanase by *K. oxytoca*. A probable arrangement of the Pul exporter responsible for the second stage of secretion is indicated. [From Pugsley, A. (1993). *Microbiol. Rev.* **57**, 89.]

ins could be assembled into a piluslike structure, which functions as a core component of the MTB exporter. While this is an attractive hypothesis and is similar to the parallelism in structure between type III exporters and flagella (see following), it currently lacks compelling evidence in its favor.

A variety of techniques have been utilized to localize the individual proteins that comprise the MTB exporter. Remarkably, 11 out of 14 proteins in the Pul system appear to be integral cytoplasmic membrane proteins. This seems strange for a system, whose function is to translocate proteins solely across the outer membrane. Several suggestions have been proffered in this regard: (i) certain inner membrane proteins may function indirectly to promote exporter assembly, (ii) a multiprotein complex may be needed for transfer of energy between the inner and outer membranes during the export cycle (see following),

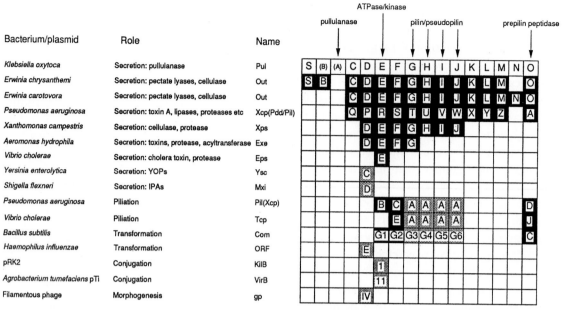

Bacterium/plasmid	Role	Name
Klebsiella oxytoca	Secretion: pullulanase	Pul
Erwinia chrysanthemi	Secretion: pectate lyases, cellulase	Out
Erwinia carotovora	Secretion: pectate lyases, cellulase	Out
Pseudomonas aeruginosa	Secretion: toxin A, lipases, proteases etc	Xcp(Pdd/Pil)
Xanthomonas campestris	Secretion: cellulase, protease	Xps
Aeromonas hydrophila	Secretion: toxins, protease, acyltransferase	Exe
Vibrio cholerae	Secretion: cholera toxin, protease	Eps
Yersinia enterolytica	Secretion: YOPs	Ysc
Shigella flexneri	Secretion: IPAs	Mxi
Pseudomonas aeruginosa	Piliation	Pil(Xcp)
Vibrio cholerae	Piliation	Tcp
Bacillus subtilis	Transformation	Com
Haemophilus influenzae	Transformation	ORF
pRK2	Conjugation	KilB
Agrobacterium tumefaciens pTi	Conjugation	VirB
Filamentous phage	Morphogenesis	gp

Fig. 9. Chart indicating homologies of components shared by MTB exporters as well as machineries promoting movement of macromolecules across bacterial envelopes. Black boxes indicate >20% overall sequence identity to the corresponding Pul counterpart, while shaded boxes indicate significant homologies that are less than this value. Known or probable functions of these components are given at the top of the figure. [From Pugsley, A. (1993). *Microbiol. Rev.* **57**, 84.]

(iii) exoproteins may be specifically concentrated and targeted for translocation by their association with an inner membrane protein complex, (iv) a large and elaborate inner membrane protein complex may be needed to properly gate the outer membrane protein channel, which must be large in order to accommodate folded protein substrates.

Central to the function of MTB exporters is a single integral outer membrane component (Pul/OutD) that exists in large multimeric complexes (10–14 monomers) that are presumed to be protein translocation channels. Accordingly, such proteins have been termed secretins. MTB secretins are homologous to an outer membrane protein encoded by filamentous phage (pIV), which presumably functions as a secretin for extrusion of phage particles across the bacterial outer membrane. A single pIV oligomer contains ~14 monomers arranged in a cylindrical fashion with a internal diameter of approximately 8 nM. Like trimeric outer membrane porins, pIV is rich in β-sheet structure and resistant to detergent-induced dissociation. Little is known about how secretin channels are gated.

Secretins possess an amino-terminal periplasmic domain that appears to determine exoprotein substrate specificity. Substrate-specific binding has been demonstrated for certain secretins, while domain exchanges between pIV homologs allows for substrate specificity switching. These experiments suggest that the amino-terminal domain of secretins may be the initial docking site of exoproteins during their transit across the outer membrane.

In vitro systems are needed to define the translocation mechanism of MTB exporters. *In vivo* systems suggest a requirement for PMF and are equivocal on an ATP requirement. If ATP hydrolysis is needed only for exporter assembly, then it is possible that such systems are driven solely by PMF. Precedent exists already for PMF-dependent import of certain small molecule ligands across the outer membrane via the TonB-dependent pathway. TonB has been suggested to "shuttle" PMF-driven conformational energy to the outer membrane in order to promote transport processes. However, it is clear that considerable future work will be required to understand the function of the many components present in MTB exporters, their overall organization and assembly, the basis of exoprotein specificity, as well as the translocation mechanism.

III. TWIN-ARGININE PROTEIN SECRETION PATHWAY

A specialized pathway for the secretion of redox enzymes across the cytoplasmic membrane of most bacteria (except mycoplasmas, *Borrelia burgdorferi,* and methanogens) and, probably, archaea exists that is independent of the GSP. Redox enzymes that employ iron–sulfur clusters, molybdopterin, polynuclear copper, tryptophan tryptophylquinone, and flavin adenine dinucleotide as cofactors appear to utilize this pathway, while enzymes that have iron porphyrins, mononuclear type I or II copper centers, dinuclear Cu_A center, and pyrrolo–quinoline quinone as cofactors utilize the GSP for their secretion. The hallmark of this recently characterized pathway is the presence of a signature motif on the signal peptide of preproteins utilizing this pathway. A consensus sequence containing an invariant arginine pair, S/T-R-R-X-F-L-K (where X represents any amino acid), is present between the N- and H-domains of the signal peptide (see Fig. 2). Aside from this feature, such signal peptides are similar to those found for GSP-dependent preproteins, although they are often somewhat longer than their GSP counterparts (up to 58 amino-acid residues). This characteristic feature has lead to the name "*t*win *a*rginine *t*ranslocation (TAT) pathway" to describe this novel pathway.

A second characteristic of the TAT pathway is that their cognate preproteins appear to be undergo extensive folding and cofactor binding in the cytoplasm prior to insertion and translocation across the membrane. This feature is in contrast to the GSP, that requires the preprotein to be in an extended or loosely folded state for engagement by the Sec machinery. Aside from the atypical

signal peptide, precisely what targeting information is also contained within the folded preprotein is not yet known. Of interest, there are metalloprotein subunits that lack any signal peptide entirely, which are secreted by association with their complementary subunit, containing a TAT-dependent signal peptide (i.e., piggyback secretion). This suggests that the TAT machinery may possess an usually large translocation channel for accommodating its substrates.

Mutational studies in *E. coli* have identified several proteins that are likely to be components of the TAT machinery. Such mutants demonstrate that the TAT machinery is nonessential in heterotrophic bacteria under appropriate conditions (e.g., in the presence of oxygen). TatA and TatE are homologs of Hcf106, which is a component of the ΔpH-dependent thylakoid import pathway of maize. Plant thylakoids possess three protein import pathways (SRP-dependent, Sec-dependent, and ΔpH-dependent), and the ΔpH-dependent pathway catalyzes the import of preproteins containing the twin-arginine motif. Thus, the TAT pathway appears to be present in all three domains of life. An additional protein unrelated to Hcf106, TatB, also functions in the bacterial TAT pathway. All three proteins are predicted to be integral cytoplasmic membrane proteins. *tatA, tatB,* and *tatE* mutants accumulate enzymatically active redox enzymes in their cytoplasm, but they are normally proficient in the secretion of proteins utilizing the GSP. This phenotype suggests that a specific mechanism(s) exist(s) that excludes redox enzymes from the GSP. By contrast, blocking the Sec machinery has no effect on the secretion of redox enzymes. Taken together, there is compelling evidence that the GSP and TAT pathways are distinct and mutually exclusive.

Little is known about the energetics of protein translocation of the TAT pathway. It is unclear whether ΔpH (which is often very minimal in bacteria such as *E. coli*) could drive protein translocation in this system, by analogy to the thylakoid pathway. The only crossover between the GSP and TAT machineries is likely to be utilization of a common signal peptidase I, which should also cleave TAT-dependent preproteins, based on the similarity of the C-domains of these signal peptides.

IV. ABC PROTEIN SECRETION PATHWAYS

A number of GSP-independent protein secretion pathways are promoted by a family of exporters that contain a highly conserved protein of the ATP-binding cassette (ABC) superfamily, and consequently, they are referred to as ABC protein secretion pathways or type I systems. ABC protein exporters are members of the ubiquitous ABC superfamily of traffic ATPases, which include importers and exporters of a large variety of small and macromolecules. There are over a dozen ABC protein exporters described currently, and this number is likely to grow significantly in the future. Hallmarks of ABC protein exporters include (i) they secrete a group of homologous proteins that can include toxins, proteases, lipases, and specific peptides, depending on exporter type, (ii) their protein substrates lack GSP-dependent signal peptides, but instead usually possess uncleaved carboxy-terminal secretion signals that vary by exporter type, (ii) they are relatively simple systems that include an ABC protein and a few accessory proteins, and (iii) protein translocation occurs in a single step across one or both bacterial membranes (without any periplasmic intermediate for gram-negative bacteria). This latter feature, combined with the ability to fuse ABC-dependent secretion signals to a large number of heterologous proteins and obtain efficient secretion, make these exporters attractive for biotechnological applications.

The nature of ABC-dependent secretion signals has been studied extensively, most notably for RTX (repeats in toxin) toxin and metalloprotease families. Deletion and gene fusion studies show that the hemolysin of uropathogenic *E. coli* contains a secretion signal in its carboxy-terminal 60 amino-acid residues, while PtrG protease of *E. chrysanthemi* contains a secretion signal encompassing only its last 29 residues. Furthermore, these minimal secretion signals are effective in promoting secretion of a large variety of naive passenger proteins by gene fusion. Such heterologous protein secretion occurs exclusively via the exporter that is homologous to the secretion signal, demonstrating the specificity of the secretion signal for a given exporter. Genetic analysis indicates that secretion signals are very tolerant to single amino-acid changes, and they suggest that such signals function in a tertiary conformation (i.e., a patch signal) that is recognized by the ABC component of the exporter.

Upstream of the secretion signals of toxins, proteases, and lipases is a conserved motif consisting of a glycine-rich sequence (GGXGSD) that is repeated 4 to 36 times. This repeated motif has been shown to form a β parallel roll structure that binds calcium ions in the case of certain metalloproteases. This arrangement may promote better separation of the secretion signal from the rest of the protein, thereby affording efficient presentation of the signal to the exporter.

Typical ABC protein exporters consist of an ABC protein and one to several accessory proteins, depending on the family (Fig. 10). All traffic ATPases possess a highly conserved structure consisting of two membrane-embedded hydrophobic domains and two hydrophilic ATP-binding domains. This architecture probably allows the ATP-binding domains to drive substrate transport across a channel formed by the hydrophobic domains (a situation akin to SecA and SecYE proteins of the GSP). Although these domains can be contained in a single polypeptide (e.g., mammalian multidrug resistance transporter or cystic fibrosis chloride channel), the most common arrangement for bacterial protein exporters is as a homodimer.

Two additional accessory proteins are found in ABC protein exporters of gram-negative bacteria that allow protein secretion across both membranes. A protein of the membrane fusion protein family (MFP) appears to functionally link the inner and outer membranes during the translocation cycle (e.g., HlyD in Fig. 10). MFP proteins have an amino-terminal hydrophobic domain anchored to the inner membrane, a central periplasmic domain, and a carboxy-terminal domain with predicted β-sheet structure that interacts with the outer membrane. The second component is an outer protein that functions as a protein channel for outer membrane traversal (e.g., TolC in Fig. 10). This latter protein is the least specific of the three, since exporters can share a common outer membrane protein (e.g., TolC for *E. coli* hemolysin, colicin V, and *S. marcesscens* hemoprotein exporters).

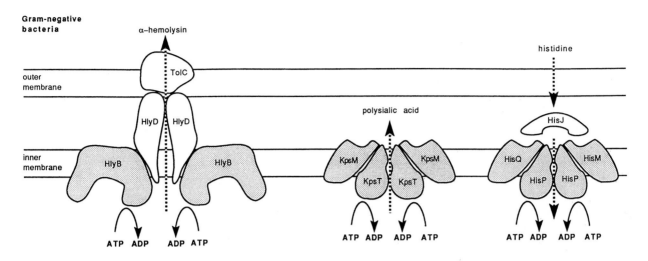

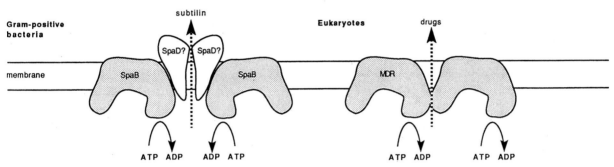

Fig. 10. Models of ABC transporters. Depicted are the *E. coli* α-hemolysin and *B. subtilis* subtilin exporters (both protein exporters), along with the *E. coli* polysialic acid and mammalian multidrug exporters and the *S. typhimurium* histidine importer. [From Fath, M., and Kolter, R. (1993). *Microbiol. Rev.* **57**, 997.]

The genes encoding ABC and MFP components are usually adjacent to the structural genes encoding the exoproteins, while the gene for the outer membrane protein may or may not be linked. Additional genes encoding cytoplasmic activators that posttranslationally modify exoproteins may be present within these genetic clusters. This arrangement allows functional transfer of these systems to other bacteria by horizontal genetic transmission.

Much less is known about the mechanism of protein translocation through ABC exporters compared to the GSP. Since secretion signals are usually carboxy-terminal, exoprotein translocation occurs in a posttranslational manner. There is no known requirement for cytoplasmic chaperones, and whether the exoprotein is translocated in a folded, loosely folded, or unfolded state is not clear. Translocation

initiates when the secretion signal interacts with the ABC protein, which defines the substrate specificity of a given system. This primary interaction leads to an ordered assembly (or stabilization) of the exporter: the ABC protein binds tightly to the MFP, which, in turn, binds to the outer membrane component. All three components must act in a concerted fashion, since the absence of the outer membrane protein leads to cytoplasmic, not periplasmic, accumulation of exoproteins. The path that the exoprotein takes through the exporter has not been defined yet.

Protein exporters require energy in the form of both ATP and PMF. PMF appears to promote the initial binding of the exoprotein to the exporter. The ABC protein drives at least a portion of the translocation cycle by ATP-dependent conformational cycling, although whether this involves membrane cycling of a mobile domain of the ABC protein akin to SecA protein is unknown. *In vitro* protein translocation systems for ABC protein exporters should clarify these and other issues relating to the translocation mechanism.

V. TYPE III PROTEIN SECRETION PATHWAYS

In recent years, it has become apparent that several gram-negative pathogenic bacteria have evolved a specialized protein secretion system, termed type III, or contact dependent, to secrete virulence determinants. This complex machinery directs the secretion and subsequent translocation into the host cell of several bacterial virulence effector proteins. These proteins, in turn, stimulate or interfere with host cell responses for the benefit of these bacterial pathogens. Remarkably, type III secretion sfiystems have been found not only in bacteria pathogenic for animals, such as *Salmonella* spp., *Yersinia* spp., *Shigella* spp., *E. coli*, *Bordetella* spp., and *Pseudomonas aeuruginosa*. but also in plant pathogenic bacteria of the genera *Erwinia, Xhanthomonas, Ralstonia,* and *Pseudomonas*.

A characteristic feature of this specialized secretion system is the absence of typical *sec*-dependent signal sequences in their substrate proteins, which are, therefore, secreted in a *sec*-independent manner without processing of their amino termini. Another property of these systems is the requirement of an activating signal for their efficient function. Certain growth conditions in the laboratory (e.g., low concentration of Ca^{2+}) have been shown to activate type III secretion systems in some bacteria. However, the physiologically relevant signals that activate these systems are not known, although it is widely believed that such signals must be derived from bacterial contact with host cells. Indeed, bacteria/host cell interactions lead to a marked increase in secretion through this pathway. Contact activation of protein secretion does not require *de novo* protein synthesis, implicating a posttranslational mechanism in the regulation of this process.

Type III secretion systems are most often encoded within chromosomal regions, known as pathogenicity islands, or on large virulence-associated plasmids. The nucleotide composition of the relevant coding regions is most often A + T rich and significantly different from that of the host chromosome, suggesting their acquisition by some mechanism of horizontal gene transfer. Consistent with this hypothesis is the frequent finding of sequences encoding remnants of transposable elements, bacteriophage proteins, or insertion sequences in the immediate vicinity of genes encoding type III secretion systems.

A. The Secretion Apparatus

A core set of at least 17 highly conserved proteins are thought to be components of the secretion machinery, based on their requirement for promotion of type III secretion. One of these components shares sequence similarity to the secretin or PulD family of protein exporters. As discussed previously, secretins are essential components of all type II secretion systems that are thought to form a channel through which the secreted substrates traverse the outer membrane. A subset of the type III secretion machinery components exhibits significant sequence similarity to components of the flagellar export apparatus. This group of proteins includes several integral membrane proteins and a membrane associated ATPase, which is thought to energize this system. This sequence homology, coupled to the architectural similarity between flagella and the type III system (see following), strongly suggests an evolutionary relationship between type III secretion and flagellar assembly. Another subset of conserved components of type III secretion systems includes several lipoproteins. The number of lipoproteins present in each system is not always the same and their primary sequence is poorly conserved also. However, some lipoproteins are highly conserved within a subset of bacterial species.

Recent electron microscopy studies have revealed the supramolecular organization of components of the type III secretion machinery of *Salmonella typhimurium*. Although structural information on type III secretion systems from other bacteria is not available, the high degree of sequence similarity among the

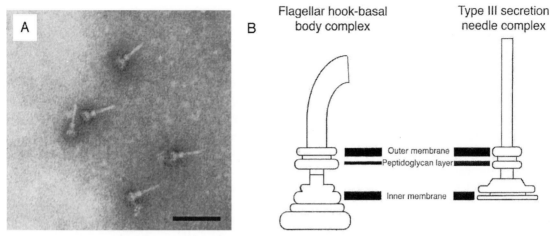

Fig. 11. Type III exporter (needle complex) of *Salmonella typhimurium*. (A) Electron micrograph of purified needle complexes from *Salmonella typhimurium*. Scale bar, 100 nm. (B) Diagram of flagellar hook basal body and needle complexes aligned to show similarities. [Courtesy of Shin-Ichi Aizawa.]

different components suggests a similar architecture for all type III protein exporters. The studies in *Salmonella typhimurium* revealed that the type III secretion machinery is organized in a supramolecular complex resembling a needle (needle complex) that spans both the inner and outer membranes of the bacterial envelope (Fig. 11). This needle complex possesses cylindrical symmetry with two clearly identifiable domains: a slender, needlelike portion, projecting outwards from the surface of the cell, and a cylindrical base, which anchors the structure to the inner and outer membranes. The base structure has some resemblance to the flagellar basal body. It is composed of two outer and two inner rings. The inner rings are 40 nm in diameter and 20 nm in thickness and interact with the cytoplasmic membrane. The outer rings interact with the outer membrane and the peptidoglycan layer and are 20 nm in diameter and 18 nm in thickness. The inner and outer rings are connected by a rod structure. The needle structure itself is 80 nm in length and 13 nm in thickness and appears hollow. The number of needle structures observed per bacterial cell varies between 10 to 100. The overall architecture of the needle complex clearly resembles that of the flagellar basal body, giving further support to the hypothesis that these two structures are evolutionary related (Fig. 11). Furthermore, the organization of this com-

plex suggests a mechanism by which proteins can be secreted through the bacterial envelope without a periplasmic intermediate. Biochemical analysis of the purified needle complexes revealed that they are composed of at least three proteins: InvG (a member of the secretin family) and two lipoproteins, PrgH and PrgK. Neither of these proteins has true homologs in the flagellar system, although FliF, a component of the flagellar MS ring structure, is also a lipoprotein. Further characterization of the needle complex will most likely yield additional structural components.

B. Substrates of Type III Secretion Systems

Despite the high degree of conservation among the components of type III secretion systems across bacteria species, there is little similarity among the substrates of the different systems, in particular, those proteins that will ultimately mediate or interfere with host cell responses. Therefore, it is evident that different pathogenic bacteria have adapted the function of the type III secretion systems for their own specific requirements. Proteins secreted through this system can be divided into at least four different categories: (i) Proteins that are required for the secretion process itself. By analogy to the flagellar system, these proteins are candidates for structural components of the needle complex; (ii) Proteins that are involved in the regulation of the secretion process. This category may include proteins that act as "plugs" of the translocation channel; (iii) Proteins that promote the process of translocation of the secreted substrates into eukaryotic host cells. These proteins have been termed translocases; (iv) Proteins that act as effectors of bacterial-induced host cellular responses. This latter category includes proteins that are agonists of cellular responses or antagonists of physiological host cell responses. Such proteins are the least conserved among the different systems, a further indication that different bacteria have assembled a set of effectors that is best suited for their specific requirements. Effector proteins are predicted to act on eukaryotic cell target molecules and include tyrosine protein phosphatases, serine/threonine protein kinases, and exchange factors for small molecular weight GTPases.

C. Specific Chaperones, Export and Translocation Signals

A characteristic feature of type III protein secretion systems is that most of their substrates are associated with specific cytoplasmic chaperones or partitioning factors. Unlike other well-characterized chaperones, such as GroEL or Hsp70, these specialized chaperones have a rather narrow binding specificity and lack nucleotide-binding or -hydrolyzing activities. While they exhibit little primary amino-acid sequence similarity, they do share a number of biochemical properties, such as a relatively small size (15–18 kDa), a low isoelectric point, and a predominantly α-helical secondary structure. They are usually, although not always, encoded in the vicinity of their cognate target proteins. Overall, the function of the type III associated chaperones is not well understood.

Deletion analysis of type III secreted proteins has established the existence of well-defined independent domains that are involved in either their secretion or translocation into the host cell. In general, the first 10 to 20 amino-acid residues of these proteins are required for their secretion, whereas an adjacent domain of 60 to 70 residues is involved in

their translocation into the host cell. Interestingly, this translocation domain overlaps with the binding site for their specific chaperone, suggesting the intriguing possibility that these chaperones are directly involved in the events that lead to the translocation of the different type III secreted proteins into the eukaryotic host cell. Recent studies carried out in *Yersinia* spp. have suggested the existence of two independent type III secretion pathways. One pathway appears to be mediated by a translationally coupled secretion signal, present in the first ~60 nucleotides of the coding mRNAs. The other pathway is chaperone-dependent and is mediated by a domain located between residues 15 and 100 of substrate proteins.

The role that this family of chaperones plays in the secretion process itself is not well understood, and it is the subject of some controversy. Most of the studies have been carried out in *Yersinia* spp. and *Salmonella typhimurium*. In these bacteria, absence of a given chaperone results in reduced secretion of the cognate substrate protein. However, it has been difficult to establish whether reduction of secretion is due to degradation of the target protein in the absence of the chaperone, the direct involvement of the chaperone in the secretion process itself, or both. It is clear that secretion of type III substrate proteins can take place in the absence of specific chaperones. However, this process requires either removal of the chaperone binding site from the cognate protein or the absence of specific secreted substrate proteins that presumably engage in direct interaction with the cognate protein. Thus, it has remained controversial whether the function of these chaperones is to deliver the secreted target proteins to the type III secretion machinery, to maintain them in a secretion competent state, to prevent premature association with other cognate secreted products, or some combination of the above. More studies will be required to clarify the physiological role, if any, of these secretion mechanisms in the different type III secretion systems, as well as to better understand the role of specific chaperones in this process.

See Also the Following Articles

ABC Transport • Cell Walls, Bacterial • Fimbriae, Pili • Flagella • Outer Membrane, Gram-Negative Bacteria

Bibliography

Bardwell, J. (1994). Building bridges: disulfide bond formation in the cell. *Mol. Microbiol.* **14**, 199–205.

Berks, B. (1996). A common export pathway for proteins binding complex redox cofactors? *Mol. Microbiol.* **22**, 393–404.

Binet, R., Letoffe, S., Ghigo, J., Delepelaire, P., and Wandersman, C. (1997). Protein secretion by gram-negative bacterial ABC exporters—a review. *Gene* **192**, 7–11.

Driessen, A., Fekkes, P., and van der Wolk, J. (1998). The Sec system. *Curr. Opin. Microbiol.* **1**, 216–222.

Fath, M., and Kolter, R. (1993). ABC transporters: bacterial exporters. *Microbiol. Rev.* **57**, 995–1017.

Hueck, C. J. (1998). Type III secretion systems in bacterial pathogens of animals and plants. *Microbiol. Molec. Biol. Rev.* **62**, 379–433.

Kubori, T., Matsushima, Y., Nakamura, D., Uralil, J., Lara-Tejero, M., Sukhan, M., Galán, J., and Aizawa, S.-I. (1998). Supramolecular structure of the *Salmonella typhimurium* type III protein secretion system. *Science* **280**, 602–605.

Pohlschroder, M., Prinz, W., Hartmann, E., and Beckwith, J. (1997). Protein translocation in the three domains of life variations on a theme. *Cell* **91**, 563–566.

Pugsley, A. (1993). The complete general secretory pathway in gram-negative bacteria. *Microbiol. Rev.* **57**, 50–108.

Russel, M. (1998). The macromolecular assembly and secretion across the bacterial cell envelope: type II protein secretion systems. *J. Mol. Biol.* **279**, 485–499.

Wandersman, C. (1996). Secretion across the bacterial outer membrane. *In* "*Escherichia coli* and *Salmonella* Cellular and Molecular Biology" F. C. Neidhardt, R. Curtiss, J. L. Ingraham, E. C. Lin. K. B. Low, B. Magasanik, W. S. Reznikoff, M. Riley, M. Schaechter, and H. E. Umbarger, (eds.), pp. 955–966. American Society for Microbiology, Washington, DC.

Protozoan Predation

Lucas A. Bouwman

Alterra, Wageningen, The Netherlands

GLOSSARY

decomposition Degradation of organic matter of all origins, vegetative, animal, and microbial.

epigrowth Organisms growing on the surface of particles.

mineralization Release of CO_2, H_2O, NH_4^+, H_2S, etc.

nematodes Invertebrate animals, worms of microscopic size (1–1.5 mm), millions found per square meter in terrestrial soils and benthic sediments, feeding there on bacteria, fungi, algae, plant roots, and on each other; include also parasitic forms, which occupy all moist ecological niches with the appropriate food in plants, animals and humans, and inside almost all organs.

protozoa Unicellular, eukaryotic, phagotrophic organisms, with or without flagella, about 5–50 μm in size, living in aquatic environments (swimming) and in soil pore water (swimming, creeping) in densities of millions per liter water and up to 10^6 per gram soil, feeding mainly on bacteria; include also forms parasitic to plants, animals, and humans.

soil food web The intimately connected complex of organisms living within the soil pore labyrinth, with bacteria and fungi feeding primarily on organic matter and protozoa and nematodes feeding on microbes and on each other.

DECOMPOSITION OF ORGANIC MATTER counterbalances primary production and releases minerals, which become available again for plant growth. Decomposition is performed by microorganisms in nature and also in the digestive tract of animals.

In all habitats where organic matter is available and physical/chemical conditions are not adverse, microorganisms, such as bacteria and fungi, decompose organic matter and release carbon dioxide (CO_2) and ammonia (NH_4^+) (Fig. 1). Besides fresh and marine waters, the labyrinth of interstitial spaces, located in the upper soil or sediment layers, is the living space where organic matter is broken down by organisms. In natural habitats, the soil is often covered by a more or less compact litter layer, consisting of plant debris in various stages of decomposition. Under agricultural conditions, organic matter is usually mixed through the upper soil layer. In all these biotopes, bacteria and fungi are grazed upon by bacterivorous and fungivorous organisms, primarily protozoa. Also metazoa, such as nematodes, microarthropods, oligochaetes, and turbellarians feed on (soil)-microbiota. As the soil ecosystem of microbiota and their predators (grazers) is hidden, its functioning as a whole has only recently been the subject of research. It appears that the role of protozoan predation in (soil) food webs is crucial. Their grazing on microbes prevents stagnation of microbial development.

I. SOIL BIOLOGICAL PROCESSES

Microbes are primarily responsible for the decomposition of primary- and secondary-produced organic matter, in aquatic as well as in terrestrial habitats. The microbes, mainly bacteria and fungi,

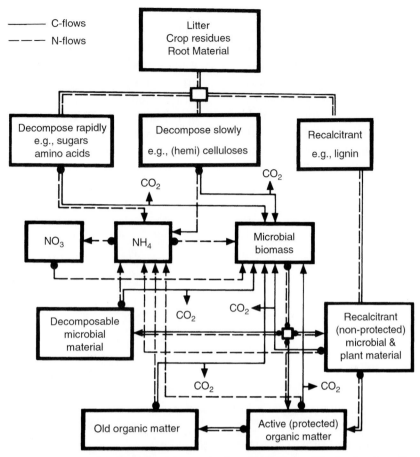

Fig. 1. Scheme of a decomposition and mineralization model of organic matter (Adapted from J. A. van Veen *et al.* (1984). *Plant and Soil* **76** 261, with kind permission of Kluwer Academic Publishers).

decompose the myriad numbers of different organic substances produced in nature which end up in the soil. Easily decomposable substances in crop debris after harvest, young roots and fresh litter, are decomposed by fast-growing generalist microbes, whereas resistant organic structures are broken down by slow-growing specialists, often fungi. In fact, a mosaic of processes with different dynamics proceeds simultaneously in the soil. The slow processes go on continuously at about the same rate, depending mainly on temperature, and are more or less regularly distributed over the soil profile. The fast processes occur incidentally and more locally at particular spots. During decomposition, minerals essential for the growth of primary producers and of microbes are released.

Due to their predation on microbes, protozoa influence the key processes in nature: decomposition of organic matter and mineralization of nutrients. The effects can be negative as the protozoan predation reduces numbers of microbes, and positive in that they stimulate microbial activity and primary production, due to the release of minerals essential for growth. Fungi, bacteria, protozoa, nematodes, and other metazoa constitute the soil food web of organisms, connected to each other by trophic relations (predation–prey), sharing the interstitial space between soil particles.

II. THE FOOD WEB OF SOIL ORGANISMS

Food webs of organisms occur in all types of biotopes: above ground, within the soil, in marine and fresh waters, and in sediments. They comprise the routes of organic matter decomposition and mineralization and the fate of the consumers through successive trophic stages. In soil, organisms occur over a continuous size spectrum and serve as food for each other, the typical length ratio between predator and prey being around ten. This allows for a number of coexisting species, each of which exploits a certain size range of prey organisms. Being aquatic animals, protozoa, in particular amoebae and flagellates, also occur in moist soil. Because they can glide over solid substrates, amoebae are especially adapted to life in soil, whereas flagellates, ciliates, and other protozoa are more prominent inhabitants of fresh and marine waters. Root growth and incorporation of above-ground produced organic matter provide the soil with the food necessary for the web of biota. In arable fields, for example, annually about 3000 kg carbon (C) is incorporated into the soil per hectare, i.e., about 30,000 kg fresh organic matter. Under conditions of equilibrium, i.e., a stable soil organic matter content, about 3000 kg C will be also decomposed

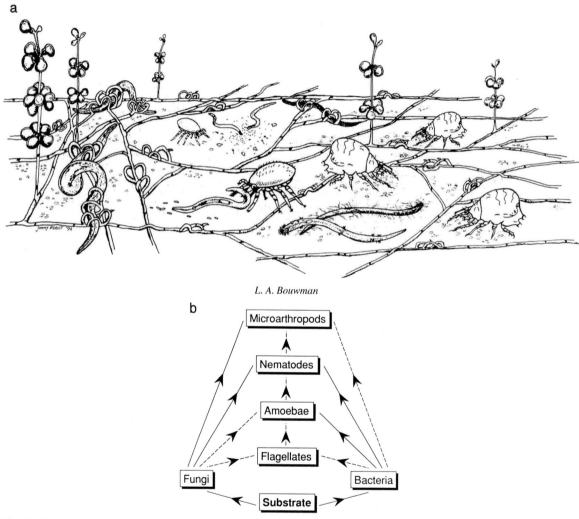

L. A. Bouwman

Fig. 2. (a) Living picture of a soil food web of organisms as it can be cultivated on an agar plate. Organisms: bacteria, flagellates, amoebae, nematodes, bacterivorous mites, nematophagous fungi (predatory and endoparasitic), nematophagous mites. (b) Scheme representing a below-ground food web, showing the major (solid lines) and minor (dashed lines) flows of nutrients (adapted from B. S. Griffiths, (1994). "Soil Protozoa" (J. F. Darbyshire, ed.), p. 880. CAB International, Oxon).

annually, largely by respiration. Microbiota, bacteria and fungi, are mainly responsible for decomposition and mineralization. They are consumed by protozoa, nematodes, microarthropods, etc. For example, the upper 25 cm of a hectare of arable soil in the Netherlands was found to contain about 250 kg bacterial C, about 15 kg protozoan C, about 5 kg fungal C, about 1 kg nematode C, and 0.5 kg arthropod C. To obtain figures for fresh weight in kilograms, amounts of C have to be multiplied by a factor of 10. This means that the weights of bacteria, protozoa, fungi, nematodes, and arthropods in the surveyed hectare arable soil corresponded to 3.5 cattle, 3 sheep, 1 calf, 5 rabbits, and 1 fat chicken, respectively. With respect to fungi, a figure of 5 kg fungal C per hectare is rather low, as often about equal amounts of bacterial and fungal C are measured (>250 kg C per hectare each), and also protozoan weights of 25 to 100 kg C per hectare are not uncommon. This food web balances enrichment and decomposition of organic matter in the soil. Figure 2 presents an imaginary living picture of this food web, as it could be cultivated on an agarplate in the laboratory (a) and also a simplified scheme of the food relations (b). Primary are the bacteria and the fungi; the bacteria are eaten by bacterivorous amoebae, flagellates, nematodes and mites. In Fig. 2a, the nematodes are eaten by nematophagous mites and fungi. With their rapid multiplication rates and small sizes, protozoa have a distinct advantage over metazoa with respect to resources that become available and, consequently, in colonizing the soil.

III. PROTOZOA

Protozoa are known as human and animal parasites, but less commonly as plant parasites. Malaria, sleeping sickness, nosema (parasite of honeybees), and club foot (parasite of *Cruciferae*) are well-known diseases caused by parasitic protozoa. Protozoa are unicellular eukaryotes, more comparable to an individual animal than to one of its constituent cells. A universal requirement for protozoa and protozoan activity is water, including for species associated with terrestrial environments. Although environments exist which harbor prokaryotes but no protozoa (hot springs with temperatures >90°C; biotopes with pH values below 1), protozoa normally occur in the presence of bacteria and free water, even at subzero temperatures and temperatures exceeding 50°C. In different ecosystems, the functional roles of protozoa are fundamentally the same; as bacterivores, they control bacterial numbers and dynamics. Having a potential for rapid growth and a high metabolic rate, relatively small protozoan biomasses may have large effects on elemental cycling. The total number of different protozoan species is not impressive when compared to other taxa, such as insects, fungi, or nematodes: 30,000 protozoan species have been described, including 8000 ciliates, 4000 flagellates, and 1000 rhizopods. In the functional sense, the free-living taxa are phagotrophic; this means they enclose food particles in a membrane-covered vacuole in which digestion takes place. In ciliates and in most flagellates, this occurs at a special site on the organism's surface, the cytostome, which is covered by a single unit membrane from which food vacuoles are formed. Amoebae engulf their prey but have no special site for that. The smallest soil protozoa are the nanoflagellates, being slightly larger than bacteria (> 1 μm). The largest are the ciliates (> 10 μm, < 1000 μm); amoebae are of intermediate size: they span a size range of about three orders of magnitude and are found at several trophic levels of food chains. Fig. 3 presents typical pictures of a flagellate, an amoeba and a ciliate.

IV. PROTOZOAN PREDATION

A. Protozoan Occurrence in Various Biotopes

In marine and in fresh water, bacteria are the primary decomposers of organic matter. Bacteria are consumed by small flagellates which, in turn, are eaten by ciliates, while planktonic metazoans constitute the next trophic level. With regard to protozoan communities, differences between fresh and marine waters are not dramatic, although the marine environment is more diverse than the fresh water. Terrestrial soils harbor amoebae and flagellates with special adaptations for interstitial life and, in this sense, a

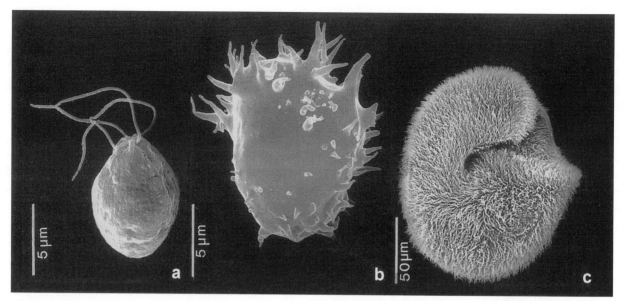

Fig. 3. Scanning electron micrographs of three main types of soil protozoa: (a) flagellate, (b) (naked) amoeba, (c) ciliate. (Adapted from W. Foissner, (1994). "Soil Protozoa" (J. F. Darbyshire, ed.) CAB International, Oxon. Front cover photographs.)

genuine terrestrial protozoan fauna exists, although specific terrestrial ciliates do not exist. In organic litter layers on top of mineral soils, fungi predominate decomposition. In particular, fungal spores can be consumed by protozoa while, contrary to the fungal-feeding nematodes and microarthropods with their specific buccal structures, protozoa are not properly equipped for consumption of fungal hyphae. In soil, naked amoebae, can glide over solid surfaces and are the most successful protozoans feeding on bacteria adhering to surfaces. Although a gram of soil contains 10^9 to 10^{10} bacteria, this impressive number does not mean that the surfaces of the solid fractions are crowded with microbial organisms. In fact, on average, not more than 0.5% of the surface is overgrown. In fine-textured soils, specific solid surfaces up to 1000 m² per gram soil occur, indicating that, on average, for each bacterium an extended area is available. However, most bacteria occur in high densities on randomly distributed aggregates of substrate. Consequently, protozoa also cluster on these "hot spots" of biological activity, which are separated by deserts without any resources. After depletion of a feeding site, amoebae need to bridge extended areas without prey to reach a new site; as amoeboid gliding is much slower than ciliary swimming, starvation is

a relatively important threat to amoebae. Because of this spatial distribution of separate spots of substrate and bacteria, it is not surprising that relatively small protozoa do not disperse by movement through the soil, in contrast to bigger bacterivorous metazoans, such as nematodes and microarthropods. Consequently, protozoa do not distribute bacteria through the soil as do nematodes, which also enhance bacterial processes in this way: they stimulate, for example, nitrification in sterilized and reinoculated soil. In soil, protozoa spread mainly by growth of populations; however, protozoan cysts have a high potential for passive distribution over large distances through the air.

B. Variation in Feeding Biology

Catching suspended bacteria (particles) requires other feeding techniques than grazing particles associated with solid surfaces. With their flat shape, easy movement through the waterfilm surrounding solid particles and epigrowth-feeding, amoebae are less vulnerable to desiccation of the soil than suspension-feeding flagellates and ciliates in the soil solution. Filter-feeding suspension protozoa retain and ingest all properly sized particles, even latex beads, after concentration of dilute food particles near the cytostome. In this way, they clear food particles from water volumes many times their own cell volume. Among protozoa feeding on suspended particles, three different techniques are observed: filter feeding, direct interception, and diffusion feeding, being comparable with trawling, spearing, and trapping of fish, respectively. Filter feeding depends on the transport of water through ciliary or pseudopodial filters. Raptorial feeders intercept food particles with a sticky surface; this interception is enhanced by electrostatic attraction (direct interception). In diffusion feeding, a motile prey is intercepted by a motionless consumer. Epigrowth feeders, which slide along the solid surfaces, intercept prey directly or spread their pseudopodia over a solid area to trap motile prey. Although all free-living protozoa and even most parasitic protozoa are phagotrophic, they are able to absorb dissolved compounds. This way of feeding is considered insignificant in nature as protozoa cannot compete with bacteria for dissolved nutrients. Raptorial amoebae, however, which capture each particle individually rather than in bulk, show a considerable degree of discrimination on the basis of qualities other than size. They also discriminate on the basis of species and physiological status, preferring actively growing specimens. Often, conditions in the soil are unfavorable to protozoan development and flagellates and ciliates are more vulnerable than amoebae. There may be lack of proper food or its physiological state may be unattractive, the soil may be too dry, or the temperature too low. Protozoa respond to these conditions by encysting and survive in a resistant, compact, low metabolic state, which is also less vulnerable to predation. (Cysts are protective, inactive forms of protozoa.) After depletion of a feeding site, at the onset of starvation, cell division of protozoa continues for one or two generations, yielding smaller cells than normal, respiration decreases, and autophagous vacuoles appear, digesting in particular mitochondria, and, finally, synthesis of macromolecules is completely stopped. Encystment and excystment of protozoa are rapid processes, enabling protozoa to promptly respond to changing physical, chemical, or feeding conditions. In soil, active and encysted protozoa can be enumerated separately and, in general, considerable numbers of encysted proto-

zoa coexist with active ones; in climates with lasting dry periods, complete populations of protozoa survive the drought encysted and rapidly excyst after the first rainfall. The occurrence of "prey deserts" and the only temporary bloom of hot spots of prey organisms explain the presence of considerable numbers of encysted protozoa in most soil samples. Particularly in clay-containing soils, large numbers of bacteria are located within more or less compact soil aggregates containing small pores. The aggregates are not accessible for all grazers and certainly not for the larger ones. Pores smaller than 6 μm diameter act as protective microhabitats against protozoan predation for bacteria. However, amoebae are flexibile and succeed in penetrating narrow spaces with bulging tentacles through pore-necks that are much narrower than their own diameter. In this way, amoebae channel confined substrates into the food web.

C. Protozoan Predation in the Aquatic Environment

Flagellates, on average, considerably smaller than amoebae, dominate soil protozoa numerically, but not in biomass. Their biomass in soil is only a small fraction of the amoebal biomass. Flagellates are well adapted to life in soil pore water, because of their shape and flagellae, where they swim around as in marine and in fresh water. In the aquatic environment, they are the dominant protozoans. In water, after organic enrichment, an initial bloom of bacteria is followed by bacterivorous flagellates and ciliates, and, after a few days, carnivorous ciliates appear, followed by small metazoa. Flagellates feed in quite another way than do amoebae, by means of filter feeding through a stoma, with the aid of their flagellae. In the aquatic environment, ciliates consume flagellates besides feeding on bacteria and algae, but in the soil ciliates, generally, are rare and do not seem to perform a quantitatively significant role. Terrestrial and benthic ciliates are neither specific and they occur in fresh and marine waters as well. In aquatic environments, flagellates (about 10^7 per liter) tend to control numbers of bacteria and often keep them at a level below 5×10^9 specimens per liter, whereas in the absence of flagellates bacterial numbers amount to 10×10^9 l^{-1}. As a consequence of protozoan grazing, the average density of bacteria in seawater is about the same as in the surface layers of oligotrophic oceans and in eutrophic coastal seas (0.5 to 2×10^9 cells per liter). The feeding by predators on the flagellates decreases the flagellate densities but this does not increase bacterial numbers. On a global scale, the enormous primary production in marine waters is mainly performed by small algal cells and cyanobacteria which are primarily consumed by protozoa; this also stresses the importance of protozoan predation in nature.

D. Protozoan Predation in the Soil

The same relationships function in the soil but, because of the decreased vulnerability of bacteria to predation, particularly in soils containing small particles (clay, organic matter), protozoa control bacterial numbers less directly. Unlike other important bacterivores such as the nematodes, generally, protozoan numbers in soil run parallel to bacterial numbers with a short time lag (Fig. 4). This parallellism between bacterial number/dynamics and protozoan number/dynamics functions only when bacteria are actively growing. Protozoan populations cease to grow when bacterial densities decrease to 10^6 cells per gram soil. Contamination of the soil, e.g., with heavy metals, can disturb the parallel growth of microorganisms and their grazers.

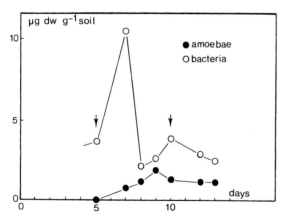

Fig. 4. Biomass of bacteria and amoebae in a forest soil (→ rainfall). (Adapted from M. Clarholm, *in* Tom Fenchel (1987). "Ecology of Protozoa: The Biology of Free-living Protists." Springer-Verlag, Berlin/Heidelberg/New York.)

V. EFFECTS OF PROTOZOAN PREDATION ON SOIL METABOLIC PROCESSES

A. Metabolic Soil Processes

Metabolic processes in the soil are largely performed by bacteria and fungi. Decomposition of organic matter leads to the disintegration of materials formed by primary producers, finally resulting in inorganic substances, CO_2, NH_4^+, H_2O, PO_4^{---}, H_2S. The steps resulting in the formation of inorganic molecules is called mineralization. Other important transformation processes of nitrogen, such as nitrification and denitrification, are also microbial in nature.

B. Protozoan Metabolism

Digestion of bacteria is the basis of direct nutrient flow through protozoa at either cell or population level. After ingestion of prey-bacteria by protozoa, up to 40% of the prey, in terms of C, is incorporated into newly formed protozoan biomass; the remaining 60% is mainly respired and, to a small extent, egested. Protozoan cells have about the same elemental ratio, C:N:P, as their prey-bacteria. Consequently, growing protozoa acquire sufficient N and P to balance the amount of C used for biomass production. This leaves 60% of the ingested nutrients to be excreted, mainly as inorganic molecules, partly as structural residues. The C:N:P ratio of bacteria is approximately 30:6:1, the C:N ratio being more stable than the C:P ratio. Bacteria can accumulate P in the form of polyphosphate, resulting in a low C:P ratio. This makes excretion of P by protozoa less predictable than excretion of N (Fig. 5).

C. Effects of Protozoa on Bacterial Mineralization of C, N, and P

Excretion of inorganic N by bacteria depends on the C:N ratio of the substrate and on its resistance to decomposition. For easily decomposable substrates, such as fresh crop debris, C:N ratios wider than 15 do not result in N mineralization, whereas decomposition of more resistant organic matter with the same C:N ratio results in inorganic N release; higher resis-

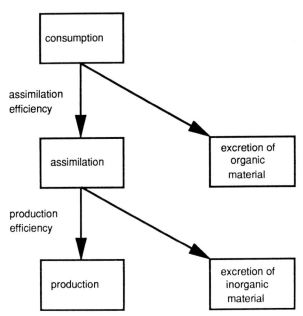

Fig. 5. Scheme relating consumption, biomass production, excretion of organic material, and excretion of inorganic material. (Adapted from P.C. de Ruiter *et al.* (1993). *J. Appl. Ecology* **30**, 95–106.)

tance is indicated by a lower production : respiration ratio of the bacteria. During wheat straw (C : N $\cong$ 95) decomposition, inorganic soil-N is immobilized, whereas during decomposition of sugar beet leaf (C : N $\cong$ 12) N mineralizes. Under conditions of N-immobilization, only predation on bacteria by bacterivores, mainly protozoa and nematodes, results in N mineralization. As mineral N is an essential nutrient for the growth of plants and microorganisms, predation on bacteria under N-immobilizing conditions is a very important process for continuous growth of plants and microbes. Effects of protozoan and nematode predation on mineralization processes in soil have been established in laboratory experiments and, with the help of food web models, in field studies. The results of both approaches support each other: both bacterivorous predators had a considerable effect on rate and pattern of CO_2 and NH_4^+ evolution, but effects on C and N did not always run parallel. The latter phenomenon is caused by the difference between the mechanisms affecting CO_2, and NH_4^+ evolution. Whereas the effects on N are mainly direct, protozoa excreting NH_4^+ after digestion of bacteria, the effects on C are mainly indirect, protozoa stimu-

lating bacterial growth. With respect to the indirect effects, a number of mechanisms are assumed:

- digesting protozoa excrete bacterial growth-limiting substances such as vitamins and minerals;
- selective grazing on "young," dividing bacteria keeps the bacterial populations in a phase of active growth and prevents their becoming senescent;
- bigger predators such as nematodes, but not protozoa, transport bacteria to new substrate sites.

Not only the activity of individual heterotrophic bacteria is enhanced in the presence of protozoa but also the activity of bacteria for specific processes, such as nitrogen fixation and nitrification. Under laboratory conditions, protozoa-stimulated activity of individual bacteria can result in increased overall soil respiration, but, as the numbers of bacteria decrease, predation can also result in unaffected, or even decreased, soil respiration. It was also shown under experimental conditions that net N mineralization in C-excess/N-limited soil only occurred in treatments with bacteria and amoebae and not in controls with only bacteria. Although protozoan predation has considerable effect on bacterial growth under N limitation, little effect results from predation under C limitation. In the aquatic environment, where phosphorus, rather than nitrogen, is the limiting nutrient for primary production, the contribution of protozoa to P mineralization is also substantial and probably essential in low-nutrient natural waters.

VI. PROTOZOAN PREDATION IN BULK SOIL AND IN THE RHIZOSPHERE

A. Organic Matter Distribution in the Soil

Soil is a heterogeneous habitat for organisms, consisting of solid, liquid, and gaseous fractions, intimately connected to each other on various scales, building a labyrinth of more or less habitable spaces and provided with variable amounts and types of food for its inhabitants. Aged organic matter is distributed more or less at random in the topsoil of tilled arable soils, whereas fresh organic matter is concentrated in local spots. Under natural conditions, organic matter is often regularly arranged in successive horizontal layers, with the litter zone covering the mineral soil. All transitional stages between fresh crop debris and aged organic matter (humus) occur, decomposing at different rates and at different growth efficiencies of the microbes. Thus, in nature, the availability of food is patchy in time, space, and quality, and microorganisms also adapt to this by changing their physiological state. Generally, growth efficiencies decrease with increasing age of the substrate and, consequently, relatively more nutrients are released. Fresh organic matter normally has its original shape and structure, whereas aged organic matter has been physically and chemically transformed and is often adsorbed to solid mineral surfaces or dissolved in the soil solution. The constitution and distribution of organic matter largely determine the taxonomic and physiological status, densities and location of bacteria and, consequently, of their grazers.

B. Hot Spots of Biological Activity

Due to their growth potential, protozoa utilize rapid fluctuations in resource levels (hot spots) and they can exploit patches of resources even in tiny habitats because of their small size. In particular, fast-growing, dense colonies of bacteria in the rhizosphere of growing plants, and in hot spots of decomposing fresh crop debris, are grazed upon by opportunistic protozoa and nematodes; these track the dynamics of the bacteria and develop fast-growing temporary, local populations of high density, which are prey for higher-order predators. Ingestion of bacteria suspended in the soil solution requires other feeding mechanisms than the ingestion of bacteria growing on surfaces or in small hardly accessible micropores. The rhizosphere is the soil layer surrounding roots, directly under the influence of the roots. It comprises up to 5% of an overgrown soil. A root passes through successive stages: colonizing, fast-growing, differentiating, and, finally, decomposing and disappearing. Thus, roots stimulate microbial growth and, consequently, protozoa and nematodes. In general, crops translocate more than 40% of the carbon fixed in the above-ground parts to the roots

(1–5 tons per hectare); Deposition of organic compounds by roots in soil is, thus, a major source of substrates for opportunistic bacteria/protozoa/nematode populations, particularly near the tips of growing roots. The rhizosphere is, thus, a very dynamic environment for food webs of organisms, tracking root development. As microbial communities in the rhizosphere are specific for different plant species, protozoan rhizosphere communities also vary between different crops. Comparable dynamics of organisms can be observed at hot spots of organic matter. Increased N mineralization due to protozoan predation in the rhizosphere of growing plants and subsequent uptake of nitrogen by plants has been observed on several occasions. There was a 20% increase in N uptake in wheat plants in the presence of a protozoan inoculum under experimental conditions. Similar results were found for shortgrass prairie due to amoebal predation of bacteria in a simulation model of the detritus food web.

C. Effects of Protozoan Predation

Generally, predation on microbes, in particular by protozoa, releases nutrients at locations with increased biological activity in the soil, where roots compete with microorganisms for nutrients. They, thus, prevent stagnation of growth of the vegetation and soil life. This mechanism is effective in soil as well as in aquatic environments. Studies of food web models of N mineralization indicate that, in soil, approximately one-third of the mineralized N originates (directly or indirectly) from faunal, largely protozoan, predation on microbes; the rest is mineralized mainly by bacteria. Protozoa are particular effective in sewage treatment plants where their presence substantially enhances decomposition of organic material.

VII. PROTOZOA UNDER STRESS CONDITIONS

A. Protozoan Encystment

After starvation, the most common stress factor for protozoa in soil is certainly drought. In soil, protozoa need at least a water film surrounding solid particles for their movement, which is necessary for actual feeding and for migration to new feeding sites. If this film is too thin or interrupted over large areas, protozoa will encyst, the basic mechanism of protozoan resistance to all types of stress. The small ciliate, *Colpoda steinii*, is only active in water films > 30 μm deep, whereas, for small amoebae, a water film depth of 3 μm is needed. There are even zonation patterns of protozoa species that can be explained by different degrees of resistance to desiccation. The inactive protozoan cyst is characterized by a decreased water content, increased cell-wall thickness, and a largely reduced metabolic activity. Obviously, encysted protozoa are less vulnerable to predation. Decrease and increase of water content are the first and the last step of en- and excystment, respectively. Among nematodes, comparable mechanisms exist to escape from adverse circumstances; juveniles can moult without removal of the epidermic remains and subsequently decrease their water content and cease moving. Favorable conditions generally stimulate excystment but sometimes specific stimuli are needed. Stress situations, caused by osmotic and pH extremes, heavy metals, organic pollutants, pesticides, and lack of food, trigger encystment in the same way as does drought. Stress specific to protozoa is unknown, as are specific biocides against protozoa. Therefore, if protozoa suffer from unfavorable conditions, they suffer amid the entire food web of soil organisms and of plants, which means that stress effects on protozoa are always direct as well as indirect. Short-term stress on soil ecosystems often causes decreased numbers of organisms and decreased or fluctuating metabolic processes. Stress for some days or weeks may cause more substantial effects.

B. Relative Vulnerability of Organisms to Heavy Metals

Lasting effects of heavy-metal contamination are increased specific bacterial respiration and decreased bacterial growth. Bacterial numbers can be increased or decreased, depending on effects of stress on protozoa and nematodes. If the heavy metal decreases numbers of bacterivores, bacteria are released from predation and, consequently, higher numbers of bacteria are counted.

Microbes are the first organisms to respond to increasing heavy-metal concentrations, but tend to continue to function under high levels of heavy metals. Plant growth is already more negatively affected at lower levels of heavy-metal contamination than the functioning of webs of soil organisms, and, within these webs, nematodes and protozoa are more vulnerable than are bacteria, and among the bacteria, specialist taxa, such as nitrifiers and denitrifiers, disappear sooner than do more general heterotrophs. The greater vulnerability of plants has consequences for the soil food web since decreased crop growth results in decreased incorporation of organic matter into the soil and, consequently, many soil organisms suffer from a lack of substrate before they are affected by toxic concentrations of heavy metals. Starved and damaged food webs in heavily contaminated bare fields recover rapidly during development of metal-tolerant vegetations.

See Also the Following Articles

Freshwater Microbiology • Heavy Metal Pollutants • Osmotic Stress • Rhizosphere • Soil Dynamics and Organic Matter, Decomposition

Bibliography

Brussaard, L. (1994). Soil ecology of conventional and integrated arable farming systems. *Agric. Ecosystems Environ.* **51**, 1–270.

Darbyshire, J. F. (1994). "Soil Protozoa." CAB International, Wallingford Oxon, U.K.

Fenchel, T. (1987). "Ecology of Protozoa: The Biology of Free-living Phagotrophic Protists." Springer-Verlag, Berlin/Heidelberg/New York.

Pseudomonas

Vinayak Kapatral, Anna Zago, Shilpa Kamath, and Sudha Chugani

University of Illinois at Chicago College of Medicine

GLOSSARY

hypersensitive response (HR) A defense response in resistant cultivar or nonhost plants, elicited by plant-pathogenic bacteria that cause disease in susceptible host plants.

lysogen A bacterial or other cell carrying a viral prophage and having the potential of producing bacteriophage under proper conditions.

operon A transcriptional unit encoding one or more genes trancribed as a single transcriptional unit.

opportunistic organism An organism that is generally harmless but becomes pathogenic in a compromised host.

sigma factor A protein that helps the RNA polymerease core enzyme recognize the promoter at the start of a gene.

septicemia Blood poisoning, associated with persistence of pathogenic organisms or their toxins in the bloodstream.

PSEUDOMONADS are ubiquitous microorganisms, found in the environments such as water, soil, plants, humans, animals, insects, sewage, and hospitals. Members belonging to this genera have remarkable nutritional versatility. In humans, they are opportunistic pathogens, found in lungs of patients suffering from cystic fibrosis (CF) or eye infections, in burn victims, and in AIDS patients. Pathogenesis is due to the secretion of a large number of toxins, which weaken or evade the host immune system, enabling the bacteria to survive. In plants, they cause disease by produc-ing hypersensitive response, resulting in leaf and root tissue damage. In soils, they detoxify environmentally hazardous compounds, such as phenols, chlorinated compounds, and pesticides. They are also useful in producing beneficial compounds, including oils, biopesticides, and nylon monomers.

I. INTRODUCTION

Pseudomonads are gram-negative, monoflagellated rods. They are obligate aerobes, but they can also grow in anaerobic conditions with arginine in the medium. The size of Pseudomonas is about 1–5 μm long and about 0.5–1.0 μm wide. The genome size of *P. aeruginosa* is about 5.9 Mb with 65% G+C content. Based on the rRNA typing and characteristic tests, the family Pseudomonadaceae is classified into five related groups. Further, distinctions among groups is based on the following seven tests: (a) colony structure, (b) odor, (c) production of phenazine pigments, (d) ability to grow at 42°C, (e) gelatin liquefaction, (f) denitrification, and (g) differential growth patterns utilizing specific carbon compounds. The genus Pseudomonas belongs to Palleroni's rRNA group I. In this group, there are three phenotypic subgroups: (a) The first subgroup, which includes *P. aeruginosa, P. fluorescens*, and *P. putida*. These species produce water-soluble pigments which fluoresce blue–green light on exposure to ultraviolet rays. (b) The second subgroup consists mainly of water and soil bacteria, such as *P. stutzeri* and *P. mendocina*. These microorganisms perform denitrification reactions under anaerobic conditions. (c) The third subgroup consists of *P. alcaligenes, P. pseudo-alcaligenes*, and *P. dentrificans*. The rRNA group II contains genus Burkholderia: group III includes genus Comamonas and Acedovorax; group IV includes Brevundemonas: and group V consists of Stenotrophomonas.

II. HUMAN PATHOGENS

P. aeruginosa, is an opportunistic human pathogen. It is commonly associated with cystic fibrosis patients, burn victims, and immunocompromised and AIDS patients. Several virulence factors are secreted by the bacteria; these factors include elastase, exotoxin A, exoenzyme S, and alginate. *P. aeruginosa* also causes osteomyelitis, urinary tract infections, and septicemia. Other life-threatening strains include *P. mallei* and *P. pseudomallei,* which cause melioidosis. Acute infections lead to septicemia, resulting in high fever, chilly pneumonia, lung abscess, and enlargement of liver and spleen, which leads to septicemia shock and coma. The subacute form causes pneumonia and lung abscesses which resemble those in pulmonary tuberculosis. These two species, in particular, have been listed as agents of biological warfare and are regulated by the Center for Disease Control (CDC, Atlanta) under Antiterrorsim and Effective Death Penalty Act of 1996. Less pathogenic strains include *P. aeruginosa keratitis,* which causes keratitis of the human eye. This infection is common in coal-miners, farmers, and people wearing contact lenses. A typical corneal damage is associated with the disease, due to the action of extracellular proteases produced by the bacteria. Some of the general characteristics of Pseudomonads are described below.

Flagella. Pseudomonas is motile by a single, polar flagellum and exhibits chemotaxis to favorable molecules, such as sugars. Pseudomonads which cause human diseases use flagella for adherence and tissue invasion at the early stages of infection. Strains of *P. aeruginosa* express either a-type or b-type of flagella. This classification is primarily based on the apparent size of the flagellin subunit (encoded by the *fliC* gene) and its antigenicity. The a-type flagellins are heterogeneous and are divided into various subgroups, whereas the b-type flagellins are homogeneous. The a-type *fliC* open reading frame (ORF)

varies between 1164 bp to 1185 bp, with the subunit size ranging from 45 to 52 kDa. The b-type *fliC* open reading frame is 1467 bp in length and encodes a 53 kDa size protein. The N- and C-terminal sequences of both these flagellins are nearly identical, whereas the central region is variable. The a-type of flagellins undergo glycosylation, whereas the b-type flagellins are phosphorylated at tryosine residue. Such modifications are unique among the prokaryotic flagella, which are often methylated at the lysine residue. The phosphorylated flagellin protein is believed to serve as a signal for intact flagellin export from cytoplasm to the flagellar assembly apparatus.

The flagellin gene is transcribed by a sigma factor called σ^{28} (encoded by *fliA*) and is similar to the σ^{28} of *E coli* and *Salmonella*. In addition to σ^{28}, there are other regulatory factors which control flagellin gene expression. Pseudomonas flagellin gene expression requires another important regulator, called the FlaR (flagellin regulator), which binds to the upstream activating sequence, also called as UAS, and transcribes the flagellin genes. The FlaR is, in turn, controlled by RpoN (σ^{54}, which regulates the expression of nitrogen assimilatory genes, and is also called NtrC) regulates both the nitrogen assimilatory genes.

Pili. The Pseudomonas pili belongs to a unique class called type IV pili, or N-methyl-phenyl-alanine (NMePhe) pili. Pili contribute significantly to the adherence capability of the bacteria to the epithelial cells and mucosal surfaces. The tip of the pili is responsible for the attachment to the host cell surface. Pili also serve as the receptors for bacteriophages. The pili are long polar filaments composed of homopolymers of a 15–18 kDa protein, called pilins, which is encoded by the *pilA* gene. PilA is first synthesized as a pre-pilin, which then undergoes processing during its export to produce final pilin subunit. After cleavage, the leader peptide of the newly generated N-terminus undergoes methylation. There are three other accessory genes, designated *pilBCD,* which are required for the biogenesis of pili. These genes are located adjacent to the pilin structural gene. The *pilD* gene encodes the prepilin peptidase, which processes the pre-pilin protein. The pilin protein is retained in the outer membrane of the cell before its assembly into an intact pilus. Pilin filaments have a diameter of ~5.2 nm, with an average length of 2.5 mm, and the subunits are arranged in a helical array which forms a hollow cylindrical structure. The pili also forms a multifunctional retractable structure, which provides the bacteria with a very unique form of motility, called twitching motility. Pili is regulated by PilR/PilS of the two-component regulatory system. In addition to PilR/PilS, an alternate sigma factor RpoN also controls the pilin gene expression. The pili bind to the glycolipids receptors which are present on the epithelial surfaces by its C-terminal region. The two types of receptors to which pili specifically bind are gangliotetraosyl ceramide (9 asiaol-GM1) and acetyl neuramin-O-sylgangliotetraosyl-ceramide (GM1).

The flagellin and pili display various analogies with respect to their structure, function, and genetic organization. Both are situated on the surface and form filamentous structures by multimerization of identical subunits, the flagellins and the pilins, respectively. Pseudomonas uses RpoN to control both flagellin motility and pili motility, especially under low nutrient condition and during initial establishment in the human host.

A. Virulence Factors Secreted by Pseudomonas

1. Elastase

Elastase is a extracellular zinc protease, which acts against eukaryotic proteins such as elastin and collagen, and destroys the proteins which maintain the structural integrity of the cell. It also degrades complement components, human immunoglobin, and serum alpha proteins. Recently, elastase is also shown to play a cellular role in Pseudomonas: the cleaving of the 16kDa nucleoside diphosphate kinase (Ndk) to generate a 12 kDa Ndk. The truncated Ndk plays a role in GTP synthesis that is required for alginate synthesis. Elastase is encoded by *lasB* gene, which codes for a pre-pro protein of 53 kDa. The pre-pro protein is translocated into the periplasm, where it undergoes autoproteolysis to generate an 18 kDa and a 33 kDa protein. The 18 kDa pro-peptide interacts noncovalently with the 33 kDa subunit. The active 33 kDa elastase is secreted outside the bacteria by the type II secretion system. The *lasB* gene is regulated by

LasR regulator, under the LasR–LasI quorum sensing system I.

2. LasA Protease

LasA protease secreted by Pseudomonas nicks elastin, hydrolyzes β-casein, and lyses *Staphylococci*. It is synthesized as a 41 kDa precursor (encoded by *lasA*) and is cleaved to produce a 22 kDa active protein. Similarly, there is another protease produced by Pseudomonas which shares similar properties to LasA protease, called LasD (encoded by the *lasD*). LasA is synthesized as a 30 kDa protein and is cleaved to generate a 23 kDa active protein. The LasA protease is active at broad pH range, whereas LasD is active only at higher pH conditions. These proteins are also considered as major virulence factors.

3. Exotoxin A

Exotoxin A (encoded by *toxA*) is the most toxic protein produced by Pseudomonas. Exotoxin A catalyzes the ADP-ribosylation of the eukaryotic translation factor (EF-2) to form ADP-ribosyl-EF-2, which results in the inhibition of the host cell protein synthesis. This mechanism of cell toxicity is identical to diphtheria toxin. The *toxA* gene is expressed at higher levels under low iron conditions in the environment and is controlled by a ferric uptake regulator called Fur (Ferric iron regulator), which also regulates siderophore production.

4. Exoenzyme S

Exoenzyme S is a 43 kDa protein encoded by *exoS* gene and is similar to the YopE of *Yersinia enterocolitica*. It has ADP phosphorylation activity toward eukaryotic proteins, such as vimentin, H-Ras and K-Ras types of GTP-binding proteins. The ADP-ribosylation requires an eukaryotic protein, called FAS (factor for activating exoenzyme S). There is another FAS-dependent ADP ribosylating enzyme, called the exoenzyme T (encoded by *exoT*), which is similar in molecular weight to ExoS. Both these proteins posess cytotoxic activity and are secreted upon contact into the eukaryotic host by the type III secretory system. Both the *exoS* and *exoT* genes are regulated by a transcriptional activator called ExsA (encoded by *exsA*), which specifically recognizes a consensus "TNAAAANA" sequence that is present upstream of the *exoS* and *exoT* genes.

5. Alginate

Alginate is a major virulence factor. It is a linear polymer composed of D-mannuroic acid linked to its C5 epimer L-guluronic acid by β (1–4) linkages. Alginate is also used in food, pharmaceutical, and chemical industries. In humans, alginate secreting *P. aeruginosa* colonize the pulmonary tracts of cystic fibrosis patients, leading to blockage of lung airways, resulting in death. Alginate biosynthetic pathway is well characterized; the genes and the enzymes are described in Fig. 1. The first enzyme in the biosynthesis is AlgA, which a bifunctional enzyme, phosphomannose isomerase–guanosine diphosphomannose pyrophosphorylase (PMI–GMP). AlgA acts at the first and the third step of the alginate synthesis. PMI scavenges the fructose 6-phosphate from the metabolic pool and diverts it to alginate synthetic pathway. AlgC as phophomannose mutase converts mannose 6-phosphate to mannose 1-phosphate, which is finally converted to GDP-mannose by AlgA. GDP-mannose is further oxidized to GDP-mannuronic acid by GDP-mannose dehydrogenase (encoded by *algD*).

Further, the gene products of *alg8*, *agl44*, and *alg60*, which are membrane associated, form an alginate polymerizing complex and catalyze alginate polymerization and secretion. The polymer gets acetylated in the periplasm by acetylase AlgF (encoded by *algF*) and, simultaneously, AlgG, an eperimase (encoded by *algG*), converts mannuronic acid to guluronic acids at random intervals of 2–3 residues of the alginate polymer. Finally, the alginate polymer is transported across the outer membrane by the porin called AlgE. Pseudomonas also produces alginate lyase (encoded by *algL*), which breaks down alginate, thereby controlling the length of alginate polymers. Alginate lyase is important for the bacteria in detaching from the cell surfaces to spread to new habitats and in generating short oligosaccharides, which are used as primers for new alginate chains. A number of environmental products, such as osmolarity, nitrogen limtation, phopshate limitation, dehydration or stress conditions, and antibiotics, stimulate the synthesis of alginate. Instablity or mutations

in the regulatory genes, such as *mucA*, *mucB*, also affect alginate synthesis.

6. Lipopolysaccharide (LPS)

LPS is another virulence factor secreted by *P. aeruginosa*. It consists of a hydrophobic lipid A, a core oligosaccharide, and an O polysaccharide. *P. aeruginosa* produces lipid A, which is endotoxic, while the O polysaccharide antigen is immunodominant. The O-antigen is encoded by the lysogenic phage called D3. Two types of LPS are produced by Pseudomonas: A type and B type. The A-type LPS is an antigenically conserved molecule with an O-polysaccharide, consisting of O-trisaccharide repeating units of D-rhamnose, organized as α1-2, α1-3, and α1-3, while B-type is serotype specific LPS and is used to distinguish nearly 20 serotypes of *P. aeruginosa*.

7. Lipases

Several different types of lipases are secreted by Pseudomonas; these have been grouped into lipolytic, hemolytic, and nonhemolytic phospholipase C.

a. Lipase

It is a 26 kDa enzyme secreted by *P. aeruginosa*, which is active against a broad range of triglycerides with fatty acyl chain lengths, varying from C6 to C8. *P. aeruginosa* lipase is stereoselective for sn-1 of the triglyceride. It is encoded by *lipA* and *lipF* genes. The LipA product is synthesized as a 29 kDa prelipase, which has a signal sequence for secretion. The pre-lipase is secreted into the periplasm by the Sec-dependent secretion pathway. Here, lipaseA protein gets in contact with the membrane-bound 37 kDa protein called LipF (encoded by the *lif* gene). LipF assists LipA protein to fold into the right conformation, whereby oxido-reductases catalyze the disulfide bond formation. Finally, the mature lipase (26 kDa in size) is secreted into the external environment by the Xcp or the type-II secretion system. Occasionally, lipases get trapped with the LPS matrix.

b. Phospholipases C

Two types of phospholipase C have been found in Pseudomonas; these are classified as hemolytic and nonhemolytic phospholipases.

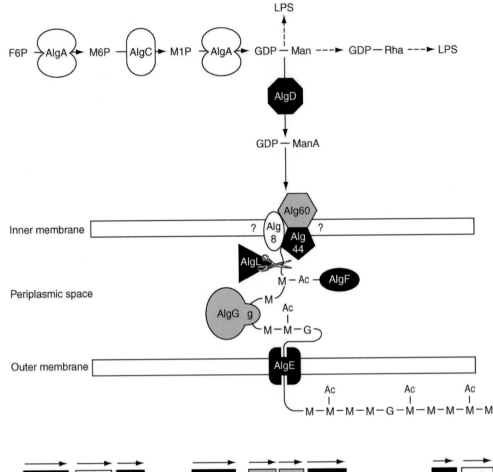

a

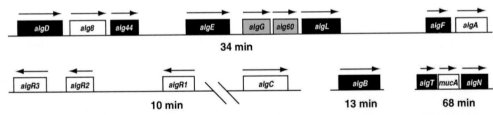

b

Fig. 1. (a) The biosynthesis of alginate by *Pseudomonas aeruginosa*. Fructose 6-phosphate (F6P), obtained from the metabolic pool, is converted to GDP-mannuronic acid (GDP-ManA), which provides mannuronate residues (M) for polymerization. Occasionally, guluronate residues (G) are incorporated via epimerization of mannuronate residues by the AlgG protein; g represents the transition stage of epimerization to guluronate. Mannuronic acid residues of bacterial alginates are partially 0-acetylated by the AlgF protein. The question marks indicate that the functions of *alg8*, *alg44*, and *alg60* gene products and the putative interactions of these proteins are hypothetical. The dashed arrows indicate enzymatic steps leading to lipopolysaccharide (LPS) synthesis. Abbreviations used: M6P, mannose 6-phosphate; M1P, mannose 1-phosphate; GDP-Man, GDP-mannose; GDP-Rha, GDP-rhamnose; and Ac, O-acetyl groups. (b) The organization of the alginate gene clusters. The alginate genes are clustered at three locations in the *P. aeruginosa* chromosome. Except for the *algC* gene, which is located at 10 min, all of the known alginate structural genes are located at 34 min. The regulatory genes map at 10 min and 13 min, and the genes responsible for the genotypic switch to alginate production are located at 68 min. The arrows above the genes represent the direction of transcription. Note that *mucA*, shown at 68 min, is different from *algA* shown at 34 min. (Reproduced from May and Chakrabarty, 1994, *TIMS.*)

(i) Hemolytic phospholipase C (PLC-H). It is a heat labile protein of about 78 kDa in size, with lipase activity on phosphatidyl choline and sphingomylein. PLC-H is encoded as part of the *plc* operon which encodes three genes. *plcR1* encodes a 23 kDa protein and *plcR2* encodes a 19 kDa protein, which modify the lipase. The *plcH* encodes a 78 kDa structural phospholipase C, which breaks down the human and mouse erythrocytes.

(ii) Nonhemolytic phospholipase (PLC-N). This is the second phospholipase produced by Pseudomonas. It is 73 kDa in size and is referred as PLC-N. It acts on phosphatidyl choline and phosphatidyl serine but not on the sheep erythrocytes. PLC-N is homologous to PLC-H. Low phosphate and aerobic conditions stimulate the PLC synthesis and activity, whereas high NaCl represses the PLC activity.

Low phosphate in the media stimulates the PLC-H by 8-fold and over 30-fold of PLC-H. Both PLC-N and PLC-H hydrolyze phosphotidylcholine to diacylglycerol and choline, resulting in the degradation of lung surfactant, leading to tissue damage and swelling. Additionally, PLC-H activity leads to hemorrhagic necrosis of the lung tissue and to the production of eicosinoids, which cause inflammation. Eicosinoids are produced due to breakdown of diacylglycerol (DAG) and arachidonic acid. DAG also activates protein kinase C of the eukaryotic cells, which results in inflammation. Choline production also leads to the accumulation of betiane, a compound which serves as osmoprotectant to the bacteria, thereby contributing to the persistence in the lungs.

8. Cytotoxin

Pseudomonas also produces a toxin against leukocytes and other eukaryotic cells. The pro-cytotoxin is encoded by the *ctx* gene, which encodes a 31 kDa protein, which undergoes a truncation to generate the 18 kDa active toxin. The CTX toxin causes pores in the lipid layers of the host cell. The CTX is encoded by a lysogenic ϕCTX phage, which is about 35 kb in size and is integrated at the *attP* site.

9. Rhamnolipids

Pseudomonas produces two types of biosurfactant, called rhamnolipids. Rhamnolipid I is a L-Rhamno-syl-L-rhamnosyl-B-hydroxy-decanoyl-β-hydroxy-decanoate and Rhamnolipid II consists of L-rhamno-syl-β-hydroxydecanoyl-β-hydroxy-decanoate. The synthesis of rhamnolipid proceeds with sequential transfer of glycosyl moiety from TDP-rhamnose to β-hydroxydecanoyl β-hydroxy-decanoate by specific rhamnosyl transferases. These compounds lyse the blood cells. Low nitrogen conditions in the presence of glucose or glycerol stimulate the production of rhamnolipids. These facilitate emulsification of water-insoluble substrates, such as alkanes.

10. Pyocins

Pseudomonads produce bacteriocins called pyocins, which kill other bacteria. Three groups of pyocins, namely, R, F, and S types, have been identified. The R- and F-types are similar to the bacteriophage tail proteins. S-types are the most abundant low molecular weight pyocins and are subdivided into S1, S2, and AP4, based on their effect on the indicator bacterial strains. S1 and S2 pyocins interfere with the phospholipid synthesis of the target bacteria, whereas the AP41 possesses endonuclease activity. The genes transcribing the S-type pyocins encode two proteins: AP41 encodes an 83 and a 10 kDa protein, S1 encodes a 64 and a 10 kDa protein, and S2 encodes a 74 and a 10 kDa protein. These are secreted into the environment and play a significant role in ecological domination by killing the bacteria. The lysed bacteria are degraded and are utilized as nutrients.

11. Lectins

Along with other virulence factors, *P. aeruginosa* produces specific sugar-binding proteins, called lectins. Two types of lectins have been characterized; PA-IL is a lectin specific to D-galactose and its derivatives, whereas PA-IIL binds specific to L-fucose and D-mannose. These lectins help the bacteria to adhere to cells and affect the immune cells of the host by inducing cytokines, such as IL-1 and IL-6, and produce cytotoxin, resulting in tissue inflammation.

12. Siderophores

Pseudomonads produce three major iron-scavenging compounds, called siderophores, under iron-deficient conditions. Siderophores bind iron and are taken in by the cell via a receptor-mediated mechanism. This is one mechanism by which Pseudomonas limits the growth of other bacteria, by competing for iron.

a. Pyochelin

It is a phenolic siderophore composed of a salicyl ring amide bonded to a thiazoline ring with a terminal N-methyl thiazolidine ring. Pyochelin binds to iron and is taken up by a 75 DA outer membrane protein. Additionally a 14 kDa protein is necessary for the uptake.

b. Pyoverdin

This is a second class of fluorescent, high affinity binding siderophore, which is made of 2,3 diamino-6-7 dihydroxyquinoline, which gives Pseudomonas a yellow–green fluorescent appearance. The structure of pyoverdin is composed of eight amino acids in the following sequence: D-serine, L-arginine, D-serine, L-dN-hydroxy-ornithine, L-threonine, l-threonine, L-lysine and a terminal L-dN-hydoxyornithine in piperdine ring. Some strains, like *P. putida* and *P. syringae*, also produce pyoveridine-like siderophores with different amino-acid sequences. The receptors for siderophores also vary in size among different species.

c. Pseudobactein

Rhizosphere colonizing *P. putida* produces a yellow–green fluorescent siderophore, called pseudobactin 358. This siderophore contains a nanopeptide linked to fluorescent dihydroxy quinoline moiety. Pseudobactin 358 is produced in iron limiting conditions, which efficiently complexes environmental iron, making it unavailable to other rhizosphere bacteria and plants.

13. Alkaline Protease

Pseudomonas infecting the cornea of the eye produce alkaline protease (encoded by the *aprA* gene), which is essential for establishment during eye infection. The secretion of the AprA requires three accessory proteins; AprD, AprE, and AprF. These form a complex spanning the periplasmic space to secrete the AprA in a single step to the external medium by Type I secretion system. Pseudomonads also produce

a inhibitor for the alkaline protease, called AprI (encoded by *aprI* gene). The secretion of AprA is similar to a hemolysin in *E. coli*. Thus, the AprDEF forms a unique secretory system, independent of the Xcp system.

14. Pigment Production

Many strains of Pseudomonas produce pigments, some which are fluorescent, while some strains produce melanin (dark brown) pigment, which have antiphagocytic properties. Some strains produce pycocyanin, which is 1-hydroxy-5-methylphenazine (pyo), and its demethylated product, 1-hydroxyphenazine (1-hp). These pigments reversibly inhibit human peripheral blood lymphocyte blastogenic properties in response to mitogen and antigens. They catalyze the NADH dependent conversion of O_2 to superoxide or water and also inhibit the proliferation of lymphocyte and nonlymphocyte cell lines.

B. Some Unique Behavior Exhibited by Pseudomonas

1. Antibiotic Resistance

Pseudomonads exhibit intrinsic resistance to a wide variety of antibiotics and chemotherapeutic agents. The resistance is either due to low permeability to antibiotics or to producing β-lactamases against β-lactam antibiotics or multidrug efflux pumps. The multidrug efflux pump consists of a inner membrane bound MexB, which is connected to outer membrane OprM by MexA, which spans the periplasmic space. Pseudomonas uses this (MexAB) efflux system to export amphipillic drugs, such as β lactams, carbanepams, tetracycline, and chloramphenicol. Pseudomonas also has another multidrug efflux system, called the MexCD–OprJ, similar in function to the MexAB–OprM system.

In recent years, there is an emergence of quinolone resistant Pseudomonas. These resistant strains have altered *gyrB* or *gyrA* genes. GyrB and GyrA subunits together form gyrase enzyme, which maintains DNA supercoiling altered forms of the enzyme, which confers quinolone resistance. Some strains exhibit low permeability to quinolones by altering LPS structure and OmpF (outer membrane protein) porins. Resistance to β-lactams is mostly due to alteration in the β lactamase gene and to low permeability.

2. Response to Oxidative Stress

Pseduomonads respond to both endogenous (aerobic growth) and exogenous (anaerobic or in macrophage) oxidative stress by producing Fe- or Mn-containing superoxide dismutase (SOD) metalloenzymes. These dismutases catalyze the diproportionation of O^- to H_2O_2 and O_2. The organism also possesses catalase, which detoxifies the H_2O_2 to form O_2 and H_2O. The superoxide dismutase A (encoded by *sodA*) is a 23 kDa dimer, which uses manganese as a cofactor, and the superoxide dismutase B (encoded by *sodB*) is Fe-dependent, which also functions as a dimer. Mucoid (alginate producing) strains have been observed to possess higher manganese superoxide dismutase activity.

3. Pseudomonas Sense Their Own Population

Like many other bacteria such as *Vibrio*, *Agrobacterium*, *Erwinia*, and *Yersinia*, Pseudomonads also sense their population by a phenomenon called "quorum sensing," where the bacteria produces low molecular weight pheromone, which accumulates in extracellular medium, allowing individual cells to sense their low population unit. Once these molecules achieve a critical concentration, a population response is stimulated. By monitoring two independent molecules in the external environment, *Pseudomonas* thus senses the environment and secretes virulence factors (antihost factors). These factors and their genes are described in the text. Here, an overall regulation of the virulence genes that are determined by cell density is described. The two quorum sensing circuits are interdependent and hierarchical. The first quorum sensing system I is the LasR–LasI type, where LasR is a regulator of LasI which produces the N-(3-oxo) dodecanoyl-L-homoserine lactone (odDHL). This low molecular weight compound is excreted by Pseudomonas in lower concentrations, correlating with little or negligible amount of transcription of critical virulence genes. When large amounts of odDHL accumulate in the external environment, the molecules enter into the bacteria and induce the transcription of several virulence genes.

Thus, this system of quorum sensing regulates the production and secretion of very important antihost factors, such as elastase, exotoxinA, LasA protease, exoenzyme S, alginate, neuraminidase, and hemolysin.

Similarly, odDHL activates a second quorum sensing system II, called VsmR–VsmI, (also called as RhlI–RhlR), which regulates the synthesis of the second low molecular weight molecule called to N-butanoyl-L-homoserine lactone (BHL). Like the system I, the system II regulates the synthesis and production of alkaline protease, elastase haemolysin, pyocyanin, hydrogen cyanide lectins, chitinase, and antistaphylolytic lipases.

4. Pseudomonas in Cystic Fibrosis Patients

Cystic fibrosis is a disease commonly found in Caucasian populations. Nearly 5% of Caucasians of European descent suffering from CF disease carry a mutation in the gene called the cystic fibrosis transmembrane conductance regulator (CFTR), which is involved in NaCl transport. The CFTR gene is located on the human chromosome VII. Nearly 80–90% of the CF patients get infected with mucoid *P. aeruginosa*, leading to the pulmonary dysfunction and early mortality. In a normal lung, the NaCl concentration is about 85 mM and the defensins are active to kill the bacteria, whereas, on a CF lung, these defensins are inactive and the bacteria survives and produce enormous amounts of alginate (mucus). Such a condition leads to blocking of the airways and death. The CF lung environment is highly conducive for Pseudomonas to produce alginate, which forms a biofilm that protects the bacteria from antibiotics and drugs. The biofilm not only protects the bacteria but also helps to escape phagocytosis by the macrophage. They also prevent complement activation. Thus, treating Pseudomonas infections in CF patients is extremely difficult.

Pseudomonas infection control. A number of chemotherapeutic agents, such as cefotaxime, benzylpenicillin, imipenem, gentamycin, tobramycin, or a gentamycin/carbenicillin combination, are administered as the drugs of choice for treating serious Pseudomonas infections. These drugs act during the new cell wall synthesis and prevent bacterial multiplication. In case of severe infections, fluorinated quinolones are also administered; these include norflaxcin, gentamycin, and ciprofloxacin. Ciprofloxacin is used in burn patients because it is highly effective against Pseudomonas. In some cases, lipid analogs are also used to limit *P. aeruginosa* infections. In case of melediosis (caused by *P. pseudomallei*), trimethoprim or sulfamethoxazole is administered.

C. Secretion System in Pseudomonas

Pseudomonas has three secretion systems, including the general secretory system (GSP), for secretion of proteins through the outer membrane. These are classified as Type I, II, and III, based on their nature of recognition of the secreted proteins, location in the membrane, and chaperone.

(i) **Type I.** Alkaline protease (*aprA*) is secreted by Type I system, which requires three gene products, i.e., AprD, AprE, AprF, to translocate the alkaline protease in a single step into the external medium.

(ii) **Type II.** This secretion system consists of the extracellular proteins secretion (xcp) gene cluster, encoding 11 gene products. A heterogeneous set of proteins is secreted, including elastase, alkaline protease, and phospholipase. These proteins have a signal peptide that directs them into the periplasm and are subsequently recognized by the Xcp system to export into the external medium. Type IV pilin biogenesis and secretion share a similar mechanism to the type II secretion.

(iii) **Type III.** This system is similar to YopE (Yersina outer proteins) secretion in Yersinia and is also present in Pseudomonas, by which Exoenzyme S and Exoenzyme A are secreted. These proteins do not have a signal sequence, but have their independent chaperones that facilitate their secretion. Plant pathogenic strains also have a similar system, which is referred to as the Hrp secretion system, for harpin secretion.

III. PLANT PATHOGENS

A large number of Pseudomonas strains cause plant disease in most important agricultural crops, such as tomato, beans, potato, and commercial crops, such as fruit crops, cherries, plums, prunes, and others. Representative species include *P. syringae*, pv. syringae, *P. syringae,* pv. tomato, and *P. syringae* pv. phaseolicola. *P. syringae* causes bacterial spots, which first appear as water-soaked, green areas on lower leaves before spreading to upper leaves, which later turn tan with reddish borders. Two related pathovars of *P. syringae* pv. syringae van Hall and P. s. pv. morsprunorum (Wormald) cause bacterial canker. Both pathogens affect sweet cherry, sour cherry, plums, and prunes in Michigan and the neighboring province of Ontario, Canada.

Disease outbreaks are sporadic and more frequent on sweet cherry than on sour cherry. *P. syringae* infections are common on peach in the southeastern United States. The disease attacks all parts of the plant and causes cankers on trunks and limbs. The branches exude gum during late spring and summer from the cankered limbs and branches wilt and die. Leaf and fruit infections occur sporadically and are of economic significance, especially in years with prolonged wet, cold weather. Leaf spots are dark brown, circular to angular, and sometimes surrounded with yellow halos. The spots coalesce to form large patches of dead tissue, along with margins of leaves, resulting in necrotic and tattered leaves. Lesions on green cherry fruit are brown with a margin of wet or water-soaked tissue. The affected tissues collapse, leaving deep, black depressions in the flesh, with margins becoming yellow to red lesions, resulting in rapid fruit destruction.

Like many other bacterial plant pathogens, such as *Erwinia* and *Xanthomonas*, *Pseudomonas* also elicit hypersensitive response (HR) in nonhost and resistant plants. In a HR condition, the plant cells produce antimicrobial compounds and kill the bacteria, accompanied by plant cell death or necrosis. These HR elicitors are also called as harpins, which are secreted by the bacteria. These harpins do not have an N-terminal signal sequence for export but are rich in glycine residues and are heat stable. The gene cluster, which encodes these proteins, is clustered in a defined region called as *hrp* loci. In *P. syringae* pv. syringae 61 pathovar, the *hrp* loci encode an operon consisting of the *hrp*AZBCDE genes. HrpB is the positive regulator of the *hrp* cluster, which regulates the Hrp gene cluster and also transcribes HprZ. HprZ elicits the HR response. The other HrpACDE proteins from a secretory apparatus for the secretion of the harpins.

Another commercially important pathogen is *P. syringae* pv. phaseolicola, which causes halo blight disease of beans. Five *avr* (avirulence genes) encoded by *P. syringae* (*avr1, avr2, avr3, avr4,* and *avr5*), which interacts with their corresponding host resistant R1, R2, R3, R4, R5 proteins, have been identified. Resistance or susceptibility of the plant depends on the presence or absence of the resistant genes. The *avr* genes cause disease in the sensitive or nonhost plants and are secreted by the *hrp* (type III secretion) and elicit the HR response.

P. solanacearum is a major plant pathogen, which infects over 200 different plants worldwide. Some of the common diseases include bacterial stripe (also caused by *P. andropogoni*) in sorghum. The disease is characterized by long narrow stripes that vary from red to black, depending on the type of sorghum. These stripes are confined between the veins and have a crusty surface as the bacterial slime dries. Similarly, *P. solanceraum* GMI1000 causes bacterial wilt of solanaceous crops by secreting a virulence protein called Pop1. Pop1 and Pop3 (93 aa truncated from Pop1) are heat stable specific HR elicitors in tobacco (nonhost plant) but not in tomato (host plant). The protein apparatus involved in the secretion of Pops is organized into six transcriptional units and encodes nearly 20 proteins. The Hrp1 unit encodes HrpA and HrpB and Hrp3 unit encodes HrpN, HrpO, and HrpP. HrpB positively regulates the *hrp* regulon. The Hrp2 unit has 8 ORFs, encoding HrpC, HrpD, HrpE, HrpF, HrpH, HrpI, HrpJ, and HrpK. Hrp4 transcription unit contains six ORFs, named as HrpQ, HrpT, HrpU, HrpV, HrpW, and HrpX. These proteins are involved in sensing the host signal, assembly, and secretion of Pop1 virulence factor to the plant by forming the type III secretion system. Transcription unit 7 is located downstream of HrpX and encodes HrpY, which not only governs the plant–bacteria interaction but also forms a part of the harpin type III secretion system. In addition to these virulence factors, some plant pathogenic *P. syringae* produce alginate.

1. Exopolysaccharide

Many plant pathogenic bacteria produce exopolysaccharides. Among the Pseudomonas species, *P. solancearum* produces exopolysaccharide (EPS), or slime. EPS chelates heavy metals, such as copper, and increases tolerance to toxic pesticides. EPS is a polymer of N-acetylglucosamine (GalNAc) units, with varying amounts of glucose, rhamnose, bacillosamine galactosa–minutonic acids. EPS is involved in the vascular system, resulting in wilting of the plants, and causes water soaked lesions as well as blight symptoms. Copper stimulates the alginate synthesis in *P. syringae*.

2. Syringomycin

P. syringae pv. syringae secretes a toxin called syringomycin, which causes necrotic lesions in host plants. This toxin disrupts the ion transport and electrical potential of the host plasma membrane, resulting in cell death. It is a lipopeptide, which is produced as a secondary metabolite by the syringomycin synthetase (encoded by *syrB* gene), similar to antibiotic synthetases of Bacillus species. SyrB secretion requires SyrD, a cytoplasmic membrane protein, which functions as an ATP driven efflux pump. SyrB is induced by plant phenolics.

3. Pectate Lyase

P. marginalis N6301 produces pectate lyase, a plant toxin. PelA encodes a polypeptide of 40 kDa. It is composed of 380 amino acids, including a 29 amino-acid signal peptide. Pectate lyase breaks down pectin present in plants and fruits and destroys the appearance and quality and commercial value of the produce. *P. viridiflava,* the second major postharvest pathogen, damages fruits and vegetable crops during storage and transit and also produces pectate lyase. In addition to these virulence proteins, Pseudomonas also secretes amidases and chitinases.

IV. ENVIRONMENTAL ASPECTS OF PSEUDOMONAS

Pseudomonads are widespread in many natural environments, where they carry out a variety of biochemical conversions and mineralize organic carbon. They metabolize a large number of natural organic compounds, including aromatic hydrocarbons and their derivatives. The enzymes involved in the degradation of these compounds are generally plasmid encoded and have low substrate specificity. These two features allow rapid evolution of new metabolic pathways for the degradation of toxic synthetic compounds (xenobiotics), such as highly chlorinated aromatics, used as pesticides, herbicides, or released into the environment by industrial processes.

1. Pseudomonads Degrade Chlorinated Aromatic Hydrocarbons

The processes involved in the degradation of these recalcitrant compounds are well studied and have been developed in pollution control. *Pseudomonas cepacia* (*Burkholderia cepacia*) AC1100 has been isolated in a chemostat from a mixed flora, obtained from a toxic dump site, for its ability to utilize 2,4,5-trichlorophenoxyacetic acid (2,4,5-T) as the only carbon and energy source. 2,4,5-T is a carcinogen and it is mixed with 2,4-dichloro-phenoxyacetic acid (2,4-D) in Agent Orange (the herbicide used as a defoliating agent in the Vietnam War). *B. cepacia* AC1100 can degrade 2,4,5-T in contaminated soil, helped by the production a surface-active compound that emulsifies this hydrophobic substrate. The pathway of degradation of the 2,4,5-T is illustrated in Fig. 2. The initial degradation of 2,4,5-T proceeds through the formation of 2,4,5-tri-chlorophenol (TCP) by two putative multicomponent dioxygenase; the oxygenase are encoded by *tftA* and *tftB* genes.

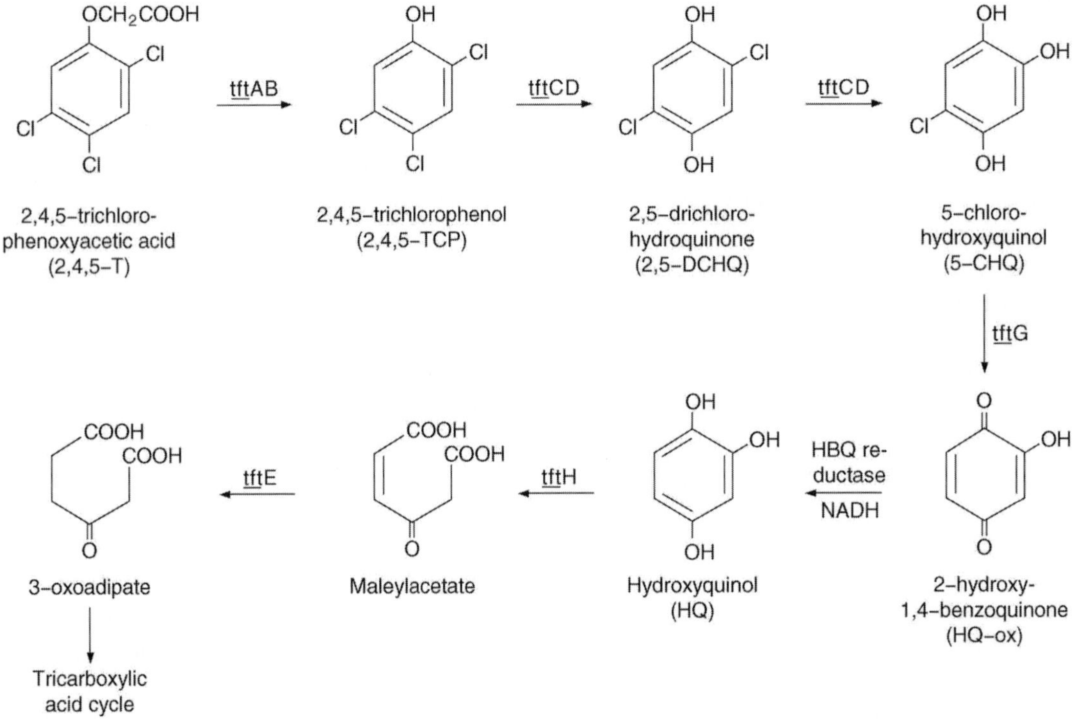

Fig. 2. The pathway of 2,4,5-trichlorophenoxyacetic acid metabolism. The *tftA* and *tftB* genes encode two subunits of the 2,4,5- tricholrophenoxyacetic acid (2,4,5-T) oxygenase enzyme responsible for the conversion of 2,4,5-T to 2,4,5-TCP (2,4,5-trichlorophenol). A two-component flavin-containing monooxygenase, encoded by the *tftC* and *tftD* genes, catalyzes the para-hydroxylation of 2,4,5-TCP to yield 2,5-DCHQ (2,5-dichlorohydroquinone). A second hydroxylation step by the same enzyme converts 2,5-DCHQ to CHQ. The intermediate 5-CHQ is further metabolized by the enzymes of the lower 2,4,5-T pathway. The 5-CHQ is dechlorinated by the *tftG* and *tftF* gene products to yield HQ (hydroquinone), the substrate for the enzyme hydroxyquinol 1,2-dioxygenase. Hydroxyquinol 1,2-dioxygenase, encoded by the *tftH* gene, catalyzes the *ortho* cleavage of HQ to yield maleylactetate. The enzyme maleylacetate reductase, encoded by the *tftE* gene, catalyzes an NADPH-dependent reduction of maleylacetate to yield 3-oxoadipate, which, ultimately, is converted to tricarboxylic acid cycle intermediates.

The conversion of 2,4,5-TCP to 2,5-dichlorohydroxyquinone and, subsequently, to 5-chlorohydroxyquinone (5-CHQ) is catalyzed by a chlorophenol 4-monooxygenase, which consists of two subunits, TftC and TftC (encoded by the *tftCD* genes). Further dissimilation of 5-CHQ requires the *tftEFGH* gene cluster. TftG is a dechlorinating enzyme that catalyzes the conversion of 5-CHQ to hydroxybenzoquinone, which is then reduced to hydroquinone (HQ) by the hydroxybenzoquinone reductase, a gene which is not a part of the *tft* operon. HQ is subsequently converted to maleylacetate by the dihydroxyquinol 1,2-dioxygenase encoded by the *tftH* gene. The *tftE* gene encodes a maleylacetate reductase. *TftF* is essential for 5-CHQ degradation, but the role in this pathway of the glutathione reductase which it encodes is still not identified. These genes, which are involved in the *ortho* cleavage of the aromatic ring between the two hydroxyl groups, are similar to other gene clusters responsible for chlorocatechol degradative pathways, such as *clcABD*, *tfdCDEF*, and *tcbCDEF*. However, in *tftEFGH*, the order of the genes is reversed and a novel enzyme, such as a 5-CHQ dechlorinase, is included. The genome of *B. cepacia* AC1100 consists of five replicons and the three 2,4,5-T degrading gene clusters are localized on different replicons. Insertion sequences (IS), which allow the genes to jump from one to the other site in the same or a different replicon, have been localized adjacent to *tftAB* and *tftEFGH* gene clusters. Such a genetic organization and the low DNA homology of *tft* genes to other members of the genus *Pseudomonas*, suggests that the 2,4,5-T degradation pathway was assembled by a series of independent gene transfers from one to another microorganisms in the consortium during selection in a chemostat for utilization 2,4,5-T.

In general, the genes that allow *Pseudomonas*, as well as other microorganisms, to degrade aromatic compounds are likely recruited from preexisting catabolic pathways. The nature of the environment dictates to a large extent the mode of evolution of the new degradative pathways in microorganisms. Environmental conditions, such as substrate availability and growth conditions, also influence the expression of catabolic genes by means of regulatory proteins. The degradation of a number of compounds by *Pseudomonas* species is regulated by LysR-type responsive transcriptional regulators (LTTRs). LTTRs are DNA binding proteins that bind approximately 50 to 60 bp upstream of the genes they regulate. The presence of an inducer molecule, which is usually a catabolic intermediate of the pathway being regulated, alters the binding pattern and results in transcriptional activation. Examples of biodegradative pathways regulated by LTTRs in *Pseudomonas* include the chromosomally encoded catecol degradative *catBCA* operon and the plasmid-encoded *pheBA* operon, which allows the growth of the *P. putida* strain PaW85 on phenol. Both the *catBCA* and the *pheBA* operons are regulated by CatR. Other examples include the 3-chlorocatechol degradative *clcABD* operon, regulated by ClcR, and the 1,2,4-trichloro-benzoate degradative *tcbCDEF* pathway in *Pseudomonas* sp. strain P51, regulated by TcbR.

2. Metabolism of Benzoate, Methylbenzoates, and Naphthalene by Pseudomonas

Several *Pseudomonas* strains degrade the BTX aromatics (benzene, toluene, and isomeric xylenes) compounds that usually occur together in gasoline and diesel oil. This genus degrades monoalkyl- and dialkyl-benzenes by different pathways, which includes the oxidative attack of the aromatic ring and formation of alkylcatecols, which are substrate for ring fission, or by the oxidation of the alkyl substituents, which leads to the formation of aromatic carboxylic acids, further oxidized to dihydroxylated ring fission substrates. The benzene ring fission occurs by *meta*-cleavage, since the ring is cleaved adjacent to one of the hydroxyls. Similarly, degradation of naphthalene occurs through the metacleavage pathway and the enzymes involved in the napthalene catabolism are encoded by the genes which are organized into the *sal* and *nah* operons, both are activated by NahR in the presence of the salicylate as inducer.

3. Polychlorinated Biphenyl Catabolism in Recombinant Pseudomonas Strains

Genes encoding polychlorinated biphenyl (PCB) degrading enzymes (*bph*) have been identified and isolated from several *Pseudomonas* sp. PCB, such as DTT and its metabolites, are pollutants present in great abundance in the ecosystem. However, most

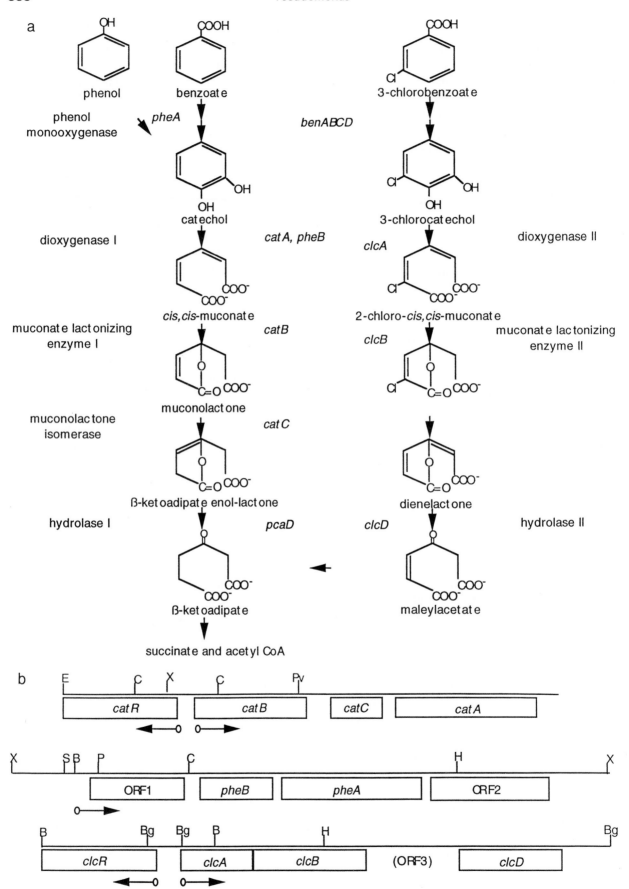

biphenyl-utilizing strains, such as the well-studied *Pseudomonas* sp. LB400, cannot degrade chlorobenzoic acids (CBAs) any further, and, therefore, they are accumulated during PCB catabolism. A general scheme of the products in the degradation of benzoate, phenol and 3-chlorocatechol is given in Fig. 3. To circumvent this limitation and to utilize *Pseudomonas* in bioremediation, recombinant strains are constructed by gene-transfer. The genes encoding the well-characterized PCB degradative pathway of LB400 are transferred into Pseudomonads capable of utilizing several CBAs.

4. Degradation of Alkanes and Cycloalkanes in Pseudomonas

P. oleovorans can grow on *n*-alkanes by virtue of the alkane hydroxylase system, which can be investigated in great detail because of its industrial application in the production of fine chemicals, such as fatty acids, alcohols, and epoxides. Another interesting degradative activity of Pseudomonads is the one toward the highly toxic and persistent insecticide, γ-hexachlorocyclohexane (γHCH). *P. paucimobilis* SS86 was isolated from an upland experimental field to which γ-HCH had been applied once a year for 12 years and it utilizes this substrate as only carbon source. Four genes are responsible for the degradation of this highly chlorinated cyclohexane, *linABCD*. *linA* encodes a γ-HCH dehydrochlorinase which is responsible for the degradation of the γ-HCH to 1,3,4,6-tetrachloro-14-cyclohexadene (1, 4-TCDN), which is then metabolized to 2,5-dichloro-2,5-cyclohexadiene-1,4 diol (2,5-DDOL) by the 1,4-TCDN halidohydrolase encoded by the *linB* gene. The *linC* gene encodes 2,5-DDOL dehydrogenase, which converts 2,5 DDOL to 2,5-dichlorohydroquinone (2,5 DCHQ), which is further degraded by LinD.

Camphor catabolism, mediated by CAM plasmid in *P. putida*, shows an example of negative regulation by the regulator CamR, which represses the *cam* hydroxylase operon (*camDCAB*) and autorepresses the *camR* gene in the absence of camphor.

5. Pseudomonas Take Part in the Natural Process of Lignin Mineralization

Several members of the Pseudomonaceae have the ability to degrade lignin and its dimers. Phenolic monomers, such as *trans*-ferulic, *p*-coumaric, and vanillic acid, which occur abundantly in the environment from the biodegradation of lignin which accomplished predominantly by white-rot fungi. These products are utilized as a unique source of carbon and energy by Pseudomonads. The investigation of their degradative ability toward these lignin monomeric components is very important for bioremediation of pollutants, such as the chlorinated form of vanillate, which are liberated in vast quantities into the environment by the wood pulp bleaching process.

6. Other Environmental Pollutants Degraded by Pseudomonas

(i) Nylon. Nylon is polymer of 6-aminohexanoate (Ahx), widely used in the textile industry. The monomers and dimers are considered waste products and are environmental pollutants. Pseudomonads can degrade such products Ahx (monomer), Ald (its linear dimer), and Acd (cyclic dimer) by producing hydrolases. The hydrolases are encoded by the *nylA* (Acd hydrolase) and *nylB* (Ald hydrolase). These genes occur on the plasmids of Pseudomonas strain NK87 and have evolved from other bacteria.

(ii) Trichloroethane. Trichloroethane (TCE) is widely used as a degreasing agent, dry cleaning fluid, fumigant, and cleanser. Such wide use of TCE has caused it to become an environmental pollutant, especially in soils and groundwater. It is known to cause anemia, kidney, and liver damage of humans. Some pseudomonas species, especially *P. putida* and *P. cepecia*, are capable of degrading TCE by produc-

---

Fig. 3. Enzymes and intermediates of benzoate, phenol, and 3-chlorobenzoate degradation (a) and their genetic organization (b) *P. putida* uses a modified β-ketoadipate pathway to degrade 3-cholrocatechol. The genes for the regulatory proteins CatR and ClcR are divergently transcribed from the *catBCA* and *clcABD* operons that they regulate. The *pheBA* operon is regulated by CatR.

ing soluble methane monooxygenases and toluene ortho monoxygenases. These enzymes convert TCE to glyoxylate, CO, and formate; finally, they are converted to CO_2 and HCl.

Metal resistance. Pseudomonas is resistant to a number of toxic metal ions, such as mercury, arsenic, cadmium, tellurium, chromium, and silver. Most of the resistance genes are plasmid encoded and, occasionally, the regulatory genes are present on the chromosome.

(i) **Mercury.** Mercury is a toxic heavy metal. Resistant *Pseudomonas* species have plasmid borne *merTP-CAD* operon. MerR, which is located upstream of the *merTPCAD* operon, is a regulator which functions as a dimer. MerR binds to *mer* operon as a inducer when Hg is present and as a repressor in the absence of Hg. The *mer* operon encodes MerT and MerP, which bind mercury and are localized in the inner membrane and periplasm of the bacteria, MerC is also an inner membrane protein. These proteins facilitate transport of mercury. MerA is a $FADH_2$ reductase but is also dependent on NADPH to convert Hg^{2+} to volatile Hg^0 and, thus, confer resistance.

(ii) **Copper.** Copper (Cu) is a major micronutrient and is a constituent of metalloenzymes and proteins involved in electron transport and redox reactions. Two forms of Cu, CuI and CuII, are normally found in bacteria. Copper is also involved in toxic free-radical formation. Copper resistance in *P. syringae* pv. tomato is well studied; resistance is due to the *cop* operon, which encodes six proteins Cop A, B, C, D, R, and S. CopR a DNA-binding protein, which activates the *cop* operon. CopD is an inner membrane protein and interacts with CopB (an outer membrane associated protein) via CopC and CopA, which are periplasmic proteins. The CopB, A, C, and D proteins form a copper transport unit. CopS is a membrane-embedded copper-sensing protein and CopR is the regulator of the *cop* operon. CopR and CopS form the two-component signal transduction system for sensing the levels of copper and regulating the *cop* operon.

(iii) **Cadmium.** Cadmium is very toxic to bacteria. *P. putida* produces a low molecular weight protein, which chelates cadmium, thereby reducing toxicity. It also confers resistance by reducing the uptake by a cadmium efflux system.

(iv) **Arsenic.** *Pseudomonas* exhibits resistance to arsenic by converting the highly toxic trivalent As III to a less toxic As IV pentavalent form by reductases.

(v) **Chromium.** Several soil-borne Pseudomonads convert hexavalent chromium (CrVI) to a less toxic trivalent (CrIII) by an NADH dependent chromate reductase. *P. putida* also has ChrA protein, which is involved in the chromate membrane transport. A membrane-bound chromate reductase has also been found in *P. fluorescens*.

V. GENETIC AND MOLECULAR TOOLS USED TO STUDY PSEUDOMONAS

1. Plasmids

The utilization of nonenteric bacteria for basic and applied molecular research has resulted in the need for well-characterized vector systems for such microorganisms. The cloning vectors used for these are generally broad host range vectors and allow the use of different species, including *E. coli*, as intermediary hosts.

a. General Type Cloning Vectors

Earlier, the general type cloning vectors were constructed from the plasmid RSF1010. This is a small plasmid (8684 bp) with a broad host range. The modified vectors were constructed by inserting DNA fragments with different antibiotic resistance as markers. In Pseudomonas species, several pMB1 (ColE1 from *E. coli*) based cloning vectors have been constructed by cloning the 1.9 Kb PstI fragment from pRO1614 into pUC18/19, pBR322, or its derivatives. These have given rise to vectors, such as pUCP18/19, the pPZ series of lacZ translational fusion vectors, and the pQFseries of vectors. The pUCP series of vectors have an ori site whose features are characteristic of the replication origins of F, R6K, pSC101, and related plasmids. The promoterless lacZ vectors are of two types, viz., transcriptional fusions which facilitate the study of promoters from Pseudomonas genes to transcribe the β galactosidase gene (*lacZ*), whereas the promoterless *lacZ* fusions are useful to generate translational in-frame fusions. Such fusions are useful to study translational control of the gene of interest.

b. Special Purpose Cloning Vectors

The expression of cloned genes in the host organism is often used to confirm the coding potential of a DNA fragment. However, large limitations are encountered while expressing cloned Pseudomonas genes in heterologous hosts, such as *E. coli*. This is primarily due to the differences in codon usage as well as variations in the structure of promoters of genes. However, recently, a T7 expression system for protein synthesis in Pseudomonas species has been developed. These vectors are constructed by ligating fragments from the parental vectors pUCP19 and pBluescriptII SK⁺/KS⁺. The resulting vectors, pUCP SK/KS, contain the stabilizing fragment (SF), which allow the maintenance of the plasmids in Pseudomonas species. It also has extended MCS (multiple cloning site, which allows the blue/white selection), ColEI ori, and T7 promoter sequence, which is driven by the T7 polymerase inducible by IPTG. The MCS has 22 different restriction enzymes, which facilitates cloning of DNA fragments and, simultaneously, allows direct gene analysis in both *E. coli* and Pseudomonas.

To study quantitatively the role and functions of promoters in Pseudomonas species, vectors containing *aphC* gene of RSF1010 or xylE gene of TOL plasmid have been constructed. The first vector allows positive selection of promoters by selecting for streptomycin-resistant transformants, while the second vector allows for screening by spraying colonies growing on plates with catechol solution. A new set of promoter probe vectors for Pseudomonas species are the ASD–LacZY or asd–mut3 (pIVPRO-2). The ASD (aspartate b-semialdehyde dehydrogenase) is an essential enzyme in the biosynthesis of diaminopimelate (DAP), a component of the peptidoglycan in the cell well of gram-negative bacteria. The product of the *asd* is a part of the biosynthetic pathway for lysine, methionine, and threonine. In absence of DAP, ASD bacteria undergo cell lysis. The LacZ or GFP expression can be assayed by plating libraries on TSA plates containing X-gal or by direct observation of colonies under ultraviolet light irradiation, respectively. Promoterless Tn–PhoA (alkaline phosphatase) vectors are available too, which randomly integrate in frame into the gene of interest. Such fusions use the target protein signal sequence to get inserted into the membrane. Tn–PhoA fusions are used to study membrane associate proteins because PhoA is active only in the periplasm.

(iii) Transposable elements. Insertional elements (IS) are a class of prokaryotic transposable elements, which have the capability of integrating into different sites on the genome in a variety of bacteria. IS are genetic entities, ranging in size from 700–2000 bp, which encode transposase necessary for transposition. During such an event, foreign genes are recruited by replicon fusion and insertional activation. Several IS elements have been identified in *B. cepacia*. Recently, a new class of IS element has been identified in *P. alcaligenes* NCIB 9867 (P25X, designated as IS1394) and has been found in both gram-positive and gram-negative bacteria. It is about 1100 bp in size, with a 27 bp inverted repeat, and belongs to the IS30 family. During transposition, it generates a 3 bp duplication of target DNA, which is a characteristic typical of the IS30 family of transposons. The transposons have been primarily used as tools for genetic analysis in bacteria. They are also used for creating gene disruptions, which are nonleaky and are linked to a selectable marker, such as antibiotic resistance or green fluorescent protein. Two such transposons which are widely used for this purpose are the Tn5 and Tn10 transposons. These belong to the composite class of transposons because their mobility is determined by two insertion sequences (IS) flanking the region of DNA encoding the selectable phenotype. Such transpositions require tranposase, which is encoded by one of the two IS elements. Recently, mini-transposons are being used for constructing stable mutations in the desired gene.

VI. BIOTECHNOLOGY

Pseudomonas strains themselves or their products have been used in large scale biotechnological applications. Pseudomonas ssp. PR3 is used in the conversion of soybean castor oil to a new value-added oxygen-containing product. This product has antifungal properties in controlling rice blast disease. Frostban is an ice minus *P. syringae*, used commercially to prevent ice nucleation in strawberry and potato fields. These include *P. fluorescens* A506, *P. fluo-*

rescens 1629RS, and *P. syringae* 742RS. *P. putida* is used as a biocontrol agent for fusarium wilt pathogen and to control black root-rot disease of tobacco. Thermostable lipases from *P. fluorescens* are used in the food and leather industries. *B. cepacia* is used for environmental cleanup of recalcitrant compounds and in oil spills. Biosurfactants are used in emulsification, phase separation, emulsion stabilization, and viscosity reductions. Polyesters produced from *P. oleoverans* are used in specialty plastics.

See Also the Following Articles

BIODEGRADATION • BIOLOGICAL WARFARE • EXOTOXINS • OXIDATIVE STRESS • PIGMENTS, MICROBIALLY PRODUCED • *XANTHOMONAS*

Bibliography

Alfano, J. R, and Collmer, A. (1996). The type III (Hrp) secretion pathway of plant pathogenic bacteria: Trafficking harpins, Avr proteins, and death. *J. Bacteriol.* 179, 5655–5662.

Chugani, S., Daubaras, D., Danganan, C., Hubner, A., McFall, S., Hendrickson, W., and Chakrabarty, A. M. (1997). Molecular and evolutionary mechanisms in the microbial degradation of synthetic chlorinated compounds. *Chemistry Today* 6, 17–22.

Daubaras, D., and Chakrabarty, A. M. (1992). The environment, microbes and bioremediation: Microbial activities modulated by the environment. *Biodegradation* 3, 125–135.

Fick, R. B., Jr. (ed.) (1993). "*Pseudomonas aeruginosa*: The Opportunist. Pathogenesis and Diseases." CRC Press. Bota Raton, FL.

May, T., and Chakrabarty, A. M. (1994). *Pseudomonas aeruginosa*: Genes and enzymes of alginate synthesis. *Trends Microbiol.* 2, 151–157.

Nakazawa, T., Furukawa, Haas, K. D., and Silver, S., (eds.) (1996). "Molecular Biology of Pseudomonads." ASM Press, Washington, DC.

Pulp and Paper

Philip M. Hoekstra

Buckman Laboratories International

GLOSSARY

biodispersant A chemical additive which will inhibit accumulation of a biofilm, or disperse a biofilm that has formed, but is itself not toxic to microorganisms at reasonable treatment levels.

biofilm Deposit on a paper machine surface caused by microorganisms attached to the surface, usually composed of a complex mixture of microorganisms, microbial products, water, and suspended and dissolved solids.

microbicide A chemical additive used to restrict the growth of microorganisms. In the paper industry, the terms *slimicide* or *microbiocide* are often used.

potentiator A nonmicrobicidal substance that enhances the effect of a microbicide, with the effect of reducing the overall toxicity of the control program. Potentiators can work synergistically with the microbicide.

slime In the paper industry, a term commonly used in place of *biofilm* (see above).

PULP AND PAPER PRODUCTS have been very important to human civilization for many centuries. Even with the growth in the use of computers, paper products remain crucial for recording, storing, and distributing information. Paper is the most widely used material for wrapping and packaging. Its other uses are myriad and continually increasing. The paper industry has major global economic importance.

Microbiological growth has a significant impact on this important industry. The papermaking process uses large volumes of water, and so there is, in the paper mill, a favorable environment for growth of many microorganisms. Paper products are made up largely of natural materials that are biodegradable, and, thus, biodeterioration of some of the products must be controlled. In these and other ways, microorganisms are the source of significant problems in the paper industry.

Much information is available on the mechanics and chemistry of papermaking. In the process of papermaking, the product is formed from a slurry of cellulose fibers suspended in water. In any particular paper machine, the cellulose raw material might be virgin fiber coming from an adjacent pulp mill. Alternatively, pulp might be purchased on the market and delivered to the paper mill, either in dry bales or as wet lap pulp. Or the fiber used may be recycled paper brought to the mill, which is pulped upon arrival. In every case, the basic raw material is a slurry of cellulose fibers in water, and there are numerous additives, purposely or inadvertently, included in the mix. These additional materials also encourage the growth of bacteria and fungi. For example, starch is one of the many additives used in this process. The whole papermaking process, including the basic materials used and many of the additives, together form an ideal matrix for microbiological growth.

The principal negative impacts of microbiological growth in the paper industry can be summarized as the following:

- Problems affecting the efficiency of the process
- Biodeterioration of the raw materials used

- Reduction in finished product quality
- Damage to process equipment.

As a result, for most paper mills, significant expenditures are required in the area of microbiological control programs to maintain efficient operations and acceptable product quality.

I. BIOFILMS IN THE PAPER MACHINE

In a paper mill, the major negative economic consequence of microbial growth is interference with production and a resulting reduction in the efficiency of the manufacturing operation. The primary cause of these runnability problems in production of paper is the formation of *biofilm*. Biofilms are composed of microbial cells and *extracellular polysaccharide* (EPS). Figure 1 is a scanning electron micrograph showing bacilli embedded in a "pure" biofilm (i.e., without the fibers and other materials that become inbedded in the matrix). Left untreated, this biofilm can become large masses of slimy material containing microorganisms, EPS, fibers, water, organic, and inorganic debris. Without effective control, this slimy material fouls surfaces in the system and can hang in masses from parts of the machine (see Fig. 2).

In papermaking, the problem-causing organisms are primarily bacterial species that colonize surfaces. If a cell contacts a surface where it finds conditions conducive to growth, it may attach there and begin secreting a protective extracellular polysaccharide

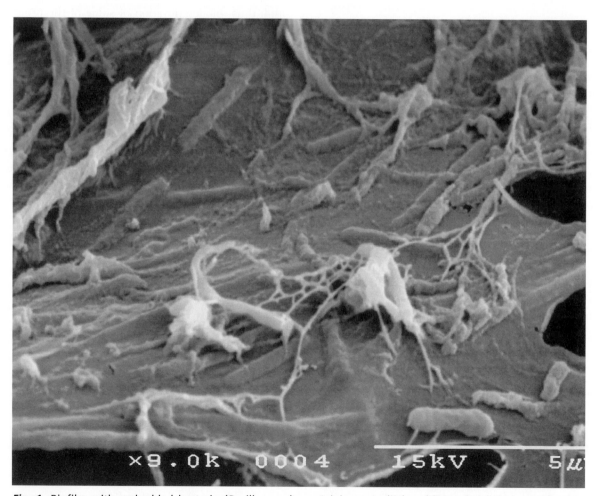

Fig. 1. Biofilm with embedded bacteria (*Bacillus* spp.), containing very little additional debris (magnification: 9000×).

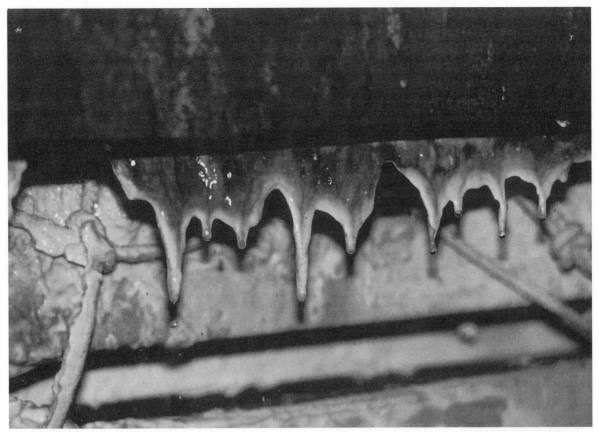

Fig. 2. Photograph of slime hanging from a support on a paper machine; in this case, the primary organisms are fungi.

layer. More rapid growth of microorganisms occurs after attachment to a surface, and most extracellular material from these organisms is produced after that attachment. This protective film allows the bacteria to control their local environment. Within this enclosure, the organisms are somewhat insulated from changes in the larger environment around them, protected from desiccation, and less likely to be affected by any chemical treatment. The biofilm is a unique niche within which a variety of microorganisms can thrive.

As often happens, the paper industry has its own vocabulary. There are a few terms that need explanation. The *wet end* of the paper machine is the area in the paper mill where the sheet is formed, just prior to pressing and drying. This is where a biofilm can cause the most serious problems. In a paper mill, biofilms are commonly referred to as *slime.* The art–science of reducing the problems caused by microbiological growth is called *slime control.* Chem-

ical additives used to reduce growth are designated *slimicides,* although the alternative term *microbicide* is preferred (and *microbiocide* is occasionally used).

The accumulation of biofilm gives rise to a number of related problems. Portions of the biofilm can break loose and cause spots or holes in the paper produced. Biofilm can cause high fluid frictional resistance losses in water pipes, reducing efficiency of transfer in heat exchangers. The organisms in the biofilm are much more difficult to kill, having the barrier of the slime layer for protection. In the special environment underneath the deposit, oxygen can become depleted. Then, in anaerobic conditions, corrosion can occur, either from the chemical microenvironment created or from microbiologically influenced processes of corrosion. In these areas, there can be proliferation of anaerobic organisms, with resulting odors and discoloration of fibers.

It is important to understand that in controlling microbial problems on the wet end of the paper machine, the primary concern is not with the total microbiological population in the circulating stock. The more important goal is to prevent the buildup of biofilms. Planktonic organisms are of little concern, in most cases, but serious problems can occur when organisms colonize the surface and produce a biofilm. As this matrix grows, fibers and other solids in the papermaking stock become entrapped, exacerbating the situation. The primary goal of slime control on the paper machine is to restrict the growth of sessile organisms on the machine surfaces. Treatments that prevent accumulation of biofilm without affecting microbial growth would not only be acceptable, but may be preferred.

The biofilms in a paper machine can be sticky, gelatinous, rubbery or stringy, or pasty. Occasionally, slime may have an appearance like tapioca. The causative organisms may be bacteria, fungi, or a mixed population of both. Since it collects solids that are in the papermaking furnish (the pulp mixture), slime can take on any color. It is not uncommon for a dreaded "pink slime" to occur in a machine. This deposit is caused by certain microscopic organisms with a typical orange, red, or pink color. Filamentous bacteria are often found in the deposit, especially if the water source is mildly treated freshwater. Growth of these filamentous bacteria or fungal hyphae can produce long iciclelike formations hanging from showers or machine frames (Fig. 2). Deposits can be found in widely varying conditions, from surfaces in overspray areas to the surfaces of rapidly rotating screens where there is very high shear.

In contrast to biofilms, there are some situations where it is important to control growth of planktonic organisms. Some examples of additives used in the papermaking process are starch slurries, polymer emulsions, clay slurries, and sizing materials. For example, after starch is cooked to produce a usable solution, it is stored before use. If microbial control is not maintained in the system, contamination will occur. In the same way, clay slurries, used in the manufacture of many grades of paper, are stored in tanks. Commonly, these tanks are not cleaned frequently, and the storage is at temperatures and pH levels perfect for growth of bacteria or yeasts. If allowed to proliferate unchecked, this growth can

quickly render the slurry unusable. A plate count may give an indication of the level of microbial contamination in the system (although this may just confirm what odors and a drop in pH have already shown). A more detailed discussion of these kinds of problems follows.

Again, the most costly problems caused by microbial activity in paper mills are related to production efficiency and product quality. For many producers of market pulp, treatment is necessary to prevent product deterioration due to fungal or bacterial growth. Foul odors caused by microbiological activity can degrade paper or additives in the mill. There have occasionally been deadly explosions, caused by flammable gases produced by biological activity. For certain paper products, there are government standards on the maximum levels of bacterial spores permitted in the paper produced. Some paper used for packaging must be treated with a fungicide to ensure that there will be no growth of mold on the package or the commercial product enclosed in that package.

II. ENVIRONMENTAL FACTORS IN THE PAPER MILL

The conditions for microbial growth vary widely from mill to mill and throughout the manufacturing process. But the situation is normally conducive to plenty of microbiological growth. In fact, in almost every mill microbicides are added to the system daily to control microbial growth. In addition, in most cases, each machine is taken off line periodically for a few hours to perform a thorough cleaning with heat and chemicals to eliminate the deposits that inevitably build up.

The growth of microorganisms in a paper machine system will fluctuate as the environment changes. Here are some of the important environmental factors in microbial growth on paper machines.

1. Water
Major amounts of water are used in a paper machine, and the quality of the water is an important factor in the microbiological problems in a mill. This

water provides the medium for growth for the organisms and carries the nutrients critical for survival. The conditions of this water—temperature, pH, dissolved oxygen, etc.—determine which species will thrive. The chemical additives used to control microbial growth are added to the water.

The amount of nutrients available will depend on the water source, the geographic location, the raw materials used in the papermaking process, and any pretreatment of the water. As water is recycled within the mill, the level of electrolytes and the level of nutrients will change. There is no typical water quality in the industry, since the amounts and types of raw materials added to the water vary so much for each type of paper or paperboard produced.

Microbial growth will occur in the bulk of the system, but it is important to keep in mind that, in most cases, we are concerned not with the population of planktonic organisms, but with the growth of biofilm on machine surfaces.

2. Nutrients

The food that enables bacteria and fungi to grow is borne by water in the system. So the quality of the water used, and the additives in the water, affect growth. Many organic materials are present in the water, including sugars, fatty acids, lignins and other components of the wood. If the fiber used is fully bleached prior to use, the amount of nutrients will be reduced. However, if recycled fiber is used, many additional contaminants are added to the process. Additives commonly used in papermaking add nutrients and stimulate microbial growth. Starch and numerous other organic materials are used to provide properties to the paper or to improve the process. Inorganic additives are also used in papermaking, such as clay, titanium dioxide, and calcium carbonate.

From the point of nutrition, there are obviously high levels of carbon available in the papermill process. But, typically, the limiting factors are low levels of potassium, nitrogen, and phosphorus, making it overall a low-nutrient system for microorganisms. Certain additives can make a significant change in microbial growth by increasing amounts of nitrogen (e.g., from retention aid polymers) or phosphorus (e.g., from potato starch).

3. Temperature

Temperature is an obvious factor in microbial growth. Figure 3 shows the typical effect of temperature on slime accretion in a paper machine. As an example, if the paper machine temperature is normally around 50°C (122°F) and, for some reason, it drops to around 45°C (112°F), there can be a large increase in slime problems. This is the expected result, since this more moderate condition is closer to the optimum growth temperature for many bacteria and fungi.

It is not uncommon for a paper machine wet end to run at 60°C (140°F). The primary reason for running at this higher temperature is to maximize drainage of water in the sheet-forming operation and, thereby, maximize production. At the higher temperature, problems from slime are substantially reduced, but there can still be considerable problems from the thermophilic organisms that can grow at this condition.

An additional effect of increasing temperature in the paper machine environment is the selection of the problem-causing organisms. Above 50°C, it is unusual to find fungi in the system. This is an important factor in slime control programs. Take, for an

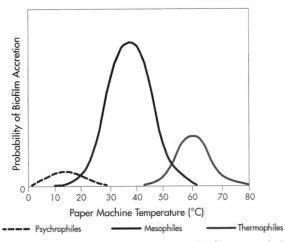

Fig. 3. The effect of temperature on biofilm growth in a paper machine; a change in temperature can make a significant change in slime problems on a paper machine.

example, a paper machine typically operating at a temperature of 55°C, with a certain microbicide being used to control the buildup of biofilm. At this condition, the problematic microbial growth is likely to be bacterial. The microbicide may have been selected to control bacteria and may not have any effect on fungal organisms. If the temperature of the system, for some reason, drops to below 50°C, the control program may no longer be effective, since the microbial population may now be a mixed population of bacteria and fungi.

Increasing the temperature of the system might reduce slime problems, but this is often not a suitable solution. It may be costly to maintain a higher temperature and might affect negatively other variables in the system.

4. pH

The pH of the system has a major effect on slime problems. Obviously, more organisms thrive at near-neutral conditions and, being near the optimum pH for survival, are more difficult to control. Most fungi grow better in acid conditions; typically, bacteria prefer neutral conditions.

In the past 15 years, there has been a major shift in the pH on the wet end of the paper machine. In the past, it was most common on most grades to operate at acid conditions (pH 4–5), but now it is much more prevalent to run at neutral to alkaline pH ranges (pH 6–8) on many grades. Slime problems are likely to be more severe at the near-neutral conditions.

There is another problem related to machine pH: most of the microbicides commonly used are not as effective at the near-neutral pH conditions. There might be several reasons for this. First, there is a different population of troublesome organisms, against which a certain chemistry may not be effective. Second, most of the microbicidal chemistries used for control are hydrolyzed more quickly at neutral and alkaline pH levels. So, in the neutral system, where the need for control is more acute, the control chemistry will break down more quickly and there will essentially be less contact time between the microbicide and the organisms.

Figure 4 illustrates the effect of increasing pH on methylenebis(thiocyanate) (MTC), a commonly

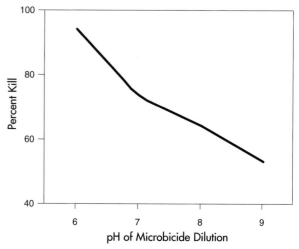

Fig. 4. The effect of pH on the efficacy of a microbicide; methylenebis (thiocyanate) was diluted at various pH levels, allowed to stand for 2 hrs, and then, at pH 5.5, a kill study was done. Typically, common microbicides are less effective at neutral and alkaline pH.

used microbicidal active. In this study, MTC was dispersed in water at pH levels ranging from 6 to 9. After 2 hrs at this condition, the efficacy of the MTC was determined in a kill study against *Enterobacter aerogenes* at pH 5.5 (2 ppm MTC used). As the pH increased, the potency of the bactericide declined rapidly.

As mentioned above, in isolated microenvironments, conditions can be dramatically different from that in the bulk of the system. Even in a system at alkaline pH, there can be significant areas within a biofilm where conditions are acidic.

5. Oxygen

Most problem-causing microorganisms in the paper mill are aerobic or faculatative species and these produce most of the problems described so far. However, special problems are caused by anaerobic organisms. Anaerobic conditions can exist at various places in the system. For example, once a deposit forms on a surface—either a microbiological deposit or some other type—underneath, oxygen becomes depleted, and growth of anaerobic organisms can occur. This will also happen where there is poor agitation or little turnover of the stock. Often, the indication is the production of foul odors that may render additives unusable or the final paper unsalable. Other anaerobic organisms, such as *Desulfovibrio* spp., can cause corrosion of the stainless steel surfaces of the paper machine in the anaerobic environment beneath a deposit. These and other organisms produce toxic gases like methane, hydrogen, or hydrogen sulfide, which are poisonous and have been the cause of explosions in some mills. These will be covered in more detail.

III. DETAILS ON MICROBIOLOGICAL PROBLEMS IN PULP AND PAPER

Following generally a typical flow through the papermaking process, here are the most common difficulties related to microbial growth.

1. Biodeterioration of Wood Fibers

The wood prepared in the pulp mill for papermaking is stored as logs and chips prior to the pulping process. The storage is in large piles exposed to the elements, so conditions are favorable for activity of decay organisms. As a result, a significant amount of fiber (estimated conservatively at 10% of the original amount) is lost in storage. The same problems occur for alternative sources of fiber, such as bagasse (fiber from crushed sugar cane), also stored unprotected outdoors. It is more common for recycled paper to be stored dry and protected from the weather, although the problem of biodeterioration exists if it is stored outdoors.

There are several reasons to store large quantities of logs or chips, other than the obvious storage of necessary raw material. It may be impossible to harvest the trees during winter or a rainy season, and so increased amounts are harvested during the best conditions. Also, extended storage reduces the levels of organic acids and fats in the wood (due in part to microbial activity). These "extractive" materials often cause spots that reduce the quality of the pulp or paper and can interfere with efficient production. Increasing storage time can reduce problems caused in the paper mill by these extractives.

Organisms involved in degradation of wood include *white rot, brown rot,* and *soft rot* fungi. The white rots degrade preferentially the lignin portion of the wood, leaving white masses or streaks in the wood, with a predominance of cellulose. Brown rots are the result of microbiological degradation, primarily of the cellulose. In a soft rot, caused by ascomycetes and some Fungi Imperfecti, both lignin and cellulose are attacked.

There are some options available to reduce the problem of wood decay. An obvious one is to minimize storage time. Careful management of the woodyard will help: wood should be used on a first-in, first-out basis. Storage of wood in such a manner as to effect good drainage and air circulation will help, as dry wood is less vulnerable to decay. The problem can be controlled with fungicides, although it may be difficult to treat the wood properly, and also not easy to verify the cost–benefit ratio of such treatment.

Considerable work is going on in the industry researching the beneficial use of fungi or isolated enzymes in *biopulping*. The wood might be treated prior to the pulping process, with the microbial enzymes eliminating some of the lignin, to expedite the subsequent pulping process.

2. Pulp Slurries in Storage

To produce the pulp used in paper manufacture, the wood is disintegrated by either chemical or mechanical means or by a combination of the two. The pulping process is done in water, and the pulp is stored and transported as an aqueous slurry. Microbial growth can become significant if the pulp slurry is stored for an extended time. The amount of nutrients available for microbial growth varies with the means of producing the pulp. Chemical pulping methods reduce the level of nutrients and bleaching diminishes this even more. The propensity of the pulp for microbiological problems depends on the nutrients available, as well as the pH, temperature, and time of storage.

The problems most likely to occur on extended storage of pulp are: (1) odors in the pulp and finished product, caused by aerobic or anaerobic organisms; (2) discoloration of pulp; (3) spots, specks, and holes in the paper produced, caused by slime accumulating on surfaces and then sloughing off; (4) increased microbiological activity in other parts of the system, downstream from the pulp storage—for example, increased biofilm problems in the paper machine.

This can give loss of production and reduced operating efficiencies on the paper machine.

These sorts of problems do not occur only when pulp is stored too long in a storage chest, but can occur in isolated areas. Often, there are areas in the system, such as unused pipes and other dead-flow spots, where microorganisms can grow unimpeded. Sometimes, chests are designed in such a manner that not every part of the system is agitated properly. A microbiological problem can occur quickly, e.g., when an impeller on an agitator has broken and some of the pulp is not mixed at all. Even if the pulp is normally treated with a microbicide to preserve it, that treatment will not contact the dead-flow areas.

Anaerobic organisms will often proliferate in these isolated areas, resulting in additional problems. Sulfate-reducing bacteria will produce discoloration from black iron sulfide precipitation. Odorous gases can be produced, like hydrogen sulfide. H$_2$S and other toxic or explosive gases can accumulate within a chest (additional discussion on this topic following). Also to be covered in detail, microbiologically influenced corrosion can damage, and even perforate, stainless steel and other metals.

Significant quantities of paper are recycled within a mill. This may be off-spec production, trim from the edges of the sheet, or material produced during a break in production on the machine. In the mill, the term *broke* is used. Broke is repulped into a slurry and stored, prior to being returned to the process. It is often necessary to add a chemical preservative to broke storage as well.

3. Water Storage

The paper industry is a leader in the reuse of water. Recycling water within the mill minimizes fresh water required and reduces the amount of water returned to the environment. Consequently, large quantities of process water (termed *white water*) are stored in the mill. Proper treatment of white water is important because it is used in many places in the mill. It is used to convey the fiber through the various stages of the process. It is common to treat this water with microbicides to control growth because, as is the case for stored pulp, if microbiological control of this water is ineffective, many problems result throughout the mill. There can be odors, pH drops, microbiologically influenced corrosion, and biofilm accretion. Some white water is clarified by removal of suspended solids and then used as shower water to clean forming fabrics and press felts during operation. Any slime that accumulates can plug showers and interfere with this important process.

Large tanks are used to store this water, and so there can be a significant incubation time. In many mills, this water is treated with microbicides as part of a control program to reduce microbiological problems. As an example of these problems, bacteria, especially of the genus *Clostridium*, will perform fermentation in anaerobic conditions and produce low molecular weight acids, such as butyric acid. At certain levels in the water, there can be sufficient concentrations of these foul-smelling acids to give a very unpleasant odor to the finished paper product. This problem is more common in mills where process water is stored in very large volumes. It is likely that suspended solids accumulate in the bottom of the tanks and provide a favorable environment for these troublesome organisms.

An example of a paper product occasionally affected by this sort of odor problem is acoustical ceiling tiles. In one case, white water from the paper machine is stored in a 100,000-gallon chest before reuse. Wet broke (off-spec production recycled within the mill) is used and can obviously exacerbate a microbiological problem. In this mill, if levels of low molecular weight organic acids (e.g., butyric acid) in the finished product exceed 100 parts per million, the consumer may notice a foul odor in a room where this tile has been installed. One short-term solution is to increase usage of fresh water. But treatment of the storage chest with a few parts per million of an effective microbicide controls the problem at this mill and maintains low levels of the troublesome acids.

Another example of problems caused by microbiological activity in water storage is buildup of hydrogen gas inside a white water storage tank. Conditions like poor agitation and stagnant flow (providing anaerobic conditions), along with no ventilation to disperse gases produced, can allow accumulation of dangerous amounts of hydrogen in a few hours.

4. Fresh Water Treatment

One primary origin of microbial contamination in the paper mill is the fresh water. For this reason it is common to treat the incoming fresh water with a microbicide, chlorination being the typical treatment. This treatment will reduce the number of organisms carried in with the water, as well as the population of organisms on the surfaces of the pipes and tanks used to transport and store this water. This will result in a lower level of organisms arriving to contaminate the paper machine.

The presence of protozoa and nematodes in a deposit (which is not commonly seen) may indicate that fresh water is the source of a problem. Algae, especially diatoms, and filamentous bacteria and iron bacteria (*Gallionella* spp., *Sphaerotilus natans*) in the deposit may indicate the same thing. If a mill uses surface water (water from a river or lake) rather than groundwater (from a well), microbiological problems will vary with the season of the year. These variations include:

- Biological loading can increase in the warm season
- Nutrients will increase during the spring and fall runoff
- Sediments and other suspended solids can increase at times (e.g., spring turnover in a lake)
- Agricultural runoff and upstream facilities can contaminate surface water supply
- Water may contain more debris during a dry season.

For this reason, the water coming to the mill is often treated with antimicrobial products, commonly chlorine.

5. Paper Machine Wet End

The major damage from microbial activity in paper mills is done on the wet end of the paper machine. The primary cause is buildup of biofilm. When a portion of this slime breaks loose, it can become embedded in the sheet, which may leave an unsightly spot. As this slime spot travels on with the sheet into the pressing and drying steps, the biofilm may be removed to leave a hole in the paper. In either case, the quality of the paper is diminished. Paper machines can operate at speeds of up to 2000 meters/minute and even higher. The sheet formed may be 10 meters across. There are significant stresses on this sheet as it is being formed. A slime hole forms a weak spot and often causes a break in the sheet. This loss of production is the most significant difficulty of all. Lost production can cost a mill upwards of $25,000/hour, depending obviously on the value of the paper the mill produces. Effective control of biofilm pays for itself quickly in improved production efficiency.

Careful cleaning and housekeeping practices, along with judicious use of effective microbicides, can get rid of the problem. Eliminating the problem does not require elimination of the organisms, but only the prevention of a troublesome amount of biofilm accretion.

Typical causative organisms for bacterial slime on a paper machine are *Pseudomonas* spp. and *Bacillus* spp. On machines at alkaline pH, actinomycetes often produce a pigmented slime. For fungal slimes, *Penicillium* spp. and *Aspergillus* spp. are common.

On similar paper machines in different parts of the world, the populations of slime producers are quite similar. From one grade of paper to another, however, typical populations are different. For example, on tissue machines around the world, fungal slime is usually predominant, probably because of the chemical additives typically used in tissue manufacture. As time passes, the typical populations of slime formers seem to be changing. A probable cause is a change in the concentration of limiting nutrients like nitrogen and phosphorus, as more recycled paper is used as the fiber source. This gradual change may be due to higher temperatures in the papermaking systems, increased use of calcium carbonate in the sheet, or higher pH ranges.

6. Machine Additives

As previously mentioned, numerous additives are used in the manufacture of paper. For example, starch is commonly added at the wet end of the paper machine or as a surface treatment. Fillers, such as calcium carbonate, clay, and titanium dioxide, are shipped as slurries to a mill and used as raw material in the sheet. Alum solutions are shipped to the mill, stored, and used in the papermaking operation. Polymer emulsions are dispersed in water at low solids and used on the machine. Blended chemicals used for sizing are diluted and stored on-site. Table I shows a list of additives commonly used.

Control of microbiological growth is crucial in these additives. In the mill, starch is cooked to produce a usable paste (typically 20–30% solids), after which it is stored and then used. A clay slurry typically contains 50–70% solids. Dispersants that are used to stabilize the clay slurry will provide additional nutrition for microbiological growth. As is the case with starch slurries, the clay slurry is stored in tanks. In many cases, these tanks are not cleaned frequently, and the storage is at temperatures and pH levels perfect for growth of bacteria or yeasts. If allowed to proliferate unchecked, this growth can quickly render these materials unusable.

It is essential to add a preservative to starch if it is stored for any length of time. Numerous organisms will proliferate in a starch slurry, especially in cooked starch. Common problem-causing organisms are *Clostridium* spp. and *Bacillus* spp. Excess microbiological activity in starch is indicated by a pH drop that results from the fermentation process. (The result of degradation of starch is usually production of organic acids.) In one typical mill, problems began when addition of warm water from storage was initiated to reduce the starch solids from 12 to 8%. The warm water was not treated and so contaminated the starch. Without addition of microbicide, the starch pH quickly dropped from 7.5 to 5 and a pink color appeared.

If the starch shows this problem with pH drop, the cause is very likely microbial activity. If one wishes to confirm this with plate counts, it is important to do these counts at conditions representative of the system itself. The problem-causing organisms are often thermophilic, since the temperature of the starch system is often high (up to 65°C). In one case, if plates were incubated at 40°C, typical counts were 0 to 500 colony-forming units per milliliter. When these were incubated at 65°C instead, the count was actually several million colonies, indicating that the problem was caused by thermophilic organisms.

When working to control microbiological problems in papermaking, proper application equipment is very important. For example, in treating starch with a preservative, the pumps that deliver the chemical should be interlocked with the pumps moving the starch and water to ensure continual treatment of every batch.

7. Toxic/Explosive Gases

The conditions needed for microbial production of explosive and/or toxic gases exist in many mills. Major explosions and fatalities have occurred. The gases produced can be hydrogen, methane, or hydrogen sulfide. Understanding the microbiology of the problem is helpful in preventing its occurrence.

Hydrogen sulfide—This gas produces the well-known offensive odor at low concentrations (above about 0.25 ppm). However, brief exposure to small quantities will desensitize the sense of smell, and high concentrations (above about 1000 ppm) are almost instantly fatal. Dangerously toxic levels will occur well before explosive levels are reached, since the explosive concentration of H_2S is above 4.3%.

Species of anaerobic sulfate-reducing bacteria,

TABLE I
Some Paper Mill Additives Affected by Microbiological Growth

Additive used	Microbiological problem	Causative organisms
Starch	Degradation, odor, pH drop	Yeast, *Pseudomonas*, *Bacillus*
Clay slurry	Odor, pH drop	Bacteria
Dilute alum solution	Deposits, screens plugged	Fungi
Polymers	Odor, screens plugged	Fungi
Dyes	Odor, screens plugged	Fungi
Calcium carbonate slurry	Odor, pH drop	Bacteria, some fungi
Sizing solutions	Odor, pH drop	Bacteria, yeast
Paper coating	Degradation, odor, pH drop	Yeast, *Pseudomonas*, *Bacillus*

such as *Desulfovibrio* and *Desulfotomaculum,* are the main producers of H_2S. These organisms are ubiquitous. They are able to survive in aerobic conditions and thrive when oxygen is depleted. Slurries and process waters that are not continually well oxygenated will provide ideal conditions. If an agitator breaks down, or the system is shut down for a time, these conditions are likely to occur. High temperatures do not solve the problem: although most sulfate reducers are mesophilic, several species are thermophiles. In addition, oxygen is much less soluble in water at warmer temperatures. As with these other gases, the major problems occur when this microbiological activity happens in a closed vessel.

Methane—Some bacteria break complex organic materials down into volatile fatty acids. Methane-producing bacteria (methanogens) use these acids to produce methane. The methanogens prefer near-neutral pH and moderate temperatures. These organisms are also strict anaerobes. Unless sulfate levels are low, it is unlikely the methanogens can compete successfully with the sulfate-reducing bacteria. In paper mills, microbial production of methane is not as common as problems with hydrogen or hydrogen sulfide. One problem with analysis is that these organisms are very difficult to grow in a culture.

Hydrogen—There are many species of hydrogen-producing bacteria, particularly of the genus *Clostridium*. Evidence gathered after serious explosions in mills has implicated hydrogen as the problem and anaerobic bacteria as the source.

In one documented example, two welders were killed when a large white water storage tank exploded during installation of a safety railing. In this mill, the storage capacity for white water had just been doubled, from 740,000 liters. The explosion occurred just one week after the new system went into service. The explosion lifted a steel shell weighing about 30 tons to a height of 20 meters, releasing about 1,000,000 liters of white water, which caused extensive flooding in the mill.

Studies showed that the cause was an accumulation of hydrogen (with a very small amount of methane) produced by microorganisms. The likely cause was poor agitation in parts of the system, where the water became stagnant. During a mill shutdown, there is little turnover of water. Laboratory studies of the causative organisms isolated from this mill did not show activity at pH 5, and at pH 5.5 produced only a small amount of hydrogen. But, as is the case more commonly in the industry, this mill operates at near-neutral pH, ideal for these bacteria.

Use of microbicides is unlikely to have a major effect in preventing problems with explosive gases. The location where the problem exists would, typically, be inaccessible to chemical additives, since these occur in spots where there is stagnation and a lack of agitation. The following solutions are more likely to be successful:

- Minimize chances for developing anaerobic conditions
- Make certain there is adequate agitation and rapid turnover of stock
- Carry out frequent cleanings
- Design chests and tanks to reduce areas of stagnation
- Fit enclosed areas with vents to reduce the chance of gases accumulating in the headspace.

8. Microbiologically Influenced Corrosion

The problem of microorganisms exacerbating corrosion is well known in industry and remains a common and costly problem. The most prevalent cause of microbiologically influenced corrosion (MIC) is probably sulfate-reducing bacteria (SRBs). The anaerobic conditions required for *Desulfovibrio* and other SRBs can be found at many sites in the mill. The paper machine stock is high in suspended solids and often contains many dissolved solids as well. The suspended solids can build up in low-flow areas. The dissolved minerals can precipitate on a surface. The organic materials in the wood, or residual organic materials in recycled paper, can form a deposit on a surface as well. Along with biofilms, all of these deposits form a protected environment where the SRBs are able to flourish. It is not uncommon for stainless steel piping in a paper mill to be perforated by the corrosion that results from this activity.

Aerobic microorganisms can also be the primary cause of corrosion. Many produce organic acids that are corrosive. Also, a dense biofilm can create a corrosion cell due to differential aeration. The region under the deposit is anoxic and becomes a small anode, paired with the cathodic area surrounding it. Pitting-type corrosion of the metal results. Manganese-fixing bacteria, which are seen infrequently in mills but more often in other industries, cause corrosion in a similar manner, although a galvanic cell can be created between the stainless steel surface and the manganese precipitated by the bacteria.

Again, the solution to microbiologically influenced corrosion is primarily keeping the system as clean as possible. Here are some factors that should be considered, many of which should minimize chances for developing anaerobic conditions:

- Make certain there is adequate agitation and rapid turnover of stock (e.g., low volumes in chests)
- Maintain a clean system with frequent boilouts
- Design of chests and tanks should reduce possibility of stagnation
- Use microbicides effectively to prevent buildup of biofilm.

9. Preservation of Paper Products

Few commercial paper products require any sort of protection from biodeterioration. In most cases, biodegradation of the paper is, in fact, preferred (after its intended use). However, there are some special products that must be protected from microbial attack. Here are some examples.

Market Pulp—Pulp is frequently produced at one location and then shipped to another location to be used in the manufacture of paper. Market pulp is an important export product for several countries. This pulp is often dried and transported in the hold of a ship. Occasionally this pulp may be treated with a fungicide to reduce mold growth. It is also common to ship pulp over shorter distances as "wet-lap," containing about 50% water. Treatment of this pulp will reduce problems like degradation of the fiber and development of odor and provide the purchaser with a higher quality product.

Reducing Spores in Food Board—For many years in the United States, there has been a special government regulation for paperboard used in milk cartons. This so-called "Dairyman's Standard" requires that there be 250 or fewer colony-forming units per gram of uncoated board. The same criterion might be required in other products used for contact with food. The rationale for the rule is that bacterial spores in

the paperboard exposed to the milk might vegetate and result in rapid spoilage. This rule is used even though the paperboard is coated with a barrier, often composed of polyethylene. There are some uncoated areas, e.g., at a seam. The troublesome organisms, in this case, are most commonly from the genus *Bacillus*.

There are various methods to control the level of spores in the milk-carton stock. Most commonly, proprietary microbicides are used in the paper machine. In systems with special problems, it is useful to determine the primary source of spore-forming species. This may indicate additional treatment is needed in the fresh water or that the source of problems is an additive.

Table II shows total bacteria counts and spore counts in samples taken at various locations in a paper machine producing paper cup stock. The data show that, in this mill, the final sheet has no detectable spores, but fiber that has dried on a surface shows numerous cells that have sporulated. The "coating color" is a starch/latex coating, used to put the desired finish on the sheet. The "couch trim" is the trimmings from the edge of the sheet, removed at a point prior to pressing and drying, which is about 80% moisture. This material is stored before being recycled back into the manufacturing process. The "sheet before press" sample is taken from the same point in the process, but has not been stored. Moisture is removed by pressing and then by drying on steam-heated drums before the final sheet is complete.

Soapwrap—Many facial soaps contain moisture when packaged, and so the packaging used must be treated to prevent unsightly fungal growth on both

TABLE II
Spore and Bacterial Counts on Paper Cup Stock

Sample	Total, CFU/g	Spore count, CFU/g
Dry fiber from chest wall	>30,000	>500
Coating color	24,000	0
Couch trim	200	130
Sheet before press	2000	0
Final sheet	200	0

soap and packaging. The organisms typically responsible are *Aspergillus* and *Penicillium* species.

This problem is solved by a surface treatment with an effective fungicide. Addition of the fungicide in the wet end of the paper machine will not work because none of the effective fungicides currently available is retained with the fiber. Instead, the fungicide is added with surface sizing solution on the paper machine or blended in lacquer used in a printing operation.

10. Miscellaneous Problems

A variety of additional microbiological problems could be listed. For example, splashing from the machine or overspray from showers can cause slime to accumulate on surfaces of floors, ladders, and catwalks, causing a safety hazard. Use of a microbicide in the system may prevent the problem, but frequent cleaning of the surface may be a more efficient solution.

There are occasional problems with algae. Cleaning and then limiting the light source can usually solve the problem of algal growth inside a mill. On rare occasions, there may be a green spot in a sheet of paper that, when analyzed, are found to be algae. Such a problem can usually be traced back to algal growth in a storage tank. Treatment with an effective algicide will eliminate the problem.

IV. CONTROL OF MICROBIOLOGICAL PROBLEMS IN THE PAPER INDUSTRY

Effective control of microbiological growth can be accomplished by a combination strategy that keeps in mind the basic factors that encourage growth. The following basics should be closely monitored:

- Minimize contamination; reduce levels of organisms entering the system where possible
- Maximize cleaning, to reduce levels of organisms by effective "housecleaning" whenever possible
- Use effective chemical treatment when necessary, safely, according to instructions on the label

- It is seldom necessary to eliminate the problem organisms; instead, growth must be controlled, and the accretion of biofilm prevented or reduced.

1. **Changing environmental conditions** is usually not feasible, since the pH and temperature of the system, for example, are set by the required properties of the paper made and capabilities of the system. However, the conditions of the system must be monitored continually, so that the cause for any sudden change in microbial growth may be resolved.

2. **Reducing contamination** can be a useful part of a control strategy. This may involve simple housekeeping, such as keeping trash from accumulating and keeping the mill clean. Effective treatment of fresh water used in the mill is often an important part of the program.

3. **Maximize cleaning**—Most paper machines periodically have a *boilout,* when a cleaning solution is circulated through the system during a shutdown. These have typically been heated solutions of caustic at pH 12–13, often with surfactants or other specialized cleaning chemicals added. An effective boilout is a very good way to reduce microbial contamination in the system and extend the running time before the growth of microorganisms becomes troublesome. On the other hand, effective use of chemical treatment can extend the production time before it is necessary to shut down the machine and run a boilout.

4. **Chemical treatment** to control microorganisms is necessary on most paper machines. (The exception might be a system where the fresh water used is quite clean, the machine is operated at a high temperature, and there are frequent cleanings of the system.) In the paper industry, the application of microbicides is usually under close supervision of the company that supplies the products used. In most of the world, the use of these materials is regulated by government agencies charged with overseeing the safety of workers and proper stewardship of the environment.

It is never necessary or prudent to attempt to sterilize a system. It is common that a paper machine might have a bacterial count of several million colony-forming units per milliliter of water in the system and still operate without any problems from microorganisms. Microbicides should be used to keep the growth of biofilm below a level where economic losses due to loss of product quality or operating efficiency become intolerable.

5. **The chemistry of microbicides** is diverse. New products are added rather infrequently, due to the major effort and the significant costs incurred well before selling any product. In addition, getting a new product approved through government bureaucracy is a slow, convoluted process. Table III gives a list of many of the major active ingredients used in the industry.

Many of the microbicidal active ingredients used are not water soluble, requiring that these actives be formulated with other ingredients that provide ready distribution into water for maximum results. Often, surfactants must be used to ensure even dispersion into the papermaking furnish. Some of these surfactants or solvents can interfere with other additives in the paper machine, produce foam, or adversely affect paper properties. The formulation is a key step to making an effective product.

6. **The typical use pattern** with microbicides varies with the application. In some cases, continuous addition of the product is required, for example, when incoming fresh water is treated with chlorine. If a preservative is used in an additive like starch, the chemical is added to every batch.

TABLE III
Microbicides Commonly Used in the Paper Industry

Class of product	Some examples	General information
Organobromines	• 2,2-Dibromo-3-nitrilopropionamide • 1,4-Bis(bromoacetoxy)butene • β-Bromo-β-nitrostyrene • 2-Bromo-4'-hydroxyacetophenone • 2-Bromo-2-nitropropane • 1,2-Dibromo-2,4-cyanobutane	• Effective bactericides • Usually not effective as fungicides • Most are solvent-based
Organosulfurs	• 3,5-Dimethyltetrahydro-1,3,5.2H-thiadiazine-2-thione • Potassium Dimethyldithiocarbamate • Potassium N-methyldithiocarbamate • Sodium Dimethyldithiocarbamate	• Effective bactericides • Usually not effective as fungicides • Odorous • Water-based
Cationics	• Alkyldimethybenzyl ammonium quats • Dialkyldimethyl ammonium quats	• May not work in high levels of suspended solids
Isothiazolinones	• 5-chloro-2-methyl-4-isothiazolin-3-one • 2-methyl-4-isothiazolin-3-one • 1,2-Benzisothiazolin-3-one	• Effective bactericides • Effective preservatives • Most water-based
Thiocyanates	• Methylenebis(thiocyanate) • 2-(Thiocyanomethylthio)benzothiazole	• Effective bactericides • Effective fungicides
Oxidizers	• Chlorine • Chlorine dioxide • Hypochlorous or hypobromous acids • Bromohydroxyacetophenone	• Must be added continuously • May damage machine fabrics • May add to AOX • Low cost per kilo
Others	• Glutaraldehyde • Bis(trichloromethyl)sulfone	

However, a much different pattern of treatment is commonly used when the purpose is to reduce the growth of biofilm in the wet end of the paper machine. The most common pattern is to feed chemicals over a certain period of time, typically 1 to 2 hr, stop the addition for a period of several hours, and then commence pumping again. This pattern, which has been used for many years, is more efficient in controlling biofilm than continuous addition.

This intermittent pattern of treatment reduces the overall cost. One reason it works well is that most of the water used on the wet end of the paper machine is continuously recycled. After the fibers form the sheet, the bulk of white water that is left is reused immediately to dilute fresh stock coming to the machine. Virtually all microbicides used commercially tend to remain in the white water and not with the fiber. So, in a typical paper machine, the final concentration of microbicide might be about 5 times higher, due to this recycling.

Figure 5 compares the concentration of a microbicide in the paper machine system, for both continuous addition and for the same amount of chemical added intermittently. In this case, we have an addition pattern of 2 hr on, 6 hr off. In this case, we apply four times the concentration for one-fourth the time. The advantage is that the biofilm is exposed (at the same cost) to a much more potent level of microbicide for a certain time. The period with no treatment is short enough that the sessile population cannot recover before another dose of control chemical is applied.

7. **Mechanisms of action** for microbicidal products have not been extensively studied. Some are known to alter the permeability of cell walls, disrupting the normal flow of nutrients into and wastes out of the cell. Oxidizers like chlorine irreversibly damage proteins and other organic constituents, resulting in a loss of enzyme activity. Some microbicides are known to chelate metals and may interfere with cell processes in that way. Others may react noncompetitively with an enzyme in place of the normal metabolite.

8. **The future for control chemicals** is likely to include innovation in the area of biofilm control with chemistries of lower toxicity. The driving forces for developing such products include: (1) reducing potential for health effects on personnel in contact with the products, (2) reducing the chance of environmental problems, and (3) increasing the supplier's probability of getting timely regulatory approval. In the future, *biodispersants* will likely be used more frequently. The aim of such products is to prevent attachment of bacteria, or possibly to break loose the biofilm that has formed, without a "cidal" effect.

For example, many bacteria attach themselves permanently to substrates by protein-based appendages (pili). Pili are almost always present in gram-negative bacteria, which dominate in microbial slimes. Protease-based biodispersants appear to have efficacy in

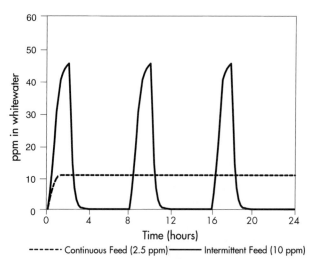

Fig. 5. The effect of recycling whitewater within a paper machine; typically, the concentration of a microbicide increases 4–5 times the initial feed rate. Also, the product can be fed continuously or intermittently: this graph compares feeding at 2.5 ppm continuously to feeding at 10 ppm for 2 hr, followed by no microbicide feed for 6 hr. The advantage is that, for the same program cost, the biofilm is treated with 4 times the level of microbicide, followed by periods of no treatment. Intermittent feed of microbicides is very successful in control of biofilms on paper machines.

TABLE IV
Potentiators

Treatment	Antibacterial efficacy	Environmental effect
A	2 ppm	30.7
B	>10,000 ppm	0.3
A + B	1 ppm	10.0

reducing biofilm. These enzymes likely keep the permanent attachment from occurring as readily.

Another path to improved effect with lower toxicity is by the use of *potentiators*. A potentiator is a nonmicrobicidal substance that enhances the effect of a microbicide, thus reducing toxicity. Potentiators can work by synergy. As shown in Table IV, Microbicide A is an effective bactericide at a dose of 2 parts per million. Potentiator B has no antibacterial effect at reasonable concentrations. On the other hand, A is much more toxic to typical microorganisms in the wastewater treatment system, 100 times as toxic as B. The benefit of the potentiator is that, with B in the system, A becomes twice as effective. At the same time, the toxic effect is reduced by 65%.

In the data from Table IV, the toxic effect is measured on a Microtox system. In this procedure, the toxic effect on a luminescent bacterium, *Vibrio fisherii,* is measured. This method is commonly used to evaluate the toxicity of outflow from the wastewater treatment of paper mills. Hundreds of studies have been done to show that there is a good reproducible relationship between the effect on this species and fish toxicity in the environment. The toxicity units used are the inverse of the concentration of toxic material that reduces the growth of *V. fisherii* by 50%, i.e., the inverse of the EC-50.

Potentiators can also work by broadening the spectrum. For instance, although Microbicide C works well against bacteria, it is not effective on molds: the minimum inhibitory concentration is 250 ppm against *Aspergillus niger,* well above any reasonable use concentration. Potentiator B has no effect at use concentrations. However, the combination is quite effective on molds, with a minimum inhibitory concentration of 25 ppm against *A. niger.*

9. **Safety and microbiological control chemistries**—Most of the products used to control microbiological growth in the paper industry are hazardous materials. It is important that these products be used safely. Important information on use of each product must be given on the Material Safety Data Sheet and the label provided with the product. These give information on any hazardous components and any health hazards associated with that particular microbicide. The supplier of the slime control chemical should provide first aid and health information and information on medical treatment of any problems that arise.

The general rules of safety for using slimicides include use of proper protective equipment. Any chances for human contact with the product must be minimized. Most microbicides can cause skin irritation and eye damage when there is direct contact. There should be chemical-resistant gloves, safety glasses with sideshields, and chemical-resistant boots available to those handling the slimicides. Adequate ventilation is important in areas where the slimicides are stored and in any area near the machine, where there might be misting of water treated with microbicide.

In most countries of the world, there is governmental oversight of the sale and use of pesticides. In the United States, for example, microbicides are regulated by the Federal Insecticide, Fungicide, and Rodenticide Act (FIFRA), which is administered by the Environmental Protection Agency (EPA). FIFRA requires that any product designed to destroy or repel any pest must be registered. EPA oversees a careful registration process that evaluates the environmental impact of each product and also ensures that these materials are reasonably safe to use.

Additional regulatory agencies may oversee other safety aspects. For example, in the United States, the activity of Food and Drug Administration (FDA) includes oversight of additives used in food packaging, since there can be migration of ingredients of the package to the foodstuff it contains. The FDA regulations include lists of ingredients specifically tested and allowed for use in the wet end of the paper machine or in coatings put on the sheet. All microbicides commonly used in the paper industry are listed as allowed for wet-end addition in the manufacture of paper for food packaging, but only a fraction of these are allowed for "dry end" addition, e.g., in a coating on the paper. In Canada, this regulation is performed by Health Canada.

For the European Community (EC), the use of microbicides is still regulated in each country by local laws. Some countries have extensive legislation (e.g., Sweden and the Netherlands), while in other countries little regulation is in place. Currently, the EC is developing the Biocidal Products Directive (BPD) to eventually replace local regulations with one international system. Once each European country approves this plan, there will be a transition period of 10 years, where microbicides will be registered according this directive, and each country will keep, in addition, its current legislation and approval process. Eventually, when a product is approved in one European country, it will also be approved in the other countries. Additionally, "ecotaxes" are levied in some European countries to further limit use of microbicides.

Allowance of materials for food contact in Europe is also regulated by each country, although the regulations of the *Bundesinstitutes fur gesundheitlichen Verbraucherrschutz und Veterinarmedizin* (BgVV) in Germany are widely consulted. There is also a movement to develop one international directive in Europe for chemical additives for food contact, but this is not likely to happen for some time.

10. **The economic benefits of effective microbiological control**—Many examples have been given of the benefits of controlling microbial problems in a paper mill. If these methods were not economically favorable to the papermaker, the industry providing the service would not exist.

The most obvious economic benefit is increased production on a paper machine. Here is an example from a paper mill. In this mill, an average break in the sheet during production results in 20–30 minutes lost time before regular production can resume. This paper machine makes 1000 tons per day of paper at a market value of $960/ton. A loss of 30 minutes of production amounts to a daily loss of over $20,000, which totals over $7 million per year. A typical slime control program might cost $5/ton of paper produced. If the control program eliminates one break per day, the benefit in this typical case is four times the cost.

V. SUMMARY

Microbiological growth has an important impact on the paper industry. The growth of fungi and bacteria causes a variety of problems resulting in significant economic loss. Typical are problems affecting the efficiency of the process, the appearance of the paper, other problems in finished product quality, biodeterioration of the materials used, and damage to process equipment.

An understanding of microbiology and the various factors that impact microbial growth enables one to find effective ways to reduce problems. By proper application of cleaning practices and microbicides, the papermaker can ensure maximum production and quality. These control programs can be costly, but the problems are much more costly. In a very competitive market, consistently high quality is crucial. The paper mill producing a sheet of paper with an odor or marred by spots and holes due to microorganisms will be quickly outclassed by its competitors.

See Also the Following Articles

BIODETERIORATION • BIOFILMS AND BIOFOULING • FRESHWATER MICROBIOLOGY • pH STRESS • TEMPERATURE CONTROL • TIMBER AND FOREST PRODUCTS • WASTEWATER TREATMENT, INDUSTRIAL

Bibliography

Biermann, C. J. (1996). "Handbook of Pulping and Papermaking" 2nd ed. Academic Press, San Diego.

Bulich, A. A. (1979). Use of luminescent bacteria for determining toxicity in aquatic environments. *In* "Aquatic Toxicology" ASTM STP 667 (L. L. Marking and R. A. Kimerie, eds.), pp. 98–106. American Society for Testing and Materials.

Characklis, W. G., and Marshall, K. C. (1990). "Biofilms." John Wiley & Sons, Inc., New York.

Hipolit, K. J. (ed.) (1992). "Chemical Processing Aids in Papermaking: A Practical Guide." Tappi Press, Chicago.

Hoekstra, P. M. (1997). Biological and freshwater deposits. *In* "Proceedings of the Tappi Deposit Analysis and Control Short Course." TAPPI, Atlanta.

King, V. M. (1990). Microbial production of explosive/toxic gases in the paper industry. *Tappi* 90, 185–188.

Lappin-Scott, H. M., Costerton, J. W., and Marrie, T. J. (1992). Biofilms and biofouling. *In* "Encyclopedia of Microbiology." Vol. 1. Academic Press, San Diego.

Ross, R. T., and Hollis, C. G. (1976). Microbiological deterioration of pulpwood, paper, and paint. *In* "Industrial Microbiology" (B. M. Miller and W. Litsky, eds.). McGraw-Hill, New York.

Sharpley, J. M. (1973). Microbiological corrosion and its control. *In* "Corrosion Inhibitors" (C. C. Nathan, ed.) National Association of Corrosion Engineers, Houston.

Smook, G. A. (1997). "Papermaking: The Process and the Product" [CD-ROM]. TAPPI, Atlanta.

ISBN 0-12-226803-2

90038